D0207467

Introductory Electronics for Scientists and Engineers

Introductory Electronics for Scientists and Engineers

Second Edition

ROBERT E. SIMPSON
University of New Hampshire

PRENTICE HALL, Englewood Cliffs, New Jersey 07632

Library of Congress Cataloging-in-Publication Data

Simpson, Robert E. (Robert Edmund)
Introductory electronics for scientists and engineers.

Includes index.
1. Electronics. I. Title.
TK7816.S545 1987 621.381 86–17499
ISBN 0-205-08377-3

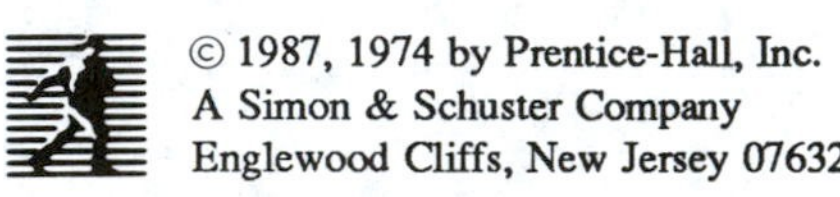

A Simon & Schuster Company
Englewood Cliffs, New Jersey 07632

Printed in the United States of America
10 9

ISBN 0-205-08377-3

Prentice-Hall International (UK) Limited, *London*
Prentice-Hall of Australia Pty. Limited, *Sydney*
Prentice-Hall Canada Inc., *Toronto*
Prentice-Hall Hispanoamericana, S. A., *Mexico*
Prentice-Hall of India Private Limited, *New Delhi*
Prentice-Hall of Japan, Inc., *Tokyo*
Simon & Schuster Asia Pte. Ltd., *Singapore*
Editora Prentice-Hall do Brasil, Ltda., *Rio de Janeiro*

TO

DONNA AND TYLER

Contents

3 Fourier Analysis and Pulses 107

4 Semiconductor Physics 151

5 The Bipolar Transistor 199

6 The Field-Effect Transistor (FET) 258

10 Operational Amplifier Circuits, 412

11 Active Filters and Regulators 469

Preface

More than half of this second edition is new. The new material essentially covers operational amplifiers and digital techniques, including an introduction to digital filters and microprocessors. The appendixes have also been changed, with the inclusion of useful reference material on digital chips and op amps. Many new problems have been added for the material carried over from the first edition, and problems have been written for the new chapters, of course.

I have kept the same general order of topics from the first edition, namely, analog electronics first and digital electronics second. Generally, this corresponds to two academic semesters of approximately 15 weeks each, with two lectures and two laboratories per week, although it may be necessary to give more than two lectures per week in some chapters. I personally find that a 15-minute talk just before each lab is very useful.

The students are usually junior physics or science majors with a few enlightened mechanical engineering majors. They have all had three semesters of calculus-based physics at the level of Weidner and Sells, three semesters of calculus, and one semester of ordinary differential equations. Calculus is assumed without apology, but Fourier analysis and some aspects of ordinary differential equations are reviewed in the text.

My general philosophy has been to emphasize the practical rather than the theoretical, but I have included some important theoretical concepts such as Fourier analysis, basic circuit analysis, and the Nyquist sampling theorem because they will never be obsolete regardless of future hardware changes.

Appendix H contains 61 suggested laboratory experiments, which are more than enough for two 15-week semesters. Generally, all component values and devices are specified in the early experiments, with less detail in the later experiments.

The first two chapters review dc and ac circuits with some new material on Thevenin's theorem, thermistors, network analysis, and bridges.

The third chapter covers the important, but often neglected, topics of pulses and Fourier analysis with new material on frequency modulation and an *RC* "speedup" network.

The fourth chapter on semiconductor physics is largely unchanged but contains new material on photovoltaic diodes.

The fifth chapter combines material on bipolar transistors from two chapters in the first edition with new material on saturated and unsaturated transistor switches. Less emphasis is placed on the equivalent circuit analysis of transistor circuits, but my years of teaching have led me to believe that the student should master the elements of a single-transistor circuit before proceeding to the "black box" approach of using integrated circuit chips.

The sixth chapter briefly covers the field-effect transistor (FET) with new material on power FETs and on comparing bipolar and field-effect devices.

Chapter 7 covers the general topic of feedback (both negative and positive) with numerous circuit examples, and several important general theorems about the effects of feedback on gain, impedances, noise, stability, the Miller effect, the gain-bandwidth product, and transistor saturation.

Chapter 8 covers the topic of noise, with the fundamentals of Johnson and shot noise, amplifier noise, bipolar versus FET noise, and flicker noise, with the addition of a section on the phase-locked loop as a modification of lock-in detection.

Chapter 9 is new and covers the basics of operational amplifiers (op amps), which have become the basic building blocks of analog electronics. A few details of internal op amp circuitry are given, along with a detailed description of the two fundamental op amp amplifier circuits: the inverting amplifier and the noninverting amplifier. The two simple rules for analyzing op amp circuits are given: the op amp voltage rule and the op amp current rule. Basic op amp parameters such as input voltage and current offsets, gain-bandwidth, compensation, and noise are covered briefly.

Chapter 10 is new and contains dozens of op amp circuits with waveforms and component values given in most cases. The circuits include amplifiers (both low power and high power), current-to-voltage converters, log amplifiers, sample-and-hold circuits, charge-sensitive amplifiers, integrating and differentiating circuits, ideal diode circuits, the Schmitt trigger, and various types of op amp multivibrators and oscillators.

Chapter 11 is new and covers active analog filters and regulators. Unlike most electronics texts, the basic driven damped mechanical oscillator is first covered before the electronic filter is introduced. Low-pass, high-pass, and bandpass filter types are covered. The mathematics is worked out rather completely in most cases to demystify the subject in the students' eyes. Voltage and current regulator circuits are treated, along with short-circuit protection, power boosting a low-current regulator, and the ubiquitous 7800 three-terminal regulator chips.

Chapter 12 is new and covers the basic concepts of digital electronics with the elements of base 2, base 8, base 10, and base 16 number systems. Boolean algebra is briefly covered with DeMorgan's theorem. The common OR, NOR, AND, NAND, XOR, and XNOR gates are covered with their truth tables and 7400 TTL chip numbers. Positive and negative logic is covered and related to the electrical behavior of the chip. The 7400 TTL

NAND gate is analyzed in detail. The various logic families are covered, including the new 74HC00 series, which will probably make most of the other families obsolete in the near future. Finally, interfacing among the various families is discussed.

Chapter 13 is new and contains a collection of commonly used digital circuits and devices, including various flip-flops, synchronous and asynchronous counters, shift registers, multiplexers and demultiplexers, the modem, the 555 chip, and various monostable multivibrators.

Chapter 14 is new and covers the subjects of binary arithmetic and memory. The arithmetic truth tables are given along with the appropriate circuits, and the various types of digital memories are discussed and compared. The lookup table technique is treated as an example of the application of inexpensive digital memory. Some useful memory chips are discussed.

Chapter 15 is new and covers the concepts of digital-to-analog conversion and analog-to-digital conversion. The various types of conversions are discussed with their advantages and disadvantages. The useful but often neglected series-parallel analog-to-digital conversion technique is presented. The Nyquist sampling theorem is carefully discussed with both a mathematical and a "seat of the pants" proof. RS-232 interfacing and parallel interfacing are discussed.

Chapter 16 is new and covers microprocessors in general and the Intel 8085 in particular as a "tried-and-true" inexpensive eight-bit microprocessor. The general architecture and the 8085 instruction set are briefly covered with references to a detailed treatment. The appropriate application areas for hard-wired circuits (composed of separate gates), microprocessor systems, and microcomputer systems are discussed. Examples using the 8085 include waveform generation, digital filtering with some elementary theory, and microprocessor-controlled measurement and control of experimental variables. Finally, several microprocessor buses are treated, including the popular S-100 bus, the Apple II bus, and the GPIB bus.

In conclusion, I would like to express my thanks to several people. My colleague, Professor Richard Kaufmann, who has also taught the electronics course at UNH, has been extremely helpful in reviewing the manuscript draft and in many discussions. Professor Robert Lambert has also been helpful in many physics discussions. Many thanks are also due Peter Hubbe, who was a superb teaching associate, and to Phil Olsen, my best electronics student. UNH Space Science Center engineers Jim Kish and Mark Widholm have also been extremely helpful in answering many questions and suggesting various neat circuits. Senior technician Ralph Varney has patiently helped in finding all sorts of electronic information and parts. The chairman of the physics department, Professor Roger Arnoldy, has provided much appreciated encouragement. Nan Collins has typed cheerfully and accurately. I would also like to express my appreciation to Donald Vincent who, as head of the UNH library, has provided a peaceful, scholarly library environment for research, writing, and contemplation. And, finally, my deepest thanks and appreciation go to my wife, Donna, and my son, Tyler, for providing encouragement and loving distraction.

R. E. S.

Introductory Electronics for Scientists and Engineers

CHAPTER 1

Direct Current Circuits

1.1 ELECTRIC CHARGE AND CURRENT

There are two kinds of electric charge, positive and negative. They are so named because they add and subtract like positive and negative numbers. All atoms contain charge; the usual picture of an atom is a small (10^{-15} m diameter) positively charged nucleus around which negatively charged electrons move in orbits of the order of 10^{-10} m diameter. Charge is measured in coulombs in mks units and in statcoulombs in the cgs system. The basic indivisible unit of charge is the charge on one electron which is -1.6×10^{-19} coulombs or -4.8×10^{-10} statcoulombs. All electric charges (positive and negative) are, in magnitude, integral multiples of the charge of the electron. However, in most electronic circuit problems the discrete nature of electric charge may be neglected, and charge may be considered to be a continuous variable. One of the most fundamental conservation laws of physics says that in any closed system the total net amount of electric charge is conserved or, in other words, is constant in time. For example, in a semiconductor if one electron is removed from a neutral atom then the atom minus the one electron (the ion) has a net electric charge of $+1.6 \times 10^{-19}$ coulombs.

The flow of electric charge, either positive or negative, is called *current*; that is, the current I at a given point in a circuit is defined as the time rate of change of the amount of electric charge Q flowing past that point.

$$I \equiv \frac{dQ}{dt} \tag{1.1}$$

The direction of the current is taken by convention to be the direction of the flow of *positive* charge. If electrons are flowing from right to left in a wire, then this electron current is electrically equivalent to positive charge flowing from left to right; hence, we say the current is to the right.

Current is measured in *amperes* (A); one ampere of current is the flow of one coulomb of charge per second, which is 6.25×10^{18} electrons per second. Other units of current are the milliampere (mA), which is 10^{-3} A, the microampere (μA), which is 10^{-6} A, the nanoampere (nA), which is

10^{-9} A, and the picoampere (pA), which is 10^{-12} A. Prefixes for various powers of ten are given in Table 1.1. Typical currents in low-power transistor electronic circuits, such as small radio receivers and amplifiers, are of the order of 1 to 10 mA; typical currents in low-power vacuum tube circuits are about 10 to 100 mA. The largest current normally encountered in vacuum tube circuits is about 500 mA, but specially designed high-current transistors carrying currents from 1 to 100 A are available.

TABLE 1.1. Powers-of-Ten Prefixes

Prefix	*Symbol*	*Meaning*	*Example*
giga	G	10^9	1 gigahertz = 1 GHz = 10^9 hertz
mega	M	10^6	1 megohm = 1 MΩ = 10^6 ohms
kilo	k	10^3	1 kilovolt = 1 kV = 10^3 volts
milli	m	10^{-3}	1 milliampere = 1 mA = 10^{-3} amperes
micro	μ	10^{-6}	1 microvolt = 1 μV = 10^{-6} volts
nano	n	10^{-9}	1 nanoampere = 1 nA = 10^{-9} amperes
pico	p	10^{-12}	1 picofarad = 1 pF = 10^{-12} farads

The two general kinds of current are direct current (dc) and alternating current (ac). In direct current the direction of charge flow is always the same. If the magnitude of the current varies from one instant of time to another, but the direction of flow remains the same, this type of current is called *pulsating direct current*. If both the direction and the magnitude of the charge flow are constant, the current is called *pure direct current* or simply *direct current*. If the charge flow alternates back and forth, the current is called *alternating current* (to be discussed further in Chapter 2). These currents are shown graphically in Fig. 1.1.

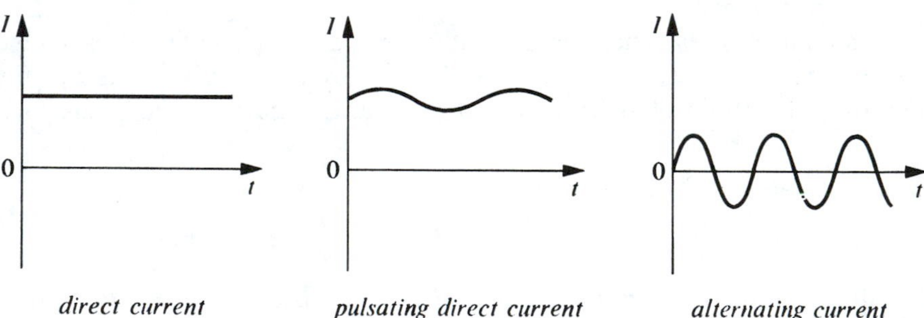

FIGURE 1.1 Types of current.

It is sometimes useful to compare electric current in a wire to water flow in a pipe, with the water molecules being analogous to the electrons in the wire. A flow of water in kilograms per second is analogous to a flow of electric charge in coulombs per second or amperes. Ignoring the individual electrons in the current is similar to ignoring the individual molecules in water flow.

1.2 VOLTAGE

The *voltage*, or electric *potential*, V at a point in a circuit is defined as the electrical potential energy W of a positive charge Q at that point divided by the magnitude of the charge; that is,

$$V \equiv \frac{W}{Q}$$

The voltage is measured relative to some other specified reference point in the circuit where we say the energy of all charges is zero. This reference point is usually called *ground* (*earth* in British literature). A good ground is a cold-water pipe or sometimes just the metal chassis or box enclosing the circuit. The point is that the circuit ground has a constant, unvarying potential, which we set equal to zero volts by convention. The earth has a constant potential or voltage simply because it is so large and is a reasonably good conductor. Any charge taken from or added to the earth through a circuit's ground wire will not appreciably change the total charge on the earth or the earth's voltage. The symbols used for ground in circuit notation are shown in Fig. 1.2.

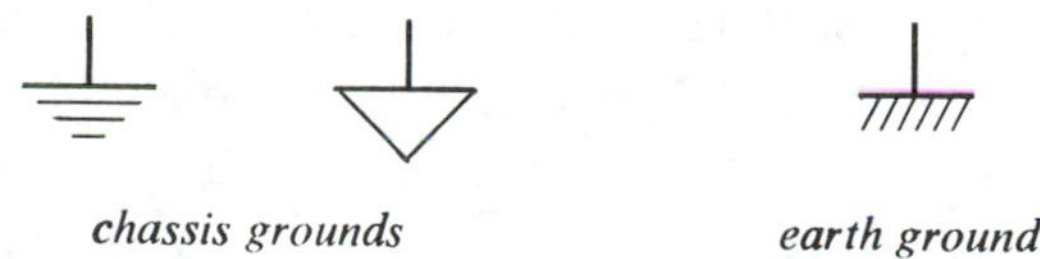

FIGURE 1.2 Ground symbols.

The unit of voltage is the volt (V); one volt is defined as one joule per coulomb (J/C). Other units of voltage are the kilovolt (kV), which is 1000 V, the millivolt (mV), which is 10^{-3} V, the microvolt (μV), which is 10^{-6} V, and the nanovolt (nV), which is 10^{-9} V. Notice that the voltage at a given point has no *absolute* meaning but only means the potential energy per unit charge *relative to ground*. This energy-per-unit-charge definition of voltage is rather difficult to understand intuitively until one considers the direction in which the charges tend to move. In all natural processes things tend to move so as to minimize the potential energy; thus, positive charges will move from points of higher voltage toward points of lower voltage (e.g., from a point with a voltage of +15 V toward a point with a voltage of +12 V). Similarly, negative charges will tend to move from points of lower voltage toward points of higher voltage. For example, negative charge (electrons) will move from a point of voltage −12 V to a point of voltage −7 V. In terms of the analogy between current and water flow, the voltage is analogous to the water pressure, because water tends to flow from points of higher pressure toward points of lower pressure. It is also sometimes useful to think of the voltage as causing or forcing the flow of current, just as one thinks of the water pressure as causing or forcing the flow of water. Thus, voltage is a "cause," and current is an "effect."

Note that the average drift speed of electrons through a copper wire is rather slow, of the order of 0.1 mm/s (10^{-4} m/s). But the speed with which electrical *energy* (the "signal") moves is normally $0.5c$ to $0.9c$, where $c = 3.0 \times 10^8$ m/s is the speed of light. The reason is that the electrons strongly repel each other by their surrounding electric fields; the electrons act, in some sense, like very hard elastic spheres. As an electron is accelerated by an applied electric field, it collides with a nearby electron, which in turn immediately collides with its neighboring electron, and so on. The situation is somewhat similar to a series of closely packed, steel ball bearings tightly packed in a pipe. If the ball bearing at one end is struck (given energy), the energy will propagate through the pipe much faster than the overall drift speed of the ball bearings themselves.†

1.3 RESISTANCE

If a current I flows through any two-terminal circuit element, then the static or dc resistance (usually referred to simply as the *resistance*) of that circuit element is defined as the voltage difference, $V_2 - V_1$, between the terminals divided by the current I (see Fig. 1.3). Strictly speaking, this

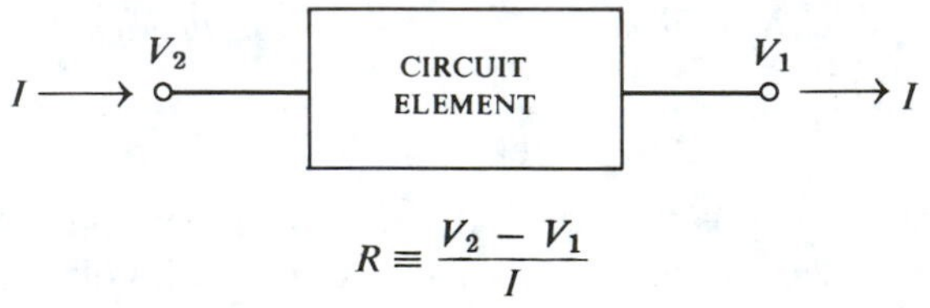

$I \longrightarrow$ V_2 CIRCUIT ELEMENT V_1 $\longrightarrow I$

$$R \equiv \frac{V_2 - V_1}{I}$$

FIGURE 1.3 Definition of resistance *R*.

definition applies only to a circuit element that converts electrical energy into heat, but this situation occurs in the overwhelming majority of cases in electronic circuitry.

$$R \equiv \frac{V_2 - V_1}{I} \tag{1.2}$$

†For a further discussion, see R. T. Weidner and R. L. Sells, *Elementary Classical Physics*, 2nd ed. (Newton, Mass.: Allyn and Bacon, Inc., 1973), Chapter 27.

The current into the circuit element must exactly equal the current leaving, from the conservation of charge law. V_2 is the voltage at the terminal where the current enters the circuit element; V_1 is the voltage where the current leaves. If the circuit element is *passive*, that is, if there is no energy given to the charge by the element, then the charge loses energy in the element. Thus, V_2 must be greater than V_1. Thus, for all passive circuit elements, the static or dc resistance is positive. The current–voltage graph for an ordinary positive resistance is shown in Fig. 1.4(a). The unit of

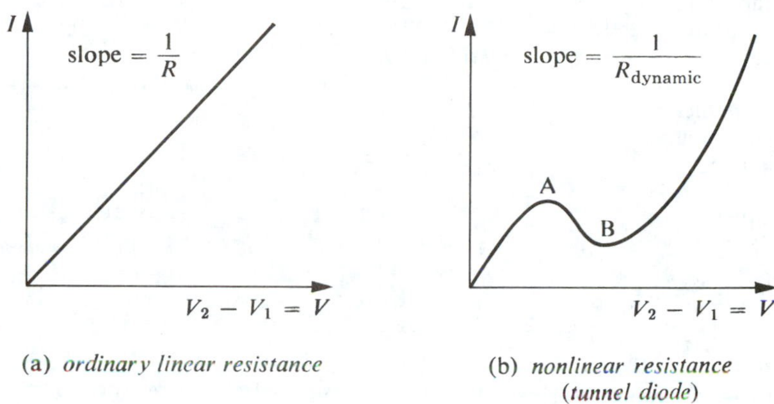

(a) *ordinary linear resistance*

(b) *nonlinear resistance (tunnel diode)*

FIGURE 1.4 Current–voltage curves.

resistance is the *ohm* (Ω); one ohm is defined as one volt per ampere. Other units of resistance are the kilohm (kΩ, sometimes written as just k), which is 10^3 Ω, and the megohm (MΩ or just M), which is 10^6 Ω.

A piece of conducting material of length L m and cross-sectional area A m^2 has a resistance of R ohms (Ω):

$$R = \rho\frac{L}{A} \tag{1.3}$$

where ρ is the *resistivity* of the material in Ω-m. The resistivity expresses the difficulty an electron has in moving through the material due to the collisions it experiences with the atoms. Resistivity varies for different materials; it depends slightly on temperature, typically increasing with increasing temperature for most metals. For carbon resistors used in most electronic circuits, resistivity increases by approximately 0.5% to 1% for a 10°C temperature increase. Resistivity does not depend on the object's size or shape but is a property of the material itself. Resistivity values for various materials are given in Table 1.2.

A *conductor* is a substance or material with a "low" resistivity. Most metals are good conductors; copper or aluminum wire is usually used in

TABLE 1.2. Resistivity Values of Various Materials

Material	*Resistivity ρ (Ω-m) at 20°C*
Conductors	
Silver	1.5×10^{-8}
Copper	1.7×10^{-8}
Aluminum	2.8×10^{-8}
Tungsten	5.5×10^{-8}
Nichrome (alloy)	100×10^{-8}
Carbon	3500×10^{-8}
Semiconductors	
Germanium	0.43
Silicon	2.6×10^{3}
Insulators	
Glass	$\approx 10^{12}$
Mica	9×10^{13}
Quartz (fused)	5×10^{16}

electronic circuits to carry current. An *insulator* is a substance or material with a "high" resistivity that is used to prevent current flow. Most plastics, rubber, air, mica, and quartz are good insulators; wire is usually covered with a plastic sheath insulator to confine the current flow to the wire. Note that germanium and silicon have resistivities that are much greater than those of metals but much less than those of insulators. Hence, they are often called *semiconductors*.

The diameter and resistance per unit length of selected copper wires are given in Table 1.3. In a practical electronic circuit, for example, a No. 22 wire 4 in. (10 cm) long would have a resistance of only 0.006 Ω.

In a formal sense a battery has a negative resistance because the battery gives energy to the charge, making V_1 greater than V_2; but this occurs because chemical energy is converted into electrical energy in a battery. Some circuit elements are said to have negative resistance; but this

TABLE 1.3. Resistance of Various Sizes of Copper Wire

AWG wire size (solid)	*Diameter (in.)*	*Resistance per 1000 ft (Ω)*
24	0.0201	28.4
22	0.0254	18.0
20	0.0320	11.3
18	0.0403	7.2
16	0.0508	4.5

always refers to *dynamic* or *ac* resistance, defined as the rate of change of voltage with respect to current:

$$R_{\text{dynamic}} \equiv \frac{dV}{dI} \tag{1.4}$$

where dV is the change in voltage across the circuit element, and dI is the change in current through the element. In Fig. 1.4(b), which is the current–voltage characteristic for a tunnel diode, the static resistance is always positive for all values of V, but between A and B the dynamic resistance is negative because I decreases as V increases. Also note that both static and dynamic resistances vary with voltage. A circuit element for which the current–voltage curve is not a straight line has *nonlinear* resistance.

A *resistor* is a two-terminal circuit element specifically manufactured to have a constant resistance. Thus, a 4.7-kΩ resistor has a resistance of 4700 Ω; a 2.2-MΩ resistor has a resistance of 2,200,000 Ω.

The resistance of a resistor usually is given by a color code (see Fig. 1.5). Bands of different colors specify the resistance according to the following rule:

R = (*first-color number*) (*second-color number*) × 10 (*raised to the third-color number*)

The first digit of the resistance is given by the color band closest to the end of the resistor. The fourth color band gives the tolerance of the resistor. Silver means ±10% tolerance, gold means ±5% tolerance, and no fourth

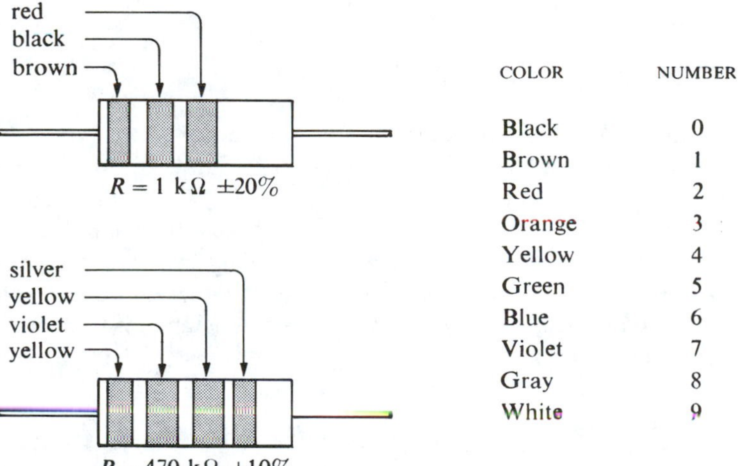

COLOR	NUMBER
Black	0
Brown	1
Red	2
Orange	3
Yellow	4
Green	5
Blue	6
Violet	7
Gray	8
White	9

FIGURE 1.5 Resistor color code.

colored band means ±20% tolerance. A 2-kΩ 10% resistor will have a resistance somewhere between 1.8 and 2.2 kΩ.

If you have difficulty remembering the color code, note that the colors follow the colors of the visible spectrum, starting with red for the number two and going through violet for the number seven. And it is extremely useful to note that if the third color is brown, the resistor is in the hundreds of ohms; if red, thousands of ohms; if orange, tens of thousands of ohms; if yellow, hundreds of thousands of ohms; if green, millions of ohms.

On certain precision metal-film resistors, the resistance value is specified to three significant figures. These resistors have five colored bands; the first three bands represent the three significant figures, the fourth band the multiplier, and the fifth band the tolerance. Certain high-reliability resistors are tested for failure rates under conditions of maximum power and voltage, and the results are expressed in percentage failure per thousand hours. The fifth colored band represents the failure rate per thousand hours according to the following scheme: brown 1%, red 0.1%, orange 0.01%, and yellow 0.001%.

The physical size of a resistor determines how much power in watts it can safely dissipate, not its electrical resistance in ohms. Characteristics of various types of resistors are given in Appendix A.

The reciprocal of resistance is called *conductance* and is usually represented by the symbols G or y:

$$G = \frac{1}{R} \tag{1.5}$$

Conductance is measured in ohms^{-1}. One ohm^{-1} is called a *siemen* or a *mho* ("ohm" spelled backwards).

1.4 OHM'S LAW

For many kinds of circuit elements it is empirically true that the resistance of the element is constant if the temperature and composition of the element are fixed. This is true over an *extremely large* range of voltages and currents. That is, changing the voltage difference between the two terminals by any factor changes the current by exactly the same factor; that is, doubling the voltage difference $V_2 - V_1$ exactly doubles the current I.

Ohm's law is simply the statement that the resistance is constant. It can be written in three ways:

$$R = \frac{V_2 - V_1}{I} \qquad V_2 - V_1 = IR \qquad I = \frac{V_2 - V_1}{R} \tag{1.6}$$

Any two-terminal circuit element of constant resistance is shown in circuit

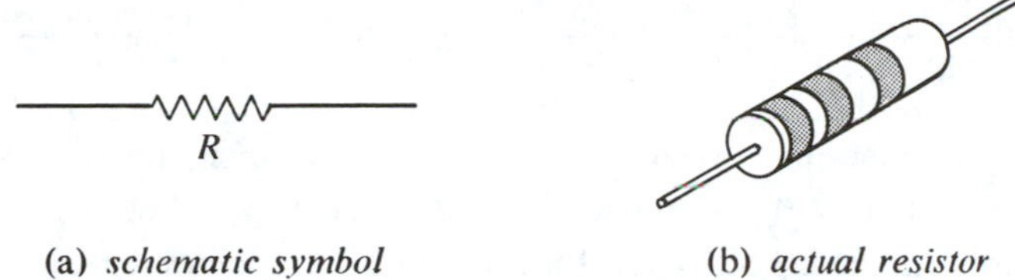

FIGURE 1.6 Resistance.

diagrams as a zig-zag line [see Fig. 1.6(a)] and is called a resistor. The larger the voltage difference, the larger the current for a fixed resistance; the larger the resistance for a fixed voltage drop, the smaller the current. Thus, "resistance" is a very appropriate name; it literally means "opposition to current flow."

These three forms of Ohm's law can be thought of in the following terms. $R = (V_2 - V_1)/I$ means that if there is a voltage difference $V_2 - V_1$ across a circuit element through which a current I is flowing, then the circuit element must have a resistance $(V_2 - V_1)/I$. $V_2 - V_1 = IR$ means that if a current I is forced through a resistance R, then a voltage difference IR will be developed between the two ends of the resistance. $I = (V_2 - V_1)/R$ means that if there is a voltage difference $V_2 - V_1$ across a resistance, then a current $(V_2 - V_1)/R$ must be flowing through the resistance. Perhaps the most important thing to remember about Ohm's law is that it is only the *difference* $(V_2 - V_1)$ in voltage across a resistor which causes current to flow. Thus a 5-kΩ resistor with one end at 35 V and the other end at 25 V will pass a current of 2 mA because

$$I = \frac{V_2 - V_1}{R} = \frac{10 \text{ V}}{5000 \ \Omega} = 0.002 \text{ A} = 2 \text{ mA}$$

A 5-kΩ resistor with one end at 1078 V and the other end at 1068 V will also pass a current of 2 mA because the voltage difference is also 10 V. It is useful to remember the shortcut that the voltage difference in volts divided by the resistance in kilohms equals current in milliamperes; in the previous example $I = 10 \text{ V}/5 \text{ k}\Omega = 2 \text{ mA}$.†

If two resistors R_1 and R_2 are connected in *series* (see Fig. 1.7), that is, if they are connected end-to-end so that the same current flows through each of them, then the total effective resistance is simply the sum of the two

†A careful analysis of free-electron drift through a conductor in response to an applied voltage difference shows that Ohm's law holds over a wide range of applied voltages because the thermal speed of a free electron (10^6 m/s at room temperature) is much larger than the drift speed (10^{-4} m/s). Thus the average time between collisions is essentially independent of the applied voltage, and the drift speed of the free electrons is proportional to the applied voltage or electric field, which is basically Ohm's law.

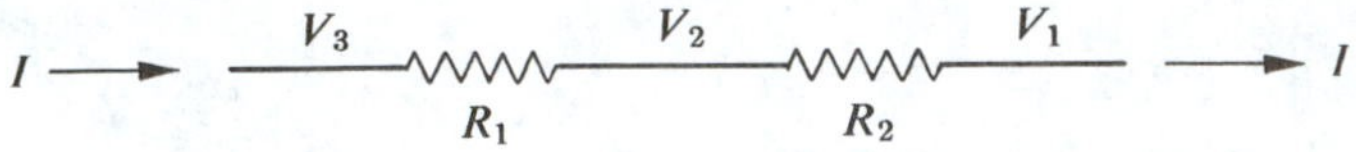

FIGURE 1.7 Two resistors in series.

individual resistances. This result follows from Ohm's law applied separately to R_1 and R_2.

$$R_{\text{total}} = \frac{V_3 - V_1}{I} = \frac{V_3 - V_2}{I} + \frac{V_2 - V_1}{I} = R_1 + R_2 \qquad \textbf{(1.7)}$$

Thus, a 1-kΩ resistor and a 3-kΩ resistor in series act like a single 4-kΩ resistor. This rule can be extended to N resistors in series, in which case the total effective resistance is equal to the sum of all the N individual resistances.

$$R_{\text{total}} = R_1 + R_2 + R_3 + \cdots + R_N \qquad \textbf{(1.8)}$$

Notice that the total resistance for a series connection always is greater than any of the individual resistances. Also notice that a straight line connecting the two resistances in a circuit diagram represents a wire or electrical connection of zero resistance. Thus all points of a straight line in a circuit diagram must be at exactly the same voltage; there is no voltage drop along a wire of zero resistance. In an actual circuit the resistance of the wire used to connect various elements is usually negligibly small; for example, a 2-in. length of No. 18 copper wire (0.04 in. diam) has a resistance of only 0.0011 Ω.

Ohms law in terms of conductance is simply

$$I = G(V_2 - V_1)$$

where $(V_2 - V_1)$ is the voltage drop across a circuit element with conductance G.

If two resistors are connected in *parallel* or side by side (see Fig. 1.8) so that the same voltage appears across each one, then the total effective

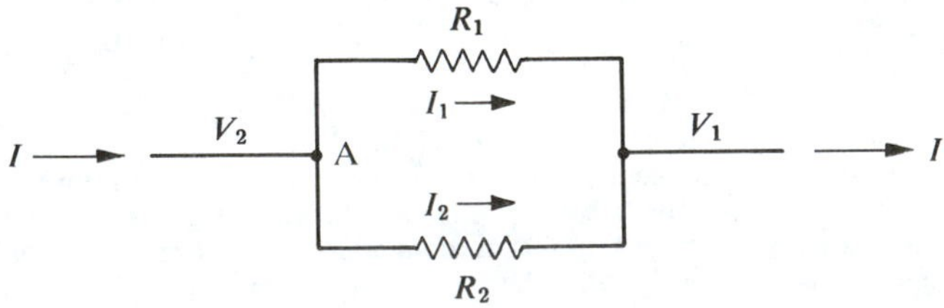

FIGURE 1.8 Two resistors in parallel.

resistance is given by

$$R_{\text{total}} = \frac{1}{\dfrac{1}{R_1} + \dfrac{1}{R_2}} = \frac{R_1 R_2}{R_1 + R_2} \tag{1.9}$$

This result follows from Ohm's law and the conservation of current. At point A, the current I entering splits up into two parts, $I = I_1 + I_2$.

$$R_{\text{total}} = \frac{V_2 - V_1}{I} = \frac{V_2 - V_1}{I_1 + I_2} = \frac{V_2 - V_1}{\dfrac{V_2 - V_1}{R_1} + \dfrac{V_2 - V_1}{R_2}} = \frac{1}{\dfrac{1}{R_1} + \dfrac{1}{R_2}} = \frac{R_1 R_2}{R_1 + R_2}$$

For example, a 6-kΩ resistor and a 4-kΩ resistor in parallel act like a single 2.4-kΩ resistor. If we have N resistors $R_1, R_2, R_3, \ldots, R_N$ all connected in parallel, the total effective resistance is given by

$$R_{\text{total}} = \frac{1}{\dfrac{1}{R_1} + \dfrac{1}{R_2} + \dfrac{1}{R_3} + \cdots + \dfrac{1}{R_N}} \tag{1.10}$$

Notice that the total resistance for a parallel connection is always less than any of the individual resistances and that the voltage drop is the same across all of the resistors in parallel. Ohm's law applied to each resistor shows that the current divides among the various resistors in such a way that the most current flows through the smallest resistance, and vice versa. For example, in the circuit of Fig. 1.8:

If $R_1 = 10\ \text{k}\Omega$ and $R_2 = 2\ \text{k}\Omega$, then $R_{\text{total}} = (10\ \text{k}\Omega)(2\ \text{k}\Omega)/12\ \text{k}\Omega = 1.67\ \text{k}\Omega$.
If $V_2 - V_1 = 8$ V, then $I = 8\ \text{V}/1.67\ \text{k}\Omega = 4.8$ mA.
The current I_1 flowing through $R_1 = 10\ \text{k}\Omega$ is $I_1 = 8\ \text{V}/10\ \text{k}\Omega = 0.8$ mA.
The current flowing through $R_2 = 2\ \text{k}\Omega$ is $I_2 = 8\ \text{V}/2\ \text{k}\Omega = 4$ mA.

In general, $I_1/I_2 = R_2/R_1$, which follows from the fact that R_1 and R_2 have the same voltage drop across them: $I_1 R_1 = I_2 R_2$.

In terms of conductance, for N resistances in series the total conductance is

$$G_{\text{total}} = \frac{1}{\dfrac{1}{G_1} + \dfrac{1}{G_2} + \cdots + \dfrac{1}{G_N}} \tag{1.11}$$

For N resistances in parallel the total conductance is

$$G_{\text{total}} = G_1 + G_2 + \cdots + G_N \tag{1.12}$$

Variable resistors, often called *potentiometers* or "pots," are also available. They come in many sizes and styles, and are usually adjusted by manually turning a shaft, as shown in Fig. 1.9. They have three terminals: one at each end of the resistor, and one for the variable position of the tap. The total resistance R_T between the two end terminals A and B is always constant and equals the resistance value of the pot. The resistance R_1 between A and the tap and the resistance R_2 between B and the tap vary as the shaft is turned. Notice that if the shaft is turned fully clockwise, the tap

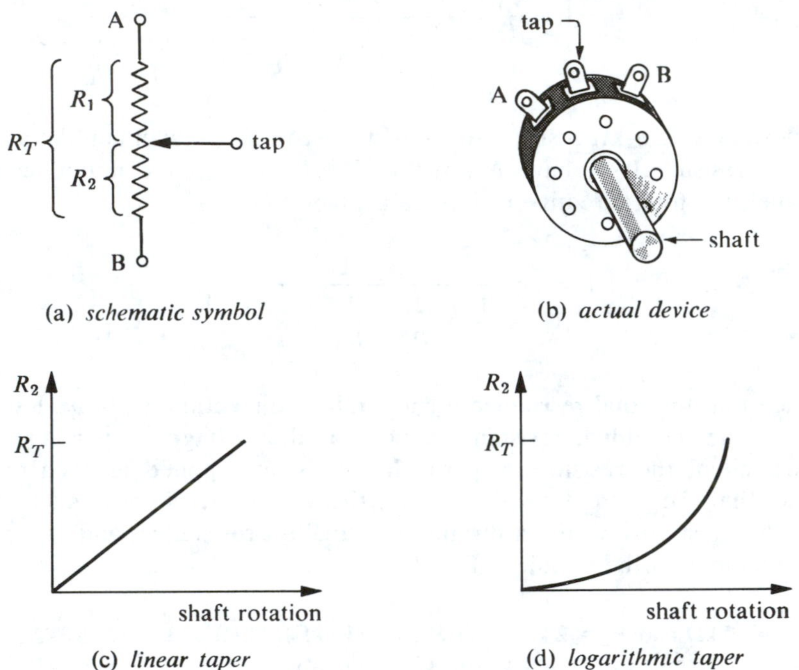

FIGURE 1.9 Variable resistor.

is electrically connected to terminal A and $R_2 = R_T$, $R_1 = 0$. The variation may be *linear taper*, with shaft rotation as shown in Fig. 1.9(c), or *logarithmic taper* as in Fig. 1.9(d). For example, a 100-kΩ pot has $R_T = R_1 + R_2 = 100$ kΩ regardless of the shaft rotation, but R_1 and R_2 depend on the shaft position—always subject to the condition $R_1 + R_2 = 100$ kΩ. A logarithmic taper is usually used in volume controls for audio equipment; a linear taper is more commonly used in scientific apparatus.

It is worthwhile to note that Ohm's law must apply to a resistance even though it is connected with another circuit element that does not obey Ohm's law. For example, a *zener diode* is a solid-state device with the nonlinear property that the voltage drop across its two terminals is essentially constant, regardless of the current through it, over a wide range of currents. Thus Ohm's law does not apply to a zener diode: in other words, the resistance of a zener diode varies with the current through the diode. Consider a resistor and a zener diode connected in series with a battery as shown in Fig. 1.10. The current through R and the zener diode must be

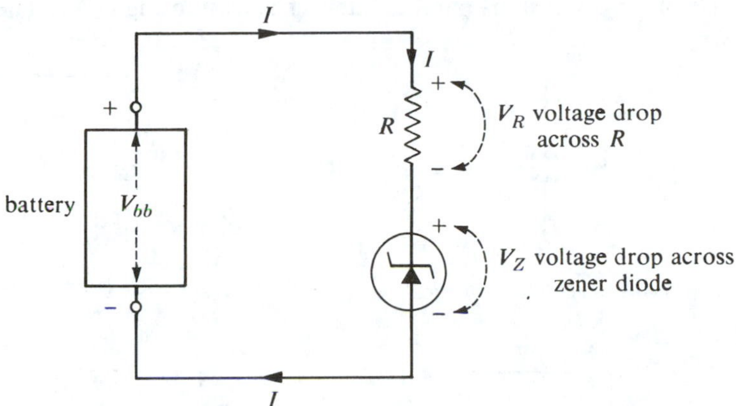

FIGURE 1.10 Zener diode circuit.

equal because they are connected in series, and the voltage of the battery, V_{bb}, must equal the voltage across R, V_R, plus the voltage across the zener, V_Z: $V_{bb} = V_R + V_Z$. Ohm's law applied to the resistor alone implies $V_R = IR$. Thus $V_{bb} = IR + V_Z$. If V_{bb} decreases, the constant zener voltage V_Z implies that I must decrease, so the IR drop across R always equals the difference between V_{bb} and V_Z.

$$IR = V_{bb} - V_Z \tag{1.13}$$

This circuit is often used to produce a constant output voltage across the zener diode. If $V_{bb} = 12\text{ V}$ and $V_Z = 6.8\text{ V}$, then $IR = 12\text{ V} - 6.8\text{ V} = 5.2\text{ V}$. If $I = 10\text{ mA}$, then $R = 5.2\text{ V}/10\text{ mA} = 520\ \Omega$ by Ohm's law. For $R = 520\ \Omega$, if V_{bb} falls to 10 V, then I must decrease from 10 mA to $(10\text{ V} - 6.8\text{ V})/520 = 3.2\text{ V}/520 = 6.16\text{ mA}$.

1.5 BATTERIES

A battery is a two-terminal device in which chemical energy is converted into electrical energy, and a voltage difference is generated between the two battery terminals. A battery tends to spew positive charge out of the positive terminal and draw it into the negative terminal. A battery also tends to spew negative charge out of the negative terminal and draw it into the positive terminal. In an actual battery negatively charged electrons are spewed out the negative terminal and are drawn in the positive terminal, even though we may, for convenience, speak of positive charge flowing. A *dry* battery has chemicals that are essentially dry or in a paste form (e.g., a flashlight battery). A *wet* battery contains liquids (e.g., an automobile battery that contains sulfuric acid solution). The circuit symbol for a battery is a series of long and short parallel lines, as shown in Fig. 1.11: The longer

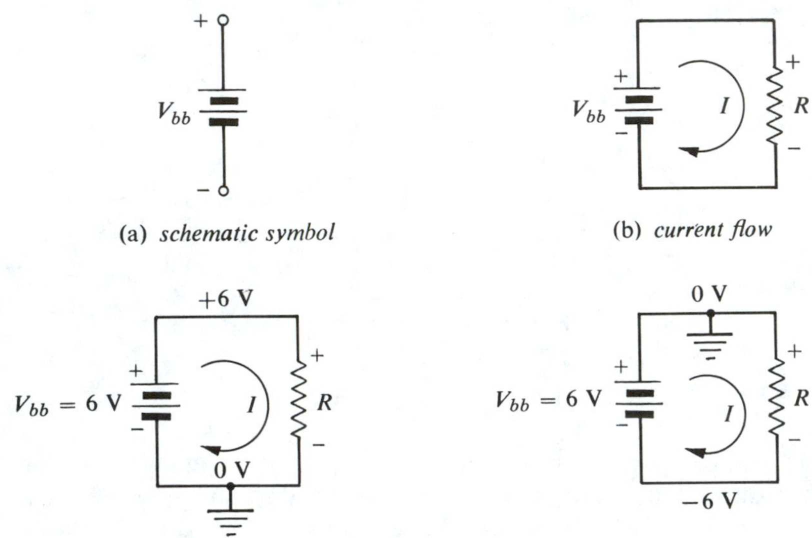

FIGURE 1.11 Batteries.

line represents the positive terminal; the shorter line, the negative terminal. Thus, if a resistor R is connected between the two battery terminals as shown in Fig. 1.11(b), then positive current will flow out of the battery "+" terminal, through the resistor, and back into the battery "−" terminal. Notice that the (positive) current flows *out* of the battery + terminal and *into* the + end of the resistor; this means the battery is *discharging*. If (positive) current flows into the battery + terminal, that battery is being *charged* by some other battery with a higher voltage.

Notice again that the straight lines drawn in the circuit diagram represent wires with *zero resistance*. Thus, there is no change in voltage along the wires represented by straight lines. The voltage at the positive battery terminal is exactly the same as the voltage at the top of the resistance in Fig. 1.11(b). To have a voltage difference between two points in a circuit, there must be some resistance between these two points (from Ohm's law, $V_2 - V_1 = IR$); that is, if $R = 0$, $V_2 - V_1 = 0$ even if $I \neq 0$. Notice also that the + and − signs represent the *relative* polarity of the voltages; that is, the top of the resistor is positive with respect to the bottom. Also notice that we may ground any *one* point in a circuit. Two such cases are shown in Fig. 1.11(c). In either case the voltage difference between the battery terminals is 6 V, and the current is the same. Notice that no current flows into the ground connection; the current flowing out of the positive terminal of the battery flows back into the negative terminal. The ground connection merely sets the zero voltage level.

A battery is rated in volts; its voltage rating V_{bb} indicates the difference in voltage that the battery will maintain between its two terminals. A perfect or ideal battery will always maintain the same voltage difference between its terminals regardless of how much current I it supplies to the rest of the circuit. However, the voltage of any real battery will decrease as more and more current is drawn from it. Thus, for a real battery, V_{bb} represents the terminal voltage when no current is drawn from the battery. V_{bb} is often called the *open-circuit* voltage. In general, the larger the battery's physical size (for the same voltage rating) the less the voltage will decrease as more current is drawn; in other words, a large battery can supply more current than a small battery at the same voltage. This behavior can be explained by the fact that a real battery has an internal resistance r, as shown in Fig. 1.12(a). The actual current must flow through both r and R, which are in series, so $I = V_{bb}/(r + R)$. Notice that the battery terminal voltage $V_A - V_B$ must equal $V_{bb} - Ir$ because of the polarity difference between the Ir voltage drop across the internal resistance r and V_{bb}.

The larger the battery for a given voltage the smaller r is ($r = 0$ for an ideal battery). The internal resistance of a battery can be determined by measuring the terminal voltage for various measured currents I drawn from the battery. The internal resistance r is then the negative slope of the graph of the terminal voltage plotted versus I, as shown in Fig. 1.12(b). Or, if the terminal voltage is measured for two currents, then r is given by [see Fig. 1.12(b)] $r = (V_1 - V_2)/(I_2 - I_1)$.

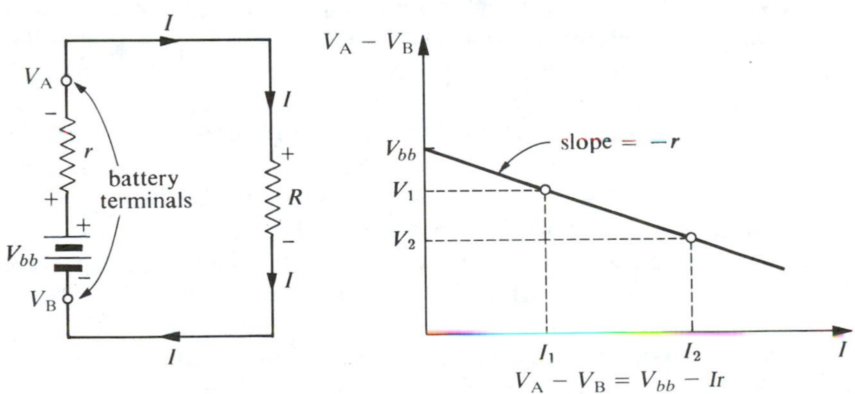

FIGURE 1.12 Internal resistance r of battery.

When a battery goes "bad," its internal resistance r increases sharply. The typical good 12-V automobile battery has an internal resistance of about 0.03 Ω. A size D 1.5-V carbon–zinc dry cell, such as is used in flashlights and in some portable transistor circuits, has $r = 0.5\ \Omega$. A 1″ × ½″ × ¼″ 9-V battery, often used to power portable transistor radios, has $r \cong 13\ \Omega$. Usually the internal resistance r is omitted from circuit diagrams, but this omission is valid only when r is much less than any other series resistance in the circuit.

In scientific circuitry a battery that very gradually goes bad (that is, whose voltage slowly decreases) is a real disadvantage, because the circuit behavior may become erratic and difficult to diagnose as the battery slowly wears out. However, with a mercury battery, which goes bad very abruptly (after a long life), the battery voltage is either all right or extremely low, in which case the circuit will usually not function at all. Thus, in most scientific instruments mercury batteries are used, particularly if proper circuit behavior depends strongly upon a certain minimum battery voltage. The nickel–cadmium battery also goes bad abruptly after a long life, and it has the additional advantage of being rechargeable. It is, however, more expensive than the mercury battery. For a brief summary of the six different types of batteries, see Appendix B.

1.6 POWER

Power is defined as the time rate of doing work or the time rate of expending energy; that is, $P \equiv dW/dt$, where W is work or energy and t is time. The units of power are thus joules per second (J/s); 1 *watt* (W) ≡ 1 J/s. We will now show that a dc current I flowing through a resistor R develops a power of I^2R or VI or V^2/R, where V is the voltage drop across the resistor.

Recalling that voltage is electrical potential energy per unit charge, we see that a charge has less electrical potential energy when it leaves a resistor than it has when it enters because of the decrease in voltage, or voltage "drop" across the resistor. The time rate at which the flowing charge gives up electrical potential energy is thus the amount of charge flowing per second multiplied by the energy lost per unit charge, which is exactly equal to the current I multiplied by the voltage drop V. Thus, $P = IV$. But, from Ohm's law we know that $V = IR$; thus the power can also be expressed as $P = I^2R$ (see Fig. 1.13). And, again from Ohm's law, $I = V/R$; thus another way of expressing the power is $P = V^2/R$. These three expressions for the power are equivalent and apply only to direct current flowing through a resistor. For alternating current the phase angle between the current and the voltage must be taken into account (more about this in Chapter 2).

The power developed in a resistor shows up as heating of the resistor.

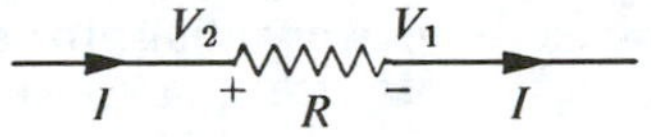

$$V \equiv V_2 - V_1$$

$$P = IV = I(IR) = I^2 R$$

$$P = IV = \left(\frac{V}{R}\right) V = \frac{V^2}{R}$$

FIGURE 1.13 Power dissipated in a resistor.

In other words, the loss of electrical potential energy (due to the *IR* voltage drop) of the charge flowing through the resistor is converted into random thermal motion of the molecules in the resistor. The kinetic energy of the flowing charges remains approximately constant everywhere in the circuit. Electrical potential energy is converted into heat energy in any dc circuit element across which there is a voltage drop and through which current flows.

If too many watts of power are converted into heat in a resistor or in any circuit element, the resistor may "burn up," in which case the resistor turns brown or black and may actually fragment, thus breaking the electrical circuit. In other words, the resistor is *open* or has an infinite resistance. Or, if the resistor is heated too much, its resistance value may increase tremendously, thus changing the operation of the circuit drastically. For these reasons resistors are rated by the manufacturer according to how much power they can safely dissipate without being damaged. Resistors are commonly available with wattage ratings of $\frac{1}{8}$ W, $\frac{1}{4}$ W (for low-power transistor circuits), $\frac{1}{2}$ W, 1 W, 2 W, 5 W, 10 W, 20 W, 50 W, 100 W, and 200 W. The larger the resistor is physically, the more power it can safely dissipate as heat. The actual sizes of several commonly used resistors rated for various power dissipations are shown in Fig. 1.14.

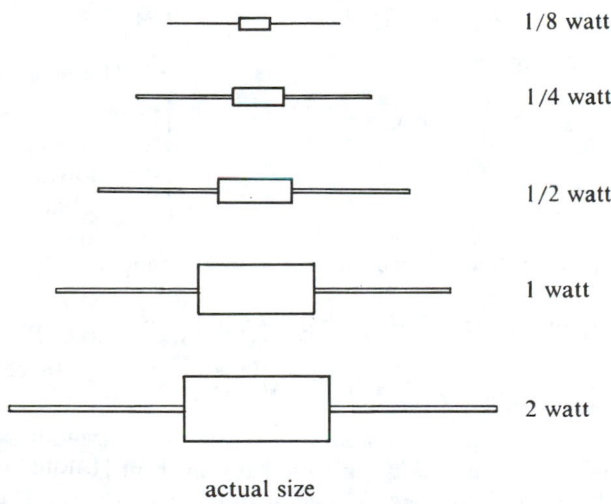

FIGURE 1.14 Resistor sizes for different power ratings.

We emphasize that the resistance does not depend upon the actual physical size. The physical size determines the power rating; for example, a $\frac{1}{2}$-W, 2.2-kΩ resistor is the same size as a $\frac{1}{2}$-W, 470-kΩ resistor. Circuit designers usually choose a power rating of at least three or four times the expected power. For example, if a 1.5-kΩ resistor is to carry 20 mA of direct current, then the power dissipated as heat in the resistor will be

$$P = I^2R = (20 \times 10^{-3}\ \text{A})^2(1.5 \times 10^3\ \Omega) = 0.6\ \text{W}$$

In the actual circuit a 1.5-kΩ, 2-W resistor would be used, or perhaps even a 1.5-kΩ, 5-W resistor if the circuit were very sensitive to heat. If a large wattage is developed in a certain part of a circuit, care should be taken to provide an adequate vertical flow path for air around the hot element so that the heat can be carried away by the resulting convection air currents. A resistor dissipating a large amount of power should never be placed in a closed chassis. Heat is an enemy of transistors as well as other circuit elements.

Notice also that two $\frac{1}{2}$-W, 1-kΩ resistors in parallel are equivalent to one 1-W, 500-Ω resistor, and two $\frac{1}{2}$-W, 1-kΩ resistors in series are equivalent to one 1-W, 2-kΩ resistor.

We will now derive an important theorem about power. If a battery has a certain fixed internal resistance r and we connect a load resistor R_L across its terminals, as shown in Fig. 1.15, how do we maximize the power

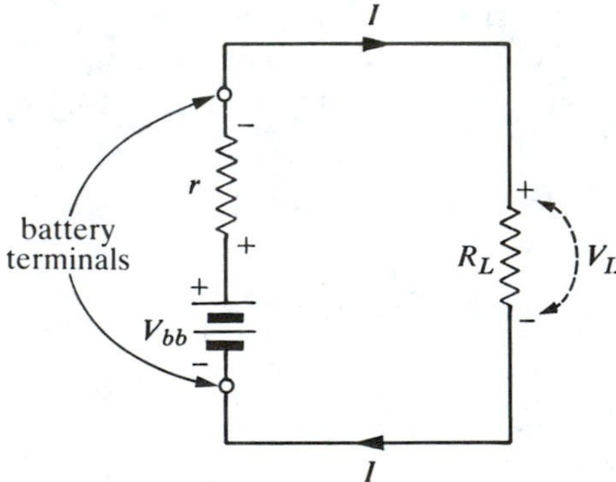

FIGURE 1.15 Circuit illustrating power transfer from source to load R_L.

$P_L = IV_L$ dissipated in R_L? If R_L is very small, I is large; but

$$V_L = IR_L$$

is small, so the power in R_L is small. If R_L is very large, the voltage V_L across R_L is nearly equal to V_{bb}; but then the current is small, so again the power in R_L is small. It seems reasonable that for some intermediate value of R_L the power dissipated in R_L is maximized. Let us set up an expression for P_L as a function of R_L and maximize it.

$$P_L = IV_L \tag{1.14}$$

But $$I = \frac{V_{bb}}{(r + R_L)} \quad \text{and} \quad V_L = IR_L = \left(\frac{V_{bb}}{r + R_L}\right) R_L$$

Therefore,

$$P_L = \frac{V_{bb}}{(r + R_L)} \cdot \frac{V_{bb}}{(r + R_L)} \cdot R_L = \left(\frac{V_{bb}}{r + R_L}\right)^2 R_L \tag{1.15}$$

$$\frac{\partial P_L}{\partial R_L} = V_{bb}^2 \frac{(r + R_L)^2 - 2R_L(r + R_L)}{(r + R_L)^4}$$

Set $\partial P_L / \partial R_L = 0$ to find the value of R_L for which P_L is an extremum.

$$\frac{\partial P_L}{\partial R_L} = 0 = \frac{V_{bb}^2}{(r + R_L)^4} [(r + R_L)^2 - 2R_L(r + R_L)]$$

$$(r + R_L)^2 = 2R_L(r + R_L)$$

$$R_L = r \tag{1.16}$$

It can be shown that P_L is a maximum when $R_L = r$ by showing that $\partial^2 P_L / \partial R_L^2$ is negative at $R_L = r$. In words, the maximum power is dissipated in the load resistance R_L when it equals the internal resistance of the battery. Under this condition the voltage V_L across the load equals one half of V_{bb}, which is the open-circuit battery voltage. One half of the total power is also dissipated in r and one half in R_L.

However, suppose we have a *fixed load resistance* R_L, V_{bb}, and variable r, and we ask the question: "How can we maximize the power dissipated in R_L?" The answer is *not* $r = R_L$, but $r = 0$! This answer is really obvious when we realize that any power dissipated in r is wasted as far as the load is concerned. It also follows mathematically from maximizing $P_L = V_{bb}^2 R_L / (r + R_L)^2$ with respect to r.

To sum up, if the load R_L is fixed, the smaller the internal resistance r the better. If we are presented with a fixed internal resistance r, then the load R_L gets maximum power when $R_L = r$. Also note that when r is fixed, the load *voltage* is very large when R_L is very large, and the load *current* is very large when R_L is very small.

1.7 TEMPERATURE VARIATIONS OF RESISTIVITY AND RESISTANCE

For most metallic conductors, the resistivity increases slowly with increasing temperature. At higher temperatures the thermal motion of the atoms increases; hence the average distance moved by a free electron between collisions decreases, producing a slightly lower drift speed. The resistance of a metallic conductor depends upon temperature according to the relation

$$R_T = R_0(1 + \alpha \Delta T) \tag{1.17}$$

where R_0 is the resistance in ohms at some reference temperature T_0, $\Delta T = (T - T_0)$ is the temperature rise in °C, and α is the linear temperature coefficient of resistance in $(°\text{C})^{-1}$. α is typically $0.005(°\text{C})^{-1}$ for most metals or about 0.5%/°C. For example, if a piece of copper has a resistance of 100 Ω at 10°C, at 300°C, it will have a resistance of

$$R = R_T = (100\ \Omega)[1 + (0.0039/°\text{C})(300°\text{C} - 10°\text{C})] = 213\ \Omega$$

Values of α for various materials are given in Table 1.4.

TABLE 1.4. Linear Temperature Coefficients of Various Materials

Material	$\alpha(°C)^{-1}$
Manganin alloy	+0.00001
Carbon	−0.0005
Iron	+0.005
Copper	+0.0039
Silver	+0.0038
Aluminum	+0.0039
Platinum	+0.0039
Tungsten	+0.0045
Iron	+0.0050
Nickel	+0.0067

The resistance of pure platinum is often used as a standard of temperature over a wide temperature range, from −190°C to over 600°C; the device is called a *resistance thermometer*. A constant known current is forced through a pure piece of platinum wire, and the voltage across the platinum wire is measured, as shown in Fig. 1.16. This voltage drop is linearly proportional to the resistance of the platinum wire, which in turn depends upon the temperature, according to equation (1.17). Notice that the voltmeter must draw zero current. Usually a potentiometer is used (this will be described later).

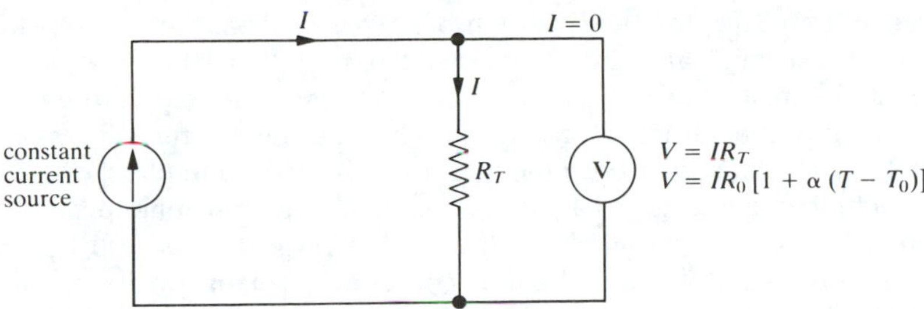

FIGURE 1.16 Resistance thermometer.

A *thermistor* is a special two-terminal device designed to have a resistance that is a strong function of temperature. The resistance, R_T, of a thermistor is given by

$$R_T = R_0 e^{A(1/T - 1/T_0)} \tag{1.18}$$

where A is a constant in kelvins (K) whose value depends upon the particular thermistor, R_0 is the resistance at temperature T_0 (K), T = the temperature of the thermistor (K), and e = 2.718 (the basis of the natural logarithms). The constant A depends slightly on temperature. Using (1.18) over a 50°C range with A constant would typically produce a ±3°C error. For greater accuracy the thermistor should be calibrated experimentally over the temperature range expected.

The resistance change for a thermistor is typically ten times the resistance change for copper for the same temperature change. The thermistor resistance *decreases* with increasing temperature, which is opposite to the temperature dependence for most metals.

Commercial thermistors are usually made from sintered mixtures of Mn_2O_3 and NiO or platinum alloys and are often encapsulated in a thin glass bead with two wire leads, as shown in Fig. 1.17. They are available in a wide range of resistance values. The thermistor resistance at 25°C (R_{25})

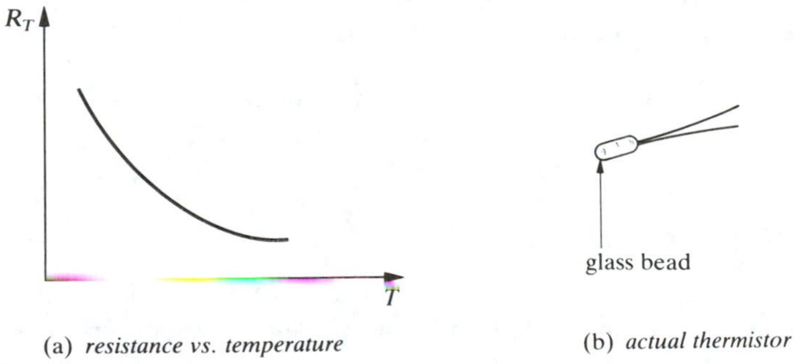

FIGURE 1.17 Thermistor.

can range from 30 Ω to 20 MΩ for various types, and the ratio of resistance at 25°C to resistance at 125°C may range from 10:1 to 100:1.

The thermistor takes a certain time to come to equilibrium if its surrounding temperature is changed. The thermistor time constant is defined as the time required for the thermistor resistance to change by 63%, with 100% being the total change in resistance for an infinite time. For example, consider a thermistor with a 10-kΩ resistance at 100°C, and a 110-kΩ resistance at 30°C, and a time constant of 100 ms. If the thermistor is initially in equilibrium at 100°C and is suddenly immersed in a 30°C environment, then its resistance will increase to

$$\begin{aligned} 110\,\text{k}\Omega + 0.63(110\,\text{k}\Omega - 10\,\text{k}\Omega) &= 10\,\text{k}\Omega + 0.63(100\,\text{k}\Omega) \\ &= 10\,\text{k}\Omega + 63\,\text{k}\Omega = 73\,\text{k}\Omega \end{aligned}$$

in the first 100 ms. Bare thermistors with no glass covering are the fastest; they are available with time constants as short as 4 ms in water and 100 ms in air. Larger thermistors encapsulated in glass would, of course, have longer time constants, perhaps 200 ms in air and 5 s in water.

Finally, it should be pointed out that the self-heating of the thermistor due to the current I flowing through it should be as small as possible. The manufacturer will specify a self-heating or dissipation constant, which gives the maximum self-heating ($I^2 R_T$) power per degree Celsius error produced. This might be 0.5 mW/°C in air and 2.5 mW/°C in water for a typical thermistor. For such a thermistor in water, the internal self-heating should be kept much less than 2.5 mW for the temperature error to be much less than 1°C.

1.8 KIRCHHOFF'S LAWS AND NETWORK ANALYSIS

The two basic laws of electricity that are most useful in analyzing circuits are Kirchhoff's current law and voltage law:

Kirchhoff's Current Law (KCL)

At any junction of wires in a circuit, the sum of all the currents entering the junction exactly equals the sum of all the currents leaving the junction. In other words, electric charge is conserved.

Kirchhoff's Voltage Law (KVL)

Around any closed loop or path in a circuit, the algebraic sum of all the voltage drops must equal zero. In other words, energy is conserved.

The current law merely says that no electric charge is being created or destroyed at the junction in question; that is, the total current entering equals the total current leaving. The voltage law says that there is no net gain or loss in electrical potential energy for any charge making a trip around any closed loop; that is, the energy the charge gains (in passing through a battery) must be all lost (as heat, radiation, etc.) in the rest of the loop. (If a changing magnetic field is present, then an induced "emf" must be placed in series with the loop just as if a battery were actually present.)

To solve for the currents and voltages in a circuit or network, using Kirchhoff's laws, we first must assume a current direction in each branch of the circuit and define a current symbol such as I_1 for that current. We use arrows to draw the currents, and we label the polarities of the voltage drops (+ and −) across the resistances, remembering that positive (conventional) current *always* flows into the "plus end" and out of the "minus end" of a resistance. It is useful to draw the current arrows in the direction the current actually flows in the circuit; if this is done, the numerical value obtained for the current at the end of the calculation will turn out to be positive. However, if we guessed wrong as to the current direction, the numerical value obtained for the current will be negative but of the same magnitude as if we had guessed the current direction correctly. In other words, a negative current at the end of a calculation is merely a "flag" that we guessed wrong when we drew the current arrow on the circuit diagram. It should also be emphasized that once the current directions are chosen (or guessed), the polarities of the voltage drops are fixed.

There are basically three methods for calculating the currents and voltages in a circuit, that is, *network analysis*. Only experience will enable you to choose the easiest method. In the following paragraphs, a *branch* is simply one path or wire through which one current can flow, a *loop* is a closed path in the circuit (an electron can flow around any loop and return to its starting point without leaving the actual circuit), and a *junction* or *node* is a point in a circuit where three or more wires come together (e.g., joined by solder). In all three methods we must have a clear circuit diagram with all the voltages and currents clearly defined. It is also useful in all three methods to look for simple series and parallel combinations of resistances and to replace them immediately by their equivalent resistances. We must also remember that *no* current flows in the ground connection.

The Branch Method

In this method we draw the current in each branch of the circuit, and we label the voltages at all batteries. Then we can write a KCL equation at each junction and a KVL equation for each closed loop in the circuit. However, a little thought will show that such a procedure applied to each junction and each closed loop will produce a number of nonindependent equations; that is, we might obtain five equations in three unknowns. Obviously, we wish to obtain n independent equations to solve for the n

unknown currents. It can be shown that if there are k junctions in the circuit, there are only $k - 1$ independent KCL equations. For example, if there are four junctions, there are three independent KCL equations.

It can also be shown that there are only $n - (k - 1)$ independent KVL equations. Thus, the total number of independent equations (KCL and KVL) is

$$(k - 1) + n - (k - 1) = n$$

We can solve these n equations for the n unknown currents. In writing the KVL loop equations, we must cover each branch in the circuit at least once.

Let us solve for the currents and voltages in the circuit of Fig. 1.18.

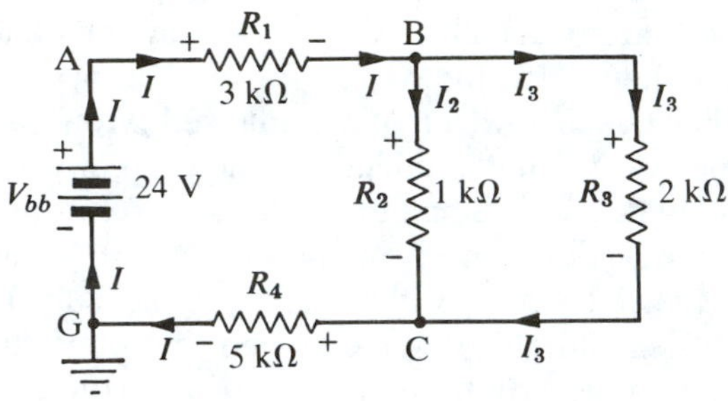

FIGURE 1.18 Circuit problem.

This problem can, of course, be solved by simply combining resistances in series and in parallel. R_2 and R_3 are in parallel, and their combined parallel resistance is 0.667 kΩ. R_1, $(R_2 \| R_3)$, and R_4 are all in series, so

$$I = \frac{V_{bb}}{R_1 + (R_2 \| R_3) + R_4} = \frac{24\ \text{V}}{8.667\ \text{k}\Omega} = 2.77\ \text{mA}$$

We can then find I_2, I_3, and the voltages by Ohm's law.

Let us illustrate the branch method by solving for the currents and voltages. First, we use arrows to draw the currents in all parts of the circuit; then we label the polarities of the voltage drops. For this circuit there are two junctions, B and C, so $k = 2$. Thus, there is $k - 1 = 1$ independent KCL equation. There are three unknown currents ($n = 3$) I, I_2, and I_3, so there are $n - (k - 1) = 3 - 1 = 2$ independent KVL equations.

The KCL says that the *total* current entering a junction equals the *total* current leaving the junction. So $I = I_2 + I_3$ at junction B and $I_2 + I_3 = I$ at junction C, which is, of course, the same equation. So we have one KCL equation:

$$I = I_2 + I_3 \tag{1.19}$$

The KVL says that starting at G, which is ground (0 V), and going around the circuit clockwise, we obtain

$$+V_{bb} - IR_1 - I_2R_2 - IR_4 = 0 \tag{1.20}$$

or

$$+V_{bb} - IR_1 - I_3R_3 - IR_4 = 0 \tag{1.21}$$

where (1.20) has been obtained by going through R_2 in our loop, and (1.21) has been obtained by going through R_3. Substituting (1.19) for I in (1.20) and in (1.21) immediately gives us two equations in the two unknowns I_2 and I_3.

$$V_{bb} - (I_2 + I_3)R_1 - I_2R_2 - (I_2 + I_3)R_4 = 0$$

$$V_{bb} - (I_2 + I_3)R_1 - I_3R_3 - (I_2 + I_3)R_4 = 0$$

or

$$V_{bb} = (R_1 + R_2 + R_4)I_2 + (R_1 + R_4)I_3 \tag{1.22}$$

$$V_{bb} = (R_1 + R_4)I_2 + (R_1 + R_3 + R_4)I_3 \tag{1.23}$$

With the resistances in kΩ and the currents in mA, we have

$$24 = 9I_2 + 8I_3 \tag{1.24}$$

$$24 = 8I_2 + 10I_3 \tag{1.25}$$

Solving equations (1.24) and (1.25) simultaneously for I_2 and I_3 yields

$$I_2 = 1.85\text{ mA} \quad \text{and} \quad I_3 = 0.92\text{ mA}$$

From (1.19)

$$I = I_2 + I_3 = 2.77\text{ mA}$$

which is the same answer we obtained from the simple series and parallel resistance analysis. However, we emphasize that the branch method and Kirchhoff's laws will always yield a solution, while the series and parallel analysis will work only for relatively simple circuits.

We can calculate the voltages at points B and C by using Ohm's law and the values for I_2 and I_3.

$$V_C = +IR_4 = (2.77\text{ mA})(5\text{ k}\Omega) = 13.85\text{ V}$$

$$V_B = V_{bb} - IR_1 = 24\text{ V} - (2.77\text{ mA})(3.0\text{ k}\Omega) = 15.7\text{ V}$$

Or,

$$V_B = +IR_4 + I_2R_2 = (2.77\text{ mA})(5\text{ k}\Omega) + (1.85\text{ mA})(1\text{ k}\Omega) = 15.7\text{ V}$$

These voltages, of course, are all relative to ground (point G), which by definition is 0 V.

Another approach would be to write down the KVL equations using the minimum number (two) of unknown currents. In the example of Fig. 1.18, going around the complete loop through the battery, R_1, R_3, and R_4, we would write

$$V_{bb} - IR_1 - (I - I_2)R_3 - IR_4 = 0 \qquad \textbf{(1.26)}$$

and

$$I_2R_2 - (I - I_2)R_3 = 0 \quad \text{(using } I_3 = I - I_2\text{)} \qquad \textbf{(1.27)}$$

from going around the small loop containing R_2 and R_3. We now have two equations that can be solved for the two unknown currents, I and I_2. There are many different ways to solve such problems, but they all eventually lead to the same solution. With a little experience, you will learn how to choose the most effective technique for a particular problem.

The Loop Current Method

In this method one loop current is drawn for each closed loop of the circuit, as shown in Fig. 1.19. It is convenient to draw all the loop currents going clockwise. Notice that a loop current, by definition, always flows *through* a

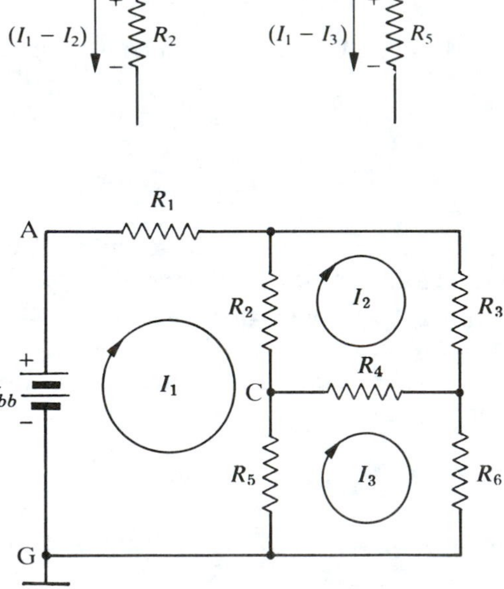

FIGURE 1.19 Circuit problem using loop currents.

junction, so the KCL is automatically satisfied. Thus we need worry only about writing the KVL equations, and we need write only one KVL for each loop. We will have an equal number of KVL equations and loop currents, so we can solve for the loop currents. In writing the KVL equations, notice that we must always use the *total* current flowing through a resistance to calculate the voltage drop across that resistance.

For the circuit of Fig. 1.19 the three KVL equations are as follows: For the I_1 loop,

$$V_{bb} - I_1R_1 - (I_1 - I_2)R_2 - (I_1 - I_3)R_5 = 0 \tag{1.28}$$

using the sign conventions shown in Fig. 1.19. For the I_2 loop, starting at point C and going clockwise,

$$(I_1 - I_2)R_2 - I_2R_3 - (I_2 - I_3)R_4 = 0 \tag{1.29}$$

For the I_3 loop, starting at point G and going clockwise,

$$(I_1 - I_3)R_5 - (I_3 - I_2)R_4 - I_3R_6 = 0 \tag{1.30}$$

Rewriting and grouping similar terms in each equation gives

$$(R_1 + R_2 + R_5)I_1 - R_2I_2 - R_5I_3 = V_{bb} \tag{1.28}$$

$$R_2I_1 - (R_2 + R_3 + R_4)I_2 + R_4I_3 = 0 \tag{1.29}$$

$$R_5I_1 + R_4I_2 - (R_4 + R_5 + R_6)I_3 = 0 \tag{1.30}$$

These three equations can now be solved for the three loop currents I_1, I_2, and I_3 either by various adroit algebraic substitutions or by the method of determinants, which will *always* yield the answer.

The Nodal Method

In this method the *voltages* are the unknowns, and the KCL is written for each junction in terms of voltages and resistances. Hence, the number of equations equals the number of independent junctions $(k - 1)$, where k is the total number of junctions. For the circuit of Fig. 1.20 there are three junctions $(k = 3)$: A, B, and C. We therefore obtain $k - 1$, or two, independent KCL equations in terms of the voltages. Let V_A, V_B, and V_C be the voltages at junctions A, B, and C, respectively. The KCL at A is

$$I_1 = I_2 + I_3$$

or, in terms of voltages,

$$\frac{V_1 - V_A}{R_1} = \frac{V_A}{R_2} + \frac{V_A - V_B}{R_3} \tag{1.31}$$

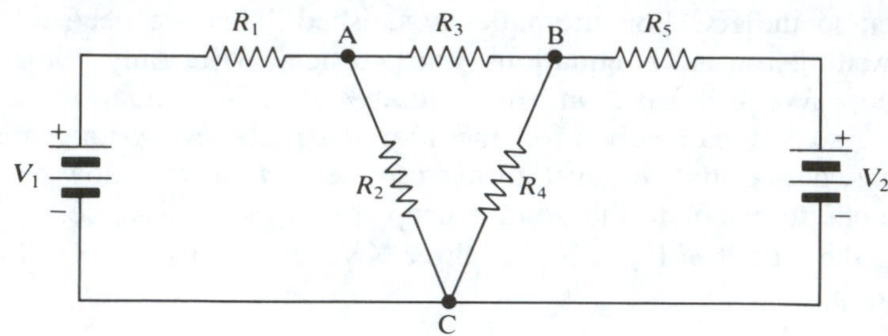

redrawn to emphasize three junctions A,B,C

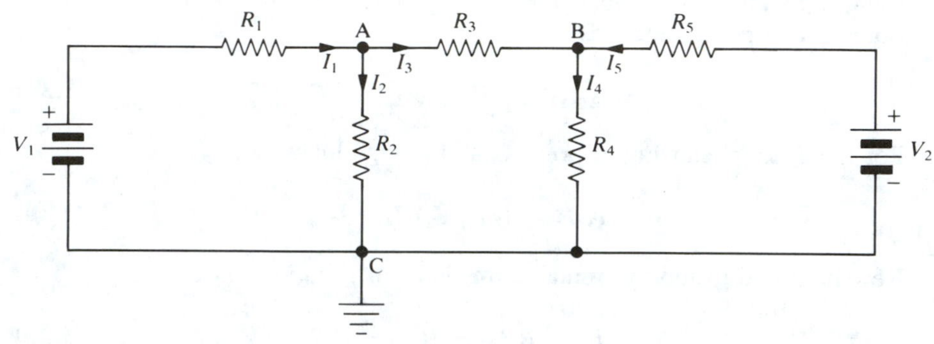

original drawing

FIGURE 1.20 Circuit problem.

The KCL at B is

$$I_3 + I_5 = I_4$$

or

$$\frac{V_A - V_B}{R_3} + \frac{V_B - V_2}{R_5} = \frac{V_B}{R_4} \qquad \textbf{(1.32)}$$

Notice that V_C does not appear in either (1.31) or (1.32) because point C is ground. (We may always eliminate any one voltage in this method by simply grounding that point.)

If we know the battery voltages V_1 and V_2 and the resistances, we can solve (1.31) and (1.32) for voltages V_A and V_B. Then we can solve for all the currents by using Ohm's law. The algebra is usually simplified by using the conductance G instead of the resistance $G_1 = 1/R_1$, etc. The nodal method is useful when the total number of junctions is less than or equal to the number of loops. In Fig. 1.20, for example, there are three loops and three junctions. The nodal method gives us *two* equations to solve, whereas the loop current method would have given us *three* loop currents and *three* KVL equations to solve.

We now provide a simple example which is very useful when designing transistor amplifiers. For this example we consider the transistor as a "black box" or circuit element with three terminals, labeled C (for collector), E (for emitter), and B (for base). Suppose that terminals C and E, a resistor R_L, and a battery V_{bb} are connected in series, as shown in Fig. 1.21.

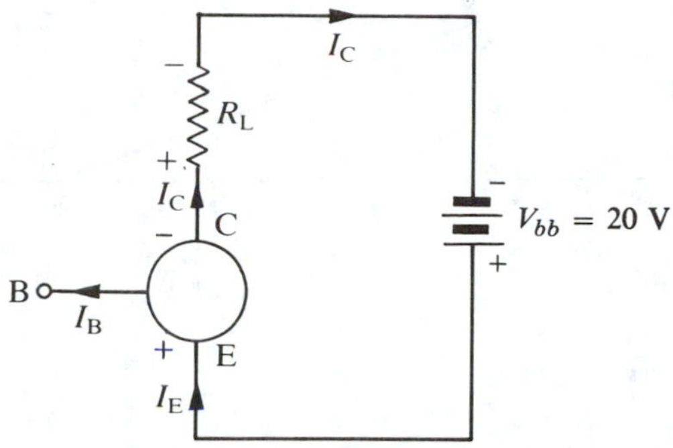

FIGURE 1.21 "Black box" transistor circuit.

If we call the transistor voltage between C and E V_{CE} (V_{CE} is negative if C is negative with respect to E), then Kirchhoff's voltage law for the loop is

$$-V_{CE} - I_C R_L + V_{bb} = 0 \tag{1.33}$$

From Kirchhoff's current law, regarding the transistor as a junction, we see that

$$I_E = I_B + I_C$$

because the transistor can create no current. As we shall see later, the base current I_B is usually about 50 or 100 times smaller than I_C, so $I_E \cong I_C$ is a good approximation. If we solve for I_C, we obtain

$$I_C = \frac{V_{bb}}{R_L} - \frac{1}{R_L} V_{CE} \tag{1.34}$$

which is of the form $y = b + mx$, where m = slope and b = y intercept. This is called the *load-line* equation and is usually graphed with V_{CE} as the independent variable and I_C as the dependent variable (see Fig. 1.22). Notice that the graph is a straight line with slope $-1/R_L$, a vertical (current) intercept of V_{bb}/R_L, and a horizontal (voltage) intercept of V_{bb}. That is, the larger R_L, the flatter the line, and the more V_{CE} changes for a given change in I_C. Also notice that the V_{bb} intercept represents the condition when there is no current flowing through the transistor and the entire battery voltage appears across the transistor. In other words, the

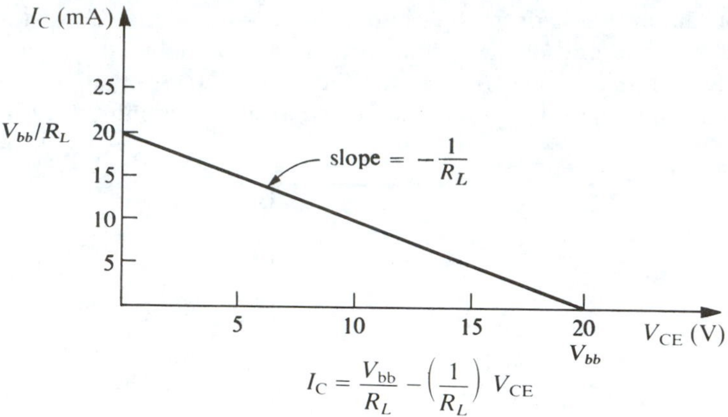

FIGURE 1.22 Transistor load line

transistor has an infinite effective resistance or is *cut off*. The V_{bb}/R_L intercept, on the other hand, represents the condition when the transistor has an effective resistance of 0 Ω or is fully conducting—*saturated* or *turned on*. In this condition the only thing limiting the current flow is the resistance R_L. This argument has neglected any resistance presented to the current flow by the transistor itself; in a real transistor V_{CE} never falls quite to zero when it is turned on.

This graph is usually called the *load line* and is completely determined by V_{bb} and R_L. Once the load line is determined, if we know I_C, we can immediately find the voltage across the transistor, V_{CE}. For example, refer to Fig. 1.22. If $I = 15$ mA, then $V_{CE} = 5$ V, and the voltage drop across R_L must be $V_{bb} - V_{CE} = 20\text{ V} - 5\text{ V} = 15$ V. The important thing to remember here is that regardless of what the transistor is doing, its collector-emitter voltage V_{CE} and its current I_c must *always* lie on the load line because of Kirchhoff's voltage law and Ohm's law.

1.9 VOLTAGE DIVIDERS

We often wish to reduce a battery or power supply voltage for a smaller value. For example, in Fig.1.18 suppose that a −4-V voltage is needed for some other part of the circuit. We have to devise some voltage divider circuit to get −4 V from the 20-V battery. One solution is shown in Fig. 1.23(a). If no current is drawn from terminal A, then the voltage at A is −4 V, the total current I drawn from the battery by R_1 and R_2 is $V_{bb}/(R_1 + R_2) = 20\text{ V}/(16\text{ k}\Omega + 4\text{ k}\Omega) = 1$ mA; thus, the voltage at A (V_A) is 1 mA × 4 kΩ = 4 V. Terminal A is negative with respect to G because the current flows from G toward A. In general, $V_A = V_{bb} \times R_2/(R_1 + R_2)$ if no current

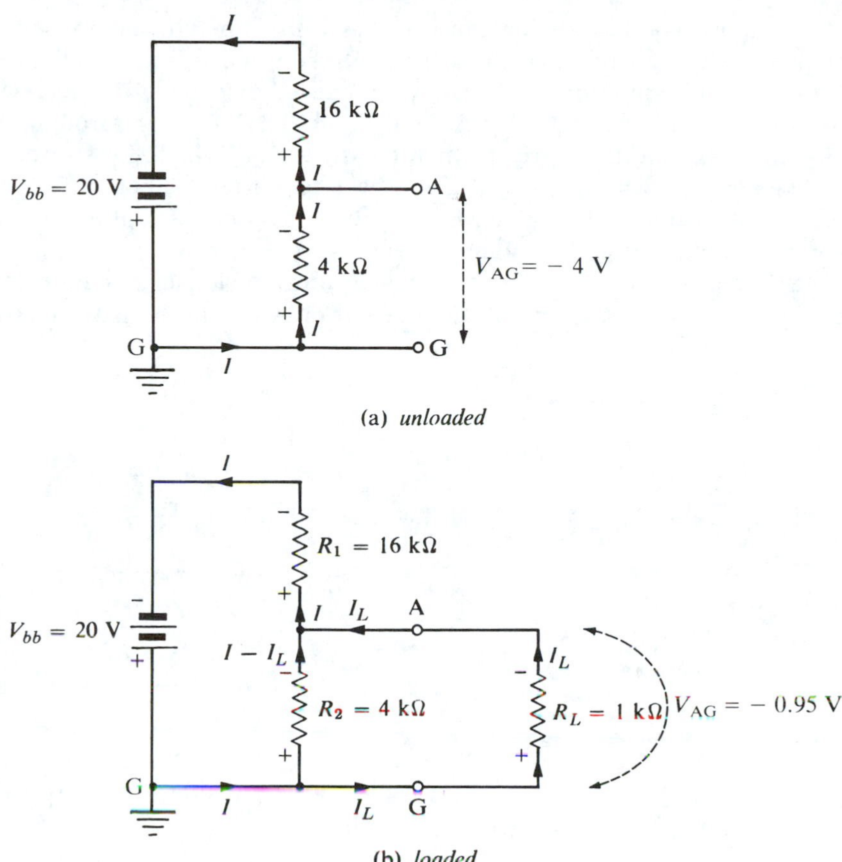

FIGURE 1.23 Voltage divider.

is drawn from A. This can be derived as follows:

$$V_A = -IR_2 = \frac{-V_{bb}}{R_1 + R_2} R_2 = -V_{bb} \frac{R_2}{R_1 + R_2} \tag{1.35}$$

There are two things to keep in mind in connection with this type of voltage divider. First, keep the total current drawn from the battery low to ensure a reasonably long battery life. Second, realize that any current drawn from terminal A may greatly affect the voltage at A. For example, if a 1-kΩ load resistor R_L were connected between A and G as in Fig. 1.23(b), then V_A would fall to only

$$V_A = I \times (R_2 \| R_L) = \frac{-V_{bb}}{R_1 + (R_2 \| R_L)} (R_2 \| R_L)$$

$$= -\frac{20\text{ V}}{16\text{ k}\Omega + 0.8\text{ k}\Omega} (0.8\text{ k}\Omega) = -0.95\text{ V}$$

In this case we say that the 1-kΩ resistor has *loaded* the voltage divider. In a similar vein, any variation in resistance R_L between A and G will cause V_A to vary. This variation in V_A can be minimized by making R_2 much less than R_L—in other words, by making the current I flowing through the R_1R_2 divider chain much greater than the current I_L which flows through R_L. Then any change in R_L will affect the total current only slightly, and thus V_G will change only slightly. But R_L then uses only a small fraction of the current drawn from the battery, which is wasteful.

If we know that R_L will remain constant, then a simpler circuit will do (see Fig. 1.24). In this circuit all the current drawn from the battery flows

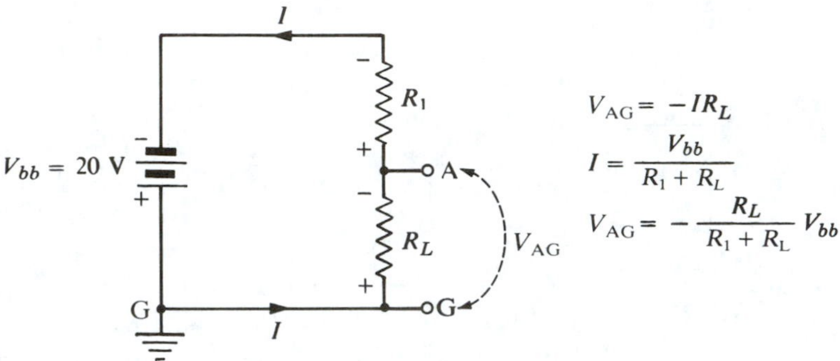

FIGURE 1.24 Resistance voltage divider.

through R_L, which may be an important advantage if the load draws a sizable current and the battery size is limited.

$$V_{AG} = -\frac{R_L}{R_1 + R_L} 20\text{ V} = -4\text{ V} \qquad \text{if } R_L = 1\text{ k}\Omega \text{ and } R_1 = 4\text{ k}\Omega \qquad \textbf{(1.36)}$$

1.10 IDEAL VOLTAGE AND CURRENT SOURCES

It is often convenient to use the concept of an ideal or perfect voltage source to describe various circuits. An *ideal voltage source* simply means a source of voltage with *zero internal resistance*, that is, a perfect battery. The circuit symbol for an ideal voltage source is a circle as shown in Fig. 1.25(a). Any battery or power supply in an actual circuit can be considered ideal if its internal resistance is small compared to any other resistances connected in series with it. As we mentioned in Section 1.5, most batteries are considered to have zero internal resistance in circuits. Such a voltage source is called ideal because it keeps on supplying the same voltage regardless of how much current we draw from it. An ideal voltage source can be approximated by a good battery. A real voltage source or battery can be represented by an ideal battery or voltage source in series with a

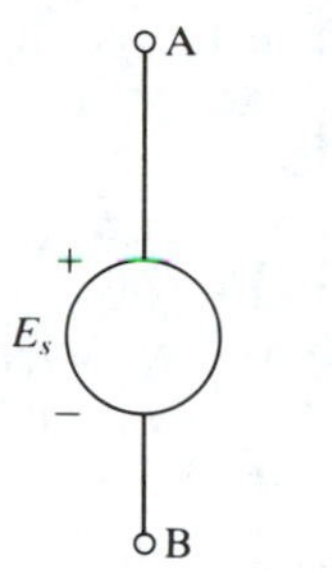

(a) *ideal voltage source*

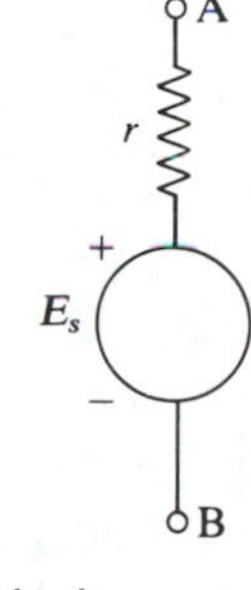

(b) *real voltage source including internal resistance r*

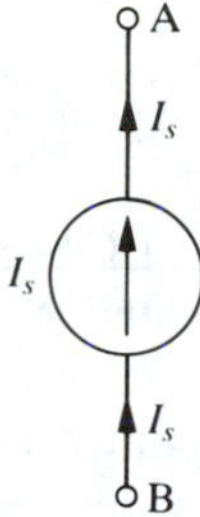

(c) *ideal current source*

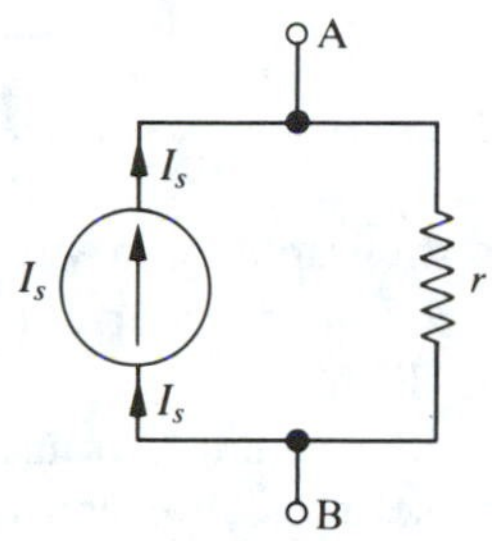

(d) *real current source including shunt resistance r*

FIGURE 1.25 Voltage and current sources.

resistance *r*, as shown in Fig. 1.25(b). The terminal voltage of a real battery, of course, falls as we draw more and more current from it because of the voltage drop across the battery's internal resistance as explained in Section 1.5.

An *ideal current source* is one that supplies a constant current regardless of what load it is connected to. Thus, an ideal current source has an *infinite internal resistance* so that changes in the load resistance will not affect the current supplied by the source. For example, a 6-V battery with a 1-MΩ internal resistance will deliver essentially 6 μA to a 1-kΩ, a 10-kΩ, or a 100-kΩ resistor connected across it. The current remains almost constant so long as the load resistance is small compared to the internal resistance. The circuit symbol for an ideal current source is shown in Fig. 1.25(c); in this circuit a current I_s *always* flows through the circle representing the ideal current source regardless of the rest of the circuit. Such a source can be approximated by a battery with a series resistance R_s much larger than any resistances in the load connected across the battery. Then the battery will always supply a current of V_{bb}/R_s regardless of any changes in the load resistances.

Any real current source can be represented by an ideal current source in parallel with a resistance *r*, as shown in Fig. 1.25(d). If a load resistance R_L is then connected to terminals A and B of the real current source, the

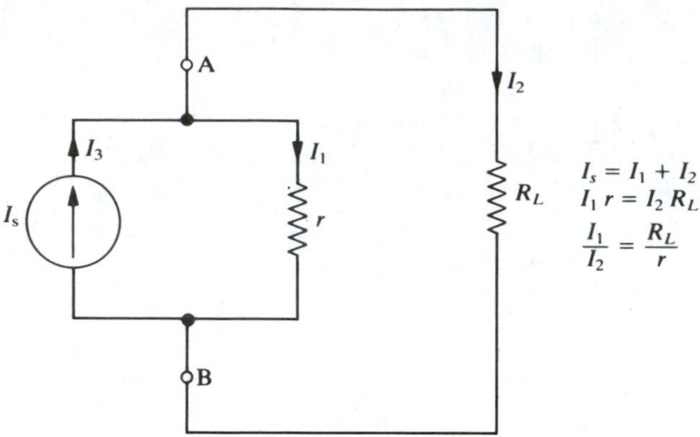

FIGURE 1.26 Current source with load R_L.

current I_s is then split between r and R_L according to Ohm's and Kirchhoff's laws; the larger current flows through the smaller resistance, as shown in Fig. 1.26.

As we will see later, transistors can be conveniently represented by ideal current sources. We can always convert from a current source to a voltage source, and vice versa. The conversion is shown in Fig. 1.27. Notice that the resistance in series with the ideal voltage source is equal to the resistance across the ideal current source, and that $E_s = I_s r$. These results follow from Thevenin's theorem, which is discussed in Section 1.12. The net result is that terminals A′B′ act electrically exactly like terminals A and B. Which circuit you choose is a matter of convenience; one may yield an easier set of circuit equations to solve.

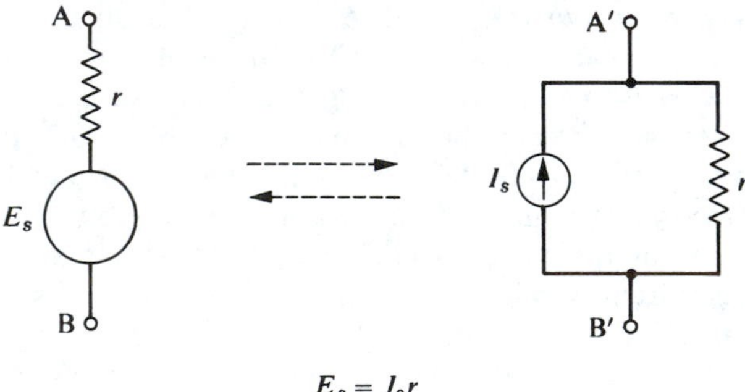

FIGURE 1.27 Equivalence of current and voltage source.

1.11 THE SUPERPOSITION THEOREM

A *linear* circuit is one in which all the KVL equations are mathematically linear in the currents and voltages. In other words, there are no terms containing products or quotients like $V_1 V_2$ or V_1/V_2. All dc circuits containing batteries and resistances are linear. Dc and ac circuits can be linear or nonlinear. In general, all circuits are linear when the voltages and currents are relatively small; they become nonlinear for larger currents and voltages. But certain circuit elements like diodes are intrinsically nonlinear for currents or voltages of any magnitude.

For any linear circuit containing more than one voltage source or current source, the *superposition theorem* states that the total current in any part of the circuit equals the algebraic sum of the currents produced by each source separately. To calculate the current due to any one particular source, replace all the other voltage sources by short circuits and all other current sources by open circuits.

In the circuit of Fig. 1.28, let us use the superposition theorem to calculate the current I through R_1. The current through R_1 due to V_1 alone is obtained by shorting out V_2, as shown in Fig. 1.28(b), and

$$I' = \frac{V_1}{R_1 + (R_2 \| R_3)}$$

The current through R_1 due to V_2 alone is obtained by shorting out V_1, as shown in Fig. 1.28(c), and

$$I'' = \frac{V_2}{R_2 + (R_1 \| R_3)} \times \frac{R_3}{R_1 + R_3}$$

where we have used

$$I'' = [R_3/(R_1 + R_3)]I_2 \quad \text{and} \quad I_2 = \frac{V_2}{R_3 + (R_1 \| R_3)}$$

By the superposition theorem the current through R_1 is just $I = I' - I''$ or

$$I = \frac{V_1}{R_1 + (R_2 \| R_3)} - \frac{V_2}{R_2 + (R_1 \| R_3)} \times \frac{R_3}{R_1 + R_3}$$

This result could be obtained from a Kirchhoff analysis with considerably more effort.

1.12 THEVENIN'S THEOREM

Thevenin's theorem is useful for analyzing many circuits. The theorem states that any combination of batteries and resistances with two terminals is electrically equivalent to an ideal battery of voltage e in series with one resistance r, as shown in Fig. 1.29. An ideal battery has a consistent voltage regardless of the rest of the circuit and has zero internal resistance. All the Kirchhoff voltage loop equations for linear circuit elements are linear; that

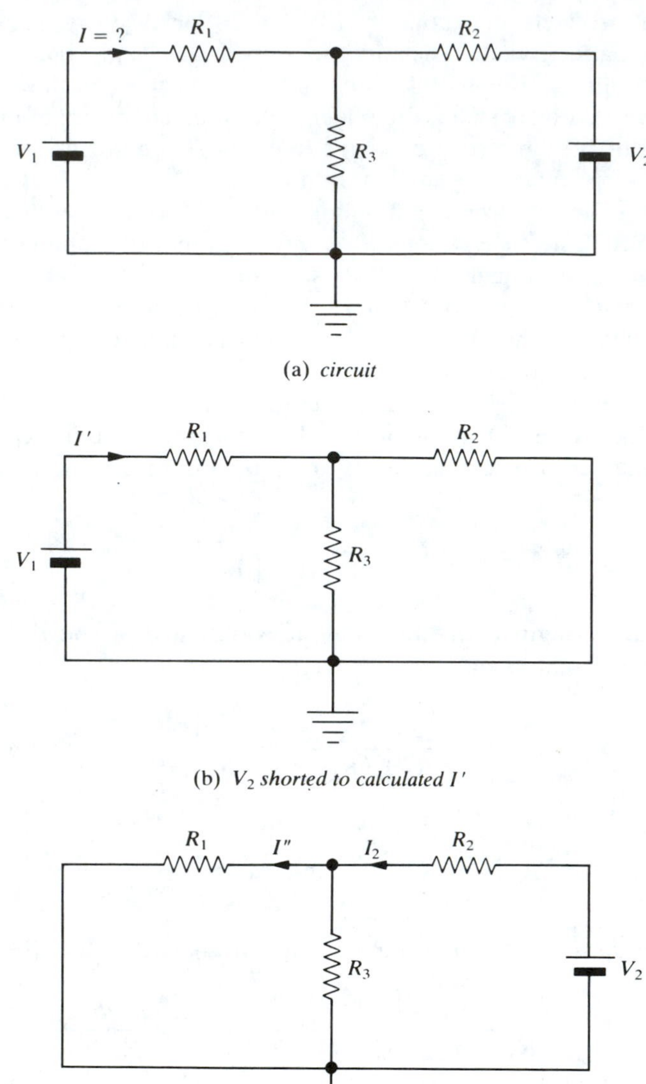

FIGURE 1.28 Superposition theorem problem.

is, the currents I and voltages V all occur raised to the first power. There are no terms containing products or quotients of I and V. The Kirchhoff equation for the Thevenin equivalent circuit is also linear. Thus for two points A and B in the circuit, the current I flowing out of A (to the "outside world") must be linearly related to the voltage difference V_{AB}. Such a

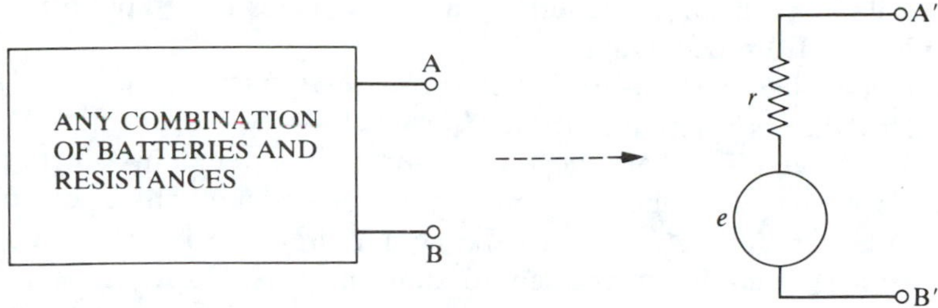

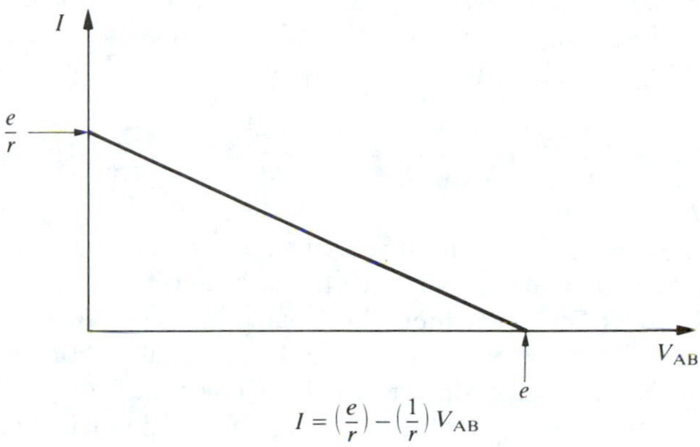

FIGURE 1.29 Thevenin equivalence circuit.

linear relationship, of course, yields a straight-line plot of V_{AB} versus I of the form

$$V_{AB} = e - Ir \tag{1.37}$$

or

$$I = \left(\frac{e}{r}\right) - \left(\frac{1}{r}\right) V_{AB} \tag{1.38}$$

But (1.37) is precisely the equation for $V_{A'B'}$ of the Thevenin equivalent circuit of Fig. 1.29. This is essentially a proof of Thevenin's theorem. It immediately follows that the Thevenin equivalent voltage e is the open-circuit voltage V_{ABoc} measured when no current flows into or out of A and B. This measurement must be made by a device that draws negligible current from the circuit. Usually an oscilloscope will do nicely. It also follows that the Thevenin resistance r is given by

$$r = -\frac{\partial V_{AB}}{\partial I} \tag{1.39}$$

Thus, once we obtain the linear relationship between V_{AB} and I, both e and

r are easily determined. Any method that determines the graph in Fig. 1.29 will uniquely determine e and r.

The equivalent resistance r is also the open-circuit voltage $V_{\mathrm{ABoc}}(I = 0)$ divided by the short-circuit current I_{sc}, which is the current flowing from terminals A and B when they are shorted together. In other words, terminals A′ and B′ of the Thevenin equivalent circuit will act electrically exactly like terminals A and B of the actual circuit. It is easy and safe to measure the open-circuit voltage to determine e. However, it is often disastrous (producing sparks, smoke, vile odors, destruction of the circuit, and embarrassment) actually to short terminals A and B to determine r. A better way to determine r experimentally is to measure the voltage V_{AB} between the two circuit terminals for a known load resistance R_L connected between A and B. Then r is given by

$$r = \frac{e - V_{\mathrm{AB}}}{I} = \frac{e - V_{\mathrm{AB}}}{V_{\mathrm{AB}}/R_L} = \frac{e - V_{\mathrm{AB}}}{V_{\mathrm{AB}}} R_L \tag{1.40}$$

where e is the open-circuit voltage measured when $R_L = \infty$. Another useful way to calculate r is to mentally short out all the batteries in the circuit being analyzed. The resistance r is then the total net resistance between output terminals A and B. However, this method of mentally shorting out the batteries to determine r is appropriate only for constant (independent) voltage sources such as batteries. It does not work for dependent voltage sources such as those in the equivalent circuit for a transistor. An example of a dependent voltage generator is one whose voltage depends on the voltage in another part of the circuit.

In any circuit

1. $e =$ the open-circuit voltage V_{ABoc}.
2. $r = V_{\mathrm{ABoc}}/I_{\mathrm{sc}}$ or $r = -(\partial V_{\mathrm{AB}}/\partial I)$ or (1.40).

In analyzing transistor amplifier circuits with a voltage divider, it is often useful to replace the actual divider with its Thevenin equivalent, as shown in Fig. 1.30. Notice that both the actual circuit and the Thevenin equivalent produce the same open-circuit voltage and the same short-circuit current V_{bb}/R_1. In Fig. 1.30, for example, if $V_{bb} = 12$ V, $R_1 = 6$ kΩ, and $R_2 = 4$ kΩ, the Thevenin equivalent circuit is that of Fig. 1.30(c).

$$e = V_{\mathrm{ABoc}} = IR_2 = \frac{V_{bb}R_2}{R_1 + R_2} = \frac{12\ \mathrm{V}}{10\ \mathrm{k\Omega}} \times 4\ \mathrm{k\Omega} = 4.8\ \mathrm{V}$$

The Thevenin resistance is simply

$$r = \frac{V_{\mathrm{ABoc}}}{I_{\mathrm{sc}}} = \frac{\left(\dfrac{V_{bb}}{R_1 + R_2}\right)R_2}{\dfrac{V_{bb}}{R_1}} = \frac{R_1 R_2}{R_1 + R_2} = \frac{(6\ \mathrm{k\Omega})(4\ \mathrm{k\Omega})}{(6\ \mathrm{k\Omega} + 4\ \mathrm{k\Omega})} = 2.4\ \mathrm{k\Omega}$$

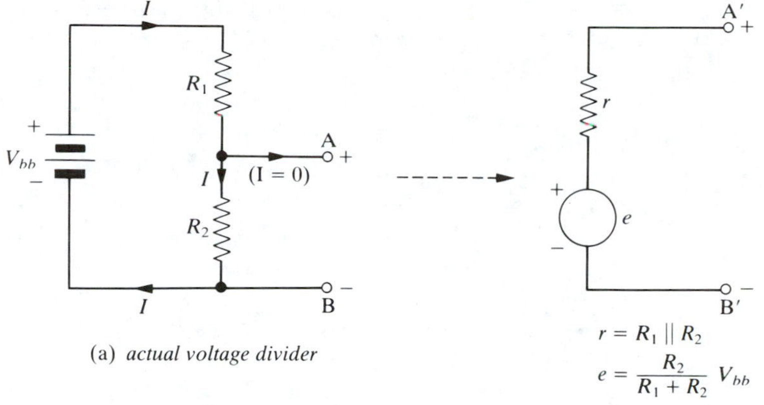

(a) *actual voltage divider*

(b) *Thevenin equivalent*

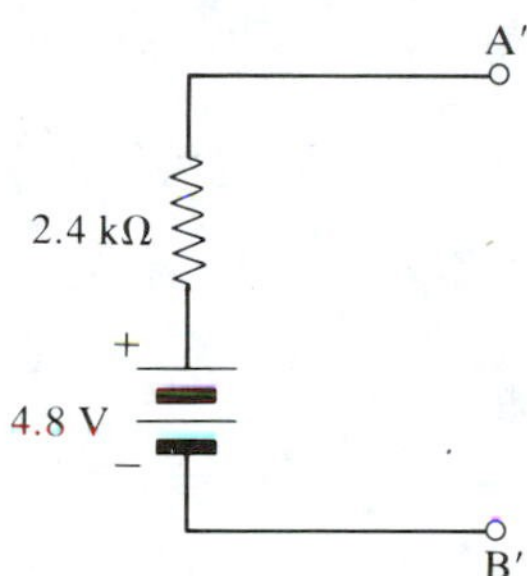

(c) *Thevenin equivalent circuit of (a) for* $R_1 = 6\ \text{k}\Omega$, $R_2 = 4\ \text{k}\Omega$, $V_{bb} = 12\ \text{V}$

FIGURE 1.30 Thevenin equivalent circuit of a voltage divider.

Or, because we have an independent voltage source (a battery), r can be calculated by mentally shorting out the battery, which puts R_1 and R_2 in parallel; then

$$r = R_1 R_2/(R_1 + R_2) = 2.4\ \text{k}\Omega$$

Another example may be helpful. Consider the circuit of Fig. 1.31. The Thevenin voltage e is the open-circuit voltage IR_4 because there is no current through (and thus no voltage drop across) R_3 due to the open-circuit condition. Thus,

$$e = IR_4 = \frac{V_{bb}}{R_2 + R_4} R_4 = \frac{12\ \text{V}}{2\ \text{k}\Omega + 4\ \text{k}\Omega}(4\ \text{k}\Omega) = 8\ \text{V}$$

The Thevenin resistance r is the open-circuit voltage e (8 V), divided by the short-circuit current I'_{sc}. When we short A and B, R_3 and R_4 are in

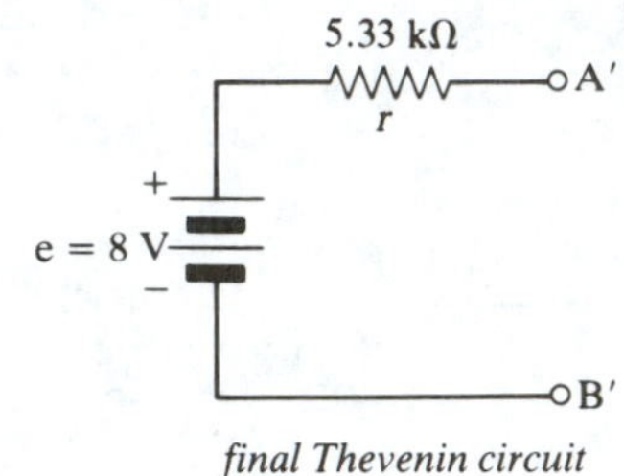

final Thevenin circuit

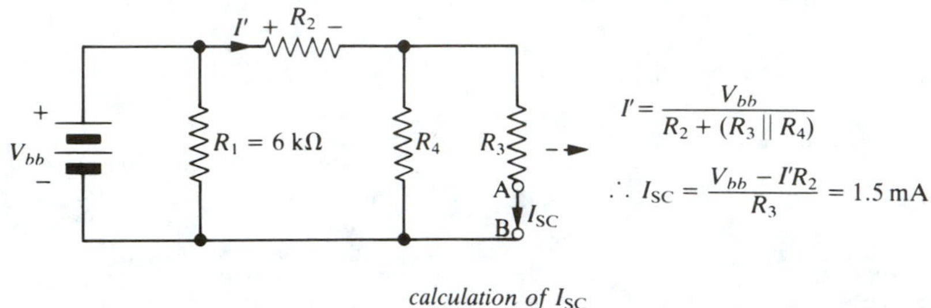

calculation of I_{SC}

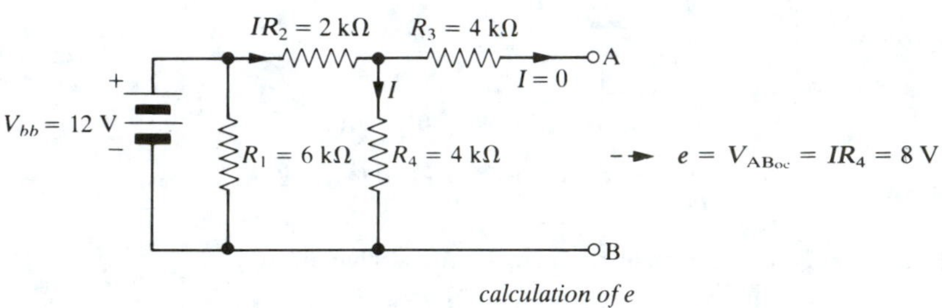

calculation of e

FIGURE 1.31 Thevenin equivalent circuit example.

parallel, and the current I' is given by

$$I' = \frac{V_{bb}}{R_2 + (R_3 \| R_4)} = \frac{12\text{ V}}{2\text{ k}\Omega + 2\text{ k}\Omega} = 3\text{ mA}$$

Thus the short-circuit current is the current through R_3.

$$I_{sc} = \frac{V_{bb} - I'R_2}{R_3} = \frac{12\text{ V} - (3\text{ mA})(2\text{ k}\Omega)}{4\text{ k}\Omega} = 1.5\text{ mA}$$

Thus,

$$r = \frac{V_{ABoc}}{I_{sc}} = \frac{e}{I_{sc}} = \frac{8\text{ V}}{1.5\text{ mA}} = 5.33\text{ k}\Omega$$

Or, mentally shorting out V_{bb}, we see that R_2 is in parallel with R_4, so r is

$$r = (R_2 \| R_4) + R_3 = 1.33\ \text{k}\Omega + 4\ \text{k}\Omega = 5.33\ \text{k}\Omega$$

Notice that R_1 does not affect r at all.

Thevenin's theorem can also be used to simplify the analysis of a complicated circuit by applying it to only one part of the circuit. Consider the circuit of Fig. 1.32. Suppose we are given V, R_1, R_2, R_3, and R_4 and

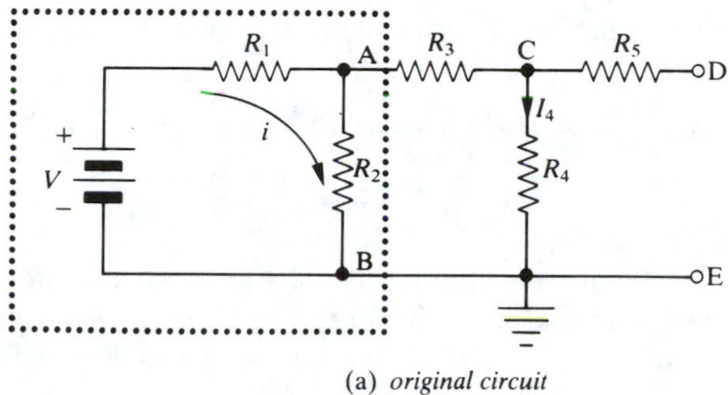

(a) *original circuit*

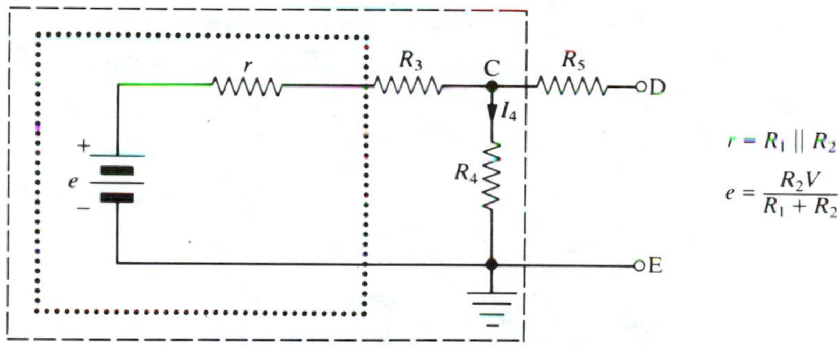

(b) *using Thevenin equivalent of circuit inside dotted line*

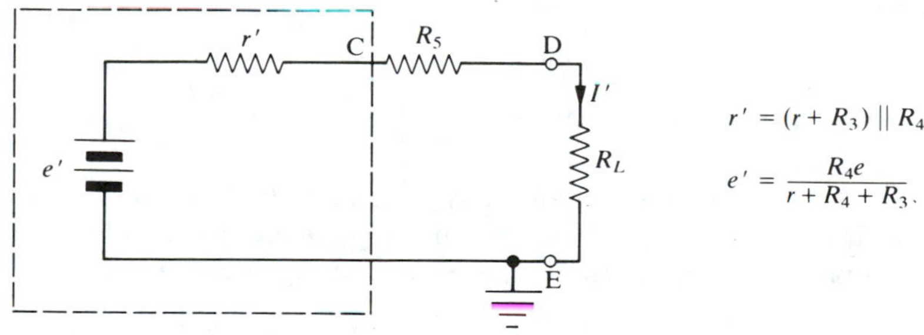

(c) *using Thevenin equivalent of circuit inside dashed line*

FIGURE 1.32 Thevenin equivalent circuit example.

are asked to find I_4. We certainly could combine R_3 and R_4 if no current were drawn from terminal D. And then R_2 and $R_3 + R_4$ are in parallel, and so on. But I_4 can also be found by replacing V, R_1, and R_2 inside the dotted line by its Thevenin equivalent, as shown in Fig. 1.32(b). V, R_1, and R_2 form a simple voltage divider looking into point A, so their Thevenin equivalent is

$$e = \left(\frac{R_2}{R_1 + R_2}\right) V \quad \text{and} \quad r = \frac{R_1 R_2}{R_1 + R_2}$$

Thus, we can immediately calculate I_4 from the circuit of Fig. 1.32(b).

$$I_4 = \frac{e}{r + R_3 + R_4} = \frac{e}{\dfrac{R_1 R_2}{R_1 + R_2} + R_3 + R_4}$$

Consider another problem using the circuit of Fig. 1.32, except put a load resistance R_L between D and E as shown in Fig. 1.32(c). We want to calculate V_D. We note that e, r, R_3, and R_4 form a voltage divider, so their Thevenin equivalent is

$$e' = \left(\frac{R_4}{r + R_3 + R_4}\right) e \qquad r' = \frac{(r + R_3)R_4}{r + R_3 + R_4}$$

Thus, we obtain the circuit of Fig. 1.32(c) and

$$I' = \frac{e'}{r' + R_5 + R_L}$$

$$V_D = I' R_L = \frac{e' R_L}{r' + R_5 + R_L}$$

This technique will be used in Chapter 15 to analyze a digital-to-analog converter.

1.13 NORTON'S THEOREM

Norton's theorem simply states that any combination of batteries and resistances with two terminals is electrically equivalent to an ideal current source in parallel with a resistance, as shown in Fig. 1.33. An ideal current source is one that puts out a constant current regardless of the rest of the circuit. It also has an infinite internal resistance, as compared to an ideal voltage source which has zero internal resistance. The equivalent resistance is the same as the equivalent resistance of the Thevenin equivalent circuit,

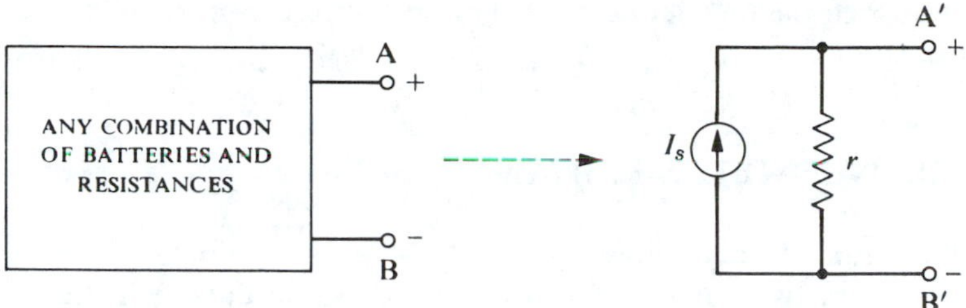

FIGURE 1.33 Norton equivalent circuit.

and the ideal constant current source supplies a current of

$$I_s = \frac{V_{\text{ABoc}}}{r} \tag{1.41}$$

Thus, a voltage divider circuit can be replaced by a Norton, or current, equivalent circuit as shown in Fig. 1.34.

We emphasize that the Thevenin equivalent is just as good as the Norton equivalent, and vice versa. Each is electrically equivalent to the original circuit being analyzed and to each other. Which equivalent circuit you use is a matter of taste and convenience, although there is a tendency to use the current equivalent circuit in describing transistor circuits and the voltage equivalent circuit for vacuum tube circuits.

Thevenin's and Norton's theorems can be extended to alternating current (ac) circuits. When applied to ac circuits, the theorem states that any combination of ac circuit elements (such as resistors, capacitors, inductors, etc.) whose equations are *linear* in current and voltage can be replaced by an ac Thevenin or Norton equivalent. Practically speaking, this means that almost any small-signal circuit can be represented by a The-

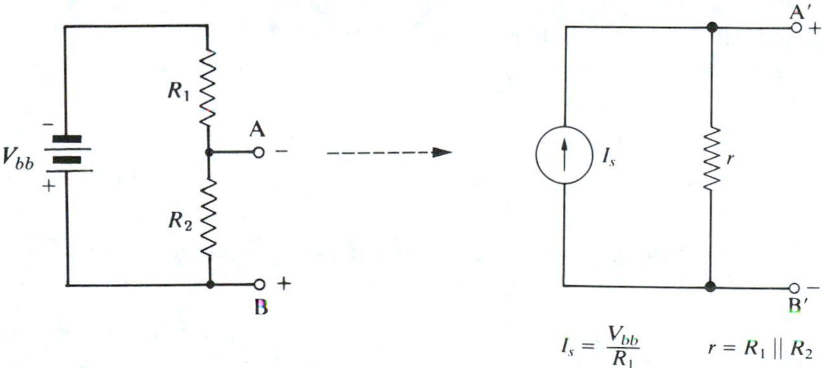

FIGURE 1.34 Norton equivalent circuit of a voltage divider.

venin or a Norton equivalent circuit. This important subject will be covered in later chapters.

1.14 THE WHEATSTONE BRIDGE

The Wheatstone bridge shown in Fig. 1.35 was originally developed to make precise measurements of resistance. If the dc current meter in Fig. 1.35(a) reads 0, then points C and D must be at the same voltage. Thus the voltage drop $V_A - V_C$ across R_1 must equal the voltage drop $V_A - V_D$ across R_2; similarly, $V_C - V_B = V_D - V_B$. Thus

$$I_C R_1 = I_D R_2 \quad \text{and} \quad I_C R_3 = I_D R_4$$

Solving for I_D/I_C, we see that

$$\frac{R_1}{R_2} = \frac{R_3}{R_4} \tag{1.42}$$

which is called the *balance* condition or equation and indicates that the bridge meter current is zero. When the bridge is balanced, then R_3, the unknown resistance, can be determined if the other three resistances are known.

$$R_3 = \frac{R_1}{R_2} R_4$$

Notice that the meter need *not* be calibrated in microamperes; it need only indicate a true "zero" reading; that is, it acts as a *null* indicator.

If the bridge is unbalanced, that is, if a current flows through the meter, or, equivalently, if a voltage difference exists between C and D, then the analysis is considerably more complicated.

Let R_M be the resistance of the meter between points C and D. The current I_M through R_M can be calculated by calculating the Thevenin equivalent of the bridge between points C and D.

The equivalent circuit of the bridge is shown in Fig. 1.35(c); the current through the meter will be

$$I_M = \frac{e}{r + R_M} \tag{1.43}$$

The Thevenin voltage e can be determined by calculating $e = V_C - V_D$ with R_M infinite (the open-circuit voltage). From the bridge circuit of Fig. 1.35(b),

$$V_C = V_1 - I_C R_1 \quad \text{and} \quad I_C = \frac{V_1}{R_1 + R_3} \quad \text{because } I_M = 0$$

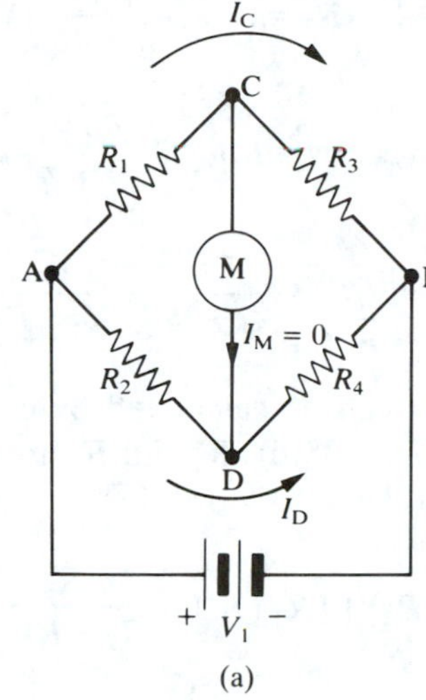

$$\frac{R_1}{R_2} = \frac{R_3}{R_4}$$

$$I_M = 0$$

(a)

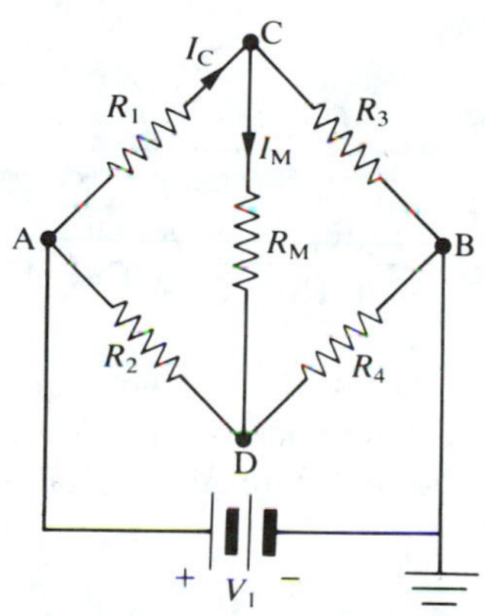

(b) *circuit showing meter resistance*

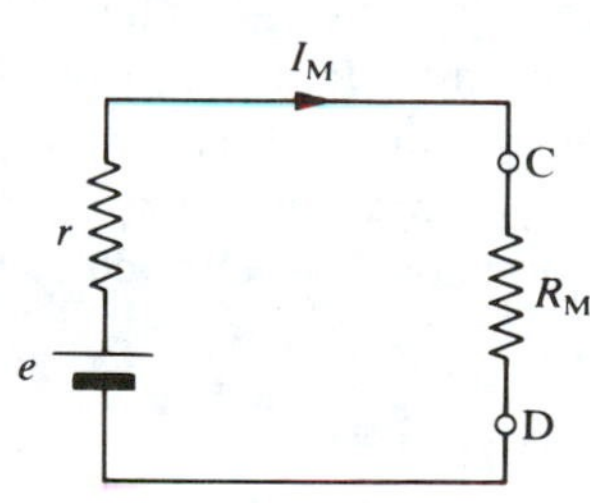

(c) *final Thevenin equivalent of bridge*

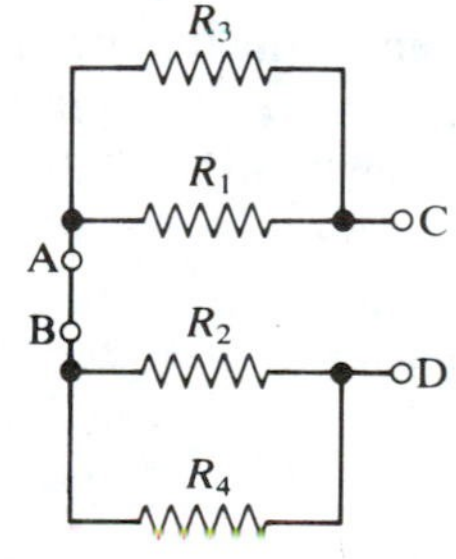

(d) *circuit for calculating r* (V_1 *shorted out*)

FIGURE 1.35 The Wheatstone bridge.

Thus $$V_C = V_1 - \frac{V_1}{R_1 + R_3} R_1 = \left(1 - \frac{R_1}{R_1 + R_3}\right) V_1 = \frac{R_3 V_1}{R_1 + R_3}$$

Similarly, $$V_D = V_1 - I_D R_2 = \frac{R_4 V_1}{R_2 + R_4}$$

Thus $$e = V_C - V_D = \left(\frac{R_3}{R_1 + R_3} - \frac{R_4}{R_2 + R_4}\right) V_1 \tag{1.44}$$

The Thevenin resistance r can be calculated by conceptually shorting the battery V_1. As shown in Fig. 1.35(d), R_1 and R_3 are then in parallel, and R_2 and R_4 are also in parallel. Hence,

$$r = (R_1 \| R_3) + (R_2 \| R_4) = \frac{R_1 R_3}{R_1 + R_3} + \frac{R_2 R_4}{R_2 R_4} \tag{1.45}$$

Substituting (1.44) and (1.45) in (1.43), we find that the current through the meter of resistance R_M is

$$I_M = \frac{\left(\dfrac{R_3}{R_1 + R_3} - \dfrac{R_4}{R_2 + R_4}\right)}{\dfrac{R_1 R_3}{R_1 + R_3} + \dfrac{R_2 R_4}{R_2 R_4} + R_M} V_1 \tag{1.46}$$

Notice that $I_M = 0$ if (1.42) holds; that is, the balance condition is the same.

There is one useful consequence of (1.46) that deserves further comment. Consider R_1, R_2, and R_4 as fixed, and allow R_3 to vary. When R_3 satisfies (1.42),

$$R_3 = \frac{R_1 R_4}{R_2} \equiv R_{3B} \tag{1.47}$$

Let R_{3B} represent this particular balance value of R_3 when $I_M = 0$. But if R_3 is greater than R_{3B}, I_M is positive; whereas if R_3 is less than R_{3B}, I_M is negative. In other words, the current I_M in the meter (or, equivalently, the voltage across the meter, $V_C - V_D = I_M R_M$) *changes sign* as R_3 passes through the balance value R_{3B}. This can be shown algebraically by noting from (1.46) that

$$I_M \propto (R_2 R_3 - R_1 R_4)$$

From (1.47) $$R_1 R_4 = R_2 R_{3B}$$

so $$I_M \propto (R_2 R_3 - R_2 R_{3B}) = R_2 (R_3 - R_{3B}) \tag{1.48}$$

Thus if $R_3 > R_{3B}$, $I_M > 0$, and if $R_3 < R_{3B}$, $I_M < 0$, as we have stated. This feature of the bridge is particularly useful because the *sign* of I_M (or, equivalently, the sign of $V_C - V_D = I_M R_M$) can be used as an indicator of whether R_3 is greater than or less than a certain value R_{3B}. For example, R_3 could be a thermistor whose resistance changes rapidly with temperature. Then R_1, R_2, and R_4 could be chosen to balance the bridge at a certain temperature, say, 37.4°C. Then if the temperature of R_3 changed from 37.4°C, the bridge unbalance current *magnitude* (I_M) would approximately indicate the *magnitude* of the temperature change, and the *sign* would indicate whether the temperature was above or below 37.4°C. Thus, the bridge unbalance current (or voltage) could be used to control a heater (or cooler) element to regulate the temperature at 37.4°C.

PROBLEMS

1. What is the electric charge on (a) one electron, (b) a He^+ ion, (c) a He^{2+} ion, (d) an As^+ ion? State why the quantization of charge can be neglected in most electrical circuit problems.
2. Doubly ionized helium ions at a concentration of 10^{13} ions/cm^3 move with a velocity of 10^5 cm/s. Calculate the current density in A/cm^2.
3. Define current.
4. Calculate how many electrons flow per second past a fixed point in a wire carrying 10 mA of current. If the current moves from left to right, which way do the electrons move?
5. Define electric potential or voltage. What are the mks units of voltage? What does "ground" mean?
6. Calculate the gain in kinetic energy for an electron moving from a point of voltage 3 V to a point of voltage 5 V. Express your answer in joules and in electronvolts.
7. Define resistance.
8. State Ohm's law, including the units of all the terms.
9. Calculate the resistance of a silver wire 2.0 m long of radius 1 mm. Repeat for an aluminum wire of the same dimensions.
10. If a wire has a resistance of 10 Ω, what would be the resistance of a wire with twice the length (made from the same material) and twice the diameter? Assume the temperatures of both wires are the same.
11. Calculate the voltage at points A, B, and C if (a) A is grounded, (b) B is grounded, (c) C is grounded.

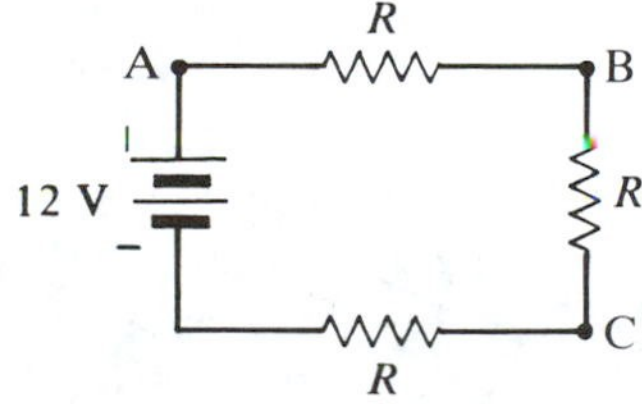

12. Calculate the voltage at points A, B, and C in the following circuit.

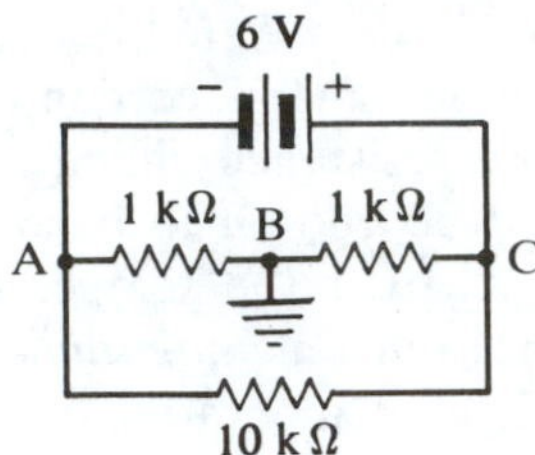

What is the current through the 10-kΩ resistor?

13. Calculate the resistance between terminals A and B.

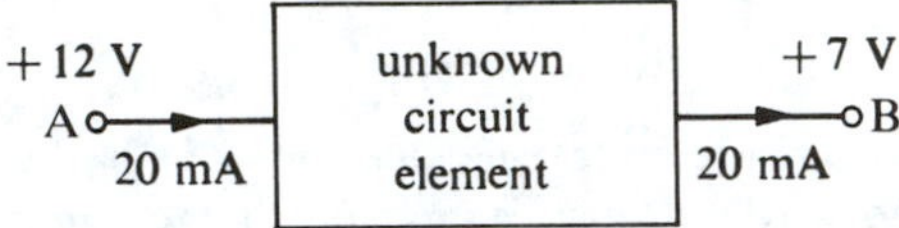

14. A fixed 1.0-kΩ resistor and a 4.0-kΩ potentiometer are connected in series across a 5-V battery of negligible internal resistance. Calculate the maximum and minimum values of V_{AB} as the potentiometer shaft is rotated.

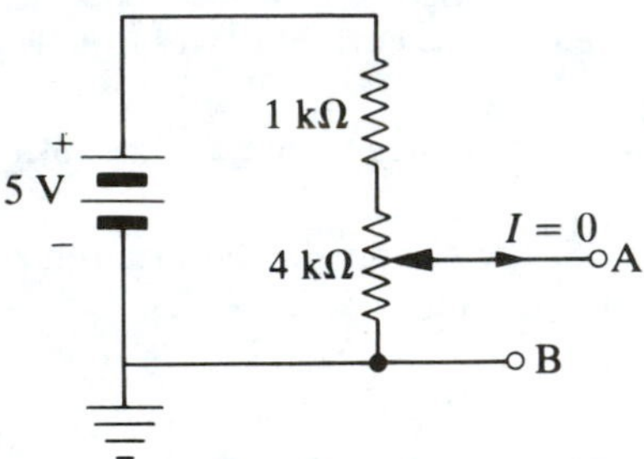

15. A gas discharge tube draws a current of 20 mA when a voltage difference of 800 V is maintained between its ends. Calculate the effective dc resistance in ohms between the ends of the tube.

16. Estimate quantitatively the static (dc) and the dynamic (ac) resistances at points A, B, and C of the device whose current–voltage curve is shown.

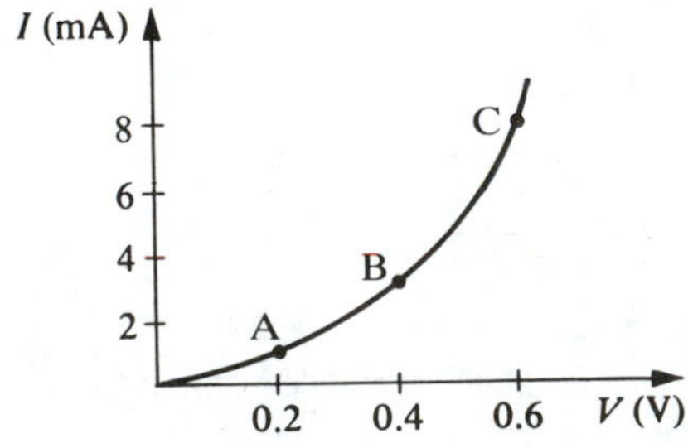

17. Calculate the voltage between A and B. What is the polarity of V_{AB}?

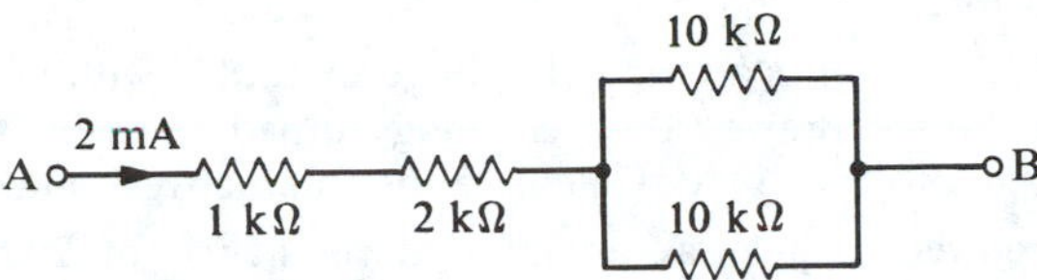

18. Calculate the current I.

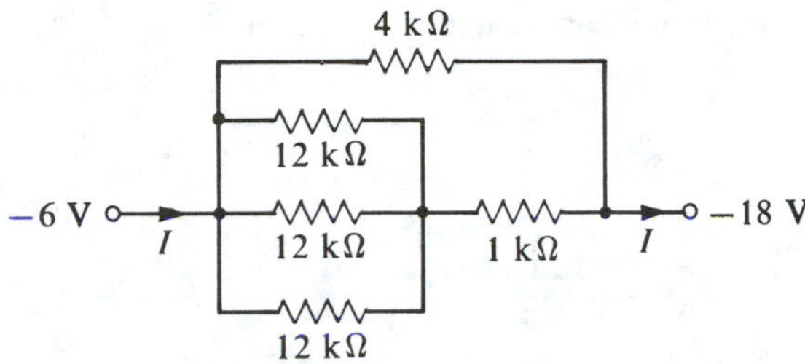

19. What happens if an ideal 0–15-V dc voltmeter is connected across a 12-V auto battery as shown? What happens if an ammeter is connected across the battery?

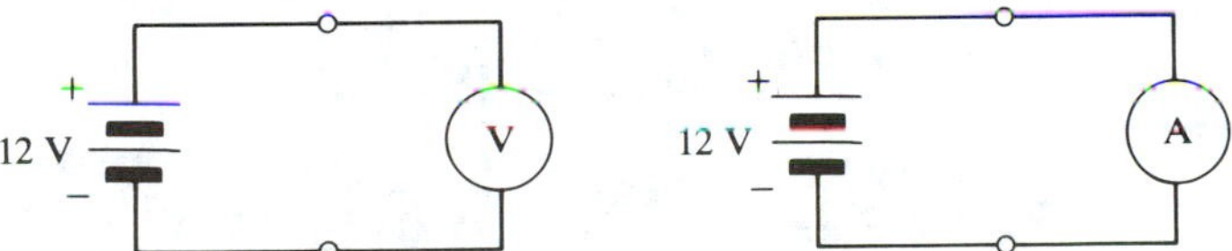

20. If the temperature of a 1000-Ω carbon resistor is 20°C, what is its resistance at 100°C?
21. If a thermistor has a resistance of 10,000 Ω at 20°C and $A = 5000$ K, what is its resistance at 40°C?
22. Describe the energy conversion process that occurs in a battery. Does the negative battery terminal draw in electrons or spew out electrons? Does the current into one battery terminal always equal the current out of the other terminal?
23. Define power, including the units.
24. Assume that a "durg" is a well-known physical quantity represented by the symbol D. What is (a) a kilodurg, (b) a megadurg, (c) a gigadurg, (d) a millidurg, (e) a microdurg, (f) a nanodurg, (g) a picodurg?
25. How would you make a crude 10-W heater–defroster for your car to run directly off the 12-V battery?
26. A 1-W, 1-kΩ carbon resistor carries a current of 30 mA. Calculate the power in watts dissipated as heat in the resistor. Would this situation be desirable in a circuit? Explain briefly.

27. Describe, on a microscopic basis, what happens when electrical energy is converted into heat in a resistor. Is heat always generated when a current flows through a resistor?

28. An automobile battery has a terminal voltage of 12.8 V with no load. When the starter motor (which draws 90 A) is being turned over by the battery, the terminal voltage drops to 11 V. Calculate the internal resistance of the battery.

29. A 30-V dc power supply has an internal resistance of 2 Ω. Calculate the terminal voltage when a current of 500 mA is being drawn from the power supply.

30. How large should the heater resistance R_h be to draw maximum power from a 12-V battery with an internal resistance of 3 Ω? Calculate the power dissipated in the heater and in the battery under such conditions.

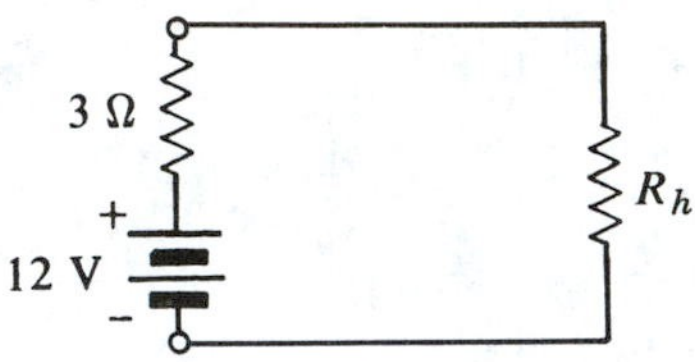

31. State Kirchhoff's laws.

32. Calculate I_1 and I_2.

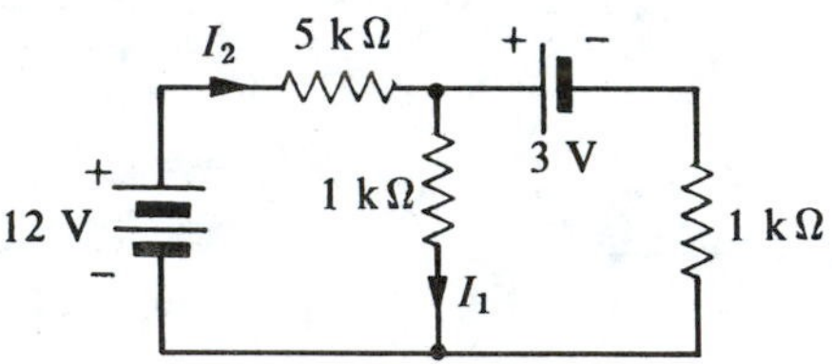

33. Calculate I_1, I_2, and I_3.

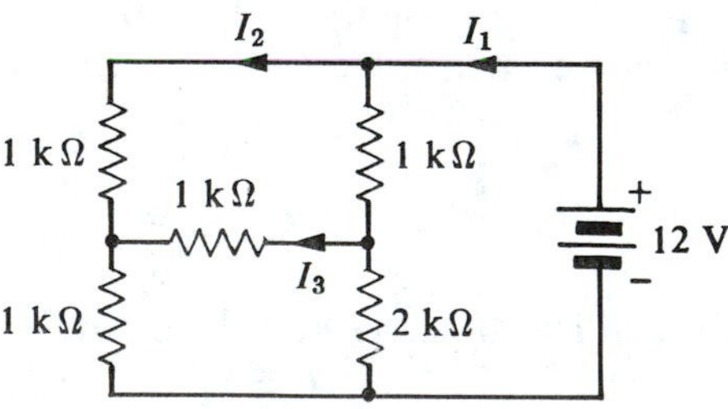

34. Describe an ideal current source and an ideal voltage source.

35. State Thevenin's theorem.

36. Calculate the Thevenin equivalent circuit.

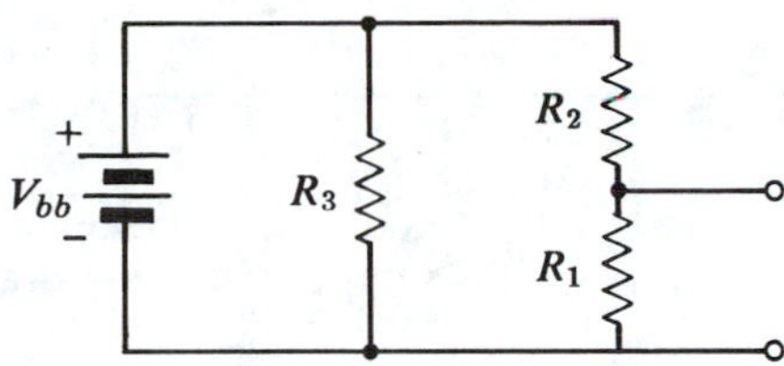

37. Calculate the Thevenin equivalent circuit.

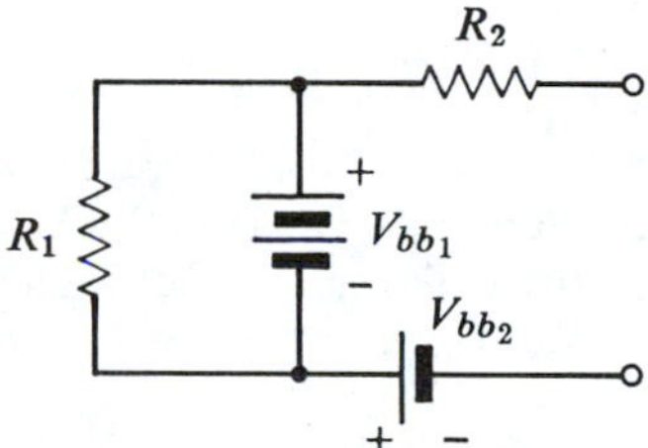

38. Calculate the Thevenin equivalent circuit.

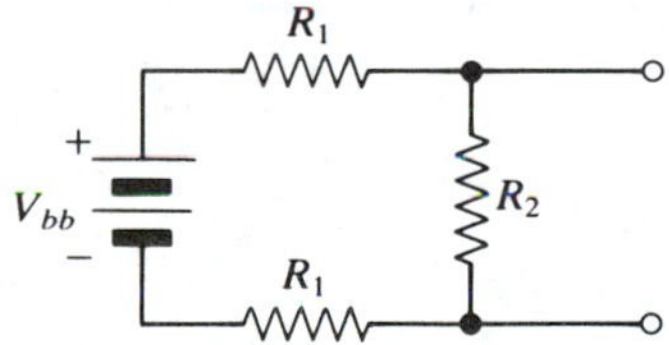

39. Calculate the Thevenin equivalent circuit.

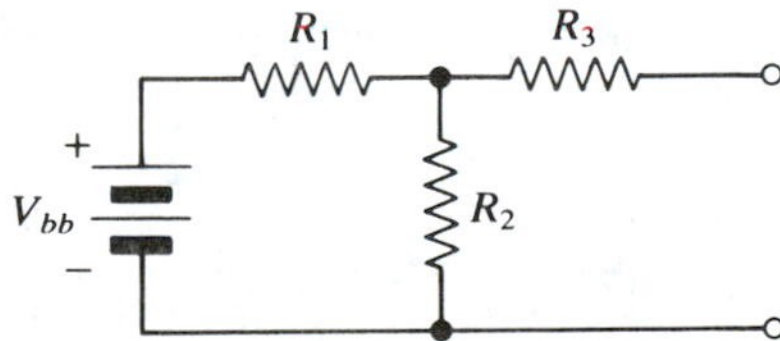

40. A "black box" with three terminals labeled E, B, and C is connected in the following circuit. (a) Calculate V_C and V_E. (b) If terminal B is 0.6 V more negative than terminal E, calculate R_1 and R_2, assuming that I_d is very large compared to the 20 μA flowing in the B lead. (This black box is a silicon npn transistor.)

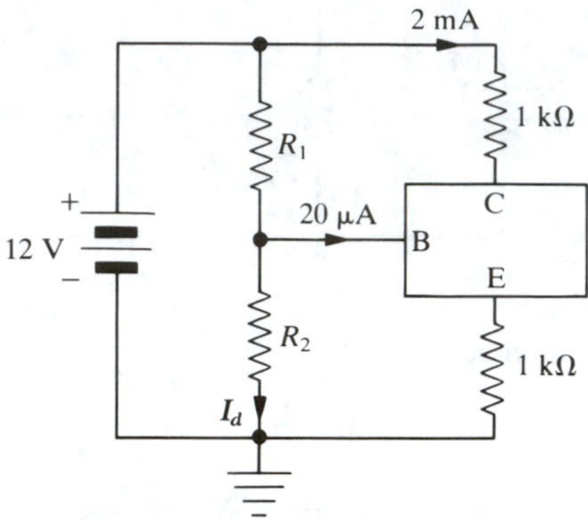

41. A black box with three terminals labeled K, G, and P is connected in the following circuit. The current flowing in the G lead is approximately 10^{-8} A. Calculate V_P, V_G, and V_K. (This black box is a vacuum tube.)

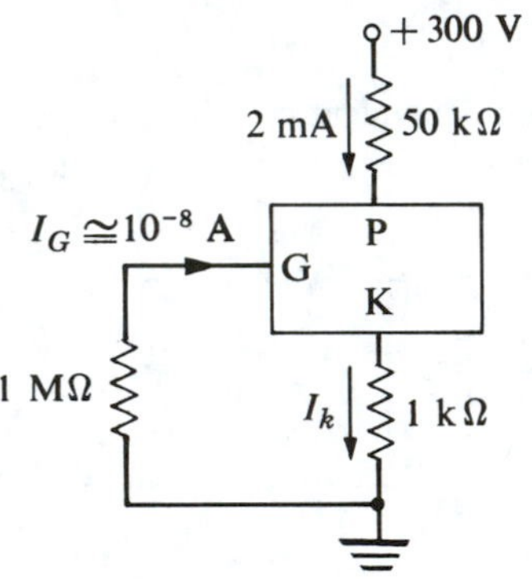

42. Calculate the voltmeter reading for (a) $R = 1\ \text{k}\Omega$ and (b) $R = 1\ \text{M}\Omega$. You may assume the voltmeter is an oscilloscope with a 1-MΩ input resistance.

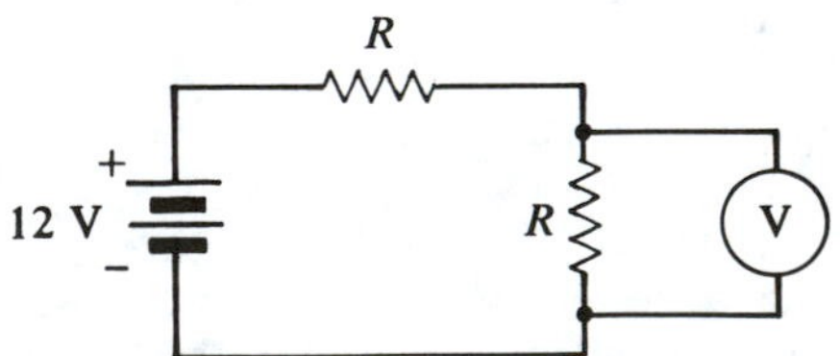

43. In the Wheatstone bridge of Fig. 1.35, calculate R_3 if $R_1 = 1.0\ \text{k}\Omega$, $R_2 = 2.0\ \text{k}\Omega$, and $R_4 = 500\ \Omega$.

44. Calculate the current I_5 through the meter ($R_5 = 10\ \Omega$) if $R_3 = 255\ \Omega$ for the bridge of Problem 43.

CHAPTER 2

Alternating Current Circuits

2.1 PERIODIC WAVEFORMS

In Chapter 1 we briefly mentioned alternating current (ac) and described it as current in which the electric charge "sloshed" back and forth along the wire. The most common type of ac is that in which the back-and-forth motion is sinusoidal; that is, if we plot a graph of the instantaneous value of the current $i(t)$ against time t, we get a sine wave as in Fig. 2.1(a). The instantaneous voltage $v(t)$ can also be sinusoidal, as shown in Fig. 2.1(b).

Such a sine wave is completely determined by specifying three things: (1) the frequency f of the wave, (2) the amplitude of the wave, and (3) the phase of the wave. The *frequency* f of the wave is the number of complete cycles that occur in one second and is expressed in cycles per second (cps). The *hertz* (Hz) is a unit of frequency now widely used. It is named after

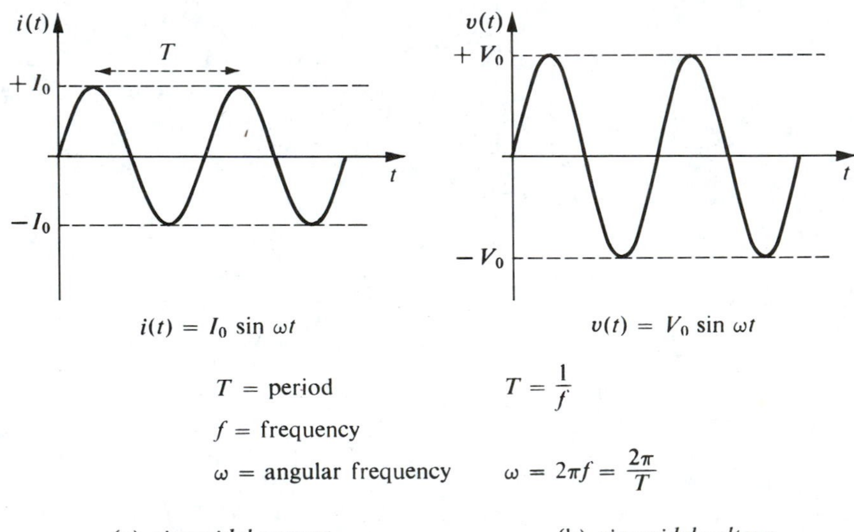

FIGURE 2.1 Sinusoidal waveforms.

Heinrich Hertz, who was the first to demonstrate electromagnetic wave propagation. One Hz = 1 cps, 1 kilohertz (kHz) = 10^3 Hz, 1 megahertz (MHz) = 10^6 Hz, 1 gigahertz (GHz) = 10^9 Hz. The *period* T of the wave is the time for one complete cycle; therefore $T = 1/f$. For example, a wave of frequency 2000 cps or 2 kHz has a period of 0.0005 s. The "60-cycle" (really 60 cps) line voltage available in most laboratories has a period of 1/60 s = 0.0167 s. Often, the *angular frequency* $\omega = 2\pi f = 2\pi/T$ is used; ω has units of radians per second (rad/s). (2π rad = 360°; 1 rad = 57°.) Note that a radian is really a pure number, so radians per second = s^{-1}. This follows from the definition of an angle θ (Fig. 2.2) in radians: $\theta \equiv s/r$, where s is the arc length and r is the radius.

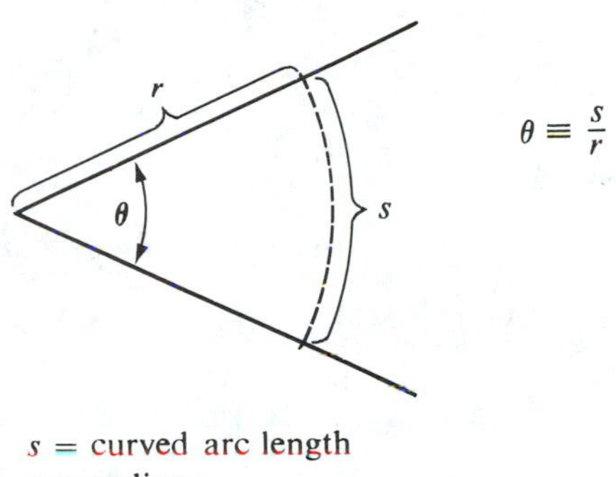

FIGURE 2.2

The amplitude of the wave is a measure of how "large" or strong the wave is. A common way to specify the amplitude is to give the change in current or voltage from the most positive value to zero; this is I_0 in Fig. 2.1(a) and V_0 in Fig. 2.1(b). The *peak-to-peak* amplitude is the change in current or voltage from the most positive value to the most negative value: $2I_0$ in Fig. 2.1(a) and $2V_0$ in Fig. 2.1(b). Another common measure of the amplitude is the *root-mean-square* (*rms*) amplitude, defined as

$$V_{\text{rms}} \equiv \left[\frac{1}{T}\int_0^T v^2(t)\,dt\right]^{1/2} \tag{2.1}$$

If $v(t)$ is sinusoidal, $v(t) = V_0 \sin \omega t$ and

$$V_{\text{rms}} = \left[\frac{1}{T}\int_0^T V_0^2 \sin^2 \omega t\,dt\right]^{1/2} = \frac{V_0}{\sqrt{2}} = 0.707V_0 \tag{2.2}$$

Rms values are useful because they occur in the expression for the power in ac circuits that will be explicitly shown later in this chapter. The 110-V,

60-Hz line voltage commonly available in laboratories has an rms voltage of 110 V; therefore, $V_0 = \sqrt{2}\,V_{rms} = 156$ V, and the peak-to-peak voltage is 312 volts! This line voltage is shown in Fig. 2.3 and can vary from about 105 V rms to 120 V rms depending on the time of day and the demands made upon the electric power company.

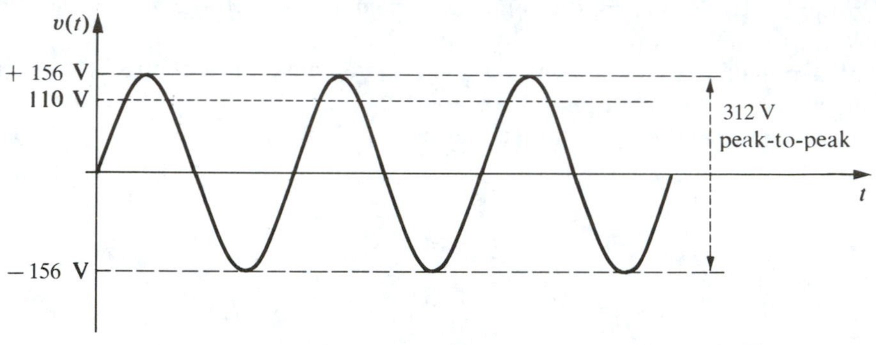

$$v(t) = V_0 \sin 2\pi f t = (154\text{ V}) \sin 2\pi(60)t$$

FIGURE 2.3 110-V-rms 60-"cycle" (60-Hz) line voltage.

The *phase* of a wave is a more subtle concept and has meaning only when specified relative to another wave of the same frequency. The phase tells us when the wave reaches its maximum value compared to the time of the maximum for the other wave. Two waves of the same frequency are *in phase* if they reach their maximum values at exactly the same time, as shown in Fig. 2.4(a). It makes no sense to compare the phase of two waves of different frequencies. The two waves of Fig. 2.4(b) are out of phase because their peaks do not occur at the same time. Their phase difference is seen to be $\frac{1}{4}T$ s or 45°, because one complete period T corresponds to 360°.

Mathematically, the initial phase is expressed as the angle ϕ in the equation $v(t) = V_0 \sin(2\pi f t + \phi) = V_0 \sin(2\pi t/T + \phi)$ which is graphed in Fig. 2.5(a), and the phase is the entire argument $(2\pi t/T + \phi)$. The equivalence between phase in degrees (or radians) and time can be seen from the fact that a change in t by one period T or a change in ϕ by 360° or 2π rad does not change $v(t)$. Expressed in words, the phase determines the wave amplitude at the arbitrary time $t = 0$, because at $t = 0$, $v = V_0 \sin \phi$. Phase is measured in degrees or radians and does not depend upon the amplitude or the frequency.

The voltage $v_1(t) = V_{01} \sin(\omega t + \phi)$ has a phase of ϕ radians relative to the voltage $v_2(t) = V_{02} \sin \omega t$; in other words $v_1(t)$ *leads* $v_2(t)$ by ϕ radians, or, less precisely, $v_1(t)$ is out of phase with respect to $v_2(t)$ by ϕ radians. If two waves (of the same frequency) are out of phase by π radians or 180°, then they appear as shown in Fig. 2.5(b). Thus two waves of equal amplitude and exactly 180° out of phase add to give exactly zero. From

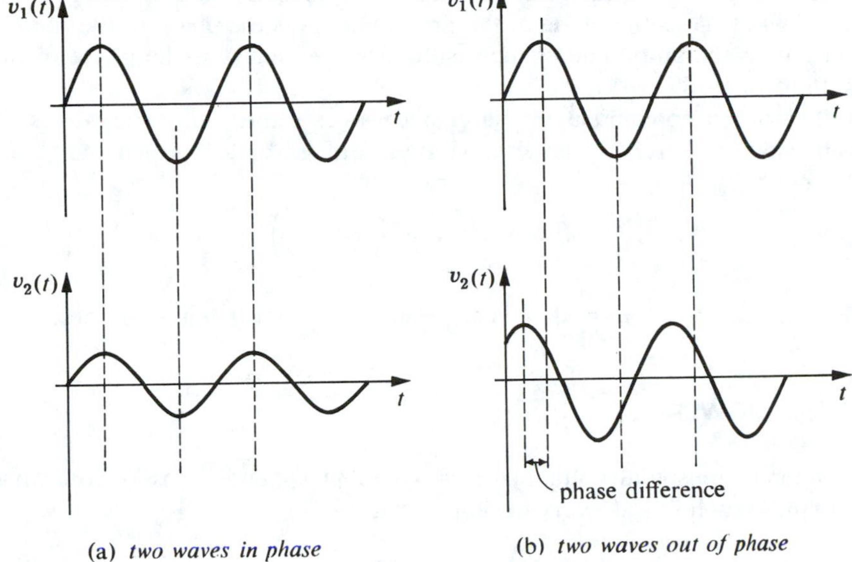

(a) *two waves in phase* (b) *two waves out of phase*

FIGURE 2.4 Phase relationships.

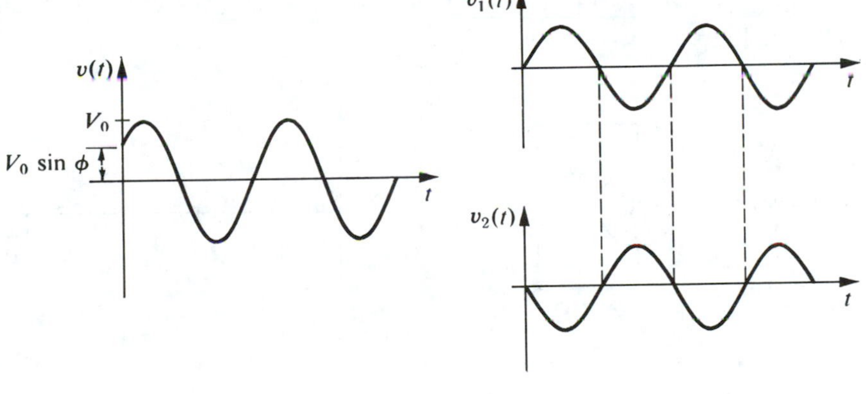

(a) $v(t) = V_0 \sin(\omega t + \phi)$ (b) *two waves 180° out of phase*

FIGURE 2.5 More phase relationships.

trigonometry any two sine waves of the same frequency and arbitrary phases will add together to give a sine wave of the same frequency ω and a definite phase ϕ. That is,

$$V_{01} \sin(\omega t + \phi_1) + V_{02} \sin(\omega t + \phi_2) = V_0 \sin(\omega t + \phi)$$

where V_{01} is the amplitude of the first wave, V_{02} is the amplitude of the second wave, ϕ_1 is the phase of the first wave, ϕ_2 is the phase of the second wave, V_0 is the amplitude of the resultant wave, and ϕ is the phase of the resultant wave.

It is often convenient to use cosine waves rather than sine waves. A cosine wave is merely a sine wave shifted in phase by 90° or $\pi/2$ rad.

$$V_0 \cos \omega t = V_0 \sin\left(\omega t + \frac{\pi}{2}\right)$$

The identity $\sin(a + b) = \sin a \cos b + \sin b \cos a$ has been used here.

2.2 AC POWER

Consider a sinusoidal voltage $v(t) = V_0 \sin \omega t$ applied across a resistance R. Ohm's law holds at every instant of time, so

$$i(t) = \frac{v(t)}{R} = \frac{V_0 \sin \omega t}{R} = \frac{V_0}{R} \sin \omega t = I_0 \sin \omega t \tag{2.3}$$

Thus the current through the resistance also varies sinusoidally with time at the same frequency as the voltage. The amplitude I_0 of the current equals V_0/R (see Fig. 2.6).

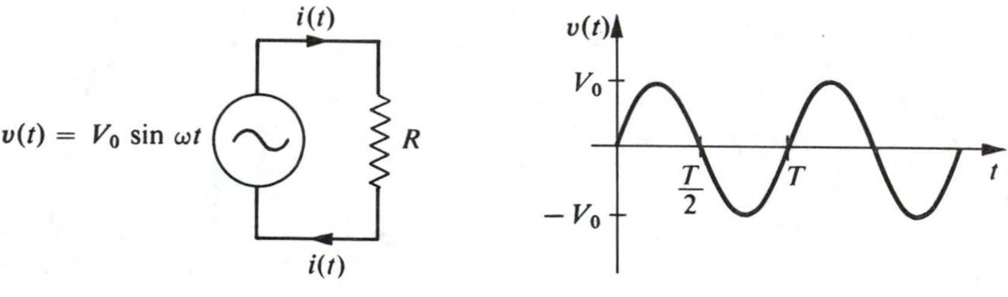

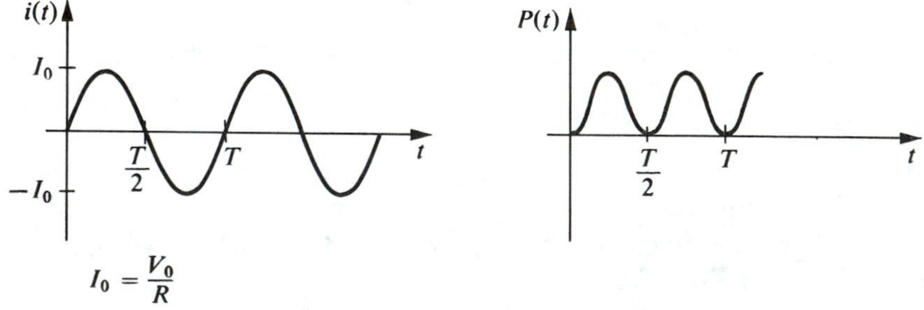

FIGURE 2.6 Alternating current and voltage through a resistance.

Notice that the current and the voltage are exactly in phase. The instantaneous power developed in the resistance is still given by $p = vi$ (the dc formula), but now v is the instantaneous value of the voltage drop across the resistance, and i is the instantaneous value of the current through the resistance. Hence, $p(t) = I_0^2 R \sin^2 \omega t$; the *instantaneous* power $p(t)$ thus varies with time, as is shown in Fig. 2.6.

Usually, however, it is much more useful to consider the *average* power, that is the power averaged over an integral number of cycles of voltages and current. The average power can be written by averaging the instantaneous power over one (or many) periods of time:

$$P_{av} = \frac{1}{T}\int_0^T p(t)\,dt = \frac{1}{T}\int_0^T v(t)i(t)\,dt$$
$$= \frac{1}{T}\int_0^T (V_0 \sin \omega t)\left(\frac{V_0}{R}\sin \omega t\right) dt \tag{2.4}$$

$$P_{av} = \frac{1}{T}\frac{V_0^2}{R}\int_0^T \sin^2 \omega t\,dt = \frac{V_0^2}{2R} \tag{2.5}$$

But $$V_{rms} = V_0/\sqrt{2}$$

so $$P_{av} = \frac{V_{rms}^2}{R} \tag{2.6}$$

And $$I_{rms} = I_0/\sqrt{2} = V_0/\sqrt{2}R = V_{rms}/R, \tag{2.7}$$

so $$P_{av} = I_{rms}^2 R$$

The result, then, for sinusoidal current flowing through a resistance is that the average power dissipated in the resistance equals $I_{rms}^2 R$ or V_{rms}^2/R. The occurrence of the rms values in the power formula is one reason rms values are useful. In words, a dc current $I = I_{rms} = I_0/\sqrt{2}$ would have the same heating effect as the ac current of amplitude I_0, and similarly for a dc voltage $V = V_{rms} = V_0/\sqrt{2}$.

If the current through a two-terminal element is sinusoidal and the voltage across it is also sinusoidal of the same frequency but differing in phase by ϕ rad, then the instantaneous power dissipated in the element is

$$p = v(t)i(t) = V_0 \sin(\omega t + \phi) I_0 \sin \omega t \tag{2.8}$$

and the average power dissipated can be shown to be

$$P_{av} = \frac{V_0 I_0}{2}\cos \phi = V_{rms} I_{rms} \cos \phi \tag{2.9}$$

2.3 CAPACITANCE

Capacitance is very important in ac circuits. A circuit element that has capacitance is called a *capacitor* or a *condenser*; it is represented in circuit diagrams by the symbol in Fig. 2.7. Capacitance is defined as the charge stored on one plate divided by the voltage difference between the two plates:

$$C \equiv \frac{Q}{V} \tag{2.10}$$

The total net charge on the two plates is always zero. (If $+Q$ is on one plate, then $-Q$ is on the other plate because the presence of positive charge

V_A V_B

(a) *schematic symbol*

$$C \equiv \frac{Q}{|V_A - V_B|}$$

(b) *definition*

FIGURE 2.7 Symbol of capacitance.

on one plate repels an equal amount of positive charge from the other plate.) The larger the capacitance, the more charge is stored on the plates for a given voltage difference; capacitance is the "amount of charge stored per volt." A capacitor has two terminals or leads, one going to each plate.

Capacitance is measured in coulombs per volt; one coulomb per volt is called a *farad* (F). A 1-F capacitor would be physically huge, so other units are used for capacitance: the microfarad (μF), which equals 10^{-6} F, and the picofarad (pF), which equals 10^{-12} F. A picofarad is often called a "puff" in informal conversation. The reason for the capacitance symbol in Fig. 2.7 is that a capacitor is actually made by placing two metal plates or foils parallel to but insulated from each other; the two terminals are connected to the two plates. Notice that if the dielectric (i.e., the insulating material) between the two plates does not break down from too high a voltage being applied across the plates, then *there is no way direct current can flow through the capacitor*. In other words, a capacitor has an infinite dc resistance in the steady state. The thicker the insulating material, the higher the voltage rating of the capacitor.

However, *alternating* current can pass through a capacitor. As a positive charge surges onto the left-hand plate, the positive charge on the right-hand plate is repelled toward the right. Thus a surge of current into one terminal results in a surge of current out of the other terminal, and, one-half cycle later, charge surges into the right-hand terminal and out of the left-hand terminal. In other words, the capacitor does pass alternating current.

There are two general types of capacitors: polarized and unpolarized. *Polarized* capacitors function properly only when a definite dc voltage polarity is maintained across the plates. One terminal is labeled + and the other −; the dc voltage on the + side must always be maintained positive with respect to the − side regardless of the ac voltage present. The most common type of polarized capacitor is the electrolytic capacitor in which the dielectric film between the two conducting plates is formed by an electrochemical reaction. The film is extremely thin; thus, large capacitances can be obtained at relatively low voltage ratings in a small volume. For example, an electrolytic capacitor as large as half a cigarette may have 100-μF capacitance with a maximum voltage rating of 25 V dc. If the dc voltage polarity across an electrolytic capacitor is reversed, the dielectric film is usually punctured, and the capacitor is permanently shorted out. Such a situation can usually be detected by measuring the dc resistance of the capacitor; a good electrolytic capacitor should yield a dc resistance of several hundred kΩ or more. In making this resistance measurement, place the red or positive ohmmeter probe on the + lead of the electrolytic capacitor.

Unpolarized capacitors can function properly with either polarity of dc voltage between the plates. There are many types. The *tubular* capacitor shown in Fig. 2.8 consists of a thin sheet of insulating material (mylar,

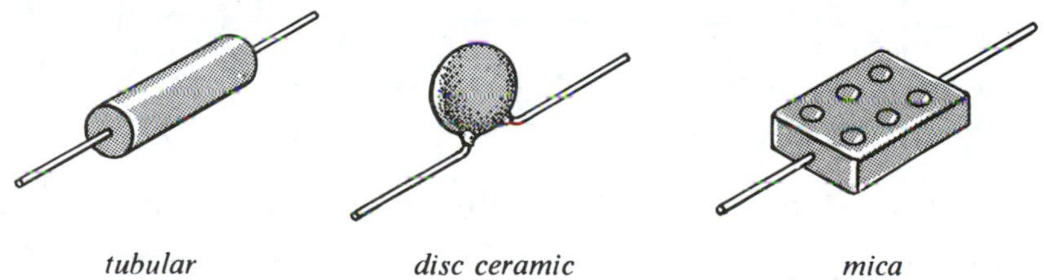

FIGURE 2.8 Various types of capacitors.

paper, etc.) rolled up in a cylinder between two thin metal foils. One wire lead is connected to each metal foil, and the entire assembly is potted in wax or some insulating plastic. Tubular capacitors are normally used for audio frequencies, and typical capacitance values range from 0.001 μF to 1 μF.

The *disc ceramic* capacitor, shown in Fig. 2.8, has a thin layer of ceramic as the insulator between the electrodes and, unfortunately, a rather large temperature coefficient; that is, the capacitance changes relatively rapidly with changing temperature. Hence, they are usually used for bypass applications where only a certain minimum capacitance is required, rather than a specific value of capacitance. They are used at frequencies up to several hundred MHz. A special type of disc ceramic, type NPO, has a zero temperature coefficient at room temperature and is widely used.

The *mica* capacitor, shown in Fig. 2.8, has a thin sheet of mica as the insulator and provides the most stable, precise value of capacitance of all capacitor types. Mica capacitors are useful up to hundreds of megahertz and are also rather expensive. Another type of expensive capacitor is the *tantalum* capacitor, which has low inductance, high capacitance per unit volume, and provides relatively high capacitance values, up to hundreds of μF. A summary of the properties of different types of capacitors is given in Appendix A.

Variable capacitors, where the capacitance is adjusted by turning a shaft, are also available. As the shaft turns, two sets of parallel metal plates mesh, without touching [Fig. 2.9(b)]: the larger the overlap, the higher the capacitance. The stationary set of plates is called the *stator*, the rotating set the *rotor*. The rotor is connected electrically and mechanically to the shaft, which is usually connected to the chassis and thus is at ground. The curved line in the schematic symbol [Fig. 2.9(a)] is always the rotor and is usually grounded.

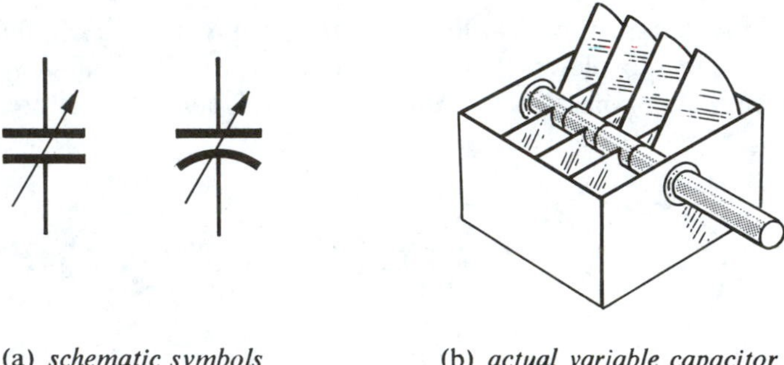

(a) *schematic symbols* (b) *actual variable capacitor*

FIGURE 2.9 Variable capacitor.

A useful way to think of a capacitor in terms of the water-flow current analogy is to regard the capacitor as an enlargement in the water pipe with a flexible membrane stretched across the enlargement, as shown in Fig. 2.10. As water surges in from the left, the membrane stretches toward the right, and water surges out of the right-hand pipe. No water actually passes completely through, from left to right, but the surge of water does flow out of the right-hand pipe. In the case of the capacitor, no direct current flows through, but alternating current does because ac is really a back-and-forth surging of electrons. Note that as the water pressure between the two sides increases, the membrane stretches and water flows out of the low-pressure side. This is analogous to a capacitor, which passes some current every time the voltage difference across the capacitor plates changes. A very stiff membrane corresponds to a small capacitance; a very flexible one to a large capacitance.

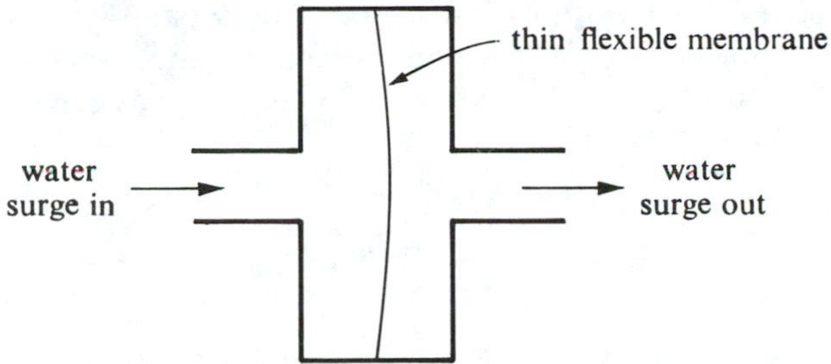

FIGURE 2.10 Water-pipe analogy of capacitor.

Capacitance depends on the area of the plates, the separation of the plates, and the material between the plates. It does not depend on current or voltage. For two parallel plates each of area A and separation d, the capacitance can be shown to be $C = \epsilon_0 A/d$, for a vacuum between the plates, and $C = k\epsilon_0 A/d$ for a material of dielectric constant k (a pure number) between the plates. If A is in square meters, d in meters, and ϵ_0 in farads per meter, then C is in farads. The constant $\epsilon_0 = 8.85 \times 10^{-12}$ F/m is the *permittivity* of free space, which occurs in so many formulae in mks units. Because C varies inversely with d, we see that if we obtain a higher voltage rating by using a thicker insulating material (larger d) between the plates, we must decrease the capacitance for a given volume capacitor. In other words, we can have a high voltage rating and a low capacitance, or a low voltage rating and a high capacitance, for a fixed physical size.

If the capacitors of capacitance C_1 and C_2 are connected in parallel as in Fig. 2.11(a), then the resultant can be regarded as one new capacitor. We would expect the new capacitor to have a capacitance larger than C_1 or C_2 because the area of the left-hand plate of the new resultant capacitor is the sum of the areas of the left-hand plates of C_1 and C_2, and similarly for the

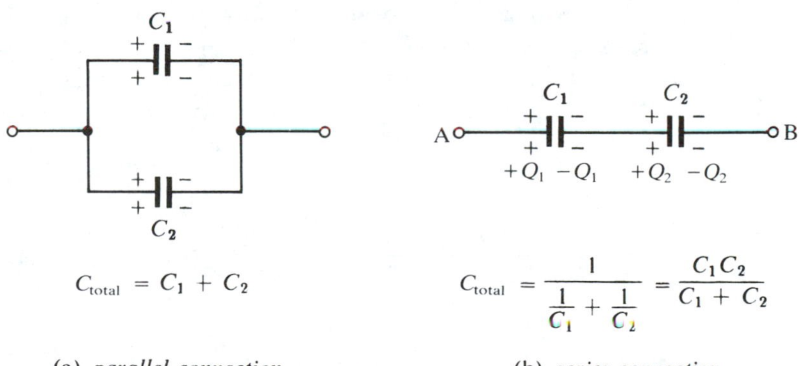

FIGURE 2.11 Combinations of two capacitors.

right-hand plates. If we do not change the plate separation of each capacitor when we connect them in parallel, we might, since the areas add, then expect simply that $C_{\text{total}} = C_1 + C_2$, which is the case, as can be seen by the following derivation. By definition

$$C_{\text{total}} = \frac{Q_{\text{total}}}{V} \tag{2.11}$$

where Q_{total} is the total charge on either plate and V is the voltage difference across the plates. Note that $Q_{\text{total}} = Q_1 + Q_2$ where $Q_1 =$ the charge on C_1 and $Q_2 =$ the charge on C_2. Therefore,

$$C_{\text{total}} = \frac{Q_1 + Q_2}{V} \tag{2.12}$$

From the definition of capacitance $Q_1 = C_1 V_1$ and $Q_2 = C_2 V_2$, and because C_1 and C_2 are in parallel, $V_1 = V_2$. Therefore

$$C_{\text{total}} = \frac{C_1 V + C_2 V}{V} = C_1 + C_2 \tag{2.13}$$

For example, a 0.01-μF capacitor and a 0.05-μF capacitor in parallel act effectively like one 0.06-μF capacitor. For N capacitors in parallel it is easily shown that

$$C_{\text{total}} = C_1 + C_2 + \cdots + C_N \tag{2.14}$$

If two capacitors C_1 and C_2 are connected in series as in Fig. 2.11(b), then the resultant can be regarded as one new capacitor. If a voltage difference V is applied between terminals A and B, with A positive with respect to B, then positive charge from the right-hand plate of C_1 will be repelled to the left-hand plate of C_2, where it must stop. Thus, $Q_1 = Q_2$, because the center wire, the right-hand plate of C_1, and the left-hand plate of C_2 must remain electrically neutral. Thus the charge on C_1 must exactly equal the charge on C_2. Let this charge be denoted by Q. By definition $C_{\text{total}} = Q/V = Q/(V_1 + V_2)$. But $V_1 = Q_1/C_1$ and $V_2 = Q_2/C_2$ and $Q_1 = Q_2 = Q$, so

$$C_{\text{total}} = \frac{Q}{\dfrac{Q}{C_1} + \dfrac{Q}{C_2}} = \frac{1}{\dfrac{1}{C_1} + \dfrac{1}{C_2}} = \frac{C_1 C_2}{C_1 + C_2} \tag{2.15}$$

Thus the total capacitance of two capacitors in series is the reciprocal of the sum of the reciprocals of the capacitances. This result is analogous to that for two resistors in parallel. For N capacitors in series

$$C_{\text{total}} = \frac{1}{\dfrac{1}{C_1} + \dfrac{1}{C_2} + \cdots + \dfrac{1}{C_N}} \tag{2.16}$$

A 0.01-μF capacitor and a 0.05-μF capacitor in series act like a 0.0083-μF capacitor.

In addition to manufactured capacitors intentionally wired in a circuit, there is always some "stray" capacitance between any two elements in the circuit: for example, between a wire and the metal chassis, or between the grid and plate of a vacuum tube, or between the base and collector of a transistor. Such stray capacitance is rarely shown on circuit diagrams but, nevertheless, can often be extremely important, particularly at high frequencies—for example, at tens or hundreds of MHz. The reason is that at high frequencies capacitors tend to act as "low resistances." In the limit as the frequency increases the capacitance acts like a short circuit. We will show this in Section 2.4

When a capacitor is charged with a voltage difference V between the plates, there is an electric field present in the region between the plates. Energy is stored in this electric field, and work must be done to create this field—that is, work must be done to charge up the capacitor. Conversely, when the capacitor discharges, the energy stored in its electric field must go somewhere; it must be dissipated as heat, or it must be stored in some other form, such as the energy of a magnetic field. The energy stored in a capacitance C with a voltage difference V between the plates is $W = \frac{1}{2}CV^2$. The derivation is short:

$$W = \int_0^V V\,dQ = \int_0^V V\,d(CV) = C\int_0^V V\,dV = \frac{1}{2}CV^2 \tag{2.17}$$

Because Q, the charge on either plate, is related to V by $Q = CV$, W can be rewritten as $W = QV/2$ or $W = Q^2/2C$.

2.4 CAPACITIVE REACTANCE

Suppose a sinusoidal voltage $v = V_0 \cos \omega t$ is applied across a capacitor as shown in Fig. 2.12(a). Then the current through the capacitor can be obtained by taking the first derivative with respect to time of the equation $Q = Cv$:

$$i = \frac{dQ}{dt} = \frac{d}{dt}(Cv) = \frac{d}{dt}(CV_0 \cos \omega t) = -\omega C V_0 \sin \omega t = \omega C V_0 \cos\left(\omega t + \frac{\pi}{2}\right) \tag{2.18}$$

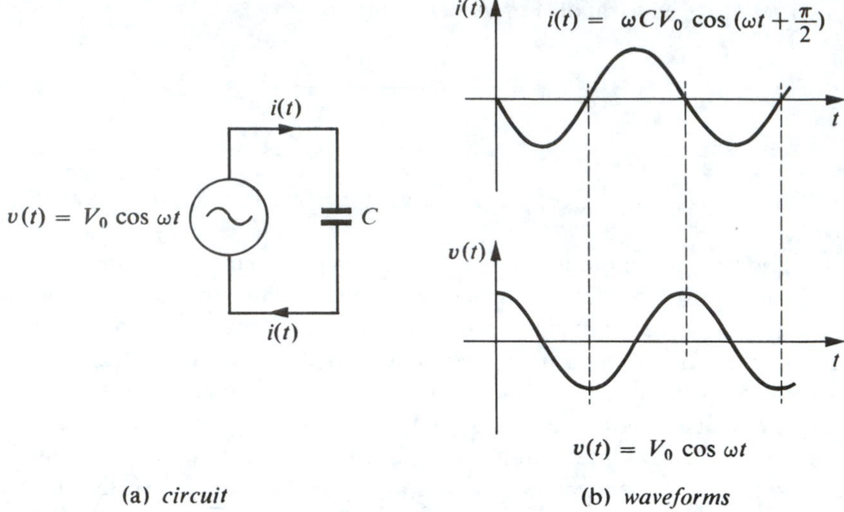

FIGURE 2.12 Current and voltage in a capacitance.

where we have used the trigonometric identity

$$\cos(\omega t + \pi/2) = -\sin \omega t$$

Thus, we see that the current i varies sinusoidally with time at the same frequency ω as does the applied voltage but is 90° or $\pi/2$ rad out of phase with respect to the voltage. The current through the capacitor leads the voltage across the capacitor by 90°, or, equivalently, the voltage lags the current by 90° or $\pi/2$ rad, as shown in Fig. 2.12(b). Notice also that the higher the frequency ω of the applied voltage, the larger the current for a given voltage; that is, at higher frequencies the capacitor presents *less* opposition to the current flow.

All these features can be consolidated very neatly by introducing complex numbers. For a brief summary of complex numbers see Appendix E. We now regard the applied voltage $V_0 \cos \omega t$ as the real part of $V_0 e^{j\omega t}$, which we can call the complex voltage (j^2 equals -1, and $e^{j\theta} = \cos\theta + j\sin\theta$):

$$v(t) = V_0 \cos \omega t = \text{real part of } [V_0 e^{j\omega t}] \tag{2.19}$$

The current can also be written as the real part of a complex number:

$$i(t) = \omega C V_0 \cos\left(\omega t + \frac{\pi}{2}\right) = \text{Re}[\omega C V_0 e^{j(\omega t + \pi/2)}] \tag{2.20}$$

where Re stands for "real part of."

Now we would like to be able to write an ac version of Ohm's law that takes into account the *phase* as well as the amplitude of the current through the capacitor. The current can be written in a form similar to Ohm's law ($i = v/R$) by writing

$$i(t) = \text{Re}[\omega C V_0 e^{j\omega t} e^{j\pi/2}] = \text{Re}[j\omega C V_0 e^{j\omega t}] \tag{2.21}$$

where we have used $e^{j\pi/2} = j$. Then we can write

$$i(t) = \text{Re}\left[\frac{V_0 e^{j\omega t}}{\dfrac{1}{j\omega C}}\right] \tag{2.22}$$

The expression says that the current through the capacitor equals the voltage divided by $1/j\omega C$, which depends only on the capacitance and the frequency of the applied voltage. In other words, the current equals the real part of the ratio of the complex voltage to the complex number $1/j\omega C$. Thus, $1/j\omega C$ plays the role of the "effective resistance" or "impedance" of the capacitor to current flow: the larger $1/j\omega C$, the smaller the current for a given voltage. The value $1/j\omega C$ is called the *capacitive reactance*, usually denoted by X_C, and is measured in ohms. The capacitive reactance can be thought of as the ac frequency-dependent resistance of a capacitor. By convention we usually omit writing "real part of":

$$i(t) = \frac{v(t)}{X_C} = \frac{V_0 e^{j\omega t}}{\dfrac{1}{j\omega C}} \tag{2.23}$$

It is important to realize that once we define capacitance by $C = Q/V$, we are forced to the conclusions that the current and voltage are 90° out of phase, and that the current through a capacitor equals the voltage across it divided by $1/j\omega C$. *The presence of the complex number j simply takes into account the phase of the current relative to the voltage*. Notice that because the current through and voltage across a capacitor differ in phase by $\phi = 90°$, the power dissipated in the capacitor is zero because $P = I_{\text{rms}} V_{\text{rms}} \cos \phi = I_{\text{rms}} V_{\text{rms}} \cos 90° = 0$. This is in complete contrast to a resistance where the current and voltage are in phase, and electrical power is dissipated as heat. Notice also that in the limiting case of zero frequency, which is direct current, the capacitive reactance goes to infinity; that is, a capacitor can pass no direct current.

It is often desired to block out the dc component of a voltage while retaining the ac component. A series capacitor called a *blocking* capacitor or a *coupling* capacitor does the trick, as shown in Fig. 2.13. If any dc current I_{dc} flowed through C, there would be a dc voltage drop across R equal to

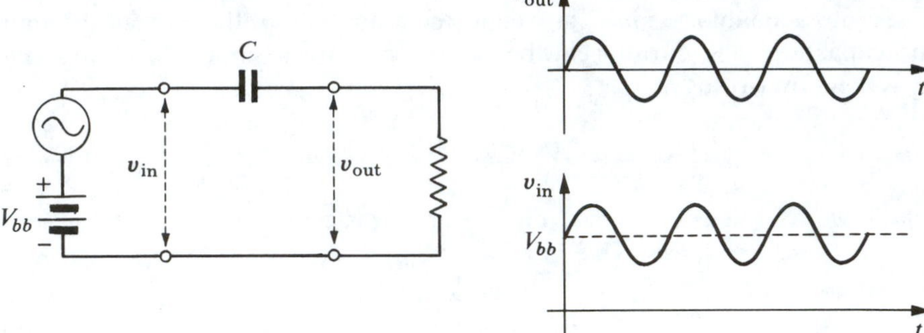

FIGURE 2.13 Coupling or blocking capacitor.

$I_{dc}R$. But the capacitor cannot pass any dc current; hence

$$I_{dc} = 0 \quad \text{and} \quad V_{dc\,out} = 0$$

2.5 INDUCTANCE

Inductance is also important in ac circuits. A circuit element that has inductance is called an *inductor*, a *choke*, or sometimes an *inductance*. It is represented in circuit diagrams as a coil (see Fig. 2.14).

Actual inductors are usually made by winding wire as a coil around some kind of core; the two terminals are the two ends of the coil. The definition of inductance is

$$L \equiv \frac{\Phi}{I} \tag{2.24}$$

the ratio of magnetic flux Φ with respect to current I. (We recall that magnetic flux equals the magnetic field induction times area, $\Phi = BA$.) The current I flowing through the coil produces a magnetic flux Φ through the cross-sectional area of the coil. The larger I is, the larger Φ is, so L is

(a) *schematic symbol* (b) *actual inductor*

FIGURE 2.14 Circuit symbol for inductance.

positive. A more easily understood definition of inductance is in terms of the voltage drop v across the inductor; L is defined from the equation $v = -L\,dI/dt$. These two definitions are, of course, equivalent, as can be seen from the Faraday law expression for the voltage produced by a changing magnetic flux. Starting with Faraday's law, $v = -d\Phi/dt = -d/dt(LI) = -L\,dI/dt$. If the current changes through the inductor at a rate of one ampere per second, and the voltage produced across the inductor is one volt, then the inductor has an inductance of one *henry* (H). Typical units used for inductance are the millihenry (mH), which equals 10^{-3} H, and the microhenry (μH), which equals 10^{-6} H.

The physical reason for the inductance is that the current flowing in the coil produces a magnetic flux through the coil. As the current changes ($dI/dt \neq 0$), the magnetic flux through the coil also changes $d\Phi/dt \neq 0$; hence a voltage is induced by Faraday's law (induced voltage or emf $= -d\Phi/dt$) between the two ends of the coil. In other words, whenever the magnetic flux lines "cut" the wire turns of the coil, a voltage is induced in the coil. The minus sign means that the voltage induced is of a polarity to oppose the change in current that produced the voltage. We can see from Faraday's law that a direct current (I = constant and $dI/dt = 0$) will produce no voltage across the coil terminals; the only voltage drop produced by direct current I through the coil will be the Ohm's law drop $V = IR$, where R is the resistance of the wire in the coil, which is usually negligibly small. For example, a low-power, 1-H iron-core inductance may have a dc resistance of about 50 Ω; hence 10 mA dc flowing through the inductance will produce a steady voltage drop across it of only $\frac{1}{2}$V. Inductance ideally depends upon the number of turns on the coil, the size of the coil, and the material inside the coil. It is usually independent of current or voltage.

Solid or powdered iron is often placed inside the coil to increase the inductance. The iron produces more magnetic flux Φ through the coil for a given number of ampere-turns. However, eddy currents always flow in the iron core, which has some resistance; thus I^2R power is dissipated as heat in the core. This power comes from the ac signal in the coil, so that ac signal is attenuated or reduced. The higher the frequency of the ac, the larger the power loss. It can be shown that the power loss is linear with frequency—doubling the frequency exactly doubles the power loss to the core. This power loss can be reduced by using a powdered iron core in which very small particles of iron are cemented in an insulating glue to reduce the eddy currents. Powdered-iron-core inductances can be used at frequencies up to several tens of MHz. For higher frequencies air-core coils are usually used. Note that iron-core inductors often change their inductance if the dc current through the coil is changed. This is due to *magnetic saturation*† of the core (see Fig. 2.15). $L = \Phi/I$, and if saturation occurs, L clearly decreases.

†Saturation means a flattening or leveling off of the Φ-vs.-I curve.

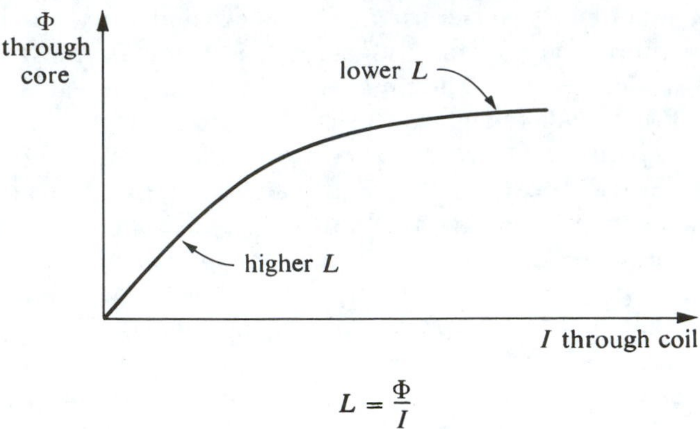

$$L = \frac{\Phi}{I}$$

FIGURE 2.15 Inductance changes with changing dc current.

If two inductors of inductance L_1 and L_2 are connected in series, as shown in Fig. 2.16(a), then we can show that the resultant can be regarded as a single inductor with an inductance equal to $L_1 + L_2$. This result follows from the argument

$$v = L\frac{dI}{dt} = v_1 + v_2 \tag{2.25}$$

But $v_1 = L_1(dI_1/dt)$ and $v_2 = L_2(dI_2/dt)$. And because L_1 and L_2 are in series, $I_1 = I_2 = I$ and $dI_1/dt = dI_2/dt = dI/dt$. Therefore

$$v = v_1 + v_2 = L_1\frac{dI_1}{dt} + L_2\frac{dI_2}{dt} = (L_1 + L_2)\frac{dI}{dt} = L\frac{dI}{dt}$$

Hence,

$$L = L_1 + L_2 \tag{2.26}$$

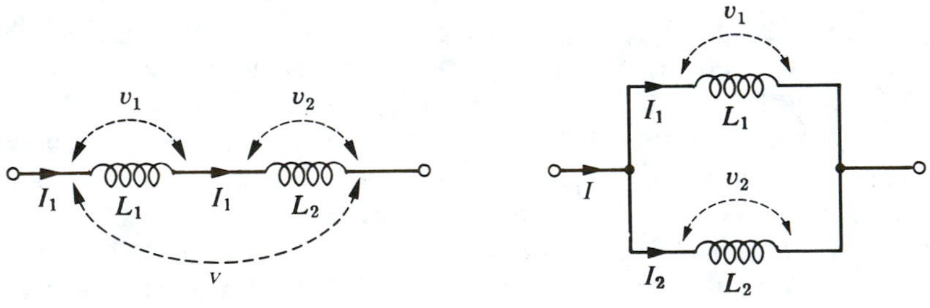

$$L_{\text{total}} = L_1 + L_2$$

(a) *series*

$$L_{\text{total}} = \frac{1}{\dfrac{1}{L_1} + \dfrac{1}{L_2}} = \frac{L_1 L_2}{L_1 + L_2}$$

(b) *parallel*

FIGURE 2.16 Series and parallel inductances.

If two inductors are connected in parallel (this is almost never done in actual practice because of magnetic coupling between the two inductors) as shown in Fig. 2.16(b), the resultant can be regarded as a single inductor of inductance $L = 1/(1/L_1 + 1/L_2)$. This result follows from $I = I_1 + I_2$. Therefore,

$$\frac{dI}{dt} = \frac{dI_1}{dt} + \frac{dI_2}{dt}$$

By the definition of inductance $dI/dt = v/L$, $dI_1/dt = v_1/L_1$, $dI_2/dt = v_2/L_2$, and, because L_1 and L_2 are in parallel, $v_1 = v_2 = v$. Therefore,

$$\frac{v}{L} = \frac{v}{L_1} + \frac{v}{L_2} \quad \text{or} \quad \frac{1}{L} = \frac{1}{L_1} + \frac{1}{L_2} \quad \text{or} \quad L = \frac{1}{\dfrac{1}{L_1} + \dfrac{1}{L_2}}$$

or

$$L = \frac{L_1 L_2}{L_1 + L_2} \tag{2.27}$$

It can be shown that an air coil of wire of length l, radius R, and N turns has an inductance of

$$L = \mu_0 \frac{N^2}{l} \pi R^2$$

It can be shown that for a given length and cross-sectional area, a wire of rectangular cross section (a flat strip) has much less inductance than a wire of circular cross section. Hence, flat metal strips are often used in very high-frequency circuits instead of ordinary round wire, particularly for ground connections. Even a perfectly straight piece of wire has an inductance because of the finite cross-sectional area of the wire. For example, a 4-in. ($\cong$ 10-cm) long piece of No. 18 (0.04-in. diameter) wire has an inductance of 0.1 μH. This inductance can be very important in determining circuit behavior at high frequencies in the tens or hundreds of MHz. Current flowing in one part of the wire creates a magnetic flux Φ in other parts of the wire, and this flux changing in time affects the current flow in the wire. The inductance also will change with frequency, because at higher frequencies the current tends to be concentrated near the surface of the wire and not in the interior; this phenomenon is called the *skin effect*. Electromagnetic fields do not penetrate into a good conductor. They decrease according to $e^{-x/\delta}$, where x is the depth of the conductor and δ is the *skin depth* of the conductor. It can be shown that δ is given by $\delta = (\pi f \sigma \mu)^{-1/2}$, where f is the frequency of the electromagnetic field, σ is the conductivity, and μ is the permeability of the conductor. The skin depth in copper at 60 Hz is 0.85 mm, and at 1 MHz it is only 0.066 mm.

When an inductance L has a certain current I flowing through it, there is a magnetic field present in the region around the inductance. Energy is stored in this magnetic field, and work must be done to create the field; that is, work must be done to increase the current flowing through the inductance from zero to I. Conversely, when the current through the inductance falls to zero, the magnetic field decreases to zero, and the energy must go somewhere. It must be dissipated as heat or stored in some other form, such as the electric field energy in a charged capacitor. The energy stored in an inductance through which a current I is flowing is given by $W = \frac{1}{2}LI^2$. The derivation is short:

$$W = \int V\,dQ = \int \left(L\frac{dI}{dt}\right)(I\,dt) = L\int_0^I I\,dI = \frac{1}{2}LI^2 \tag{2.28}$$

2.6 MUTUAL INDUCTANCE

In the previous discussion of inductors in series and parallel, we have tacitly assumed that the magnetic flux lines, which cut the wires of an inductor and induce a voltage, come from the inductor itself and not from any other coil. But magnetic flux lines from *any* source will induce a voltage in a wire if they cut the wire. Thus we expect a changing current in one conductor to produce a voltage in a second conductor if the magnetic flux lines from the first conductor cut through the second.

This concept of changing magnetic flux in one conductor being produced by changing current in a different conductor leads us to the concept of *mutual inductance* between the two conductors. Consider the circuit of Fig. 2.17 where we have two coils of inductances L_1 and L_2 with N_1 and N_2 turns, respectively. Suppose a time-varying current I_1 is flowing in L_1. Let Φ_{12} be the magnetic flux through coil 2 caused by the current I_1 flowing in coil 1. The voltage v_{12} induced in L_2 will be given by Faraday's law:

$$v_{12} = -N_2\frac{d\Phi_{12}}{dt} \tag{2.29}$$

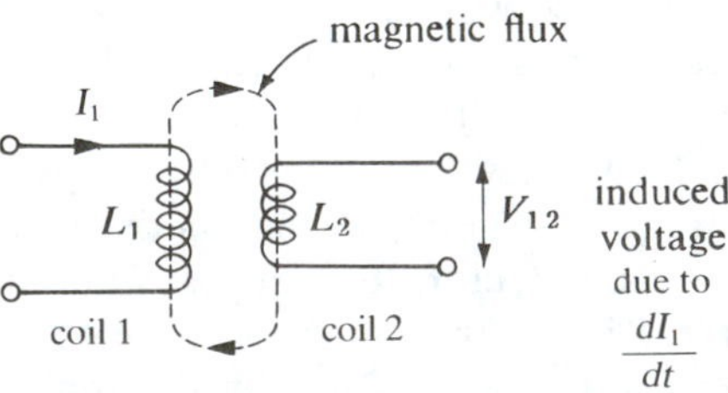

FIGURE 2.17 Mutual inductance between two coils.

But we know from elementary electricity and magnetism that the flux in L_2 must be proportional to I_1. That is, $\Phi_{21} = KI_1$, where K is a constant depending on the distance between the two coils, the number of turns in coil 1, and the relative orientation of the coils. Therefore the voltage induced in L_2 is

$$v_{12} = -N_2K\frac{dI_1}{dt} \tag{2.30}$$

We define $-N_2K$ as M_{12}, the *mutual inductance* between coil L_1 and L_2. In terms of M_{12},

$$v_{12} = M_{12}\frac{dI_1}{dt} \tag{2.31}$$

M_{12} clearly has units of henrys, just like self-inductance. The mutual inductance depends on the number of magnetic flux lines from L_1 that pass through L_2 (i.e., on the geometry—the distance between the coils, the relative orientation of the two coils, and the medium between L_1 and L_2). A little thought will show that the further apart the two coils are, the smaller the mutual inductance is, and if the coils are oriented at right angles, the mutual inductance will be minimum. If the coil axes are parallel, the mutual inductance will be maximum. Any substance between the coils with a high magnetic permeability, such as iron, increases the mutual inductance.

The self-inductance L_2 also creates an induced voltage in coil 2 due to the changing I_2. The total voltage v_2 across coil 2 will therefore by given by

$$v_2 = M_{12}\frac{dI_1}{dt} - L_2\frac{dI_2}{dt} \tag{2.32}$$

A changing current in L_2 produces magnetic flux lines that cut L_1; hence by a similar argument

$$v_1 = M_{21}\frac{dI_2}{dt} - L_1\frac{dI_1}{dt} \tag{2.33}$$

The mutual inductance M_{21} can be shown to equal M_{12}. Two coils wound close together so that most of the flux lines from one pass through the other produces a *transformer*, the circuit symbol for which is shown in Fig. 2.18. If both coils are wound on a common iron core, then almost all the magnetic flux lines produced by a current in one coil will pass through the other coil, as shown in Fig. 2.18(d). The practical details of transformers are discussed in Appendix A.

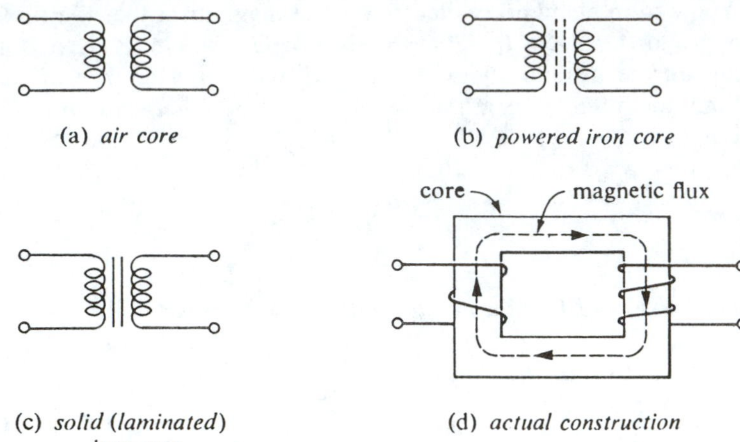

FIGURE 2.18 Transformers.

2.7 INDUCTIVE REACTANCE

Suppose a sinusoidal voltage $v = V_0 \cos \omega t$ is applied across an inductance, as shown in Fig. 2.19. Then the current i through the inductance can be obtained by integrating with respect to time the $v = L(di/dt)$ equation that

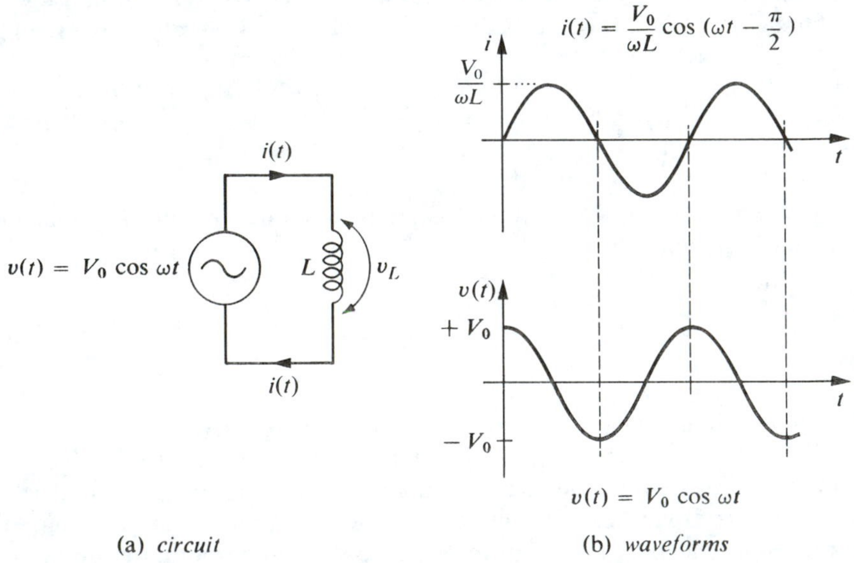

FIGURE 2.19 Current and voltage in an inductance.

defines inductance:

$$v = L\frac{di}{dt} \quad \text{so} \quad di = \frac{v}{L}\,dt \tag{2.34}$$

Now
$$\int_0^t di = i(t) - i(0)$$

So
$$i(t) - i(0) = \frac{1}{L}\int_0^t v\,dt = \frac{1}{L}\int_0^t V_0 \cos \omega t\,dt$$

and
$$i(t) - i(0) = \frac{V_0}{\omega L}\sin \omega t \tag{2.35}$$

If we choose $t = 0$ when the current equals zero, then $i(0) = 0$ and

$$i(t) = \frac{V_0}{\omega L}\sin \omega t = \frac{V_0}{\omega L}\cos\left(\omega t - \frac{\pi}{2}\right) \tag{2.36}$$

where we have used the trigonometric identity $\cos(\omega t - \pi/2) = \sin \omega t$. The current i varies sinusoidally with time at the same frequency ω as does the applied voltage and is 90° or $\pi/2$ rad out of phase with respect to the voltage. The current in the case of the inductance *lags* the voltage by $\pi/2$ rad or 90° as shown in Fig. 2.19. Thus, the power dissipated in an inductance as heat is zero because of the 90° phase difference between the current and the voltage. Notice that the higher the frequency of the applied voltage, the smaller the current; that is, at higher frequencies the inductance presents more "opposition" to the current flow. The use of complex numbers can be used to consolidate these features. Again, we regard the applied voltage $V_0 \cos \omega t$ as the real part of $V_0 e^{j\omega t}$ and write

$$i = \frac{V_0 \cos\left(\omega t - \dfrac{\pi}{2}\right)}{\omega L} = \text{Re}\left(\frac{V_0 e^{j(\omega t - \pi/2)}}{\omega L}\right)$$

$$i = \text{Re}\left(\frac{-jV_0 e^{j\omega t}}{\omega L}\right) \quad \text{using } e^{-j\pi/2} = -j$$

$$i = \text{Re}\left(\frac{V_0 e^{j\omega t}}{j\omega L}\right) \tag{2.37}$$

where we have used $-j = 1/j$.

This expression for the current is in a form similar to Ohm's law; the current equals the real part of the complex voltage divided by the complex number $j\omega L$. Thus $j\omega L$ plays the role of the effective resistance of the inductance to current flow—the larger $j\omega L$, the smaller the current for a given voltage. The *inductive reactance* X_L is defined as $j\omega L$ and can be thought of as the ac frequency-dependent resistance of an inductance. The presence of the complex number j simply takes into account the *phase* of the current relative to the voltage. By convention, we usually omit writing Re and simply write

$$i = \frac{V_0 e^{j\omega t}}{j\omega L} = \frac{v(t)}{X_L} \tag{2.38}$$

Notice that in the limiting case of zero frequency, which is direct current, the inductive reactance goes to zero, which means that inductance offers *no* opposition to the flow of direct current. Direct current flowing through an inductance is limited only by the resistance of the wire in the coil.

The reciprocal of reactance is sometimes called *susceptance* and is measured in siemens or ohms^{-1}. The susceptance of a capacitance C is

$$B = j\omega C \tag{2.39}$$

The susceptance of an inductance is

$$B = \frac{1}{j\omega L} \tag{2.40}$$

Thus the current through a capacitor C across which a voltage difference v exists is

$$i = B_C v = j\omega C v \tag{2.41}$$

and the current through an inductance L is

$$i = B_L v = \frac{v}{j\omega L} \tag{2.42}$$

2.8 THE COMPLEX VOLTAGE PLANE

Let us now consider a few points about the representation of sinusoidal voltages by complex numbers. The voltage $V_0 \cos \omega t$ can certainly be written as the real part of the complex exponential $V_0 e^{j\omega t}$ because $V_0 e^{j\omega t} = V_0 \cos \omega t + jV_0 \sin \omega t$. We can represent $V_0 e^{j\omega t}$ in the complex voltage plane by a point that moves counterclockwise in a circle of radius V_0 with an angular frequency ω. At any time t the actual voltage $V_0 \cos \omega t$ is the projection of the complex voltage $V_0 e^{j\omega t}$ on the real voltage axis. Such a diagram is often referred to as a rotating vector diagram or *phasor* diagram (see Fig. 2.20). The vector drawn from the origin out to the point $V_0 e^{j\omega t}$ is called the *voltage vector* or *phasor*. The voltage found in the actual circuit is the projection of the rotating complex voltage vector on the real axis.

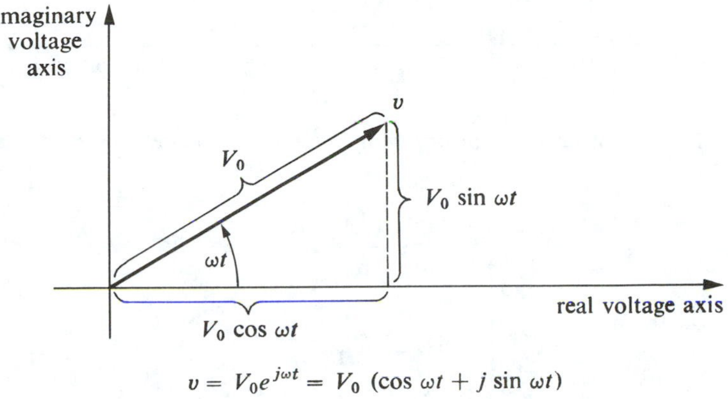

$$v = V_0 e^{j\omega t} = V_0 (\cos \omega t + j \sin \omega t)$$

FIGURE 2.20 Basic rotating vector diagram.

One of the main advantages of this kind of diagram is that the phase difference between two voltages (of the same frequency) is simply the geometrical angle between their two rotating vectors. This will be particularly useful in analyzing phase-shifter circuits in which the output voltage has been shifted in phase with respect to the input voltage. That is, if we have two voltages

$$v_1 = V_{01} \cos(\omega t + \phi_1) \text{ and } v_2 = V_{02} \cos(\omega t + \phi_2)$$

then at $t = 0$ they would be represented by the diagram in Fig. 2.21. The phase difference between v_2 and v_1 is $\phi_2 - \phi_1$, the angle between v_1 and v_2; v_2 leads v_1 by $\phi_2 - \phi_1$. $v_1 + v_2$ can be calculated very easily from the rotating vector diagram by adding v_1 and v_2 using the standard parallelogram method. The phase and amplitude of $v_1 + v_2$ can be read off the diagram immediately.

An example will be useful in analyzing *RC* and *RL* circuits. First, consider two sinusoidal voltages v_1 and v_2 90° out of phase, with v_1 leading v_2. They could be drawn as in Fig. 2.22(a), and their sum would be drawn as the hypotenuse. Notice that the sum $v_1 + v_2$ leads v_2 by θ and lags v_1 by $(90° - \theta)$. They could also be drawn as in Fig. 2.22(b) with $v_1 + v_2$ along the real voltage axis, or as in Fig. 2.22(c) with v_1 along the real voltage axis.

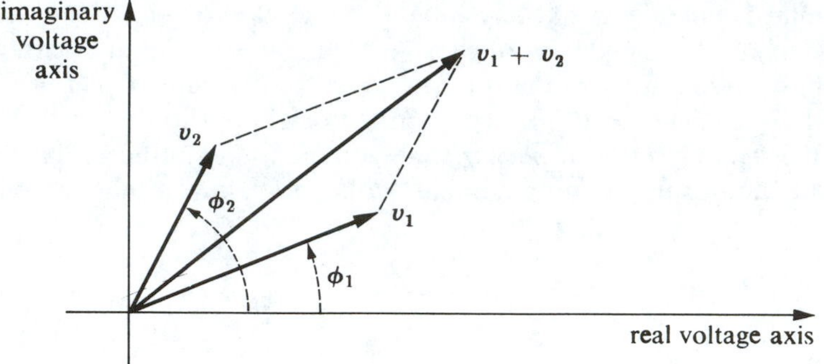

FIGURE 2.21 Two sinusoidal voltages with different phases and their sum at a certain instant of time.

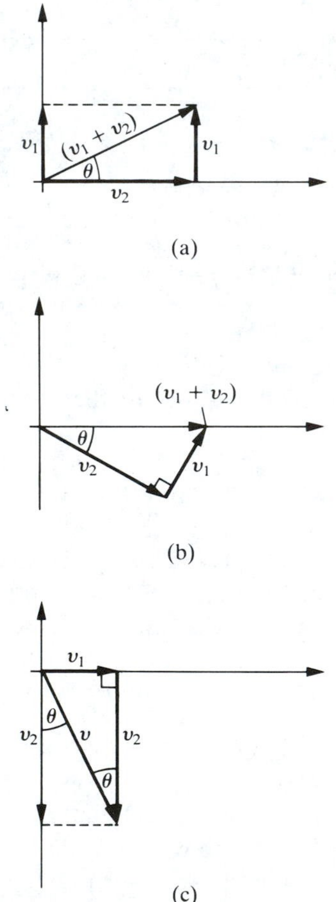

FIGURE 2.22 Voltage addition of v_1 and v_2 90° out of phase.

One voltage obviously can be drawn at any angle in any quadrant, but the voltage triangle and the phases between any two voltages are the same. In all cases the relative phase of the voltages is the same: v_2 lags $v_1 + v_2$ by θ, and v_1 leads $v_1 + v_2$ by $90° - \theta$.

2.9 *RC* HIGH-PASS FILTER

Consider the circuit of Fig. 2.23 in which the output is taken across the resistor. The *gain* of any circuit is defined as the output divided by the input. The *attenuation* of any circuit is defined as the reciprocal of the gain. The voltage gain of the *RC* high-pass filter equals v_2/v_1 and clearly depends on the frequency of the input. At very low frequencies the capacitor presents a very high reactance, thus giving a small output; and at very high frequencies the capacitor is essentially a short circuit, thus making the output nearly equal to the input. In other words, low frequencies will be attenuated, and high frequencies will be passed without much loss in amplitude, with the gain approaching 1.0 as the frequency increases without limit ($\omega \to \infty$). For this reason the circuit is called a *high-pass* circuit or filter. The gain is zero for dc (zero frequency) because the capacitor passes no direct current. If a dc voltage is applied at the input, once the transients have died out all the dc input voltage will appear across C and none across R, thus giving zero output.

Let us calculate the gain, assuming that a negligible amount of current is drawn from the output terminals and that the input voltage is $v_1 = V_{01} \cos \omega t = \mathrm{Re}(V_{01} e^{j\omega t})$. Then the ac current i will be as shown in Fig. 2.24(a). For this sinusoidal input of angular frequency ω we need only replace the capacitor by its complex capacitive reactance $X_C = 1/j\omega C$ and treat the circuit as follows [see Fig. 2.24(b)].

$$\frac{v_2}{v_1} = \frac{iR}{i(X_C + R)} = \frac{iR}{i\left(\dfrac{1}{j\omega C} + R\right)} = \frac{R}{\dfrac{1}{j\omega C} + R} = \frac{1}{1 + \dfrac{1}{j\omega RC}} = \frac{j\omega RC}{1 + j\omega RC} \tag{2.43}$$

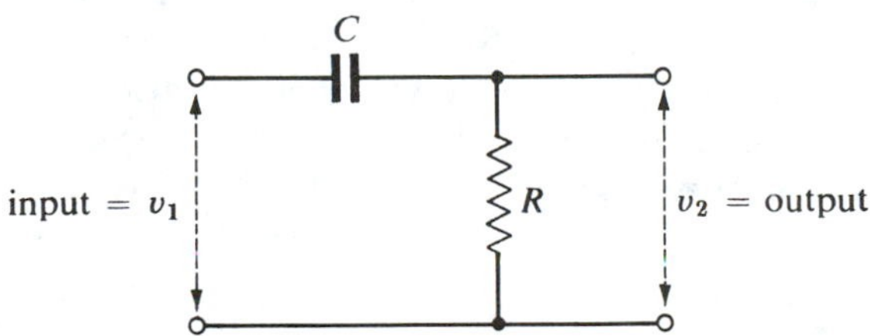

FIGURE 2.23 *RC* high-pass filter.

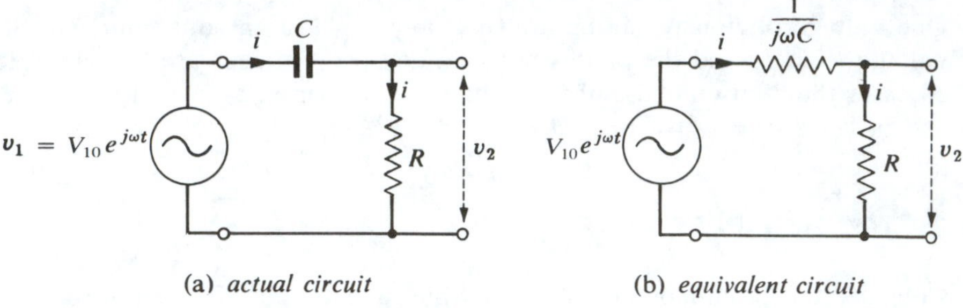

(a) *actual circuit*

(b) *equivalent circuit*

FIGURE 2.24 *RC* high-pass filter with source.

The fact that the voltage gain is complex merely means that the output differs in *phase* from the input. The magnitude or absolute value of the gain will tell us the magnitude of the output divided by the magnitude of the input:

$$\left|\frac{v_2}{v_1}\right|^2 = \left(\frac{v_2}{v_1}\right)^*\left(\frac{v_2}{v_1}\right) = \left(\frac{-j\omega RC}{1 - j\omega RC}\right)\left(\frac{j\omega RC}{1 + j\omega RC}\right) = \frac{\omega^2 R^2 C^2}{1 + \omega^2 R^2 C^2} \qquad \textbf{(2.44)}$$

$$\left|\frac{v_2}{v_1}\right| = \frac{\omega RC}{\sqrt{1 + \omega^2 R^2 C^2}} \qquad \textbf{(2.45)}$$

Notice that this expression for the gain approaches 1.0 as the angular frequency ω of the input approaches infinity, which agrees with our intuitive feeling that the capacitor acts like a short circuit at very high frequencies.

Also notice from (2.45) that for very "low" frequencies ($\omega RC \ll 1$) the gain decreases as the first power of the frequency or by a factor of 2 each time the frequency decreases by a factor of 2.

$$\left|\frac{v_2}{v_1}\right| \cong \omega RC \propto \omega \qquad \textbf{(2.46)}$$

If we plot a graph of the magnitude of the gain versus frequency on log-log paper, we obtain Fig. 2.25. The curved graph [Fig. 2.25(a)] of the actual gain can be approximated quite well by two straight lines [Fig. 2.25(b)], giving a sharp break or "knee" in the graph. The frequency at which this break occurs is given by $\omega_B = 1/RC$. This is called the *break-point* frequency or usually simply the *breakpoint*. For example, a high-pass

*Indicates complex conjugate.

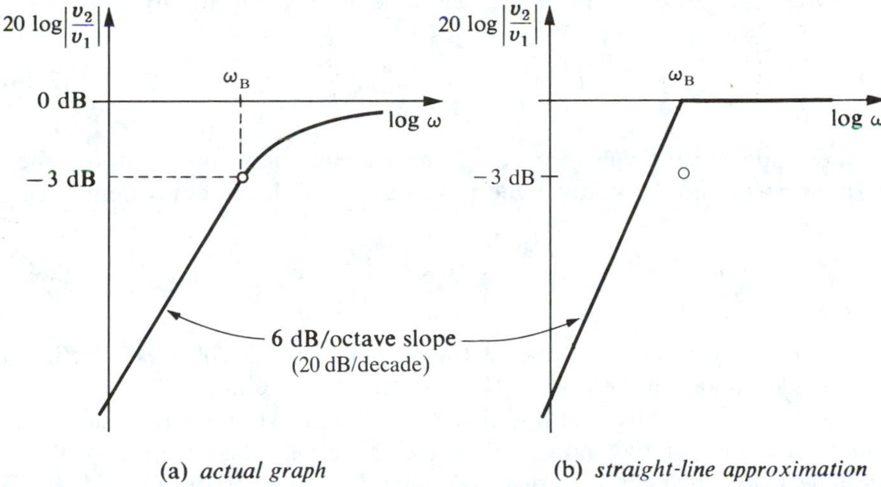

FIGURE 2.25 Gain vs. frequency for *RC* high-pass filter.

filter with $R = 10\ \text{k}\Omega$ and $C = 0.1\ \mu\text{F}$ will have a breakpoint at

$$\omega_B = 1/RC = 1/(10^4\ \Omega)(10^{-7}\ \text{F}) = 10^3\ \text{rad/s, or } f_B = \omega_B/2\pi = 160\ \text{Hz}$$

At the breakpoint the voltage gain $|v_2/v_1| = 0.707$, as can be seen by substituting $\omega = 1/RC$ in the expression for $|v_2/v_1|$. As a gross simplification, the *RC* high-pass filter can be thought of as passing frequencies above $\omega_B = 1/RC$ and attenuating those below ω_B.

It is common practice to express gain (or attenuation) in decibels (dB). The voltage gain v_2/v_1 in decibels is defined as

$$\left(\frac{v_2}{v_1}\right)_{\text{in dB}} \equiv 20 \log_{10}\left(\frac{v_2}{v_1}\right) \quad \textbf{(2.47)}$$

The phase difference (if any) between v_2 and v_1 does not affect the gain expressed in decibels. Note that a negative gain in decibels merely means the output is *less* than the input; a positive gain means, of course, that the output is greater than the input.

$\left\|\frac{v_2}{v_1}\right\|$	*in dB*	$\left\|\frac{v_2}{v_1}\right\|$	*in dB*
0.01	−40	0.5	−6
0.1	−20	0.707	−3
1.0	0	1.414	+3
10.0	+20	2.0	+6
100.0	+40	4.0	+12

Often the power gain is expressed in decibels according to

$$\left(\frac{p_2}{p_1}\right)_{\text{in dB}} \equiv 10 \log_{10}\left(\frac{p_2}{p_1}\right) \tag{2.48}$$

If the impedance levels are the same at the input and output, then $p_2/p_1 = v_2^2/v_1^2$ and the voltage and power gains in dB are equivalent.

$$\left(\frac{p_2}{p_1}\right)_{\text{dB}} = 10 \log_{10}\left(\frac{p_2}{p_1}\right) = 10 \log_{10}\left(\frac{v_2}{v_1}\right)^2 = 20 \log_{10}\left(\frac{v_2}{v_1}\right) \tag{2.49}$$

Thus a power gain of 20 dB means the output power is 100 times the input power and the output voltage is 10 times the input voltage.

Notice that because the log of the product of two gains is equal to the sum of the logs of the individual gains, the total dB gain of two filters, amplifiers, or whatever in series is equal to the *sum* of the individual dB gains. A 10-dB amplifier driving a 30-dB amplifier has a net gain of 40 dB, provided the second amplifier does not "load" the first one. In other words, the second amplifier should not draw too much current from the output of the first amplifier. More on this when input and output impedance are discussed in Chapter 5.

Expressed in decibels, the voltage gain or power gain at the breakpoint for an *RC* high-pass filter is −3 dB or 3 dB "down," relative to the gain of 1.0 at high frequencies. If the gain is down 3 dB at the breakpoint, the voltage gain is down by a factor of 0.707, and the power gain is down by a factor of 2. At lower frequencies the voltage gain continues to fall off at a constant rate of 6 dB per octave frequency change or, equivalently, 20 dB per decade (×10) frequency change according to the straight-line approximation to the gain curve. For example, if the gain at 1000 Hz for an *RC* high-pass filter is 10 dB down, then the gain at 500 Hz will be 16 dB down; at 250 Hz it will be 22 dB down; and so forth.

The phase of the output is different from the phase of the input. This can be seen most easily from a rotating vector diagram. Choose the phase of the current as the reference phase, and recall that the current leads the voltage in a capacitor by 90°. Therefore, the voltage $v_R = iR$ must be 90° more counterclockwise (ahead of v_C). Thus, we have the diagram of Fig. 2.26 because $v_1 = v_R + v_C$. Thus, the output voltage across the resistor leads the input voltage v_1 by θ, which is given by $\theta = \tan^{-1}(|v_C|/|v_R|) = \tan^{-1}(1/\omega RC)$. Notice that as the frequency increases, the phase shift θ goes to zero, and that as θ increases, the output voltage decreases. Sometimes this circuit is used to shift the phase of a sinusoidal signal; the amount of phase shift can be varied by using a variable resistor for R. When $R = 1/\omega C$, $\theta = 45°$ and $|v_2/v_1| = 1/\sqrt{2} = 0.707$. However, as a phase shifter this circuit has two disadvantages: The amplitude of the output voltage changes as R is changed, and also the maximum phase shift possible with this circuit is 90° in the limit as the output voltage goes to zero.

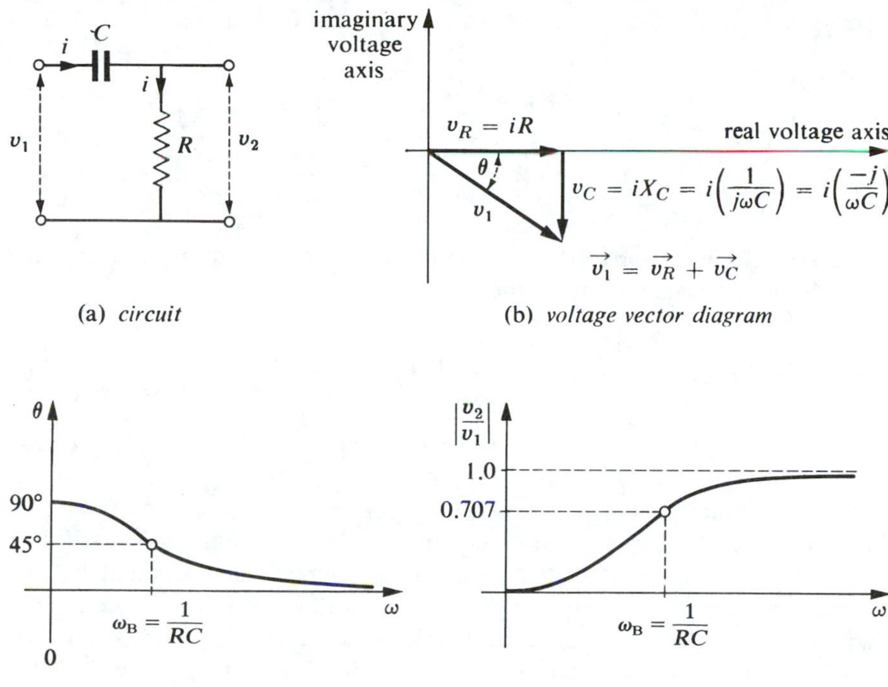

FIGURE 2.26 *RC* phase shifter.

2.10 *RC* LOW-PASS FILTER

Consider the circuit of Fig. 2.27, in which the output is taken across the capacitor. The voltage gain v_2/v_1 clearly will go to zero at very high frequencies because the capacitor acts like a short circuit at high frequencies. At zero frequency (dc) the gain will be unity if we assume no current is drawn from the output. The circuit is therefore called a *low-pass filter* because it passes the low frequencies and attenuates the high frequencies. For a sinusoidal input of angular frequency ω, the voltage gain (assuming

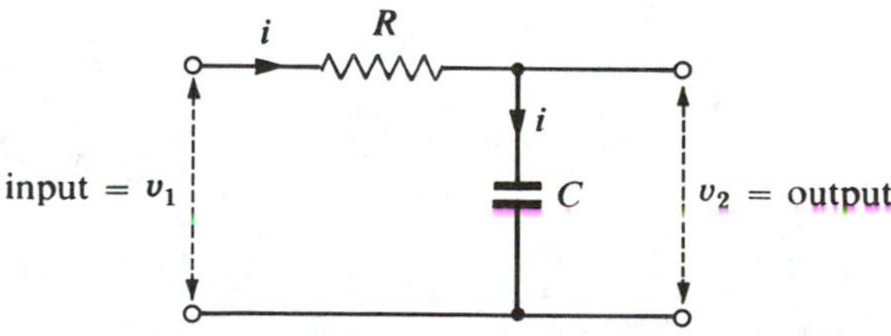

FIGURE 2.27 *RC* low-pass filter.

no output current) is

$$\frac{v_2}{v_1} = \frac{iX_C}{i(R + X_C)} = \frac{i\left(\dfrac{1}{j\omega C}\right)}{i\left(R + \dfrac{1}{j\omega C}\right)} = \frac{-\dfrac{j}{\omega C}}{R - \dfrac{j}{\omega C}} = \frac{1}{1 + j\omega RC} \qquad \textbf{(2.50)}$$

The complex gain means that the output has been shifted in phase relative to the input. The magnitude of the gain is

$$\left|\frac{v_2}{v_1}\right| = \left[\left(\frac{v_2}{v_1}\right)^* \left(\frac{v_2}{v_1}\right)\right]^{1/2} = \left[\left(\frac{1}{1 - j\omega RC}\right)\left(\frac{1}{1 + j\omega RC}\right)\right]^{1/2} = \frac{1}{\sqrt{1 + \omega^2 R^2 C^2}} \qquad \textbf{(2.51)}$$

which goes to unity as $\omega \to 0$, and which goes to zero as $\omega \to \infty$. If we plot the magnitude of the gain versus frequency on log-log paper, we obtain Fig. 2.28. The curved graph of the actual gain can be approximated by two straight lines, giving a sharp break or knee in the curve at the breakpoint frequency $\omega = 1/RC = \omega_B$. At the breakpoint the voltage gain is down by 0.707 or 3 dB down relative to the gain of unity at zero frequency. You see this by substituting $\omega = 1/RC$ in the expression for the gain. The slope of the straight line is 6 dB per octave. That is, if the voltage gain at 1000 Hz is down 12 dB, then the voltage gain at 2 kHz is down 18 dB.

The *RC* circuit basically accounts for the decrease in gain with increasing frequency for *all* amplifier circuits, so it is worth studying in detail. The capacitance *C* is often the stray capacity between the circuit wiring and the chassis, or it may be an inherent capacity built in a transistor

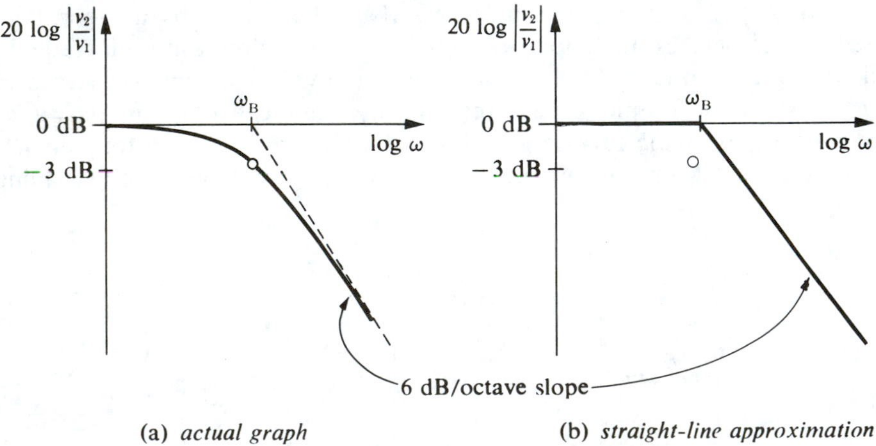

FIGURE 2.28 Gain vs. frequency for *RC* low-pass filter.

or tube. Thus, we can see that if we are "stuck" with a certain minimum capacitance between a signal-carrying wire and ground, and if we wish to maximize the high-frequency response of the circuit, then we should take care that the effective series resistance R is as small as possible. We will discuss this in more detail in Chapter 7.

The difference in phase between the input and the output can be calculated most easily from a rotating vector diagram. Choose the phase of the current as the reference phase and recall that the current leads the voltage in a capacitor by 90°. Thus we have, assuming $i_{out} = 0$, the diagrams of Fig. 2.29. The output is seen to lag the input in phase by an angle

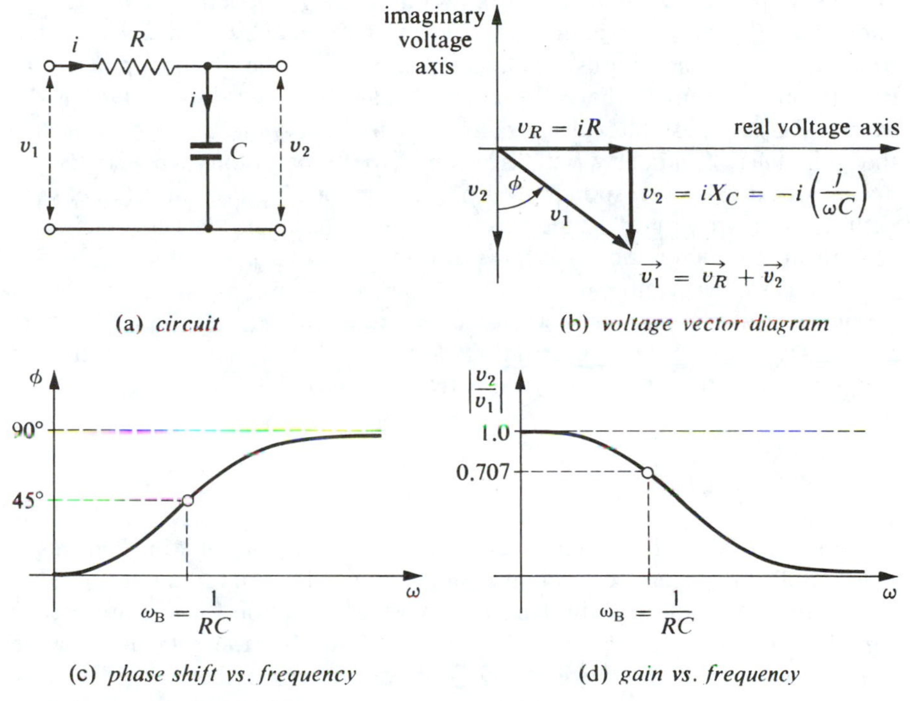

(a) *circuit*

(b) *voltage vector diagram*

(c) *phase shift vs. frequency*

(d) *gain vs. frequency*

FIGURE 2.29 *RC* phase shifter.

$\phi = \tan^{-1}(v_R/v_C) = \tan^{-1}\omega RC$. Notice that as the frequency increases, the phase shift ϕ goes to 90° and the output voltage goes to zero in amplitude. If either the resistance or the capacitance is made variable, the phase shift can be varied, but the amplitude of the output varies with the phase shift. The maximum phase shift is 90°.

Most four-terminal networks composed of simple combinations of R, L, and C with a resistive source and load have gains that vary with frequency and contribute a frequency-dependent phase shift; that is, the phase difference between the output and the input is a function of the frequency. If the network contributes the minimum possible phase shift for a given gain-versus-frequency behavior, the network is termed a *minimum phase-shift network*, and for such networks the gain characteristics and the phase characteristics are not independent. Knowing one characteristic enables you to calculate the other. Fortunately, most simple four-terminal networks fall into this class. Examples are the simple *RC* and *LR* low- and high-pass filters.

The exact mathematical theory of minimum phase-shift networks is beyond the scope of this book, and the interested reader is referred to Bode's theory in the *Radio Engineer's Handbook* (first edition) by Fred. E. Terman, McGraw-Hill (1943, p. 218). The essence of the theory is that the phase shift produced by the network at a frequency f depends on the rate of *change* of the network gain with respect to frequency evaluated at that frequency f. In other words, the network introduces the greatest phase shift in a frequency range where the network gain is rapidly changing. In the simple *RC* low-pass filter, for example, at high frequencies ($f \gg 1/2\pi RC$) the gain falls off linearly with frequency (6 dB per octave or 20 dB per decade), and this corresponds to a phase shift of 90°. For an *LC* network with the gain varying at 12 dB per octave, the phase shift is 180°. Gain and phase shift characteristics are shown for various networks in Fig. 2.30.

It takes considerable ingenuity to produce a simple network that is not a minimum phase-shift network. One such network is the phase shifter in Fig. 2.31, in which the gain is absolutely constant from dc to infinite frequency, and the phase shift varies from 0° to 180°

2.11 *RLC* CIRCUITS

In general, practical electronic circuits consist of combinations of resistance, inductance, and capacitance as well as tubes and transistors. By a "brute-force" solution of the Kirchhoff voltage equation for a simple series combination of R, L, and C, we will now show that the total impedance ("net opposition to the current flow") offered by R, L, and C is given by the magnitude of the complex sum of the resistance R, the inductive reactance $j\omega L$, and the capacitive reactance $1/j\omega C$. The *impedance* Z of any collection of R, L, and C is the complex algebraic sum of R, $j\omega L$, and $1/j\omega C$, treating each term as a "resistance" and combining terms in series or parallel as the circuit implies. Z is measured in ohms but is complex. Thus, the impedance of R and L in series is $Z = R + j\omega L$, and the impedance of R and C in parallel is $R \times (1/j\omega C)/(R + 1/j\omega C)$. The magnitude I_0 of the current through R, L, and C in series is given by $I_0 = V_0/|Z|$, where $Z = R + j\omega L + 1/j\omega C$ is the total impedance, $|Z| = \sqrt{R^2 + (\omega L - 1/\omega C)^2}$, and V_0 equals the magnitude of the applied voltage.

Consider the series *RLC* circuit of Fig. 2.32. The applied voltage is $V_0 \cos \omega t$; we are given R, L, and C and are asked to calculate the current

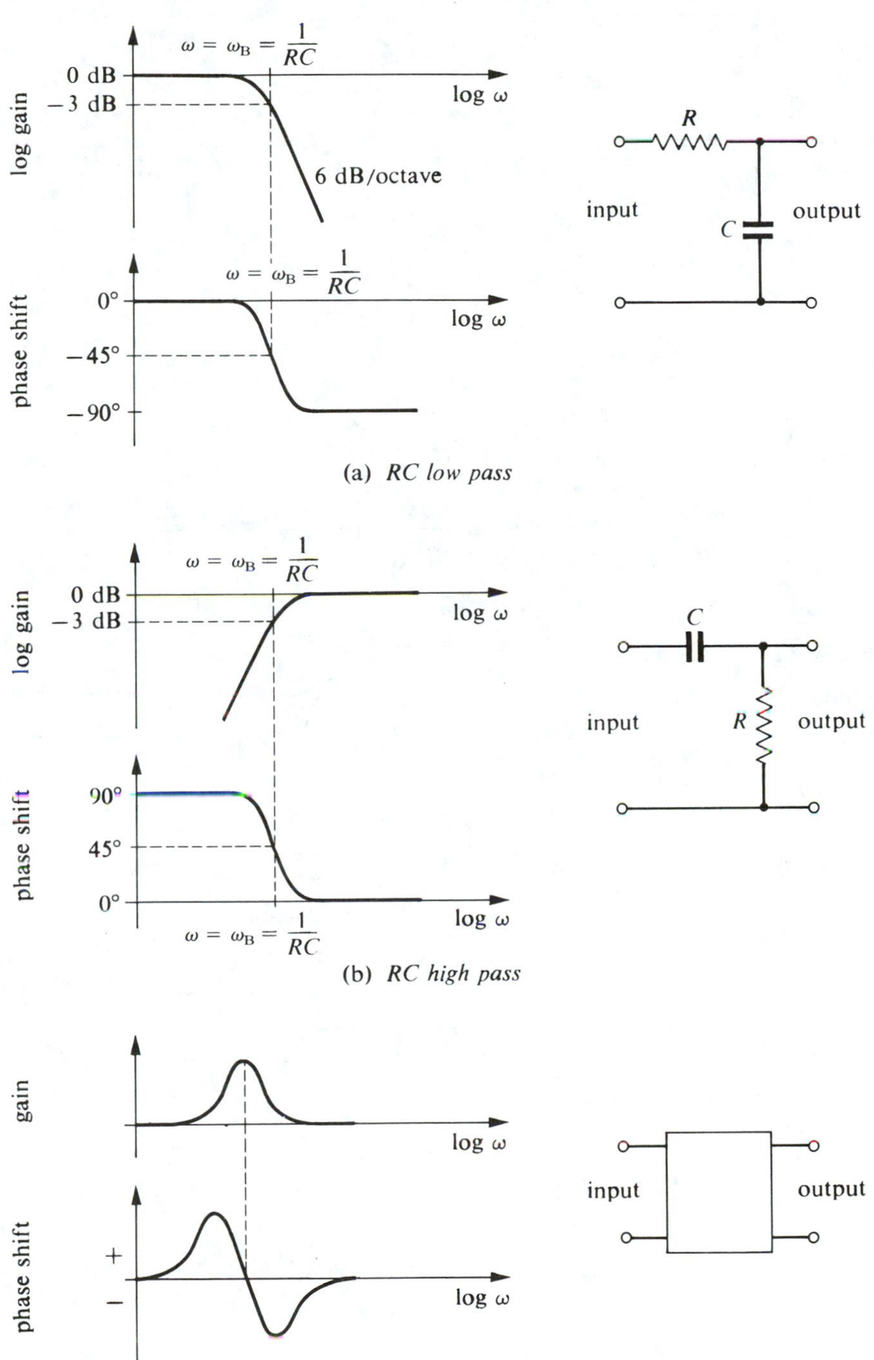

FIGURE 2.30 Gain and phase shift for three circuits.

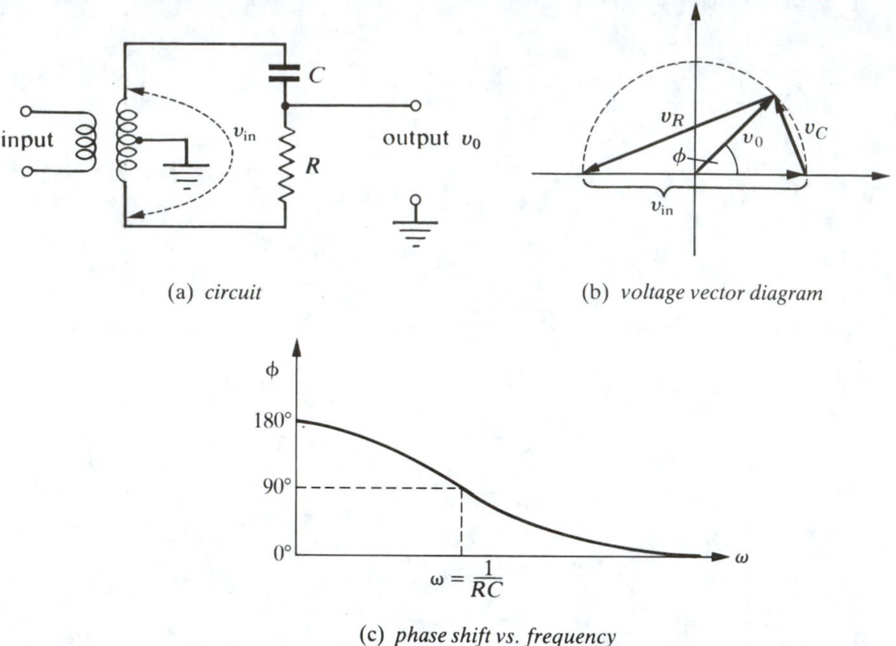

(a) *circuit*

(b) *voltage vector diagram*

(c) *phase shift vs. frequency*

FIGURE 2.31 Phase shifter.

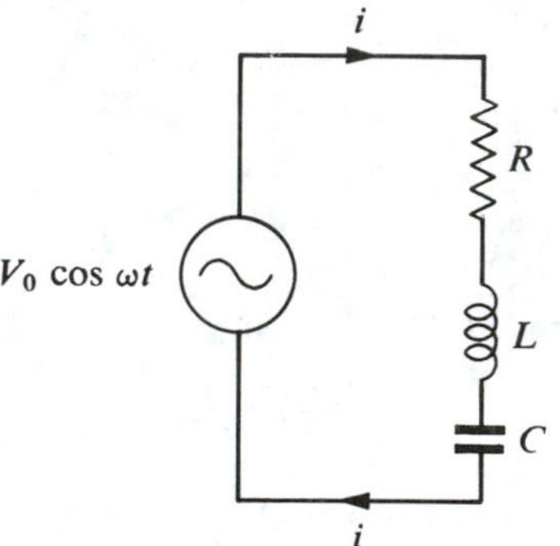

FIGURE 2.32 Series *RLC* circuit.

i. The Kirchhoff voltage law is

$$V_0 \cos \omega t - v_R - v_L - v_C = 0 \qquad \text{or} \qquad V_0 \cos \omega t = iR + L\frac{di}{dt} + \frac{Q}{C} \tag{2.52}$$

which when differentiated once with respect to time yields a second-order differential equation

$$L\frac{d^2i}{dt^2} + R\frac{di}{dt} + \frac{i}{C} = -\omega V_0 \sin \omega t$$

As we have seen, the current has the same frequency as the applied voltage for the capacitor alone and for the inductance alone; it is therefore a reasonable guess in the *RLC* circuit in Fig. 2.32 that the current has the same frequency as the applied voltage. However, the phase of the current may be different from the phase of the voltage. To take this possibility into account, we will assume a solution of the form $i = I_0 \cos(\omega t - \theta)$ and solve for I_0 and θ. I_0 and θ are both assumed to be constants. I_0 is the magnitude of the current; θ is the phase of the current relative to the voltage. If θ is positive, then the current lags the voltage. Of course, the final justification for the form of the assumed solution lies in our being able to find the solution that satisfies the differential equation and the boundary conditions. Now we substitute the assumed solution $i = I_0 \cos(\omega t - \theta)$ in the second-order differential equation in i and try to solve for I_0 and θ. We obtain one $\cos \omega t$ term and one $\sin \omega t$ term after a little algebra:

$$I_0\left[\left(\omega L - \frac{1}{\omega C}\right)\cos\theta - R\sin\theta\right]\cos\omega t + \left[I_0\left(\omega L - \frac{1}{\omega C}\right)\sin\theta + I_0 R\cos\theta - V_0\right]\sin\omega t = 0 \quad \textbf{(2.53)}$$

Because the functions $\cos \omega t$ and $\sin \omega t$ are *orthogonal*, the coefficient of each must equal zero. Another way of seeing this is to note that equation (2.53) must hold for all values of t. In particular, when $t = 0$, $\sin \omega t = 0$, so the coefficient of $\cos \omega t$ must equal zero; and when $t = \pi/2\omega = T/4$, $\cos \omega t = \cos(\pi/2)$ is zero, so the coefficient of $\sin \omega t$ must also equal zero. By either reasoning we obtain two equations that can be solved for I_0 and θ.

$$\left(\omega L - \frac{1}{\omega C}\right)\cos\theta - R\sin\theta = 0$$

and

$$I_0\left(\omega L - \frac{1}{\omega C}\right)\sin\theta + I_0 R\cos\theta - V_0 = 0$$

Hence

$$\tan\theta = \frac{\omega L - \dfrac{1}{\omega C}}{R} \quad \textbf{(2.54)}$$

$$I_0 = \frac{V_0}{R\cos\theta + \left(\omega L - \dfrac{1}{\omega C}\right)\sin\theta} \tag{2.55}$$

From equation (2.54) we find $\theta = \tan^{-1}(\omega L - 1/\omega C/R)$. That is, if $\omega L > 1/\omega C$, then the current lags behind the applied voltage by θ, or, equivalently, the applied voltage leads the current by θ. From equation (2.55) and the impedance triangle shown in Fig. 2.33, we obtain

$$I_0 = \frac{V_0}{\left[R^2 + \left(\omega L - \dfrac{1}{\omega C}\right)^2\right]^{1/2}} \tag{2.56}$$

which is very interesting. I_0, the magnitude of the current, equals V_0, the magnitude of the voltage, divided by $[R^2 + (\omega L - 1/\omega C)^2]^{1/2}$, *which is precisely the magnitude of the complex number* $[R + j\omega L + 1/j\omega C]$. Thus, by

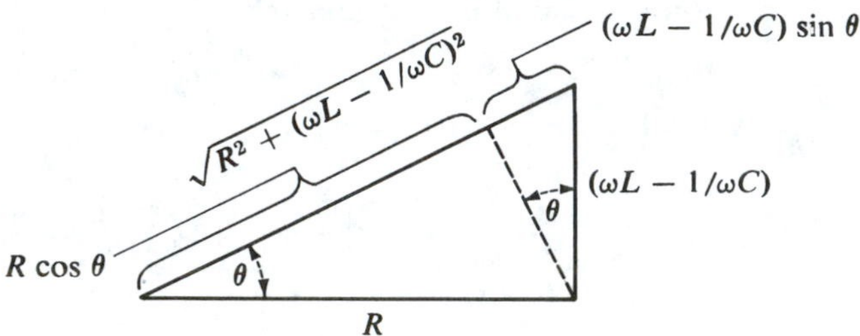

FIGURE 2.33 Impedance triangle for series *RLC* circuits.

treating the problem as a series connection of three complex reactances R, $j\omega L$, and $1/j\omega C$, we can get the magnitude of the current without even writing down the differential equation, much less solving it! In other words, using the complex inductive and capacitive reactances provides us with a quick shortcut to solving the differential equation for the current amplitude.

Let us show explicitly how the use of complex impedance simplifies the job of finding the amplitude of the current. We replace the voltage $V_0 \cos\omega t$ by $V_0 e^{j\omega t}$, L by its inductive reactance $j\omega L$, and C by its capacitive reactance $1/j\omega C$, as shown in Fig. 2.34. The total impedance is thus

$$Z = R + j\omega L + \frac{1}{j\omega C} \tag{2.57}$$

and

$$|Z| = \sqrt{R^2 + \left(\omega L - \frac{1}{\omega C}\right)^2} \tag{2.58}$$

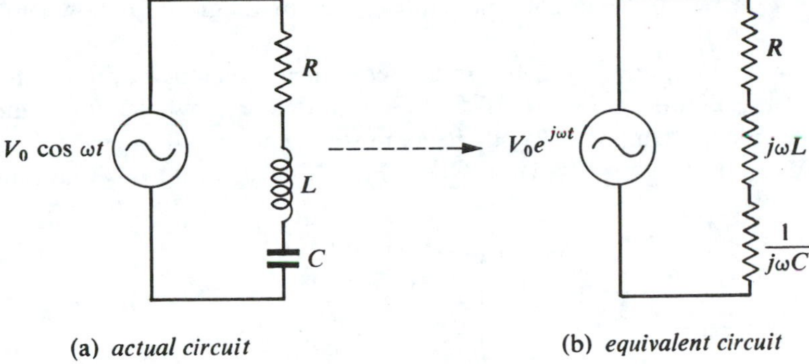

FIGURE 2.34 Series *RLC* circuit and equivalent.

The reciprocal of the impedance Z is called the *admittance* Y,

$$Y \equiv \frac{1}{Z} \tag{2.59}$$

and is measured in siemens or ohms^{-1} or mhos.

The current flowing in such a series circuit, Fig. 2.34(b), can be written down from Ohm's law:

$$i = \frac{V_0 e^{j\omega t}}{R + j\omega L + \dfrac{1}{j\omega C}} \tag{2.60}$$

The magnitude of the current is obtained merely by taking the magnitude or absolute value of both sides of the above equation. If we let I_0 be the magnitude of the current, then we see that

$$I_0 = \frac{|V_0 e^{j\omega t}|}{\left|R + j\omega L + \dfrac{1}{j\omega C}\right|} = \frac{V_0}{\left[R^2 + \left(\omega L - \dfrac{1}{\omega C}\right)^2\right]^{1/2}} = \frac{V_0}{|Z|} \tag{2.61}$$

which is the same result as was obtained by the relatively laborious procedure of solving the differential equation.

To obtain the phase of the current, we will draw a rotating voltage vector diagram and start by picking a reference phase. Draw $v_R = iR$ along the real voltage axis. Because R, L, and C are in series, the current through each of them is the same in magnitude and phase. Thus the voltage v_L across the inductance is given by $v_L = iX_L = ij\omega L$ which leads the current by 90°. The voltage across the capacitance is given by $v_C = iX_C = i/j\omega C$. The rotating voltage vector v_C will then be $v_C = -ji/\omega C$ which lags the current by 90°. The rotating voltage vector diagram is shown in Fig. 2.35.

Notice that the voltage across the resistance is exactly in phase with the current through the resistance. We now note that, at any instant of time, by Kirchhoff's voltage law the voltage v_{in} must exactly equal the instantaneous sum of v_R, v_L, and v_C. Thus we have Fig. 2.35(c). From this diagram the

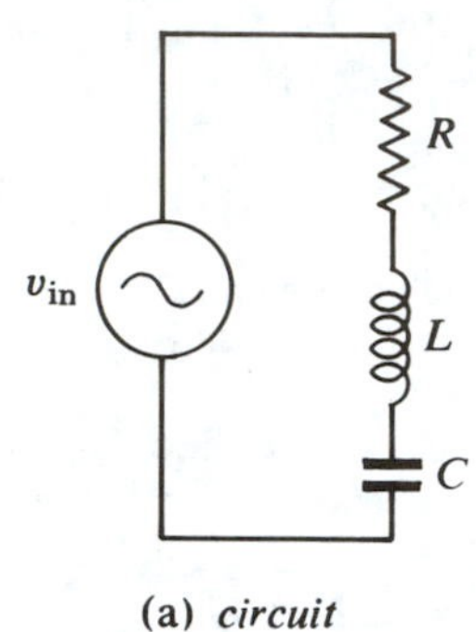

(a) *circuit*

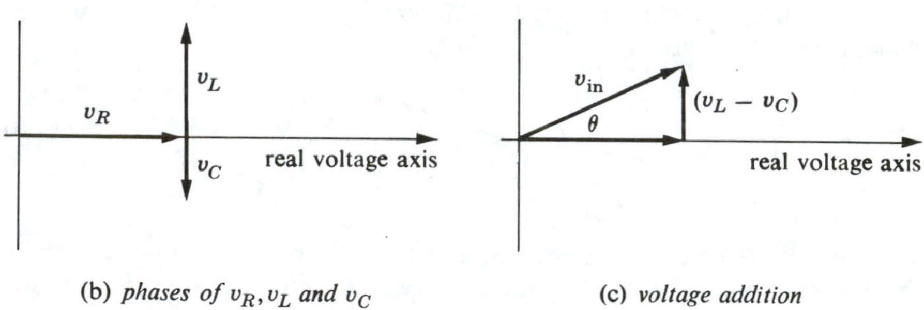

(b) *phases of* v_R, v_L *and* v_C

(c) *voltage addition*

FIGURE 2.35 Phases of voltages in a series *RLC* circuit.

phase θ of the current relative to the applied voltage is given by $\theta = \tan^{-1}(\omega L - 1/\omega C)/R$. If ωL is greater than $1/\omega C$, the current lags the applied voltage by θ; if ωL is less than $1/\omega C$, the current leads the applied voltage by θ. For later times ($t > 0$) all the voltage vectors rotate counterclockwise at the constant angular frequency ω, with θ remaining constant. And, as usual, the actual voltages measured in the circuit will be the real parts of the complex voltages we have diagrammed, that is, the projection of the voltage on the horizontal real voltage axis.

To sum up, complex numbers provide a shortcut to finding the currents in circuits without having to solve differential equations, and they automatically take care of the phase differences among the various voltages and currents.

2.12 SERIES AND PARALLEL RESONANCE

In the series *RLC* circuit $I_0 = V_0/[R^2 + (\omega L - 1/\omega C)^2]^{1/2}$; the current magnitude I_0 is maximum when $\omega L = 1/\omega C$, and the maximum value of i is given by $I_{0\,max} = V_0/R$. A graph of I_0 as a function of angular frequency is shown in Fig. 2.36. The frequency ω_0 at which $\omega L = 1/\omega C$ is called the *resonant* frequency and is given by $\omega_0 = 1/\sqrt{LC}$ in angular frequency (rad/s), or by $f_0 = 1/(2\pi\sqrt{LC})$ in Hz or cps. Also, v_L and v_C are equal in

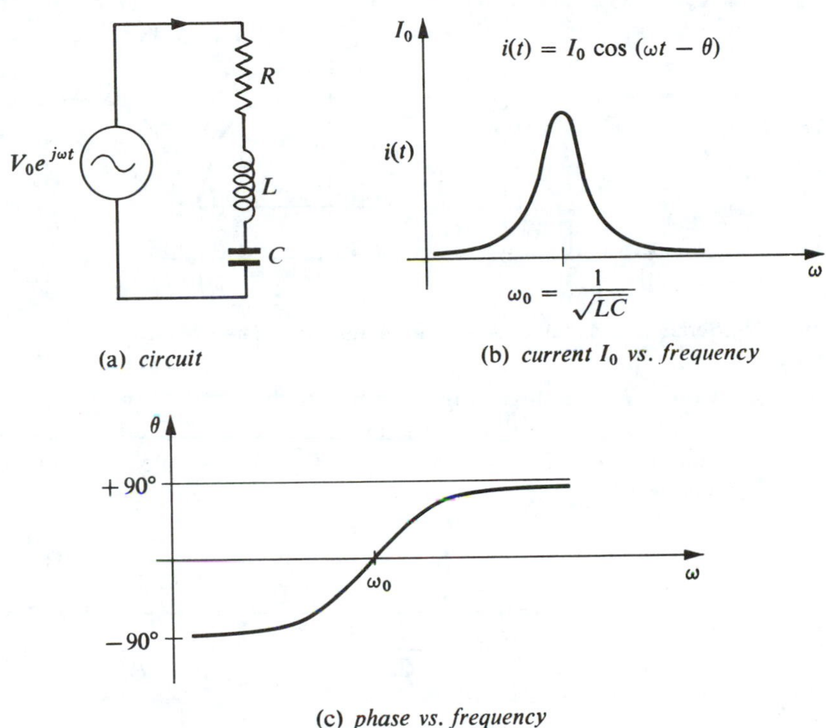

FIGURE 2.36 Series *RLC* circuit.

magnitude and 180° out of phase at resonance, and $Z = R$ at resonance, which means that the only limiting agent to the current at resonance is the resistance. Thus, if R is very small, the current at resonance will be very large. The voltage at resonance across the inductance $v_L = I_{0\,max}(\omega L)$ may be extremely high (even greater than V_0), perhaps high enough to cause arcing in the inductance. Similarly, $v_C = I_{0\,max}(1/\omega C)$ at resonance may be large enough to destroy the capacitor. Notice also that the phase of the current relative to the applied voltage changes from a lag to a lead as ω passes through $\omega_0 = 1/\sqrt{LC}$, because $\tan\theta = (\omega L - 1/\omega C)/R$, as shown in Fig. 2.36(a).

One application of the series *RLC* circuit is to attenuate a narrow band of frequencies around ω_0 by the circuit shown in Fig. 2.37. The voltage gain versus frequency has a relative minimum at ω_0 because the impedance between A and B is a minimum at ω_0 from equation (2.58). At ω_0 the gain equals $R/(R + R_1) = 1/(1 + R_1/R)$, which can be much less than one if $R_1 \gg R$.

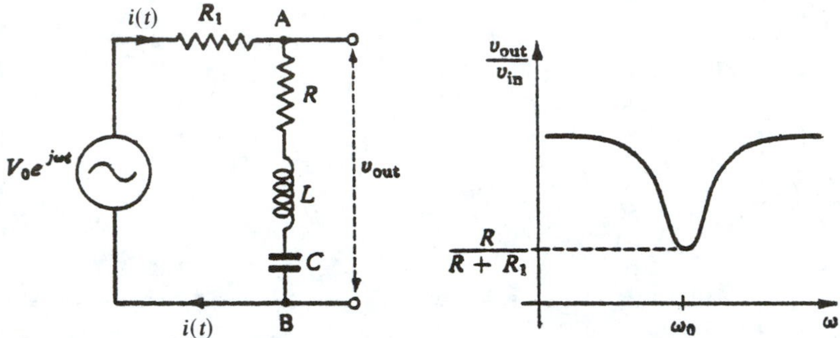

FIGURE 2.37 Series *RLC* circuit to attenuate frequencies near ω_0.

Consider the *RLC* circuit in Fig. 2.38 in which L and C are in parallel; this circuit behaves rather differently from the series *RLC* circuit. Let us calculate the current as a function of frequency. The parallel combination of L and C has an impedance of

$$Z = \frac{j\omega L\left(\dfrac{1}{j\omega C}\right)}{j\omega L + \dfrac{1}{j\omega C}} = \frac{-\dfrac{jL}{C}}{\omega L - \dfrac{1}{\omega C}} = \frac{j\omega L}{1 - \omega^2 LC} = \frac{j\omega L}{1 - \dfrac{\omega^2}{\omega_0^2}} \qquad \textbf{(2.62)}$$

which approaches *infinity* as ω approaches the resonant frequency ω_0.

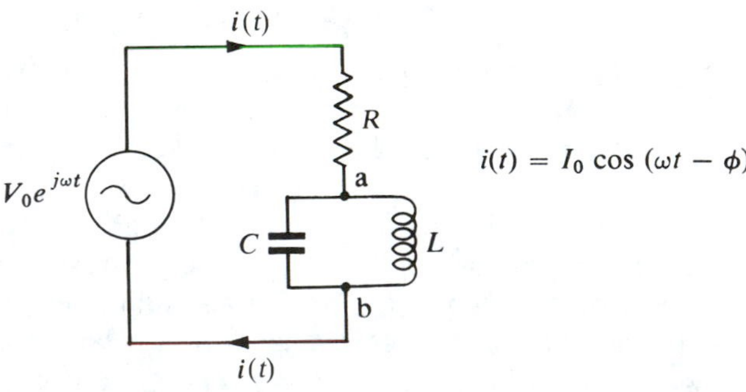

FIGURE 2.38 Parallel *LC* circuit.

Therefore, at resonance, the current $i = v/Z$ goes to zero, and the voltage v_{ab} goes through a maximum. Thus we have just the opposite behavior from the series circuit.

Notice that the impedance changes from inductive to capacitive as ω passes through ω_0. For $\omega < \omega_0$, $Z = j\omega L/\gamma$, and for $\omega > \omega_0$, $Z = -j\omega L/\gamma$, where $\gamma = |(1 - \omega^2/\omega_0^2)|$. Because the voltage across the parallel LC circuit is given by $v = iZ$, the phase of the output voltage relative to the input voltage $V_0e^{j\omega t}$ changes by 180° as ω passes through ω_0, as shown in Fig. 2.39(c).

$$i = \frac{v}{Z} = \frac{V_0 e^{j\omega t}}{R + \left(\dfrac{-\dfrac{jL}{C}}{\omega L - \dfrac{1}{\omega C}}\right)} \tag{2.63}$$

$$v_{ab} = \frac{Z_{ab}}{R + Z_{ab}} \cdot V_0 e^{j\omega t} \to V_0 e^{j\omega t} \qquad \text{as } Z_{ab} \to \infty$$

The parallel LC circuit is sometimes called a *tank* circuit and is most often used to select one desired frequency from a signal containing many

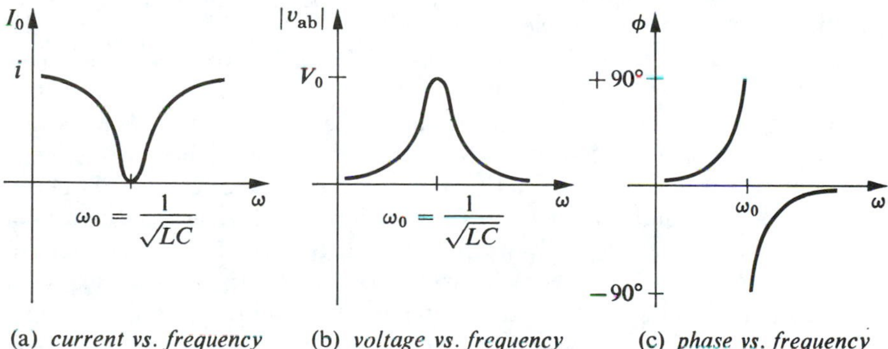

(a) *current vs. frequency* (b) *voltage vs. frequency* (c) *phase vs. frequency*

FIGURE 2.39 Current and voltage as functions of frequency in a parallel LC circuit.

different frequencies. This selection is accomplished by the basic circuit in Fig. 2.40 in which the voltage gain (v_{out}/v_{in}) is maximum $(=1)$ at $\omega_0 = 1/\sqrt{LC}$. This circuit is said to be *tuned* to the frequency ω_0. Often either L or C is made variable so that different frequencies may be selected or tuned. This is precisely what you do when tuning in a radio station; the tuning knob is usually a variable capacitor in a parallel LC circuit.

The parallel LC circuit will tend to oscillate. This can be seen by the following physical argument. Suppose at $t = 0$ the capacitor C is fully

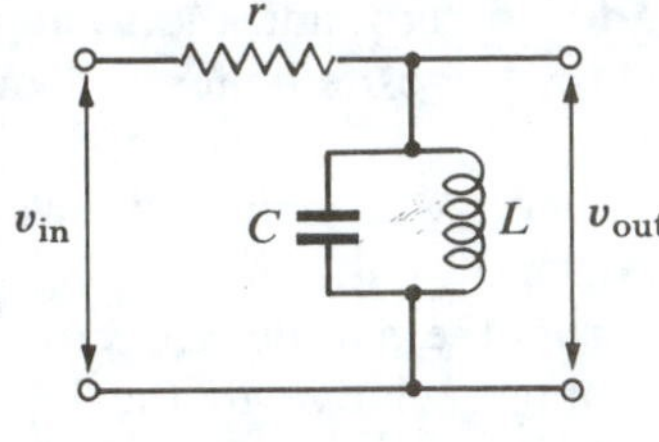

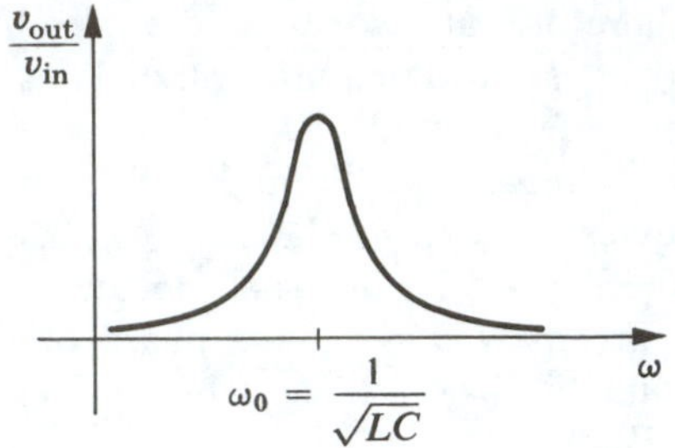

FIGURE 2.40 Parallel *LC* circuit used to select frequencies near ω_0.

charged and the current is zero. The energy of the circuit is then all in the electric field between the capacitor plates. The capacitor C will start to discharge through the inductance, as shown in Fig. 2.41(b). The current flowing through L creates a magnetic field around L, and energy is stored in this magnetic field. This increase in energy will be exactly balanced by the decrease in the energy stored in the electric field between the plates of C, the total energy of the LC circuit remaining constant. But by Lenz's law the magnetic field lines cutting the turns in L induce a voltage across the terminals of L of such a polarity as to *oppose* the increasing current. Eventually the current will reverse direction because of this induced voltage and will recharge C with the opposite polarity, as in Fig. 2.41(f). The capacitor C will now discharge through L in the opposite direction, as in 2.41(f), eventually returning the circuit to the state of Fig. 2.41(a) when the entire cycle starts over again. This back-and-forth flow of current is called *oscillation*, and the oscillations for an LC circuit will in principle last forever with a constant peak-to-peak amplitude. If, however, there is any *resistance* at all in the circuit, there will be a conversion of electrical energy to heat energy in the resistance at a rate i^2R. This loss of electrical energy occurs regardless of the *direction* of the current flow. Thus the electrical energy of the circuit gradually decreases, and the oscillations die out. In actual practice, of course, there is always some resistance associated with any LC circuit, mainly due to the resistance of the wire in the inductance. The only exception would be a circuit constructed entirely from superconducting material, which, at this writing, must be kept at liquid helium temperatures (≈ 4 K) and, hence, is somewhat impractical. Thus, the oscillations in a real circuit die out eventually. If constant amplitude oscillations are desired, then some circuit must be devised to continuously replenish the energy lost to heat in each cycle. This is precisely what one does in building an oscillator, which will be covered in Chapter 7.

It can be shown that the current in a pure LC circuit (no resistance) is exactly sinusoidal, constant in amplitude, and of angular frequency

$$\omega_0 = \frac{1}{\sqrt{LC}}$$

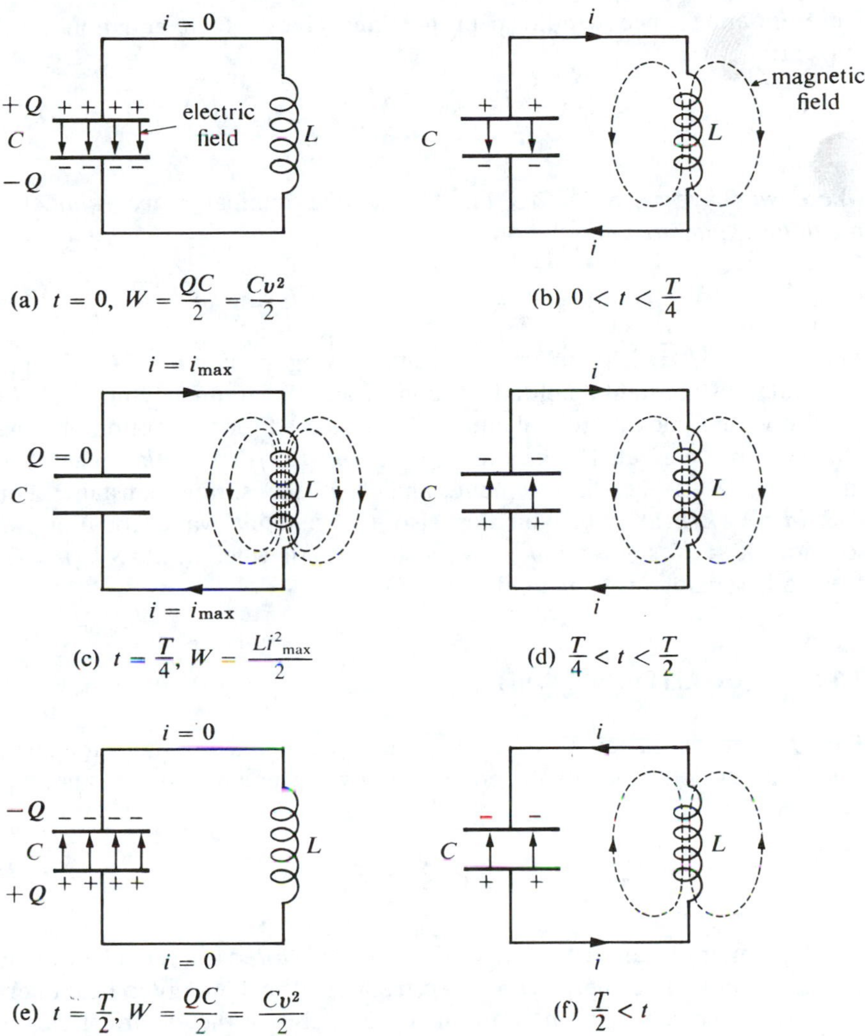

FIGURE 2.41 Oscillation of current in parallel *LC* circuit.

To show this, we simply write the Kirchhoff voltage law for the closed loop containing L and C.

$$v_C - v_L = 0$$

$$v_C + L\frac{di}{dt} = 0$$

$$\frac{Q}{C} + L\frac{di}{dt} = 0 \qquad \textbf{(2.64)}$$

Differentiating once with respect to time gives us an equation in the current:

$$\frac{i}{C} + L\frac{d^2 i}{dt^2} = 0 \qquad \frac{d^2 i}{dt^2} + \frac{1}{LC}i = 0 \tag{2.65}$$

where we have used $i = dQ/dt$. This is the familiar *wave equation* or *oscillator equation*. The solution is

$$i = I_0 \cos(\omega_0 t + \phi) \qquad \text{or} \qquad i = I_0 \sin(\omega_0 t + \theta) \tag{2.66}$$

where $\omega_0 = 1/\sqrt{LC}$. Thus, the current flowing in a pure LC circuit is sinusoidal, of constant amplitude I_0 and of angular frequency $\omega_0 = 1/\sqrt{LC}$. Precisely the same differential equation is obtained for the motion of a mass attached to a perfect, Hooke's law spring: $m(d^2X/dt^2) = -kX$ where m is the mass, X is the displacement, and k is the spring constant. Thus, $d^2X/dt^2 + (k/m)X = 0$, which is also the familiar wave equation with solution $X = X_0 \sin(\omega_0 t + \theta) = X_0 \cos(\omega_0 t + \phi)$, where $\omega_0 = \sqrt{k/m}$. L is seen to be analogous to m, and C to $1/k$.

2.13 Q (QUALITY FACTOR)

The Q or quality factor of any circuit can be defined as 2π times the energy stored per cycle divided by the energy lost or dissipated (usually as heat) per cycle.

$$Q \equiv 2\pi\frac{W_S}{W_L} \tag{2.67}$$

The energy stored refers to the energy stored in the electric and magnetic fields. The energy converted from electrical energy to heat energy is "lost" in the sense that it is no longer available in the electrical current. The higher the Q, the less energy is converted into heat per cycle. An ideal circuit that dissipates no energy would have an infinite Q.

Notice that the energy has been converted from a more *ordered* form (i.e., the coherent average drift of electrons in the current flow in the wire) to the more *disordered* random thermal motion of the molecules. This conversion corresponds to an increase in entropy and is consistent with the second law of thermodynamics, which says that the reverse process will never occur with 100% efficiency—that is, the thermal energy can never be completely converted back into the more ordered electrical energy. Thus, some (and usually all in most circuits) of the energy converted into heat is permanently lost from the electrical current.

We will now show from the above definition that the Q of a series RLC circuit can be expressed as $Q = \omega L/R$. From the definition, equation (2.67), we can write $Q = 2\pi W_S/W_H$, where W_S is the energy stored in one cycle and W_H is the energy converted to heat in one cycle. Assume $i = I_0 \cos \omega t$ is the current flowing in the circuit. The energy stored in the magnetic field of L when a current i is flowing is $W = \frac{1}{2}Li^2$. W will therefore be maximum when i is maximum. Because i and v_C are 90° out of phase, when i is maximum, $v_C = 0$. Therefore when $i = i_{max} = I_0$, *all* the energy is stored in the magnetic field. $W = \frac{1}{2}Li_{max}^2$, and none is stored in the electric field of the capacitor because $v_C = 0$. $W_S = \frac{1}{2}LI_0^2$ and $W_H = \int_0^T p\,dt$, where p is the instantaneous power converted into heat, $p = i^2(t)R$, and the period $T = 2\pi/\omega$. Thus,

$$\begin{aligned} Q &= 2\pi \frac{W_S}{W_H} = 2\pi \frac{\frac{1}{2}LI_0^2}{\int_0^T i^2(t)R\,dt} = \frac{\pi L I_0^2}{R\int_0^T I_0^2 \cos^2 \omega t\,dt} \\ &= \frac{\pi L}{R} \frac{1}{\int_0^T \cos^2 \omega t\,dt} = \frac{\pi L}{R} \cdot \frac{1}{T/2} = \frac{\omega L}{R} \end{aligned} \tag{2.68}$$

Note that at the resonance frequency ω_0, Q can be written as $Q = 1/R\sqrt{L/C}$. Thus the higher the ratio of L to C, the higher the Q at resonance. However, it is very difficult to make a large inductance without also increasing R. And, as expected, the larger R is, the lower the Q, because a larger R will dissipate more energy as heat.

For the parallel RLC circuit of Fig. 2.42, we show that the Q is also given by $Q = \omega L/R$. Here we explicitly draw the resistance of the wire in the inductance by including R in series with L. But we can always write $i_1(t)$ in the form $i_{1\,max} \cos(\omega t - \theta)$, so then Q becomes

$$\begin{aligned} Q &= \frac{\pi L i_{1\,max}^2}{\int_0^T i_{1\,max}^2 \cos^2(\omega t - \theta) R\,dt} \\ &= \frac{\pi L}{R} \frac{1}{\int_0^T \cos^2(\omega t - \theta)\,dt} = \frac{\pi L}{R} \cdot \frac{1}{\frac{T}{2}} = \frac{\omega L}{R} \end{aligned} \tag{2.69}$$

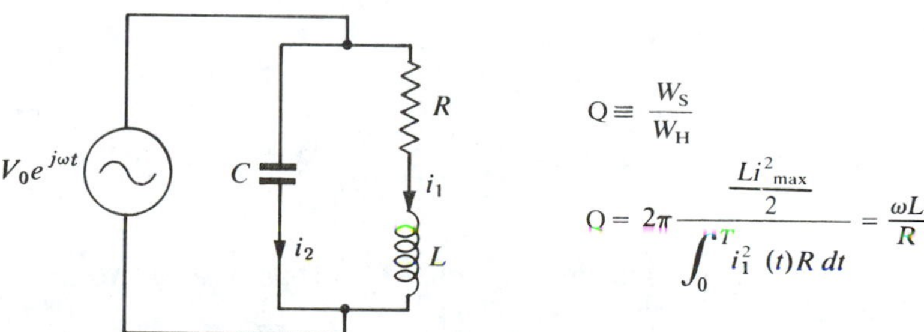

FIGURE 2.42 Q calculation for a parallel *RLC* circuit.

which is the same as for the series RLC circuit. Note that we could have solved explicitly for $i_1(t)$ by writing

$$i_1(t) = \frac{V_0 e^{j\omega t}}{R + j\omega L} = \frac{V_0 e^{j(\omega t - \theta)}}{\sqrt{R^2 + \omega^2 L^2}}$$

where $\theta = \tan^{-1}(\omega L/R)$. Then $i_{1\,\max} = V_0/\sqrt{R^2 + \omega^2 L^2}$, but this procedure is not necessary to calculate the Q because $i_{1\,\max}$ cancels out in the expression for Q.

The resistance R in an actual RLC circuit is usually the resistance of the wire in the inductance L. Practically speaking, we usually can neglect any power dissipated as heat in the capacitor, which we have done by not drawing any resistance associated with C. If we draw the parallel LC circuit as in Fig. 2.43, with R representing the leakage resistance of the capacitor,

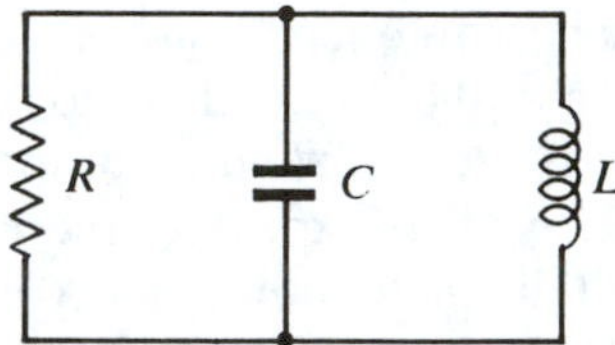

FIGURE 2.43 Parallel *LC* circuit with capacitor leakage resistance *R*.

then an ideal circuit with zero power dissipated as heat would have an infinite R. Thus we would expect Q to be proportional to R. A calculation from the definition of Q yields the result $Q = R/\omega L$ or $Q = \omega_0 RC$ or $Q = R\sqrt{C/L}$ at resonance.

We will now show that for a series RLC circuit, the sharpness of the current-versus-frequency curve depends on the Q: A high Q corresponds to a narrow, sharp curve and a low Q corresponds to a broad curve. The current as a function of frequency is given by

$$i = \frac{V_0 e^{j\omega t}}{R + j\omega L + \dfrac{1}{j\omega C}} = I_0 e^{j(\omega t - \theta)}$$

$$I_0 = \frac{V_0}{\left[R^2 + \left(\omega L - \dfrac{1}{\omega C}\right)^2\right]^{1/2}} = \frac{V_0}{|Z|} \tag{2.70}$$

Clearly I_0 reaches a maximum when $\omega L = 1/\omega C$ (i.e., at resonance), and $I_{0\,\max} = V_0/R$, as shown in Fig. 2.44.

Let us now define the width $\Delta\omega$ of the I_0-versus-ω curve as the width

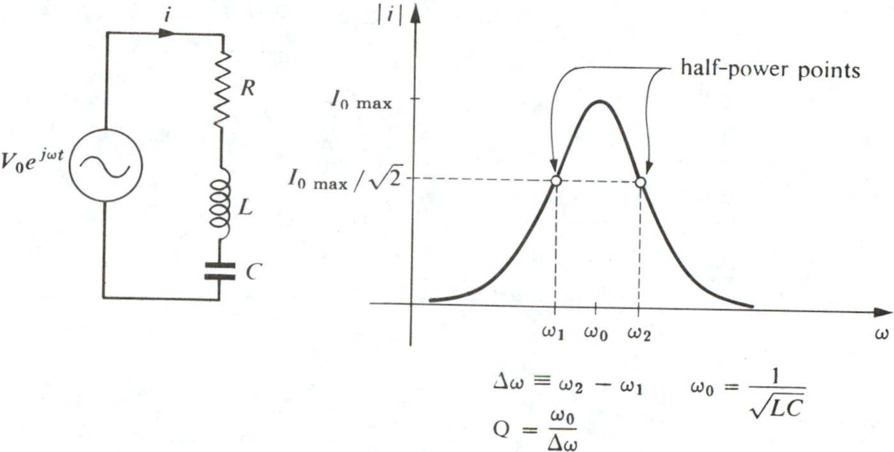

FIGURE 2.44 Frequency response of series *LC* circuit.

from ω_1 to ω_2, where $i(\omega_1) = i(\omega_2) = I_{0\max}/\sqrt{2}$. That is, $\Delta\omega$ is the width of the curve where the current has fallen to $1/\sqrt{2} = 0.707$ of its value at resonance. At ω_1 and ω_2 the power dissipated in R is one-half of the power at resonance. Hence, the points ω_1 and ω_2 are sometimes called the *half-power* points. To find $\Delta\omega$, we note that $I_{0\max} = V_0/R$ and i has fallen to $I_{0\max}/\sqrt{2}$ when $Z = \sqrt{2}R$. Therefore at $\omega = \omega_1$ and at $\omega = \omega_2$, $Z = \sqrt{2}R$ or $|Z(\omega_1)| = |Z(\omega_2)| = \sqrt{2}R$.

$$\sqrt{R^2 + \left(\omega_1 L - \frac{1}{\omega_1 C}\right)^2} = \sqrt{R^2 + \left(\omega_2 L - \frac{1}{\omega_2 C}\right)^2} = \sqrt{2}R \qquad \textbf{(2.71)}$$

Thus, ω_1 and ω_2 are the two roots of $(\omega L - 1/\omega C)^2 = R^2$. If we assume a symmetrical curve about $\omega = \omega_0$ and also that $\Delta\omega \ll \omega_0$, then substituting $\omega_2 = \omega_0 + \Delta\omega/2$ and $\omega_1 = \omega_0 - \Delta\omega/2$, we obtain $\Delta\omega = \omega_0/\mathrm{Q}$. In other words,

$$\mathrm{Q} = \frac{\omega_0}{\Delta\omega} \qquad \textbf{(2.72)}$$

Thus a high-Q circuit has a narrow width $\Delta\omega$ for a given frequency ω_0, and a low-Q circuit has a broad width $\Delta\omega$. A high-Q circuit is said to be *sharp* or to have a high selectivity because it can be used to select a narrow band of frequencies around ω_0 and to reject other frequencies (see Fig. 2.45). The best intuitive way to think of Q is with the relationship $\mathrm{Q} = \omega_0/\Delta\omega$.

For the parallel *RLC* circuit we can also show that, if $\omega_0 L \gg R$ (the resistance associated with L), $\mathrm{Q} \cong \omega_0/\Delta\omega = \omega_0 L/R$, where ω_0 is the angular resonant frequency and $\Delta\omega$ is the full width between the half-power

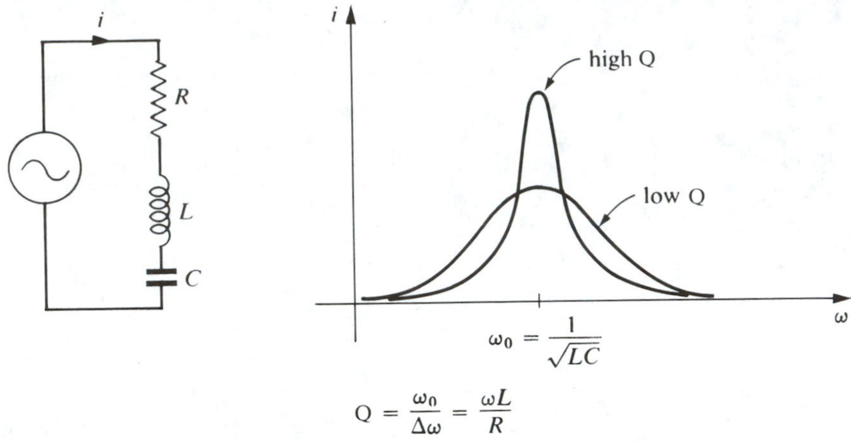

FIGURE 2.45 Frequency response of high-Q and low-Q series *RLC* circuit.

points. The impedance of the parallel tank at resonance is $Z_0 = Q\omega_0 L = Q/\omega_0 C$, and at the half-power points the impedance is $Z = Z_0/\sqrt{2}$. Notice that at resonance the Q effectively multiplies the impedance of the inductance or of the capacitance.

If an *RLC* circuit (either series or parallel) is excited with a step function voltage (or indeed with any rapidly changing voltage), the current will tend to oscillate or "ring" with an exponentially decreasing amplitude, as shown in Fig. 2.46. The differential equation from the KVL for a step-function input is

$$V_{in} = V_0 = V_L + V_R + V_C \quad \text{for } t \geqq 0$$

$$V_0 = L\frac{di}{dt} + Ri + \frac{Q}{C}$$

or

$$V_0 = L\frac{d^2Q}{dt^2} + R\frac{dQ}{dt} + \frac{Q}{C} \tag{2.73}$$

with the boundary condition $Q = 0$ at $t = 0$. The solution of (2.73) can be written as

$$Q = Q_h + Q_p$$

where Q_p is any particular solution and Q_h is the solution to the homogeneous equation.

$$0 = L\frac{d^2Q_h}{dt^2} + R\frac{dQ_h}{dt} + \frac{Q_h}{C} \tag{2.74}$$

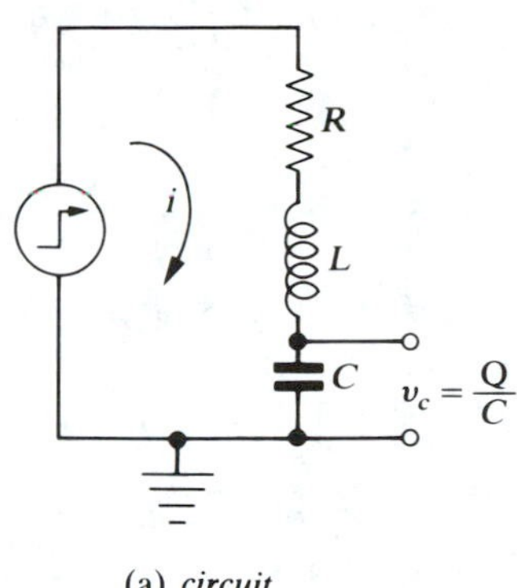

(a) *circuit*

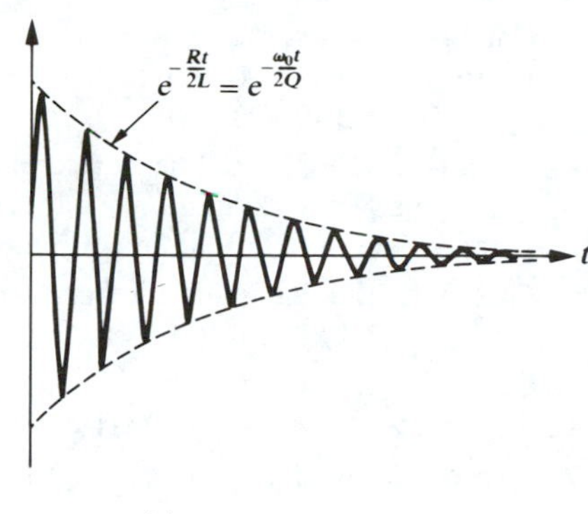

(b) *voltage across c*

FIGURE 2.46 Ringing excited by step input.

By inspection $Q_0 = CV_0$ is a constant particular solution of (2.73). The solution to (2.74) can be shown to be

$$Q_h = Ae^{-Rt/2L}e^{\pm j\omega_0 t} \tag{2.75}$$

where A is a constant, and provided $R^2/4L^2 \ll 1/LC = \omega_0^2$ or $4Q^2 \gg 1$. Using $Q = \omega_0 L/R$, we can write the solution as

$$Q = CV_0(1 - e^{-\omega_0 t/2Q}e^{\pm j\omega_0 t}) \tag{2.76}$$

Thus we see that for a high-Q circuit ($4Q^2 \gg 1$), the charge on the capacitor oscillates at approximately ω_0, and the amplitude of the oscillations decreases exponentially according to

$$e^{-\omega_0 t/2Q} \tag{2.77}$$

In other words, the *envelope* of the oscillations or ringing decreases according to (2.77) as shown in Fig. 2.46(b). The current and voltage also decrease according to (2.77). Thus a measurement of the voltage across the *RLC* circuit versus time can yield the circuit Q.

PROBLEMS

1. Calculate the period in seconds of a 400-Hz sinusoidal voltage. If the zero-to-peak amplitude is 5 V, calculate the maximum instantaneous rate of change of the voltage in volts per second.
2. Prove that the root-mean-square (rms) value of

$$V(t) = V_0 \cos \omega t$$

is equal to $V_0/\sqrt{2}$. What is the rms value of the voltage of Problem 1?

3. Calculate the average power (in watts) dissipated as heat in a device passing a current of $i(t) = 4 \cos \omega t$ if the voltage across the device is (a) $v(t) = 10 \cos(\omega t + 30°)$, (b) $v(t) = 10 \cos \omega t$.
4. Calculate the capacitance in microfarads between two 1-cm^2 conducting plates 1 mm apart in a vacuum. Repeat if the space between the plates is filled with a plastic dielectric with a dielectric constant of 8.
5. Calculate the capacitive reactance in ohms of a 0.01-μF capacitor at (a) 100 Hz, (b) 1 kHz, (c) 100 kHz, (d) 1 MHz.
6. Calculate the inductive reactance in ohms of a 2.5-mH choke at (a) 100 Hz, (b) 1 kHz, (c) 100 kHz, (d) 1 MHz.
7. Express L dimensionally in terms of ohms and seconds. Repeat for C.
8. Calculate the energy in joules stored in a 2000-μF capacitor charged to 5 V. Physically, how is the energy stored?
9. A 1-H inductance carries a current of 500 mA. The wire breaks, and in 10^{-3} s the current drops to zero. What would happen?
10. Calculate the impedance Z_{AB} in the form $a + jb$ and $|z|e^{j\theta}$ for

11. Calculate the impedance Z_{AB} in the form $a + jb$ and $|z|e^{j\theta}$ for

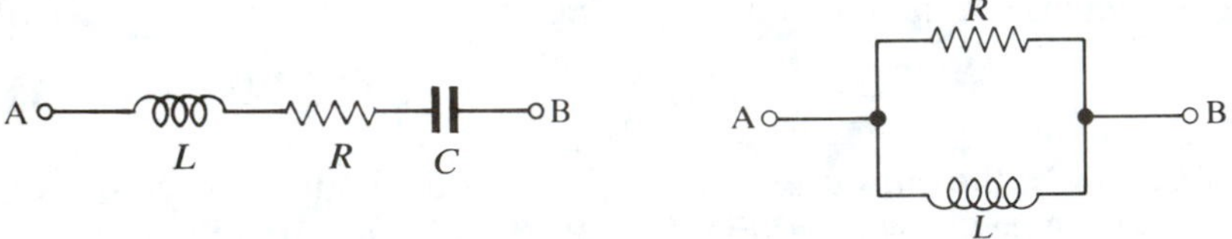

12. Calculate the impedance Z_{AB} in the form $a + jb$ and $|z|e^{j\theta}$ for

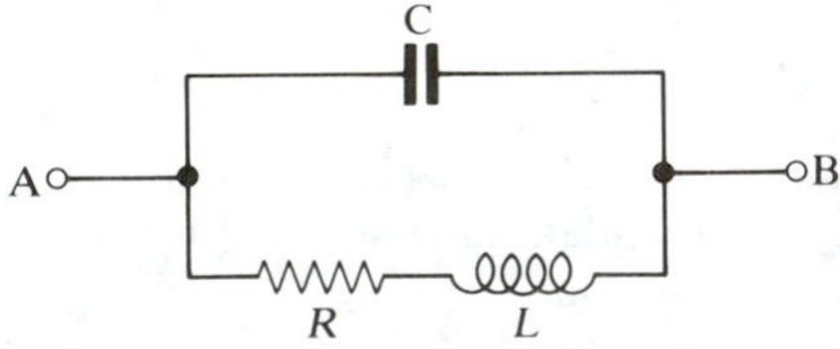

13. Design a high-pass RC filter with a breakpoint at 100 kHz. Use a 1-kΩ resistance. Explain in words why the high-pass filter attenuates the low frequencies.

14. Design a low-pass RC filter that will attenuate a 60-Hz sinusoidal voltage by 12 dB relative to the dc gain. Use a 100-Ω resistance. Explain in words why the low-pass RC filter attenuates the high frequencies.

15. For a low-pass RC filter prove that (a) at the frequency $\omega = 1/RC$ the voltage gain equals $0.707 = 1/\sqrt{2}$; (b) the rise time of the output pulse equals $2.2RC$ for a zero rise time input pulse.

16. Calculate the slope of the gain versus frequency curve for a low-pass RC filter in dB per octave and also in dB per decade for high frequencies ($\omega \gg 1/RC$). [One decade frequency change is a factor of ten (e.g., from 10 to 100 Hz, or from 50 to 500 kHz).]

17. Carefully sketch the voltage vector diagram for a high-pass RC filter and calculate the phase of the output voltage relative to the phase of the input voltage.

18. Derive an expression for the voltage gain and phase shift for the following LR circuit.

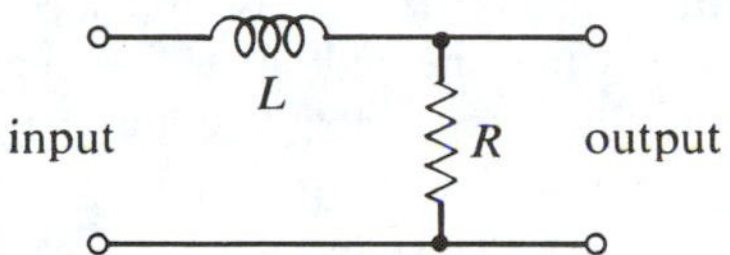

19. Design a parallel LC resonant circuit or tank to resonate at 1 MHz. Assume the inductance $L = 100\ \mu$H and has a dc resistance of 10 Ω. What is the Q of this circuit at resonance?

20. Carefully sketch the rotating voltage vector diagram for a series RLC circuit at resonance. If the circuit has a Q of 100, calculate the voltage ratings L and C must have.

21. Sketch a graph of the magnitude of the impedance versus frequency for series and parallel RLC circuits. State the change in phase of the impedance as the frequency passes through resonance.

22. Sketch the approximate gain-versus-frequency curve for the following circuit. You may treat the circuit as being composed of two independent RC filters.

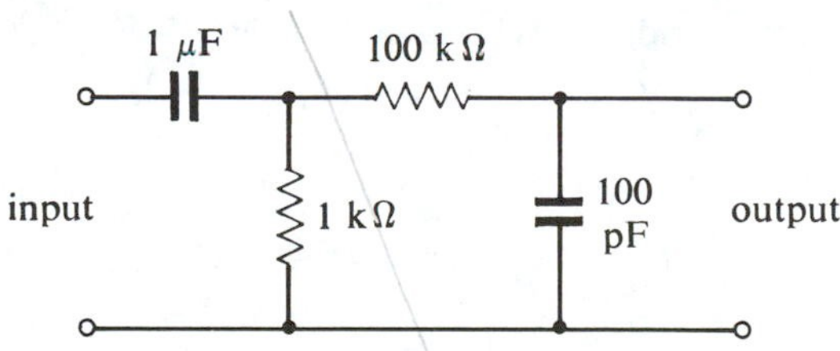

23. For a high Q parallel RLC circuit prove that $Q = \omega_0/\Delta\omega$, where ω_0 is the (angular) resonant frequency and $\Delta\omega$ is the width at the half-power points.

24. For a high-Q parallel RLC circuit prove that at resonance the impedance equals the Q times the inductive reactance at resonance. Calculate the im-

pedance at resonance for

$$L = 100\ \mu\text{H} \quad C = 0.001\,\mu\text{F} \quad R = 5\,\Omega$$

25. For a series RLC circuit show that the angular frequency for the charge on C (or the current) is approximately

$$\omega \cong \omega_0\left(1 - \frac{1}{8Q^2}\right)$$

26. Derive (2.75) for the series RLC circuit excited by a step-function voltage. [*Hint*: Solve (2.74) by assuming a solution of the form $Q_h = e^{\lambda t}$ where λ is an unknown constant independent of time but depending upon R, L, and C. Assume $R^2/4L^2 \ll 1/LC$.]

27. A general ac bridge can be made from the Wheatstone bridge of Fig. 1.35 by replacing each of the four resistances by a general impedance Z, that is, $R_1 \rightarrow Z_1$, etc. The battery of Fig. 1.35 is also replaced by a sinusoidal ac source of angular frequency ω, and the null meter is replaced by a sensitive ac voltmeter such as an oscilloscope. Show that the general balance condition is

$$\frac{Z_1}{Z_2} = \frac{Z_3}{Z_4}$$

28. With $Z_1 = R_1$, $Z_2 = R_2 + j\omega L_2$, $Z_3 = R_3$, and $Z_4 = R_4 + j\omega L_4$, show that the two balance conditions are

$$\frac{R_1}{R_2} = \frac{R_3}{R_4} \quad \text{and} \quad \frac{R_1}{L_2} = \frac{R_3}{L_4}$$

This is an inductance bridge. Also sketch the bridge.

29. With $Z_1 = R_1$, $Z_2 = R_2 + 1/j\omega C_2$, $Z_3 = R_3$, and $Z_4 = R_4 + 1/j\omega C_4$, show that the two balance conditions are

$$\frac{R_1}{R_2} = \frac{R_3}{R_4} \quad \text{and} \quad R_1 C_2 = R_3 C_4$$

This is a capacitance bridge. Also sketch the bridge.

30. With $Z_1 = R_1$, $Z_2 = R_2 \| C_2$, $Z_3 = R_3$, and $Z_4 = R_4 + 1/j\omega C_4$, show that the two balance conditions are

$$\frac{R_4}{R_2} + \frac{C_2}{C_4} = \frac{R_3}{R_1} \quad \text{and} \quad \omega^2 R_2 C_2 R_4 C_4 = 1$$

This is the Wien bridge. If $R_2 = R_4 = R$ and $C_2 = C_4 = C$, find the two balance conditions. Also sketch the bridge.

CHAPTER 3

Fourier Analysis and Pulses

3.1 INTRODUCTION

In many situations we encounter voltage changes that are not sinusoidal in shape. Any relatively rapid change in voltage is usually referred to as a *pulse*. For example, every time an ionizing particle such as a proton or electron passes through a Geiger Mueller tube, a small voltage pulse several microseconds long will be produced at the output of the tube. This pulse must then be amplified and recorded in some way to count the protons or electrons. And every time an ionizing particle passes through a scintillation crystal such as NaI (sodium iodide), a light flash of very short duration is produced in the crystal or scintillator. This light flash is converted to a voltage pulse lasting about 10^{-6} s by a photomultiplier tube, and this voltage pulse must then be amplified and counted. Voltage pulses may also be used to indicate the beginning and the end of a time interval and, therefore, must be amplified and recorded. The electronic signals in modern computers also consist of voltage pulses.

3.2 DESCRIPTION OF A PULSE

An ideal single pulse is rectangular in shape, as shown in Fig. 3.1(a). This pulse can be described by giving its amplitude V_0 and its width or duration τ. An actual pulse encountered in the laboratory will always appear somewhat rounded if observed with a sufficiently fast oscilloscope, as shown in Fig. 3.1(b). The steepness of the sides of this pulse is described by giving the *rise time*, which is defined as the time for the pulse to rise from 10% to 90% of its maximum value. The *fall time* of the pulse is defined as the time for the pulse to fall from 90% to 10% of its maximum value. An ideal pulse has zero rise and fall times.

If a series of pulses occurs regularly in time as shown in Fig. 3.2, we can describe the pulses completely by giving the number of pulses per second (called the *frequency* or repetition rate), the pulse width τ, the pulse amplitude V_0, and the phase. In Fig. 3.2 the frequency or repetition rate is 200 kHz, the pulse width is 1 μs, and the amplitude is 10 V. Notice that if

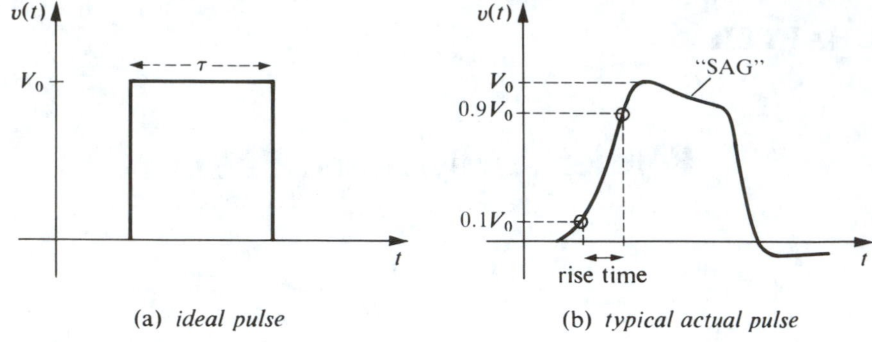

FIGURE 3.1 Pulses.

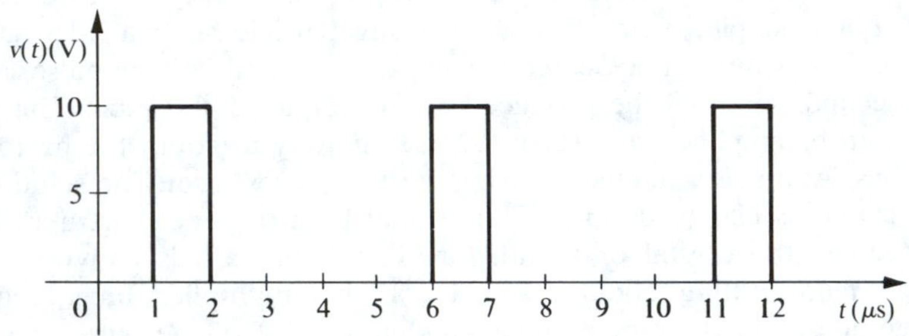

FIGURE 3.2 Train of regular pulses.

the zero voltage level were at the top of the 1-μs pulses, we would say that we had 10-V negative pulses of pulse width 4 μs with the same frequency of 200 kHz.

3.3 FOURIER ANALYSIS

In this section we outline the general idea of Fourier analysis and apply it to the problem of analyzing a single rectangular pulse. Fourier analysis basically says that any function can be written as a sum of sine and cosine functions of different frequencies and amplitudes. In this book we are interested in analyzing voltage waveforms, so from now on we shall speak about the problem of analyzing a voltage $v(t)$. The Fourier analysis theorem says that if the voltage $v(t)$ is defined from $t = -T/2$ to $t = T/2$, then $v(t)$ can be written as the following sum of sines and cosines:

$$v(t) = \frac{a_0}{2} + \sum_{n=1}^{\infty} a_n \cos n\omega t + \sum_{m=1}^{\infty} b_m \sin m\omega t \qquad \omega = \frac{2\pi}{T} \tag{3.1}$$

Or

$$v(t) = \frac{a_0}{2} + a_1 \cos\frac{2\pi t}{T} + a_2 \cos\frac{4\pi t}{T} + \cdots + b_1 \sin\frac{2\pi t}{T} + b_2 \sin\frac{4\pi t}{T} + \cdots$$

T is the minimum time in which the voltage waveform repeats itself, and $1/T$ is the *fundamental* frequency. In equation (3.1) a_0, a_n ($n = 1, 2, 3, \ldots$), and b_m ($m = 1, 2, 3, \ldots$) are constants that can be calculated from $v(t)$ with the following equations:

$$a_n = \frac{2}{T}\int_{-T/2}^{T/2} v(t)\cos n\omega t\, dt \qquad b_m = \frac{2}{T}\int_{-T/2}^{T/2} v(t)\sin m\omega t\, dt \qquad \textbf{(3.2)}$$

If we start with the voltage $v(t)$ and proceed to express it as the sum of the sines and cosines, then we say this process is one of "analysis"; if we start with the sines and cosines and add them together to form the voltage $v(t)$, we say this process is one of "synthesis."

Usually we know the response of any circuit to an input of sines and/or cosines. Thus we can determine the response of a circuit to a complicated input voltage by analyzing the input waveform into sines and cosines and then calculating the response of the circuit to the sine and cosine components. Strictly speaking, this procedure will work only for a linear circuit, but many circuits encountered are linear, and any circuit at all will act in a linear manner if the input voltage is sufficiently small.

Let us now consider the meaning of the a and b Fourier coefficients. The first term, $a_0/2$, represents simply the average dc level of the voltage from $t = 0$ to $t = T$, since $a_0/2 = (1/T)\int_{-T/2}^{T/2} v(t)\, dt$. Thus if the area of the $v(t)$ curve above the $v = 0$ horizontal axis equals the area below the axis, then $a_0/2 = 0$. This will be the case, for example, whenever $v(t)$ has been passed through a series capacitor, because a capacitor can pass no direct current. a_1 is the coefficient of the term $\cos 2\pi t/T = \cos 2\pi f t$; thus a_1 is the amplitude of the Fourier cosine component of frequency f. Similarly, a_2 is the amplitude of the cosine component of frequency $2f$, and so on. The point to remember here is that the shape of the voltage $v(t)$ determines just what the coefficients a_n and b_m are. The voltage can be described equally well by giving its shape versus time $v(t)$, or by giving all the a_n and b_m Fourier coefficients.

If we assume the Fourier series expansion exists and converges to the voltage waveform $v(t)$, then it is relatively easy to show mathematically that the Fourier coefficients a_n and b_m must be given by equation (3.2). Starting with the Fourier expansion

$$v(t) = \frac{a_0}{2} + \sum_{n=1}^{\infty} a_n \cos\frac{2\pi n t}{T} + \sum_{m=1}^{\infty} b_m \sin\frac{2\pi m t}{T} \qquad \textbf{(3.3)}$$

we multiply by $\cos(2\pi kt/T)$, where k is some positive integer, and integrate with respect to time from $t = -T/2$ to $t = T/2$.

$$\int_{-T/2}^{T/2} v(t) \cos \frac{2\pi kt}{T}\, dt$$

$$= \int_{-T/2}^{T/2} \left(\frac{a_0}{2} + \sum_{n=1}^{\infty} a_n \cos \frac{2\pi nt}{T} + \sum_{m=1}^{\infty} b_m \sin \frac{2\pi mt}{T} \right) \cos \frac{2\pi kt}{T}\, dt \tag{3.4}$$

There are three types of integrals:

$$\int_{-T/2}^{T/2} \frac{a_0}{2} \cos \frac{2\pi kt}{T}\, dt = \frac{a_0}{2} \int_{-T/2}^{T/2} \cos \frac{2\pi kt}{T}\, dt = 0 \tag{3.5}$$

$$\int_{-T/2}^{T/2} \left(\sum_{m=1}^{\infty} b_m \sin \frac{2\pi mt}{T} \right) \cos \frac{2\pi kt}{T}\, dt$$

$$= \sum_{m=1}^{\infty} b_m \int_{-T/2}^{T/2} \sin \frac{2\pi mt}{T} \cos \frac{2\pi kt}{T}\, dt = 0 \tag{3.6}$$

$$\int_{-T/2}^{T/2} \sum_{n=1}^{\infty} a_n \cos \frac{2\pi nt}{T} \cos \frac{2\pi kt}{T}\, dt = \sum_{n=1} a_n \int_{-T/2}^{T/2} \cos \frac{2\pi nt}{T} \cos \frac{2\pi kt}{T}\, dt \tag{3.7}$$

All the integrals involving the product of a sine and a cosine are zero. The only integral that is not zero is the one for which $n = k$ in the product of two cosine terms. This integral works out to be

$$\int_{-T/2}^{T/2} \cos \frac{2\pi kt}{T} \cos \frac{2\pi kt}{T}\, dt = \frac{T}{2} \tag{3.8}$$

Thus we have shown that

$$\int_{-T/2}^{T/2} v(t) \cos \frac{2\pi kt}{T}\, dt = a_k \frac{T}{2}$$

or

$$a_k = \frac{2}{T} \int_{-T/2}^{T/2} v(t) \cos \frac{2\pi kt}{T}\, dt \tag{3.9}$$

which is the expression for the a coefficient of equation (3.2) except that the subscript k has been used in place of n. A similar argument will yield the result

$$b_k = \frac{2}{T} \int_{-T/2}^{T/2} v(t) \sin \frac{2\pi kt}{T}\, dt \tag{3.10}$$

A few mathematical points might be in order here: There are certain restrictions on the function $v(t)$ to ensure that the expansion exists. Namely, $v(t)$ can contain only a finite number of discontinuities and must remain bounded, but any voltage waveform encountered in electronics will satisfy these conditions. Also, if we plot the Fourier expansion versus time for times outside the $-T/2$ to $T/2$ interval, we will merely get a periodic extension of $v(t)$ with period T. Thus, a periodic voltage, such as a square wave or a sawtooth wave, can be Fourier analyzed for t from 0 to ∞. At a point of discontinuity in $v(t)$, the Fourier expansion "splits the difference"; for example, if $v(t)$ jumps discontinuously at the instant of time t_1 from 2 V to 4 V, then the Fourier expansion will give 3 V as the value of $v(t)$ at t_1.

The Fourier series can be expressed in a simpler form if we allow the coefficients to be complex. If we define (complex) Fourier coefficients c_n according to

$$
\begin{aligned}
c_n &\equiv \frac{a_n - jb_n}{2} \qquad \text{for } n > 0 \\
&\equiv \frac{a_0}{2} \qquad \text{for } n = 0 \\
&\equiv \frac{a_n + jb_n}{2} \qquad \text{for } n < 0
\end{aligned}
\tag{3.11}
$$

then the Fourier series (3.3) becomes

$$v(t) = \sum_{n=-\infty}^{\infty} c_n e^{jn\omega t} \tag{3.12}$$

where

$$c_n = \frac{1}{T}\int_{-T/2}^{T/2} v(t)\, e^{-jn\omega t}\, dt \tag{3.13}$$

This can be proven by substituting the Euler identities

$$\cos\theta = \frac{1}{2}(e^{j\theta} + e^{-j\theta}) \quad \text{and} \quad \sin\theta = \frac{1}{2j}(e^{j\theta} - e^{-j\theta})$$

into (3.3).

If $v(t)$ is real, $v^*(t) = v(t)$, and, consequently, $c_n = c^*_{-n}$. This complex version of the Fourier series will be used in Chapter 15 in deriving the Nyquist sampling theorem, which is so important in analog-to-digital conversion.

Consider a specific example; let us analyze the square wave shown in Fig. 3.3(a) with pulse width τ and frequency $1/T$. The voltage waveform to be analyzed equals V_0 from $-\tau/2$ to $+\tau/2$ and equals zero from $\tau/2$ to $T/2$

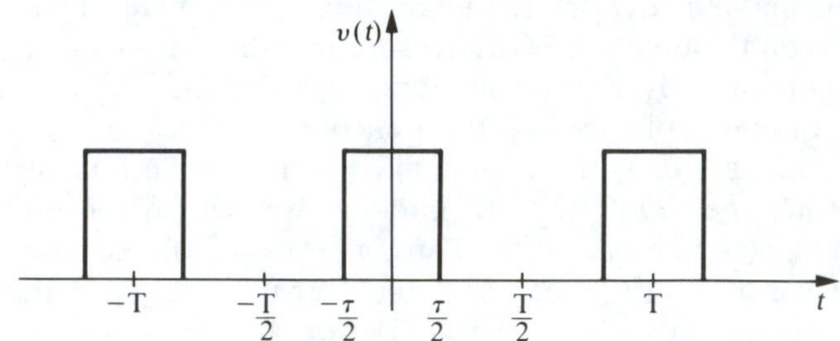

(a) *asymmetrical square wave with pulse width τ and period T*

(a) *a*

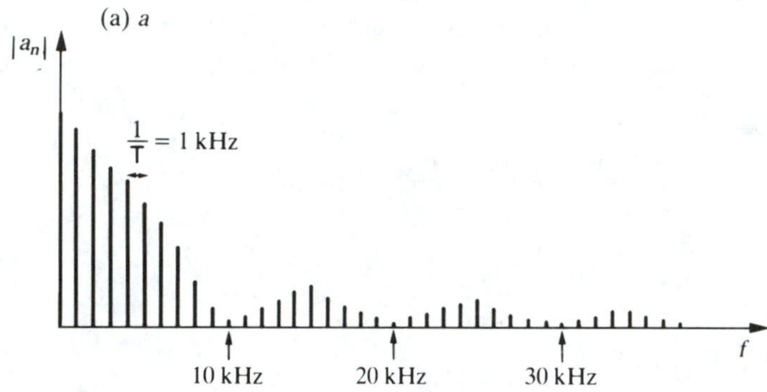

(b) *Fourier spectrum for* τ = 100 μs, T = 1 ms

FIGURE 3.3 Square wave and its Fourier spectrum.

and $-\tau/2$ to $-T/2$. Note that $v(t)$ is even, so all the b_m sine coefficients equal zero.† The a_n cosine coefficients are given by

$$a_n = \frac{2}{T}\int_{-T/2}^{T/2} v(t)\cos n\omega t\,dt = \frac{4}{T}\int_0^{T/2} v(t)\cos n\omega t\,dt$$

since $v(t)\cos n\omega t$ is *even*.

$$a_n = \frac{4}{T}\int_0^{\tau/2} V_0 \cos n\omega t\,dt = \frac{4\,V_0}{T}\left(\frac{\sin n\omega t}{n\omega}\bigg|_0^{\tau/2}\right) = \frac{4\,V_0}{n\omega T}\sin\frac{n\omega\tau}{2}$$

---

†1. A function is even if $v(-t) = v(t)$.
2. A function is odd if $v(-t) = -v(t)$.
3. $\int_{-T}^{T} v(t)f(t)\,dt = 0$ if the integrand is odd; i.e., if $v(t)$ is even and $f(t)$ is odd, or if $v(t)$ is odd and $f(t)$ is even.
4. $\int_{-T}^{T} v(t)f(t)\,dt = 2\int_0^T v(t)f(t)\,dt$ if the integrand is even; i.e., if both $v(t)$ and $f(t)$ are even or both are odd.
5. If $v(t)$ is even, all the b_m coefficients = 0; if $v(t)$ is odd, all the a_n coefficients = 0. If $v(t)$ is neither even nor odd, then, in general, we will have both a_n and b_m coefficients in the Fourier expansion of $v(t)$.

Using $\omega \equiv (2\pi/T)$

$$a_n = \frac{2V_0}{\pi n}\sin\frac{n\pi\tau}{T} \tag{3.14}$$

$$a_0 = \frac{2}{T}\int_{-T/2}^{T/2} v(t)\,dt = \frac{4}{T}\int_0^{\tau/2} V_0\,dt = 2V_0\frac{\tau}{T}$$

Thus, the Fourier coefficients depend upon the ratio of the pulse width to the period, τ/T. Notice that a_n is nonzero for *all* values of n ($n = 0, 1, 2, 3, \ldots$) unless $\sin n\pi\tau/T = 0$. This means that the square wave contains very-high-frequency Fourier components.

If $\tau/T = \frac{1}{2}$, which corresponds to a symmetrical square wave, then we have

$$a_n = \frac{2V_0}{\pi n}\sin\frac{n\pi}{2} = \frac{2V_0}{\pi n}(-1)^{(n-1)/2} \qquad \text{for } n = 1, 3, 5, \ldots$$

$$a_n = 0 \qquad \text{for } n = 2, 4, 6, \ldots$$

$$\frac{a_0}{2} = \frac{V_0}{2} \tag{3.15}$$

Thus the symmetrical square-wave expansion is

$$v(t) = \frac{V_0}{2} + \sum_{n\ \text{odd}}^{\infty} \frac{2V_0}{\pi n}(-1)^{(n-1)/2}\cos\frac{2\pi n t}{T} \tag{3.16}$$

In words, the symmetrical square wave equals an average dc level of one half the pulse amplitude plus a series of cosine waves having frequencies of odd integral multiples of $2\pi/T$, the fundamental frequency. Notice that the amplitude a_n of the odd harmonics decreases with increasing frequency, as shown in Fig. 3.3.

When are the Fourier coefficients $a_n = 0$? From (3.14) we see that $a_n = 0$ when the argument of the sine equals an integral multiple of π:

$$\frac{n\pi\tau}{T} = k\pi \qquad \text{where } k = 1, 2, 3, \ldots$$

or

$$\frac{n}{T} = \frac{k}{\tau} \tag{3.17}$$

But $1/T = f_0$ is the fundamental frequency of the pulse train; for example, if $T = 1$ ms, then $f_0 = 1$ kHz. All the Fourier component frequencies must be integral multiples of f_0 from the basic Fourier expansion (3.1) of any periodic function. In other words, the Fourier components are all spaced f_0 apart.

The $1/\tau$ term in (3.17) is the reciprocal of the pulse width; for example, if $\tau = 100\ \mu s$, then $1/\tau = 10\ \text{kHz}$. For $k = 1$, (3.17) gives us the first Fourier coefficient that is zero:

$$nf_0 = \frac{1}{\tau}$$

If $T = 1$ ms and $\tau = 100\ \mu s$, then $n = 10$, and $a_{10} = 0$. For $k = 2$, then $a_{20} = 0$, and so on. In general, the Fourier coefficients are zero for

$$nf_0 = \frac{1}{\tau}, \quad \frac{2}{\tau}, \quad \frac{3}{\tau}, \quad \ldots$$

For our $f_0 = 1$-kHz pulse frequency with pulse width 100 μs, $a_n = 0$ for 10 kHz, 20 kHz, 30 kHz, ... This specific example is shown in Fig. 3.3(b).

Finally, notice that for a symmetrical "square" wave ($\tau = T/2$), (3.17) implies all the even Fourier coefficients are zero, in agreement with (3.15).

Fourier analysis can be extended to cover the problem of analyzing a voltage waveform extending from $t = -\infty$ to $t = +\infty$. The argument will not be given here, but the result is expressed most concisely in complex notation as follows:

$$v(t) = \frac{1}{\sqrt{2\pi}} \int_{-\infty}^{\infty} g(\omega)e^{-j\omega t}\, d\omega \tag{3.18}$$

$$g(\omega) = \frac{1}{\sqrt{2\pi}} \int_{-\infty}^{\infty} v(t)e^{+j\omega t}\, dt \tag{3.19}$$

The $g(\omega)$ is the Fourier coefficient; it is the amplitude of the Fourier component of angular frequency ω. The integral over ω is analogous to the summation over the different frequencies $n2\pi t/T$. $v(t)$ and $g(\omega)$ are often called a Fourier transform pair.

The above is referred to as the Fourier integral theorem, and the interested reader is referred to *Mathematical Methods for Physicists* by G. Arfken.† Notice that the basic idea is still the same; an arbitrary voltage $v(t)$ is expressed as a sum (an integral is just a sum of an infinite number of infinitesimally small terms) of waves of different frequencies. In this case the waves are written in the form of complex exponentials, but their real and imaginary parts are the familiar trigonometric cosines and sines.

$$e^{j\omega t} = \cos \omega t + j \sin \omega t \tag{3.20}$$

†(New York: Academic Press, 1985), Chapter 14.

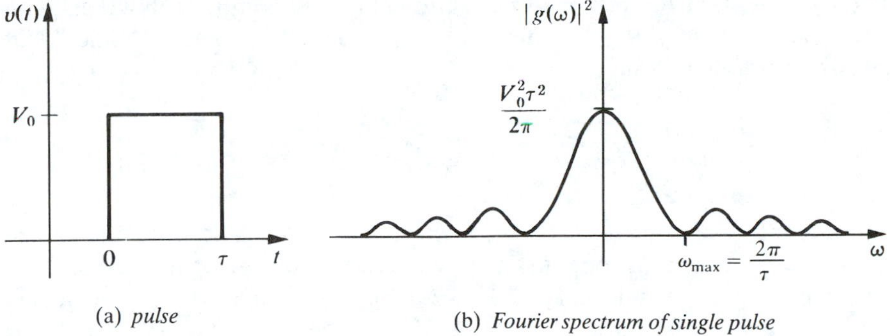

(a) *pulse*

(b) *Fourier spectrum of single pulse*

FIGURE 3.4 Single rectangular voltage pulse and Fourier components.

Let us now analyze a single rectangular pulse of amplitude V_0, width τ, and zero rise and fall times as shown in Fig. 3.4(a). The voltage pulse being analyzed is zero for t less than 0 and for t greater than τ, and it equals V_0 for $0 \leqq t \leqq \tau$. Therefore, the Fourier coefficient $g(\omega)$ is given by

$$g(\omega) = \frac{1}{\sqrt{2\pi}} \int_{-\infty}^{\infty} v(t) e^{j\omega t}\, dt = \frac{1}{\sqrt{2\pi}} \int_{0}^{\tau} V_0 e^{j\omega t}\, dt = \frac{V_0}{\sqrt{2\pi}} \frac{e^{j\omega t}}{j\omega} \bigg|_0^{\tau}$$

$$g(\omega) = \frac{V_0}{j\omega\sqrt{2\pi}} (e^{j\omega\tau} - 1) \tag{3.21}$$

The coefficient $g(\omega)$ is the amplitude in volts per rad/s bandwidth of the Fourier component of angular frequency ω. It is more convenient to consider the *power* of the Fourier component, which is proportional to the amplitude squared, so we will calculate $|g(\omega)|^2$.

$$|g(\omega)|^2 = g^*(\omega) g(\omega) = \frac{V_0}{-j\omega\sqrt{2\pi}} (e^{-j\omega\tau} - 1) \frac{V_0}{j\omega\sqrt{2\pi}} (e^{j\omega\tau} - 1)$$

$$= \frac{V_0^2}{2\pi\omega^2} [2 - (e^{j\omega\tau} + e^{-j\omega\tau})] = \frac{V_0^2}{2\pi\omega^2} [2 - 2\cos\omega\tau]$$

$$= \frac{2V_0^2}{\pi\omega^2} \sin^2 \frac{\omega\tau}{2}$$

using $1 - \cos\omega\tau = 2\sin^2(\omega\tau/2)$ or, rewriting,

$$|g(\omega)|^2 = \frac{V_0^2\tau^2}{2\pi} \left[\frac{\sin\frac{\omega\tau}{2}}{\frac{\omega\tau}{2}} \right]^2 \tag{3.22}$$

Thus, we see that the power in the various Fourier components varies with frequency as $[\sin^2(\omega\tau/2)]/\omega^2$, which is plotted in Fig. 3.4(b). Notice that $|g(\omega)|^2$ falls to zero when

$$\omega = \frac{2\pi}{\tau} \equiv \omega_{\text{max}} \quad \text{or} \quad \frac{1}{\tau} = f_{\text{max}}$$

Subsequent zeros occur when $\omega = 4\pi/\tau,\ 6\pi/\tau, \ldots$. The area under the $|g(\omega)|^2$ curve is proportional to the power in the Fourier components.

It can be seen by inspecting Fig. 3.4(b) that most of the power occurs for frequencies *from zero to* $f_{\text{max}} = 1/\tau$, the frequency of the first zero. Thus we would expect a circuit capable of passing frequencies from zero to $f_{\text{max}} = 1/\tau$ to pass the voltage pulse without serious distortion of the pulse shape, since most of the energy of the pulse lies in this range of frequencies. This result is extremely important, practically speaking, because it is impossible to build a circuit that will pass all frequencies from zero to infinity without attenuation. The gain of any real circuit will decrease if the signal frequency is increased enough. The range of frequencies that a circuit will pass with a certain gain is called the *bandwidth*. The preceding result tells us that a bandwidth of approximately $1/\tau$ (or more) is necessary to pass a pulse of pulse width τ. For example, a bandwidth of 1 MHz is necessary to pass a 1-μs pulse. The narrower the pulse width (the "faster" the pulse), the larger is the bandwidth required. In actual electronic circuit design many ingenious techniques are used to increase the bandwidth of circuits to pass fast pulses with minimum distortion.

In order to pass a perfectly rectangular pulse (zero rise and fall times) with no distortion, a circuit must have an infinite bandwidth, from dc to infinity. No real circuit can have an infinite bandwidth, so we are led to the question: What will the output pulse look like for a circuit of finite bandwidth? One's physical intuition (properly trained) can answer this question to a surprisingly high degree of accuracy. It is intuitively reasonable (and correct) that any rapidly changing part of a voltage waveform, such as the steep leading or trailing edges of a pulse, will contain *high*-frequency Fourier components, while a slowly changing portion of the waveform, such as the flat top of a pulse, will contain *low*-frequency Fourier components. Thus, if a zero-rise-time pulse is fed into a circuit of finite bandwidth, we would expect the leading and trailing edges of the pulse to be rounded off because the high-frequency components in these edges are not passed by the circuit. This is illustrated in Fig. 3.5. The smaller the bandwidth, the more rounding of the pulse will occur; if the bandwidth is much less than from 0 to $f = 1/\tau$, then the amplitude of the output pulse will be appreciably less than that of the input pulse.

It is a general feature of Fourier analysis that any rapid change in the voltage waveform being analyzed contains high-frequency Fourier com-

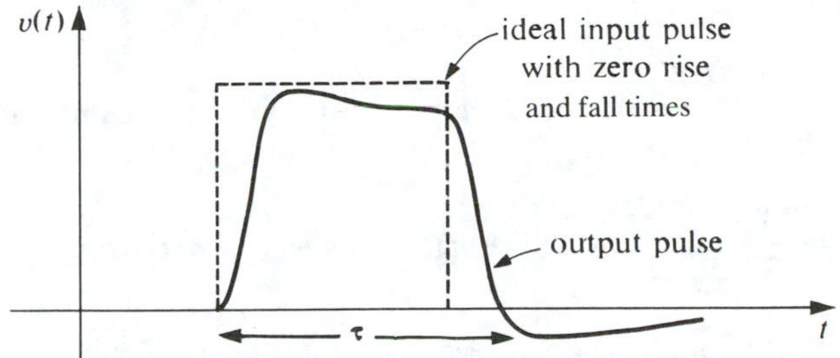

FIGURE 3.5 Rounding of output pulse due to finite bandwidth.

ponents. For example, note the leading and trailing edges of pulses, the sharp "corners" of the triangular or sawtooth waves, and the discontinuity of the slope of a rectified sine wave as indicated in Fig. 3.6.

Thus, a circuit incapable of passing the high-frequency Fourier components will round off any sharp corners or rapid changes in the input waveform; the output will be distorted and may even be much less in amplitude than the input.

FIGURE 3.6 Examples of rapid voltage changes (circled) which contain high-frequency Fourier components.

Let us Fourier analyze the *rectified* or "half" sine wave of Fig. 3.7(a). (As we will see in the next chapter, this waveform is produced when a sine wave is rectified by an ideal diode.) Notice that we have placed $t = 0$ at the center of the peak, thus making v(t) an *even* function of t.

$$v(t) = 0 \qquad \text{for } \frac{-T}{2} \leq t < \frac{-T}{4}$$

$$= V_0 \cos \omega t \qquad \text{for } \frac{-T}{4} \leq t \leq \frac{T}{4} \quad \left(\omega = \frac{2\pi}{T}\right)$$

$$= 0 \qquad \text{for } \frac{T}{4} \leq t \leq \frac{T}{2}$$

Because $v(t)$ is even, all the b_m coefficients are zero, and we have for the a_n

coefficients:

$$a_n = \frac{2}{T}\int_{-T/4}^{T/4} v(t)\cos n\omega t\, dt = \frac{4V_0}{T}\int_0^{T/4} \cos \omega t \cos n\omega t\, dt$$

$$= \frac{2V_0}{T}\int_0^{T/4} [\cos \omega(1+n)t + \cos \omega(1-n)t]\, dt$$

$$= \frac{2V_0}{T}\left[\frac{\sin \omega(1+n)t}{\omega(1+n)}\bigg|_0^{T/4} + \frac{\sin \omega(1-n)t}{\omega(1-n)}\bigg|_0^{T/4}\right]$$

$$= V_0\left[\frac{\sin\left[\frac{(1+n)\pi}{2}\right]}{(1+n)\pi} + \frac{\sin\left[\frac{(1-n)\pi}{2}\right]}{(1-n)\pi}\right] \tag{3.23}$$

We immediately see that $a_n = 0$ for all odd values of $n > 1$ and that

$$a_0 = V_0\left[\frac{1}{\pi} + \frac{1}{\pi}\right] = \frac{2V_0}{\pi} \tag{3.24}$$

Evaluating a_1 is a bit tricky because of the $(1 - n)$ factor in the denominator. Recalling that $\lim_{x\to 0}[(\sin x)/x] = 1$, we can write the $(1 - n)$ sine terms as

$$\frac{\sin\left[(1-n)\frac{\pi}{2}\right]}{(1-n)\pi} = \frac{1}{2}\frac{\sin x}{x} \qquad \text{where } x = \frac{(1-n)\pi}{2}$$

Thus
$$a_1 = V_0\left[0 + \frac{1}{2}\right] = \frac{V_0}{2}$$

Similarly
$$a_2 = \frac{2V_0}{3\pi} \qquad a_4 = \frac{-2V_0}{15\pi} \tag{3.25}$$

and so on. An experimental Fourier spectrum of a rectified sine wave is shown in Fig. 3.7(b).

An interesting example of Fourier analysis occurs in ordinary AM radio broadcasting. The AM stands for "amplitude modulation" and refers to the fact that the amplitude of the high-frequency electromagnetic wave broadcast by the radio station is changed or modulated according to the type of audible sound transmitted. For example, an AM station at 930 on your dial broadcasts an electromagnetic wave at a frequency $f_c = 930$ kHz. This wave is called the *carrier* wave, and if it were of constant amplitude no audible sound would be heard through the receiver. If a sound wave of

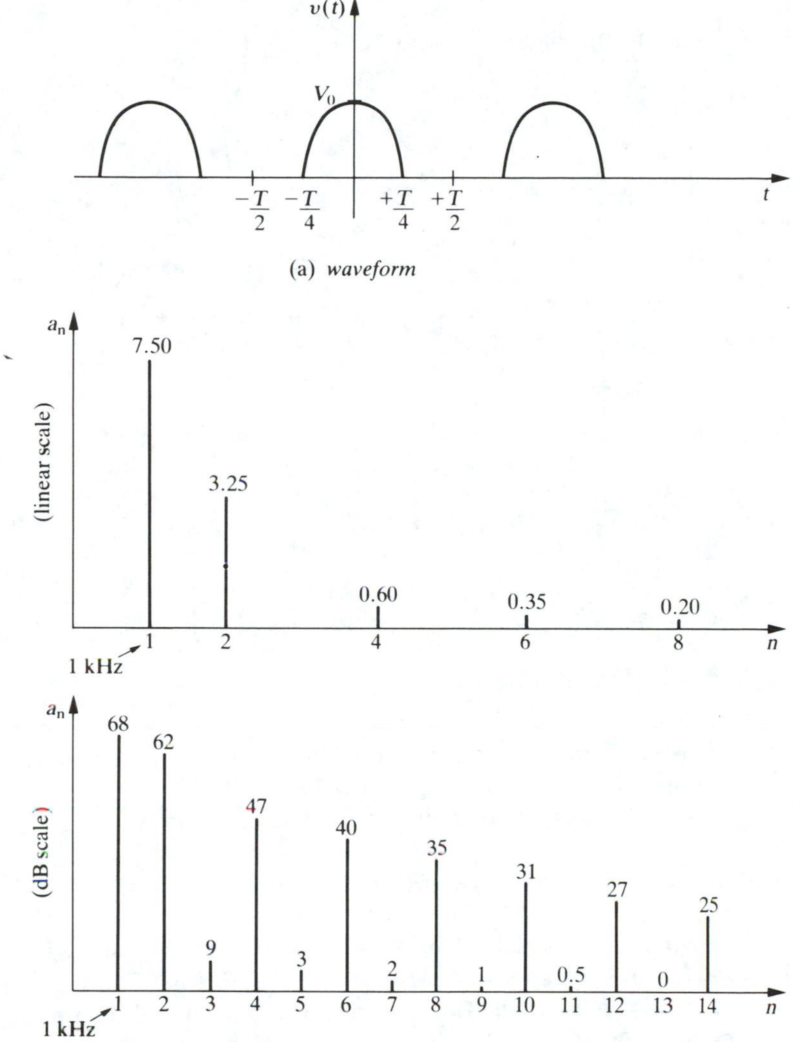

(b) *experimental Fourier spectrum of* (a) *[using Tektronix 5L4N Spectrum Analyzer]*

FIGURE 3.7 Rectified sine wave.

frequency $f_m = 400$ Hz is to be transmitted via this high-frequency carrier, then the 400-Hz sound wave is first converted into a 400-Hz voltage by a microphone, and the 400-Hz voltage is used to change the amplitude of the carrier (i.e., to *modulate* the carrier) as shown in Fig. 3.8. The receiver now must somehow extract the 400-Hz modulation envelope from the modulated carrier. The interesting thing that Fourier analysis reveals is that the receiver bandwidth must be at least 400 Hz wide in order to receive the 400-Hz audio signal. That is, if the receiver bandwidth were too narrow,

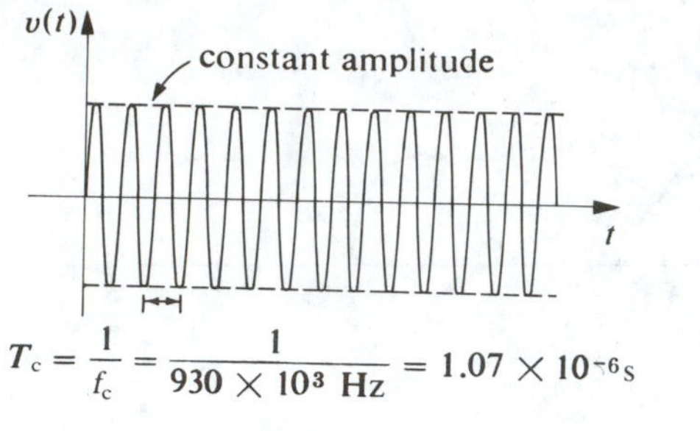

(a) *unmodulated 930 kHz carrier wave*

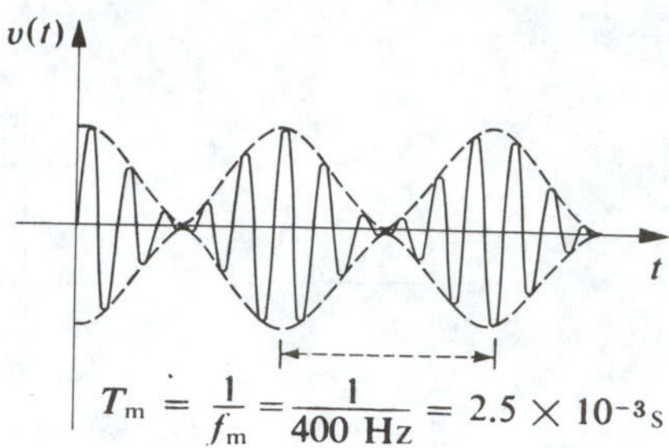

(b) *carrier with 100% 400 Hz amplitude modulation*

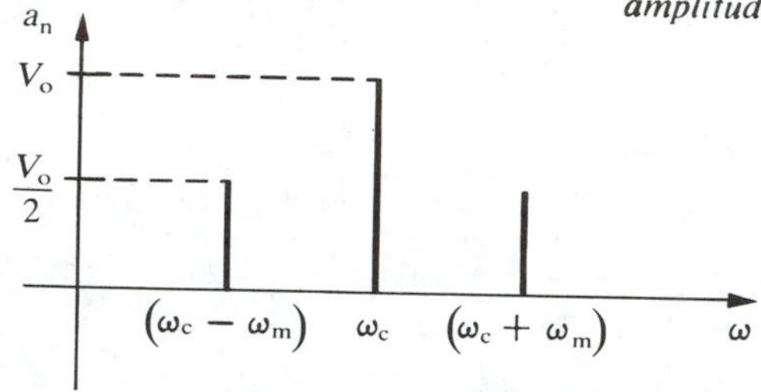

(c) *Fourier components of 100% modulated carrier of* (b)

FIGURE 3.8 Amplitude modulation.

say, from 929.9 to 930.1 kHz, a bandwidth of 0.2 kHz or 200 Hz, no 400-Hz signal would be present in the receiver.

Let us Fourier analyze the amplitude-modulated carrier.

$$v(t) = V_0(1 + M \cos \omega_m t) \cos \omega_c t \tag{3.26}$$

where $\omega_m = 2\pi f_m$, $f_m = 400$ Hz, $\omega_c = 2\pi f_c$, $f_c = 930$ kHz, and M is the degree of modulation, $0 \leqq M \leqq 1$. M is 1.0 in Fig. 3.8(b). The modulated carrier wave $v(t)$ is an even function, so only cosine terms will be present in the Fourier expansion. Also, we can see that $a_0 = 0$, because the average dc level is zero. Thus

$$v(t) = \sum_{n=1}^{\infty} a_n \cos n\omega t \tag{3.27}$$

where

$$a_n = \frac{2}{T} \int_{-T/2}^{T/2} v(t) \cos n\omega t \, dt \tag{3.28}$$

Choose $T = 2\pi/\omega_m \equiv T_m$; this is the shortest time interval over which $v(t)$ is periodic. Substituting (3.26) in (3.28) yields, with $\omega = \omega_m$,

$$a_n = \frac{4}{T_m} \int_0^{T_m/2} V_0(1 + M \cos \omega_m t) \cos \omega_c t \cos n\omega_m t \, dt$$

The trigonometric identity

$$\cos\alpha\cos\beta = \frac{1}{2}[\cos(\alpha+\beta) + \cos(\alpha-\beta)] \tag{3.29}$$

is used to change the integrand product of three cosine factors into a product of two cosines: Let $\alpha = \omega_c t$ and $\beta = \omega_m t$. Thus a_n becomes

$$\begin{aligned} a_n &= \frac{4V_0}{T_m}\int_0^{T_m} \cos\omega_c t \cos n\omega_m t\, dt \\ &+ \frac{4V_0 M}{T_m}\int_0^{T_m/2} \cos(\omega_c+\omega_m)t \cos n\omega_m t\, dt \\ &+ \frac{4V_0 M}{T_m}\int_0^{T_m/2} \cos(\omega_c-\omega_m)t \cos n\omega_m t\, dt \end{aligned} \tag{3.30}$$

Each integral equals zero unless the arguments of the two cosine factors are equal. Thus, there are only three nonzero Fourier components, one for each integral of (3.30). The frequencies $n\omega_m$ of these three components are $n\omega_m = \omega_c$ (the carrier frequency), $n\omega_m = \omega_c - \omega_m$ (ω_m lower than the carrier frequency) and $n\omega_m = \omega_c + \omega_m$ (ω_m higher than the carrier frequency).

The Fourier coefficients work out to be

$$a_n = V_0 \qquad \text{for } n\omega_m = \omega_c \tag{3.31}$$

and

$$a_n = \frac{MV_0}{2} \qquad \text{for } n\omega_m = \omega_c \pm \omega_m \tag{3.32}$$

Thus the Fourier spectrum of the amplitude-modulated carrier consists of a central component at the carrier frequency ω_c and two other equal components called *sidebands* spaced equally on either side of the carrier—one ω_m above and the other ω_m below the carrier frequency. For $M = 1$ or 100% modulation, as shown in Fig. 3.8(b), the sidebands are each one half the amplitude of the carrier, or are 6 dB down with respect to the carrier. For the receiver to "hear" the 400-Hz modulation signal, it must receive from ω_c to $\omega_c + \omega_m$ or from ω_c to $\omega_c - \omega_m$; that is, the receiver bandwidth must be at least ω_m wide.

If the carrier waveform is not modulated to zero (i.e., if $M < 1$), as shown in Fig. 3.9(a), then the sidebands are smaller as shown in Fig. 3.9(b).

In actual practice the modulation signal is not usually a pure sine wave but is a complicated audio waveform as shown in Fig. 3.9(c) with frequency components ranging from 20 Hz to about 20 kHz, the range of response of the human ear. A Fourier analysis of a carrier whose envelope is amplitude

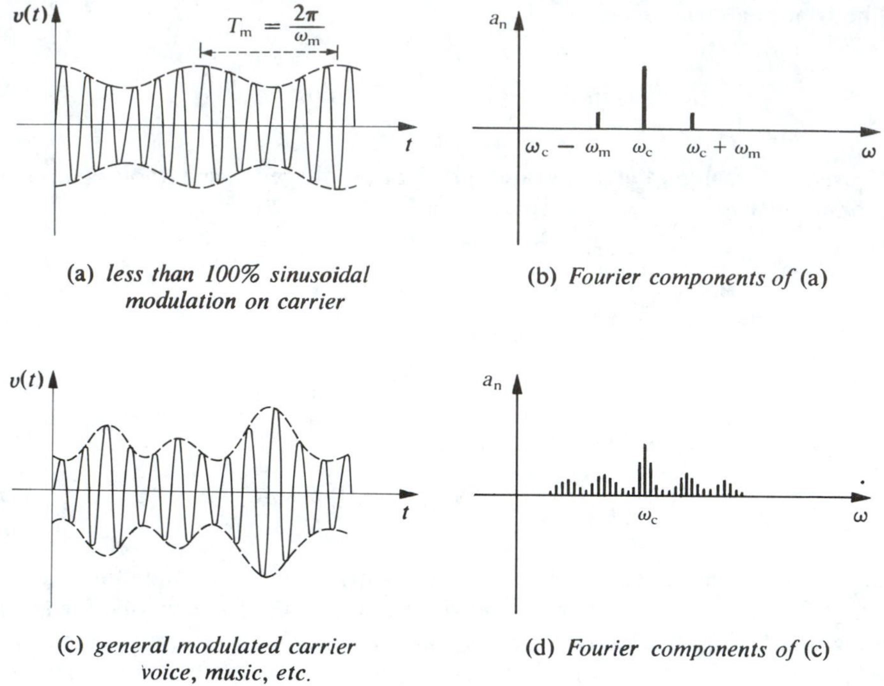

(a) *less than 100% sinusoidal modulation on carrier*

(b) *Fourier components of* (a)

(c) *general modulated carrier voice, music, etc.*

(d) *Fourier components of* (c)

FIGURE 3.9 Amplitude modulation waveforms.

modulated by such a complicated audio waveform will show that the sidebands (Fourier components) lie in the range from $\omega_c - \omega_{max}$ to $\omega_c + \omega_{max}$, as shown in Fig. 3.9(d). ω_{max} is the maximum frequency component of the audio modulation signal. For high-fidelity sound, we need

$$\omega_{max} = 2\pi \times 20 \text{ kHz}$$

It is necessary to receive only one set of sidebands. Thus the receiver must have a bandwidth from ω_c to $\omega_c + \omega_{max}$, 20 kHz for high fidelity. To pack more AM stations into the AM band, the maximum modulation frequency $\omega_{max}/2\pi$ is limited by law to 10 kHz, so an AM receiver bandwidth must be a maximum of 10 kHz. The reason AM radio music is not high fidelity is simply that only audio frequencies up to 10 kHz can be broadcast. The crisp, high-pitched sounds from 10 kHz to 20 kHz are simply not present in AM broadcasts.

An AM receiver with a diode detector is shown in Fig. 3.10. The amplitude-modulated carrier frequency ω_c and the sidebands from the antenna are selected by the tuned LC circuit and amplified by the rf amplifier. The diode then passes only the positive half-waves, yielding the waveform of Fig. 3.10(b). This waveform is then passed through a low-pass

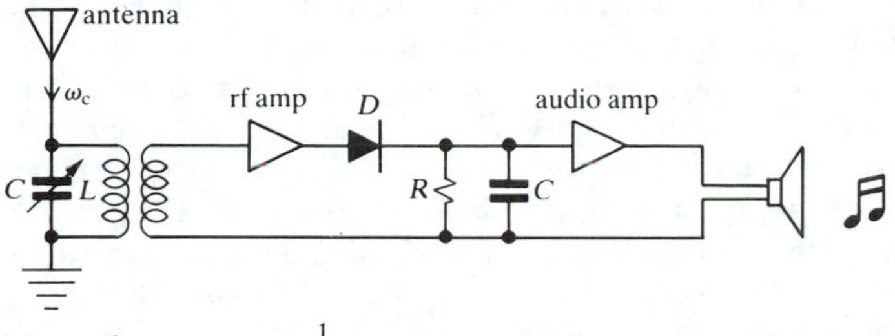

(a) *circuit*

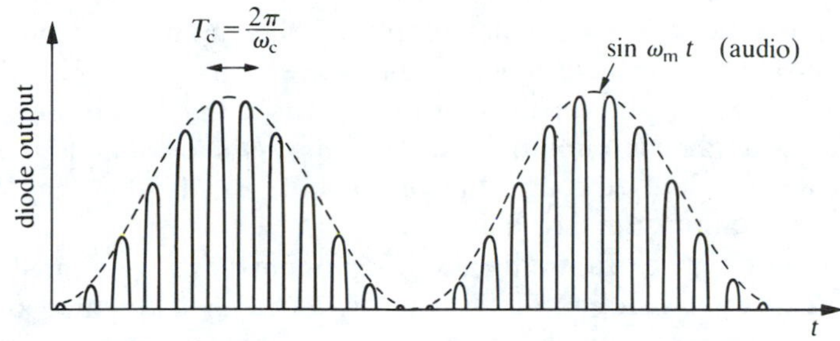

(b) *output of diode before passing through the RC low-pass filter*

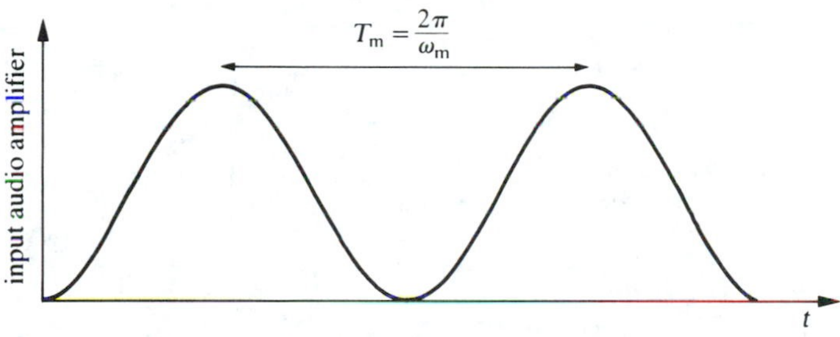

(c) *filtered diode output (original audio signal)*

FIGURE 3.10 Simple diode AM receiver.

RC filter that passes only the audio envelope at ω_m, provided

$$T_m \gg RC \gg T_c \qquad \text{where } T_m = \frac{2\pi}{\omega_m} \quad \text{and} \quad T_c = \frac{2\pi}{\omega_c}$$

Thus, the filter output is the same as the original audio modulation of the transmitted wave.

Another type of amplitude modulation widely used in long-range noncommercial radio communication is *single sideband* (SSB). In this technique only one set of sidebands is actually transmitted, not the carrier or the other set of sidebands. Both sets of sidebands contain the same information (the modulating audio signal), so only one set need be broadcast. The carrier itself, being a wave of constant amplitude, really contains no information, so it is wasteful to broadcast it, especially because the carrier power is many times the power in one set of sidebands. At the receiver the carrier is generated at a low power level and used to demodulate the sidebands that have been received. That is, the carrier at ω_c is beat against the received sidebands at $\omega_c + \omega_m$, and the difference beat frequency created is the desired audio signal at ω_m.

It should be pointed out that any waveform amplified (or attenuated) by a *nonlinear* circuit will have an entirely different Fourier spectrum at the output as compared to its spectrum at the input. A linear circuit is one in which the output is simply a multiple of the input. A nonlinear circuit essentially changes the shape of the waveform and, thus, the Fourier spectrum. The output-versus-input graph for a linear circuit is a straight line; for a nonlinear circuit, a curved line, as shown in Fig. 3.11. Such graphs need not go through the origin but usually do; that is, there is usually zero output for zero input.

To see how extra frequency components are generated, we write $v_{out} = Av_{in}$, and we think of the gain A as being a function of the input voltage. Then we expand the gain A in a power series in v_{in}, using a standard Taylor series expansion.

$$A = A(0) + \left.\frac{dA}{dv_{in}}\right|_0 v_{in} + \frac{1}{2!}\left.\frac{d^2A}{dv_{in}^2}\right|_0 v_{in}^2 + \cdots \tag{3.33}$$

Thus,

$$v_{out} = \left[A(0) + \left(\left.\frac{dA}{dv_{in}}\right|_0\right) v_{in} + \frac{1}{2!}\left(\left.\frac{d^2A}{dv_{in}^2}\right|_0\right) v_{in}^2 + \cdots\right] v_{in} \tag{3.34}$$

which can be rewritten as

$$v_{out} = a_1 v_{in} + a_2 v_{in}^2 + a_3 v_{in}^3 + \cdots = \sum_{n=1}^{\infty} a_n v_{in}^n \tag{3.35}$$

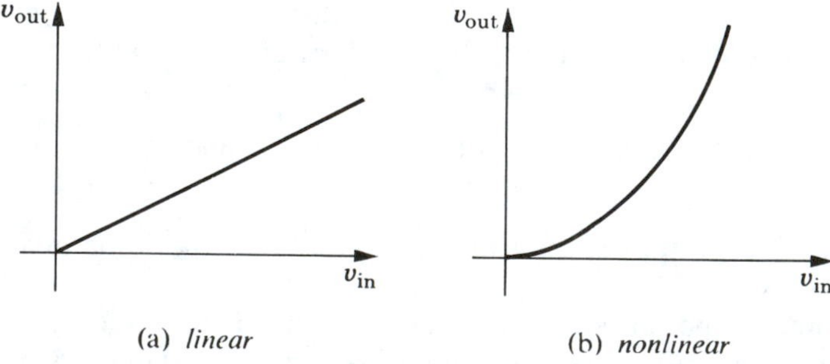

FIGURE 3.11 Input–output graphs.

where $a_1, a_2, a_3, \ldots$ are constants:

$$a_1 = A(0) \qquad a_2 = \left.\frac{dA}{dv_{\text{in}}}\right|_0 \qquad \text{and so on}$$

Thus the constants a_n depend on the exact shape of the output-versus-input graph. Any particular device or amplifier will have a certain set of constants a_n.

If the input contains the sum of two different frequencies $v_{\text{in}} = V_{01}\cos\omega_1 t + V_{02}\cos\omega_2 t$, then the $a_2 v_{\text{in}}^2$ term will generate *new* frequencies of $\omega_1 + \omega_2$ and $\omega_1 - \omega_2$. Let us neglect a_3 and higher a_k's, which is a good approximation for small inputs. Then

$$v_{\text{out}} \cong a_1 v_{\text{in}} + a_2 v_{\text{in}}^2 \tag{3.36}$$

$$\begin{aligned} v_{\text{out}} &\cong a_1(V_{01}\cos\omega_1 t + V_{02}\cos\omega_2 t) + a_2(V_{01}\cos\omega_1 t + V_{02}\cos\omega_2 t)^2 \\ &\cong a_1 V_{01}\cos\omega_1 t + a_1 V_{02}\cos\omega_2 t \\ &\quad + a_2(V_{01}^2\cos^2\omega_1 t + 2V_{01}V_{02}\cos\omega_1 t\cos\omega_2 t + V_{02}^2\cos^2\omega_2 t) \end{aligned} \tag{3.37}$$

Using the trigonometric identity

$$\cos\omega_1 t\cos\omega_2 t = \frac{1}{2}[\cos(\omega_1 + \omega_2)t + \cos(\omega_1 - \omega_2)t]$$

we see that the two frequencies $\omega_1 \pm \omega_2$ are indeed present in the output. If we also use the identities $\cos^2\omega t = \frac{1}{2}(1 + \cos 2\omega t)$, we see that the new frequencies $2\omega_1$ and $2\omega_2$ and a dc component are also present in the output:

$$\begin{aligned} v_{\text{out}} &= \overbrace{a_1 V_{01}\cos\omega_1 t}^{\omega_1} + \overbrace{a_1 V_{02}\cos\omega_2 t}^{\omega_2} + \overbrace{\frac{a_2 V_{01}^2}{2}\cos 2\omega_1 t}^{2\omega_1} \\ &\quad + \overbrace{a_2 V_{01}V_{02}\cos(\omega_1 + \omega_2)t}^{\omega_1+\omega_2} + \underbrace{a_2 V_{01}V_{02}\cos(\omega_1 - \omega_2)t}_{\omega_1-\omega_2} \\ &\quad + \underbrace{\frac{a_2 V_{02}^2}{2}\cos 2\omega_2 t}_{2\omega_2} + \underbrace{\frac{a_2 V_{01}^2}{2} + \frac{a_2 V_{02}^2}{2}}_{\text{dc component}} \end{aligned} \tag{3.38}$$

If the $\omega_1 \pm \omega_2$ frequencies are undesirable, they are called *intermodulation distortion*.

A nonlinear circuit specifically built to generate the sum and difference frequencies $\omega_1 \pm \omega_2$ for two inputs is usually called a *mixer*. The two inputs at ω_1 and ω_2 are said to be mixed together. In superheterodyne radio receivers the signal picked up from the antenna at the carrier frequency ω_c plus any sidebands due to the audio modulation of the carrier is mixed or "beat against" a frequency ω_{LO} generated by the local oscillator in the receiver (see Fig. 3.12). The difference frequency or *beat* frequency $\omega_{LO} - \omega_c'$

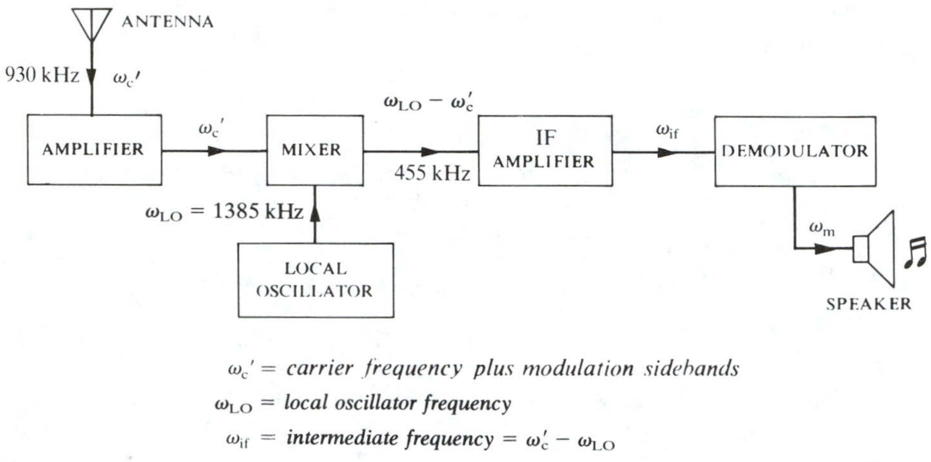

FIGURE 3.12 Superheterodyne radio receiver block diagram.

contains the modulation sidebands and is amplified and demodulated to recover the audio modulation. Usually the local oscillator is tuned so that the difference frequency (called the *intermediate* frequency or i.f.) $\omega_{LO} - \omega_c$ is constant. Usually $(\omega_{LO} - \omega_c)/2\pi = 455$ kHz for AM radios. Thus, if $\omega_c/2\pi = 930$ kHz, $\omega_{LO}/2\pi$ must equal 1385 kHz. Details of the detection process are beyond the scope of this book; the interested reader is referred to any book on commercial radio.

The sum beat frequency $\omega_c' + \omega_{LO}$ is usually well outside the bandpass of the mixer and i.f. amplifier; hence it is strongly attenuated. In the microwave region of thousands of megahertz, a circuit designed to generate the difference frequency $\omega_1 - \omega_2$ is often called a *down converter*, because the output frequency $\omega_1 - \omega_2$ has been shifted down or reduced compared to the input frequency ω_1.

In FM radio the information is transmitted by changing or modulating the *frequency* of the carrier wave; FM stands for frequency modulation. If we let ω_c be the fixed central carrier frequency, then ω_c is varied or modulated above and below its fixed central value by the audio modulation signal. If the modulation signal is $v_m = V_a \cos \omega_m t$, then the FM carrier-wave frequency is

$$\omega = \omega_c + \Delta\omega \cos \omega_m t \tag{3.39}$$

where $\Delta\omega$ is the maximum frequency swing (the *deviation*) of the carrier. The FM carrier wave is then

$$v_c(t) = V_0 \cos \omega t = V_0 \cos[\omega_c t + \Delta\omega \cos \omega_m t] \tag{3.40}$$

The carrier frequency swings from $\omega_c + \Delta\omega$ to $\omega_c - \Delta\omega$, as shown in Fig. 3.13. In commercial FM radio $\Delta\omega/2\pi$ is limited to 75 kHz, so the maximum carrier-frequency swing is ±75 kHz around $\omega_c/2\pi$; for example, if $\omega_c/2\pi = f_c = 100$ MHz, then the swing is from 100.075 to 99.925 MHz. Commercial FM stations are assigned frequency f_c every 200 kHz, so their signals do not overlap. Notice that the frequency swing of the carrier is determined by the *amplitude* ($\Delta\omega$) of the audio modulation, whereas the audio modulation *frequency* (ω_m) determines how often the carrier frequency swings up and

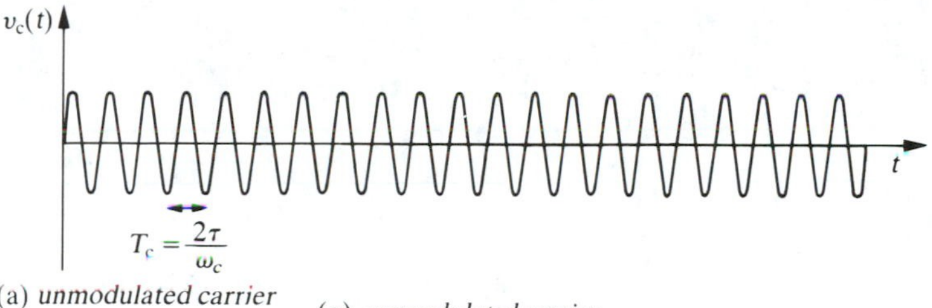

(a) unmodulated carrier

(a) unmodulated carrier

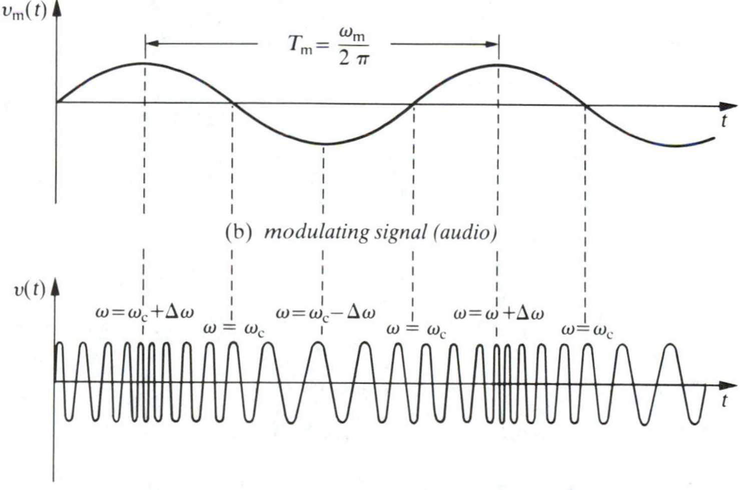

(b) modulating signal (audio)

(c) frequency modulated carrier

FIGURE 3.13 FM radio waveforms.

down. For example, if the audio modulation signal is a 2-kHz wave, and if $f_c = 100$ MHz, the carrier frequency will equal 100.075 MHz every 0.5 ms. The modulation frequency $\omega_m/2\pi = f_m$ in commercial FM is not limited to 10 kHz as in the case of AM. It can range up to 20 kHz, thus covering the full range of frequencies audible to the healthy human ear: 20 Hz to 20 kHz. Thus FM music sounds "crisper" and has higher fidelity than AM music because of the presence of the higher-frequency components from 10 kHz to 20 kHz.

A Fourier analysis of (3.40) is rather complicated and leads to an infinite number of Fourier components or sidebands separated by the modulation frequency for the case of a purely sinusoidal modulation signal. The sideband amplitudes depend upon Bessel functions of the first kind and also on the ratio of the maximum carrier-frequency swing to the modulation frequency, $\Delta\omega/\omega_m$, which is usually called the *modulation index*. Generally, if the modulation index is much less than 1, only the two sidebands at $\omega_c \pm \omega_m$ are important, as shown in Fig. 3.14(a), and the bandwidth is approximately $2\omega_m$, which is the same as for AM. If the modulation index is much greater than 1, then many sidebands are present, as shown in Fig. 3.14(b).†

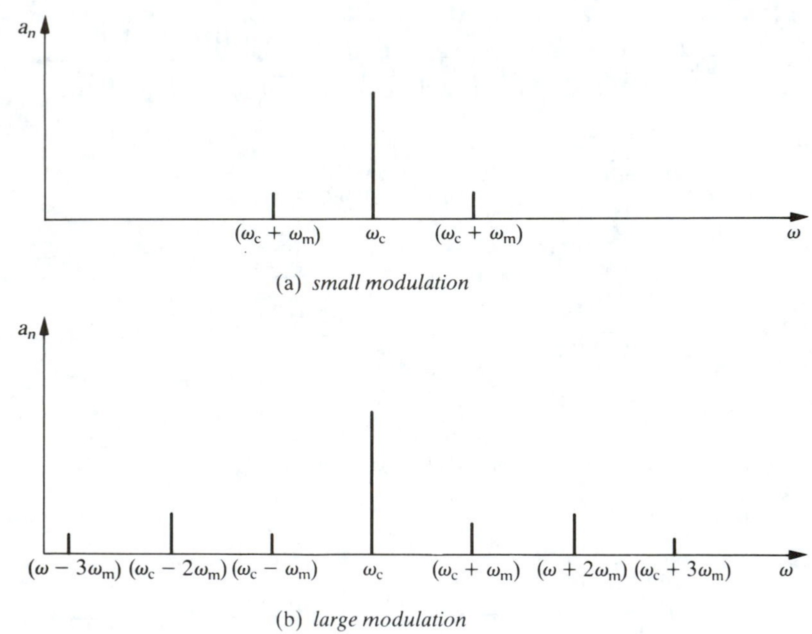

FIGURE 3.14 Fourier components of sinusoidally modulated FM signal.

†*Electronic & Radio Engineering*, F. Terman. McGraw-Hill, New York, 1955, Chap. 17. See *Information Transmission, Modulation & Noise*, M. Schwartz. McGraw-Hill, New York, 1959, Chap. 3.

One important difference between AM and FM systems is their ability to reject or discriminate against noise. In AM a change in the amplitude of the AM carrier (due to a lightning flash or any extraneous voltage spike introduced into the receiver) changes the information transmitted, because the amplitude of the carrier determines the amplitude of the audio output. However, in FM the amplitude of the FM carrier does not affect the audio output at all in theory. The audio output frequency is determined by how rapidly the *frequency* of the carrier changes, and the audio output amplitude is determined by the magnitude of the frequency swing of the carrier. Thus, an FM receiver usually contains a *limiter* circuit that limits or "chops off" all voltage fluctuations larger than some fixed amplitude. Hence any large voltage spikes are chopped off, and very little noise appears in the FM audio output.

3.4 INTEGRATING CIRCUIT (LOW-PASS FILTER)

The RC filter with the output taken across the capacitor is a low-pass filter, as we have already seen in Section 2.10. Let us now consider the effect of such a filter on a rectangular pulse of width τ. The RC low-pass filter, its gain–frequency curve, and the Fourier power spectrum of the pulse are shown in Fig. 3.15.

Without doing any calculations at all, we can predict the qualitative effect of the RC low-pass filter on a rectangular pulse of width τ by comparing the breakpoint frequency of the filter ω_B with the ω_{max} frequency of the pulse. This is true because most of the energy of the pulse lies in the Fourier frequency components below ω_{max}, and the filter will pass frequencies below its breakpoint frequency ω_B. Thus, if $\omega_{max} < \omega_B$, the pulse will be passed by the filter with little rounding off or distortion. If $\omega_{max} > \omega_B$, the pulse will be appreciably rounded off and attenuated. Because $f_{max} = 1/\tau$ and $\omega_B = 1/RC$, $\omega_{max} \ll \omega_B$ implies $\tau \gg RC$, and the pulse will suffer little rounding off. $\omega_{max} \gg \omega_B$ implies $\tau \ll RC$, and the pulse will be seriously rounded off. These results are shown in Fig. 3.16. This is what we would expect intuitively; a large capacitor is slow to charge up and discharge, and thus a large capacitor (large RC) would result in a rounded output pulse.

The RC low-pass filter circuit is often known as an *integrating circuit*, because if $RC \gg \tau$ the output voltage is essentially the integral (with respect to time) of the input voltage pulse. This result can be shown by writing the Kirchhoff voltage equation for the circuit for $0 \leqq t \leqq \tau$:

$$v_{in} = V_0 = v_R + v_C = iR + \frac{Q}{C} \tag{3.41}$$

Then we substitute $i = dQ/dt$ and solve the resulting differential equation for $Q(t)$ because we wish to obtain an expression for v_{out}, which equals the

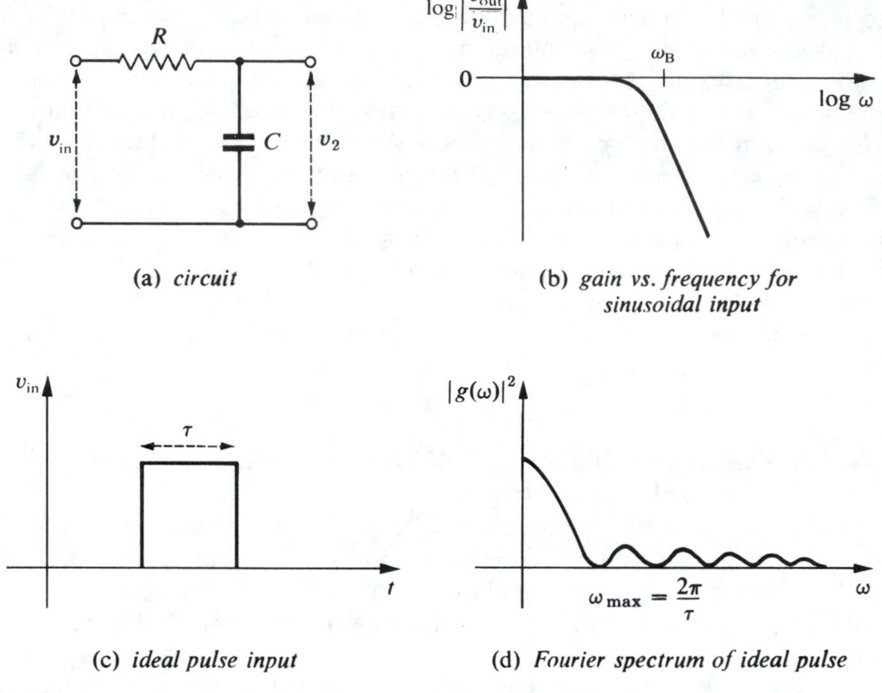

(a) *circuit*

(b) *gain vs. frequency for sinusoidal input*

(c) *ideal pulse input*

(d) *Fourier spectrum of ideal pulse*

FIGURE 3.15 *RC* low-pass filter or integrating circuit.

voltage across the capacitor, Q/C.

$$v_{in} = V_0 = \frac{dQ}{dt}R + \frac{1}{C}Q \quad \text{or} \quad \frac{dQ}{dt} + \frac{1}{RC}Q = \frac{V_0}{R} \tag{3.42}$$

whose solution is

$$Q = CV_0(1 - e^{-t/RC}) \tag{3.43}$$

The exact solution for the output voltage on the capacitor is thus $v_{out} = Q/C$ or

$$v_{out} = V_0(1 - e^{-t/RC}) \tag{3.44}$$

which satisfies the boundary condition $Q = 0$ when $t = 0$. Now if $RC \gg \tau$, then $t/RC \ll 1$, and in the above expression for v_{out} we may approximate $e^{-t/RC}$ by $1 - t/RC$ because $e^{-x} \cong 1 - x$ if $x \ll 1$. Thus the output voltage can be written as

$$v_{out} \cong \frac{V_0 t}{RC}$$

or

$$v_{out} \cong \frac{1}{RC}\int_0^t V_0\,dt = \frac{1}{RC}\int_0^t v_{in}\,dt \qquad \textbf{(3.45)}$$

Thus the output voltage is approximately the integral of the input voltage, for $0 \leqq t \leqq \tau$, that is, for times less than the pulse width τ. For times greater than τ (still with $RC \gg \tau$), it can easily be shown that the output voltage decays exponentially to zero according to the formula $v_{out} = (V_0\tau/RC)e^{-(t-\tau)/RC}$. The smaller τ/RC is, the more nearly linear the output voltage is from $t = 0$ to $t = \tau$, but the output amplitude is smaller. With $\tau \ll RC$ an integrating circuit is often used to generate an approximately linear output voltage or *sawtooth* from an input square pulse.

If an ideal rectangular voltage pulse with zero rise time is fed into an RC low-pass filter circuit with $RC \ll \tau$ as shown in Fig. 3.16(c), then the output pulse can be shown to have a rise time of $2.2RC$, from $v_{out} = V_0(1 - e^{-t/RC})$ and the definition of rise time. Thus, in agreement with our previous results, the time constant RC should be as small as possible for the sharpest possible output pulse. It can be shown that for any circuit with a bandwidth B Hz the output rise time in seconds (for a zero-rise-time input) will be given by R.T. $= 0.35/B$. This result is intuitively reasonable because the larger the bandwidth, the smaller the rise time; for an ideal amplifier with an infinite bandwidth the output rise time would be zero. The relationship between rise time and bandwidth can be derived approximately by the following argument. A sinusoidal wave goes from zero to peak amplitude in

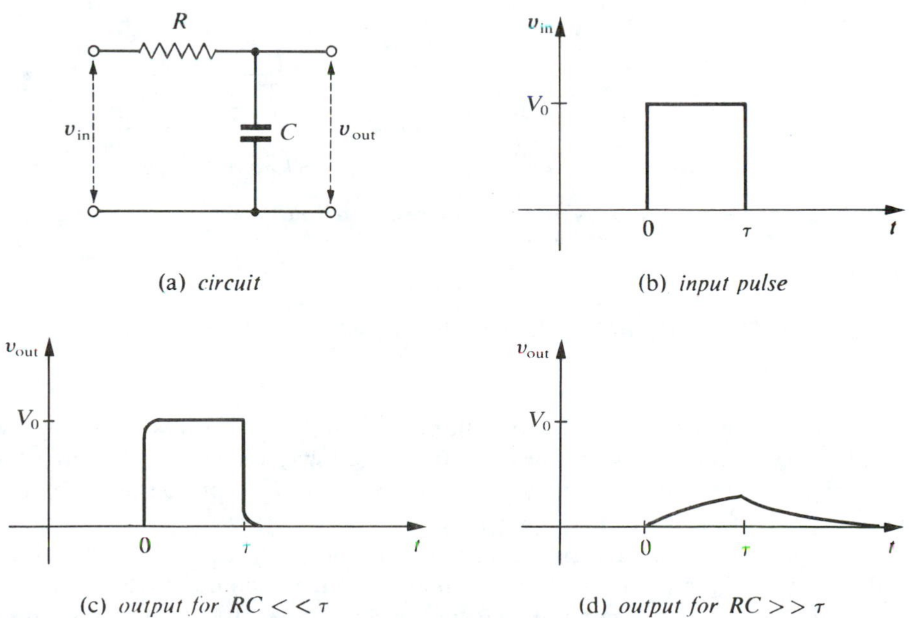

FIGURE 3.16 Pulse distortion in an *RC* low-pass filter.

one quarter of a period. Thus for a circuit with bandwidth B, capable of passing frequencies up to B Hz, a rectangular voltage pulse could rise from zero to its final amplitude in a time no shorter than one quarter of the period of the *maximum* frequency B (i.e., $1/4B$). Thus we would expect an output rise time of

$$\text{R.T.} = \frac{1}{4B} = \frac{0.25}{B} \tag{3.46}$$

A more accurate calculation gives R.T. = $0.35/B$ (see Fig. 3.17).

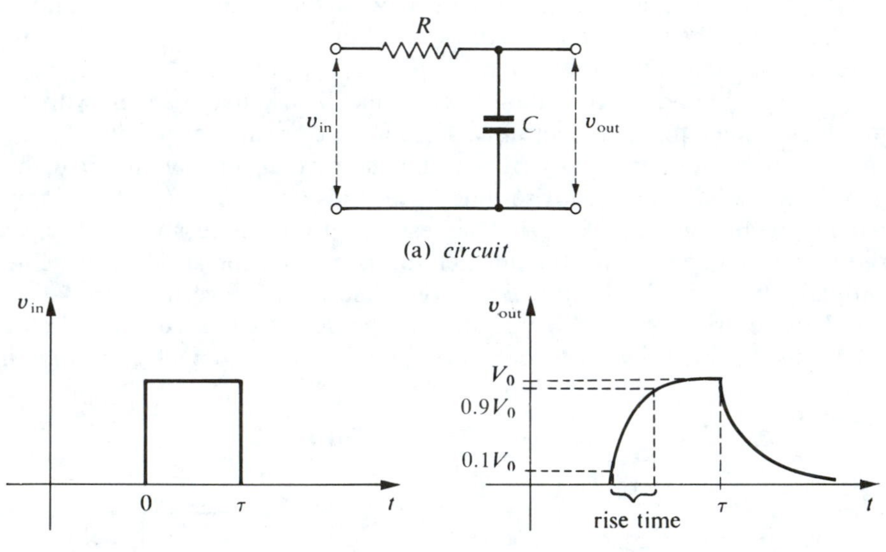

FIGURE 3.17 Effect of RC low-pass filter on rise time if $RC \ll \tau$.

3.5 DIFFERENTIATING CIRCUIT (HIGH-PASS FILTER)

The RC high-pass filter with the output taken across the resistance and a pulse input behaves quite differently from the low-pass filter. Let us first try to predict qualitatively the output waveform from a consideration of the Fourier components of the input pulse. The gain-versus-frequency curve of the high-pass RC filter is shown in Fig. 3.18; $\omega_B = 1/RC$. Because most of the energy of a pulse of width τ occurs in frequency components from zero (dc) to $f_{max} = 1/\tau$, we can see that for the high-pass filter to pass the pulse without appreciable distortion, the breakpoint must be much less than

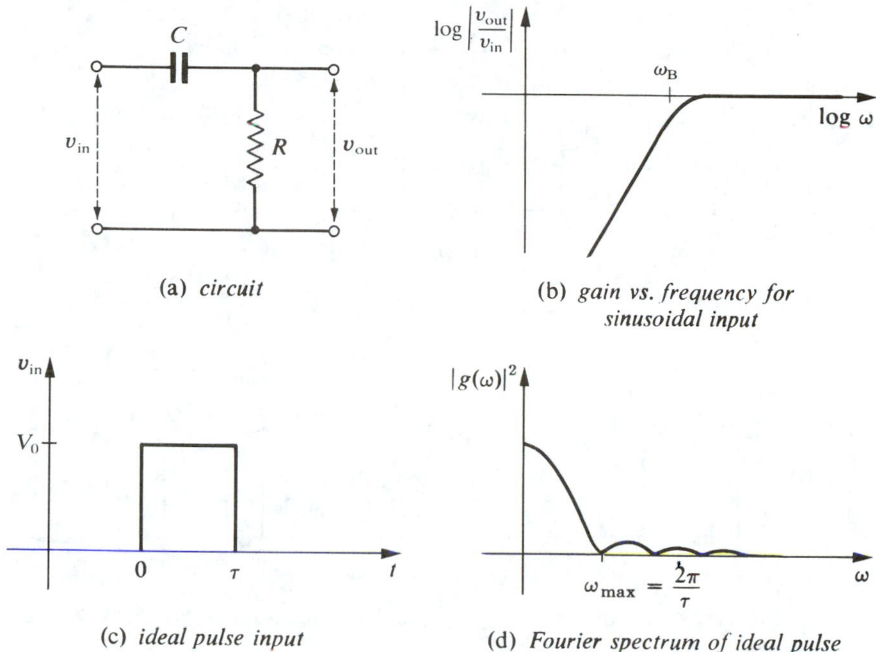

FIGURE 3.18 High-pass *RC* filter or differentiating circuit.

$f_{max} = 1/\tau$. Even then, the extremely-low-frequency components of the pulse (less than $\omega_B = 1/RC$) will be attenuated. And because of the series capacitor, there will be no dc passed by the RC high-pass filter. Hence, the output pulse will be lacking the dc and the low-frequency components below $\omega_B = 1/RC$. Because of the lack of the low-frequency components, the top of the pulse will "sag" somewhat—it takes low-frequency components to maintain the output voltage near V_0 for any length of time. Because of the zero gain of the filter at dc, the net charge passed by the filter must be zero, which means that the total area under the output voltage–versus–time curve must be exactly zero.

$$v_{out} = iR$$

$$\int_0^\infty v_{out}\,dt = \int_0^\infty iR\,dt = R\int_0^\infty i\,dt = RQ \tag{3.47}$$

But $Q = \int i\,dt =$ *the total net charge passed through the capacitor C*, which must be exactly zero. Therefore, $\int_0^\infty v_{out}\,dt = 0$. Thus, for $\omega_B \ll \omega_{max}$ or $1/RC \ll 1/\tau$ or $RC \gg \tau$, the input and output pulses will be as shown in Fig. 3.19. Because $\int_0^\infty v_{out}\,dt = 0$, the area of the pulse above the $v = 0$ axis must exactly equal the area below the axis. Thus there is *always* a long

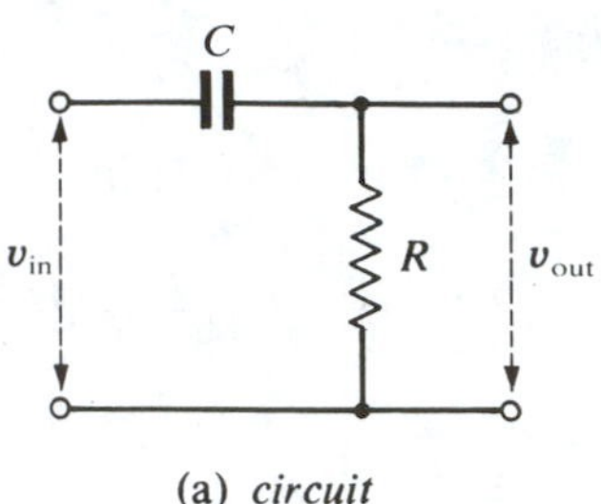

(a) *circuit*

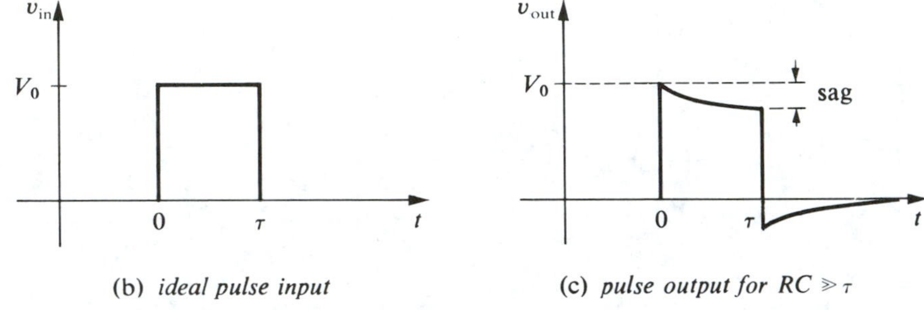

(b) *ideal pulse input* (c) *pulse output for RC* $\gg \tau$

FIGURE 3.19 Input and output pulses for differentiating circuit when $RC \gg \tau$.

negative tail accompanying every positive pulse passed through an *RC* high-pass filter. The presence of this long negative tail may give rise to severe problems if a large number of positive pulses are fed into the input. This situation is commonly encountered in nuclear pulse amplifiers when high nuclear radiation levels produce many pulses per second (i.e., a high counting rate). Each positive pulse then will appear at the output of the high-pass *RC* filter superimposed on top of the negative tails of all the preceding pulses. This may result in the baseline of the pulses falling well below 0 V; the pulses may go from −2 to +1 V instead of from 0 to +3 V. The addition of a diode across the resistance is a simple remedy and is discussed in Chapter 4. Techniques to counteract this are beyond the scope of this book but are generally referred to as *baseline restoration* or *dc restoration*.

If, on the other hand, ω_B is close to or higher than f_{max}, then only a small fraction of the pulse energy will be passed by the filter, namely, the energy in the highest-frequency components of the pulse. We would thus expect to get a narrow spike or pulse out—a positive spike out when the input pulse jumps from 0 to V_0, and a negative spike out when the input pulse jumps from V_0 back to 0, because the high-frequency components occur when the input voltage changes quickly. This result can also be arrived at intuitively by saying that the capacitor will pass only rapid changes in voltage because

$$Q = Cv_C \qquad i = C\frac{dv_C}{dt} \qquad v_{out} = iR = RC\frac{dv_C}{dt} \tag{3.48}$$

where i is the current passed through the capacitor. dv_{out}/dt is large and positive when the input pulse jumps from 0 to V_0, and is large and negative when the input jumps from V_0 to 0. A sketch of the expected output is shown in Fig. 3.20. Again, notice that because $\int_0^\infty v_{out}\,dt$ must equal 0, the area under the positive spike must exactly equal the area under the negative spike.

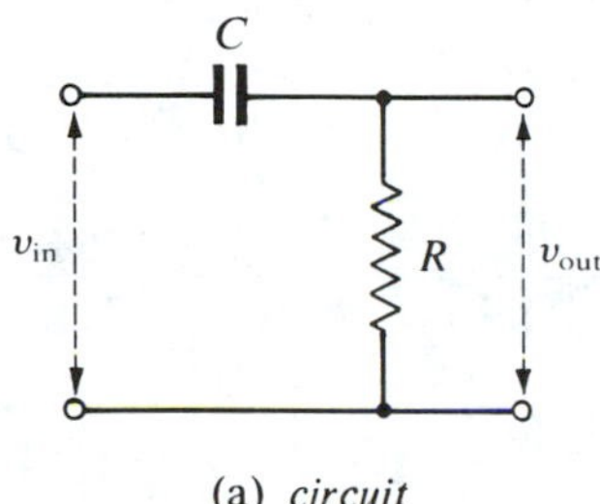

(a) *circuit*

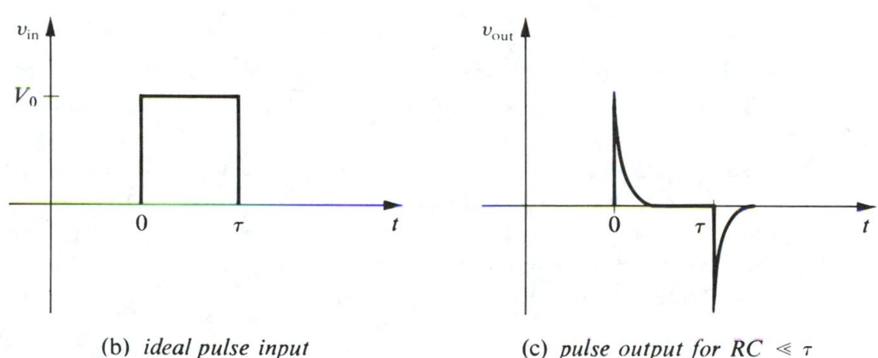

(b) *ideal pulse input* (c) *pulse output for* $RC \ll \tau$

FIGURE 3.20 Input and output pulses for differentiating circuit when $RC \ll \tau$.

To determine the output more quantitatively, we note that *if* $RC \ll \tau$, then $v_R \ll v_C$. This follows from (neglecting any current drawn from the output)

$$\frac{|v_R|}{|v_C|} = \frac{|iR|}{|iX_C|} = \frac{R}{\dfrac{1}{\omega C}} = \omega RC \tag{3.49}$$

For an input pulse of width τ the maximum Fourier component angular frequency $\omega \cong 1/\tau$. Therefore, the *maximum* value for v_R/v_C is

$$\left(\frac{|v_R|}{|v_C|}\right)_{\max} = \omega_{\max} RC = \frac{RC}{\tau} \ll 1 \tag{3.50}$$

Thus, $v_R/v_C = RC/\tau \ll 1$ for $\omega_{\max}$. If $\omega < \omega_{\max}$, the ratio v_R/v_C is even smaller. Therefore the Kirchhoff voltage law can be written with the approximation $v_R \ll v_C$.

$$v_{\text{in}} = v_C + v_R \cong v_C = \frac{Q}{C} \tag{3.51}$$

This equation can be solved easily for i by differentiating with respect to time.

$$\frac{dv_{\text{in}}}{dt} \cong \frac{i}{C} \qquad i \cong C\frac{dv_{\text{in}}}{dt} \tag{3.52}$$

The output voltage is simply $v_{\text{out}} = iR$.

$$v_{\text{out}} = iR \cong RC\frac{dv_{\text{in}}}{dt} \tag{3.53}$$

The output is therefore approximately equal to the time constant RC multiplied by the time derivative of the input. Hence, the circuit is called a *differentiating* circuit.

The result that the output is the derivative of the input can be shown quantitatively by considering the circuit response to an input pulse with a finite slope, as shown in Fig. 3.21.

The Kirchhoff voltage equation is, as usual (again neglecting any current drawn from the output),

$$v_{\text{in}} = v_C + v_R = \frac{Q}{C} + iR \tag{3.54}$$

Differentiating yields

$$\frac{dv_{\text{in}}}{dt} = \frac{i}{C} + R\frac{di}{dt}$$

Thus, for $0 \leq t \leq t_1$, $v_{\text{in}} = KT$, $dv_{\text{in}}/dt = K$. The differential equation is then

$$K = \frac{i}{C} + R\frac{di}{dt} \quad \text{or} \quad \frac{di}{dt} + \frac{1}{RC}i = \frac{K}{R}$$

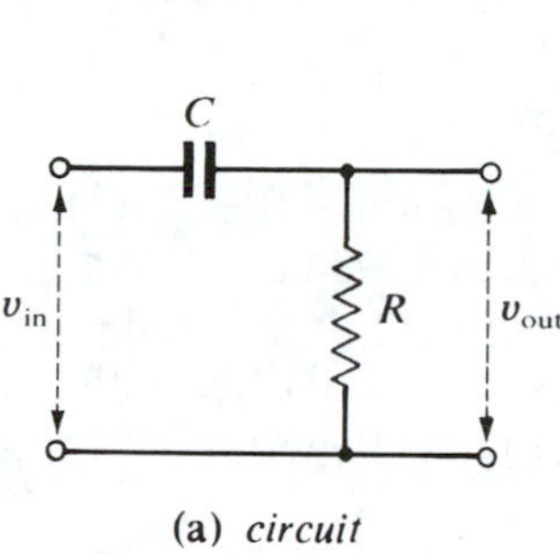

(a) *circuit*

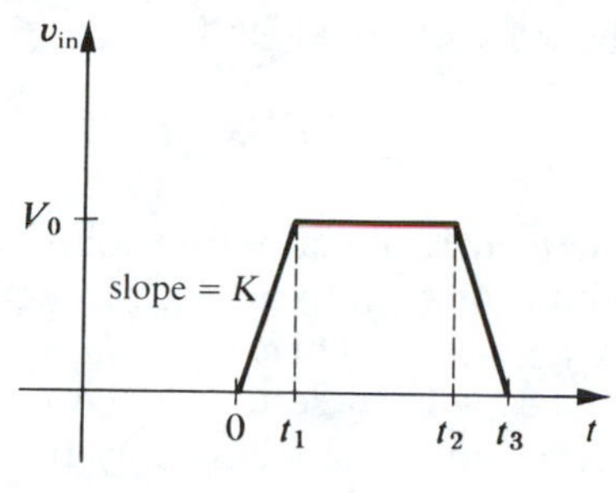

(b) *input pulse with finite slopes*

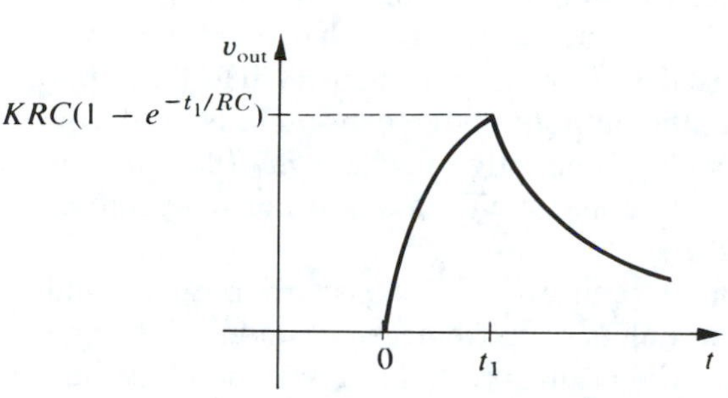

(c) *output leading edge (magnified)*

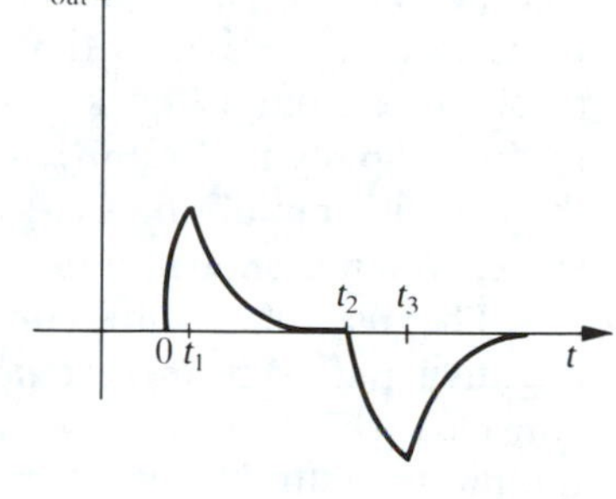

(d) *complete output*

FIGURE 3.21 Differentiating circuit waveforms.

which has the solution

$$i = KC(1 - e^{-t/RC})$$

The output voltage across the resistor is then given by

$$v_{out} = iR = KRC(1 - e^{-t/RC}) \tag{3.55}$$

which is shown greatly magnified in Fig. 3.21(c). The output voltage thus rises exponentially toward $RCK = RC(dv_{in}/dt)$ during the time interval $0 < t < t_1$. If $RC \ll t_1$, then the output will rise to a voltage very close to RCK in a time less than t_1. The rise time of the input pulse is equal to $0.8t_1$, so if RC is much less than the pulse rise time, the output will rise to RCK in the time interval from 0 to t_1. Finally, because $dv_{in}/dt = K$ for $0 < t < t_1$, v_{out} rises to $RC(dv_{in}/dt)$, hence, the name *differentiating* circuit.

For the time interval $t_1 < t < t_2$ the input voltage is equal to V_0, and the differential equation is

$$v_{in} = V_0 = \frac{Q}{C} + iR \quad \text{or} \quad \frac{di}{dt} + \frac{1}{RC} i = 0$$

the solution to which is

$$i = i_1 e^{-(t-t_1)/RC} \quad \text{or} \quad v_{\text{out}} = iR = Ri_1 e^{-(t-t_1)/RC} \tag{3.56}$$

where i_1 is the current at time $t = t_1$. The output voltage thus falls to zero with a time constant RC. In equation (3.56) t is in the interval $t_1 < t < t_2$, or $t - t_1 < \tau$. Thus, if τ/RC is large (in other words if $RC \ll \tau$), the output falls rapidly to zero. For the negative-going portion of the pulse the same sort of thing happens; only the sign of the output voltage is changed [see Fig. 3.21(d)].

We summarize our results: If RC is much less than the rise time, then the output voltage across the resistor will rise rapidly to $RC(dv_{\text{in}}/dt)$ and will then fall rapidly to zero as $e^{-t/RC}$, thus producing a sharp positive spike whose amplitude is RC times the derivative of the input. When the input pulse drops from V_0 back to zero, the output voltage can similarly be shown to drop rapidly to $RC(dv_{\text{in}}/dt)$ (which is negative because $dv_{\text{in}}/dt < 0$) and then to rise rapidly back to zero, thus producing a sharp negative output spike, which is shown in Fig. 3.21.

The use of a differentiating circuit to produce sharp positive and negative pulses from rectangular pulses is extremely common. The time constant RC of the circuit is usually chosen to be at least ten times less than the pulse width. The smaller RC is, the sharper the output spike is but the smaller is its amplitude. For example, a pulse with a 1-μs pulse width would be differentiated into sharp positive and negative spikes by a differentiating circuit with $RC = 0.1$ μs, for example, with $R = 10$ kΩ and $C = 10$ pF. It is best to choose C no smaller than 10 pF. This is because the typical stray capacitance in a circuit may be about 5 pF, and the desired capacitance should always be greater than the (uncertain) stray capacity of the circuit, particularly if many circuits are to be made.

Sometimes to achieve extremely sharp voltage spikes, a rectangular pulse is differentiated twice; the output of the first differentiating circuit is fed into a second differentiating circuit as shown in Fig. 3.22.

The output from R_2C_2 will be sharper than from just one differentiating circuit because the derivative has been taken twice. Also, the output from R_2C_2 will go negative because of the negative slope of the output from R_1C_1. The net result is an output pulse that is narrower or "sharper," smaller in amplitude, and that has a long negative tail. This technique is commonly used in pulse amplifiers in nuclear physics.

3.6 PULSE-SHARPENING CIRCUIT

Whenever a capacitance C is charged or discharged through a series resistance R, the charge or discharge is always exponential with a time constant $\tau = RC$. If a step-function voltage is applied to a simple RC

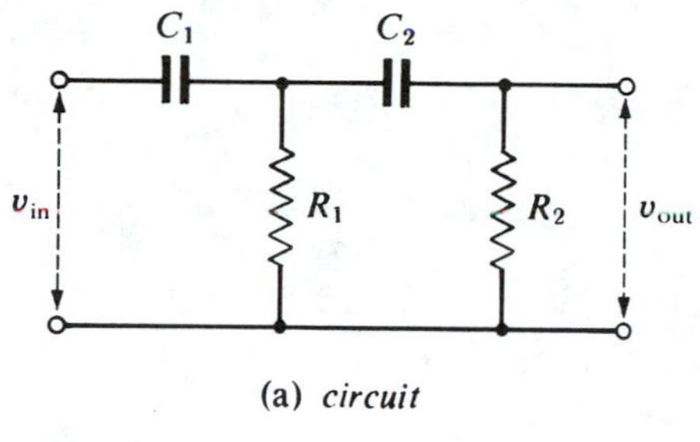

(a) *circuit*

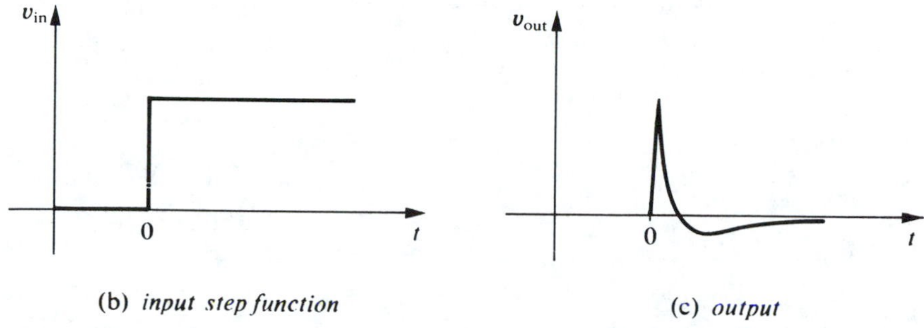

(b) *input step function*

(c) *output*

FIGURE 3.22 Double differentiation.

low-pass filter, the rise time (10% to 90%) and fall time (90% to 10%) each equal 2.2RC.

An ingenious circuit shown in Fig. 3.23 allows us to decrease the rise time at the expense of attenuating the entire signal. Using the method of nodal analysis at junction A, we have from the KCL

$$i + i' = i_2$$

$$\frac{v_{in} - v_{out}}{R_1} + \frac{d}{dt} C_1(v_{in} - v_{out}) = \frac{v_{out}}{R_2}$$

$$C_1 \frac{dv_{out}}{dt} + \left(\frac{1}{R_1} + \frac{1}{R_2}\right) v_{out} = C_1 \frac{dv_{in}}{dt} + \frac{v_{in}}{R_1} \tag{3.57}$$

Letting $R_{12} = R_1 \| R_2 = R_1 R_2/(R_1 + R_2)$, we have

$$\frac{dv_{out}}{dt} + \frac{1}{R_{12}C_1} v_{out} = \frac{dv_{in}}{dt} + \frac{1}{R_1 C_1} v_{in} \tag{3.58}$$

Assuming the input v_{in} is of the form

$$v_{in} = V_0(1 - e^{-t/\tau_{in}})$$

then

$$\frac{dv_{in}}{dt} = \frac{V_0}{\tau_{in}} e^{-t/\tau_{in}} \tag{3.59}$$

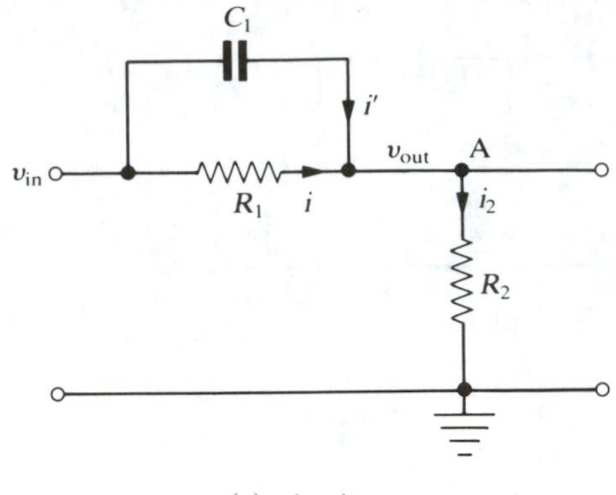

(a) *circuit*

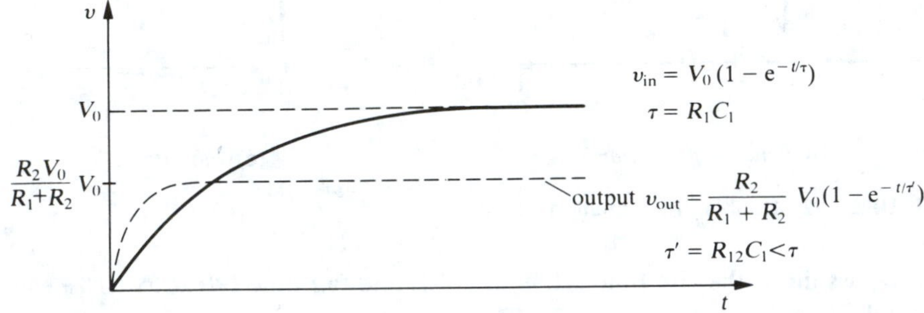

(b) *output waveform compared to input waveform*

FIGURE 3.23 Pulse-sharpening circuit.

Substituting (3.59) in (3.58), we obtain

$$\frac{dv_{out}}{dt} + \frac{1}{R_{12}C_1} v_{out} = V_0\left(\frac{1}{\tau_{in}} - \frac{1}{R_1 C_1}\right) e^{-t/\tau_{in}} + \frac{V_0}{R_1 C_1} \tag{3.60}$$

If we choose R_1 and C_1 such that $R_1 C_1 = \tau_{in}$, the (constant) characteristic time of the input step, then we simplify (3.60) to

$$\frac{dv_{out}}{dt} + \frac{1}{R_{12}C_1} v_{out} = \frac{V_0}{R_1 C_1} \tag{3.61}$$

The right side of (3.61) is now simply a constant term, so we can solve for v_2 by writing $v_{out} = v_h + v_p$, where v_p is any *particular* solution to (3.61) and v_h is the solution of the homogeneous equation

$$\frac{dv_h}{dt} + \frac{1}{R_{12}C_1} v_h = 0 \tag{3.62}$$

Clearly

$$v_h = DE^{-t/R_{12}C_1} \tag{3.63}$$

where D is a constant. And by inspecting (3.61), we see that a particular solution is

$$v_p = \frac{R_{12}C_1}{R_1C_1} V_0 = \frac{R_2}{R_1 + R_2} V_0 \tag{3.64}$$

Thus, the general solution for the output v_{out} is

$$v_{out} = v_h + v_p = De^{-t/R_{12}C_1} + \frac{R_2}{R_1 + R_2} V_0 \tag{3.65}$$

Using the boundary condition $v_{out}(0) = 0$, we can evaluate D:

$$0 = D + \frac{R_2}{R_1 + R_2} V_0 \quad \text{or} \quad D = -\frac{R_2}{R_1 + R_2} V_0 \tag{3.66}$$

Substituting (3.66) in (3.65), we obtain the final result for v_{out}:

$$v_{out} = \frac{R_2 V_0}{R_1 + R_2}(1 - e^{-t/R_{12}C_1}) = \frac{R_2 V_0}{R_1 + R_2}(1 - e^{-t/\tau'}) \tag{3.67}$$

The important point is that the characteristic time for the output rise is $\tau' = R_{12}C_1$, which is less than the characteristic time for the input rise, $\tau_{in} = R_1C_1$.

$$\tau' = R_{12}C_1 = \frac{R_2}{R_1 + R_2} \tau_{in} \tag{3.68}$$

Expressed in terms of the 10% to 90% rise time,

$$\frac{R.T._{out}}{R.T._{in}} = \frac{2.2\tau'}{2.2\tau_{in}} = \frac{2.2R_{12}C_1}{2.2R_1C_1} = \frac{R_2}{R_1 + R_2} \tag{3.69}$$

The output rise time is thus less than the input rise time by the factor $R_2/(R_1 + R_2)$, which is very desirable. The price paid is that the amplitude of the output pulse is less by the same factor. Thus we see that the input amplitude divided by the input rise time is equal to the output amplitude divided by the output rise time:

$$\frac{V_0}{\tau_{in}} = \frac{\frac{R_2}{R_1 + R_2} V_0}{\tau'} \tag{3.70}$$

For example, if the input rise time is 1 μs and the input step amplitude is $V_0 = 1$ V, then we choose $R_1C_1 = 1\ \mu$s; for example, $R_1 = 1\ \text{k}\Omega$ and $C_1 = 10^{-9}$ F = 0.001 μF. Then if $R_2 = 1\ \text{k}\Omega$, $R_{12} = 500\ \Omega$, and the output rise time is $R_{12}C_1 = (500\ \Omega)(10^{-9}\ \text{F}) = 0.5\ \mu\text{s} = 500$ ns, but the output amplitude will be only $R_2V_0/(R_1 + R_2) = 0.5$ V.

Qualitatively, we can think of this circuit as a filter that passes the higher-frequency Fourier components with less attenuation than the lower-frequency Fourier components; this effect shortens the rise time. However, the greater attenuation of the lower-frequency components reduces the ultimate amplitude of the output. The capacitor C_1 is sometimes called a *speed-up* capacitor. It can also be shown that the output fall time is also shortened by the same factor compared to the input fall time.

3.7 COMPENSATED VOLTAGE DIVIDER

We recall from Chapter 1 that the simple two-resistor network of Fig. 3.24(a) acts as a voltage divider with a division ratio of

$$\frac{v_{out}}{v_{in}} = \frac{R_1}{R_1 + R_2}$$

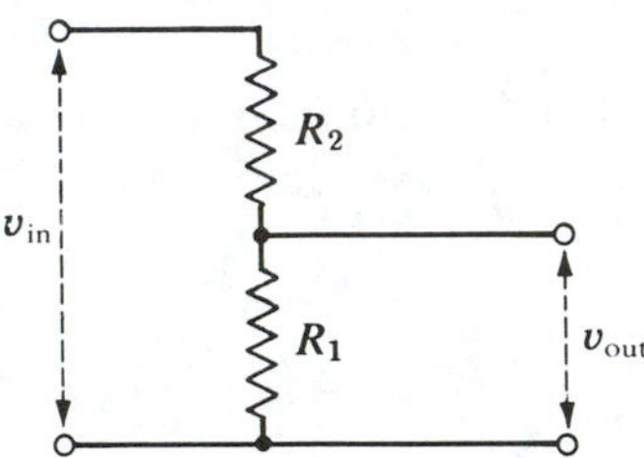

(a) *ideal voltage divider with gain equal to* $R_1/(R_1 + R_2)$

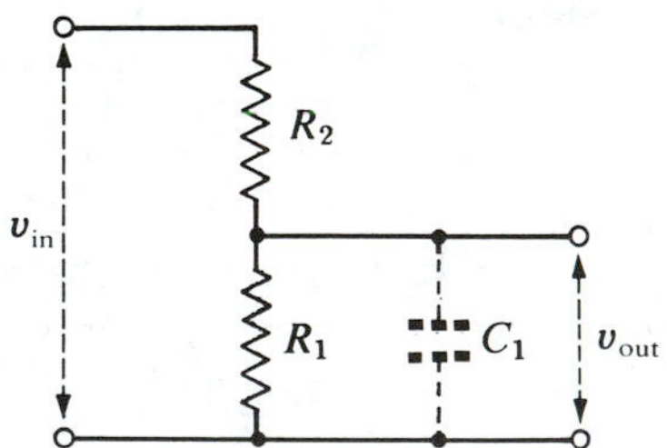

(b) *actual voltage divider with gain less than* $R_1/(R_1 + R_2)$ *at high frequencies*

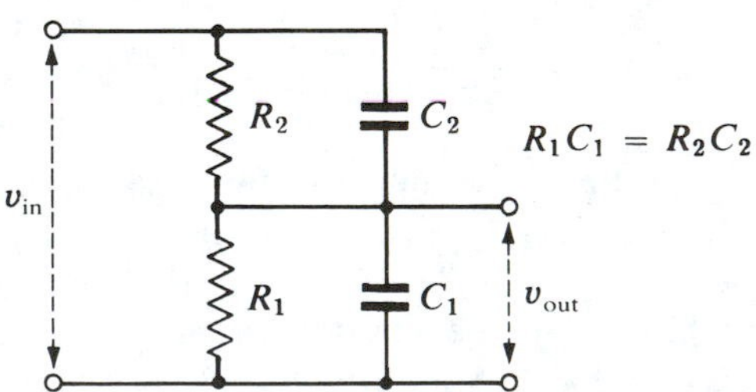

(c) *compensated voltage divider with gain equal to* $R_1/(R_1 + R_2)$

FIGURE 3.24 Voltage divider.

if negligible current is drawn from the output. However, if very narrow or "fast" pulses are fed into such a divider, they are severely rounded off as well as attenuated; that is, the output pulse rise time is substantially larger than the input pulse rise time, which is generally an undesirable situation. The reason for this rounding off of the input pulses is that there is always some stray or unavoidable circuit capacitance C_1 across R_1; this capacitance is shown as a dashed capacitor in Fig. 3.24(b). The very-high-frequency Fourier components of the input pulse are simply shorted out by C_1, thus rounding off the output.

A calculation of the division ratio including C_1 (again assuming negligible output current) yields the result:

$$\frac{v_{\text{out}}}{v_{\text{in}}} = \frac{(R_1 \| C_1)}{R_2 + (R_1 \| C_1)} = \frac{\dfrac{R_1 X_{C_1}}{R_1 + X_{C_1}}}{R_2 + \dfrac{R_1 X_{C_1}}{R_1 + X_{C_1}}}$$

$$= \frac{R_1}{R_1 + R_2\left(1 + \dfrac{R_1}{X_{C_1}}\right)} = \frac{R_1}{R_1 + R_2(1 + j\omega R_1 C_1)} \tag{3.71}$$

The division ratio is seen to equal $R_1/(R_1 + R_2)$ only when the frequency is low enough such that $R_1 \ll X_{C_1} = 1/\omega C_1$, or $\omega R_1 C_1 \ll 1$. And as the frequency ω increases, the division ratio becomes much smaller (i.e., there is more attenuation). The frequency at which the division ratio begins to differ appreciably from the dc value of $R_1/(R_1 + R_2)$ is approximately that frequency at which $R_1 \cong X_{C_1} = 1/\omega C_1$ or $\omega = 1/R_1 C_1$. The larger R_1 or C_1 is, the lower the frequency at which this takes place. For example, a typical value of C_1 is 10 pF, so if $R_1 = 100\ \text{k}\Omega$, then frequency components of about $\omega = 1/(10^5)(10^{-11}) = 10^6$ rad/s or $f = 160$ kHz or higher will be seriously attenuated. In terms of pulse width τ the principal Fourier components are from 0 to $1/\tau$, so pulses narrower than $\tau = 1/f = 1/160\ \text{kHz} = 6.3\ \mu\text{s}$ would be seriously rounded off.

The way to avoid this rounding off is to make the division ratio a constant value independent of frequency. The capacitance C_1 cannot be eliminated, so we must add a frequency-dependent impedance to the R_2 part of the divider. Because C_1 tends to short out the high-frequency components, we wish to add a component to R_2 to make the impedance of the top half of the divider decrease as frequency increases. A little thought will show that adding another capacitance C_2 in parallel with R_2 will accomplish this as shown in Fig. 3.24(c). The capacitance C_2 tends to short out R_2 partially as the frequency increases.

The division ratio will be independent of frequency if the ratio of R_1 to

R_2 is exactly the same as the ratio of X_{C_1} to X_{C_2}; that is, if

$$\frac{R_1}{R_2} = \frac{X_{C_1}}{X_{C_2}} = \frac{\dfrac{1}{\omega C_1}}{\dfrac{1}{\omega C_2}} = \frac{C_2}{C_1}$$

which implies that

$$R_1 C_1 = R_2 C_2 \tag{3.72}$$

If we desire a 10:1 division ratio, then a suitable choice could be

$$R_1 = 1\ \text{M}\Omega \qquad R_2 = 9\ \text{M}\Omega \qquad C_1 = 72\ \text{pF} \quad \text{and} \quad C_2 = 8\ \text{pF}$$

Then C_1 is the total capacitance across the output; for example, if the stray capacitance is 10 pF, an additional 62 pF would be added in parallel to give a total of 72 pF for C_1. Usually C_2 is made adjustable; it is commonly a small ceramic trimmer capacitor that is tuned to R_1C_1/R_2 by a screwdriver adjustment. If C_2 is too large (compared to R_1C_1/R_2), the high-frequency Fourier components will be attenuated less than the low-frequency ones, and the output will appear slightly differentiated as shown in Fig. 3.25. If C_2

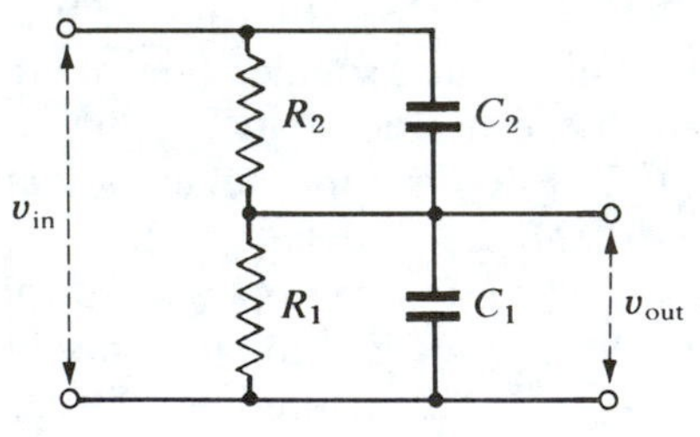

(a) *compensated voltage divider*

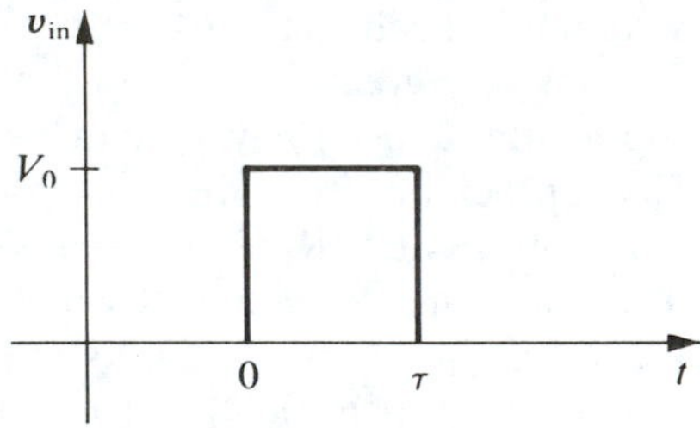

(b) *ideal pulse input*

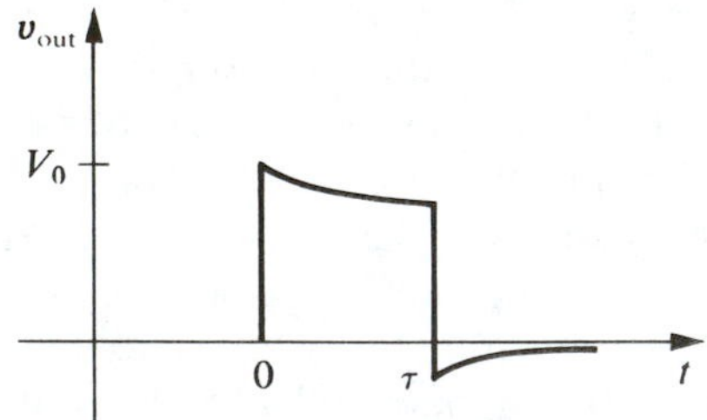

(c) *pulse output for* $C_2 > (R_1/R_2)\ C_1$

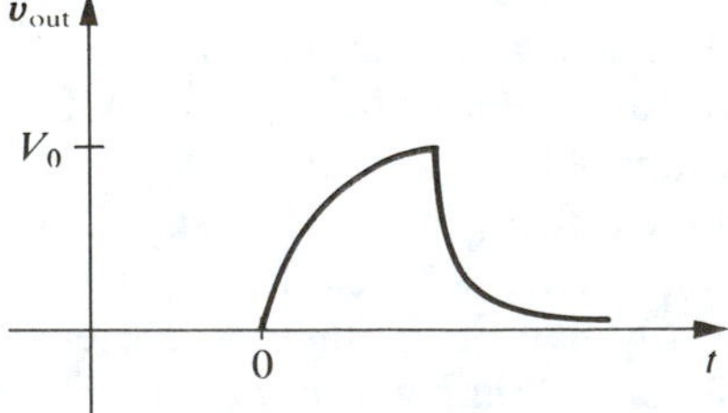

(d) *pulse output for* $C_2 < (R_1/R_2)\ C_1$

FIGURE 3.25 Pulse output from compensated voltage divider.

is too small (compared to R_1C_1/R_2), the high-frequency Fourier components will be attenuated more than the low-frequency ones, and the output will appear rounded as in Fig. 3.25.

The principal advantage of such a compensated divider is that it presents a very high impedance to the input. It draws little current into the input and therefore loads the circuit being measured very little. It can be shown that the compensated divider is equivalent to a single RC parallel combination as shown in Fig. 3.26, with

$$R_{\text{eff}} = R_1 + R_2 \quad \text{and} \quad C_{\text{eff}} = \frac{R_1}{R_1 + R_2} C_1 \tag{3.73}$$

Thus the 10:1 compensated divider with $R_2 = 9\ \text{M}\Omega$, $R_1 = 1\ \text{M}\Omega$, $C_1 = 72\ \text{pF}$, $C_2 = 8\ \text{pF}$ is electrically equivalent to a 10-MΩ resistor in parallel

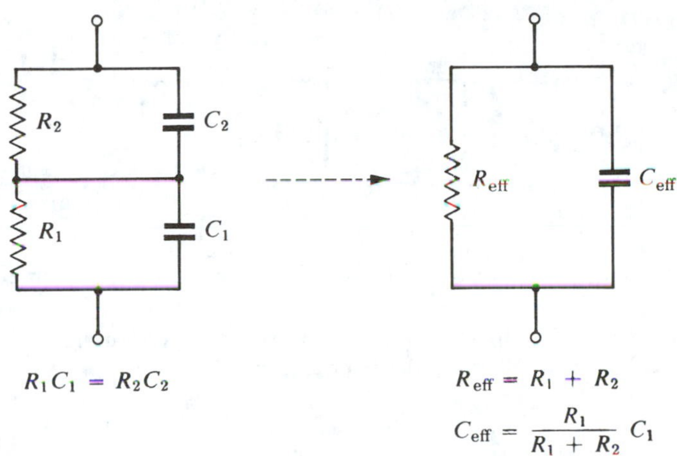

FIGURE 3.26 Compensated voltage divider equivalent.

with only a 7.2-pF capacitor. Not only is the dc input impedance of the compensated divider high (10 MΩ), but the effective capacitance presented to the input pulse is only 7.2 pF. Thus, the effective shunt capacitance of the divider as seen by the input signal is less than the original stray capacitance C_1 (10 pF) present. The price one pays for this is the reduced gain, but this can be overcome by feeding the output into a suitable high-frequency amplifier.

Almost all measurements of fast pulses with oscilloscopes are made with such a compensated voltage divider scope probe. The input to the divider is the pulse taken from the circuit under investigation, and the output from the divider is fed directly into the oscilloscope. The oscilloscope input capacitance is usually about 47 pF, and this appears across R_1.

Thus, to make C_1 total 72 pF, only 25 pF need be added across R_1. The effective shunt capacitance of the divider is still only 7.2 pF, which is less by a factor of about 7 than the input capacitance of the oscilloscope. The only thing you must remember in using such a probe is that all signals are attenuated by a factor of 10; that is, a 2.0-V pulse in the circuit under investigation will appear as a 0.2-V pulse on the oscilloscope. Also, a *good short* ground lead to the oscilloscope probe is essential for fast pulses.

PROBLEMS

1. Given the pulses shown below, determine the (a) pulse width, (b) pulse frequency or repetition rate, (c) period, (d) amplitude.

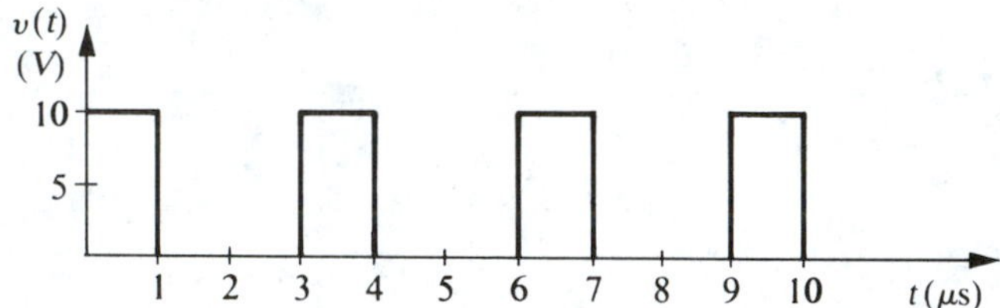

2. Explain briefly, yet clearly, in words the meaning of the Fourier analysis theorem.
3. Fourier analyze the square wave shown. Without performing any integrals, calculate the first (dc) term in the Fourier expansion.

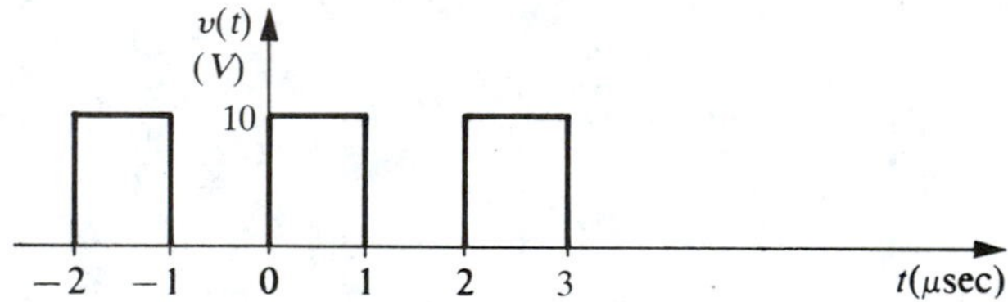

4. Fourier analyze the sawtooth wave shown.

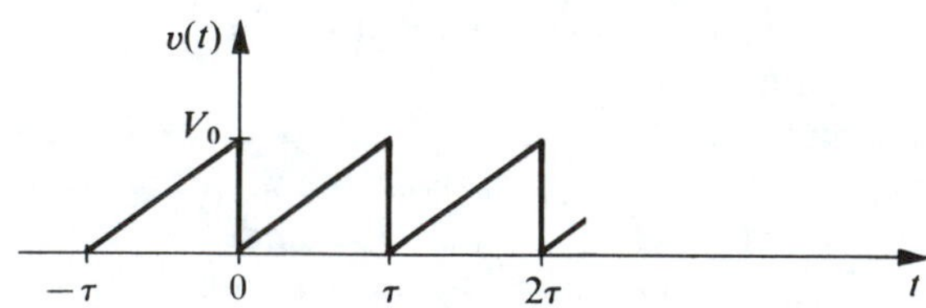

5. What approximate bandwidth would be required to amplify randomly occurring pulses of 50-ns width?

6. Calculate how long after the switch S is closed it will take the condenser to charge up from 3 V to 3.78 V. The capacitor is initially uncharged.

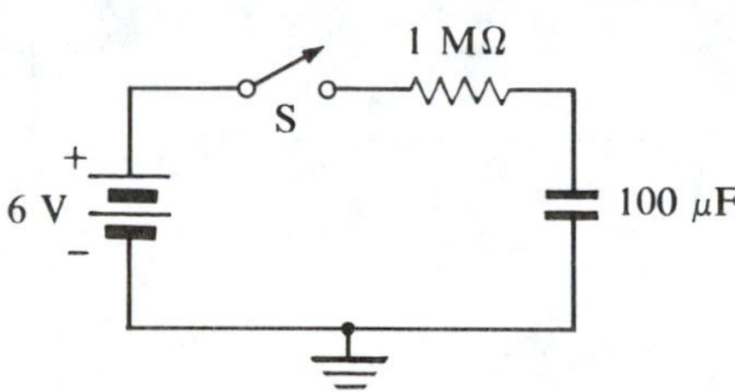

7. Prove that the 10% to 90% rise time for the RC low-pass filter is $2.2RC$ for a perfect step-function input.

8. The rise time is independent of the input pulse amplitude—true or false? The output voltage will reach a specific voltage (say, +3 V) in a shorter time if the amplitude of the input step function is increased—true or false?

9. The audio signal V_a to be carried by an AM radio station is shown below. (a) Sketch the modulated carrier wave. (b) Sketch the expected sidebands. [*Hint*: What harmonics would give this wave shape?]

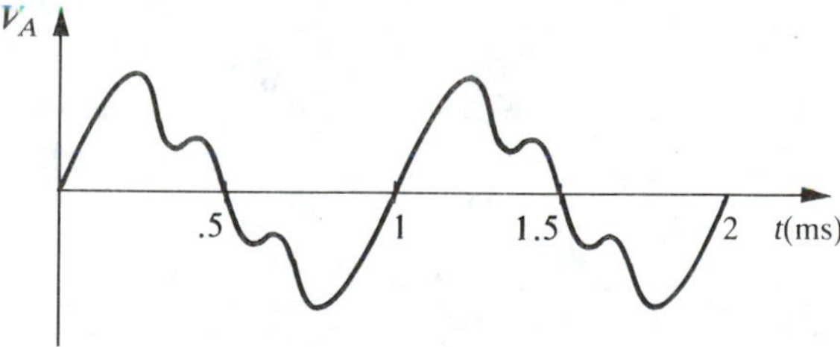

10. Compensate the voltage divider shown.

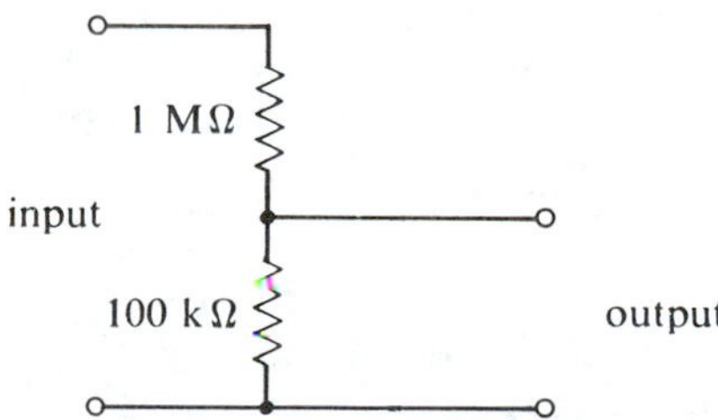

11. Sketch the approximate output from a differentiating circuit with $R = 10\ \text{k}\Omega$, $C = 0.01\ \mu\text{F}$ for the following inputs:

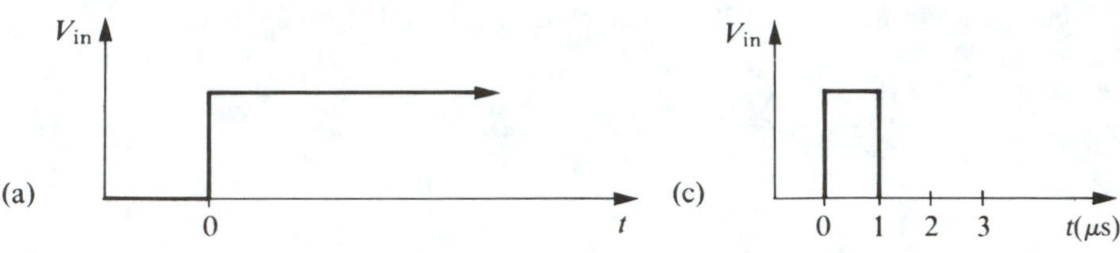

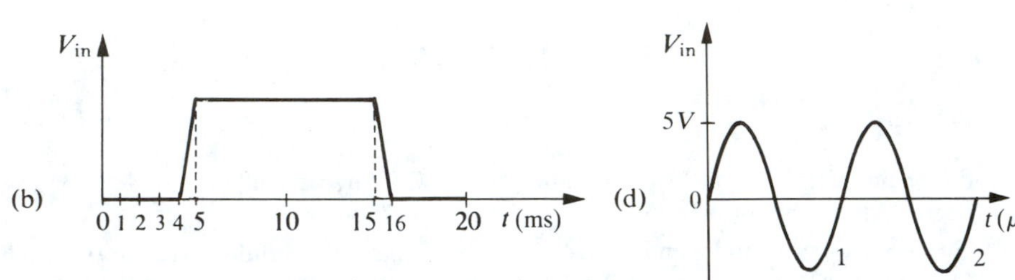

12. Sketch the approximate output from an integrating circuit with $R = 10\ \text{k}\Omega$, $C = 0.001\ \mu\text{F}$ for the inputs of Problem 11.

13. Derive the expression $v_{out} \cong RC(dv_{in}/dt)$ for a differentiating circuit. Be careful to state your assumptions.

14. Sketch the approximate scope trace you would see for the following input if the scope is on ac coupling.

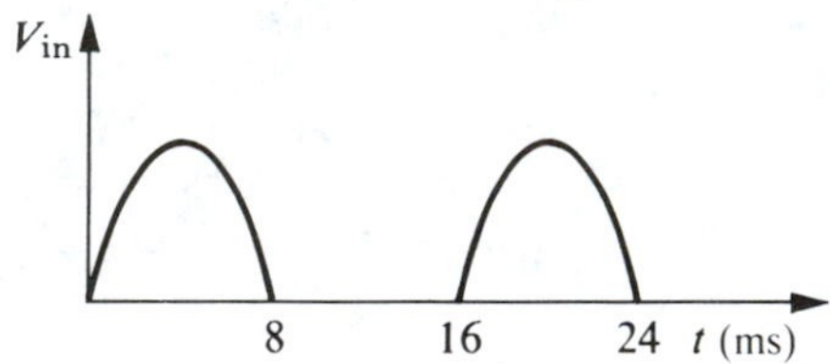

15. Derive (3.24).

16. Derive (3.25).

17. Sketch the output pulse for each of the two input pulses. The amplifier has a gain of 10 from dc out to 2 MHz.

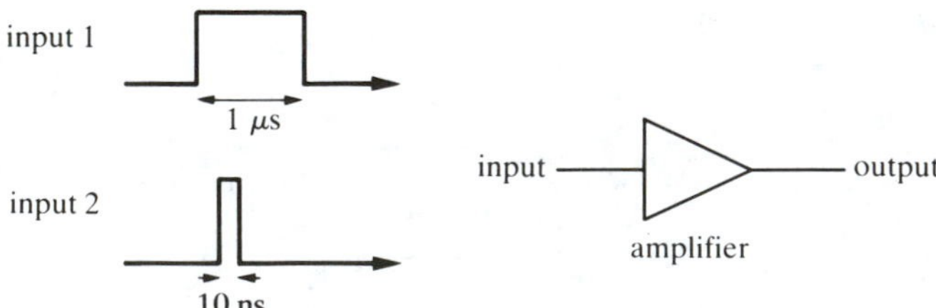

18. Sketch the output waveform from the RC low-pass filter for (a) $RC = 1\ \mu s$, (b) $RC = 100\ \mu s$.

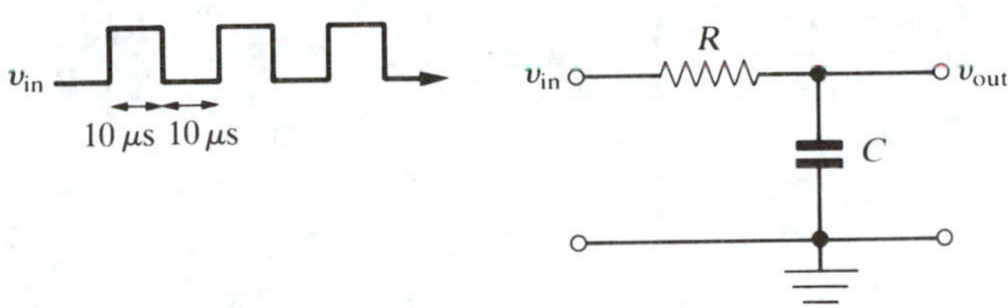

19. Repeat Problem 17 for the same input waveform to an RC high-pass filter for (a) $RC = 1\ \mu s$, (b) $RC = 100\ \mu s$.

20. Fourier analyze the following waveform.

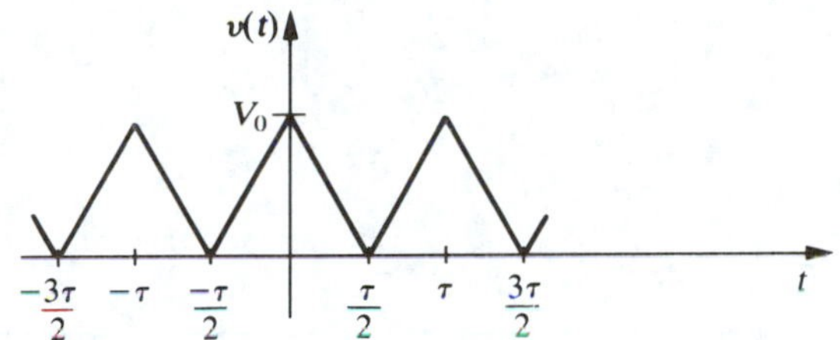

21. Sketch the AM radio wave for the following modulation waveform.

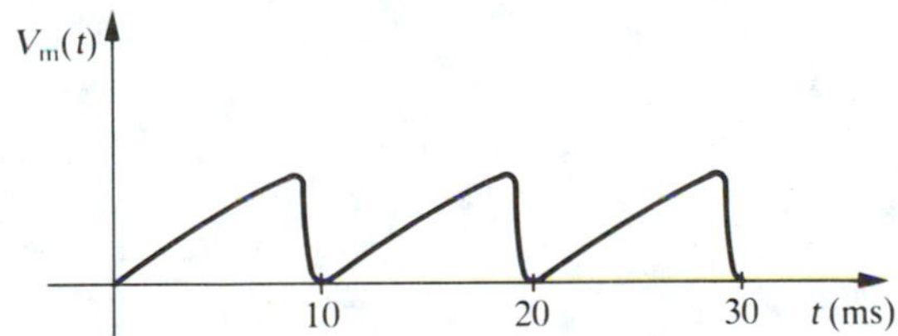

22. In a 100-MHz FM radio wave, the modulation is a 400-Hz pure sine wave. Sketch a graph of the frequency of the carrier wave versus t.

23. What is the output pulse rise time?

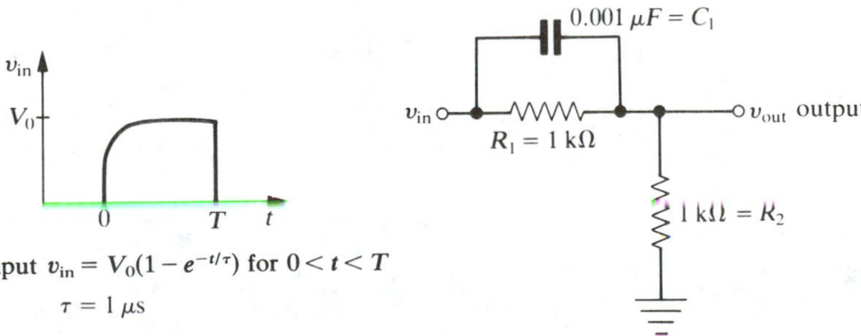

24. In the circuit of Problem 23, if the input v_1 is a negative step, $v_1 = v_0 e^{-t/\tau}$. Show that the output will be

$$v_2 = \frac{R_2}{R_1 + R_2} e^{-t/R_{12}C} \quad \text{where } R_{12} = R_1 \| R_2 = \frac{R_1 R_2}{R_1 + R_2}$$

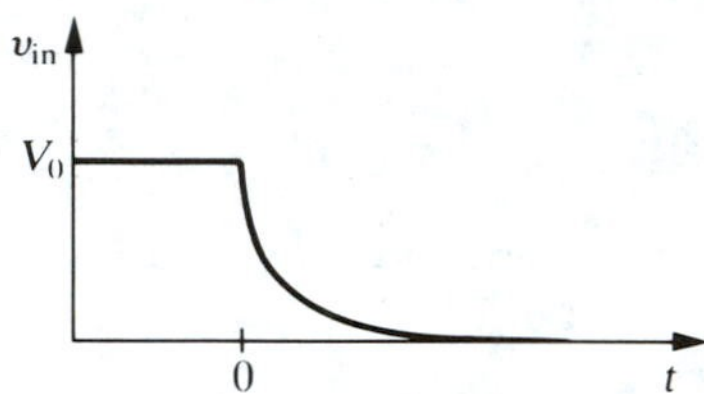

CHAPTER 4

Semiconductor Physics

4.1 INTRODUCTION

In this chapter we consider some of the basic features of the physics of semiconductors that will enable us to understand physically why a transistor can be made to amplify signals, why transistor parameters depend so strongly upon temperature and the magnitudes of the various resistances, voltages, and currents in transistor circuits. Most semiconductor devices are made from either germanium or silicon crystals; hence particular emphasis will be placed upon these elements.

4.2 ENERGY LEVELS

In classical physics energy can be thought of as the ability to do work. It has units of joules (in mks units) or ergs (in cgs units) or electron volts. One electron volt (eV) is equal to 1.6×10^{-12} erg or 1.6×10^{-19} joule and is defined as the energy an electron gains when it falls through a potential drop (or voltage drop) of one volt. The electron volt is commonly used to measure energy in nuclear, atomic, and semiconductor physics. A useful number to remember is $1/40\,\text{eV} = 0.025\,\text{eV}$, which is the approximate average translational kinetic energy an atom or free electron has in thermal equilibrium at room temperature.

Classically, there is no restriction on the amount of energy an object may have. However, in quantum physics where we are dealing with very small objects such as nuclei, atoms, electrons, or molecules, the energy an object may have is often restricted by the basic quantum-mechanical laws that apply in such situations. If we calculate the energy possible for a single isolated atom according to the quantum theory, we find that there are a number of sharp or discrete energy values or *levels* possible. These energy levels depend mostly on the orbital electronic properties, not on the nuclear properties, because the nuclear states are usually so stable. The higher levels are usually more closely spaced than the lower levels, but the main point to remember is that the energy levels are discrete. It is impossible for the atom to have any energy between these discrete energy levels, which

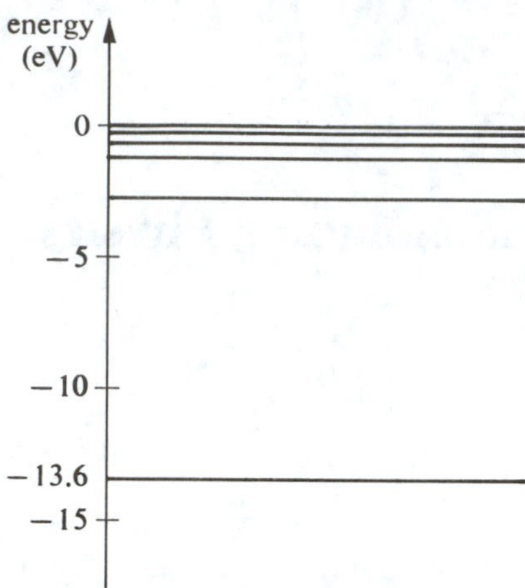

FIGURE 4.1 Energy levels for a single hydrogen atom.

are usually spaced of the order of electron volts apart. The energy level diagram for a hydrogen atom, for example, is given in Fig. 4.1. The horizontal axis is essentially meaningless. Larger atoms containing more electrons in general have more complicated energy level diagrams, but the levels are still discrete with forbidden gaps between them and look basically like those in Fig. 4.1.

4.3 CRYSTALS

A *crystal* is a solid in which the atoms are arranged in a regular periodic way. Most elements, if extremely pure, will form crystals when cooled carefully from the liquid. The exact geometry of the atoms depends on the chemical bonds formed among the outer or *valence* electrons of the atoms. The electrons in the inner shells ordinarily take no part in bonds between atoms because they are so tightly bound to their own nucleus. Thus, the crystal structure depends on the atomic number and the valence of the atom involved. Some compounds, particularly the simpler ones, also form crystals easily: for example, NaCl, KCl, GaAs, and others. If the regular, periodic arrangement of atoms is unbroken throughout the entire piece of material, then we have a perfect "single" crystal. A solid is called *polycrystalline* if it consists of a large number of smaller single crystals oriented randomly with respect to one another. In general, a substance takes on a polycrystalline form upon solidifying from the liquid and special experimental techniques and extremely pure samples must be used to make a single crystal of an appreciable size. In fact, much of the recent progress in transistor and chip technology has come from improvements in the purification techniques for silicon.

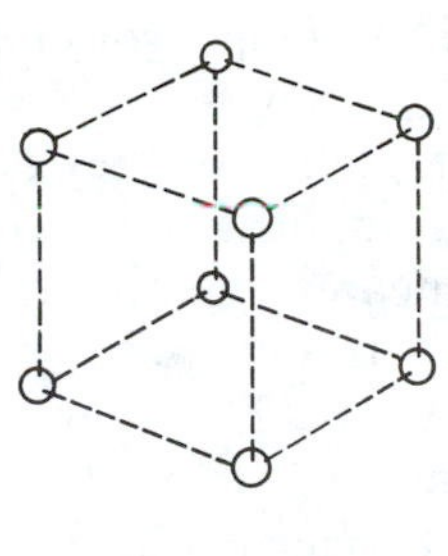

(a) *unit cell*

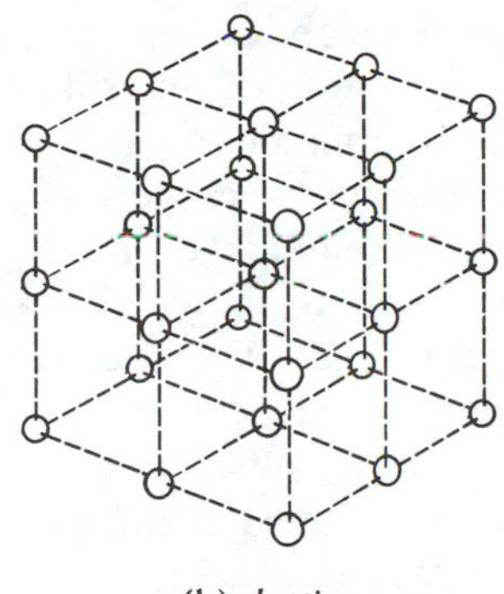

(b) *lattice*

FIGURE 4.2 Cubic crystal lattice.

The regular geometrical arrangement of atoms is called a crystal *lattice*. One example of a crystal is a lattice in which all the atoms occur at the corners of cubes of identical size, as shown in Fig. 4.2. Such an arrangement is called a *cubic* lattice. The cube is called a unit cell of this crystal, because the entire crystal lattice can be built up conceptually by a regular "stacking-up" of cubes.

There are many types of imperfections or defects present in any real single crystal: There can be mechanical breaks or discontinuities in the crystal structure extending over many unit cells, atoms can be missing, an extra atom can be present—in an unusual position in a unit cell, or an atom of a different element (with the "wrong" number of valence electrons) may be present in a regular lattice position. The type and number of such imperfections play an extremely important role in determining the mechanical and electrical properties of the crystal. Most transistors and chips today are made from extremely small single crystals of silicon, although some special-purpose semiconductor devices are made from single crystals of compounds such as GaAs. The unit cell of germanium and silicon is a face-centered cubic cell, as shown in Fig. 4.3, in which the atoms

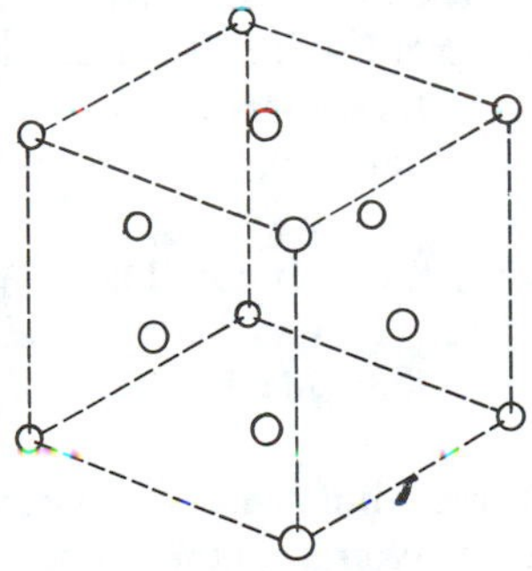

FIGURE 4.3 Face-centered cubic cell (Ge and Si).

occur at the eight corners of a cube and in the centers of the six faces of the cube. The length of one side of the cube is called the *lattice constant*; it is 5.66 Å in germanium and 5.43 Å in silicon (1 Å = 10^{-8} cm = 10^{-10} m).

If several face-centered cubic cells are drawn together, as they occur in either a germanium or a silicon crystal, a very complicated interlocking structure results in which each atom has four nearest neighbors. Both germanium and silicon are Group IV elements in the chemical periodic table and hence have four outer or valence electrons. The four valence electrons enter into covalent electronic bonds with the four nearest-neighbor atoms in the lattice. The resulting covalent chemical bonds literally hold the lattice together, and energy must be supplied from some source to break a valence electron loose from such a bond. The inner electrons are much more tightly bound to the atom's nucleus and are never free to move in the lattice.

4.4 ENERGY LEVELS IN A CRYSTAL LATTICE

If several identical atoms are brought together, the energy levels of the individual isolated atoms split into a closely spaced set of energy levels, as shown in Fig. 4.4(a) and (b).

If we calculate, from quantum theory, the possible energy levels for an electron of an atom in a crystal containing a large number of atoms (e.g., 10^{20} atoms for a tiny crystal composing a transistor), we find that the allowed energy levels are very closely spaced together and that there are large forbidden energy gaps between groups of closely spaced energy levels. A group of many closely spaced energy levels is called an energy *band*. The most important two bands for the purpose of understanding transistors are the *valence* band and the *conduction* band shown in Fig. 4.4. The gap between the valence and conduction bands is 0.72 eV in germanium and 1.09 eV in silicon. Note that these energy gaps are much greater than kT, which is equal to 0.025 eV at room temperature. The electron energy levels in either band are so closely spaced in energy that the band may be regarded as a continuum for most purposes.

The electrons in the conduction band are bound only very weakly to individual atoms in the lattice; hence they are essentially free to move throughout the lattice and take part in conduction of electrical current. Electrons in the valence band are those electrons that form the covalent bonds between atoms of the lattice. They are "valence electrons" in chemical terminology, hence the name "valence" band. Electrons in atomic shells inside the valence electrons lie in energy bands of the order of electron volts below the valence band. These electrons play no part in conduction at temperatures under several thousand degrees, and so from now on we will consider only the valence and conduction band electrons.

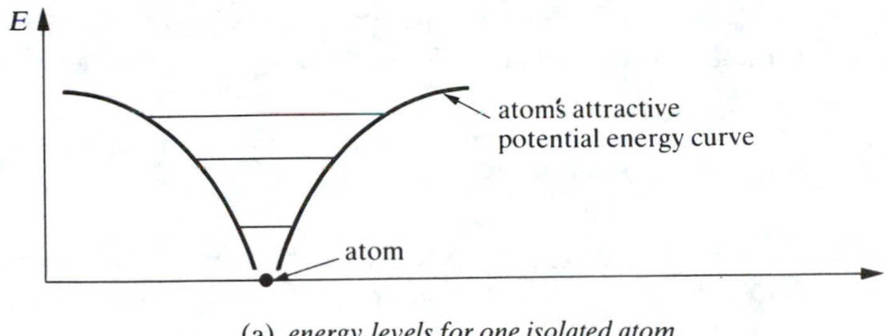

(a) *energy levels for one isolated atom*

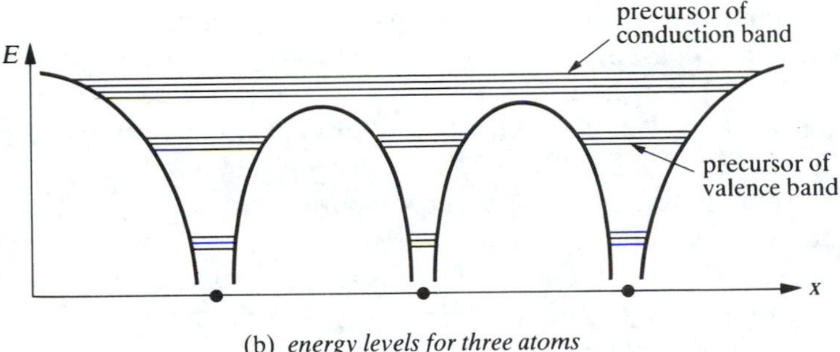

(b) *energy levels for three atoms*

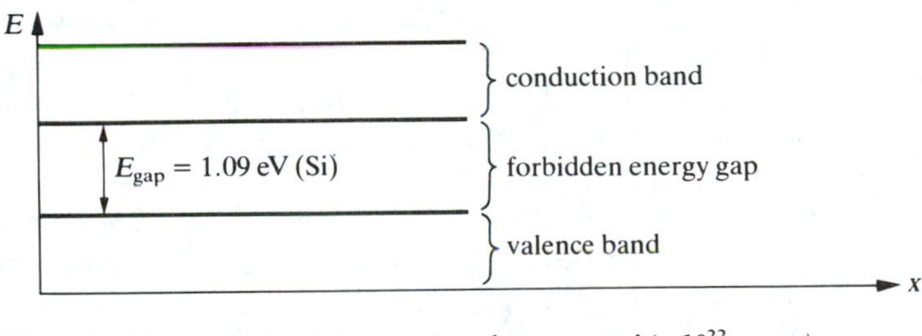

(c) *energy bands in a semiconductor crystal (~10^{22} atoms)*

FIGURE 4.4 Atomic electron energy levels.

The physical reason for the energy gap is simply that it takes a certain amount of energy or work, E_{gap}, to pull an electron loose from a bond between lattice atoms and make it free to move through the crystal. Thus, the energy gap is physically the amount of energy that must be given an electron in a bond to free it from that bond. We can now see that at absolute zero, when all the electrons must fall into the bonds between the lattice atoms, all the electrons must lie in the valence band, and *none* lies in the conduction band.

The problem of conduction in a crystal is more complicated than we have just stated because of two quantum-mechanical laws we have not yet considered: the Pauli exclusion principle and the Fermi–Dirac statistics.

4.5 PAULI EXCLUSION PRINCIPLE

An energy level is characterized by a set of quantum numbers that come from a quantum-mechanical calculation of the physical system's allowed energies. In the simplest case of a hydrogen atom each electron has four quantum numbers: n, l, m_l, and m_s. It can be shown from quantum mechanics that the energy of a hydrogen atom depends only upon n, the principal quantum number according to the formula $E = -\mu e^4/2\hbar^2 n^2$. The electronic charge is e, μ is the reduced mass of the electron and nucleus, $\hbar = h/2\pi$, where h is Planck's constant, l is called the azimuthal quantum number. The electron's orbital angular momentum L is given by $L = \sqrt{l(l+1)}\hbar$, where $l = 0, 1, 2, 3, \ldots, n-1$. The symbol m_l is called the magnetic quantum number; the z component L_z of the orbital angular momentum is given by $L_z = m_l\hbar$, where $m_l = -l, \ldots, +l$. The value of m_s determines the z component of the electron spin angular momentum according to $S_z = m_s\hbar$, and $m_s = -\frac{1}{2}$ or $+\frac{1}{2}$. Hence the $n = 2$ energy level, for example, can contain electrons with quantum numbers n, l, m_l, and m_s equaling 2, 0, 0, $+\frac{1}{2}$ or 2, 0, 0, $-\frac{1}{2}$ or 2, 1, 1, $+\frac{1}{2}$ or 2, 1, 1, $-\frac{1}{2}$ or 2, 1, 0, $+\frac{1}{2}$ or 2, 1, 0, $-\frac{1}{2}$ or 2, 1, -1, $+\frac{1}{2}$ or 2, 1, -1, $-\frac{1}{2}$, a total of eight possible states or sets of quantum numbers.

The Pauli exclusion principle says that no two (or more) electrons can have exactly the same set of quantum numbers. Thus, in an energy level characterized by the quantum numbers n, l, m_l, m_s, if $n = 2$ there can be at most eight electrons. We say that the $n = 2$ level is "filled" with eight electrons. A hydrogen atom never has eight electrons, but a hydrogen-like energy level can exist in many heavy atoms. For an energy band in a crystal, each of the closely spaced energy levels has its own set of quantum numbers. Electrons in the crystal will naturally tend to fill up the lowest energy levels first but always subject to the restriction that no two electrons can occupy the same state; that is, no two can have the same set of quantum numbers.

The most important implication of the exclusion principle is that a completely full energy band of electrons cannot conduct. This can be seen by realizing that conduction implies that an electron is raised from a lower to a slightly higher energy state. However, in a completely full band the higher energy state is already occupied by an electron, and the exclusion principle prevents a second electron from jumping up to this higher state. Thus for conduction to occur, an energy band must be only partially full; there must be empty higher energy states for the electron to jump up to. Another implication of the exclusion principle is that electrons for conduction come mainly from near the top of the partially filled band because these electrons need only a small increase in energy to jump up to an unfilled level in the band.

4.6 FERMI–DIRAC STATISTICS

To understand the electrical behavior of a semiconductor, we clearly must know how many electrons are in the conduction band and how many are in the valence band, because only these electrons in the conduction band can move and create a current flow in response to an applied voltage. The basic problem is then to calculate the distribution of the electrons in the crystal among the various allowable energy levels. We have already stated that a quantum-mechanical calculation of the energy levels in a crystal lattice yields the result that the allowable energy levels lie in two bands (conduction and valence), separated by a forbidden energy region or gap. The next thing to calculate is the probability $F(E)$ that a given energy level E somewhere in a band is actually occupied by an electron. If we let $N(E)\,dE$ equal the number of electrons per unit volume between E and $E + dE$, then we write $N(E)\,dE$ as the product of the number $\rho(E)\,dE$ of allowable or available energy states per unit volume in the range E to $E + dE$ [where $\rho(E)$ is often called the *density of states*] and the probability $F(E)$ that the energy level E is actually occupied.

$$N(E)\,dE = \rho(E)\,dE\,F(E) \tag{4.1}$$

The $\rho(E)$ and therefore $N(E)\,dE$ will be zero in the energy gap and positive in the conduction and valence bands. The exact form of $\rho(E)\,dE$ comes from the quantum-mechanical solution to the problem of calculating the allowable energy levels for electrons in a periodic crystal lattice. The result is

$$\rho(E) = \frac{2^{7/2} m^{3/2} \pi}{h^3} E^{1/2} \tag{4.2}$$

for electrons not too near the top of the band. $E = 0$ represents the energy of the bottom of the band,† m is the mass of the electron, and h is Planck's constant.

The Fermi function $F(E)$ tells us the probability that the energy level E is actually occupied by an electron. It seems intuitively reasonable that $F(E)$ should depend on temperature, because the higher the temperature, the more thermal energy is available to be distributed among the electrons

†Near the top of the band electrons undergo Bragg reflections from atoms in the lattice, and the net result is that fewer energy levels are allowable than are predicted from equation (4.2).

and the nuclei of the lattice atoms. And the more energetic the electrons, the greater the probability that a higher energy state is occupied. At absolute zero when there is zero† thermal energy available, all vibrational motion of the atoms in the lattice should cease; and all the electrons should be lying in the lowest possible energy levels, subject only to the restriction of the Pauli exclusion principle. Thus we would expect the Fermi function $F(E)$ to be temperature dependent and, at absolute zero, to be a constant value from $E = 0$ up to some maximum energy, which is the energy of the highest filled state.

$F(E)$ is calculated by an involved statistical argument. The problem is to calculate the probability that an energy level E is occupied subject to four conditions: (1) The total number of electrons is constant (conservation of charge). (2) The total energy of all the electrons remains constant (conservation of energy). (3) No two electrons can be distinguished from one another. (4) No two (or more) electrons can lie in the same quantum state. This total energy depends on the temperature, which is assumed to be constant (i.e., the particles are assumed to be in thermal equilibrium). Once the probability is known, the maximum probability is calculated, because we know nature in equilibrium always assumes the most probable configuration. There are clearly many ways to distribute a fixed total amount of energy among N electrons; this calculation gives us the most *likely* or most probable distribution of electrons among the various energy levels.

The net result of the calculation is that the Fermi function is given by

$$F(E) = \frac{1}{e^{(E-E_F)/kT} + 1} \tag{4.3}$$

where E_F is a constant energy called the Fermi energy, k is Boltzmann's constant equal to 1.38×10^{-16} ergs/deg $= 1.38 \times 10^{-23}$ J/deg, and T is the temperature in kelvins. To see the physical meaning of $F(E)$, let us graph it for various temperatures (see Fig. 4.5). At absolute zero the Fermi function

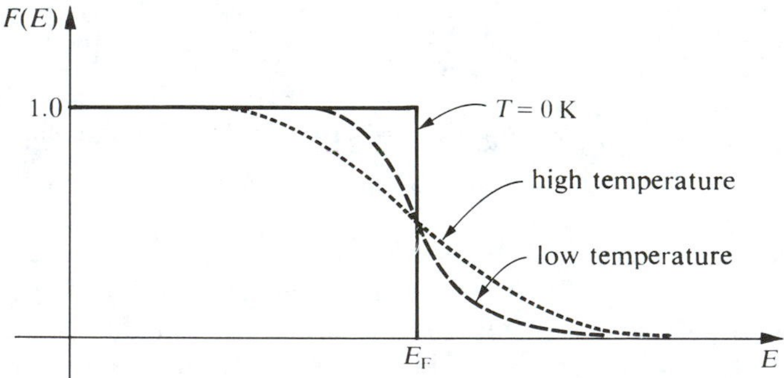

FIGURE 4.5 The Fermi function vs. *E*.

†Actually at absolute zero, there is a nonzero, fixed minimum value of energy available to the electrons and atoms; this is called the *zero-point* energy and comes from a careful consideration of the uncertainty principle of quantum mechanics.

$F(E)$ is 1.0 from $E = 0$ up to $E = E_F$; for energies above E_F, $F(E) = 0$. This says that at absolute zero all the allowable energy levels from 0 up to E_F are filled and that no levels above E_F are filled, which is just what we expect physically. E_F represents the highest energy level filled at absolute zero.

The physical meaning may become clearer if we refer to Fig. 4.6,

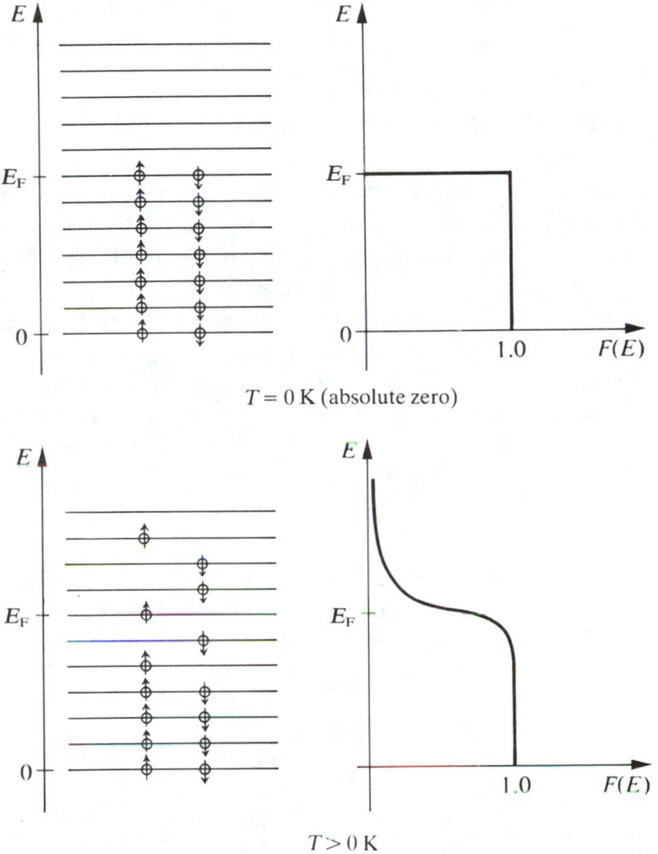

FIGURE 4.6 Simple energy level picture and Fermi function.

which is a simple energy level picture of a hypothetical solid with evenly spaced, allowed energy levels. The density of states $\rho(E)$ gives the number of these energy levels per unit energy. At absolute zero the electrons have cascaded down to the lowest energy states possible, subject to the restriction of the Pauli exclusion principle, namely, that no more than one electron can occupy any one energy level with a specific set of quantum numbers. In Fig. 4.6 we have assumed that the spin quantum number does not affect the energy; that is, the energy does not depend upon the quantum number m_s. Thus two electrons, one with spin "up" ($m_s = +\frac{1}{2}$) and one with spin "down" ($m_s = -\frac{1}{2}$), can occupy the same energy level. The Fermi energy will be at the topmost occupied energy level at 0 K. Notice that the only way to change the Fermi energy would be to add more electrons or to change the spacing of the energy levels. Figure 4.6(b) shows the distribution of the electrons at a temperature above absolute zero. The density of the electrons per unit energy increases for energies above the Fermi level as the temperature is raised.

4.7 ELECTRON ENERGY DISTRIBUTION

Before we can multiply the density of states $\rho(E)\,dE$ by the Fermi function to obtain the number of electrons actually between E and $E + dE$, we must first know E_F. If we note that at absolute zero all the electrons in the lattice will be bound in the covalent bonds between the atoms, we see that the conduction band must be empty and the valence band full. Therefore, since the Fermi level is the maximum energy an electron can have at absolute zero, the Fermi energy must be somewhere above the valence band and below the bottom of the conduction band. An exact treatment gives the result that

$$E_F = \frac{E_{gap}}{2} \tag{4.4}$$

where $E = 0$ is the top of the valence band and E_{gap} is the energy gap. In other words, the Fermi energy lies exactly halfway between the valence and conduction bands. If we take $E = 0$ to be the top of the valence band, then from (4.2) the density of states in the conduction band must be

$$\rho(E) = \frac{2^{7/2} m^{3/2} \pi}{h^3} (E - E_{gap})^{1/2} \tag{4.5}$$

Thus the number $N(E)\,dE$ of electrons in the conduction band between E and $E + dE$ will be given by $N(E)\,dE = \rho(E)\,dE\,F(E)$. $F(E)$, $\rho(E)$, and $N(E)$ are graphed on an energy level diagram in Fig. 4.7.

Substituting for $\rho(E)$ and $F(E)$, we obtain

$$N(E)\,dE = \frac{2^{7/2} m^{3/2} \pi}{h^3} (E - E_{gap})^{1/2} \frac{1}{e^{(E-E_F)/kT} + 1}\,dE \tag{4.6}$$

For pure silicon or germanium with E in the conduction band, $E - E_F \gg kT$. Hence,

$$e^{(E-E_F)/kT} + 1 \cong e^{(E-E_F)/kT}$$

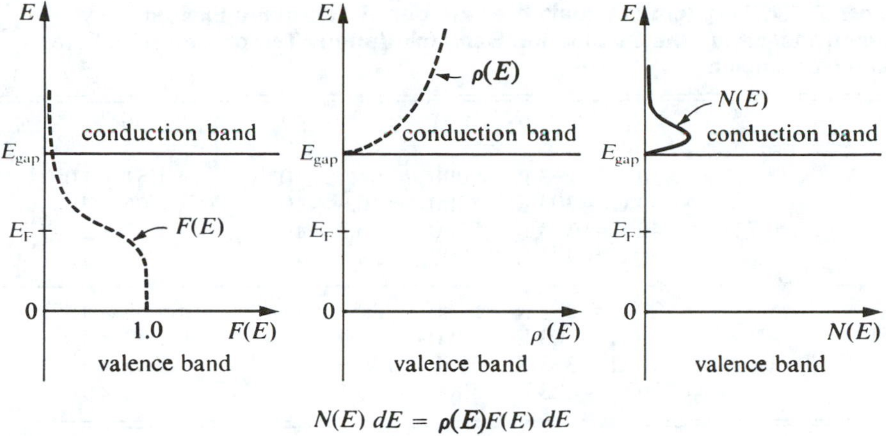

FIGURE 4.7 Density of electrons in conduction band.

So we may write

$$N(E)\,dE \cong \frac{2^{7/2} m^{3/2} \pi}{h^3} (E - E_{\text{gap}})^{1/2}\, e^{-(E-E_F)/kT}\, dE \tag{4.7}$$

The total number N_{cb} of electrons in the conduction band can be obtained by integrating $N(E)\,dE$ from the bottom of the conduction band to the top, which can be taken to be $E = \infty$ for all practical purposes. The result is

$$N_{cb} = \int_{E_{\text{gap}}}^{\infty} N(E)\,dE = \frac{2^{5/2}(m\pi kT)^{3/2}}{h^3}\, e^{-E_{\text{gap}}/2kT} \tag{4.8}$$

where we have set $E_F = E_{\text{gap}}/2$. It is useful to rewrite N_{cb} as

$$N_{cb} = AT^{3/2} e^{-E_{\text{gap}}/2kT} \tag{4.9}$$

where $$A \equiv \frac{2^{5/2}(m\pi k)^{3/2}}{h^3} = 4.6 \times 10^{15} \frac{\text{electrons}}{\text{cm}^3} (\text{deg})^{-3/2}$$

The most important thing to remember is that the total number of electrons in the conduction band depends on the temperature according to $T^{3/2}$ times $e^{-E_{\text{gap}}/2kT}$. Table 4.1 gives some numerical values for various temperatures. Note that the exponential factor $e^{-E_{\text{gap}}/kT}$ is much smaller for silicon because of the larger value of E_{gap}. Therefore, the number of electrons thermally excited to the conduction band is much less for silicon than for germanium.

Also notice that the exponential factor $e^{-E_{\text{gap}}/2kT}$ is a much stronger

TABLE 4.1. N_{cb} (cm^{-3}) Equals the Number of Thermally Excited Electrons/cm^3 in the Conduction Band for Various Temperatures in Silicon and Germanium

	T	$T^{3/2}$	$e^{-E_{gap}/2kT}$	$AT^{3/2}e^{-E_{gap}/2kT} = N_{cb}$
	20°C = 293 K	5000	6.21×10^{-7}	1.51×10^{13}
Ge	30°C = 303 K	5280	10.3×10^{-7}	2.65×10^{13}
($E_{gap} = 0.72$ eV)	40°C = 313 K	5560	16.2×10^{-7}	4.39×10^{13}
	50°C = 323 K	5800	28.8×10^{-7}	8.15×10^{13}
	20°C = 293 K	5000	3.52×10^{-10}	0.858×10^{10}
Si	30°C = 303 K	5280	7.09×10^{-10}	1.82×10^{10}
($E_{gap} = 1.1$ eV)	40°C = 313 K	5560	13.56×10^{-10}	3.67×10^{10}
	50°C = 323 K	5800	26.1×10^{-10}	7.36×10^{10}

function of temperature than $T^{3/2}$. Therefore, the gap energy E_{gap} mainly determines the number of electrons in the conduction band for pure germanium and silicon.

From the graphs in Fig. 4.8 we see that, starting at 20°C, the electron

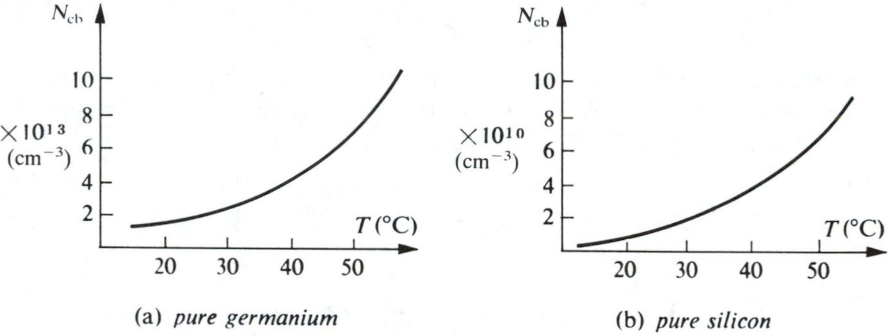

(a) *pure germanium* (b) *pure silicon*

FIGURE 4.8 Conduction band electron density N_{cb} as a function of temperature.

density in the conduction band N_{cb} doubles for a 13°C temperature rise for germanium; for silicon N_{cb} doubles for only an 8°C rise. However, there are far fewer electrons in absolute number in silicon for the same temperature.

4.8 CONDUCTION IN SEMICONDUCTORS

The Fermi level in pure semiconductors lies halfway between the top of the valence band and the bottom of the conduction band. Hence, at absolute zero all the electrons are in the valence band, which is then full and thus can carry no current. At higher temperatures some electrons acquire enough thermal energy to break loose from a valence bond in the lattice, thus jumping from the valence band up to the conduction band. The conductivity of a semiconductor then increases with increasing temperature.

In a metal the Fermi level lies *in* the conduction band, which is only partially full. Hence the metal is a good conductor at any temperature because of the many empty energy levels in the conduction band above the filled levels. There is no energy gap that the electrons must jump to get into the conduction band.

In a crystalline insulator the energy level picture is qualitatively the same as for a semiconductor, but the energy gap is much larger. For example, in diamond the energy gap is 5 eV, which is so large that an enormous temperature (approximately 60,000°C!) is required to thermally excite an electron from the valence band all the way up to the conduction band, where it may then contribute to conduction (see Fig. 4.9).

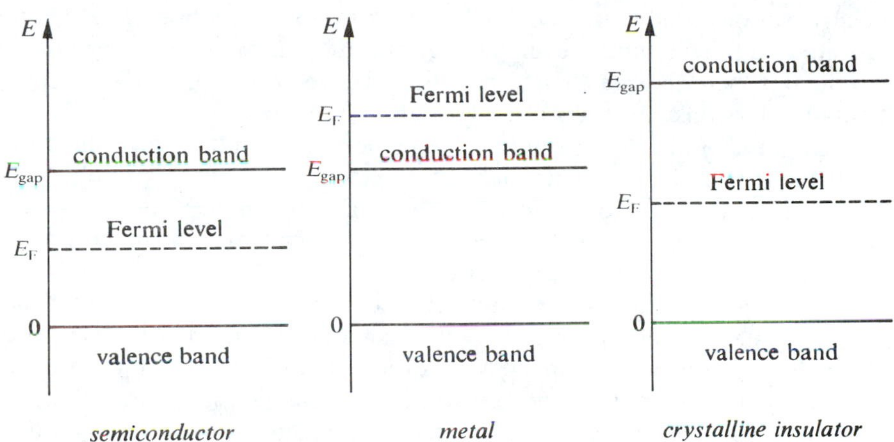

FIGURE 4.9 Location of Fermi level in various materials.

If we are to use semiconductors as practical components in electrical circuits, then they must be able to carry a reasonable amount of current. From the preceding section the total number N_{cb} of electrons per unit volume in the conduction band of a pure semiconductor was given by

$$N_{cb} = AT^{3/2}e^{-E_{gap}/2kT}$$

Substituting numerical values gives us

1. $N_{cb} = 1.5 \times 10^{13}$ electrons/cm^3 for Ge at room temperature.
2. $N_{cb} = 8.6 \times 10^{9}$ electrons/cm^3 for Si at room temperature.

Are there enough electrons in either pure germanium or pure silicon to carry a practical amount of current? The answer is no, for the following reasons. The maximum current I that can be passed is the number of electrons flowing per second multiplied by the charge per electron. The number flowing per second is the number per unit volume times the area times the speed. Thus, $I = N_{cb}Aev$, where e is the electronic charge. It can be shown that the average speed v of electrons flowing in germanium is of the order of 4×10^4 cm/s for a 10 V/cm electric field (voltage gradient) applied.[†] If we assume a generous cross-sectional area of 1 mm × 1 mm, then the current I is

$$\begin{aligned} I = N_{cb}Aev &= (1.5 \times 10^{13}\ \text{cm}^{-3})(0.1\ \text{cm})^2(1.6 \times 10^{-19}\ \text{C})(4 \times 10^4\ \text{cm/s}) \\ &= 9.6 \times 10^{-4}\ \text{C/s} = 960\ \mu\text{A} = 0.96\ \text{mA} \end{aligned}$$

A value of 0.96 mA is too small a current for many practical applications. In addition, the actual current will be less than that because not every electron in the conduction band will contribute to conduction. The current passed by a silicon cube is even less because the number of electrons in the conduction band is smaller and also because, for a given applied electric field the electrons move more slowly than in germanium.

Because of the temperature dependence of the number of electrons available for conduction, one might try to increase the number of electrons by warming the semiconductor. However, a quick numerical calculation shows that even for a temperature of 100°C, the maximum current passed is still too small for practical use in circuits.

There is another way of exciting electrons up into the conduction band, and that is by the absorption of electromagnetic radiation. An electron in the valence band can jump up to the conduction band if it absorbs an electromagnetic photon of energy greater than the gap, $h\nu \geqq E_{gap}$, where ν is the frequency of the photon. This process is called *photoconduction*. Some devices (photocells) detect light using this process, and the photovoltaic cell converts sunlight into electrical energy by photoconduction.

The practical way of increasing the number of electrons in the conduction band so that the semiconductor can carry a reasonable amount of current is to effectively add more electrons by introducing "impurity" or "donor" atoms to the lattice. The process of adding impurity atoms is called *doping* the crystal. Consider a silicon lattice at absolute zero. All the valence electrons are locked in covalent bonds between the atoms, so none

---

[†]The speed is calculated from the electron mobility μ, which is defined as the electron drift speed per unit electric field.

$$\mu \equiv \frac{v(\text{cm/s})}{|\mathbf{E}|(\text{V/cm})}$$

	Ge	Si	
$\mu =$	3600	1200	for electrons
$\mu =$	1700	250	for holes

is available for conduction. Each atom has four valence electrons that bond to the four nearest atoms in the lattice. If we now introduce a donor atom with *five* valence electrons in a regular lattice site in place of a silicon atom, then there clearly will be one electron left over (i.e., not bound in a covalent bond with the four nearest silicon neighbors). This one electron is only very *loosely* bound to its atom and therefore can easily be pulled loose. Once freed it can migrate through the crystal under the influence of an applied electric field and contribute to conduction. In other words, only a small amount of energy is necessary to free this fifth electron from its donor atom and raise it up into the conduction band.

On the energy level diagram for silicon the fifth electron when loosely attached to its donor atom then lies in an energy level in the gap only slightly below the bottom of the conduction band (see Fig. 4.10). The

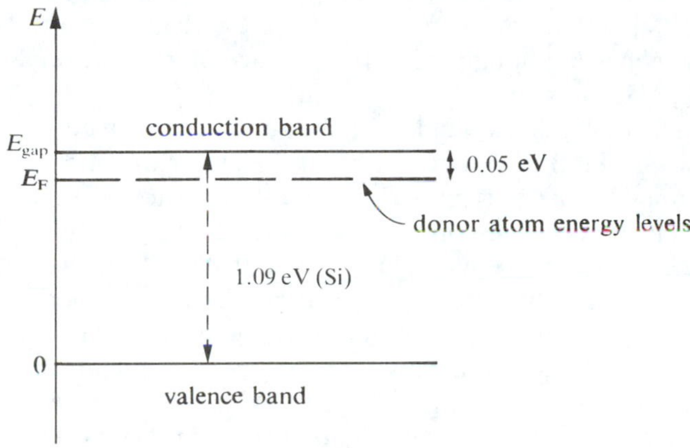

FIGURE 4.10 n-type semiconductor energy level diagram.

concentration of these donor atoms is such that they are spaced of the order of 30 Å apart in the lattice. Thus, each donor atom has four silicon atoms as its nearest neighbors in the lattice. Notice that the donor atom, once it loses its fifth electron, is a positively charged ion and is still locked in the crystal lattice by the four remaining valence electrons—being bonded to the adjacent four silicon atoms. Thus, it is only the electrons, not the positive ions, that contribute to conduction. The atom that gives up its electron is called a *donor* atom because it *donates* an electron to the conduction band. The donor energy level in silicon lies only about 0.05 eV below the conduction band, so at room temperature (300 K), where the average thermal agitation energy $kT = 0.025$ eV, almost all the donor atoms are ionized and contribute an electron to the conduction band. A typical concentration of donor impurity atoms is $10^{16}/\text{cm}^3$, so there will be of the order of 10^{16} electrons/cm^3 in the conduction band from the ionized donor atoms. Thus a current of the order of amperes can be carried by such a semiconductor. In silicon for every electron in the conduction band that has been thermally excited up from the valence band, there will be of the order of 10^6 electrons in the conduction band from the donor atoms.

In the conduction band of silicon at 300 K, there will thus be roughly 8.6×10^9 electrons/cm^3 from the valence band and about 10^{16} electrons/cm^3 from the ionized donor atoms. There will be 10^{16} positively charged donor atoms/cm^3 fixed in the lattice. There will also be 8.6×10^9 singly positive ionized silicon atoms/cm^3 in the valence band, because breaking one valence bond and lifting one electron up to the conduction band must leave behind exactly one singly ionized silicon atom. This positively ionized silicon atom is fixed in the lattice, but it can attract an electron from an adjacent silicon atom. Thus, the positive charge appears to move in the opposite direction from the motion of the electron it attracted. These positive vacancies or charges are called *holes* and move in the valence band quite freely, although not quite so rapidly as do electrons in the conduction band. A hole may be thought of as the absence of an electron, or something like an empty spot in a nearly full parking lot.

If an external electric field is applied to such a doped semiconductor, most of the current flow will be due to the electrons in the conduction band, and only roughly *one* part in 10^6 will be due to the positive holes in the valence band. Because the current is primarily carried by the negatively charged electrons, the semiconductor is called an *n-type* semiconductor. The electrons are called the *majority* carriers, and the positive holes are called the *minority* carriers. Any impurity atom having five valence electrons, such as arsenic, antimony, or phosphorus, will produce an n-type semiconductor when added to either pure germanium or pure silicon.

If, on the other hand, we add an impurity atom with only *three* valence electrons, such as boron, aluminum, gallium, or indium, in a regular lattice site in place of a silicon atom, then there will be one too few valence electrons to make the required four bonds to the four nearest silicon atoms. The impurity atom will pull an electron over from an adjacent silicon atom to fill the empty bond, thus leaving the adjacent silicon atom positively charged. The impurity atoms are called *acceptor* atoms because they accept electrons. In other words, a positive hole has been created. The impurity atom is now negatively charged, but it remains locked in the lattice by its four bonds to adjacent silicon atoms. The positively charged adjacent silicon atom may attract a valence electron from another silicon atom, and then the positive charge is transferred to the new atom, and so on. In this way the positive charge or hole will move through the lattice as the valence electrons are attracted from one atom to another. If the concentration of the impurity atoms is about 10^{16} cm^{-3}, then the hole concentration will greatly exceed the concentration of conduction band electrons which have been thermally excited across the energy gap. To give some numbers for silicon: There will be 10^{16} holes/cm^3 in the valence band from the impurity atoms, 8.6×10^9 electrons/cm^3 in the conduction band, and 8.6×10^9 holes/cm^3 in the valence band from thermally broken Si–Si bonds. Thus, conduction will take place primarily due to motion of the positively charged holes in the valence band. Hence, this type of doped semiconductor is called a *p-type* semiconductor. The holes are called the *majority* carriers, and the electrons are called the *minority* carriers.

On the energy level diagram of silicon (see Fig. 4.11) the acceptor

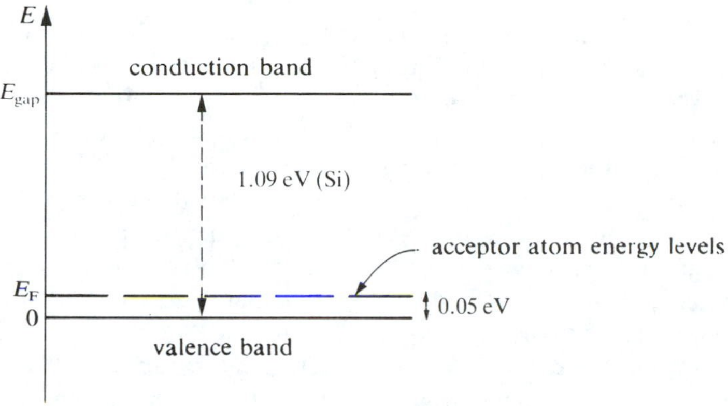

FIGURE 4.11 p-type semiconductor energy level diagram.

atoms have energy levels about 0.05 eV above the top of the valence band. When a valence electron from a neighboring silicon lattice atom jumps over to the acceptor atom, this corresponds to an electron jumping up from the valence band to the acceptor energy level. This process is very probable at room temperature, because the average thermal agitation energy $kT = 0.025$ eV is greater than the energy the electron needs to make the jump.

The presence of the impurity atoms, either donor or acceptor, radically changes the value of the Fermi energy. In n-type material the Fermi level is just above the donor atom energy levels and below the bottom of the conduction band. This position follows from the fact that the Fermi energy physically means the maximum occupied energy level at absolute zero. At absolute zero the valence band and all the donor levels must be filled, because as the crystal is cooled to absolute zero all the electrons in the conduction band must fall to the lowest possible energy levels, thereby filling the donor levels and the valence band. In p-type material the Fermi level can be shown to be below the acceptor energy levels and above the valence band. A more refined treatment shows that the Fermi level is slightly temperature dependent. An increase in temperature will slightly lower the Fermi level in n-type material and raise the Fermi level in p-type material.

It is worth pointing out explicitly that if the electric field **E** applied to a

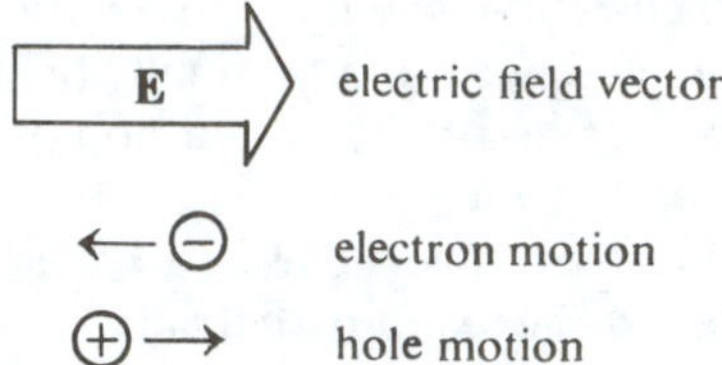

FIGURE 4.12 Electron and hole motion.

semiconductor points from left to right, as shown in Fig. 4.12, then electrons will move from right to left against the electric field, and holes will move from left to right. A positive hole moving from left to right actually involves electrons moving from right to left, but from now on we will speak of the holes as if they were positively charged particles capable of moving in the direction of the applied electric field. It can be shown that one can treat the motion of the holes exactly by treating them as positively charged particles with an effective mass slightly larger than the electron mass. The larger hole mass simply means that for a given applied electric field, a hole will accelerate less rapidly than will an electron.

To sum up: In an n-type semiconductor the impurity or donor atoms have one more valence electron than the atoms composing the crystal lattice. These donor atoms readily donate their extra valence electron to the conduction band, thus producing mobile electrons in the conduction band to carry current and positively charged donor atoms or ions locked in the lattice. In a p-type semiconductor the impurity or acceptor atoms have one less valence electron than the atoms composing the lattice. These acceptor atoms readily attract electrons up out of the valence band, thus producing mobile positive holes in the valence band to carry current and negatively charged acceptor atoms or ions locked in the lattice. In both n- and p-type semiconductors, at room temperature, there are always present a few ($10^{13}\ \text{cm}^{-3}$ in Ge and $10^{10}\ \text{cm}^{-3}$ in Si) holes in the valence band and electrons in the conduction band caused by thermal breaking of lattice bonds.

4.9 p-n JUNCTIONS

If a p-type semiconductor is joined to an n-type semiconductor to form a "good" junction, that is, a junction at which there are few breaks or imperfections in the lattice structure, then this junction will act as a rectifier—it will conduct current readily in one direction and only very poorly in the other direction. Such a junction cannot be formed by merely pressing together a p-type semiconductor and an n-type semiconductor; this procedure would produce a poor junction with gaps at the junction larger than the lattice spacing. A good junction is usually made by changing over

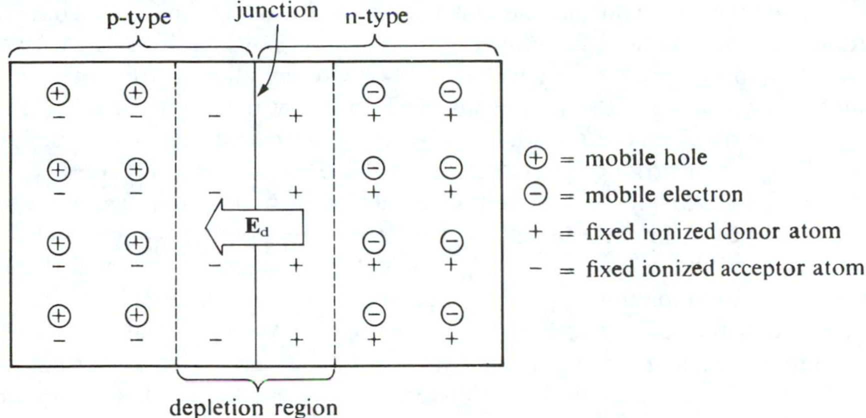

FIGURE 4.13 p-n junction (only majority carriers shown).

the type of impurity from p- to n-type while the crystal is being grown. We will consider a junction in which the type of impurity changes abruptly as we cross the junction; this type of junction is called an *abrupt* junction.

An abrupt p-n junction is shown in Fig. 4.13. Let us first consider the behavior of the majority carriers on both sides of the junction. Majority carriers are mobile and are continually diffusing around in the lattice, due to thermal motion. If holes in the p-type material diffuse away from the junction back into the p-type material, then a concentration gradient will build up, thus tending to diffuse the holes back toward the junction. Diffusion always results in a flow from regions of higher concentration towards regions of lower concentration. Similarly, if electrons in the n-type material diffuse away from the junction, the concentration gradient thus established will tend to diffuse the electrons back towards the junction. Near the junction some holes will diffuse across the junction from the p-type material into the n-type material where they will recombine with the mobile electrons of the n-type material. Similarly some electrons will diffuse across the junction from the n-type material into the p-material and will recombine with the mobile holes of the p-type material. Recombination of a hole and an electron merely means that the electron drops from the conduction band down into the hole in the valence band. Notice that after recombination neither the electron nor the hole can conduct because the electron is no longer in the conduction band, and the valence band hole is filled.

The result of this recombination is the formation of a thin region at the junction where there are essentially no mobile charge carriers. This is called the *depletion region* because the mobile charge carriers have been depleted here. The only charges remaining in the depletion region are the ionized acceptor and donor impurity atoms that are fixed in the crystal lattice and, of course, the electrons in the filled valence band. These fixed charges produce an electric field $\mathbf{E}_d$ in the depletion region pointing from the n-type toward the p-type material. This field tends to sweep any electrons near the junction back toward the n-type material, and any holes back toward the p-type material. Thus, the depletion region has no free charge carriers, and the electric field produced tends to keep the depletion region free of any mobile charge carriers that are either created in it by thermal excitation or that may diffuse into it. In effect, the depletion region is a thin slab of insulator sandwiched between the p- and n-type materials. A typical thickness d for the depletion region is about one micron, which is 10^{-4} cm or 10,000 Å. If the doping of the p- and n-type materials is equal—that is, if the number of acceptor atoms per cubic centimeter, [p], equals the number of donor atoms per cubic centimeter, [n]—then the depletion region extends equally into the p- and n-type materials on either side of the p-n interface. However, if $[p] \neq [n]$, then the depletion region extends unequally into the p- and n-type materials. It can be shown that $d_n/d_p = [p]/[n]$, where d_n is the depletion region thickness in the n-type material, and d_p is the depletion region thickness in the p-type material. The total depletion region thickness $d = d_n + d_p$. This result is reasonable because the electrons and holes *diffuse* across the p-n interface and then recombine. If there are many more donor atoms per cm^3 in the n-type material than there are acceptor atoms per cm^3 in the p-type material ($[n] \gg [p]$), then the electrons diffusing into the p-type material will have to diffuse a long way before they all recombine with the holes. Thus we expect a large d_p. Conversely, the few holes from the p-type material that diffuse over into the n-type material very quickly recombine with the plentiful electrons there; thus we get a small d_n.

The effective voltage V_c developed between the p- and n-type materials is called the *contact potential* and, if we assume parallel plate geometry, is given by $V_c = E_d d$, where d is the thickness of the depletion region. V_c is about 0.2 V for a germanium junction and 0.5 V for a silicon junction. The electric field in the depletion region points from the n-type material toward the p-type material. This field thus tends to keep the majority carriers *out* of the depletion region, but notice that any *minority* carriers thermally generated in the depletion region will be swept across the depletion region by this same electric field.

When equilibrium is established for an isolated p-n junction, there is no net current flowing across the junction. For every majority hole from the p-type material diffusing across against the electric field to the n-type material, there will be a minority hole from the n-type material accelerated back by the electric field to the p-type material. A similar two-way flow will occur for electrons.

If a battery is connected to a p-n junction with a polarity to make [see Fig. 4.14(a)] the p-type material negative with respect to the n-type

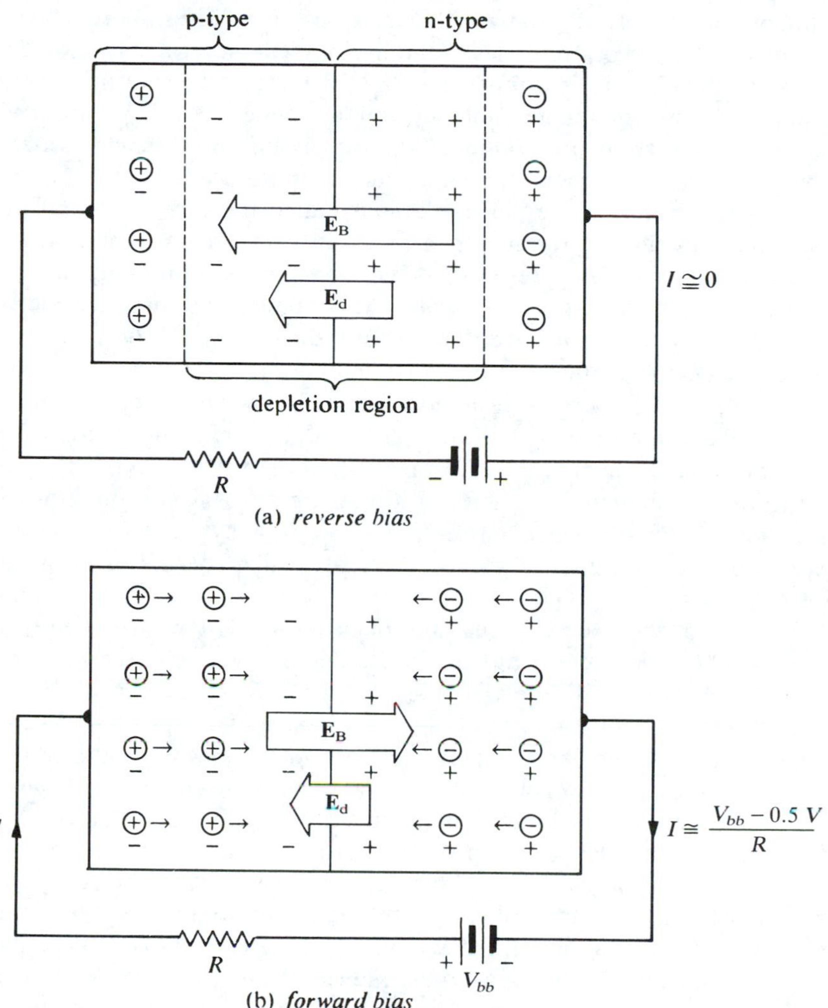

FIGURE 4.14 Biased p-n junction.

material, only a very small current will flow, and the junction is said to be *reverse* biased.

It is only in the depletion region that the p-n junction has a high resistance due to the absence of mobile charge carriers; hence the battery will apply an electric field $\mathbf{E_B}$ across the depletion region only. In the reverse bias configuration $\mathbf{E_B}$ is in the same direction as $\mathbf{E_d}$ and will tend to pull the mobile charge carriers away from the junction, thus increasing the thickness of the charge depletion region, which acts like an insulating slab, and little current will flow through the junction. Actually the preceding argument applies only to the majority charge carriers. The minority carriers (electrons in the p-type and holes in the n-type) will be attracted *toward* the junction by the electric field introduced by the battery. Thus at the junction the minority carriers will recombine, and the junction will pass current. More minority carriers are continually being created by thermal excitation on both sides of the junction, so this current passed due to the minority carriers will continue to be passed. This is called the *reverse current*. In a germanium junction at room temperature the reverse current is on the order of several microamperes (10^{-6} A), whereas in a silicon junction it is on the order of 10 nA (10^{-8} A). Because the number of minority carriers created depends on thermal excitation across the energy gap, we expect the reverse current of a p-n junction to be temperature dependent. At room temperature a 10°C increase in temperature will approximately double the reverse current in a germanium junction; a 6°C increase will approximately double the reverse current in a silicon junction. At 70°C a typical reverse current for a germanium junction is 100 μA, but for a silicon junction only 1 μA. Thus, if low reverse currents are important, particularly at high temperatures, one *must* use silicon. Virtually all modern transistors and chips are silicon.

Because of the fixed charges present in the depletion layer due to the ionized donor and acceptor atoms, the depletion layer acts like a charged capacitor. Even though the depletion region as a whole is electrically neutral, there is positive charge on the n-type side of the p-n interface due to ionized donor atoms and negative charge on the p-type side due to the ionized acceptor atoms. Thus we have two layers of charge in a parallel plate configuration, which makes up a capacitor. The situation is somewhat different from a parallel plate capacitor where the insulating region between the plates contains no charge. In the reverse biased junction the charges are distributed throughout the volume of the depletion region. As the reverse bias increases in magnitude, the depletion region grows thicker (the capacitor "plates" become farther apart), and the capacitance decreases. An exact calculation shows that the junction capacitance for an abrupt junction is given by

$$C_J = \frac{\mathcal{K}}{\sqrt{V}} \tag{4.10}$$

where $\mathcal{K}$ is a constant depending on the transistor material and geometry, and V is the reverse bias voltage across the junction. C_J may vary from several pF for a low-power transistor especially designed to operate at high frequencies, to several hundred pF for a high-power transistor. In general, with transistors, as with any device, the higher the power handling capability is, the worse the high-frequency response is, because higher powers mean physically larger structures for heat dissipation, which in turn mean larger capacitances.

If a battery is connected across the p-n junction with a polarity to make the p-type material positive with respect to the n-type material [see Fig. 4.14(b)], a large current will flow, and the junction is said to be *forward biased*. The resistance R is present only as a precaution to limit the current flow. In this case, the electric field from the battery $\mathbf{E}_B$ set up across the depletion layer is in a direction to pull the mobile charge carriers *across* the depletion layer; that is, electrons will be attracted from the n-type material over into the p-type material, and holes from the p-type material over into the n-type material. Recombination will take place when the electrons reach the p-type material and when the holes reach the n-type material; thus a current will flow through the junction. More electrons will continually be injected into the n-type material by the wire connected to the negative terminal of the battery, and electrons will continually be taken out of the p-type material by the wire connected to the positive terminal of the battery. The taking out of electrons from the p-type material is electrically equivalent to injecting positive holes into the p-type material. Thus, a continual flow of electrons and holes moves toward the junction from the n-type and p-type materials, respectively, and the junction passes current. Notice that the electric field $\mathbf{E}_d$ in the depletion layer due to the contact potential opposes the forward bias electric field of the battery, so no current will flow through the junction until the battery voltage applied to the junction exceeds the contact potential. In other words, it takes a finite voltage to "turn on" a forward biased p-n junction, about 0.5 V for a silicon junction. Once the turn-on voltage is exceeded, the forward biased junction presents essentially a short circuit to the flow of current; that is, $I \cong (V_{bb} - V_{\text{turn on}})/R$, where R is the external resistance in series with the forward biased junction.

The turn-on voltage $V_{\text{turn on}}$ is slightly temperature dependent because of the weak temperature dependence of the Fermi energy. In a pure or *intrinsic* semiconductor with no doping, the Fermi level is in the energy gap halfway between the valence and the conduction band. In a doped n-type semiconductor the Fermi level is between the donor levels and the bottom of the conduction band (see Fig. 4.10).

Thus, an increase in temperature lowers the Fermi energy for n-type material, because at higher temperatures a larger fraction of the mobile charge carriers are from electrons thermally excited across the gap from the valence band to the conduction band. A similar argument shows that the Fermi level is raised for an increase in temperature in p-type material. The net result is that if the temperature increases, a *smaller* forward bias voltage is required to maintain a constant current flowing through a forward biased p-n junction. If the temperature rises and the forward bias voltage is maintained constant by the circuit, then the current flowing through the junction will increase. In other words, a temperature rise will lower the effective turn-on voltage. Empirically, for silicon p-n junctions near room temperature, the turn-on voltage *decreases* by approximately 2.5 mV for every degree Celsius temperature rise.

In the reverse biased configuration the electric field within the semiconductor exists only in the depletion region. Thus, the motion of the mobile charge carriers in the rest of the semiconductor is governed mainly by diffusion—there simply is little or no electric field present to "hurry" the charges along. Charge will flow through the semiconductor because of the concentration gradients set up by the depletion layer absorbing charge carriers and the wire contacts injecting more carriers. The charge carriers will then diffuse from regions of greater concentration toward regions of lesser concentration. Thus, there is a certain definite lag in the propagation of a current through a semiconductor, but this lag or *transit time* does not become important until frequencies of the order of tens or hundreds of megahertz or higher are considered. And the diode manufacturers make the physical size of the semiconductor material used as small as possible to minimize the time necessary for a charge carrier to diffuse from the connecting wire to the depletion layer. In some diodes the conductivity of the semiconductor material is deliberately made less so as to have an electric field exist inside the material; this field moves the charge carriers faster than by diffusion alone.

In summary, we have seen that a p-n semiconductor junction will conduct current very well in one direction (the forward direction) and very poorly in the other direction (the backward direction). This one-way type of conduction is called *rectification*, and the device is called a *diode*.

An ideal or perfect diode would present zero resistance in the forward direction and infinite resistance in the backward direction. If we plot a graph of current I passed by an ideal diode versus V the voltage difference across it, we would get a 90° break in the curve as is shown in Fig. 4.15(a).

An equivalent circuit for a forward biased diode is thus a resistance R_f,

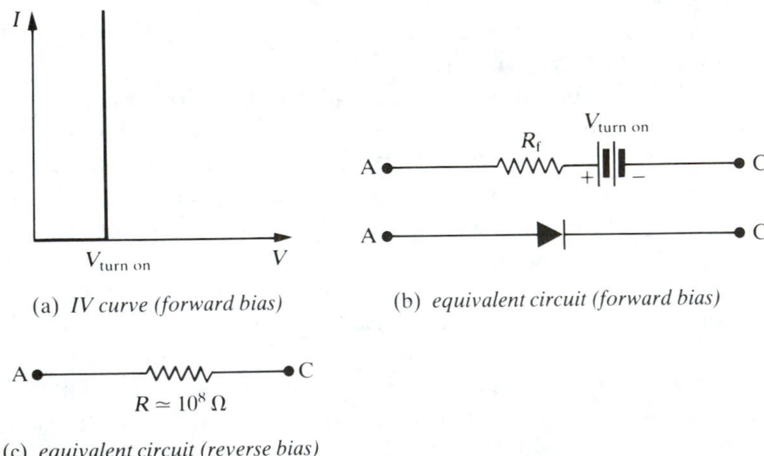

(a) *IV curve (forward bias)*

(b) *equivalent circuit (forward bias)*

(c) *equivalent circuit (reverse bias)*

FIGURE 4.15 Ideal diode.

representing the forward dc resistance of the diode, in series with a battery representing the turn-on voltage $V_{\text{turn on}}$, as shown in Fig. 4.15(b).

The equivalent circuit for a reverse biased diode is simply a large resistance about $10^8\ \Omega$. This simply means that if a reverse bias of several volts exists across a diode, it will conduct a current of about 0.01 μA.

The value of R_f depends on how strongly the diode is conducting. A small Si signal diode might conduct 10 mA with a 0.7-V drop across it. If $V_{\text{turn on}} = 0.6$ V then $R_f = 0.1\text{ V}/10\text{ mA} = 10\ \Omega$. If it conducted 20 mA with a 0.75-V drop across it, then $R_f = 0.15\text{ V}/20\text{ mA} = 7.5\ \Omega$.

The current–voltage curve for a real semiconductor diode differs from the ideal diode curve in several ways: The reverse current is not exactly zero because the thermally generated minority carriers will pass current even when the diode is reverse biased; the forward current is finite because the semiconductor material composing the diode has some resistance; and a minimum voltage (the turn-on voltage) must exist across the diode before it will pass appreciable current. The reverse current increases with increasing temperature because there are more minority carriers at higher temperatures. Typical curves for germanium and silicon diodes are shown in Fig. 4.16. It can be shown that the current–voltage equation for a p-n junction is

$$I = I_0(e^{eV_B/kT} - 1) \tag{4.11}$$

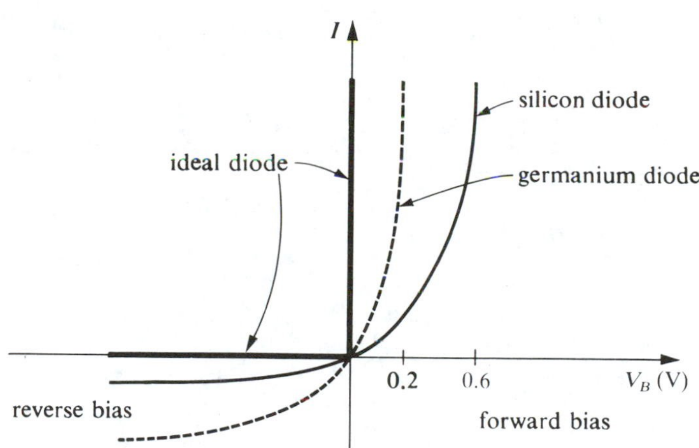

FIGURE 4.16 Current–voltage curves for real diodes.

where V_B is the applied bias voltage, e is the electronic charge, k is Boltzmann's constant, and T is the absolute temperature. V_B is positive for forward bias and negative for reverse bias. I_0 is the reverse current for large reverse bias. Notice that the silicon diode passes much less reverse current than does the germanium diode, and that the turn-on voltage differs appreciably between the two types. Therefore, in applications requiring an extremely small reverse current, silicon is used, and if one has only a small voltage to turn on the diode, germanium is used. Silicon diodes can be used at temperatures up to about 200°C, whereas germanium diodes can be used only up to about 85°C. Silicon diodes are now almost universally used instead of germanium.

The explanation for the turn-on voltage and the equation for the current–voltage curve for a p-n junction can be obtained from a more detailed energy level picture of both sides of the junction. We recall that

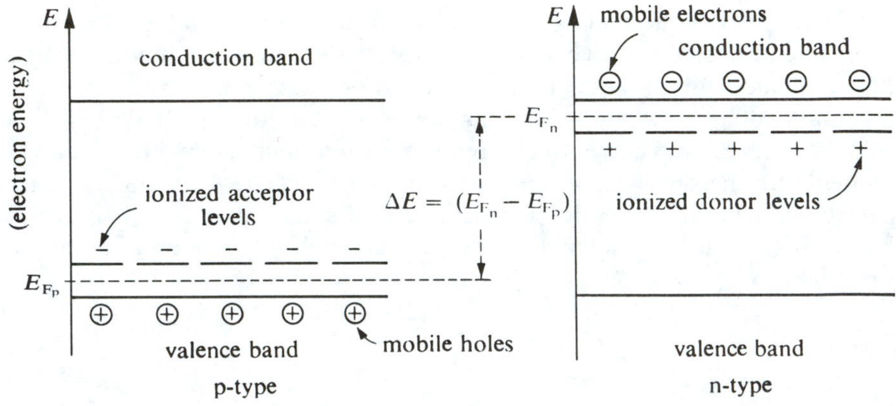

(a) *separate p-type and n-type materials*

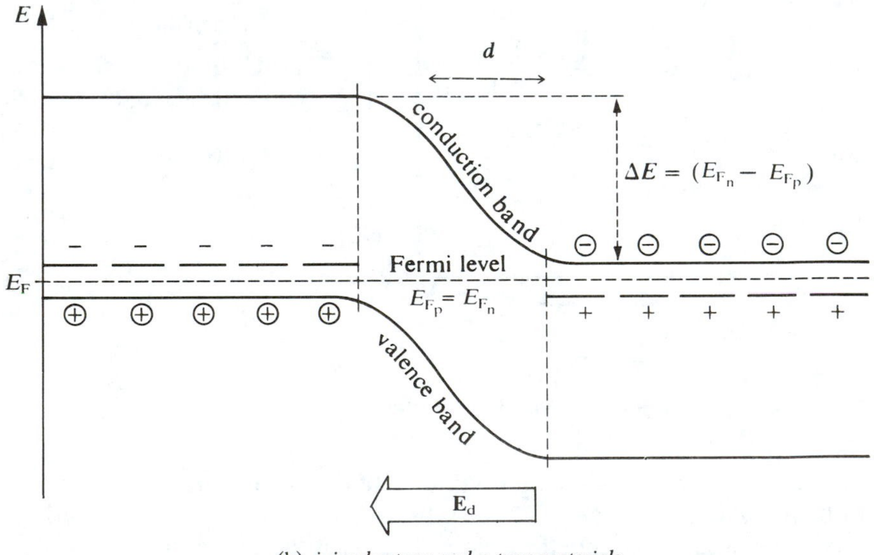

(b) *joined p-type and n-type materials*

FIGURE 4.17 Energy levels in an unbiased p-n junction.

the Fermi energy for n-type material lies just above the donor levels and in p-type material lies just below the acceptor levels. When any two materials—metal, semiconductor, or insulator—are brought together, their Fermi levels must equalize. If the Fermi levels are not initially equal, then electrons will flow from the material having the higher Fermi level into the other material until the Fermi levels are equalized, as shown in Fig. 4.17(b).

Referring to Fig. 4.17, we can draw some conclusions. The conduction band of the p-type material is higher in energy than that of the n-type material by an amount ΔE equal to the difference in the Fermi energies, which is approximately equal to the gap energy E_{gap}. The electric field $\mathbf{E}_{\text{d}}$ in the depletion region (pointing from n-type to p-type) simply arises from the slope of the bottom of the conduction band as we go across the junction, $E_{\text{d}} \cong \Delta E/ed$. Thus, an electron at the bottom of the conduction band in the n-type material must somehow get ΔE energy to climb up the potential hill and arrive in the conduction band of the p-type material. The turn-on voltage $V_{\text{turn on}}$ of the junction is approximately given by $eV_{\text{turn on}} = \Delta E$, because for any electrons to flow from the n-side to the p-side, they must attain a minimum of ΔE energy, which means a forward bias of at least $V_{\text{B}} = \Delta E/e$ volts must be applied. The reason for the larger turn-on voltage of silicon as compared to germanium p-n junctions is simply that the energy gap (and therefore ΔE) is larger for silicon than for germanium.

There are four types of current which can flow across the junction: two from majority carriers and two from minority carriers. Let J represent current densities due to majority carriers and j current densities due to minority carriers.

Any electrons in the conduction band of the n-type material are majority carriers and are due to the ionized donor atoms. They will flow over to the p-type material only if their total energy is greater than E_V, the energy of the bottom of the conduction band in the p-type material. The number of such electrons is represented by the shaded area in Fig. 4.18 and will provide a current density J_{e}, which is given by

$$J_{\text{e}} \propto \int_{E_V}^{\infty} N_{\text{cb}}(E)\, dE \tag{4.12}$$

where $N_{\text{cb}}(E)$ is the density of electrons in the n-type conduction band as a function of energy. If the junction is forward biased (see Fig. 4.18) by an applied voltage V_B, then all energy levels on the n-type side will be raised by an amount eV_B if we assume the p-side is grounded. (Grounding the p-side merely fixes all the energy levels in the p-type material.) Therefore,

$$J_{\text{e}} \propto \int_{E_V - eV_B}^{\infty} N_{\text{cb}}(E)\, dE \propto e^{-(E_V - eV_B)/kT} \propto e^{eV_B/kT} \tag{4.13}$$

Let J_{e_0} be the current density when no bias is applied ($V_B = 0$). Then J_{e} can

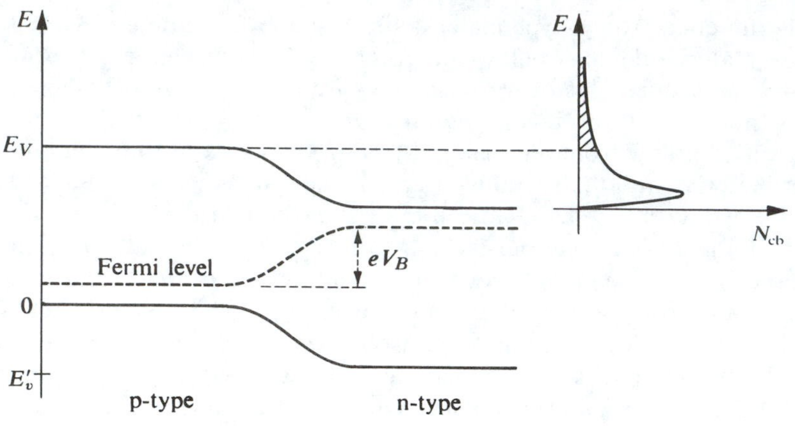

V_B = *applied forward bias voltage*
N_{cb} = *number of electrons/cm³ in conduction band*

FIGURE 4.18 Forward biased p-n junction energy level diagram.

be written simply as

$$J_e = J_{e_0} e^{eV_B/kT} \tag{4.14}$$

Any electrons in the p-type material thermally excited from the valence band to the conduction band will be minority carriers and will tend to "fall down" the potential hill, thus flowing from the p-type to the n-type material. The resulting current density j_e will be proportional only to the number of minority carriers and, hence, to the temperature so long as there is a downhill potential slope over to the n-type material. The net current flow from these two *electron* currents will then be $J_e - j_e$, where positive current means effective positive charge flowing from the p-type side to the n-type side of the junction.

Any holes in the valence band of the p-type material are majority carriers and are due to the ionized acceptor atoms. They will flow over to the n-type material only if their energy is less than E'_v of Fig. 4.18. An argument similar to that given for the majority carrier electrons in the n-type material shows that the resulting current density J_h from these majority carrier holes is given by

$$J_h = J_{h_0} e^{eV_B/kT} \tag{4.15}$$

Any holes in the valence band of the n-type material are minority carriers and are produced by thermal excitation. They give rise to a voltage-independent current density j_h because the holes spontaneously flow up the potential hill from the n-type side to the p-type side of the junction. The net current flow from these two *hole* currents is then $J_h - j_h$ with the same sign convention.

The net current density flowing across the junction is the algebraic sum of these four current densities:

$$J_{net} = J_h - j_h + J_e - j_e \tag{4.16}$$

If we now realize that the net current J must equal zero when no bias is applied ($V_B = 0$), we must have

$$J_{h_0} - j_h + J_{e_0} - j_e = 0 \tag{4.17}$$

It can be shown that the hole and electron currents must separately equal zero in equilibrium, so

$$J_{h_0} = j_h \quad \text{and} \quad J_{e_0} = j_e \tag{4.18}$$

Therefore,

$$J_{net} = J_{h_0}e^{eV_B/kT} - J_{h_0} + J_{e_0}e^{eV_B/kT} - J_{e_0} \tag{4.19}$$

Rewriting yields

$$J = (J_{h_0} + J_{e_0})(e^{eV_B/kT} - 1) \tag{4.20}$$

Letting $J_0 = J_{h_0} + J_{e_0}$, we have

$$J = J_0(e^{eV_B/kT} - 1) \tag{4.21}$$

Equation (4.21) is called the *diode equation* or the *rectifier equation* and tells us that as the forward bias $V_B > 0$ is increased, the total current passed through the junction increases rapidly with V_B. And for large reverse bias ($V_B < 0$) the current will decrease to a value of $-J_0$. For large forward bias the increase of J will be somewhat slower than implied by equation (4.21) because of the inherent resistance of the semiconductor material, which we have neglected in this treatment.

Diodes can also be made with a sharp junction between a metal and a semiconductor, the metal acting like a p-type material. Such diodes are called *point contact* diodes. They are superior to junction diodes at high frequencies but can carry far less current than either Ge or Si junction diodes.

A Schottky barrier diode is another example of a metal-semiconductor diode. It is usually made with n-type silicon and aluminum in integrated circuits and is the basis of Schottky and low-power Schottky integrated circuits, which will be discussed in Chapter 12.

The aluminum is evaporated on the n-type silicon, and a depletion region or *Schottky barrier* region is formed (in the silicon), containing only positive ionized donor atoms. The Fermi levels of the aluminum and the n-type silicon equalize, and a rectifying metal-semiconductor junction is formed with a low turn-on voltage of 0.3 V. If the doping of the silicon is very heavy, the junction is ohmic—it conducts almost equally well for both polarities of bias voltages. This technique is used in manufacturing integrated circuits.

All metal-semiconductor contacts do not act as rectifying junctions. Only extremely sharp p-n junctions will rectify. The gradual junctions found between the metal leads and the semiconductor material of diodes and transistors do not rectify; they pass current equally well in both directions. These contacts are usually made by depositing a thin metal coating on the semiconductor and soldering the thin wire to the metal coating.

Another type of diode, called the *zener* or *avalanche* or *reference* diode, can be made to break down at a specified reverse voltage V_z. The breakdown process occurs, briefly, because an electron or hole obtains enough energy (from the reverse bias electric field) between collisions to break a covalent bond in the crystal lattice and thereby create two new charge carriers, which in turn are accelerated and create more charge carriers. This process occurs very sharply at a certain voltage V_z. If one tries to increase the voltage drop across the diode above V_z, then enough additional charge carriers are produced in the diode to increase the voltage drop across whatever resistance is in series with the diode. The net result is that the zener diode draws just enough current to keep the voltage drop across it constant at essentially V_z volts. The zener diode circuit symbol and current–voltage curve are shown in Fig. 4.19. Its principal use is in regulating voltages, for if its reverse current changes from I_A to I_B, the voltage across the diode will remain essentially constant at the zener

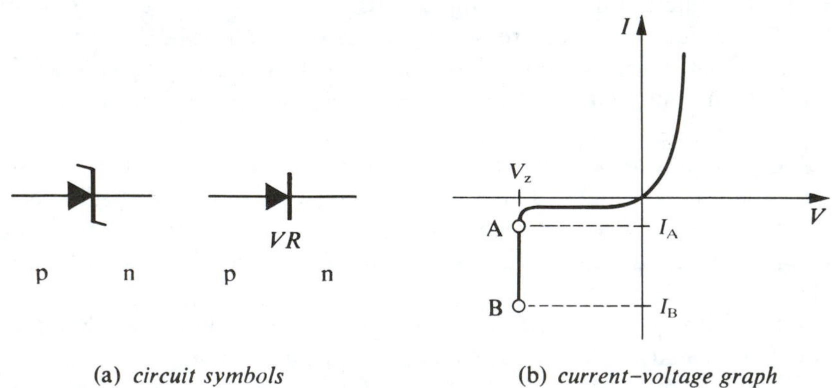

(a) *circuit symbols*　(b) *current–voltage graph*

FIGURE 4.19 The zener or voltage regulator diode.

voltage V_z. In a typical zener diode a 1-mA change in current will result in only a 1-mV change in the voltage across the diode, corresponding to a dynamic resistance of only about 1 Ω. Special zener diodes can be made with a very low temperature coefficient to provide an extremely stable reference voltage.

If the amount of impurities is increased to the order of 10^{19} cm^{-3}, then the impurity energy levels merge to form a small band, and the diode can conduct due to a quantum-mechanical "tunneling" process for small forward biases on the order of several tenths of a volt. The tunneling falls to zero for larger forward biases, yielding the current–voltage curve shown in Fig. 4.20. This device is, appropriately, called a *tunnel* diode and is useful

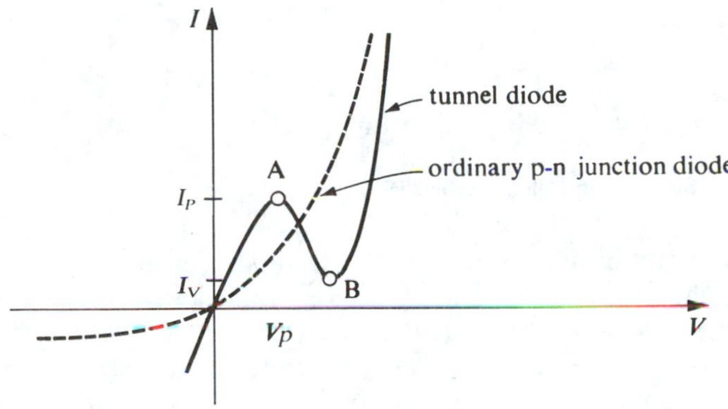

FIGURE 4.20 Tunnel diode current–voltage curve.

not as a rectifier, but because it exhibits a negative dynamic resistance from point A to point B on the current–voltage curve. The peak current I_P is typically from 1 to 10 mA; the valley current I_V is 1 mA or less, and V_P is only about 50 mV. Oscillators and amplifiers utilizing this negative resistance phenomenon can be made. Because the tunneling occurs with majority rather than minority carriers, it can be shown that the tunnel device will operate at extremely high frequencies—up to 10^{10} Hz (10 GHz), which is well up in the so-called microwave or radar region. The tunnel diode is also useful because from points 0 to A its current increases more rapidly with voltage than for ordinary p-n diodes.

Another useful semiconductor device is the *silicon controlled rectifier* (SCR). The SCR is a four-layer device (p-n-p-n or n-p-n-p) with three terminals called the *anode*, the *cathode*, and the *gate*, as shown in Fig. 4.21. The SCR will not conduct in the forward direction until the anode–cathode voltage V_{AC} exceeds a certain value V_C, which depends on the (small) gate current I_G. Once the SCR conducts or "fires," V_{AC} drops to a very small value, and the SCR current I_A is independent of the gate current—the SCR acts just like a forward biased diode. The SCR current I_A drops to zero only

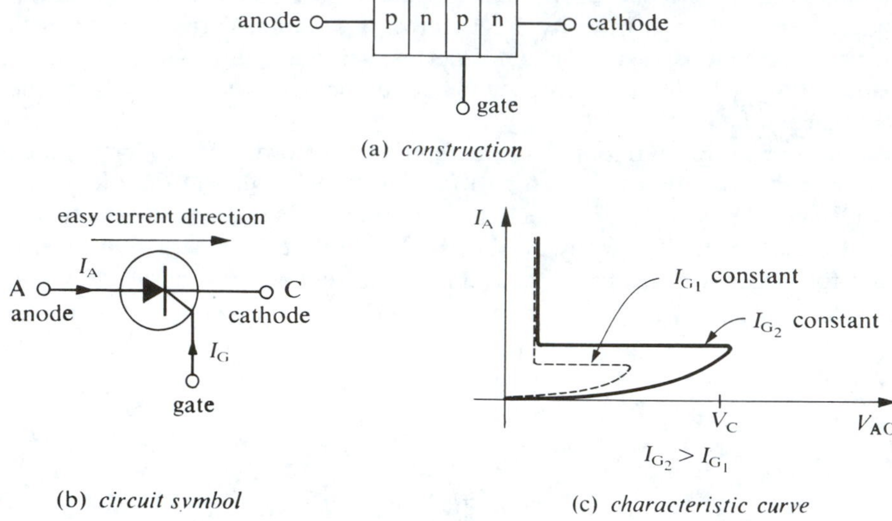

FIGURE 4.21 Silicon controlled rectifier (SCR).

when the polarity of V_{AC} is reversed. The diode can be made to conduct again only when the anode is made positive with respect to the cathode; the exact value that will fire the diode depends on the gate current. The SCR is commonly used to control high-power ac circuits such as electric motors. One SCR can control currents of tens of amperes. The SCR is a solid-state

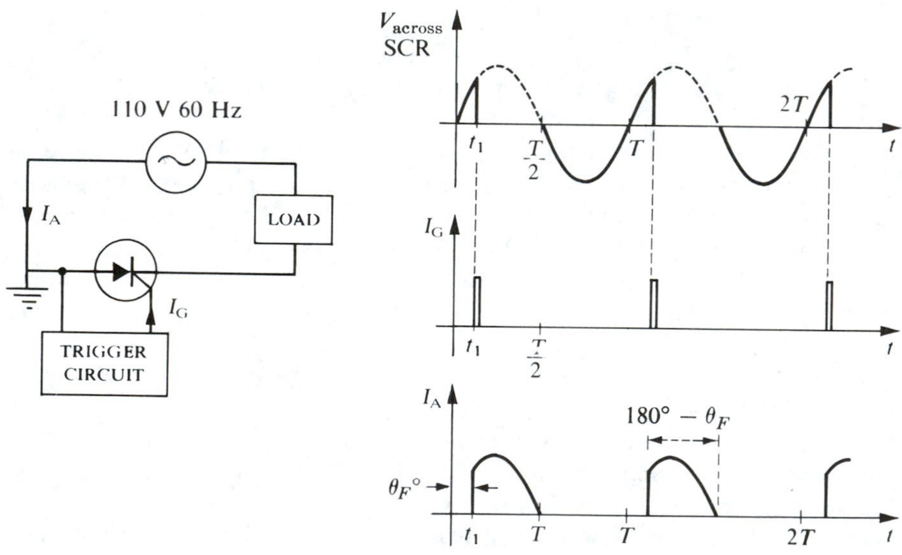

FIGURE 4.22 SCR operation.

equivalent of a thyratron vacuum tube. A gate current of 50 mA can turn on an SCR, which can then carry a current of 20 A.

The basic operation of an SCR in controlling an ac current is to fire the SCR at various phases of the ac voltage across the SCR, as shown in Fig. 4.22. If a trigger circuit supplies a pulse of current to the SCR gate at t_1, the SCR will fire and conduct so long as its anode voltage remains positive with respect to its cathode. The phase angle corresponding to t_1 is often called the *firing angle*, θ_F; the number of electrical degrees during which the SCR conducts equals $180° - \theta_F$. The trigger pulse must last approximately 30 μs or more. The time t_1 of the trigger pulse will determine at what voltage the SCR will fire. The current passed by the SCR will then last from t_1 to $T/2$ on each cycle. By varying the time t_1 at which the trigger pulses occur, the average current passed by the SCR can be adjusted. SCR trigger circuits commonly use small neon lamps that will not conduct until the voltage across the lamp terminals reaches approximately 80 V. One such circuit is shown in Fig. 4.23. The capacitor C is charged up

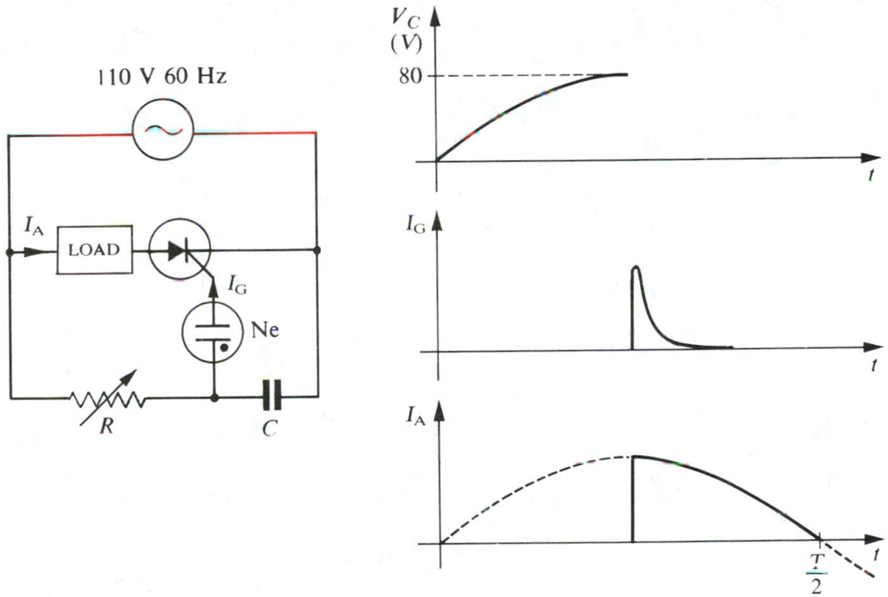

FIGURE 4.23 SCR controlled load.

positively through R by the 110-V, 60-Hz supply. When V_C reaches the firing voltage of the neon lamp, it fires and C is discharged through the SCR gate, thus providing a pulse of gate current to turn on the SCR. The capacitor should be large enough so that the I_G pulse is large enough to turn on the SCR even for a very low anode–cathode voltage, which would be the situation when the firing angle is close to 180°. When the 110-V ac reverses polarity, a gate pulse is again applied to the SCR, but the SCR does not fire because its anode is now negative with respect to its cathode. As R is increased, it takes a longer time for C to charge up to the neon bulb firing voltage, and θ_F is larger, resulting in a smaller average current through the load.

The *triac* is useful to control ac current. It is basically two SCRs connected so that it can conduct current in either direction, unlike the SCR. The triac symbol and characteristic curves are shown in Fig. 4.24. The two triac terminals are called main terminal #1 (MT1) and main terminal #2 (MT2) instead of anode and cathode. When the triac gate is triggered by a brief current pulse I_C (longer than approximately 50 μs), it

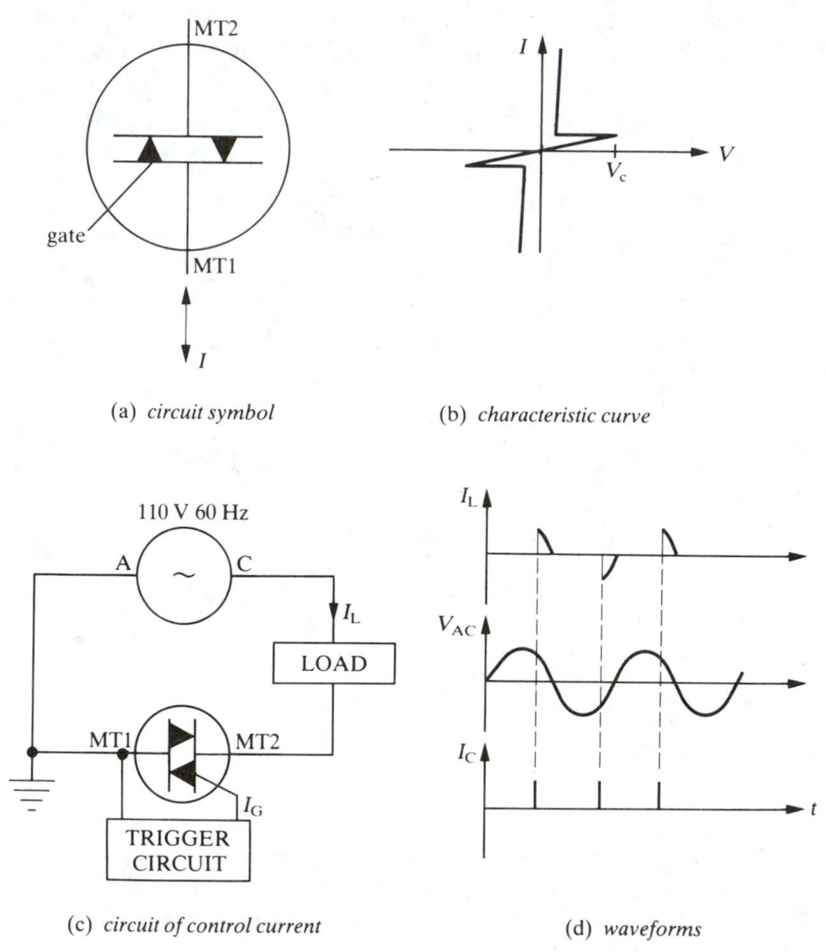

(a) *circuit symbol* (b) *characteristic curve*

(c) *circuit of control current* (d) *waveforms*

FIGURE 4.24 The triac.

conducts current regardless of the polarity of the voltage between MT1 and MT2. The current drops to zero only when this voltage drops to zero, as shown in Fig. 4.24(d). Thus, the phase of the gate current pulse relative to the phase of the ac voltage determines the average current through the load.

4.10 THE PHOTOVOLTAIC DIODE

A p-n junction can be used to convert the solar energy in sunlight directly into electrical energy; the device is called a *photovoltaic diode* or a *photovoltaic cell* or a *solar cell*. Many of the space satellites presently in orbit use arrays of photovoltaic cells to supply power to their electronic circuits.

The basic operation of a solar cell is simply that a photon is absorbed by a semiconductor atom, and an electron is raised across the energy gap from the valence band to the conduction band, leaving a hole behind in the valence band. For silicon the energy gap is 1.09 eV, so the minimum photon energy for this type of absorption is 1.09 eV, corresponding to a photon wavelength of 1140 nm, which is in the near infrared. All visible photons are more energetic than this because the visible range of wavelengths is from approximately 350 to 700 nm.

The electric field in the depletion region of the junction then accelerates the electron and hole and thereby produces a *photocurrent* I', which is a *reverse* current, flowing out of the p material, as shown in Fig. 4.25(a). Neglecting recombination of the electrons and holes (which is appreciable in real cells), we determine the cell current to be

$$I' = I_0(e^{eV/kT} - 1) - I_L$$

where I_L is the light-induced current.

The maximum voltage of a silicon cell is 1.09 V, the gap energy in volts, because any electron excited by photon absorption to a level above the bottom of the conduction band quickly returns to the bottom of the conduction band by emitting phonons (quanta of vibrational energy). These low-energy phonons produce heat not electrical energy. A practical silicon solar cell has an open-circuit voltage V_{oc} of approximately 0.7 V.

The maximum theoretical power efficiency of a silicon single-crystal solar cell is 27%; real commercial cells have a 12%–14% efficiency. The incident solar-power density at the earth's surface on a clear sunny day (winter or summer) is approximately 900 W/m^2, so a 12% efficient cell would give approximately 108 W power for each square-meter area. Processes contributing to the inefficiency are reflection from the silicon surface (30% from untreated silicon, much less from a surface with an antireflection coating), recombination of electron-hole pairs, I^2R losses due to bulk silicon resistance and contact resistance, phonon production, blocking of

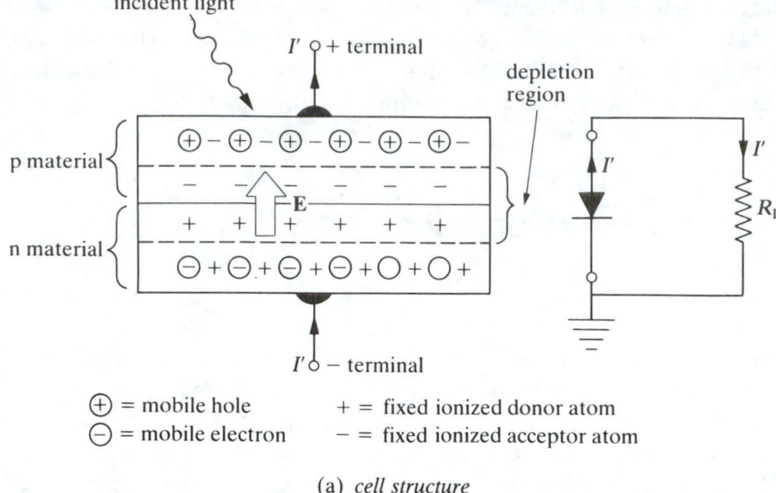

(a) *cell structure*

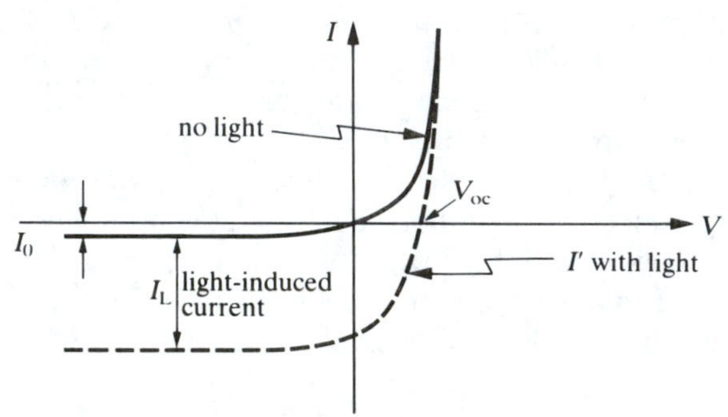

(b) *cell current* 1-*voltage curve without and with illumination*

FIGURE 4.25 Photovoltaic cell.

some incident light by the electrical connections at the top surface of the cell, and thinness of the cell (i.e., too thin to absorb all the incident light).

The physics of such a cell is rather complicated. The incoming photons are exponentially absorbed, so a thick cell is better. But a thick cell is much more expensive because of the expense of the high-purity single-crystal material, and recombination can be higher in thick cells. There is a continuous recombination of the electron-hole pairs, so an equilibrium is reached between the photoproduction of the pairs and the recombination. The electrons and holes move by diffusion outside the depletion region, and they drift in response to the electric field inside the depletion region. Finally, Poisson's equation must be satisfied.

An interesting recent development is the manufacture of inexpensive *amorphous* silicon solar cells. The amorphous silicon does not have the long-range crystalline order of a single crystal, and small microcrystals are separated from each other by small *microvoids*. The result of this structure is that there are many (unwanted) energy levels in the forbidden 1.09-eV gap, thus rendering the material useless for a good p-n junction solar cell.

However, when an amorphous silicon film from a glow discharge of SiH_4 is deposited on a substrate, some of the hydrogen is trapped in the amorphous silicon film and effectively quenches or saturates the unwanted energy levels in the gap. The resultant energy level picture is similar to that of single-crystal silicon. Such amorphous silicon can be doped in the usual way to form a p-n junction that can be made into a solar cell. Amorphous silicon solar cells have recently been made with about the same efficiency as single-crystal silicon cells and are widely used to power small solar-powered pocket calculators. Considerable research in amorphous cell technology is presently being conducted.

4.11 DIODE APPLICATIONS

One of the most common uses of the diode is in power supplies where the standard 110-V, 60-Hz ac line voltage is converted into a dc voltage. A simple diode-resistance circuit shown in Fig. 4.26(a) will convert the input ac voltage $v_{in} = V_{AB} = V_0 \sin \omega t$ into a pulsating dc voltage. The explanation is simply that the diode conducts only on the positive half-cycles and

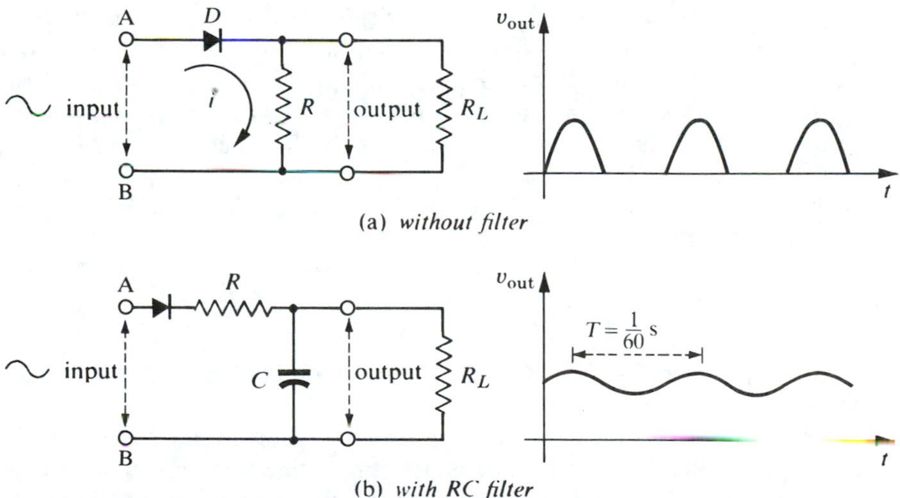

FIGURE 4.26 Half-wave diode rectification.

not on the negative half-cycles of the input. We assume in this section that the diode has an infinite resistance in the reverse direction and has zero turn-on voltage. Thus the current i passed by the diode is unidirectional; that is, it is pulsating direct current always flowing in the direction shown in Fig. 4.26(a). The positive output voltage is the iR voltage drop across R. Notice that when the 60-Hz input voltage makes point A negative with respect to point B, the diode is reverse biased, and essentially all the 60-Hz voltage appears across the diode because $i = 0$. For the circuit to function properly the diode must be capable of withstanding this reverse voltage without breaking down. The appropriate diode rating is the *peak reverse voltage* (PRV) or *peak inverse voltage* (PIV). Modern silicon rectifier diodes come with PIV ratings up to 500 V or more. Diodes are also rated according to how much current they may safely pass when forward biased. The maximum current rating usually refers to the average dc current that the diode may safely pass. The peak or surge current rating is generally much higher and refers to the maximum peak current the diode may safely pass in a very short interval of time. Typical modern silicon rectifier diodes cost less than $1 each, can carry average currents of one ampere and peak currents of tens of amperes, and have peak inverse ratings of 500 V.

If a steady dc output voltage is desired, an RC low-pass filter is added, as shown in Fig. 4.26(b). The resistance R also limits the current surge through the diode when the circuit is turned on, when C is completely uncharged. The capacitance C is charged and serves to smooth out the amplitude variations in the output. Another view of the filter is that it passes dc and the lower-frequency Fourier components of the pulsating voltage of Fig. 4.26(a) and attenuates the higher-frequency components. The lower the breakpoint frequency $\omega_B = 1/RC$, the more effective is the filtering. Thus, the larger RC is, the better the filtering will be. The remaining ac variation in the output voltage is called *ripple*. Too large a resistance R cannot be used because the output dc voltage will fall if an appreciable dc current is drawn from the output. For example, if $f = \omega/2\pi = 60$ Hz, then we desire $\omega_B = 1/RC \ll 2\pi \times 60$ or $1/RC \ll 120\pi = 377$. The maximum dc output current usually fixed an upper limit for R. Thus, if we desire the output voltage to fall by less than 1 V as I_{out} increases from 0 to 100 mA, then the maximum voltage drop across R is $(I_{out\,max})R = 1$ V or $R_{max} = 1\text{ V}/100\text{ mA} = 10\ \Omega$. Thus C is determined from

$$C \gg \frac{1}{377R} = \frac{1}{3770} = 2.65 \times 10^{-4}\text{ F} = 265\ \mu\text{F}$$

We would therefore try to use $C = 1000\ \mu$F or $2000\ \mu$F if the budget permits.

It should also be noted that a larger dc output current implies a smaller load resistance R_L. When the diode is not conducting, the capacitance C is discharging through R_L. Thus, the smaller R_L, the more rapid the discharge of C and the more the output voltage falls before the next positive

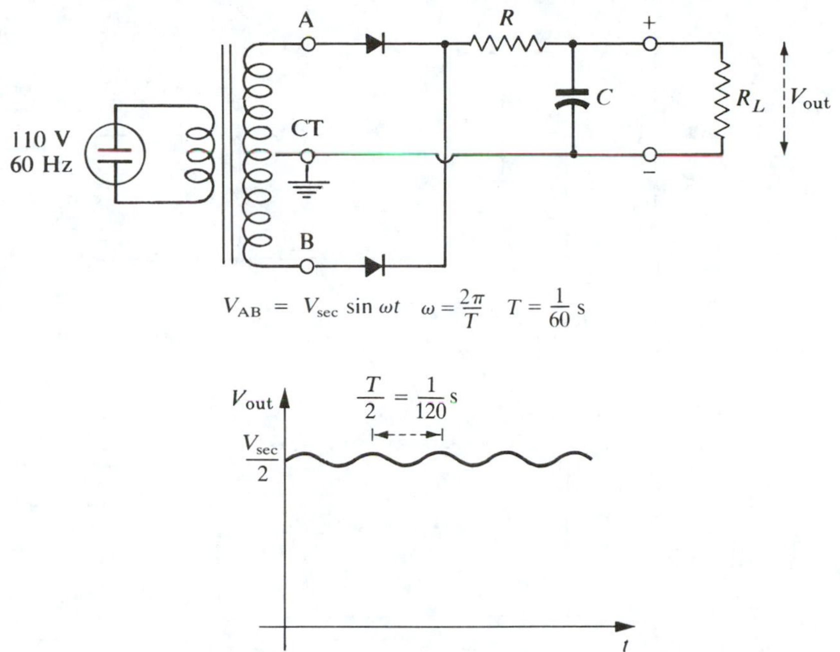

FIGURE 4.27 Full-wave rectifier.

half-cycle, when D conducts and charges C again. In other words, the larger the dc output current, the larger the ripple.

Practical power supply circuits are considerably more complicated than the half-wave circuit of Fig. 4.26. Usually, both halves of the input ac voltage are utilized, either in a full-wave circuit shown in Fig. 4.27 or in the full-wave bridge circuit of Fig. 4.28. Usually a transformer is used to change the 110-V, 60-Hz line voltage to the approximate desired output voltage. V_{AB} is the secondary voltage of the transformer. Notice that the half-wave circuit ripple is 60 Hz, while the full-wave ripple is 120 Hz. Also, the full-wave circuit requires a center tap "CT" on the transformer, whereas the full-wave bridge circuit does not. For the same transformer secondary voltage the half-wave and the full-wave bridge circuits give twice the output voltage that the full-wave circuit does. More current can be drawn from the half-wave circuit than the full-wave circuit without overheating the transformer because current flows through the half-wave circuit transformer only during 50% of the time; that is, the *duty cycle* is 50%. A complete schematic for a practical regulated power supply is given in Appendix A.

A simple AM detector can be made with a diode as shown in Fig. 4.29. The diode rectifies the amplitude-modulated input so that only the positive peaks appear at the input to the low-pass RC filter. If the time constant of the filter is chosen so that $T_c \ll RC \ll T_m$, the carrier oscillations will be

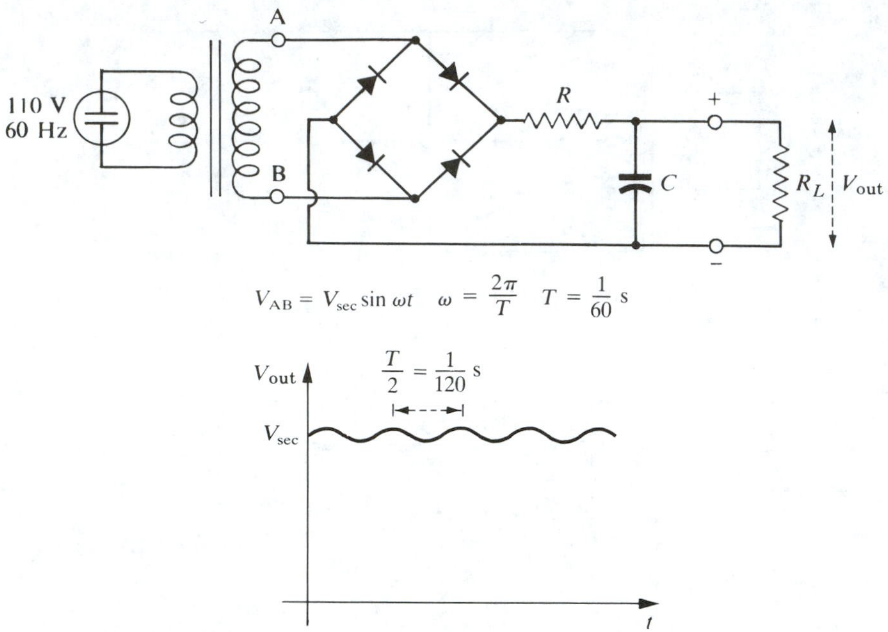

FIGURE 4.28 Full-wave bridge rectifier.

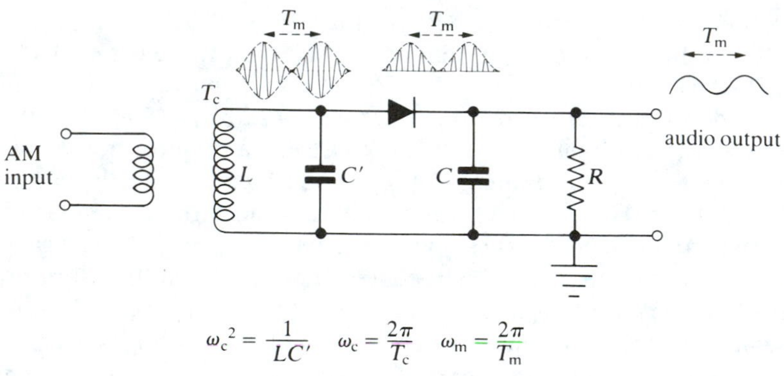

FIGURE 4.29 AM diode detector.

filtered out but not the modulation. Thus, the output voltage across C will follow the audio modulation envelope of the AM wave; the output will be the desired audio. For AM radio the carrier frequency is from 550 to 1500 kHz, so $T_c \cong 1\ \mu s$. The audio modulation is from 20 Hz to 10 kHz, so T_m is from 50 ms to 100 μs. Thus, a reasonable choice for the time constant would be $RC \cong 10\ \mu s$.

A simple FM (frequency modulation) detector can be constructed from diodes using the circuit of Fig. 4.30. The top half of the transformer

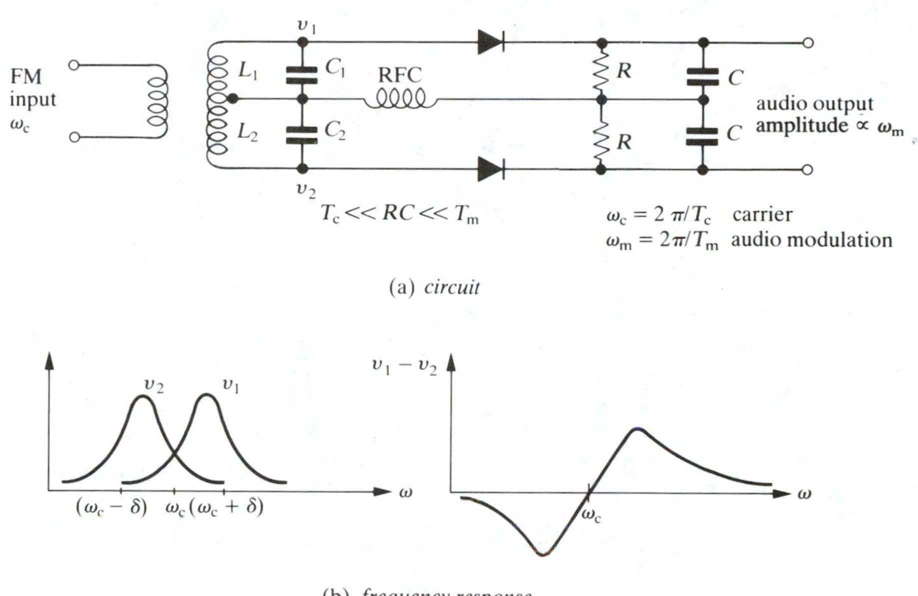

FIGURE 4.30 FM discriminator or detector.

secondary (L_1 and C_1) is turned to a frequency slightly higher than the carrier frequency ($\omega_c + \delta$), and the bottom half (L_2 and C_2) is tuned to a slightly lower frequency ($\omega_c - \delta$). The response of the tuned transformer secondary is shown in Fig. 4.30(b). The difference voltage ($v_1 - v_2$) across the output terminals is proportional to ($\omega - \omega_c$), that is, to how far the frequency has been modulated from the unmodulated carrier frequency ω_c. Thus, if RC is large enough to filter out the carrier ($RC \gg T_c = 2\pi/\omega_c$) and small enough to pass the audio modulation ($RC \ll T_m = 2\pi/\omega_m$), the output ($v_1 - v_2$) will be proportional to the audio modulation signal. In other words, the FM signal will have been *demodulated*.

Another common use of diodes is in clipping circuits where the amplitude of a signal must be limited or clipped. In Fig. 4.31(b) the diode conducts only on the negative half-cycle of the input; thus for the negative portion of the input the output is limited to the turn on voltage. The positive portion of the input is passed through to the output, provided only that $R \ll R_r$, where R_r is the reverse resistance of the diode. If a battery V_{bb} is added, as shown in Fig. 4.31(c), then the diode will not start to conduct

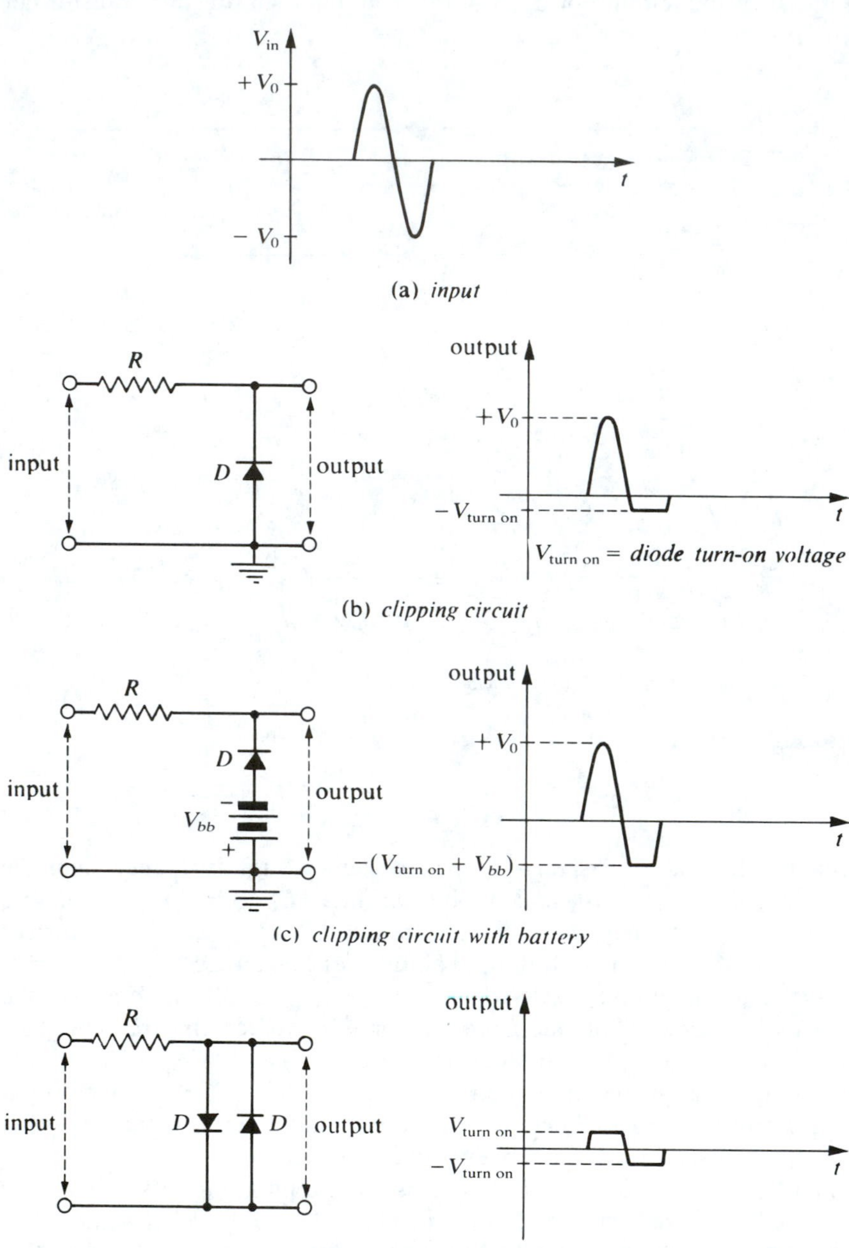

FIGURE 4.31 Diode clipping circuits.

until the input is more negative than $-(V_{bb} + V_{\text{turn on}})$. Thus the negative excursion of the output is clipped off at $-(V_{bb} + V_{\text{turn on}})$ volts. The circuit in Fig. 4.31(d) will pass voltages of either polarity only if the amplitude is less than approximately the diode turn-on voltage; that is, this circuit rejects large-amplitude inputs and passes small-amplitude inputs.

High-voltage transformers are expensive and impractical at voltages much above several thousand volts. A useful diode power supply circuit with no high-voltage transformer is a voltage doubler, shown in Fig. 4.32.

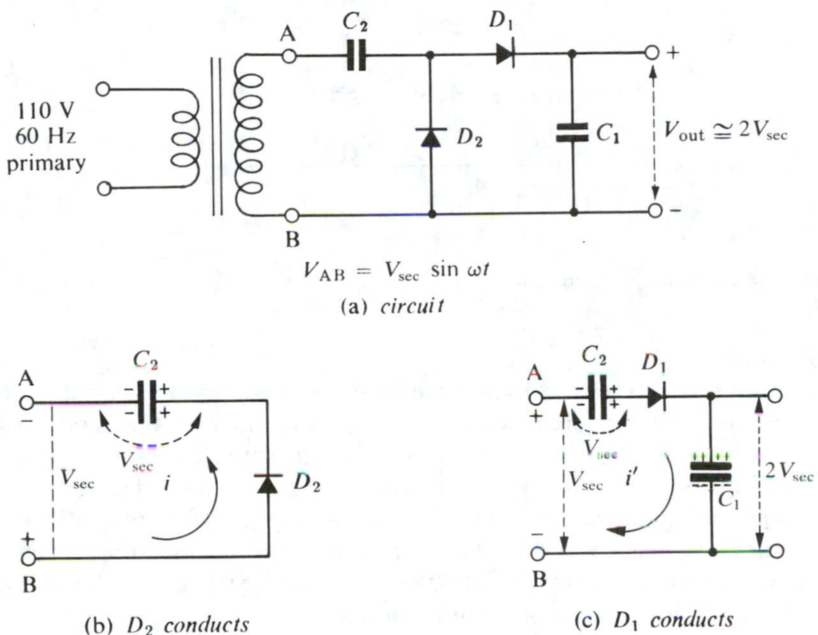

FIGURE 4.32 Voltage doubler.

The dc output voltage will be equal to twice the zero-to-peak input voltage between terminals A and B, which is usually the voltage from the secondary of a transformer. This can be seen by the following argument. If A is *negative* with respect to B, then diode D_2 conducts and diode D_1 does not. Current i flows through D_2 to charge up C_2, as shown in Fig. 4.32(b). The voltage across C_2 will be V_{sec}, the peak secondary transformer voltage. On the next half-cycle when A is *positive* with respect to B, D_1 conducts and D_2 does not. Thus current i' flows as shown in Fig. 4.32(c). C_1 is thus charged up to a voltage $2V_{sec}$ because the peak transformer secondary voltage $V_{AB} = V_{sec}$ and the voltage on $C_2 = V_{sec}$ *add*. C_2 can be an electrolytic capacitor because the voltage drop across it is always of the same polarity. The value of V_{sec} depends on the turns ratio of the transformer; it can be up to several kilovolts.

Higher output voltages can be achieved with similar circuits. A voltage tripler is possible, and a voltage quadrupler is shown in Fig. 4.33. In

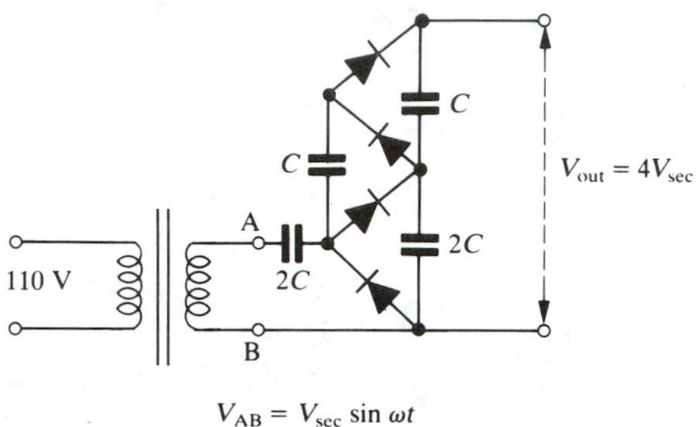

FIGURE 4.33 Voltage quadrupler.

general, these circuits can be extended to yield higher output voltages limited only by the breakdown of the capacitors and the diodes (when reverse biased). The first successful atom-smashing experiment was performed in 1932 by Cockroft and Walton in the Cavendish Laboratory, England, using a voltage multiplier containing vacuum tube diodes and capacitors as the source of high dc voltage. Protons were accelerated through a voltage drop of 250,000 V dc and struck a lithium target, producing disintegration of the lithium nuclei.

A simple diode clipping circuit provides for dc baseline restoration as mentioned in Chapter 2. Recall that when a positive pulse is passed through an RC coupling network (an RC high-pass filter), the output pulse area above and below the zero voltage axis must be equal. For negligible distortion of the pulse shape, the time constant of the circuit should be much larger than the pulse width T. The output pulse, shown in Fig. 4.34(a), then has a long negative tail that decays with the characteristic time constant RCs. If many pulses come along at the input in a time of the order of RCs or less, then the cumulative effect of all the negative pulse tails is to appreciably depress the dc voltage level or baseline of the output as shown in Fig. 4.34(b). The addition of a diode across the output as shown in Fig. 4.34(c) and (d) will essentially cure the problem by providing a low-impedance path to ground for negative outputs. There are many other possible diode clipping or limiting circuits. Several diode circuits useful in computers are discussed in Chapter 12.

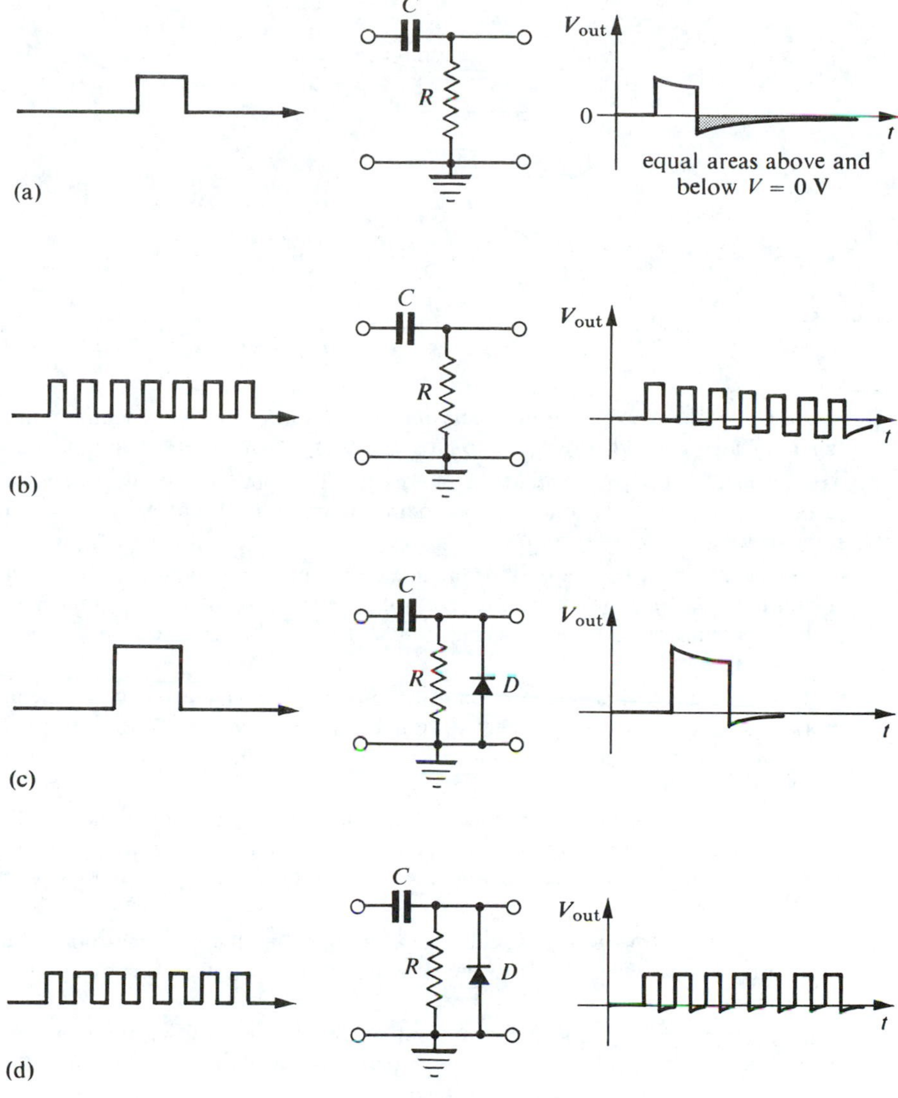

FIGURE 4.34 Elementary diode baseline restoration circuit.

PROBLEMS

1. A new atom, confusium, is discovered with the following energy level diagram. How much work must be done (how much energy must be expended) to ionize a confusium atom in the ground state?

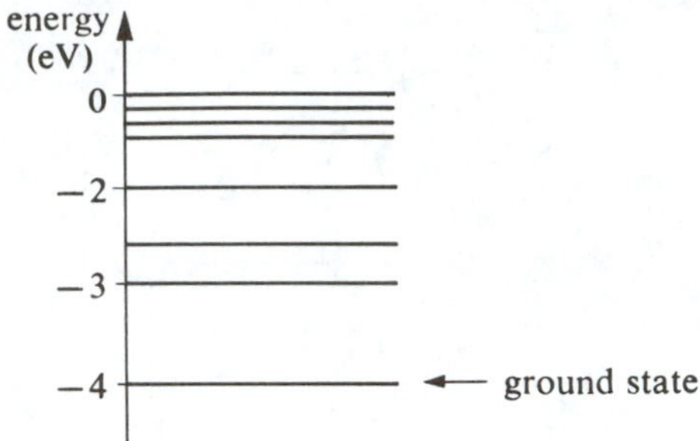

2. In the energy level diagram of confusium in Question 1, at what temperature (approximately) would you expect to have the energy level at −3 eV populated?
3. Approximately how large would the energy gap in a pure semiconductor have to be for it to be a nonconductor at room temperature? What is the physical significance of the energy gap?
4. Calculate an approximate speed for (a) an electron in thermal equilibrium with a silicon lattice at room temperature, and (b) an electron in thermal equilibrium with a germanium lattice at room temperature. The electron mass $m = 9 \times 10^{-31}$ kg.
5. Calculate an approximate speed for the random thermal motion of a silicon atom (atomic weight = 28) in thermal equilibrium at room temperature. Repeat for germanium (atomic weight = 73) at room temperature. 1 amu = 1.7×10^{-27} kg.
6. Calculate the total number of mobile charge carriers in a pure semiconductor if 10^{10} valence bonds/cm^3 are broken on the average due to thermal agitation. Does one distinguish between majority and minority charge carriers in a pure semiconductor?
7. At roughly what temperature would you expect large numbers of electrons to be excited from the valence band to the conduction band in pure silicon? In pure germanium?
8. A silicon crystal is doped with pentavalent arsenic. Is the resulting semiconductor p-type or n-type? If the crystal is doped with trivalent phosphorus, is the resulting semiconductor p-type or n-type?
9. Calculate the average distance between arsenic doping atoms in a silicon crystal if there are 10^{16} As atoms per cm^3.
10. Briefly explain physically why the donor atom energy levels in an n-type semiconductor lie so close to the conduction band.
11. Briefly give a physical interpretation of the Fermi energy. Why can't the Fermi energy lie in the conduction band for a pure semiconductor? What does the Fermi factor mean physically?
12. Briefly explain why the Fermi energy cannot lie below the donor energy levels in an n-type semiconductor.

13. Calculate an approximate temperature at which most of the donor atoms would be ionized in a doped n-type semiconductor with the following energy level diagram.

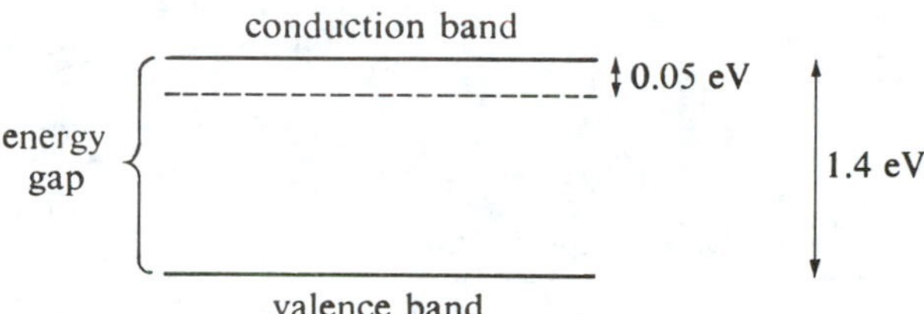

14. At roughly what temperature would you expect a conventional silicon transistor to cease operating as it is cooled down? Explain briefly.
15. Explain why a positively ionized donor atom does not contribute to conduction in an n-type semiconductor.
16. In an abrupt p-n junction with no applied bias, why will there be a small depletion region formed at the junction?
17. Why does an n-type doped semiconductor have many more mobile negative charge carriers than positive carriers? Give typical approximate numbers for silicon and germanium.
18. Carefully state the Pauli exclusion principle. Why does this principle imply that a completely filled band cannot conduct?
19. Describe in words and sketch a diagram of the motion of minority charge carriers in a reverse biased abrupt p-n junction. Include the depletion region.
20. Distinguish among a metal, a semiconductor, and an insulator on the basis of the location of the Fermi level.
21. Set up (do not evaluate) an expression for the number of electrons/cm^3 in the conduction band of an n-type semiconductor with energies more than 0.1 eV above the bottom of the conduction band. Define all symbols carefully.
22. Sketch the energy level diagram (to scale) for an abrupt p-n silicon junction with a 0.3-V forward bias applied. Include the positions of the Fermi levels and the various bands.
23. Sketch a full-wave diode rectifier circuit to produce a −12-V dc output voltage relative to ground. Include approximate numerical values for the transformer and the other components used.
24. Repeat Problem 23, using a full-wave bridge rectifier circuit.
25. Design a diode clipping circuit to clip off negative-going pulses so that the output is never more negative than −3 V.
26. Sketch the output waveform to scale for the following.

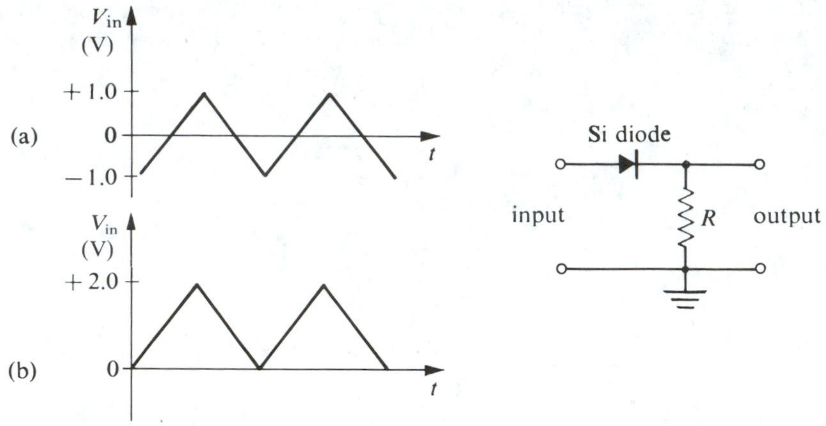

27. Sketch the output waveform to scale for the following.

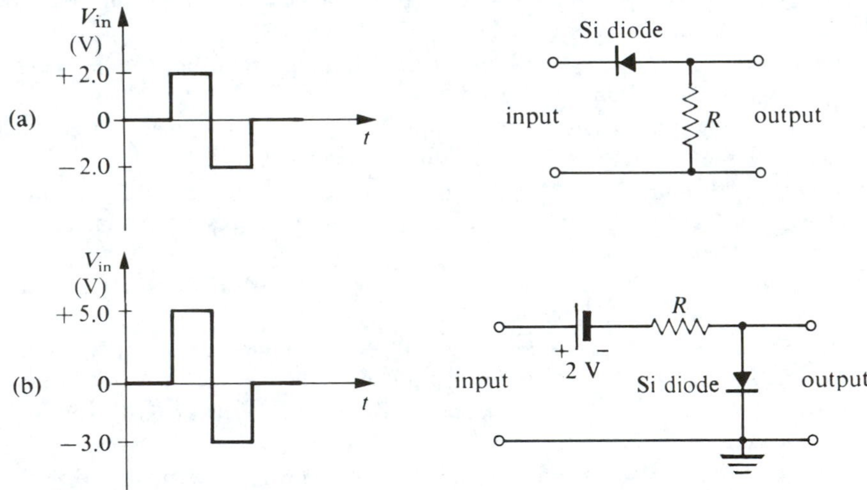

28. Explain why the ripple amplitude in a power supply increases as the load current drawn from it is increased.

29. Diagram a simple half-wave power supply with *RC* filter to supply an output voltage of approximately −5 V. Specify the transformer secondary voltage and the *RC* values.

30. Repeat Problem 29 for a full-wave power supply to supply an output voltage of +5 V.

31. (a) Approximately how large an area of solar cells would be required to charge up a 12-V battery with a charging current of 1.0 A? Assume a typical solar cell power efficiency of 12% and 0.7 V per solar cell. [*Hint*: You will need a charging voltage of approximately 14 V to charge a 12-V battery.] (b) Would the solar cells be hooked up in series or in parallel? How much area would each cell have?

CHAPTER 5

The Bipolar Transistor

5.1 INTRODUCTION

In this chapter we will consider the physical construction of the bipolar transistor, explain transistor properties on the basis of how forward biased and reverse biased p-n junctions work, and design a common emitter transistor amplifier circuit.

5.2 TRANSISTOR CONSTRUCTION

A transistor consists of three layers of doped semiconductor material in the order p-n-p, which is called a *pnp* transistor, or in the order n-p-n, which is called an *npn* transistor. Figure 5.1 shows the construction and the circuit symbols used for pnp and npn transistors. The center layer is always called the *base*. The outside layers are called the *emitter* and the *collector*. They look identical in Fig. 5.1, but they are not. Briefly, the collector is thermally connected to the outside world for better heat dissipation, whereas the emitter is not; also, the emitter is more heavily doped.

The boundary between the different types of semiconductors must be abrupt; that is, the type of impurity atom must suddenly change from donor to acceptor as we go from n-type to p-type and from p-type to n-type. Also, the crystal lattice structure must be undistorted or unbroken as we go from p-type to n-type material, or vice versa. One cannot make a transistor by gluing or clamping together separate pieces of doped semiconductor because there would be too many imperfections in the crystal structure at the interface. The entire transistor must be grown from one single crystal with the type of impurity changed abruptly to create the p-n junctions.

Starting with an n-type single crystal of silicon for the collector, a group III p impurity atom such as boron is diffused downward from the top to form the p-type base. Then a group V n impurity atoms such as phosphorus is diffused downward to form the n-type emitter. A thin layer of insulating SiO_2 is formed on the top by exposing the silicon to oxygen gas at a high temperature. A stencil of organic polymer deposited directly on the top surface of the silicon determines where the impurity atoms diffuse

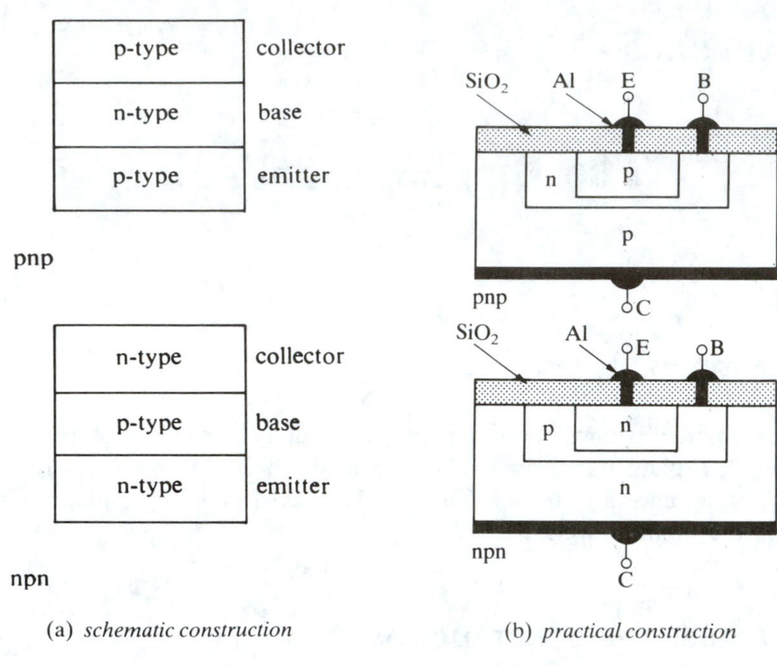

(a) *schematic construction* (b) *practical construction*

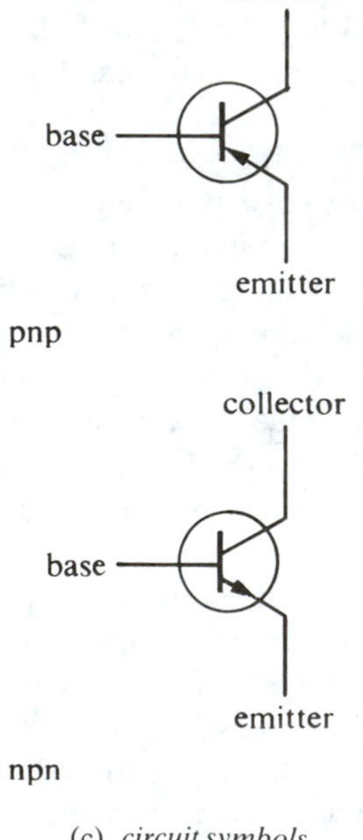

(c) *circuit symbols*

FIGURE 5.1 Transistor construction and circuit symbols.

into the silicon crystal. Finally, metallic aluminum is deposited to form the connections to the emitter, base, and collector. The resultant npn bipolar transistor is called a *vertical diffused planar transistor* and is shown in Fig. 5.1(b).

An alternate technique is to grow a thin epitaxial p-type layer on top of the n-type collector crystal. Epitaxial means there is no discontinuity in the crystal lattice structure as we go from the n-type to the p-type material. Epitaxial growth can be controlled more precisely than diffusion, so an epitaxial base can be made thinner than a diffused base. The base thickness in an epitaxial transistor can be as small as $20\ \text{nm} = 2 \times 10^{-8}\ \text{m}$, producing a high-gain, high-speed transistor.

An ohmmeter can be used to distinguish between pnp and npn transistors by measuring the resistance between the base and the emitter for different polarities of the ohmmeter probes. In most ohmmeters the red probe is positive in voltage with respect to the black probe. Hence for a pnp transistor placing the red lead on the emitter lead and the black lead on the base lead should forward bias the base-emitter junction and yield a low resistance, somewhere between 1 and $10\ \Omega$. If the leads are reversed, the measured resistance should be roughly a thousand times higher; on the $R \times 1$ scale the resistance should appear infinite. For an npn transistor the black lead on the emitter lead and the red lead on the base lead should yield a low resistance; reversing the leads should give a high resistance. Similar arguments apply to the base-collector and collector-emitter resistances.

A transistor "burned out" or damaged by excessively high voltage can also be identified with an ohmmeter by a similar procedure. Measure *six* resistances to be absolutely sure: the base-emitter, base-collector, and collector-emitter resistances, each with two polarities of the ohmmeter leads on the $R \times 1$ scale. A good transistor will have the readings of Table 5.1. R_{BE} means the base positive with respect to the emitter, R_{EB} the emitter positive with respect to the base; the other junctions have similar designations. Both R_{CE} and R_{EC} are high for any transistor because in either case there is one reverse biased junction.

Because the operation of the transistor depends upon the nonlinear

TABLE 5.1. R × 1 Ohmmeter Scale Readings for a Good Transistor

	R_{BE}	R_{EB}	R_{BC}	R_{CB}	R_{CE}	R_{EC}
pnp	high	low	high	low	high	high
npn	low	high	low	high	high	high

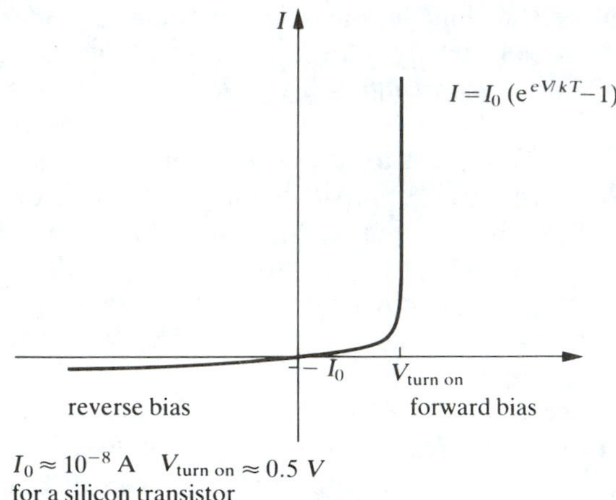

FIGURE 5.2 p-n junction current–voltage graph.

characteristics of the p-n junctions, we repeat the p-n current–voltage curve here for convenience in Fig. 5.2.

5.3 BIASING AND CURRENT FLOW INSIDE A TRANSISTOR

From bottom to top in Fig. 5.1(a), the three layers of semiconductor material are called the *emitter*, the *base*, and the *collector* for both the pnp transistor and the npn transistor. We will now confine our attention to a pnp silicon transistor. The two p-type regions are not exactly identical, as we will see later; hence they go by different names.

Let us first consider the problem of biasing the junctions. In this section *bias* or *bias voltage* means a dc voltage applied to the transistor at all times; this voltage is necessary for the transistor to function properly. The base-emitter junction is biased in a *forward* direction; that is, the p-type emitter is positive in voltage with respect to the n-type base. Thus, a large current flows from the emitter into the base. This could be accomplished by connecting a battery of voltage V_{bb1} and a resistor R_1 as shown in Fig. 5.3(a). The base-collector junction is biased in a *reverse* direction; that is, the n-type base is positive with respect to the p-type collector. A battery of voltage V_{bb2} (typically from 3 to 30 V) and a resistor R_2, also shown in Fig. 5.3(a), would be suitable for this bias. A more common and convenient arrangement is to connect the battery V_{bb2} and resistor R_2 between the emitter and the collector as shown in Fig. 5.3(b) and (c). This arrangement is more convenient because the two batteries have a common point that can

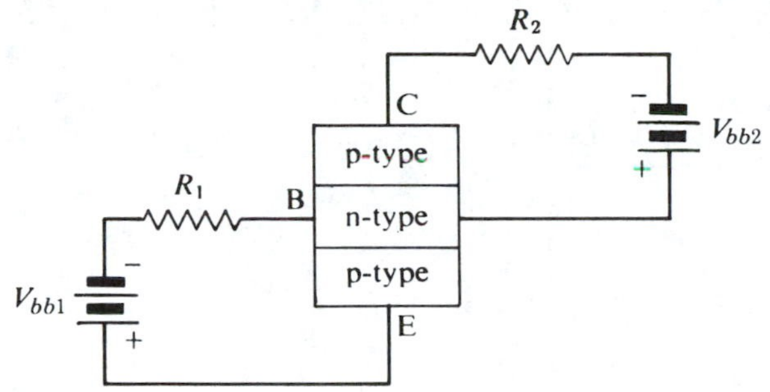

(a) pnp *transistor bias circuit*

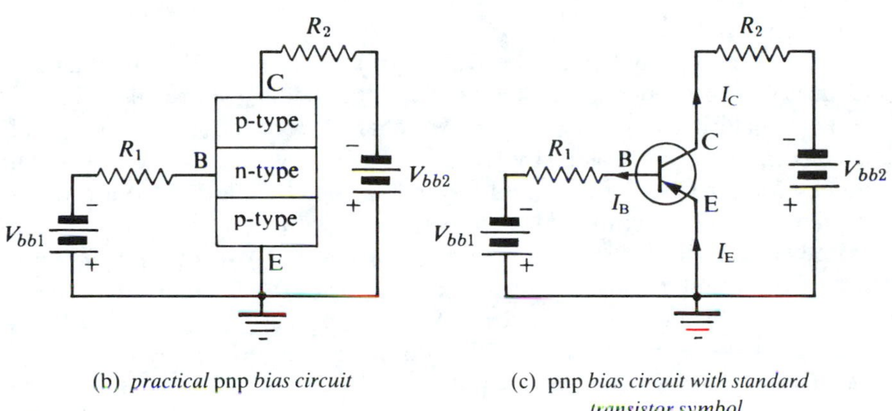

(b) *practical* pnp *bias circuit*

(c) pnp *bias circuit with standard transistor symbol*

FIGURE 5.3 Transistor biasing.

be grounded. The two circuits provide essentially the same base-collector reverse bias voltage because the base-emitter junction is forward biased, so the voltage difference between the base and emitter is only 0.5 V to 0.6 V. In other words, the base-emitter junction has a very low resistance compared to the high resistance of the reverse biased base-collector junction. The contacts between the external wire leads and the transistor are not rectifying; they conduct current equally well in both directions.

Because of the reverse bias between base and collector, there is a depletion region formed between the base and the collector with an electric field $\mathbf{E}_d$ pointing from the base *toward* the collector as shown in Fig. 5.4. The magnitude of the electric field is simply $\mathbf{E}_d \cong V_{CB}/d \cong V_{CE}/d$, where d is the thickness of the depletion region if we assume that the base-collector junction resistance is large compared to the base-emitter junction resistance. The point here is that most of the collector-emitter voltage V_{CE} appears across the reverse biased base-collector junction. Notice that the electric field tends to sweep majority carriers in the base and

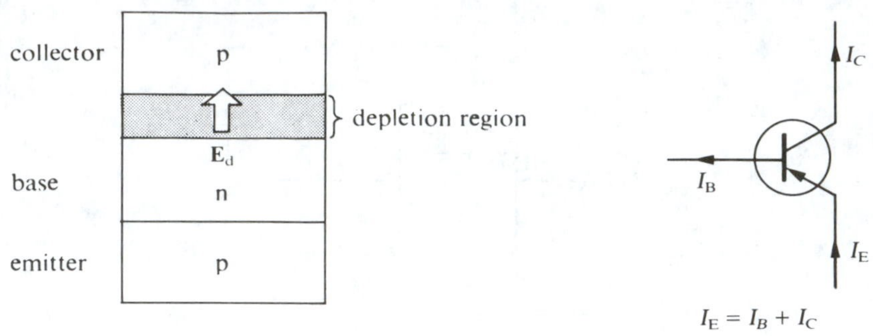

FIGURE 5.4 Depletion regions and field at base-collector junction of pnp transistor.

collector material away from rather than across the base-collector junction. Outside the depletion region there is no appreciable electric field in the transistor material, and the motion of charge carriers, both holes and electrons, is governed solely by diffusion—which is simply the statistical tendency for particles with random thermal motion to migrate from regions of higher concentration toward regions of lower concentration.

Because of the conservation of current, we must have $I_E = I_B + I_C$, that is, the current I_E flowing in the emitter must exactly equal the current I_B leaving via the base plus the current I_C leaving via the collector. Again, remember that we are talking about positive currents consisting of the flow of positively charged holes. We would expect the effective resistance of the base-emitter junction to be very small because of its *forward* bias; and similarly, we expect the resistance of the base-collector junction to be relatively high because of its *reverse* bias. One might be tempted at this point to conclude that the base current I_B would be large compared to the collector current, but this conclusion is absolutely wrong. Exactly the opposite situation occurs: $I_B \ll I_C$. To explain this, we must consider the current flow inside the transistor more carefully.

Let us consider in detail the motion of positive holes injected from the emitter into the base. The base is essentially a field-free region, so the holes move around by diffusion. These holes in the base can do one of three things: (1) They can recombine with the mobile electrons in the base (the majority carriers of the n-type base material). (2) They can diffuse through the base and recombine with electrons injected into the base region from the wire going to R_1. (3) They can diffuse across the base region into the depletion layer of the reverse biased base-collector junction where the electric field $\mathbf{E}_d$ sweeps them over into the collector. Once in the collector the holes diffuse around and eventually recombine with the electrons injected into the collector from the wire connected to R_2. In other words, a positive current I_C flows out of the collector.

Positive holes are continuously created in the emitter because the wire going to the positive terminal of the battery V_{bb2} continuously draws electrons out of the emitter, thus creating holes in the emitter. The recombination of the holes and electrons in the base region forms the base current I_B, and the holes swept into the collector region form the collector current I_C.

The single most important feature of a transistor that makes it a useful device is the thin design of the base region, so process (3) is much more probable than processes (1) or (2). That is, very few of the holes injected from the emitter into the base recombine in the base; most of them are swept into the collector region by the electric field $\mathbf{E}_d$ existing at the base-collector junction. This is done by making the thickness of the base region small compared to the "mean free path" or the "diffusion length" of the holes in the base. In such a case the holes in the base simply do not have much time to recombine with electrons before they are swept into the collector by the electric field $\mathbf{E}_d$ at the base-collector junction. In some modern bipolar transistors the base is only 50 nm = 5×10^{-8} m thick, about one-tenth the wavelength of visible light!

In an actual transistor 98% or 99% of the holes injected into the base will be swept into the collector. Thus, the base current will be only 2% or 1% of the emitter current. If we let $\alpha = I_C/I_E \cong 0.99$, then the conservation of current equation $I_E = I_B + I_C$ becomes

$$\frac{I_C}{\alpha} = I_B + I_C$$

or

$$\frac{I_C}{I_B} = \frac{\alpha}{1 - \alpha} \tag{5.1}$$

Also,

$$I_B = (1 - \alpha)I_E \quad \text{and} \quad I_C = \alpha I_E \tag{5.2}$$

If $\alpha = 0.99$, then

$$\frac{I_C}{I_B} = \frac{0.99}{1 - 0.99} = 99 \qquad I_B = 0.01 I_E \quad \text{and} \quad I_C = 0.99 I_E$$

If $\alpha = 0.98$, then

$$\frac{I_C}{I_B} = \frac{0.98}{1 - 0.98} = 49 \qquad I_B = 0.02 I_E \quad \text{and} \quad I_C = 0.98 I_E$$

Notice that the ratio I_C/I_B is a very strong function of α. A 1% change in α will double the I_C/I_B ratio. It is common notation to let $\beta \equiv \alpha/(1 - \alpha)$. In terms of β we then have

$$I_C = \beta I_B \tag{5.3}$$

and

$$I_E = (\beta + 1) I_B \tag{5.4}$$

The β is usually on the order of 20 to 200 for most transistors. Notice also that the collector current is much larger than the base current and that the collector and emitter currents are essentially equal. For example, a low-power transistor may have an emitter current of 5.0 mA, a base current of 50 μA, and a collector current of 4.95 mA if $\alpha = 0.99$. Also notice that if α remains constant, then we will obtain amplification if we can use I_B as an input and I_C as an output, because $I_C/I_B = \beta \gg 1$. More on this later.

The value of α, and therefore β, varies with emitter current. At low emitter currents an appreciable fraction of the holes injected from the (p-type) emitter into the (n-type) base may recombine. At higher emitter currents a smaller fraction of the emitter current is lost to recombination because the number of electrons in the base is now much less than the number of holes injected from the emitter. Thus α and β increase with increasing emitter current: β may double as the emitter current is increased from 0.5 mA to several mA for a typical transistor. α and β decrease slightly (by about 10% to 20%) as the emitter current is increased from several mA to around 100 mA, due to increased base conductivity resulting from the larger number of charge carriers in the base.

As we have just mentioned, the main feature of a transistor is the thinness of the base region, which enables most of the holes injected from the emitter into the base to reach the collector and form the collector current. Another physical feature of transistor construction is that the doping of the emitter is made larger than that of the base. This is done so that the concentration of holes injected from emitter to base is larger than the concentration of electrons in the base, thus producing minimum recombination of holes and electrons in the base and thus higher values of α and β for the transistor.

A third feature of transistor construction is that the collector is physically larger than the emitter. There is more power dissipated as heat in the collector than in the emitter because the base-collector voltage V_{BC} is much larger (typically several volts) than the base-emitter voltage V_{BE} (typically 0.2 V for germanium and 0.5 V to 0.6 V for a silicon transistor). This is simply because the base-collector junction is *reverse* biased and the emitter-base junction is *forward* biased. The power dissipated as heat in the base-collector region is $I_C V_{BC}$, whereas the power dissipated in the emitter-base region is only $I_E V_{BE}$. For example, a silicon transistor with $\alpha = 0.99$, $I_E = 5$ mA, and $V_{BC} = 10$ V has a collector current of $I_C = \alpha I_E =$ 4.95 mA and will have $I_E V_{BE} = (5 \text{ mA})(0.6 \text{ V}) = 3.0$ mW and $I_C V_{BE} =$ $(4.95 \text{ mA})(10 \text{ V}) = 49.5$ mW. A typical low-power transistor can safely dissipate approximately several hundred milliwatts. Because of these differences between the emitter and the collector, one cannot blithely interchange the emitter and collector leads even though both emitter and collector are p-type material. The dopings of the collector and emitter differ, and the breakdown voltage of the emitter-base junction is *much* less (typically 5 V) than the breakdown voltage for the base-collector junction (typically 30 V). The collector is often connected to the metal case of the transistor for better heat dissipation. The circuit symbol distinguishes between the emitter and collector; the arrow of the symbol in Fig. 5.1 always denotes the emitter in either pnp or npn transistors.

5.4 AMPLIFICATION

Amplification occurs in a device when the output is greater than the input. The amplification or gain A usually is defined as the output divided by the input; that is, the voltage gain is defined as $A_v \equiv V_{out}/V_{in}$, the current gain as $A_i \equiv I_{out}/I_{in}$, and the power gain as $A_p = P_{out}/P_{in}$. If we regard the base current as the input and the collector current as the output, the transistor will amplify so long as most of the holes injected into the base from the emitter are collected in the collector region. This can be seen by the following argument and by referring to Fig. 5.5.

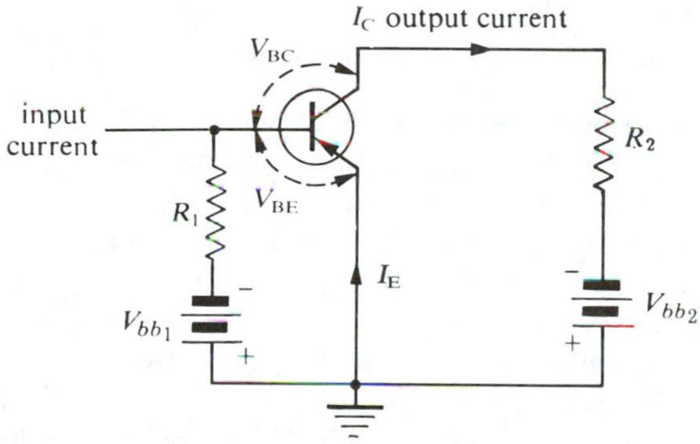

FIGURE 5.5 Transistor amplification.

A change in the input current changes the base current and therefore changes the base-emitter voltage V_{BE}. The emitter-base junction is forward biased, so a very *small* change in V_{BE} will result in a very *large* change in the current flowing from the emitter into the base. This follows from the steepness of the current–voltage curve (Fig. 5.2) for a forward biased junction. A fraction α (on the order of 0.99) of this large *increase* in emitter-base current will then flow in the collector lead. Thus, a small change ΔI_B in input current I_B produces a large change ΔI_C in output current I_C. And every hole passing from the base to the collector is accelerated by the electric field $\mathbf{E}_d$ in the depletion region of the base-collector junction. In other words, every hole going from the base to the collector falls through a voltage drop V_{BC} with $V_{BC} = E_d d$, where d is the width of the depletion region. The power output is the power dissipated in the external resistance R_2: $P_{out} \cong I_C^2 R_2$. The input power is much smaller; it is given approximately by $P_{in} \cong I_{in} V_{BE}$. Thus the power gain A_p would be equal to $I_C^2 R_2 / I_{in} V_{BE}$. For a typical low-power transistor with $\alpha = 0.99$, a 10-μA change in input current will produce a 1-mA change in collector current; R_2 is several kilohms, and $V_{BE} = 0.6$ V for a silicon transistor. Hence

$$A_p \cong \frac{(1 \text{ mA})^2 (2 \text{ k}\Omega)}{(10 \ \mu\text{A})(0.6 \text{ V})} = 333$$

If the output voltage is measured between the collector and ground V_{CE}, it will change according to $\Delta V_{CE} = \Delta I_C R_2$. The change in the input voltage V_{BE} will be the very small voltage change across the forward biased base-emitter junction, on the order of 10 mV at most. Hence the voltage gain will be approximately

$$A_v \cong \frac{\Delta V_{CE}}{\Delta V_{BE}} = \frac{\Delta I_C R_2}{\Delta V_{BE}} \cong \frac{(1 \text{ mA}) R_2}{10 \text{ mV}} = 200$$

if $R_2 = 2 \text{ k}\Omega$.

The net result of the transistor construction and biasing is that a very small change in the base-emitter voltage produces a large increase in the current flowing from the emitter to the base because the base-emitter junction is *forward* biased. In other words, we are working on the very steep forward bias portion of the p-n junction current–voltage curve of Fig. 5.2. The bias is necessary to ensure we are on the steep portion of the curve rather than on the flatter portion near the origin. In other words, the bias is necessary to turn the base-emitter junction on. The large fraction of holes collected by the depletion electric field $\mathbf{E}_d$ at the base-collector junction ensures that most of the large increase in the emitter-base current shows up as collector current.

5.5 BIASING AND GRAPHICAL TREATMENT

A transistor has three terminals, so if we have two input and two output terminals one of the three terminals must be common between the input and the output. These three possible configurations are called the "common emitter," the "common collector," and the "common base" configurations. In this section we will restrict ourselves to the npn common emitter configuration shown in Fig. 5.6, because it is the most useful configuration for most amplifiers. The base-emitter junction is forward biased, so it is easiest to control this junction by controlling and measuring the base current I_B rather than the voltage V_{BE}, because V_{BE} is essentially constant at 0.6 V

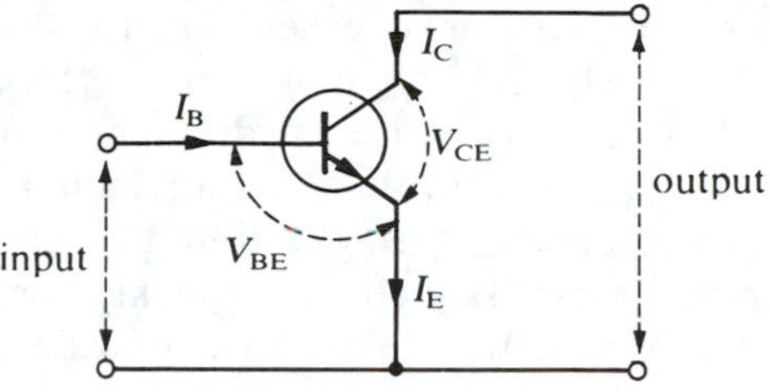

FIGURE 5.6 Common emitter configuration.

for a silicon transistor—whereas I_B may vary by a factor of ten or more from 5 μA to 50 μA. Because it is reverse biased, it is easiest to control the collector-base junction by controlling the collector-base voltage V_{BC}, which is approximately equal to the collector-emitter voltage V_{CE}. If we measure the current–voltage curves of these two junctions for a typical medium-power silicon npn transistor, we obtain the curves shown in Fig. 5.7.

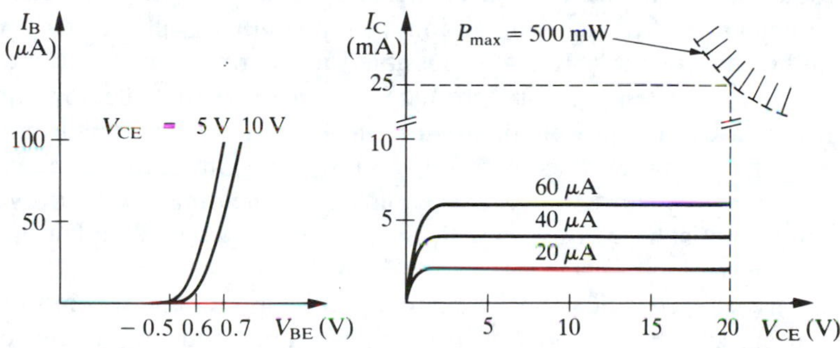

FIGURE 5.7 Input and output graphs for a 2N2222, low-cost (@$0.25) silicon medium-power npn transistor in the common emitter configuration.

The input graph of V_{BE} versus I_B is merely the graph of the forward biased p-n junction similar to Fig. 5.2. It can be seen that the base-emitter voltage remains approximately constant at 0.6 V if the base current exceeds about 20 μA. For a germanium transistor the input graph would be similar except that the base-emitter voltage would be approximately constant at 0.2 V.

The output graph of I_C versus V_{CE} is the current–voltage graph for a reversed biased p-n junction; the collector current is approximately independent of V_{CE}. However, the collector current increases drastically if the base current increases, because an increase in base current means the forward biased base-emitter junction has been turned on more.

The equation $P_{max} = I_C V_{CE} = 500$ mW has been plotted on the output graph to indicate the maximum allowable values for the collector current and collector-emitter voltage. Every transistor is rated for the maximum power it may dissipate without overheating; for the 2N2222, $P_{max} =$ 500 mW. The product of I_C and V_{CE} must *never* exceed 500 mW, even for a very short time; this means that the shaded region on the output graph corresponding to $I_C V_{CE} > 500$ mW is a forbidden region for transistor operation. If the power dissipated in the transistor exceeds the maximum power rating, the transistor will quietly and quickly burn out.

At this point it might be well to ask exactly what would happen if the transistor were overheated. The answer is that as the temperature increases, the impurity atoms tend to diffuse through the semiconductor from points of high concentration toward points of low concentration; that is, donor atoms from the p-type emitter and collector would diffuse into the n-type base, and acceptor atoms would diffuse from the base into the emitter and collector. This diffusion destroys the sharpness of the p-n junctions and produces gradual junctions that have current–voltage graphs totally different from those of Fig. 5.2. If the temperature is lowered, the impurity atoms will *not* diffuse back into their original positions. Such a process goes from a less ordered (more probable) to a more ordered (less probable) state and would violate the second law of thermodynamics. Thus, a transistor (or any semiconductor device containing p-n junctions) will be irreversibly damaged by overheating. It is also possible for "thermal runaway" to occur, in which a rise in temperature produces an increase in collector current, which increases the power dissipated in the transistor, which in turn increases the temperature still more, and so on. This may burn out the transistor in a very short time (milliseconds). Overheating can be prevented by placing a resistance in series with the emitter, as we will explain later in this chapter.

Let us now consider the problem of biasing a common emitter amplifier, using a 2N2222 npn transistor whose input and output curves are shown in Fig. 5.7. To make the transistor function as an amplifier, we must (1) bias the base-emitter junction in the *forward* direction, (2) bias the base-collector junction in the *reverse* direction, and (3) arrange the circuit so that neither temperature changes nor the input signal disturbs these bias conditions. Also, if we are going to make many copies of the same circuit, we want the circuit behavior to be essentially independent of transistor parameter changes we might encounter when we replace transistors. In general, there is a wide sample-to-sample variation of transistor parameters. The parameter $\beta = \alpha/(1 - \alpha)$ may vary by ±50% or more for inexpensive transistors. Specially selected and more expensive transistors are available with smaller tolerances on β and other parameters. The β for the transistor whose curves are shown in Fig. 5.7 is approximately equal to 100 ($\beta = I_C/I_B$).

The base-collector junction can be biased in the reverse direction by hooking up a resistance R_C and a battery of voltage V_{bb} as shown in Fig. 5.8(a). Remember for an npn transistor the base is p-type and the collector is n-type, so connecting the collector to the positive terminal will reverse

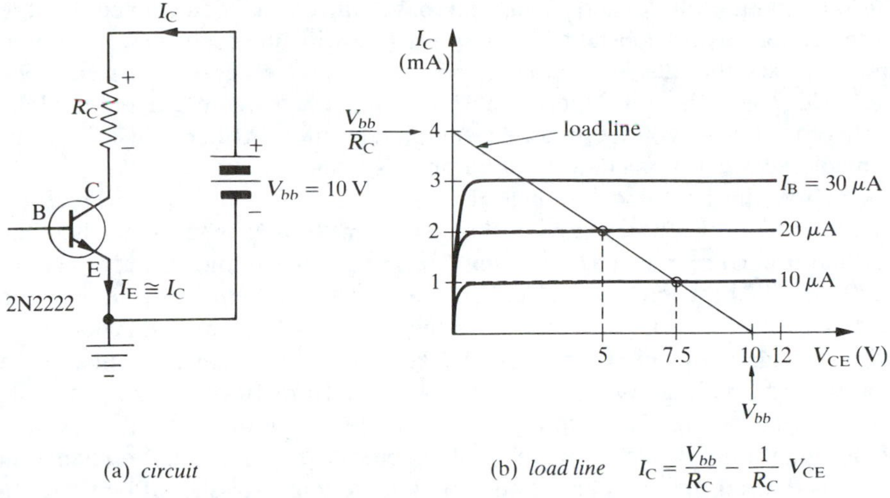

FIGURE 5.8 Common emitter amplifier circuit.

bias the base-collector junction. The Kirchhoff voltage law for V_{bb}, R_C, and V_{CE} is

$$V_{bb} - V_{CE} - I_C R_C = 0 \tag{5.5}$$

which can be solved for I_C,

$$I_C = \frac{V_{bb}}{R_C} - \frac{1}{R_C} V_{CE} \tag{5.6}$$

Equation (5.6) can be plotted on the output graph for the transistor, as shown in Fig. 5.8(b). The plot is a straight line with slope $-1/R_C$, a vertical collector current intercept of V_{bb}/R_C, and a horizontal voltage intercept of V_{bb}. This straight line is called the "load line," and its significance is that regardless of the behavior of the transistor the collector current I_C and the collector-emitter voltage V_{CE} must *always* lie on the load line. In other words, the linear relation of I_C and V_{CE} is forced upon us by Kirchhoff's voltage law and Ohm's law. In a practical circuit the load line should be kept below the P_{max} graph to avoid burning out the transistor. Notice also that the intercepts have a simple physical meaning: The vertical intercept $I_C = V_{bb}/R_C$ corresponds to all the battery voltage appearing across the resistor R_C and none across the transistor. This means the transistor has a resistance of zero ohms; it is fully conducting or "turned on." The horizontal intercept $V_{CE} = V_{bb}$ means that all the battery voltage appears across the transistor and none across R_C; thus, no current I_C can flow through R_C. The transistor has an effective resistance of infinity; in other words, it is completely "turned off" or nonconducting. For a point on the load line with certain values of I_C and V_{CE}, the base current is determined by the intersection of the constant base current line with the load line. As a final point, notice that this is a dc load line relating dc voltages and currents. The ac load line, which is appropriate for the ac collector current and ac collector-emitter voltage, may be completely different because of various capacitances and inductances. More on this later.

Once the load line is drawn, the current amplification can be read off the graph. If the base current changes from 10 μA to 20 μA, then the collector current must change from 1 mA to 2 mA, giving a current gain of 100. If the base current varies sinusoidally from 10 μA to 20 μA, the collector current will vary approximately sinusoidally from 1 mA to 2 mA. The voltage gain will also be large because only a very small change in the base-emitter voltage is necessary to change I_B from 10 μA to 20 μA while V_{CE} changes from 7.5 V to 5.0 V. If a large change in I_C is desired, the load line should be steep, meaning R_C should be small; if a large change in V_{CE} is desired, the load line should be flat, meaning R_C should be large. In other words, for a large current gain choose R_C small; for a large voltage gain choose R_C large.

It can also be seen that the magnitude of the change in collector current or collector-emitter voltage for a given change in base current will depend on what the initial base current is with no input. The "operating" point or the "quiescent" point on the load line is the point that represents I_C, V_{CE}, and I_B for no input signal.

The problem of fixing the operating point is basically the problem of setting the base-emitter forward bias so that the dc base current has the desired value with no signal input. If the input signal that is to be amplified has roughly equal positive and negative swings, then the operating point should be chosen approximately in the center of the load line. To be specific, suppose $V_{bb} = 10$ V and $R_C = 2.5$ kΩ. Then the load line has a vertical intercept of 10 V/2.5 kΩ = 4 mA, and a horizontal intercept of 10 V. If we choose the operating point to be $I_C = 2$ mA, then $V_{CE} = 5$ V, which implies that $I_B = I_C/\beta = 2\text{ mA}/100 = 20\ \mu\text{A}$.

Thus, our biasing problem is to supply a steady current of 20 μA to the base even when there is no input signal. The value of base current either can be read off the graph of I_C versus V_{CE}, or it can be calculated from $I_B = I_C/\beta$. This biasing could be done by connecting up a 1.5-V battery and a 45-kΩ resistor, as shown in Fig. 5.9(a). The Kirchhoff voltage law for the base-emitter junction bias circuit is

$$+1.5\text{ V} - I_B R - V_{BE} = 0$$

But $V_{BE} \cong 0.6$ V for any silicon transistor, so

$$0.9\text{ V} - I_B R = 0$$

Thus $R = 0.9\text{ V}/I_B = 0.9\text{ V}/20\ \mu\text{A} = 45\text{ k}\Omega$.

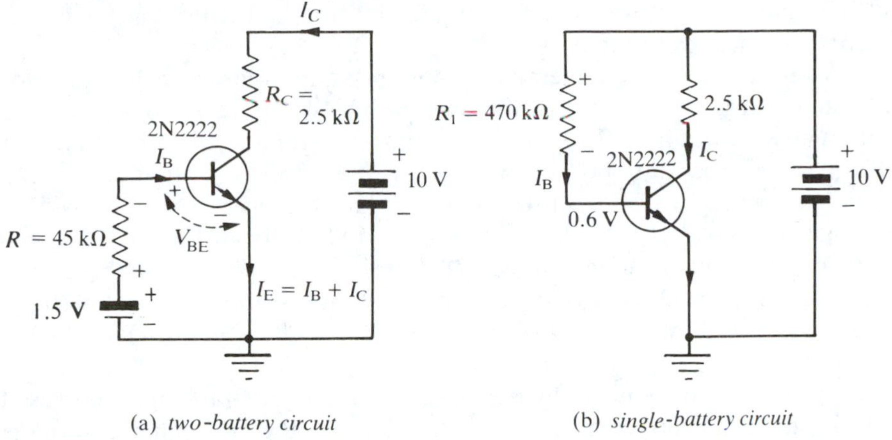

FIGURE 5.9 Common emitter biasing circuits.

However, this biasing circuit of the battery and the 45-kΩ resistor has three disadvantages: (1) It requires a separate battery in addition to the 10-V battery biasing the collector-base junction. A second battery is expensive, heavy, and bulky. (2) This circuit is unstable with respect to changes in temperature. (3) If a different transistor of the same type (2N2222) but with a different β is put in the same circuit, the operating point may change appreciably; that is, the operating point is a sensitive function of the transistor β. One possible circuit using only one battery is shown in Fig. 5.9(b).

We recall that $\beta = 100$ from the graph of Fig. 5.7; if the graph is not available, we can look up the value of β for the particular transistor in a transistor manual. The value of R_1 can be calculated as follows. For the transistor to be "turned on," that is, to conduct an appreciable (>1 mA) amount of collector current, the base-emitter junction must be forward biased. If the transistor is npn silicon, the base must be approximately 0.6 V positive with respect to the emitter. In Fig. 5.9(b) the emitter is grounded, so its voltage is exactly zero. Therefore the base should be at +0.6 V. At the desired operating point the transistor will draw 2 mA of current with no signal input. Thus $I_C = 2$ mA, and $I_B = I_C/\beta = 2\text{ mA}/100 = 20\ \mu\text{A}$. R_1 is now calculated from Ohm's law. We know the current through R_1 and the voltage at each end:

$$R_1 = \frac{V_{bb} - V_B}{I_B} = \frac{10\text{ V} - 0.6\text{ V}}{20\ \mu\text{A}} = \frac{9.4\text{ V}}{20\ \mu\text{A}} = 470\text{ k}\Omega$$

The biasing arrangement for the base shown in Fig. 5.9(b) is certainly simple; it involves only one resistor R_1. However, it still has two disadvantages: (1) Its operating point will change drastically if another transistor of the same type but with a different β is substituted. (2) It is unstable with respect to changes in temperature.

There can be a considerable sample-to-sample variation in transistor characteristics, especially β, which can vary by up to ±50% or more. The biasing circuit of Fig. 5.9(b) is designed to work with a transistor with a β of 100, and the operating point is $V_{CE} = 5$ V, $I_C = 2$ mA. If another 2N2222 transistor with a β of 50 is used in this circuit, the operating point will be changed to $V_{CE} = 7.5$ V, $I_C = 1$ mA. This follows from the fact that the base-emitter voltage of both 2N2222 transistors is 0.6 V, and the circuit provides the same 20-μA base current to both transistors. Thus $I_C = \beta I_B = (50)(20\ \mu\text{A}) = 1$ mA and $V_{CE} = 10\text{ V} - I_C R_C = 10\text{ V} - (1\text{ mA})(2.5\text{ k}\Omega) = 7.5$ V

A two-resistor voltage divider as shown in Fig. 5.10(a) can also be used to bias the transistor with one battery, and it has the additional advantage

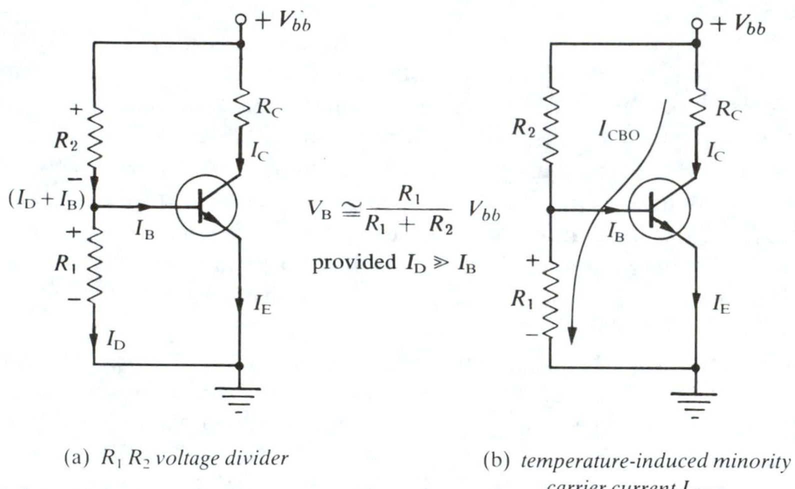

FIGURE 5.10 Two-resistor bias circuit (temperature unstable).

that the bias voltage tends to change less due to temperature-induced changes in R_1 and R_2. That is, the bias in Fig. 5.10(a) depends upon the ratio of R_1 to R_2, which is less sensitive to temperature changes than the single-resistor bias circuit of Fig. 5.9. Also R_1 and R_2 in Fig. 5.10 are usually much less than R_1 in Fig. 5.9(b), thus reducing the change in the operating point caused by temperature-induced changes in the base current. This is covered in the next several paragraphs. The voltage at the base V_B is given by $V_B = I_D R_1$ where I_D is the current flowing through R_1. And

$$V_{bb} = (I_D + I_B)R_2 + I_D R_1 \quad \text{or} \quad I_D = \frac{V_{bb} - I_B R_2}{R_1 + R_2}$$

Thus, $V_B = I_D R_1$ becomes

$$V_B = \frac{V_{bb} - I_B R_2}{R_1 + R_2} R_1 \tag{5.7}$$

Usually R_1 and R_2 are made sufficiently small so that $I_B \ll I_D$ or $I_B R_2 \ll V_{bb}$. Then $I_D \cong V_{bb}/(R_1 + R_2)$, and

$$V_B \cong \frac{R_1}{R_1 + R_2} V_{bb} \tag{5.8}$$

However, the bias circuit of Fig. 5.10 is still unstable with respect to temperature fluctuations. *The temperature instability arises from the thermal generation of minority carriers in the semiconductor material of the transistor.* Any increases in temperature, from whatever source, will increase the number of covalent bonds broken in the semiconductor lattice and thereby will increase the number of minority carriers in all three regions of the transistor. The relative increase of majority carriers will be small because almost all the donor or acceptor atoms are already ionized at room temperature. Let us focus our attention on the thermally generated *minority* carriers in the base region and the collector region. A higher temperature means more mobile electrons in the base and holes in the collector for an npn transistor.

The voltage drop between the base and collector is a reverse voltage only for the majority carriers but is a *forward* voltage for minority carriers. Therefore, the thermally generated electrons in the base and holes in the collector will flow across the base-emitter junction and constitute a current from base to collector. This reverse current is usually denoted by I_{CBO}, where the subscripts stand for collector to base with the emitter open. This current flows even when the emitter of the transistor is open and it depends strongly upon the temperature, because the flow consists of *thermally* generated holes and electrons. For a silicon transistor I_{CBO} will increase by a factor of 2 (from 5 nA to 10 nA) for a 6°C rise in temperature. At 70°C, I_{CBO} is approximately equal to 50 nA.

As the temperature increases, the transistor will turn on more (I_C will increase) because of two factors: (1) the increasing I_{CBO} will increase the forward bias of the base-emitter junction, and (2) the base-emitter voltage required for a given collector current will decrease. Let us now consider these two factors in more detail.

As the temperature increases, the number of minority carriers in the base and collector increases and I_{CBO} increases. From the direction of I_{CBO} [see Fig. 5.10(b)], it can be seen that an increase in I_{CBO} will make the base voltage more positive, thus increasing the forward bias voltage between base and emitter. In other words, an increasing temperature turns the transistor on more. This increases the emitter-base current and also the collector current. The increase in collector current increases the power dissipated in the transistor and thereby raises the transistor temperature. This temperature rise generates more minority carriers, and the cycle starts over again. If not checked, the transistor can eventually destroy itself from overheating.

A quantitative explanation of this I_{CBO} overheating cycle is as follows. Including the effect of I_{CBO}, the collector current is

$$I_C = \alpha I_E + I_{CBO} \tag{5.9}$$

and the base current is

$$I_B = (1 - \alpha)I_E - I_{CBO} \tag{5.10}$$

The Kirchhoff current law for the transistor is still

$$I_E = I_B + I_C \tag{5.11}$$

When we eliminate I_E and use $\beta = \alpha/(1 - \alpha)$ and $\alpha = \beta/(\beta + 1)$, (5.9) becomes

$$I_C = \beta I_B + (\beta + 1)I_{CBO} \tag{5.12}$$

and

$$I_E = (\beta + 1)I_B + (\beta + 1)I_{CBO} \tag{5.13}$$

The immediate conclusion we can draw is that a small temperature-induced change in I_{CBO} is multiplied by the large factor $\beta + 1$. Thus, the collector current can change appreciably with temperature, and thermal runaway and transistor burnout can occur.

The turn-on voltage or V_{BE} for a transistor tends to change with temperature—the transistor turning on more as the temperature increases because of the thermal generation of minority carriers in the base—and the resulting change in the Fermi level as discussed in Chapter 4. Equivalently, the base-emitter voltage required for a given collector current decreases with increasing temperature. For both germanium and silicon transistors, near room temperature, the base-emitter voltage decreases at approximately -2.5 mV/°C, although one would expect slightly less temperature dependence for silicon because of the larger energy gap.

The only link that can be broken in the chain just mentioned is the change in the base-emitter voltage. There is absolutely no way to prevent an increase in temperature from increasing the number of minority carriers. However, if we can devise a bias circuit for the base that will hold the base-emitter voltage constant at the desired voltage, regardless of the increase in I_{CBO}, then the collector current will remain constant except for the small increase due to the increasing I_{CBO}. There is no way to prevent

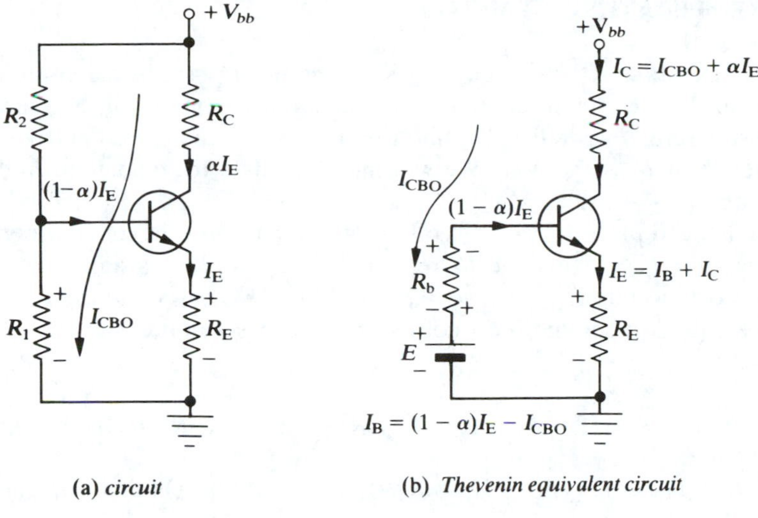

(a) *circuit* (b) *Thevenin equivalent circuit*

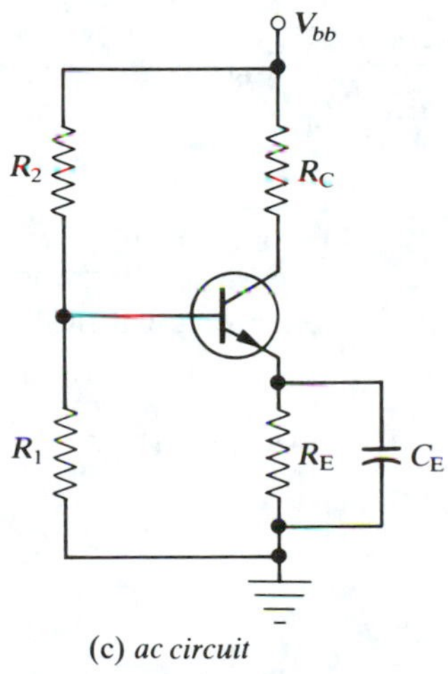

(c) *ac circuit*

FIGURE 5.11 Stable bias circuit.

V_B from becoming more positive due to the increase in I_{CBO}, but if we also can make the emitter voltage go more positive at the same time then the forward bias V_{BE} will not increase nearly so much. A little thought will show that a resistor R_E placed in series with the emitter, as shown in Fig. 5.11, will do the trick. For then, as the temperature increases, I_{CBO} increases, V_B goes more positive, and I_C and I_E increase. But now $V_E = + I_E R_E$ also goes more positive, so $V_B - V_E = V_{BE}$ remains much more nearly constant.

5.6 TEMPERATURE STABILITY

To calculate exactly how large a resistor one must place in the emitter lead to stabilize the common emitter amplifier against changes in temperature, we must carefully write the Kirchhoff equations for the circuit and solve for I_C in terms of I_{CBO} and V_{BE}. We will then calculate the change in I_C due to a change in I_{CBO} and in V_{BE}.

First we replace the R_1R_2 voltage divider by its Thevenin equivalent, as shown in Fig. 5.11(b). The R_b represents the parallel resistance of R_1 and R_2; in most circuits $R_1 \ll R_2$, so $R_b \cong R_1$. Now we can write the Kirchhoff voltage equation for the loop containing the base-emitter junction of the transistor:

$$E - (1 - \alpha)I_E R_b + I_{CBO}R_b - V_{BE} - I_E R_E = 0 \qquad \textbf{(5.14)}$$

Using $I_E = (I_C - I_{CBO})/\alpha = [(\beta + 1)/\beta](I_C - I_{CBO})$ to eliminate I_E and $\alpha = \beta/(\beta + 1)$ to eliminate α, we obtain

$$I_C = \frac{R_b + R_E}{R_E + \dfrac{R_b}{\beta + 1}} I_{CBO} + \frac{E - V_{BE}}{\dfrac{\beta + 1}{\beta} R_E + \dfrac{R_b}{\beta}} \qquad \textbf{(5.15)}$$

We now can see how changes in I_{CBO} and in V_{BE} will affect the collector current I_C. A small change ΔI_{CBO} and a small change ΔV_{BE} will produce, to first order, a change ΔI_C in the collector current:

$$\Delta I_C = \left(\frac{\partial I_C}{\partial I_{CBO}}\right)_{V_{BE}} \Delta I_{CBO} + \left(\frac{\partial I_C}{\partial V_{BE}}\right)_{I_{CBO}} \Delta V_{BE} \qquad \textbf{(5.16)}$$

Using (5.15) we obtain

$$\Delta I_C = \frac{R_b + R_E}{R_E + \dfrac{R_b}{\beta + 1}} \Delta I_{CBO} - \frac{\beta}{(\beta + 1)R_E + R_b} \Delta V_{BE} \qquad \textbf{(5.17)}$$

Equation (5.17) can be simplified by noting that usually $R_E \gg R_b/(\beta + 1)$ and β typically is about 100. Thus

$$\Delta I_C \cong \left(1 + \frac{R_b}{R_E}\right) \Delta I_{CBO} - \frac{1}{R_E} \Delta V_{BE} \qquad \textbf{(5.18)}$$

At this point we should explicitly point out that for a temperature increase ΔI_{CBO} is positive, but ΔV_{BE} is negative. Thus, both terms in (5.12) tend to *increase* I_C for a temperature rise. The value of ΔI_{CBO} is about 2 nA for a

10°C temperature rise in silicon. For both germanium and silicon devices, $\Delta V_{BE} \cong -2.5\ \text{mV/°C}$ as discussed in Chapter 4. R_B/R_E is usually about 10 in most circuits.

It is now clear that the larger R_E is, the better the temperature stability is. For a 10°C rise, with $R_E = 500\ \Omega$, $R_b/R_E = 10$, and $\Delta I_{CBO} = 2\ \text{nA}$, (5.18) implies

$$\Delta I_C \cong (11)(2\ \text{nA}) + \frac{25\ \text{mV}}{500\ \Omega} = 22\ \text{nA} + 50\ \mu\text{A} \cong 50\ \mu\text{A} = 0.05\ \text{mA}$$

Notice, however, that if $R_E = 0$, the situation is disastrously different. Setting $R_E = 0$ in (5.17), we obtain

$$\Delta I_C \cong (\beta + 1)\Delta I_{CBO} - \frac{\beta}{R_b}\Delta V_{BE} \qquad \textbf{(5.19)}$$

For a 10°C rise (5.19) implies

$$\Delta I_C \cong (101)(2\ \text{nA}) + \frac{(100)(2.5\ \text{mV})}{10\ \text{k}\Omega} = 0.45\ \text{mA}$$

As we will see in Chapter 8, the presence of R_E in the emitter provides negative feedback, which stabilizes the circuit operation against changes in temperature, supply voltage, and so on, but which also decreases the voltage gain. For amplifying ac signals at the frequency $f = \omega/2\pi$, R_E is usually connected in parallel with a large capacitance C_E as shown in Fig. 5.11(c). We say R_E is "bypassed" by C_E so that $1/\omega C_E \ll R_E$ at the ac frequencies to be amplified. Thus, R_E is effectively shorted out or bypassed at the signal frequencies, and a larger ac gain is obtained. But, because C_E has an infinite reactance for dc, R_E is present for very slow changes due to temperature changes, and the preceding arguments about temperature stability still apply.

The temperature-induced change in collector current caused by the change in the base-emitter voltage can also be minimized by adding diodes, as shown in Fig. 5.12 for an npn transistor. (For a pnp transistor the diodes should be reversed.) The principle of operation of the circuit is that as the temperature rises the transistor base-emitter voltage required for a constant collector current decreases, so if the base bias circuit supplied a constant voltage to the base the collector current would increase. But if the temperature rises, the voltage drops across the two diodes will also decrease, thus biasing the transistor a bit more off and tending to stabilize the transistor collector current.

A quantitative analysis of the circuit is as follows. The Kirchhoff voltage law for the loop containing the two diodes and the transistor

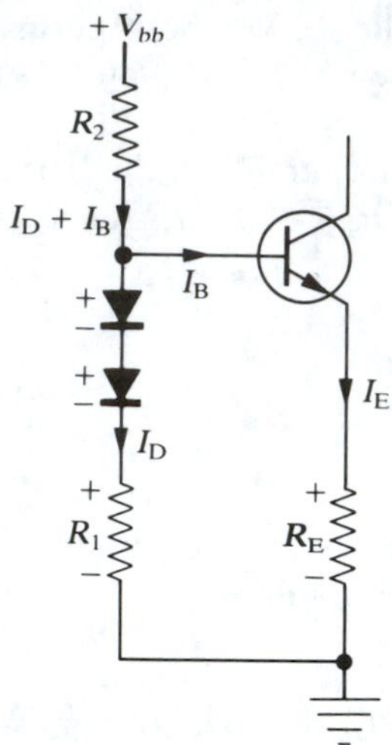

FIGURE 5.12 Diode temperature stabilization.

base-emitter junction is

$$I_D R_1 + 2V_D - V_{BE} - I_E R_E = 0 \tag{5.20}$$

where V_D is the forward voltage drop across each voltage. And provided $I_D \gg I_B$, which is good design, I_D is given by

$$I_D \cong \frac{V_{bb} - 2V_D}{R_1 + R_2} \tag{5.21}$$

where V_D is the forward voltage drop across each diode. Eliminating I_D, we can write the emitter current as

$$I_E \cong \frac{1}{R_E}\left[\left(\frac{V_{bb} - 2V_D}{R_1 + R_2}\right)R_1 + 2V_D - V_{BE}\right]$$

But $V_D \cong V_{BE}$ if the diodes and the transistor are both made from silicon. This equality is particularly accurate if the diodes and the transistor are both made on the same silicon chip, which is the case with integrated circuits. Using $V_D = V_{BE}$, we obtain

$$I_E = \frac{1}{R_E}\left[\frac{V_{bb}R_1 + V_D(R_2 - R_1)}{R_1 + R_2}\right] \tag{5.22}$$

The only strongly temperature-dependent term in (5.22) is V_D, and the V_D term can be eliminated by simply making $R_1 = R_2$. The emitter current (and also the collector current) is then essentially independent of the temperature and is given by

$$I_E = \frac{V_{bb}}{2R_E} \tag{5.23}$$

This type of circuit is necessary whenever a transistor encounters widely varying temperatures or whenever the transistor collector current changes appreciably in circuit operation—for example, when the collector current of a power transistor changes from 1 to 10 A. The diodes and the transistor should be mounted close together so that they are at the same temperature. One good technique is to mount the diodes in a gob of heat-sink compound (a good thermal conductor but also a good electrical insulator) right on the transistor case.

5.7 COMMON EMITTER AMPLIFIER DESIGN

Let us now quickly go through a "seat of the pants" design of a common emitter amplifier stage and illustrate the assumptions and approximations that go into an actual design problem. We first choose a transistor. For frequencies up to several MHz and an environment under 70°C, we may choose an inexpensive silicon transistor such as the 2N2222. For higher-frequency operation we need a more expensive high-frequency transistor such as the 2N2369 and/or negative feedback, which will be discussed in Chapter 7.

Let us now focus our attention on the 2N2222, an npn silicon transistor costing approximately $0.30 in small quantities. We will choose the voltage divider bias circuit [Fig. 5.11(c)] for reasons of stability discussed in the previous section. The 2N2222 characteristics are:

Maximum power in 25°C air	500 mW
Maximum voltages	$V_{CB} = 60$ V
	$V_{CE} = 30$ V
	$V_{EB} = 5$ V
Maximum junction temperature	125°C
Minimum β	35 @ $I_C = 0.1$ mA
	50 @ $I_C = 1.0$ mA
	75 @ $I_C = 10$ mA

We first choose a dc power supply voltage V_{bb} less than the 30-V maximum for V_{CE}, say, $V_{bb} = 20$ V. Referring to Fig. 5.11(c), we see that the KVL implies

$$+V_{bb} - I_C R_C - V_{CE} - I_E R_E = 0$$

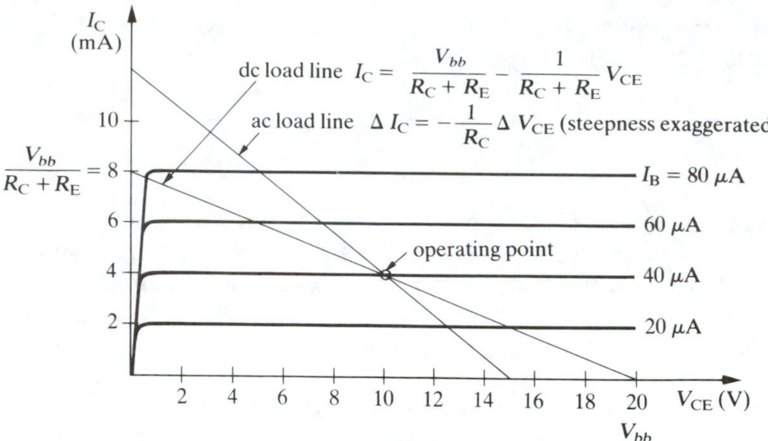

FIGURE 5.13 Load line.

If $\beta \gg 1$, $I_E \cong I_C$, and we can solve for I_C:

$$I_C = \frac{V_{bb}}{(R_E + R_C)} - \frac{V_{CE}}{(R_E + R_C)} \tag{5.24}$$

which is the load-line equation and which is drawn on the 2N2222 curves in Fig. 5.13. The two intercepts are $(0, V_{bb}/(R_E + R_C))$ and $(V_{bb}, 0)$. Choosing $V_{bb} = 20$ V fixes the intercept on the V_{CE} axis. Thus, a choice of $R_E + R_C$ fixes the other intercept and determines the load line. We must, of course, keep the load line below the P_{max} curve to avoid burning out the transistor. The choice of load line is determined by the kind of amplifier we desire. A steep load line will produce a high current gain and a low voltage gain; a flatter load line will produce a high voltage gain and a low current gain.

Let us choose the I_C intercept at 8 mA. Then

$$8\text{ mA} = \frac{V_{bb}}{R_E + R_C} \quad \text{and} \quad R_E + R_C = \frac{20\text{ V}}{8\text{ mA}} = 2.5\text{ k}\Omega$$

Thus, the sum of R_E and R_C must equal 2.5 kΩ for this particular load line.

The operating point must now be chosen. This is the point on the load line representing the transistor's dc collector current and dc collector-emitter voltage when there is no input signal applied at the base. In general, transistors function best if I_C is 1 mA or greater and if V_{CE} is 1 V or larger. If these conditions are not met, the transistor β is usually greatly reduced. We choose the operating point near the center of the load line, $I_C = 4$ mA and $V_{CE} = 10$ V, so that the transistor can respond to either positive or negative input signals on its base. In other words, the (npn) transistor can be turned on more (by a positive signal on its base), and the collector current could increase from 4 mA to nearly 8 mA, or from a negative signal on its base, its collector current could decrease from 4 mA to nearly 0.

Our problem is now to "bias" the transistor so that it conducts 4 mA collector current with $V_{CE} = 10$ V when there is no external signal on the base. We now show that a choice of R_E, R_1, and R_2 will accomplish this. We first choose R_E because once R_E is fixed the 4-mA emitter current will fix the emitter voltage, and the base voltage must be 0.6 V higher than the emitter because the base-emitter junction is turned on. The base voltage and V_{bb} then determine the values of R_1 and R_2. However, there are usually two conflicting requirements to consider in choosing R_E: First, for a large voltage gain R_C should be large; since $R_E + R_C$ is fixed at 2.5 kΩ, this means R_E should be small. Second, for temperature stability R_E should be large from (5.18). Let us assume the V_{BE} term in (5.18) is the dominant term (this will be true for silicon transistors at temperatures below approximately 50°C), and let us assume that the circuit will be exposed to temperature fluctuations of 25° to 45°C at the most (i.e., $\Delta T = +20$°C). Then if we require that the temperature-induced change in collector current be less than 5% of the dc value at 25°C, we have from (5.18) that

$$\Delta I_C \cong \frac{-\Delta V_{BE}}{R_E} \leqq 0.05 I_C \quad \text{or} \quad R_E \geqq \frac{-\Delta V_{BE}}{0.05 I_C}$$

Thus $$R_E \geq \frac{-(-2.5\ \text{mV/°C})(+20\text{°C})}{(0.05)(4\ \text{mA})} = 250\ \Omega$$

and $$R_C = 2.5\ \text{k}\Omega - 250\ \Omega = 2.25\ \text{k}\Omega$$

From Ohm's law the emitter voltage is then $V_E = I_E R_E = (4\ \text{mA})(250\ \Omega) = 1$ V. The base voltage is then $V_B = V_E + 0.6\ \text{V} = 1\ \text{V} + 0.6\ \text{V} = 1.6$ V. We must now choose R_1 and R_2 to produce a base voltage of 1.6 V. From Fig. 5.14 we see that if we make $R_1 + R_2$ small enough, the "divider" current I_d will be much larger than the base current I_B. This is a very desirable situation because it makes the operating point independent of the transistor's base current—independent of the transistor's β, which may vary ±50% among various samples of the same transistor type. Let us then choose I_d to be 10 times I_B. Because $I_B = I_C/\beta = 4\ \text{mA}/75 = 53\ \mu\text{A}$, we have $I_D = 530\ \mu\text{A} = 0.53$ mA. Thus, from Ohm's law

$$V_B = I_D R_1 \quad \text{or} \quad R_1 = \frac{V_B}{I_D} = \frac{1.6\ \text{V}}{0.53\ \text{mA}} = 2.8\ \text{k}\Omega$$

R_2 is determined by

$$V_{bb} - V_B = I_D R_2 \quad \text{or} \quad R_2 = \frac{V_{bb} - V_B}{I_D} = \frac{20\ \text{V} - 1.6\ \text{V}}{0.53\ \text{mA}}$$

$$R_2 = 35\ \text{k}\Omega$$

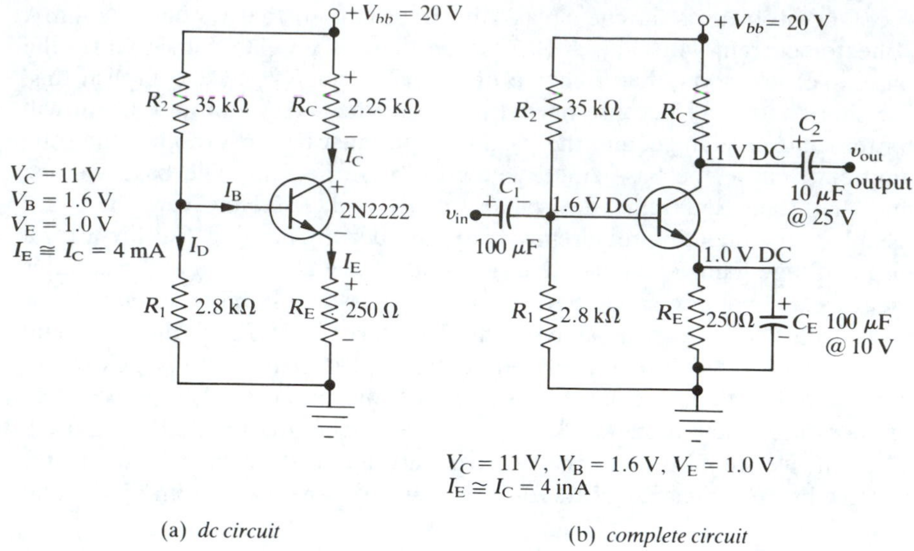

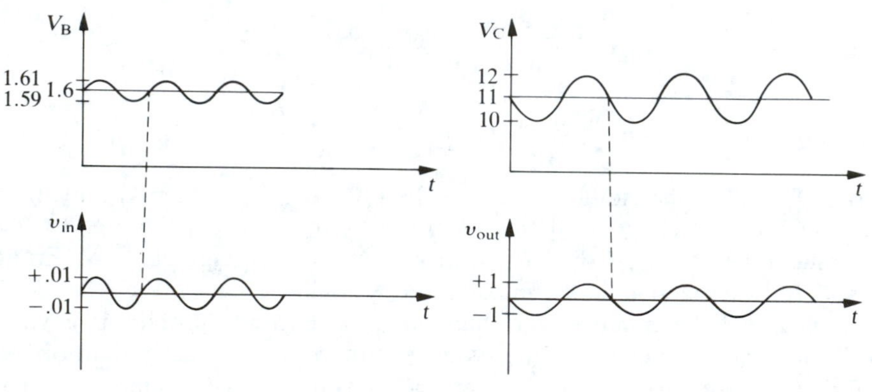

(c) *waveforms with sinusoidal input*

FIGURE 5.14 Common emitter amplifier circuit.

The complete circuit is shown in Fig. 5.14 with all the component values.

It is generally desirable to have as large a value for R_1 as possible to avoid drawing too much current from whatever signal source is connected to the base. In other words, there are two conflicting requirements as to the choice of R_1: We need a large value of R_1 to avoid loading down the signal source as just mentioned, but we need a small value of R_1 to make I_D much larger than I_B. The preceding example illustrates a reasonable compromise for choosing R_E and R_L.

We will now show why we must add C_E in parallel with the emitter resistor R_E, to avoid decreasing the gain of the amplifier. We assume that the transistor has the proper dc bias voltages applied. Suppose an input signal of +2 mV is applied to the base of the transistor. This signal makes the base 2 mV more positive, thus turning the npn transistor on more and increasing I_E. But, the voltage V_E at the emitter equals $+I_E R_E$, so as I_E increases, V_E becomes more positive. Hence the change in voltage between the base and the emitter due to the signal is *less* than 2 mV. The net voltage V_{BE} present across the base-emitter junction is the input the transistor "sees" and thus determines how much the transistor turns on (i.e., how large the output is). Because V_E also goes positive, as V_B does, V_{BE} is less than the signal input. Hence, the gain is decreased. This is an example of negative feedback, which will be discussed in a later chapter. To prevent this, we must prevent the voltage of the emitter from changing as the base voltage changes due to the signal input. The addition of a capacitor C_E from the emitter to ground as shown in Fig. 5.14(b) will accomplish this by making the emitter an ac ground—by smoothing out any variations in the emitter voltage due to the signal at the base. The dc bias is not affected because C_E will pass no direct current. If we make the capacitance large enough so that its reactance is small compared to R_E, then the emitter will be an ac signal ground, that is, $X_C = 1/\omega C_E \ll R_E$. This inequality is hardest to satisfy at the lowest frequency of operation. For an audio amplifier, for example, the lowest frequency the human ear can hear is about 20 Hz, so we want $X_C \ll R_E$ at 20 Hz. Therefore,

$$C_E \gg \frac{1}{\omega_{min} R_E} = \frac{1}{2\pi f_{min} R_E} = \frac{1}{(2\pi)(20\ \text{Hz})(250\ \Omega)} = 32\ \mu\text{F}$$

A 100-μF electrolytic capacitor of a low voltage rating (5 to 10 V) will serve nicely, because the dc voltage drop across R_E is only $I_E R_E = (4\ \text{mA})(250\ \Omega) = 1.0\ \text{V}$. The plus terminal of C_E should be at the emitter because the emitter dc voltage is always positive with respect to ground for our npn transistor. Notice that if the minimum frequency to be amplified is larger, a smaller capacitance will suffice.

As far as the input ac signal is concerned, R_E is shorted out so the emitter is a good ac ground. We also note that the $+V_{bb}$ terminal of the power supply is also an ac ground (remember the large filter capacitor to ground inside the power supply). Thus, the ac equivalent circuit as seen by the input signal is as shown in Fig. 5.15.

We now can calculate an approximate expression for the voltage gain. The voltage gain A_v is defined as the change in the output voltage divided by the change in the input voltage:

$$A_v = \frac{v_{out}}{v_{in}} = \frac{\Delta V_C}{\Delta V_B} \tag{5.25}$$

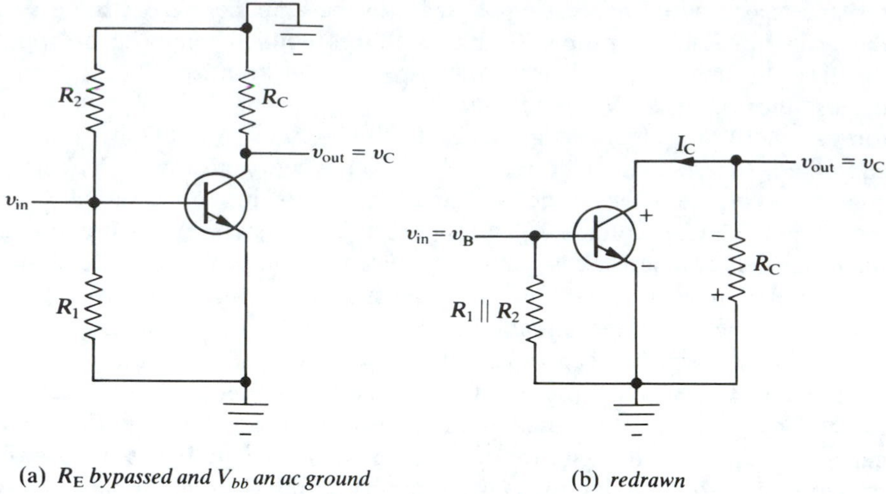

FIGURE 5.15 Common emitter amplifier ac equivalent circuit.

Let us now represent changes in voltage and current by lowercase letters, and use capital letters for dc values; that is, $i_C = \Delta I_C$, $v_C = \Delta V_C$, $i_B = \Delta I_B$, and so on. The Kirchhoff voltage law implies

$$v_C + i_C R_C = 0$$

or

$$v_C = -i_C R_C \tag{5.26}$$

The minus sign merely means that if the collector current I_C increases ($\Delta I_C = i_C$ positive) the dc collector voltage V_C will decrease ($\Delta I_C = i_C$ negative). In other words, a positive-going input voltage at the base will cause a negative-going voltage at the collector. For example, if V_B changes from 1.60 V to 1.61 V, V_C might change from 9 V to 8 V.

Using $v_C = v_{out}$, and $i_B = i_C/\beta$, the voltage gain can be written as

$$A_v = \frac{v_C}{v_B} = -\frac{i_C R_C}{v_B} = -\frac{\beta i_B R_C}{v_B}$$

But from Ohm's law the ac signal base current i_B is

$$i_B = \frac{v_B}{R_{BE}}$$

where R_{BE} is the effective transistor input resistance at the base. Thus the voltage gain is

$$A_v = -\frac{\beta R_C}{R_{BE}} \tag{5.27}$$

Notice that we expect R_{BE} to be a fairly small resistance because it is the effective ac resistance of the *forward* biased base-emitter junction. R_{BE} is the inverse slope of the steep I_B-versus-V_{BE} graph. As we shall shortly see, R_{BE} is typically 2.6 kΩ for $I_C = 1$ mA, and it varies inversely with I_C ($R_{BE} = 260\ \Omega$ for $I_C = 10$ mA). For most transistor amplifier circuits R_C and R_{BE} are in the same "ball-park," so we can say that the approximate voltage gain of the common emitter amplifier circuit of Fig. 5.14(b) is equal to the transistor β.

Notice also that (5.26) implies that the ac load line seen by the signal is steeper than the dc load line, as shown in Fig. 5.13.

The only problem remaining is to feed the signal to be amplified into the base terminal and take the amplified output from the collector terminal. To avoid changing the dc bias voltages shown in Fig. 5.14(a), we must not allow any dc current to flow in the input or out of the output. The simplest way to do this is to place capacitors, called dc "blocking" or "coupling" capacitors, in series with the input and output. The resulting circuit is shown in Fig. 5.14(b). However, C_1 and C_2 form high-pass RC filters (see Section 2.8). We recall that the breakpoint for a high-pass RC filter is at $\omega_B = 2\pi f_B = 1/RC$, as shown in Fig. 5.16. At frequencies higher than f_B

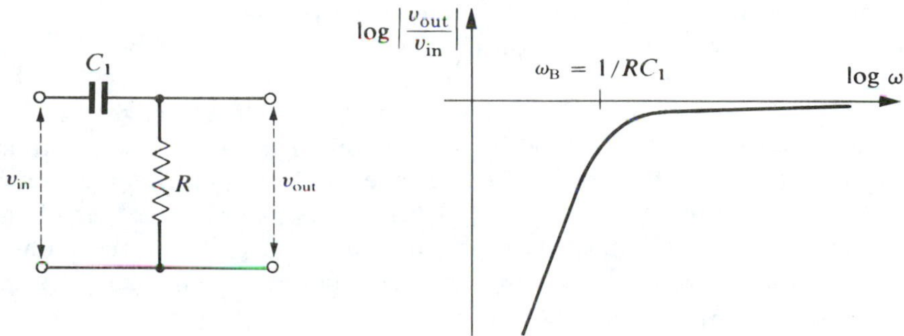

FIGURE 5.16 High-pass *RC* filter.

the filter gain is essentially unity; that is, a negligible signal voltage drop occurs across the capacitor. Thus, we choose the capacitance such that the breakpoint is appreciably less than the lowest frequency the amplifier will be required to amplify—20 Hz for our audio amplifier. The lower the frequency to be amplified, the larger C_1 and C_2 must be.

The effective resistance R in the output high-pass filter is merely R_L, the load resistance that the amplifier is driving, as shown in Fig. 5.17. This load may be the input of another amplifier or it may be a loudspeaker or it

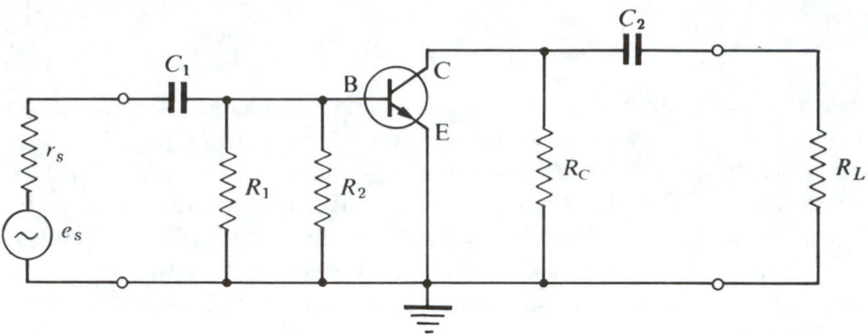

FIGURE 5.17 Ac equivalent circuit of the common emitter amplifier of Fig. 5.14(b).

may be an actual resistance. To be specific, suppose $R_L = 10\ \text{k}\Omega$. Then we must choose

$$\frac{1}{R_L C_2} = \omega_B \ll 2\pi \times 20\ \text{Hz}$$

or $$C_2 \gg \frac{1}{2\pi 20 R_L} = \frac{1}{(2\pi)(20\ \text{Hz})(10^4\ \Omega)} = 1.6 \times 10^{-6}\ \text{F}$$

or $C_2 \gg 1.6\ \mu\text{F}$. A 10-μF, 25-V capacitor would be adequate. Unless there is some voltage at the top of R_L more positive than $V_C = +11$ V, the polarity of C_2 should be as shown in Fig. 5.14(b) with the positive side connected to the collector.

A glance at Fig. 5.17 shows that if capacitor C_2 is effectively a short circuit, then R_C and R_L will be in parallel. This will be true at all frequencies for which $1/(\omega C_2) \ll R_L$—above the breakpoint of the output high-pass filter formed by C_2 and R_L. Thus, at these frequencies the voltage gain expression (5.27) must be changed by replacing R_C by the parallel combination of R_C and R_L. Thus for the complete circuit, including the load resistance R_L, the voltage gain is

$$A_v = -\frac{\beta(R_C \| R_L)}{R_{BE}} \tag{5.28}$$

The measured gains and impedances for the common emitter amplifier circuit of Fig. 5.14(b) are

Voltage gain	$A_v = -180$
Circuit input impedance	$Z_{in} \cong 1\ \text{k}\Omega$
Circuit output impedance	$Z_{out} \cong 2\ \text{k}\Omega$
3-dB bandwidth	$= 400$ kHz

The resistance R in the input high-pass filter containing C_1 is not just R_1 but the total effective ac resistance (or impedance really) seen by the ac signal. That is, we must draw the ac equivalent circuit, which is obtained by realizing that the power supply terminal at 20 V dc is an ac ground. Thus the ac equivalent circuit is as shown in Fig. 5.17. Notice that R_E is not present in the ac equivalent circuit because we have $X_{C_E} \ll R_E$; the capacitance C_E is an ac short circuit. The input high-pass filter therefore consists of C_1 and the parallel combination of R_1, R_2, and the effective input resistance R_{BE} between the transistor base and emitter terminals. If we assume $R_{BE} = 1\ \text{k}\Omega$, then the effective resistance in the filter is equal to $R_1 \| R_2 \| R_{BE} = 740\ \Omega$, as shown in Fig. 5.18. Thus the input high-pass filter

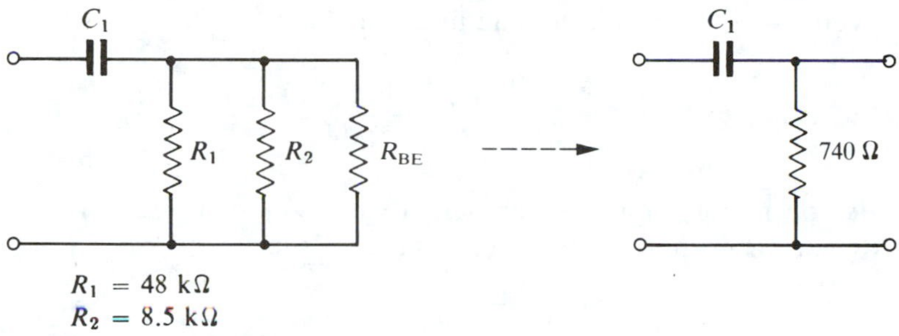

FIGURE 5.18 Input high-pass *RC* filter circuit.

consists of C_1 and a 740-Ω resistor. Therefore to pass frequencies down to 20 Hz we should have

$$\omega_B = 2\pi f_B = \frac{1}{R'C_1} \ll 2\pi(20\ \text{Hz}) \qquad \text{where } R' = R_1 \| R_2 \| R_{BE} = 740\ \Omega$$

which implies

$$C_1 \gg \frac{1}{R' 2\pi(20\ \text{Hz})} = \frac{1}{(2\pi)(20\ \text{Hz})(740\ \Omega)} \cong 10\ \mu\text{F}$$

Hence a 50- or 100-μF capacitor is needed for C_1. Fortunately, the dc voltage C_1 must withstand is not too high, so we may use a low-voltage electrolytic capacitor for C_1, which will be inexpensive. If the dc voltage of the input is more positive than +1.6 V, we hook up C_1 as shown in Fig. 5.14(b). If we expected a normal dc voltage of approximately 5 V to exist across C_1, we would choose a 10-V or a 25-V rating for C_1. The final amplifier circuit is shown in Fig. 5.14 with the dc voltages given for various points in the circuit.

To sum up, the common emitter amplifier has a large voltage gain, 180° phase difference between input and output, a large current gain, and medium input and output impedances. It is the most widely used transistor configuration.

5.8 COMMON COLLECTOR AMPLIFIER DESIGN

In the common collector configuration, shown in Fig. 5.19, the collector terminal is common to both the input and the output. The input is at the base, and the output is taken off the emitter, that is, across the emitter resistor R_E.

The dc bias design is similar to that for the common emitter amplifier. Assume $V_{bb} = 20\text{ V}$ and $\beta = 100$. A dc load line is chosen and drawn on the I_C-versus-V_{CE} curves. The load line equation is

$$I_E = \frac{V_{bb}}{R_E} - \frac{V_{CE}}{R_E} \tag{5.29}$$

The dc operating point is chosen (e.g., $V_{CE} = V_{bb}/2 = 10\text{ V}$, and $I_E = 10\text{ mA}$). R_E is now determined from Ohm's law:

$$R_E = \frac{V_E}{I_E} = \frac{V_{bb} - V_{CE}}{I_E} = \frac{10\text{ V}}{10\text{ mA}} = 1\text{ k}\Omega$$

(R_E could also be calculated from the load line I_E intercept where $V_{CE} = 0$.) $V_E = 10\text{ V}$, so $V_B = V_E + 0.6\text{ V} = 10.6\text{ V}$. If we now choose $I_D \gg I_B$, R_1 and R_2 are determined. $I_B = I_E/(\beta + 1) = 10\text{ mA}/101 \cong 0.1\text{ mA}$, so we choose $I_D = 10 I_B = 1\text{ mA}$. Then

$$R_1 = \frac{V_B}{I_D} = \frac{10.6\text{ V}}{1\text{ mA}} = 10.6\text{ k}\Omega$$

$$R_2 = \frac{V_{bb} - V_B}{I_D} = \frac{20\text{ V} - 10.6\text{ V}}{1\text{ mA}} = 9.4\text{ k}\Omega$$

The ac design is simply adding two coupling or blocking capacitors C_1 and C_2. C_1 and the parallel combination of R_1, R_2, and R_{BE} form a high-pass RC filter at the input. But R_{BE} is usually very large because when V_B becomes more positive, so does V_E, thereby making V_{BE} small. If V_{BE} is small, the base draws little current and R_{BE} is large. Thus $R_1 \| R_2 \| R_{BE} \cong R_1 \| R_2$ and we choose C_1 so that the filter breakpoint or knee is less than the lowest signal frequency ω_L: $C_1 \gg 1/(\omega_L R_1 \| R_2)$. Similarly, C_2 and the load form a high-pass RC filter at the output, so we choose C_2 so that the filter breakpoint or knee is less than the lowest signal frequency ω_L: $C_2 \gg 1/\omega_L R_L$.

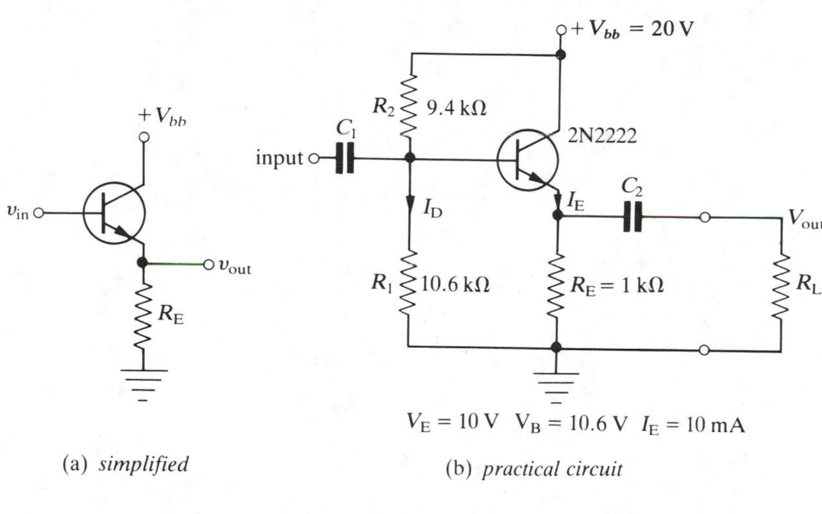

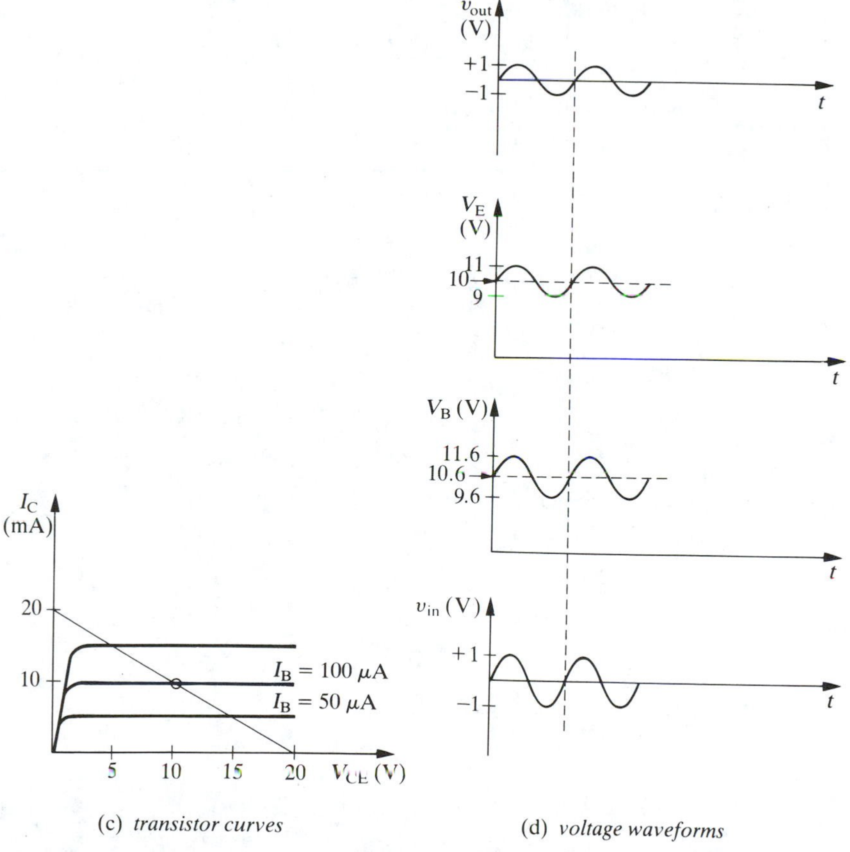

FIGURE 5.19 Common collector amplifier.

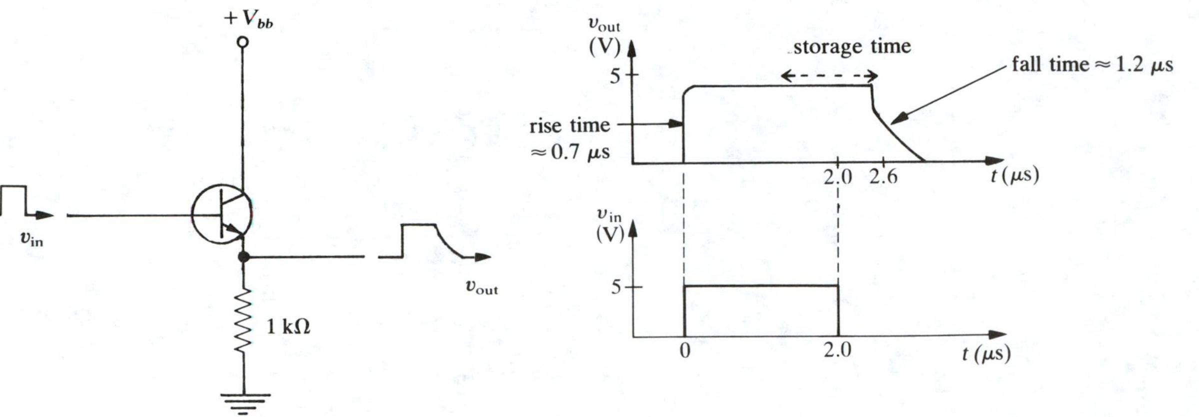

(e) *turn-on and turn-off speed*

FIGURE 5.19 Continued.

The measured gains and impedances of the complete circuit of Fig. 5.19 are

Voltage gain	$A_v = 1$
Circuit input impedance	$Z_{in} = 5\ \text{k}\Omega$
Circuit output impedance	$Z_{out} = 5\ \Omega$
3-dB bandwidth	> 10 MHz

Notice that the voltage gain of essentially 1 is reasonable because if the base-emitter junction is forward biased, V_{BE} will remain essentially constant at 0.6 V. Thus, the Kirchhoff voltage law for the input loop $V_{in} - V_{BE} - V_{out} = 0$ implies that $\Delta V_{in} = \Delta V_{out}$, which means a gain of 1.0. Here we have assumed that X_{C_1} is negligible. Also notice that as the input goes more positive, the transistor turns on and I_E increases, which means that the output *also* goes more positive. In other words, the output voltage swing is *in phase* with the input voltage swing; the common collector amplifier is thus often called the "emitter follower" because the output voltage on the emitter "follows" the input voltage at the base.

The large current gain is intuitively reasonable because of the steepness of the I_B (or I_E)-versus-V_{BE} curve; a very small change in V_{BE} (small change in I_B) will produce a very large change in I_E.

The high input impedance is essentially a result of the emitter voltage following the base voltage, thus tending to minimize the base-emitter difference voltage and, hence, the base current. The small base current means, of course, that the input resistance R_{BE} looking into the base is large.

The low output impedance occurs because the ac emitter current swing is much larger than the ac base current swing, and because the input (base) and output (emitter) voltage swings are nearly equal.

All of these properties of the emitter follower can be explained in terms of negative feedback, which we will do in Chapter 8.

An approximate voltage gain expression can be derived by assuming the transistor base presents an effective resistance R_{BE} to the base current (i.e., $V_{BE} = I_B R_{BE}$). Then the KVL implies

$$V_{in} - V_{BE} - V_{out} = 0$$

or

$$V_{in} = V_{BE} + V_{out} = I_B R_{BE} + V_{out}$$

Thus

$$A_v = \frac{\Delta V_{out}}{\Delta V_{in}} = \frac{\Delta V_{out}}{\Delta I_B R_{BE} + \Delta V_{out}} = \frac{i_E R_E}{i_B R_{BE} + i_E R_E}$$

Using $i_B \cong i_E/(\beta + 1)$, we get

$$A_v = \frac{R_E}{\dfrac{R_{BE}}{\beta + 1} + R_E} = \frac{1}{1 + \dfrac{R_{BE}}{(\beta + 1)R_E}}$$

From the transistor curves of Fig. 5.19(c), for the operating point $I_C = 10\text{ mA}$, $V_{CE} = 10\text{ V}$, $I_B = 100\ \mu\text{A}$, so $R_{BE} = V_{BE}/I_B \cong 0.6\text{ V}/100\ \mu\text{A} = 6\text{ k}\Omega$. Thus for $R_E = 1\text{ k}\Omega$ and $\beta = 100$,

$$A_v \cong \frac{1}{1 + \dfrac{6\text{ k}\Omega}{101\text{ k}\Omega}} = 0.94$$

To sum up, the common collector amplifier has a unity voltage gain, a large current gain, no phase inversion between the input and output, high input impedance, and low output impedance. Its principal use is in driving low-impedance loads such as long cables or loudspeakers.

It should be mentioned that if a rectangular pulse is fed into an emitter follower, the output rise time will differ significantly from the output fall time. For the npn circuit of Fig. 5.19, the negative-going edge of the input pulse will turn the transistor *off*, and the output voltage at the emitter can fall only by the stray capacitance C_{stray} (between the output terminal and ground) discharging through the emitter resistor R_E. Thus, the output time constant for the negative-going edge will be $R_E C_{stray}$. However, for the positive-going edge of the input pulse, the npn transistor will be turned on and the output voltage will rise by the stray capacitance C_{stray} charging through the transistor, which, being turned on, presents a very small output resistance (usually about $100\ \Omega$ or less). Thus the time constant for the positive-going edge of the output pulse will be an order of magnitude less than for the negative-going edge. For a pnp transistor the reverse situation holds. The time constant for the negative-going edge will be less than for the positive-going edge of the output. To sum up, it is harder (slower) to turn a transistor off than to turn one on. This is shown in Fig. 5.19(e) along with the storage time which will be covered in 5.11.

An interesting and useful variation of a common collector amplifier or emitter follower is the "double emitter follower" or "Darlington" amplifier shown in Fig. 5.20. The Darlington circuit consists of two transistors connected together such that the emitter current of the first transistor provides the base current of the second transistor. The advantage of this arrangement is that the two transistors may be regarded as one "supertransistor" with a β essentially equal to the *product* of the β's of the two individual transistors. This can be seen from the following approximate argument:

$$I_{E_1} = I_{B_2} \quad \therefore \quad I_{B_1} \cong \frac{I_{E_1}}{\beta_1} = \frac{I_{B_2}}{\beta_1} = \frac{I_{E_2}/\beta_2}{\beta_1} = \frac{I_{E_2}}{\beta_1\beta_2} \tag{5.30}$$

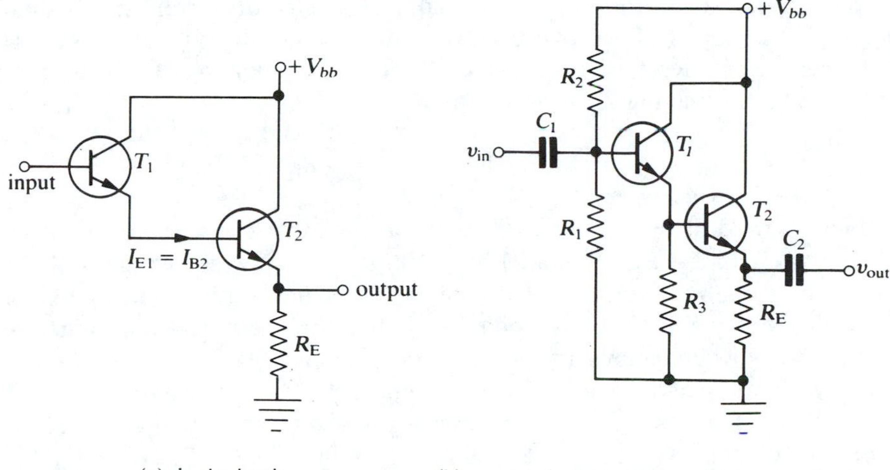

(a) *basic circuit* (b) *practical circuit with two discrete transistors*

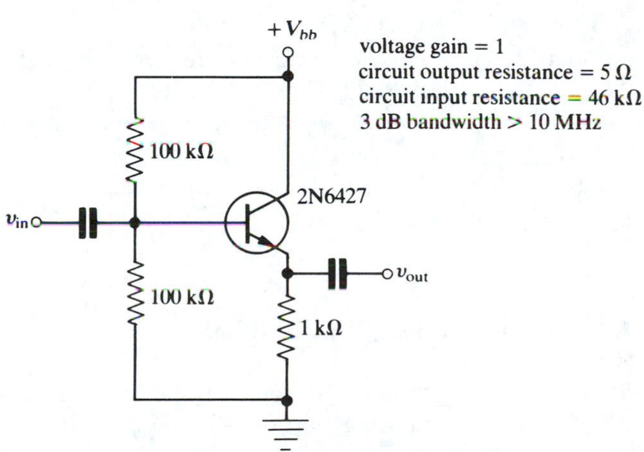

(c) *emitter follower circuit with a 2N6427 single package Darlington transistor*

FIGURE 5.20 Darlington emitter amplifier.

Thus the base current of the input transistor T_1 is the emitter current of the output transistor T_2 divided by $\beta_1\beta_2$, the product of the β's of the two transistors. The same equations hold approximately for the ac signal currents: $i_{B_1} \cong i_{E_2}/\beta_1\beta_2$. An overall β of approximately $100 \times 100 = 10{,}000$ may be achieved, thus producing a circuit with an extremely high input impedance and an extremely low output impedance when connected as an emitter follower.

In practice the principal difficulty is to make the emitter current of T_1

large enough to turn on T_2 and to make the collector current of T_2 small enough to keep T_2 from overheating. Thus we would probably never use the same transistor type for T_1 and T_2 because, assuming $\beta = 100$, if we turn on T_1 by making $I_{E_1} = 1$ mA, then

$$I_{B_2} = I_{E_1} = 1\text{ mA} \quad \text{and} \quad I_{C_2} = \beta_2 I_{B_2} = 100\text{ mA}$$

which might destroy T_2. Or, if we choose $I_{C_2} = 10$ mA to avoid overheating T_2, then $I_{B_2} = I_{C_2}/\beta_2 = 10\text{ mA}/100 = 100\ \mu\text{A} = I_{E_1}$, which would probably not be enough to turn on T_1. And if T_1 is not turned on, its gain is extremely low. Hence, T_1 should be a transistor especially designed to operate at low collector currents of approximately 100 μA, or else T_2 must be capable of handling large collector currents of approximately 100 mA.

A practical circuit including biasing is shown in Fig. 5.20(b). Resistances R_1 and R_2 set the dc operating point: $V_{B_1} = R_1/(R_1 + R_2)\,V_{bb}$, $V_{E_2} = V_{B_1} - 1.2$ V, and $I_{E_2} = V_{E_2}/R_E$. Resistance R_3 deserves some comment. If the temperature rises, I_{E_1} may increase enough (from thermally generated minority carriers) to turn on T_2. R_3 prevents this by diverting some of I_{E_1} away from the base of T_2.

Darlington transistors are available commercially in a single package with three leads: the emitter (of T_2), the base (of T_1), and the collector (both T_1 and T_2). They are widely used to control large currents (I_{E_2}) with small currents (I_{B_1}). Examples are the TIP-122 (npn) and the TIP-127 (pnp) power Darlingtons at \$1.00 each. They will carry up to $I_C = 5$ A and can dissipate up to 65 W. A low-power Darlington is the npn 2N6427 which has a β of 5000 or more and costs \$0.35 each.

5.9 COMMON BASE AMPLIFIER DESIGN

In the common base configuration, shown in Fig. 5.21, the base terminal is common to both the input and the output. The input is at the emitter, and the output is taken off the collector, that is, across the collector resistor R_C.

The dc bias design is straightforward. Assume $V_{bb} = 20$ V and $\beta = 100$. The load line equation is

$$I_C = \frac{V_{bb}}{R_C + R_E} - \frac{1}{R_C + R_E} V_{CE} \tag{5.31}$$

The dc operating point is chosen, say, $V_{CE} = 10$ V and $I_C = 2$ mA. The sum $R_C + R_E$ is determined by the load line I_C intercept where $V_{CE} = 0$ or from

$$I_C(R_C + R_E) = V_{bb} - V_{CE}$$

Thus,

$$R_C + R_E = \frac{10\text{ V}}{2\text{ mA}} = 5\text{ k}\Omega$$

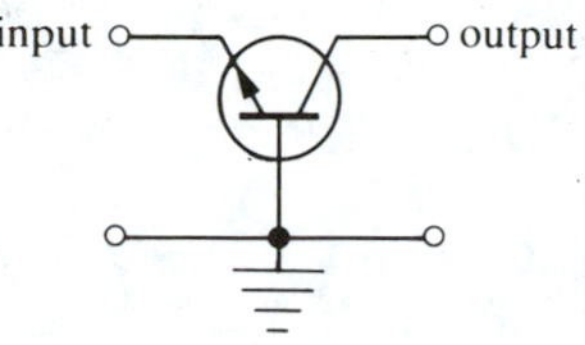

(a) *basic circuit*

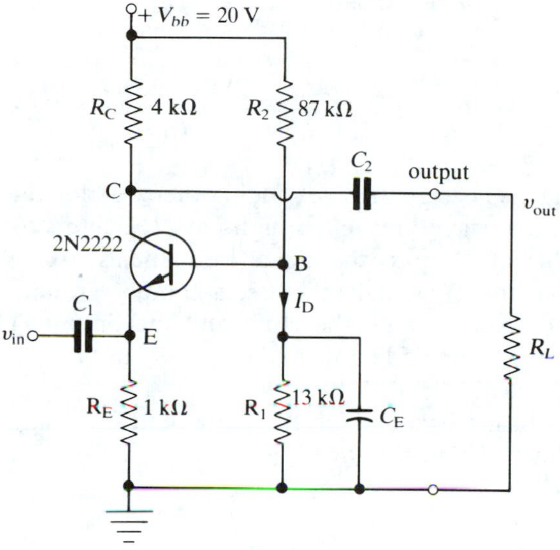

(b) *practical circuit*

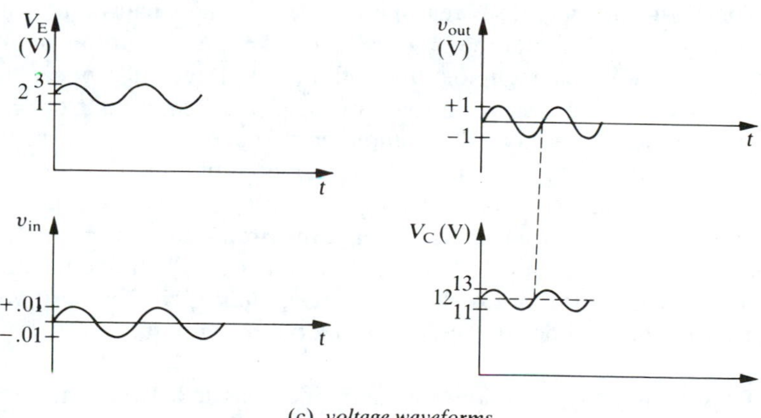

(c) *voltage waveforms*

FIGURE 5.21 Common base amplifier.

For high-voltage gain we want a large R_C, so we choose $R_E = 1\ \text{k}\Omega$ and $R_C = 4.0\ \text{k}\Omega$. Then

$$V_E = I_E R_E = I_C R_E = (2\ \text{mA})(1\ \text{k}\Omega) = 2\ \text{V}$$

and

$$V_B = V_E + 0.6\ \text{V} = 2.6\ \text{V}$$

$I_B = I_C/\beta = 2\ \text{mA}/100 = 0.02\ \text{mA}$, so we choose $I_D = 10 I_B = 0.2\ \text{mA}$.

Then

$$R_1 = \frac{V_B}{I_D} = \frac{2.6\ \text{V}}{0.2\ \text{mA}} = 13\ \text{k}\Omega$$

$$R_2 = \frac{V_{bb} - V_B}{I_D} = \frac{20\ \text{V} - 2.6\ \text{V}}{I_D} = 87\ \text{k}\Omega$$

The ac bias design involves adding the two blocking capacitors C_1 and C_2 in the usual way to make the high-pass RC filters formed by C_1 and R_E and by C_2 and R_L pass the signal frequencies. But we must also add a bypass capacitor C_E to make the base a good ac ground so that the base will be truly common to both the input and the output. This means that the reactance of C_E should be much less than R_1 at the lowest signal frequency ω_L: $1/\omega_L C_E \ll R_E$.

The measured values for the gains and impedances of the common base amplifier of Fig. 5.20 are

Voltage gain	$A_v = 250$
Circuit input impedance	$Z_{in} = 17\ \text{k}\Omega$
Circuit output impedance	$Z_{out} = 3.5\ \text{k}\Omega$
3-dB bandwidth	$= 150\ \text{kHz}$

The high-voltage gain is reasonable because the small input voltage is applied to the emitter, and the base voltage is kept constant (an ac ground). Thus, provided X_{C_1} is negligible, the full input voltage will appear between the base and emitter, just as in the case of the common emitter amplifier. However, in the common base amplifier when the input goes positive, the output *also goes positive*. A positive input makes the emitter more positive than the base, or, equivalently, the base is made more negative than the emitter, thus turning the transistor off more (i.e., the collector current decreases). This decrease in the collector current causes the output voltage at the collector to rise, thus giving a positive-going output. In other words, the output is *in phase* with the input but much larger in magnitude.

To sum up, the common base amplifier has a high voltage gain, no phase inversion between input and output, a relatively low input impedance, and a relatively high output impedance. Its principal use is at very high frequencies (50 MHz and higher), because the transistor base physically separates the input (emitter) and the output (collector). Thus, there is minimal feedback possible from output to input, which minimizes unwanted oscillations in practical circuits.

An example of a practical common base transistor 400-MHz amplifier is shown in Fig. 5.22. The common base configuration provides excellent

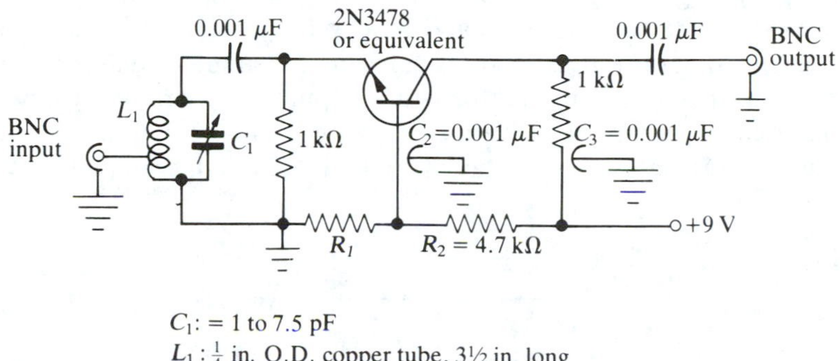

FIGURE 5.22 Common base 400-MHz amplifier.

isolation between the input and the output, thus producing an amplifier circuit with less tendency to oscillate than the common emitter configuration. The input "coil" L_1 resonates with C_1 at 400 MHz. L_1 is the small inductance of a $\frac{1}{4}$-in. diameter copper tube, approximately 0.05 μH. The base is an ac ground because of C_2. Notice the extra filtering of the supply voltage from C_3. Both C_2 and C_3 are "feed-through" bypass capacitors, so R_1 and R_2 are on the outside of the metal chassis, while all the other components are on the inside where they are well shielded. R_1 is adjusted to set the dc operating point (I_C) for the best measured signal-to-noise ratio.

Finally, for all three transistor configurations—the common emitter, common collector (emitter follower), and common base—the following checks should be made in debugging a circuit designed to amplify a small input signal (i.e., analog amplifiers, as opposed to a "switching" circuit that just turns on or off):

1. Is the base-emitter junction forward biased at approximately 0.6 V? For an npn transistor the base should be positive with respect to the emitter.
2. Is the base-collector junction reverse biased by at least 2 or 3 V? For an npn transistor the collector should be positive with respect to the base.
3. Is the dc collector current at least 1 mA? Check by measuring the dc voltage drop across R_C or R_E.
4. Is the divider current I_D large compared to I_B?

5.10 TRANSISTOR EQUIVALENT CIRCUITS

A transistor equivalent circuit is basically a circuit consisting of ideal voltage and/or current "sources" or "generators" and passive components (R, L, and C), which acts electrically exactly like the transistor. In other words, the transistor can be replaced by an appropriate collection of generators and passive components of the equivalent circuit. The advantage of the equivalent circuit is that one can predict the transistor's exact behavior (gain, etc.) by writing down the Kirchhoff voltage and current laws for the various loops and junctions and solving for the desired quantities by using only algebra and Ohm's law. A simple example will show how the calculations are performed. Suppose a certain transistor's equivalent circuit is simply an ideal voltage generator of magnitude Av_{in} in series with a resistance R as shown in Fig. 5.23. Algebraically, A is a

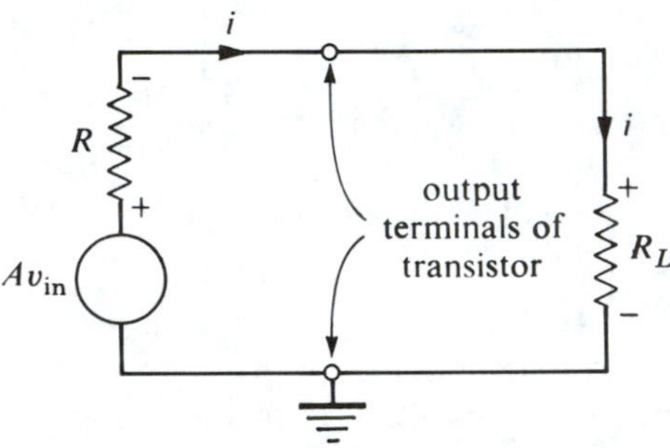

FIGURE 5.23 Simple equivalent circuit.

positive number, and v_{in} is the amplitude of the input voltage to the transistor. The R_L is the load resistance connected to the output terminals. The voltage generator produces a voltage A times as large as the input, so one might intuitively think of A as the voltage gain at this stage of the calculation. However, the voltage gain is

$$A_v = \frac{v_{\text{out}}}{v_{\text{in}}} = \frac{iR_L}{v_{\text{in}}} \tag{5.32}$$

and the current i can be obtained from the Kirchhoff voltage equation for the loop containing the generator, R, and R_L.

$$Av_{\text{in}} - iR - iR_L = 0 \quad \text{or} \quad i = \frac{Av_{\text{in}}}{R + R_L} \tag{5.33}$$

Thus the voltage gain becomes

$$A_v = \frac{v_{out}}{v_{in}} = \frac{iR_L}{v_{in}} = \frac{Av_{in}R_L}{(R + R_L)v_{in}}$$

or

$$A_v = A\left(\frac{R_L}{R + R_L}\right) \tag{5.34}$$

We see that the voltage gain depends on A, R, and R_L, and only as R_L becomes very large compared to R do we get $A_v \cong A$.

Let us now develop a perfectly general equivalent circuit that will apply to *any* four-terminal device, including a transistor, as shown in Fig. 5.24. I_1 and I_2 are the currents flowing into the input and output, respec-

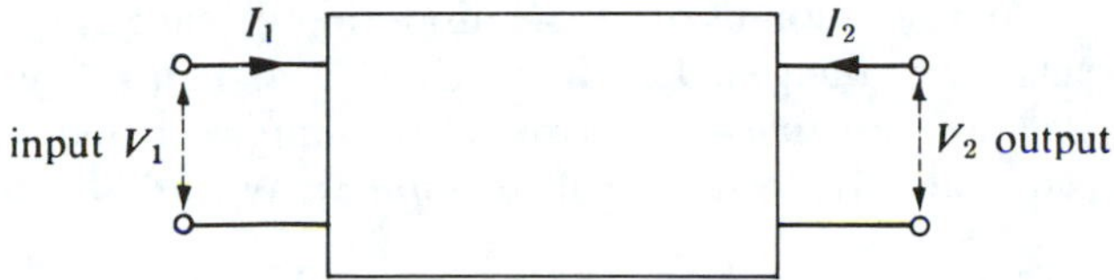

FIGURE 5.24 General four-terminal black box.

tively; and V_1 and V_2 are the voltage differences across the input and the output terminals, respectively. We have four variables: I_1, V_1, I_2, and V_2; they represent the total, instantaneous values of the currents and voltages. There are two Kirchhoff voltage equations we can write—one for the input loop and one for the output loop. In general, they can be expressed as

$$f(I_1, V_1, I_2, V_2) = 0 \tag{5.35}$$

and

$$g(I_1, V_1, I_2, V_2) = 0 \tag{5.36}$$

where f and g are mathematical functions whose exact form depends on the internal structure of the black box. We now choose (arbitrarily) to solve for the input voltage V_1 and the output current I_2 in terms of the other two variables I_1 and V_2. In mathematical language we are treating I_1 and V_2 as the independent variables and I_2 and V_1 as the dependent variables. This can always be done by solving (5.35) for V_1 and substituting the resulting expression for V_1 into (5.36), thus obtaining an equation not involving V_1.

$$g(I_1, I_2, V_2) = 0$$

This equation can then be solved for I_2 in terms of I_1 and V_2:

$$I_2 = I_2(I_1, V_2) \tag{5.37}$$

Similarly, (5.36) can be solved for I_2 and substituted in (5.35), which can then be solved for V_1:

$$V_1 = V_1(I_1, V_2) \tag{5.38}$$

In general, we are interested in the response of the transistor to ac signals, so we will take the differential of (5.37) and of (5.38) to obtain expressions for the change in I_2, dI_2, and the change in V_1, dV_1.

$$dI_2 = \left(\frac{\partial I_2}{\partial I_1}\right)_{V_2} dI_1 + \left(\frac{\partial I_2}{\partial V_2}\right)_{I_1} dV_2 \tag{5.39}$$

$$dV_1 = \left(\frac{\partial V_1}{\partial I_1}\right)_{V_2} dI_1 + \left(\frac{\partial V_1}{\partial V_2}\right)_{I_1} dV_2 \tag{5.40}$$

Let us change to a notation useful for considering ac signals or any change in the currents and voltages. Let $i_2 = dI_2$, $i_1 = dI_1$, $v_1 = dV_1$, and $v_2 = dV_2$. That is, lowercase v's and i's refer to *changes* in the voltages and currents or, equivalently, to *ac* signal *amplitudes*. With this notation,

$$i_2 = \left(\frac{\partial I_2}{\partial I_1}\right)_{V_2} i_1 + \left(\frac{\partial I_2}{\partial V_2}\right)_{I_1} v_2 \tag{5.41}$$

$$v_1 = \left(\frac{\partial V_1}{\partial I_1}\right)_{V_2} i_1 + \left(\frac{\partial V_1}{\partial V_2}\right)_{I_1} v_2 \tag{5.42}$$

We now define the h parameters for our four-terminal black box in terms of the partial derivatives:

$h_{21} \equiv \left(\frac{\partial I_2}{\partial I_1}\right)_{V_2}$ *a pure number* $\qquad h_{22} \equiv \left(\frac{\partial I_2}{\partial V_2}\right)_{I_1}$ *a conductance in mhos or siemens* (1 mho = 1 ohm^{-1} = 1 siemen)

$h_{11} \equiv \left(\frac{\partial V_1}{\partial I_1}\right)_{V_2}$ *a resistance in ohms* $\qquad h_{12} \equiv \left(\frac{\partial V_1}{\partial V_2}\right)_{I_1}$ *a pure number*

(5.43)

The h parameters have a variety of dimensions; hence, the name "hybrid" parameters. With this notation we have

$$i_2 = h_{21} i_1 + h_{22} v_2 \tag{5.44}$$

$$v_1 = h_{11} i_1 + h_{12} v_2 \tag{5.45}$$

Equations (5.44) and (5.45), relating the dependent variables i_2 and v_1 to the independent variables i_1 and v_2 via the h parameters, determine the equivalent circuit. The term $h_{21}i_1$ means there is a current generator of magnitude h_{21} times i_1. The term $h_{22}v_2$ means the voltage v_2 appears across a conductance h_{22} (or equivalently across a resistance of $1/h_{22}$ ohms). The term $h_{11}i_1$ means the current i_1 flows through an effective resistance of h_{11} ohms. The term $h_{12}v_2$ means there is a voltage generator of magnitude $h_{12}v_2$. Therefore, we can draw the equivalent circuit of Fig. 5.25. Equation (5.44) is seen to be merely the Kirchhoff current equation for the output, and equation (5.45) is merely the Kirchhoff voltage equation for the input. The important point here is that the perfectly general mathematical treatment that led to equations (5.44) and (5.45) implies the equivalent circuit of Fig. 5.25.

Some physical feeling for the h parameters can be obtained from the equivalent circuit of Fig. 5.25. The h_{11} parameter is a resistance in the input circuit, usually called the "input resistance." The term $h_{12}v_2$ is the amplitude of a voltage generator in the input; it represents how much of the output voltage v_2 is transferred or fed back to the input, and h_{12} is called the "reverse voltage transfer ratio." The word "reverse" is used to denote the transfer from the output back to the input. The h_{21} parameter represents how much of the input current i_1 is transferred to the output; h_{21} is called the "forward current transfer ratio." The higher the value of h_{21} is, the larger is the change in output current for a given input current change. We call h_{22} the "output admittance" because it is an admittance or conductance directly across the output terminals.

The preceding development is entirely mathematical and is exact; that is, no approximations have been made except that v and i must refer to small signals because equations (5.41) and (5.42) hold exactly only for infinitesimal i's and v's. However, we have not yet shown that the equivalent circuit of Fig. 5.25 is, in fact, a representation of a *real* transistor. To do so, we must look at the *experimental* input and output curves for a transistor and see if we can accurately represent the transistor with the equivalent circuit by choosing numerical values for the h parameters. For a useful equivalent circuit we would like a constant set of h parameters to represent the experimental curves over a *wide* range of currents and

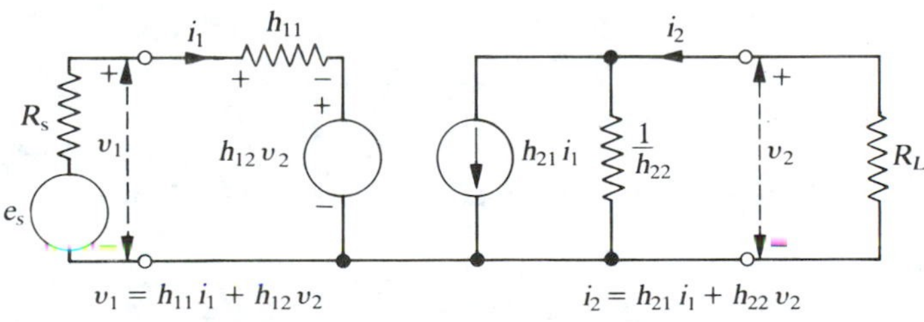

FIGURE 5.25 Transistor h parameter equivalent circuit.

voltages. The equivalent circuit is simply not useful if the h parameters are strong functions of current and voltage.

A transistor has three terminals, whereas our black box from which the equivalent circuit was developed has four. Hence, for the equivalent circuit to be applied to a transistor, one transistor terminal must be common between the input and the output. This can be either the emitter, the collector, or the base, called, respectively, the "common emitter" (CE), the "common collector" (CC), or the "common base" (CB) configuration.

For the widely used common emitter configuration, we have:

$$h_{21} = \left(\frac{\partial I_2}{\partial I_1}\right)_{V_2} = h_{fe} = \left(\frac{\partial I_C}{\partial I_B}\right)_{V_{CE}} = \beta \tag{5.46}$$

The h_{fe} parameter is often loosely called the "current gain." Capital letter subscripts refer to dc values; thus

$$h_{FE} = \frac{I_C}{I_B}$$

In most cases, $h_{fe} \cong h_{FE}$.

$$h_{11} = \left(\frac{\partial V_1}{\partial I_1}\right)_{V_2} = h_{ie} = \left(\frac{\partial V_{BE}}{\partial I_B}\right)_{V_{CE}} \tag{5.47}$$

But $I_E = I_0(e^{eV_{BE}/kT} - 1)$ for the base-emitter junction from Chapter 4, and $I_E = (\beta + 1)I_B$, so

$$I_B = \frac{I_0}{(\beta + 1)}(e^{eV_{BE}/kT} - 1) \cong \frac{I_0}{(\beta + 1)} e^{eV_{BE}/kT}$$

Thus $$h_{ie} = \left(\frac{\partial V_{BE}}{\partial I_B}\right)_{V_{CE}} = \left(\frac{\partial I_B}{\partial V_{BE}}\right)^{-1} = \left[\frac{\partial}{\partial V_{BE}}\left(\frac{I_0}{(\beta + 1)} e^{eV_{BE}/kT}\right)\right]^{-1}$$

$$h_{ie} = \frac{(\beta + 1)kT}{eI_0} e^{-eV_{BE}/kT}$$

or for $T = 300$ K and $\beta = 100$

$$h_{ie} \cong \frac{\beta kT}{eI_C} \cong \frac{2.6 \text{ V}}{I_C} \tag{5.48}$$

Thus, if $I_C = 100\ \mu$A, then $h_{ie} \cong 26$ kΩ. If $I_C = 1$ mA, then $h_{ie} \cong 2.6$ kΩ. A useful formula is

$$h_{ie}(\text{in k}\Omega) \cong \frac{2.6 \text{ k}\Omega}{I_C(\text{in mA})} \tag{5.49}$$

The effective ac signal resistance of the base-emitter junction R_{BE} which we introduced earlier in this chapter is equal to h_{ie}.

The h parameters will be different for the three configurations, and letter subscripts are usually used to distinguish one from another according to Table 5.2. The second letter of the subscript denotes the common terminal of the configuration, e.g., h_{fe} is h_{21} for the common emitter configuration and h_{fc} is h_{21} for the common collector configuration.

TABLE 5.2. *h* Parameter Values for the Three Transistor Configurations

h parameter	*CE*	*CC*	*CB*
h_{11}	$h_{ie} \approx 1\ k\Omega$	$h_{ic} \approx 1\ k\Omega$	$h_{ib} \approx 10\ \Omega$
h_{12}	$h_{re} \approx 10^{-4}$	$h_{rc} \approx 1$	$h_{rb} \approx 2 \times 10^{-4}$
h_{21}	$h_{fe} \approx 100$	$h_{fc} \approx -100$	$h_{fb} \approx -0.99$
h_{22}	$h_{oe} \approx 2 \times 10^{-5}$ siemens	$h_{oc} \approx 2 \times 10^{-5}$ siemens	$h_{ob} \approx 2 \times 10^{-7}$ siemens
Δh	$\Delta_e h \approx 2 \times 10^{-2}$	$\Delta_c h \approx 100$	$\Delta_b h \approx 4 \times 10^{-4}$
$\Delta h \equiv h_{11}h_{12} - h_{12}h_{21}$			

CE = common emitter
CC = common collector
CB = common base

From the general h parameter equivalent circuit of Fig. 5.25 and (5.44) and (5.45), we can calculate the voltage gain, the current gain, the input impedance, and the output impedance of the transistor. The results are

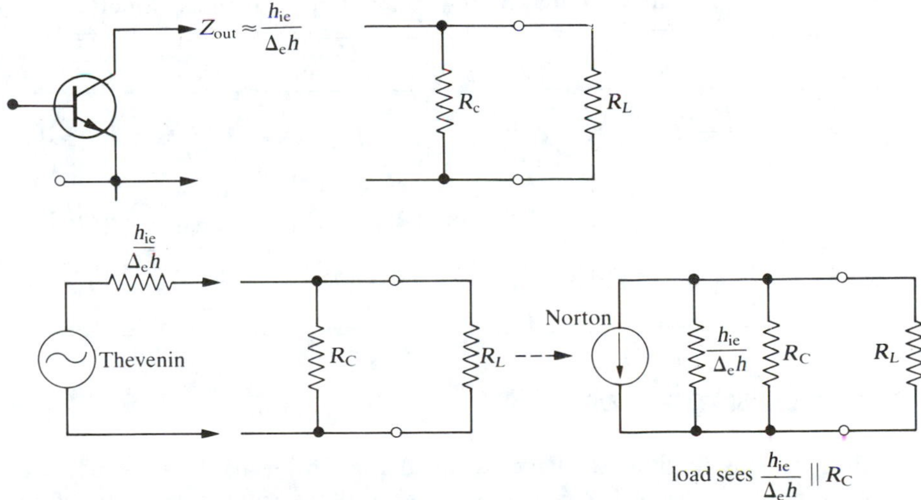

FIGURE 5.26 Transistor "sees" $R_C \parallel R_L$.

voltage gain: $$A_v = \frac{-h_{21}R_L}{\Delta h R_L + h_{11}} \tag{5.50}$$

current gain: $$A_i = \frac{h_{21}}{1 + h_{22}R_L} \tag{5.51}$$

input impedance: $$Z_{in} = \frac{\Delta h R_L + h_{11}}{1 + h_{22}R_L} \tag{5.52}$$

output impedance: $$Z_{out} = \frac{h_{11} + R_s}{\Delta h + h_{22}R_s} \tag{5.53}$$

where $$\Delta h = h_{11}h_{22} - h_{12}h_{21}$$

Equations (5.50)–(5.53) apply to the *transistor alone*, with no input biasing resistors, collector resistor, or load resistor.

For the common emitter configuration, R_C and R_L (see Fig. 5.26) are in parallel in terms of ac, so R_L in (5.50)–(5.52) must be replaced by $R_C \| R_L$. The output impedance expression (5.53) for the transistor alone refers to the impedance looking into the transistor collector terminal. Therefore the load R_L sees the *transistor* output impedance in parallel with R_C. The output impedance can never be larger than R_C.

The input impedance (5.52) is that of the transistor alone. A signal source sees $R_1 \| R_2$ in parallel with (5.52). Thus the source sees $Z_{in} \| R_1 \| R_2$. The input impedance can never be larger than $R_b = R_1 \| R_2$.

Similar arguments apply to the CB and CC configurations.

The approximate gains and impedances for the transistor *circuits* are given in Table 5.3 in terms of the widely used common emitter parameters.

TABLE 5.3. Approximate Gains and Impedances for Transistor Circuits

	CE	*CC*	*CB*
A_v	$-\dfrac{h_{fe}(R_C \| R_L)}{h_{ie}} \approx -h_{fe} \approx 100$	1	$\dfrac{h_{fe}R_L}{h_{ie}} \approx h_{fe} \approx 100$
A_i	$h_{fe} \approx 100$	$-h_{fe} \approx -100$	1
Z_{in}	$(h_{ie} \| R_1 \| R_2)$	$\sim h_{fe}(R_E \| R_L \| R_1 \| R_2)$	$\left(\dfrac{h_{ie}}{h_{fe}} \| R_1 \| R_2\right)$
Z_{out}	$\dfrac{h_{ie}}{\Delta_e h} \| R_C \cong 5\ \text{k}\Omega \| R_C$	$\sim \dfrac{h_{ie}}{h_{fe}} \| R_E$	$\dfrac{h_{fe}R_s}{\Delta_e h} \| R_C$

5.11 TRANSISTOR SWITCHES

In the previous sections we have assumed that the transistors are used to amplify relatively small sinusoidal signals or other small pulses. In other words, the purpose of the circuit was to produce an enlarged replica of the input with higher voltage, current, and power. Such an amplifier is often called a "linear" amplifier. But in some applications the transistor is merely used to turn something completely on or off; that is, it acts like a switch. Thus the two states of the transistor are "full on" or "full off."

Figure 5.27 shows a transistor, with a load R_L in series with its collector, and the dc load line. As long as the base voltage is kept below

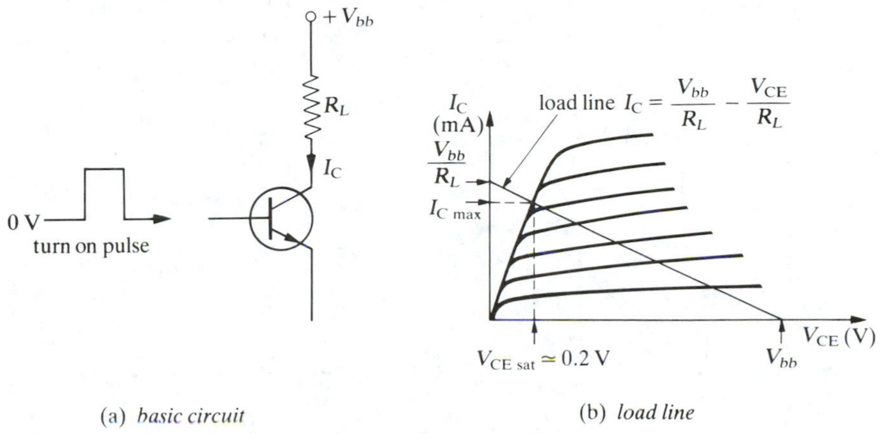

FIGURE 5.27 Transistor switch.

approximately 0.5 V (ground is fine), the transistor is "full off," $I_C = 0$, and $V_{CE} = V_{bb}$. If the base voltage is raised to 0.6 V or 0.7 V, the transistor will turn on, and I_C will depend on I_B: $I_C = \beta I_B$. But the load line implies that regardless of how large I_B is, V_{CE} cannot fall below the value $V_{CE\,sat}$ and I_C cannot exceed $I_{C\,max}$. $V_{CE\,sat}$ is the "saturation" collector emitter voltage and is typically 0.2 V for $I_C = 10$ mA, increasing for larger collector currents. $V_{CE\,sat}$ may be as high as 1 V for $I_C = 5$ A for a high-power transistor. From Kirchhoff's voltage law and Ohm's law,

$$I_{C\,max} = \frac{V_{bb} - V_{CE\,sat}}{R_L} \cong \frac{V_{bb}}{R_L}$$

When the (npn) transistor is "full on" ($I_C = I_{C\,max}$), its collector will be at approximately 0.2 V and its base at 0.6 or 0.7 V. Thus, the base-collector junction is *forward* biased instead of reverse biased, which was true for all the linear amplifier circuits we considered in Sections 5.7–5.9. When the base-collector junction is forward biased, the transistor is said to be in "saturation"—increasing the base current will no longer increase the collector current. In other words, $I_B > I_{C\,max}/\beta$ for saturation. The situation for linear amplifiers is $I_B = I_C/\beta$, with the base collector junction reverse biased.

Let us now go through a quick switching design problem. Suppose we

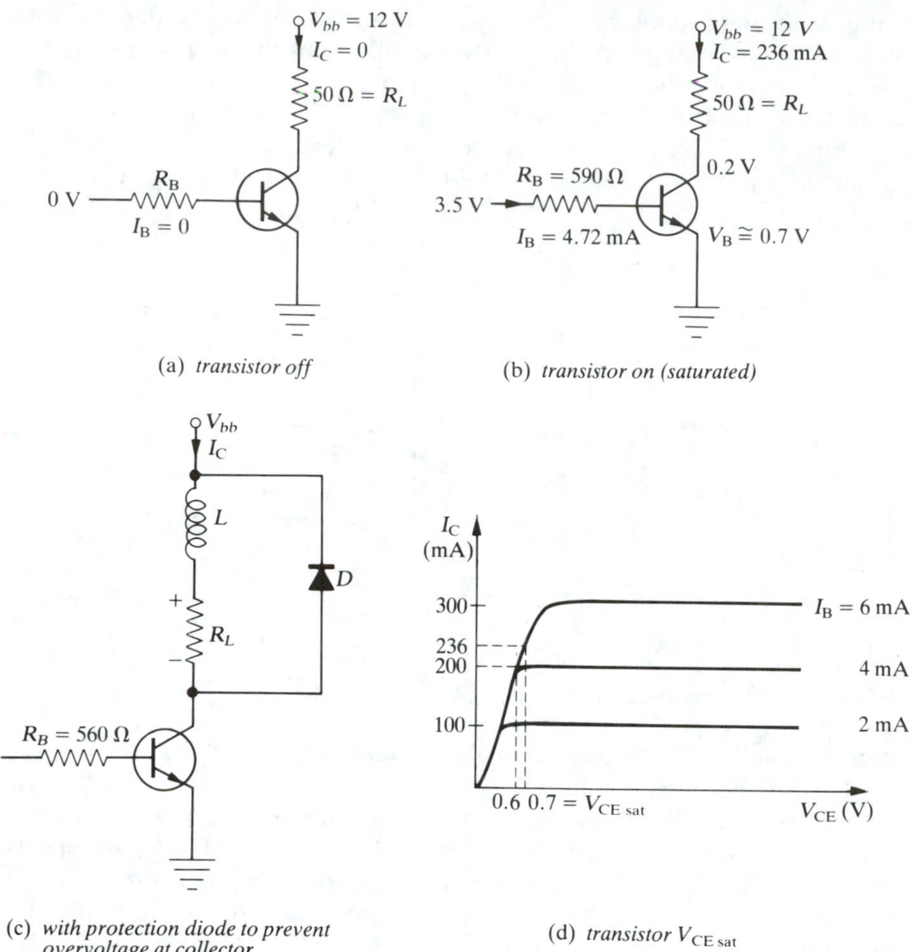

FIGURE 5.28 Switching problem circuit.

have a solenoid-operated valve with 50 Ω resistance that requires a current of 200 mA to open, and we wish to hold this valve open with a positive 3.5-V pulse coming from a microcomputer that can supply up to 5 mA output current when its output is +3.5 V. The circuit is shown in Fig. 5.28. We clearly must choose a transistor capable of carrying a 200-mA collector current. Assume the transistor $\beta = h_{FE} = 50$. The voltage drop across the solenoid valve will be (50 Ω)(200 mA) = 10 V, so the supply voltage must be at least 10 V. Let us choose $V_{bb} = 12$ V.

When the input signal from the microcomputer is 0 V, the base current will be zero and the collector current through the 50-Ω solenoid coil will be essentially zero. (It will equal $I_{CO} \simeq \mu$A due to the thermally generated minority carriers in the transistor and leakage current.)

When the input signal is 3.5 V, base current will flow into the base, and the transistor will be on with $V_{BE} \cong 0.7$ V. The base current will be

$$I_B = \frac{3.5\text{ V} - V_B}{R_B} \cong \frac{3.5\text{ V} - 0.7\text{ V}}{R_B}$$

The smaller R_B the larger I_B, and if the transistor is not saturated $I_C = h_{FE} I_B$ will also be larger. If the transistor is saturated, $V_{CE} = 0.2$ V, so $V_C = 0.2$ V and $I_C = I_{C\,max}$ is fixed by Ohm's law *regardless* of I_B:

$$I_{C\,max} = \frac{V_{bb} - V_C}{R_L} = \frac{12\text{ V} - 0.2\text{ V}}{50\ \Omega} = 236\text{ mA}$$

If we assume $V_{BE} = 0.7$ V for a saturation (this value depends upon the transistor type), then for the transistor to be at the edge of saturation,

$$I_B = \frac{I_{C\,max}}{h_{FE}} = \frac{236\text{ mA}}{50} = 4.72\text{ mA}$$

The microcomputer input can supply up to 5 mA, so this value of I_B is okay. Thus, R_B must be

$$R_B = \frac{3.5\text{ V} - 0.7\text{ V}}{4.72\text{ mA}} = 590\ \Omega$$

In other words, $R_B = 590\ \Omega$ will allow the 3.5-V input to drive the transistor to the edge of saturation.

If we use a lower value of R_B, I_B will be increased but I_C will remain locked at 236 mA because the transistor is saturated. For example, if $R_B = 560\ \Omega$, then

$$I_B = \frac{3.5\text{ V} - 0.7\text{ V}}{560\ \Omega} = 5.0\text{ mA}$$

This is the maximum current the microcomputer can supply at 3.5 V, so R_B must be greater than 560 Ω. For R_B between 560 Ω and 590 Ω the transistor will be saturated and I_C will be 236 mA. If R_B is greater than 590 Ω, the transistor will not be saturated, V_C will be above 0.2 V, and $I_C = h_{FE} I_B$ will hold. If we assume $V_{BE} = 0.6$ V for the nonsaturated transistor, $R_B = 725\ \Omega$ will produce $I_B = 4$ mA and $I_C = 200$ mA. In this nonsaturation condition any change in the 3.5-V input will produce a change in I_C, which is bad.

A comment about the load: If the load inductance is appreciable, as would be the case for a coil of many turns, any sudden change in the load

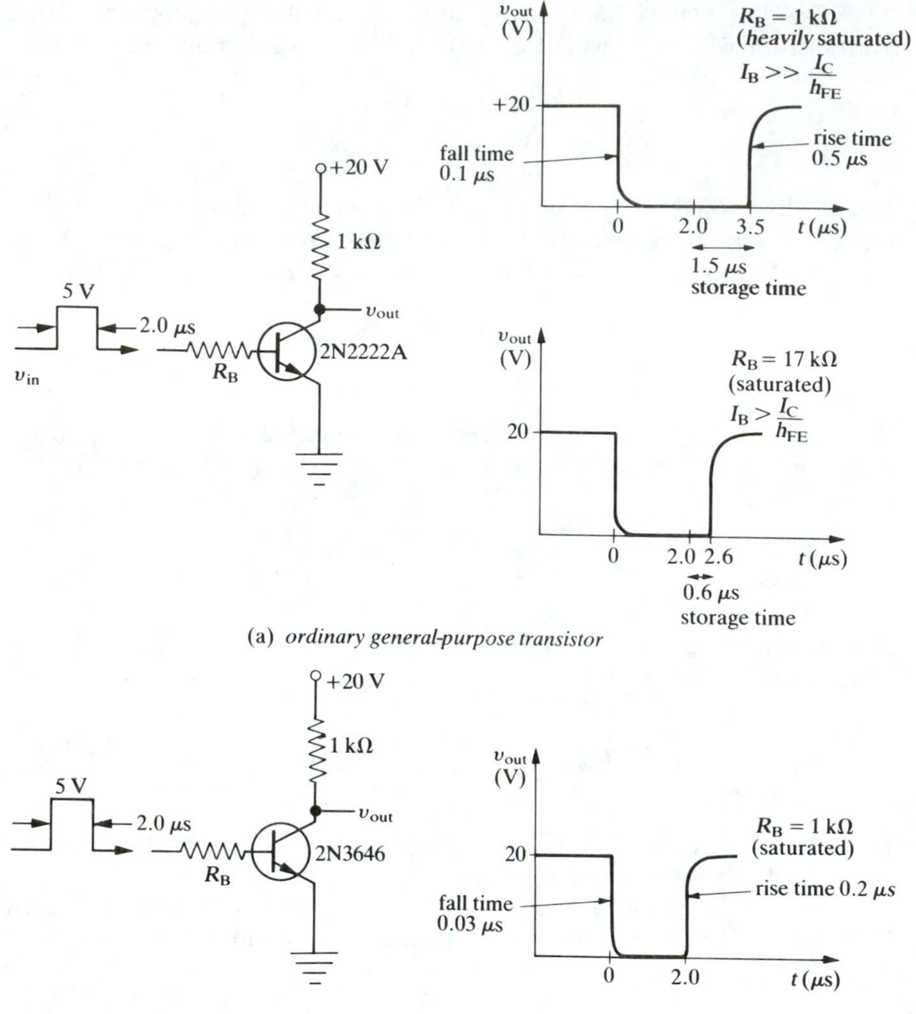

FIGURE 5.29 Storage time in standard transistor.

current would produce a large-amplitude voltage oscillation due to Faraday's law, $V_L = -L(dI/dt)$. The voltage swing might be enough to damage the transistor. To prevent this, connect a diode across the load, as shown in Fig. 5.28(c). The diode is reverse biased for normal circuit operation, but if a voltage oscillation has occurred, the diode will conduct on the reverse polarity oscillation and will limit the voltage across the load to 0.6 V. The oscillation will also be quickly damped out.

Finally, we should mention that although saturating the transistor provides a margin for error in I_B to keep the transistor fully on, a saturated transistor is slower to turn off than an unsaturated one with the same collector current, because in a saturated transistor the lack of reverse bias at the base collector junction means there is no depletion region there. Thus, the base is thicker, and because there is no electric field in the base the excess charge carriers in the base (due to the excess I_B) must slowly *diffuse* out of the base before the transistor can turn off. During the time it takes the excess electrons to diffuse out of the base, the transistor is still on, and this results in a "storage time" which lengthens the time the transistor is on. This is shown in Fig. 5.29(a). Notice the larger the base current the more saturation and the longer the storage time. Using a special switching transistor will eliminate the storage time as shown in Fig. 5.29(b).

To sum up, if it is important to make sure the transistor is fully on, use excess I_B to achieve saturation. Then small changes in I_B will not change I_C. The disadvantage is that the saturated transistor will be slow to turn off. If it is important to turn the transistor off quickly, then the transistor should not be saturated when it is conducting; this is achieved by using less base current. The disadvantage is that if the base current even slightly decreases, the collector current will decrease also.

PROBLEMS

1. (a) Describe the construction of a pnp transistor and an npn transistor. (b) What is the physical meaning of α?
2. Describe how to test an npn transistor with an ohmmeter.
3. Label the polarities of the two voltage supplies:

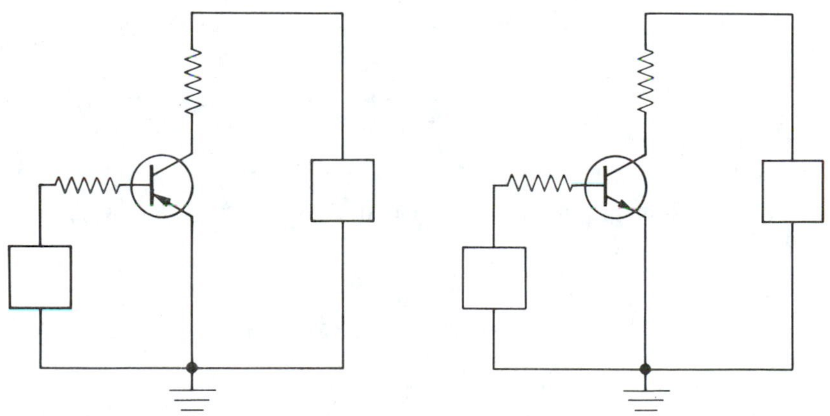

4. (a) The base voltage of an "on" silicon pnp transistor is always approximately ______V more ______ than the emitter. (b) The base voltage of an "on" silicon npn transistor is always approximately ______V more ______ than the emitter.
5. Carefully sketch the depletion region in a properly biased npn transistor. Show the mobile charge carriers, the fixed ionized impurity atoms, and any electric field vector present.

6. (a) How are I_C, I_B, and I_E related? (b) How are I_C and I_B related in terms of α? In terms of β? (c) How are I_C and I_E related in terms of α? In terms of β? (d) How are I_B and I_E related in terms of α? In terms of β?
7. Explain why the base of a transistor is purposely made thin. Would increasing the doping concentration of the base tend to increase or decrease the α? Explain.
8. Sketch a graph of the collector current versus base-emitter voltage for a silicon transistor. Repeat for the base current versus base-emitter voltage. Include approximate numerical values for the voltages and currents. Assume the transistor $\beta = 50$.
9. Explain briefly why a transistor amplifies when connected in the common emitter configuration.
10. Consider a transistor with a maximum power dissipation of 200 mW and a 20-V power supply. On a graph of I_C versus V_{CE} sketch the maximum power curve and shade in the forbidden region of operation. Also draw the dc load line for a 2-kΩ collector resistor. Is this a safe load line? Repeat for a 400-Ω collector resistor. Is this load line safe?
11. Calculate I_C and V_{CE}. The transistor is silicon and has a β of 100.

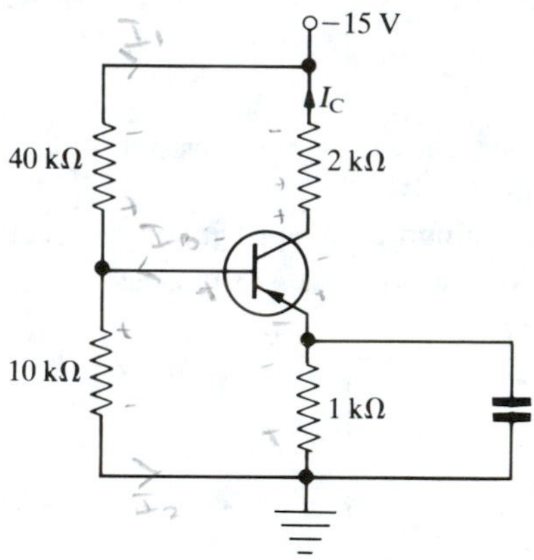

12. Calculate R_1 and R_2 if $I_D = 20I_B$. The transistor is silicon and has a β of 100. $V_E = 1$ V.

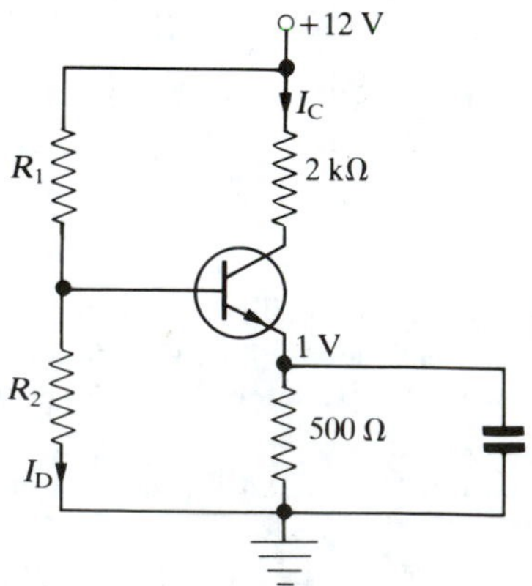

13. (a) Calculate R_1 and R_2 if $I_E = 1.5\ \text{mA}$, $\beta = 100$. (b) Calculate the voltage gain. (c) Estimate the input and output impedances.

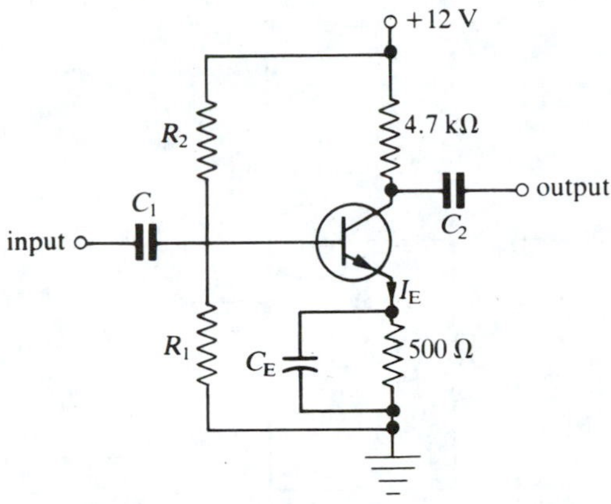

14. Calculate C_1 and C_2 if the desired bandwidth is (a) 20 Hz to 20 kHz, (b) 300 Hz to 3 kHz ($R_L = 100\ \text{k}\Omega$).

15. (a) Calculate R_1 and R_2 if $I_E = 5\ \text{mA}$, $\beta = 100$. (b) Estimate the input and output impedances.

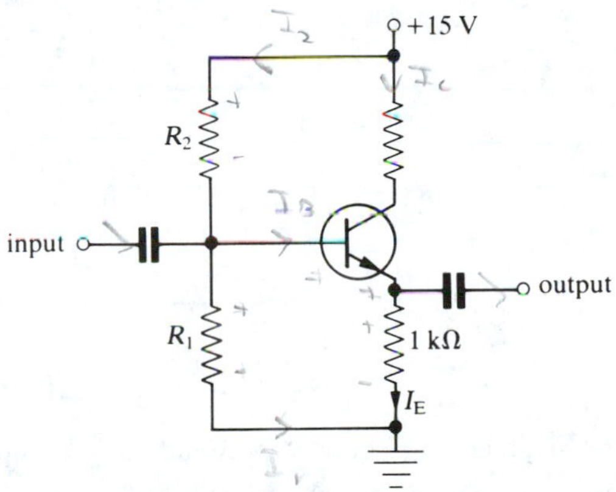

16. (a) Calculate R_E if $I_E = 4\ \text{mA}$, $\beta = 100$. (b) Carefully sketch the output voltage to scale, for the 1-V input pulses shown.

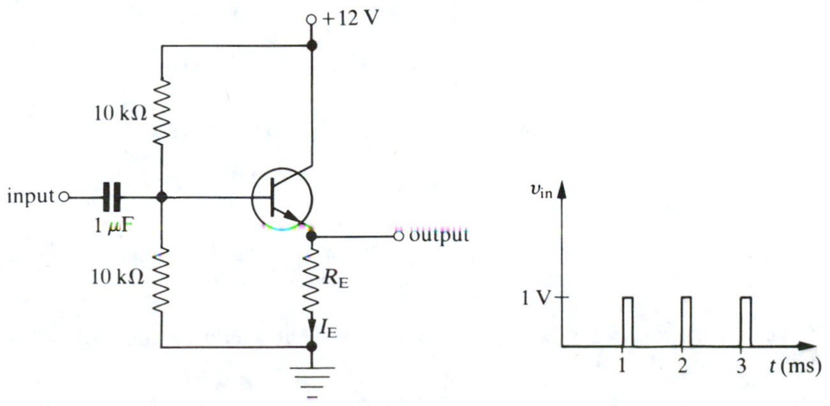

17. The output dc voltage (with no input) is exactly zero, and $I_E = 6$ mA. Calculate (a) R_E and (b) R_1 and R_2.

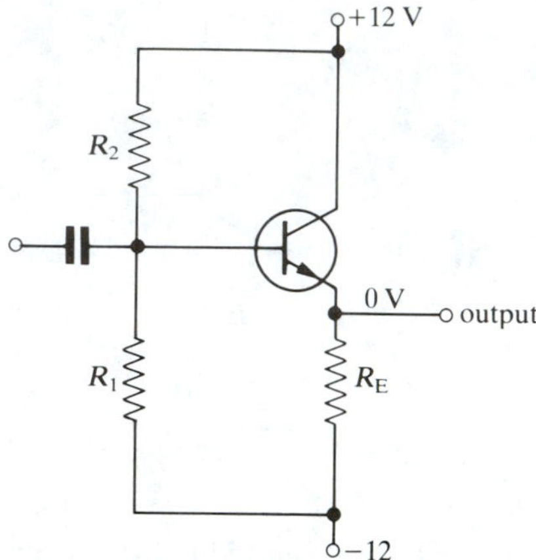

18. In the simple equivalent circuit shown calculate (a) the current gain, (b) the voltage gain, (c) the input impedance, and (d) the output impedance.

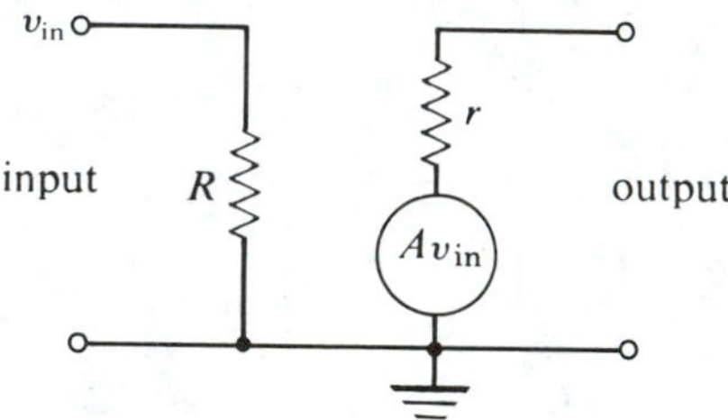

19. Explain why the h parameter equivalent circuit can be applied to any four-terminal device so long as small signals are concerned.
20. State the units of the four h parameters.
21. Derive the h parameter equivalent circuit of the following resistive "black box."

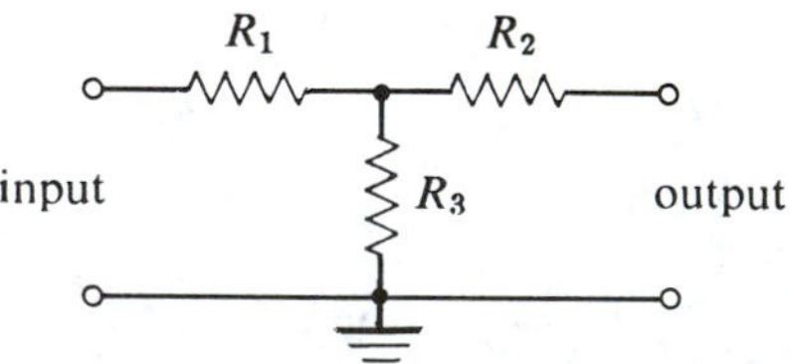

22. Give typical values for h_{ie}, h_{re}, h_{fe}, and h_{oe}. What would be the values for an ideal transistor?

23. (a) Calculate the four h parameters and β for the transistor whose input and output characteristic curves are shown below if the dc operating point is $I_C = 2$ mA, $V_{CE} = 10$ V. (b) As the collector current is increased, how do h_{fe} and h_{oe} change? (c) Compare h_{fe} with $\beta = \alpha/(1 - \alpha)$. (d) Explain why V_{CE} at the operating point should be greater than (1 mA, 1 V).

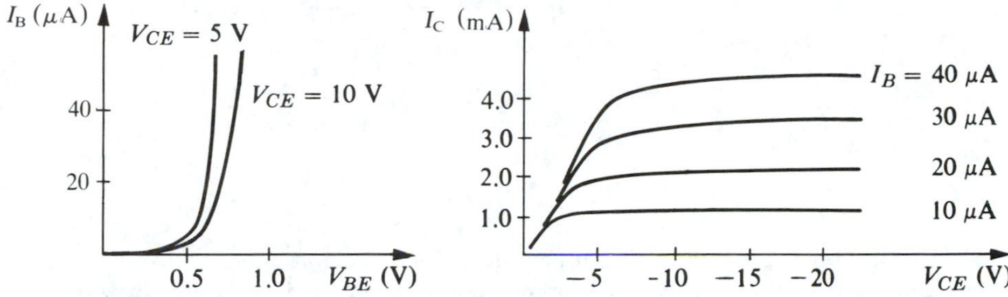

24. Show that the input impedance for a transistor connected in the common emitter configuration is equal to h_{ie} if $h_{re} = 0$. [*Hint*: Use the h parameter equivalent circuit.]

25. Calculate an approximate value for h_{ie} for a transistor at room temperature with $h_{fe} = 100$ and an operating point $V_{CE} = 10$ V, $I_C = 1$ mA. Repeat if the operating point is changed to $V_{CE} = 5$ V, $I_C = 100$ μA.

26. From the definition of h_{ie} and the current–voltage graph for the emitter-base junction, show (by a graphical argument) that the value of h_{ie} decreases for increasing emitter current. Sketch a rough graph of h_{ie} versus I_E.

27. Calculate the approximate 1-kHz ac input impedance to the *circuit* (not the transistor alone) between terminals A and B. $h_{ie} = 2$ kΩ, $h_{re} = 10^{-4}$, $h_{fe} = 100$, $h_{oe} = 10^{-5}$ mhos.

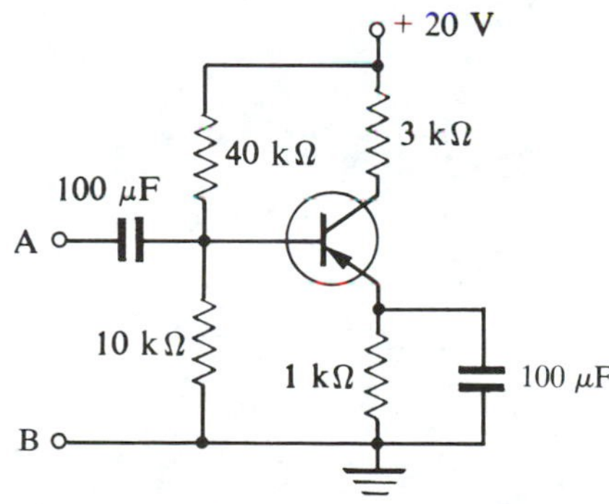

28. Calculate the voltage and current gains for the amplifier of Problem 27.

29. Calculate the output impedance for the amplifier of Problem 27.

30. Explain why R_1 should not be much less than h_{ie} for a common emitter amplifier. (R_1 is the external resistance from base to ground.)

31. What is the effective output impedance of an amplifier driven by a constant input voltage whose output versus load resistance is given in the following table?

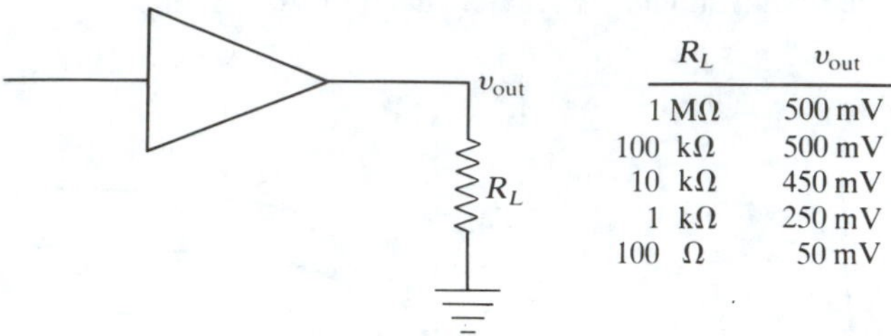

R_L	v_{out}
1 MΩ	500 mV
100 kΩ	500 mV
10 kΩ	450 mV
1 kΩ	250 mV
100 Ω	50 mV

32. Explain briefly, without using *h* parameters, why one would expect the voltage gain of an emitter follower or common collector amplifier to be less than unity.

33. Design an emitter follower for the input pulses shown. The transistor should have $I_E = 1$ mA *only* when the input pulse is at 0 V. [*Hint*: This involves choosing an appropriate dc operating point for the transistor.]

34. For the Darlington circuit shown, calculate (a) V_{E2}, (b) I_{E2}, and (c) I_{E1}. The transistors are silicon and $\beta_1 = 100$, $\beta_2 = 50$.

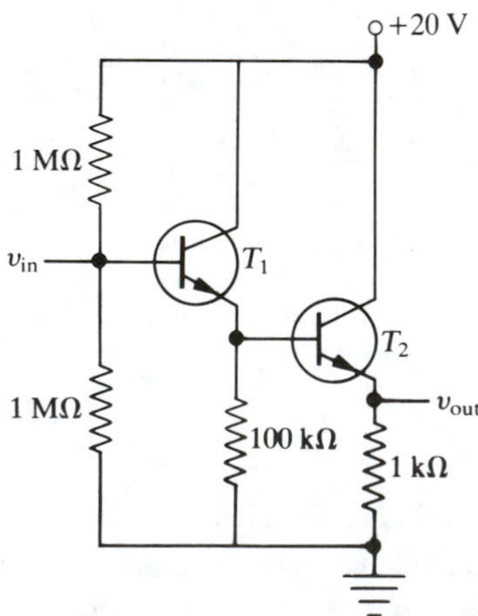

35. Calculate C_3, R_3, and R_4. The transistor is silicon. What would you estimate the output impedance to be?

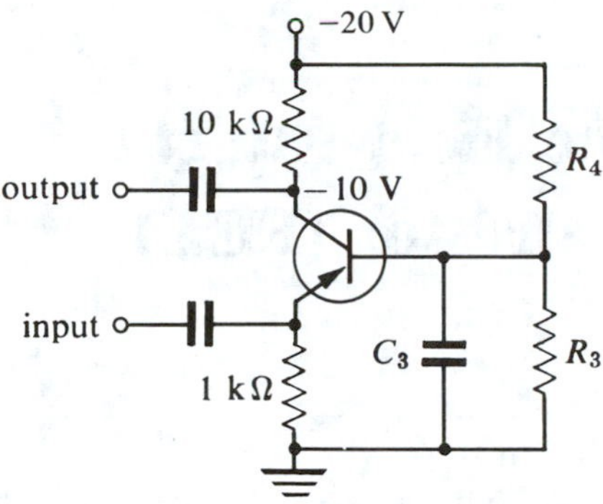

36. What is wrong with the common base amplifier shown below? Why? Correct the mistakes.

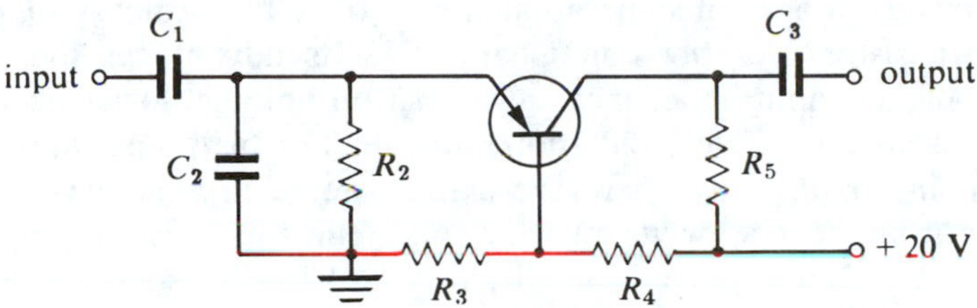

37. Calculate (a) V_{bb} if I_L must be 400 mA when the transistor is saturated "on," (b) the base current that must be supplied to turn the transistor on, and (c) R_B so that the transistor is just saturated when it is "on." $V_{CE\,sat} = 0.3$ V and $h_{FE} = 100$ for $I_C < 0.4$ A for the 2N3055.

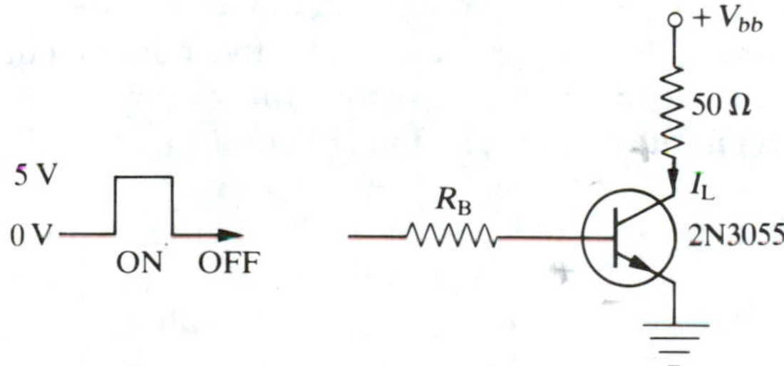

CHAPTER 6

The Field-Effect Transistor (FET)

6.1 INTRODUCTION

The field-effect transistor (FET) is a special type of transistor that offers a number of advantages in some applications over the ordinary bipolar pnp or npn transistors described in Chapter 5. Its advantages include considerably higher input impedance, zero turn-on voltage, lower noise, and a higher resistance to nuclear radiation damage. The high input impedance is important in applications where the signal source driving the FET has a high output impedance or where very small currents must be amplified. If the FET input impedance is much higher than the source output impedance, then little signal is lost due to impedance mismatch; that is, the source is not appreciably "loaded." In general, however, the gain available from a FET is less than from a carefully selected bipolar transistor. Because a FET carries current only by the flow of majority carriers, it is sometimes termed a "unipolar" transistor, although the term FET is more common. An ordinary transistor is often termed a "bipolar" transistor to distinguish it from a FET, because it carries current by the flow of both majority and minority carriers. In an npn transistor, for example, the electrons are majority carriers in the emitter and the collector but are minority carriers in the base.

There are two general types of field-effect transistors: the junction field-effect transistor (JFET), and the metal-oxide-semiconductor field-effect transistor (MOSFET). In this chapter we will discuss the construction of various types of FETs and the y parameter equivalent circuit, and we will give examples of several practical FET circuits.

6.2 JFET CONSTRUCTION

A junction field-effect transistor is basically a small slab of doped semiconductor, either p- or n-type, with the opposite type impurity diffused in either side of the slab, as shown in Fig. 6.1. The main slab is called the *channel*,

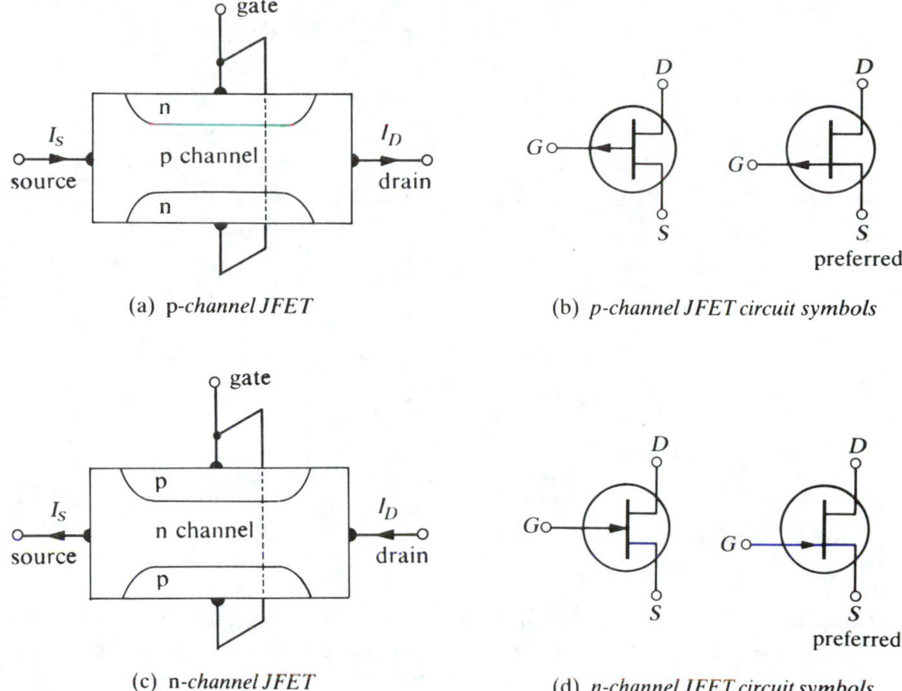

(a) p-*channel JFET*

(b) *p-channel JFET circuit symbols*

(c) n-*channel JFET*

(d) *n-channel JFET circuit symbols*

FIGURE 6.1 JFET construction.

and the two small regions of opposite type semiconductor material on the sides of the channel are connected together and called the *gate*.

In Fig. 6.1(a), the channel is made of p-type material and the gate of n-type material; this device is termed a *p-channel FET* or a *p-channel junction FET*, to be more precise. An n-channel junction FET is shown in Fig. 6.1(c). Most, but not all, JFETs are n-channel because of the greater conductivity of n-type material.

The direction of the arrow in the gate lead of the circuit symbol is the direction of "easy" current flow if the gate-channel junction were forward biased. It should be emphasized that the gate-channel junction in a JFET is *always reverse biased* so that gate current I_G *never* flows in the direction of the arrow in the JFET symbol. Because of this reverse bias, the gate current is the reverse or leakage current in the gate-channel junction, which is of the order of 1 to 10 nA. In other words, the input impedance looking into the gate is extremely high, of the order of 100 MΩ or even higher. This is the principal difference between the JFET and the bipolar transistor. Finally, note that $I_S + I_G = I_D$ and $I_S \cong I_D$.

Consider the n-channel JFET of Fig. 6.2. Ohmic contacts are made on each end of the channel, and one is called the *source* and the other the

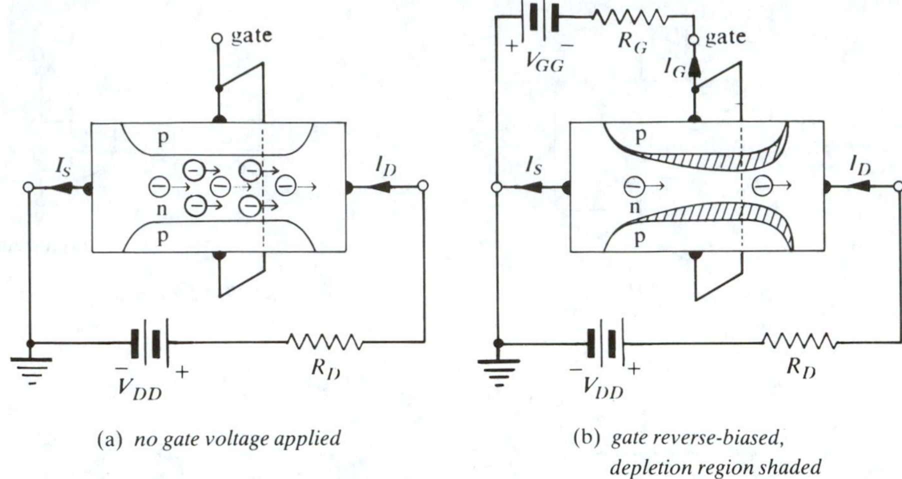

(a) *no gate voltage applied*

(b) *gate reverse-biased, depletion region shaded*

FIGURE 6.2 Current flow in an n-channel JFET.

drain. That is, the source and drain terminals are merely the two ends of the channel. If a voltage difference is applied between the source and the drain, with the drain positive with respect to the source, then the majority carrier electrons in the n-type channel will flow from the source towards the drain. A battery V_{DD} and a resistor R_D will accomplish this as shown in Fig. 6.2. If the gate terminal is not connected to anything, then the current I_D flowing into the drain must exactly equal the current I_S leaving the source. In this situation with the gate floating, the current is limited only by the resisitivity of the channel and, of course, the external resistance R_D. Recall that the resistance of a slab of material of cross-sectional area A, length l, and resistivity ρ is given by

$$r = \rho \frac{l}{A} \tag{6.1}$$

We can see that the drain current I_D will be given by

$$I_D = \frac{V_{DD}}{r + R_D} = \frac{V_{DD}}{\dfrac{\rho l}{A} + R_D} \tag{6.2}$$

In other words, the FET is acting just like a resistor, with its resistance being the resistance r of the channel material with a constant length l and cross-sectional area A.

However, if the gate material is reverse biased with respect to the channel, as shown in Fig. 6.2(b), then the situation changes drastically.

Recall from Chapter 4 that a reverse biased p-n junction will have a depletion region formed at the junction, which contains no mobile charge carriers but only fixed ionized impurity atoms. The region is shaded in Fig. 6.2(b): the larger the reverse bias voltage applied to the gate, the thicker the depletion region and the smaller the cross-sectional area of the channel through which the hole current I_D can flow. Notice that the electron current in the n-channel continues to flow only through n-type material, unlike the flow of electrons in an npn transistor where the mobile electrons find themselves minority carriers in the field-free p-type base material. Also notice that because of the ohmic voltage drop along the channel material as we go from the source toward the drain, the gate-channel reverse bias voltage is larger near the drain end than near the source end. Hence, the depletion layer is thicker nearer the drain than near the source.

If the voltage V_{DS} between the drain and the source is kept fixed, then as the gate reverse bias is increased, that is, as the gate is made more negative, the drain current will decrease as the depletion regions enlarge and narrow the cross-sectional area of the channel available for drain current flow. Eventually, the two depletion regions will meet and effectively block the channel, thereby reducing the drain current to zero. In a very real physical sense the gate terminal, when reverse biased, acts like a "gate" in restricting and cutting off the flow of current through the channel. When the two depletion regions meet and reduce the drain current to zero, this is referred to as *pinchoff*, because the channel has literally been pinched off by the two depletion regions. Typical values of gate-source voltage required to achieve pinchoff are several volts. The exact value of the pinchoff voltage depends on the dielectric constant ϵ of the channel material, the number of impurity atoms n per unit volume in the channel, and the physical thickness H of the channel measured perpendicular to the channel current (i.e., from gate to gate). An exact analysis yields

$$V_{\text{pinchoff}} = \frac{enH^2}{8\epsilon} \qquad \textbf{(6.3)}$$

As expected, a thicker channel implies a larger pinchoff voltage. The pinchoff voltage is often called $V_{GS(\text{off})}$.

A detailed analysis shows that I_D decreases as $|V_{GS}|$ increases (V_{GS} is always negative for an n-channel JFET) according to

$$I_D = I_{DSS}\left[1 - \frac{V_{GS}}{V_p}\right]^2 \qquad \textbf{(6.4)}$$

where V_p is the pinchoff voltage and I_{DSS} is the maximum drain current when $V_{GS} = 0$.

The triple subscripts in I_{DSS} deserve comment. The first two refer to the current flowing from the drain to the source. The third subscript S

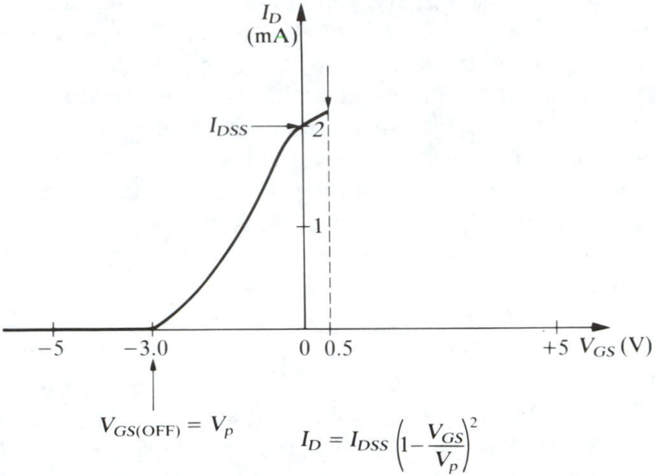

FIGURE 6.3 I_D versus V_{GS} for n-channel depletion MOSFET.

refers to the condition of the third terminal (the gate); the gate is "shorted" to the source; that is, $V_{GS} = 0$.

A graph of (6.4) is shown in Fig. 6.3. The pinchoff voltage is −3.0 V, and I_{DSS} is 2.0 mA. Notice that the rate of change of I_D with respect to V_{GS} is larger for larger I_D (smaller V_{GS}). The *transconductance* of the FET is defined as the slope of the graph in Fig. 6.3.

$$g_m = y_{fs} = \left(\frac{\partial I_D}{\partial V_{GS}}\right)_{V_{DS}} \tag{6.5}$$

Typical values for the transconductance are from 1000 to 5000 μmhos. (1 mho = 1 ohm^{-1} = 1 siemen.) The transconductance can be thought of as telling us how much the drain current I_D will change in response to a change in the gate source voltage V_{GS}. For example, a 2000-μmho transconductance means 2 mA/V.

The relationship between I_D and V_{GS} also depends on the temperature. At low values of I_D the temperature coefficient is positive: An increase in temperature produces an increase in I_D. But at higher values of I_D the temperature coefficient is negative: An increase in temperature produces a *decrease* in I_D. Thus at higher values of I_D, FETs can be safely connected in parallel to carry large currents. If one of the FETs heats up more than the others, it will draw less current and thus will cool down. If it had a positive temperature coefficient, any increase in temperature would make it draw more current and heat up even more; that is, the FET would self-destruct.

If the voltage on the gate is kept constant and the drain-source voltage V_{DS} is increased, then the drain current increases. But as V_{DS} is increased, the depletion region near the *drain* end of the gate will increase in thickness even though the gate remains at a constant voltage, because of the increasing negative voltage at the drain end of the channel caused by the *IR* drop along the channel. In other words, as V_{DS} is increased, the cross-sectional area of the channel is decreased, which increases the channel resistance. Thus the drain current will *not* increase linearly with V_{DS}, as for an ordinary resistor obeying Ohm's law, but will increase at a slower and slower rate as V_{DS} is increased. The resulting graph of I_D versus V_{DS} for a fixed gate-source reverse bias voltage is shown in Fig. 6.4. As V_{DS} is

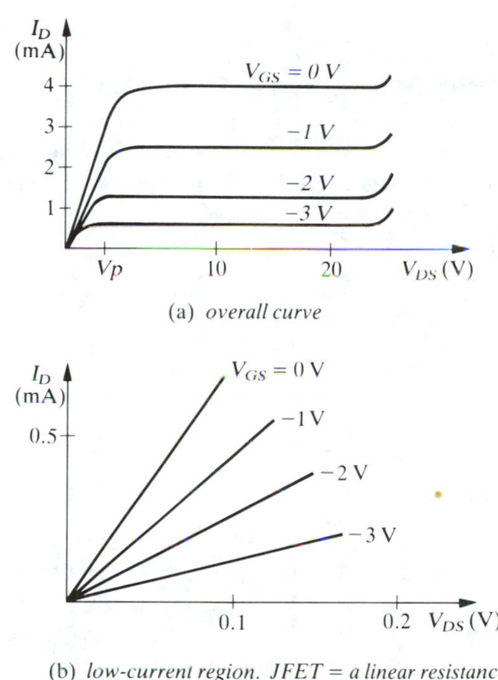

FIGURE 6.4 Drain current versus drain-source voltage.

increased, the depletion regions nearly meet, and the cross-sectional area of the channel available to pass drain current is very small. The dynamic resistance of the channel is then very large (of the order of 10^4 to 10^5 Ω), and further increases of V_{DS} result in only a very small increase in drain current. This region of operation is on the flat portion of the I_D-versus-V_{DS} graph in Fig. 6.4(a), and is sometimes referred to as the *constant current*, *saturation* region, or the *pinchoff* region. If V_{DS} is greater than the pinchoff voltage, then I_D is essentially constant for constant V_{GS}. For $V_{DS} < V_p$, both V_{GS} and V_{DS} determine I_D.

If the drain-source voltage is increased far enough, the depletion region blocking the channel breaks down and the drain current rises sharply, possibly destroying the FET. Breakdown usually occurs for drain-source voltages of the order of 20 to 50 V. In practical FET circuits the drain-source voltage is kept well below the breakdown voltage.

For low values of V_{DS} the drain current I_D increases approximately linearly with V_{DS} as shown in Fig. 6.4(b). In this region the FET acts like an ohmic resistance that can be controlled by varying the gate-source voltage V_{GS}, a higher value of V_{GS} resulting in a higher value of the FET channel resistance. When operated like this, the FET is termed a *voltage-controlled resistance*. More on this in 6.5.5.

6.3 FET *y* PARAMETER EQUIVALENT CIRCUIT

We will now treat the FET as a four-terminal "black box" and derive an equivalent circuit. Consider the four-terminal black box or network of Fig. 6.5.

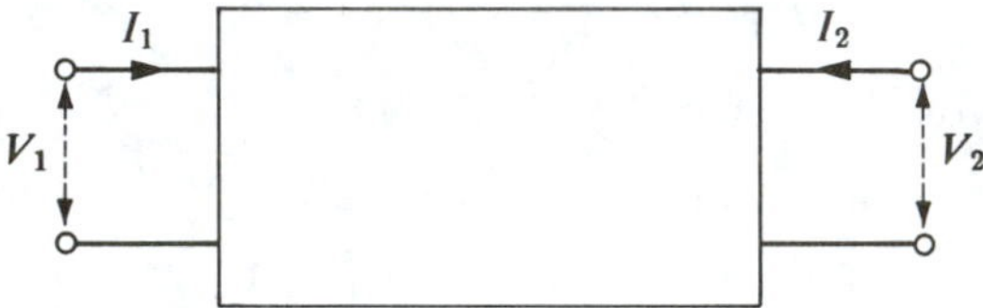

FIGURE 6.5 FET four-terminal network.

By the same argument as was used in Chapter 5 in developing the *h* parameter bipolar transistor equivalent circuit, we can choose any two of the four variables I_1, I_2, V_1, V_2 as our independent variables (which we think of intuitively as "causes") and the other two as dependent variables (which we think of as "effects"). For an FET we choose V_1 and V_2 as the independent variables and I_1 and I_2 as the dependent variables. We then can write the appropriate Kirchhoff equations and express I_1 and I_2 each in terms of V_1 and V_2:

$$I_1 = I_1(V_1, V_2) \tag{6.6}$$

$$I_2 = I_2(V_1, V_2) \tag{6.7}$$

The functions $I_1(V_1, V_2)$ and $I_2(V_1, V_2)$ depend on the type of FET.

We now consider the effect on I_1 and I_2 of small changes ΔV_1 and ΔV_2:

$$\Delta I_1 = \left(\frac{\partial I_1}{\partial V_1}\right)_{V_2} \Delta V_1 + \left(\frac{\partial I_1}{\partial V_2}\right)_{V_1} \Delta V_2 \tag{6.8}$$

$$\Delta I_2 = \left(\frac{\partial I_2}{\partial V_1}\right)_{V_2} \Delta V_1 + \left(\frac{\partial I_2}{\partial V_2}\right)_{V_1} \Delta V_2 \tag{6.9}$$

The notation is now changed: $i_1 = \Delta I_1$, $i_2 = \Delta I_2$, $v_1 = \Delta V_1$, $v_2 = \Delta V_2$. The lowercase i's and v's now represent *changes* in current and voltage, respectively; one can think of them as signal amplitudes, that is, peak-to-peak current or voltage swings. In terms of the new notation

$$i_1 = \left(\frac{\partial I_1}{\partial V_1}\right)_{V_2} v_1 + \left(\frac{\partial I_1}{\partial V_2}\right)_{V_1} v_2 \tag{6.10}$$

$$i_2 = \left(\frac{\partial I_2}{\partial V_1}\right)_{V_2} v_1 + \left(\frac{\partial I_2}{\partial V_2}\right)_{V_1} v_2 \tag{6.11}$$

We now define four "y parameters" as the four partial derivatives occurring in (6.10) and (6.11):

$$\begin{aligned} y_{11} &\equiv \left(\frac{\partial I_1}{\partial V_1}\right)_{V_2} & y_{12} &\equiv \left(\frac{\partial I_1}{\partial V_2}\right)_{V_1} \\ y_{21} &\equiv \left(\frac{\partial I_2}{\partial V_1}\right)_{V_2} & y_{22} &\equiv \left(\frac{\partial I_2}{\partial V_2}\right)_{V_1} \end{aligned} \tag{6.12}$$

In terms of the four y parameters, (6.10) and (6.11) become

$$i_1 = y_{11}v_1 + y_{12}v_2 \tag{6.13}$$

$$i_2 = y_{21}v_1 + y_{22}v_2 \tag{6.14}$$

Notice that all the y parameters have dimensions of conductance or admittance; they are all measured in siemens or mhos or micromhos.

Equations (6.13) and (6.14) imply the equivalent circuit of Fig. 6.6(a). This equivalent circuit can apply to *any* four-terminal network, not just to an FET. Of course, as with any equivalent circuit, the problem is to find or choose the circuit that accurately describes the actual four-terminal network behavior over a wide range of currents and voltages for one *fixed* set of the y parameters.

Because an FET, like an ordinary bipolar transistor, has only three

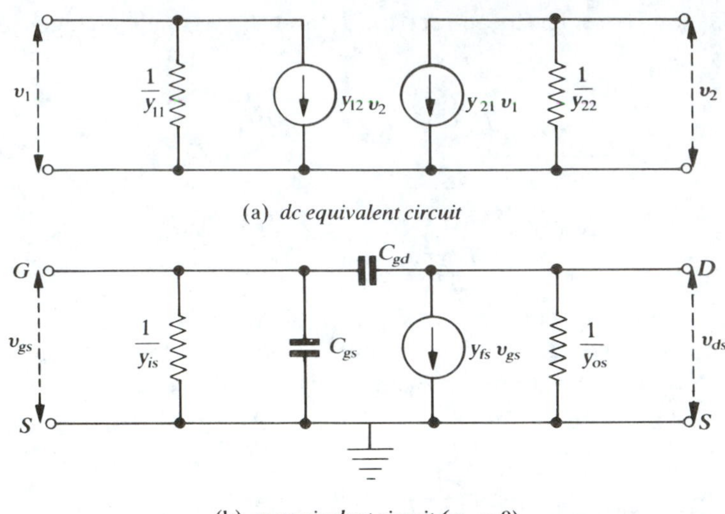

(a) *dc equivalent circuit*

(b) *ac equivalent circuit* ($y_{12} = 0$)

FIGURE 6.6 *y* parameter JFET equivalent circuit.

terminals, one terminal must be common between the input and the output of Fig. 6.5. The three possibilities are the source terminal common, the drain terminal common, or the gate terminal common. This gives rise, respectively, to the "common source," the "common drain," or the "common gate" configurations. These three configurations are shown in bare outline without any dc biasing in Fig. 6.7.

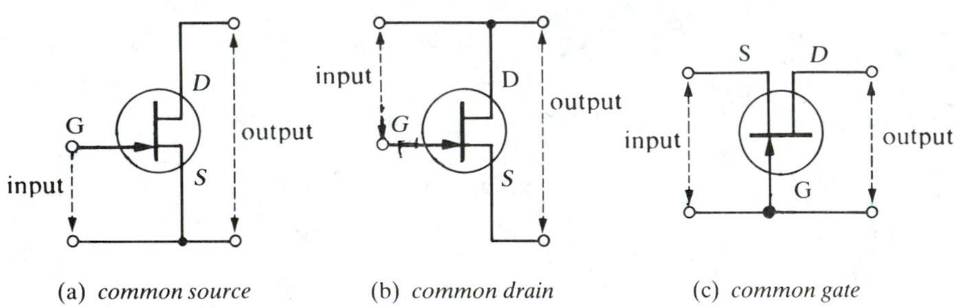

(a) *common source* (b) *common drain* (c) *common gate*

FIGURE 6.7 FET configurations (n-channel).

The gate of an FET corresponds to the base of a bipolar transistor; the source and drain correspond to the emitter and collector, respectively. Hence, the FET common source is analogous to the common emitter configuration, the FET common drain to the common collector or emitter follower configuration, and the FET common gate to the common base configuration.

Typical values for the y parameters for a low-power JFET in the common source configuration are

$$y_{11} = y_{is} \cong 10^{-8} \text{ siemens or mhos} \qquad \frac{1}{y_{11}} \cong 100\ \text{M}\Omega$$

$$y_{12} = y_{rs} \cong 10^{-10} \text{ siemens or mhos} \qquad \frac{1}{y_{12}} \cong 10{,}000\ \text{M}\Omega$$

$$y_{21} = y_{fs} = g_m \cong 4000\ \mu\text{siemens or } \mu\text{mhos} = 4 \times 10^{-3}\ \Omega^{-1} = 4\ \text{mA/V}$$

$$y_{22} = y_{os} \cong 10^{-5} \text{ siemens or mhos} \qquad \frac{1}{y_{22}} \cong 100\ \text{k}\Omega$$

A good approximation for almost all JFETs is $y_{is} = 0$. Thus, the input impedance of the JFET between gate and source is essentially $1/y_{is} \cong 100\ \text{M}\Omega$ or more. The physical reason for this high input resistance is that the input is the *reverse* biased gate-channel junction. The only current flowing out of the gate (n-channel device) is the reverse current I_G due to the thermal generation of minority carriers in the depletion region between the gate and the channel. $I_G \cong 10$ nA, and approximately doubles for each 5°C rise in temperature. Thus the JFET input resistance will decrease with increasing temperature (e.g., from 100 MΩ at 25°C to perhaps 50 MΩ at 30°C). To prevent this, we usually connect a $1\ \text{M}\Omega = R_G$ external resistance from gate to ground. The JFET input resistance will then be $R_G \| R_{\text{in}} \cong R_G$, which is independent of temperature if $R_G \ll R_{\text{in}}$.

An important JFET parameter is the capacitance between the gate and the channel due to the reverse biased gate-channel junction. Any reverse biased p-n junction has such a capacitance, as we saw in Chapter 4: $C \propto V^{-1/2}$, where V is the reverse bias voltage across the junction. This capacitance depends on the thickness of the depletion layer and thus on V_{GS} and V_{DS} as well as the geometry. The discrete capacitances C_{gs} and C_{gd} of the equivalent circuit of Fig. 6.6(b) represent this junction capacitance. Typical values of C_{gs} and C_{gd} are 1 to 5 pF. Other notations used are $C_{gss} = C_{iss} = C_{gs} + C_{gd}$, the total gate-source capacitance with the drain and source an ac short circuit; $C_{dss} = C_{oss} \cong C_{gd} = C_{rss}$ with the gate and source an ac short circuit.

6.4 THE METAL-OXIDE-SEMICONDUCTOR FIELD-EFFECT TRANSISTOR (MOSFET)

The ordinary junction FET that we have considered up to now contained only p- and n-type semiconductor material. For an n-channel FET, for example, the n-type channel is bounded on each side by two p-type regions connected together to form the gate. The interface between the gate and the channel is a p-n junction—hence the name "junction" FET. The gate is reverse biased with respect to the channel, thereby producing a high gate-channel impedance of approximately 10^8 to $10^{10}\ \Omega$. This large impedance arises from the small reverse leakage current (on the order of nanoamperes) in silicon that flows from the channel to the gate. We recall from Chapter 4 that this leakage current is temperature dependent because it arises from the thermal generation of minority charge carriers in the depletion region between the gate and the channel.

The metal-oxide-semiconductor field-effect transistor (MOSFET), sometimes called the insulated-gate field-effect transistor (IGFET), is similar to the JFET but differs in two important ways: First, the MOSFET has smaller internal parasitic capacitances and hence will operate at higher frequencies than the ordinary JFET. Second, the gate of a MOSFET is metal and separated from the channel by a thin layer of extremely good insulating material, usually silicon dioxide (SiO_2) for a silicon FET. The SiO_2 is a far better insulator than the reverse biased gate-channel junction of an ordinary FET, so the MOSFET gate-channel impedance is far larger. Impedances of 10^{12} to $10^{15}\ \Omega$ can be achieved with modern MOSFETs. Because the gate terminal is usually connected to the input, we usually say that the input impedance of a MOSFET is about 10^{12} to $10^{15}\ \Omega$. Because of the SiO_2 layer between the gate and the channel, there is no p-n junction and the gate can be *either* polarity with respect to the channel.

The MOSFET gate-to-channel capacitance is very small (several pF), so large voltages can be created between the gate and the channel by the accumulation of only small amounts of charge ($V = Q/C$). Hence, great care must be taken to avoid building up electrostatic charge on the gate of a MOSFET to prevent high-voltage breakdown and permanent damage to the SiO_2 layer. It is standard practice for MOSFETs to come from the manufacturer with a thin shorting wire wound around the leads. This shorting wire is left in place until the MOSFET is installed in the circuit. Soldering irons used on MOSFET circuits should also be externally grounded, with a ground wire attached to the tip itself to keep 60-Hz leakage voltages off the tip.

There are two types of MOSFETs: the depletion type and the enhancement type. In the depletion type, shown in Fig. 6.8, the channel is open when the gate-source voltage is zero, whereas in the enhancement type, shown in Fig. 6.9, the channel is normally closed when $V_{GS} = 0$. Thus with $V_{GS} = 0$ the depletion MOSFET is "ON" but the enhancement MOSFET is "OFF." The curves of I_D versus V_{GS} and I_D versus V_{DS} are shown in Fig. 6.10 for an n-type depletion MOSFET.

A negative gate-source voltage of several volts (the gate negative with respect to the source) will cut off the channel and force $I_D = 0$. This value of V_{GS} is usually referred to as $V_{GS(\text{off})}$. Physically, the reason I_D decreases as V_{GS} becomes more negative is that a negative charge on the gate repels electrons out of the n-type channel and reduces the channel conductivity.

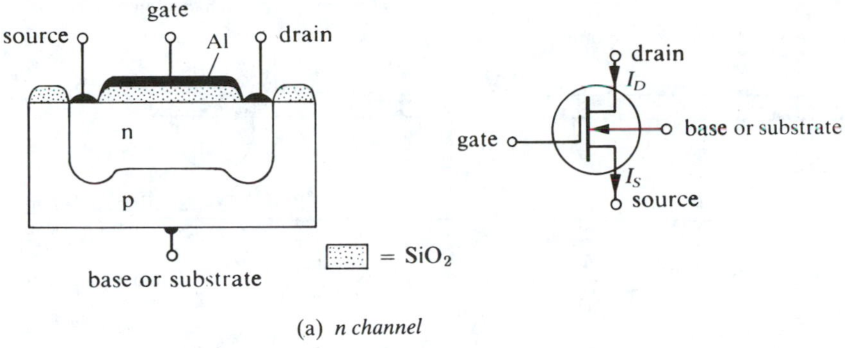

(a) *n channel*

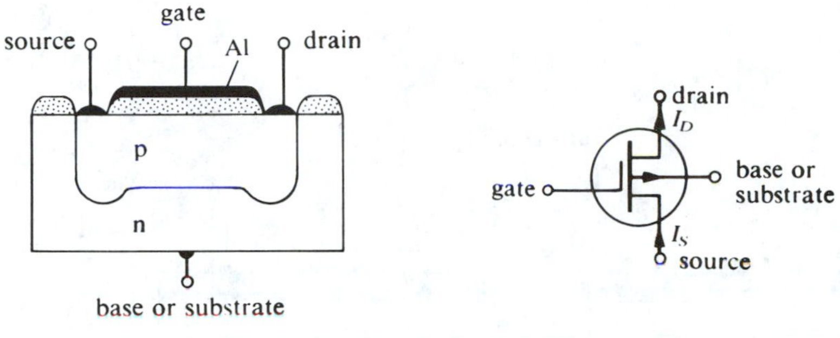

(b) *p-channel (rarely made)*

FIGURE 6.8 Depletion MOSFET construction and schematic symbols.

This repulsion is due to the electric field from the negative charge on the gate penetrating through the SiO_2 layer into the channel—hence the name *field*-effect transistor.

The MOSFET I_D-versus-V_{GS} curve is quite similar to the JFET curve except that V_{GS} must be kept $\leqq 0.5$ V for the JFET; otherwise the gate-channel junction will be forward biased. But in contrast to the JFET, V_{GS} can be of *either* polarity for the MOSFET because of the presence of the SiO_2 layer insulating the gate from the channel.

The enhancement MOSFET can be either n-channel or p-channel, as shown in Fig. 6.9. A dc bias must be applied to the gate-source junction to open up the channel: V_{GS} positive in an n-channel enhancement MOSFET, and V_{GS} negative in a p-channel enhancement MOSFET. For an n-channel enhancement MOSFET, making the gate positive repels positive mobile charge carriers in the substrate away from the gate and attracts mobile electrons from the n-type regions near the source and drain into the region near the gate, thus creating an n-type channel joining the source and drain, as shown in Fig. 6.9.

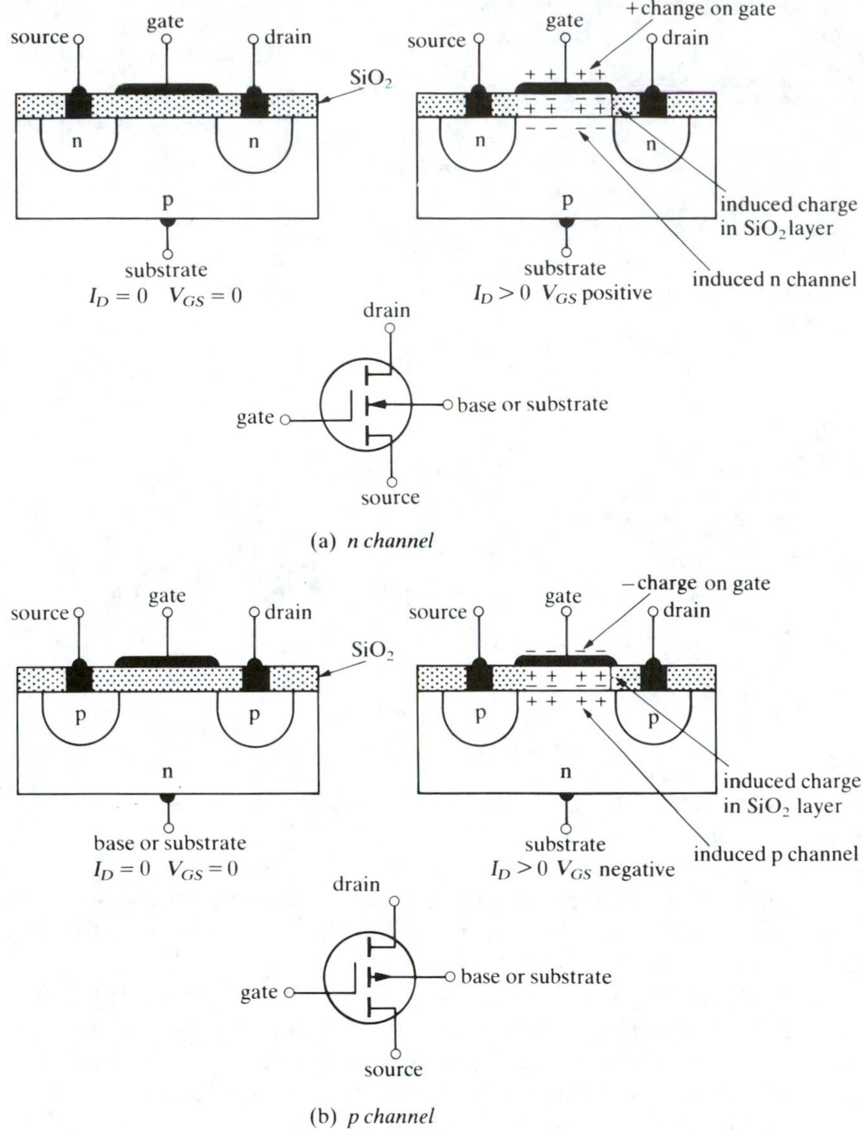

FIGURE 6.9 Enhancement MOSFET construction and schematic symbols.

The base or substrate is not an "active" terminal of a MOSFET. The substrate is merely the mechanical crystal "foundation" on which the MOSFET is built. The substrate is p-type material in an n-channel MOSFET. *The channel-substrate junction must be kept reverse biased at all times in any MOSFET*. This reverse bias is usually produced by connecting the substrate directly to the source. Then at the source end of the channel the

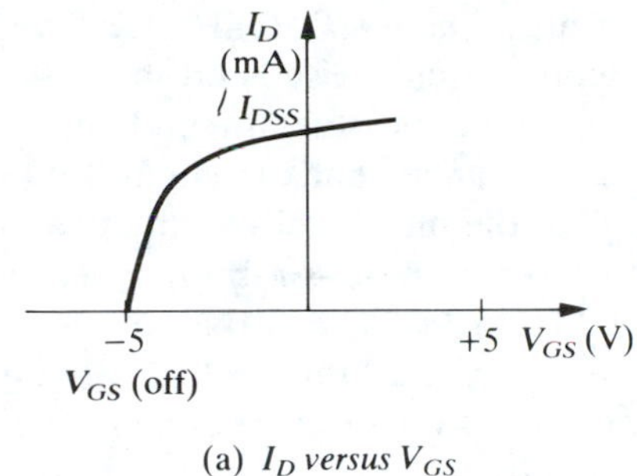

(a) I_D *versus* V_{GS}

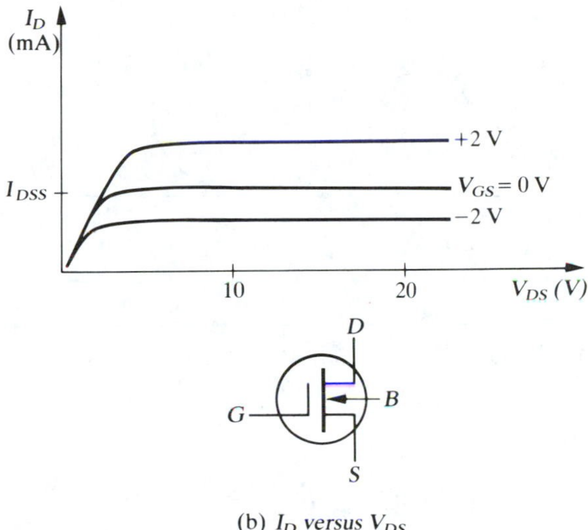

(b) I_D *versus* V_{DS}

FIGURE 6.10 I_D versus V_{GS} for n-channel depletion MOSFET.

channel-substrate junction voltage is zero, and it cannot conduct. As we go along through the channel from the source toward the drain, the voltage in the n-type channel becomes more *positive* (because of the I_SR drop, where R = channel resistance), thus *increasing* the channel-substrate reverse bias.

If this source-substrate connection is made internally, the MOSFET will have only *three* leads: source, gate, and drain. If the substrate is not internally connected, the MOSFET will have *four* leads: source, gate, drain, and substrate. When in doubt, wire the substrate to the source.

A similar argument applies to a p-channel MOSFET; connecting the gate to the source again makes the channel-substrate junction reverse biased at all points. Generally, p-channel enhancement MOSFETs are more common than n-channel.

If the channel of an n-channel MOSFET is ever driven negative in a circuit (e.g., in a switching circuit, discussed in 6.5.4), the substrate must be biased *more negative* than the most negative channel voltage. Similarly, the n-type substrate of a p-channel enhancement MOSFET must always be biased *more positive* than the most positive channel voltage.

The graphs of I_D versus V_{GS} for both types of enhancement MOSFETs are shown in Fig. 6.11. The value of V_{GS} required to produce $I_D = 10\ \mu A$ is denoted by V_T, a "threshold" voltage. It can be considered as a "turn-on threshold" voltage for enhancement MOSFETs and is approximately several volts.

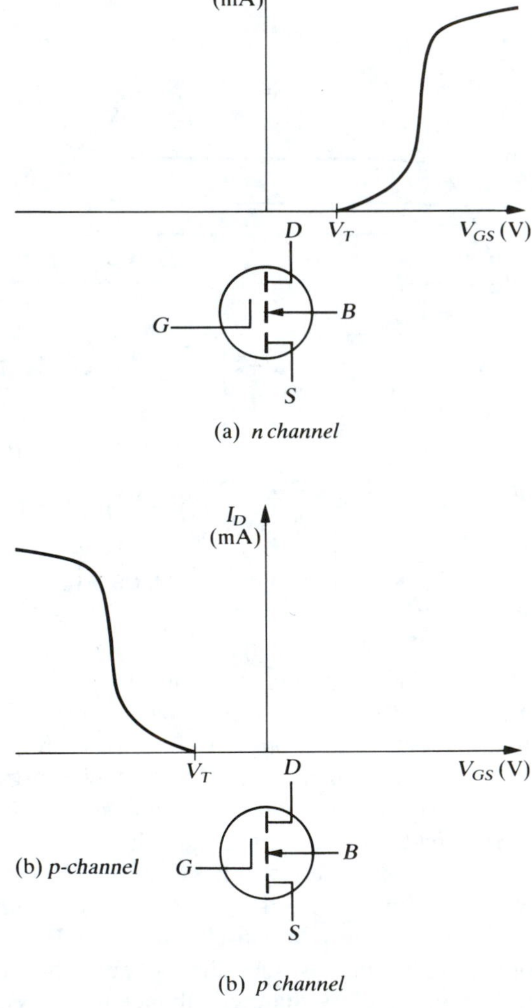

FIGURE 6.11 I_D versus V_{GS} for enhancement MOSFETs.

In summary, the MOSFET differs from the JFET by having a much higher input impedance, operating with either polarity of gate-source voltage, and being much more susceptible to permanent damage from electrostatic voltages generated by synthetic clothing or rugs. Precautions for handling MOSFETs include the following:

1. Ground soldering iron tip.
2. Never remove or insert a MOSFET in a circuit with power on.
3. Do not wear synthetic clothing.
4. Leave protective metal ring around MOSFET until MOSFET is in circuit.
5. Store MOSFET in protective (black) foam.
6. Never insert MOSFET into ordinary (white) foam.

6.5 SAMPLE FET CIRCUITS

In this section we will give several FET circuits for JFETs and MOSFETs.

6.5.1 The Common Source FET Amplifier

The common source FET amplifier is analogous to the common emitter bipolar transistor amplifier. Hence we expect moderate voltage and current gains and a relatively high output impedance. We expect, however, an input impedance much higher than for a bipolar transistor circuit because the FET input (gate-channel) junction is reverse biased rather than forward biased as for a bipolar transistor.

A practical common source FET amplifier is shown in Fig. 6.12 for an inexpensive (less than $1) n-channel FET. Perhaps the first difference that is apparent between the FET common source amplifier characteristics and those of the bipolar transistor common emitter amplifier is that the FET voltage gain is substantially less, and the input impedance is much higher.

There is no voltage drop across R_G because the gate current I_G is essentially zero. This means the dc gate voltage is zero because the bottom of R_G is grounded. Thus, the gate-source junction is reverse biased by the $I_S R_S$ dc voltage drop across the source resistance R_S, as shown in Fig. 6.12(b). Also notice that because the input impedance is so large ($Z_{in} \cong R_G = 10\ \text{M}\Omega$), we can use a small value of C_1 and still obtain decent low-frequency gain. In other words, the breakpoint frequency $\omega_B = 1/R_G C_1$ of the input high-pass filter formed by C_1 and R_G is very low because R_G is so large.

A y parameter analysis yields

$$A_v \cong -y_{21}(R_D \| R_L) \qquad Z_{in} \cong R_G \qquad Z_{out} \cong R_D$$

A practical n-channel enhancement MOSFET amplifier is shown in Fig. 6.12(c). Notice that the two bias resistors are necessary to set $V_{GS} = +4$ V dc to turn on the MOSFET.

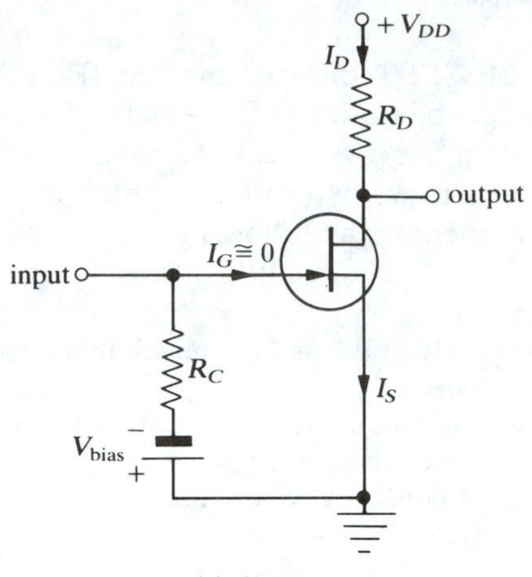

(a) *basic circuit*

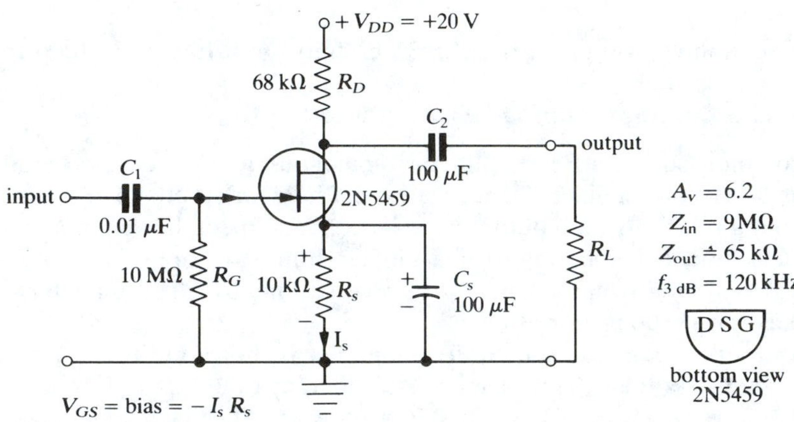

(b) *practical n-channel JFET circuit (or n-channel depletion MOSFET with $R_S = 0$)*

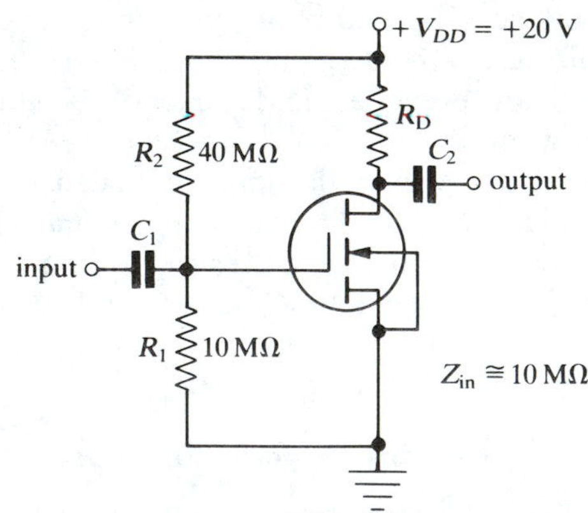

(c) *practical n-channel enhancement MOSFET circuit*

FIGURE 6.12 Common source JFET amplifier.

6.5.2 The Common Drain FET Amplifier

The common drain FET amplifier or source follower is analogous to the common collector or emitter follower amplifier made with bipolar transistors. Hence, we expect a low output impedance, a broad bandwidth, a voltage gain close to unity, and a high current gain.

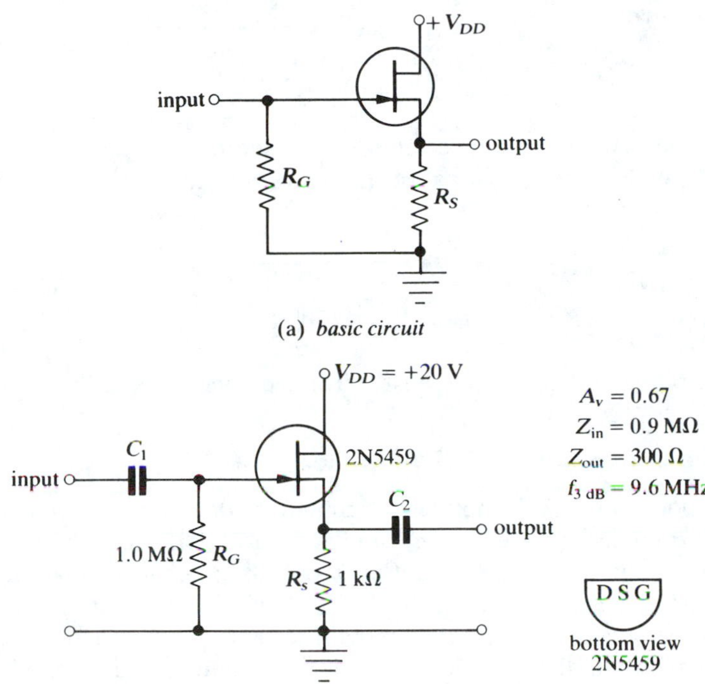

(b) *practical n-channel JFET circuit (or n-channel depletion MOSFET)*

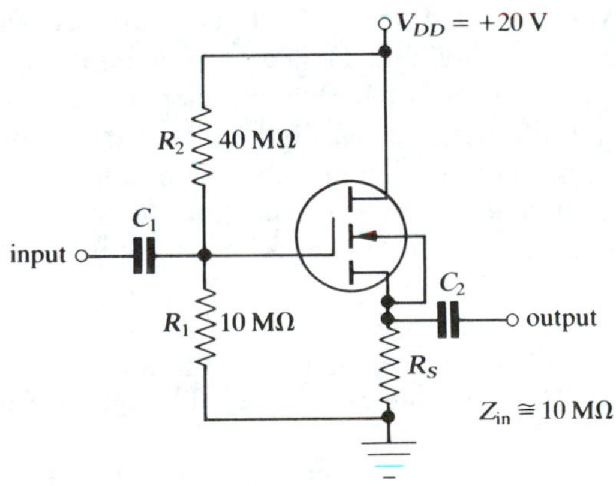

(c) *practical n-channel enhancement MOSFET circuit*

FIGURE 6.13 Common drain JFET amplifier.

A practical common drain amplifier is shown in Fig. 6.13 along with measured circuit parameters. A higher voltage gain of 0.9 can be obtained at the expense of a higher output impedance of 1 kΩ by increasing the source resistance R_S from 1 kΩ to 10 kΩ. A y parameter analysis of the circuit yields

$$Z_{in} \cong R_G \qquad Z_{out} \cong \frac{1}{\dfrac{1}{R_S} + y_{21}}$$

which is the parallel combination of the source resistance R_S and a resistance equal to the reciprocal of the forward transconductance, and

$$A_v \cong \frac{y_{21}}{y_{21} + \dfrac{1}{R_S \| R_L}}$$

The gate-source dc bias is $I_S R_S$, just as for the common source amplifier.

6.5.3 The Common Gate FET Amplifier

The common gate FET amplifier is similar to the common base bipolar transistor amplifier, so we expect similar characteristics: a high voltage gain, a low input impedance, and a high output impedance. The circuit of such an amplifier is shown in Fig. 6.14. The high (56-kΩ) output impedance makes for very poor high-frequency gain if there is even a small capacitance C_o between the output terminal and ground. For example, with $C_o = 50$ pF, the upper 3-dB frequency is only 300 kHz.

The common gate FET amplifier provides excellent isolation between the input at the source and the output at the drain, as does the common base bipolar transistor amplifier. Hence, the common gate amplifier is mainly used at frequencies of 100 MHz or higher where unwanted feedback from output to input often causes oscillations. A sample circuit is shown in Fig. 6.14(c) with tuned circuits in the source and drain.

Finally, for all three amplifier configurations (common source, common drain, and common gate) the following checks should be made in debugging a circuit.

1. In a JFET circuit the gate-channel junction must always be reverse dc biased. For an n-channel JFET the gate must always be negative with respect to the source.
2. The drain current must be approximately 1 mA or more for most JFETs; otherwise the JFET is cut off or has a very low gain.
3. In a MOSFET circuit the channel-substrate junction must *always be reverse biased*. When in doubt, connect the substrate to the source.

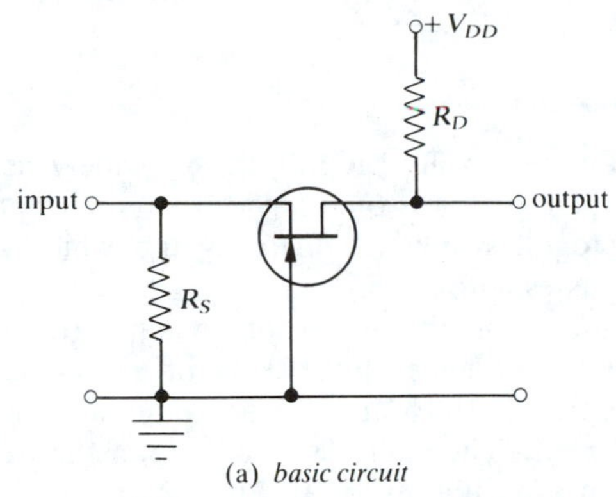

(a) *basic circuit*

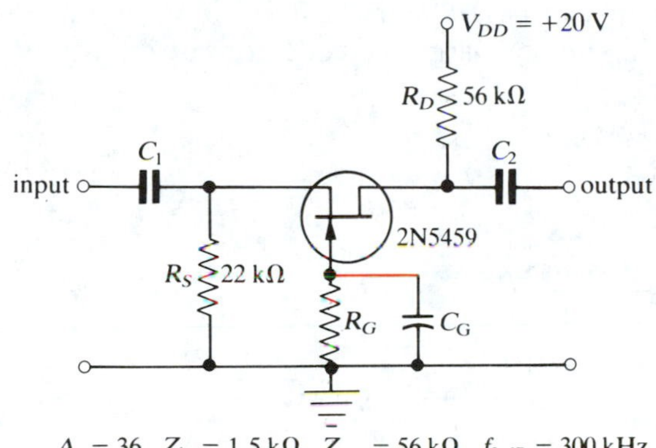

$A_v = 36,\ Z_{in} = 1.5\ \text{k}\Omega,\ Z_{out} = 56\ \text{k}\Omega,\ f_{3\ dB} = 300\ \text{kHz}$

(b) *practical circuit*

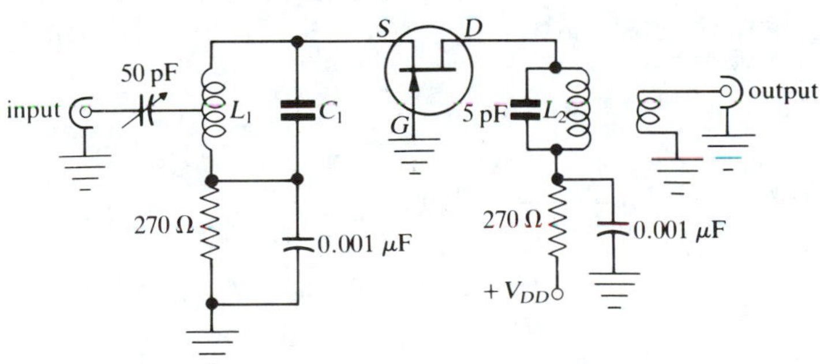

C_1 may be C_{SG} of the JFET

(c) 144-*MHz common gate amplifier*

FIGURE 6.14 Common gate JFET amplifier.

4. In an enhancement MOSFET be sure the channel is opened; in other words, be sure that drain current is flowing. Check by measuring the voltage drop across the drain resistor.

6.5.4 FET Switches

FETs, JFETs, and especially MOSFETs are widely used to make analog switches. An *analog* switch is one that will pass current in either direction like an ordinary toggle switch. (A *digital* switch, which we will consider in a later chapter, passes current in only one direction.) FET switches are generally superior to bipolar transistor switches because (1) there is no 0.6-V turn-on voltage for the FET, (2) the off resistance of a FET switch is much larger, typically 100 MΩ or more, and (3) the gate input draws essentially no current. The main disadvantage of the FET switch is that its "on" resistance is typically 50 to 100 Ω, whereas the "on" resistance of a bipolar transistor can be 10 Ω or less ($R_{on} \approx V_{CE\,sat}/I_C \approx 0.2\text{ V}/20\text{ mA} = 10\ \Omega$), depending on how large the collector current is.

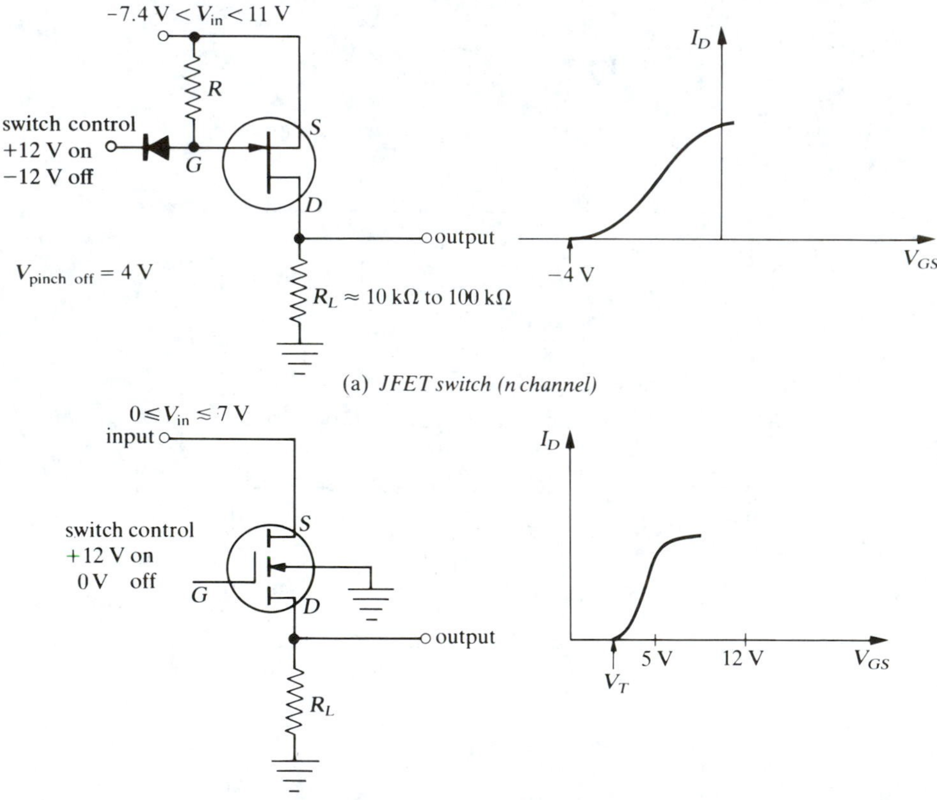

FIGURE 6.15 FET switches.

Figure 6.15 shows two switch circuits—one for a MOSFET and one for a JFET. The main disadvantage with any JFET switch is that the gate-source junction must *always* be kept reverse biased. Thus, in the JFET circuit of Fig. 6.14(a), when the switch control signal is −12 V, the diode is on, $V_G = -11.4$ V, and the JFET is off if V_{in} is more positive than −7.4 V because the pinchoff voltage is −4 V. The diode is to isolate the switch control circuit from the signal when the JFET is on. Thus, when the switch control signal is +12 V, the JFET is turned on, the diode is off, and the maximum signal input should be +11 V to keep a minimum 1-V reverse bias on the diode.

The enhancement MOSFET switch circuit of Fig. 6.15(b) is simpler because, as for any MOSFET, the gate can be either polarity with respect to the source. However, the channel-substrate junction must *always be reverse biased*. With the gate at ground, $I_D = 0$ (the drain-source resistance ≈ $10^{10}\ \Omega$ or more), the switch is off, and $V_{out} = 0$. With the gate at $+12\text{ V} \gg V_T$, the MOSFET is on (the drain-source resistance is 50 to 100 Ω), and $V_{out} = V_{in}$. The only requirement for the switch control "on" voltage level is that it be much larger than the threshold voltage V_T: $V_{on} \gtrsim 5$ V is usually fine. With the switch on and $V_{on} = +12$ V, any signal more positive than +7 V would start to turn off the MOSFET by making the gate less than 5 V more positive than the source. Thus the maximum signal level is $V_{\text{max signal}} < V_{on} - 5$ V. Similarly, with the switch off and $V_G = 0$ V, if the signal went negative the channel (source)-substrate junction would be forward biased, which would shunt the signal input to ground. Thus the signal input must never be negative.

If the substrate were kept at −12 V, positive and negative input signals could be switched with $V_{on} = +12$ V, $V_{off} = -12$ V, provided

$$-12\text{ V} < V_{in} < +7\text{ V} \quad \text{or} \quad -|V_{off}| < V_{in} < +7\text{ V}$$

An enhancement MOSFET can also be used to switch a load R_L on or off, as shown in Fig. 6.16. A zero-voltage switch control will keep the

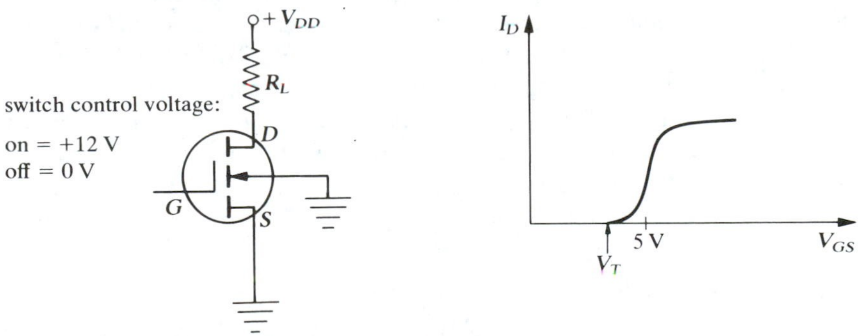

FIGURE 6.16 MOSFET switch.

MOSFET off, and a positive switch control input of approximately 5 V or more ($\gg V_T$) will turn on the MOSFET. Power MOSFETs, such as the VMOS type, can carry drain currents up to 10 A when on, so this switch can control 10-A circuits from a control source that supplies essentially zero current to the MOSFET gate.

6.5.5 The JFET Variable Resistance or Voltage-Controlled Resistance

If the JFET drain-source voltage is kept under approximately 0.5 V, the I_D-versus-V_{DS} curves are essentially linear, as shown in Fig. 6.17(b). Thus, in this region the JFET is acting like an ordinary, linear, "ohmic" resistance, with its resistance being determined by the value of V_{GS}. Such a JFET is often called a *voltage-controlled resistance* or VCR. Notice that both the dc resistance (V_{DS}/I_D) and the ac resistance ($\partial V_{DS}/\partial I_D$) are constant and equal, right down to $V_{DS} = 0$.

The circuit of Fig. 6.17(c) shows a JFET VCR used in a simple voltage divider. The minimum drain-source resistance is $r_{DS\,\text{on}}$ with the JFET fully on, and the maximum is essentially infinite with the JFET off. Thus, the maximum output voltage is v_{in}, and the minimum is $r_{DS(\text{on})}v_{\text{in}}/(R + r_{DS(\text{on})})$.

In Fig. 6.17(d) the JFET VCR is used as the resistive load for a photomultiplier tube whose maximum anode current is approximately 1 μA. Thus, the JFET will always be in the linear region, and r_{DS} will always be an ohmic (linear) resistance.

6.5.6 The JFET Constant Current Source

If the drain-source voltage V_{DS} of a JFET is greater than 1 or 2 V and if V_{GS} is held constant, then the drain current is essentially constant, independent of V_{DS}, as can be seen from the I_D-versus-V_{DS} curve of Fig. 6.18(a).

If the gate is connected to the source as shown in Fig. 6.18(b), $V_{GS} = 0$ and the drain current will be constant: $I_D = I_{DSS}$. Even if R_L varies, as long as $V_{DS} > 2$ V, $I_D = I_L$ will be constant. These devices are sometimes called *constant current diodes* or *current regulator diodes* and are available commercially. Values of I_{DSS} range from approximately 0.5 to 5 mA. They are simply JFETs with the gate internally connected to the source.

If a source resistance R_S is added, $V_{GS} = I_S R_S$, as shown in Fig. 6.18(c). By adjusting R_S, we can obtain any value of I_D less than I_{DSS}; the larger R_S is, the smaller I_D is.

6.6 POWER MOSFETS

There are two general types of power MOSFETs: a low-voltage VMOS type and a high-voltage DMOS type. All types are constructed on a body or

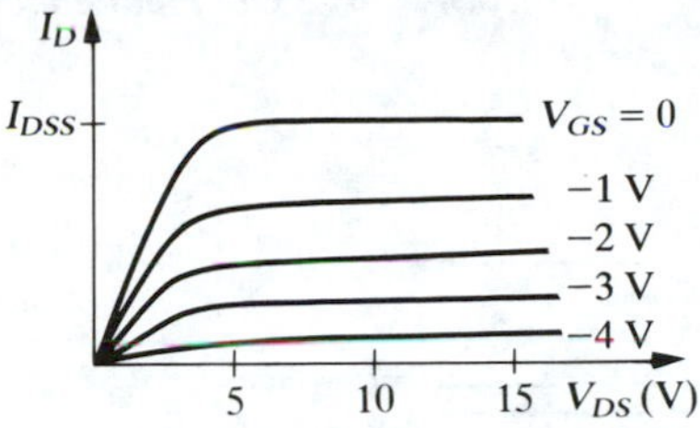

(a) *general I_D vs. V_{DS} curves*

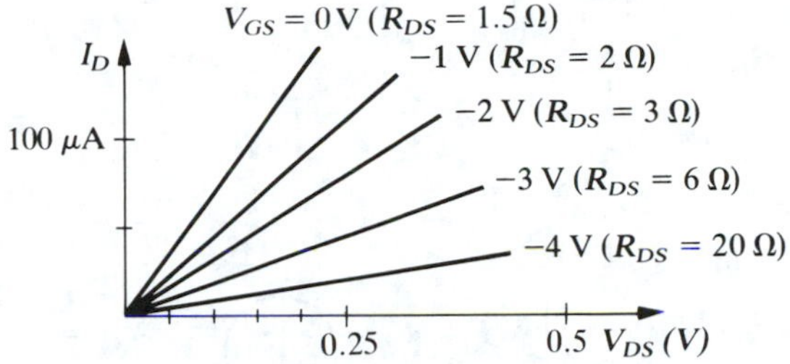

(b) *expanded curves near $I_D \approx 0$. I_D vs. V_{DS} is linear*

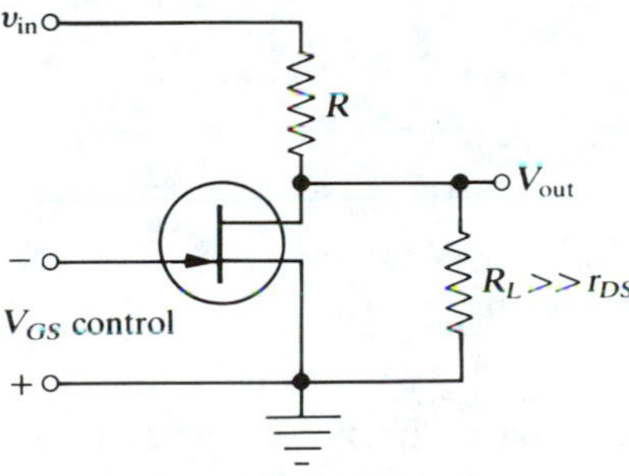

(c) *voltage divider*

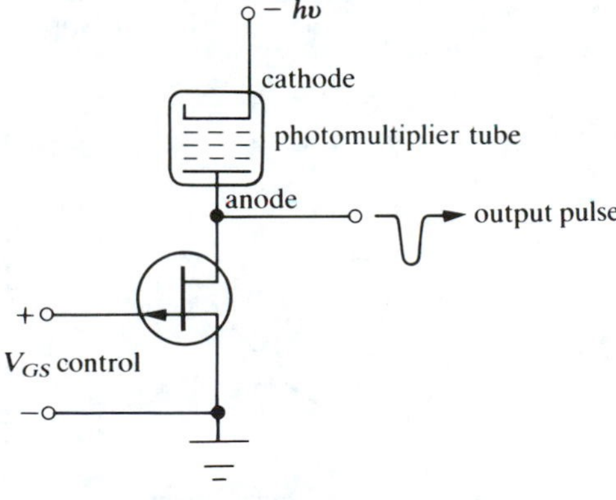

(d) *photomultiplier load*

FIGURE 6.17 JFET output curves.

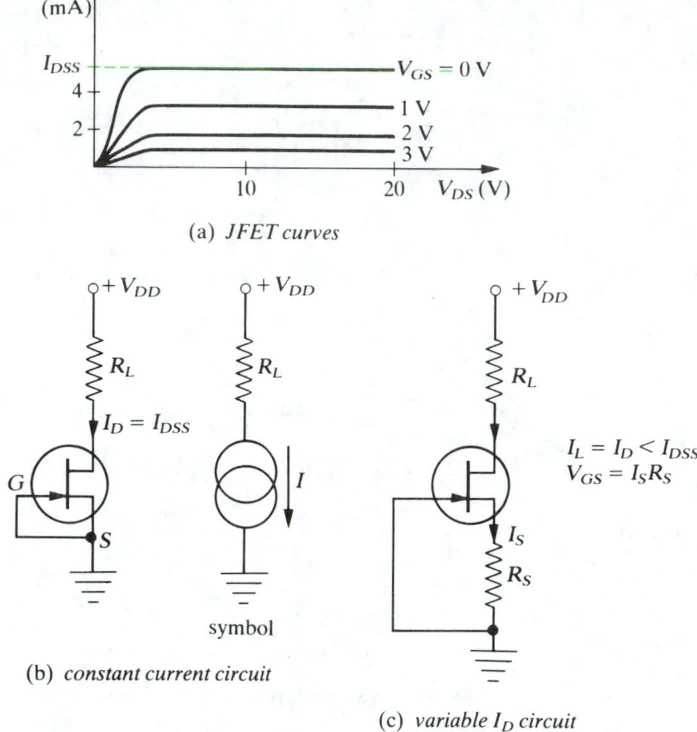

FIGURE 6.18 JFET constant current source.

substrate that produces a parasitic diode between the drain and source, as shown in Fig. 6.19(a). Most power MOSFETs are n-channel devices because the mobility of p-type silicon is less than that of n-type silicon. They also have a relatively large gate-to-source internal capacitance (≈100

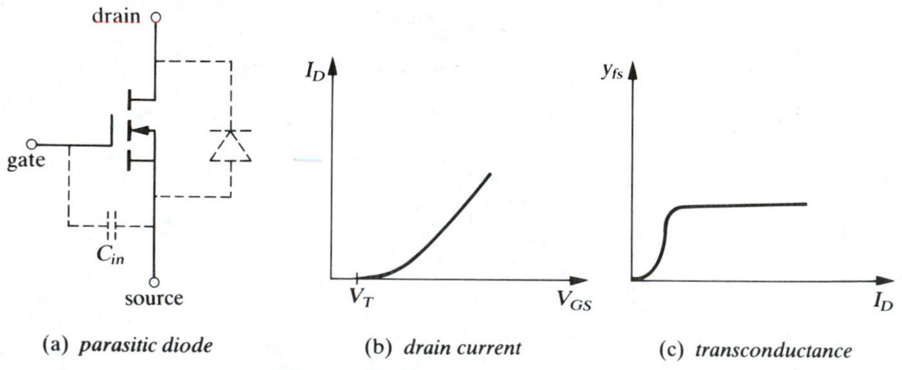

FIGURE 6.19 Power MOSFET.

to 1000 pF), so they do *not* need zener diode protection against electrostatic voltages as low-power MOSFETs do. Because of the absence of the internal zener diode, the gate can swing *either* polarity with respect to the source without damage. The gate-source voltage typically can be up to ±30 V, and large MOSFETs can carry up to 10-A drain current. However, to turn the power MOSFET on or off rapidly, C_{in} must be charged or discharged rapidly, thus requiring a low-impedance driving circuit.

The principal advantage is that its drain current is almost exactly *linearly* proportional to the gate-source voltage, as shown in Fig. 6.19(b). And, of course, the input gate current is negligibly small in most applications, the main exception being when the MOSFET is switched on and off rapidly, thus drawing ac current pulses to charge and discharge its gate-source capacitance C_{in}.

Further advantages of power MOSFETs over power bipolar transistors are as follows: (1) They are more resistant to second breakdown when running at high powers. (2) They have a negative temperature coefficient, which means that as a MOSFET heats up, its resistance decreases; thus it dissipates less I^2R power and automatically cools down. (3) They have lower input capacitances. (4) They have no minority carrier storage time.

Power MOSFETs are useful in controlling electric motors, in switching power supplies, and in the output stages of power audio amplifiers.

A high-voltage CMOS driver can turn a p-channel MOSFET off and on as shown in Fig. 6.20.

Another useful application is in the modified Darlington circuit shown in Fig. 6.20(b). The input MOSFET has an extremely high input impedance. The switching time is faster than in the conventional Darlington with two bipolar transistors, and the output transistor is never driven to saturation.

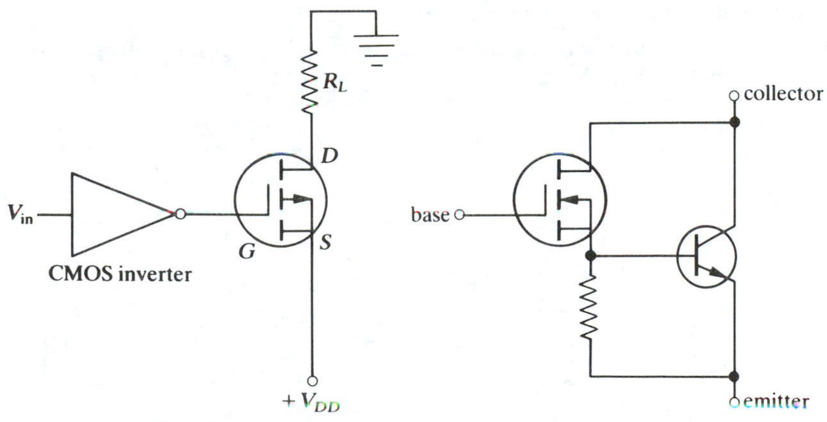

(a) *CMOS Inverter Driven Switch* (b) *modified Darlington*

FIGURE 6.20 Power MOSFET applications.

6.7 THE FET VERSUS THE BIPOLAR TRANSISTOR

In Table 6.1 we summarize the important qualitative differences between the FET and the bipolar transistor.

TABLE 6.1. Comparison of Bipolar Transistor and FET Parameters

Parameter	*Bipolar Transistor*	*FET*
Input impedance	Low ≈ 1 kΩ	Very high ≳ 10 MΩ
Input turn on	0.6 V	Zero (good for switches)
Gain	Higher ≈ $100 v_{out}/v_{in}$	Lower ≈ 3000 μmhos i_{out}/v_{in}
ON resistance	Lower ≈ 10 Ω	Higher ≈ 100 Ω
Temperature coefficient	Lower ≈ 0.6 μV/°C	Higher ≈ 5 μV/°C
Input offset voltage	Lower ≈ 25 μV	Higher ≈ 500 μV
Sample-to-sample spread	Tighter	Wider (in I_{DSS})
Noise	Lowest for low R_S	Lowest for high R_S

R_S = source resistance.

PROBLEMS

1. Identify the majority carriers in the channel for n-channel and for p-channel FETs.
2. Explain why the depletion region between the gate and the channel is thicker at the drain end than at the source end.
3. Fill in the polarities of the power supplies and the voltage drops across the source and drain resistors.

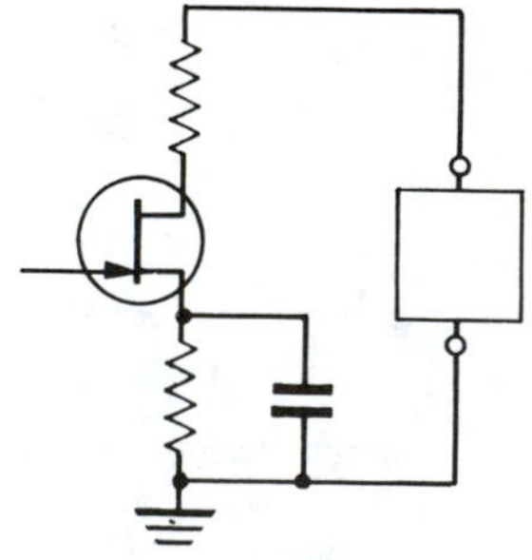

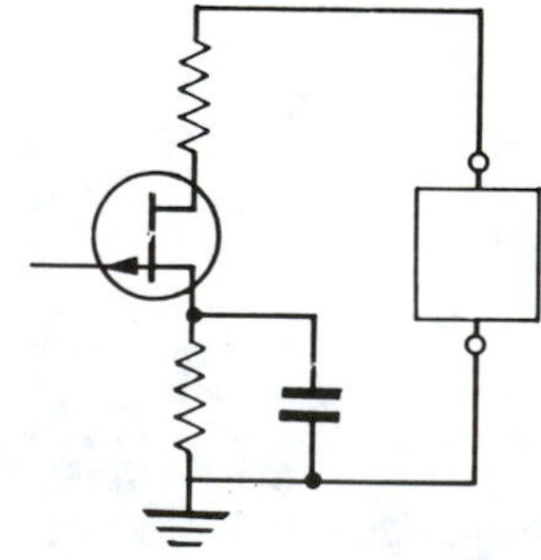

4. (a) Sketch the load line. (b) Determine V_S, V_D, and I_D at the operating point.

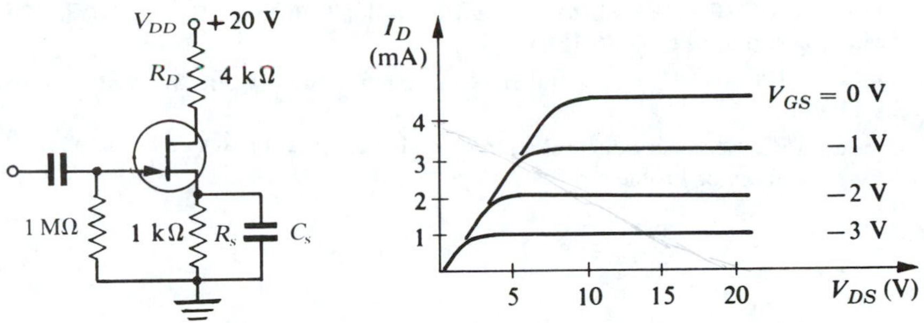

5. Explain the physical meaning of the pinchoff voltage.
6. What is I_{DSS}?
7. Explain how to connect the substrate lead in a four-lead MOSFET.
8. Explain why the input impedance of a MOSFET can be even higher than that of a JFET.
9. Calculate $y_{21} = g_m = y_{fs}$ and y_{22} for the FET whose output curves are given in Problem 4. Does it depend upon I_D?
10. State the expected JFET resistance measurements taken with an ohmmeter on the $R \times 1$ scale for (a) gate-source, (b) source-gate, (c) source-drain, (d) drain-source.
11. For the FET amplifier shown below, (a) explain why the dc gate voltage is very close to 0 V; (b) calculate the operating point (I_D, V_{DS}). Use the curve of Problem 4.

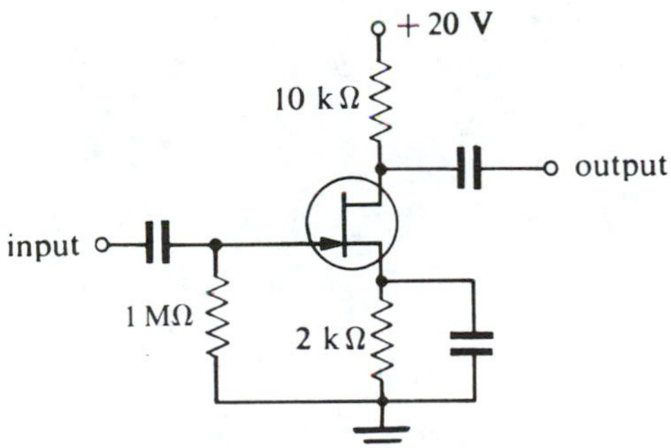

12. For a FET whose curves are given in Problem 4, (a) calculate R_D and R_S if the desired operating point is $I_D = 0.5$ mA, $V_{DS} = 10$ V; (b) calculate C_S if the amplifier must amplify signals from 1 to 30 kHz. $V_{DD} = 20$ V.
13. Explain why the output impedance of a common source FET amplifier is almost exactly equal to the drain resistance R_D. Is it slightly higher or lower than R_D?
14. Design a common drain or source follower FET amplifier using a FET with

$y_{21} = y_{fs} = 1000\ \mu$mhos when the drain current is 3 mA. The FET curve is that of Problem 4, and the desired output impedance is 250 Ω or less.

15. Design a MOSFET switch to turn a 10-mA load on and off with a 0-to-5-V switching signal (see Fig. 6.16).

16. Design a 100-MHz JFET amplifier. What configuration would be best to avoid oscillation?

17. Design a FET constant current source to give (a) $I = 4.0$ mA, (b) $I = 2.0$ mA. Use the curves of Problem 4.

CHAPTER 7

Feedback

7.1 INTRODUCTION

In almost all practical circuits "feedback" is employed. *Feedback* may be defined as the taking of a portion of the output of a circuit and coupling or feeding it back into the input. If the portion of the output that is fed back is *in* phase with respect to the input, then the feedback is termed *positive feedback*. If the output fed back is *out* of phase with respect to the input, then the feedback is termed *negative feedback*. Positive feedback is usually used in oscillators and to increase the gain of a circuit as in super-regenerative receivers. With positive feedback a circuit can be made to generate an output with no external input; a random noise voltage or voltage transient (when the circuit is turned on) at the input creates an output, part of which is fed back to reinforce the input, and the cycle repeats itself until some nonlinearity of the circuit results in a constant or repetitive output.

Negative feedback is used in almost all amplifiers. Its only disadvantage is that the gain of the amplifier is reduced because the fed-back voltage subtracts from the input. The gain can usually be increased to the desired value by merely adding more stages of amplification, however. The advantages of negative feedback are numerous: increased circuit stability against almost *any* type of disturbance (e.g., power supply or temperature fluctuation and component aging), increased input impedance, and decreased output impedance. It also provides an increased frequency bandwidth for a constant gain and decreased distortion introduced by the circuit. Negative feedback is sometimes referred to as *bootstrapping*, because the voltage fed back, usually to the emitter of the first stage, goes in the same direction as the input voltage at the base. For example, if the signal input voltage at a transistor base becomes more positive, negative feedback will result from a positive-going feedback voltage being applied to the emitter, thus decreasing the base-emitter voltage difference, which is the effective input to the amplifier. The first-stage amplifier "reaches down and pulls up" its emitter voltage like a person pulling up on his or her bootstraps.

7.2 NEGATIVE VOLTAGE FEEDBACK

Let us consider negative feedback in which a fraction $B < 1$ of the output voltage is subtracted from the input voltage, so the effective input to the amplifier is

$$v' = v - Bv_{out}$$

This type of feedback is called *negative voltage feedback* and is shown schematically in Fig. 7.1.

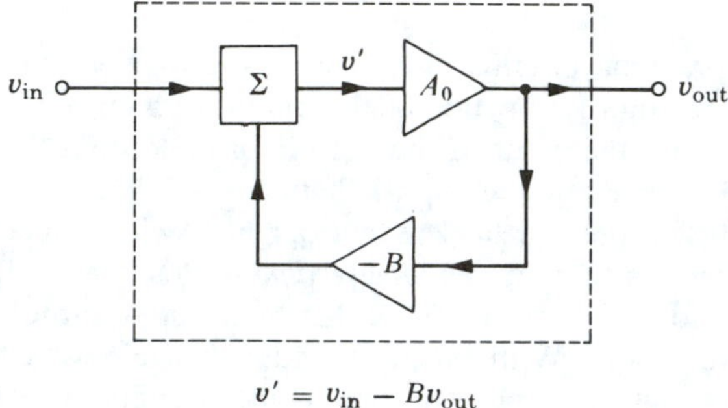

FIGURE 7.1 Generalized negative voltage feedback amplifier.

The triangle A_0 represents an amplifier with a gain A_0; that is, $A_0 = v_{out}/v'$, where v' is the input. A_0 is called the *open-loop gain*. The triangle $-B$ represents a circuit that is an amplifier with a fractional gain of $-B$; that is, its output is $-B$ times its input. B is sometimes called the *feedback factor*. The minus sign means the output is out of phase with respect to its input. The square Σ represents a summing or addition circuit whose output $v' = v_{in} - Bv_{out}$. The entire amplifier is the system in the dotted rectangle, and its gain with the negative voltage feedback is, by definition,

$$A_f \equiv \frac{v_{out}}{v_{in}} \tag{7.1}$$

A_f is called the closed-loop gain and can be calculated from the two equations

$$A_0 = \frac{v_{out}}{v'} \tag{7.2}$$

$$v' = v_{in} - Bv_{out} \tag{7.3}$$

Eliminating v' between (7.2) and (7.3) yields the following expression for the amplifier gain:

$$A_f = \frac{v_{out}}{v_{in}} = \frac{A_0}{1 + A_0 B} \tag{7.4}$$

As we expect, the overall or closed-loop gain A_f of the amplifier depends on A_0 and B. We also see from (7.4) that the gain A_f with negative feedback is *less* than the gain A_0 without negative feedback. Usually, the fraction B of the output voltage that is fed back is much less than unity, and the gain without negative feedback A_0 is made so large that $A_0B \gg 1$. For example, $B = 0.1$ and $A_0 = 1000$ would make $A_0B = 100$. With the approximation $1 + A_0B \cong A_0B$, (7.4) becomes simply

$$A_f = \frac{A_0}{1 + A_0 B} \cong \frac{1}{B} \tag{7.5}$$

We thus have an astounding result: The gain of the entire amplifier depends *only upon B*, the fraction of the output voltage fed back, and not at all upon the gain A_0 of the amplifier without feedback. For example, if $B = 0.01$, corresponding to only 1% of the output voltage being fed back to the input, and if $A_0 = 10{,}000$, then $A_0B = (10{,}000)(0.01) = 100$, which is much greater than unity. Therefore the overall gain $A_f \cong 1/0.01 = 100$.

The beauty of this situation is that B can be made extremely constant. For example, B can be determined by the ratio of two extremely stable resistors, R_1 and R_2, with the circuit of Fig. 7.2. Thus the amplifier gain is

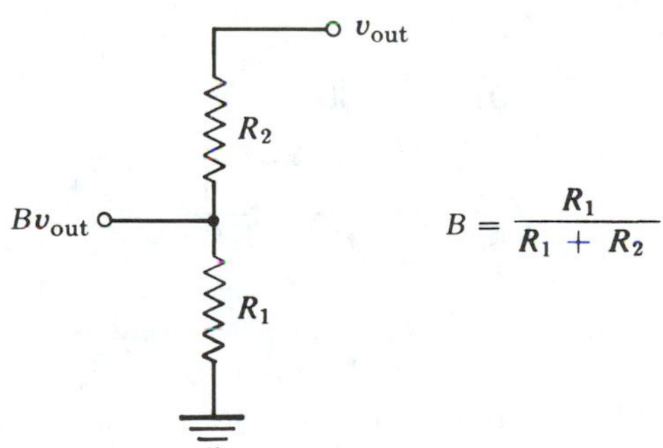

FIGURE 7.2 Resistor divider to obtain feedback voltage.

given by $A_f \cong 1/B = (R_1 + R_2)/R_1 = 1 + R_2/R_1$. If a gain of $A_f = 100$ is desired, typical values might be $R_1 = 100\ \Omega$, $R_2 = 9{,}900\ \Omega \cong 10\ \text{k}\Omega$. Even if the gain A_0 varies widely (e.g., from 5,000 to 20,000) the gain of the amplifier, A_f, will remain essentially fixed at $1/B$. Substituting $A_0 = 5{,}000$, 10,000, and 20,000 into the exact gain expression (7.4) shows how constant

TABLE 7.1

	A_0	$A_f = A_0/(1 + A_0B)$
$B = 0.01$ *fixed*	5,000	98.3
	10,000	99.0
	20,000	99.6

the gain A_f is (see Table 7.1). Thus, if A_0 changes by a factor of 2, A_f changes by less than 1%. Notice that this argument holds regardless of *what* factor causes the change in A_0—temperature changes, power supply voltage fluctuations, aging of transistors, or any other factors. The gain with feedback A_f remains constant so long as A_0B remains very large compared to 1. A gain change of a factor of 2 in A_0 is, of course, rather unusual. A change in A_0 of, say, 10% would be more likely. If A_0 changes by 10%, from 10,000 to 11,000, A_f changes only from 99.0 to 99.1, a change of only 0.1%. A quick calculation shows that an absolute change in A_0 results in a change in A_f smaller by approximately a factor of $(1/A_0B)^2$, since

$$\frac{\partial A_f}{\partial A_0} = \frac{\partial}{\partial A_0}\left(\frac{A_0}{1 + A_0B}\right) = \frac{(1 + A_0B) - A_0B}{(1 + A_0B)^2} = \frac{1}{(1 + A_0B)^2} \cong \frac{1}{(A_0B)^2} \tag{7.6}$$

In the case just discussed,

$$A_0B = (10{,}000)(0.01) = 100$$

so

$$\frac{\partial A_f}{\partial A_0} \cong \frac{1}{(A_0B)^2} = \frac{1}{(100)^2} = 10^{-4}$$

which is indeed small. It is easy to show that a relative or percentage change in A_0 results in a relative change in A_f smaller by a factor of $1/A_0B$. Notice also that so long as $A_0B \gg 1$, the gain of the amplifier, $A_f = 1/B$, is completely independent of the h parameters of the individual transistors. This is an exceedingly comforting thought, because the usual transistor manuals do not contain all four h parameters; they usually contain only $h_{21} = h_{fe}$ along with the maximum power, currents, and voltages.

The input impedance of an amplifier without negative feedback is *increased* by adding negative feedback. This effect seems intuitively reasonable if we realize that the part of the output voltage fed back to the input opposes the input voltage. Thus, the net effective input voltage applied to the amplifier is reduced. Hence, we expect a smaller current to flow into the input terminals of the amplifier, which means that the amplifier input effectively presents a higher impedance to the source. We will now show that the input impedance with feedback Z_{inf} is larger than the input impedance without feedback Z_{ino} by a factor of $(1 + A_0B)$; that is,

$$Z_{inf} = (1 + A_0B)Z_{ino}$$

Consider an amplifier with gain A_0 without feedback, an input impedance Z_{ino}, and an output impedance Z_{out}, as shown in Fig. 7.3(a). The output of the amplifier is represented by an ideal Thevenin voltage generator of magnitude A_0v_{in} in series with the output impedance Z_{out}. The input terminals have an impedance of Z_{ino}; we have assumed there is negligible coupling of the output back to the input in the amplifier of Fig. 7.3(a), by assuming there is no voltage generator in series with the input terminals. In short, the amplifier of Fig. 7.3(a) is ideal.

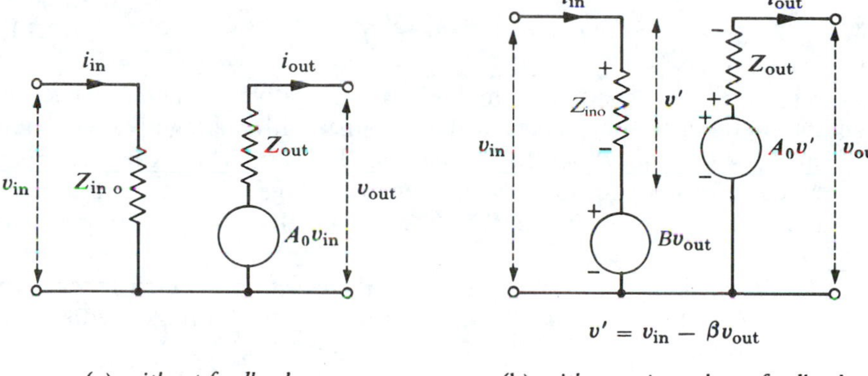

(a) *without feedback* (b) *with negative voltage feedback*

FIGURE 7.3 Feedback amplifier.

If we now add negative voltage feedback by adding a voltage generator Bv_{out} in series with the input, as shown in Fig. 7.3(b), we can write the following equations for the amplifier with negative feedback:

$$v_{in} = v' + Bv_{out} \tag{7.7}$$

$$v' = i_{in}Z_{ino} \tag{7.8}$$

Because of the negative voltage feedback, the overall amplifier gain is given by

$$A_f = \frac{v_{out}}{v_{in}} = \frac{A_0}{1 + A_0B} \tag{7.9}$$

Using the definition $Z_{inf} = v_{in}/i_{in}$, we can find the input impedance of the amplifier with feedback by eliminating v' and v_{out} from equations (7.7) and (7.8). Substituting (7.8) in (7.7) to eliminate v' gives

$$v_{in} = i_{in}Z_{ino} + Bv_{out} \tag{7.10}$$

And solving (7.9) for v_{out} and substituting in (7.10) to eliminate v_{out} gives

$$v_{in} = i_{in}Z_{ino} + B\frac{A_0}{1 + A_0B}v_{in}$$

Thus the input impedance is given by

$$Z_{inf} = \frac{v_{in}}{i_{in}} = \frac{Z_{ino}}{1 - \dfrac{A_0B}{1 + A_0B}}$$

$$Z_{inf} = (1 + A_0B)Z_{ino} \tag{7.11}$$

So the input impedance of the amplifier with negative voltage feedback is *greater* than the input impedance of the amplifier without negative feedback by a factor of $(1 + A_0B)$, which is usually about 10 to 100. A higher input impedance is, of course, almost always an advantage, because as the input impedance increases less current is drawn from the source supplying the input.

The output impedance of an amplifier is *decreased* by adding negative feedback. Let's look at the ideal amplifier with negative voltage feedback in Fig. 7.3(b); we have

$$v' = v_{in} - Bv_{out} \tag{7.12}$$

$$v_{out} = A_0v' - i_{out}Z_{out} \tag{7.13}$$

The output impedance with feedback is

$$Z_{outf} = -\frac{\partial v_{out}}{\partial i_{out}}$$

so we eliminate v' between (7.12) and (7.13) and solve for v_{out} in terms of i_{out}:

$$v_{out} = \frac{A_0v_{in} - i_{out}Z_{out}}{1 + A_0B} \tag{7.14}$$

Thus the output impedance with negative voltage feedback is

$$Z_{outf} = \frac{Z_{out}}{1 + A_0 B} \quad \textbf{(7.15)}$$

Z_{outf} is less than the output inpedance without feedback by a factor $1/(1 + A_0B)$, which is usually from 1/10 to 1/100 in most circuits. A low output impedance is generally a significant advantage in a circuit: The voltage drop is smaller across Z_{outf} internal to the circuit, and a faster rise time is possible when charging an external capacitive load; that is, the high-frequency response of the circuit is improved.

Negative feedback also decreases the distortion in an amplifier. *Distortion* here means any change in the shape of the input signal introduced by the amplifier. A high-fidelity amplifier is one in which the output is a faithful replica, only larger, of the input. A high-fidelity amplifier has low distortion, and all high-fidelity amplifiers have large amounts of negative feedback.

We can regard any distortion of the input signal by the amplifier as an "extra" voltage generated within the amplifier. It is a general rule, as we have seen, that any change inside the feedback loop (inside the amplifier)—for example, a change in A_0—results in a very small change in the gain with feedback A_f. So we should not be too surprised to find that any distortion voltage generated *within* the amplifier is effectively decreased by the presence of negative feedback. Suppose we represent the distortion voltage simply by an ideal voltage generator v_D in series with the output as in Fig. 7.4. The equations are

$$v_{out} = A_0 v' - i_{out} Z_{out} + v_D \quad \textbf{(7.16)}$$

$$v_{in} = v' + B v_{out}$$

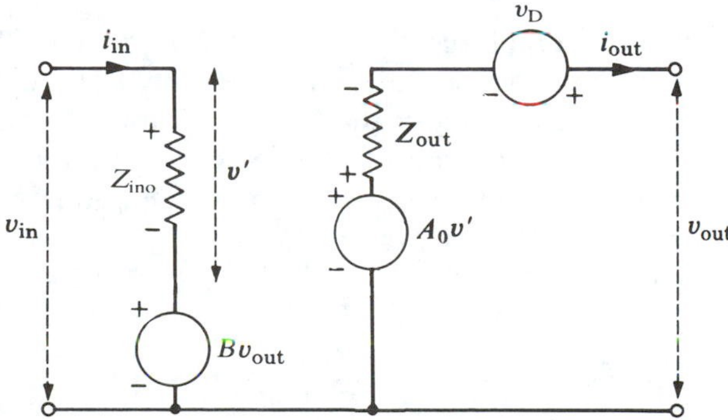

FIGURE 7.4 Negative-feedback amplifier containing distortion voltage v_D.

Eliminating v' gives us

$$v_{out} = A_0(v_{in} - Bv_{out}) - i_{out}Z_{out} + v_D$$

Neglecting the $i_{out}Z_{out}$ term (i.e., assuming negligible loading of the output), we obtain

$$v_{out} \cong A_0(v_{in} - Bv_{out}) + v_D$$

$$v_{out}(1 + A_0B) = A_0v_{in} + v_D$$

or

$$v_{out} = \frac{A_0}{1 + A_0B} v_{in} + \frac{1}{1 + A_0B} v_D \tag{7.17}$$

We see that the gain for the input signal v_{in} is larger, by a factor of A_0, than the gain for the distortion voltage v_D.

In an actual amplifier, of course, the distortion is generated not in series with the output but at several points closer to the input. Consider the two-stage amplifier in Fig. 7.5. The v_{D1} represents a distortion voltage

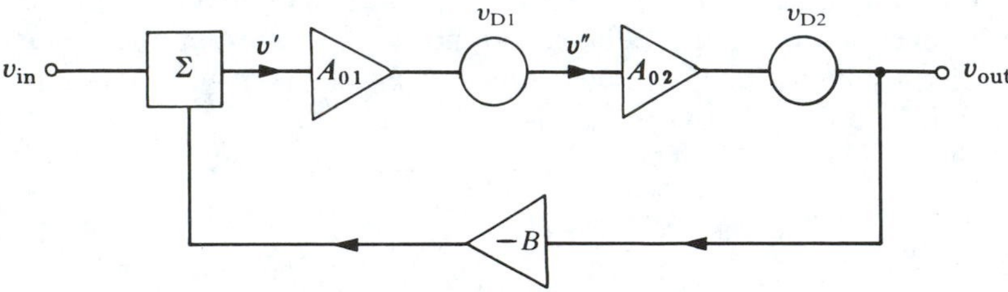

FIGURE 7.5 Two-stage, negative-feedback amplifier containing distortion at two points.

introduced at the output of the first stage whose gain is A_{01}, and v_{D2} represents a distortion voltage introduced at the output of the second stage, whose gain is A_{02}. It seems clear that the distortion v_{D1} will affect the output more than v_{D2}, because v_{D1} occurs closer to the input and is therefore subject to more amplification than v_{D2}. For this two-stage amplifier we can write

$$v' = v_{in} - Bv_{out} \tag{7.18}$$

$$v'' = A_{01}v' + v_{D1} \tag{7.19}$$

$$v_{out} = A_{02}v'' + v_{D2} \tag{7.20}$$

Eliminating v' and v'', we obtain

$$v_{\text{out}} = A_{02}A_{01}(v_{\text{in}} - Bv_{\text{out}}) + A_{02}v_{\text{D1}} + v_{\text{D2}}$$

$$v_{\text{out}}(1 + A_{02}A_{01}B) = A_{02}A_{01}v_{\text{in}} + A_{02}v_{\text{D1}} + v_{\text{D2}}$$

or
$$v_{\text{out}} = \frac{A_{02}A_{01}}{1 + A_{02}A_{01}B}v_{\text{in}} + \frac{A_{02}}{1 + A_{02}A_{01}B}v_{\text{D1}} + \frac{1}{1 + A_{02}A_{01}B}v_{\text{D2}} \tag{7.21}$$

The overall gain of the amplifier with negative feedback is $A_{02}A_{01}/(1 + A_{02}A_{01}B)$, as we would expect, but the gain applied to the distortion voltage v_{D1} is only $A_{02}/(1 + A_{02}A_{01}B)$, or a factor of A_{01} less. And the gain for the distortion v_{D2} is only $1/(1 + A_{02}A_{01}B)$, or a factor of $A_{02}A_{01}$ less. As we expected, the distortion voltage v_{D1} generated near the front end (input) of the amplifier is amplified more. In general, any unwanted voltage generated near the front end of an amplifier presents a more serious problem than those unwanted voltages introduced nearer the output regardless of whether these voltages are a distortion of the input, random noise voltages generated by transistors, resistors, or faulty contacts, or voltages picked up from external sources. We conclude that most efforts to reduce noise, pickup, and distortion should go into the *first* stage of the amplifier.

In power amplifiers most of the distortion is generated in the last stage, so negative feedback is very effective in reducing the distortion.

7.3 EXAMPLES OF NEGATIVE-FEEDBACK AMPLIFIER CIRCUITS

We now illustrate the use of negative voltage feedback by giving several actual amplifier circuits. First, consider the emitter follower circuit of Fig. 7.6. The input voltage is

$$v_{\text{in}} = v_{\text{BE}} + i_{\text{E}}R_{\text{E}}$$

But the output voltage is $v_{\text{out}} = i_{\text{E}}R_{\text{E}}$, so

$$v_{\text{in}} = v_{\text{BE}} + v_{\text{out}}$$

The voltage v' amplified by the transistor is v_{BE} and

$$v_{\text{BE}} = v_{\text{in}} - v_{\text{out}} \tag{7.22}$$

which is of the form for a negative-voltage-feedback amplifier:

$$v' = v_{\text{in}} - Bv_{\text{out}}$$

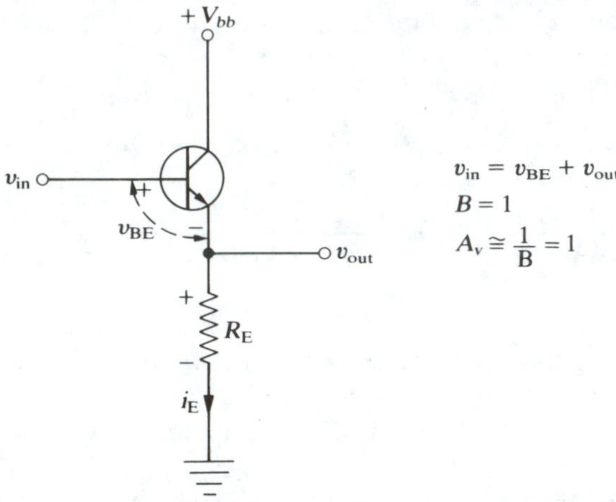

FIGURE 7.6 Negative feedback in emitter follower.

with $B = 1$. And the gain with the negative feedback is

$$A_f \cong \frac{1}{B} = 1 \tag{7.23}$$

Thus we see that the unity gain of the emitter follower is simply due to all the output voltage being fed back and subtracted from the input—that is, due to strong negative feedback.

The common emitter amplifier with an unbypassed emitter resistor is another example of a negative-voltage-feedback amplifier as shown in Fig. 7.7. The ac input and output voltages are

$$v_{in} = v_{BE} + i_E R_E \qquad v_{out} = -i_C R_C$$

When $v_{in} = v_B$ goes positive, i_E increases and v_E also goes positive, thus making v_{BE} less than v_{in}; that is, we have *negative* feedback. Using

$$I_C = \frac{\beta}{\beta + 1} I_E \quad \text{and} \quad i_C = \frac{\beta}{\beta + 1} i_E$$

we obtain

$$v_{in} = v_{BE} - \frac{(\beta + 1)R_E}{\beta R_C} v_{out}$$

The voltage amplified by the transistor is v_{BE}, where

$$v_{BE} = v_{in} + \frac{(\beta + 1)R_E}{\beta R_C} v_{out}$$

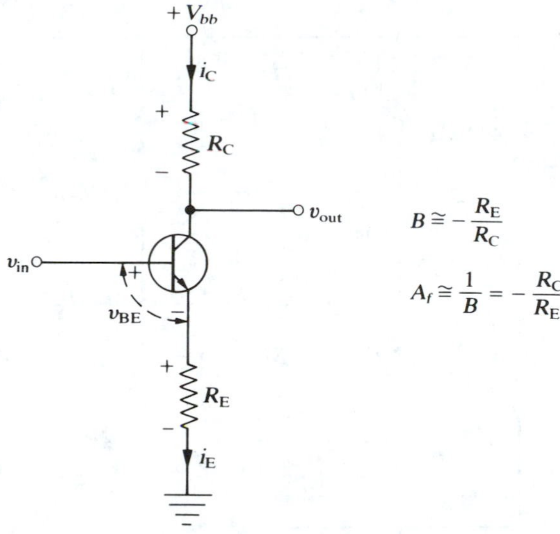

FIGURE 7.7 Negative feedback in the common emitter amplifier with R_E not bypassed.

which is of the form for a negative-voltage-feedback amplifier ($v' = v_{BE}$),

$$v' = v_{in} - Bv_{out}$$

with $B = -(\beta + 1)R_E v_{out}/\beta R_C$. Thus the gain with the negative feedback is

$$A_f \cong \frac{1}{B} = \frac{-\beta R_C}{(\beta + 1)R_E} \cong -\frac{R_C}{R_E} \tag{7.24}$$

The minus sign represents the 180° phase inversion in the common emitter amplifier.

The gain expression (7.24) can also be obtained from a simple approximate argument. A small change in the input voltage at the base will appear almost undiminished at the emitter: $v_{in} \cong v_E$. Thus the emitter current change will be $i_E = v_E/R_E \cong v_{in}/R_E$ and $i_E \cong i_C$. Therefore the voltage gain will be

$$A_v = \frac{v_{out}}{v_{in}} = \frac{v_C}{v_{in}} \cong \frac{-i_C R_C}{v_E} \cong \frac{-i_E R_C}{v_E} = -\frac{R_C}{R_E}$$

which is (7.24).

A two-stage, dc coupled amplifier with negative voltage feedback is shown in Fig. 7.8. The closed loop or gain with feedback is

$$A_f \cong \frac{1}{B} = \frac{R_1 + R_2}{R_1} = 1 + \frac{R_2}{R_1} \tag{7.25}$$

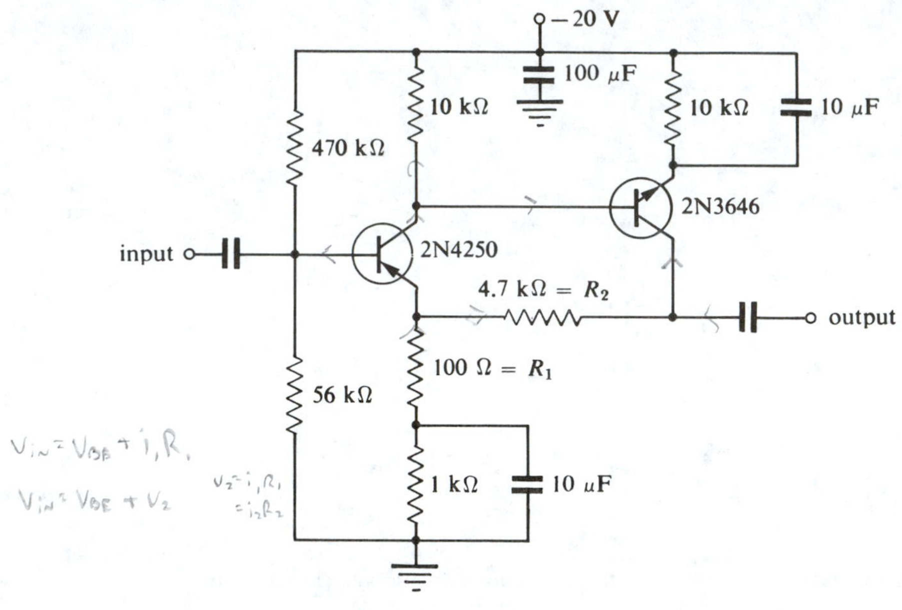

Measured Circuit Performance

100-kHz voltage gain	42
100-kHz input impedance	43 kΩ
100-kHz output impedance	200 Ω
upper 3-dB frequency	>10 MHz

FIGURE 7.8 Negative-feedback amplifier.

Because there is no coupling capacitor between the two stages, the negative feedback extends right down to dc. A low-power Darlington transistor such as the 2N6427 can be substituted for the 2N3646 for lower output impedance (about 80 Ω) at the expense of less bandwidth.

The basic idea of negative voltage feedback is extensively used in op amp amplifier circuits, as explained in Chapters 9 and 10.

7.4 NEGATIVE CURRENT FEEDBACK

If the voltage fed back and subtracted from the input is proportional to the output *current* (rather than the output voltage, as in Section 7.2), then the result is a feedback amplifier whose output current I_0 is constant: that is, I_0 depends only on the input voltage, not on the amplifier gains, load resistance, supply voltage, or anything else. Figure 7.9 shows the circuit.

The voltage fed back equals BI_0, where B is the *feedback factor*, and

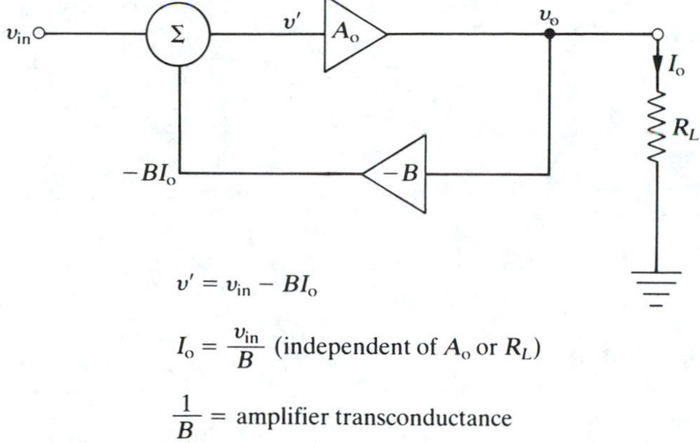

FIGURE 7.9 Negative-current-feedback amplifier.

has the dimensions of resistance. The equations are

$$v' = v_{in} - BI_0 \qquad v_0 = I_0 R_L = A_0 v'$$

Eliminating v', we obtain

$$I_0 = \frac{A_0 v_{in}}{R_L + A_0 B}$$

If $A_0 B \gg R_L$,

$$I_0 \cong \frac{v_{in}}{B} \quad \text{or} \quad \frac{I_0}{v_{in}} = \frac{1}{B} \tag{7.26}$$

The current output per volt input is *constant* and equal to $1/B$. This is often called the *transconductance* of the amplifier and could be expressed in milliamperes per volt or similar units. This basic idea can be used with op amps to construct almost ideal constant current sources (see Chapters 9 and 10).

7.5 POSITIVE FEEDBACK

For the negative-feedback amplifier of Fig. 7.10(a), the feedback voltage Bv_{out} *subtracts from* the input signal v_{in}. The net input to the amplifier of gain A_0 is $v' = v_{in} - Bv_{out}$, and the overall gain is

$$A_v = \frac{v_{out}}{v_{in}} = \frac{A_0}{(1 + A_0 B)}$$

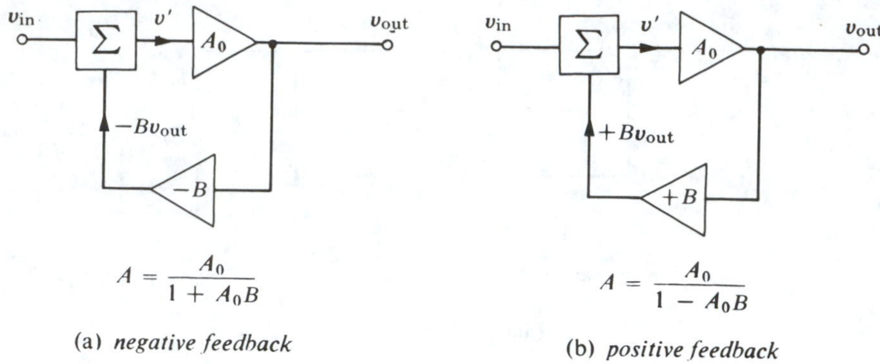

FIGURE 7.10 Feedback amplifiers.

As B is increased from 0, the overall gain drops from A_0 and levels out at $A_v = 1/B$ in the limit as $A_0B \gg 1$. However, in the positive-feedback circuit of Fig. 7.10(b), the feedback voltage Bv_{out} ***adds to*** the input signal v_{in}. The net input to the amplifier of gain A_0 is $v' = v_{in} + Bv_{out}$, and the overall gain is

$$A_v = \frac{v_{out}}{v_{in}} = \frac{A_0}{1 - A_0B}$$

Notice that as B increases from 0 in the case of positive feedback, the gain starts at A_0 and blows up to infinity when $A_0B = 1$.

What does an infinite gain mean physically? There are no infinities in practical electronics. What the mathematical result $A_v \to \infty$ as $A_0B \to 1$ means is that *if* the gain A_0 remains constant, then the overall gain goes to infinity. However, as the input $v' = v_{in} + Bv_{out}$ keeps increasing, the gain A_0 falls off in all practical circuits; in other words, A_0 is the gain only for small inputs v'. Stated another way, as the net input v' increases in magnitude, the inherent nonlinearities in the amplifier become significant and the output tends to approach a constant magnitude. (These non-linearities are usually negligible in an amplifier designed for small signal inputs.) The final result is that if $A_0B \geqq 1$, the output of the circuit has a constant magnitude and is often quite independent of the input v_{in}. If v_{in} is removed, the circuit still can generate an output—that is, we have an oscillator. Another way of looking at an infinite gain is to say that a finite output is generated with zero input.

Positive feedback is used in two ways: first, and most common, to make an oscillator or signal generator—that is, to produce oscillating waveforms (sine or square waves, etc.) with no input. For this purpose $A_0B \geqq 1$. The output waveform depends on the type of feedback network used, as will be explained in detail later. The second purpose is to increase the gain of an amplifier by having $A_0B < 1$. However, if either B or A_0 increases to the point where $A_0B = 1$, the amplifier will break into oscillation.

A sinusoidal oscillator of constant amplitude can be made by taking a physical system that tends naturally to oscillate when perturbed and adding just enough energy to the system to compensate for the energy lost (by friction, radiation, etc.) in each cycle of oscillation. There are two requirements: (1) The amount of energy added per cycle must exactly equal the amount lost per cycle to produce a constant magnitude oscillation. (2) This energy must be added in phase with the oscillation, that is, at just the right time to increase rather than decrease the amplitude of the oscillation. In other words, the feedback must be positive.

Let us now consider a mechanical oscillator consisting of a simple pendulum mounted on a frictionless bearing and swinging back and forth so that the bob of mass m passes through a viscous oil bath, as shown in Fig. 7.11. During each cycle as the bob passes back and forth through the oil

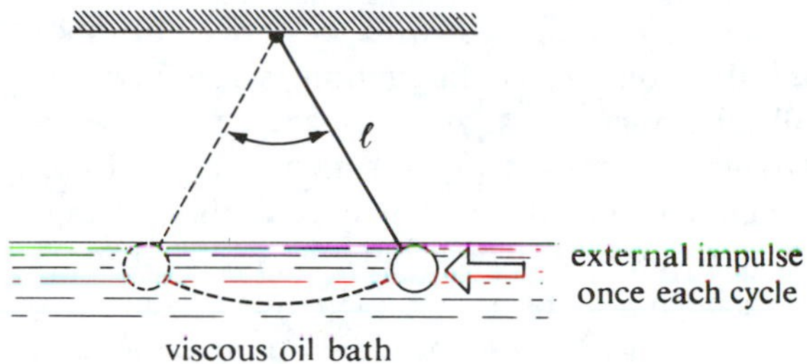

FIGURE 7.11 Mechanical oscillator.

bath, mechanical energy of oscillation is lost (as heat) due to the viscous retarding force exerted by the oil on the bob. If this energy loss is not too large, the frequency of the pendulum is given by $f = (1/2\pi)\sqrt{g/\ell}$, and the amplitude of the oscillations will slowly fall to zero.† To keep the amplitude constant, we must add by some external agency exactly the same amount of energy to each cycle as is lost each cycle to heat in the oil bath. This can be done by applying an impulse every $T = 1/f$ seconds; that is, the impulse must be periodic and must have exactly the same frequency as the pendulum oscillation. For the impulse to *add* energy to the pendulum, it must be applied to the left when the bob is at the extreme right or when it is moving from right to left. If the impulse is applied too soon in the cycle

†An exact analysis will show that the amplitude of the oscillations is given by

$$x(t) = x_0 \sin\left\{\omega_0 t\left[1 - \left(\frac{1}{2\omega_0\tau}\right)^2\right]^{1/2}\right\} e^{-t/2\tau}$$

where $\tau \equiv m/\gamma$ is called the *relaxation time*, γ is the viscous retarding force constant: $-\gamma(dx/dt) = F_{\text{viscous}}$, and ω_0 is the natural oscillation frequency with no friction ($\gamma = 0$).

(i.e., when the bob is moving toward the right), energy will be taken *from* the pendulum and the amplitude will decrease. In other words, the point in the oscillation cycle when the impulse is applied (i.e., the phase) is important.

If the applied impulse supplies more energy to the pendulum each cycle than is lost to heat in the oil bath, then the amplitude of oscillation increases. The velocity of the bob through the oil also increases, and the energy loss as heat in the oil bath increases until a new steady-state situation is reached when the energy input from the external impulse each cycle exactly equals the energy loss in the oil bath each cycle. The amplitude of the oscillations grows until nonlinearities in the amplifier decrease the gain A_0 so that A_0B falls to unity. However, it should be emphasized that an oscillator "pushes itself" by means of the positive feedback, coupling some of the amplified output back into the input with the proper phase.

Another example of oscillation can be found in the wild snowshoe hare and Canadian lynx populations in the North American Arctic. The hare population naturally oscillates with a period of about ten years. The amplitude of the population fluctuation is roughly a factor of four from maximum to minimum years. As the hare become more numerous due to natural reproduction, they overgraze their feed and the lynx also begin to increase in number. The Canadian lynx is the principal predator for the hare. The larger number of lynx plus the reduced food available per hare reduces the hare population; and with fewer hare the lynx decrease in number, and the hare browse becomes more abundant. These factors in turn lead to an increase in the hare population, and the cycle begins again. As might be expected, the lynx population also oscillates at the same frequency, and slightly out of phase with respect to the hare population—a peak in hare population precedes a peak in the lynx population. Five- and ten-year oscillations are actually common in many forms of wildlife. These oscillations tend to be most violent in northern latitudes, because the arctic ecosystems tend to be simpler—there are fewer predator species that prey on only one species. The harsh arctic environment also tends to reduce greatly the diversity of species able to survive.

If the feedback signal is precisely in phase with the input of an amplifier, oscillation will occur if $A_0B \geqq 1$. Often the phase shift depends on the frequency, due to RC or LC networks in the circuit. For example, in an RC coupled amplifier an RC high-pass filter is used at the input and at the output; and at very low frequencies, when $\omega < 1/RC$, there is an appreciable phase shift as well as attenuation. The phase difference between the output across R and the input is

$$\phi = \tan^{-1}\left(\frac{1}{\omega RC}\right) \qquad \textbf{(7.27)}$$

If $\omega = 1/RC$, then $\phi = 45°$. If the RC network is designed to pass frequencies down to $f = 20$ Hz, then typical values might be: $C = 1\ \mu$F, $R = 30\ \text{k}\Omega$. Then $RC = 0.03$ s; and when $f = 5$ Hz, $\omega = RC$ and we have a 45° phase shift. The phase shift approaches 90° as $\omega \to 0$, but the gain $\to 0$, so a *single RC* network can never produce a phase shift near 90°. Several networks, however, may produce a 180° phase shift and oscillation.

At higher frequencies *shunt* capacitance becomes important in determining the gain and phase shift. Usually we have a shunt capacitance C_s between the signal lead and ground in parallel with a resistance R_s. The parallel impedance of C_s and R_s is

$$Z = \frac{X_{C_s} R_s}{X_{C_s} + R_s} = \frac{R_s}{1 + j\omega R_s C_s} \qquad |Z| = \frac{R_s}{\sqrt{1 + \omega^2 R_s^2 C_s^2}} \tag{7.28}$$

Z falls to zero as ω approaches infinity. When $\omega = 1/R_s C_s$, the phase of Z changes by 45° compared to the low-frequency phase. For $C_s = 10$ pF and $R_s = 10\ \text{k}\Omega$, $R_s C_s = 10^{-7}$ s, and the 45° phase shift occurs at $f = 1.6$ MHz. The phase shift approaches 90° as $\omega \to \infty$, but $Z \to 0$ as $\omega \to \infty$.

If a number of phase-shifting networks exist in an amplifier, it is possible to have an accidental phase shift that produces positive feedback. If the amplifier gain at this frequency exceeds $1/B$, the amplifier will oscillate. It is customary to consider the amplifier gain and phase shift as a function of frequency by plotting the complex value of the loop gain $A_0 B$ as a function of frequency in a polar plot. Such a plot is called a *Nyquist diagram* (see Fig. 7.12). The magnitude of the loop gain $A_0 B$ at a

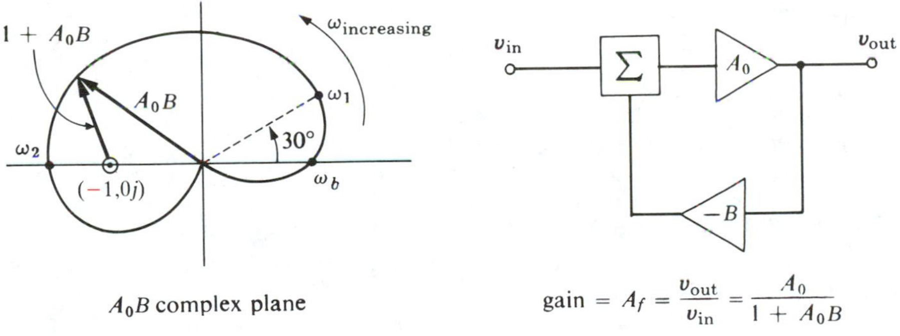

FIGURE 7.12 Nyquist diagram.

frequency ω is the distance from the origin out to the point of the $A_0 B$ graph corresponding to the frequency ω. The desired phase of the amplifier for negative feedback and stable operation is $A_0 B$ positive and purely real; this is termed the *midband* frequency ω_b and corresponds to the point of the $A_0 B$ vector lying on the positive real axis. As the frequency differs from ω_b, both the magnitude and the phase of $A_0 B$ change, and this is indicated by the tip of the $A_0 B$ vector lying off the positive real axis. For example, at the frequency $\omega_1 > \omega_b$ in Fig. 7.12, the phase of A_0B is 30° different from the phase at the midband.

If $A_0B = -1$, then the gain with feedback $A_f = A_0/(1 + A_0B)$ diverges to infinity, and the system is unstable (i.e., oscillation results). Oscillation also results if A_0B is negative and $|A_0B| > 1$. This situation occurs in Fig. 7.12 at the frequency ω_2. It can be shown that if the complex vector drawn from the point $(-1, 0j)$ to the point A_0B (i.e., the vector $1 + A_0B$) does *not* enclose the point $(-1, 0j)$ then the circuit will be stable. This is the famous Nyquist criterion for stability.

Examples of stable and unstable Nyquist plots are shown in Fig. 7.13.

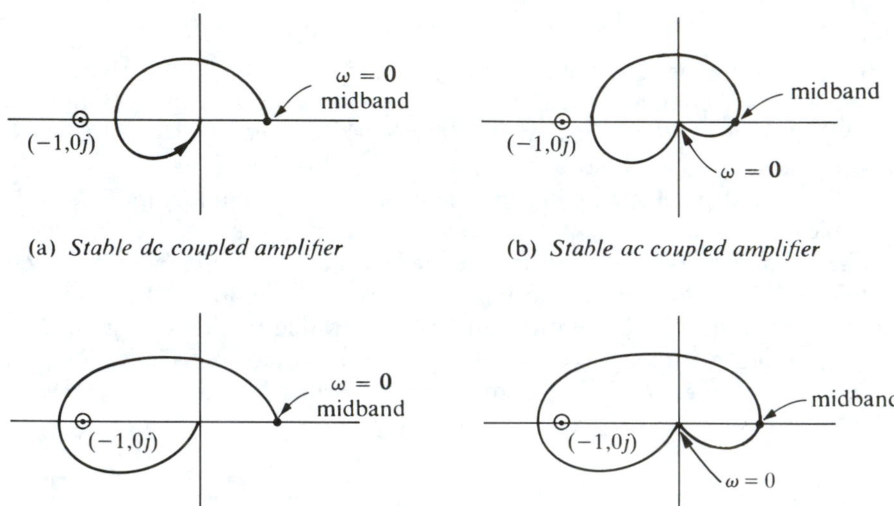

FIGURE 7.13 Nyquist diagrams A_0B complex plane.

The physical meaning of the vector $1 + A_0B$ enclosing the point $(-1, 0j)$ is simply that A_0B has a magnitude greater than 1 ($A_0 > 1/B$), and the feedback is positive rather than negative.

From Chapter 2 the phase shift of a network is often determined from the slope of the gain-versus-frequency plot. Thus, a clue that instability and oscillations may be encountered is often apparent from a rapidly changing gain-versus-frequency graph, especially if the gain is large at these frequencies. An example is shown in Fig. 7.14. The hump in gain near ω_1 for curve 2 corresponds to the A_0B loop gain approaching -1 and the $1 + A_0B$ vector almost encircling the point $(-1, 0j)$. If the gain hump is high enough, the A_0B curve will enclose the point $(-1, 0j)$ and the circuit will be unstable and will oscillate at ω_1. It is an unpleasant and common feature of such circuits that they will always oscillate at ω_1 if A_0B is

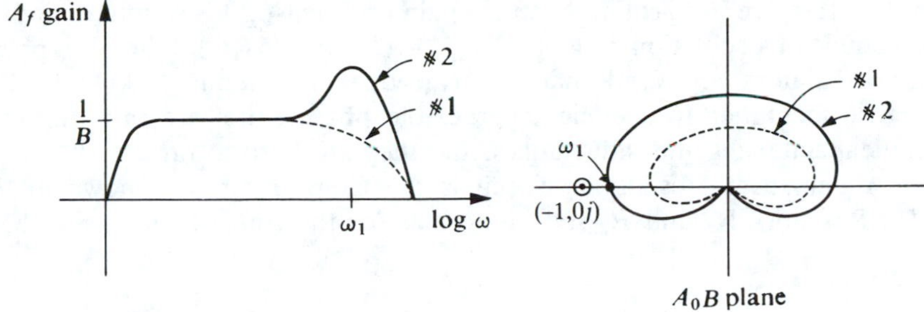

FIGURE 7.14 Almost unstable amplifier.

negative and exceeds 1 at ω_1. They will not amplify signals benignly at frequencies different from ω_1. The reason is basically that there are always some Fourier components of noise voltages at the frequency ω_1; these voltages initiate the positive feedback, and oscillations result because the gain is essentially infinite at ω_1.

An oscillator circuit in which the 180° phase shift in the feedback from collector to base is easily seen is the *RC* phase-shift oscillator shown in Fig. 7.15. In this circuit the 180° phase shift comes from the three *RC* high-pass

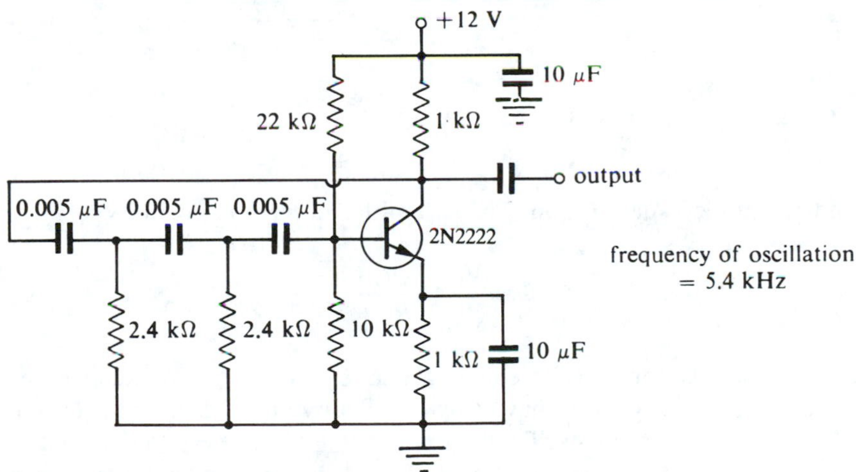

FIGURE 7.15 *RC* phase-shift oscillator.

filters in series between the collector and the base. The resistance to ground in the third *RC* filter is essentially h_{ie} of the transistor, which is about 2 kΩ. Each filter contributes about 60° of phase shift and also attenuates the signal. For three identical *RC* filters, where the loading effects of one filter on the other are important, a voltage gain of at least 29 is required, and the frequency of oscillation is given by $f_0 = (1/2\pi)(1/6RC)$. The RC phase-shift oscillators are used mainly for frequencies about 10 kHz or less, because at higher frequencies appreciable phase shifts are introduced by stray capacitances and inductances, and they are hard to tune.

A very useful oscillator circuit is the Colpitts circuit, shown in Fig. 7.16. Resistors R_1 and R_2 set the dc bias for the transistor along with R_E

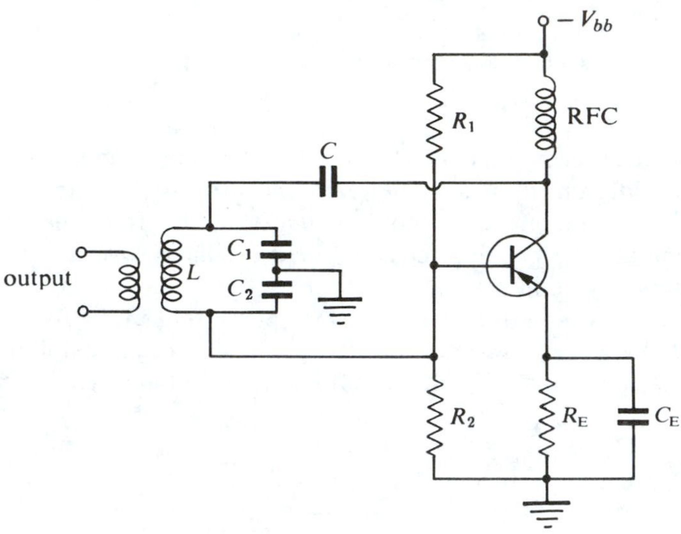

FIGURE 7.16 Colpitts oscillator.

and C_E in the usual fashion. The dc emitter current is given by

$$I_E = \frac{V_E}{R_E} \cong \frac{R_2 V_{bb}}{R_1 + R_2} \frac{1}{R_E}$$

"RFC" stands for radio frequency choke and is an inductance in the collector lead designed to have a large inductive reactance at the frequency of operation. For an oscillation frequency of 500 kHz, a 2.5-mH RFC would have an inductive reactance of $|X_L| = 79\text{ k}\Omega$, which would provide adequate ac voltage gain. The ac voltage on the collector is coupled through the coupling capacitor C back to the base resonant circuit consisting of L and C_1 and C_2. Thus the reactance of C should be small at the operating frequency. For example, at $f = 500$ kHz, $C = 0.1\ \mu$F, the capacitive reactance $|X_C| = 3\ \Omega$.

The feedback from collector to base can be seen to be positive by the following argument. The emitter is an ac ground, so we have a common emitter configuration. In this configuration there is a 180° phase change from base to collector; that is, a positive-going voltage on the base produces a negative-going voltage on the collector. The base and collector are connected by the tank circuit consisting of L and $C_1 + C_2$, and there is a 180° phase change between the voltages on opposite ends of an LC tank. Hence, a positive-going voltage on the base produces an amplified negative-going voltage on the collector, which is coupled back to the base as a positive-going voltage by the tank.

At high frequencies where air-core transformers suffice, the output is usually transformer coupled from the inductance L. At lower, audio frequencies the output can be taken off the collector through a series coupling capacitor.

The frequency of oscillation is the resonant frequency of the tank. Because C_1 and C_2 are in series, their effective capacitance is $C_1C_2/(C_1 + C_2)$. Hence the resonant frequency is

$$f_0 = \frac{1}{2\pi}\left[\sqrt{L\left(\frac{C_1C_2}{C_1 + C_2}\right)}\right]^{-1} \tag{7.29}$$

The amount of feedback is determined by the ratio of C_1 to C_2. C_2 is connected between the base and emitter (the emitter is an ac ground because of C_E): hence the smaller C_2 and the larger its capactive reactance $1/\omega C_2$, the larger the feedback voltage to the base. Therefore, to change the oscillation frequency without changing the feedback ratio, the ratio C_1/C_2 should be kept constant while the effective series capacitance $C = C_1C_2/(C_1 + C_2)$ is changed. The Colpitts oscillator is very popular and with proper choice of transistor can operate up to 30 MHz or higher.

The junction of C_1 and C_2 can be thought of as a tap that takes a fraction of the ac voltage across the tank and applies it to the base as the feedback voltage. This fraction could equally well be obtained by tapping off the inductor L in the tank circuit. Such an oscillator is called a Hartley oscillator and is shown in Fig. 7.17.

No discussion of oscillators would be complete without mention of the crystal oscillator. This oscillator uses a precisely ground, thin quartz crystal to determine the oscillation frequency through the piezoelectric effect. The piezoelectric effect is the mechanical vibration of the crystal in response to an applied electric field, and the generation of an oscillating electric field as a result of its vibration. The quartz crystal is ground to the proper dimension so that one of its mechanical resonant frequencies is in the region of interest. The Q of the crystal's mechanical vibration is several orders of magnitude higher than that of a conventional LC tank, whose Q is usually 100 at most. The crystal is, in effect, a very high-Q LC resonant circuit. The crystal is mounted between two metal plates and has two terminals, one connected to each plate. Crystals can be purchased for \$5–\$10 each in frequencies from several kHz to nearly 100 MHz. Crystal oscillators find

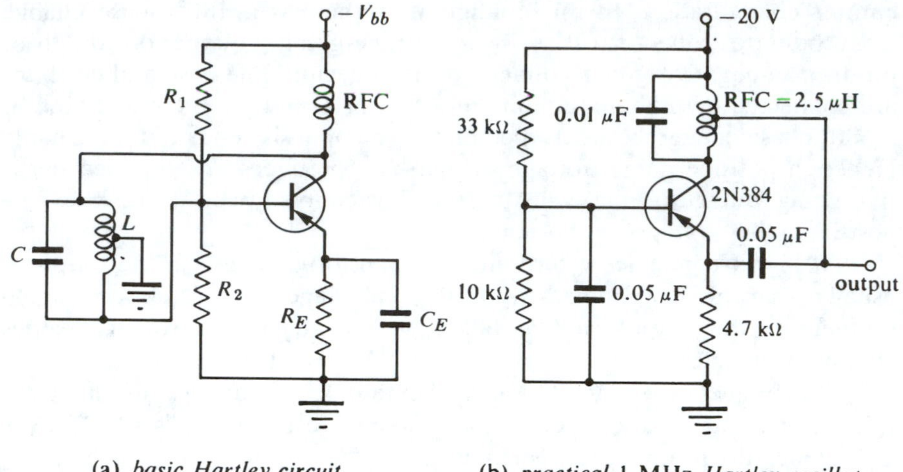

(a) *basic Hartley circuit*

(b) *practical* 1 MHz *Hartley oscillator (from a Handbook of Selected Semiconductor Circuits NAVSHIPS* 93484. *U.S. Government Printing Office*)

FIGURE 7.17 Hartley oscillator.

their principal application in low-power, fixed-frequency circuits (tens of mW power) where frequency stability is the main consideration. Stability of the order of $\Delta f/f = 10^{-6}$ or 10^{-7} is easily obtainable with inexpensive crystals. The crystal can be used in almost any type of oscillator circuit. Greater frequency stability can be obtained by placing the crystal in a

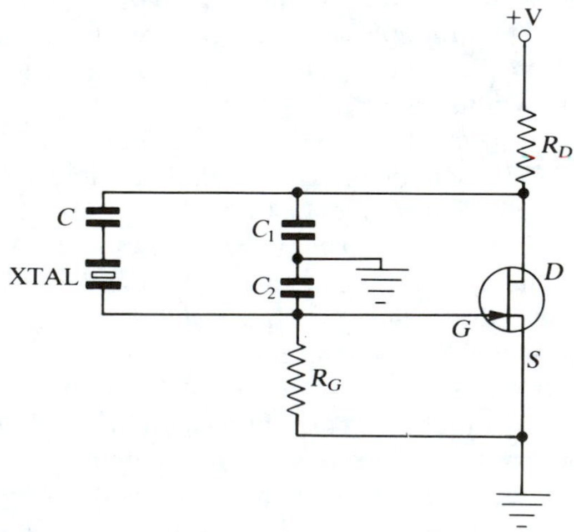

FIGURE 7.18 Crystal Colpitts oscillator.

temperature-controlled oven, because temperature changes produce slight changes in the crystal size, which in turn results in small changes in the crystal frequency. Figure 7.18 shows a simple JFET crystal Colpitts oscillator.

7.6 PRACTICAL COMMENTS AND NEUTRALIZATION

As anyone who has ever tried to build an amplifier or oscillator knows, many things can go wrong. Oscillators don't oscillate, and amplifiers tend to oscillate at strange frequencies, thereby obliterating the desired signal to be amplified. When an oscillator will not oscillate, the usual cure is to increase the amount of positive feedback and/or to increase the gain of the amplifier section to achieve the condition $A_0B \geqq 1$. For example, to increase B, you can vary the ratio of the capacitances in the resonant circuit of the Colpitts oscillator. The gain can be increased by changing transistors, adjusting the impedance in the collector, or changing the dc operating point to obtain a high value of h_{fe} or g_m for the transistor or FET, respectively.

Unwanted oscillations in an amplifier can often be eliminated by adding relatively large bypass capacitors from the V_{bb} connection to ground at each transistor. This procedure ensures that the V_{bb} supply for each transistor is a good ac ground and eliminates unwanted ac feedback from transistor to transistor through the V_{bb} power supply leads. In fact, it is good practice to include such bypass capacitors in the initial design of any oscillator or amplifier. Often, too, a small resistance, on the order of 1 kΩ, is put in series with the V_{bb} lead to each transistor along with the bypass capacitor, to increase the isolation between stages as is shown in Fig. 7.19.)

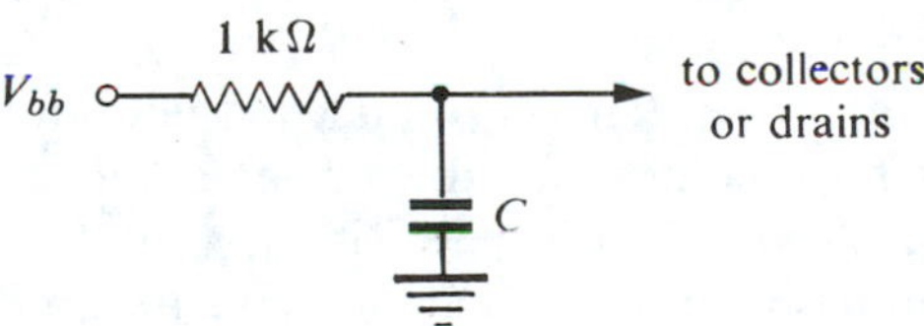

FIGURE 7.19 Filtering power supply lead to prevent unwanted coupling.

The lower the operating frequency of the circuit, the larger the bypass capacitance must be. $C = 100$ pF is adequate for frequencies over 10 MHz, whereas $C = 1\ \mu$F might be required for audio frequencies.

Another source of unwanted oscillations in an amplifier is excessive lead length. As a teenager I once laboriously and lovingly constructed an eight-tube superheterodyne AM receiver with three stages of i.f. amplification on what seemed like a chassis of reasonable size—approximately 8″ × 15″ × 3″. The receiver oscillated at a large number of frequencies and sounded somewhat like a cage full of fighting tomcats. The cure is to build as compactly as possible and use minimum lead lengths.

Particularly at frequencies above 1 MHz it is desirable to use a tuned *LC* circuit in both the base and collector in transistor circuits and for the gate and drain in FET circuits. Such a circuit is often called a tuned base-tuned collector or a tuned gate-tuned drain circuit. The FET circuit is shown in Fig. 7.20. This circuit will oscillate because of the inherent

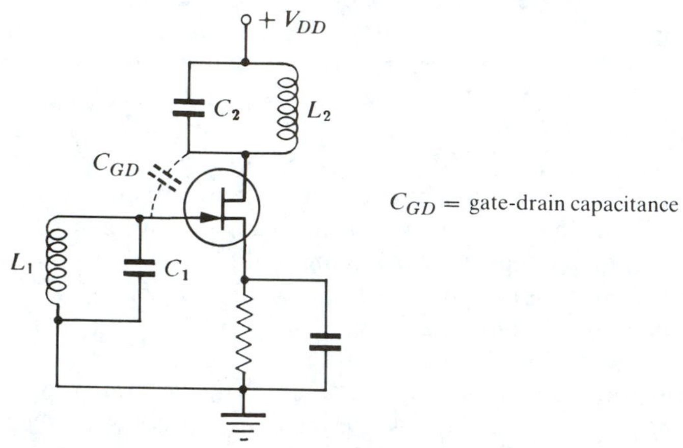

FIGURE 7.20 Tuned gate–tuned drain FET oscillator.

capacitive coupling between the input and the output, that is, due to the base-collector or gate-drain capacitances.

The gate-drain capacitance is usually *small*, of the order of several pF, and thus has a large impedance; for example, if $C_{GD} = 2$ pF, then $|X_{C_{GD}}| = 1/\omega C_{GD} = 80\ \text{k}\Omega$ at 1 MHz. The C_{GD} and the gate-ground impedance R_G form a high-pass filter connecting the drain to the gate. This filter will introduce a phase shift of nearly +90° if $X_{C_{GD}} \gg R_G$. If the L_2C_2 tank circuit in the drain has a natural resonant frequency $\omega_D = 1/\sqrt{L_2C_2}$ higher than the operating frequency, then the impedance of the L_2C_2 tank is inductive and introduces a +90° phase shift. Thus the total phase shift introduced by L_2C_2 and C_{GD} is approximately 180°, which added to the 180° phase shift between gate and drain inherent in the common source configuration gives *positive* feedback: If the gate goes positive, the signal fed back to the gate through C_{GD} also goes positive. Such an oscillator is called a *tuned-gate–tuned-drain* oscillator, as mentioned above.

An amplifier made from a bipolar transistor in the common emitter configuration with a tuned *LC* circuit in the input and the output will tend to oscillate by this same argument. The common base transistor amplifier and the common gate FET amplifier will *not* oscillate in this circuit configuration because of the excellent input–output isolation; that is, the emitter-collector and the source-drain capacitances are extremely low, well under 1 pF. Such circuits are used mostly for amplifiers at frequencies above 50 MHz for precisely this reason.

Unwanted coupling or feedback between stages of a multistage amplifier can sometimes result from an overlap of rf grounding currents in the chassis, which serves as a common ground for the various amplifier stages. This overlap of currents and consequent coupling is greater when the various circuit grounds are connected to the chassis at different locations. Minimum overlap is thus obtained by connecting all the bypass capacitor grounds and other grounds to one point on the chassis, taking care to use the shortest ground leads possible. A certain amount of experimentation is usually necessary. Figure 7.21 shows the proper and improper connection of an rf bypass capacitor.

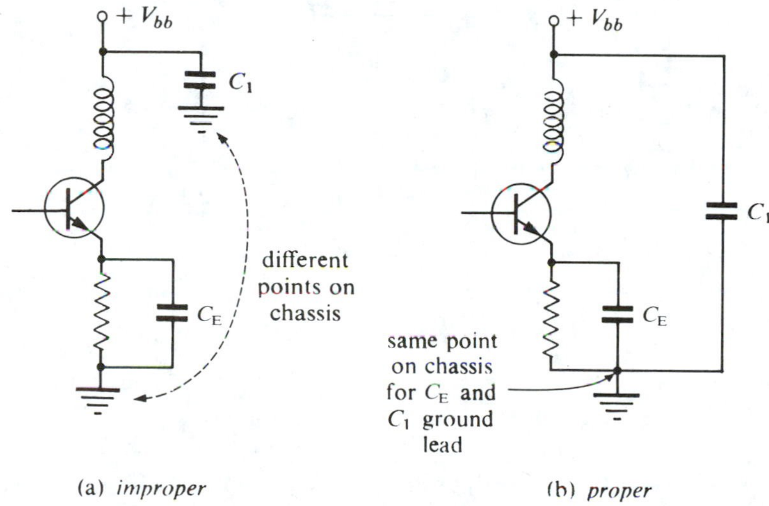

FIGURE 7.21 Ground connections.

7.7 THE MILLER EFFECT

As we have just seen, the inherent capacitance between the base and collector of a bipolar transistor (or the gate and drain of an FET) can lead to unwanted oscillations in the tuned-base–tuned-collector amplifier because some of the output voltage on the collector is fed back in phase with the input. This capacitance and its feedback can also lead to a significant decrease in the high-frequency response of an amplifier.

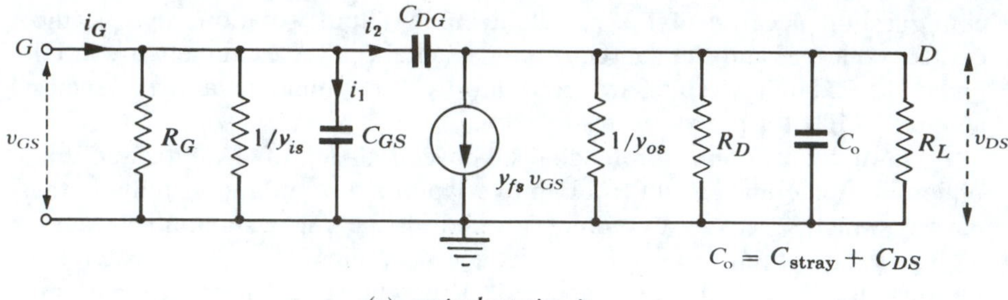

(a) *equivalent circuit*

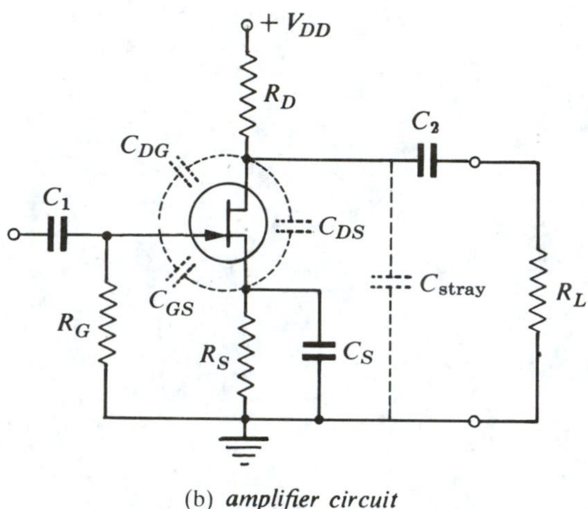

(b) *amplifier circuit*

FIGURE 7.22 High-frequency FET circuit.

Consider a JFET whose high-frequency equivalent circuit is shown in Fig. 7.22(a). This circuit differs from the simpler equivalent circuit in Chapter 6 only by the addition of three capacitances: C_{GS} between the gate and source, C_{DG} between the gate and drain, and C_o between the drain (output) and ground. C_o is the capacitance between the drain and ground plus any stray or shunt capacitances between the output lead and ground. We will now show that C_{DG} plays a major role in attenuating the high-frequency gain. In effect, C_{DG} is multiplied by the voltage gain of the amplifier and appears electrically to be connected across the input. The proof is as follows.

The input impedance is

$$Z_{in} = \frac{v_{GS}}{i_G}$$

and the ac gate current is

$$i_G \cong i_1 + i_2 = \frac{v_{GS}}{\dfrac{1}{j\omega C_{GS}}} + \frac{v_{GS} - v_{DS}}{\dfrac{1}{j\omega C_{DG}}}$$

We have assumed here that $1/j\omega C_{GS} \ll R_G$ or $1/y_{is}$. We recall from Chapter 6 that the voltage gain A_v of the FET common source amplifier is given by

$$A_v = \frac{v_{DS}}{v_{GS}} = \frac{-y_{fs}R_{DL}}{1 + y_{os}R_{DL}} \cong -y_{fs}R_{DL}$$

where $R_{DL} = R_D \| R_L$. Eliminating v_{DS}, we find that

$$i_G \cong \frac{v_{GS}}{\dfrac{1}{j\omega C_{GS}}} + \frac{v_{GS} + y_{fs}R_{DL}v_{GS}}{\dfrac{1}{j\omega C_{DG}}}$$

or

$$Z_{in} = \frac{v_{GS}}{i_G} = \{j\omega C_{GS} + j\omega C_{DG}(1 + y_{fs}R_{DL})\}^{-1}$$

$$\cong \{j\omega[C_{GS} + (1 + |A_v|)C_{DG}]\}^{-1} \quad \textbf{(7.30)}$$

The coefficient of $j\omega$ in equation (7.30) is the effective capacitance ω across the input:

$$C_{in} \cong C_{GS} + (1 + |A_v|)C_{DG} \quad \textbf{(7.31)}$$

If the voltage gain A_v is large, then the input capacitance can be many times larger than the gate-source capacitance C_{GS} of the FET. This enhancement of the input capacitance is called the *Miller effect* and also holds for bipolar transistors in the common emitter configuration. For bipolar transistors $C_{in} \cong C_{BE} + (1 + |A_v|)C_{CB}$, where C_{BE} is the base-emitter capacitance and C_{CB} is the collector-base capacitance.

It is important to emphasize that the Miller effect holds only for FETs and bipolar transistors in the common source and common emitter configurations, respectively. The common gate or common drain FET amplifiers and the common base or common collector bipolar transistor amplifiers do not exhibit the Miller effect and thus will amplify at high frequencies.

Capacitance values for an inexpensive FET (2N5459) and an inexpensive bipolar transistor (2N4250) are given in Table 7.2. Notice that the "interelectrode" capacitance, which is multiplied by the voltage gain in equation (7.31), is about 2 to 4 pF. If the voltage gain is about 50, then a capacitance of 100 to 200 pF will appear across the input. The impedance

TABLE 7.2

Device	Capacitance	
2N5459 FET	$C_{GS} = 4.5$ pF	$C_{DG} = 1.5$ pF
2N4250 Bipolar transistor (2N5087)	$C_{BE} = 4.8$ pF	$C_{CB} = 3.5$ pF

of a 100-pF capacitor is only 167 Ω at 10 MHz, so it is easy to see how the Miller effect reduces the high-frequency gain. Even at the relatively low frequency of 10 kHz, the impedance of a 100-pF capacitor is 167 kΩ, which can have a drastic effect if it is in parallel with a high resistance. For example, in an FET common source amplifier, the gate-to-ground resistor may be 1 MΩ. Thus, at 10 kHz, the gate-to-ground impedance would be 1 MΩ in parallel with 167 kΩ, which is only 143 kΩ. Thus the input impedance can be significantly less than the value of the gate-to-ground resistor even in the audio frequency range.

It is worth pointing out that the interelectrode or junction capacitance, which is multiplied by the gain in a bipolar transistor or an FET, is the capacitance inherent in the reverse biased p-n junction and therefore depends on the voltage across the junction. The junction capacitance is given by $C = KV^{-1/2}$, where K is a constant depending on the type of junction and V is the reverse bias voltage across the junction. For the 2N5087 bipolar transistor, for example, C_{CB} is 4.8 pF when $V_{CB} = 1$ V dc and falls to 1.7 pF when $V_{CB} = 10$ V dc. Thus, some improvement in high-frequency operation can be obtained by increasing the reverse bias voltage between the base and collector for a bipolar transistor or between the gate and drain for an FET.

7.8 THE CASCODE AMPLIFIER

The Miller effect for an FET can be reduced by approximately a factor of 100 by connecting two FETs in a *cascode* arrangement shown in Fig. 7.23. Q_1 is an FET in a conventional common source configuration, and Q_2 acts as the drain load of Q_1. Q_2 can also be thought of as a common gate amplifier. The relatively low input impedance of the common gate amplifier effectively reduces the voltage gain between the gate and the drain of Q_1. Thus C_{DG} for Q_1 is not multiplied by a large factor. The voltage gain comes in Q_2, which is isolated from the input because the gate of Q_2 is an ac ground. Q_2 can be a bipolar transistor to achieve higher gain.

The bipolar circuit of Fig. 7.23(b) is similar; Q_2 is a common base amplifier in a low input impedance. The differential amplifier, shown with bipolar transistors in Fig. 7.23(d), has no Miller effect, although, strictly speaking, it is not a cascode circuit. Q_1 is an emitter follower that drives the common base amplifier Q_2.

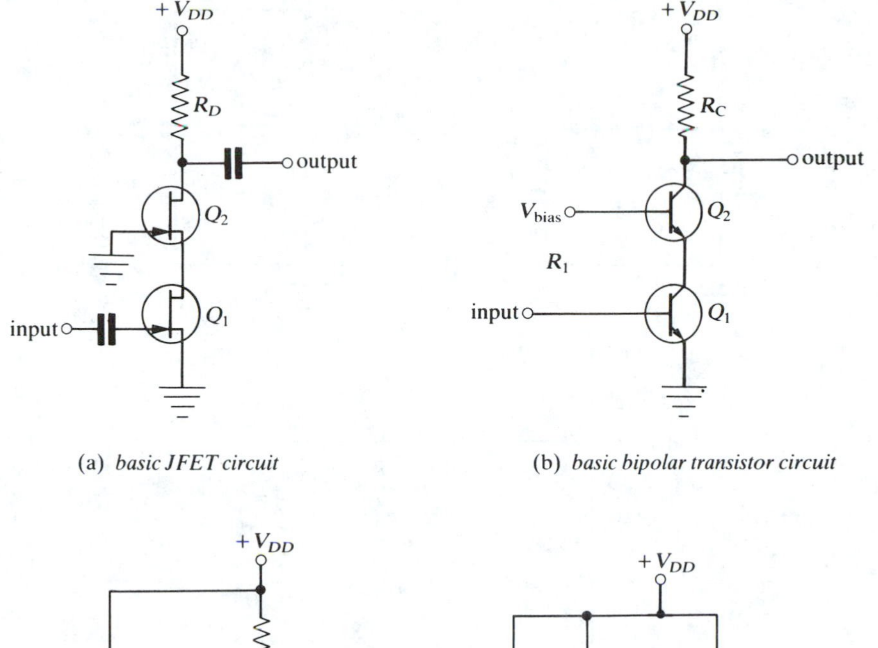

(a) *basic JFET circuit*

(b) *basic bipolar transistor circuit*

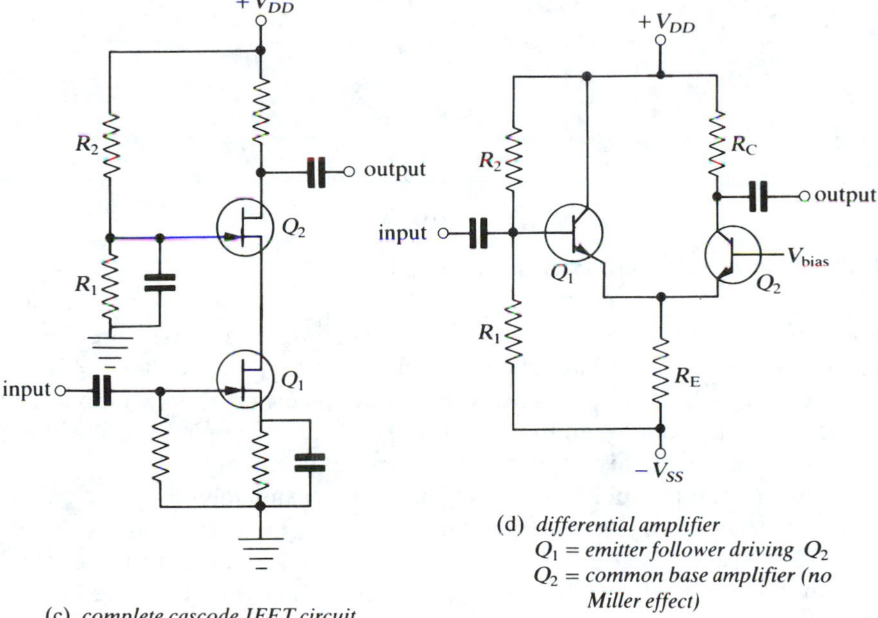

(c) *complete cascode JFET circuit*

(d) *differential amplifier*
Q_1 = emitter follower driving Q_2
Q_2 = common base amplifier (no Miller effect)

FIGURE 7.23 Cascode amplifier.

7.9 THE GAIN–BANDWIDTH PRODUCT

Whenever a shunt capacitance is driven by an amplifier output, as shown in Fig. 7.24, we have a low-pass RC filter in which R is the output resistance of the amplifier and C_{shunt} is the total shunt capacitance (i.e., the capaci-

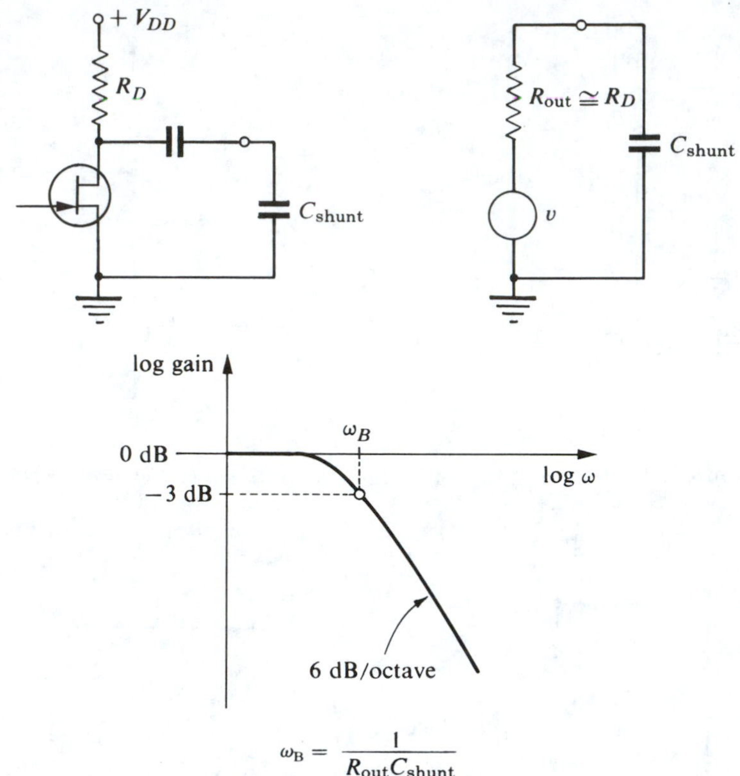

FIGURE 7.24 Amplifier driving capacitive load.

tance between the output terminal and ground). The upper breakpoint where the voltage gain has fallen by 3 dB or a factor of $0.707 = 1/\sqrt{2}$ is given by $\omega_B = 1/R_{out}C_{shunt}$. The situation is somewhat more complicated but basically the same when the output impedance is complex (Z_{out}) so long as the real part of Z_{out} is large compared to the imaginary part.

For an FET the mid-frequency gain is approximately given by $|A_v| \cong g_m R_D$, where g_m is the transconductance ($g_m = y_{fs}$) and R_D is the value of the drain resistor. The output impedance of an FET is equal to the parallel combination of R_D and $1/y_{22}$. Usually $R_D \ll 1/y_{22}$, so that the output impedance is approximately R_D. Thus we can write for the upper 3-dB angular frequency ω_B

$$\omega_B = \frac{1}{R_{out}C_{shunt}} \cong \frac{1}{R_D C_{shunt}}$$

Eliminating R_D with $|A_v| = g_m R_D$ and rearranging, we obtain

$$|A_v|\omega_B = \frac{g_m}{C_{shunt}}$$

In words, the product of the voltage gain $|A_v|$ and the upper 3-dB frequency equals the transconductance of the device divided by the total shunt capacitance between the output signal lead and ground. The bandwidth of the amplifier is conventionally taken to mean the frequency in hertz at which the gain has fallen 3 dB, or 0.707 times the mid-frequency gain. Thus, if the lower 3-dB frequency is small compared to ω_B, we may say the bandwidth f is essentially equal to $\omega_B/2\pi$. Thus the voltage gain–bandwidth product is

$$|A_v|\Delta f = \frac{g_m}{2\pi C_{\text{shunt}}} \tag{7.32}$$

The significance of this gain–bandwidth product is that it depends only on characteristics of the *device*, g_m, and the *wiring* shunt capacitance, not on the particular circuit used. The gain–bandwidth product is often denoted by f_T. For an FET with $g_m = 4000\ \mu$mhos and $C_{\text{shunt}} = 5$ pF, the gain–bandwidth product is

$$A_v\Delta f = \frac{4 \times 10^{-3}\ \text{mhos}}{(2\pi)(5 \times 10^{-12}\ \text{F})} = 127\ \text{MHz}$$

At a frequency of 127 MHz the voltage gain will have fallen to unity. If $R_D = 2.2$ kΩ, the mid-frequency gain is $|A_v| = g_m R_D = (4 \times 10^{-3})(2.2 \times 10^3) = 8.8$, and the upper 3-dB frequency is 14.4 MHz. Attempts to raise the gain at higher frequencies by increasing R_D while keeping the bandwidth are doomed to failure by equation (7.32). For example, if R_D is raised to 4.7 kΩ to increase the mid-frequency gain to 18.8, then the upper 3-dB frequency falls to 6.8 MHz. For a given device (g_m) and wiring (C_{shunt}) you can have a high gain over a small bandwidth or a low gain over a large bandwidth, as illustrated in Fig. 7.25. However, amplifiers can be cascaded in series to achieve a much higher gain at a slightly reduced bandwidth so

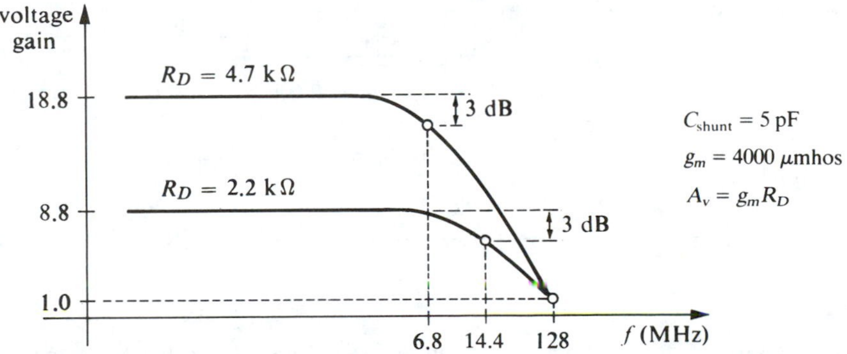

FIGURE 7.25 Constancy of gain–bandwidth product.

long as the output impedance of each stage is much less than the input impedance of the next stage. For example, two identical FET amplifiers in series, each with a gain–bandwidth product of 127 MHz, an upper 3-dB frequency of 14.4 MHz, and a mid-frequency gain of 8.8, would together have a gain of (8.8)(8.8) = 77.3 and an upper 3-dB frequency or bandwidth of approximately 9.3 MHz. The new upper 3-dB frequency for two identical amplifiers (of any type) is

$$\omega_{B2} = (\sqrt{2} - 1)^{1/2}\omega_B = 0.643\,\omega_B \tag{7.33}$$

where ω_B is the upper 3-dB frequency of one amplifier.

PROBLEMS

1. What is the principal disadvantage of negative feedback? What are the advantages?
2. (a) Derive the voltage gain expression $A_v = A_0/(1 + A_0 B)$ for an amplifier with negative-feedback factor B and gain A_0 without feedback. (b) Show by numerical substitution that if $A_0 = 10^6$ and $B = 10^{-2}$, the gain with feedback, A_v, varies approximately 0.002% for a 20% change in A_0, the gain without feedback.
3. Prove that the percentage change ΔA_0 in the gain without feedback produces a change $\Delta A'_f \cong \Delta A'_0/A_0 B$ in the gain with feedback. $\Delta A'_0$ = percentage change in A_0.
4. Explain why an unbypassed emitter resistor in a common emitter amplifier results in decreased gain.
5. Give the amplifier characteristics if 1% of the output is fed back out of phase (i.e., $B = 0.01$).

	$B = 0$	$B = 0.01$
Voltage gain	10,000	
Input impedance	20 kΩ	
Output impedance	5 kΩ	

6. Negative feedback affects distortion, gain, and impedance levels but only for the circuit actually *within* the feedback loop. The bias divider chain, $R_1 R_2$, is

outside the feedback loop. The amplifier inside the dashed line without feedback has a gain $A_0 = 10{,}000$ and input impedance 1 kΩ.

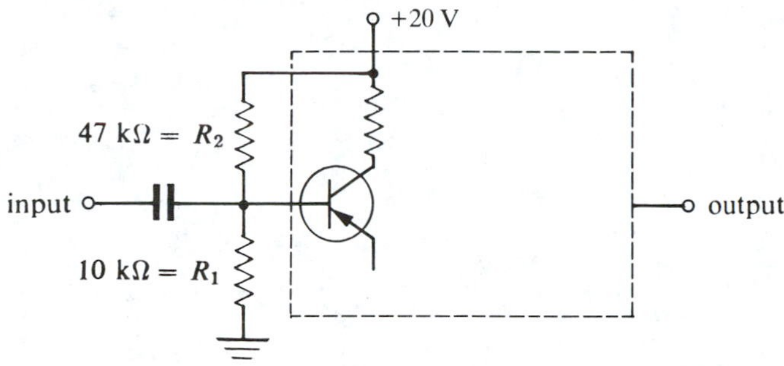

Calculate (a) the input impedance of the circuit if there is no negative feedback and (b) the input impedance of the circuit if there is negative feedback within the dashed line with $B = 0.05$.

7. Explain (a) why negative feedback will increase rather than decrease the input impedance of an amplifier, and (b) why a large input impedance is usually desirable.

8.

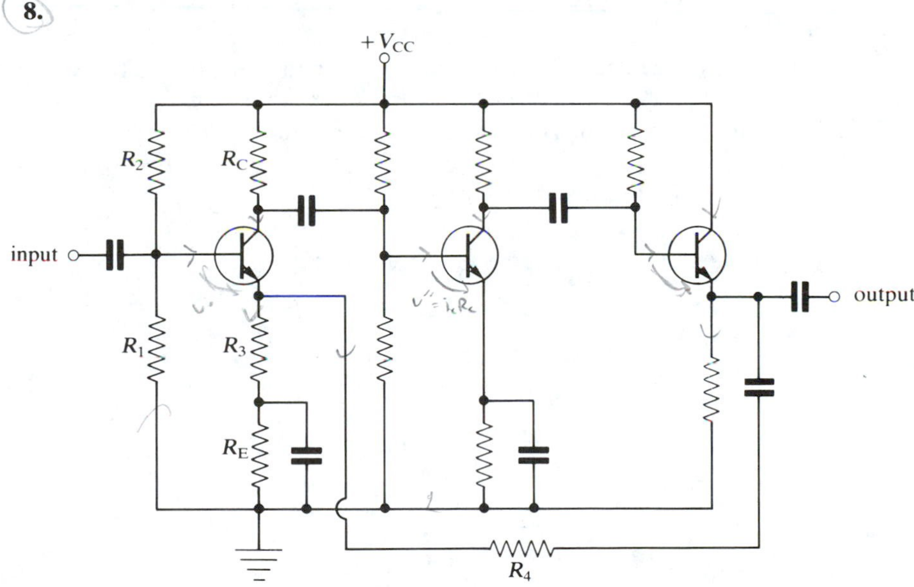

(a) Is the feedback in the above amplifier positive or negative? Explain.
(b) What is the voltage gain of the above amplifier?
[*Hint*: First calculate B.]
(c) What effect would a capacitance in parallel with R_4 have on the gain?

9.

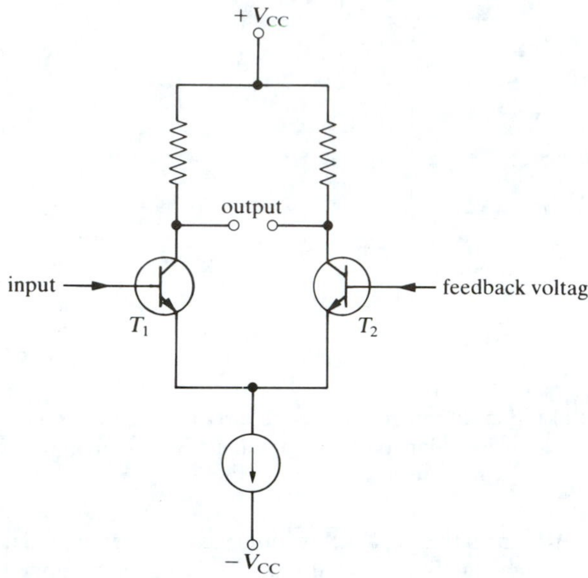

To obtain negative feedback in the above differential amplifier circuit, should the feedback voltage at the base of T_2 be in phase or out of phase with respect to the input?

10. Prove that a voltage gain of at least 29 is necessary for the phase shift oscillator of Fig. 7.15 and that the frequency of oscillation is $f_0 = (1/2\pi\sqrt{6})(1/RC)$.

11. Calculate the feedback for B for the Colpitts oscillator of Fig. 7.18.

12. In the oscillator circuit shown, what determines the phase of the feedback? How could the amount of feedback be changed?

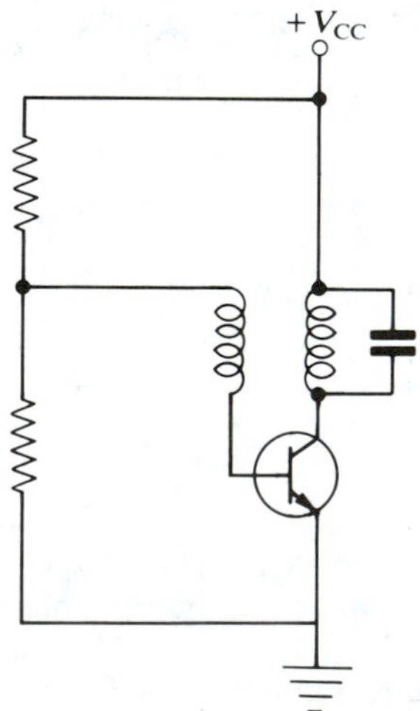

13. A relaxation oscillator can be made with a neon tube as shown. The neon tube is filled with low-pressure neon gas, indicated by the solid dot inside the tube symbol. The tube will not conduct until the voltage across the tube, V_{AB}, exceeds a critical breakdown voltage of approximately 100 V. Then the tube suddenly conducts heavily, and the voltage drop V_{AB} across the tube during conduction V_C is much less than the breakdown voltage. Typically $V_C \cong 10$ V. The effective tube resistance during conduction is very low, approximately 100 Ω. (a) Sketch the output waveform versus time if $R = 500$ kΩ and $C = 1\ \mu$F. (b) How could you adjust the frequency of the waveform? (c) How could you adjust the amplitude of the output? (d) If the current flowing through the neon tube during the discharge of C must be limited, how would you change the circuit to accomplish this?

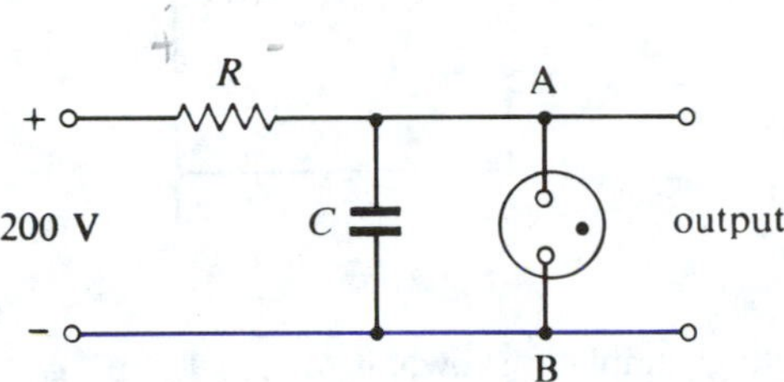

14. A relaxation oscillator can be made with a silicon-controlled rectifier (SCR) as shown. (a) Sketch the output waveform versus time if the SCR breaks down when its anode–cathode voltage reaches 20 V. (b) How could you adjust the frequency of the output? (c) If R_1 increased so as to increase the gate current, how will the output be changed? Sketch. (d) Explain how r affects the peak discharge current of the capacitor.

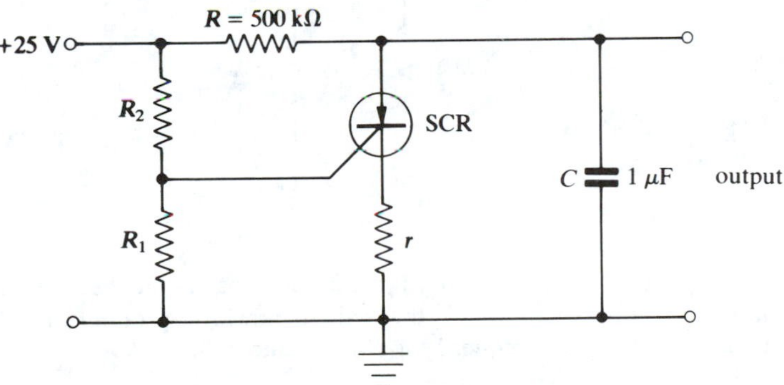

15. Briefly discuss the appropriate remedies for an oscillator that won't oscillate. You may assume that the components are all good and that the circuit has been properly wired.

16. What effect would the capacitance C have on the gain of the amplifier? Would the effect depend on the voltage gain of the amplifier?

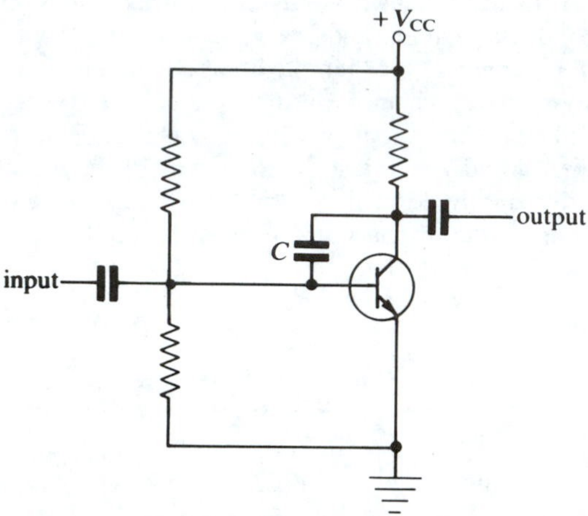

17. In the JFET voltage amplifier shown, if $C_{GD} = 1.5$ pF, $g_m = y_{fs} = 4000$ μmhos. (a) What is the approximate voltage gain? (b) What will be the input impedance at 1 kHz? At 1 MHz?

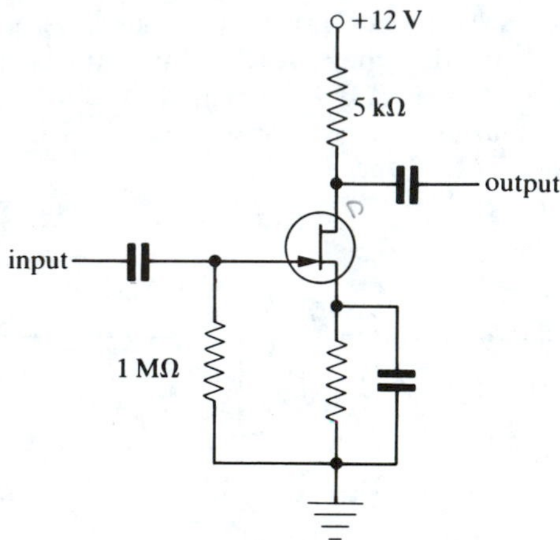

18. In the amplifier of Problem 17, if the total capacitance between drain and ground is 5.0 pF, what is (a) the gain–bandwidth product, (b) the 3-dB frequency, (c) the 3-dB frequency if R_D is reduced to 2.2 kΩ?

CHAPTER 8

Noise

8.1 INTRODUCTION

Noise, to the layman, is an unwanted or unpleasant acoustical disturbance such as a cough during the pianissimo symphony passage or the garbage man's clanging the cans outside the bedroom at 4 A.M. In this chapter *noise* shall mean any unwanted electrical disturbance that tends to obscure or otherwise hinder the observation of the electrical signal of interest.

The first point that must be made is that the magnitude of the noise alone is not important. The noise magnitude is important only *compared* to the signal amplitude. In other words, the signal-to-noise *ratio* is important, not the size of the signal alone or the noise alone. The *signal-to-noise ratio* is usually defined as the ratio in dB of the signal power to the noise power; it is occasionally defined as the ratio of the rms signal voltage to the rms noise voltage.

$$\left(\frac{\mathrm{S}}{\mathrm{N}}\right)_{\text{in dB}} = 10\log_{10}\left(\frac{P_{\mathrm{S}}}{P_{\mathrm{N}}}\right) = 10\log_{10}\left(\frac{V_{\mathrm{S}}}{V_{\mathrm{N}}}\right)^2 = 20\log_{10}\left(\frac{V_{\mathrm{S}}}{V_{\mathrm{N}}}\right)$$

Thus a 20-dB signal-to-noise ratio (20 dB S/N) means the signal is 20 dB "above" the noise or the signal amplitude is ten times the noise amplitude, for example, a 1.0-V signal and 0.1-V noise or a 10-mV signal and 1-mV noise.

With a high signal-to-noise ratio and a small signal, a usable, large signal can be obtained by amplification with a decent amplifier that itself does not introduce too much noise. However, with a low signal-to-noise ratio, amplification is useless. The amplifier cannot tell the difference between a signal voltage and a noise voltage, and so it amplifies each impartially. Thus, even if the amplifier introduces a negligible amount of noise, the signal-to-noise ratio at the output will be the *same* as at the input. And, of course, all amplifiers introduce some noise of their own in addition to the noise present in the input signal. Hence, we come to the initially discouraging conclusion that the signal-to-noise ratio at the output of *any* amplifier is *always* less than the signal-to-noise ratio at the input.

Noise, as we have defined it, is any *unwanted* electrical disturbance; so noise can be regular or random in nature. Noise in a circuit can be classified roughly into four categories: (1) *interference*, either regular or random, from sources outside the circuit; (2) inherent *thermal noise* generated within resistances in the circuit (Johnson noise); (3) inherent *shot noise*, which is the statistical fluctuation of current due to the discrete nature of the charge on the electron; and (4) transistor and tube *flicker noise*, which is a random, low-frequency noise increasing with decreasing frequency.

8.2 INTERFERENCE

Many electrical sources outside a circuit can introduce noise voltages into the circuit. Near and distant lightning flashes are a source of broadband electromagnetic radiation ranging from a few kHz to hundreds of MHz in frequency. Many man-made devices are also sources of electromagnetic radiation: automobile ignition systems, medical diathermy machines, and any machine that contains a spark gap, such as electric motors with sparking between the commutator and the brushes. Noisy electric power transformers and high-voltage power lines are sources of 60-Hz and multiples of 60-Hz electromagnetic fields. The cure is to shield the circuit, particularly the input stage or the "front end," by metal plates or screens. If bolts or screws are used to fasten the metal shield to the chassis, the spacing of the bolts should be small compared to the wavelength of the interfering radiation. As a rule of thumb, one should be extremely careful in shielding against interference above several MHz; the shield should preferably be soldered rather than bolted in place. Double-shielded cable might also be used at the input of high-gain amplifiers to reduce the interference. If the source of the interference can be eliminated, fine, but if shielding is the only possibility the shielding should be at the source as well as at the circuit.

Interference at 60 Hz, the power line frequency, often is caused by *ground loops*, which are closed loops of wire all points of which are supposed to be at ground potential. Because of the very low resistance of these ground loops, appreciable 60-Hz currents may be induced in them by the 60-Hz magnetic fields present around the power lines. These induced 60-Hz currents in the ground loops may then induce small 60-Hz voltages in the input leads of an amplifier or receiver. The cure is to avoid creating ground loops by bringing all the ground connections together at one point in the circuit. A certain amount of experimentation is usually necessary.

Mechanical vibration can often be converted into electrical noise. This noise is often called *microphonic* noise. Vibration may cover a broad range of frequencies in aircraft and rockets, or it may be relatively narrowband. For example, the natural vibration frequency of most floors lies between 10 Hz and 20 Hz—the larger the object, of course, the lower the vibration frequency. Transistors are, of course, essentially immune to this type of noise because they are literally solid devices. However, in any circuit containing cables, vibration of the cable produces noise voltages in the braided outer shield due to piezoelectric effects, flexing of the dielectric, and the flexing and intermittent contacts between the wires in the braid. This may be a serious problem at the input of a high-gain amplifier where the signal itself may only be millivolts in amplitude. The cure for this cable noise is to eliminate the vibration by shock-mounting the circuit or, to stop the source of the vibration, to use a special low-noise cable. This cable has a cylindrical layer of *conducting* plastic extruded over the dielectric surrounding the central wire. A second concentric metal shield surrounds this conducting plastic, then a thin layer of polyethylene, and finally the standard outer braided metal shield. The conducting plastic and the inner shield serve to attenuate the noise generated by flexing of the braid.

If the interference frequencies differ substantially from the signal frequencies, then frequency selective filters can be used to eliminate the interference. These filters should themselves introduce little noise in the circuit. They are often put in the circuit after an initial stage of amplification—after the "preamp"—but not at the final output. If the interference consists of occasional spikes that are much larger than the signal, these spikes can be attenuated and the signal retained by using two diodes as shown in Fig. 8.1. Any voltage less than the turn-on voltage of the diodes (0.2 V for Ge, 0.5 V for Si) will pass through unaffected, but a noise spike of either polarity larger than the turn-on voltage will cause one of the diodes to conduct, thus limiting the output voltage to the diode turn-on voltage. This circuit is particularly useful when a large noise spike will disable the following circuit for an appreciable time (e.g., by charging a capacitor that may cut off the circuit). The capacitor then takes a finite time to discharge, and during this time the circuit may be inoperative.

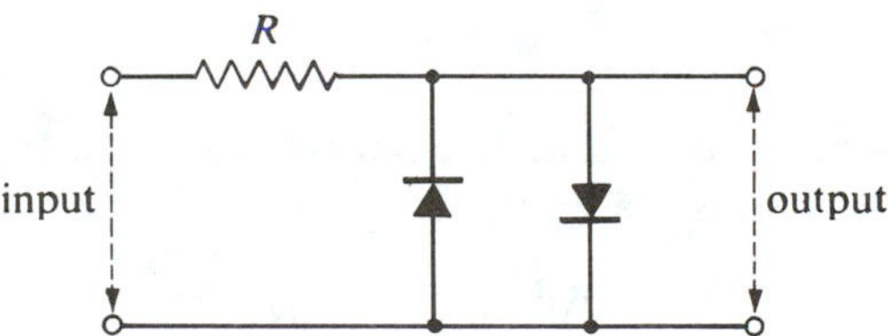

FIGURE 8.1 Diode limiter.

8.3 THERMAL NOISE OR JOHNSON NOISE

The thermal energy of matter is basically the random vibrational energy of the atoms: the higher the temperature, the more violent the motion and the larger the thermal energy per atom. The famous equipartition theorem of statistical mechanics says that for every mathematical degree of freedom of a physical system in equilibrium at T K or absolute (°C + 273° = K) there is associated an average energy of $kT/2$, where k = Boltzmann's constant = 1.38×10^{-23} J/K, and T is the absolute temperature in K. Thus a point mass with three degrees of freedom for motion in the x, y, and z directions will on the average have $3kT/2$ energy. Examples are free atoms and free electrons. A free electron at room temperature, $T = 20°\text{C} = 293$ K, will have

$$E = \frac{3}{2}kT = \frac{3}{2}(1.38 \times 10^{-23}\ \text{J/K})293\ \text{K} = 6.1 \times 10^{-21}\ \text{J}$$

However, this is the *average* energy; some electrons will have more, some less. According to classical theory, this random thermal energy will completely die out as the temperature approaches absolute zero. Quantum theory, however, predicts a small residual or *zero-point* vibrational energy even at absolute zero.

Because many of the electrons in a resistance are essentially free and in constant, random, vibrational motion, the voltage difference between the two ends of any resistance will fluctuate randomly. J. B. Johnson, in 1928, showed that the power associated with such fluctuations varied linearly as the bandwidth B_W of the measuring instrument. For example, a sensitive rms power meter with a response from dc to 1000 Hz placed across a resistance might measure 0.01 μW noise power, but a power meter with a response from dc to 2000 Hz would measure 0.02 μW noise power. Johnson also found that the square of the noise voltage varied linearly with the resistance R. Noise power is proportional to the mean-square noise voltage, and the expression for the mean-square noise voltage generated by a resistance R at T degrees absolute in a bandwidth B_W is

$$\overline{e_n^2} = 4kTRB_W \quad \textbf{(8.1)}$$

where k is Boltzmann's constant. The bar over e_n^2 indicates an average. The root-mean-square or rms noise voltage is

$$(\overline{e_n^2})^{1/2} = (4kTRB_W)^{1/2} \quad \textbf{(8.2)}$$

This noise is commonly referred to as *Johnson noise*, *thermal noise*, or *resistor noise*.

An approximate derivation follows from the equipartition theorem. The random statistical fluctuations of the electrons in the resistor produce a fluctuating difference in the electron density at the two ends of the resistor. This difference in electron density produces a voltage difference between the two ends of the resistor, which, like all circuit elements, has an effective shunt capacitance C, as shown in Fig. 8.2(a). The energy associated with the voltage fluctuations is stored in the electric field of the shunt capaci-

(a) *equivalent circuit for resistance and its inherent shunt capacitance*

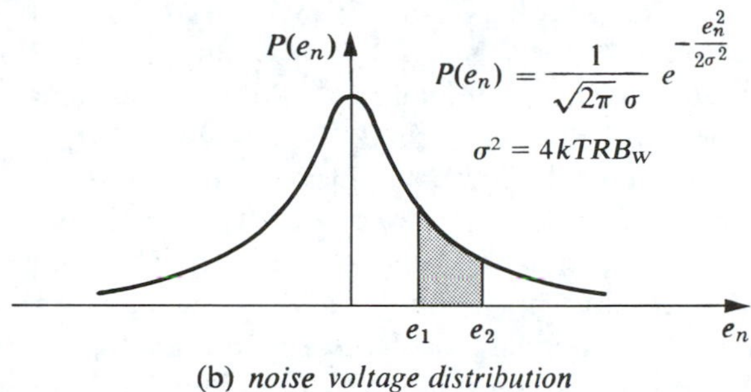

(b) *noise voltage distribution*

FIGURE 8.2 Johnson noise.

tance C. Thus, the energy of the voltage fluctuations is given by $E = CV^2/2$, where V is the instantaneous voltage difference across the capacitor. By the equipartition theorem, if the resistor and capacitor are in thermal equilibrium at T K, the average energy stored in the capacitor must equal $kT/2$. Thus

$$\frac{kT}{2} = \frac{C\overline{V^2}}{2} \quad \text{or} \quad \overline{V^2} = \frac{kT}{C} \tag{8.3}$$

where $\overline{V^2}$ is the average-squared noise voltage. The bandwidth B_W of the parallel RC circuit is from dc out to the frequency where the total parallel impedance is 3 dB down from R.

$$B_W = \frac{1}{2\pi RC} \tag{8.4}$$

Solving equation (8.4) for C and substituting in (8.3), we obtain an expression for the average-squared noise voltage in terms of R:

$$\overline{V^2} = 2\pi kTRB_W \tag{8.5}$$

which is close to the exact equation, (8.1).

The noise voltage e_n developed across R can be of either polarity, and its magnitude varies statistically according to a Gaussian distribution:

$$P(e_n) = \frac{1}{\sqrt{2\pi}\,\sigma} e^{-e_n^2/2\sigma^2} \tag{8.6}$$

where
$$\sigma^2 = 4kTRB_W$$

The expression $P(e_n)de_n$ is the probability that the noise voltage is between e_n and $e_n + de_n$. The probability that the noise voltage is between e_{n1} and e_{n2} is $\int_{e_{1n}}^{e_{2n}} P(e_n)\,de_n$, shown as the shaded area in Fig. 8.2(b). Notice that small noise voltages are more probable than large ones. In statistical language, $\sigma = \sqrt{4kTRB_W}$ is the *standard deviation* of the random noise. The probability that the noise voltage is between $-\sqrt{4kTRB_W}$ and $+\sqrt{4kTRB_W}$ is 0.68 or 68%, and thus there is a 32% probability that the noise is greater in magnitude than $\sqrt{4kTRB_W}$. In noise calculations it is common practice to assume that the noise voltage e_n exactly equals $\sqrt{4kTRB_W}$, but we should always keep in mind that the noise amplitude varies randomly with time.

An important mathematical consequence due to the Gaussian or random character of Johnson noise is that the total noise voltage from two or more independent random noise sources adds as the square root of the sum of the squares of the individual noise voltages. For example, two resistances R_1 and R_2 in series generate independent Johnson voltages of $e_{n1} = \sqrt{4kTR_1B_W}$ and $e_{n2} = \sqrt{4kTR_2B_W}$, respectively. Thus the total noise voltage is given by

$$e_{n\,\text{total}} = \sqrt{e_{n1}^2 + e_{n2}^2} = \sqrt{4kTR_1B_W + 4kTR_2B_W} = \sqrt{4kT(R_1 + R_2)B_W}$$

which is simply the noise voltage that would be generated by a single resistance $R_1 + R_2$. In other words, noise *powers* simply add, since power is proportional to the square of the voltages. This method of addition of independent noise sources is used repeatedly in analyzing the noise generated by different components in a circuit.

It also should be emphasized that it is only a *resistance* that generates noise; a pure reactance, either capacitive or inductive, generates no Johnson noise. For example, a parallel LC resonant circuit as shown in Fig. 8.3

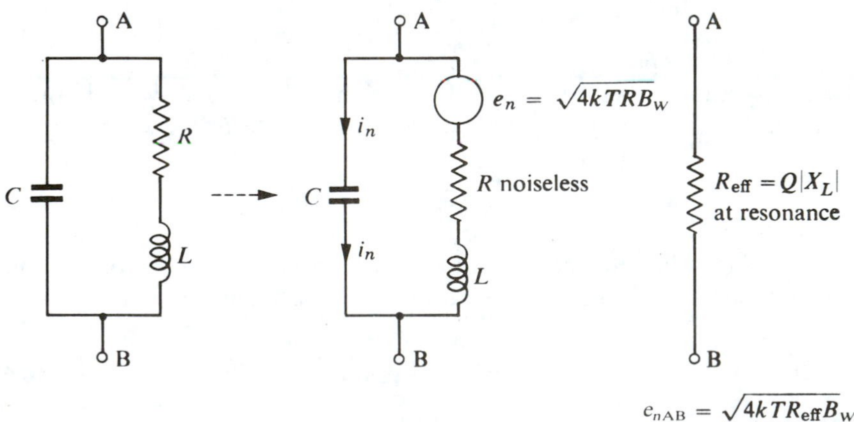

FIGURE 8.3 Noise generated in a parallel *LC* circuit.

will generate Johnson noise only due to the resistance R of the inductance. The noise voltage $e_n = \sqrt{4kTRB_W}$ appears across the resistance R, *not* across the tank terminals A and B. The noise voltage appearing between A and B, e_{nAB}, is considerably larger: e_{nAB} equals the resulting noise current i_n times the capacitive reactance:

$$e_{nAB} = i_n \times \frac{1}{j\omega C} = \frac{e_n}{j\omega L + R + \dfrac{1}{j\omega C}} \times \frac{1}{j\omega C}$$

At resonance, $\omega_0^2 = 1/LC$, so $e_{nAB} = e_n/j\omega_0 RC$. Eliminating ω_0 by using $\omega_0^2 = 1/LC$, we have

$$|e_{nAB}| = \frac{e_n}{R\sqrt{\dfrac{C}{L}}} = \frac{\sqrt{4kTRB_W}}{R\sqrt{\dfrac{C}{L}}}$$

But the Q of the circuit at resonance is given by $Q = \omega_0 L/R = (1/R)\sqrt{L/C}$, so $e_{nAB} = Q\sqrt{4kTRB_W}$. Thus, the noise voltage generated by the resistance R appears across the tank terminals Q times larger. The same result can be obtained by realizing that the tank has a purely resistive impedance $R_{\text{eff}} = Q|X_L| = \omega^2 L^2/R = Q^2 R$ at resonance. Thus, at resonance

$$|e_{nAB}| = \sqrt{4kTR_{\text{eff}}B_W} = Q\sqrt{4kTRB_W} \qquad \textbf{(8.7)}$$

Johnson noise is often called *white noise* because its power density depends only on the bandwidth B_W, not on the frequency f. That is, *any* 1-kHz bandwidth, whether from 1 to 2 kHz or from 1 to 1.001 MHz, of Johnson noise will contain the same average noise power. The word "white" is used because white noise contains "all frequencies," just as white visible light contains a mixture of all of the colors of the spectrum.

A little thought will show that the formula $e_n^2 = 4kTRB_W$ cannot hold indefinitely as the frequency increases; for if it did, one resistance would develop an *infinite* noise power because $\int_0^\infty e_n^2\, df \to \infty$. It is clear that the noise power per unit bandwidth must decrease as the frequency increases. The Johnson noise formula holds up to a frequency of about $f = kT/h$, where h (Planck's constant) $= 6.67 \times 10^{-34}$ Js, at which point the equipartition theorem of classical statistical mechanics breaks down and quantum statistical mechanics takes over. In classical theory the average energy per degree of freedom is kT ($kT/2$ kinetic and $kT/2$ potential); and as the frequency increases, the number of possible modes of oscillation increases, thus yielding an infinite power for high frequencies. In quantum theory the average energy per degree of freedom is

$$\overline{E} = \frac{hf}{e^{hf/kT} - 1} \tag{8.8}$$

which approaches kT for low frequencies $f \ll kT/h$, and which decreases monotonically as the frequency increases above $f = kT/h$. For room temperature, $T \cong 300$ K, $f = kT/h \cong 10^{13}$ Hz, which is in the infrared region of the electromagnetic spectrum; so we may safely use the Johnson noise formula up into the microwave region of 10 GHz $= 10^{10}$ Hz. For a different perspective on the infinite power implied by the classical equipartition theorem, consult any modern physics book on "ultraviolet catastrophe" and the Rayleigh–Jeans law of blackbody radiation.

Most practical resistors used in circuits exhibit *more* noise than the Johnson noise formula, $e_n^2 = 4kTRB_W$, so the Johnson noise expressions should be regarded as a *lower limit* on the noise generated with a resistor. Variations in noise power of up to a factor of 100 have been observed in resistors of the same resistance value made by the same company. Also, different type resistors exhibit different noise characteristics. In general, wirewound resistors exhibit the least noise, and standard composition carbon resistors the most, with metal-film resistors occupying an intermediate position. The least noisy resistors are also the most expensive, and wirewound resistors have considerably more inductance than other types, so, in general, they cannot be used in high-frequency circuits. Also, the higher the voltage across and the current through a resistor, the more excess noise is generated.

8.4 SHOT NOISE

In most electronic circuit problems the basic quantization of the electronic charge can be neglected (just as we neglect the existence of water molecules in a water pipe), and charge and current can be treated as continuous variables. However, all currents consist of the flow of electrons or ions (or holes) that are discrete charges, and random fluctuations in the number of charge carriers flowing past a given point in the circuit produce random fluctuations or noise in the current. This occurs in transistors and in all photoemission devices, such as photomultiplier tubes and phototubes. Random fluctuations will occur in the collector or drain current simply because the number of electrons arriving per second fluctuates. The resultant noise is termed *shot* noise because the fluctuations are similar to those occurring when a hail of shot strikes a target. For a vacuum diode where the cathode emission current is limited only by the cathode temperature. that is, where each electron emitted from the cathode reaches the plate, the rms current noise is given by

$$i_{\hat{n}} = (\overline{i_n^2})^{1/2} = \sqrt{2eI_{dc}B_W} \tag{8.9}$$

where e is the electronic charge, 1.6×10^{-19} C, B_W is the bandwidth, and I_{dc} is the average or dc emission current. Another way of looking at this shot noise current is that $i_n = (\overline{i_n^2})^{1/2}$ is the random fluctuating ac current superimposed on the dc current. The exact expression for the shot noise for other devices will vary, but for most devices the general feature remains that the noise current is proportional to the *square root* of the dc current. Semiempirical techniques are still used for determining the noise characteristics of actual circuits. Also, like Johnson noise, shot noise is "white"; that is, there is a constant power density per unit bandwidth, independent of frequency.

The shot noise expression can be derived from a basic statistical argument as follows. Consider a device emitting discrete charged particles from terminal A to terminal B that flow through a resistance R, as shown in Fig. 8.4. This device may be a temperature-limited diode or a transistor or a

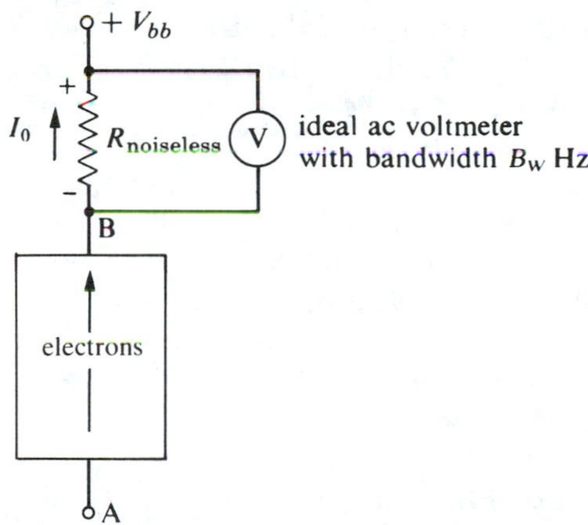

FIGURE 8.4 Shot noise.

vacuum tube. If the average dc current flowing is I_0, then the average number $\bar{r}$ of charge carriers flowing per second is given by $\bar{r} = I_0/e$, where e is the net charge on each carrier. The current may be different from I_0, however, due to random fluctuations in the number of charge carriers emitted from the device. If we let I be the instantaneous current flowing corresponding to $r = I/e$ charge carriers flowing per second, then the fluctuating noise voltage developed across R will be given by $v_n = (I - I_0)R = (r - \bar{r})eR$. A more meaningful way of looking at the fluctuations in the rate of charge carrier flow is to look at the absolute number n of charge carriers flowing in a time interval Δt: $n = r\,\Delta t$. In terms of n the noise voltage is

$$v_n = \frac{(n - \bar{n})eR}{\Delta t} \tag{8.10}$$

where $\bar{n} = \bar{r}\,\Delta t$ is the number of charge carriers flowing in Δt corresponding to the average dc current I_0. The mathematical theory of random processes can now be invoked to tell us a typical value for the fluctuation $(n - \bar{n})$ in the number of charge carriers flowing in a time Δt. If the average number of charge carriers flowing in Δt is $\bar{n}$, then the probability of n, the actual number flowing in Δt, is given by the standard Gaussian distribution:

$$P(n) = \frac{1}{\sqrt{2\pi}\,\sigma}\, e^{-(n-\bar{n})^2/2\sigma^2} \tag{8.11}$$

Strictly speaking, the probability of having a number between n_1 and n_2 flow in Δt is given by $\int_{n_1}^{n_2} P(n)\,dn$. $P(n)$ is normalized to unity: $\int_0^{\infty} P(n)\,dn = 1$; that is, it is certain that some value of n occurs between 0 and ∞. Notice that it is most likely that $n = \bar{n}$—that the average number of charge carriers flows; that is, $P(n)$ is maximum when $n = \bar{n}$. Notice also that if n is substantially different from $\bar{n}$, the probability of getting that value of n is extremely small.

The difference $n - \bar{n}$ between the actual number n of charge carriers and the *average* number $\bar{n}$ represents the fluctuation or the noise. To calculate the magnitude of the fluctuation or noise, we must obtain a representative value for $n - \bar{n}$. From the statistical nature of $P(n)$ we see that the larger $(n - \bar{n})^2/2\sigma^2$, the less probable that value of n. It is conventional to say that a representative or typical value of the variable in a Gaussian distribution is that value that makes the exponent equal to *minus one*. That is, $(n - \bar{n})^2/2\sigma^2 = 1$ or $(n - \bar{n}) = \pm\sqrt{2}\sigma$. This value of the fluctuation means that $P(n) = P_{max}e^{-1} = 0.37P_{max}$ and that there is a probability of $\int_{n'}^{\infty} P(n)\,dn = 0.08$ that the fluctuation is larger in value than $n' = \bar{n} + \sqrt{2}\sigma$, and a probability $\int_0^{n'} P(n)\,dn = 0.08$ that the fluctuation is less than $n' = \bar{n} - \sqrt{2}\sigma$. That is, there is a 16% chance that $|n - \bar{n}|$ exceeds $\sqrt{2}\sigma$.

It can be shown for most physical processes that $\sigma = \sqrt{\bar{n}}$; the standard deviation is approximately equal to the square root of the average number. Substituting $n - \bar{n} = \sqrt{2}\sigma = \sqrt{2\bar{n}}$ into equation (8.10) gives

$$v_n = \frac{(n - \bar{n})eR}{\Delta t} = \frac{\sqrt{2\bar{n}}eR}{\Delta t}$$

But $\bar{n}$ can be expressed in terms of the average dc current flowing,

$\bar{n} = I_0\Delta t/e$. Thus the noise voltage is given by

$$v_n = \sqrt{\frac{2I_0\Delta t}{e}\frac{eR}{\Delta t}} = \sqrt{\frac{2I_0e}{\Delta t}}R \tag{8.12}$$

Notice that the noise voltage increases as the time interval Δt, during which we count the number of charge carriers, *decreases*. In other words, if we average the noise voltage over a long time interval, Δt, the noise voltage will go to zero. If we identify this time interval with the inverse of the bandwidth of the voltmeter we are using to measure the noise voltage across R, $\Delta t = 1/B_W$; then $v_n = \sqrt{2I_0eB_W}R$. This is not unreasonable, because $1/B_W$ is the fastest response time of the voltmeter; any fluctuations occurring in times less than $1/B_W$ simply do not affect the voltmeter. Thus, the square of the noise voltage developed across R is

$$v_n^2 = 2I_0eB_WR^2 \tag{8.13}$$

The noise power developed in R is

$$P_n = \frac{v_n^2}{R} = 2I_0eB_WR \tag{8.14}$$

and the noise current squared is

$$i_n^2 = \frac{v_n^2}{R^2} = 2I_0eB_W \tag{8.15}$$

which is the shot noise expression. The two important factors are that i_n varies as the square root of I_0 and as the square root of B_W.

Some numerical values for the shot noise voltage developed across R are instructive. If $R = 1\ \text{k}\Omega$ and $I_0 = 1$ mA, then for various bandwidths B_W we have $v_n = 1.3 \times 10^{-7}\sqrt{B_W}$; see Table 8.1. Thus for an audio amp-

TABLE 8.1

B_W	v_n (*shot noise*)	e_n (*Johnson noise*)
1 Hz	0.013 μV	0.0041 μV
10 kHz	1.3 μV	0.41 μV
1 MHz	13 μV	4.1 μV
100 MHz	130 μV	41 μV

lifier with a bandwidth of 10 Hz–20 kHz, the noise voltage would be several microvolts. The noise voltage becomes increasingly bothersome as the bandwidth increases. Again it should be pointed out that this noise source (as with any noise source) is most serious when it occurs at the front end of a high-gain amplifier, where the noise will be amplified by all the following amplifier stages. Another way of looking at the seriousness of the problem is that at the front end of an amplifier the signal voltage is usually very small; thus the signal-to-noise ratio will be lowered appreciably by the presence of even a small amount of noise.

8.5 CALCULATING AMPLIFIER NOISE

An exact calculation of all the Johnson noise and shot noise generated within a practical amplifier would be impossibly difficult, but a few comments can be made. First, the *noise figure* (N.F.) is used to describe how much the amplifier degrades or decreases the signal-to-noise ratio. The noise figure is defined in dB as ten times the logarithm of the signal-to-noise power ratio at the input divided by the signal-to-noise power ratio at the output:

$$\text{Noise Figure} = \text{N.F.} = 10\log_{10}\left[\frac{(\text{S/N})_{\text{in}}}{(\text{S/N})_{\text{out}}}\right]_{\text{power}} = 20\log_{10}\left[\frac{(\text{S/N})_{\text{in}}}{(\text{S/N})_{\text{out}}}\right]_{\text{voltage}} \tag{8.16}$$

The noise figure will always be greater than 0 dB because the output S/N ratio is always less than the input S/N ratio, due to the additional noise contributed by the amplifier. A perfect amplifier that adds no noise whatsoever thus has a N.F. of 0 dB. An amplifier that decreases the S/N power ratio by a factor of 2 would have a 3-dB N.F. In general, most rf amplifiers in the 10–500-MHz frequency region tend to have noise figures in the vicinity of 5 to 10 dB. High-quality audio-frequency low-noise amplifiers can have a noise figure of less than 1 dB.

For a transistor amplifier, two noise parameters are necessary to represent the transistor noise. All the noise in the entire transistor amplifier is represented by one equivalent noise voltage generator, e_n, and one equivalent current generator, i_n, at the input as shown in Fig. 8.5. $e_n^2 B$ is the mean-square voltage generated in a bandwidth B_W, so e_n has units of V/$\sqrt{\text{Hz}}$. Similarly, i_n has units of A/$\sqrt{\text{Hz}}$. The input signal source is represented as a voltage source e_s in series with a source resistance R_s. The Johnson noise generated by R_s is represented by a voltage generator $e_{sn} = \sqrt{4kTB_W R_s}$. B_W refers to the bandwidth of the amplifier, since we will be referring all noise calculations to the amplifier output. The amplifier voltage and current noise generators e_n and i_n are assumed to be

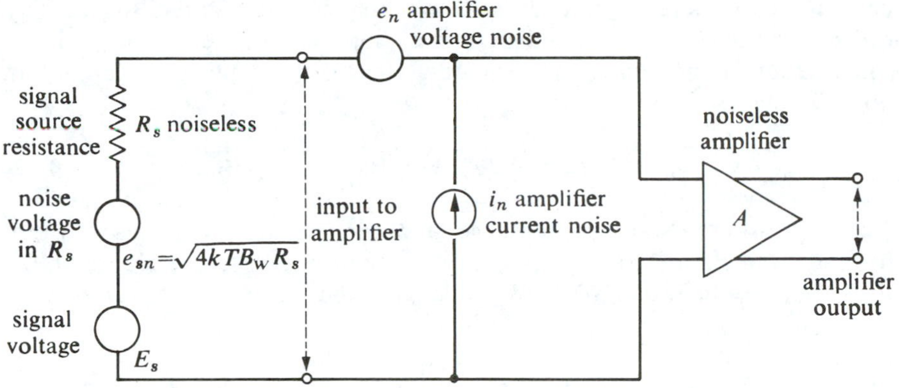

FIGURE 8.5 Transistor amplifier and source noise equivalent circuit.

completely independent of each other and to generate perfectly white noise (i.e., constant noise power per unit bandwidth).

The current noise i_n is mainly due to the shot noise in the base current, and the voltage noise e_n is due to the Johnson noise of the base resistance and the collector shot noise current flowing through the emitter resistance.

The amplifier current noise i_n flows through the source resistance R_s and produces a noise voltage of $i_n R_s$. (We assume an infinite amplifier input impedance.) Since e_n and $i_n R_s$ are in series, their noise powers add quadratically, assuming e_n and i_n are not correlated. The total amplifier input noise is then

$$e_{\text{in}} = \sqrt{e_n^2 B_W + i_n^2 R_s^2 B_W + e_{sn}^2}$$

Thus we immediately see that the *source* resistance R_s partially determines the input noise to the transistor amplifier. If R_s is very small, the main noise input is due to the transistor voltage noise e_n; if R_s is very large, the main noise input is due to the transistor current noise i_n and to the Johnson noise of R_s itself. The voltage noise e_n tends to decrease as the collector current I_C increases, whereas i_n tends to increase with increasing I_C. Thus, in principle, we can choose an intermediate value for I_C to minimize the total amplifier noise.

We can now calculate the amplifier output signal-to-noise ratio from analyzing the equivalent circuit of Fig. 8.5. The signal input voltage to the amplifier is E_s, assuming negligible loading of the signal source (i.e., assuming the amplifier input impedance is large compared to R_s). Thus the signal output voltage from the amplifier is AE_s. The noise voltage in the output of the amplifier is simply the amplifier gain A times the total effective input noise voltage to the amplifier. There are three noise sources at the input: the source noise e_{sn}, the effective amplifier voltage, and current noise sources, e_n and i_n. If we assume these three sources are random in nature and independent of one another, then the total noise is the square root of the sum of the squares. Hence, the total amplifier output noise voltage is given by

$$N_{out} = Ae_{in} = A\sqrt{e_{sn}^2 + e_n^2 B_W + (i_n R_s)^2 B_W} \quad \textbf{(8.17)}$$

The expression $i_n R_s$ is the noise voltage generated by the noise current flowing through the source resistance R_s. Putting the expressions for the three noise voltages in N_{out}, we obtain for the output voltage signal-to-noise ratio:

$$(\text{S/N})_{out} = \frac{AE_s}{A\sqrt{4kTB_W R_s + e_n^2 B_W + i_n^2 R_s^2 B_W}} \quad \textbf{(8.18)}$$

It is immediately obvious that the signal-to-noise ratio has been decreased by the amplifier, for the signal-to-noise ratio of the input signal source is

$$(\text{S/N})_{in} = \frac{E_s}{\sqrt{4kTB_W R_s}} \quad \textbf{(8.19)}$$

The better the amplifier is, the smaller e_n and i_n will be, and the degradation of the signal-to-noise ratio will be less. The noise figure is a measure of how much the signal-to-noise ratio is reduced by the amplifier—the better the amplifier, the smaller the noise figure. The noise figure can be expressed in terms of the noise generators; from (8.16), (8.18), and (8.19) we have

$$\text{N.F.}_{\text{dB}} = 10\log_{10}\left[\frac{4kTR_s + e_n^2 + i_n^2 R_s^2}{4kTR_s}\right] = 10\log_{10}\left[1 + \frac{e_n^2 + i_n^2 R_s^2}{4kTR_s}\right] \quad \textbf{(8.20)}$$

The *noise factor F* is defined as the ratio of the power signal-to-noise ratios:

$$F \equiv \frac{(\text{S/N})\text{ power input}}{(\text{S/N})\text{ power output}} = 1 + \frac{e_n^2 + i_n^2 R_s^2}{4kTR_s} \quad \textbf{(8.21)}$$

A perfect amplifier introduces no noise of its own ($e_n = i_n = 0$) and has a noise figure of 0 dB and a noise factor of 1.0: the lower the noise factor the better.

By the preceding definition the noise figure of an amplifier depends not only on the noise generated by the amplifier (e_n and i_n) but also on the source resistance R_s. One really should speak of the noise figure of the *amplifier–source combination* rather than of the amplifier alone. Two amplifiers cannot be compared on the basis of their noise figures; one must also take into consideration the source resistance. For example, two identical amplifiers with the same equivalent noise generators, e_n and i_n, will have different noise figures if used with different source resistances: the smaller the source resistance, the larger (and worse) the noise figure. This may seem puzzling, but a little thought will show that for a small R_s the source contributes very little noise, and the amplifier contributes most of the noise. Hence, the signal-to-noise ratio is degraded substantially by the amplifier, and the noise figure is high. However, for a *fixed* value of source resistance, the lower the noise figure the better the output signal-to-noise ratio.

What really matters is the signal-to-noise ratio at the amplifier output; to maximize this, we obviously should use an amplifier that contributes as little noise, e_n and i_n, as possible. However, the output signal-to-noise ratio also depends on R_s and T, the source temperature—the cooler the source and the smaller the source resistance, the less the output noise and the greater the output signal-to-noise ratio.

Let's assume that we can vary R_s, and let's see how this freedom affects the noise factor. Inspection of the noise factor expression (8.21) shows us that the noise factor F becomes infinitely large as $R_s \to \infty$ because then $e_n^2 \ll i_n^2 R_s^2$ and

$$\lim_{R_s\to\infty} F = \lim_{R_s\to\infty}\left[1 + \frac{e_n^2 + i_n^2 R_s^2}{4kTR_s}\right] = \lim_{R_s\to\infty}\left[\frac{i_n^2 R_s}{4kT}\right] \qquad \textbf{(8.22)}$$

The noise factor *also* becomes infinitely large as $R_s \to 0$ because then $e_n^2 \gg i_n^2 R_s^2$ and

$$\lim_{R_s\to 0} F = \lim_{R_s\to 0}\left[1 + \frac{e_n^2 + i_n^2 R_s^2}{4kTR_s}\right] = \lim_{R_s\to 0}\left[\frac{e_n^2}{4kTR_s}\right] \qquad \textbf{(8.23)}$$

Thus we immediately see that there must be a relative minimum in the noise factor as R_s varies, as shown in Fig. 8.6.

Let us now minimize the noise factor by varying R_s in (8.21) by setting $\partial F/\partial R_s = 0$.

$$\frac{\partial F}{\partial R_s} = \frac{\partial}{\partial R_s}\left[1 + \frac{e_n^2 + i_n^2 R_s^2}{4kTR_s}\right]$$

$$= \frac{4kTR_s(2i_n^2 R_s) - (e_n^2 + i_n^2 R_s^2)4kT}{(4kTR_s)^2} = 0$$

which implies

$$R_s = \frac{e_n}{i_n} \qquad \textbf{(8.24)}$$

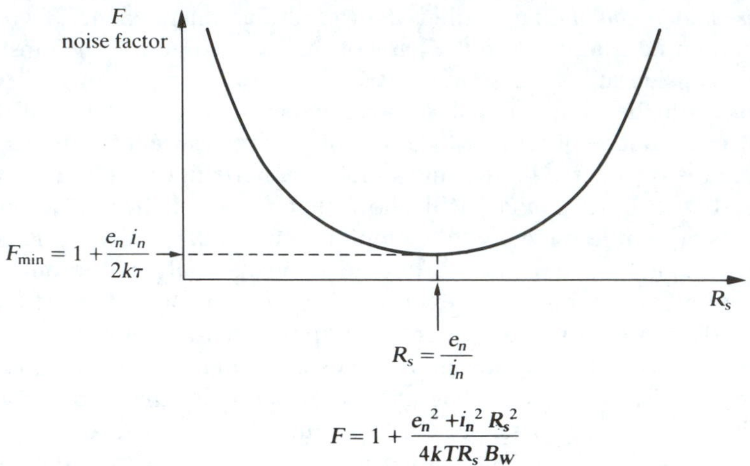

FIGURE 8.6 Noise factor versus source resistance.

In words, for a given amplifier with a fixed e_n and i_n, the noise factor will be minimized when the source resistance $R_s = e_n / i_n$. From (8.21), the minimum value of the noise factor is then

$$F_{\min} = 1 + \frac{e_n i_n}{2kT} \tag{8.25}$$

As a general rule, e_n is approximately the same for bipolar transistors and JFETs, about $10\,\text{nV}/\sqrt{\text{Hz}} = 10^{-8}\,\text{V}/\sqrt{\text{Hz}}$. However, i_n is *much* smaller for JFETs because it is essentially the shot noise in the extremely small gate current; for FETs i_n is about $0.001\,\text{pA}/\sqrt{\text{Hz}} = 10^{-15}\,\text{A}/\sqrt{\text{Hz}}$. Thus, $e_n / i_n \approx 10\,\text{k}\Omega$ for bipolar transistors, and $e_n / i_n \approx 10\,\text{M}\Omega$ for JFETs. Therefore, to achieve the minimum noise factor and the highest output signal-to-noise ratio, we would generally use a bipolar transistor amplifier for low source resistances of several kΩ but a JFET amplifier for a high source resistance of several megohms or more.

The preceding calculation may seem irrelevant, because it doesn't seem easy to change the source resistance. However, for ac signals from tens of hertz to tens of kilohertz, a transformer will change R_s as seen by the amplifier input. The transformer must be extremely well shielded and should have negligible primary winding resistance to avoid increasing the total source resistance. It should also have negligible power loss in the core material, which means a high-quality low-loss laminated iron core for frequencies up to 30 kHz and a powdered-iron, ferrite, or air core for higher frequencies.

We recall that for a transformer with a turns ratio of $N = N_{\text{sec}} / N_{\text{pri}}$, the voltage is changed by a factor of N and the current by a factor of $1/N$, so the impedance is changed by a factor of N^2.

$$V_{sec} = NV_{pri} \qquad I_{sec} = \frac{I_{pri}}{N} \qquad Z_{sec} = \frac{V_{sec}}{I_{sec}} = N^2 Z_{pri} \tag{8.26}$$

For example, a step-up transformer with $N = 10$ will increase the primary impedance by a factor of 100. The transformer output terminals will appear to have a voltage generator of 10 times higher amplitude and an output impedance 100 times larger. This impedance transformation is illustrated in Fig. 8.7.

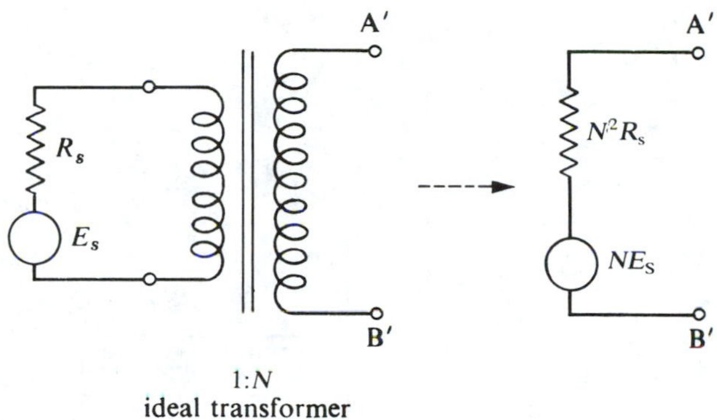

FIGURE 8.7 Transformer impedance transformation.

The circuit containing the "impedance matching" transformer is shown in Fig. 8.8. A calculation of the noise factor for the transformer source with voltage NE_s and source resistance N^2R_s yields

$$F = \left[1 + \frac{e_n^2 + i_n^2 N^4 R_s^2}{4kTN^2R_s}\right] \tag{8.27}$$

which is identical to (8.21) if we replace the transformed source resistance N^2R_s by R_s. The noise factor (8.21) was minimized for $R_s = e_n/i_n$; hence the noise factor (8.27) for the transformer circuit will be a minimum for

$$N^2R_s = \frac{e_n}{i_n} \tag{8.28}$$

Thus the desirable turns ratio for the transformer is

$$N = \sqrt{\frac{e_n}{i_nR_s}} \tag{8.29}$$

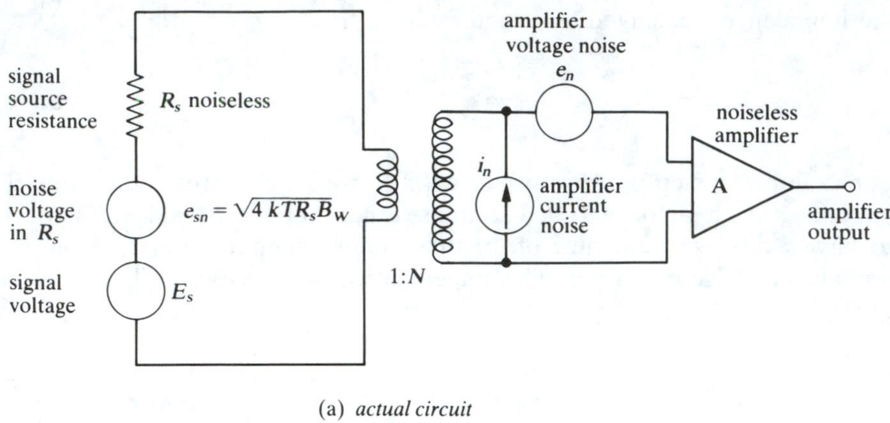

(a) *actual circuit*

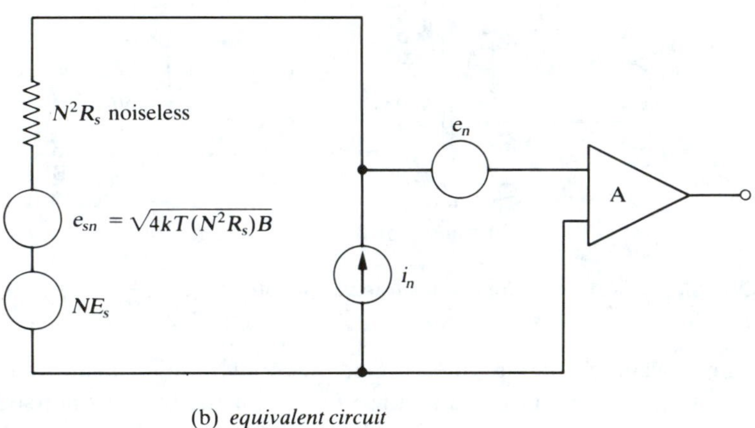

(b) *equivalent circuit*

FIGURE 8.8 Amplifier with impedance matching transformer.

This result can also be obtained by minimizing (8.27) with respect to N or by maximizing the voltage S/N ratio for the output with respect to N.

$$(\text{S/N})_{\text{out}} = \frac{ANE_s}{A\sqrt{4kTN^2R_sB_W + e_n^2B_W + (i_nN^2R_s)^2B_W}} \tag{8.30}$$

A common use of this transformer technique is when the signal source has a low impedance and the desired source impedance e_n/i_n is much higher. A step-up transformer is then used to achieve a better signal-to-noise ratio. This situation occurs, for example, when a crystal diode is used to detect a microwave frequency; the diode produces a small dc output voltage proportional to the incident microwave power. The diode output impedance is typically about 50 Ω, and e_n/i_n for the amplifier is usually about 10^5 to $10^6\ \Omega$. Hence a step-up transformer with turns ratio $N = (e_n/i_n)/R_s = \sqrt{10^6/50} = 141$ should be used. If, on the other hand, the source resistance is larger than the optimum value, e_n/i_n, a step down of the source resistance is called for. One example of such a source is a photomultiplier tube, which acts essentially as a constant current source (i.e., it has a very high output impedance). In this case the source noise is almost entirely shot noise and would be represented by a *current* noise generator i_{sn}. However, a step-down transformer is not a practical solution because its primary resistance is too high, thereby increasing the effective source noise. Thus the only alternative is to choose an amplifier with as large an e_n/i_n ratio as possible, which usually means a high-input-impedance FET amplifier.

Another technique for low-noise counting of photons from weak sources uses modern amplitude discriminators and fast amplifiers to convert each photomultiplier voltage pulse resulting from one photon to a standardized voltage pulse. These pulses are then fed into a digital-to-analog converter, the output of which is proportional to the incident photon intensity. In this way low-amplitude noise pulses and baseline drift are prevented from affecting the output.

Note that a *transformation* of impedance is necessary to improve the signal-to-noise ratio. If R_s is less than the optimum value of e_n/i_n, it is not possible to just add another resistance R_A in series with R_s to make $R_s + R_A = e_n/i_n$. This will *lower* the signal-to-noise ratio, as can be seen by referring to the signal-to-noise expression (8.30). More noise will be generated due to the increased thermal noise voltage generated in R_A and to the amplifier current noise flowing through R_A, which produces an additional noise voltage in R_A.

8.6 FLICKER NOISE

In the preceding calculation of amplifier signal-to-noise ratio we have assumed that the amplifier voltage and current equivalent noise generators e_n and i_n were independent of one another and frequency. In general, however, e_n and i_n do depend on frequency because at *low* frequencies a new type of noise, called *flicker noise*, is larger than either Johnson noise or shot noise. Flicker noise empirically has a power dependence of f^{-1}, although the exponent can vary from about 0.9 to 1.4. That is, the flicker noise power is given by $P_{fn} = kf^{-x}$, where $x \cong 1.0$. The f^{-1} frequency dependence is 3 dB per octave, or a factor of 10 increase in power for each decade of frequency decrease. Flicker noise usually is the major noise source below several hundred hertz in most electronic amplifiers. The frequency at which flicker noise equals the other noise in a device is often called the *corner frequency*. A general graph of noise versus frequency is

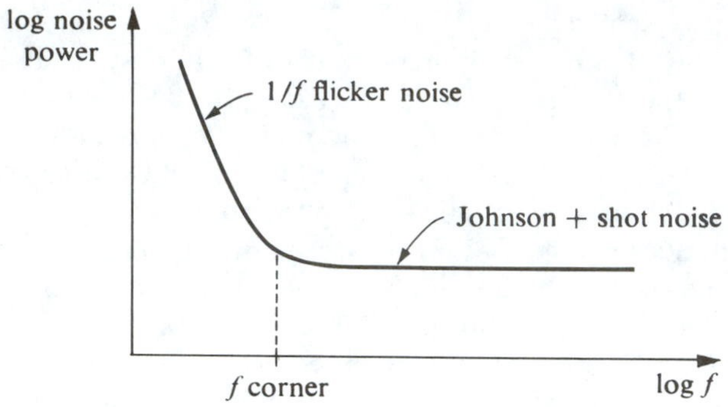

FIGURE 8.9 Noise spectrum.

shown in Fig. 8.9. For semiconductor diodes used to detect microwave frequencies around 10 GHz, the corner frequency is about 10 kHz. Schottky barrier diodes have a corner frequency around 1 kHz.

Flicker noise is also occasionally referred to as *pink noise* to distinguish it from white noise. Because of the $1/f$ dependence, there is the same pink noise power in each octave or frequency decade, that is, from 10 to 20 Hz or from 50 to 100 Hz. The exact cause of flicker noise is not well understood, but it appears to depend on surface characteristics. Measurements that verify the $1/f$ frequency dependence have been made on some devices down to frequencies of approximately 10^{-4} Hz. It is clear that the $1/f$ dependence cannot extend down to dc because the total noise power would then be infinite. All transistors and vacuum tubes exhibit flicker noise, as do some resistors. Carbon composition resistors carrying substantial current exhibit $1/f$ noise, but metal-film or wirewound resistors are much less noisy.

For minimal noise the amplifier bandwidth should also be the minimum necessary to amplify the Fourier frequency components of the signal. Any amplifier bandwidth in excess of this merely amplifies noise and thus degrades the signal-to-noise ratio. The bandwidth is usually limited by putting in appropriate low-pass and high-pass filters.

The optimal value of the source resistance, $R_{\text{opt}} = e_n/i_n$, varies with frequency. Typically, e_n/i_n ranges from a high value of about 10 MΩ at 100 Hz to a low value of about 100 kΩ at 100 kHz. Hence, great care must be taken in choosing the impedance matching ratio depending on the frequency of the signal.

The flicker noise will also depend slightly on the power supply voltage, so some experimentation in this respect is worthwhile. Most amplifier noise generators e_n and i_n are substantially independent of temperature, but in FET amplifiers the shot or current noise varies as the square root of the gate-channel leakage current, which in turn varies with temperature. The net effect is for the FET shot noise to double for approximately every 15° of temperature rise. This effect is only noticeable, of course, above the corner frequency—where the shot noise is not masked by the low-frequency flicker noise.

8.7 NOISE TEMPERATURE

The noise figure of an amplifier is a somewhat ambiguous concept. Intuitively one would like the noise figure to depend only on the amplifier's characteristics, but, as we saw in Section 8.5, it depends on the bandwidth, the source resistance, and the temperature of the source resistance. A different parameter for describing an amplifier's noise is the *noise temperature*. Consider an amplifier driven by a source with resistance equal to the input impedance of the amplifier. If the source resistance were at 0 K, it would generate no Johnson noise and all the noise in the amplifier output would be generated within the amplifier. If we now increase the source resistance temperature until the output noise power exactly doubles, the resistance is contributing an amount of noise power equal to that generated within the amplifier. Thus, the temperature *increase* of the resistance is a measure of the amplifier noise. Noise temperature is independent of the amplifier bandwidth, so it can be used to compare amplifiers of different bandwidths; however, one cannot make such a comparison on the basis of the amplifier noise figures.

The noise temperature is also linearly proportional to the noise power in the output, so the signal-to-noise ratio is inversely proportional to the noise temperature. An amplifier with a noise temperature of 100 K will have twice the output signal-to-noise power of an amplifier with a noise temperature of 200 K.

A noise temperature can be defined for circuits other than amplifiers—for example, an antenna or a transmission line. An antenna driving a transmission line feeds a certain amount of noise into the line. The antenna impedance should be matched to the line impedance for optimal power transfer. Hence, if we replace the antenna by a resistance equal to the antenna impedance and heat the resistance up until the noise at the output of the line is the same as with the antenna connected, then the temperature of this resistance is the noise temperature of the antenna. The noise temperature is a measure of how much noise power the antenna produces; it is not a measure of the actual physical temperature of the antenna material. Some of the noise in the antenna output is blackbody electromagnetic radiation received by the antenna from external sources. Hence, the antenna noise temperature often depends on the direction the antenna points (i.e., on what the antenna "sees").

A noise temperature for a transmission line can be defined similarly. The line is replaced by a resistance equal to the characteristic impedance of the line. The temperature this resistance must have in order to generate a noise power equal to that generated by the line is the noise temperature of the line.

Because the noise temperature is always proportional to the actual noise power, the total noise power of a system is proportional to the sum of the noise temperatures of the various parts of the system. The same addition technique cannot be used to calculate the total system noise from the noise figures of the various parts.

8.8 LOCK-IN DETECTION

Lock-in detection or *phase-sensitive* detection is probably the single most useful experimental technique for increasing the signal-to-noise ratio of a noisy signal. The basis of the technique is to compress all the signal information into a very narrow bandwidth Δf and amplify only frequencies in this bandwidth, thus rejecting all noise outside Δf.

Consider the phase-sensitive detector in Fig. 8.10. If a sinusoidal signal $V_{\text{sin}} = V_0 \sin \omega_0 t$ is "chopped" or sampled by a switch at the same frequency as the signal (ω_0), then the switch output V_s' will be a half-wave rectified wave provided the switch is on from exactly 0 to $T/2$ and off from exactly $T/2$ to T. A Fourier analysis of V_s' yields

$$V_s' = \frac{a_0}{2} + \sum b_n \sin n\omega_0 t \qquad \omega_0 = \frac{2\pi}{T} = 2\pi f_0 \tag{8.31}$$

with

$$\frac{a_0}{2} = \frac{V_0}{\pi} \qquad \text{(average dc value)}$$

$$b_1 = \frac{V_0}{2} \qquad (\text{at } f_0)$$

$$b_2 = \frac{2V_0}{3\pi} \qquad (\text{at } 2f_0)$$

$$b_3 = 0 \qquad (\text{at } 3f_0)$$

$$\vdots$$

(All higher odd coefficients are zero.) Thus a low-pass filter with a unit gain at dc whose break frequency is much less than f_0 (i.e., $1/RC \ll \omega_0$) will pass only the dc component; the filter output will be V_0/π.

If the signal and the switch are still at the same frequency but out of phase by ϕ ($0 < \phi < 2\pi$), then, if V_{sin} leads the switch,

$$V_{\text{sin}} = V_0 \sin(\omega_0 t + \phi)$$

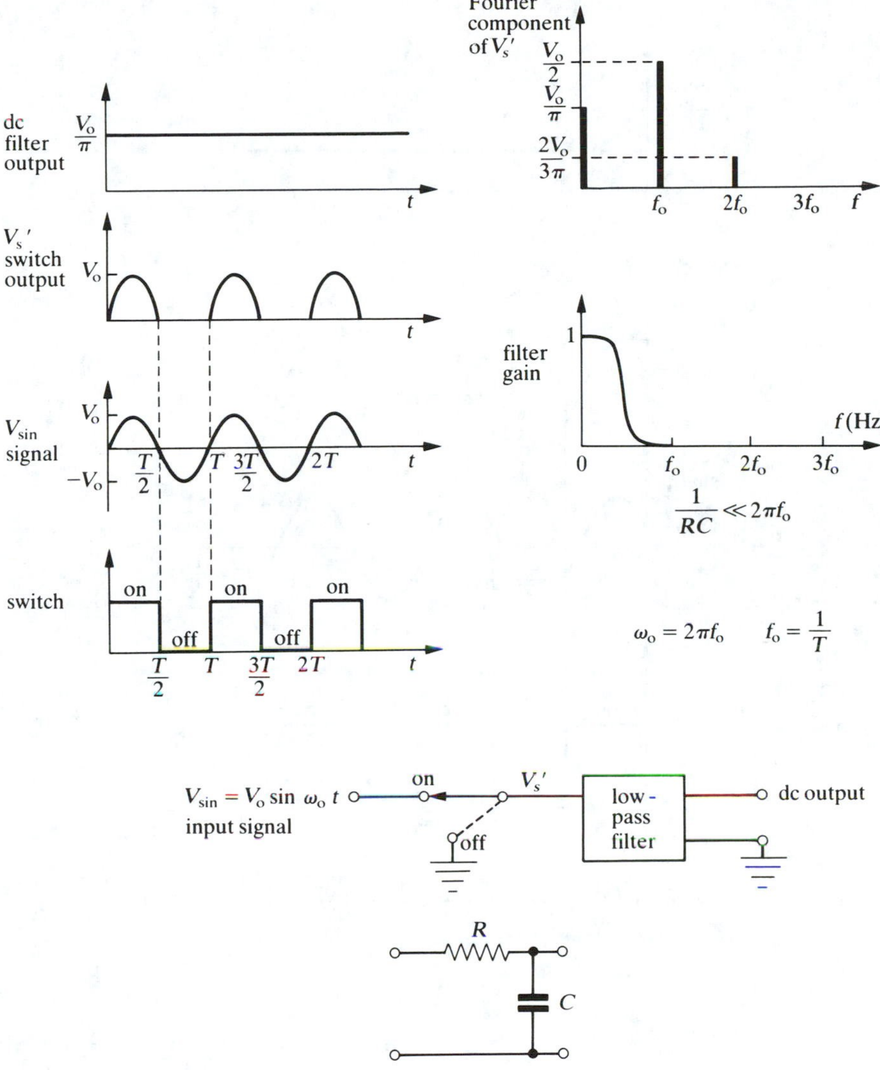

FIGURE 8.10 Phase detector with signal and switch exactly in phase.

The first (dc) term of the Fourier expression of V_s' is now

$$\frac{a_0}{2} = \frac{1}{T}\int_0^{T/2} V_{\sin}(t)\,dt = \frac{V_0}{T}\int_0^{T/2} \sin(\omega_0 t + \phi)\,dt = \frac{V_0}{\pi}\cos\phi \quad \textbf{(8.32)}$$

Again, if $1/RC \ll \omega_0$, the filter output will equal $a_0/2$; the dc level of the filter output will be proportional to the cosine of the *phase difference*

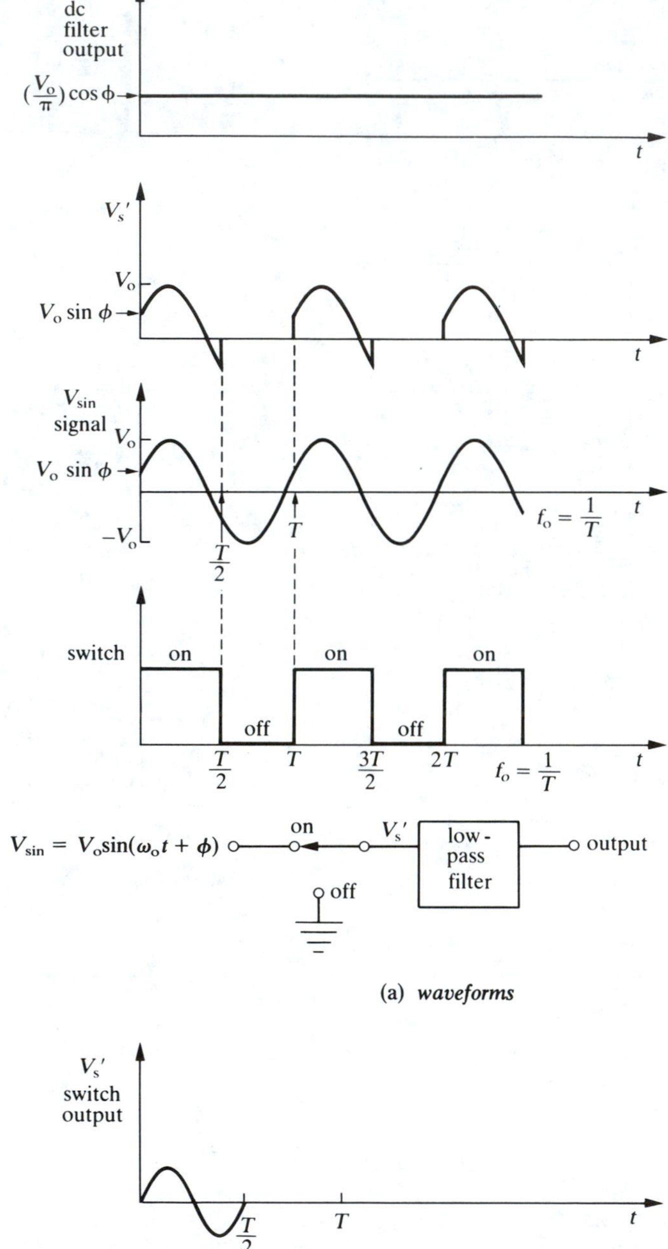

(a) *waveforms*

(b) *signal component at* $2f_o$

FIGURE 8.11 Phase detector with signal and switch out of phase by ϕ.

between the signal and the switch. The waveforms are shown in Fig. 8.11(a). Clearly the dc output is less than in Fig. 8.10, where $\phi = 0$; if $\phi = 90°$, the dc output will be zero.

If the input signal is not purely sinusoidal but still has the same fundamental frequency f_0 as the switch, then we can see that all the even harmonics of f_0 will not contribute to the dc filter output. For example, the second harmonic at $2f_0$ will have one positive peak and one negative peak from 0 to $T/2$ (when the switch is on) that will average to zero exactly as shown in Fig. 8.11(b). Any odd harmonic of f_0 will contribute to the dc output.

A final modification to the circuit is to use a *full*-wave rectification of the signal as shown in Fig. 8.12. This doubles the dc filter output because

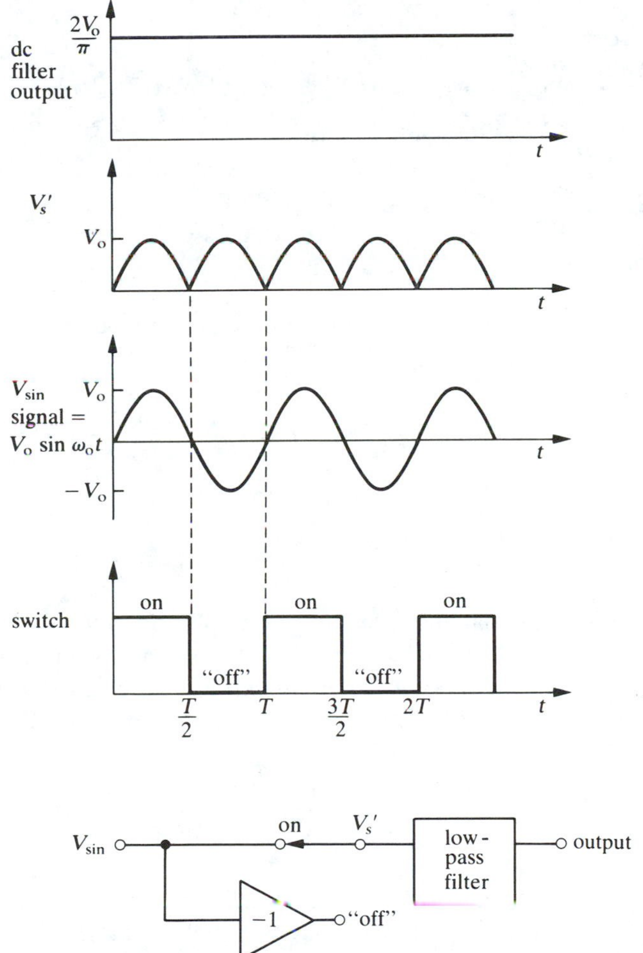

FIGURE 8.12 Full-wave phase detector.

the negative signal peaks are also used. The dc filter output is

$$\frac{a_0}{2} = \frac{2V_0}{\pi}\cos\phi \tag{8.33}$$

All the previous general comments apply: Any *odd* harmonics of the switch frequency in the input will contribute to the dc output, and the even harmonics will all average to zero.

An important point about the phase detector is that only the *odd multiples* of the switch frequency contribute to the dc filter output. Let f_0 = the switch frequency; that is, the switch is in the "on" position from 0 to $T/2$ and in the -1 gain "off" position from $T/2$ to T, where $f_0 = 1/T$. Suppose the input signal frequency differs from the switch frequency by Δf.

$$f_{\text{signal}} = f_0 + \Delta f$$

$$\omega_{\text{signal}} = \omega_0 + \Delta\omega \qquad \omega_0 = 2\pi f_0,\ \Delta\omega = 2\pi\Delta f$$

Then $$V_{\sin} = V_0 \sin(\omega_0 + \Delta\omega)t$$

which can be written as

$$V_{\sin} = V_0 \sin(\omega_0 t + \phi) \tag{8.34}$$

where $\phi = \Delta\omega t$. The dc filter output v_{out} will then be (for the full-wave circuit)

$$v_{\text{out}} = \frac{2}{T}\int_0^{T/2} V_0 \sin(\omega_0 t + \phi)\, dt = \frac{2V_0}{\pi}\cos\phi$$

or $$v_{\text{out}} = \frac{2V_0}{\pi}\cos(\Delta\omega t) \tag{8.35}$$

Thus, (8.35) implies that the filter output amplitude will slowly vary at an angular frequency $\Delta\omega$ or a period of $T' = 2\pi/\Delta\omega$, as shown in Fig. 8.13.

Now comes the main point: If the breakpoint ω_B of the low-pass filter is much less than $\Delta\omega$, the filter output will be essentially zero. That is, if

$$\omega_B = \frac{1}{RC} \ll \Delta\omega \tag{8.36}$$

the output is zero. If $\omega_B = 1/RC \gg \Delta\omega$, the filter output will vary at $\Delta\omega$. In other words, the low-pass filter output is equal to $2V_0/\pi$ when $\omega_{\text{signal}} = \omega_{\text{switch}}$ ($\Delta\omega = 0$) and falls off rapidly to zero if $|\omega_{\text{signal}} - \omega_{\text{switch}}| = \Delta\omega \gg 1/RC$. Thus the low-pass filter output is sharply peaked at $\omega_{\text{signal}} = \omega_{\text{switch}}$

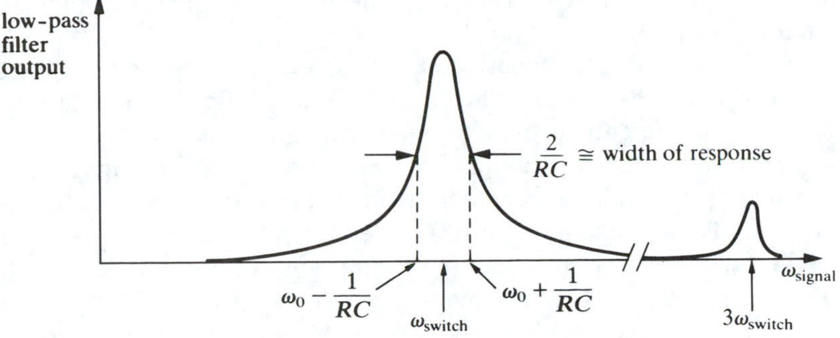

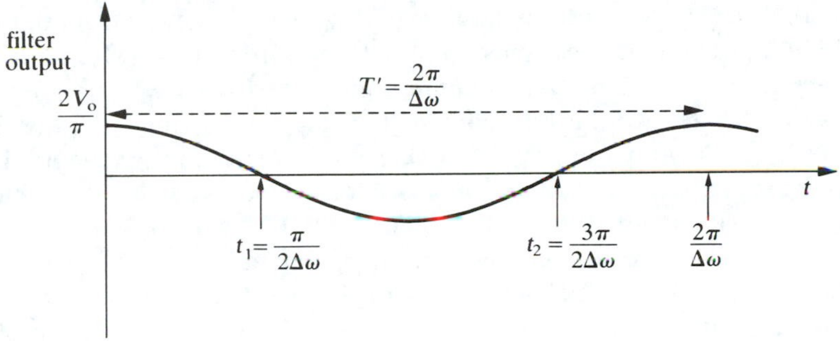

FIGURE 8.13 Phase detector output for $f_{signal} = f_{switch}$.

with a width of approximately $1/RC$, so the lock-in bandwidth is

$$\text{Bandwidth} \cong \frac{1}{RC} \tag{8.37}$$

This bandwidth can be made very narrow by simply using a large time constant RC; the phase detector is thus a narrowband voltmeter whose gain is sharply peaked at $\omega_{signal} = \omega_{switch}$.

The phase detector will also response to any *odd* multiple of the switch frequency but with a lower overall gain.

The lock-in technique for signal detection involves doing two things. First, we must convert the signal to ac form, if it is not already ac, and feed it into the phase detector. The signal must be converted to ac at *precisely* the switch frequency; this means we must use the *same* frequency source to drive the switch and to convert the signal to ac (i.e., to modulate the signal). Second, we must slowly sweep through the signal in a time that is long compared to RC.

Consider a specific experiment, electron spin resonance (ESR) or nuclear magnetic resonance (NMR). In ESR or NMR the sample contains magnetic dipoles (electron spins or nuclear magnetic moments) that absorb energy only at one (or several discrete) values of an external magnetic field if a fixed-frequency electromagnetic field of the proper polarization is fed into the sample. For example, in ESR free electrons bathed in 10,000-MHz (10-GHz) electromagnetic radiation in a microwave waveguide will absorb 10 GHz energy only when the magnetic field equals 0.3751 T (1 gauss = 10^{-4} tesla). In NMR protons bathed in 42.6-MHz electromagnetic radiation inside a coil in a resonant LC tank will absorb 42.6 MHz energy only when the external magnetic field equals 1.000 T. The sample always absorbs energy over a very small but finite range of magnetic field known as the *line width*, as shown in Fig. 8.14. The resonance signal is the energy absorption curve of Fig. 8.14(a). We could observe it by detecting and amplifying the absorbed energy as we increase or "sweep" the magnetic field from 0.3570 T to 0.3572 T. However, if we sweep the field upward very slowly, the signal will vary slowly in time and contain only very-low-frequency Fourier components. This would be difficult to amplify because of the low-frequency $1/f$ flicker noise present in all crystal diodes and amplifiers. If we sweep more rapidly, the signal Fourier components will be at higher frequencies, say, from 100 Hz to 100 kHz; but this means an amplifier bandwidth of 100 Hz to 100 kHz, or approximately 100 kHz would be necessary to reproduce the signal with negligible distortion. (Leaving out any of the signal's frequency components by an inadequate amplifier bandwidth will distort the signal.) If the source resistance is 50 Ω and is at 300 K, then the source noise voltage is

$$
\begin{aligned}
e_{sn} &= \sqrt{4kTR_sB_{\hat{W}}} = \sqrt{4 \times 1.38 \times 10^{-23} \times 300 \times 50 \times 10^5} \\
&= 0.29\ \mu\text{V}
\end{aligned}
$$

which, practically speaking, is enough to obscure many small resonance signals of great scientific interest.

What we actually do is convert the resonance signal to essentially a single frequency by sinusoidally modulating the magnetic field up and down at a frequency f_m, which may range from 100 Hz to 100 kHz, while simultaneously sweeping the magnetic field slowly upward from 0.3570 to 0.3572 T. The amplitude of the magnetic field modulation is made much less than the line width (e.g., 10^{-5} T). The energy absorbed by the sample is then amplitude modulated *at the modulation frequency* f_m, and the *amplitude* of the fluctuating energy absorbed is proportional to the *slope* of the energy absorption curve of Fig. 8.14. This amplitude is zero off to the left of the line, around 0.3500 T, increases and then decreases to zero at the center of the line at 0.3571 T, and then increases above 0.3571 T

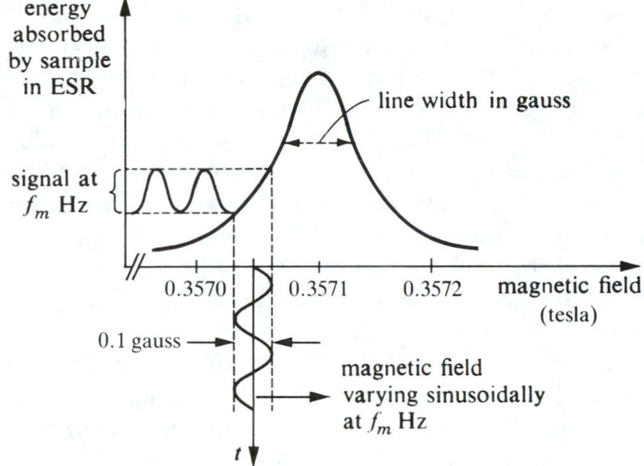

(a) *energy absorption curve*

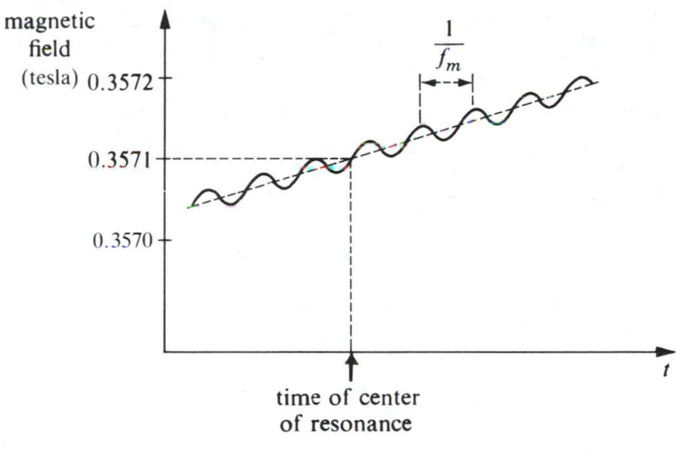

(b) *magnetic field sweep*

FIGURE 8.14 Electron spin resonance.

but with a 180° phase shift relative to the signal on the left side of the line. The point is that the amplitude must not change appreciably in the time RC, which is essentially the minimum time during which the dc output voltage can change. A typical value for RC is 1 s, so the energy absorption curve should be swept through no faster than once in 10 or 20 s. The effective frequency bandwidth of the switch and RC output filter is approximately $1/RC$, because it is a low-pass filter with a breakpoint frequency $\omega_B = 1/RC$.

Any noise voltage present along with the signal will be random in phase or "incoherent" with respect to the modulation frequency driving the switch. Therefore, short positive and negative noise spikes are equally likely during the time the switch is closed. And any low-frequency noise, such as $1/f$ flicker noise or 60-Hz or 120-Hz interference, tends to give the capacitor a net charge of zero, because the positive and negative noise cycles will each be sampled equally by the switch if the output is averaged over many noise points—that is, if the breakpoint frequency of the output filter is less than the noise frequency: $1/2\pi RC < f_{\text{noise}}$. This is equivalent to the earlier statement that the bandwidth of the switch–RC filter combination is essentially equal to $1/RC$.

In practical lock-in detectors the switch is rarely mechanical. It is usually an amplifier that is gated on and off by a square wave of frequency f_m. For example, a square wave of 10 V amplitude could be applied to the collector of a transistor amplifier.

In summary, only a signal frequency coherent or in phase with f_m, the

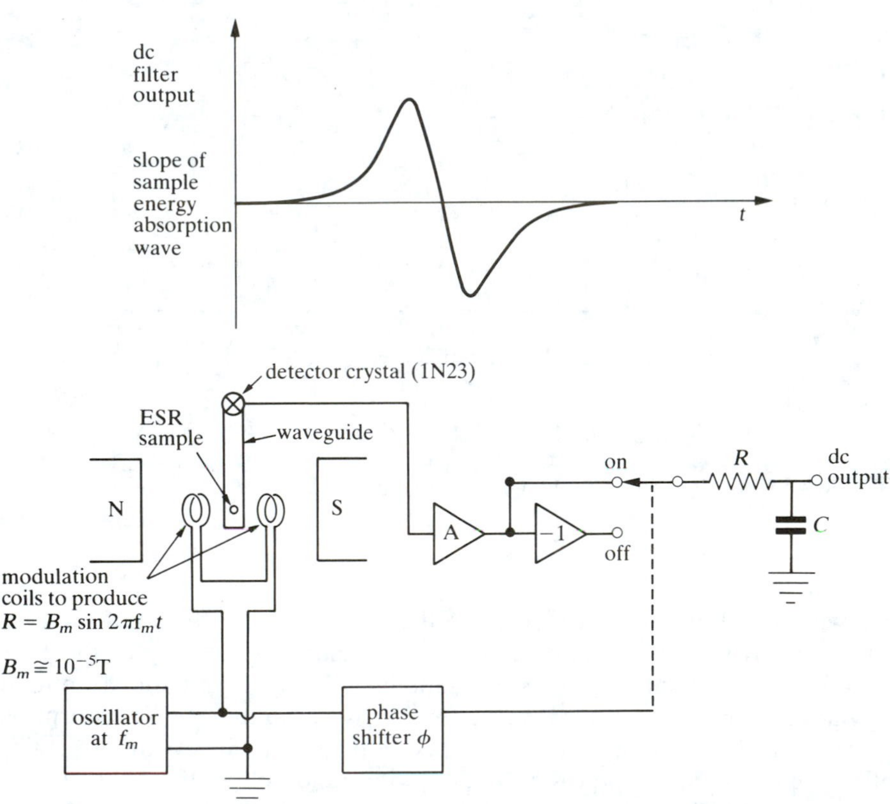

FIGURE 8.15 Lock-in detection of ESR.

modulation frequency driving the switch, will be passed through the lock-in detector.

In microwave ESR experiments the detector might be a point-contact silicon crystal diode, the 1N23 family for 10 GHz, for example, as shown in Fig. 8.15. These diodes exhibit $1/f$ flicker noise as well as white noise, and their corner frequency is approximately 50 to 100 kHz. The crystal noise is usually far larger than the amplifier noise, so a considerable increase in signal-to-noise ratio is achieved by using a modulation frequency above the corner frequency. $f_m = 100$ kHz is a typical choice. Schottky diodes have a corner frequency down around 1 kHz, enabling one to use lower modulation frequencies, which are experimentally more convenient because of their greater skin depth.

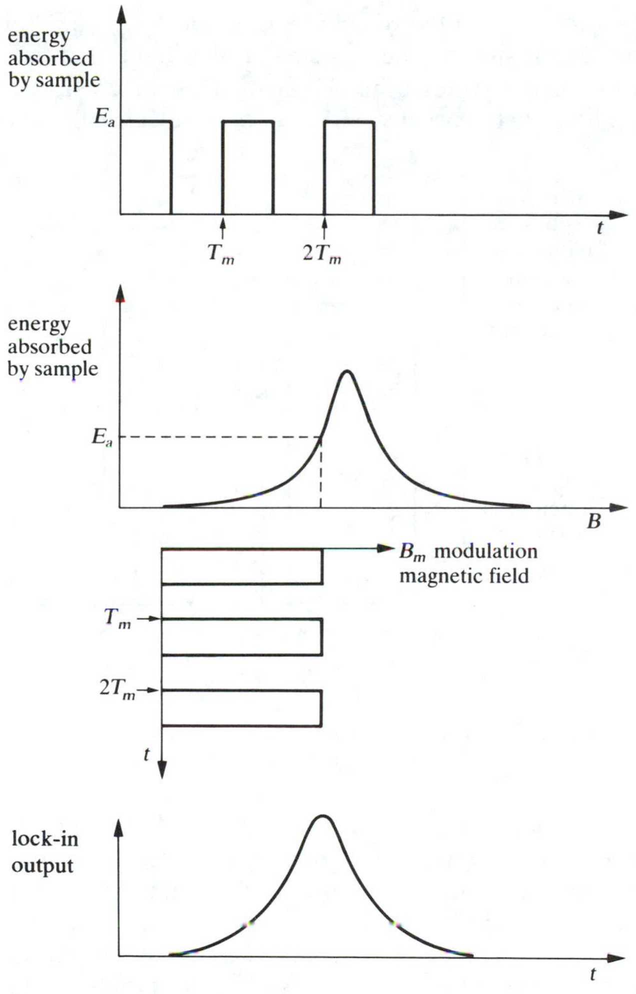

FIGURE 8.16 Large square-wave modulation.

Another type of modulation is possible in which the desired signal is turned completely on and completely off as shown in Fig. 8.16. If the modulation amplitude is larger than the signal line width, then the energy absorbed will vary from E_a to 0 at the modulation frequency $f_m = 1/T_m$. Thus the energy absorbed is amplitude modulated at f_m. This signal is fed into the lock-in detector, and the output of the lock-in is proportional to the energy absorption curve, not to its slope as in the previously discussed case of small sinusoidal modulation. A slow sweep of the magnetic field is also necessary.

Lock-in detection is not confined to ESR and NMR experiments. It is useful whenever the signal is at dc or a low frequency where $1/f$ amplifier flicker noise is large. For example, if the response of a complicated biological system to a light stimulus is to be measured, then the light can be chopped by a rotating slotted wheel at a frequency f_m. The desired signal is the electrical response of the system to the light and is taken off a transducer of some sort, for example, electrodes attached to the skull (see Fig. 8.17). The signal is usually buried in other electrical signals and noise

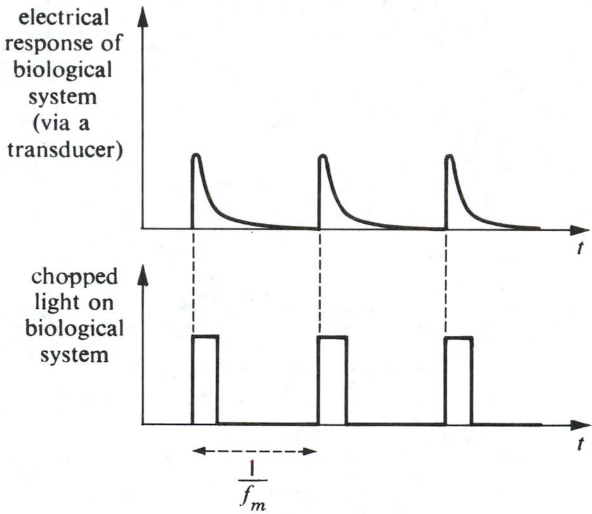

FIGURE 8.17 Light stimulus and electrical response.

and therefore difficult to detect. But if the transducer output is fed into a lock-in detector with a reference frequency of f_m, the lock-in output will be proportional to the system response to the light and not to any other stimulus that may produce electrical signals in the transducer. The essential point is to "chop" or amplitude modulate the stimulus at a frequency f_m and to use the same chopping frequency source to supply the switch in the lock-in detector.

8.9 THE PHASE-LOCKED LOOP

The phase-locked loop is basically a lock-in detector with the dc output of the low-pass filter fed back to adjust the frequency f_s of the oscillator that drives the switch, as shown in Fig. 8.18(a). The switch oscillator is a

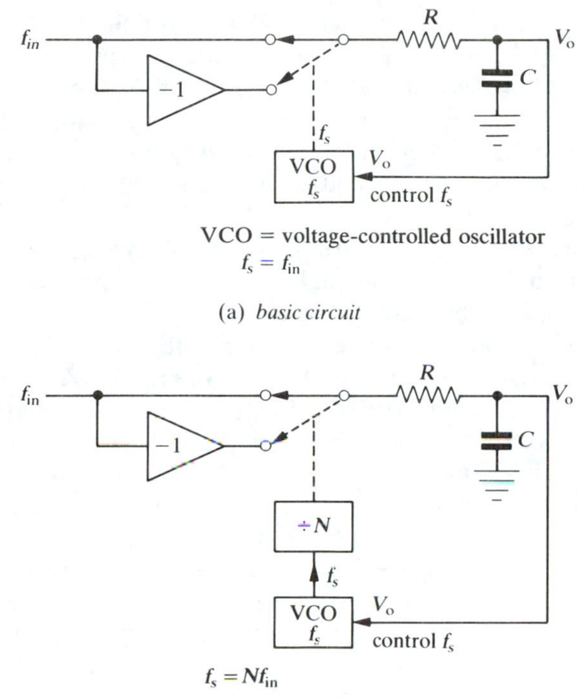

FIGURE 8.18 Phase-locked loop.

voltage-controlled oscillator whose frequency is proportional to its dc "control" voltage, V_0, which is the output of the low-pass filter. If the input frequency $f_{in} > f_s$, the filter output V_0 increases and causes f_s to increase until $f_s = f_{in}$. If $f_{in} < f_s$, then V_0 decreases and causes f_s to decrease until $f_s = f_{in}$. In all cases V_0 is proportional to f_{in}; the circuit acts as a frequency-measuring device and can be used to generate a dc voltage V_0 proportional to f_{in}. Obvious applications are demodulation of FM signals and the generation of a pure frequency f_s, which is frequency locked to a noisier input f_{in}.

If a divide-by-N counter is placed at the output of the VCO as shown in Fig. 8.18(b), then f_s will be locked to Nf_{in}. With counter techniques of digital electronics, N can be a very large number, $N = 2^{10} = 1024$ or higher.

8.10 SIGNAL AVERAGING TECHNIQUES

Noise is random not only in frequency but also in polarity. A noise voltage averaged over a long time will average to zero, because there will be as many positive noise pulses as negative ones. A signal, on the other hand, has a definite, nonrandom polarity. Hence, if we can repeatedly average over a signal plus noise, then the noise contributions ought to get smaller and eventually average to zero, and each average should add to the signal amplitude.

There are several types of instruments available, some digital and some analog, to perform such signal averaging. The signal must be repetitive, not necessarily periodic. The analog signal averager known as a *waveform educator* works as follows. At the beginning of each sweep through the signal, a trigger pulse or command of some sort must start the sweep. The time during which the signal exists is divided up into many intervals called *channels*, say, 100; the voltage in each interval is measured and stored on a separate capacitor, one for each interval. Thus on every sweep each capacitor receives a voltage proportional to the signal value plus the noise value during that interval for that particular sweep. The same process is repeated when the next trigger pulse comes along, and the capacitors accumulate more voltage each sweep. After N sweeps each capacitor should have a voltage equal to N times the signal voltage v_s plus the sum of N noise voltages, which add as the square root of the sum of the squares because the noise is random in time. Thus the voltage v stored after N sweeps is given by

$$v = Nv_s + \sqrt{e_{n1}^2 + e_{n2}^2 + \cdots + e_{nN}^2} \tag{8.38}$$

where v_s is the signal voltage present at each sweep and e_{n1} is the noise voltage present on the first sweep, and so on. If the noise is constant in character for all the sweeps (i.e., if the experimental conditions do not change), then the square of the noise should be the same for each sweep:

$$e_{n1}^2 = e_{n2}^2 = \cdots = e_{nN}^2 = e_n^2$$

Thus

$$v = Nv_s + \sqrt{Ne_n^2}$$

or

$$v = Nv_s + \sqrt{N}\sqrt{e_n^2} = Nv_s + \sqrt{N}e_{\text{rms}n} \tag{8.39}$$

where $e_{\text{rms}n}$ is the root mean square noise. The signal-to-noise ratio after N

sweeps is then

$$S/N = \frac{Nv_s}{\sqrt{N}e_{rmsn}} = \sqrt{N}\frac{v_s}{e_{rmsn}} \tag{8.40}$$

Hence, the signal-to-noise ratio S/N increases as the square root of the number of sweeps through the signal. There is no theoretical upper limit on the improvement in the signal-to-noise ratio that can be obtained by this technique. The practical limits are as follows: the stability and reproducibility of the trigger pulse—it must occur at precisely the same point in time relative to the signal; the ability of the capacitor storage to keep the stored signal and noise for the duration of the N sweeps; and, of course, the noise generated within the instrument.

Signal averagers are generally used in pulse experiments where the signal occurs once after some sort of stimulus. Lock-in techniques are usually used in steady-state experiments where the signal can be sinusoidally modulated. Pulse experiments are widely used in biology. The stimulus could be a light or sound pulse that produces a response that is usually buried in noise. The stimulus is usually also used to provide a trigger pulse for the signal averager, as shown in Fig. 8.19. The signal averaging

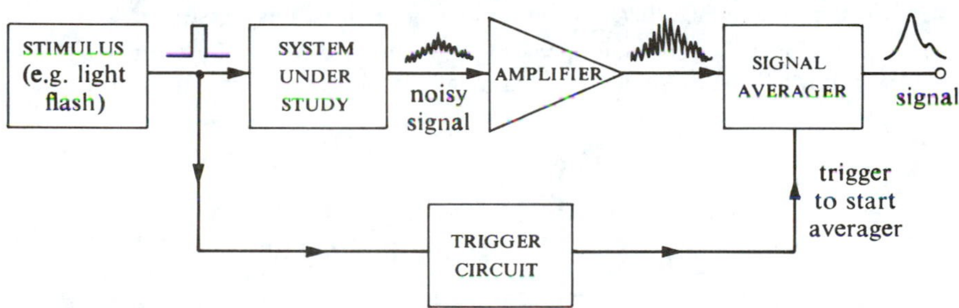

FIGURE 8.19 Signal averages in pulse experiment.

technique is not quite as good as the lock-in technique in extracting signals buried in a great deal of noise, but it does have the advantage of presenting the true signal shape as the output rather than the approximate first derivative as in the lock-in technique.

The digital signal averager is sometimes called the *computer of average transients* (CAT) and, in general, is considerably more expensive than the analog waveform eductor. However, most general-purpose computers with the proper program can be used as a CAT. In theory the CAT can be used for an infinite time to improve the signal-to-noise ratio without limit because of the permanent memory and the absence of drift in its digital circuitry, but there will always be some noise and drift in the input transducer and the input analog-to-digital converter. The CAT has the advantages that its memories in which the signal is stored are more permanent, and its output is already binary coded for further computer processing if desired. The intervals or channels in a waveform eductor can be made as short as 1 μs, whereas the shortest CAT interval is about 30 μs.

Signal averaging can also be used in pulsed resonance experiments such as NMR or ESR spin echo experiments. Modern instruments have a time resolution of about 5 to 10 ns and so can be used with extremely fast signals, such as in photoluminescence and fluorescence.

PROBLEMS

1. Which would be a better source for a high-fidelity music system: (a) a signal voltage of 10 mV along with a noise voltage of 1 μV, or (b) a signal voltage of 1 V along with a noise voltage of 1 mV? Briefly explain.
2. Calculate the Johnson noise voltage generated in a 1-MΩ resistance at 300 K as measured by a scope with a 10-MHz bandwidth.
3. For Johnson noise generated in a resistance R at $T°$, what is the percentage probability that the noise voltage is (a) greater than $+\sqrt{4kTB_WR}$, (b) less than (more negative than) $-\sqrt{4kTB_WR}$, (c) between 0 and $+\sqrt{4kTB_WR}$, and (d) between 0 and $-\sqrt{4kTB_WR}$?
4. How can one reduce the Johnson noise generated within a resistance?
5. Calculate the noise voltage appearing at resonance across the terminals AB of the following tank circuit, assuming negligible current is drawn from the tank.

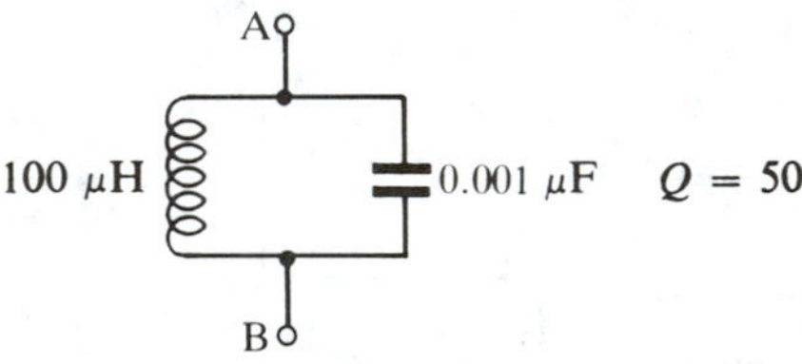

6. Sketch a rough but highly magnified graph of the current flowing through a diode carrying 112 mA current as seen by an oscilloscope with a bandwidth of 10 MHz.
7. Explain why the relative or percentage fluctuations of current due to shot noise are smaller at higher dc currents.
8. For a bipolar transistor amplifier with an input of $E_s = 1$ mV, $e_n = 2 \times 10^{-8}$ V/$\sqrt{\text{Hz}}$, $i_n = 2$ pA/$\sqrt{\text{Hz}}$, calculate (a) the optimal source resistance for the best output signal-to-noise ratio, (b) the best noise factor, (c) the noise factor for $R_s = 100\ \Omega$.
9. Show that the noise figure for a transistorized amplifier-source combination diverges to infinity for zero source resistance and also for infinite source resistance.

10. (a) Calculate the noise figure for a 50-Ω source driving an FET low-noise amplifier with an effective input voltage noise generator, $e_n = 5 \times 10^{-8}\ \text{V}/\sqrt{\text{Hz}}$, and an effective input current noise generator, $i_n = 4 \times 10^{-14}\ \text{A}/\sqrt{\text{Hz}}$. (b) Calculate the input transformer turns ratio to maximize the output signal-to-noise ratio.

11. Explain why a step-down transformer is not a practical solution to the problem of maximizing the signal-to-noise ratio for a larger source impedance.

12. Briefly explain the meaning of the words "lock-in" in a lock-in detector.

13. Briefly explain how the lock-in detector output signal-to-noise ratio depends upon the output time constant.

14. Why does a larger output time constant in a lock-in detector require a slower passage through the signal?

15. Explain why noise of random phase and polarity will not appear in the output of a lock-in detector, but noise with frequency an odd multiple of the reference frequency will contribute to the output.

16. Two equal resistances initially at different temperatures are suddenly connected in parallel. Discuss the noise voltages produced across the two resistors, the power flow, and the final equilibrium situation.

17. Calculate the approximate mean-square voltage expected at the output if the input is a 10-MΩ resistance at 70°C. You may neglect the noise generated in the 10-kΩ resistor. [*Hint*: What is the approximate bandwidth of the 10-kΩ, 100-pF low-pass filter?]

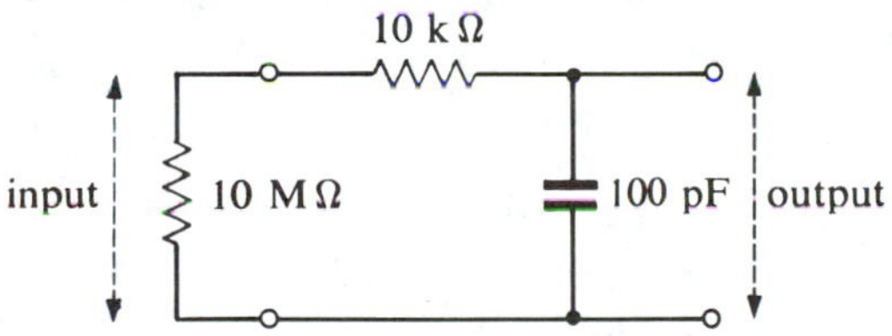

18.

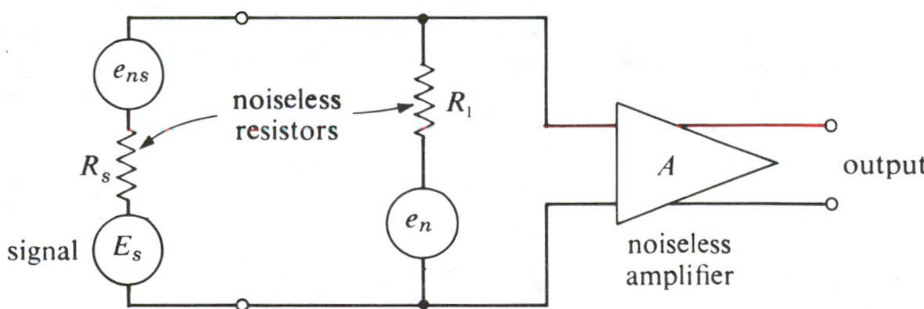

The source noise is pure Johnson noise; $e_{ns} = \sqrt{4kTB_W R_s}$. If all the effective input amplifier noise is that of a voltage noise source $e_n = \sqrt{4kTB_W R_1}$ (no current noise), (a) show that the total input noise to the amplifier is

$$e_{nT} = \sqrt{\left(\frac{R_s}{R_1 + R_2}\right)^2 e_n^2 + \left(\frac{R_1}{R_1 + R_s}\right)^2 e_{ns}^2}$$

[*Hint:* Use superposition to calculate the noise input; remember that the noise voltages add quadratically.] (b) Show that the output signal-to-noise power ratio is

$$(\text{S/N})_{\text{power}} = \frac{E_s^2}{4kTB_W} \times \frac{R_1}{R_s(R_1 + R_s)}$$

(c) Sketch a graph of the output signal-to-noise power ratio versus R_s for R_1 and E_s fixed.

19. Show that the noise temperature of two resistances R_1 and R_2 in series is equal to their common temperature. [*Hint*: Noise voltages add as the square root of the sum of the squares.]

20. Calculate the output from the lock-in low-pass filter if $v_{in} = V_0 \sin(3\omega t)$, where ω is the switch frequency. Repeat for $v_{in} = V_0 \sin(3\omega t + \phi)$.

21. In the phase-locked loop shown, what is the VCO frequency? If the $\div 2^{10}$ is replaced by a multiplier $\times 10$, what is the VCO frequency?

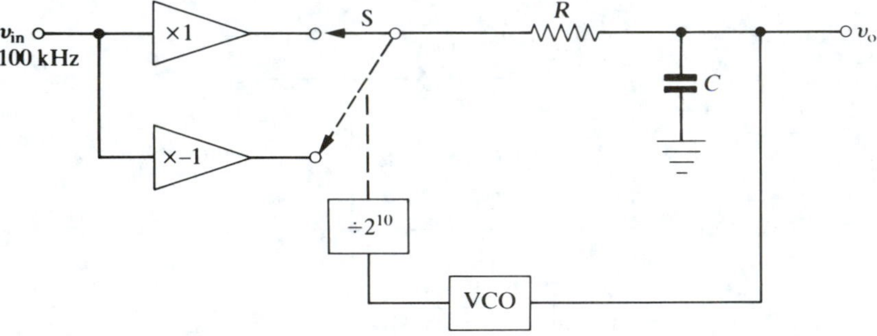

22. If the signal-to-noise voltage average is 10:1 without signal averaging, what would be the best signal-to-noise voltage you would expect if the signal were swept through and averaged 1000 times?

CHAPTER 9

Operational Amplifiers

9.1 INTRODUCTION

An *operational amplifier* or *op amp* is a high-gain dc coupled amplifier with two inputs; it is the basic building block in most analog electronic systems and is also used in some digital electronic systems. It is usually used with negative feedback, so its properties are independent of variables such as temperature, supply voltage, and so on. Operational amplifiers are one of the truly great bargains in electronics because of their excellent electrical performance, their low price (less than $1 each in many cases), and the wide range of types available.

The name "operational amplifier" originally meant an amplifier that could be made to perform various mathematical operations, such as addition, multiplication, differentiation, and so on. These operational amplifiers were widely used in analog computers, where an analog voltage input represented a mathematical variable and the output of the operational amplifier represented a function of the variable. Analog computers have largely been replaced by digital computers, but the name "operational amplifier" remains because the op amp is such an extremely useful electronic building block.

A modern op amp is usually made as an integrated circuit (IC), which means the transistors, resistors, diodes, and other components composing the op amp are all formed on a single silicon chip. (Strictly speaking, this is a *monolithic* IC.) Modern electronic manufacturing techniques can produce thousands of transistors, resistors, and diodes on a chip of 1 mm^2 area or less. Thus, extreme miniaturization is possible with very small shunt capacitance and series inductance, which gives excellent high-frequency performance. The resulting op amp not only is inexpensive but also has far better electrical characteristics than an op amp made from discrete components soldered together. The one limitation of such ICs is in handling high power and high voltage. Because of the close spacing of the components on the chip, most op amps (but not all) and other ICs cannot handle powers greater than approximately hundreds of milliwatts or voltages of more than 50 V. With the addition of an external *heat sink* to dissipate power to the surrounding air, op amps can safely handle watts of power and tens of amperes of current. Special high-voltage op amps are available that can operate at hundreds of volts. However, following the general rule of "you never get something for nothing," op amps that operate at high voltages or high powers generally sacrifice some other property, usually high-frequency operation. As a general rule, if one parameter of an op amp is unusually good, some other parameter must be unusually poor.

A modern op amp means an integrated circuit op amp, and we will use the term "op amp" in this sense from now on.

9.2 INTEGRATED CIRCUIT CONSTRUCTION

At this writing most ICs are made from silicon. A small, carefully selected, ultrapure (< 0.01 part per million impurities typically) silicon single crystal is used as a seed crystal to grow a larger ultrapure silicon single crystal. The seed is dipped into p-type molten silicon and slowly raised up into a cooler region. A large p-type silicon crystal about 5 cm in diameter and up to 50 cm long can be grown in this way. This large crystal is sawed into "wafers," which are polished and chemically etched to form a very smooth, chemically clean silicon surface. (Considerable silicon is wasted as sawdust.) The p-type wafers form the *substrate* and are typically 120 μm thick. A thin layer ~ 10 μm) of n-type silicon is grown on top of the p-type substrate; this n-type layer is *epitaxial*, which means it continues the crystal structure of the substrate. The epitaxial n-type layer eventually forms the collectors of the npn transistors in the IC. The wafer is then oxidized at very high temperatures (about 1000°C) in a furnace to form a thin (~ 0.5-μm) insulating layer of SiO_2 covering the entire epitaxial n-type surface of the wafer. In some processes we start with an n-type substrate and grow an epitaxial p-type layer on it.

A complicated series of photoetching and diffusing processes follows to build up the desired npn transistor structure, the p-n diode structure, and the appropriate resistances. If we desire to remove certain areas of the SiO_2 surface layer and to leave other areas, we cover the entire wafer with a uniformly thin film of photosensitive emulsion, called *photoresist*. The wafer is illuminated with ultraviolet light through a precisely made mask or stencil, which exposes only certain areas. The exposed areas of the photoresist layer become polymerized while the unexposed areas remain unpolymerized. A chemical solvent is then used that dissolves only the unpolymerized photoresist. Thus we are left with only certain areas of the SiO_2 layer covered with polymerized photoresist. The entire wafer is then bathed in hydrofluoric acid to etch away the exposed SiO_2, exposing the substrate, but not attacking the SiO_2 covered by the polymerized photoresist. The remaining SiO_2 acts as a mask or stencil for diffusing n-type impurities partly into the exposed p-type substrate, the impurities diffusing much more readily into the exposed silicon than into the SiO_2.

To make a horizontal transistor as shown in Fig. 9.1(a), a p-type impurity is diffused into the n-type epitaxial layer to form the base, and a final n-type diffusion into the p region forms the emitter. The resultant transistor is a *horizontal integrated transistor*. A common modification to the horizontal integrated transistor is to diffuse an n^+ layer on the p substrate before growing the n epitaxial layer. (n^+ means extra heavy n doping.) This reduces the resistivity of the long horizontal collector region.

A similar series of photoetching processes is used to build up the desired circuit, which may consist of dozens of transistors, diodes, resistors, and perhaps one or two capacitors. The resistances are formed from strips of doped silicon etched to the proper width and thickness to yield the desired resistance. It is difficult to manufacture high resistances by such techniques; few ICs contain resistances greater than 100 kΩ. Capacitances of approximately 1–10 pF can be made with SiO_2 as the dielectric and a vapor deposited layer of aluminum as one of the plates. The final network of "wires" connecting the components with one another is formed by vacuum depositing a thin layer of aluminum over the entire wafer and etching away the undesired areas.

Many transistors can be made on one chip; this is called *monolithic* construction. All the transistors, resistors, and other components of a monolithic circuit are thus based on the same substrate.

If the p-type substrate of the chip is kept more negative in voltage than any of the epitaxial n-collector "islands," then *many* npn transistors can be made on the same substrate because each npn transistor will be isolated from the others by the reverse biased collector-substrate junction.

If the n^+-type substrate is used as the collector, the resulting transistor is called a "vertical" transistor as shown in Fig. 9.1(b). Vertical transistors are more expensive than horizontal transistors, but they are intrinsically faster devices because they can be made with thinner base regions. We clearly cannot make many vertical transistors on the same substrate.

With these techniques a complete complicated circuit containing hundreds or even thousands of horizontal transistors can be made on one small chip, perhaps only 1 mm × 1 mm in area. A two-transistor monolithic circuit is shown in Fig. 9.1(c). A "two-emitter" transistor is shown in Fig. 9.1(d). This is widely used as the input stage for a digital NAND ("not and") gate.

One monolithic transistor can occupy as little as 60 μm × 60 μm surface area on the chip. In addition to the small circuit size made possible by monolithic construction, the transistors have very similar characteristics and are always at the same temperature. The dc current gain h_{FE}, for example, may vary by a factor of 2 in a production run of discrete transistors, but two transistors on a monolithic chip will probably have h_{FE} matched within 10% and will have the same V_{BE} within 4 mV for the same collector current. Also, their characteristics will stay matched over a wide temperature range.

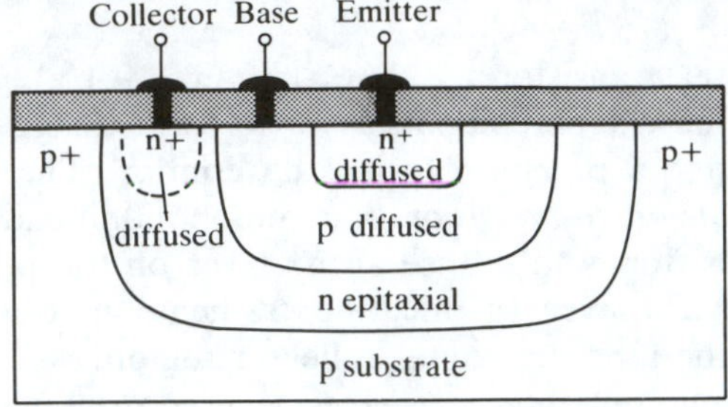

(a) *horizontal integrated npn transistor (used in IC chips)*

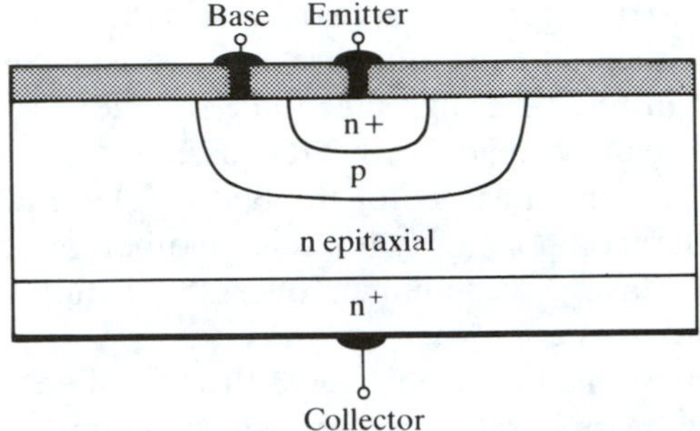

(b) *vertical planar epitaxial transistor (used for single transistors)*

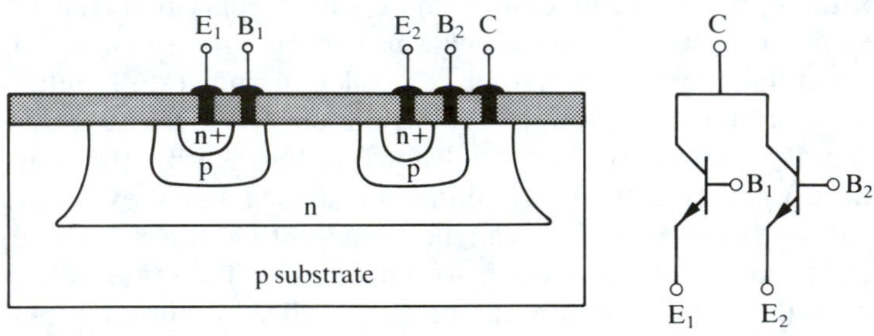

(c) *two transistor circuit on common substrate*

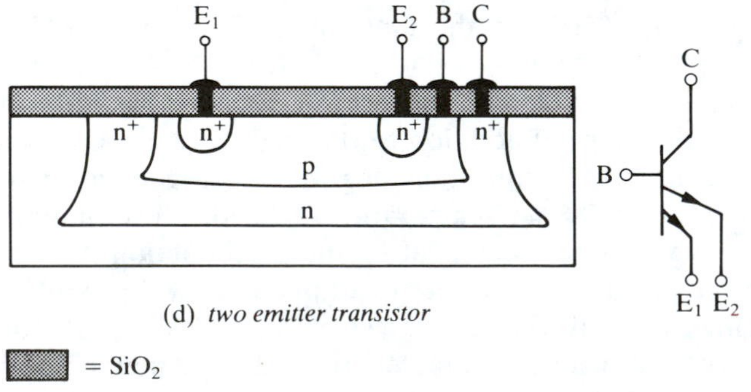

(d) *two emitter transistor*

= SiO_2

FIGURE 9.1 Transistor construction.

The monolithic construction with horizontal transistors has several disadvantages, however. First, there is a small leakage current between the different transistors through the reverse biased collector-substrate junctions, but this usually can be neglected. Second, there is a parasitic capacitance between the collector and substrate that is usually several pF. Third, the collector resistance is slightly larger for a monolithic transistor because the current path from the base to the collector terminal is larger in the monolithic device. Thus the $V_{CE\,sat}$ value is larger for the monolithic device.

Both npn and pnp transistors can be made with this technique, but most transistors tend to be npn. The reason is that the p-type dopant is usually B in B_2H_6 diborane gas, and the n-type dopant is usually P in PH_3 phosphine gas. The normal isotopes are ^{11}B and ^{31}P. Thus the n-type impurity ^{31}P has a smaller diffusion coefficient than the p-type ^{11}B impurity because it is heavier. The collector region must be heated to allow the base and emitter impurities to diffuse in, and thus it is desirable to minimize the diffusion of the collector impurities into the base region. Thus we get a better transistor when the collector impurity diffusion is less, which means an n-type collector.

Finally, the n^+ region is necessary because the Al contact is a weak p-type impurity. Thus if some Al diffuses into the n-type emitter or collector, a p-n junction will be formed between the Al lead and the emitter or collector, which is obviously undesirable—we want an ohmic junction at the Al leads. The extra n-impurity concentration in the n^+ region swamps out the p-type Al impurities in the collector or emitter and makes the junctions ohmic.

Most modern op amps and many other ICs are monolithic in construction.

The completed IC is cut away from the wafer and mounted in some sort of case with wire leads. Four case types are common (see Fig. 9.2): a hermetically sealed metal case (TO-5, etc.), which looks somewhat like an individual transistor except for the greater number of leads; the "flat-pak," which looks like a small rectangular solid with leads protruding from two sides; and the dual-in-line package, which is perhaps the most common. Other dual-in-line packages are available: 4-pin, 8-pin, 16-pin, 20-pin, 40-pin, and so on. Great care is taken to form good electrical and mechanical connections between the small IC aluminum conductors and the wire leads, which are typically 0.019-in.-diameter gold-plated kovar (an alloy of iron, nickel, and cobalt with a low coefficient of thermal expansion).

The absence of large capacitors in an IC makes it impossible to stabilize a common emitter bipolar transistor amplifier stage with a large bypass capacitor across the emitter resistor. In an IC it is much easier to make several additional transistors and diodes to stabilize a common emitter amplifier stage than to make one large capacitor. Hence, a basically simple common emitter amplifier may look rather complicated in an IC. Ad-

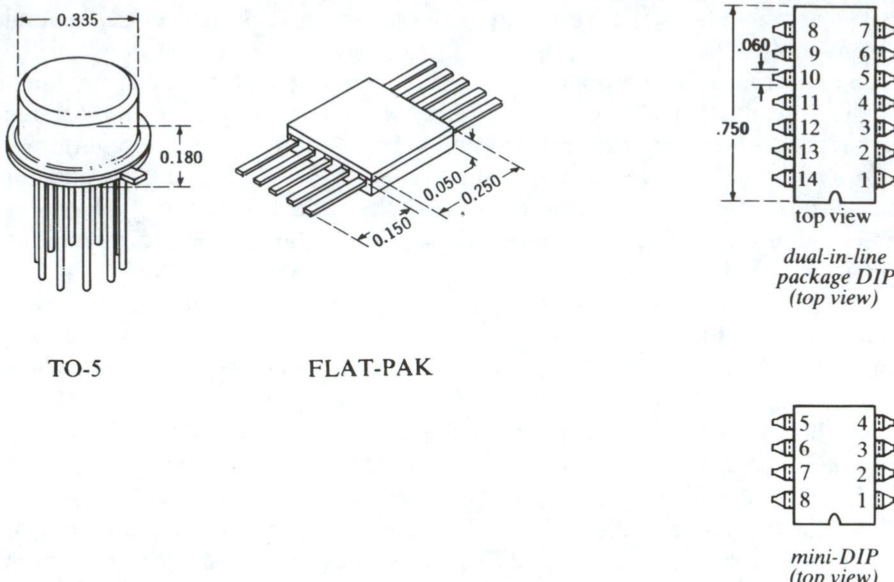

FIGURE 9.2 Three common integrated circuit cases.

vantage is usually taken in the IC design of the tight thermal coupling between the various components on the chip; that is, all transistors and resistors will be locked at essentially the same temperature because they are close together on the same semiconductor chip.

9.3 IC HEAT DISSIPATION

Like a transistor, an IC chip has a certain thermal resistance Θ (in °C/W) that depends on the physical construction of the IC chip. The temperature inside the transistor, T_J (the junction temperature), equals the ambient temperature T_A plus the thermal resistance times the electrical power P dissipated in the chip.

$$T_J = T_A + \Delta T_J = T_A + \Theta P \tag{9.1}$$

For example, if the ambient air temperature is $T_A = 25°C$, the chip thermal resistance is $\Theta = 100°C/W$, and the chip draws 10 mA from a 5-V power supply, the junction temperature is

$$T_J = T_A + \Theta T_J = 25°C + (100°C/W)(5\ V)(0.01\ A) = 30°C$$

The better the thermal coupling between the chip and the outside air, the smaller Θ and the better the cooling of the chip. For example, a metal-can power transistor mounted in a heat sink with large metal fins may have Θ ≅ 1–2°C/W, whereas a small low-power transistor or IC may have Θ ≅ 100°C/W.

The important thing to remember in evaluating manufacturers' specifications is that the *junction* temperature is always higher (sometimes by as much as 50°C) than the ambient temperature.

The insides of a medium-power 2N2222 transistor and a 2N3771 power transistor are shown in Fig. 9.3. Notice that most of the volume of the transistor is due to the heat sink, not to the npn semiconductor material.

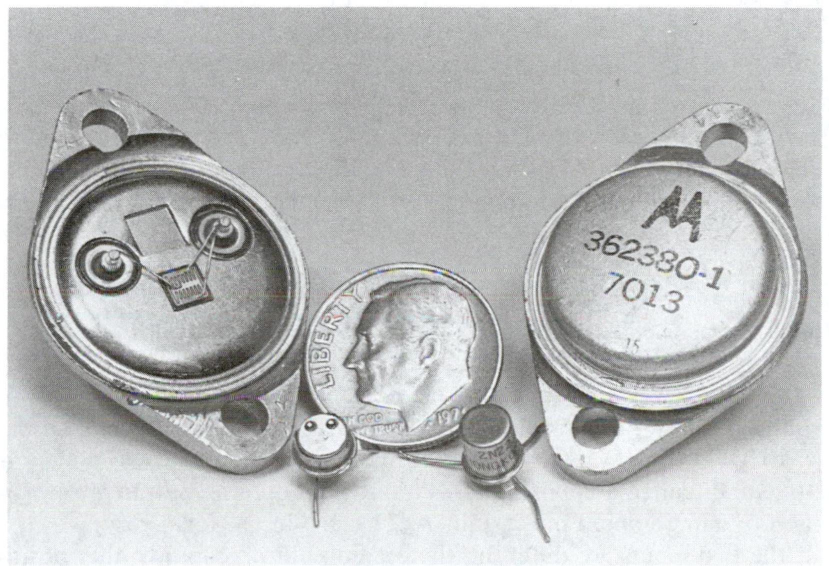

FIGURE 9.3 Small transistor and power transistor with case cut away.

9.4 THE IDEAL VERSUS THE ACTUAL OP AMP

An op amp is a high-gain dc coupled amplifier with two inputs, high input impedance, and low output impedance. The two inputs are the *inverting* input (usually labeled −) and the *noninverting* input (usually labeled +). A positive-going input at the noninverting input produces a positive-going output, while a positive-going input at the inverting input produces a negative-going output. In other words, the output voltage is in phase with the noninverting input and 180° out of phase with the inverting input. The output voltage of any op amp is equal to the op amp *open-loop* gain A_o (the voltage gain with no feedback—i.e., with the feedback loop open) times the differences between the noninverting input v_2 and the inverting input v_1 for negligible output current.

$$v_{out} = A_o(v_2 - v_1) \quad \textbf{(9.2)}$$

A_o is usually 10^5 or higher at very low frequencies. It is important to note that v_2 and v_1 are the voltages *directly at* the two input terminals of the op amp.

The standard circuit symbol for an op amp and an ideal equivalent circuit for an ideal op amp are shown in Fig. 9.4. R_{in} is the input resistance

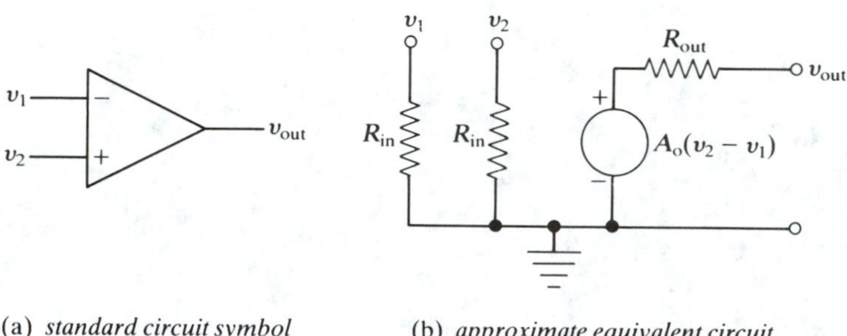

(a) *standard circuit symbol* (b) *approximate equivalent circuit*

FIGURE 9.4 Operational amplifier.

to ground (usually 1 MΩ or higher) and R_{out} is the output resistance (usually 100 Ω to 1000 Ω). We immediately see that if the two inputs v_2 and v_1 are exactly equal (called a *common mode* input), the op amp output is zero. We say the op amp common mode gain A_{cm} is zero in the ideal case.

If the two inputs are different, the op amp output can be either positive or negative depending on whether v_2 or v_1 is larger. The op amp differential gain is very high; a small difference between v_2 and v_1 will produce a large output. The op amp usually operates from two power supply voltages (typically +15 V and −15 V), and the output voltage can range from near the positive supply voltage to near the negative supply voltage (typically +13 V to −13 V for an op amp made from bipolar transistors, and +15 V to −15 V for an FET op amp).

In Fig. 9.4, R_{in} is the input resistance, assumed to be equal for the two inputs, and R_{out} is the output resistance. If no current is drawn from the output terminal, then there is no voltage drop across R_{out}, and the output terminal voltage will equal $A_o(v_2 - v_1)$. Remember, A_o is the *open*-loop gain of the op amp or, in other words, the "bare bones" gain of the op amp alone with no external feedback circuitry connected. Op amps are almost neven run as amplifiers without negative feedback applied with external resistors, capacitors, and so on, and, as we have seen in Chapter 8, the gain of *any* amplifier with negative voltage feedback depends only on the feedback factor B according to $A_v = 1/B$. Usually B depends only on the passive components (resistors, etc.) in the feedback network. Specific feedback networks are discussed in Sections 9.6.1 and 9.6.2.

An ideal op amp would, of course, have an infinite gain, infinite input resistance (so as to draw no current from the input signal source), a zero output resistance (so as to be able to supply any amount of output current), and infinite bandwidth. Naturally no such op amp exists; the ideal and actual op amp parameters for a typical inexpensive op amp are shown in Table 9.1.

TABLE 9.1. Op Amp Parameters

Parameter	*Ideal Op Amp*	*Typical Op Amp*
Differential voltage gain A_o at low frequencies	∞	10^5–10^9
Common mode voltage gain	0	10^{-5}
Gain bandwidth product f_T	∞	1–20 MHz
Input resistance R_{in}	∞	$10^6\ \Omega$ (bipolar) 10^9–$10^{12}\ \Omega$ (FET)
Output resistance R_{out}	0	100–1000 Ω

Notice that the parameters for the actual op amp are pretty close to ideal. They certainly are *far* better than could be obtained from an amplifier constructed from discrete components. Also notice that the input resistance is many orders of magnitude larger than the output resistance, which means that the output of one op amp can be fed into the input of another op amp without losing any voltage gain. If the input stage of the op amp is made from bipolar transistors, the input impedance is approximately 1 MΩ, but for an FET input stage the input resistance can be 1 GΩ or higher. Op amps are almost always used with external negative feedback, which will further increase the input resistance and decrease the output resistance according to the feedback theory of Chapter 8. Also, we point out that the open-loop gain A_o does decrease with increasing frequency, as will be discussed in Section 9.8.1; A_o typically falls off as $1/f$ (6 dB/octave or 20 dB/decade) for frequencies above several hundred Hz.

9.5 OP AMP CIRCUITRY

The first stage of an op amp is basically a two-transistor (either bipolar or FET) amplifier with a common emitter (or source) resistance as shown in Fig. 9.5. The input voltages are applied to the two bases, which are usually

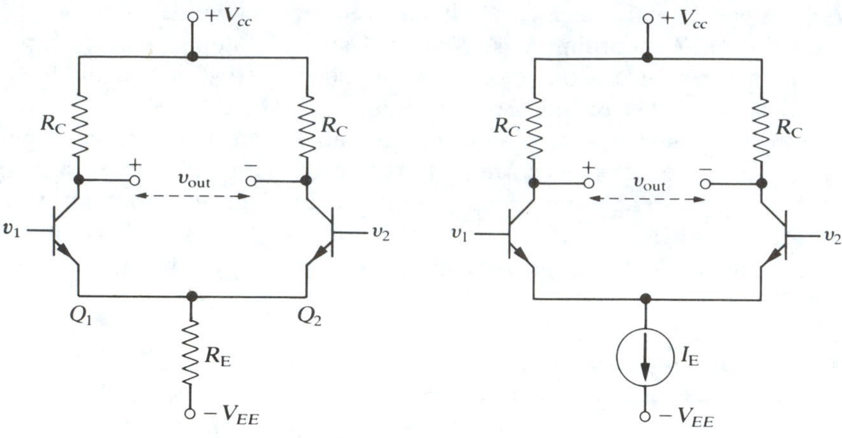

(a) *basic differential amplifier* (b) *with constant current source in emitter*

FIGURE 9.5 Op amp input circuitry.

near dc ground. The output voltage is taken between the two collectors and is fed to the next stage of the op amp for further amplification. A standard h parameter analysis shows that the differential voltage gain is given by

$$v_{out} = \frac{h_{fe} R_C}{h_{ie}} (v_2 - v_1) \tag{9.3}$$

Thus, the gain should be zero for any common mode inputs, $v_2 = v_1$. (Such common mode inputs can easily occur from 60-Hz pickup, where the pickup voltage is the same at both inputs.) This analysis assumes that the two input transistors Q_1 and Q_2 are exactly equal, with identical values for h_{fe} and h_{ie}, and that the two collector resistances R_C are also exactly equal. Such is not the case, of course, and so for a real op amp, the common mode gain is not zero.

It can be shown that the common mode gain is inversely proportional to the magnitude of the emitter resistance R_E. Thus, to minimize the common mode gain, we must use a large value of R_E or, equivalently, a constant current source in place of R_E, as shown in Fig. 9.5(b). The circuitry of Fig. 9.6(a) is designed to make the emitter current I_{E3} constant and independent of the temperature. This is very important because the op amp is dc coupled, and thus any temperature drift that causes a slow small shift in the output voltage of the first stage produces a slow large change in the output dc voltage because of the amplification of all the subsequent stages.

The basic problem is that the turn-on voltage of any silicon p-n junction (in particular the base-emitter junction of Q_3) changes with

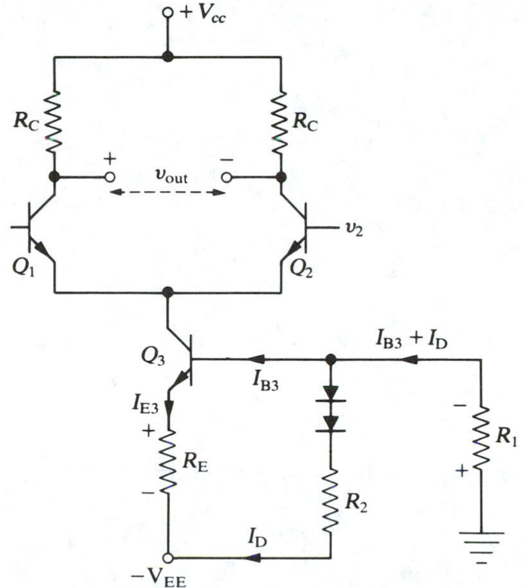

(a) *constant current source for* $(I_{E1} + I_{E2})$

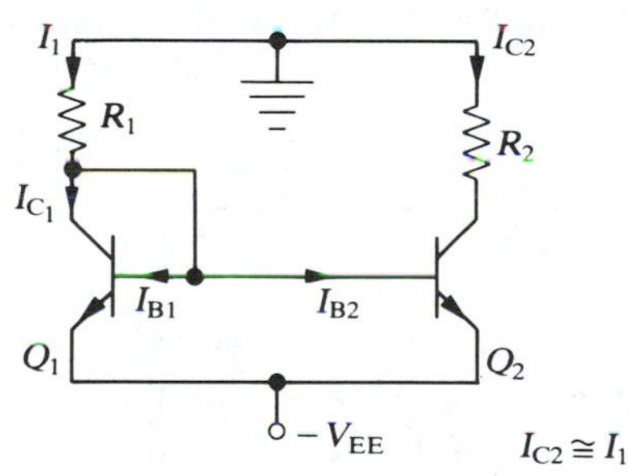

(b) *current mirror*

FIGURE 9.6 Op amp circuitry.

temperature at an approximate rate of −2.5 mV/°C (turning on more as the temperature rises). The circuitry of Fig. 9.6 makes the emitter current I_E depend only on V_{EE} and R_E, as can be seen from the following argument. As the temperature increases, the base-emitter junction of Q_3 tends to conduct more, but the voltage drop across the diodes decreases, thus applying a smaller forward bias to Q_3.

The exact analysis is as follows. The base emitter voltage of Q_3 is

$$V_{BE3} = V_{B3} - V_{E3}$$

But $$V_{B3} = -(I_D + I_{B3})R_1 \cong -I_D R_1 \qquad \text{if } I_D \gg I_{B3}$$

and $$V_{E3} = -|V_{EE}| + I_{E3}R_E$$

Thus, $$V_{BE3} = -I_D R_1 + |V_{EE}| - I_{E3}R_E$$

The emitter current is then

$$I_{E3} = \frac{|V_{EE}| - I_D R_1 - V_{BE3}}{R_E}$$

But $$I_D \cong \frac{|V_{EE}| - 2V_D}{R_1 + R_2} \qquad \text{if } I_D \gg I_{B3}$$

where V_D is the voltage drop across each diode. Eliminating I_D from the expression for I_{E3} yields

$$I_{E3} = \frac{|V_{EE}|}{R_E} - \frac{|V_{EE}|R_1 - 2V_D R_1 + V_{BE3}(R_1 + R_2)}{R_E(R_1 + R_2)} \tag{1}$$

Assuming V_D and V_{BE} are the most temperature-sensitive terms, we obtain the rate of change of emitter current with respect to temperature as

$$\frac{dI_{E3}}{dT} = \frac{2R_1 \dfrac{\partial V_D}{\partial T} - (R_1 + R_2)\dfrac{\partial V_{BE3}}{\partial T}}{R_E(R_1 + R_2)} \tag{2}$$

where we have neglected the temperature dependence of the various resistances. But the base-emitter junction of Q_3 and the two diodes are all made from the same silicon chip with exactly the same impurities, and so forth, and they are all at the same temperature because they are so close together on the chip. Thus the base-emitter junction of Q_3 and the diodes will have almost exactly the same temperature dependence:

$$\frac{\partial V_D}{\partial T} = \frac{\partial V_{BE3}}{\partial T} \equiv \frac{\partial V}{\partial T}$$

Thus the temperature dependence of the emitter current of Q_3 is

$$\frac{dI_{E3}}{dT} = \left[\frac{2R_1 - (R_1 + R_2)}{R_E(R_1 + R_2)}\right]\frac{\partial V}{\partial T} \tag{3}$$

which can be made zero by simply choosing $R_1 = R_2$. In other words, choosing $R_1 = R_2$ makes I_{E3} independent of the temperature.

If Q_3 and the diodes carry approximately the same current, then $V_{BE3} = V_D$, and, from (1), I_{E3} becomes

$$I_{E3} \cong \frac{V_{EE}}{2R_E} \tag{4}$$

Thus, we conclude that the circuit of Fig. 9.6 provides an extremely constant current I_{E3} for the two input transistors Q_1 and Q_2.

Another way of generating a constant current is to use the "current mirror" circuit of Fig. 9.6(b). It can be shown I_{C2} is given by

$$I_{C2} = \frac{\beta}{\beta + 2} I_1 \tag{9.4}$$

where $\beta = h_{FE} = I_C/I_B$ for Q_1 and Q_2.

9.5.1 A DC Level Shifter

In any common emitter amplifier the dc output voltage at the collector is quite different from the dc input voltage at the base because of the large base-collector voltage drop necessary for amplification. Thus if there are several such common emitter stages in a dc coupled op amp, something must be done to shift the output level of one stage to match the dc input level of the next stage. In an ac coupled amplifier large series capacitors can be used, but large capacitors are not practical in ICs. (The largest capacitance possible in an IC op amp is about 30 pF.)

One level-shifting circuit is shown in Fig. 9.7. The voltage drop across R_1 equals the base emitter voltage of the transistor; thus, the value of R_1 determines I_D by $I_D = V_{BE}/R_1$, which in turn determines the total voltage

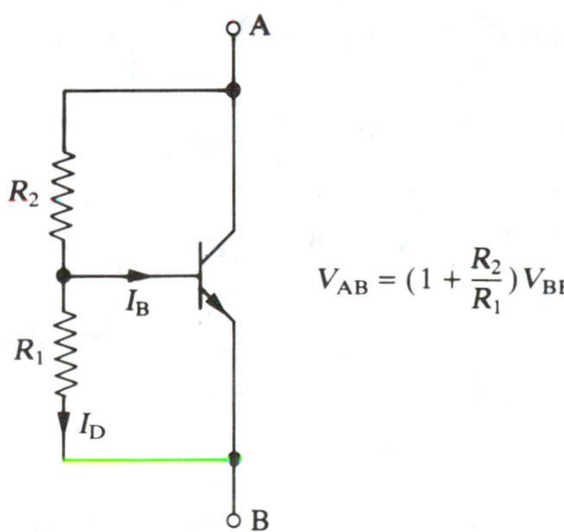

FIGURE 9.7 Dc level shifter.

drop between A and B from $V_{AB} \cong I_D(R_1 + R_2)$. The exact analysis is

$$V_{BE} = I_D R_1 \quad \text{and} \quad I_D \cong \frac{V_{AB}}{R_1 + R_2} \qquad (I_B \ll I_D)$$

Thus

$$V_{BE} \cong \frac{V_{AB} R_1}{R_1 + R_2}$$

or

$$V_{AB} \cong \left(1 + \frac{R_2}{R_1}\right) V_{BE} \tag{9.5}$$

The voltage shift V_{AB} then is determined solely by the ratio of R_1 and R_2, which can be made very stable. The level shifter is used to bias the push-pull output stage in many op amps, as will be explained in Chapter 10. V_{BE}, of course, does vary with temperature but by only 2.5 mV/°C. For example, if $R_1 = 1.0\ \text{k}\Omega$ and $R_2 = 2.2\ \text{k}\Omega$,

$$V_{AB} = \left(1 + \frac{2.2}{1.0}\right) 0.6\ \text{V} = 1.92\ \text{V}$$

The main thing to keep in mind from Section 9.5 is that the circuitry used in integrated circuits is different from that used in circuits constructed from discrete components. The lack of high-value resistances and capacitances and the ease of using transistors (some connected as diodes) dictate a different design philosophy, and thus IC circuit diagrams look quite different from diagrams for discrete component circuits.

9.6 TWO SIMPLE OP AMP AMPLIFIERS

We will now describe the two most common op amp amplifier circuits: the noninverting amplifier and the inverting amplifier.

9.6.1 The Noninverting Amplifier

A noninverting single-input amplifier is shown in Fig. 9.8(a). We will now derive an expression for the voltage gain. The KCL at the junction of R_1 and R_2 states

$$i_2 = i_1 + i'$$

Using Ohm's law, we can rewrite i_2 as

$$i_2 = \frac{v_{out} - v_1}{R_2} = \frac{v_1}{R_1} + i' \tag{9.6}$$

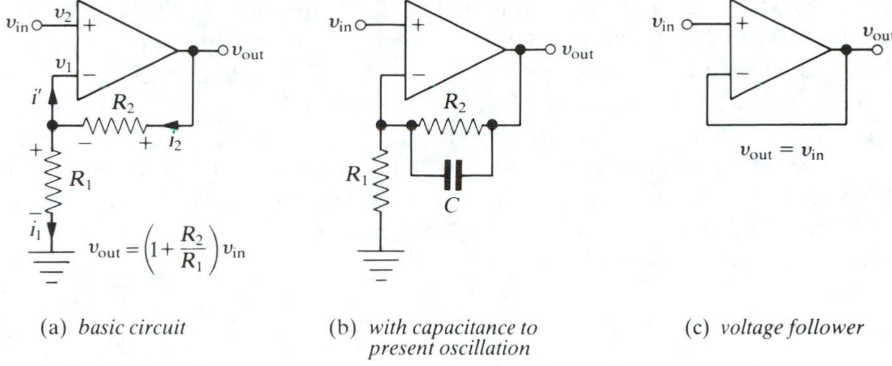

(a) *basic circuit* (b) *with capacitance to present oscillation* (c) *voltage follower*

FIGURE 9.8 The noninverting amplifier.

Because the input resistance of the amp is so large, we can neglect i' compared to i_1, and (9.6) becomes

$$\frac{v_{out} - v_1}{R_2} \cong \frac{v_1}{R_1} \tag{9.7}$$

The output voltage v_{out} is always related to v_1 and $v_2 = v_{in}$ by

$$v_{out} = A_o(v_2 - v_1) = A_o(v_{in} - v_1) \tag{9.8}$$

We now can obtain an expression for the voltage gain v_{out}/v_2 by eliminating v_1 from (9.7) and (9.8). The result is

$$A_v = \frac{v_{out}}{v_{in}} \cong \frac{A_o(R_1 + R_2)}{(A_o + 1)R_1 + R_2} \tag{9.9}$$

If the open-loop op amp voltage gain A_o is very large (usually $A_o > 10^5$), then

$$(A_o + 1)R_1 \gg R_2 \quad \text{and} \quad A_o + 1 \cong A_0$$

Thus (9.9) simplifies to

$$A_v \cong 1 + \frac{R_2}{R_1} \tag{9.10}$$

For example, if $R_1 = 1.0\ \text{k}\Omega$ and $R_2 = 4.7\ \text{k}\Omega$, then $A_v = 5.7$. Notice that the gain is positive, which means that the output is in phase with the input. In other words, a positive input produces a positive output, and a negative input produces a negative output. Also, the gain depends only upon the *ratio* of R_2 to R_1, which is essentially independent of temperature because both resistors usually have the same temperature coefficient of resistance. The gain is also independent of the op amp open-loop gain and any other op amp parameters, such as supply voltage.

The voltage gain expression (9.10) could have been derived immediately by using the feedback theory of Chapter 8. Neglecting i' because of the high input resistance of the op amp, we get $i_1 \cong i_2$, which, from Ohm's law, can be written as

$$\frac{v_1}{R_1} \cong \frac{v_{out} - v_1}{R_2} \quad \text{or} \quad \frac{v_1}{v_{out}} \cong \frac{R_1}{R_1 + R_2}$$

But v_1/v_{out} is the fraction of the output voltage fed back in series with the input; in other words, $v_1/v_{out} = B$, the feedback factor or fraction. And, from Chapter 8, the voltage gain with negative feedback is

$$A_v \cong \frac{1}{B} = \frac{1}{\dfrac{R_1}{R_1 + R_2}} = 1 + \frac{R_2}{R_1}$$

which is (9.10).

Often a high-speed op amp (one with good high-frequency gain) may oscillate at a high frequency when connected as in Fig. 9.8(a). The solution is to place a small capacitance C (generally about 10 to 100 pF) in parallel with R_2, as shown in Fig. 9.8(b). The effect of the capacitance is to increase the amount of negative feedback at high frequencies and thereby decrease the high-frequency gain and prevent the oscillation. The gain will start to fall below (9.10) when the reactance of C becomes comparable to R_2—that is, at frequencies where

$$\frac{1}{\omega C} \lessapprox R_2 \quad \text{or} \quad \omega \gtrapprox \frac{1}{R_2 C} \quad \text{or} \quad f \gtrapprox \frac{1}{2\pi R_2 C}$$

For example, if the circuit tended to oscillate at 1 MHz and $R_2 = 4.7\ \text{k}\Omega$, we would try a capacitance C such that $1/\omega C = R_2$ at 1 MHz, or

$$C = \frac{1}{\omega R_2} = \frac{1}{(2\pi \times 10^6\ \text{Hz})(4.7 \times 10^3\ \Omega)} = 34\ \text{pF}$$

Another useful form of the noninverting amplifier is shown in Fig. 9.8(c). This is called a *voltage follower* and has a voltage gain of unity. The circuit can be thought of as a limiting case of the circuit of Fig. 9.8(a) in which $R_1 \to \infty$ and $R_2 = 0$. By inspecting (9.10) we see the gain is 1.0. We can also think of this circuit as having a feedback factor of $B = 1$, much like the emitter follower circuit of Chapter 5. With a modern op amp the voltage follower of Fig. 9.8(c) is much better than the single-transistor emitter follower; the op amp voltage follower has a gain of almost exactly 1.0, its input impedance is very high (at least 1 MΩ for a bipolar op amp and 1 GΩ for an FET op amp), and its output impedance is very low, typically less than 1 Ω. The voltage follower does have a power gain and a current gain, of course. It is thus often used to drive a low-impedance load, such as a long coaxial cable, or it is used as a "buffer" amplifier to isolate two different parts of a system—it is almost an ideal "one-way" device for signals. It draws almost no current from the circuit supplying the input because of its high input resistance, and it can supply a large output current to low-impedance loads.

Notice that the inequality used in deriving (9.10) can be written as

$$(A_o + 1)R_1 \gg R_2 \quad \text{or} \quad A_o + 1 \gg \frac{R_2}{R_1} \cong A_v$$

In other words, (9.10) holds if the open-loop gain A_o (without feedback) is much greater than the gain A_v with feedback (the *closed-loop* gain); in other words, the gain is *greatly reduced* by negative feedback. The term *loop gain* means A_o/A_v or the open-loop gain minus the closed-loop gain in dB. For example, if $A_o = 100\text{ dB }(10^5)$ and $A_v = 40\text{ dB }(10^2)$, the loop gain is $A_o/A_v = 10^5/10^2 = 10^3$ or $100\text{ dB} - 40\text{ dB} = 60\text{ dB}$: the larger the loop gain, the more accurate is the gain expression (9.10).

The open-loop gain of any op amp decreases with increasing frequency, as shown in Fig. 9.9. For the graph of Fig. 9.9 the loop gain is

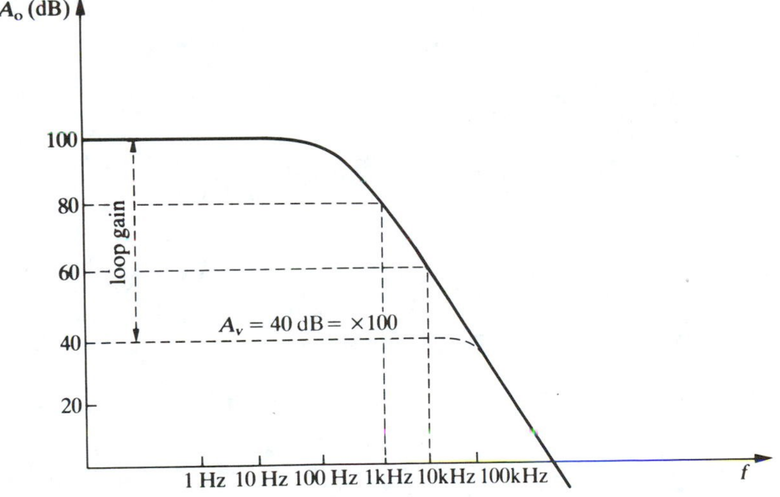

FIGURE 9.9 Open-loop gain.

60 dB from dc out to 100 Hz, but it decreases as $1/f$ for frequencies above 100 Hz. For example, at 1 kHz the loop gain is only 40 dB and at 10 kHz is only 20 dB. Thus, we conclude that the approximate gain expression (9.10) becomes less and less accurate as the frequency increases, and that the lower the closed loop gain A_v the better the accuracy of (9.10) at high frequencies. As a general rule-of-thumb, the percentage error in the approximate gain expression (9.10) [compared to the exact gain expression (9.9)] is approximately equal to the reciprocal of the loop gain times 100. For example, if the loop gain = 40 dB = 100, the percent error is 1%; a 20-dB loop gain produces an error of 10%.

9.6.2 The Inverting Amplifier

An inverting single-input amplifier is shown in Fig. 9.10. The voltage gain expression can be derived by using the Kirchhoff current law and Ohm's

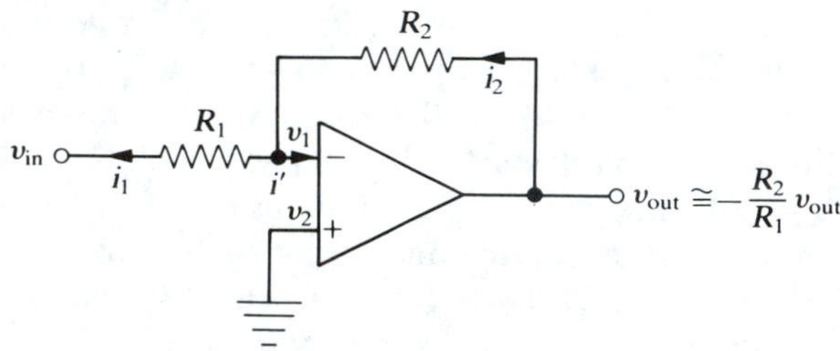

FIGURE 9.10 The inverting amplifier.

law. The KCL at the inverting input is

$$i_2 = i_1 + i' \qquad (1)$$

From Ohm's law, (1) becomes

$$\frac{v_{out} - v_1}{R_2} = \frac{v_1 - v_{in}}{R_1} + i' \qquad (2)$$

The output voltage v_{out} is always related to the input voltage by

$$v_{out} = A_o(v_2 - v_1) \qquad (3)$$

$v_2 = 0$ because the noninverting input is grounded, so (3) becomes

$$v_{out} = -A_o v_1 \qquad (4)$$

Neglecting i' in (2) as usual and eliminating v_1 using (2) and (4) yield an expression for the voltage gain:

$$A_v = \frac{v_{out}}{v_{in}} \cong -\frac{A_o R_2}{(A_o + 1)R_1 + R_2} \tag{9.11}$$

For a large op amp open-loop gain, $(A_o + 1)R_1 \gg R_2$ and $A_o + 1 \cong A_o$, so (9.11) becomes

$$A_v \cong -\frac{R_2}{R_1} \tag{9.12}$$

The minus sign merely means that the output is out of phase with respect to the input; a positive input produces a negative output, and a negative input produces a positive output. Also, the gain depends only on the ratio of the two resistances R_2 and R_1, which means the gain is essentially independent of temperature and other parameters (such as the op amp open-loop gain, etc.) as long as $(A_o + 1)R_1 \gg R_2$. Notice that the inequality to be satisfied can be written as

$$(A_o + 1) \gg \frac{R_2}{R_1} = |A_v|$$

In other words, (9.12) holds (and the gain A_v depends only on the ratio of R_2 to R_1) if the open-loop gain A_o (without feedback) is much greater than the gain with feedback. In other words, (9.12) holds if the gain is greatly reduced by negative feedback or if the loop gain is large. It should be clear now why such large open-loop gains are useful; with a large A_o we can have a large and stable A_v because $A_o \gg A_v$. Then the gain with feedback will be extremely stable (and adjustable) because it depends only on the passive components (R_1, R_2) in the feedback network. This conclusion applies to the noninverting amplifier of Section 9.6.1 also.

The larger the loop gain the more accurate (9.12) is; the same discussion in Section 9.6.1 on accuracy applies here to the inverting amplifier circuit. For example, a loop gain of 60 dB $= 10^3$ means a 0.1% error in (9.12).

Aside from the phase inversion, the main difference between the noninverting and the inverting amplifier circuits is in the input impedance. The input impedance of the noninverting circuit is very high because of the small current flowing into the noninverting input. The input impedance of the inverting circuit is R_1, as we will see in Section 9.7. Another point is that the high impedance noninverting amplifier input is *much* more susceptible to pick up than is the low-impedance input to the inverting amplifier. Consequently, if you don't need a high input impedance, the inverting amplifier is usually the better choice. We often wire an external $R' \cong$

100 kΩ to 1 MΩ resistor from the noninverting input to ground to stabilize the input resistance at R' for the noninverting amplifier. (See Section 10.3.)

9.7 TWO SIMPLE RULES FOR ANALYZING OP AMP CIRCUITS

Careful inspection of the derivations of the gain expressions for the two op amp circuits of the preceding sections shows that two approximations were made: (1) The input current i' flowing into the op amp input terminal was neglected compared to the other currents; and (2) the open-loop op amp gain A_o was assumed to be very large compared to the gain with feedback (A_v). The following extensions of these two rules will enable us to analyze any op amp circuit with very little algebra.

I. The current into (or out of) each op amp input terminal is approximately zero. We will call this the *op amp current rule* (OACR).
II. The voltage difference between the two op amp input terminals is approximately zero. We will call this the *op amp voltage rule* (OAVR).

The current rule basically says that the input impedance to the op amp at either input terminal is much higher than the external impedance from the input terminal to ground. Because the input resistance is typically 1–10 MΩ for a bipolar op amp and 10^9 Ω (1 GΩ) to 10^{12} Ω for an FET op amp, this approximation is usually excellent. This approximation might break down only, for example, when R_1 or R_2 in the op amp circuits of Figs. 9.8 and 9.10 were 1 MΩ or higher for a bipolar op amp or 100 MΩ or higher for an FET op amp. Such resistance values are rarely used in practice.

The voltage rule is equivalent to saying that the open-loop op amp gain A_o is essentially infinite. The output voltage of an op amp is always given by $v_{out} = A_o(v_2 - v_1)$. Thus, because v_{out} can never be more positive than $+V_{cc}$ or more negative than $-V_{cc}$ (typically $V_{cc} = 15$ V), it follows that if A_o is infinite, $v_2 - v_1$ must be zero. More precisely, the maximum value of $v_2 - v_1$ is given by

$$(v_2 - v_1)_{max} = \frac{v_{out\,max}}{A_o} = \frac{15\text{ V}}{10^5} = 1.5 \times 10^{-4}\text{ V} = 150\ \mu\text{V}$$

Thus, the maximum difference between v_2 and v_1 is only 150 μV, and it will be less if the supply voltage is less or if the open-loop gain of the op amp is higher. The op amp is dc coupled, so this approximation holds from dc to the upper operating frequencies of the op amp. This approximation would break down only when the inequality $A_v \ll A_o$ breaks down—that is, only when A_v is too high (corresponding to a small amount of negative feedback). Another way to think of this rule in an actual op amp circuit is that the negative feedback plus the high gain of the op amp effectively nulls out the voltage difference between the two inputs. The OAVR applies only when the op amp is not saturated (i.e., only when its output is not at $+V_{cc}$ or $-V_{cc}$ or, in other words, only for small input signals).

We will now quickly show that the two op amp circuits of Section 9.6 can be easily analyzed by using the two rules just stated.

For the noninverting amplifier of Fig. 9.8 the OAVR implies $v_2 = v_1$. The OACR implies that no current flows into the inverting input of the op amp, so the same current flows through R_1 and R_2 and thus they form a simple voltage divider. Thus, we can immediately write

$$v_1 = \frac{R_1}{R_1 + R_2} v_{\text{out}} = v_2 = v_{\text{in}}$$

The voltage gain is then

$$A_v = \frac{v_{\text{out}}}{v_{\text{in}}} = 1 + \frac{R_2}{R_1}$$

which is (9.10).

For the voltage follower of Fig. 9.8(c) the OAVR implies that $v_2 = v_1$, but $v_1 = v_{\text{out}}$ because the inverting input is directly connected to the output; so $v_{\text{in}} = v_2 = v_{\text{out}}$, and $A_v = v_{\text{out}}/v_{\text{in}} = 1$.

For the inverting amplifier of Fig. 9.10, the OAVR implies that $v_1 = v_2$, and v_2 is zero because the noninverting input is grounded. Thus, $v_1 = 0$; the inverting input is a "virtual" ground. The OACR implies that $i' = 0$, so $i_1 = i_2$. By Ohm's law, $i_1 = i_2$ becomes

$$\frac{v_1 - v_{\text{in}}}{R_1} = \frac{v_{\text{out}} - v_1}{R_2} \quad \text{or} \quad \frac{-v_{\text{in}}}{R_1} = \frac{v_{\text{out}}}{R_2}$$

Thus, the voltage gain is

$$A_v = \frac{v_{\text{out}}}{v_{\text{in}}} = -\frac{R_2}{R_1}$$

which is (9.12).

9.8 OP AMP PARAMETERS

We will now discuss a number of op amp parameters that determine the circuit gain, frequency response, impedance levels, and stability.

9.8.1 Gain and Bandwidth

The open-loop gain A_o of an op amp is extremely high at dc and low frequencies ($10^4 = 80\,\text{dB}$ to $10^6 = 120\,\text{dB}$) but decreases with increasing frequency. A typical graph is shown in Fig. 9.11 with the gain flat from dc

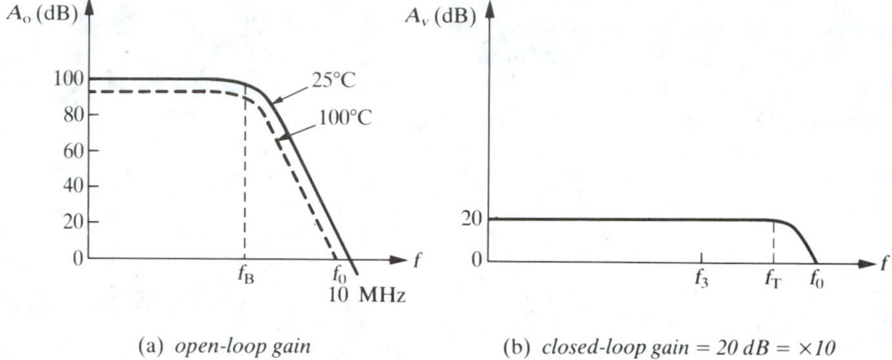

(a) *open-loop gain* (b) *closed-loop gain = 20 dB = ×10*

FIGURE 9.11 Op amp gain versus frequency.

to f_B and then decreasing linearly with frequency at 20 dB/decade. Negative feedback, of course, always reduces the gain, and the resulting gain of the op amp circuit can never be larger than the open loop gain. Thus, the gain of the circuit with negative feedback is basically the open-loop gain "lopped off" to a lower gain as shown in Fig. 9.11(b).

The gain also depends upon the temperature (decreasing with increasing temperature), and upon the supply voltage (increasing with increasing supply voltage). The supply voltage greatly affects the gain (typically 50% less for a 5-V decrease in V_{cc}), whereas the temperature effect is usually smaller. The manufacturer's specification sheets should be consulted for the details.

The gain of the inverting and the noninverting amplifiers can be rewritten in a form useful in evaluating how accurate the approximate gain expressions (9.10) and (9.12) are. The point is that as the frequency increases, the open-loop gain decreases steadily, which makes the approximations $A_oR_1 \gg R_2$ or $A_oB \gg 1$ or $A_o \gg R_2/R_1 = A_v$ less and less accurate. The gain expressions are usually accurate for "low" frequencies (below f_B) where the open-loop gain is large.

For the noninverting amplifier the exact closed-loop gain (9.9) can be rewritten as

$$A_v = \frac{1 + \dfrac{R_2}{R_1}}{1 + \dfrac{1 + R_2/R_1}{A_o}} \cong 1 + \frac{R_2}{R_1} \tag{9.13}$$

For the inverting amplifier the exact closed-loop gain (9.11) can be rewritten as

$$A_v = \frac{-\dfrac{R_2}{R_1}}{1 + \dfrac{1 + R_2/R_1}{A_o}} \cong -\frac{R_2}{R_1} \tag{9.14}$$

The conclusion we draw is that the gain error, that is, the difference between the approximate expressions (9.10) and (9.12) and the exact expressions (9.13) and (9.14), depends on $(1 + R_2/R_1)/A_o$: the smaller $(1 + R_2/R_1)/A_o$, the less the gain error. In other words, for a given open-loop gain A_o, the more negative feedback (lower gain) we have, the better. For example, at frequencies below f_B for the op amp of Fig. 9.11, the open-loop gain $A_o = 100\,\text{dB} = 10^5$, and if the desired gain of a noninverting op amp circuit $1 + R_2/R_1 = 20\,\text{dB} = 10$, the exact actual gain will be, from (9.13),

$$A_v = \frac{10}{1 + \dfrac{10}{10^5}} = 9.999$$

which is only 0.01% less than the approximate gain $1 + R_2/R_1 = 10.000$.

However, for higher frequencies, A_o decreases substantially. It usually decreases at a rate of 20 dB/decade or 6 dB/octave; that is, it varies as $1/f$. This is the same variation with frequency as that of the low-pass RC filter discussed in Chapter 2. For frequencies well above the break frequency f_B there is a 90° phase shift in the open-loop gain with respect to the open-loop gain below f_B (the dc open-loop gain). For this frequency range the exact gain expression for the noninverting amplifier is

$$|A_v| = \frac{A_o\left(1 + \dfrac{R_2}{R_1}\right)}{\left[A_o^2 + \left(1 + \dfrac{R_2}{R_1}\right)^2\right]^{1/2}} \tag{9.15}$$

The exact gain expression for the inverting amplifier is

$$|A_v| = \frac{A_o\dfrac{R_2}{R_1}}{\left[A_o^2 + \left(1 + \dfrac{R_2}{R_1}\right)^2\right]^{1/2}} \tag{9.16}$$

For example, if at $f = 1\,\text{MHz} > f_B$, A_o has decreased to 30 dB (×31.6), then if $R_2/R_1 = 10$, from (9.16), the actual gain for an inverting amplifier is

$$A_v = \frac{(31.6)(10)}{[(31.6)^2 + (11)^2]^{1/2}} = 9.44$$

The conclusion is that the open-loop gain A_o should be much larger than $1 + R_2/R_1$ to get the expected gain.

The *bandwidth* (BW) of an op amp can be defined in various ways. One way is to define it as the frequency at which the open-loop gain has fallen to unity or 0 dB (f_o in Fig. 9.11). For the op amp of Fig. 9.11, the bandwidth or gain bandwidth product would be 10 MHz. The open-loop gain is extremely difficult to measure, so the bandwidth is usually determined from a measurement of the rise time (R.T.) of a rectangular pulse fed into the op amp connected as a voltage follower, as shown in Fig. 9.12. The bandwidth

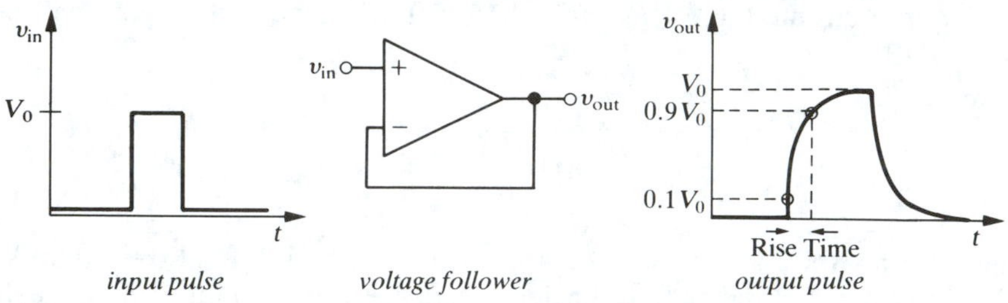

FIGURE 9.12 Rise time measurement.

is related to the rise time (10%–90%) of the output pulse by

$$\text{BW} = \frac{0.35}{\text{R.T.}} \tag{9.17}$$

For example, if the rise time were measured to be 0.50 μs, then the bandwidth would be $\text{BW} = 0.35/(0.50 \times 10^{-6}\text{ s}) = 7 \times 10^5\text{ Hz} = 0.7\text{ MHz}$. Of course, the rise time of the input pulse must be short compared to the op amp rise time, and the oscilloscope used to measure the output rise time must be fast enough. Also, a *small* input pulse must be used for this measurement because the frequency response is also limited by the *slew rate* of the op amp (see Section 9.8.3). For large-amplitude pulses (volts) the maximum slew rate of the amplifier is usually the limiting factor.

Notice that if the open-loop gain A_o decreases with frequency as $1/f$ (which is true for most op amps), then the product of the gain with feedback A_v and the frequency [f_T in Fig. 9.11(b)], where A_v meets the open-loop gain, is constant for any amount of negative feedback, that is, for any circuit gain. In other words, the gain bandwidth product is constant. We can have a high gain and a small bandwidth or a low gain and a large bandwidth. For example, if the gain bandwidth is 10 MHz, we could have a gain of 1.0 and a bandwidth of 10 MHz or a gain of 4.0 and a bandwidth of 2.5 MHz.

9.8.2 Op Amp Compensation

For any amplifier as the frequency increases, the gain decreases and the phase shift between the input and the output increases. The rate at which the gain decreases with frequency determines the ultimate phase shift, 6 dB/octave (20 dB/decade) for 90° and 12 dB/octave (40 dB/decade) for 180°, and so on, just like the *RC* filters in Chapter 2. Because of the phase shift in the amplifier gain, a zero-phase-shift feedback network (two resistors, for example) between the output and the *inverting* input may result in *positive* feedback because of a 180° shift from a 12 dB/octave rolloff in the amplifier gain and another 180° shift from the *inverting* op amp input. If the gain is high enough at the frequency where the feedback is positive, the amplifier will oscillate. The point of this section is to show how to prevent op amps from oscillating.

We recall from Chapter 8 that the overall gain of a negative feedback amplifier is

$$A_v = \frac{A_o}{1 + A_o B} \tag{9.18}$$

which reduces to $A_v = 1/B$ if $A_o B \gg 1$ or $A_o \ll 1/B$. B is the fraction of the output voltage fed back and subtracted from the input signal. In op amps negative feedback is achieved by taking a fraction of the output and applying it to the *inverting* input terminal, but this assumes there is no phase shift in the amplifier gain. Oscillations will occur if $A_o B = -1$, because this makes the gain infinite, i.e., a finite output for zero input. The minus sign simply represents another 180° phase shift, which changes the negative feedback into positive feedback. In op amps the extra 180° of phase shift can come from phase shifts inside the op amp, which become larger at higher frequencies. To sum up, there are two conditions for oscillation: (1) positive feedback (the feedback signal being in phase with the input), and (2) enough gain so that $|A_o B| \geqq 1$.

The gain of a simple *RC* low-pass filter of Fig. 9.13 is

$$A_v = \frac{v_{\text{out}}}{v_{\text{in}}} = \frac{1}{1 + j\omega RC} \quad \text{or} \quad |A_v| = \frac{1}{[1 + \omega^2 R^2 C^2]^{1/2}}$$

The gain falls off as $1/\omega$, that is, at 6 dB/octave or 20 dB/decade. The phase shift of the low-pass filter is $\phi = \tan^{-1} \omega RC$. The filter output lags the input by ϕ, and $\phi \to 90°$ as $\omega \gg 1/RC$. Notice that the phase shift cannot exceed

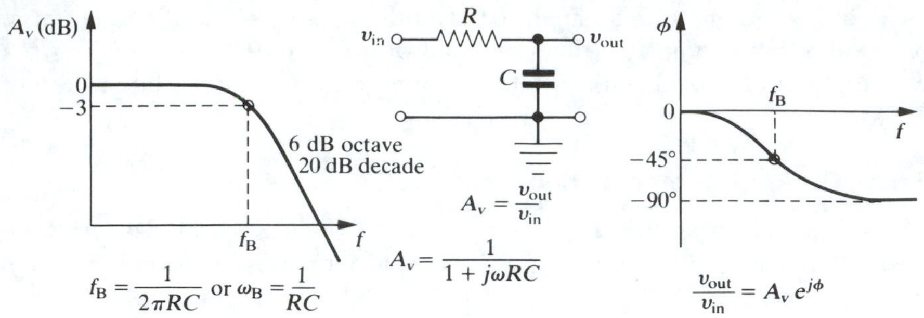

FIGURE 9.13 Simple low-pass *RC* filter.

90° for any frequency. In general, the more rapidly the gain falls off with increasing frequency, the greater the phase shift. For example, for two *RC* low-pass filters in series, the gain falls off as $1/\omega^2$ and the phase shift approaches 180° as $\omega \gg 1/RC$. Every additional low-pass *RC* filter will introduce an additional 90° of phase shift at high frequencies.

An op amp generally has a number of low-pass *RC* filters inherent in its circuitry, and the open-loop gain and phase shifts typically look like Figs. 9.14(a) and 9.14(b).

Consider two amplifiers [Figs. 9.14(c) and 9.14(d)] constructed with the op amp whose open-loop gain versus frequency is that shown in Fig. 9.14(a). The first amplifier has a gain of $A_{v1} = 100 = 40$ dB ($B_1 = 0.01$); the second amplifier, $A_{v2} = 10 = 20$ dB ($B_2 = 0.1$). The amplifier gains A_v are drawn as dashed lines in Fig. 9.14(a). The first amplifier with a gain of 100 will not oscillate, whereas the second amplifier with a gain of 10 will oscillate, as the following argument shows.

For the first amplifier, $A_{v1} = 100$ or 40 dB, and thus $B_1 = 0.01$. From Fig. 9.14(a), we see that the open-loop op amp gain has fallen to 40 dB at ω_1; in other words, $A_o = 1/B_1$ at ω_1. Thus for $\omega \leqq \omega_1$, $A_oB_1 \geqq 1$. But the gain-versus-frequency curve is falling off at only 6 dB/octave or less for $\omega \leqq \omega_1$, so the amplifier phase shift must be 90° or less. Thus, no oscillation is possible because at ω_1, where $A_{oB1} = 1$, the phase shift is 90° or less. But for amplifier 2, $A_{v2} = 20$ dB and $B_2 = 0.1$. The open-loop op amp gain has fallen to 20 dB at ω_2, so $A_o = 1/B_2$ at ω_2. Thus, for $\omega \leqq \omega_2$, $A_oB_2 \geqq 1$. But for $\omega \geqq \omega_2$, the phase shift in A_o is 180° because of the 12-dB/octave gain versus frequency, so near ω_2 we have the two conditions for oscillation: positive feedback, and $A_oB_2 = 1$. Thus, we see that the important point in determining whether the op amp circuit will oscillate is where the straight-line extrapolation of the $A_v = 1/B$ gain curve meets the A_o open-loop op amp curve. If the intersection occurs where the A_o slope is less than 12 dB/octave, then no oscillations will result. If the intersection occurs where the slope is 12 dB/octave or greater, the circuit will oscillate. Notice

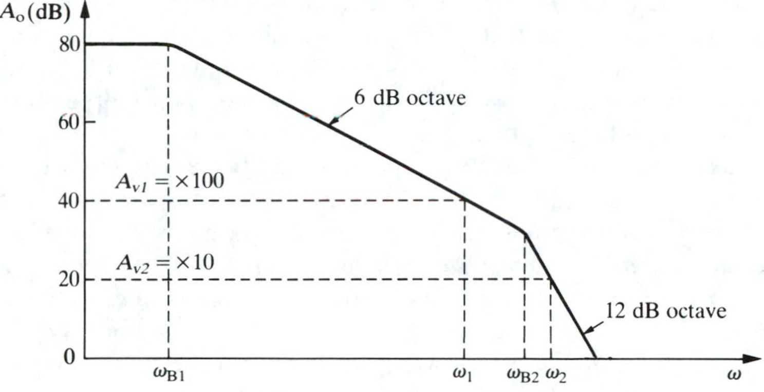

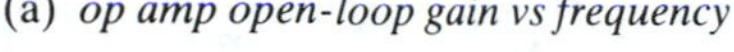

(a) *op amp open-loop gain vs frequency*

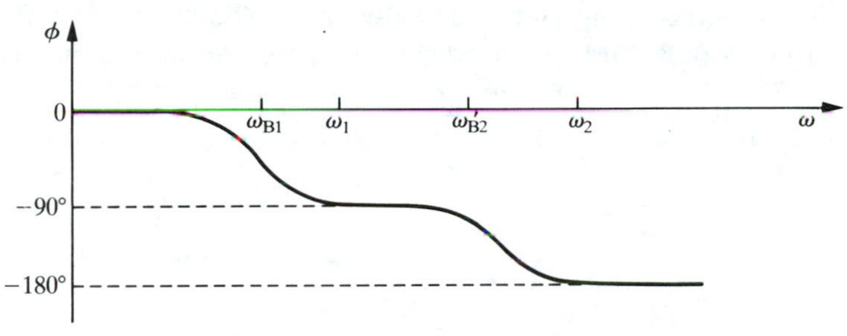

(b) *op amp phase shift vs frequency*

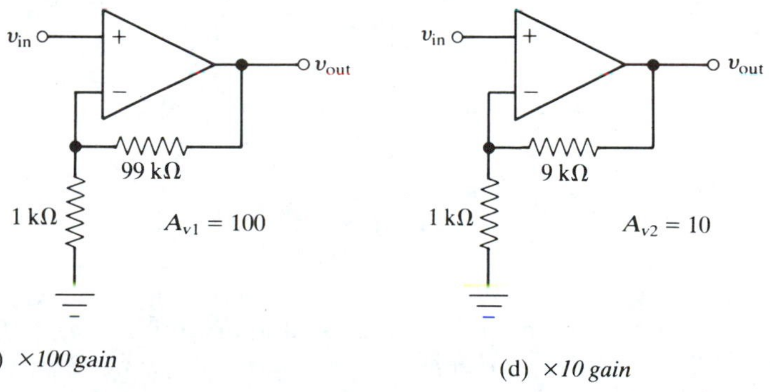

(c) *×100 gain*

(d) *×10 gain*

FIGURE 9.14 Op amp gain and phase shift.

that the *lower* A_v is, the more the circuit is apt to oscillate. Thus the voltage follower is more apt to oscillate than is an amplifier with a higher gain.

We can therefore conclude that the way to make an op amp unconditionally stable against oscillation is to have the open-loop gain A_o fall off at a rate of only 6 dB/octave where the circuit gain A_v intersects it. Many op amps contain an internal capacitor that forces the open-loop gain to fall off at 6 dB/octave down to 0 dB (a gain of unity). Such op amps are unconditionally stable against oscillation for circuit gains A_v down to 0 dB and are called *internally compensated* op amps. They can be used for any A_v gain greater than or equal to unity without oscillation as long as there is no extra phase shift in the feedback network. The 741 is one of the most popular internally compensated bipolar op amps, but there are many others.

A second type of op amp is called *uncompensated* and has an open-loop gain curve typically like that of Fig. 9.15. External compensation capacitors can be added to roll off the gain with a break frequency determined by the capacitor—the larger the capacitance, the lower the break frequency. The manufacturer's spec sheet should be consulted for details, but it should be clear that a larger bandwidth can be obtained with an uncompensated op amp and a carefully chosen compensation capacitor. Figure 9.15 illustrates this. The bandwidth for the internally compensated op

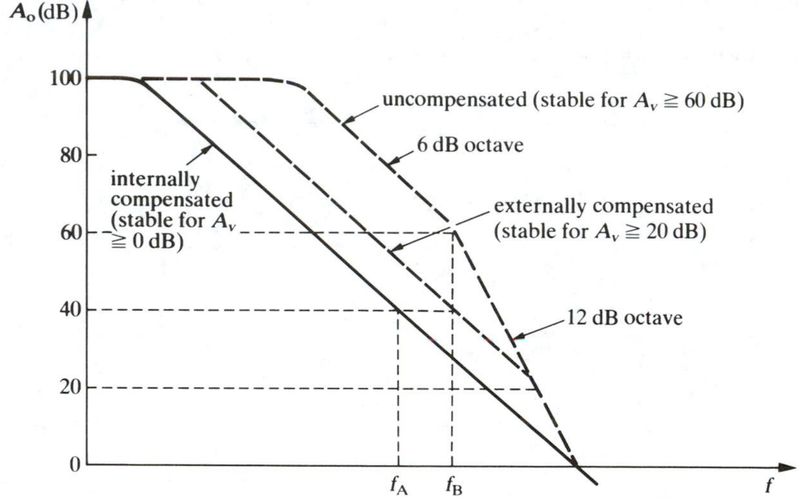

FIGURE 9.15 Op amp gain curves.

amp of gain $A_v = 40$ dB would be f_A, whereas the bandwidth for the externally compensated amplifier with equal gain would be $f_B > f_A$. Careful choice of the external compensation components can also increase the large-signal frequency response.

TABLE 9.2. Compensated and Equivalent Uncompensated Op Amps

	Compensated	*Uncompensated Version*
GP[†] bipolar	741	748 ($A_v > 5$)
GP bipolar	307	301
"Super beta" (low input bias current)	308	312
MOSFET	3160	3130
High voltage	343	344

[†]GP = general purpose.

Many widely used compensated op amps are available in an uncompensated or decompensated version. A few are listed in Table 9.2.

9.8.3 Slew Rate

Slew rate (S.R.) is the maximum rate at which the output voltage of an op amp can change; it is usually expressed in volts per microsecond. This limitation of op amps comes from the constant current charging of a capacitor, either inside the op amp (a compensation capacitor) or a load capacitor. In either case the rate at which the voltage across the capacitor changes is

$$\frac{dV}{dt} = \frac{d}{dt}\left(\frac{Q}{C}\right) = \frac{1}{C}\frac{dQ}{dt} = \frac{I}{C}$$

The capacitance C is fixed, so the maximum rate at which the capacitor voltage can change is limited by the maximum charging current I, which depends upon the op amp. There is always some maximum value of the charging current due to the limitations of the circuitry, so there is an upper limit on dV/dt; this upper limit is called the *slew rate* of the op amp. It ranges from typically $0.5\ \mathrm{V}/\mu\mathrm{s}$ for inexpensive op amps to $500\ \mathrm{V}/\mu\mathrm{s}$ for special high-speed op amps.

If a sinusoidal voltage is amplified, the voltage changes most rapidly with respect to time as the waveform passes through zero amplitude. The rate of change of voltage is

$$\frac{dV}{dt} = \frac{d(V_0 \sin \omega t)}{dt} = \omega V_0 \cos \omega t$$

Thus, the maximum rate of change of voltage is

$$\left(\frac{dV}{dt}\right)_{\max} = \omega V_0$$

The important point here is that the maximum rate of change of voltage depends on both the amplitude *and* the frequency. Letting S.R. (slew rate) stand for $(dV/dt)_{max}$, we have

$$\text{S.R.} = \omega V_0 = 2\pi f V_0 \qquad \textbf{(9.19)}$$

For example, if an op amp has a slew rate of 4 V/μs, then the maximum frequency at which the op amp output will be free of slew rate distortion, for a 1-V peak-to-peak output ($V_0 = 0.5$ V), will be

$$f = \frac{\text{S.R.}}{2\pi V_0} = \frac{4\ \text{V}/\mu\text{s}}{(2\pi)(0.5\ \text{V})} = 1.3\ \text{MHz}$$

Notice that as the amplitude of the op amp output increases, the frequency for no slew rate distortion decreases.

Slew rate distortion thus first shows up on the zero crossings of sinusoidal waveforms; the op amp simply can't change its output voltage fast enough to faithfully reproduce the sinusoidal waveform at its steepest points where it crosses zero. If the frequency and the amplitude of the input signal continue to increase, the output waveform will become triangular, as shown in Fig. 9.16.

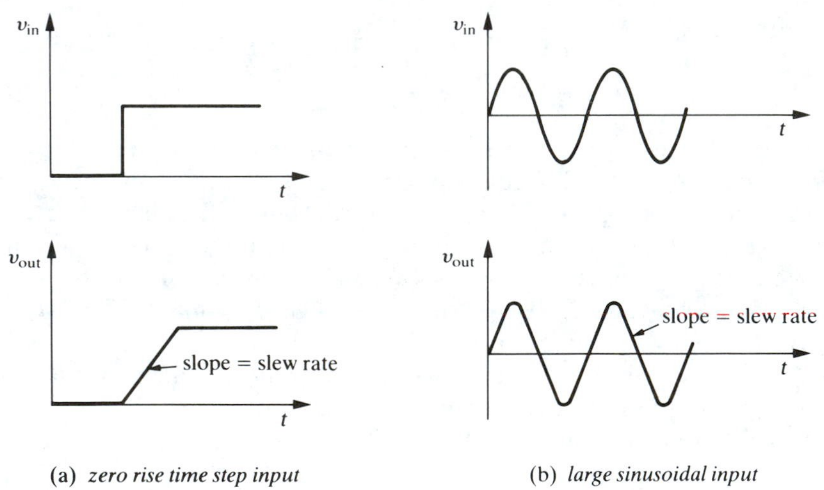

FIGURE 9.16 Slew rate distortion of sinusoidal input.

If a higher slew rate is required, an op amp with a higher slew rate can be obtained, or if the limiting slew rate is due to the charging of the load capacitance, an output buffer amplifier can be used to supply a higher charging current to the load, that is, a lower output impedance. In some op amps the slew rate can be increased by the cautious use of positive feedback (*feed forward compensation*). With the 318 op amp, for example, the slew rate can be increased from 70 V/μs to approximately 200 V/μs. Too much positive feedback will, of course, result in continuous oscillation. The spec sheets should be consulted for details.

In general, the slew rate is proportional to the gain bandwidth product for the op amp; for internally compensated op amps, we can do nothing except get an op amp with a higher slew rate. But for externally compensated op amps, we can try using a smaller compensation capacitance at the risk of producing unwanted oscillations. The manufacturer's data sheet should be consulted for details of the external compensation.

9.8.4 Input Bias Current

In a bipolar op amp both input stage transistors must be forward biased, and thus some dc input current must flow into each input terminal. Thus there *must* be a dc path to ground for both inputs in *any* op amp circuit. This input bias current is typically from 10 to 100 nA for bipolar op amps and from 0.001 nA (1 pA) to 0.50 nA (500 pA) for JFET op amps; it is much lower for MOSFET op amps—approximately 1 to 10 pA. The input FETs are, of course, reverse biased; the gate is reverse biased with respect to the channel, so the input JFET bias currents are merely the reverse currents in the gate-channel reverse biased junctions. In the MOSFETs the input bias currents are the leakage currents through the SiO_2 insulating layers. These bias currents are all functions of temperature; for increasing temperature they decrease for bipolar op amps and increase for FET op amps. Typically, the bias current may decrease from 100 to 50 nA as T increases from 25° to 100°C for bipolar op amps, and it may double for every 10°C temperature rise for FET op amps.

A word of caution is in order with regard to op amp bias current specifications. If the manufacturer specifies the bias current at a *junction* temperature of 25°C, this means the junctions in the actual chip must be at 25°C. With power applied the chip junction temperature is *always higher* than ambient (room) temperature because of the i^2R "joule heat" dissipated in the chip. The actual junction temperature depends on the ambient temperature, the power P dissipated in the chip, and the thermal coupling between the chip and the outside environment (i.e., the type of package, type of heat sink, etc.). For a typical eight-pin plastic-case IC the thermal resistance $\Theta \cong 150°\text{C/W}$, so if $I = 10\text{ mA}$ from ±15-V power supplies, the junction temperature T_J will be

$$\begin{aligned} T_J &= T_A + \Theta P \\ &= 25°\text{C} + (150°\text{C/W})(0.01\text{ A})(30\text{ V}) = 70°\text{C} \end{aligned}$$

Thus the bias current will be determined by the 70°C junction temperature not the 25°C ambient temperature. For a bipolar device the 70°C bias current will be approximately 50% less than the 25°C bias current, and for an FET device it will be approximately 30 times larger.

Because the two input transistors differ slightly in an actual op amp, the input bias current I_B is defined as the *average* value of the two bias currents. The input *offset* current I_{io} is defined as the *difference* between the two bias currents and is typically 1–20 nA for bipolar op amps and much less for FET op amps (1–10 pA). I_B is always larger than I_{io}.

Generally, the most serious cause of a nonzero output voltage with a zero input voltage is that the bias currents in the two inputs of the op amp flow through different resistances, thus producing a differential input voltage. We will now show, for the inverting amplifier circuit, that the output voltage with zero external input equals the bias current of the op amp times the feedback resistance R_2 plus the usual $-(R_2/R_1)v_{in}$ term. In Fig. 9.17(a)

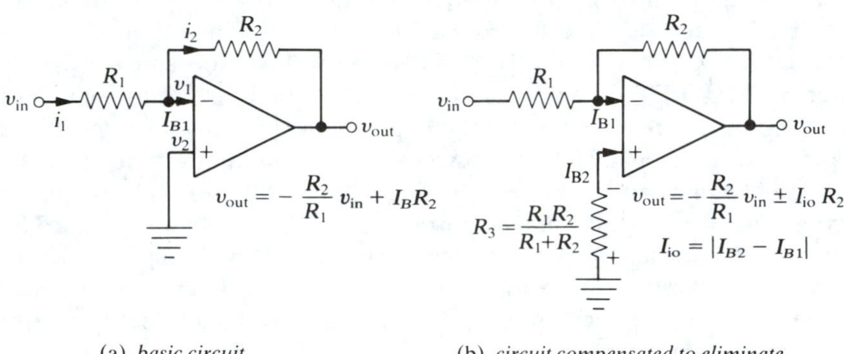

FIGURE 9.17 Bias current effects in the inverting op amp circuit.

the Kirchhoff current law at the inverting input implies

$$i_1 = I_{B1} + i_2 \tag{1}$$

Using Ohm's law (1) becomes

$$\frac{v_{in} - v_1}{R_1} = I_{B1} + \frac{v_1 - v_{out}}{R_2} \tag{2}$$

But the OAVR implies $v_1 = v_2$ and $v_2 = 0$. Thus (2) becomes

$$\frac{v_{in}}{R_1} = I_{B1} - \frac{v_{out}}{R_2} \tag{3}$$

Solving (3) for the output voltage, we obtain

$$v_{out} = -\left(\frac{R_2}{R_1}\right) v_{in} + I_{B1}R_2 \qquad \textbf{(9.20)}$$

Thus the output voltage has a dc shift equal to $I_{B1}R_2$; this shift is independent of the input voltage. The output (9.20) can also be written as

$$v_{out} = -\frac{R_2}{R_1}[v_{in} - I_{B1}R_1] \qquad \textbf{(9.21)}$$

which tells us that the bias current term is negligible only if $I_{B1}R_1 \ll v_{in}$.

In an ideal op amp $I_{B1} = 0$, and (9.20) reduces to the gain expression for an inverting op amp. Because I_{B1} is so small (especially for FET op amps), the $I_{B1}R_2$ term is often negligible unless R_2 is extremely large (megohms or more). For example, if $R_2 = 100\,\text{k}\Omega$ and $I_{B1} = 100\,\text{nA}$, the $I_{B1}R_2$ term equals $(10^{-7}\,\text{A})(10^5\,\Omega) = 0.01\,\text{V}$.

To eliminate the bias current term $I_{B1}R_2$ in the output voltage, we must add a voltage at the noninverting terminal such that this voltage times the circuit voltage gain produces an output voltage that cancels the $I_{B1}R_2$ term. Because a bias current of $I_{B2} \cong I_{B1}$ will also flow into the noninverting terminal, we simply place a resistance R_3 in series with the noninverting terminal, as shown in Fig. 9.17(b). Then I_{B2} flowing through R_3 will produce a voltage $v_2 = -I_{B2}R_3$ at its noninverting terminal. To cancel the $I_{B2}R_2$ term in the output voltage, we require

$$v_2 A_v + I_{B1}R_2 = 0$$

But the circuit gain A_v for a voltage at the noninverting input voltage is $A_v = 1 + R_2/R_1$. Using $v_2 = -I_{B2}R_3$, we have

$$-I_{B2}R_3\left(1 + \frac{R_2}{R_1}\right) + I_{B1}R_2 = 0$$

Solving for R_3, we obtain (assuming $I_{B1} = I_{B2}$)

$$R_3 = \frac{R_1R_2}{R_1 + R_2} \qquad \textbf{(9.22)}$$

With R_3 equal to the parallel combination of R_1 and R_2, the output voltage should not depend on I_B at all. It is common practice to add R_3 to inverting op amp circuits at the first design stage.

Because the bias currents in the two op amp input terminals are never *exactly* equal, R_3 is often made adjustable. The difference in the two bias currents is called the *input bias offset current* I_{io}, and the output with R_3 is

$$v_{out} = -\left(\frac{R_2}{R_1}\right) v_{in} \pm I_{io} R_2 \tag{9.23}$$

Typically for bipolar op amps $I_B \cong 60$ nA and $I_{io} \cong 10$ nA, whereas for JFET input op amps $I_B \cong 30$ pA and $I_{io} \cong 5$ pA. Thus we see that adding R_3 replaces the $I_B R_2$ term by the much smaller $I_{io} R_2$ term.

For the noninverting circuit shown in Fig. 9.18, the Kirchhoff current

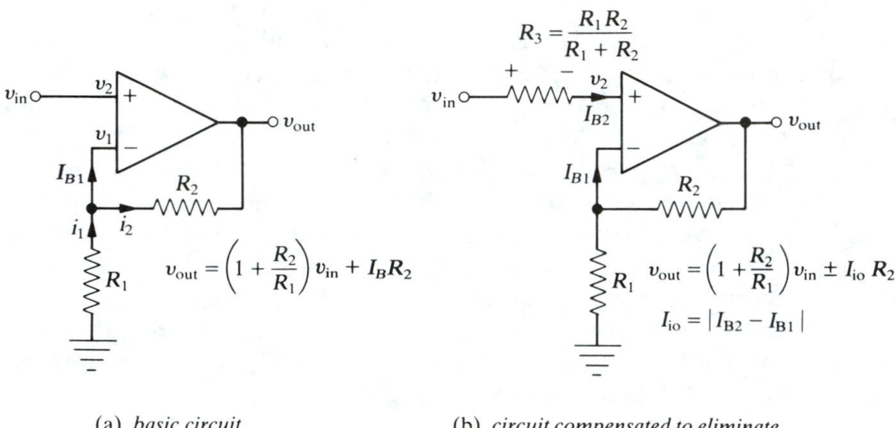

(a) *basic circuit*

(b) *circuit compensated to eliminate bias current term in output*

FIGURE 9.18 Bias current effects in the noninverting op amp circuit.

law at the inverting terminal implies

$$i_1 = I_{B1} + i_2 \tag{1}$$

and Ohm's law implies

$$\frac{-v_1}{R_1} = I_{B1} + \frac{v_1 - v_{out}}{R_2} \tag{2}$$

But the OVAR implies $v_1 = v_2$ and $v_2 = v_{in}$. Thus

$$-\frac{v_{in}}{R_1} = I_{B1} + \frac{v_{in} - v_{out}}{R_2} \tag{3}$$

Solving (3) for v_{out} yields

$$v_{out} = \left(1 + \frac{R_2}{R_1}\right) v_{in} + I_{B1} R_2 \tag{9.24}$$

which is very similar to (9.20) for the inverting circuit.

Thus we see that the bias current term in the output voltage is $I_{B1}R_2$ for the noninverting amplifier as well as for the inverting amplifier. To cancel out the $I_{B1}R_2$ term in v_{out}, we add a resistance R_3 in series with the noninverting terminal, as shown in Fig. 9.18(b). Then the bias current in the noninverting terminal is

$$I_{B2} = \frac{v_{in} - v_2}{R_3} \quad \text{or} \quad v_2 = v_{in} - I_{B2} R_3$$

In other words, the input voltage at the op amp has been decreased by $I_{B2}R_3$. The circuit gain for a voltage at the noninverting input is $A_v = 1 + R_2/R_1$. Thus to cancel out the $I_B R_2$ term in the output in (9.24), we require.

$$A_v(-I_{B2}R_3) + I_{B1}R_2 = 0$$

or

$$\left(1 + \frac{R_2}{R_1}\right)(-I_{B2}R_3) + I_{B1}R_2 = 0$$

Solving for R_3, we obtain (assuming $I_{B2} = I_{B1}$)

$$R_3 = \frac{R_1 R_2}{R_1 + R_2}$$

which is the same resistance required for the inverting amplifier.

To sum up, in both the inverting and the noninverting amplifier circuits the $I_B R_2$ term in the output is replaced by $I_{io} R_2$ if we add a resistance $R_3 = R_1 R_2/(R_1 + R_2)$ in series with the noninverting input.

9.8.5 Input Offset Voltage

The two input transistors in an op amp never have exactly the same base emitter characteristics, so even with both inputs connected together, there generally will be a nonzero output voltage. We can think of this output voltage as being caused by a small voltage V_{io} in series with one of the inputs shown in Fig. 9.19(a). This input voltage is called the *input offset voltage* and typically ranges from 0.1 mV to several mV and changes with temperature with a typical temperature coefficient of approximately 5 μV/°C. The input offset voltage problem exists for *any* type of op amp circuit.

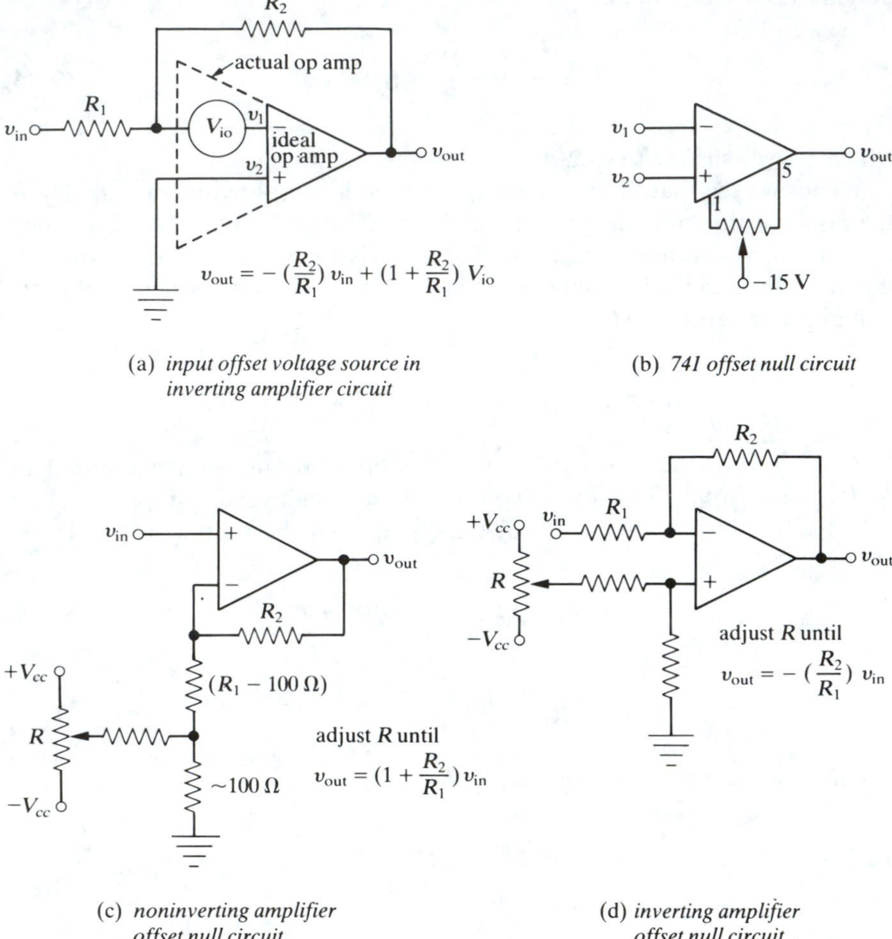

FIGURE 9.19 Input offset voltage effects.

The output voltage due to the input offset voltage for either the noninverting or the inverting amplifier circuits can be shown to be (with $v_{in} = 0$)

$$v_{out} = \pm \left(1 + \frac{R_2}{R_1}\right) V_{io} \tag{9.25}$$

Thus, for example, in a noninverting amplifier with a gain of 100 and an input offset voltage of 4 mV, the output voltage with zero input would be

$$v_{out} = (100)(4 \times 10^{-3}\text{ V}) = 0.4\text{ V}$$

Notice that if the gain of the circuit is increased, the output voltage will be increased. Thus, offset voltage problems are worse in high-gain amplifiers than in low-gain amplifiers.

An offset null circuit is usually suggested by the op amp manufacturers to cancel the V_{io} term in the output; one such circuit for the popular 741 op amp is shown in Fig. 9.19(b). The potentiometer is adjusted with zero input until the output is "nulled" or set to zero.

If the op amp does not have offset null terminals on the chip, the offset null circuits of Figs. 9.19(c) and 9.19(d) can be used. The general idea is to add a small adjustable dc voltage of either polarity to the op amp input terminal that is not connected to the signal input in order to zero the dc output.

Finally, even when the appropriate offset null circuitry has been used to set the output dc voltage to zero with $v_{in} = 0$, the output dc voltage will still vary with temperature because the (input) offset voltage is temperature dependent with a typical temperature coefficient of 5 μV/°C. We summarize the effects of the input bias current I_B, the input offset current I_{io}, and the input offset voltage V_{io} as follows:

For the inverting amplifier of Fig. 9.17(a) the output is

$$v_{out} = -\left(\frac{R_2}{R_1}\right) v_{in} \pm \left(1 + \frac{R_2}{R_1}\right) V_{io} + I_B R_2 \qquad \textbf{(9.26)}$$

With the addition of $R_3 = R_1 R_2/(R_1 + R_2)$ in series with the noninverting input as shown in Fig. 9.17(b), the output is

$$v_{out} = -\left(\frac{R_2}{R_1}\right) v_{in} \pm \left(1 + \frac{R_2}{R_1}\right) V_{io} \pm I_{io} R_2 \qquad \textbf{(9.27)}$$

For the noninverting amplifier of Fig. 9.18(a), the output is

$$v_{out} = \left(1 + \frac{R_2}{R_1}\right) v_{in} \pm \left(1 + \frac{R_2}{R_1}\right) V_{io} + I_B R_2 \qquad \textbf{(9.28)}$$

With the addition of R_3 as shown in Fig. 9.18(b) the output is

$$v_{out} = \left(1 + \frac{R_2}{R_1}\right) v_{in} \pm \left(1 + \frac{R_2}{R_1}\right) V_{io} \pm I_{io} R_2 \qquad \textbf{(9.29)}$$

The input offset voltage and the input bias and offset currents are all temperature dependent. The input offset voltage can always be nulled out at any fixed temperature by the circuits of Fig. 9.19. The resistance $R_3 = R_1 R_2/(R_1 + R_2)$ can be adjusted to null out the offset $I_{io} R_2$ current term at any fixed temperature.

We make a final comment on the inverting amplifier versus the noninverting amplifier. The inverting amplifier generally has a lower input impedance because its inverting input is a virtual ground, so $R_{in} \cong R_1$. But it is not as susceptible to pickup because of its lower R_{in}. The noninverting amplifier has a much higher (and temperature-dependent) input resistance, but because of this it is much more susceptible to pickup. Unless a high input impedance is absolutely necessary—for example, for a high-impedance input voltage source—the inverting amplifier is probably the better choice.

9.8.6 Input and Output Resistance

The equivalent circuit of an op amp with the input offset voltage and the input bias currents represented by voltage and current generators is shown in Fig. 9.20. R_{id} is the input resistance between the two input terminals (the

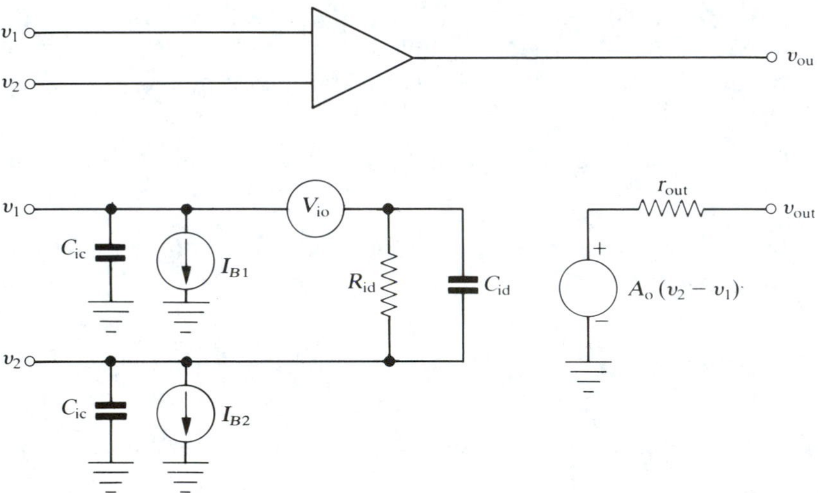

FIGURE 9.20 Op amp equivalent circuit showing input and output resistance.

differential input resistance). We have assumed that the resistance between the two input terminals and ground is infinite. R_{id} is typically 1 MΩ for inexpensive bipolar op amps, but it could be as high as 100 MΩ for special bipolar op amps whose input stages contain high-beta transistors that operate on collector currents of approximately 1 μA. For FET op amps R_{id} might be as high as 10^{12} Ω. There is also a small capacitance C_{id} of approximately 1–3 pF in parallel with R_{id} and a small capacitance C_{ic} between each input terminal and ground. C_{ic} is also several pF.

The input resistance will reduce the gain by a voltage divider action with the input circuitry. The effect of the input impedance is to add a term

R_2/A_oR_{id} to the denominator of the voltage gain expression. For example, the noninverting op amp circuit gain (9.9) becomes

$$A_v = \frac{1 + \dfrac{R_2}{R_1}}{1 + \dfrac{1}{A_o}\left(1 + \dfrac{R_2}{R_1}\right) + \dfrac{R_2}{A_oR_{id}}}$$

Usually this effect of the input resistance term on the gain is small, unless the open-loop gain A_o is rather low, which would be the case at high frequencies.

As we have seen in Chapter 8, negative feedback has a profound effect on input and output resistances, and op amps are invariably used with negative feedback. The effect of the negative feedback is to raise the input resistance and to lower the output resistance. For series negative voltage feedback (which is used in the noninverting amplifier circuit of Fig. 9.8), we recall that

$$R_{in} = R_{ino}(1 + A_oB)$$

where R_{ino} is the input resistance with no negative feedback. Thus, we immediately see that the input resistance of the noninverting amplifier is extremely large, probably 100 MΩ (assuming $R_{ino} = 1$ MΩ, $A_oB = 100$) or even higher.

However, for the inverting amplifier circuit of Fig. 9.10, the situation is entirely different. First, the negative feedback is in parallel, not in series, and its effect is basically to null the voltage at the inverting input to the same value as at the noninverting input. If the noninverting input is grounded, as it is for the simple circuit of Fig. 9.10, then the voltage at the inverting input, v_1, is nulled to almost zero. Thus, the input signal sees the resistance R_1 effectively connected to ground. We say the inverting input of the op amp is a ***virtual*** ground. Thus, the input resistance of the inverting amplifier circuit is equal to R_1, which is usually 1 to 10 kΩ in most circuits.

The output resistance r_{out} of the op amp with no negative feedback is called the *open-loop output resistance* and is shown in Fig. 9.21. The magnitude of r_{out} typically ranges from 10 Ω (for higher-power op amps) to 500 Ω (for lower-power op amps) and is approximately 100 Ω for a typical general-purpose op amp. It increases as the frequency increases.

Adding negative feedback, of course, makes the output impedance of the op amp circuit much lower, typically 1 Ω or less, depending on the circuit gain. We recall from negative feedback theory that the output resistance R_{out} with feedback is given by

$$R_{out} = \frac{r_{out}}{1 + A_oB} \cong \frac{r_{out}}{A_oB}$$

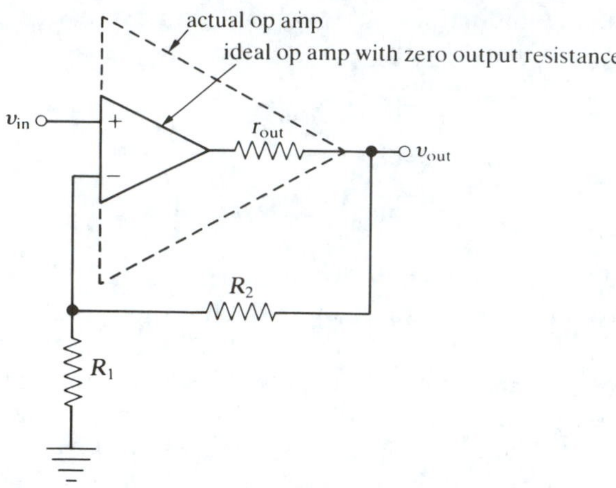

FIGURE 9.21 Op amp output resistance.

or, using $A_v = 1/B$,

$$R_{out} \cong \frac{A_v r_{out}}{A_o}$$

This result holds for inverting and noninverting amplifiers. A quick calculation shows that if $A_v = 10$, $r_{out} = 100\ \Omega$, and $A_o = 10^4 = 80$ dB (at low frequencies), then the amplifier circuit output resistance R_{out} at low frequencies is low indeed—$0.1\ \Omega$. However, if the frequency is high enough, the open-loop gain is drastically reduced and R_{out} is increased. For example, if A_o is only $100 = 40$ dB, R_{out} is $10\ \Omega$. A low output resistance, of course, is required when large currents have to be supplied to a load or when the load resistance is small; thus it is usually a problem only in high-power circuits. To avoid problems caused by an output resistance that is too high, we should use an op amp whose open-loop gain at the signal frequency is large enough (much larger than A_v, the desired gain), or else we should add a current booster circuit at the output of the op amp, as described in Chapter 10.

9.8.7 Common Mode Rejection Ratio (CMRR)

For an ideal op amp the output voltage is $v_{out} = A_o(v_2 - v_1)$, where v_2 and v_1 are the voltages *directly* at the op amp terminals. Thus, the *differential* voltage gain is high, but the gain is zero for equal input voltages at the noninverting and the inverting input terminals. For a real op amp equal voltages at the two input terminals $v_1 = v_2$ (a common mode input) produce a small voltage output. The gain for this situation is called the *common mode gain*, A_{cm}, and is zero for an ideal op amp. The *common mode rejection ratio* (CMRR) of an op amp is defined as the ratio of the differential gain to the common mode gain:

$$\text{CMRR} = \frac{A_{\text{diff}}}{A_{\text{cm}}} \tag{9.30}$$

The CMRR is infinite for an ideal op amp and is typically 70 to 100 dB for a general-purpose op amp at dc and low frequencies and tends to decrease with increasing frequency at a rate of 20 dB/decade. Thus the CMRR may be as low as 30 dB at several hundred kilohertz. A high CMRR is important when there is a large amount of pickup (usually 60 Hz) at the op amp input terminals. This is especially troublesome with circuits with high input resistance.

9.8.8 Power Supply Rejection Ratio (PSRR)

For an ideal op amp, changes in the power supply voltage will have no effect on the output, but for an actual op amp power supply voltage changes (usually 120-Hz ripple) will produce a nonzero 120-Hz ripple voltage at the output. The effect of the power supply voltage change ΔV_{cc} is referred to the effective change ΔV_{io} in the input offset voltage it produces; that is, the power supply rejection ratio is defined as

$$\text{PSRR} = \frac{\Delta V_{\text{io}}}{\Delta V_{cc}} \tag{9.31}$$

and is expressed in μV/V or in dB (1 μV/V = − 120 dB). An ideal op amp has negative infinite PSRR; a typical value for a general-purpose op amp is −70 to −100 dB. Since the output voltage (for the inverting or noninverting op amp circuits) equals $(1 + R_2/R_1)(V_{\text{io}})$, the output change due to power supply voltage fluctuation is

$$\Delta v_{\text{out}} = \left(1 + \frac{R_2}{R_1}\right)\Delta V_{\text{io}} = \left(1 + \frac{R_2}{R_1}\right)(\text{PSRR})(\Delta V_{cc}) \tag{9.32}$$

For example, an op amp with a PSRR = −80 dB = 10^{-4} = 100 μV/V in a noninverting circuit with gain $1 + R_2/R_1 = 1000$ hooked up to a power supply with a 5.0-mV_{rms} 120-Hz ripple would produce an output 120-Hz ripple of

$$\Delta v_{\text{out}} = (1000)(10^{-4})(5.0\ \text{mV}) = 0.5\ \text{mV}_{\text{rms}}$$

9.8.9 Noise in Op Amps

Electrical noise is generated in many locations in an op amp, but the total effective noise is assumed to be generated by a voltage noise source and a

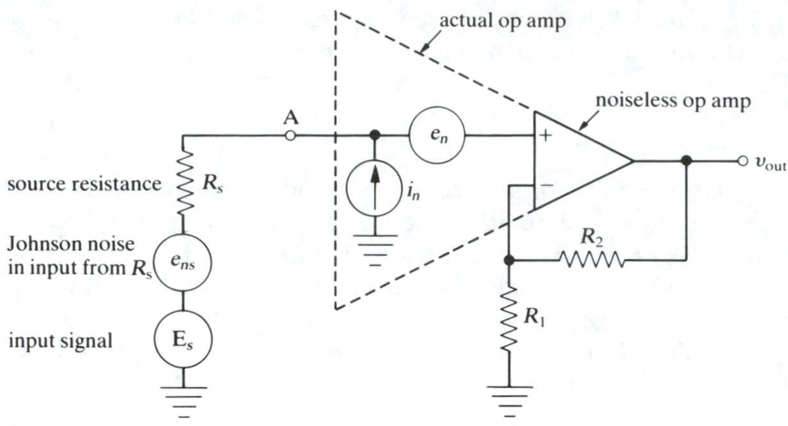

FIGURE 9.22 Op amp noise sources.

current noise source at the input to the (noiseless) op amp, as shown in Fig. 9.22 where we have neglected the Johnson noise generated by R_1 and R_2. e_n is called the *equivalent input noise voltage*, and i_n the *equivalent input noise current*. e_n and i_n are assumed to be ideal voltage and current sources, respectively: that is, e_n has zero internal resistance, and i_n has infinite internal resistance. In other words, e_n supplies a constant noise voltage regardless of what load impedance it sees, and i_n supplies a constant noise current regardless of its load impedance. e_{ns} is the noise voltage in the incoming signal, E_s is the signal voltage, and R_s the source resistance. By inspecting Fig. 9.22, we see that the noise current i_n will flow through the source resistance R_s (negligible noise current flows into the op amp because of its high input resistance), so the source resistance will partially determine the noise in the op amp output.

As explained in Chapter 8, the rms noise voltage produced by the noise source e_n depends on the bandwidth of the measuring instrument: the larger the bandwidth, the larger the rms noise voltage measured. The total mean-square noise voltage produced by the source e_n in a bandwidth B_W is

$$e_t^2 = e_n^2 B_W \quad \text{or} \quad e_t = e_n \sqrt{B_W} \tag{9.33}$$

e_n is thus measured in V/$\sqrt{\text{Hz}}$. Similarly, the total rms noise current is

$$i_t = i_n \sqrt{B_W} \tag{9.34}$$

The noise input voltage e_n can be measured by shorting the op amp input (point A in Fig. 9.22) to ground. This shorts the noise current i_n to ground so that the only voltage input to the op amp is e_n. Thus, if the op amp gain is A_v and its bandwidth is B_W, the output noise voltage v_{on} will be

$$v_{out} = e_n A_v \sqrt{B_W} \qquad \left(A_v = 1 + \frac{R_2}{R_1}\right)$$

which can be solved for the equivalent input noise voltage e_n:

$$e_n = \frac{v_{out}}{A_v \sqrt{B_W}}$$

To measure the equivalent input current noise i_n, we must allow the noise current to flow through some resistance to produce a voltage input to the op amp. If we assume that the op amp input resistance is large compared to the source resistance R_s, all the noise current will flow through R_s, thus producing an effective input noise voltage of $i_n R_s$. But there is a third noise source, due to R_s alone. From Chapter 8 any resistance R generates a random Johnson noise voltage in a bandwidth B_W of

$$e_{ns} = (4kTRB_W)^{1/2}$$

The resistor noise per unit bandwidth is thus $(4kTR)^{1/2}$ and is measured in $V/\sqrt{\text{Hz}}$.

It can be shown that for noncorrelated (independent) noise voltage sources in series the voltages add as the square root of the sum of the squares, basically because there is some cancellation when a positive noise voltage from one noise source occurs at the same time as a negative noise voltage from the other noise source. Thus, the total input noise voltage to the op amp e_{nt} is the square root of the sum of the squares of all three noise voltages: e_n, $i_n R_s$, and e_{ns}:

$$e_{nt} = e_n^2 + (i_n R_s)^2 + e_{ns}^2 = [e_n^2 + (i_n R_s)^2 + 4kTR_s]^{1/2} \tag{9.35}$$

The op amp output noise voltage is then

$$v_{out} = e_{nt} A_v \sqrt{B_W} = [e_n^2 + (i_n R_s)^2 + 4kTR_s]^{1/2} A_v \sqrt{B_W} \tag{9.36}$$

and the output signal-to-noise ratio is

$$(\text{S/N})_{out} = \frac{A_v E_s}{[e_n^2 + (i_n R_s)^2 + 4kTR_s]^{1/2} A_v \sqrt{B_W}} \tag{9.37}$$

The signal-to-noise ratio of the input signal is

$$(\text{S/N})_{in} = \frac{E_s}{[4kTR_s]^{1/2} \sqrt{B_W}} \tag{9.38}$$

TABLE 9.3. Useful Popular Op Amps

Op Amp		A_o	f_T (*MHz*)	*S.R.* ($V/\mu s$)	I_B (*nA*)	I_{io} (*nA*)	V_{io} (*mV*)	I_{max} (*mA*)
741	GP[†] bipolar (standard)	2×10^5	1.2	0.5	80	20	2	20
4136C	Quad bipolar	3×10^5	3.0	1.0	40	5	0.5	20
TI071C	Low-noise BIFET (FET input stage plus bipolar)	2×10^5	3.0	13	0.03	0.005	3	10
CA3140A	GP MOSFET	1×10^5	3.7	7	0.01	0.0005	2	+10/−1
LM318	GP fast bipolar	2×10^4	15	70	150	30	4	10

[†]GP = general purpose.

The noise figure of the amplifier is defined as the input signal-to-noise ratio divided by the output signal-to-noise ratio, expressed in decibels:

$$\text{N.F.} = \log_{10}\frac{(\text{S/N})_{\text{in}}}{(\text{S/N})_{\text{out}}}$$

Remember $(\text{S/N})_{\text{out}} < (\text{S/N})_{\text{in}}$ for any real circuit, so N.F. > 1. (An ideal noiseless amplifier thus has a noise figure of 0 dB.) Therefore the noise figure is given by

$$\text{N.F.} = 10\log_{10}\frac{[e_n^2 + (i_nR_s)^2 + 4kTR_s]^{1/2}}{[4kTR_s]^{1/2}} \tag{9.39}$$

The best amplifier is the one with the minimum noise figure, which is obtained when the source resistance equals the ratio e_n/i_n, called the *amplifier noise resistance*.

$$R_s = \frac{e_n}{i_n} \tag{9.40}$$

For most bipolar op amps e_n/i_n is about 10 kΩ, whereas for most FET input op amps e_n/i_n is about 10 MΩ. Hence we have the important general rule-of-thumb: For small source resistances use bipolar input op amps; for large source resistances use FET input op amps. Table 9.3 lists the characteristics of several popular op amps.

PROBLEMS

1. An op amp draws 20 mA from the +15-V and the −15-V supplies. Calculate (a) the power dissipation and (b) the approximate internal ("junction") temperature if the chip thermal resistance $\Theta = 50°\text{C/W}$ and the ambient temperature is 20°C.

2. What is an approximate value for f_B and for A_o for a typical inexpensive op amp? A_o is the open-loop gain, and f_B is the break frequency, where A_o is 3 dB down ($0.707A_o$) from the dc value. What is A_o for $f = 4f_B$? For $f = 10f_B$?

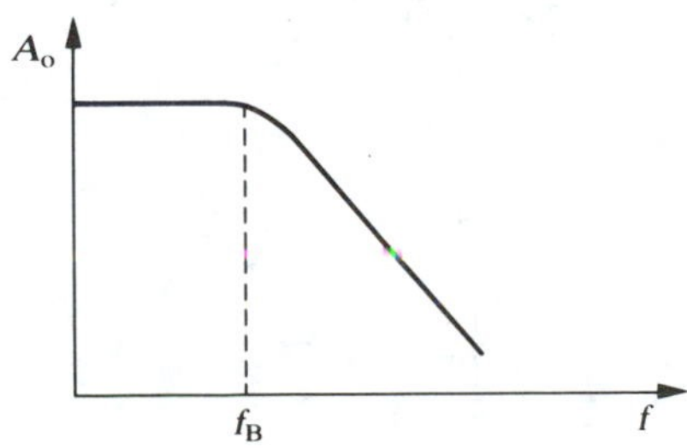

3. In the differential amplifier stage shown, with zero input, $I_{C_1} = I_{C_2} = 2$ mA. Suppose a positive input v_1 increases I_{C_1} to 2.5 mA. v_2 remains constant. (a) Calculate I_{C_2} and v_{out} for this positive input v_1. (b) What would v_{out} be for only one transistor Q_1? (The emitter of Q_1 is grounded and $v_{out} = v_{C_1}$.)

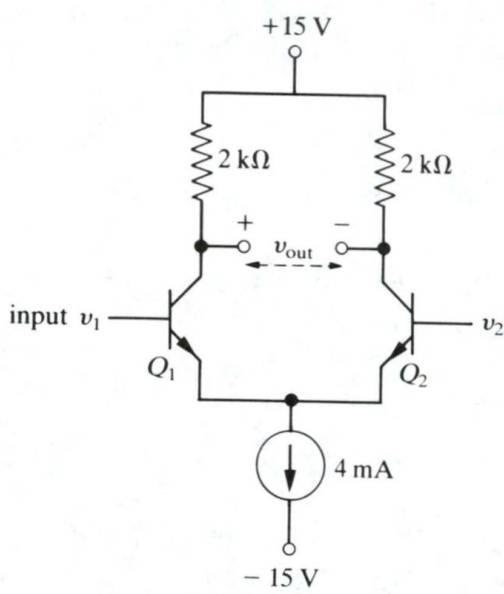

4. Show that $I_E \cong V_{bb}/2R_E$ independent of the temperature if $R_1 = R_2$. You may assume the transistor base emitter voltage and each diode voltage are the same at any given temperature.

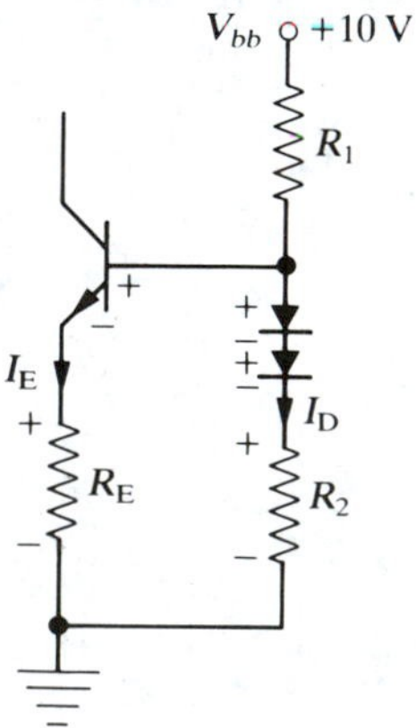

5. Q_1 is connected as a diode. Show that (a) $I_1 = (V_{bb} - V_D)/R_1$ and (b) $I_{C_2} \cong I_1$.

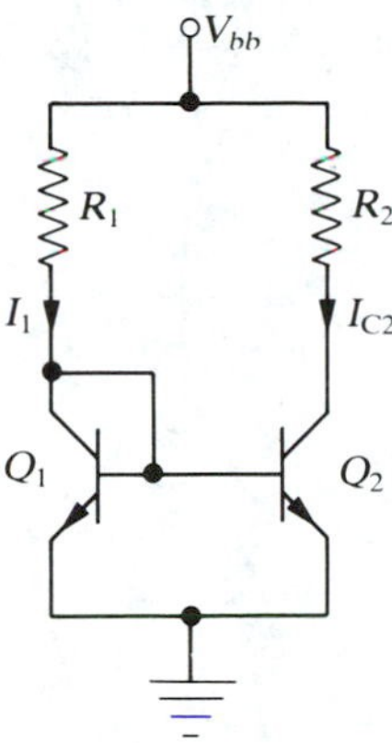

6. If $I_{C_1} = 1$ mA, what is the dc voltage at the base of Q_3? [*Hint*: Q_2 acts as a level shifter.]

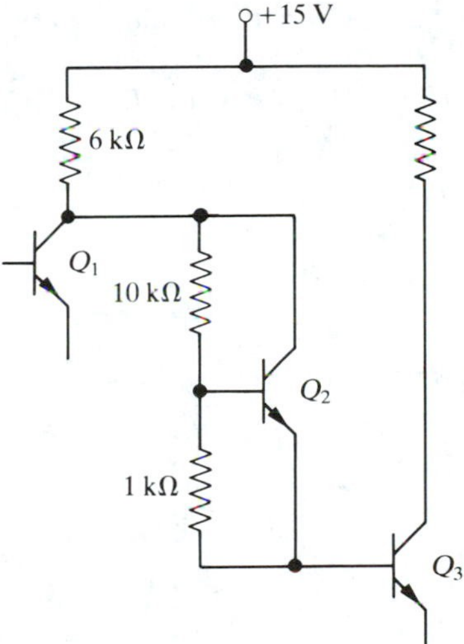

7. Calculate: (a) the voltage gain, (b) the upper 3-dB frequency for a 741 and for an 071, (c) the input impedance, and (d) the approximate effect of wiring a 1600-pF capacitor in parallel with the 10-kΩ resistor.

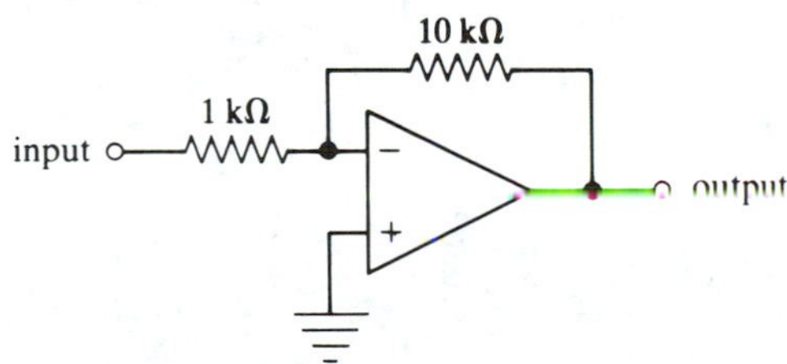

8. Calculate: (a) the voltage gain, (b) the upper 3-dB frequency for a 741 and for an 071, (c) the input impedance for a 741 (approximate), and (d) the approximate effect of adding a 3200-pF capacitor in parallel with the 4.7-kΩ resistance.

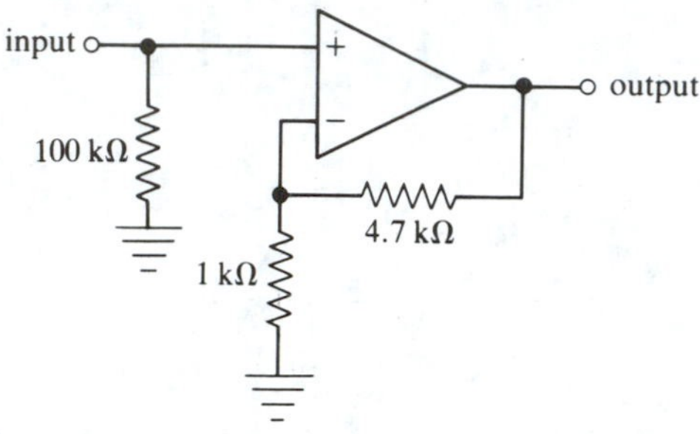

9. Briefly state the op amp current rule and the op amp voltage rule.
10. Briefly list the important parameters for an inexpensive (a) bipolar op amp and (b) FET input op amp.
11. If the upper 3-dB frequency is 300 kHz for $R_2 = 10$ kΩ, what will it be for (a) $R_2 = 50$ kΩ and (b) $R_2 = 100$ kΩ?

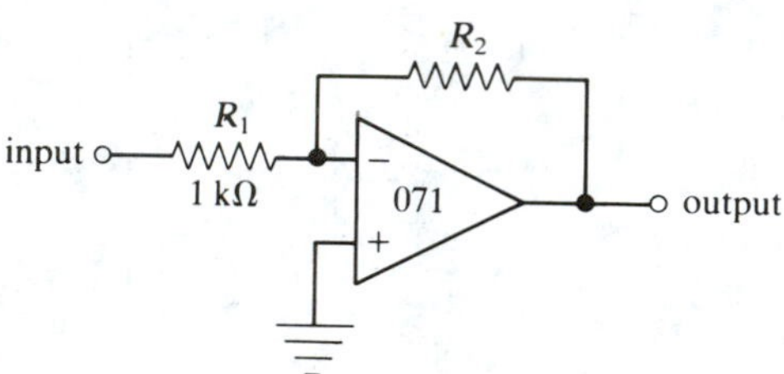

12. Repeat Problem 11 for the noninverting amplifier shown below.

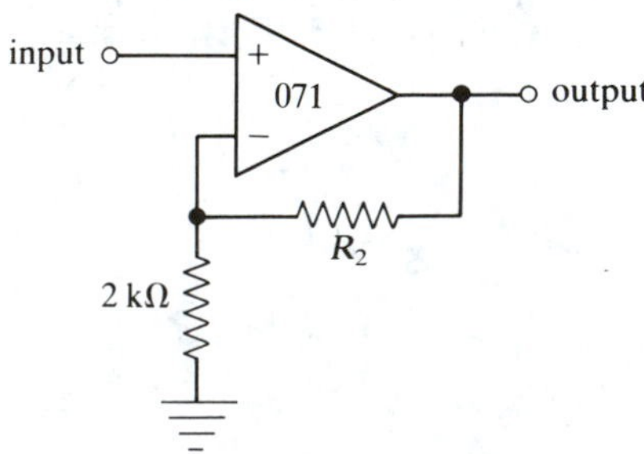

13. The open-loop gain versus frequency for an uncompensated op amp is shown below. Would the inverting amplifier be stable for (a) $R_2 = 10\ \text{k}\Omega$, (b) $R_2 = 100\ \text{k}\Omega$? $R_1 = 1\ \text{k}\Omega$.

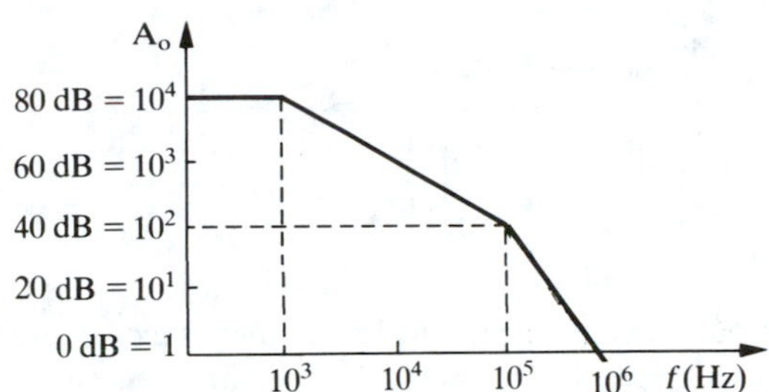

14. For a 741 op amp with a slew rate of $0.5\ \text{V}/\mu\text{s}$, for $R_2 = 20\ \text{k}\Omega$, what is the maximum input sinusoidal frequency to avoid slew rate distortion in the output? The input peak-to-peak amplitude is 10 mV. Repeat for $R_2 = 200\ \text{k}\Omega$.

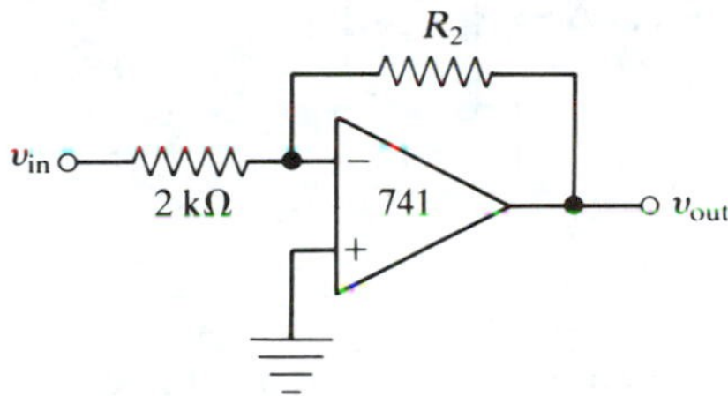

15. If the bias current $I_B = 200$ nA in the inverting amplifier circuit of Problem 14 with $R_1 = 2\ \text{k}\Omega$ and $R_2 = 200\ \text{k}\Omega$: (a) Calculate the dc output voltage for zero input. (b) Diagram how to null out this voltage. (c) Compare bipolar and FET input op amps with respect to bias current. (d) If the input offset current $I_{io} = 30$ nA, what will the output be for zero input?

16. Repeat Problem 15 for the noninverting amplifier shown below if $I_B = 400$ nA and $I_{io} = 20$ nA.

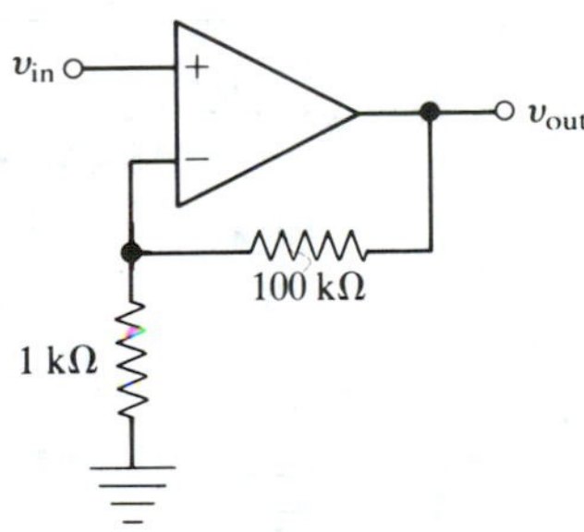

17. If the temperature coefficient of the input offset voltage is 6 μV/°C and the offset voltage is 2 mV at 20°C, calculate the dc output voltage at 20°C and 70°C.

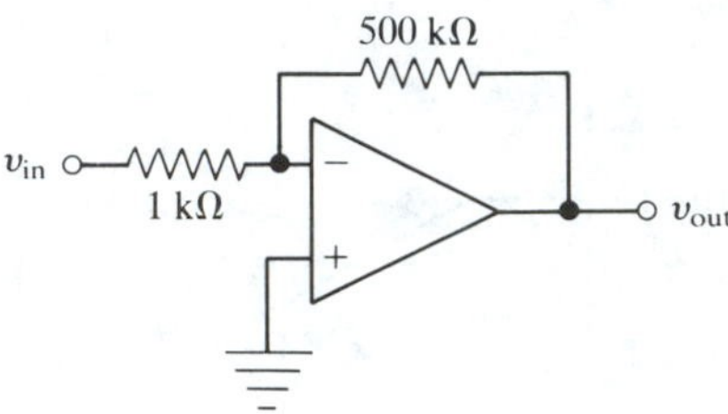

18. Although the output resistance of most op amps is extremely low when negative feedback is used, a typical inexpensive op amp can supply only a maximum current of approximately 25 mA. Calculate the maximum amplitude sinusoidal input (at low frequencies ~1 kHz) that will produce a nondistorted output across $R_L = 100\ \Omega$. Repeat for $R_L = 1\ \text{k}\Omega$.

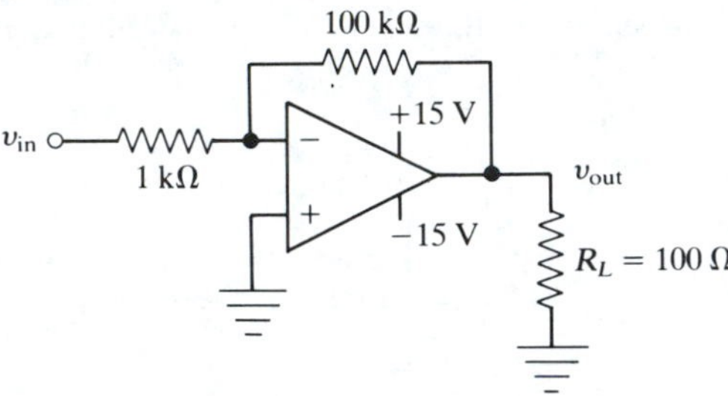

19. Briefly discuss the differences between the input noise in bipolar and in FET op amps.

20. (a) If a noisy signal source has a source impedance $R_s = 5\ \text{M}\Omega$, what would be the preferred type of op amp to maximize the signal-to-noise at the op-amp output? (b) Repeat for $R_s = 5\ \text{k}\Omega$. (c) Repeat for $R_s = 100\ \Omega$. [*Hint*: What circuit component would you have to add between the source and the op amp input?]

21. With $v_{in} = 0$, $v_{out} = 24$ mV for $R_1 = 10\ \text{k}\Omega$, $R_2 = 100\ \text{k}\Omega$, and $R_3 = R_1 \| R_2$. With $v_{in} = 0$, $v_{out} = 14$ mV for $R_1 = 20\ \text{k}\Omega$, $R_2 = 100\ \text{k}\Omega$, with R_3 again equal

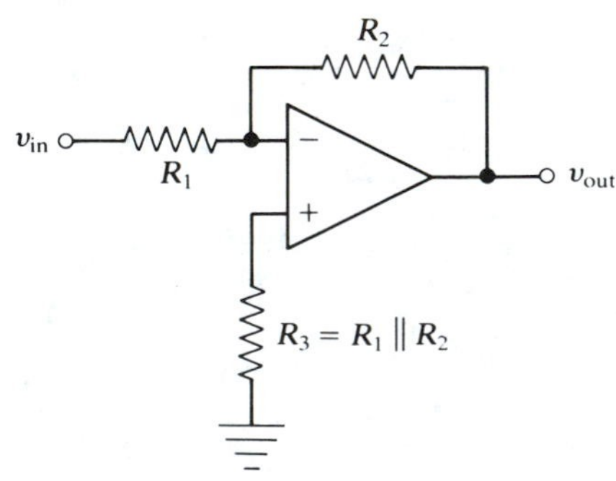

to $R_1 \| R_2$. Calculate V_{io} and I_{io}, the input offset voltage and input offset current of the op amp.

22. Repeat Problem 21 for the noninverting amplifier below.

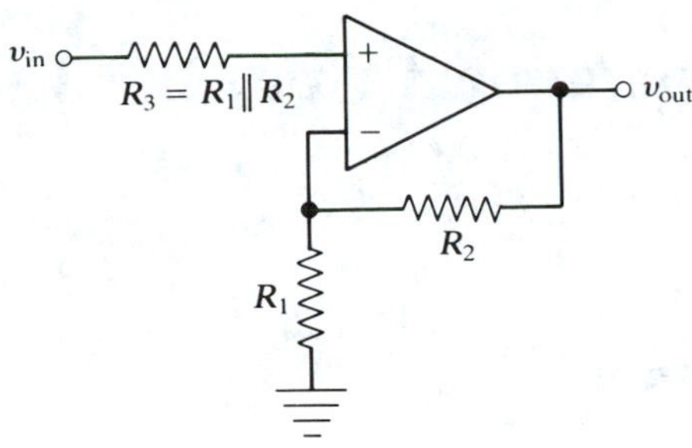

CHAPTER 10

Operational Amplifier Circuits

10.1 INTRODUCTION

In this chapter we present a number of linear and nonlinear op amp circuits to show the wide range of possible applications. A brief explanation is given for each circuit along with pertinent waveforms, impedances, and some specific applications.

10.2 THE INVERTING AMPLIFIER

Perhaps the simplest of all op amp circuits is the inverting amplifier shown in Fig. 10.1. As we have seen in Chapter 9, the exact voltage gain expression is

$$A_v = \frac{-\dfrac{R_2}{R_1}}{1 + \dfrac{R_1 + R_2}{A_o R_1}} \tag{10.1}$$

If $A_o R_1 \gg (R_1 + R_2)$, which is almost always the case, we have the simple gain expression

$$A_v \cong -\frac{R_2}{R_1} \tag{10.2}$$

(10.2) can also be derived immediately by using the two op amp rules of Chapter 9, which we repeat here for convenience:

I. Op amp current rule (OACR): The current flowing into (or out of) each op amp terminal is zero.
II. Op amp voltage rule (OAVR): The voltage difference between the inverting and the noninverting terminals is zero.

As a review, we repeat the voltage gain derivation, using the two rules.

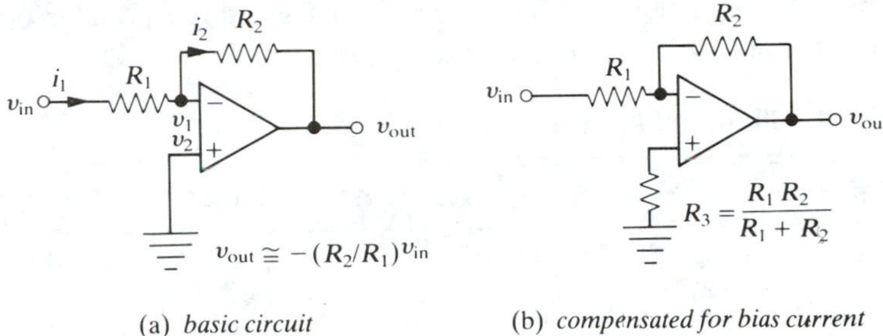

FIGURE 10.1 The inverting amplifier.

The OACR implies $i_1 = i_2$, and the OAVR implies $v_2 = v_1$. Thus, using Ohm's law to evaluate i_1 and i_2, we have $v_{in}/R_1 = -v_{out}/R_2$, or $v_{out}/v_{in} = -R_2/R_1$.

The input resistance is simply equal to R_1, because by the OAVR the voltage at the inverting input must be zero. The inverting input is termed a *virtual ground*. Notice that this is true regardless of the value of R_1 or R_2—that is, regardless of the circuit gain.

The output resistance of the amplifier circuit of Fig. 10.1 is reduced to below that of the bare (no feedback) op amp by the negative feedback and is given by

$$R_{out} = \frac{r_{out}}{1 + \dfrac{R_1 A_o}{R_1 + R_2}} \cong \frac{1 + \dfrac{R_2}{R_1}}{A_o} r_{out} \cong \frac{A_v}{A_o} r_{out} \qquad \textbf{(10.3)}$$

where r_{out} is the output resistance of the bare op amp. Typical values of the output resistance R_{out} are extremely low—a few ohms or less.

The circuit bandwidth (BW) (see 9.8.1) depends on the gain with feedback, A_v: the smaller the gain the larger the bandwidth. For most op amps the open-loop op amp voltage gain decreases as $1/f$ (20 dB/decade or 6 dB/octave) for $f > f_B$; f_B is the break frequency where the op amp open-loop gain A_o is 3 dB down from its dc value. f_B is usually 100 Hz or less. Thus,

$$\text{BW} = f_B\left(\frac{A_o}{A_v}\right) \qquad \textbf{(10.4)}$$

This is simply another way of writing the constant gain bandwidth product: $(\text{BW})A_v = f_B A_o$. For example, if the open-loop gain break frequency is $f_B = 500$ Hz, and the open-loop gain is 10^5, then for a gain of 20 the bandwidth would be BW = (500 Hz)(10^5)/20 = 2.5×10^6 Hz.

Actual measurements of these parameters for three popular op amps are given in Table 10.1 for $R_1 = 1\ \text{k}\Omega$ and $R_2 = 100\ \text{k}\Omega$ in Fig. 10.1(a).

TABLE 10.1. Inverting Amplifier Parameters

Op Amp	A_v	R_{in} ($k\Omega$)	R_{out} (Ω)†	*BW* (*kHz*)	*GBW* (*MHz*)
741 General-purpose bipolar	90	1	2	11	1.0
318 Fast bipolar	90	1	140	120	11
071 BIFET	90	1	1	35	3.1

†at 1 kHz.

The dc output voltage with zero input is

$$v_{oo} = \pm V_{io}\left(1 + \frac{R_2}{R_1}\right) + I_B R_2 \qquad \textbf{(10.5)}$$

where V_{io} is the input offset voltage and I_B is the bias current of the particular op amp used. Notice that the larger R_2 is (the larger the gain), the larger the output offset voltage v_{oo}. The polarities of the input offset current term and the input offset voltage term can be either positive or negative, depending on the particular op amp. Also, recall from Chapter 9 that the bias current term $I_B R_2$ is replaced by the smaller offset current term $I_{io} R_2$ if we add a resistance $R_3 = R_1 R_2/(R_1 + R_2)$ in series with the noninverting input.

With $R_3 = R_1 \| R_2$ in series with the noninverting input as shown in Fig. 10.1(b), the dc output voltage with zero input will be

$$v_{oo} = \pm V_{io}\left(1 + \frac{R_2}{R_1}\right) \pm I_{io} R_2 \qquad \textbf{(10.6)}$$

where I_{io} is the input *offset* current, which is the difference between the two input bias currents at the two op amp input terminals. I_{io} is always less than I_B, so adding R_3 will always reduce v_{oo}. With the appropriate offset voltage compensation circuit as discussed in 9.8.5, the V_{io} term can be canceled. But V_{io}, I_B, and I_{io} are all temperature dependent, so the output voltage will always drift in response to temperature changes.

10.3 THE NONINVERTING AMPLIFIER

The noninverting amplifier circuit is shown in Fig. 10.2. The input is connected directly to the noninverting terminal of the op amp, and R_1 and R_2 form a voltage divider that applies a fraction $B = R_1/(R_1 + R_2)$ of the output voltage to the inverting input terminal. Thus, a fraction Bv_{out} is

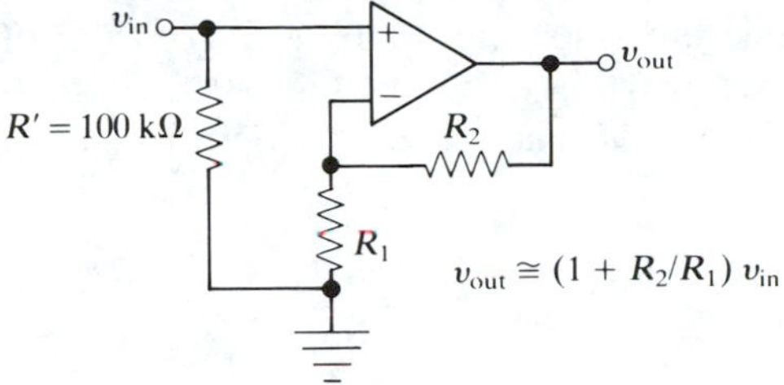

(a) *basic circuit*

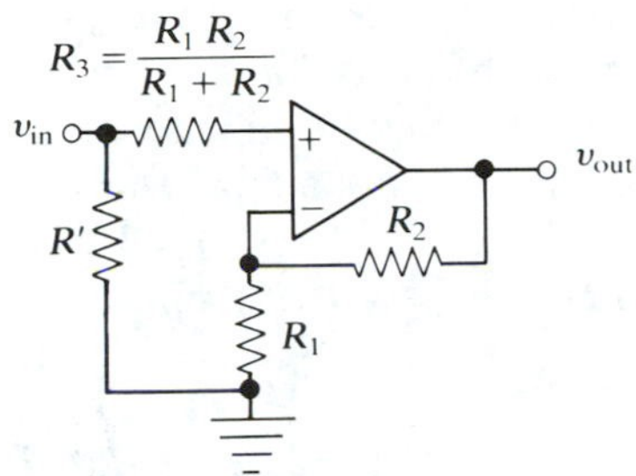

(b) *compensated for bias current*

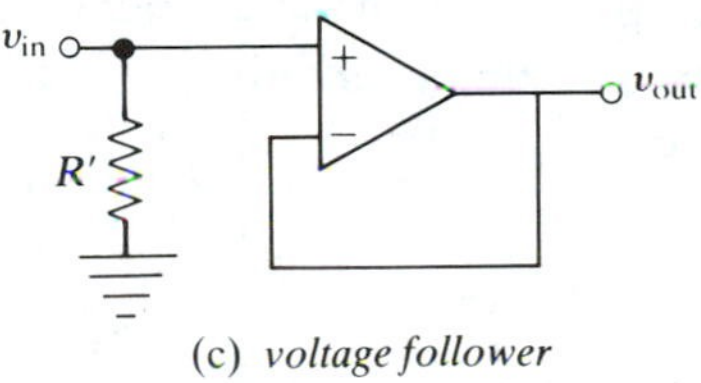

(c) *voltage follower*

FIGURE 10.2 The noninverting amplifier.

subtracted from the input, and we have the classical negative voltage feedback amplifier.

As we have seen in Chapter 9, the exact voltage gain expression is

$$A_v = \frac{1 + \dfrac{R_2}{R_1}}{1 + \dfrac{R_1 + R_2}{A_o R_1}} \tag{10.7}$$

and if $A_o R_1 \gg R_1 + R_2$, which is usually the case, we have the simple gain expression

$$A_v \cong \frac{1}{B} = 1 + \frac{R_2}{R_1} \tag{10.8}$$

(10.8) can also be quickly derived from the two op amp rules.

The input resistance for the noninverting circuit of Fig. 10.2 is much larger than for the inverting circuit of Fig. 10.1. From either direct analysis or the feedback theory of Chapter 8, the input resistance of the amplifier is

$$R_{\text{in}} = (1 + A_o B)R_{\text{ino}} \cong \frac{A_o}{A_v} R_{\text{ino}} \tag{10.9}$$

where R_{ino} is the input resistance of the bare op amp (usually very large) and can be estimated from the bias current I_B given in the op amp spec sheet. For example, if the bias current is 1 nA, then $R_{\text{ino}} \approx 1\ \text{V}/I_B \cong 1\ \text{V}/10^{-9}\ \text{A} = 10^9\ \Omega$. In most cases the input resistance of the noninverting amplifier is so large it can be considered infinite for all practical purposes. Thus, it is common practice to wire in a resistance $R' \cong 100\ \text{k}\Omega$ from the noninverting input to ground to stabilize the input impedance. (Some dc resistance R' to ground is absolutely essential if the input is ac coupled.)

The output resistance R_{out} of the noninverting amplifier circuit of Fig. 10.2 is

$$R_{\text{out}} = \frac{R_o}{1 + A_o B} \cong R_o \frac{A_v}{A_o} \tag{10.10}$$

where R_o is the output resistance of the base op amp with no feedback. Typical values of R_{out} are extremely low—several ohms or less.

The bandwidth is the same as for the inverting amplifier:

$$\text{BW} = f_{\text{B}}\left(\frac{A_o}{A_v}\right) \tag{10.11}$$

where f_{B} is the frequency where the op amp open-loop gain is 3 dB less than the dc gain, A_o is the dc open-loop gain, and $A_v = 1 + R_2/R_1$ is the circuit gain with feedback.

The dc output with zero input is

$$v_{\text{oo}} = \pm V_{\text{io}}\left(1 + \frac{R_2}{R_1}\right) + I_B R_2 \tag{10.12}$$

With the addition of $R_3 = R_1 R_2/(R_1 + R_2)$ in series with the noninverting input, as shown in Fig. 10.2(b), the dc output with zero signal input is

$$v_{\text{oo}} = \pm V_{\text{io}}\left(1 + \frac{R_2}{R_1}\right) \pm I_{\text{io}} R_2 \tag{10.13}$$

which is the *same* as for the inverting amplifier.

A special case of the noninverting amplifier is the voltage follower shown in Fig. 10.2(c). It is the noninverting amplifier with $R_2 = 0$, and

$R_1 \to \infty$; in other words, all the output voltage is fed back. The voltage follower thus has a voltage gain of 1, but a very high input resistance and a very low output resistance, and a large bandwidth. Thus, it is useful as a buffer amplifier when it is desirable to isolate two parts of a system. It is almost an ideal one-way device because of its high input and low output resistances—changes at the output will have almost no effect on its input, and it will draw almost no current from its input. In other words, it acts as an impedance transformer; it allows us to couple a signal from a source with a high output resistance (such as a glass pH electrode, a wick electrode in neurophysiology, or a photomultiplier tube) to a lower impedance circuit.

Actual measurements of these parameters for some popular op amps are shown in Table 10.2 for $R_1 = 1\ \text{k}\Omega$, $R_2 = 100\ \text{k}\Omega$ in Fig. 10.2(a).

TABLE 10.2. Noninverting Amplifier Parameters

Op Amp	A_v	R_{out} (Ω)†	*BW* (*kHz*)	*GBW* (*MHz*)
741	94	2	10	0.94
318*	92	140	200	18
071	96	1	34	3.3

*5 pF in parallel with $R_2 = 100\ \text{k}\Omega$.
†at 1 kHz.

10.4 THE DIFFERENTIAL AMPLIFIER

A difference or differential amplifier is one whose output is proportional to the difference between two input voltages $v_{\text{out}} = A_v(v_B - v_A)$. The amplifier is shown in Fig. 10.3(a). Such an amplifier is useful when both signals contain large (and equal) amounts of noise, such as 60-Hz pickup. This is often the case when the input signals come fron transducers that have a high impedance to ground (biological wick electrodes, for example). Another useful application is when the dc level of the two inputs is high (several volts) and when the signal change is small (millivolts). This is the case for strain gauges.

Applying the OACR and the KCL to the junction at the inverting input, we have

$$i_1 = i_2 \qquad \frac{v_A - v_1}{R_1} = \frac{v_1 - v_{\text{out}}}{R_2} \tag{1}$$

The OAVR implies $v_1 = v_2$, but v_2 is given by

$$v_2 = \frac{R_4}{R_3 + R_4} v_B = v_1 \tag{2}$$

i_2 R_2

i_1 R_1

v_A v_1

R_3 v_2

v_B v_{out}

R_4

if $R_1/R_2 = R_3/R_4$

$$v_{out} = \frac{R_2}{R_1}(v_B - v_A)$$

(a) *basic circuit*

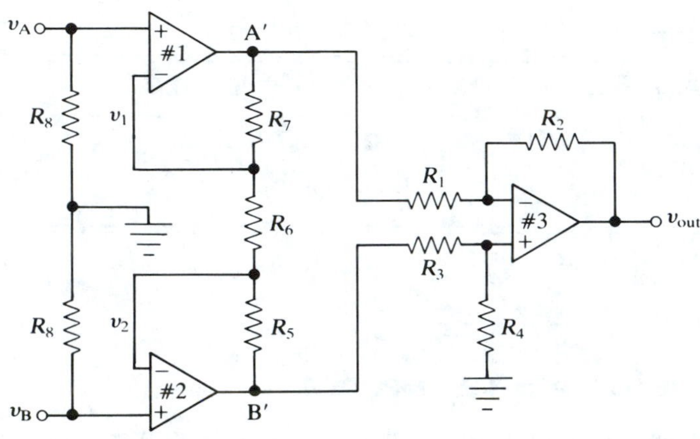

(b) *instrumentation amplifier*

FIGURE 10.3 The difference amplifier.

Eliminating v_1 between (1) and (2), we get

$$v_{out} = \frac{R_2}{R_1}\frac{1 + \dfrac{R_1}{R_2}}{1 + \dfrac{R_3}{R_4}}(v_B - v_A)$$

If we choose the resistances such that $R_1/R_2 = R_3/R_4$, then the output is given by

$$v_{out} = \frac{R_2}{R_1}(v_B - v_A) \tag{10.14}$$

which indicates that the output is proportional to the difference between the two input voltages v_B and v_A.

The resistances must be chosen carefully to satisfy $R_1/R_2 = R_3/R_4$. Resistances of 1% or better tolerance are generally required to get a high CMRR. If (10.14) holds exactly, the CMRR = ∞; that is, the gain for $v_A = v_B$ is zero. Also, notice that the input impedances for v_A and v_B are different, being much higher for v_B. This can be a very serious source of noise pickup.

The usual instrumentation amplifier consists of three op amps and is shown in Fig. 10.3(b). Op amp #3 and R_1, R_2, R_3, and R_4 comprise the standard basic circuit of Fig. 10.3(a). Thus, the output is given by

$$v_{out} = \frac{R_2}{R_1}(v_{B'} - v_{A'}) \qquad \textbf{(1)}$$

if $R_1/R_2 = R_3/R_4$.

The input impedance for both v_A and v_B is R_8 because both inputs go directly to the high impedance noninverting input terminals. From the OAVR we have

$$v_1 = v_A \quad \text{and} \quad v_2 = v_B$$

Thus, for a common mode input ($v_A = v_B$) there is zero voltage drop across R_6, and therefore no current flows through the R_5, R_6, R_7 resistor chain. Thus, $v_{A'} = v_{B'}$, and there is zero differential input to op amp #3, and v_{out} is zero. But for a differential input ($v_A \neq v_B$) the same current must flow through the R_5, R_6, R_7 chain because zero current flows into the inverting input terminals of op amps #1 and #2. Thus,

$$\frac{v_{A'} - v_1}{R_7} = \frac{v_1 - v_2}{R_6} = \frac{v_2 - v_{B'}}{R_5} = \frac{v_{A'} - v_{B'}}{R_5 + R_6 + R_7} \qquad \textbf{(2)}$$

Solving for $v_{B'} - v_{A'}$; we obtain

$$v_{B'} - v_{A'} = -\frac{R_5 + R_6 + R_7}{R_6}(v_B - v_A)$$

It is common practice to choose $R_5 = R_7 = R$, so the gain expression becomes

$$v_{out} = -\frac{R_2}{R_1}\left(1 + \frac{2R}{R_6}\right)(v_B - v_A) \qquad \textbf{(10.15)}$$

The overall differential gain can be set by adjusting R_6.

Many different commercial instrumentation amplifiers are available at prices from less than a dollar to hundreds of dollars. One such (expensive) instrumentation FET amplifier is the 4253, which is especially

TABLE 10.3. Selected Instrumentation Amplifier Specifications

Parameter	*4253*	*LF152*	*AD624A*
Circuit gain	1–10,000	1 to 1000	1 to 1000
Output impedance ($A_v = 1$)	0.1 Ω max @ dc 1.0 Ω max @ 3 kHz	1.2 Ω	
Output swing	±10 V @ ±5 mA	±9 V	±10 V
Output capacitive load	max 0.01 μF		
Input impedance	10^{13} Ω‖3 pF diff. or cm	2 × 10^{12} Ω‖2.5 pF diff. 2 × 10^{12} Ω‖5.0 pF cm	10^9 Ω
Input swing	±36 V max	±12 V (*must* not exceed $-V_{CC}$)	±10 V max
CMRR	76 dB dc to 60 Hz A = 1, 10 114 dB dc to 100 Hz A = 1000 108 dB dc to 500 Hz A = 1000	85 dB A = 1 105 dB A = 10 125 dB A = 100 125 dB A = 1000	70 dB A = 1 100 dB A = 10 110 dB A = 1000
Input offset voltage	±1 mV RTI max ± 4 μV/°C ±15 mV RTO max ± 100 μV/°C ±3 μV/% power supply drift ±5 μV/day shift	8 mV 10 μV/°C ±100 μV/V power supply	200 μV 2 μV/°C
Input noise voltage	1.6 to 160 Hz 0.3 μV_{rms} RTI 1.6 to 160 Hz 30 μV_{rms} RTO	10 Hz to 10 kHz 8 + 450/A μV_{rms}	0.2 μV 0.1 to 10 Hz A = 1000
Input bias current	−10 pA max @ 25°C(×2/10°C) −55 pA @ 50°C 0.05 pA/V cm input 0.05 pA/V signal input 0.02 pA/% power supply drift	3 pA @ 25°C (0.5 pA offset current)	±50 nA ±50 pA/°C
Slew rate		1 V/μS	5 V/μs
Input current noise	1.6 to 160 Hz 1 pA_{rms}	10 Hz to 10 kHz 0.01 pA_{rms}	
Power supply	±15 V @ ±21 mA_{max} (full load)	±22 V 0.7 mA	±5 to 18 V @ 3.5 mA

suited for amplifying small differential signals in the presence of large common mode voltages. It has a very high input impedance, low drift with time and temperature, low input noise, low bias current, adjustable gain with one external resistor, and output short circuit protection. The 4253 specifications are listed in Table 10.3. A less expensive instrumentation amplifier is the AD624A.

10.5 THE OP AMP POWER BOOSTER

Most inexpensive op amps can put out a maximum current of approximately 10 to 20 mA before the output waveform is seriously distorted. The circuit shown in Fig. 10.4(a) will boost the output current (and power) to the limits of the transistor, which can be chosen to be a high-power transistor. The circuit of Fig. 10.4(a) has one disadvantage, however; its output voltage is always positive (or negative if a pnp transistor and negative supply are used). Thus, this circuit can only "source" current; it cannot "sink" current. Notice that the output transistor is *inside* the feedback loop and is acting essentially as an emitter follower. From basic negative feedback theory $B = R_1/(R_1 + R_2)$, and the overall voltage gain of the op amp and transistor of Fig. 10.4(a) must be

$$A = \frac{1}{B} = 1 + \frac{R_2}{R_1} \tag{10.16}$$

If both positive and negative output voltage swings are desired, the bottom of R_E can be tied to a negative supply voltage or the push-pull circuit of Fig. 10.4(b) can be used. If the op amp output swings positive, then the npn transistor Q_1 is turned on and Q_2 is turned off; hence the output voltage goes positive. If the op amp swings negative, then the pnp transistor Q_2 is turned on and Q_1 is turned off; hence the output goes negative. You might recall that the turn-on voltage for any silicon transistor is approximately 0.5 V and thus might be tempted to conclude that the op amp output would have to be at least ±0.5 V to get a nonzero output and that therefore the output will be seriously distorted. This is not the case because the transistors are *inside* the feedback loop. The OAVR states that the voltage at the noninverting input terminal (v_{in}) must equal the voltage at the inverting terminal, which is $R_1/(R_1 + R_2)$ times the *output* voltage:

$$v_{in} = \frac{R_1}{R_1 + R_2} v_{out}$$

Thus, the overall voltage gain of the op amp plus transistors is

$$A = \frac{v_{out}}{v_{in}} = \frac{R_1 + R_2}{R_1} = 1 + \frac{R_2}{R_1} \tag{10.17}$$

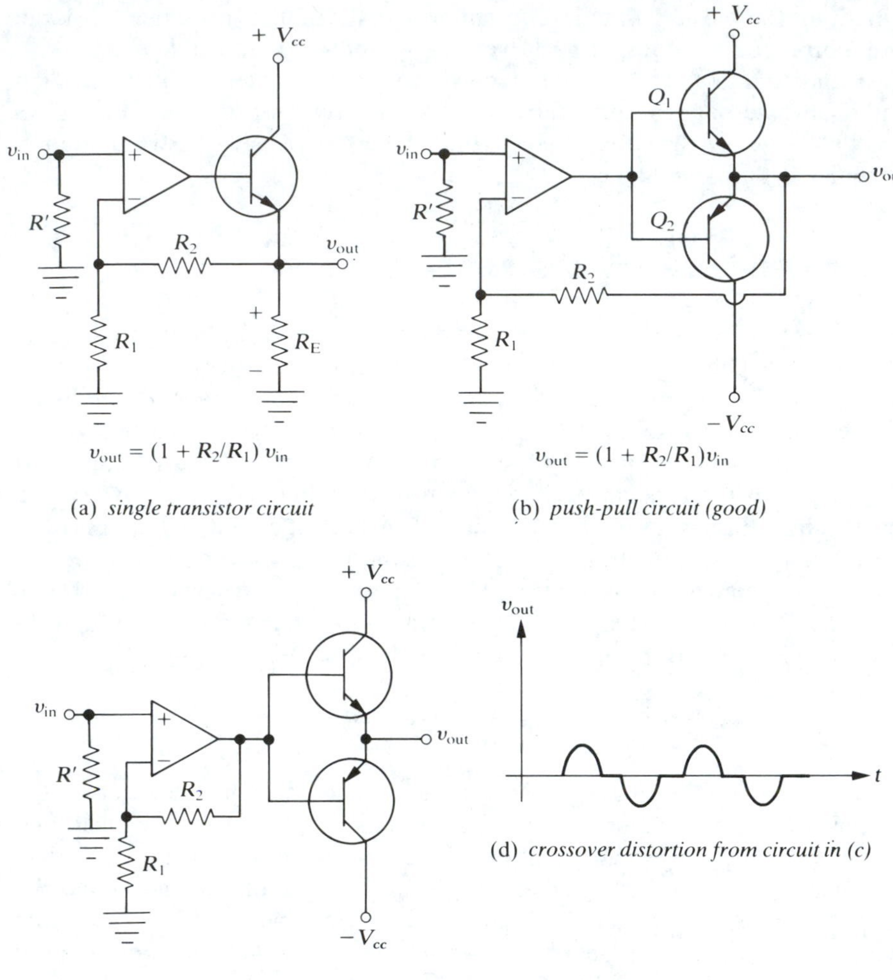

FIGURE 10.4 Power booster.

Push-pull transistors are available on a single chip that will supply up to several hundred mA of output current. Examples are the MC1438 and the LH0063.

It is instructive to investigate what would happen if the output transistors were outside the feedback loop, as shown in Fig. 10.4(c). In this case the output of the op amp would have to be greater than approximately 0.5 V (either + or −) to turn on one of the output transistors. Thus, the output would "miss" the first 0.5 V of signal and would look like Fig. 10.4(d) for a pure sinusoidal input. Such distortion is called *crossover* distortion, because it occurs as the input signal crosses or changes polarity.

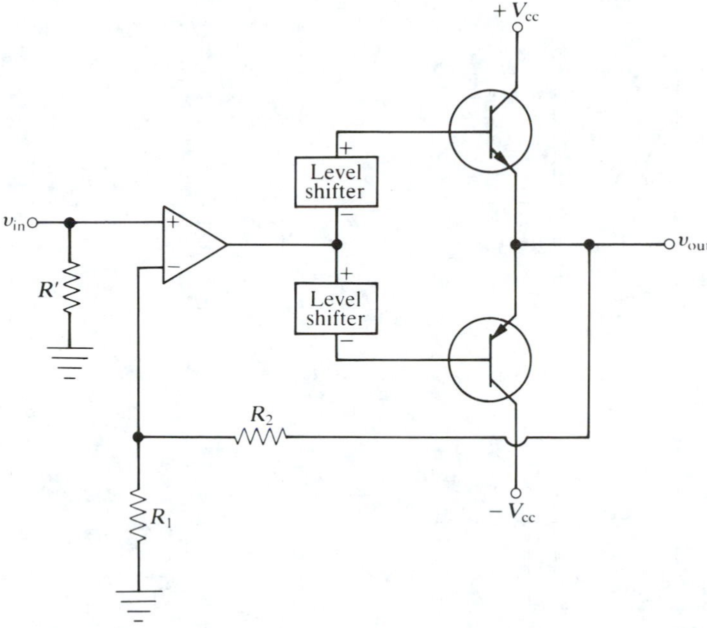

FIGURE 10.5 Push-pull power booster with level shifter.

If minimum distortion is required, the crossover distortion can be decreased even more by adding level-shifter circuits (see 9.5.1), as shown in Fig. 10.5. Both the npn and the pnp transistors are biased slightly on by the level shifter even with zero signal from the output of the op amp. The price paid is the higher power dissipation in the transistors with zero input signal.

10.6 COMPENSATION OR EQUALIZATION AMPLIFIERS

The conventional noninverting op amp of Fig. 10.6(a) has a flat frequency response out to f_2; it will amplify signals from dc up to f_2 with a gain of $1 + R_2/R_1$. The gain at f_2 will be 3 dB less than the dc gain.

To avoid having the dc output level depend on the slowly drifting (due to aging or temperature changes) op amp voltage offset and bias current, it is often desirable to decrease the amplifier gain at dc and low frequencies. A capacitance C_1 in series with R_1 will accomplish this, as shown in Fig. 10.6(b). There must be some dc path from the inverting output to ground through v_{out}, shown in Fig. 10.6(b) as R_L. The voltage gain will be

$$A = 1 + \frac{R_2}{Z_1}$$

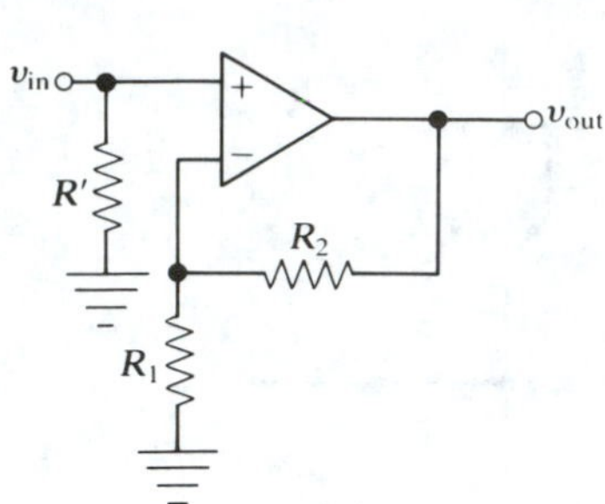

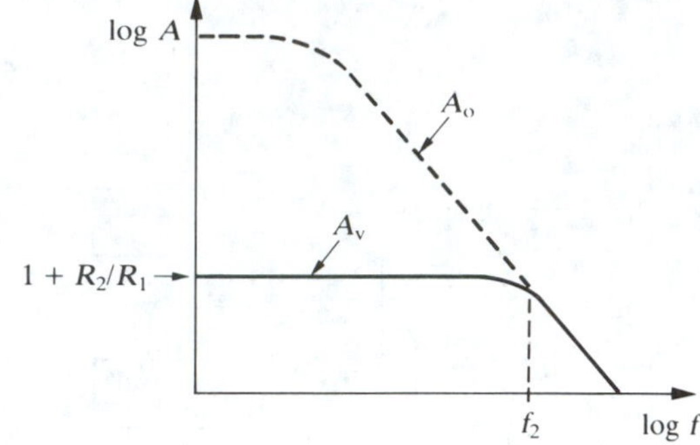

(a) *dc coupled*

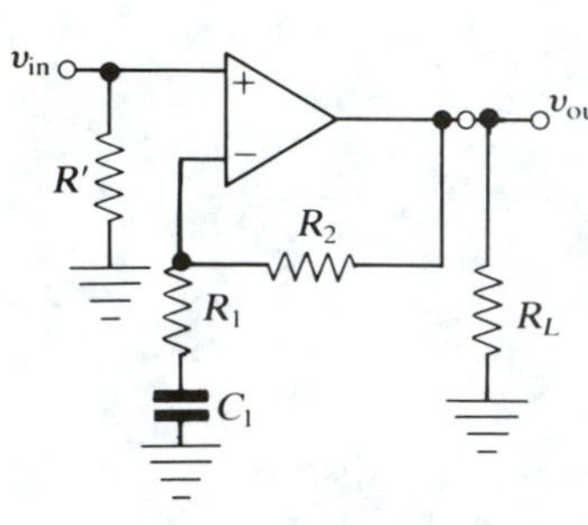

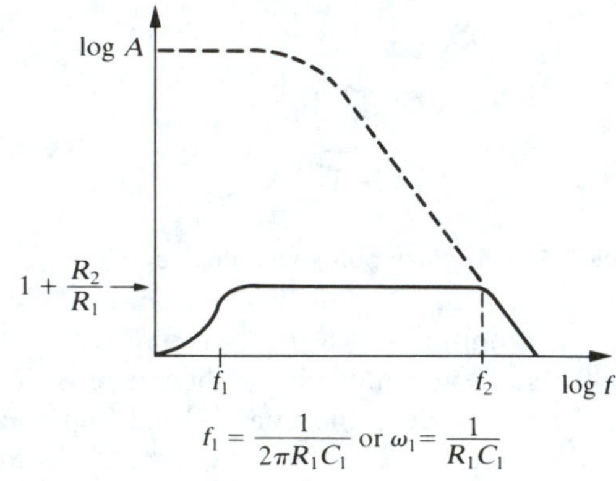

$$f_1 = \frac{1}{2\pi R_1 C_1} \text{ or } \omega_1 = \frac{1}{R_1 C_1}$$

(b) *ac coupled to reduce dc gain*

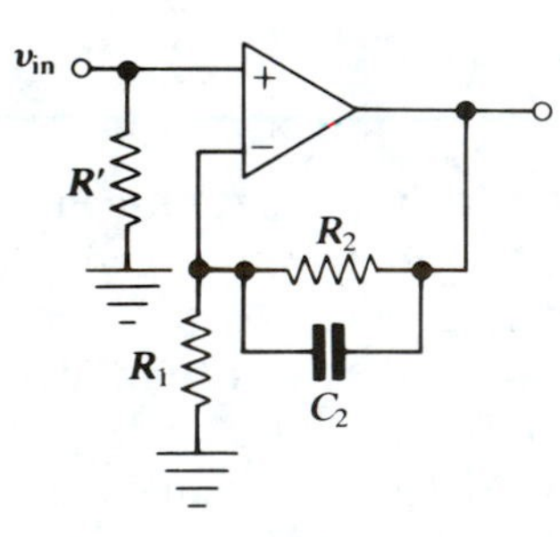

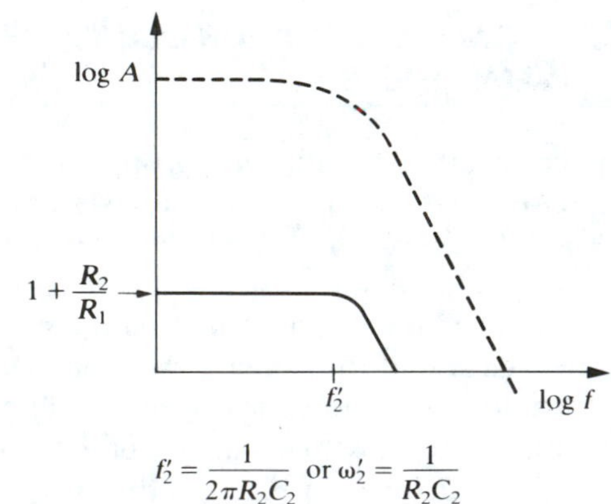

$$f_2' = \frac{1}{2\pi R_2 C_2} \text{ or } \omega_2' = \frac{1}{R_2 C_2}$$

(c) *reduced high-frequency gain*

FIGURE 10.6 Equalization amplifiers.

where Z_1 is the series impedance of R_1 and C_1: $Z_1 = R_1 + 1/j\omega C_1$. Thus, the amplifier gain is

$$A = 1 + \frac{R_2}{R_1 + \dfrac{1}{j\omega C_1}} = 1 + \frac{j\omega R_2 C_1}{1 + j\omega R_1 C_1}$$

If we let $\omega_1 = 1/R_1 C_1$, then we can write the gain as

$$A = 1 + \frac{\dfrac{j\omega R_2}{\omega_1 R_1}}{1 + \dfrac{j\omega}{\omega_1}} \tag{10.18}$$

If $\omega \gg \omega_1$, then $A \to 1 + R_2/R_1$, which is the midfrequency gain. If $\omega \ll \omega_1$, then $A \cong 1 + j(\omega/\omega_1)R_2/R_1$, which decreases (to 1) with decreasing frequency. At $\omega = \omega_1$, $A \cong 1/\sqrt{2}(R_2/R_1)$ if $R_2/R_1 \gg 1$, that is, A is 3 dB less than the midfrequency gain. The gain-versus-frequency curve is shown in Fig. 10.6(b).

Similarly, to reduce the gain at high frequencies (but less than f_2), we can add a capacitance C_2 in parallel with R_2, as shown in Fig. 10.6(c). The voltage gain will then be

$$A = 1 + \frac{Z_2}{R_1}$$

where Z_2 is the parallel impedance of R_2 and C_2:

$$Z_2 = \frac{R_2\left(\dfrac{1}{j\omega C_2}\right)}{R_2 + \dfrac{1}{j\omega C_2}} = \frac{R_2}{1 + j\omega R_2 C_2}$$

Thus,

$$A = 1 + \frac{\dfrac{R_2}{R_1}}{1 + j\omega R_2 C_2}$$

If we let $\omega_2 = 1/R_2 C_2$, then the gain can be written as

$$A = 1 + \frac{\dfrac{R_2}{R_1}}{1 + \dfrac{j\omega}{\omega_2}} \tag{10.19}$$

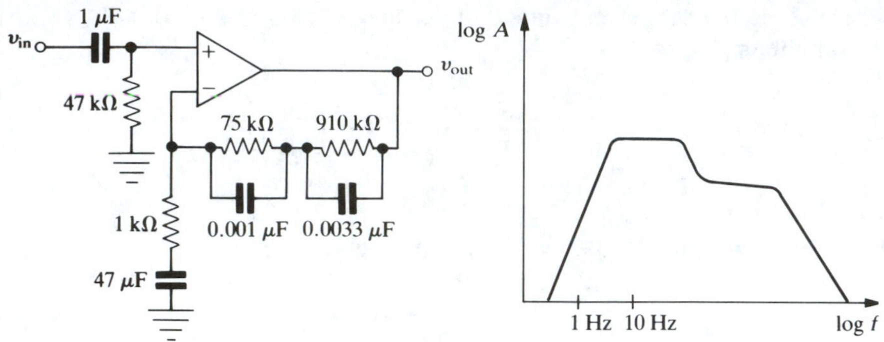

FIGURE 10.7 RIAA phonograph compensation amplifier.

Thus if $\omega \ll \omega_2'$, $A \to 1 + R_2/R_1$, which is the midband gain; if $\omega \gg \omega_2'$, $A \to 1 + R_2\omega_2/j\omega R_1$, which decreases (to 1) with increasing frequency. At $\omega = \omega_2'$, $A \cong 1/\sqrt{2}(R_2/R_1)$ if $R_2/R_1 \gg 1$, that is, A is 3 dB less than the midfrequency gain. The gain-versus-frequency curve is shown in Fig. 10.6(c).

The popular RIAA compensation curve for LP phonograph records can be created with three capacitors, as shown in Fig. 10.7. The 1-μF, 47-kΩ high-pass RC filter at the input has a break frequency of 3.4 Hz.

10.7 THE SUMMING AMPLIFIER

An op amp can easily be hooked up to act as an adder; that is, the output voltage will be the sum of a number of independent input voltages (with each input multiplied by its own gain if desired). The circuit is shown in Fig. 10.8; it uses the fact that for the inverting configuration the inverting input is a virtual ground, which means that the various inputs are electrically isolated from each other. No input should affect any other input.

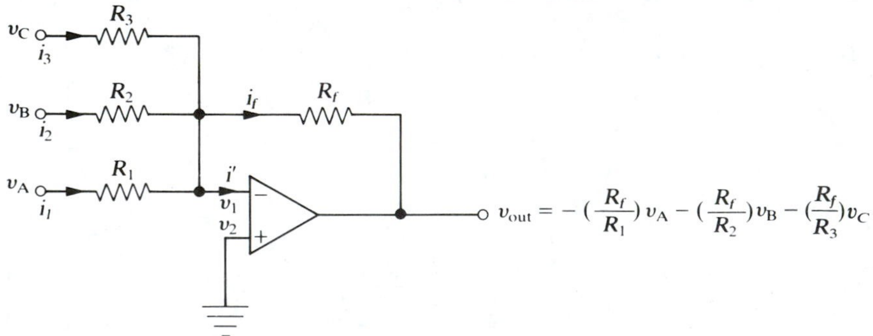

FIGURE 10.8 The summing amplifier.

The output voltage expression can be obtained easily by writing the KCL at the inverting input:

$$i_1 + i_2 + i_3 = i_f + i'$$

Neglecting i' from the OACR and using Ohm's law give

$$\frac{v_A - v_1}{R_1} + \frac{v_B - v_1}{R_2} + \frac{v_C - v_1}{R_3} = \frac{v_1 - v_{out}}{R_f}$$

But $v_1 = v_2 = 0$ from the OAVR, so the output is

$$v_{out} = -\left(\frac{R_f}{R_1}\right)v_A - \left(\frac{R_f}{R_2}\right)v_B - \left(\frac{R_f}{R_3}\right)v_C \tag{10.20}$$

Thus we see that the output is a linear sum (with negative coefficients) of the inputs. If $R_1 = R_2 = R_3 = R_f$, then the output is $v_{out} = - v_A - v_B - v_C$. The output can easily be made positive by simply adding an inverting amplifier to the output.

This summing circuit can be made into a digital-to-analog converter very simply—v_A is the least significant bit, v_B the next more significant bit, and so on. We then choose the resistances so that $R_2 = R_1/2$, $R_3 = R_2/2$, and so on, so that v_B will be weighted twice as heavily in the output as v_A, and so forth. This will be explained at greater length in Chapter 15.

10.8 THE CURRENT-TO-VOLTAGE CONVERTER (THE TRANSCONDUCTANCE AMPLIFIER)

It is often desirable to convert a small input current to an output voltage for further processing. Examples of input current sources are a phototube, a photovoltaic cell, a photomultiplier tube, a photodiode, and a phototransistor. The op amp circuit of Fig. 10.9 does the trick; the output voltage equals the negative of the input current times the feedback resistance R. The KCL at the inverting input implies

$$i_{in} = i_2 + i'$$

The OACR implies $i' = 0$. But Ohm's law implies

$$i_2 = \frac{v_1 - v_{out}}{R}$$

But the OAVR implies $v_1 = v_2$, and $v_2 = 0$, so

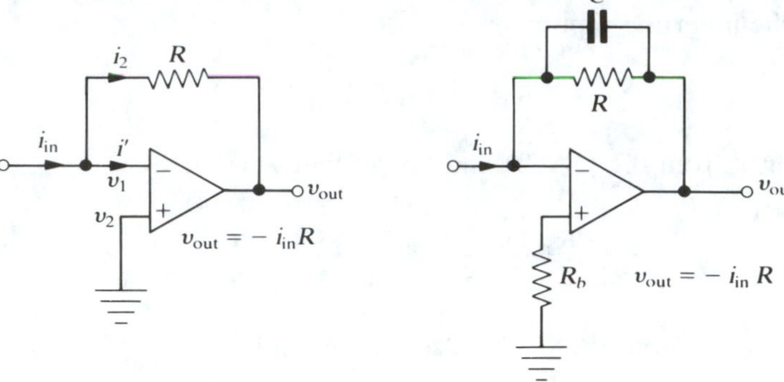

(a) *basic circuit*

(b) *reduced high-frequency gain from C, reduced output offset from R_b*

FIGURE 10.9 Current-to-voltage converter.

$$i_{in} = \frac{-v_{out}}{R}$$

or

$$v_{out} = -i_{in}R \tag{10.21}$$

For example, if $R = 1\ \text{M}\Omega$, then a 1-μA input current will produce a 1-V output. We say the output will be $1\ \text{V}/\mu\text{A}$. If $R = 5\ \text{M}\Omega$, we would have $5\ \text{V}/\mu\text{A}$. The reader might well ask, Why not just run the current directly through the resistance R and avoid the expense of the op amp? After all, the output voltage will still be $1\ \text{V}/\mu\text{A}$ for $R = 1\ \text{M}\Omega$. The answer is that the op amp provides *power* gain; in other words, the output impedance of the op amp is much less than R and thus can drive a wide variety of circuits. If the output were taken off R directly (without the op amp), there would be no power gain and the output resistance would be high. The op amp circuit also provides much better isolation between the input and output.

If only low-frequency or slow dc variations in the input current are of interest, it is desirable to connect a capacitance in parallel with R as shown in Fig. 10.9(b) to reduce the high-frequency gain so that high-frequency pickup and possible oscillations are prevented. The maximum frequency response of the circuit will then be approximately

$$2\pi f_{max} = \omega_{max} \cong \frac{1}{RC}$$

because the output voltage cannot change in a time shorter than the RC time constant.

A resistance R_b in series with the noninverting input might be necessary to decrease the dc output voltage due to the offset bias current of the op amp, as shown in Fig. 10.9(b). R_b can be adjusted to zero the output voltage at any fixed temperature, but temperature changes will always change the output dc voltage due to the temperature coefficient of the bias current.

10.9 THE VOLTAGE-TO-CURRENT CONVERTER

The circuit of Fig. 10.10(a) will supply a current through the "load" resistance R_L, which is independent of R_L; the current equals the input voltage divided by R_1.

Because the op amp inputs draw negligible current (OACR), there is no voltage drop across R_3 and the input voltage equals the voltage at the noninverting terminal: $v_{in} = v_2$. The OAVR implies $v_2 = v_1$, and there is no current through or voltage drop across R_2 from the OACR, so $v_1 = i_1 R_1$. Thus

$$v_{in} = v_2 = v_1 = i_1 R_1$$

But $i_1 = i_L$ because no current flows through R_2 into the inverting input. Thus

$$v_{in} = i_L R_1$$

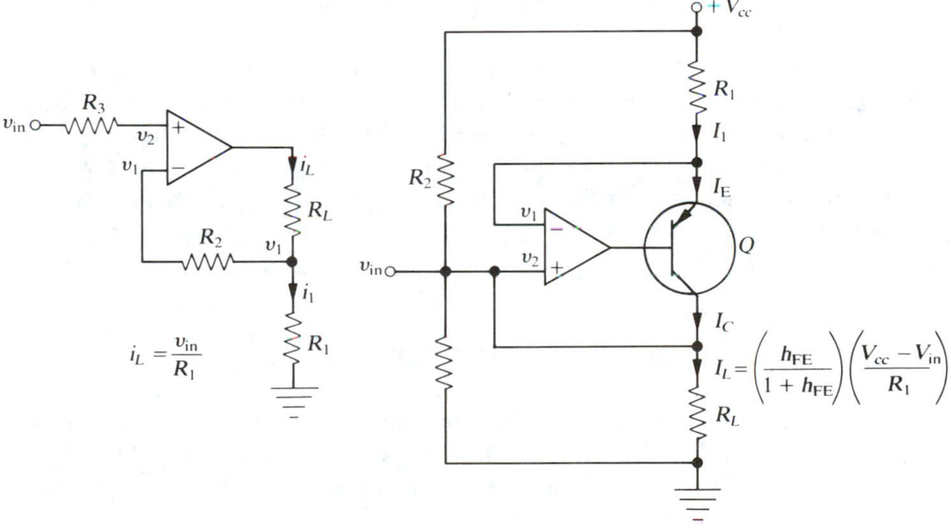

(a) *"floating" load* (b) *one end of load grounded*

FIGURE 10.10 Voltage-to-current converter.

or the current through the load R_L is simply

$$i_L = \frac{v_{\text{in}}}{R_1} \tag{10.22}$$

Notice:

1. The load current can be changed by changing the input voltage.
2. No current (or power) is drawn from the v_{in} source.
3. The load current is independent of the load resistance R_L.
4. The maximum load current is the maximum current output of the op amp, which is typically 20 mA for an inexpensive general-purpose op amp.
5. Both ends of the load resistance are above ground.
6. R_2 and R_3 can be adjusted to minimize op amp bias current efforts on the output.

If one end of the load resistance must be grounded, the circuit of Fig. 10.10(b) can be used.

By the OACR, $I_1 = I_E$ and $I_C = I_L$. And

$$I_L = I_C = \alpha I_E = \left(\frac{h_{FE}}{1 + h_{FE}}\right) I_E = \left(\frac{h_{FE}}{1 + h_{FE}}\right)\left(\frac{V_{cc} - v_1}{R_1}\right)$$

From the OAVR $v_1 = v_2 = v_{\text{in}}$. Thus

$$I_L = \left(\frac{h_{FE}}{1 + h_{FE}}\right)\left(\frac{V_{cc} - v_{\text{in}}}{R_1}\right) \tag{10.23}$$

The load current depends on the input voltage, V_{cc}, R_1, and h_{FE} of the transistor. We want I_L to be controlled by only v_{in}, so we want V_{cc}, R_1, and h_{FE} to be constant. It is easy to keep V_{cc} and R_1 constant, but h_{FE} will depend slightly on $I_C = I_L$. The best solution is to pick a transistor with a value of h_{FE} much greater than 1, so $h_{FE}/(1 + h_{FE}) \cong 1$. This means using a Darlington transistor for Q.

10.10 THE CURRENT-TO-CURRENT CONVERTER

An op amp can be used to multiply a current with the circuit of Fig. 10.11. The input voltage v_{in} is zero from the OAVR. Because the inverting input of the op amp draws negligible current (the OACR), the input current and the voltage v'_{out}, are simply related by Ohm's law:

$$i_{\text{in}} = \frac{-v'_{\text{out}}}{R_2}$$

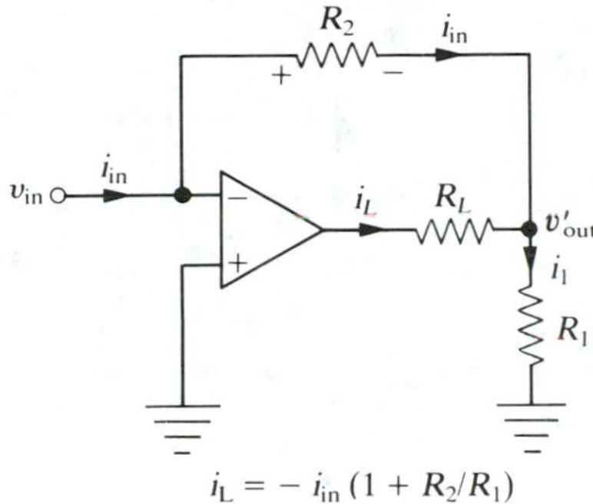

FIGURE 10.11 Current-to-current converter.

The KCL at the junction of R_L and R_1 implies

$$i_{in} + i_L = i_1 = \frac{v'_{out}}{R_1}$$

Using $v'_{out} = -i_{in}R_2$ yields

$$i_{in} + i_L = \frac{-i_{in}R_2}{R_1}$$

or

$$i_L = -i_{in}\left(1 + \frac{R_2}{R_1}\right) \tag{10.24}$$

Thus we see that the current through the load resistance is independent of R_L and depends only on the input current and the ratio R_2/R_1. Notice also that both ends of the load resistance must be floating above ground. A resistance can be placed in series with the noninverting input to minimize bias current effects on the output.

10.11 THE LOGARITHMIC AMPLIFIER

With the circuit of Fig. 10.12, we can use an op amp to produce an output voltage that is proportional to the logarithm of its input voltage. The qualitative explanation of the logarithmic amplifier is that as the input voltage goes more positive, the output of an op amp goes negative, which tends to turn on the npn transistor (the emitter going negative is equivalent to the base going positive). Thus, there is a smaller and smaller resistance (of the transistor) between the op amp output and the inverting input, which lowers the gain of the op amp. Thus, the larger the input signal voltage, the smaller the op amp gain, which effectively "compresses" the gain—a factor

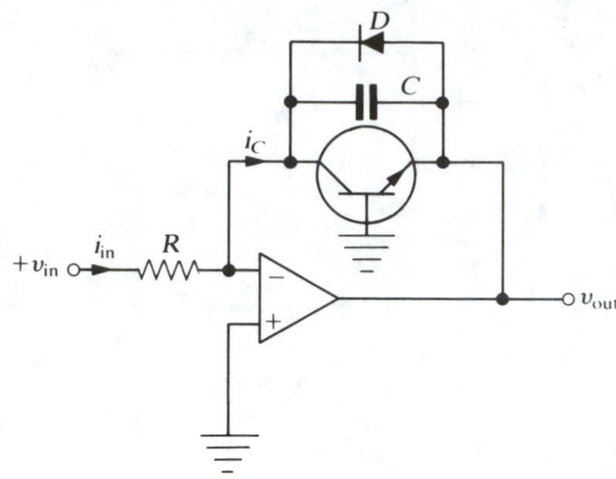

FIGURE 10.12 The logarithmic amplifier.

of 10 or 100 in the input will produce a much smaller factor in the output.

The KCL and the OACR applied to the junction at the inverting terminal imply $i_{in} = i_C$. The input current i_{in} can be expressed with Ohm's law, and the collector current of the transistor can be expressed in terms of its base emitter voltage.

$$i_C = \frac{v_{in} - 0}{R} = I_0\, e^{eV_{BE}/kT}$$

But the base emitter voltage of the transistor is simply the negative of the output voltage: $V_{BE} = -v_{out}$. Thus,

$$\frac{v_{in}}{R} = I_0 e^{-ev_{out}/kT}$$

Solving for the output voltage, we obtain

$$v_{out} = -\frac{kT}{e} \ln\left(\frac{v_{in}}{I_0 R}\right) \tag{10.25}$$

Using $v_{in} = +i_{in}R$, we can rewrite the output as

$$v_{out} = -\frac{kT}{e} \ln\left(\frac{i_{in}}{I_0}\right)$$

Notice:

1. The input voltage v_{in} must be positive (otherwise no current would flow through the transistor).

2. The output voltage range is limited to approximately −0.2 V to −0.7 V, because $v_{out} = V_{EB}$.
3. I_0 is usually about 100 pA or less because it is the reverse current for the transistor base-emitter junction and is temperature dependent.
4. The factor kT/e at room temperature is 0.026 V.
5. For small input currents i_{in} it is necessary to use an op amp with a bias current that is much smaller than the smallest i_{in}; this usually means an op amp with an FET input stage. For small input currents the input voltage v_{in} is also small, which means the output voltage due to input voltage offset may swamp the desired output; this means we must carefully null out the offset voltage with appropriate circuitry.
6. It is often necessary to add a small capacitance across the transistor to prevent oscillation as shown in Fig. 10.12.
7. When the op amp output goes positive (from a negative input), the npn transistor is turned off and the maximum collector emitter voltage may be exceeded. A diode is sometimes placed across the transistor to protect it as shown in Fig. 10.12.
8. If the input is a current source (with high impedance), the resistance R can be omitted—the high output resistance of the current source takes its place.

10.12 THE IDEAL DIODE

An ordinary silicon p-n junction diode does not pass an appreciable forward current until its forward voltage exceeds approximately $v_D = 0.5$ V, as shown in Fig. 10.13(a). However, if the diode is put in the feedback loop of an op amp, as shown in Fig. 10.13(b), the turn-on voltage of the diode is effectively reduced to $0.5\text{ V}/A_o$, where A_o is the open-loop gain of the op amp. The qualitative explanation is that if the input voltage goes positive by only a slight amount (approximately v_D/A_o, where v_D is the diode turn-on voltage), the output of the op amp will go strongly positive and will turn on the diode. With the diode turned on the op amp circuit is essentially a voltage follower (see Section 10.3) with a gain of unity. Thus, the output

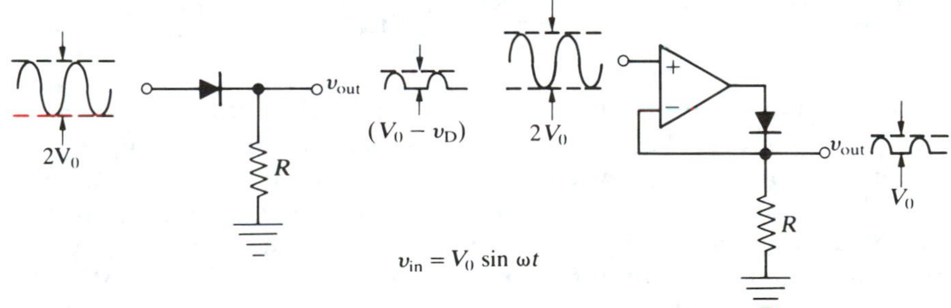

(a) *real diode half-wave rectifier* (b) *ideal diode half-wave rectifier*

FIGURE 10.13 The ideal diode.

equals the input, and the circuit acts like an ideal forward biased diode with almost zero turn-on voltage. But for a small negative input the op amp output will go strongly negative, thus turning the diode off. There is then no negative feedback, and the op amp output will saturate at close to the negative supply voltage $-V_{cc}$. The diode is then strongly reverse biased and the output of the circuit will be zero, because no current flows through R. Thus, the circuit acts like an ideal reverse biased diode.

An ideal diode clamp can be made with the circuit of Fig. 10.14(a). The

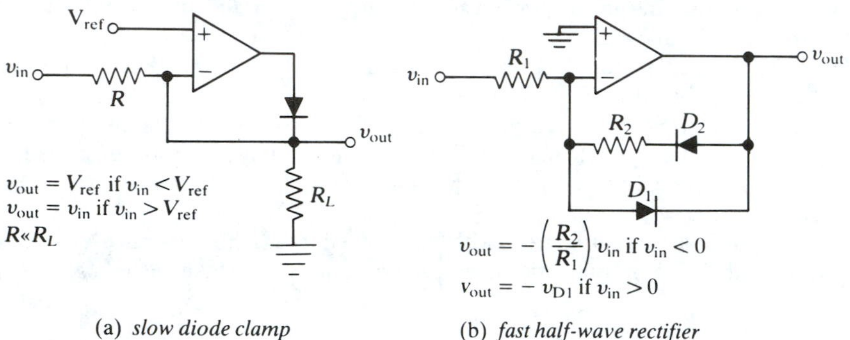

FIGURE 10.14 Diode clamp circuits.

noninverting input of the op amp is connected to a reference voltage. If the input voltage is less than the reference voltage, the op amp output goes positive, which turns on the diode. The output voltage is then equal to the reference voltage by the OAVR because the two inputs to the op amp must be the same. We say the output is *clamped* to the reference voltage, regardless of the input voltage, as long as the input is *less* than the reference voltage. But if the input voltage is greater than the reference voltage, then the op amp output goes strongly negative (approximately $-V_{cc}$), which turns off the diode. Thus, the output is proportional to the input because of the voltage divider action of R and R_L:

$$v_{out} = \frac{R_L v_{in}}{R + R_L} \cong v_{in}$$

If $R \ll R_L$, the output approximately equals the input. Notice also that as the input voltage goes below the reference voltage, the op amp output must go from $-V_{cc}$ to near V_{ref}, which can be 10 or 20 V, depending on V_{cc} and V_{ref}. Thus, the slew rate of the op amp may limit how fast the circuit can make this transition. For example, if the slew rate of the op amp is 2 V/μs, the op amp will take 9 μs for its output to swing from -13 V to $+5$ V. Thus to avoid distortion in the output, the input signal cannot change appreciably in 9 μs or a shorter time.

The circuit of Fig. 10.14(b) is a fast half-wave rectifier. If the input is negative, the op amp output is positive and D_2 is on and D_1 is off. Then we have a standard inverting op amp, and $v_{out} = -(R_2/R_1)v_{in}$. If the input goes positive, the op amp output swings negative but only to -0.5 V (the drop across D_1); D_2 is off and D_1 is on. Then $v_{out} = -v_{D1}$ because the voltage at the inverting op amp input equals 0 by the OAVR. The circuit is thus a half-wave rectifier whose output is scaled by the factor $-R_2/R_1$. This circuit is faster because the output voltage swing is less.

An ideal full-wave rectifier can be made with the circuit of Fig. 10.15.

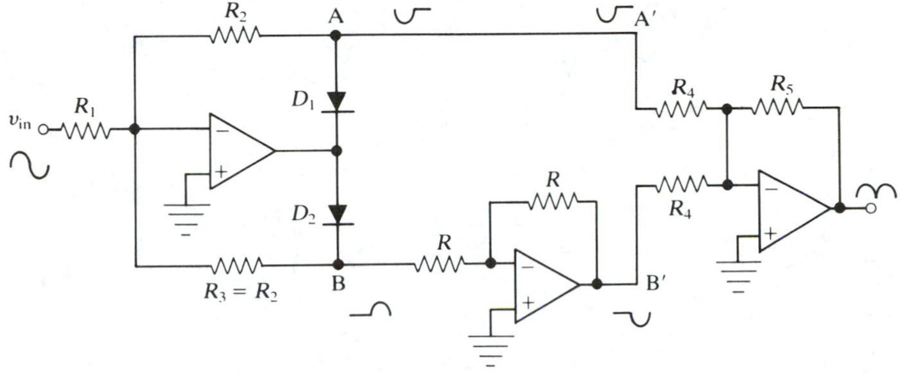

FIGURE 10.15 The ideal full-wave rectifier.

If the input is positive, the op amp output is negative; then D_1 is on and D_2 is off. Then we have a standard inverting op amp circuit, and the output is $v_A = -(R_2/R_1)v_{in}$. The output v_B is zero because there is no current through R_3. But if the input is negative, then the op amp output is positive and D_1 is off and D_2 is on. Again, we have a standard inverting op amp circuit, and the output is $v_B = -(R_3/R_1)v_{in}$, which is positive because v_{in} is negative. The output v_A is zero because there is no current through R_2. Thus, the positive part of the input appears as a negative voltage at output A, and the negative part of the input appears as a positive voltage at the output B.

An inverter amplifier with a gain of -1 in series with B will change the polarity of v_B. Then the two outputs at A′ and B′ are summed and inverted in sign by using a summing amplifier to produce the full-wave positive output. Then the output of the summer will be R_2R_5/R_1R_4 times the absolute value of the input; that is, the output will be that of a full-wave rectifier.

10.13 THE PEAK DETECTOR

If we add a high-quality capacitor to the output of the ideal half-wave rectifier of Fig. 10.13(b), let $R \to \infty$, and add a voltage follower, we have a peak detector as shown in Fig. 10.16(a). The output voltage will be stored

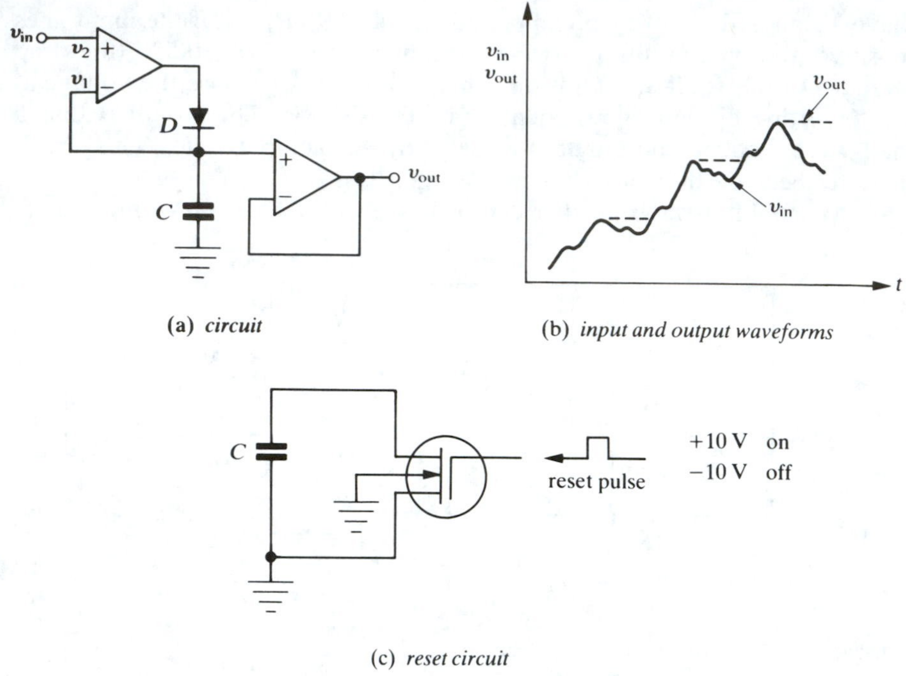

FIGURE 10.16 Peak detector.

on the capacitor, and the voltage at the inverting input of the op amp will equal the output voltage $v_1 = v_{out}$. Thus if the input rises above v_{out}, the first op amp output will go more positive and turn on D, thus producing a voltage follower—the output voltage will follow the input voltage. The capacitance C will be charged through the diode by the op amp to the new more positive input voltage. But if the input voltage drops below v_{out}, the op amp output will fall and turn off D. Then the capacitance will be isolated and will hold its charge and voltage. The voltage on the capacitance is then fed into the input of a voltage follower. The discharge current from the capacitor will be the sum of the small reverse diode current and the bias current into the inverting input of the first op amp and the voltage follower.

If the input varies with time, then the output will rise to the most positive value of the input and hold there at the peak value until a more positive input comes along, as shown in Fig. 10.16(b). Notice:

1. The output voltage will droop (gradually decrease in time) as the capacitance discharges. To minimize this droop, use an op amp with a small input bias current (i.e., an FET op amp). The capacitance should also have a very high leakage resistance and low polarization, which usually means a polystyrene or silver mica capacitor must be used. The voltage follower op amp should also have a low input bias current, which means an FET op amp. The maximum droop rate is $dv_{out}/dt = I_B/C$, so a large capacitance is desirable.

2. The maximum rate of output change is limited by the slew rate of the op amp and by C.
3. The circuit can be reset by shorting out the capacitance. An FET switch can be used as shown in Fig. 10.16(c), but the switch must have an extremely high resistance when off. Thus, a MOSFET device would be appropriate. The positive reset pulse turns the MOSFET on and discharges the capacitor.
4. The output voltage change per unit time is limited by the current supplied by op amp 1.

$$\frac{dv_{out}}{dt} = \frac{d}{dt}\left(\frac{Q}{C}\right) = \left(\frac{1}{C}\right)I_{max}$$

where I_{max} is the maximum output current of the first op amp.

10.14 THE SAMPLE-AND-HOLD CIRCUIT

A circuit similar to the peak detector (Section 10.13) can be used to sample a waveform or pulse and store it for any desired length of time. This is particularly useful in analog-to-digital converters (covered in Chapter 15) where a voltage level at a certain time in an input analog waveform must be sampled and held for a period of time during which time the voltage level is converted into digital form. The circuit, shown in Fig. 10.17, uses a high-

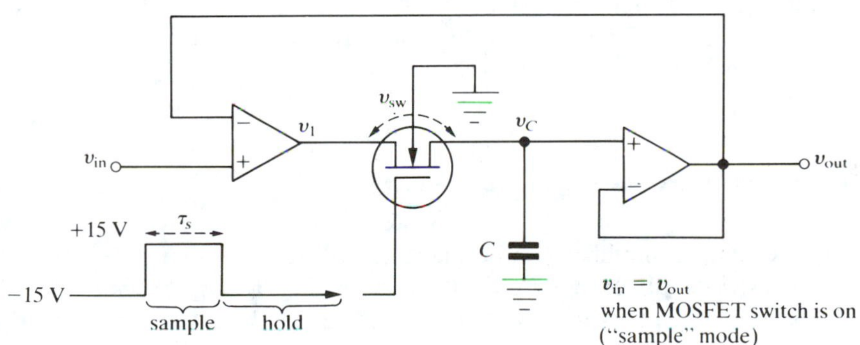

FIGURE 10.17 Sample-and-hold circuit.

input-impedance op amp at the input. When the MOSFET switch is "on" (from a positive voltage on its gate), the circuit is sampling the input, and $v_{out} = v_C \to v_{in}$ because the whole circuit acts like a voltage follower; that is, the capacitance is charged to the input voltage through the MOS-FET. When the MOSFET is "off" the input voltage is "held" on the capacitor. The second voltage follower presents a high impedance across the capacitor, so the capacitor voltage does not decay appreciably before the next sample is taken. Notice:

1. The circuit samples the input for a time τ_s when the signal on the gate of the MOSFET is +15 V, thus turning the MOSFET on. The sampling time τ_s should be short enough so that the input signal does not change appreciably during τ_s. When the MOSFET gate voltage is at −15 V, the MOSFET is off and the circuit holds the input.
2. The bias current I_B of the second voltage follower must be small so that the capacitor does not discharge too quickly. The rate at which the capacitor voltage decreases or droops is

$$\frac{dV}{dt} = \left(\frac{1}{C}\right) I_B$$

so it is desirable to have I_B small and C large.
3. The capacitor charges through the series resistance of the "on" MOSFET, which is typically 50–100 Ω. Thus, the charging time constant is $RC \cong (100\ \Omega)(C)$, which should be smaller than the time during which the input signal changes appreciably. Thus, a small C is best. (The MOSFET resistance and C form a low-pass RC filter.)
4. The first voltage follower supplies the current to charge C. Because the current must be large enough to charge C quickly, the op amp output impedance should be small:

$$\frac{dV}{dt} = \frac{I_{out}}{C}$$

where I_{out} = the output current of the first op amp.
5. C should be high quality, and its leakage resistance should be large to avoid droop (polystyrene, teflon, or mica).

10.15 THE OP AMP DIFFERENTIATOR

An operational amplifier can be made to differentiate by using a capacitor and resistor feedback network, as shown in Fig. 10.18. Assume $C' = 0$ in the following explanation. Applying the KCL at the inverting input of the op amp yields

$$i_{in} = i_2 + i'$$

If we let Q be the charge on the capacitor C then $i_{in} = dQ/dt$. Neglecting i' as usual from the OACR and writing i_2 from Ohm's law yield

$$\frac{dQ}{dt} = \frac{v_1 - v_{out}}{R}$$

But $v_1 = v_2 = 0$ from the OAVR, which implies $Q = Cv_{in}$. Thus,

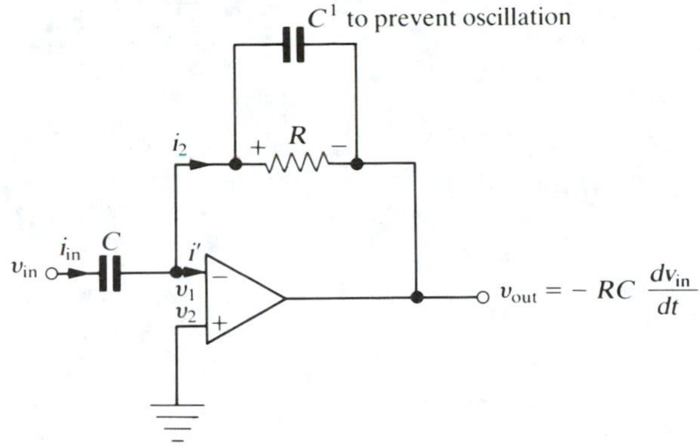

FIGURE 10.18 Op amp differentiator.

$$\frac{d(Cv_{in})}{dt} = -\frac{v_{out}}{R}$$

or

$$v_{out} = -RC\left(\frac{dv_{in}}{dt}\right) \tag{10.26}$$

Notice that, unlike the passive RC differentiator of Chapter 3, we have made no assumptions about the input signal period compared to RC. In theory the op amp differentiator will work over a frequency range from dc to infinity, unlike the passive integrator of Chapter 3. In practice, of course, things are not so ideal. The differentiator tends to oscillate because its gain is so large at high frequencies. A practical cure for oscillation is a small capacitor C' connected across R to decrease the gain at high frequencies. The reactance of C' should be comparable to R at the frequency of oscillation. Sometimes a resistance in series with C is necessary.

10.16 THE OP AMP INTEGRATOR

An op amp can be made to integrate by using the circuit of Fig. 10.19. Applying the KCL at the inverting terminal of the op amp yields

$$i_{in} = i_2 + i'$$

If Q is the charge on the capacitor, then $i_2 = dQ/dt$. By the OACR we can neglect i'. By Ohm's law $i_{in} = (v_{in} - v_1)/R$, and $v_1 = v_2$ from the OAVR, and $v_2 = 0$. Thus, we have

$$\frac{v_{in}}{R} = \frac{dQ}{dt}$$

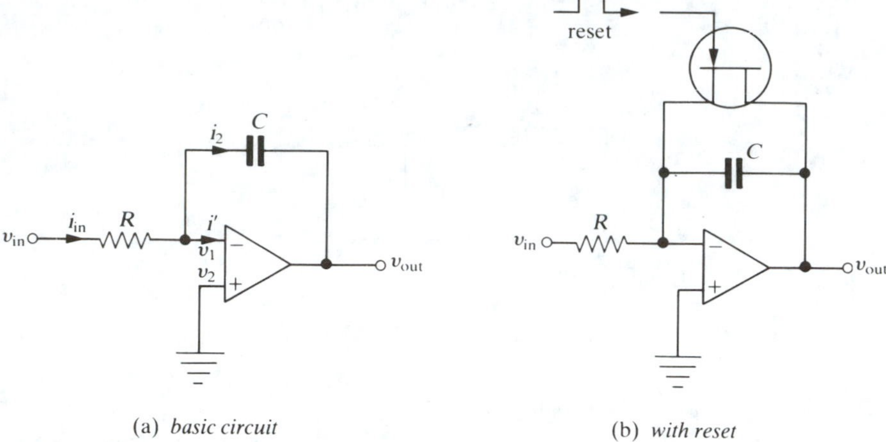

FIGURE 10.19 Op amp integrator.

But $Q = CV$, where V is the voltage across the capacitor, and $V = v_1 - v_{out} = -v_{out}$. Thus,

$$\frac{v_{in}}{R} = \frac{d(-Cv_{out})}{dt} = -C\left(\frac{dv_{out}}{dt}\right)$$

or

$$v_{out} = \frac{-1}{RC}\int v_{in}\,dt \qquad \textbf{(10.27)}$$

As in the case of the differentiator, notice that we made no assumptions about the input signal period compared with RC. Thus, in theory the op amp integrator will integrate over a frequency range from dc to infinity, unlike the passive RC integrator of Chapter 3. Notice:

1. At the start of the integration we want the output to be zero, which requires that we start with no charge on the capacitor. Thus, there must be a shorting switch across C that should be opened when the integration is to start (i.e., when the input signal is applied). An FET switch can be used, as shown in Fig. 10.19(b). The signal to the gate that turns on the FET and shorts out C is called the *reset* signal. The time required to reset the integrator is about $R'C$, where R' is the "on" resistance of the FET channel, because C discharges through the FET.
2. If the input voltage is constant, $v_{in} = V_0$, then the output will be a negative-going linear ramp: $v_{out} = (-V_0/RC)t$.
3. $i_{in} = v_{in}/R$, so the output is proportional to the time integral of the input current: $v_{out} = -(1/C)\int i_{in}\,dt$.
4. There usually will be a long-term dc drift in the output of the integrator due to input offset effects, both voltage and current. For example, if V_{io} is the input offset voltage of the op amp, then V_{io} will be integrated by the integrator (if the capacitor is not shorted) and the output will be

$$v_{out} = \frac{-1}{RC}\int V_{io}\,dt$$

This drift also depends on the temperature because V_{io} depends on the temperature, typically 1–10 μV/°C. One cure is to use a large time constant RC, but this makes the output small for a real signal input. Another widely used cure is a resistance R_2 connected across C to bleed off the charge on C in a time of approximately R_2C. But this requires integrating the signal input in a time short compared to R_2C. Bias current effects can be reduced by using a low-bias-current FET op amp. Integrators, in general, are much better behaved than differentiators.

10.17 THE CHARGE-SENSITIVE AMPLIFIER

In some cases the signal source voltage remains essentially constant, and the source capacitance changes, such as in a capacitance microphone. Or the signal source emits a certain amount of charge dQ that must be detected, as in an X-ray proportional counter. What is required is a circuit to convert this charge dQ into an output voltage change; the amplifier to do this is called a *charge-sensitive amplifier*. The circuit of Fig. 10.20(a)

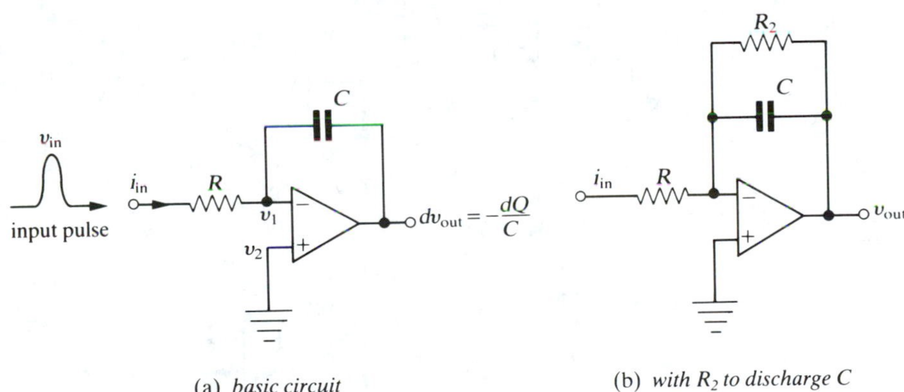

FIGURE 10.20 Charge-sensitive amplifier.

produces an output voltage change dv_{out} proportional to dQ. The circuit is basically an integrator. The op amp voltage rule implies $v_1 = v_2 = 0$. And the op amp current rule implies the input charge dQ all piles up on the capacitor plate. Thus

$$dQ = C(-\,dv_{out})$$

or

$$dv_{out} = -\frac{dQ}{C} \tag{10.28}$$

Thus the output voltage change is linearly proportional to the change in dQ. The smaller is C, the larger is the output voltage change for a given change in charge.

A little thought will show that as more and more input increments of charge come along, the output voltage steadily gets more and more negative until the op amp is saturated and its output voltage is near $-V_{cc}$. Thus, we must discharge the capacitor between the dQ inputs; to do so, we add a resistance R_2 in parallel with C, as shown in Fig. 10.20(b). R_2C limits the maximum counting rate, which must be less than $1/R_2C$. For example, if $R_2C = 1\ \mu s$ ($C = 10$ pF, $R_2 = 100$ kΩ), the maximum rate of dQ events is approximately 1 MHz. For example, a plastic scintillator and photomul-

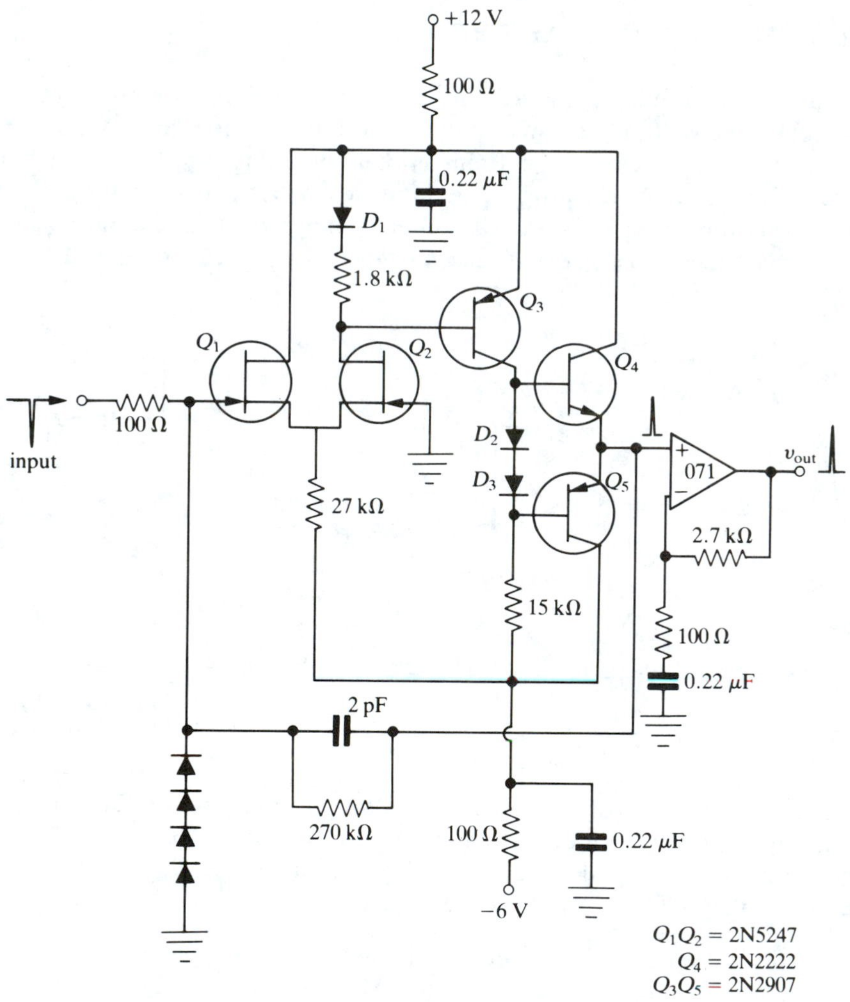

(a) *discrete component circuit*

FIGURE 10.21 Charge-sensitive amplifiers.

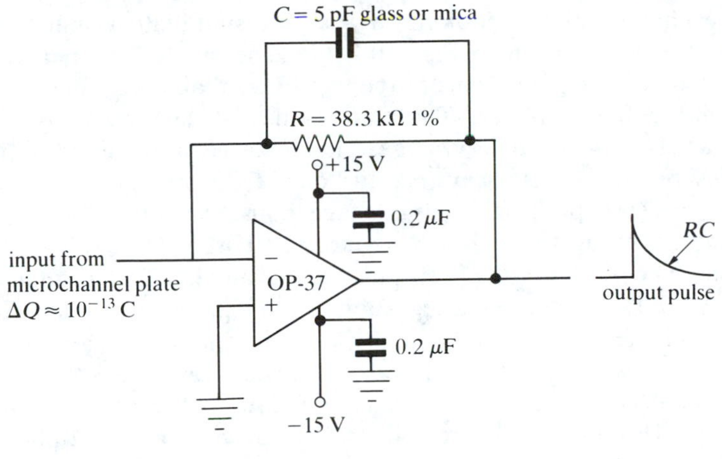

(b) *op amp circuit*

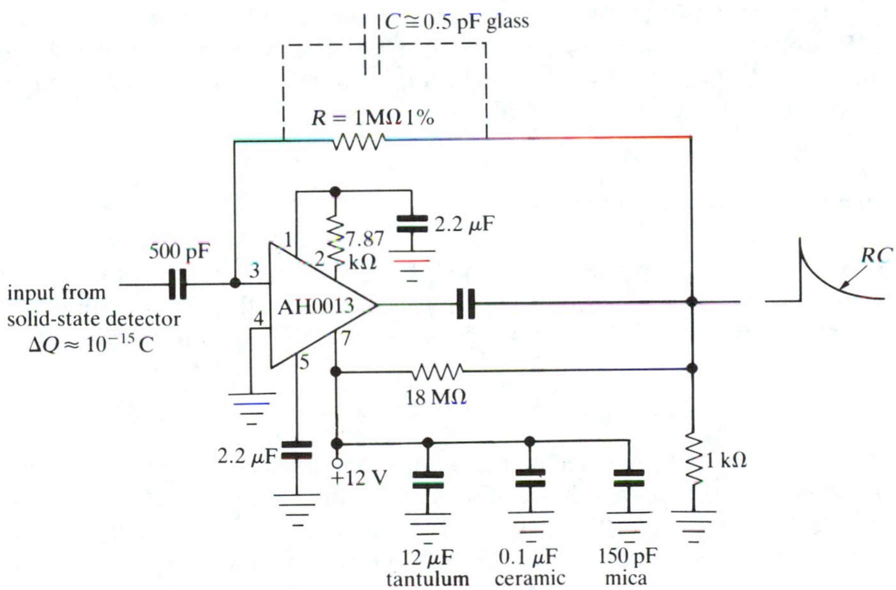

(c) *low-noise pre amp circuit*

FIGURE 10.21 Continued.

tiplier tube will produce a 50-ns pulse input with $dQ = 24 \times 10^{-12}$ C = 24 pC. The output will then be a 4-V pulse with $C = 6$ pF. Commercial charge-sensitive amplifiers can detect charge increments $dQ = 0.16$ to 10 pC.

A charge-sensitive preamplifier used to amplify fast (50–ns wide) pulses from a photomultiplier (looking at a plastic scintillator) built from discrete components is shown in Fig. 10.21(a). The small, fast, negative input pulse is fed into the gate of one of the two FETs that are connected as a standard differential amplifier. The four diodes at the input provide protection against excessive input voltage. The diode D_1 in the drain of the other FET ensures that Q_3 is always biased "on." Q_3 is a common emitter amplifier and drives a push-pull output stage consisting of Q_4 and Q_5. Diodes D_2 and D_3 ensure that Q_4 and Q_5 are just barely conducting with no signal input. The output of the push-pull stage is amplified by a standard noninverting BIFET op amp. The integration of the input is due to the 2-pF capacitance from the emitters of Q_4 and Q_5 to the input and the 100-Ω resistance in series with the input. Two high-speed charge-sensitive amplifier circuits are shown in Figs. 10.21(b) and 10.21(c). The OP-37 op amp circuit is designed to amplify current pulses from a "microchannel plate." The AH0013 circuit is designed to amplify very small pulses from a solid-state electron detector. Notice the very careful bypassing of the +12-V supply; this is necessary for low noise operation.

The LeCroy TRA1000 charge-sensitive preamplifier comes as a 16-pin DIP (dual-in-line-package) and can also be used as a current-to-voltage converter. Its gain is 2.7 to 270 mV/μA or 100 to 500 mV/pC. Its noise can be as low as 0.1 fC (1 fC = 10^{15} C).

10.18 OP AMP COMPARATORS

The output of any op amp is always equal to the very large open-loop gain times the difference between the input voltage at the noninverting and inverting terminals of the op amp, provided the op amp is not saturated.

$$v_{out} = A_o(v_2 - v_1)$$

It is important to remember that v_2 and v_1 are the voltages at the op amp *terminals*, not the input voltages elsewhere in the circuit. The open-loop op amp gain is typically 10^5 to 10^6, and the maximum output voltage is the supply voltage V_{cc}, typically 15 V. Most bipolar op amp outputs can swing within 1–2 V of $\pm V_{cc}$; thus if $V_{cc} = 15$ V, the maximum output voltage might be 13 V. For discussion purposes we will assume the swing is to $\pm V_{cc}$. Thus, if the *differential* input is greater than 15 V/10^6 = 15 μV, the op amp output will be saturated at either +15 V or −15 V, depending on the polarity of the input. In other words, as the differential input slowly increases from a large negative value to a large positive value, the output suddenly changes from $-V_{cc}$ to $+V_{cc}$ as the input changes from −15 μV to +15 μV [see Fig. 10.22(a)]. An ideal comparator's response is shown in Fig. 10.22(b).

The time for the op amp to change from $-V_{cc}$ to $+V_{cc}$ is usually very short but is always limited by the slew rate of the op amp. Any op amp thus

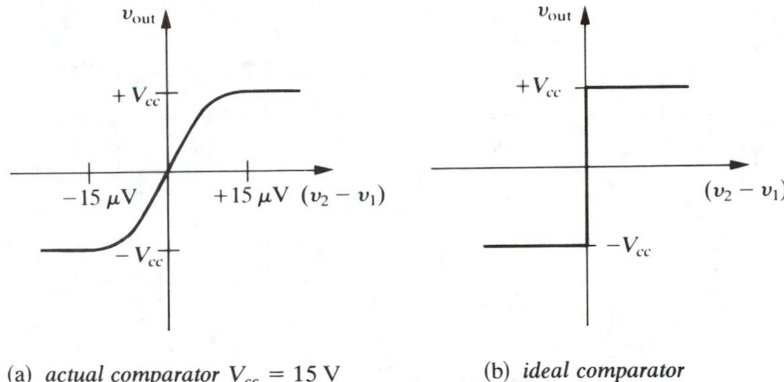

(a) *actual comparator* $V_{cc} = 15$ V (b) *ideal comparator*

FIGURE 10.22 Comparator waveforms.

will essentially "compare" the two inputs at its two terminals and give either a large positive or a large negative output, provided its differential input exceeds approximately 15 μV. An op amp specially made to do this is called a *comparator*; it may switch its output from low to high in a very short time (as short as 50–100 ns) and may have a slew rate of up to 500 to 1000 V/μs. (An ordinary general-purpose op amp slew rate is approximately 1–5 V/μs.) This time to switch from $-V_{cc}$ to $+V_{cc}$ is called the *response time* of the comparator. One should bear in mind, however, that the faster the comparator, the more prone it is to unwanted oscillations, so unless a very short response time is essential to the circuit application, it may be better to use an ordinary op amp.

The comparator output can be restricted or clipped to values less than $\pm V_{cc}$ by connecting two zener diodes to the output as shown in Fig. 10.23(a). The maximum positive output voltage will be $V_D + V_{Z2}$, and the maximum negative voltage will be $-(V_D + V_{Z1})$, where V_D is the drop across the forward biased zener diode and V_Z is the zener voltage. One of the comparator inputs can be a fixed reference voltage, so the output will indicate when the input is greater than or less than the reference voltage, as shown in Fig. 10.23(b).

Some comparators have *open collector* outputs, which means one must wire an external resistor from the output to a positive supply voltage, as shown in Fig. 10.23(c). Then the output is either $+V_{cc}$ when Q is off or approximately 0.2 V when Q is strongly on (saturated). This circuit is useful in digital applications where the two logic levels are 3.5 V and 0.2 V.

If the input changes slowly, the comparator output should flip suddenly when the differential input voltage goes through zero. But if the input is noisy, the differential input may change sign several times in a short time interval as the overall dc input level changes sign; in other words, the comparator output will flip back and forth several times from $+V_{cc}$ to $-V_{cc}$ instead of making just one flip. The output will then contain a brief burst of

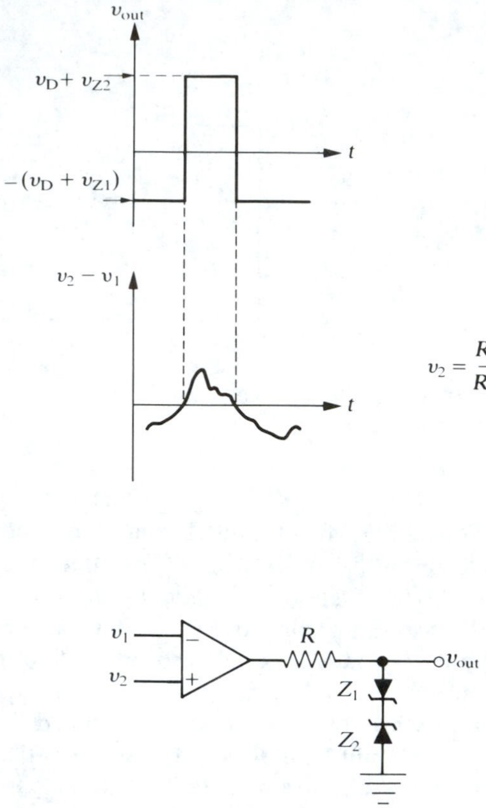

(a) *clipped output*

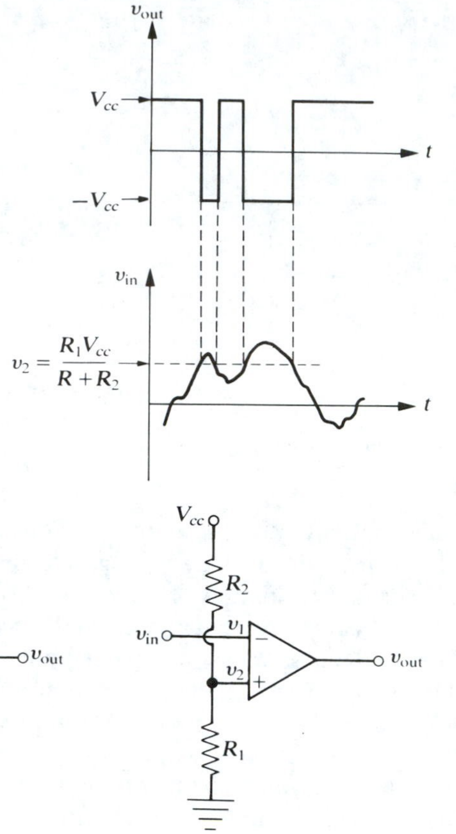

(b) *input compared to reference voltage*

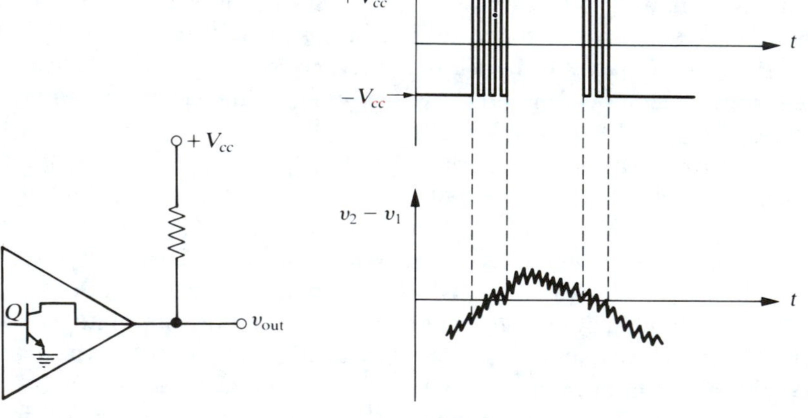

(c) *open collector output*

(d) *noisy input*

FIGURE 10.23 Comparators.

large "noise" pulses, as shown in Fig. 10.23(d), that may wreak havoc with the rest of the circuit. To avoid this, we need less noise on the input signal, a slower comparator, or a circuit that flips its output up when the differential input goes 15 μV positive but flips back when the differential input goes several tenths (or more) of a volt negative—a circuit that has a different input voltage threshold for a positive output flip compared to a negative output flip. This behavior is called *hysteresis*, and the Schmitt trigger circuit of Section 10.19 solves the problem.

A simple comparator application is shown in Fig. 10.24; when the

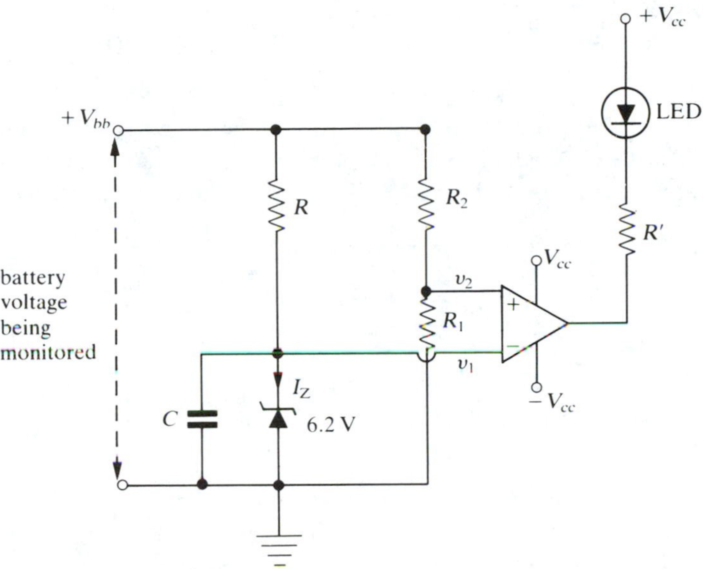

FIGURE 10.24 Low-battery indicator.

battery voltage drops below a certain value determined by R_1 and R_2, the op amp comparator output goes low and turns on the LED. The comparator inputs are

$$v_2 = \frac{R_1}{R_1 + R_2} V_{bb} \quad \text{and} \quad v_1 = 6.2 \text{ V (the zener voltage)}$$

Under normal operation with a good battery $v_2 > v_1$, the op amp output is high (V_{cc}), and the LED is out. When V_{bb} decreases to the value where $v_2 < v_1$, then the op amp output goes low ($-V_{cc}$), which turns on the LED. For example, if the battery voltage is $V_{bb} = 13.2$ V when fully charged (a "12"-V auto battery) and if we desire the LED to go on when V_{bb} falls to 12 V, then we choose R_1 and R_2 so that $v_2 = 6.2$ V when $V_{bb} = 12$ V. Thus

$$\frac{v_2}{V_{bb}} = \frac{6.2\text{ V}}{12\text{ V}} = \frac{R_1}{R_1 + R_2} = 0.517$$

Hence,
$$1 + \frac{R_2}{R_1} = \frac{1}{0.517} = 1.934 \quad \text{or} \quad \frac{R_2}{R_1} = 0.934$$

R is chosen so that the zener diode conducts approximately $I_Z = 1\text{–}2$ mA. Thus

$$I_Z R \cong V_{bb} - V_Z$$

or
$$R \cong \frac{V_{bb} - V_Z}{I_Z} = \frac{13.2\text{ V} - 6.2\text{ V}}{2\text{ mA}} \cong 3.5\text{ k}\Omega$$

R' is chosen so that the LED is brightly on when the op amp comparator output is low ($-V_{cc}$), which occurs for a current of 10 mA for most LEDs. The voltage drop across a lighted LED is approximately 1.5 V, so

$$1.5\text{ V} + I_{\text{LED}} R' = V_{cc} - (-V_{cc})$$

$$R' = \frac{2V_{cc} - 1.5\text{ V}}{I_{\text{LED}}}$$

$$R' = \frac{28.5\text{ V}}{I_{\text{LED}}} = \frac{28.5\text{ V}}{10\text{ mA}} \cong 2.8\text{ k}\Omega$$

A capacitance C across the zener diode might be necessary to prevent noise from flipping the comparator. Finally, we note that a slow comparator (i.e., an ordinary op amp) would work fine in this application.

10.19 THE SCHMITT TRIGGER

If we tie the output of the comparator to the noninverting input, as shown in Fig. 10.25(a), then the voltage at this input v_2 will depend on the output. If the output is high, v_2 will be higher than it will be when the output is low; this produces hysteresis. The voltage at which the comparator flips depends on the "history" of the circuit—whether the circuit output was previously high or low.

A quick numerical calculation should make the point clear. From Fig. 10.25(b), the KCL implies

$$i_2 + i_3 = i_1$$

$$\frac{V_{\text{ref}} - v_2}{R_2} + \frac{v_{\text{out}} - v_2}{R_3} = \frac{v_2}{R_1}$$

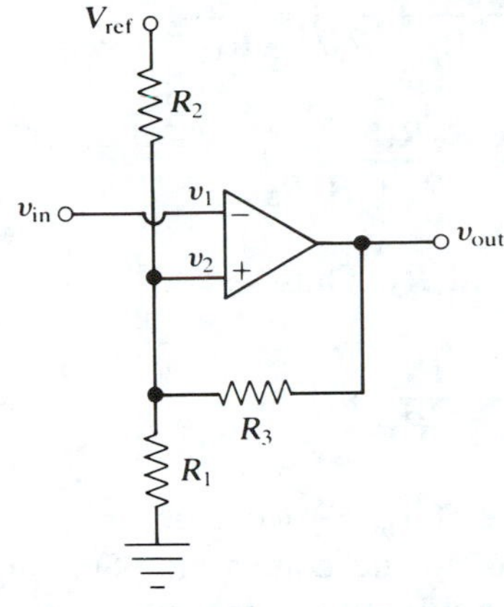

(a) *basic circuit*

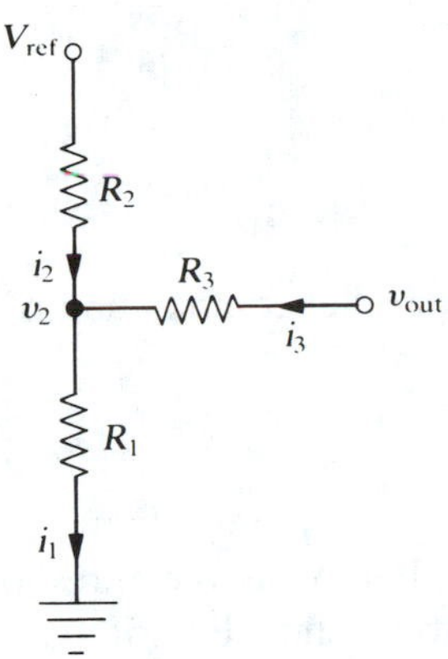

(b) *voltage divider circuit*

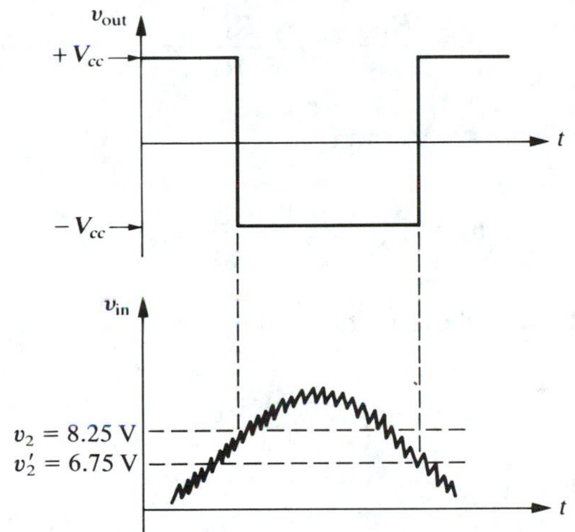

(c) *waveforms for* $R_1 = R = 1\ \text{k}\Omega$, $R_3 = 10\ \text{k}\Omega$, $V_{cc} = 15$ V

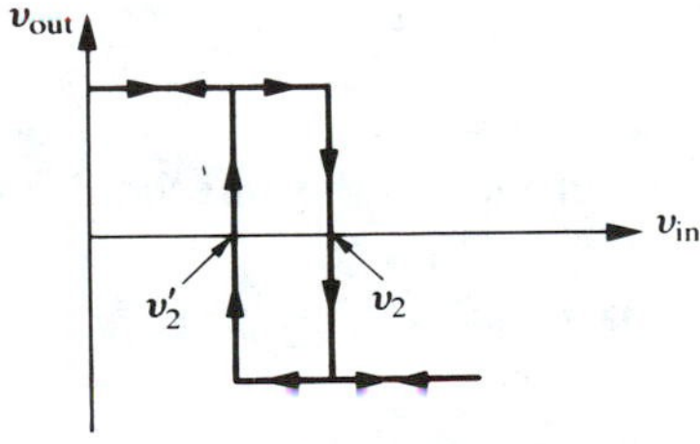

(d) *input-output hysteresis curve*

(e) *simple Schmitt trigger circuit symbol*

FIGURE 10.25 The Schmitt trigger.

or
$$\frac{V_{ref}}{R_2} + \frac{v_{out}}{R_3} = v_2\left(\frac{1}{R_1} + \frac{1}{R_2} + \frac{1}{R_3}\right) = \frac{v_2}{R_{123}}$$

where
$$R_{123} = \frac{R_1 R_2 R_3}{R_1 R_2 + R_1 R_3 + R_2 R_3}$$

is the parallel combination of R_1, R_2, and R_3. Thus,

$$v_2 = \frac{R_{123}}{R_2} V_{ref} + \frac{R_{123}}{R_3} v_{out} \tag{1}$$

Remember, v_2 is the threshold voltage; if $v_{in} = v_1$ is greater than v_2, the comparator flips low; if v_{in} is less than v_2, the comparator flips high. We immediately see from (1) that the threshold voltage depends on v_{out}. If the output is high, $v_{out} = V_{cc}$, and

$$v_2 = \frac{R_{123}}{R_2} V_{ref} + \frac{R_{123}}{R_3} V_{cc} \tag{2}$$

If the output is low, $v_{out} = -V_{cc}$ and

$$v_2' = \frac{R_{123}}{R_2} V_{ref} - \frac{R_{123}}{R_3} V_{cc} \tag{3}$$

The difference in the threshold voltages is

$$v_2 - v_2' = \frac{2R_{123}}{R_3} V_{cc} \tag{4}$$

If $R_1 = R_2 = R$, then

$$v_2 = \frac{R_3}{R + 2R_3} V_{ref} + \frac{R}{R + 2R_3} V_{cc} \cong \frac{V_{ref}}{2} + \frac{R}{2R_3} V_{cc}$$

and
$$v_2' = \frac{R_3}{R + 2R_3} V_{ref} - \frac{R}{R + 2R_3} V_{cc} \cong \frac{V_{ref}}{2} - \frac{R}{2R_3} V_{cc}$$

where we have assumed $R \ll R_3$. The difference in the threshold voltage is then

$$v_2 - v_2' \cong \frac{R}{R_3} V_{cc} \tag{10.29}$$

If $V_{ref} = V_{cc} = 15\text{ V}$, $R_1 = R_2 = R = 1\text{ k}\Omega$, and $R_3 = 10\text{ k}\Omega$, then

$$v_2 = 8.25\text{ V} \qquad v_2' = 6.75\text{ V} \qquad v_2 - v_2' = 1.5\text{ V}$$

Thus, if the output is already high, $v_2 = 8.25$ V, and the input voltage is low and increasing, it must exceed 8.25 V to flip the comparator output to low. If the output is already low, $v_2' = 6.75$ V, and the input voltage is high and decreasing, it must fall below 6.75 V to flip the comparator output to high. If the input goes up to 8.3 V (the output will flip low) and then a noise pulse decreases the input below 8.25 V, the comparator will *not* flip back high. The net effect is to make the comparator insensitive to small amplitude noise in the input, provided the peak-to-peak noise amplitude is less than the difference in the triggering levels. Thus the noise pulses in the output are eliminated as the input passes through the threshold voltage. The waveforms are shown in Fig. 10.25(c) for a comparator with hysteresis. Without hysteresis the waveform is that of Fig. 10.23(d). The input-output voltage graph is shown in Fig. 10.25(d), and a simple Schmitt trigger circuit symbol is shown in Fig. 10.25(e). The open circle at the output means the output is inverted compared to the input. The use of a Schmitt trigger is extremely common in "cleaning up" noisy signals.

10.20 THE SQUARE-WAVE GENERATOR OR ASTABLE MULTIVIBRATOR

If the output of a comparator is fed back to its two inputs, we can make a square-wave oscillator or *astable* multivibrator, as shown in Fig. 10.26(a). Because of the feedback from the output to the noninverting input through R_1 and R_2, the voltage at the noninverting input of the op amp is always equal to

$$v_2 = \frac{R_1}{R_1 + R_2} v_{out} \equiv \lambda v_{out}$$

Thus, the circuit is like the Schmitt trigger circuit in that the "reference" voltage v_2 at the noninverting input depends on the output. Thus, as the capacitor C charges and discharges, the comparator action of the op amp will flip the output at a voltage depending on the output voltage. Let us assume that C is uncharged ($v_1 = 0$) and $v_{out} = +V_{cc}$. C will then charge up from i flowing through R ("toward" V_{cc}) with a time constant RC. The output will flip from $+V_{cc}$ to $-V_{cc}$ when C charges to $v_1 = v_2 = \lambda V_{cc}$. When the output flips to $-V_{cc}$, v_2 will immediately change to $-\lambda V_{cc}$ and C will start discharging from i' flowing through R ("toward" $-V_{cc}$) with a time constant RC. When v_1 reaches $v_2 = -\lambda V_{cc}$, the output will flip from $-V_{cc}$ to $+V_{cc}$. The waveforms for v_1 (on C) and v_{out} are shown in Fig. 10.26(b). The output is a square wave.

An analysis of the charging and discharging waveforms shows that the period is given by

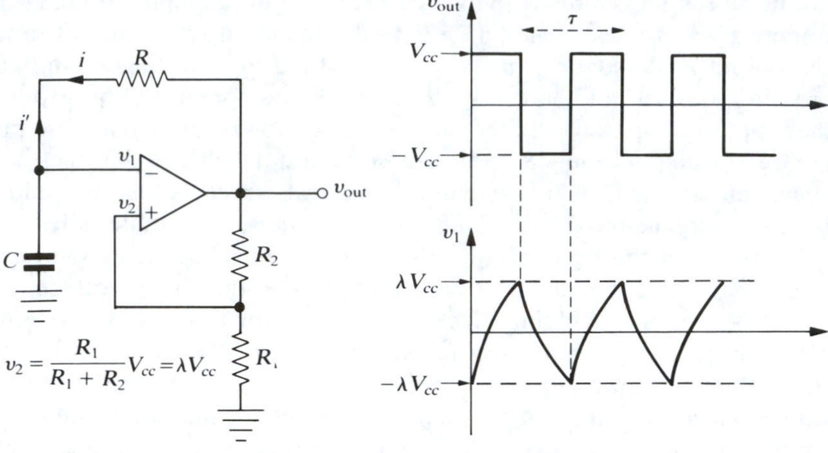

(a) *circuit for symmetric output* (b) *waveforms*

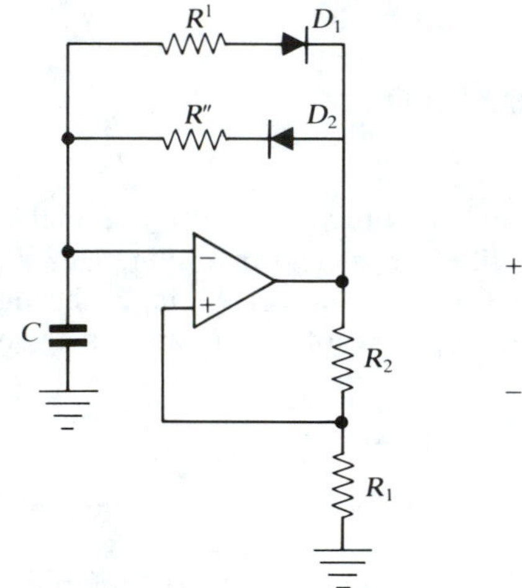

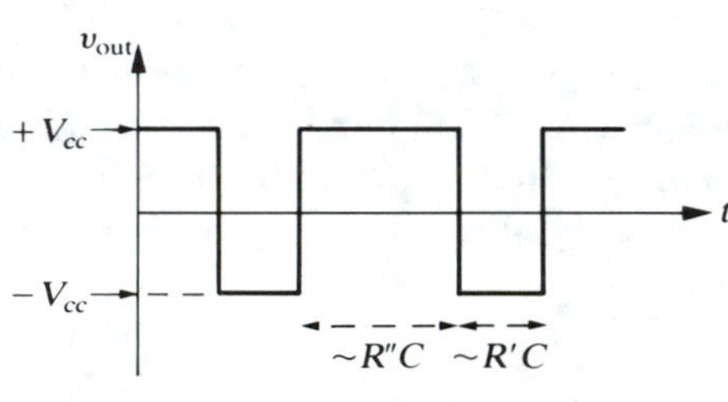

(d) *output waveform for $R'' > R'$*

(c) *circuit for asymmetric output*

FIGURE 10.26 Square-wave generator.

$$T = 2RC \ln\left[\frac{1+\lambda}{1-\lambda}\right] \qquad \text{where } \lambda = \frac{R_1}{R_1 + R_2} \qquad \textbf{(10.30)}$$

Notice:

1. The period is independent of the output amplitude; the period depends only on R, C, R_1, and R_2.
2. The bias current of the op amp should be less than the charging and discharging current of C.

The output square wave can be made asymmetrical by having the capacitor C charge and discharge with different time constants. The circuit of Fig. 10.26(c) accomplishes this with the two diodes. When the output is high, diode D_2 will be on and diode D_1 will be off. Thus, C will charge through R'' with a time constant $R''C$, and the output will be high for approximately $R''C$ seconds. When the output is low, diode D_1 will be on and diode D_2 will be off. Then C will discharge through R' with a time constant $R'C$, and the output will be low for approximately $R'C$ seconds. The waveform output for $R'' > R'$ is shown in Fig. 10.26(d).

For both the symmetrical and the asymmetrical circuits the output has to change from a large positive voltage (V_{cc} or near V_{cc}) to a large negative voltage ($-V_{cc}$ or near $-V_{cc}$) in a time short compared to the period of the square wave. Thus, the slew rate of the op amp must be large enough to accomplish this. If the slew rate is not large enough, the output will not be square—there will be appreciable rise and fall time, and at higher frequencies the output will be rounded off into an approximately triangular shape. For an inexpensive op amp with a slew rate of $1.0\,\text{V}/\mu\text{s}$, the maximum square-wave output frequency is approximately 10 kHz.

10.21 THE TRIANGLE-WAVE GENERATOR OR RAMP GENERATOR

A linearly increasing voltage or *ramp* can be produced if a capacitance C is charged with a constant current, as shown in Fig. 10.27(a). If i_0 is the constant charging current, then the (output) voltage on the capacitance will be $V_C = Q/C = i_0 t/C$, where t is the time. If we add a constant current source to the square-wave generator circuit of Fig. 10.26 so that it supplies the charging current to the capacitor, then the capacitor voltage will increase linearly with time as it charges and will decrease linearly with time as it discharges. The basic circuit is shown in Fig. 10.27(b) for C charging, and the constant current sources are realized with n-channel JFETs as shown in Fig. 10.27(c). Assume the op amp (comparator) output is high (V_{cc}) and there is zero charge on C. Then $v_2 = \lambda V_{cc}$ [$\lambda = R_1/(R_1 + R_2)$], and $v_1 = v_C$ increases as C charges through the two FETs. With a high op amp output, Q_2 conducts strongly because its gate is positive with respect to its source; that is, Q_2 acts like a low resistance of several hundred ohms. But Q_1 acts like a constant current source because its gate is negative with respect to its source (see Fig. 6.4 for typical FET curves). C continues to charge linearly until $v_C = v_2 = \lambda V_{cc}$, at which point the output flips to $-V_{cc}$. Then C will discharge linearly through the two FETs with a constant discharge current. As C discharges, Q_1 is strongly on and Q_2 acts like a constant current source. When v_C decreases to $-\lambda V_{cc}$, the output flips high and C starts to charge linearly. Thus, the output on the capacitor will be a ramp or a sawtooth waveform as shown in Fig. 10.27(d).

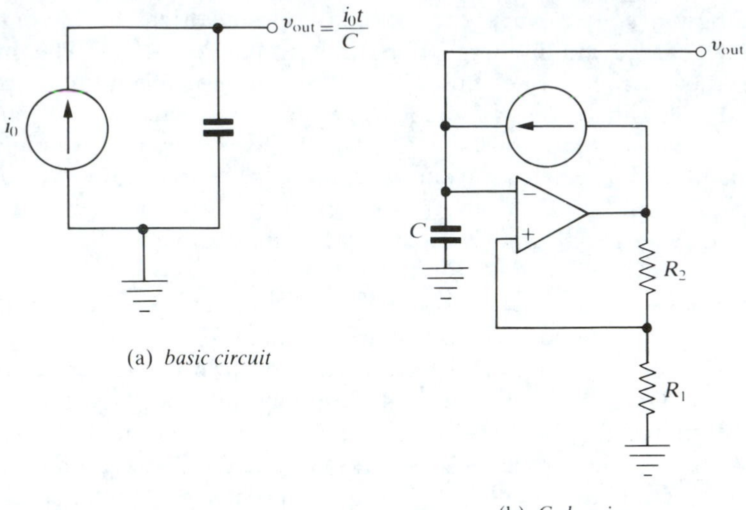

(a) *basic circuit*

(b) *C charging*

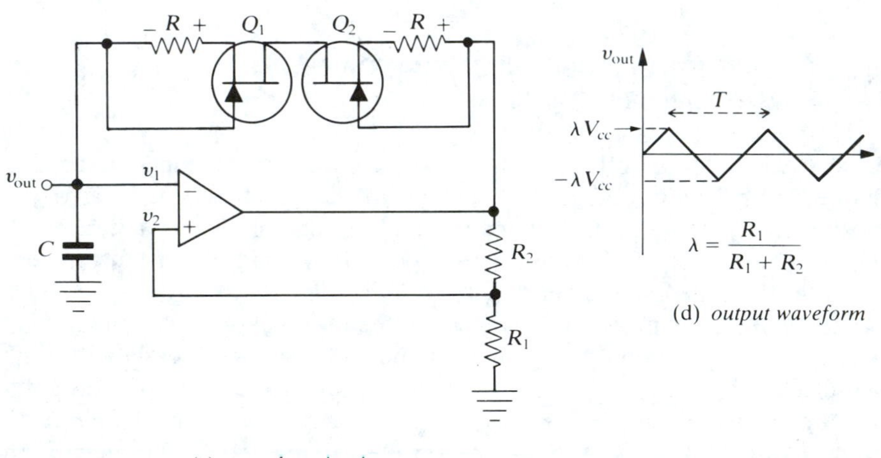

(c) *complete circuit*

(d) *output waveform*

FIGURE 10.27 Ramp generator.

One-half the period of the sawtooth will be the time it takes C to charge from $-\lambda V_{cc}$ to $+\lambda V_{cc}$. If we let i_0 be the charging current, then

$$i_0 = \frac{dQ}{dt} = \frac{d}{dt}(CV_C) = C\frac{dV_C}{dt} = C\left(\frac{2\lambda V_{cc}}{\frac{T}{2}}\right)$$

Solving for T gives

$$T = \frac{4\lambda C V_{cc}}{i_0} \quad \text{and} \quad f = \frac{1}{T} = \frac{i_0}{4\lambda C V_{cc}} \tag{10.31}$$

Notice:

1. The period depends not only on the capacitance C, the resistances R_1 and R_2, and the charging current (determined by the FETs), but also on the magnitude of the supply voltage V_{cc}.
2. For equal positive and negative slopes on the sawtooth output waveform, the two FETs should be matched.
3. The i_0R drop must be large enough to pinch off the FET—that is, large enough to ensure that the FET is operating on the constant current portion of its I_D-versus-V_{DS} curve.
4. The bias current of the op amp should be less than i_0.

10.22 THE MONOSTABLE MULTIVIBRATOR OR ONE SHOT

A monostable multivibrator has only one stable state and one unstable or transient state. The circuit normally sits in the stable state until a trigger pulse comes along and flips it into its unstable state. It remains in the unstable state for a fixed time, depending on the circuit time constant, and then it automatically flips back into its stable state. The output voltages for the stable and unstable states are different, so the output voltage changes for a time interval depending on the circuit time constant. Thus, in a monostable multivibrator one input trigger pulse (usually short) produces one output pulse whose width depends on the circuit time constant RC.

If we add a diode in parallel with the capacitance of the square-wave generator of Fig. 10.26, we have the monostable multivibrator circuit in Fig. 10.28. The voltage v_1 at the inverting input of the op amp can never be more positive than the diode turn-on voltage, about 0.6 V. If we choose R_1 and R_2 so that $v_2 = \lambda V_{cc} = (R_1/(R_1 + R_2))V_{cc} > 0.6$ V, then the noninverting input will be more positive than the inverting input and the op amp output will be $+V_{cc}$. This is the stable state: $v_1 = v_C = 0.6$ V, and $v_{out} = +V_{cc}$.

If we feed in a negative trigger pulse through a capacitor to the noninverting input, then, if v_2 is driven below v_1, the op amp will flip the output to the low state: $v_{out} = -V_{cc}$. For the op amp to do this, the trigger pulse must be larger in amplitude than $\lambda V_{cc} - 0.6$ V. When the op amp flips its output low, v_2 is changed from $+\lambda V_{cc}$ to $-\lambda V_{cc}$. The capacitor, which was initially at 0.6 V, will now discharge through R towards the $-V_{cc}$ voltage at the output, and v_1 will become more negative, thus reverse biasing the diode. C will discharge until v_1 reaches $v_2 = -\lambda V_{cc}$, at which point the output will

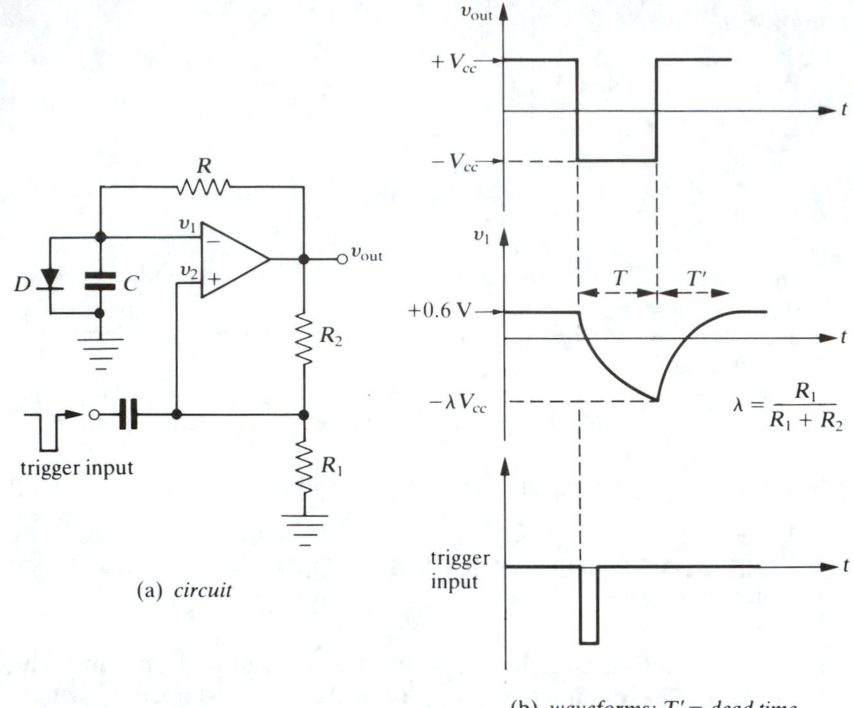

FIGURE 10.28 Negative output monostable multivibrator.

flip back to the stable state of $v_{\text{out}} = +V_{cc}$. The time it takes C to discharge depends on the time constant RC. As soon as the output flips up to $+V_{cc}$, the capacitor starts charging until $v_1 = 0.6$ V. During this charging time T' (or recovery time or *dead time*), the circuit is generally unable to respond to a trigger pulse unless the trigger pulse is much larger than $\lambda V_{cc} - 0.6$ V.

It can be shown that the time the output remains low is given by

$$T = RC \ln \frac{1 + \dfrac{V_{\text{D}}}{V_{cc}}}{1 - \lambda} = RC \ln\left[\frac{V_{cc} + V_{\text{D}}}{V_{cc}(1 - \lambda)}\right] \qquad \textbf{(10.32)}$$

and the recovery time is

$$T' = RC \ln \frac{V_{cc}(1 + \lambda)}{V_{cc} - V_{\text{D}}}$$

where V_{D} is the diode forward voltage. If we make $\lambda \ll 1$ ($R_1 \ll R_2$), then $T' \ll T$, and the recovery time is short compared to the length of the output pulse.

A monostable multivibrator with a positive output pulse and no dead time is shown in Fig. 10.29(a). In the stable state the FET is normally biased off by $-V_{bb}$, and the voltage at the noninverting input is $v_2 = (R_1/(R_1 + R_2))\,V_{cc} = \lambda V_{cc}$. The output is normally low because the invert-

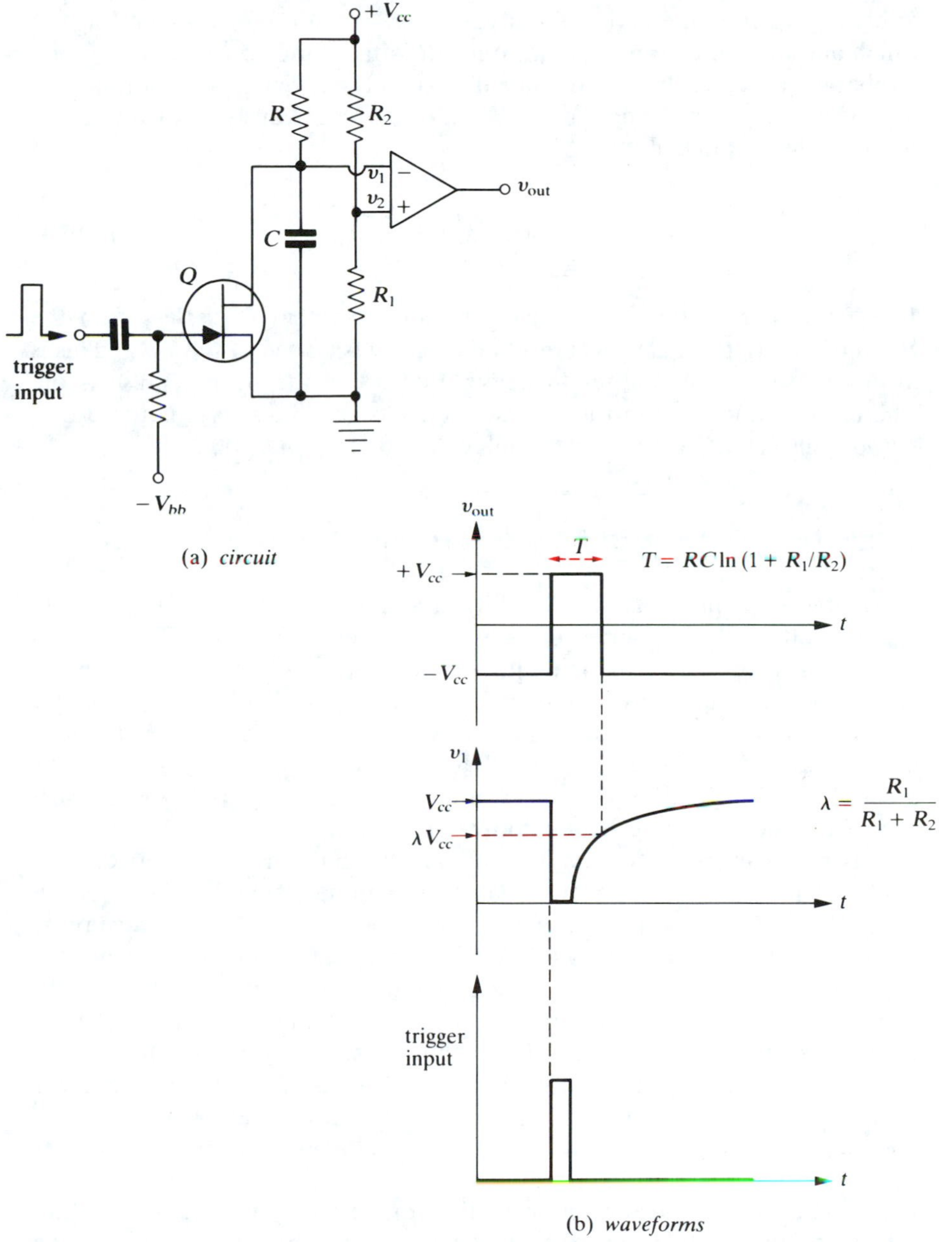

FIGURE 10.29 Positive output monostable multivibrator.

ing input voltage v_1 equals V_{cc} because C has charged up to V_{cc} through R. When a positive trigger pulse comes in, the FET is turned on briefly, which quickly discharges C through the FET to ground, thus lowering the voltage v_1 at the inverting input from V_{cc} to near ground. Thus, the trigger pulse makes v_1 less than v_2, which causes the comparator to flip the output to high ($+V_{cc}$). At the end of the trigger pulse, the FET is turned off again, which allows C to charge again through R with a time constant RC. When C charges to λV_{cc}, the comparator flips the output back to its normal low state. The waveforms are shown in Fig. 10.29(b). It can be shown that the width of the output pulse is

$$T = RC \ln\left(1 + \frac{R_1}{R_2}\right) \qquad \textbf{(10.33)}$$

The FET must be turned on strongly enough so that rC is less than the width of the trigger pulse, where r is the "on" resistance of the FET. This is to ensure that C discharges to below λV_{cc} while the trigger pulse is on. There is no dead time in this circuit; a new trigger pulse will cause a new output pulse at any time after the fall of the first output pulse.

10.23 THE VOLTAGE-CONTROLLED OSCILLATOR

A *voltage-controlled oscillator* (VCO) is a device that converts an input analog voltage into a series of pulses or waveforms whose frequency is linearly proportional to the magnitude of the input voltage. Another name is a *voltage-to-frequency converter* (V/F). The output waveform is typically a series of narrow pulses; the pulse repetition rate or pulse frequency is linearly proportional to the input voltage. VCOs are presently available on a chip for several dollars. The basic circuit, shown in Fig. 10.30(a), consists of an integrator followed by a comparator.

The circuit operation is as follows: Assume that the capacitor is uncharged and that a positive input voltage is applied. The input is integrated, and the output v_A of the integrator goes negative according to $v_A = -(1/RC) \int v_{in}\, dt$, as shown in Fig. 10.30(b). The noninverting input of the comparator is held constant at a negative reference voltage, so the comparator output is initially low. When the integrator output v_A drops below this negative reference voltage, the comparator output will flip high. But this high comparator output lasts only a very short time, because it closes the switch (a transistor) across C, which immediately discharges C and causes v_A to drop to zero, thus flipping the comparator output back to its initial low state. The resulting comparator output is a narrow positive pulse. The switch opens as soon as the comparator output goes low, the integrator starts integrating the input once more, and the process repeats itself. The larger the input voltage, the less time is taken for the integrator

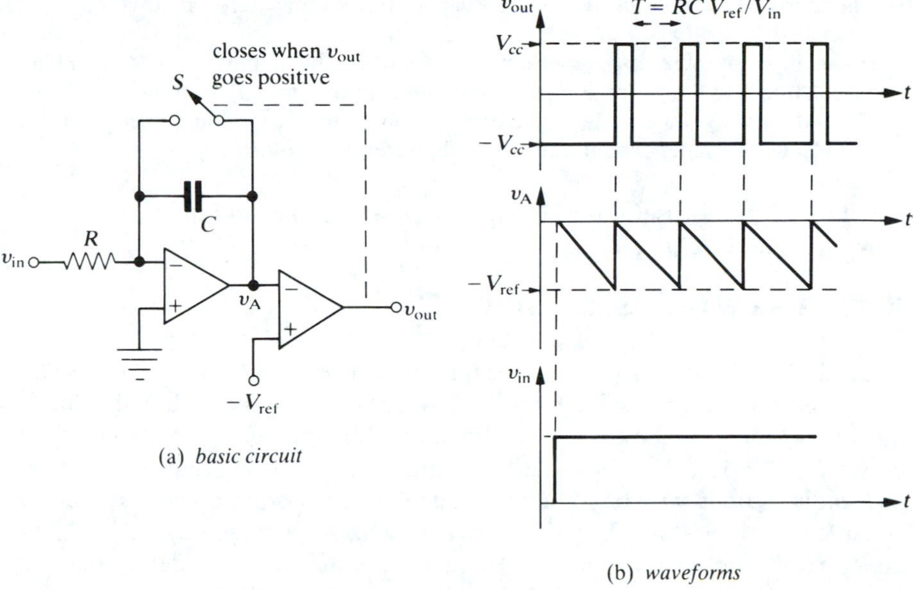

FIGURE 10.30 Voltage-controlled oscillator.

output to fall from zero to the negative reference voltage, and the shorter the time between the output pulses—in other words, the output pulse frequency is higher.

If we let T be the time for the integrator output to go from 0 to $-V_{ref}$, then, since the integrator output is the voltage on C, we have

$$-V_{ref} = \frac{-1}{RC}\int_0^T v_{in}\,dt = \left(\frac{-1}{RC}\right) v_{in} T$$

and

$$T = \frac{RCV_{ref}}{v_{in}}$$

Thus the output frequency is

$$f = \frac{1}{T} = \frac{v_{in}}{RCV_{ref}} \tag{10.34}$$

which is linearly proportional to the input voltage. This assumes that the integrator is a true integrator. Notice:

1. The integrator output must go below the negative reference voltage to flip the comparator.
2. The time to discharge the capacitor through the transistor switch should be negligible compared to the pulse period T.

3. The narrow output pulses can be shaped by a monostable multivibrator as described in Section 10.22.
4. If the input voltage changes during the time T, the output frequency will be proportional to the average input voltage during the time T.
5. If the dc input voltage is derived from the same reference voltage, then the ratio v_{in}/V_{ref} will be independent of drifts in the reference voltage.

This circuit is useful for converting an analog signal into digital form, as we will see in Chapter 15.

10.24 SINE-WAVE OSCILLATORS

A number of different sine-wave oscillators can be made with op amps. The basic idea behind any oscillator is to provide positive feedback and to have the open-loop gain times the feedback fraction greater than or equal to unity: $A_o B \geqq 1$. If the feedback is positive for many frequencies, a nonsinusoidal output will be produced, such as in a square-wave oscillator (astable multivibrator). But if positive feedback occurs only at one frequency (or over a narrow band of frequencies), then a sinusoidal output will result at that one frequency if $A_o B \geqq 1$.

10.24.1 Phase-Shift Oscillator

To produce positive feedback, we must make the phase of the feedback signal equal to the phase of the input signal. If the input signal is at the inverting input of an op amp, the output will be 180° out of phase with respect to the input. Thus, for positive feedback we must produce another 180° of phase shift of the op amp output before feeding it back to the inverting input. One way to do this is to feed the op amp output into three simple RC filters, each of which produces an average phase shift of 60°; this produces a total 180° phase shift in the RC network and a total phase shift of 360°, which means positive feedback. The circuit is shown in Fig. 10.31.

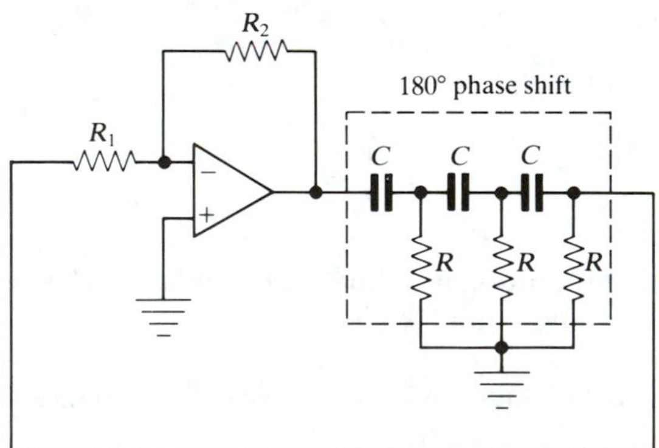

FIGURE 10.31 Phase-shift oscillator.

It can be shown that the RC network produces a total 180° phase shift at only one frequency: $f = 1/2\pi\sqrt{6}\ RC$. At this frequency the feedback fraction $B = 1/29$. Thus, the gain of the circuit must be at least 29 for oscillation. The gain is determined by R_2 and R_1, so these resistances usually are chosen to produce oscillation, and the three capacitors are tuned to vary the frequency of oscillation. These phase-shift oscillators generally are useful for frequencies under 100 kHz.

A similar oscillator is the *dual-integration oscillator*, which uses two integrators in series to produce the required 360° of phase shift—each integrator producing 180°.

10.24.2 Wien Bridge Oscillator

One of the most commonly used oscillator circuits is the *Wien bridge oscillator*. It is much more easily tunable over a wide range of frequencies than is the phase-shift oscillator of the preceding section, and it produces a low-distortion sinusoidal output. One way of drawing the circuit is shown in Fig. 10.32(a); we have a simple inverting amplifier circuit with feedback taken from the output back to the noninverting input of the op amp. At very high frequencies there is zero positive feedback because the lower capacitance acts as a short circuit, whereas at very low frequencies there is also essentially zero positive feedback because the upper capacitor acts like an infinite impedance. Thus, we immediately expect positive feedback at some finite nonzero frequency that depends on the values of R and C.

Redrawing the circuits as in Fig. 10.32(b) shows us that we have a bridge (the Wien bridge) with the two inputs to the op amp connected where the meter of the analogous *Wheatstone bridge* [Fig. 10.32(c)] would be. The output of the op amp drives the bridge. If the op amp is not saturated, we can apply the OAVR and conclude that the bridge is balanced—the voltage at points A and B must be the same. Thus, we can immediately write the balance condition

$$\frac{R_2}{R_1} = \frac{Z_4}{Z_3}$$

where Z_4 is the impedance of the series RC arm, $Z_4 = R + 1/j\omega C$, and Z_3 is the impedance of the parallel RC arm, $Z_3 = 1/(1/R + j\omega C)$. Substituting in for Z_4 and Z_3 in the balance equation yields

$$\frac{R_2}{R_1} = 2 + j\left(\omega RC - \frac{1}{\omega RC}\right)$$

Equating the real and the imaginary parts implies

$$\frac{R_2}{R_1} = 2 \quad \text{and} \quad \omega RC = \frac{1}{\omega RC} \quad \text{or} \quad \omega = \frac{1}{RC} \tag{10.35}$$

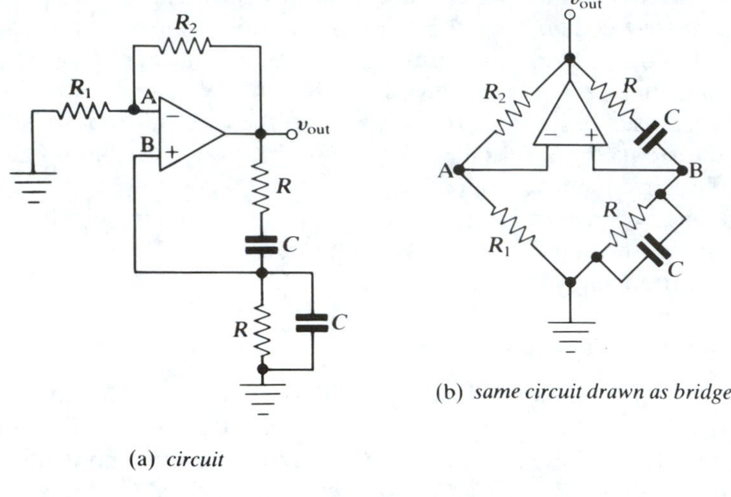

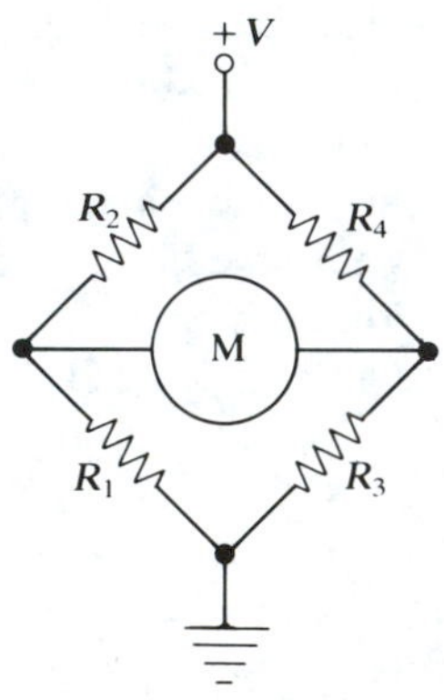

FIGURE 10.32 The Wien bridge oscillator.

Thus we see that the bridge is balanced only at the frequency $\omega_0 = 1/RC$ and that the gain of the op amp inverting amplifier must be 2.

If there is too little gain (R_2/R_1 is too small), the circuit will not oscillate. If there is too much gain, the output will be a distorted sine wave. Thus, the resistance ratio R_2/R_1 is usually made adjustable. The amplitude of the output can be regulated by changing the R_2/R_1 ratio with some long time-constant feedback. If the output amplitude increases, we should make the ratio smaller; if the output decreases, we should make the ratio larger. The principal mechanism that changes the output amplitude is a *thermal* one—if any factor increases the output amplitude, both R_2 and R_1 heat up due to the larger current flowing through them, whereas if the output amplitude decreases, both R_2 and R_1 cool down. Thus, to stabilize the output amplitude, we want R_2/R_1 to decrease with increasing temperature and to increase with decreasing temperature. This can easily be accomplished by having a negative temperature coefficient for R_2 and/or a positive temperature coefficient for R_1. A thermistor can be used for R_2 because it has a negative temperature coefficient; an incandescent lamp is often used for R_1 because it has a positive temperature coefficient.

The oscillator frequency can be adjusted by varying R or C. C is usually varied by using air capacitors "ganged" together, which means the same shaft changes both capacitances; R can be changed by a ganged selector switch.

10.24.3 *LC* Oscillators

At frequencies above several hundred kHz, most sine-wave oscillators are made with *LC* resonant circuits. At lower frequencies the inductance would be too large physically to use an *LC* circuit. Two of the most popular are the Colpitts and the Hartley shown in Fig. 10.33. These circuits can operate

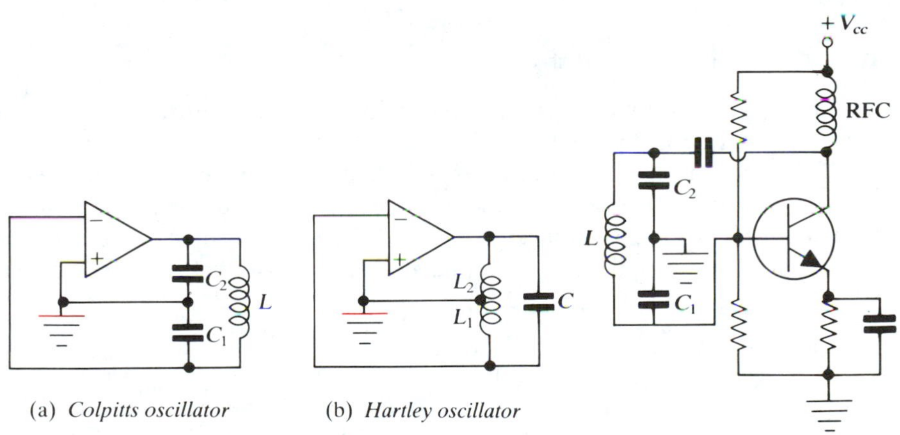

(a) *Colpitts oscillator* (b) *Hartley oscillator* (c) *transistor Colpitts oscillator*

FIGURE 10.33 *LC* oscillator circuits.

at frequencies up to many MHz, and with proper component choice they can have a frequency stability of ±0.01% or better. Both the Colpitts and the Hartley circuits feed back a fraction of the op amp output to the inverting input; the 180° phase shift across the *LC* network plus the 180° phase shift at the output of the op amp produces the positive feedback necessary for oscillation.

The two series capacitors in the Colpitts oscillator have a total capacitance of $C = C_1C_2/(C_1 + C_2)$, so the resonant frequency is

$$\omega_0 = \frac{1}{\sqrt{\dfrac{LC_1C_2}{C_1 + C_2}}} \tag{10.36}$$

The oscillation frequency may be considerably less in the op amp oscillator circuit than in the bare LC_1C_2 circuit: the smaller C_1 is, the larger the amplitude of oscillation is and the more distortion produced. A similar argument for the Hartley circuit shows that its oscillation frequency is

$$\omega_0 = \frac{1}{\sqrt{(L_1 + L_2)C}} \tag{10.37}$$

Very few op amps can operate at frequencies much over 10 MHz, so at higher frequencies the Colpitts and Hartley circuits are usually used with individual high-frequency transistors that can operate up to 100 MHz or higher. A Colpitts oscillator with a discrete transistor is shown in Fig. 10.33(c).

10.24.4 Crystal Oscillators

Better frequency stability is obtained when the LC resonant circuit is replaced by a quartz piezoelectric crystal. A *piezoelectric crystal* is one in which a mechanical strain produces a voltage difference, and vice versa. The crystal basically acts as a very high-Q LC resonant circuit; its equivalent circuit is shown in Fig. 10.34(a). The capacitance C_2 is between the metal plating applied to the opposite faces of the crystal; the capaci-

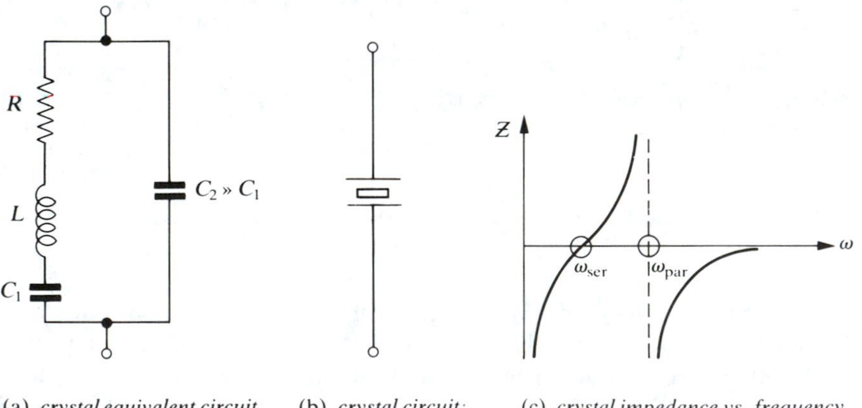

(a) *crystal equivalent circuit* (b) *crystal circuit symbol* (c) *crystal impedance vs. frequency*

FIGURE 10.34 Piezoelectric crystal.

tance C_1 is inherent in the crystal. C_2 is invariably much larger than C_1. There are two resonant frequencies: the series resonant frequency of L and C_1, and the resonant frequency of L in parallel with the series combination of C_1 and C_2:

$$\omega_{ser} = \frac{1}{\sqrt{LC_1}} \quad \text{(series)} \tag{10.38}$$

and

$$\omega_{par} = \frac{1}{\sqrt{L\left(\dfrac{C_1C_2}{C_1 + C_2}\right)}} \quad \text{(parallel)} \tag{10.39}$$

Because $C_2 \gg C_1$, the series and parallel resonant frequencies are very nearly equal, typically within 1%. The reactance of the crystal versus frequency is shown in Fig. 10.34(c). The crystal is inductive for frequencies between the two resonant frequencies (for $\omega_{ser} < \omega < \omega_{par}$), and it is capacitive for frequencies below ω_{ser} and above ω_{par}. The crystal frequency can be tuned by adding a variable capacitor in parallel with the crystal, which effectively changes C_2. Only a slight change in the crystal resonance can be produced by this technique because C_1, not C_2, is the primary determinant of the frequency. The crystal oscillator in a quartz clock or watch oscillates at precisely 32.768 kHz and is divided by 2^{15} to yield a 1-Hz frequency. A piezoelectric crystal can be used in any oscillator circuit, in a Colpitts or a Hartley, for example, to replace the LC tank.

A crystal Colpitts oscillator circuit is shown in Fig. 10.35(a) with an

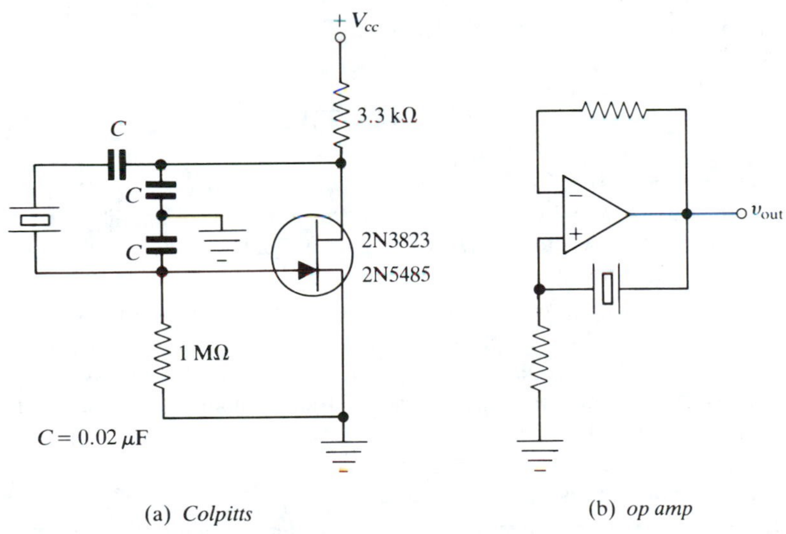

FIGURE 10.35 Crystal oscillators.

FET instead of an op amp, and an op amp oscillator circuit is shown in Fig. 10.35(b).

The frequency stability of a typical inexpensive crystal depends on the temperature changes and the power supply voltage changes. It is easy to obtain an overall stability of 0.1 ppm (one part in 10^7) or better with no special precautions. If the temperature is regulated (in a constant-temperature oven) and the power supply voltage is regulated, stabilities better by factors of 10 or 100 are obtainable. Commercial *temperature-compensated crystal oscillators* (TCXO) are available with a frequency stability of 1 to 0.1 ppm over a 50°C range. Over a relatively short time (hours) a frequency drift of only a few parts in 10^{11} is obtainable from a commercial crystal oscillator. For a frequency stability of one part in 10^{12}, we must use an atomic beam apparatus.

PROBLEMS

1. Derive the voltage gain expression (10.14) for a differential amplifier.
2. Derive the voltage gain expression (10.15) for the instrumentation amplifier.
3. Design a one-transistor emitter follower power-booster circuit to amplify the output from an op amp. The output should be able to swing both positive and negative with respect to ground.
4. For the amplifier of Fig. 10.6(c), prove that the voltage gain

$$A = 1 + \frac{R_2}{(1 + j)R_1}$$

when $\omega = \omega_2 = 1/R_2C_2$.

5. For the amplifier of Fig. 10.6(b) prove that the voltage gain

$$A = 1 + \frac{j}{1 + j}\left(\frac{R_2}{R_1}\right)$$

when $\omega = \omega_1 = 1/R_1C_1$.

6. Design an op amp amplifier (noninverting) that will amplify signals from approximately 50 Hz to 10 kHz with a gain of 20. Frequencies below 50 Hz and above 10 kHz should be attenuated.
7. Design a summing amplifier circuit to sum from inputs v_A, v_B, v_C, v_D and to produce an output of $v_{out} = v_A + 2v_B + 4v_C + 8v_D$.
8. Design a current-to-voltage converter to convert a 1-μA input to a 2-V output. The 1-μA current input has (signal) Fourier components up to 100 Hz, and higher frequency noise is present.
9. Design a voltage-to-current converter to convert a 1-mV input voltage to a 1-mA current through a 1-kΩ load resistor.
10. Design a current-to-current converter (or current multiplier) to produce a 10-mA output for a 100-μA input.

11. For the logarithmic amplifier show that if an input v_1 produces an output V_1 and v_2 produces V_2,

$$V_2 - V_1 = \left(\frac{kT}{e}\right)\ln\left(\frac{v_2}{v_1}\right)$$

12. If in the logarithmic amplifier, the input $v_1 = 10\,\text{mV}$ for an output $V_1 = -0.04\,\text{V}$, calculate R if $I_0 = 10^{-10}$ A. Calculate the output for an input $v_2 = 100\,\text{mV}$.

13. Design an ideal diode op amp circuit to produce a *negative* half-waveform from a sinusoidal input.

14. Design an ideal diode (op amp) clamp circuit to produce an output voltage clamped to +5 V; that is, $v_{\text{out}} \geqq 5$ V. Sketch the input and output waveforms for an arbitrary input.

15. Design an ideal (op amp) half-wave rectifier to produce a half-wave output that is twice the amplitude of the input. Repeat for an ideal (op amp) full-wave rectifier.

16. Design a (op amp) peak detector circuit. Is a bipolar or an FET op amp better? Calculate the approximate droop rate in volts per second if $C = 0.001\ \mu$F and the op amp input impedance is $10^{10}\ \Omega$.

17. Design a sample-and-hold circuit. Calculate the droop rate if $I_B = 100$ pA and $C = 0.01\ \mu$F. What are the two conflicting requirements for C? If $\Delta\tau$ is the time between successive samples, explain why we desire

$$R_{\text{MOSFET}}C \ll \Delta\tau$$

18. Design an op amp differentiator whose output is

$$v_{\text{out}} = -(10^{-6})\frac{dV_{\text{in}}}{dt}$$

If an input step function rises from 0 to 1 V in 2 μs and then is constant at 1 V, sketch the output.

19. Design an op amp integrator whose output is

$$v_{\text{out}} = -10^4 \int v_{\text{in}}\, dt$$

20. The comparator is bipolar. (a) Sketch the output waveform if $V_{bb} = +5$ V. (b) Repeat for $V_{bb} = -5$ V. (c) How would you change the circuit to produce output voltages of +5 V and −0.6V?

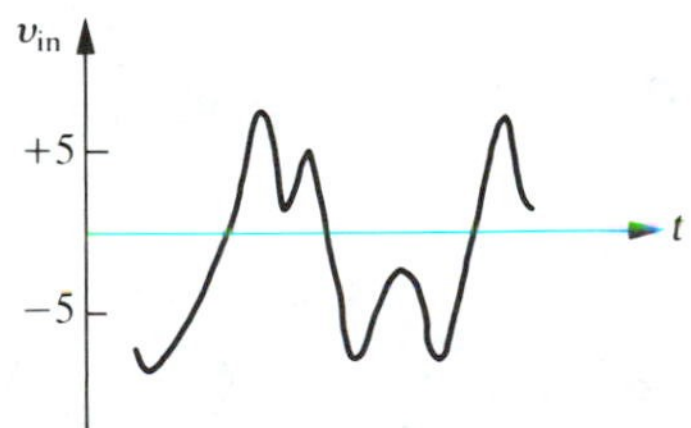

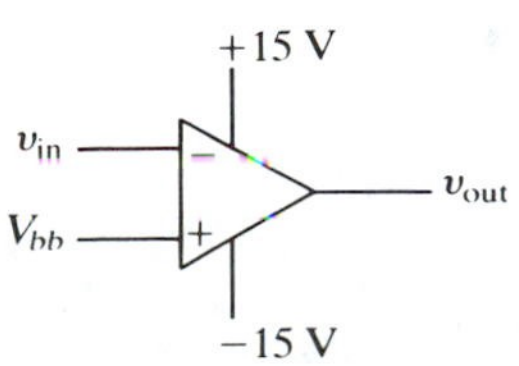

21. Derive the threshold voltage equation (10.29) for the Schmitt trigger.

22. Sketch the output waveform if the two threshold voltages are 3.0 V and 3.5 V.

23. Derive (10.30), the period for the astable multivibrator.

24. Sketch the output waveform for the astable multivibrator of Problem 23.

25. Derive (10.31), the period for the ramp or sawtooth generator.

26. Sketch the output waveform in Fig. 10.27 if $R_1 = 1\ \text{k}\Omega$, $R_2 = 2.2\ \text{k}\Omega$, and $\pm V_{cc} = \pm 15$ V.

27. Derive (10.32), the period for the negative output monostable multivibrator.

28. Sketch the output waveform in Fig. 10.28 if $\pm V_{cc} = \pm 15$ V, $R = 10\ \text{k}\Omega$, $C = 0.01\ \mu\text{F}$, $R_1 = 1\ \text{k}\Omega$, $R_2 = 1\ \text{k}\Omega$. What is the current through the diode?

29. Derive (10.33), the period for the positive output monostable multivibrator.

30. Derive (10.34), the frequency of the VCO.

31. Sketch the output waveform in Fig. 10.30 if $\pm V_{cc} = \pm 12$ V, $V_{ref} = 6$ V, $R = 1\ \text{k}\Omega$, $C = 1\ \mu\text{F}$.

32. Derive (10.35), the balance conditions for the Wien bridge oscillator.

CHAPTER 11

Active Filters and Regulators

11.1 INTRODUCTION TO FILTERS

Next to amplifiers, filters are probably the most common analog circuit element. A filter is basically any input–output circuit whose gain depends on frequency. For example, a low-pass filter passes low frequencies and attenuates high frequencies. There are four common types of filters: low-pass, high-pass, bandpass, and band reject; their gain magnitude $|A|$-versus-frequency graphs are shown in Fig. 11.1(a).

There are two general classes of filters: *passive*, which consist of the passive components R, L, and C; and *active*, which contain active elements such as amplifiers as well as the usual passive components. The availability of inexpensive high-quality op amps has revolutionized active filter design in recent years, and this section will be devoted to the basics of op amp active filter design. Op amp active filters can amplify and filter the incoming signal, and they completely eliminate the need to use a real inductor with all its attendant problems of size, weight, strong magnetic fields, tendency to vibrate, expense, fragility, and so on. Also, both the tuning and the Q of an active op amp filter can be varied by changing a resistance, in contrast to passive filters. At frequencies below 100 Hz passive filters are impractical because of the large inductors and capacitors required, whereas active filters are practical. The only application where passive filters are superior is high-frequency filtering above approximately 500 kHz. Finally, an op amp active filter can be made with a very high input impedance and a very low output impedance, so impedance matching problems in the sources and loads are almost eliminated. Thus it is quite simple to cascade active filters in series.

Recall from Chapter 2 that a simple (passive) low-pass RC filter as shown in Fig. 11.2(a) has a gain versus frequency or *transfer function* of

$$A = \frac{1}{1 + j\omega RC} \quad \text{or} \quad |A| = (AA^*)^{1/2} = \frac{1}{\sqrt{1 + \omega^2 R^2 C^2}} = \frac{1}{\sqrt{1 + \left(\frac{\omega}{\omega_B}\right)^2}} \tag{11.1}$$

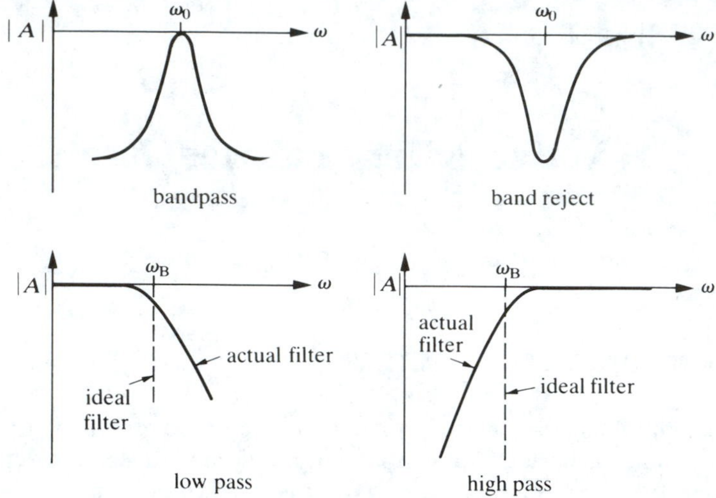

FIGURE 11.1 General filter curves.

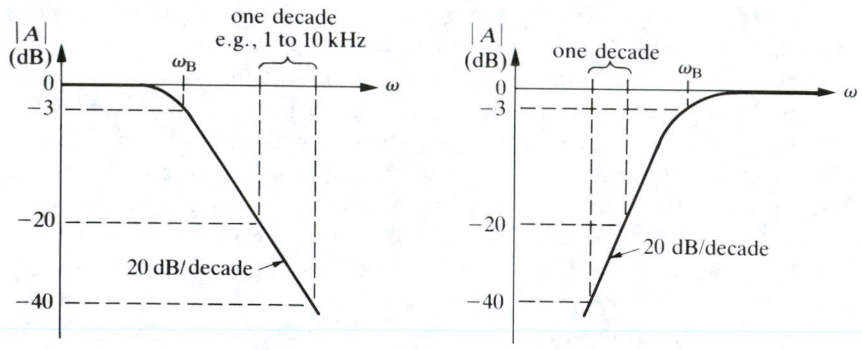

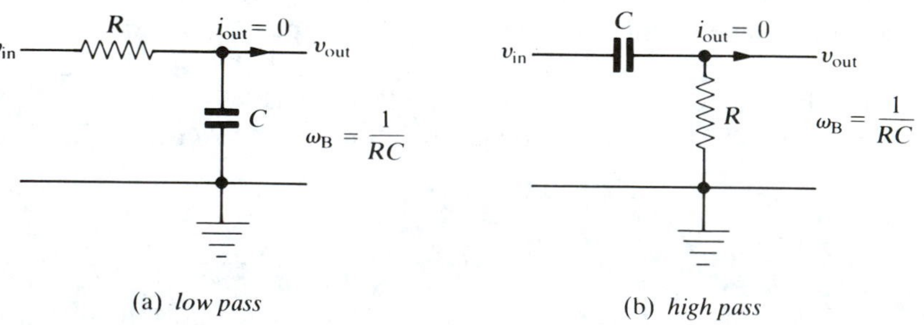

(a) *low pass* (b) *high pass*

FIGURE 11.2 Passive *RC* filters.

where $\omega_B = 1/RC$ is the break frequency. As shown in Fig. 11.2(a), the gain is 1 (0 dB) at dc, 0.707 (−3 dB) at $\omega = \omega_B$, and falls off at a rate of 20 dB/decade or 6 dB/octave (as ω^{-1}) above the break frequency. A passive RC high-pass filter [Fig. 11.2(b)] has a transfer function of

$$A = \frac{j\omega RC}{1 + j\omega RC} = \frac{\dfrac{j\omega}{\omega_B}}{1 + \dfrac{j\omega}{\omega_B}} \quad \text{or} \quad |A_v| = \frac{\omega RC}{\sqrt{1 + \omega^2 R^2 C^2}} = \frac{\dfrac{\omega}{\omega_B}}{\sqrt{1 + \left(\dfrac{\omega}{\omega_B}\right)^2}} \tag{11.2}$$

Its gain is 0 dB for very high frequencies, −3 dB at $\omega = \omega_B$, and falls off at 20 dB/decade for frequencies below ω_B.

An ideal filter passes only the desired frequencies and completely attenuates the unwanted frequencies. Thus, an ideal low-pass filter has a rectangular gain versus frequency as shown in Fig. 11.1 with a sharp "knee" at the break frequency. If two actual passive low-pass RC filters are cascaded in series, as shown in Fig. 11.3(a), the gain above the break frequency falls off at a rate of 40 dB/octave (as ω^{-2}), but the knee is quite rounded. To prevent the two RC filters from affecting one another, we must put some sort of buffer circuit between them (with a high input impedance and a low output impedance), as shown in Fig. 11.3(a).

The overall gain of the two low-pass RC filters interconnected with the buffer is simply the product of the gain of the individual filters:

$$A = \frac{1}{1 + j\omega RC}\frac{1}{1 + j\omega RC} = \frac{1}{(1 + j\omega RC)^2}$$

or

$$A = \frac{1}{1 + 2j\omega RC - \omega^2 R^2 C^2} \tag{11.3}$$

If we define a new frequency in relation to the break frequency $\omega_B = 1/RC$, we can simplify the transfer function. Let $\omega' = \omega/\omega_B$ be the new frequency variable; that is, $\omega' = \omega RC$. Then the gain or transfer function is

$$A = \frac{1}{1 - \omega'^2 + 2\omega' j} \quad \text{and} \quad |A| = \frac{1}{1 + \omega'^2} \tag{11.4}$$

Thus, we see that the magnitude of the gain will decrease at a rate of 40 dB/decade or 12 dB/octave ($|A| \propto \omega^{-2}$). This is called a *second-order* filter.

A second-order low-pass passive filter can be constructed from R, L, and C, as shown in Fig. 11.3(b). The output is taken across the capacitor, and we intuitively expect the high frequencies to be attenuated because the

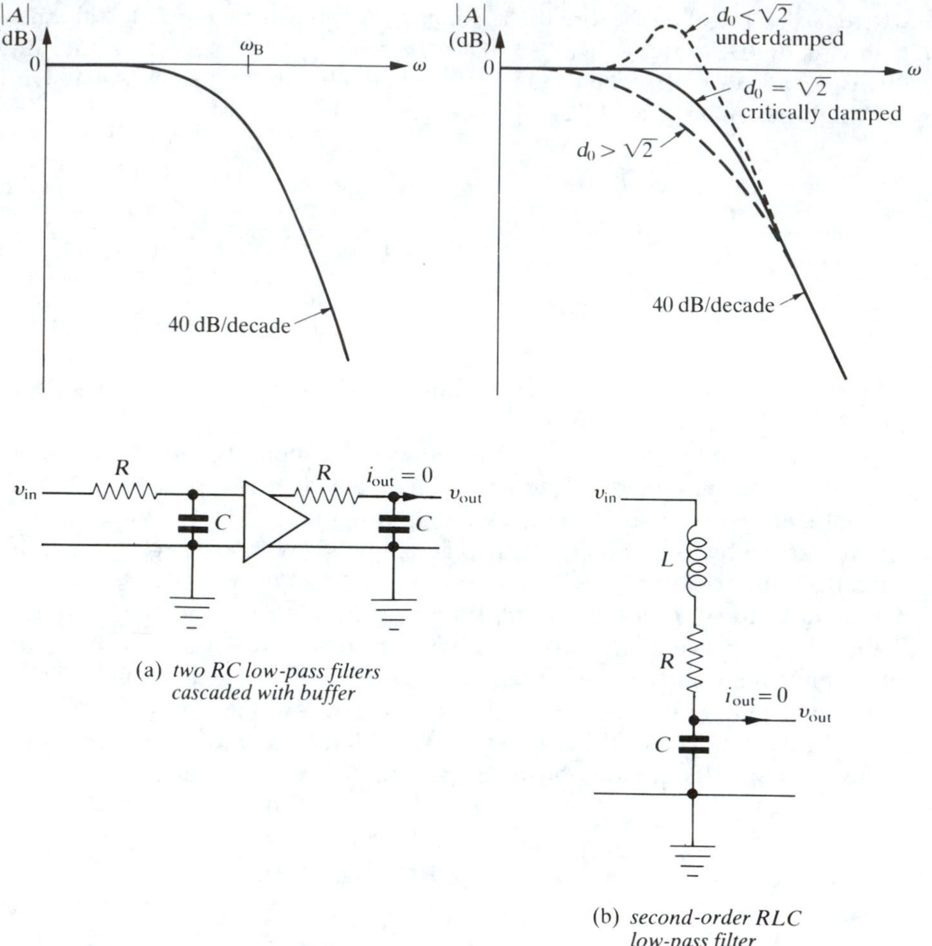

FIGURE 11.3 Second-order low-pass filters.

capacitance will tend to act like a short and the inductor will have a high impedance at high frequencies. The transfer function of this filter is

$$A = \frac{X_C}{X_L + R + X_C} = \frac{\dfrac{1}{j\omega C}}{j\omega L + R + \dfrac{1}{j\omega C}} = \frac{1}{1 + j\omega RC - \omega^2 LC}$$

which is very similar to (11.3). Again, if we define a new frequency in relation to the resonant frequency $\omega_0 = 1/\sqrt{LC}$, we can simplify the transfer function. Let $\omega' = \omega/\omega_0 = \omega\sqrt{LC}$. Then

$$A = \frac{\omega_0^2}{\omega_0^2 - \omega^2 + \dfrac{j\omega R}{L}} = \frac{1}{1 - \omega'^2 + j\dfrac{\omega' R}{\omega_0 L}} \tag{11.5}$$

But we recall that the Q of an RLC circuit is given by $Q = \omega L/R$; thus the Q at resonance is $Q_0 = \omega_0 L/R$. In terms of Q_0 the transfer function is

$$A = \frac{1}{1 - \omega'^2 + \dfrac{j\omega'}{Q_0}} \tag{11.6}$$

which is very similar to (11.4). It is customary to describe the filter in terms of the *damping factor d* rather than the Q. We define the damping factor (at resonance) as $d_0 = 1/Q_0$.[†] Thus, the transfer function is

$$A = \frac{1}{1 - \omega'^2 + j\omega' d_0} \tag{11.7}$$

The magnitude of the transfer function is then

$$|A| = \frac{1}{[(1 - \omega'^2)^2 + \omega'^2 d_0^2]^{1/2}} \tag{11.8}$$

We see that for frequencies well above the resonant frequency, $|A|$ will vary as ω^{-2} (remember $\omega' = \omega/\omega_0$), thus producing a gain rolloff of 40 dB/decade or 12 dB/octave—that is, a "second"-order filter.

However, the passive LRC low-pass second-order filter differs in one important respect from the buffer-cascaded RC low-pass filter. The shape of the transfer function of the LRC filter depends strongly on the magnitude of R—in other words, on the Q or the damping factor d. For zero R the Q is infinite, the damping factor is zero, and the gain will be infinite at the resonant frequency $\omega_0 = 1/\sqrt{LC}$ ($\omega' = 1$). As R is increased, the gain at resonance decreases, and there is a certain value of R (and Q_0 and d_0) above which there is no relative maximum in the $|A|$-versus-frequency curve. We can calculate this value of R (or Q_0 or d_0) by looking for a relative minimum in the denominator of $|A|$ in (11.8). Let $D = (1 - \omega'^2)^2 + \omega_0'^2 d^2$. When D has a relative minimum, $|A|$ will have a relative maximum, as shown by the dotted line in Fig. 11.3(b). We calculate the first derivative of D with respect to the frequency ω' and set it equal to

[†]Some references define the damping factor as $d = 1/2Q_0$.

zero to locate the minimum:

$$\frac{\partial D}{\partial \omega'} = \frac{\partial}{\partial \omega'}[(1-\omega'^2)^2 + \omega'^2 d_0^2] = 2(1-\omega'^2)(-2\omega') + 2\omega' d_0^2$$

$$= 4\omega'^3 + 2(d_0^2 - 2)\omega'$$

For the minimum $\partial D/\partial\omega' = 0$, and thus

$$2\omega'^2 = 2 - d_0^2 \quad \text{or} \quad \omega'^2 = 1 - \frac{d_0^2}{2}$$

so

$$\omega' = \left(1 - \frac{d_0^2}{2}\right)^{1/2} = \left(1 - \frac{1}{2Q_0^2}\right)^{1/2}$$

This is a reasonable result, because in the limit as $d_0 \to 0$ ($Q_0 \to \infty$), $\omega' \to 1$, which means $\omega = \omega_0$. The second derivative of D with respect to ω' turns out to be positive, indicating a relative minimum in D and, thus, a relative maximum in $|A|$. So there is a real solution for ω' only if $1 - d_0^2/2$ is positive; that is, if $d_0^2/2 < 1$ or if $d_0 < \sqrt{2}$. Therefore only if $d_0 < \sqrt{2}$ will there be a relative minimum in D and a relative maximum in $|A|$. If $d_0 = \sqrt{2}$, there is no real solution for ω', and the $|A|$-versus-frequency curve must have no relative maximum, as shown by the solid line in Fig. 11.3(b). In other words, *for* $d_0 = \sqrt{2}$, *there is maximum flatness in the gain-versus-frequency curve*. For $d_0 > \sqrt{2}$ there is also no relative maximum, but the curve is more rounded near ω_B as shown by the dashed line in Fig. 11.3(b). In terms of the Q, $d_0 < \sqrt{2}$ is the same as $Q_0 > 1/\sqrt{2}$ or $\omega_0 L/R > 1/\sqrt{2}$ or $R/L < \sqrt{2}\omega_0$, and $d_0 > \sqrt{2}$ is the same as $R/L > \sqrt{2}\omega_0$.

It is conventional to call the case $d_0 < \sqrt{2}$ *underdamped*, the case $d_0 = \sqrt{2}$ *critically damped*, and the case $d_0 > \sqrt{2}$ *overdamped*. This nomenclature comes from a transient analysis of a damped oscillator, either electrical or mechanical. This nomenclature for the shape of the filter response curves is similar to (but not exactly the same as) the nomenclature for the mechanical free damped oscillator covered in the next section. We point out that the mechanical analogue of the *LRC* circuit of Fig. 11.3(b) is the driven mechanical simple harmonic oscillator with a viscous retarding force proportional to the *speed* of the mass. This is true only for low speeds of the mass in a viscous medium; at higher speeds the fluid retarding force is proportional to the square of the speed.

11.2 THE DAMPED DRIVEN MECHANICAL OSCILLATOR

The damped driven mechanical oscillator is the basic mechanical model for all second-order filters and is shown in Fig. 11.4(a). The solution is as

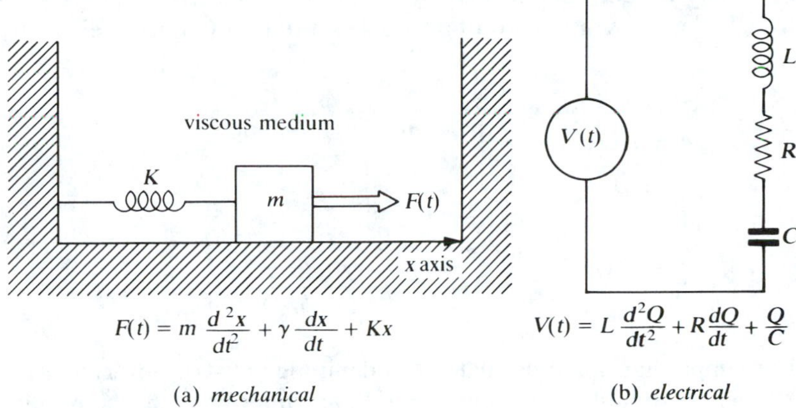

FIGURE 11.4 Damped driven oscillator.

follows. From Newton's second law the sum of all the x forces on the mass m must equal ma_x. Thus

$$\sum F_x = ma_x$$

$$F(t) - Kx - \gamma\frac{dx}{dt} = m\frac{d^2x}{dt^2}$$

or

$$F(t) = m\frac{d^2x}{dt^2} + \gamma\frac{dx}{dt} + Kx \tag{11.9}$$

where $F(t)$ is the external driving force, K the spring constant, and $\gamma\, dx/dt$ the viscous retarding force on m resulting from the motion of m through the surrounding viscous medium.

The electrical equation for the driven RLC circuit of Fig. 11.4(b) is

$$V(t) = L\frac{d^2Q}{dt^2} + R\frac{dQ}{dt} + \frac{Q}{C}$$

so $Q \leftrightarrow x$, $L \leftrightarrow m$, $R \leftrightarrow \gamma$, and $1/C \leftrightarrow K$. Let us assume a sinusoidal driving force $F(t) = F_0 \cos \omega t = F_0\text{Re}(e^{j\omega t})$. Then (11.9) becomes

$$\frac{F_0 e^{j\omega t}}{m} = \frac{d^2x}{dt^2} + \frac{\gamma}{m}\frac{dx}{dt} + \frac{K}{m}x \tag{11.10}$$

An obvious trial solution is

$$x = x_0 e^{j\omega t}$$

where x_0 may be complex if x is out of phase with $F(t)$. ($x_0 = Xe^{j\phi}$, and $x = Xe^{j(\omega t+\phi)}$ with X a real constant.) Substituting in (11.10) yields

$$\left(-\omega^2 + j\frac{\omega\gamma}{m} + \frac{K}{m}\right)x_0e^{j\omega t} = \frac{F_0e^{j\omega t}}{m}$$

or

$$x_0 = \frac{\dfrac{F_0}{m}}{-\omega^2 + j\dfrac{\omega\gamma}{m} + \dfrac{K}{m}} \tag{11.11}$$

The simple harmonic oscillator (no damping: $\gamma = 0$; no driving force: $F_0 = 0$) has a resonant frequency $\omega_0 = \sqrt{K/m}$. In terms of ω_0 the oscillation amplitude x_0 of a driven damped oscillator be written as

$$x_0 = \frac{\dfrac{F_0}{m}}{\omega_0^2 - \omega^2 + \dfrac{j\omega\gamma}{m}} = \frac{\dfrac{F_0}{m\omega_0^2}}{1 - \omega'^2 + j\dfrac{\omega'\gamma}{m\omega_0}} \tag{11.12}$$

where $\omega' = \omega/\omega_0$. The velocity is

$$\frac{dx}{dt} = \frac{d}{dt}(x_0e^{j\omega t}) = j\omega x_0e^{j\omega t} = v_0e^{j\omega t}$$

The velocity amplitude v_0 is thus

$$v_0 = \gamma\omega x_0 = \frac{\dfrac{j\omega F_0}{m}}{\omega_0^2 - \omega^2 + \dfrac{j\omega\gamma}{m}} = \frac{\dfrac{j\omega' F_0}{m\omega_0}}{1 - \omega'^2 + j\dfrac{\omega'\gamma}{m\omega_0}} \tag{11.13}$$

The acceleration is

$$\frac{d^2x}{dt^2} = j\omega\left(\frac{dx}{dt}\right) = j\omega v_0e^{j\omega t} = a_0e^{j\omega t}$$

The acceleration amplitude a_0 is thus

$$a_0 = j\omega v_0 = -\omega^2x_0 = \frac{\dfrac{-\omega^2F_0}{m}}{\omega_0^2 - \omega^2 + \dfrac{j\omega\gamma}{m}} = \frac{\dfrac{-\omega'^2F_0}{m}}{1 - \omega'^2 + j\dfrac{\omega'\gamma}{m\omega_0}} \tag{11.14}$$

We immediately see that the oscillation amplitude x_0 (11.12) is similar in form to the transfer function (11.6) for the second-order *RC low-pass* filter. Plots of $|x_0|$, $|v_0|$, and $|a_0|$ are shown in Fig. 11.5. Notice that the $|x_0|$

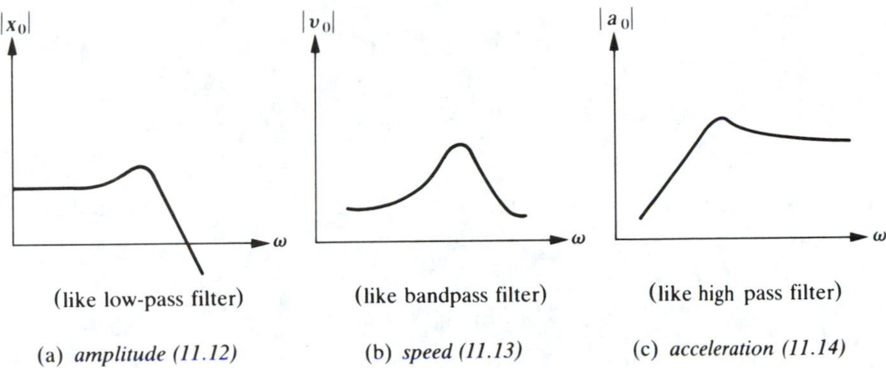

FIGURE 11.5 Damped driven mechanical oscillator.

plot is like the gain plot of a low-pass filter, the $|v_0|$ plot is like a bandpass filter, and the $|a_0|$ plot is like a high-pass filter.

For the free (not driven) damped oscillator, the differential equation is

$$\sum F_x = m\frac{d^2x}{dt^2}$$

$$-Kx - \gamma\frac{dx}{dt} = m\frac{d^2x}{dt^2}$$

Hence,

$$\frac{d^2x}{dt^2} + \frac{\gamma}{m}\frac{dx}{dt} + \frac{K}{m}x = 0 \qquad \textbf{(11.15)}$$

Letting $\gamma' \equiv \gamma/m$ and $\omega_0^2 \equiv K/m$, we have

$$\frac{d^2x}{dt^2} + \gamma'\frac{dx}{dt} + \omega_0^2 x = 0 \qquad \textbf{(11.16)}$$

Note that ω_0 is the angular frequency of the undamped ($\gamma = 0$) simple harmonic oscillator. Assuming a solution of the form $x = Ae^{\alpha t}$ and substituting in (11.16), we obtain

$$\alpha^2 + \gamma'\alpha + \omega_0^2 = 0$$

Thus α is given by the quadratic formula as

$$\alpha = \frac{-\gamma' \pm \sqrt{\gamma'^2 - 4\omega_0^2}}{2}$$

and the general solution for x is

$$x = A_1 e^{\alpha_1 t} + A_2 e^{\alpha_2 t} \tag{11.17}$$

where $\alpha_1 = -\dfrac{\gamma'}{2} + \sqrt{\dfrac{\gamma'^2}{4} - \omega_0^2}$ and $\alpha_2 = -\dfrac{\gamma'}{2} - \sqrt{\dfrac{\gamma'^2}{4} - \omega_0^2}$

The constants A_1 and A_2 are determined by the initial conditions, that is, by the values of x and dx/dt at $t = 0$. Letting

$$\omega_1^2 \equiv \omega_0^2 - \frac{\gamma'^2}{4}$$

x can be written as

$$x = A_1 e^{-(\gamma'/2)t} e^{j\omega_1 t} + A_2 e^{-(\gamma'/2)t} e^{-j\omega_1 t} \tag{11.18}$$

We will now take the special case of $A_2 = 0$ to simplify the algebra. (This involves no loss of generality; it merely means choosing a particular set of initial conditions.) It can be shown the Q of this mechanical oscillator is

$$Q = \frac{\omega_0}{\gamma'}$$

We now note that there are three possible cases, depending on the relative magnitude of γ' and ω_0.

Case I: ω_1^2 is positive or $\omega_0 > \gamma'/2$ or $Q = \omega_0/\gamma' > 1/2$. In this case, x will slowly damp out to zero amplitude and will simultaneously oscillate at the frequency ω_1.

$$x = A_1 e^{-(\gamma'/2)t} e^{j\omega_1 t} \tag{11.19A}$$

This damped oscillation is shown in Fig. 11.6(a). This case is called *underdamped*. Note that the oscillation frequency ω_1 is always less than ω_0, the frequency of the undamped oscillator. In fact, we can rewrite ω_1 in terms of the Q:

$$\omega_1 = \sqrt{\omega_0^2 - \frac{\gamma'^2}{4}} = \omega_0\sqrt{1 - \frac{\gamma'^2}{4\omega_0^2}} = \omega_0\sqrt{1 - \frac{1}{4Q^2}}$$

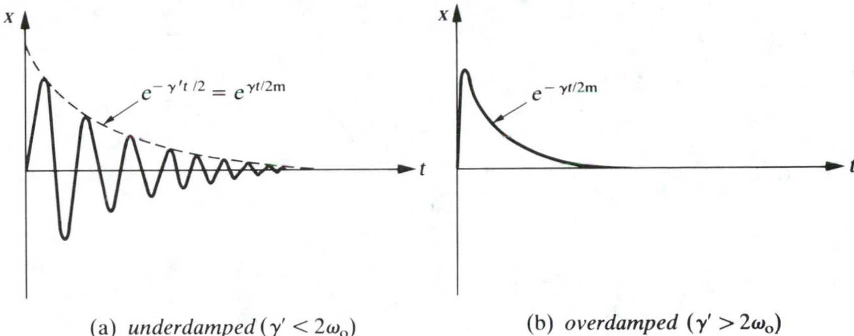

(a) *underdamped* ($\gamma' < 2\omega_o$)

(b) *overdamped* ($\gamma' > 2\omega_o$)

FIGURE 11.6 Underdamped oscillation amplitude versus time.

For $Q = 2$, $\omega_1/\omega_0 = 0.9354$; and for $Q = 10$, $\omega_1/\omega_0 = 0.9975$. Thus as the Q increases, ω_1 becomes very nearly equal to ω_0.

Case II: ω_1^2 is zero or $\omega_0 = \gamma'/2$ or $Q = 1/2$. In this case, x will decay exponentially with no oscillations:

$$x = A_1 e^{-(\gamma'/2)t} \tag{11.19B}$$

This case is called *critically damped*. It is the case of the minimum possible damping (smallest γ) without oscillations. Note that the "critically damped" filter in Section 11.1 had $Q = 1/\sqrt{2}$.† This means that a filter with $Q = 1/\sqrt{2} = 0.707$ will have no gain peak near ω_0, but it will oscillate or ring slightly as in Fig. 11.6(a) when the input is removed.

Case III: ω_1^2 is negative or $\omega_0 < \gamma'/2$ or $Q < 1/2$. In this case, the $e^{j\omega_1 t}$ term in (11.18) becomes a real exponential and x decays exponentially even more rapidly than in case II:

$$x = A_1 e^{-(\gamma'/2+|\omega_1|)t} \tag{11.19C}$$

This case is called *overdamped* and is shown in Fig. 11.6(b).

11.3 THE NEGATIVE-IMPEDANCE CONVERTER AND GYRATOR

At this point we briefly indicate how an op amp can substitute for an inductor in a circuit. An op amp can convert any impedance into its

†In the general case of a critically damped oscillator, there is a second possible solution: $x = Bte^{-(\gamma'/2)t}$. This is because the original second-order equation (11.16) must have two independent constants in the general solution in order to be able to satisfy the two initial conditions x and dx/dt at $t = 0$.

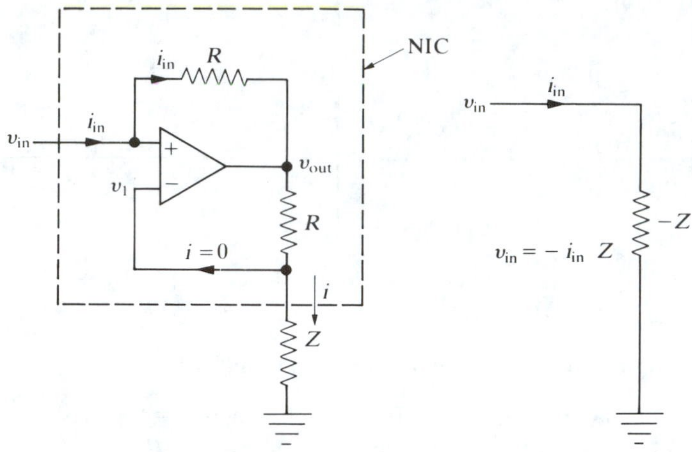

FIGURE 11.7 Negative-impedance converter.

negative with the circuit of Fig. 11.7. The input impedance looking into the noninverting input of the op amp is equal to the negative of the impedance Z. The argument is simple. The input impedance is

$$Z_{in} = \frac{v_{in}}{i_{in}}$$

The OAVR implies $v_{in} = v_1 = iZ$, and $i = v_{out}/(R + Z)$, so

$$v_{in} = \frac{v_{out}Z}{R + Z} \quad \textbf{(1)}$$

The OACR implies i_{in} must all flow through R. Thus from Ohm's law

$$i_{in} = \frac{v_{in} - v_{out}}{R} \quad \textbf{(2)}$$

Eliminating v_{out} from (1) and (2) yields

$$Z_{in} = \frac{v_{in}}{i_{in}} = -Z \quad \textbf{(11.20)}$$

For example, if $Z = j\omega L$, then $Z_{in} = -j\omega L$; if $Z = 1/j\omega C$, $Z_{in} = -1/j\omega C = j/\omega C$. In the latter case Z_{in} is inductive in a "phase" sense because of the $+j$, but Z_{in} *decreases* with increasing frequency; for an inductor, however, the impedance should *increase* with increasing frequency. This circuit is called a *negative-impedance converter* (NIC).

The NIC can be used to make a true inductive impedance out of a capacitive impedance with the circuit of Fig. 11.8(a). The input impedance of this circuit is given by

$$Z_{\text{in}} = \frac{R^2}{Z}$$

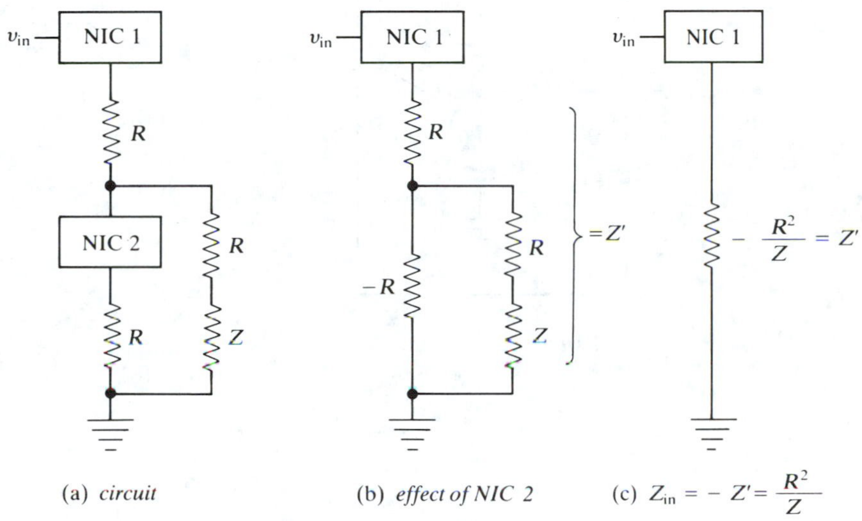

FIGURE 11.8 The gyrator.

The argument is based on the action of the NIC. In Fig. 11.8(b) NIC #2 acts on R and converts it to $-R$. NIC #1 produces an input impedance of $-Z'$, where Z' is the total impedance of R in series with the parallel combination of $-R$ and $R + Z$. Thus

$$Z' = R + \frac{(R + Z)(-R)}{(R + Z) - R} = -\frac{R^2}{Z}$$

The input impedance of NIC #1 is then

$$Z_{\text{in}} = -Z' = \frac{R^2}{Z} \qquad \textbf{(11.21)}$$

Thus, if $Z = 1/j\omega C$, then $Z_{\text{in}} = j\omega R^2 C$, which has the proper phase *and* frequency dependence for an inductor. This circuit is called a *gyrator*. The effective inductance $L_{\text{eff}} = R^2C$. For example, if $R = 100\ \Omega$, $C = 0.01\ \mu\text{F}$, $L_e = (100\ \Omega)^2(10^{-8}\ \text{F}) = 10^{-4}\ \text{H} = 100\ \mu\text{H}$.

11.4 THE SALLEN KEY ACTIVE FILTER

If two low-pass RC filters are connected to an op amp as shown in Fig. 11.9, the overall transfer function of the circuit is exactly the same as

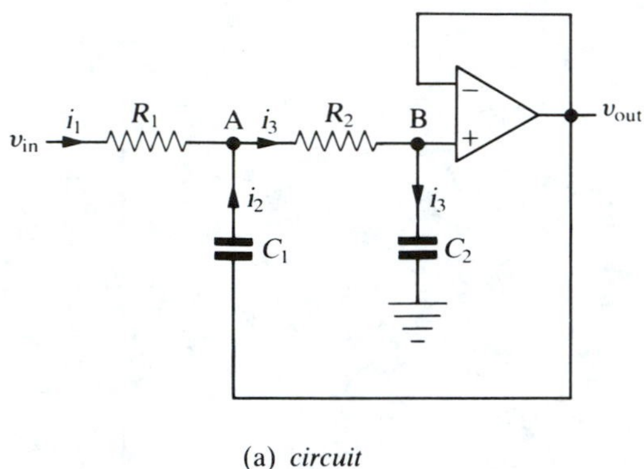

(a) *circuit*

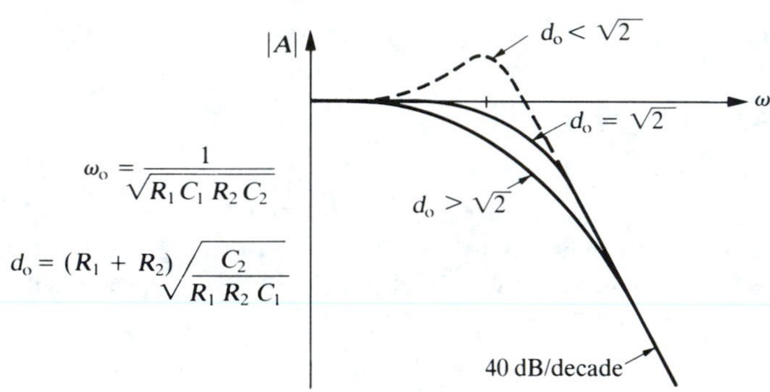

(b) *gain vs. frequency*

FIGURE 11.9 Low-pass Sallen Key active filter.

(11.7), the transfer function of a passive low-pass RLC filter. It can be shown that the transfer function of the low-pass *Sallen Key filter* is

$$A = \frac{1}{1 - \omega^2 R_1 C_1 R_2 C_2 + j\omega(R_1 + R_2)C_2} \tag{11.22}$$

For convenience we repeat the transfer function of the passive *RLC* low-pass filter:

$$A = \frac{1}{1 - \omega'^2 + j\omega' d_0} \tag{11.7}$$

where $\omega' = \omega/\omega_0$, $\omega_0 = 1/\sqrt{LC}$, and $d_0 = 1/Q_0$ is the damping factor that determines the size of the peak in the gain-versus-frequency curve.

If we define the break frequency of the Sallen Key filter as

$$\omega_0 = \frac{1}{\sqrt{R_1 C_1 R_2 C_2}} \tag{11.23}$$

and the new frequency variable $\omega' = \omega/\omega_0$, then (11.22) becomes

$$A = \frac{1}{1 - \omega'^2 + j\omega(R_1 + R_2)C_2} \tag{11.24}$$

Rewriting the $j\omega$ term in A to express it in terms of ω' gives

$$j\omega(R_1 + R_2)C_2 = j\frac{\omega}{\omega_0}(R_1 + R_2)C_2\omega_0 = j\omega'(R_1 + R_2)C_2\frac{1}{\sqrt{R_1 C_1 R_2 C_2}}$$

$$= j\omega'(R_1 + R_2)\sqrt{\frac{C_2}{R_1 C_1 R_2}}$$

Thus by comparing this with (11.7), we see that the damping factor for the Sallen Key circuit is

$$d_0 = (R_1 + R_2)\sqrt{\frac{C_2}{R_1 R_2 C_1}} \tag{11.25}$$

Thus, with two resistances, two capacitors, and one op amp, we have a circuit that behaves exactly like the passive *RLC* low-pass filter. Its break frequency and damping factor can be adjusted by varying the *RC* components, and because of the op amp it has a high input impedance and a low output impedance. Its gain falls off as f^{-2} (40 dB/decade or 12 dB/octave) for frequencies well above the break frequency, and we do not have all the serious disadvantages of an inductor. It is also very easy to vary the break frequency and the damping factor *independently*. This can be seen by inspecting (11.23) and (11.25). It becomes even clearer if we take the common case where $R_1 = R_2 = R$. Then the break frequency is

$$\omega_0 = \frac{1}{R\sqrt{C_1 C_2}} \tag{11.26}$$

and the damping factor is

$$d_0 = 2\sqrt{\frac{C_2}{C_1}} \tag{11.27}$$

Hence, the ratio of C_1 and C_2 determines the damping factor, and the break frequency can be set independently by choosing R. For example, if we desire critical damping, $d_0 = \sqrt{2}$, then from (11.27)

$$\sqrt{2} = 2\sqrt{\frac{C_2}{C_1}} \quad \text{or} \quad \frac{C_2}{C_1} = \frac{1}{2}$$

Thus we might choose $C_2 = 0.10\ \mu\text{F}$ and $C_1 = 0.20\ \mu\text{F}$. If we desire a break frequency of 1 kHz, then from (11.26)

$$R = \frac{1}{\omega_0\sqrt{C_1 C_2}} = \frac{1}{(2\pi)(10^3\ \text{Hz})\sqrt{(2 \times 10^{-7}\ \text{F})(1 \times 10^{-7}\ \text{F})}} = 1130\ \Omega$$

If the damping factor is equal to $\sqrt{2}$, we have the maximum flatness of the gain-versus-frequency curve, critical damping; if $d < \sqrt{2}$, we have an underdamped filter with a hump in the gain curve; if $d > \sqrt{2}$, we have an overdamped filter with a more rounded knee. For the special case $R_1 = R_2 = R$, the damping factor $d_0 = 2\sqrt{C_2/C_1}$; thus we have

Underdamped	$d_0 < \sqrt{2}$	$\frac{C_2}{C_1} < \frac{1}{2}$
Critically damped	$d_0 = \sqrt{2}$	$\frac{C_2}{C_1} = \frac{1}{2}$
Overdamped	$d_0 > \sqrt{2}$	$\frac{C_2}{C_1} > \frac{1}{2}$

If we try the special case $C_1 = C_2$, we find that $d_0 \geqq 2$ for any R_2 and R_1; thus only the overdamped case is possible.

The transfer function $A = v_{out}/v_{in}$ of the Sallen key unity gain filter can be derived from elementary circuit analysis: From the KCL

$$i_1 + i_2 = i_3 \tag{1}$$

From Ohm's law

$$i_1 = \frac{v_{\text{in}} - v_{\text{A}}}{R_1} \qquad \textbf{(2)}$$

$$i_2 = \frac{v_{\text{out}} - v_{\text{A}}}{\dfrac{1}{j\omega C_1}} = j\omega C_1(v_{\text{out}} - v_{\text{A}}) \qquad \textbf{(3)}$$

There is no current flowing into the op amp noninverting terminal, so

$$i_3 = \frac{v_{\text{A}}}{R_2 + \dfrac{1}{j\omega C_2}} \qquad \textbf{(4)}$$

But the OAVR implies $v_{\text{B}} = v_{\text{out}}$, so

$$i_3 = \frac{v_{\text{B}}}{\dfrac{1}{j\omega C_2}} = \frac{v_{\text{out}}}{\dfrac{1}{j\omega C_2}} = j\omega C_2 v_{\text{out}} \qquad \textbf{(5)}$$

We thus have five equations in six unknowns: i_1, i_2, i_3, v_{in}, v_{out}, and v_{A}.
Solving for $v_{\text{out}}/v_{\text{in}}$, we obtain the transfer function

$$A = \frac{v_{\text{out}}}{v_{\text{in}}} = \frac{1}{1 - \omega^2 R_1 R_2 C_1 C_2 + j\omega(R_1 + R_2)C_2} \qquad \textbf{(11.28)}$$

A second-order *high*-pass Sallen Key filter can be made by simply interchanging the resistances and capacitors. The circuit is shown in Fig. 11.10. The same expressions as for the low-pass circuit apply for d_0 and ω_0.

An interesting and useful variation of the unity gain Sallen key filter can be made by varying the gain of the op amp from unity; this is done by varying the feedback factor with R_3 and R_4, as shown in Fig. 11.11. We have set $R_1 = R_2 = R$ and $C_1 = C_2$. A straightforward analysis of the circuit yields the transfer function:

$$A = \frac{A_o}{1 - \omega^2 R^2 C^2 + (3 - A_o)j\omega RC} = \frac{A_o}{1 - \omega'^2 + j\omega'(3 - A_o)} \qquad \textbf{(11.29)}$$

where $\omega_0 = 1/RC$, $\omega' = \omega/\omega_0$, and $A_o = (1 + R_4/R_3)$ is the gain of the op amp for inputs at the noninverting input terminal.

By comparing this transfer function with the standard transfer function for the LRC low-pass filter, (11.7), we immediately see that the damping

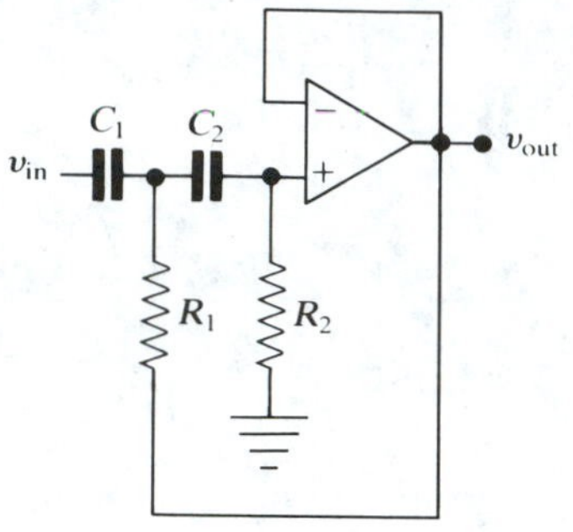

(a) *circuit*

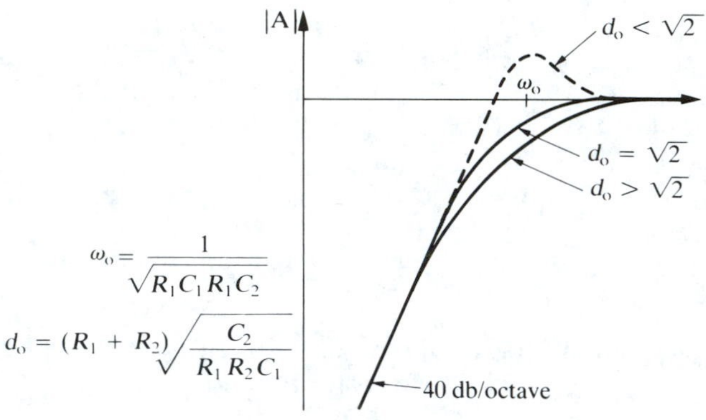

(b) *gain vs. frequency*

FIGURE 11.10 High-pass Sallen Key active filter.

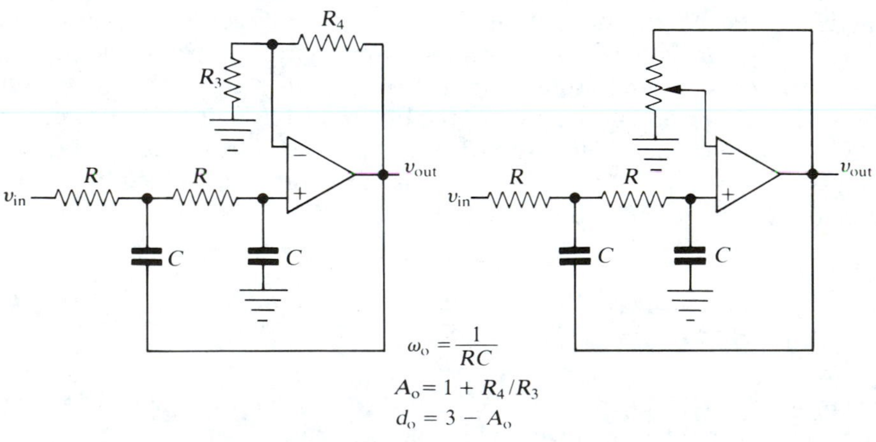

(a) *fixed-gain circuit* (b) *variable-gain circuit*

FIGURE 11.11 Variable-gain low-pass Sallen Key filter.

factor is determined by the *gain* of the op amp, which is totally independent of the break frequency:

$$d_0 = 3 - A_o \tag{11.30}$$

$$\omega_0 = \frac{1}{RC} \tag{11.31}$$

The damping factor depends only on the ratio of R_4/R_3:

$$d_0 = 3 - A_o = 3 - \left(1 + \frac{R_4}{R_3}\right) = 2 - \frac{R_4}{R_3} \tag{11.32}$$

The main advantage of this circuit is that the damping factor and the break frequency can be adjusted completely independently. For example, if a critically damped filter is desired, $d_0 = \sqrt{2}$, which means $2 - R_4/R_3 = \sqrt{2}$ or $R_4/R_3 = 0.586$. We might choose $R_4 = 1.586\ \text{k}\Omega$ and $R_3 = 1.000\ \text{k}\Omega$. Or we could use a potentiometer and make the R_4/R_3 ratio continuously adjustable, as shown in Fig. 11.11(b). This is very convenient for varying the damping.

Underdamped	$d_0 < \sqrt{2}$	$\frac{R_4}{R_3} > 0.586$
Critically damped	$d_0 = \sqrt{2}$	$\frac{R_4}{R_3} = 0.586$
Overdamped	$d_0 > \sqrt{2}$	$\frac{R_4}{R_3} < 0.586$

Notice that if $R_4/R_3 \geqq 2$, the damping factor will be negative, which means the filter will oscillate!

A final point about the low-pass active filter is that there must be a low-resistance path from the input to ground (through the signal source) to allow the small (but nonzero) bias current from the noninverting input of the op amp to flow to ground. This is merely an example of the general op amp rule that there must always be a low-resistance path from each op amp input to ground because the op amp inputs go directly to the bases of the input transistor.

A high-pass variable gain Sallen Key filter can be made by interchanging R and C, as shown in Fig. 11.12. The same expressions for d_0 and ω_0 apply to the low-pass and high-pass circuits.

The principal advantages of the Sallen Key filter are the following:

1. The damping and break frequency are independently adjustable.
2. It uses only one op amp.
3. Interchanging R and C changes the filter from low-pass to high-pass.

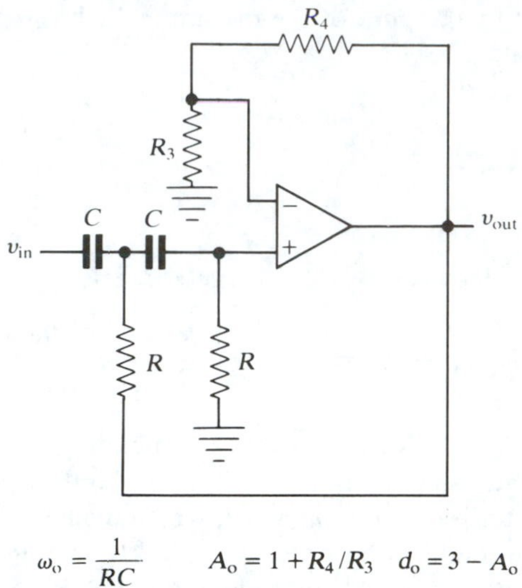

$$\omega_o = \frac{1}{RC} \qquad A_o = 1 + R_4/R_3 \quad d_o = 3 - A_o$$

FIGURE 11.12 High-pass variable-gain Sallen Key filter.

However, the Sallen Key filter is sensitive to small changes (due to temperature, etc.) in its component values and to slight mismatches in component values.

11.5 FILTER BEHAVIOR IN THE TIME AND FREQUENCY DOMAINS

Filters can be characterized by two general descriptions: (1) the description of the gain and phase shift as a function of *frequency*, and (2) the description of the output waveform as a function of *time*. In the frequency domain the effect of underdamped, critically damped, and overdamped circuits can easily be seen, as shown in Fig. 11.13(a). The phase shift of the output of the filter relative to the input for an ideal filter is shown in Fig. 11.13(b).

Filter behavior in the time domain can most easily be understood by considering a step function input as shown in Fig. 11.13(c). For a perfect low-pass filter the output will be a rounded step function with no overshoot or ringing. The higher the filter break frequency, the shorter the rise time.

We already know that an underdamped mechanical oscillator will "ring" after being struck; common examples are a tuning fork, which is very underdamped, or a bell or a gong. From (11.19), a critically damped or overdamped oscillator will not ring or oscillate at all, corresponding to a damping factor of $d_0 \geqq 2$. A ringing in the filter output corresponds to overshoot. The output rise time is also of interest—in general, the shorter

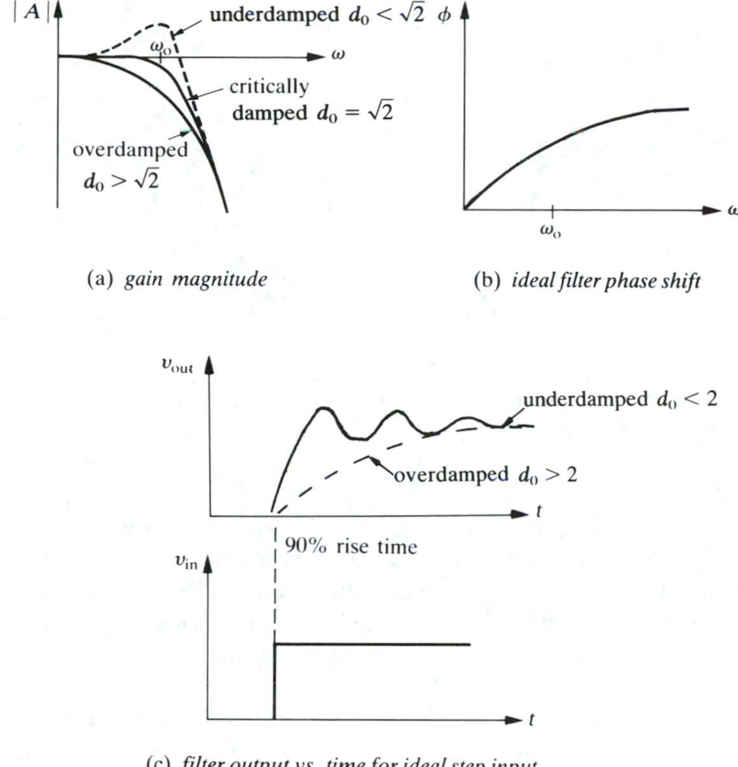

(a) *gain magnitude*

(b) *ideal filter phase shift*

(c) *filter output vs. time for ideal step input*

FIGURE 11.13 Frequency- and time-domain filter characteristics.

the rise time, the better. Also, the *settling* time is of interest—the time for the filter output to settle down to its steady-state value—in other words, the time for the ringing to damp out to within a certain percentage of the steady-state value.

The value of the damping factor determines not only the shape of the filter gain versus frequency but also the presence and amount of overshoot, ringing, and rise time in the time domain. In general, a large degree of underdamping ($d_0 \ll 2$) produces lots of ringing in the time domain and a large peak in the gain-versus-frequency curve at the break frequency. However, an underdamped filter does not, surprisingly, produce the shortest rise time. The most common underdamped filter is called a *Chebyshev* filter with a damping factor $d_0 = 0.767$. It has the sharpest knee in the frequency domain—but at the cost of ripple in the passband gain—as shown in Fig. 11.14(a). It also has the worst overshoot in the time domain, as shown in Fig. 11.14(b).

The critically damped filter is called the *Butterworth* filter with a damping factor $d_0 = 1.414$. It has the flattest pass band—the flattest gain

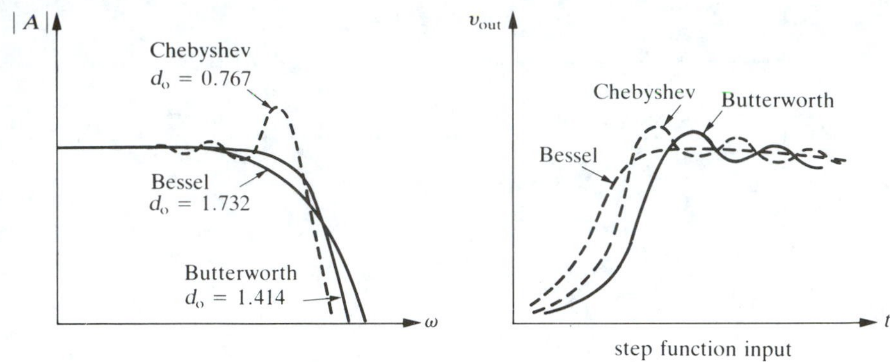

FIGURE 11.14 Three common low-pass filter types and their characteristics in the frequency and the time domains.

versus frequency with no ripples in the frequency domain—but it has overshoot in the time domain in response to step inputs.

The most common overdamped filter is the *Bessel* filter with $d =$ 1.732. It has the roundest knee of the gain versus frequency, but it has the best transient response, with the shortest rise time and no overshoot or ringing.

For frequencies well above the break frequency for all three types of low-pass second-order filters, the gain falls off at the same rate, namely, 40 dB/decade or 12 dB/octave (as f^{-2}). All of these op amp active filters have high input impedances and low output impedances, so they can be cascaded to yield higher-order filters.

Cascading several second-order active filters is possible because of their high input and low output impedances, but it is not quite as simple as it might appear. For example, consider two low-pass second-order 1-kHz Butterworth filters, each with a damping factor of 1.414 and a 3-dB point of 1.0 kHz. If they are cascaded in series, the 3-dB point of the resultant fourth-order filter is less than 1.0 kHz, and the gain is down by 6 dB at 1.0 kHz; in other words, the knee of the fourth-order filter is rounded off, even though its gain drops off at 24 dB/octave (f^{-4}) for frequencies well above 1.0 kHz. To make a true Butterworth fourth-order filter with a 3-dB cutoff frequency at 1.0 kHz requires that we cascade an overdamped second-order 1.0-kHz filter with an underdamped second-order 1.0-kHz filter as shown in Fig. 11.15. Careful gain calculations indicate each filter should have a cutoff frequency of 1.0 kHz, and the damping factor of the first filter should be 1.858 and that of the second filter 0.765. In general, the filter nearest the input should have the highest damping factor.

For cascaded Bessel or Chebyshev filters the cutoff frequencies of the two filters and also the damping factors are different. For example, a fourth-order 1.0-kHz cutoff Chebyshev 3-dB deep filter consists of two

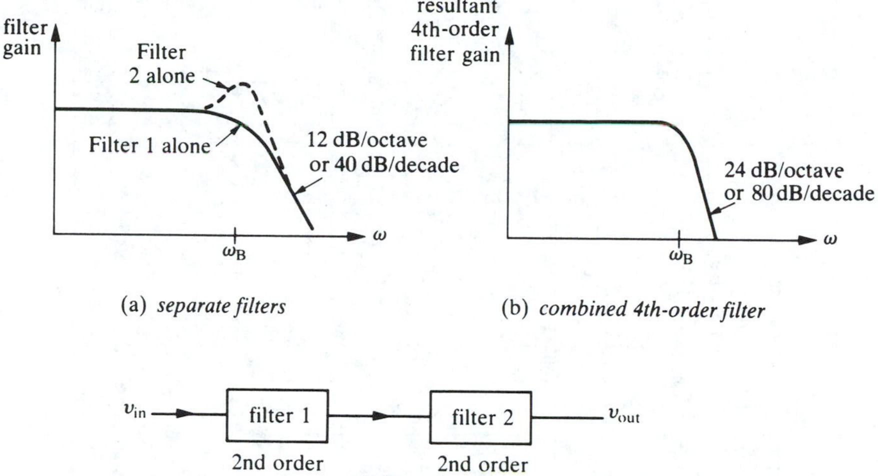

(a) *separate filters* (b) *combined 4th-order filter*

FIGURE 11.15 The fourth-order Butterworth filter.

second-order filters—the first with a cutoff frequency of 440 Hz and a damping factor of 0.923, and the second with a cutoff of 947 Hz and a damping factor of 0.176, as shown in Fig. 11.16.

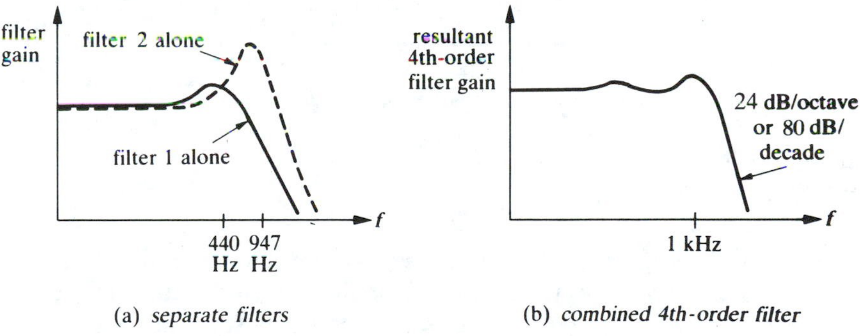

(a) *separate filters* (b) *combined 4th-order filter*

FIGURE 11.16 True fourth-order Chebyshev filter.

As a general rule, the more underdamped the filter (the smaller d_0), the higher the precision of the R and C components to avoid the possibility of filter oscillation. For example, 10% tolerance components may be adequate for the overdamped Bessel filter, whereas 1% or 2% tolerance might be required for the underdamped 3-dB Chebyshev filters.

Figure 11.17 shows graphs of the gain versus frequency for the Butterworth, Bessel, and the 3-dB Chebyshev filter, all normalized to a 1-kHz cutoff. Notice that the sharper knee of the Chebyshev filter is

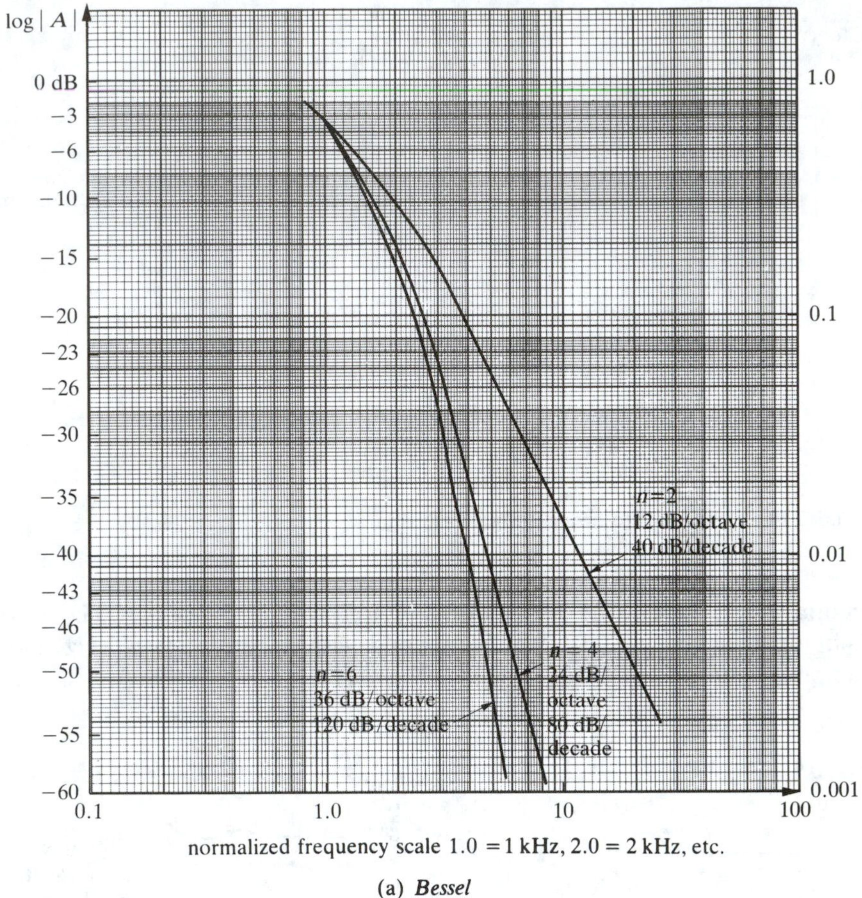

(a) *Bessel*

FIGURE 11.17 Filter gain curves.

obtained at the expense of ripple in the pass-band, whereas the Bessel filter has the rounded knee but the best transient behavior in the time domain (minimum ringing). To sum up: the Bessel filter is best for minimum ringing in response to sudden changes in the input amplitude; the Chebyshev filter has the sharpest knee but the worst transient behavior; and the Butterworth filter is a compromise between a sharp knee and minimum ringing.

Table 11.1 gives the cutoff frequencies and damping factors for the three low-pass filter types. f_1, d_1 are the cutoff frequency and damping factor of the first (second-order) filter section, and so on.

To illustrate the use of Fig. 11.17 and Table 11.1 in designing filters, let's design a Bessel filter with a cutoff frequency of 2 kHz and an attenuation at 6 kHz of at least 20 dB. A glance at Fig. 11.17 shows that a second-order filter has an attenuation of only about 16 dB at a normalized frequency of 3.0 (6.0 kHz = 3.0 × 2.0 kHz), but a fourth-order filter has an

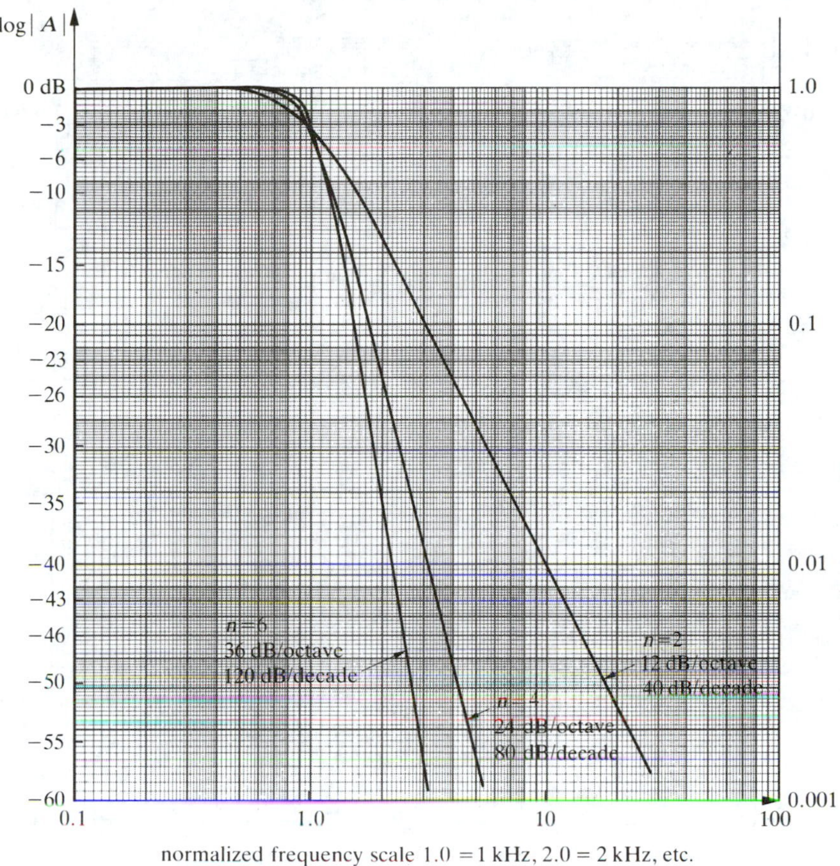

(b) *Butterworth*

FIGURE 11.17 Continued.

attenuation of 23 dB, and a sixth-order filter 29 dB. Thus we can use a fourth-order Bessel filter as shown in Fig. 11.18. (Notice that we could get 20 dB from only a second-order Butterworth filter, and 24 dB from a second-order 3-dB Chebyshev filter.)

For a fourth-order Bessel filter the first stage has $f_1 = 1.436$ and $d_1 = 1.916$. Thus the first-stage cutoff frequency should be $f_1 = 1.436 \times 2.0\text{ kHz} = 2.872\text{ kHz}$, and thus, from (11.31), $2\pi f_1 = 1/R_1C_1$.

If $C_1 = 0.01\ \mu\text{F}$, $R_1 = 5.542\text{ k}\Omega$. If $C_1 = 0.001\ \mu\text{F}$, $R_1 = 55.42\text{ k}\Omega$. The first-stage damping factor is from (11.32) $d_1 = 2 - R_{f_1}/R_{g_1}$. Thus

$$\frac{R_{f_1}}{R_{g_1}} = 2 - d_1 = 2 - 1.916 = 0.084$$

If $R_{g_1} = 100\text{ k}\Omega$, $R_{f_1} = 8.4\text{ k}\Omega$; if $R_{g_1} = 10\text{ k}\Omega$, $R_{f_1} = 840\ \Omega$. Usually R_1,

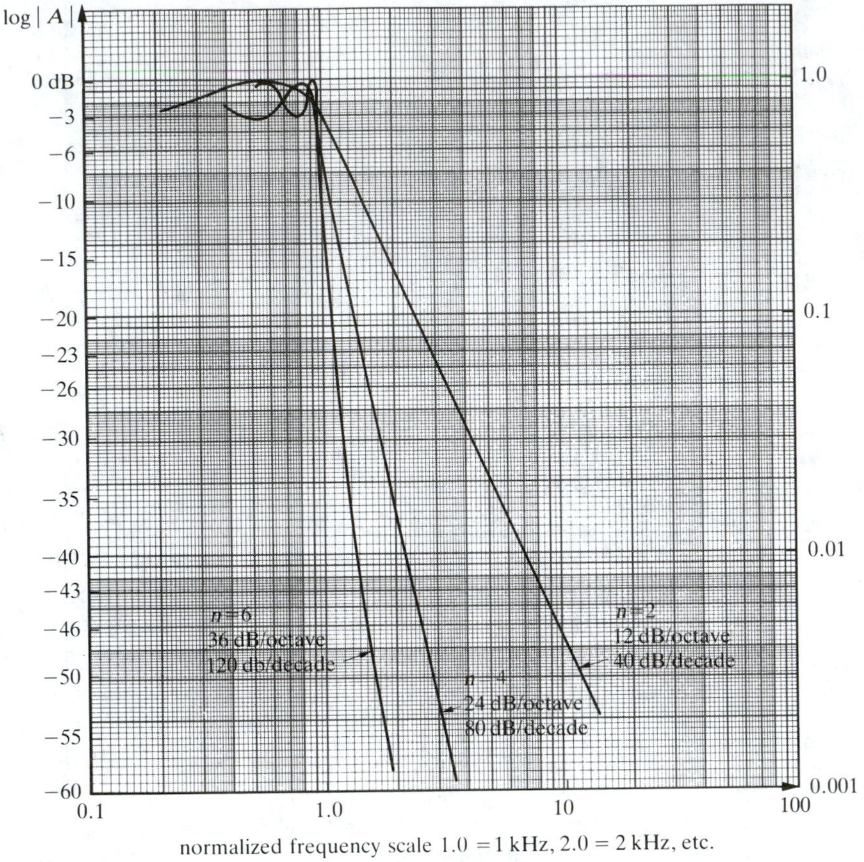

log | A |

0 dB, −3, −6, −10, −15, −20, −23, −26, −30, −35, −40, −43, −46, −50, −55, −60

1.0, 0.1, 0.01, 0.001

0.1, 1.0, 10, 100

n = 6
36 dB/octave
120 db/decade

n = 4
24 dB/octave
80 dB/decade

n = 2
12 dB/octave
40 dB/decade

normalized frequency scale 1.0 = 1 kHz, 2.0 = 2 kHz, etc.

(c) *Chebyshev*

FIGURE 11.17 Continued.

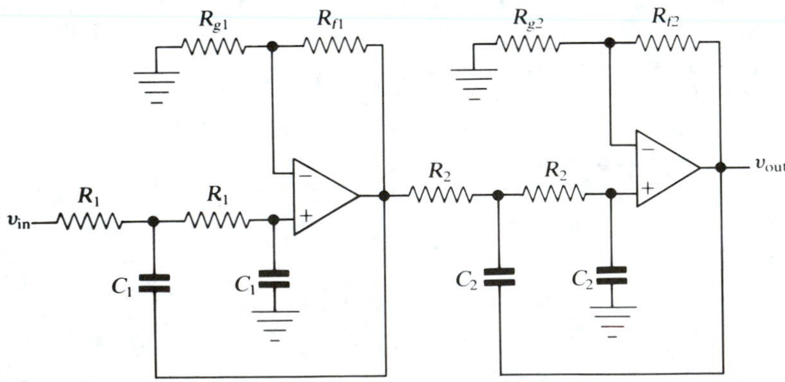

FIGURE 11.18 Fourth-order variable-gain low-pass Sallen Key active filter.

TABLE 11.1. Low-Pass Variable Gain Sallen Key Filters

Filter Type	*Order*	f_1	d_1	f_2	d_2	f_3	d_3	*dB/octave*	*RC Tolerance (%)*
Bessel	2	1.274	1.732					12	10
	4	1.436	1.916	1.610	1.241			24	10
	6	1.609	1.959	1.694	1.636	1.910	0.977	36	10
Butterworth	2	1.000	1.414					12	10
	4	1.000	1.848	1.000	0.765			24	5
	6	1.000	1.932	1.000	1.414	1.000	0.518	36	2
3-dB Chebyshev	2	0.841	0.767					12	5
	4	0.443	0.929	0.950	0.179			24	1
	6	0.298	0.958	0.722	0.289	0.975	0.0782	36	1

TABLE 11.2. High-Pass Variable Gain Sallen Key Filters

Filter Type	*Order*	f_1	d_1	f_2	d_2	f_3	d_3	*dB/octave*	*RC* *Tolerance (%)*
Bessel	2	0.785	1.732					12	0.1
	4	0.696	1.916	0.621	1.241			24	0.1
	6	0.621	1.959	0.590	1.636	0.524	0.977	36	0.1
Butterworth	2	1.000	1.414					12	0.1
	4	1.000	1.848	1.000	0.765			24	0.05
	6	1.000	1.932	1.000	1.414	1.000	0.518	36	0.02
3-dB Chebyshev	2	1.189	0.767					12	0.05
	4	2.257	0.929	1.053	0.179			24	0.01
	6	3.356	0.958	1.385	0.289	1.026	0.0782	36	0.01

R_{f_1}, and R_{g_1} are chosen so that each op amp input sees the same input impedance.

Similarly, for the second section we find if $C_2 = 0.01\ \mu F$, $R_2 = 4.943\ k\Omega$; if $C_2 = 0.001\ \mu F$, $R_2 = 49.43\ k\Omega$. And, if $R_{g_2} = 10\ k\Omega$, $R_{f_2} = 7.59\ k\Omega$; if $R_{g_2} = 100\ k\Omega$, $R_{f_2} = 75.9\ k\Omega$.

The values of the products R_1C_1 and R_2C_2 are fixed by the cutoff frequencies of the two stages. Whether we use a large R and a small C or vice versa depends on the desired filter input impedance. In general, the first filter input impedance should be large compared to the input signal source output impedance. Similarly, the second filter input impedance should be large compared to the first filter output impedance. It can be shown that the filter input impedance is approximately given by $Z = 5/\omega_0 C = 5|X_C|$. Thus, choosing C_1 determines the first filter input impedance and choosing C_2 determines the second filter input impedance. Then the break frequency $\omega_0 = 1/R_1C$ determines R_1. Resistance values greater than approximately 100 kΩ tend to be noisy and unreliable, so a good rule is to use resistances from 1 to 100 kΩ.

Table 11.2 gives the cutoff frequencies and damping factors for the three high-pass filter types. Notice that the damping factors are the same as for the low-pass filters of Table 11.1, and the frequencies are the reciprocals of those in Table 11.1.

11.6 THE BANDPASS FILTER

A *bandpass filter* is one that passes only a relatively narrow band of frequencies (called the *passband*) and attenuates all other frequencies both above and below the passband. Its general gain-versus-frequency curve is shown in Fig. 11.19(a). The passband is generally defined as those frequen-

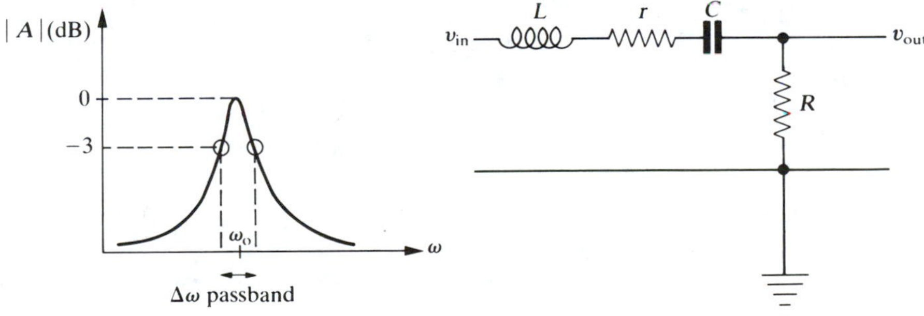

(a) *gain vs. frequency* (b) *passive LRC bandpass filter circuit*

FIGURE 11.19 Bandpass filter.

cies that are attenuated less than 3 dB; in other words, if the filter gain is A_o at the maximum, it is $0.707A_o$ (down 3 dB) at each edge of the passband. The frequency range both above and below the passband is sometimes called the *stopband*.

The simplest passive *LRC* passband filter is shown in Fig. 11.19(b). At the resonant frequency of L and C ($\omega_0 = 1/\sqrt{LC}$), the gain is 1 if $r \ll R$; at all other frequencies above and below resonance the gain is less than unity. For frequencies well above and well below resonance, the gain falls off at only 6 dB/octave—as $1/f$. The width of the passband is determined by Q_0 according to $\Delta\omega = \omega_0/Q_0$, and Q_0 is determined by the L/R ratio: $Q_0 = \omega_0 L/R$. With passive components it is difficult to obtain a Q much larger than 25, and for low resonant frequencies large (and expensive) inductors are required. Thus passive bandpass filters are limited to high-frequency low-Q applications.

One of the simplest ways to realize a bandpass filter is to incorporate a frequency selective element in the feedback loop of an op amp circuit, as shown in Fig. 11.20(a). We recall that a twin T filter attenuates only one frequency ω_0 and passes other frequencies above and below ω_0, as shown in Fig. 11.20(b). Thus, with the twin T in the feedback loop there is negative feedback for all frequencies except the resonant frequency of the twin T. Hence, the circuit is a bandpass filter with maximum gain at the twin T resonant frequency, as shown in Fig. 11.20(c). If we assume a perfect twin T filter (with perfectly matched components), then the impedance of the

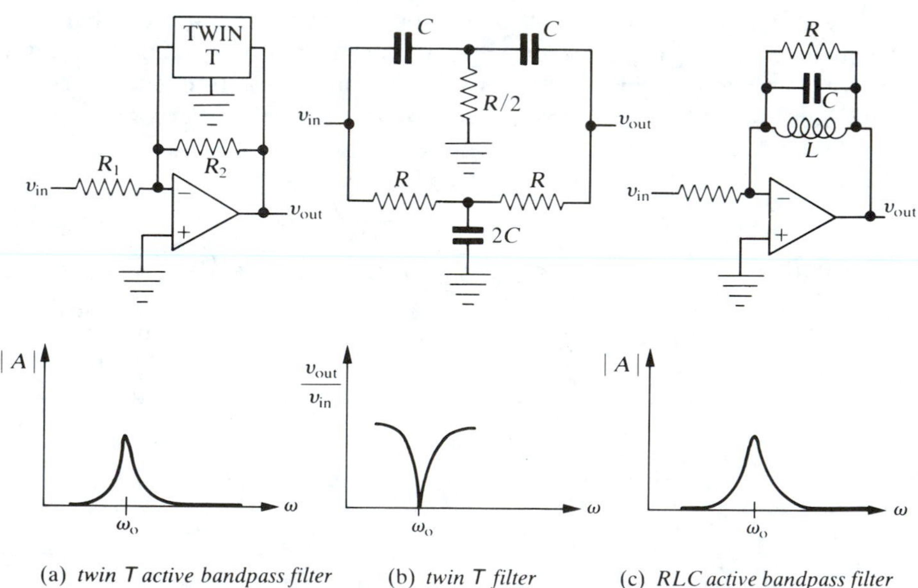

(a) *twin T active bandpass filter* (b) *twin T filter* (c) *RLC active bandpass filter*

FIGURE 11.20 Twin T feedback active filter.

twin T will be infinite at ω_0, and the circuit will be a simple inverting op amp with maximum gain $A_o = -R_2/R_1$. This bandpass filter can have a Q up to 50, depending on the component matching in the twin T. Fortunately, twin T filters are commercially available for resonant frequencies from 1 Hz to approximately 50 kHz, so it is easy to build such a bandpass filter by purchasing the appropriate twin T. However, such a filter cannot be easily tuned, and this is its main disadvantage.

Any feedback element will do as long as it has a maximum impedance at its resonant frequency—a parallel *RLC* circuit, for example, as shown in Fig. 11.20(c). The state variable filter of Section 11.6.3 also can act as a bandpass filter.

11.6.1 The Multiple-Feedback Active Bandpass Filter

We can make a more sophisticated and versatile active bandpass filter with an op amp, using the circuit shown in Fig. 11.21(a). Qualitatively, the

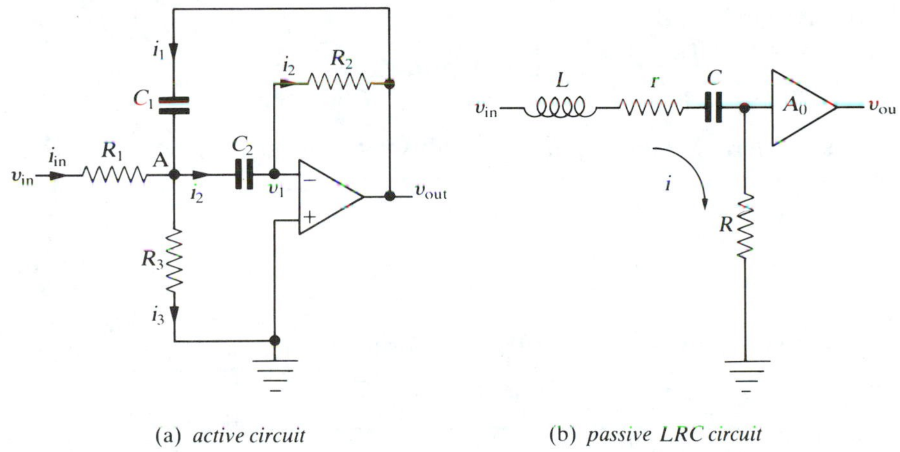

(a) *active circuit* (b) *passive LRC circuit*

FIGURE 11.21 Bandpass filter.

circuit can be thought of as a combination of two first-order *RC* filters: one a high-pass, and one a low-pass. The op amp and R_2 and C_2 form a high-pass filter (a differentiator) with a break frequency of $1/R_2C_2$. But the op amp and R_1 and C_1 form a low-pass filter (an integrator) with a break frequency of $1/R_1C_1$. Thus, by the proper choice of the resistance and capacitances, we can make a bandpass filter by overlapping the two filter passbands slightly.

The transfer function of the passive *LRC* bandpass filter of Fig.

11.21(b) is easily calculated:

$$A = \frac{iRA_o}{i\left(j\omega L + r + \dfrac{1}{j\omega C} + R\right)} = \frac{RA_o}{R + r + j\left(\omega L - \dfrac{1}{\omega C}\right)} \cong \frac{j\omega RCA_o}{1 - \omega^2 LC + j\omega RC}$$

where we have assumed $r \ll R$. Rewriting, we have

$$A = \frac{\dfrac{j\omega}{\omega_0 Q_0} A_o}{1 - \dfrac{\omega^2}{\omega_0^2} + \dfrac{j\omega}{\omega_0 Q_0}} = \frac{\dfrac{j\omega'}{Q_0} A_o}{1 - \omega'^2 + \dfrac{j\omega'}{Q_0}} \qquad \textbf{(11.33)}$$

where $Q_0 = \omega_0 L/R$, $\omega' = \omega/\omega_0$, and $\omega_0 = 1/\sqrt{LC}$.

The op amp circuit of Fig. 11.21(a) is a realization of a bandpass filter. We can see this by the following argument. The KCL at point A implies

$$i_{in} + i_1 = i_3 + i_2$$

or

$$\frac{v_{in} - v_A}{R_1} + (v_{out} - v_A)j\omega C_1 = \frac{v_A}{R_3} + v_A j\omega C_2 \qquad \textbf{(1)}$$

where we have set $v_1 = 0$ from the OAVR. By the OACR the current through C_2 must equal the current through R_2. Thus,

$$v_A j\omega C_2 = -\frac{v_{out}}{R_2} \qquad \textbf{(2)}$$

Eliminating v_A and solving for the transfer function give

$$\frac{v_{out}}{v_{in}} = \frac{-1}{R_1\left[j\omega C_1 + \left(\dfrac{1}{R_1} + \dfrac{1}{R_3}\right)\dfrac{1}{j\omega R_2 C_2} + \dfrac{C_1 + C_2}{R_2 C_2}\right]} \qquad \textbf{(11.34)}$$

Let $C_1 = C_2 = C$ because most circuits use fixed and equal capacitances (they are difficult to tune). Let $R_{13} = R_1 \| R_3 = R_1R_3/(R_1 + R_3)$. After some algebra, we obtain

$$\frac{v_{out}}{v_{in}} = \frac{-j\omega \dfrac{R_2 R_3}{R_1 + R_3} C}{1 - \omega^2 R_{13} R_2 C^2 + 2j\omega R_{13} C} \qquad \textbf{(11.35)}$$

which is of the same general form as the transfer function (11.33) for the simple passive LRC circuit. [Notice that if we let $R_1 = R_3 = R$ and $R_2 = 2R$, (11.35) becomes identical to the transfer function (11.33) for the passive LRC bandpass filter.] To simplify things, we now set $R_1 = R_3 = R$.

$$\frac{v_{\text{out}}}{v_{\text{in}}} = \frac{\dfrac{-j\omega R_2 C}{2}}{1 - \dfrac{\omega^2 R R_2 C^2}{2} + j\omega RC} \tag{11.36}$$

Comparing (11.36) for the op amp active filter with (11.33) for the passive LRC filter, we have, from equating coefficients of ω^2 in the denominator,

$$\frac{RR_2C^2}{2} = \frac{1}{\omega_0^2} \quad \text{or} \quad \omega_0 = \sqrt{\frac{2}{RR_2C^2}} \tag{11.37}$$

From equating coefficients of $j\omega$ in the denominator, we obtain

$$RC = \frac{1}{\omega_0 Q_0} \quad \text{or} \quad Q_0 = \frac{1}{\omega_0 RC} = \sqrt{\frac{R_2}{2R}} \tag{11.38}$$

From equating coefficients of $j\omega$ in the numerator, we get

$$\frac{-R_2C}{2} = \frac{A_o}{\omega_0 Q_0} \quad \text{or} \quad A_o = -\frac{1}{2}\frac{R_2}{R} \tag{11.39}$$

The op amp active filter of Fig. 11.21(a) is a bandpass filter because it has the same transfer function as the passive LRC bandpass filter of Fig. 11.21(b). However, the op amp filter has two distinct advantages over the passive filter: First, it has no inductor; it uses only an op amp and inexpensive easily available resistors and capacitors. Second, its Q can be changed by changing the resistances.

The disadvantages of this op amp bandpass filter are that the circuit cannot produce a Q much greater than 10; a high Q usually means a high gain because the Q and the gain expression are similar; and the op amp bias current term I_BR_2 produces a dc offset output voltage.

Notice that the Q and the gain expression each depend only on the ratio R_2/R. If we desire a Q of 5 and a resonant frequency of 1 kHz, then

$$Q_0 = 5 = \sqrt{\frac{R_2}{2R}} \quad \text{or} \quad 25 = \frac{R_2}{2R} \quad \text{or} \quad R_2 = 50R$$

$$A_o = -\frac{1}{2}\frac{R_2}{R} = -\frac{1}{2}(50) = -25$$

$$\omega_0^2 = \frac{2}{RR_2C^2} = \frac{2}{(R)(50R)C^2} = \frac{1}{25R^2C^2} \quad \text{or} \quad \omega_0 = \frac{1}{5RC}$$

Thus $$RC = \frac{1}{5\omega_0} = \frac{1}{(5)(2\pi)(10^3 \text{ Hz})} = 3.18 \times 10^{-5} \text{ s}$$

We might choose a convenient capacitance value, say, $C = 0.01\ \mu\text{F} = 10^{-8}$ F. Then

$$R = \frac{3.18 \times 10^{-5} \text{ s}}{10^{-8} \text{ F}} = 3180\ \Omega \quad \text{so} \quad R_2 = 50R = 159 \text{ k}\Omega$$

Notice that if $R_1 \neq R_3$, we could vary Q and A independently. If $C_1 \neq C_2$ and $R_1 \neq R_3$, it can be shown that

$$\omega_0 = \frac{1}{\sqrt{R_2 C_1 C_2 R_{13}}} \tag{11.40}$$

$$Q_0 = \frac{\sqrt{\dfrac{R_2}{R_{13}}}}{\sqrt{\dfrac{C_2}{C_1}} + \sqrt{\dfrac{C_1}{C_2}}} \tag{11.41}$$

$$A_o = \frac{-R_2 C_2}{R_1(C_1 + C_2)} \tag{11.42}$$

11.6.2 Active Inductor (Gyrator) Bandpass Filter

In general, an inverting op amp can be made into a bandpass filter by using an impedance Z that has a maximum at a certain "resonant" frequency in the negative feedback loop, as shown in Fig. 11.22. Then there

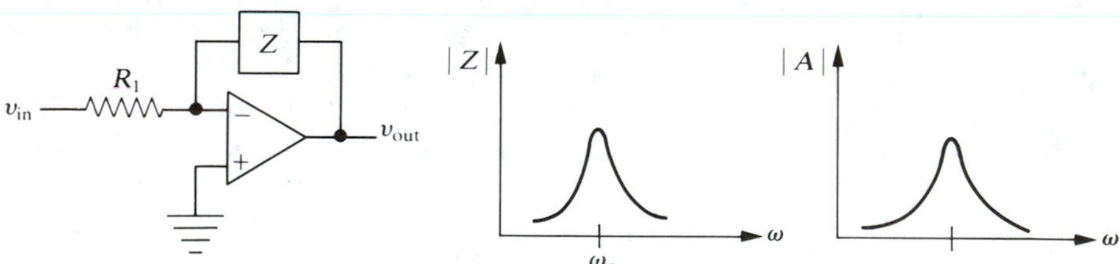

FIGURE 11.22 Active bandpass filter.

will be strong negative feedback at all frequencies except near the resonant frequency, and we will have a bandpass filter. If we use a parallel LRC circuit for Z, then the circuit will pass frequencies near $\omega_0 = 1/\sqrt{LC}$.

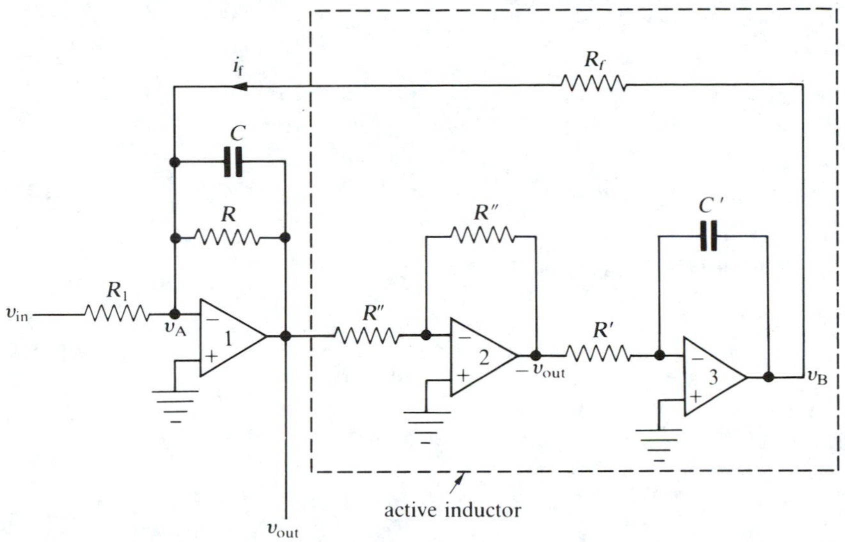

FIGURE 11.23 Active inductor bandpass filter.

If we use an op amp circuit to make an "artificial" inductor from a capacitor, then we have the active op amp bandpass filter of Fig. 11.23. The circuit inside the dashed lines acts like an inductor with an effective inductance of $L = R_f R'C'$. The argument is simple. Op amp #2 has a gain of -1, and op amp #3 is a simple integrator. Therefore

$$v_B = -\frac{1}{R'C'}\int(-v_{out})\,dt$$

or

$$\frac{dv_B}{dt} = \frac{1}{R'C'}v_{out}$$

But

$$i_f = \frac{v_B - v_A}{R_f} = \frac{v_B}{R_f}$$

from the OAVR. Thus

$$R_f\frac{di_f}{dt} = \frac{1}{R'C'}v_{out}$$

or

$$v_{out} = R_f R'C'\frac{di_f}{dt}$$

which is of the form

$$v_{out} = L_{eff} \frac{di_f}{dt}$$

Thus

$$L_{eff} = R_f R' C' \tag{11.43}$$

The resonant frequency is therefore

$$\omega_0 = \frac{1}{\sqrt{L_{eff} C}} = \frac{1}{\sqrt{R_f R' C' C}} \tag{11.44}$$

The Q is (remember R is the *parallel* resistance of the LC circuit)

$$Q_0 = \omega_0 RC = \frac{R}{\omega_0 L} = \frac{\sqrt{R_f R' C' C} R}{R_f R' C'} = \sqrt{\frac{C}{R_f R' C'}} R \tag{11.45}$$

Thus we see that once the resonant frequency has been chosen, Q_0 can be *independently* chosen by choice of R. In other words, if the circuit is tuned by changing R_f or R' or C', the *absolute* bandwidth ($\Delta\omega$)

$$\Delta\omega = \frac{\omega_0}{Q} = \frac{\omega_0}{\omega_0 RC} = \frac{1}{RC} \tag{11.46}$$

remains constant if the filter is tuned by adjusting R_f and R'. A Q of up to 100 can generally be obtained with this circuit. At resonance the gain is

$$A_v = -\frac{R}{R_1} \tag{11.47}$$

because the impedance of the parallel LRC circuit at resonance is R. Thus the gain can be set independently by choosing R_1.

11.6.3 The State Variable Active Bandpass Filter

The most useful bandpass filter with high Q and ease of adjustment is probably the state *variable* filter shown in Fig. 11.24(a). It can yield a Q up to 100, is easy to adjust, and has low-pass and high-pass outputs as well as a bandpass output. The state variable filter is basically an analog computer with two integrators and a summing amplifier that will solve the differential equation for the driven damped simple harmonic oscillator discussed in Section 11.2.

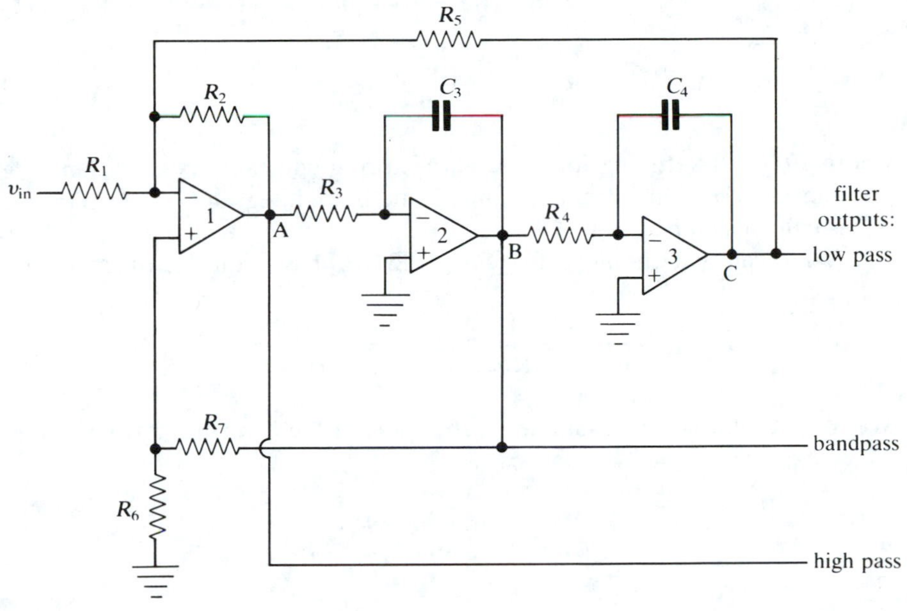

(a) *circuit*

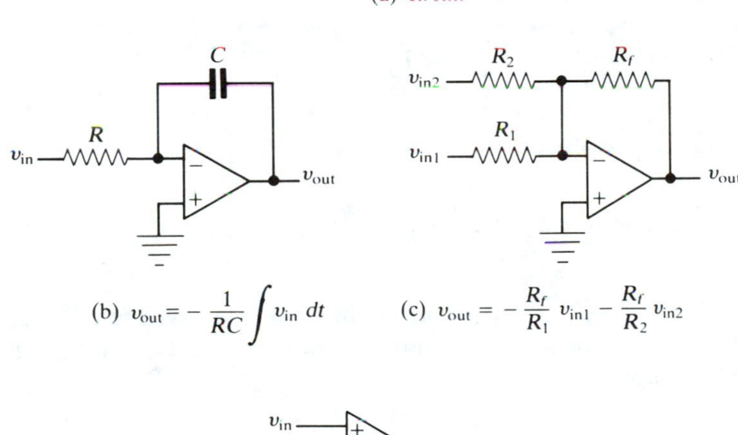

(b) $v_{out} = -\frac{1}{RC}\int v_{in}\,dt$

(c) $v_{out} = -\frac{R_f}{R_1}v_{in1} - \frac{R_f}{R_2}v_{in2}$

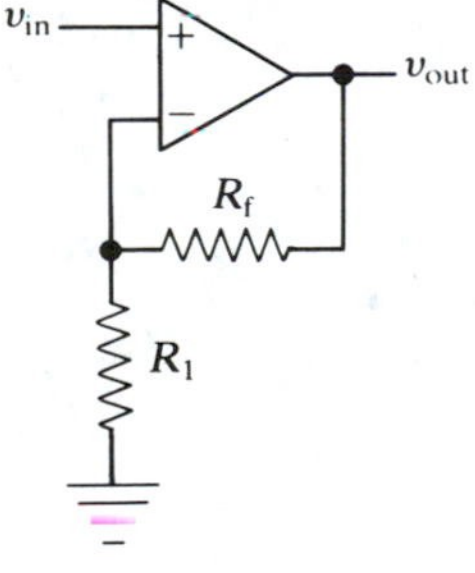

(d) $v_{out} = (1 + \frac{R_f}{R_1}\; v_{in}$

FIGURE 11.24 State variable filter.

$$F(t) - Kx - \gamma\dot{x} = m\ddot{x} \tag{1}$$

or

$$\ddot{x} = \frac{F(t)}{m} - \frac{K}{m}x - \frac{\gamma}{m}\dot{x} \tag{2}$$

where $F(t)$ is the driving force, K is the spring constant, γ is the damping constant, and m is the mass. Op amps 2 and 3 are integrators, and op amp 1 is basically a summer.

We recall that an integrator [Fig. 11.24(b)] has a transfer function of

$$\frac{v_{\text{out}}}{v_{\text{in}}} = -\frac{1}{RC}\int v_{\text{in}}\,dt = -\frac{v_{\text{in}}}{j\omega RC}$$

We also recall that for a summer with inputs at the inverting terminal [Fig. 11.24(c)],

$$v_{\text{out}} = -\frac{R_f}{R_1}v_{\text{in1}} - \frac{R_f}{R_2}v_{\text{in2}}$$

and for an input to the noninverting terminal [Fig. 11.24(d)],

$$v_{\text{out}} = \left(1 + \frac{R_f}{R_1}\right)v_{\text{in}}$$

Thus for op amp 1 of the state variable filter

$$v_{\text{A}} = -\frac{R_2}{R_1}v_{\text{in}} - \frac{R_2}{R_5}v_{\text{C}} + Gv_{\text{B}} \tag{3}$$

where G is the (positive) gain for the input at the noninverting terminal. Using the principle of superposition, we can calculate G. Set $v_{\text{C}} = v_{\text{in}} = 0$. The OAVR implies

$$\frac{R_6}{R_6 + R_7}v_{\text{B}} = \frac{R_{15}}{R_{15} + R_2}v_{\text{A}} \qquad \text{where } R_{15} = \frac{R_1 R_5}{R_1 + R_5}$$

Thus

$$G = \frac{v_{\text{A}}}{v_{\text{B}}} = \frac{R_6}{R_6 + R_7}\,\frac{R_2 + R_{15}}{R_{15}}$$

Finally, (3) becomes

$$v_{\text{A}} = -\frac{R_2}{R_1}v_{\text{in}} - \frac{R_2}{R_5}v_{\text{C}} + \left[\frac{R_6}{R_6 + R_7}\,\frac{R_2 + R_{15}}{R_{15}}\right]v_{\text{B}} \tag{4}$$

But because op amp 2 is an integrator,

$$v_B = -\frac{1}{R_3C_3}\int v_A\,dt$$

or

$$v_A = -R_3C_3\frac{dv_B}{dt} = -R_3C_3\dot{v}_B$$

For op amp 3,

$$v_C = -\frac{1}{R_4C_4}\int v_B\,dt$$

or

$$v_B = -R_4C_4\frac{dv_C}{dt} = -R_4C_4\dot{v}_C \tag{5}$$

Thus

$$v_A = R_3C_3R_4C_4\ddot{v}_C \tag{6}$$

Using (5) and (6) in (4) gives

$$v_A = R_3C_3R_4C_4\ddot{v}_C = -\frac{R_2}{R_1}v_{in} - \frac{R_2}{R_5}v_C - \left[\frac{R_6}{R_6+R_7}\frac{R_2+R_{15}}{R_{15}}\right]R_4C_4\dot{v}_C$$

or

$$\ddot{v}_C = -\frac{R_2}{R_1R_3C_3R_4C_4}v_{in} - \frac{R_2}{R_3C_3R_4C_4R_5}v_C - \left[\frac{R_6}{R_6+R_7}\frac{R_2+R_{15}}{R_{15}}\right]\frac{\dot{v}_C}{R_3C_3} \tag{7}$$

which is the same form as (2) for the damped driven mechanical simple harmonic oscillator. Thus we make the comparisons:

$$v_C \leftrightarrow x$$

$$-\frac{R_2}{R_1R_3C_3R_4C_4}v_{in} \leftrightarrow \frac{F(t)}{m}$$

$$-\frac{R_2}{R_3C_3R_4C_4R_5} \leftrightarrow -\frac{K}{m}$$

$$-\left[\frac{R_6}{R_6+R_7}\frac{R_2+R_{15}}{R_{15}}\right]\frac{1}{R_3C_3} \leftrightarrow -\frac{\gamma}{m}$$

Thus we immediately can conclude that the resonant frequency is given by

$$\omega_0^2 = \frac{K}{m} = \frac{R_2}{R_3C_3R_4C_4R_5} \tag{11.48}$$

The Q is given by

$$Q_0 = \omega_0 \frac{m}{\gamma}$$

which works out to be

$$Q = \sqrt{\frac{R_2 R_3 C_3}{R_4 C_4 R_5}} \frac{(R_6 + R_2)R_{15}}{R_6(R_2 + R_{15})} \tag{11.49}$$

It is common to choose $R_3 = R_4 \equiv R$, $C_3 = C_4 \equiv C$, and $R_1 = R_2 = R_5 \equiv R'$. Then $R_{15} = R'/2$ and

$$\omega_0 = \frac{1}{RC} \tag{11.50}$$

and

$$Q = \frac{R_6 + R_7}{3R_6} \tag{11.51}$$

Thus we see that ω_0 and Q are completely independent of one another.

From the treatment of the mechanical oscillator in 11.2, we found the x, the $\dot{x}$, and the $\ddot{x}$ response as functions of the driving frequency (see Fig. 11.5). In other words, v_A corresponds to a high-pass output, v_B to a bandpass output, and v_C to a low-pass output. Thus we have the significant advantage of three filters in one as well as the convenience of being able to adjust ω_0 and Q independently.

Several bandpass filters can be cascaded in series to obtain a variety of output-versus-frequency curves. For example, if two bandpass filters with slightly different resonant frequencies, ω_{10} and ω_{20}, are cascaded in series, the output will look like the curve of Fig. 11.25(a). The depth of the dip in the center of the passband depends on the Q of the two filters and the

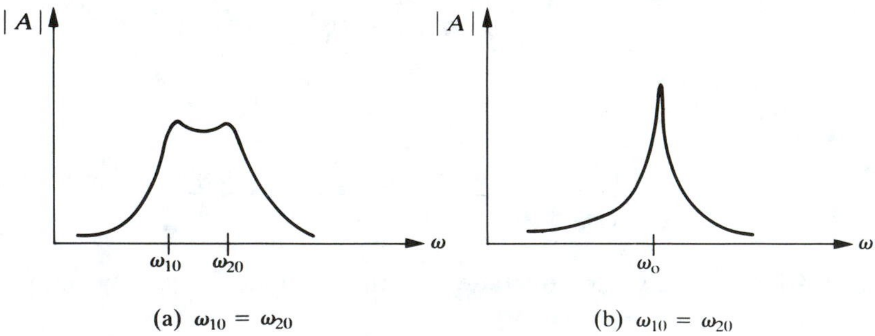

FIGURE 11.25 Output curves for two cascaded bandpass filters.

difference between their resonant frequencies. If the bandpass filters have exactly the same resonant frequency, the output will be sharply peaked at the resonant frequency as shown in Fig. 11.25(b).

In summary, then, the state variable filter is a "three-in-one" filter with high-pass, low-pass, and bandpass outputs. A Q of up to 100 can be obtained. The resonant frequency and the Q can be varied independently. As with any integrator, the bias currents for the integrator op amps should be low, which means using an FET input or a *superbeta* op amp usually. Because three op amps are required, they should all be on the same chip—that is, a *quad* op amp should be used.

State variable filters are available on a chip and are often called *universal active filters*. The chip contains the three op amps for the summer and the two integrators. Some chips contain a fourth, uncommitted, op amp that can be used for additional gain. Several resistors must be added to the chip to make a complete active filter. In the EPSCO Model 941 filter shown in Fig. 11.26, R_7 and R_8 determine the frequency and R_9 determines the Q.

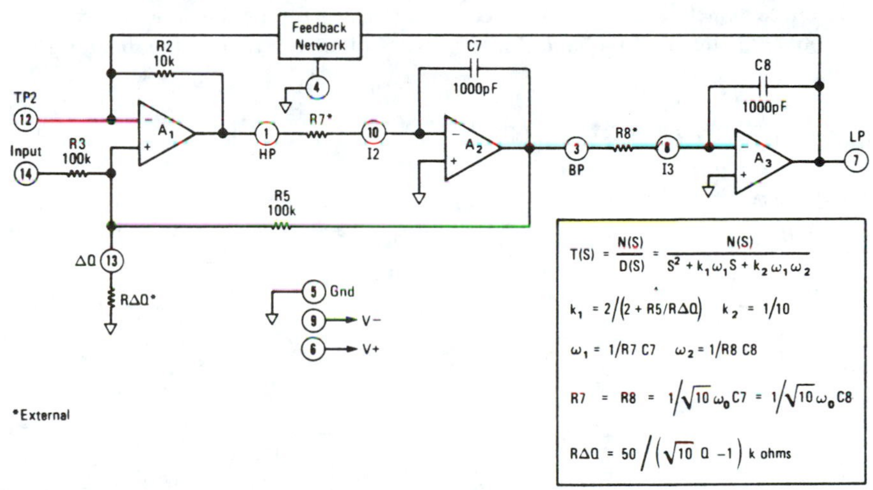

FIGURE 11.26 Commercial universal active filter. (Courtesy EPSCO, Inc., Westwood, Mass.)

The f_0Q product is a maximum of 2×10^6 for the 941. Thus if $f_0 = 20$ kHz, the maximum Q is 100. At this writing the price per chip is approximately $100.

The following comments apply to any active filters:

1. If the bandwidth is 80% or less than the resonant frequency, it is best to use a bandpass filter—for example, if $f_0 = 2.00$ kHz and the desired passband is 800 Hz. However, if the desired passband is greater than this, then it is best to

use separate low-pass and high-pass filters to achieve the desired wide passband. For example, if the desired passband is from 300 to 3000 Hz, we would use a high-pass filter with a break frequency of 300 Hz and a low-pass filter with a break frequency of 3000 Hz.

2. To achieve the desired Q, we must match the components as carefully as possible; we can't just grab any two 5% 1.0-kΩ resistors, for example. Precision metal-film resistors are the best. Sometimes it is necessary to measure several resistors and to select those that are closest to the desired resistance values.
3. The capacitors should have low temperature coefficients and be well matched like the resistors. This means generally NPO ceramic, mica, or tantalum capacitors—never use ordinary ceramic, mylar, or polypropylene capacitors.
4. For filters that use three op amps, the op amps should all be on the same chip to ensure they will be at the same temperature; that is, quad op amps should be used.
5. The op amp should have sufficient bandwidth and slew rate. One rule-of-thumb is that the gain bandwidth product should be at least 5 or 10 times greater than the center frequency of the filter times the Q. For example, with a bandpass filter with $f_0 = 2$ kHz and a Q of 20, the gain bandwidth product of the op amp should be at least 400 kHz.
6. If an extremely steep falloff is required (e.g., greater than fourth order), it is probably better to use a *digital* filter; digital filters are treated in Chapter 16.

11.7 THE BANDSTOP FILTER

If a filter passes all frequencies except those in a narrow region, it is called a *bandstop* filter. Its gain-versus-frequency curve is shown in Fig. 11.27(a).

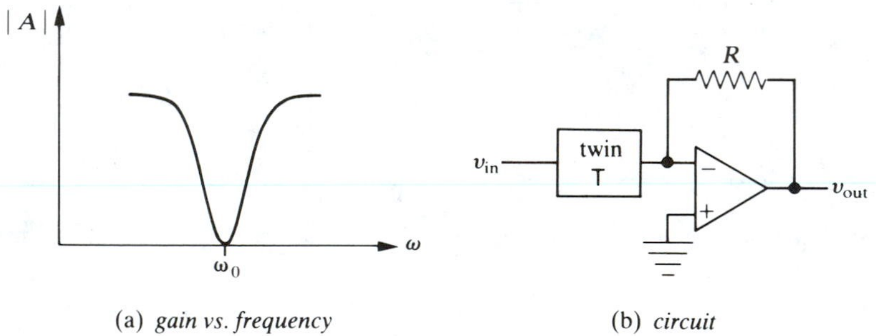

(a) *gain vs. frequency* (b) *circuit*

FIGURE 11.27 Bandstop filter.

Perhaps the simplest bandstop filter contains a twin T filter in place of the input resistance in the inverting op amp circuit, as shown in Fig. 11.27(b). At the resonant frequency of the twin T, its impedance is maximum (infinite for a perfectly matched T) and thus the gain is a minimum. Any circuit whose impedance is a maximum at a certain frequency could be used—for example, a parallel *LRC* circuit with an active inductor. The references at the end of the book should be consulted for further details.

11.8 REGULATED POWER SUPPLIES

A dc power supply operating from an ac line is, in a sense, a "super" low-pass filter. An ideal dc power supply attenuates all frequencies except dc ($\omega = 0$) and will deliver a constant output voltage over a wide range of currents, due to a varying load and in spite of temperature or line-voltage fluctuations [see Fig. 11.28(a)]. A battery with zero internal resistance would be such an ideal supply except that battery voltage tends to decrease with decreasing temperature. If the battery internal resistance r is greater than zero, the output or terminal voltage will decrease as the load current I_L drawn from the supply increases, as shown in Fig. 11.28(b). Thus, an ideal power supply would have a zero output impedance, which leads us to the thought of using a dc coupled emitter follower circuit with the output taken off the emitter. The basic circuit is shown in Fig. 11.28(c).

Notice that a constant voltage, the *reference voltage*, must be supplied to the base to produce an output that is approximately constant: $V_{out} = V_{ref} - V_{BE}$. The output voltage can therefore be no more stable than the reference voltage. The output can, however, supply much more current than the reference supply, which need only supply enough current to drive the base of the transistor. Recall that the unbypassed emitter resistance R_E provides the negative feedback. If the emitter current increases because the unregulated voltage on the collector increases, then the emitter becomes more positive, which decreases the base emitter forward bias and thereby decreases the emitter current. The simple emitter follower of Fig. 11.28(c) can be improved greatly by amplifying the feedback signal from the output to the base. An op amp with open gain A_o can be added to the feedback to improve the constancy of the output, as shown in Fig. 11.28(d). If A_o is large enough, the dc output voltage is $V_{out} \cong [(R_1 + R_2)/R_1]V_{ref}$. The reference voltage is usually supplied from a very stable low-current zener diode. Let the output voltage of the error amplifier be V_B, since it is the base voltage of the power pass transistor. We have $V_B = V_{BE} + V_{out}$, $V_B = A_o(V_{ref} - v_1)$, and $v_1 = [R_1/(R_1 + R_2)]V_{out}$. Eliminating v_1 yields

$$V_{out} = \frac{A_o V_{ref} - V_{BE}}{1 + A_o \dfrac{R_1}{R_1 + R_2}}$$

$V_{BE} \cong 0.6\text{ V}$, and usually $V_{ref} > 1\text{ V}$ and $A_o \geq 1000$, so $A_o V_{ref} \gg V_{BE}$.

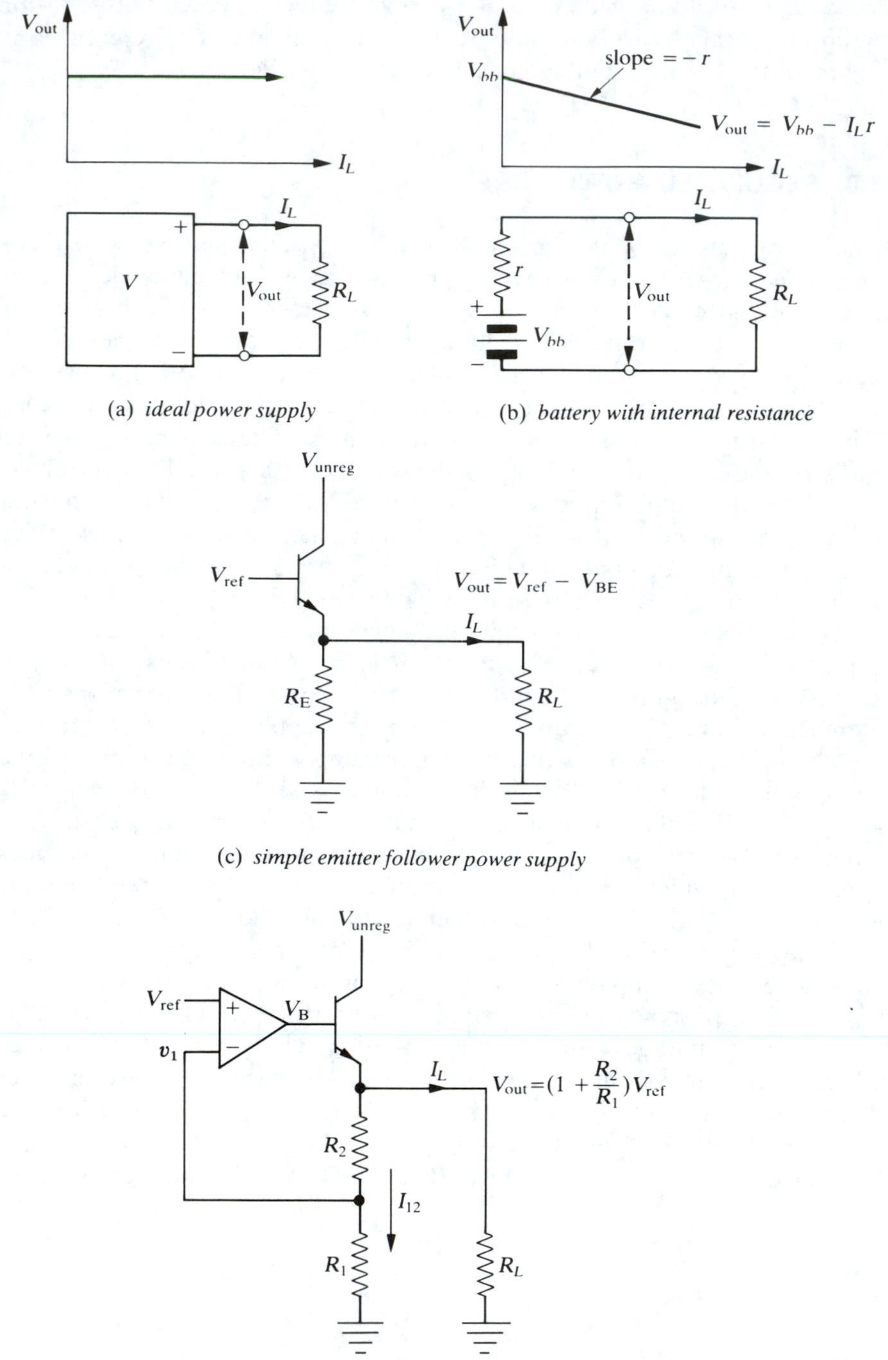

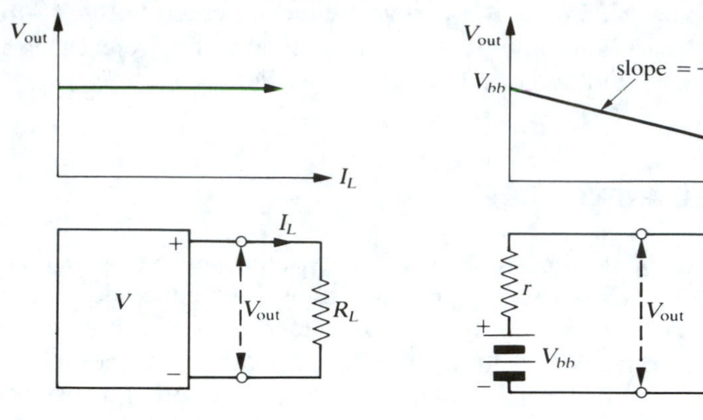

(a) *ideal power supply*

(b) *battery with internal resistance*

(c) *simple emitter follower power supply*

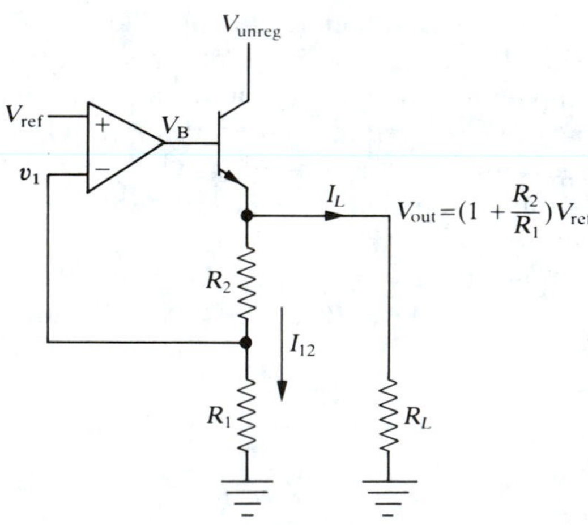

(d) *emitter follower with op amp*

FIGURE 11.28 Power supply.

Thus

$$V_{out} \cong \frac{A_o V_{ref}}{1 + A_o \dfrac{R_1}{R_1 + R_2}} \cong \frac{R_1 + R_2}{R_1} V_{ref} \qquad \textbf{(11.52)}$$

provided $A_o R_1/(R_1 + R_2) \gg 1$, which is always the case. This is a special case of the negative-feedback amplifier gain expression $A_f = A_o/(1 + A_o B) \cong 1/B$. Here $B = R_1/(R_1 + R_2)$ is the fraction of the output voltage fed back to the input. Notice that the output voltage is independent of the unregulated voltage and the error amplifier gain. The stability of the output voltage depends only on the stability of the reference V_{ref} and of the resistors R_1 and R_2.

The expression for the regulated output voltage can also be obtained from the OAVR in just one step. The voltage at the noninverting input must equal the voltage at the inverting input:

$$V_{ref} = \frac{R_1}{R_1 + R_2} V_{out}$$

Thus the output voltage is

$$V_{out} = \frac{R_1 + R_2}{R_1} V_{ref} \qquad \textbf{(11.53)}$$

Notice that the output transistor must carry a current of $I_L + I_{12}$. Also, the output voltage of the op amp can never be more than approximately $V_{cc} - 2$ V for the usual bipolar op amp, and therefore the maximum output voltage of the regulator will be $V_{cc} - 2\text{ V} - V_{BE} \cong V_{cc} - 2.6$ V.

The output voltage can never be more stable than the reference voltage at the noninverting input to the op amp. Thus, considerable attention is paid to creating a stable reference voltage. The reference voltage supplies almost zero current because it feeds the input to the op amp. A zener diode is usually used to supply the reference voltage. Zener diodes are available in voltages from several volts to several hundred volts and have positive temperature coefficients of approximately 0.1%/°C for voltages above 6 V and negative temperature coefficients for voltages below 6 V. Their dynamic resistance is usually only several ohms, which means their voltage changes by only 1 μV for a 1-μA change in current. Usually 6-V zener diodes have the smallest temperature coefficient and dynamic resistance and are thus used for the reference voltage.

A combination of a positive-temperature-coefficient zener diode and a negative-temperature-coefficient base-emitter junction transistor is available commercially as a reference diode or as part of a regulator chip, as shown in Fig. 11.29(a). The overall temperature coefficient of the regulated

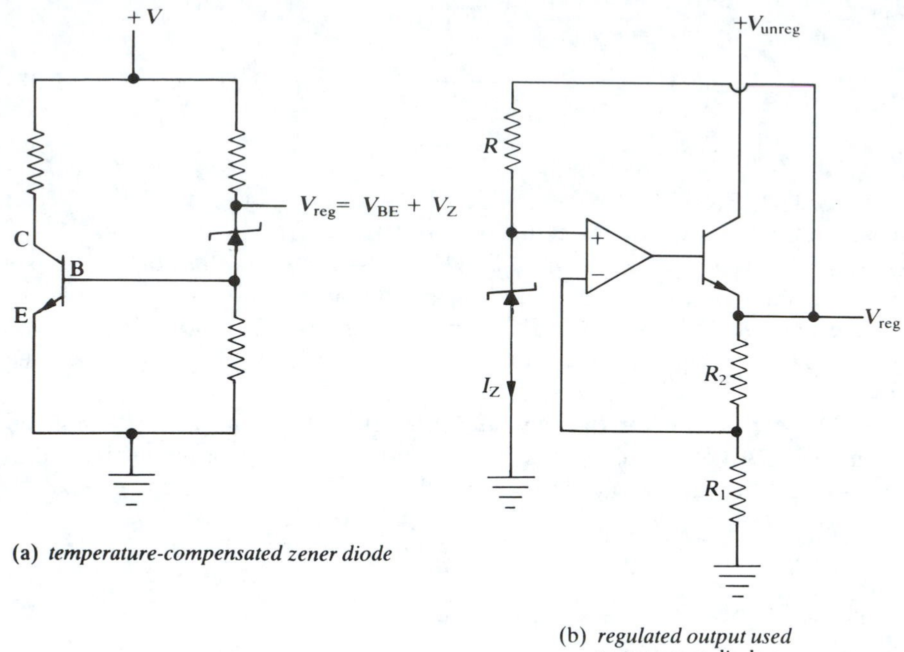

FIGURE 11.29 Reference diode with minimum temperature coefficient.

voltage can be made very small with such a reference diode, about 0.001%/°C. The circuit operates as follows: As the temperature increases, V_{BE} decreases (at approximately 2.5 mV/°C) because of the negative temperature coefficient of the forward biased base-emitter junction of the transistor, but the voltage across the zener diode increases because of its positive temperature coefficient. The regulated output $V_{reg} = V_Z + V_{BE}$ thus has a temperature coefficient equal to the difference in the temperature coefficients of the transistor base-emitter junction and the zener diode.

The zener diode is often run off the regulated output voltage of the supply rather than off the unregulated input voltage, as shown in Fig. 11.29(b). The resistance R is chosen so that the zener diode draws the optimum current for its minimum dynamic resistance (typically 10–20 mA, depending on the particular zener diode used). For example, if $V_{reg} = 10$ V, $V_Z = 6.2$ V, $I_Z = 10$ mA, $R = (10\text{ V} - 6.2\text{ V})/10\text{ mA} = 380\ \Omega$.

Fortunately, many different low-priced regulator chips are available, which include the temperature-compensated zener diode reference, the op amp, the output transistor, and built-in short-circuit protection so that the output terminals can be shorted together without damaging the chip. The simplest regulator chips have only three terminals and require only an unregulated supply voltage and two filter capacitors, as shown in Fig. 11.30.

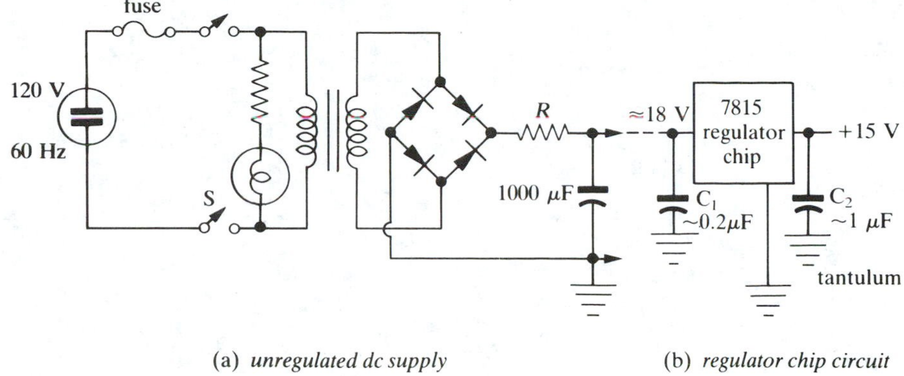

FIGURE 11.30 Three-terminal voltage regulator voltage supply.

The input unregulated voltage comes from a transformer–rectifier circuit as described in Chapter 4; the input unregulated voltage should be at least 2 or 3 V more than the regulated output voltage but not more than 35 V for the popular 7800 series. If the unregulated supply has a large filter capacitor (about 1000 μF), then C_1 is typically 0.2 μF ceramic or larger and C_2 is 1 μF tantalum. The output regulated voltage is specified by the chip (e.g., 15 V for a 7815, 5 V for a 7805, etc.). Its output voltage will vary no more than about 5 mV against line voltage changes, and its output impedance is about 0.02 Ω, which means its output voltage will change by only 2 mV for a 100-mA change in output current. The 7815 will supply output currents up to 1 A with proper heat sinking but up to only 0.15 A (150 mA) with no external heat sink.

The 7900 series regulator chips supply a negative voltage; for example, the 7915 supplies −15 V.

11.8.1 Heat Sinks

All circuits drawing power must be capable of dissipating heat to their environment to avoid overheating and permanently destroying the semiconductor junctions. If the temperature of the semiconductor chip is constant, the input dc electrical power to the chip must equal the output electrical power plus the heat power output to the environment. A *heat sink* is the name given to the device to efficiently transfer the heat generated within the chip to the environment, usually the surrounding air. An ordinary transistor or IC with no heat sink can usually dissipate approximately 10–100 mW of power to the surrounding air without overheating. A heat sink is basically something thermally connected to the IC case or transistor case; it has a large surface-area-to-volume ratio to enhance the transfer of heat to the surrounding air. As the word "sink" implies, the heat is not consumed but merely transferred to the air: the larger the surface area of

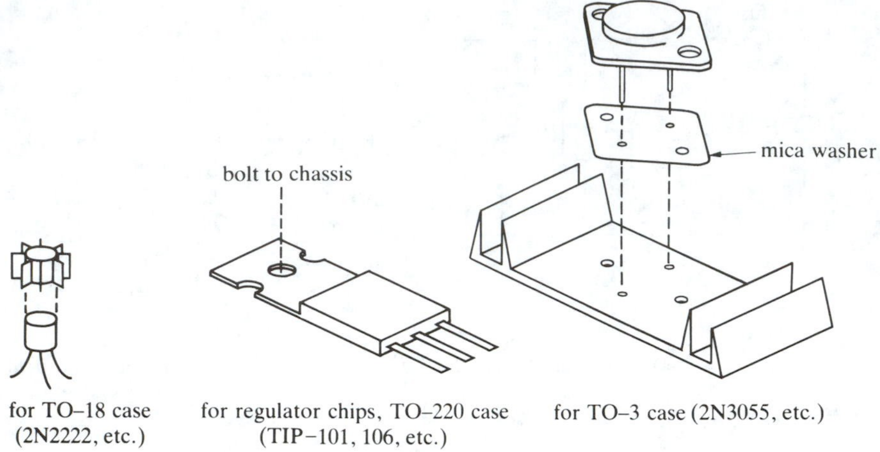

FIGURE 11.31 Heat sinks.

the heat sink, the faster the heat flow to the air. Thus, heats sinks usually have some kind of fins; several common types are shown in Fig. 11.31.

The important point is to have a good conductive heat path from the transistor or chip case right through to the air. Good mechanical contact between the device case and the heat sink is essential, and a thin layer of *heat-sink compound* is usually smeared on the device case before the heat sink is attached. This compound is unusual in that it has a high thermal conductivity but a low electrical conductivity. It looks somewhat like white toothpaste and, incidentally, is very difficult to remove from clothing. Often a thin insulating washer (usually mica) is placed between the device case and the heat sink. The heat sink will be most effective if air can move freely past the fins; the best configuration is when the fins are vertical. Forced airflow (from fans) may be necessary. The heat-sink fins are often black to increase their ability (their *emissivity*) to radiate infrared radiation. The metal chassis can sometimes be used as a heat sink.

In certain critical cases a small diamond-chip heat sink (approximately 1 mm^3) can be bonded to the heat-producing chip. This is very effective in carrying away heat because the thermal conductivity of diamond is five times that of copper.

11.8.2 Short-Circuit Protection

The simple regulator circuit of Fig. 11.29 with the op amp and the single output transistor will self-destruct if the output is shorted to ground. The probability of this happening at least once is almost 100% with inexperienced persons and breadboarded circuits—even with professors! If the output is shorted, in Fig. 11.29 the output voltage drops and the voltage at the inverting input of the op amp also drops. The op amp output then goes strongly positive in an attempt to turn on the transistor and thereby raise the output voltage so that

$$\frac{R_1}{R_1 + R_2} V_{out} = V_{ref}$$

This increases the output current almost without limit, which burns out the transistor because the power dissipated in the transistor equals $V_{unreg} \times I_{out}$ with the output shorted to ground.

The internal short-circuit protection built into commercial IC regulator chips involves negative dc feedback that turns off the pass transistor if the output current exceeds a certain value. A simple example is shown in Fig. 11.32. Two silicon diodes are connected in series between the base of the

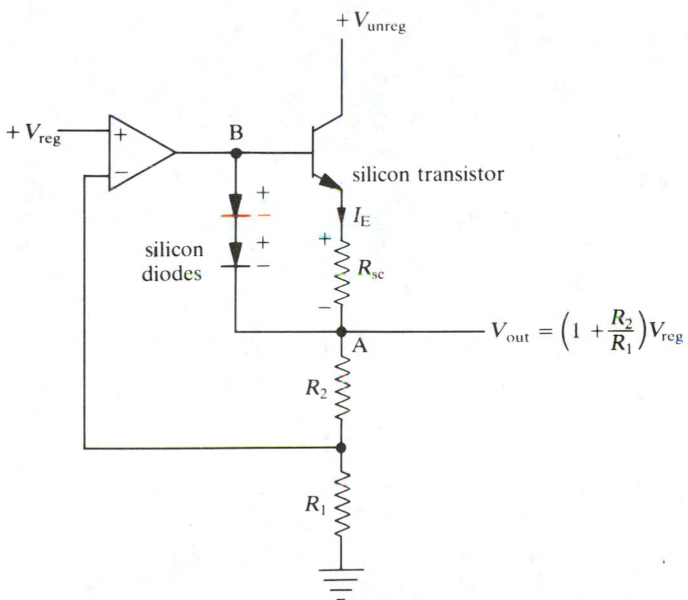

FIGURE 11.32 Simple short-circuit protection circuit.

pass transistor and one end of a resistance R_{sc} in series with the emitter. If $R_{sc} = 0.06\ \Omega$ and I_E exceeds 10 A, then the $I_E R_{sc}$ drop across R_{sc} is $(10\ \text{A})(0.06\ \Omega) = 0.6\ \text{V}$ and the two diodes will be turned on. Any additional current from the op amp output flows through the two diodes instead of into the base of the pass transistor; thus the transistor current is not turned on more by the increased op amp output current. This can be seen because

$$V_B - V_A = V_{BE} + I_E R_{sc}$$

The *maximum* possible value for $V_B - V_A = 2V_D = 2(0.6\text{ V}) = 1.2\text{ V}$. Thus

$$1.2\text{ V} \geq V_{BE} + I_E R_{sc} = 0.6\text{ V} + I_E R_{sc}$$

Hence,
$$0.6\text{ V} \geqq I_E R_{sc} \quad \text{or} \quad I_{E\,max} = \frac{0.6\text{ V}}{R_{sc}}$$

For $R_{sc} = 0.6\ \Omega$ the transistor will be limited to $I_{E\,max} = 1$ A.

11.8.3 Power-Boosting Regulators

More than 1 A of regulated output current can be obtained from a 7800 chip that can supply only 1 A by itself with a heat sink. The circuit of Fig. 11.33 does the trick. The circuit operation is as follows: For small output

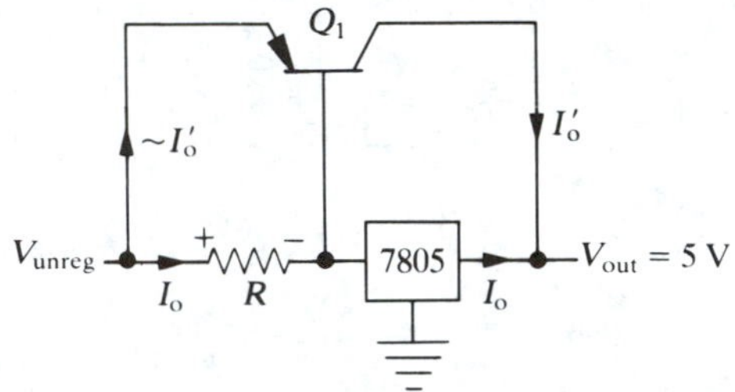

(a) *simple circuit;* $I_{out} = I_o + I_o'$

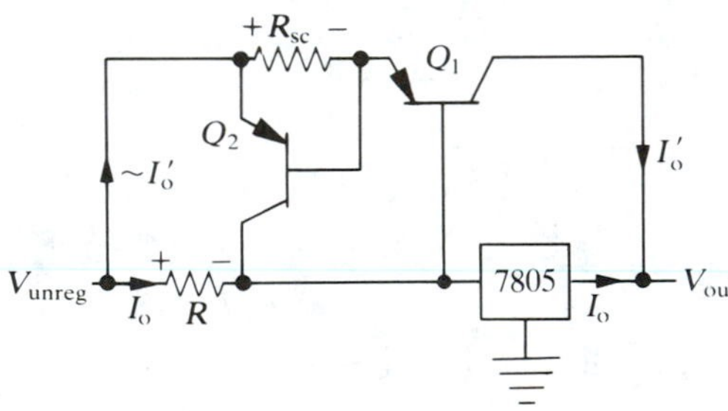

(b) *with short-circuit protection*

FIGURE 11.33 Power-boosting circuit for regulator chip.

currents, I_oR is less than the turn-on voltage of Q_1; Q_1 is off and $I_o' = 0$; that is, the chip does the regulating. When I_oR reaches the turn-on voltage of Q_1 (about 0.6 V), Q_1 will begin to conduct and shunt current around the regulator chip. For example, if we wish to limit the current I_o supplied by the 7805 chip to 0.2 A, then we choose $R = 3\ \Omega$. Then for currents above 0.2 A, Q_1 will be turned on because $V_{BE} = I_o R = (0.2\ A)(3\ \Omega) = 0.6\ V$, and all current above 0.2 A will pass through Q_1 not through the regulator chip. The regulator chip will still regulate the output voltage. If the output voltage increases for any reason (unregulated voltage input increase due to line voltage fluctuation, etc.), the regulator chip will decrease the output current I_o, which will decrease the current through R, thereby turning Q_1 off slightly and decreasing I'_o, which decreases V_{out}. Q_1 must be a power transistor capable of handling the output current I'_o. Because we need at least 2 V across the regulator chip, the unregulated voltage must be approximately 0.6 V (the maximum value of $I_o R$) higher than $V_{out} + 2\ V$: for example, if $V_{out} = 5\ V$, $V_{unreg} \geqq 7.6\ V$.

However, this circuit is not short-circuit protected. If the output is shorted to ground, V_{out} decreases and the chip increases I_{out} in an attempt to keep the output voltage constant, which will increase the current through R, thereby turning on Q_1 more, thus increasing I'_{out} and the output current. Thus, more output current is supplied for the output short, and we have positive dc feedback that ends when Q_1 (or the chip) is destroyed by heat. The addition of another transistor to limit the base emitter voltage of the pass transistor Q_1 provides short-circuit protection, as shown in Fig. 11.33(b). In normal operation $I_o R$ is less than 0.6 V, so Q_1 is off because $I_o R = I'_o R_{sc} + V_{BE1}$. When $I_o R$ exceeds 0.6 V, Q_1 turns on and $I'_o > 0$. Normally, $I'_o R_{sc}$ is less than 0.6 V, so Q_2 is off. The current through Q_1 is limited when $I'_o R_{sc}$ exceeds 0.6 V, for then Q_2 is turned on, which limits the current through Q_1. Q_2 must be able to carry the maximum current of the regulator chip. The unregulated voltage must be approximately 1.2 V higher than $V_{out} + 2\ V$.

A final note on the power transistors: For a negative supply Q_1 must be npn; for a positive supply, it must be pnp. Good choices are 2N3055 (npn) and 2N3740 (pnp).

11.8.4 Foldback Current Limiting

There is one serious problem with the short-circuit protection of the preceding section: When the output is shorted, the power dissipation in the chip and the pass transistor is *maximum*, because the *entire* unregulated voltage exists across the chip and pass transistor, and the current through is at the maximum limit set by the short-circuit protection circuit. If the output short persists for a long time, the temperature rise in the power supply may be excessive. In the simplified circuit of Fig. 11.34(a), the pass transistor's emitter current $I_E = I_{out} + I_{12} \cong I_{out}$. When the output is shorted to ground, $V_{CE} = V_{unreg}$. If the current is limited to 2 A by some circuit and $V_{unreg} = 20\ V$, then the power in Q_1 will be $P_1 = I_E V_{CE} = (2\ A)(20\ V) = 40\ W$. This is much larger than the normal power. For example, if $I_{out} = 1\ A$ under normal conditions with $V_{out} = 15\ V$, then $P_1 = I_E V_{CE} = (1\ A)(5\ V) = 5\ W$. The point is, we would like a short-circuit

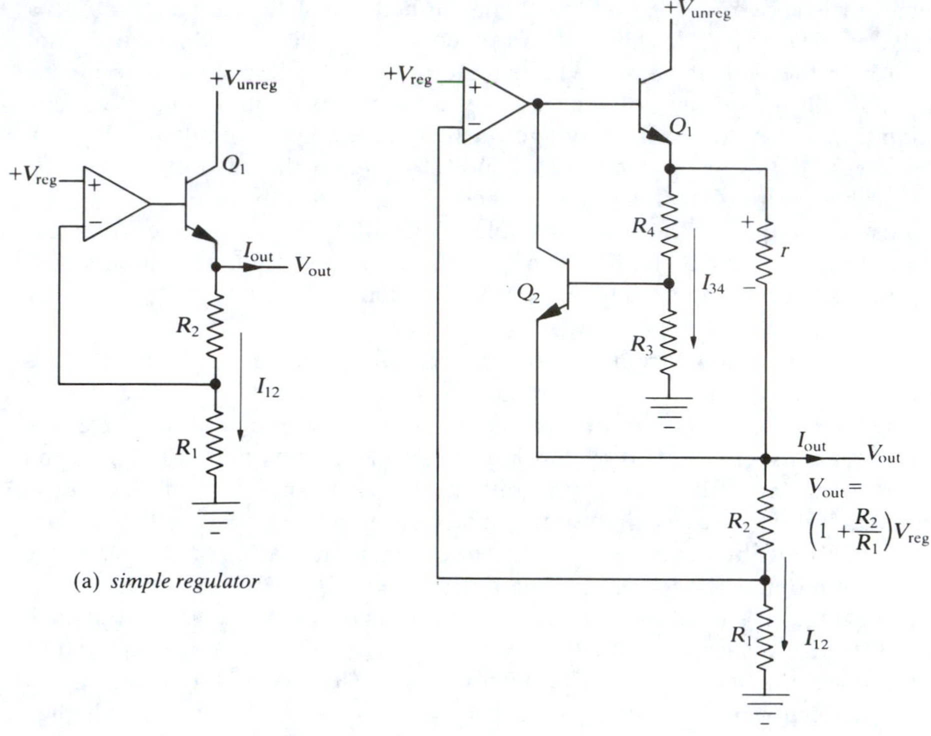

FIGURE 11.34 Foldback short-circuit protection.

protection circuit that will protect the circuit by limiting the power dissipation in the pass transistor Q_1. In other words, we need the current limited to a smaller value only when the output is shorted. *Foldback* current limiting is designed to do just that.

The circuit in Fig. 11.34(b) shows a simplified form of a foldback protection circuit. The circuit works by turning on Q_2 when the output is shorted, and this turns Q_1 off by "robbing" it of its base current. Let us calculate the base emitter voltage for Q_2, to see when it turns on.

$$V_{BE2} = V_{B2} - V_{E2}$$

But

$$V_{E2} = V_{out} \quad \text{and} \quad V_{B2} = \frac{R_3}{R_3 + R_4} V_{E1} = \frac{R_3}{R_3 + R_4}[V_{out} + (I_{out} + I_{12})r]$$

so

$$V_{BE2} = \frac{R_3}{R_3 + R_4}[V_{out} + (I_{out} + I_{12})r] - V_{out}$$

Neglecting I_{12} compared to I_{out}, we have

$$V_{BE2} \cong \frac{I_{out} r R_3 - V_{out} R_4}{R_3 + R_4}$$

If I_{out} increases to the point where V_{BE2} equals the turn-on voltage for Q_2, then the output current will be limited. But the expression for V_{BE2} also contains a (negative) term dependent on the output *voltage* V_{out}. Thus, if V_{out} goes to zero (from a short to ground), V_{BE2} will *increase*, thereby turning Q_2 on and Q_1 off.

When $V_{BE2} = 0.6$ V, Q_2 will be turned on and Q_1 will be turned off, and the output current will be limited. This will occur when

$$V_{BE2} = 0.6\text{ V} = \frac{I_{out} r R_3 - V_{out} R_4}{R_3 + R_4}$$

or when the output current is

$$I_{out} = \frac{(0.6\text{ V})(R_3 + R_4) + V_{out} R_4}{r R_3}$$

Taking $V_{out} = 15$ V, $r = 1\ \Omega$, $R_3 = 10\text{ k}\Omega$, $R_4 = 1\text{ k}\Omega$, the output current will be limited to

$$I_{out} = \frac{(0.6\text{ V})(11\text{ k}\Omega) + (15\text{ V})(1\text{ k}\Omega)}{(1\ \Omega)(10\text{ k}\Omega)} = 2.16\text{ A}$$

But when the input is shorted, $V_{out} = 0$ and Q_2 will turn on when the output current is

$$I_{out} = \frac{(0.6\text{ V})(11\text{ k}\Omega)}{(1\ \Omega)(10\text{ k}\Omega)} = 0.66\text{ A}$$

Thus, under normal operation the power supply will supply up to $I_{out} =$ 2.16 A before the current-limiting action of Q_2 occurs. But if the output is shorted, the output current will be limited to $I_{out} = 0.66$ A, thus producing far less power dissipated in the pass transistor. Qualitatively, this occurs because with the output shorted to ground, the emitter of Q_2 is also grounded (instead of being at the positive voltage V_{out}), and thus Q_2 is easier to turn on with its emitter less positive.

11.8.5 Adjustable Voltage Regulators

Adjustable output voltages can be obtained from three-terminal or four-terminal regulator chips. In addition to the unregulated input, ground, and

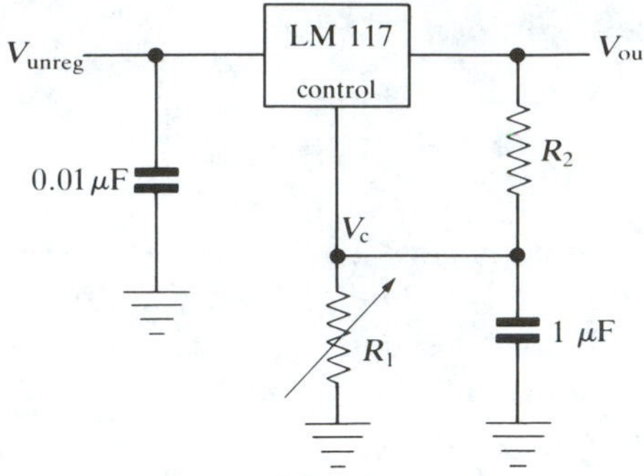

FIGURE 11.35 Adjustable voltage regulator.

regulated output terminals, there is a fourth "control" terminal, as shown in Fig. 11.35. The numerical dc voltage V_c at the control terminal is specified for the particular chip, and the regulated output voltage is

$$V_{reg} = \left(1 + \frac{R_2}{R_1}\right) V_c$$

Thus, the regulated output voltage can be controlled by adjusting the R_2/R_1 ratio. For example, if we desire $V_{reg} = 13$ V, and if $V_c = 5$ V for our particular regulator chip, then

$$1 + \frac{R_2}{R_1} = \frac{V_{reg}}{V_c} = \frac{13\text{ V}}{5\text{ V}} = 2.6 \quad \text{and} \quad \frac{R_2}{R_1} = 1.6$$

The regulated output voltage can be adjusted by using a potentiometer for R_1 and R_2. Finally, as with all regulators, the input unregulated voltage must be at least 2 or 3 V larger than the regulated output voltage.

Table 11.3 lists four common fixed-voltage three-terminal regulator chips for output voltages of ± 5 V and ± 15 V. All of these regular chips contain internal thermal overload protection and internal short-circuit protection. Solid tantalum capacitors are best to ensure adequate bypassing at high frequencies, although ceramic capacitors can be used in many applications.

11.8.6 Current Regulators

If we have a constant voltage, we can make a constant current supply by using the concept of a voltage-to-current converter (Section 10.9). The general circuit is shown in Fig. 11.36. Using the OAVR, we see that

$$V_{ref} = \frac{R_3}{R_3 + R_L}\, v_O = R_3 I_L$$

TABLE 11.3. Common Regulator Chips

Type	*No. of Terminals*	*Output (V)*	*Input (V)*	*Output Current (A) (w/heat sink)*	*Load Regulation (mV)*	*Line Regulation (mV)*	*Ripple Rejection (dB)*
7805C	3	+5	7–35	1.0	10	50	80
7905	3	−5	7–35	1.5	5–15	2–8	66
7815C	3	+15	17–40	1.0	10	50	80
7915C	3	−15	17–40	1.5	5–15	3–5	70

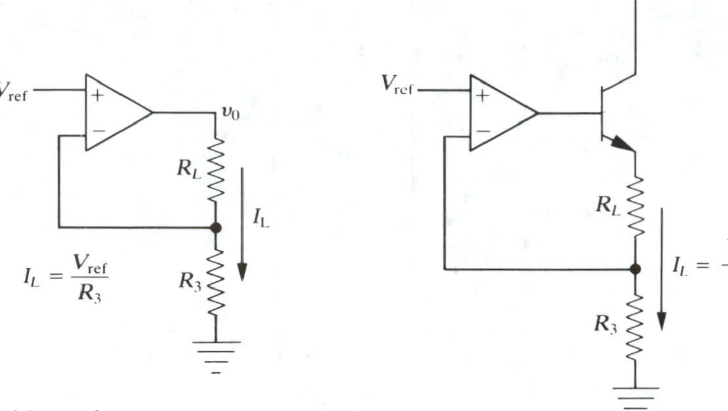

(a) *simple op amp circuit*

(b) *with current booster*

FIGURE 11.36 Constant current source.

Thus
$$I_L = \frac{V_{ref}}{R_3}$$

which means the load current I_L is constant so long as the reference voltage and R_3 are constant, even if the load resistance R_L varies. One disadvantage of this circuit is that both ends of the load are above ground. Notice that the maximum load current is limited to the maximum output current of the op amp. The load current can be boosted by adding a current booster transistor as shown in Fig. 11.36(b).

With some three-terminal regulator chips (the 317 family), we can easily make a constant current source with one end of the load grounded, as shown in Fig. 11.37. The 117 chip maintains a constant 1.25-V difference between the output terminal and the control terminal, which draws less than 100 μA. Thus $1.25\text{ V} = (I_L + I_C)R$. If $I_C \ll I_L$—that is, if $I_L \gg 100\ \mu$A, then

$$I_L = \frac{1.25\text{ V}}{R}$$

Thus, the load current is constant even if R_L varies. If the load current is not large compared to the 100-μA control current, then a voltage follower circuit can be used, as shown in Fig. 11.37(b).

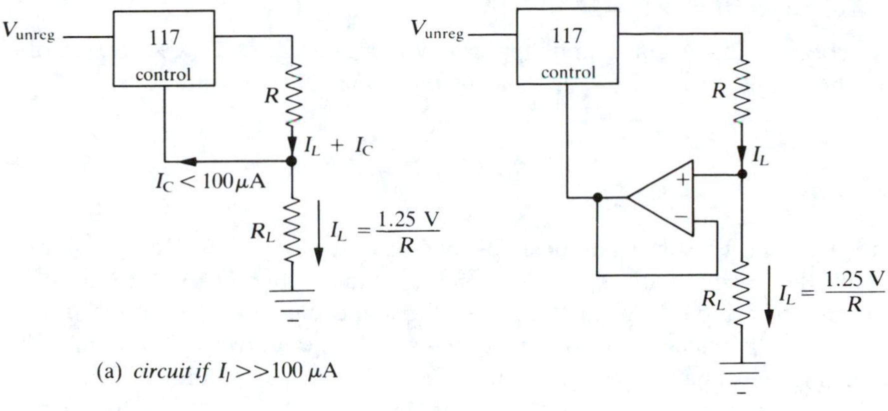

FIGURE 11.37 Constant current source with regulator chip.

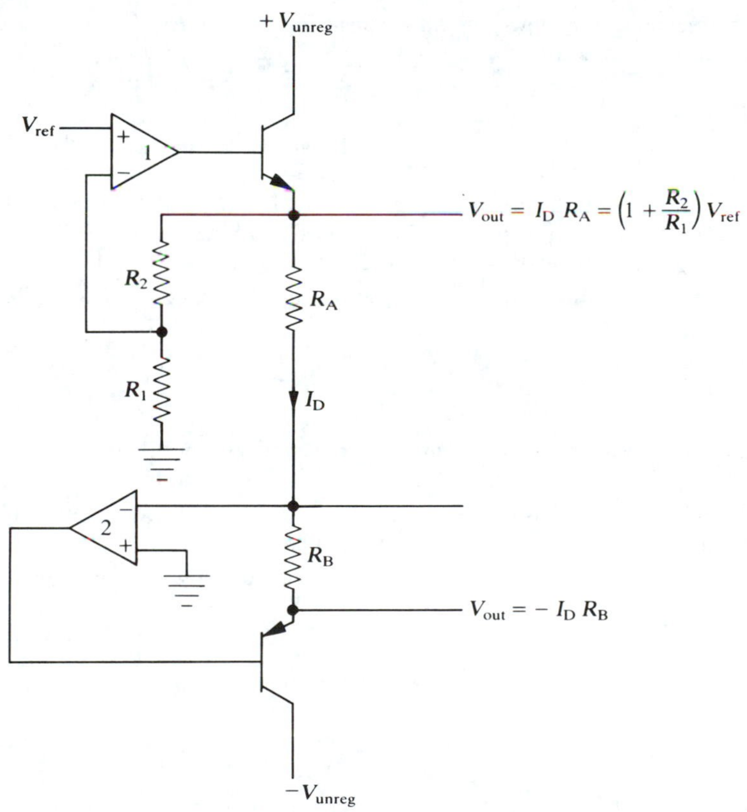

FIGURE 11.38 Split (equal and opposite) voltage supply.

11.8.7 The Split Voltage Supply

To obtain exactly equal and opposite voltages (e.g., ±15 V), we use the circuit of Fig. 11.38. Op amp 1 keeps the upper output voltage at

$$V_{\text{out}} = \left(1 + \frac{R_2}{R_1}\right) V_{\text{ref}}$$

and V_{out} can be set at any desired voltage by choice of R_2/R_1. Op amp 2 keeps the junction of R_A and R_B at ground by the OAVR because its noninverting input is grounded. And the same current must flow through R_A and R_B because of the OACR—negligible current flows into the inverting terminal of op amp 2. Thus, $V_{\text{out}} = I_D R_A$, and the negative output voltage must be $-I_D R_B$. Thus, if we choose $R_A = R_B$, we have equal and opposite output voltages.

PROBLEMS

1. For a second order low-pass filter consisting of two RC low-pass filters connected by a buffer amplifier, $A = 1/(1 - \omega'^2 + 2\omega' j)$. Show that

$$|A| = \frac{1}{1 + \omega'^2}$$

2. For a second-order low-pass filter consisting of R, L, and C in series, show that at resonance $A = -jQ_0 = -j/d_0$. What does $-j$ represent physically in terms of phase? Sketch $|A|$ versus ω' for $d_0 = 2, \sqrt{2}$, and 1.
3. Derive (11.17), the solution of the damped mechanical harmonic oscillator

$$\frac{d^2x}{dt^2} + \frac{\gamma}{m}\frac{dx}{dt} + \frac{K}{m}x = 0$$

4. Show that the "critically damped" mechanical harmonic oscillator $\omega_1 = \sqrt{\omega_0^2 - \gamma'^2/4} = 0$ corresponds to the electrical case $d_0 = 1/Q_0 = 2$.
5. Design a simple Sallen Key second-order, critically damped, low-pass filter with a break frequency $\omega_0/2\pi = 2$ kHz. Sketch $|A|$ versus f.
6. Show that for the simple Sallen key second-order low-pass filter, if $C_1 = C_2$, then

$$d_0 = \frac{1 + \dfrac{R_2}{R_1}}{\sqrt{\dfrac{R_2}{R_1}}}$$

Also show that the minimum value of d_0 for this case is 2 when $R_1 = R_2$. Is it then possible (with $C_1 = C_2$, $R_1 = R_2$) to obtain an underdamped or critically damped filter?

7. Design a simple Sallen Key second-order, critically damped, high-pass filter with a break frequency of 2 kHz.
8. Design a variable gain, Sallen Key, second-order, underdamped ($d_0 = 1.00$), low-pass filter with a break frequency of 500 Hz.
9. Explain why a negative damping factor is disastrous in a filter.
10. Design a fourth-order low-pass Butterworth filter with a break frequency at 4 kHz.
11. Prove that the gain for the *LRC* bandpass filter of Fig. 11.19(b) is

$$A = \frac{j\omega' d_0}{1 - \omega'^2 + j\omega' d_0} = j\omega' d_0 A_{LP}$$

where $d_0 = 1/Q_0 = R/\omega_0 L$, $\omega' = \omega/\omega_0$, $\omega_0^2 = 1/LC$, and $r \ll R$. A_{LP} = the gain of the *RLC* low-pass filter. Show that $A = 1$ when $\omega = \omega_0$.

12. For the *LRC* bandpass filter of Problem 11, prove that $|A| = 1/\sqrt{2}$ when $\omega = \omega_0 \pm \Delta\omega$, where $\Delta\omega = \omega_0/2Q_0$; that is, $Q_0 = \omega_0/2\Delta\omega$. Assume $\Delta\omega/\omega_0 \ll 1$.
13. Show that the transfer function (11.35) for the active bandpass filter of Fig. 11.21 is identical in magnitude to the transfer function (11.33) for the passive *LRC* bandpass filter if $R_1 = R_3 = R$ and $R_2 = 2R$.
14. Design an active bandpass filter (Fig. 11.21) with a Q of 5 and a resonant frequency of 10 kHz.
15. Repeat Problem 14 for the circuit of Fig. 11.23 and a Q of 50.
16. Design a state variable filter (Fig. 11.24) with a Q of 40 and $\omega_0 = 1$ kHz. Sketch the gain versus frequency for the low-pass, the high-pass, and the bandpass outputs.
17. You desperately need a 10-V, 50-mA power supply. Design a simple power supply using a 5-V zener diode, a 741 op amp, and a 2N2222 transistor. Assume you have a +15-V unregulated supply and the power supply for the 741.
18. Design a 5-V @ 1-A regulated power supply (useful for most digital circuits using transistor-transistor logic) "from scratch." Start with the 110-V, 60-Hz line voltage, and use a 7805 3-pin regulator chip.
19. Design a circuit to supply a constant 250-mA current to a 10-Ω load using an LM117 regulator chip. Suppose the current is only 200 μA to a 10-kΩ load. What modifications are required?

CHAPTER 12

Basic Digital Concepts

12.1 INTRODUCTION

There are fundamentally two kinds of electronics—analog and digital. In *analog* electronics the signals vary continuously in amplitude. Thus, there are an infinite number of signal amplitudes that must be processed, and it is difficult to distinguish between a small change in the signal and a small change due to noise or a drift in temperature, power supply voltage, and so on. Sinusoidal waves and exponentially decaying voltages are two simple examples of analog waveforms.

Digital electronics, by contrast, deals with digital signals that have only two possible amplitude values, say, +5 V and 0 V. Technically, we should say *binary* digital electronics, because there are only *two* amplitude values, but the term "digital" usually means two possible discrete amplitude values. Digital techniques can do almost anything that analog techniques can at present, with the significant advantage of greater noise immunity. Any signal or piece of information can be converted into digital form by using a digital language, and it can be transmitted, amplified, and so forth, without any significant distortion as long as the *distinction* between the two binary states (5 V and 0 V) remains clear. A large noise pulse of 5 V would change a 0-V level to 5 V, and this would constitute distortion, but the smaller noise pulses so common to analog circuits usually will not blur the distinction between the two voltage levels.

The most commonly used digital language is binary; it has only two symbols or digits: "0" and "1." Each binary digit, whether a 0 or a 1, is termed a *bit*. A series of bits grouped together to represent a single number is termed a ***word***. For example, 10110 is a five-bit word. A four-bit word is called a ***nibble***; an eight-bit word, a ***byte***. Any number (e.g., a signal amplitude in a short time interval) can be expressed in binary form in the usual way as shown in Table 12.1. Many more digits are required to represent a number in binary form than in decimal form, but this is not a serious practical disadvantage because of the high speed of digital circuits.

The advantage of the binary digital system is that the information can be rapidly processed by inexpensive digital circuits in such a way as to minimize errors due to voltage drifts, component aging, distortion, or noise, because the circuits need only distinguish between two voltage levels.

TABLE 12.1. Counting in Decimal, Binary, Octal, and Hexadecimal

Decimal	*Binary*	*Octal*	*Hexadecimal*
0	00000	0000	00
1	00001	0001	01
2	00010	0002	02
3	00011	0003	03
4	00100	0004	04
5	00101	0005	05
6	00110	0006	06
7	00111	0007	07
8	01000	0010	08
9	01001	0011	09
10	01010	0012	0A
11	01011	0013	0B
12	01100	0014	0C
13	01101	0015	0D
14	01110	0016	0E
15	01111	0017	0F
16	10000	0020	10
17	10001	0021	11
18	10010	0022	12
19	10011	0023	13
20	10100	0024	14
21	10101	0025	15
22	10110	0026	16
23	10111	0027	17
24	11000	0030	18
25	11001	0031	19
26	11010	0032	1A
27	11011	0033	1B

These two voltage levels need not be precisely defined for modern digital circuitry; a typical allowable range for logical 0 may be from 0.1 to 0.4 V, and for logical 1 from 2.4 to 5.0 V. Thus, a change of 0.1 V or 0.2 V will not blur the distinction between a 0 and a 1 in a digital circuit, but such a change might have a drastic effect on an analog circuit.

Analog signals can be converted into digital form with an *analog-to-digital* (A/D) converter, processed digitally, and then converted back to analog form with an *digital-to-analog* (D/A) converter. The spectacular pictures from the recent space probes were converted into digital form in the satellite and transmitted in digital form to earth where they were converted to analog form. The important point is that once the pictures are in digital form it is relatively easy to transmit them over long distances in the presence of considerable electromagnetic interference (noise) without appreciable distortion.

A digital *computer* basically is a device that can perform a wide variety of mathematical operations (addition, subtraction, multiplication, comparison, etc.) on numbers in digital form at extremely high speeds. A typical modern digital computer, for example, can add two eight-bit numbers in approximately 10^{-7} s. This high speed more than compensates for the large number of bits required to represent a number in binary form. The numbers are fed in with a keyboard or from a magnetic tape, and the output usually appears on a cathode ray tube or a printer.

Modern digital hardware is truly impressive. There is available a wide variety of digital circuits that will perform mathematical operations on digital numbers and that will store digital information as well as instructions. The technology of modern solid-state digital electronics has all been developed in the last two decades, with progress continuing at a rapid pace at this writing. The devices get more and more powerful each year, and the costs continue to decrease as more and more applications result in higher production.

Modern computers, including minicomputers and large computers, all are composed almost entirely of *digital* electronics, with the exception of the keyboard input and the output printer or cathode ray tube. The *microprocessor* is essentially a very small computer on one chip with limited memory and no input–output devices. The ubiquitous portable electronic calculator also consists of a few very small digital ICs with a keyboard input, and a light emitting diode (LED) or liquid crystal output. The majority of the weight and volume is from the battery, keyboard, and package, not from the digital circuitry.

At this writing, digital computers appear to be useful in two main areas: (1) the rapid processing of large quantities of data (often called *number crunching*) (this includes doing mathematical computations such as numerical integrations, etc.); and (2) the storing of large quantities of data in such a way that *selective rapid* retrieval is possible. Social security records, tax records, and data banks in almost any area are examples.

Finally, we should mention the famous computer user's maxim: "garbage in, garbage out" (GIGO). A computer is an electronic circuit and thus will do precisely what you tell it to do in the program with the data—no more and no less. If inaccurate information is fed in, the output will be inaccurate. If incorrect instructions are fed in, the output will be incorrect. The accuracy of a computer depends on the accuracy of the data *and* instructions fed in. In working with any computer it is usually a good idea to "calibrate" it by feeding in a simple set of data and a problem to which we already know the correct answer to see if the computer indeed gives us the correct answer.

12.2 NUMBER SYSTEMS

The familiar decimal (base ten) system uses the ten digits 0–9. Any decimal number can be expressed as a sum of various powers of ten with the coefficients being the digits. For example,

$$378 = 3 \times 10^2 + 7 \times 10^1 + 8 \times 10^0 = 300 + 70 + 8$$

For the general case,

$$N_2N_1N_0 = N_2 \times 10^2 + N_1 \times 10^1 + N_0 \times 10^0$$

N_2 is the *most significant bit* (MSB) corresponding to the highest power of the base ten, and N_0 is the *least significant bit* (LSB). We usually write the MSB on the left and the LSB on the right. To specify the decimal base, we write a subscript 10 after the number (e.g., 378_{10}).

For any number system with a base B and digits N_0 (LSB), N_1, N_2, N_m (MSB), the decimal equivalent N_{10} of the number $N_mN_{m-1} \ldots N_3N_2N_1N_0$ is given by

$$N_{10} = N_m \times B^m + \cdots + N_3 \times B^3 + N_2 \times B^2 + N_1 \times B^1 + N_0 \times B^0 \tag{12.1}$$

Because digital circuits have two stable electrical states, the binary number system is appropriate for their analysis. Digital electronic systems also use the octal system (base eight), and the hexadecimal system (base 16). Counting in the four number systems is shown in Table 12.1.

Any binary number can be expressed as a sum of various powers of 2, with the coefficients being the digits. For example,

$$1011_2 = 1 \times 2^3 + 0 \times 2^2 + 1 \times 2^1 + 1 \times 2^0 = 8 + 0 + 2 + 1 = 11_{10}$$

The subscript 2 means the number is in binary. The coefficient of the highest power of 2 is the MSB, and the coefficient of the lowest power of 2 is the LSB.

Any octal number can be expressed as a sum of various powers of 8, with the coefficients being the digits. There are eight digits (0, 1, 2, 3, 4, 5, 6, 7) in the octal number system. For example.

$$253_8 = 2 \times 8^2 + 5 \times 8^1 + 3 \times 8^0 = 128 + 40 + 3 = 171_{10}$$

The subscript 8 means the number is in octal.

Any hexadecimal number can be expressed as a sum of various powers of 16, with the coefficients being the digits. There are 16 digits in the hexadecimal number system (0, 1, 2, 3, 4, 5, 6, 7, 8, 9, A, B, C, D, E, F). For example,

$$3D7_{16} = 3 \times 16^2 + 13 \times 16^1 + 7 \times 16^0 = 768 + 208 + 7 = 983_{10}$$

Notice that we have to use the decimal equivalents of the letters A, B, . . . , F for the coefficients of the powers of 16. The subscript 16 means the number is in hexadecimal or "hex" for short. Another common notation for hexadecimal numbers is to add the capital letter "H" (e.g., 3D7H = $3D7_{16}$).

The conversion of a decimal number to a binary number is a little harder. We divide by 2 repeatedly and note the remainders. The remainders are the bits of the binary number. For example, to convert 13_{10} to binary, we perform the following divisions:

$$
\begin{array}{rll}
13 \div 2 = 6 & \text{with a remainder of } 1 & \text{LSB} \\
6 \div 2 = 3 & \text{with a remainder of } 0 & \\
3 \div 2 = 1 & \text{with a remainder of } 1 & \\
1 \div 2 = 0 & \text{with a remainder of } 1 & \text{MSB}
\end{array}
$$

The *last* remainder is the MSB, and the *first* remainder is the LSB. Thus, $13_{10} = 1101_2$.

The conversion of a decimal number into an octal number involves dividing by 8 and noting the remainder. For example,

$$
\begin{array}{rll}
23_{10} \div 8 = 2 & \text{with a remainder of } 7 & \text{LSB} \\
2 \div 8 = 0 & \text{with a remainder of } 2 & \text{MSB}
\end{array}
$$

Thus, $$23_{10} = 27_8$$

The conversion of a decimal number into a hexadecimal number involves dividing by 16 and noting the remainders. For example,

$$
\begin{array}{rll}
46_{10} \div 16 = 2 & \text{with a remainder of } 14 = \text{E} & \text{LSB} \\
2 \div 16 = 0 & \text{with a remainder of } 2 & \text{MSB}
\end{array}
$$

Thus, $$46_{10} = 2\text{E}_{16}$$

Notice that the remainder is always less than the base if the division is correct; thus the remainder can be represented by one of the symbols 0, . . . , B', where B' is the "largest" symbol; $B' = 1$ in binary, 7 in octal, F in hex.

The conversion between binary and octal is relatively easy. From octal to binary we simply express each octal digit in terms of a three-bit binary number. (Three binary bits can represent each of the eight octal digits 0, . . . , 7.) For example,

$$36_8 = (011)(110) = 011110_2$$

From binary to octal we start with the LSB and group the binary bits in groups of three. Each group of three is replaced by an octal digit. For example,

$$10111101_2 = (10)(111)(101) = 275_8$$

The conversion between binary and hexadecimal is also relatively easy. From hexadecimal to binary we simply express each hex digit in terms of a four-bit binary number. (Four binary bits can represent each of the 16 hex digits 0, . . . , F.) For example,

$$2B5_{16} = (0010)(1011)(0101) = 001010110101_2$$

From binary to hexadecimal we start with the LSB and treat the binary bits in groups of four. Each four-bit group is replaced by a hex digit. For example,

$$011011110100 = (0110)(1111)(0100) = 6F4_{16}$$

Conversions between octal and hexadecimal and vice versa can probably best be done by converting to binary as an intermediate step.

12.3 NUMBER CODES

A *number code* is a relationship between the binary digits and the number represented. All number systems are codes and the decimal equivalent is given by (12.1). But there are other useful relationships or codes that relate decimal numbers and groups of bits that do not obey (12.1). These relationships are called *codes*, and we will now discuss several useful codes.

12.3.1 The 8421 BCD Code

BCD stands for *binary coded decimal*. The 8421 BCD code represents each of the ten decimal digits by the usual four-bit binary word. For example,

$$63_{10} = (0110)(0011)_{\text{BCD}}$$

From BCD to decimal we start at the LSB, treat the bits in groups of four, and convert each group of four bits into its decimal equivalent. For example,

$$01000111_{\text{BCD}} = (0100)(0111) = 47_{10}$$

Thus, we see that it is quite easy to convert from decimal to BCD and from BCD to decimal. It is much easier to convert from BCD to decimal than from straight binary to decimal, because we only have to count up to 9 in binary to do so. However, it takes more bits to represent a number in BCD than in straight binary. Notice that a BCD number is *not* the same as the number in base two; for example, $01100011_{BCD} = 63_{10}$, but $01100011_2 = 99_{10}$.

Inexpensive hardware chips are available to convert from one code to another; these chips are called *encoders*, *coders*, or *decoders*. One such example is a "BCD to seven-segment LED† decoder" chip. It takes a number in BCD code on four separate input lines and transforms it into the appropriate electrical output to light up the correct segments of a seven-segment LED display. There are other BCD codes, such as the 4221 and the 5421 BCD codes, where the numbers refer to the weighting factors of the bits. The 8421 BCD is the most common, and "BCD" usually means 8421 BCD.

12.3.2 The Excess-Three Code (XS3)

The excess-three (XS3) code is similar to the BCD code. Each four-bit group in the BCD code is increased by three (0011) to make the XS3 code. Thus, $00010010_{BCD} = (0001 + 0011)(0010 + 0011) = 01000101_{XS3}$. The XS3 code is useful in arithmetic operations on binary numbers (e.g., in addition and subtraction). Counting in XS3 code is shown in Table 12.2.

TABLE 12.2. The XS3 Code

Decimal	*Binary*	*BCD*	*XS3*	*Gray*
0	0000	0000 0000	0011 0011	0000
1	0001	0000 0001	0011 0100	0001
2	0010	0000 0010	0011 0101	0011
3	0011	0000 0011	0011 0110	0010
4	0100	0000 0100	0011 0111	0110
5	0101	0000 0101	0011 1000	0111
6	0110	0000 0110	0011 1001	0101
7	0111	0000 0111	0011 1010	0100
8	1000	0000 1000	0011 1011	1100
9	1001	0000 1001	0011 1100	1101
10	1010	0001 0000	0100 0011	1111
11	1011	0001 0001	0100 0100	1110
12	1100	0001 0010	0100 0101	1010
13	1101	0001 0011	0100 0110	1011
14	1110	0001 0100	0100 0111	1001
15	1111	0001 0101	0100 1000	1000

12.3.3 The Gray Code

The Gray code is not a BCD code, although a four-bit group is used to represent each decimal digit. Counting in the Gray code is shown in Table 12.2. Notice that *only one* bit changes in the Gray code for each increment

†LED stands for light emitting diode.

in the decimal count. The rule in counting in the Gray code is to change the least significant bit to produce the next higher number. The Gray code is useful in some high-speed A/D conversion circuits, which will be discussed later. The basic experimental fact here is that some high-speed circuits do not work well if many bits change simultaneously.

12.3.4 The ASCII Code

An *alphanumeric* code is one that represents both numbers and letters. The American Standard Code for Information Interchange (ASCII) is the single most common alphanumeric code and is used in most keyboards. In the ASCII code each number, letter, or command is represented by a seven-bit word. The ASCII code is shown in Table 12.3 (partial listing). Notice that there are no Greek letters, subscripts, or superscripts in the ASCII code.

Most keyboards generate an ASCII binary output for each character. The first bit generated is usually a start bit or a *strobe* bit and is a 1, followed by the appropriate seven-bit ASCII word from Table 12.3, followed by one or two 0 stop bits. Note that at the beginning of each ASCII character there is a 0 to 1 transition (from the stop bit to the start bit). The bits are generated at different times and come out one after the other, with each bit lasting a fixed time. Such data are called *serial* data; the bits in serial data are all on the same wire but occur at different times. Two examples are shown in Fig. 12.1.

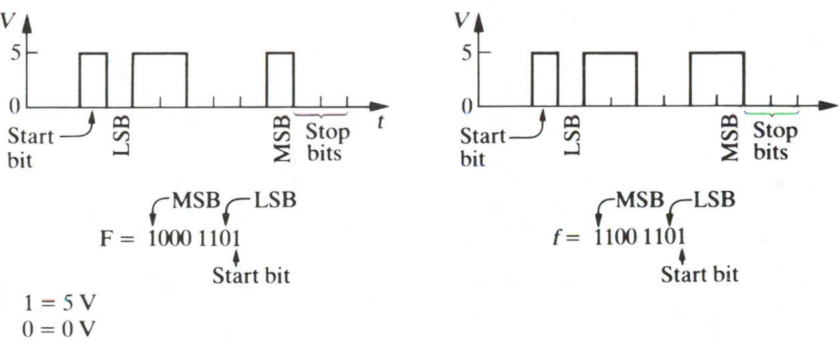

FIGURE 12.1 ASCII serial data for the letters F and f.

Another common alphanumeric code is the extended binary coded decimal interchange code (EBCDIC), an eight-bit code shown in Table 12.4. It is generally used in larger IBM computers.

12.3.5 Parity-Error Detecting Code

Any error in a digital system involves changing a 1 bit to a 0 bit, or vice versa. The parity detecting code is designed to determine when such a bit

TABLE 12.3. The ASCII Code

Keyboard Character	*ASCII*	*Keyboard Character*	*ASCII*
A	100 0001	S	101 0011
a	110 0001	s	111 0011
B	100 0010	T	101 0100
b	110 0010	t	111 0100
C	100 0011	U	101 0101
c	110 0011	u	111 0101
D	100 0100	V	101 0110
d	110 0100	v	111 0110
E	100 0101	W	101 0111
e	110 0101	w	111 0111
F	100 0110	X	101 1000
f	110 0110	x	111 1000
G	100 0111	Y	101 1001
g	110 0111	y	111 1001
H	100 1000	Z	101 1010
h	110 1000	z	111 1010
I	100 1001	0	011 0000
i	110 1001	1	011 0001
J	100 1010	2	011 0010
j	110 1010	3	011 0011
K	100 1011	4	011 0100
k	110 1011	5	011 0101
L	100 1100	6	011 0110
l	110 1100	7	011 0111
M	100 1101	8	011 1000
m	110 1101	9	011 1001
N	100 1110	,	010 1100
n	110 1110	.	010 1110
O	100 1111	+	010 1011
o	110 1111	−	010 1101
P	101 0000	(	010 1000
p	110 0000	)	010 1001
Q	101 0001		
q	111 0001		
R	101 0010		
r	111 0010		

TABLE 12.4. EBCDIC Code (Partial)

A	1100 0001	B	1100 0010	C	1100 0011
D	1100 0100	E	1100 0101	F	1100 0110
G	1100 0111	H	1100 1000	I	1100 1001
J	1101 0001	K	1101 0010	L	1101 0011
M	1101 0100	N	1101 0101	O	1101 0110
P	1101 0111	Q	1101 1000	R	1101 1001
S	1110 0010	T	1110 0011	U	1110 0100
V	1110 0101	W	1110 0110	X	1110 0111
Y	1110 1000	Z	1110 1001	0	1111 0000
1	1111 0001	2	1111 0010	3	1111 0011
4	1111 0100	5	1111 0101	6	1111 0110
7	1111 0111	8	1111 1000	9	1111 1001

change has occurred. An extra bit called the parity bit is added to each word. The parity bit (either a 0 or a 1) is set to make the total number of 1's (including the parity bit) in the word even—0, 2, 4, ... For example, if the correct word at the transmitter contains two 1's, the parity bit is set to 0. If the correct word at the transmitter contains five 1's, the parity bit is set to 1. At the receiver each incoming word is checked for the number of 1's. If the total number of 1's in the word is even, the word is assumed to be correct and the word is processed. But if the number of 1's is found to be odd, then a *parity error* is indicated in some way. This system only *detects* parity errors; it does not correct them. Also, if two bits are changed in the word so as to keep the number of 1's constant, there will be no parity error. Words containing an even number of 1's are said to have *even* parity, and words with an odd number of 1's are said to have *odd* parity. A parity-error detection scheme can also operate with the parity of each word (including the parity bit) odd.

12.4 BOOLEAN ALGEBRA

Boolean algebra is a very general algebra that is different from ordinary algebra in several ways. For example, it has only three operations: OR, usually indicated by a + sign; AND, usually indicated by a dot · ; and the *complement* or NOT, usually indicated by an overline. If A and B are Boolean variables, then "A or B" is written A + B, "A AND B" is written A · B, and "NOT A" is written $\bar{A}$. Boolean algebra does not contain the operations of subtraction or division. From now on we will restrict ourselves to two-valued Boolean algebra; that is, there are only two elements (usually called 0 and 1) in the set to which the algebra applies. In 1938 Shannon showed that such a two-valued Boolean algebra can describe the operation of two-valued electrical switching circuits. With the development of inexpensive, fast, and versatile digital binary solid-state circuits, Boolean algebra has taken on great practical significance.

Let us consider a single binary variable A. We define a function called the *complement of* A and write it as $F = \bar{A}$. If $A = 0$, $\bar{A} = 1$; if $A = 1$, $\bar{A} = 0$. Thus the complement function is also called the NOT function: $\bar{A}$ = NOT A. Technically, A and $\bar{A}$ are the only two possible functions of the single binary variable A. Physically, A and $\bar{A}$ could represent a switch being on or off, or a voltage being 5 V or 0 V.

Let us consider two binary variables A and B. Each variable can be 0 or 1. Thus, there are four possible combinations of A and B: 0,0; 0,1; 1,0; or 1,1. Any function of A and B must be expressed in terms of 0 and 1 (the only elements of the algebra), so there are 16 possible binary functions F of the two binary variables A and B. These functions are shown in Table 12.5.

TABLE 12.5. Truth Table for the 16 Possible Binary Functions of Two Binary Variables

A	B	F_0	F_1	F_2	F_3	F_4	F_5	F_6	F_7	F_8	F_9	F_{10}	F_{11}	F_{12}	F_{13}	F_{14}	F_{15}
0	0	0	0	0	0	0	0	0	0	1	1	1	1	1	1	1	1
0	1	0	0	0	0	1	1	1	1	0	0	0	0	1	1	1	1
1	0	0	0	1	1	0	0	1	1	0	0	1	1	0	0	1	1
1	1	0	1	0	1	0	1	0	1	0	1	0	1	0	1	0	1

All this may seem terribly abstract and unrelated to electronics, but remember that digital electronic circuits can have only *two* states, which are equivalent to the 0 and 1 symbols. Usually the 0 state is represented in the actual circuitry by a low voltage level (e.g., 0.2 V for TTL chips); the 1 state is represented by a high voltage level (e.g., 3.5 V for TTL chips). (TTL stands for transistor-transistor logic which is a very common type of digital chip.) We can think of Table 12.5 as listing all the possible inputs A and B and all the possible outputs F for a digital circuit. For example, the output function F_1 is 1 only if both $A = 1$ and $B = 1$. Then the function F_1 (output) is 1; for any other combination of A and B, $F_1 = 0$. We can think of each function F as a special type of digital circuit or gate; such gates will be described in detail in Section 12.5.

Two of the most important functions of two binary variables are F_1 and F_7. F_1 is usually called the AND operation; we say F_1 = A AND B and usually write it as $F_1 = A \cdot B$. WE DO NOT SAY A "TIMES" B; THE "·" SYMBOL MEANS "AND"! Other notations for AND are: AB, $A \wedge B$, and $A \cap B$. Mathematicians call $A \cap B$ the *intersection* of A and B. The possible output values for F_1 for the four possible input values A and B can be written in a table, usually called a *truth table*, as shown in Table 12.6(a).

TABLE 12.6. Truth Tables for the AND and the OR Functions

A	B	$F_1 = A \cdot B$	A	B	$F_7 = A + B$
0	0	0	0	0	0
0	1	0	0	1	1
1	0	0	1	0	1
1	1	1	1	1	1
(a) AND truth table			(b) OR truth table		

F_7 is usually called the OR operation; we say F_7 = A OR B and usually write it as $F_7 = A + B$. WE DO NOT SAY A "PLUS" B: THE "+" SYMBOL MEANS "OR"! Other notations for OR are $A \vee B$ and $A \cup B$. Mathematicians call $A \cup B$ the *union* of A and B. The possible values for F_7 for the four possible input values of A and B can be written in a truth table as shown in Table 12.6(b). The other functions are named in Table 12.7.

There are 256 possible functions of three binary variables and 65,536 possible functions of four binary variables. For N binary variables, there are 2^{2^N} possible functions. Thus for $N \geqq 3$ it is impractical to write the truth tables for all the possible functions. But for any one function it is practical.

TABLE 12.7. Names of the 16 Possible Functions of Two Variables

Function Symbol	*Name of Function*
$F_0 = 0$	Null
$F_1 = A \cdot B \; (= A \wedge B = A \cap B)$	AND
$F_2 = A \cdot \bar{B}$	A AND NOT B
$F_3 = A$	A
$F_4 = \bar{A} \cdot B$	NOT A AND B
$F_5 = B$	B
$F_6 = A \cdot \bar{B} + \bar{A} \cdot B = A \oplus B$	EXCLUSIVE OR (XOR)
$F_7 = A + B \; (= A \vee B = A \cup B)$	OR
$F_8 = \overline{(A + B)} = \bar{F}_7$	NOT OR (NOR)
$F_9 = A \cdot B + \bar{A} \cdot \bar{B} = \overline{A \oplus B} = \bar{F}_6$	EXCLUSIVE NOR (XNOR)
$F_{10} = \bar{B} = \bar{F}_5$	NOT B
$F_{11} = A + \bar{B} = \bar{F}_4$	A OR NOT B
$F_{12} = \bar{A} = \bar{F}_3 = A'$	NOT A
$F_{13} = \bar{A} + B = \bar{F}_2$	NOT A OR B
$F_{14} = \overline{(A \cdot B)} = \bar{F}_1$	NOT A AND B (NAND)
$F_{15} = 1 = \bar{F}_0$	Identity

For example, one possible function of three binary variables is

A	B	C	F
0	0	0	0
0	0	1	0
0	1	0	0
0	1	1	0
1	0	0	0
1	0	1	1
1	1	0	1
1	1	1	1

$$F = A \cdot (B + C) = A \cdot B + A \cdot C$$

Some useful theorems of Boolean algebra are listed in Table 12.8.

TABLE 12.8. Useful Boolean Algebra Theorems

$0 \cdot 0 = 0$
$1 \cdot 1 = 1$
$1 \cdot 0 = 0 \cdot 1 = 0$
$0 \cdot A = A \cdot 0 = 0$
$0 + A = A + 0 = A$
$1 \cdot A = A \cdot 1 = A$
$1 + A = A + 1 = 1$
$A \cdot A = A$
$A + A = A$
$\bar{0} = 1$
$\bar{1} = 0$
$\bar{\bar{A}} = A$
$A \cdot \bar{A} = 0$
$A + A = 1$
$A + B = B + A$
$A \cdot B = B \cdot A$
$A + A \cdot B = A$
$A + (B + C) = (A + B) + C$
$A + (B \cdot C) = (A + B) \cdot (A + C)$
$A \cdot (B \cdot C) = (A \cdot B) \cdot C$
$A \cdot (B + C) = A \cdot B + A \cdot C$
$\left.\begin{array}{l} \overline{A + B} = \bar{A} \cdot \bar{B} \\ \overline{A \cdot B} = \bar{A} + \bar{B} \end{array}\right\}$ DeMorgan's theorem

All of the preceding theorems can be proved by writing out the truth tables. For example, consider the theorem A + (A · B) = A. We write:

A	B	A · B	A + (A · B)
0	0	0	0
0	1	0	0
1	0	0	1
1	1	1	1

Thus we see that the column for A + (A · B) is identical to the A column; in other words, A + (A · B) = A.

The theorems $\overline{A+B} = \bar{A}\cdot\bar{B}$ and $\bar{A}\cdot\bar{B} = \overline{A+B}$ are examples of DeMorgan's theorem, which is probably the most important theorem for digital electronics. In words, DeMorgan's theorem says that any logical Boolean binary expression remains unchanged if we do four things to it:

1. Change all variables to their complements; for example, A to $\bar{A}$, $\bar{A}$ to A, and so on.
2. Change all OR operations to AND operations; that is, replace "+" signs by "·"
3. Change all AND operations to OR operations; that is, replace "·" by "+".
4. Take the complement of the entire expression.

Consider the following examples:

$$\bar{A}+\bar{B} \xrightarrow{1} \bar{\bar{A}}+\bar{\bar{B}} = A+B \xrightarrow{2} A\cdot B \xrightarrow{4} \overline{A\cdot B} \qquad \text{thus} \qquad \bar{A}+\bar{B} = \overline{A\cdot B}$$

$$\overline{A\cdot B} \xrightarrow{1} \overline{\bar{A}\cdot\bar{B}} \xrightarrow{3} \overline{\bar{A}+\bar{B}} \xrightarrow{4} \overline{\overline{\bar{A}+\bar{B}}} = \bar{A}+\bar{B} \qquad \text{thus} \qquad \overline{A\cdot B} = \bar{A}+\bar{B}$$

$$\overline{A+B} \xrightarrow{1} \overline{\bar{A}+\bar{B}} \xrightarrow{2} \overline{\bar{A}\cdot\bar{B}} \xrightarrow{4} \overline{\overline{\bar{A}\cdot\bar{B}}} = \bar{A}\cdot\bar{B} \qquad \text{thus} \qquad \overline{A+B} = \bar{A}\cdot\bar{B}$$

$$\bar{A}\cdot\bar{B} \xrightarrow{1} \bar{\bar{A}}\cdot\bar{\bar{B}} = A\cdot B \xrightarrow{3} A+B \xrightarrow{4} \overline{A+B} \qquad \text{thus} \qquad \bar{A}\cdot\bar{B} = \overline{A+B}$$

$$A\cdot\bar{B} \xrightarrow{1} \bar{A}\cdot\bar{\bar{B}} = \bar{A}\cdot B \xrightarrow{3} \bar{A}+B \xrightarrow{4} \overline{\bar{A}+B} \qquad \text{thus} \qquad A\cdot\bar{B} = \overline{\bar{A}+B}$$

$$\overline{\bar{A}\cdot(B+C)} \xrightarrow{1} \overline{\bar{\bar{A}}\cdot(\bar{B}+\bar{C})} = \overline{A\cdot(\bar{B}+\bar{C})} \xrightarrow{2\,3} \overline{A+(\bar{B}\cdot\bar{C})} \xrightarrow{4} \overline{\overline{A+(\bar{B}\cdot\bar{C})}}$$

$$= A+(\bar{B}\cdot\bar{C})$$

12.5 BINARY GATES

A *gate* is a circuit that produces a digital output for one or more digital inputs. The output is a function of the input(s), and the type of function determines the name of the gate. Thus, there are AND gates, OR gates, and so on. The gate is the standard building block of all digital electronic systems. The revolution in digital circuitry has resulted from the availability of many types of high-speed gates and combinations of gates at low prices. One inexpensive monolithic chip can contain four or six electrically independent gates. For example, a 74LS00 chip contains four independent two-input NAND gates, and a 74LS04 chip contains six independent inverters. Other chips can contain a more complicated combination of transistors, diodes, and gates to perform more elaborate operations. Such chips are called *medium-scale integration* (MSI) and *large-scale integration* (LSI). In this sense, "integration" means packing more hardware into one chip.

The logical symbols 0 and 1 are represented by different voltage levels in the gate circuitry. In *positive logic* 0 is a low voltage and 1 is a high voltage. For example, in standard transistor-transistor logic (TTL), 0 is represented by a voltage level of +0.2 V, and 1 by +3.5 V. In *negative logic*, 0 is the higher voltage, and 1 is the lower voltage. Positive and negative logics will be discussed in Section 12.6.

12.5.1 The OR Gate

For a two-input OR gate the output is given by F = A + B. That is, the output is 1 if either or both of the two inputs A and B are 1. The truth table and the standard OR gate schematic symbol are shown in Fig. 12.2(a) and

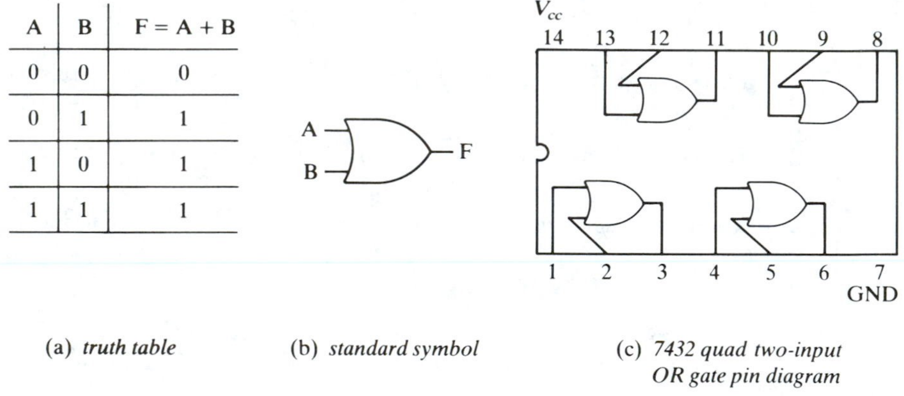

A	B	F = A + B
0	0	0
0	1	1
1	0	1
1	1	1

(a) *truth table* (b) *standard symbol* (c) *7432 quad two-input OR gate pin diagram*

FIGURE 12.2 The OR gate.

(b). A standard 7432 TTL quad-two-input OR gate is shown in Fig. 12.2(c). The magnitude of the voltage output for two 1 inputs is the same as for the case of only one 1 input.

An OR gate can have more than two inputs. A three-input OR gate truth table and schematic symbol are shown in Fig. 12.3. Notice that the

A	B	C	F = A + B + C
0	0	0	0
0	0	1	1
0	1	0	1
0	1	1	1
1	0	0	1
1	0	1	1
1	1	0	1
1	1	1	1

(a) *OR truth table*

A
B
C
F = A + B + C

(b) *standard symbol*

FIGURE 12.3 Three-input OR gate.

output is 1 when any one or more of the inputs are 1. The magnitude of the output is independent of the number of 1 inputs.

A simple positive-logic two-input OR gate can be constructed from diodes and resistors, as shown in Fig. 12.4. (In positive logic, 0 is a low

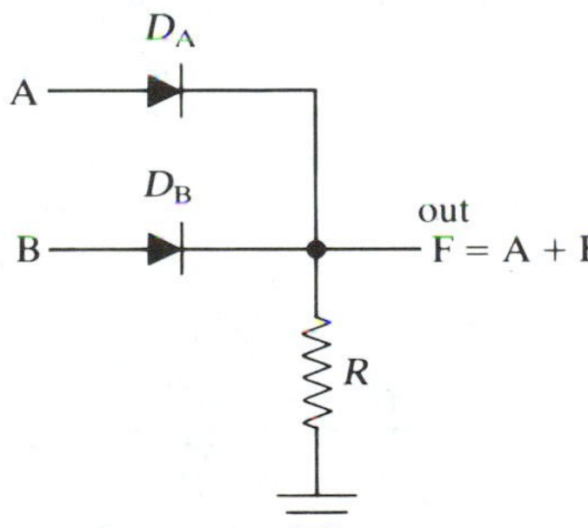

V_A(V)	V_B (V)	V_{out} (V)
0	0	0
0	5.0	4.4
5.0	0	4.4
5.0	5.0	4.4

FIGURE 12.4 Diode resistor logic (DRL) OR gate.

voltage and 1 is a high voltage.) We assume the diode turn-on voltages are exactly the same: 0.6 V. If both inputs are zero or below the turn-on voltage of the diodes, then the output will be zero because both diodes will be off. A positive-voltage input V_A at A (A = 1) and $V_B = 0$ V (B = 0) will forward bias diode D_A and reverse bias diode D_B, so the output voltage is $V_{out} = V_A - V_D$, where V_D is the turn-on voltage for diode D_A. Usually, $V_A \gg V_D$, so $V_{out} \cong V_A$. A similar argument shows that for A = 0 and B = 1, $V_{out} \cong V_B$. If both inputs are positive, $V_A = V_B = V_{in} \gg V_D$, then the output will also be positive, $V_{out} \cong V_{in}$, which follows from the following argument. From the KVL we have

$$V_{in} = V_{D_A} + V_{out}$$

and

$$V_{in} = V_{D_B} + V_{out}$$

Adding yields

$$2V_{in} = V_{D_A} + V_{D_B} + 2V_{out}$$

or

$$V_{out} = V_{in} - \frac{1}{2}(V_{D_A} + V_{D_B})$$

If $V_{in} \gg$ the diode turn-on voltage, then

$$V_{out} \cong V_{in}$$

If the two diodes have different turn-on voltages, the diode with the higher turn-on voltage will not conduct when both inputs are high. The circuit still works as an OR gate, assuming V_{in} is much larger than the diode turn-on voltages.

Modern digital OR gates are made from transistors and resistors, not diodes and resistors. A modern TTL OR gate circuit will be discussed in Section 12.7.

12.5.2 The AND Gate

For a two-input AND gate the output is given by $F = A \cdot B$. That is, the output is 1 only if *both* inputs are 1. If either or both inputs are 0, then the output is 0. The truth table and the standard schematic symbol are shown in Fig. 12.5. A standard 7408 TTL quad-two-input AND gate is shown in Fig. 12.5(c). Notice that the output is 1 only if *all* the inputs are 1.

An AND gate can have more than two inputs. A three-input AND gate truth table and schematic symbol are shown in Fig. 12.6.

A simple positive-logic two-input AND gate can be constructed from diodes and resistors as shown in Fig. 12.7. (In positive logic, 0 is a low voltage and 1 is a high voltage.) If A is 0, diode D_A is forward biased regardless of B, and if B is 0, diode D_B is forward biased regardless of A. Thus if *either* A or B is 0, the output equals the diode turn-on voltage (approximately 0.6 V). Thus, the output will go positive to V_{cc} if and only if *both* inputs A and B go positive enough to reverse bias both diodes. Thus, each input must be more positive than $V_{cc} - 0.6$ V to produce a 1 (positive) output. When both inputs are more positive than $V_{cc} - 0.6$ V, the output will approximately equal V_{cc}. Modern digital AND gates are made from

A	B	F = A • B
0	0	0
0	1	0
1	0	0
1	1	1

(a) *AND truth table*

(b) *standard symbol*

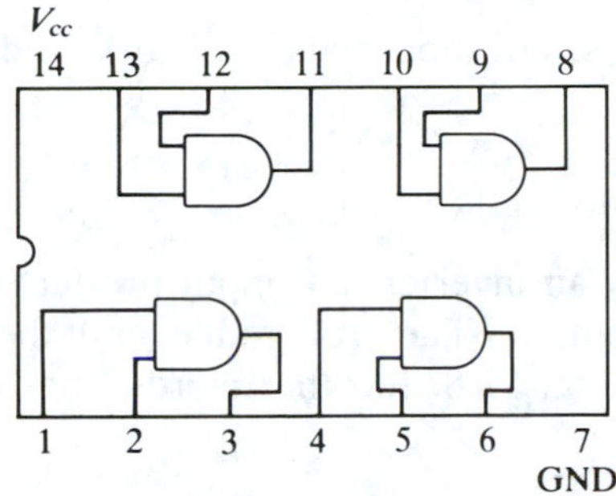

(c) *74LS08 or 7408 quad two-input AND gate pin diagram*

FIGURE 12.5 The AND gate.

A	B	C	F = A • B • C
0	0	0	0
0	0	1	0
0	1	0	0
0	1	1	0
1	0	0	0
1	0	1	0
1	1	0	0
1	1	1	1

(a) *truth table*

(b) *standard symbol*

FIGURE 12.6 Three-input AND gate.

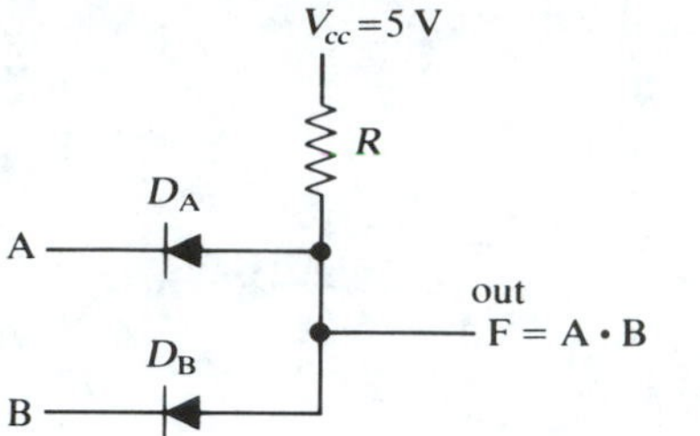

V_A (V)	V_B (V)	V_{out} (V)
0	0	0.6
0	5.0	0.6
5.0	0	0.6
5.0	5.0	5.0

FIGURE 12.7 Diode resistor logic (DRL) AND gate.

transistors and resistors, not diodes and resistors, and will be described in Section 12.7.

12.5.3 The NOT Gate

A NOT gate is simply an inverter; a 0 input produces a 1 output, and a 1 input produces a 0 output. The truth table and the standard schematic symbol are shown in Fig. 12.8. In other words, the output is the comple-

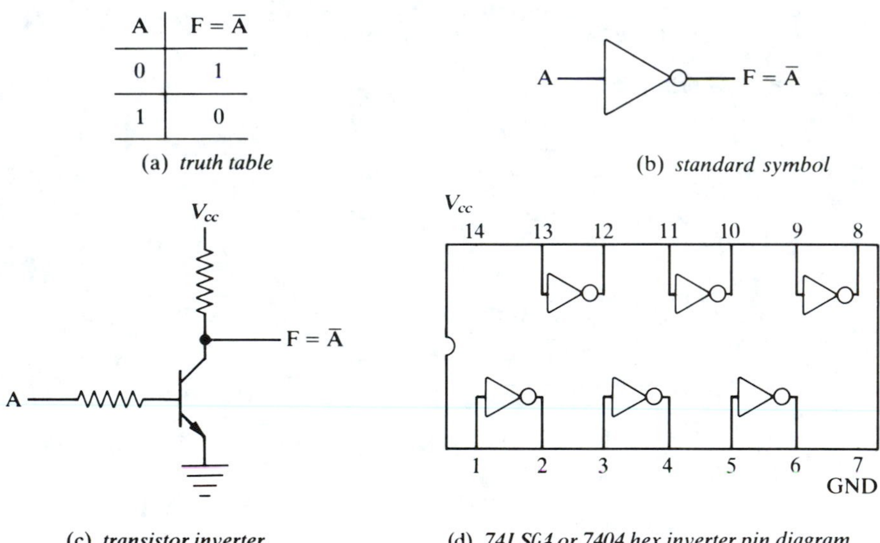

FIGURE 12.8 The NOT gate or inverter.

ment or the inverse of the input. The Boolean algebra symbol for the complement of A is $\bar{A}$, or sometimes A′. A common emitter amplifier is an obvious choice for a NOT gate because of the 180° phase inversion between the input at the base and the output at the collector. For positive logic the npn transistor of Fig. 12.8(c) is normally biased off. Thus, for a 0 logic input the output at the collector is at a high positive voltage (V_{cc}) corresponding to a logic 1. With a 1 logic input the input voltage on the base is positive, which turns on the npn transistor. Thus, the collector voltage drops to a low positive voltage, corresponding to a logic 0 output. A standard 7404 TTL hex inverter in a DIP package is shown in Fig. 12.8(d).

The OR, AND, and NOT gates form the theoretical basis of all modern digital systems. It can be shown that any mathematical operation can be performed by appropriate combinations of OR, AND, and NOT gates. The NOT gate can be combined with the OR and the AND gates to form a NOT-OR or NOR gate and also a NOT-AND or NAND gate. The NOR and NAND gates are actually the most commonly used simple gates because of their availability, performance, and low cost, and we will now describe them.

12.5.4 The NOR Gate

The NOR gate is an OR gate followed by a NOT gate. The truth table and standard schematic symbol are shown in Fig. 12.9. Notice that a 1 output is

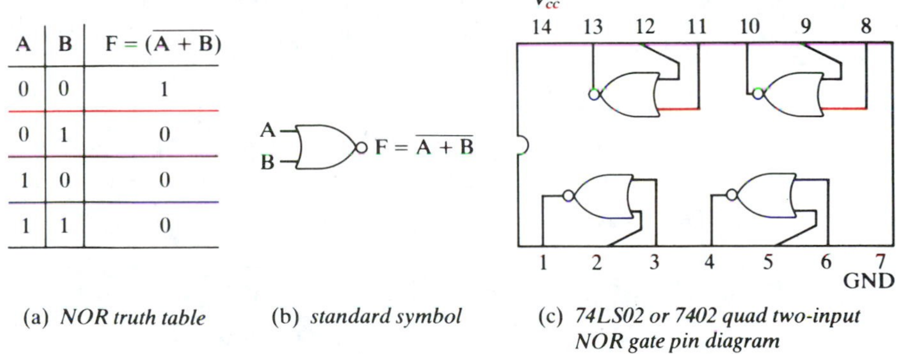

A	B	$F = \overline{(A + B)}$
0	0	1
0	1	0
1	0	0
1	1	0

(a) *NOR truth table* (b) *standard symbol* (c) *74LS02 or 7402 quad two-input NOR gate pin diagram*

FIGURE 12.9 The NOR gate.

produced only when none of the inputs is 1. In other words, any 1 input produces a 0 output. Only the small open circle at the output distinguishes the NOR gate from the OR gate symbol. A standard 7402 TTL quad two-input NOR gate in a 14-pin DIP package is shown in Fig. 12.9(c).

AND, OR, and NOT gates can all be made from NOR gates alone as shown in Fig. 12.10. The realization of the AND gate is an example of DeMorgan's theorem. The operation of all three circuits can be verified by writing out the truth tables. It is common to use a "leftover" NOR gate to make an inverter, as in Fig. 12.10(a).

(a) *NOT gate* (b) *OR gate* (c) *AND gate*

FIGURE 12.10 Realization of NOT, OR, and AND gates, using only NOR gates.

12.5.5. The NAND Gate

The NAND gate is an AND gate followed by a NOT gate. The truth table and the standard schematic symbol are shown in Fig. 12.11. Notice that any

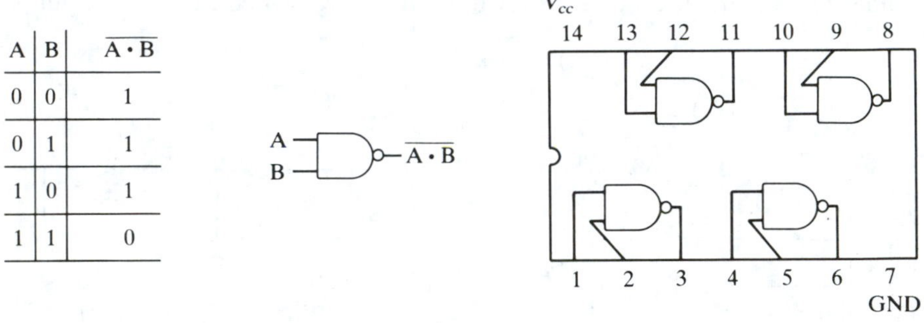

A	B	$\overline{A \cdot B}$
0	0	1
0	1	1
1	0	1
1	1	0

(a) *NAND truth table* (b) *standard symbol* (c) *74LS00 or 7400 quad two-input NAND gate pin diagram*

FIGURE 12.11 The NAND gate.

0 input produces a 1 output; only two 1 inputs will produce a 0 output. Only the small circle distinguishes the NAND gate from the AND gate symbol. A standard 7400 TTL quad two-input NAND gate in a 14-pin DIP package is shown in Fig. 12.11(c).

Three- and four-input NAND gates are available. A truth table and the standard schematic symbol for a three-input 7410 NAND gate are shown in Fig. 12.12.

AND, OR, and NOT gates can all be made from NAND gates alone, as shown in Fig. 12.13. The realization of the OR gate is an example of DeMorgan's theorem. The operation of all three circuits can be verified by writing out the truth table.

In general, circuit designers use NOR and NAND gates rather than

A	B	C	$\overline{A \cdot B \cdot C}$
0	0	0	1
0	0	1	1
0	1	0	1
0	1	1	1
1	0	0	1
1	0	1	1
1	1	0	1
1	1	1	0

(a) *truth table*

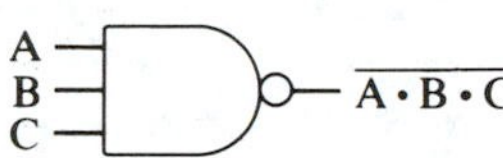

(b) *standard symbol*

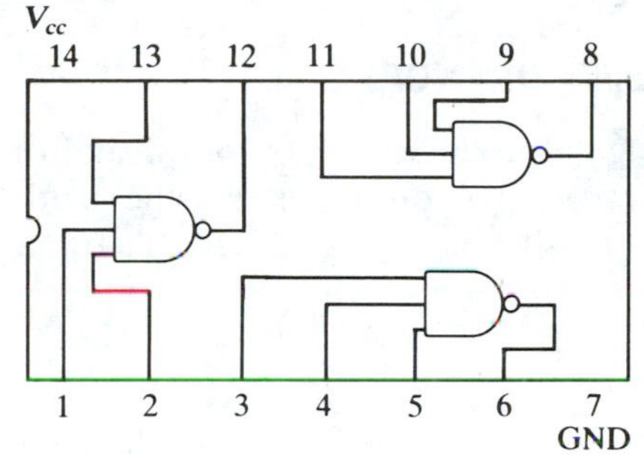

(c) *7410 triple-three-input NAND gate pin diagram*

FIGURE 12.12 Three-input NAND gate.

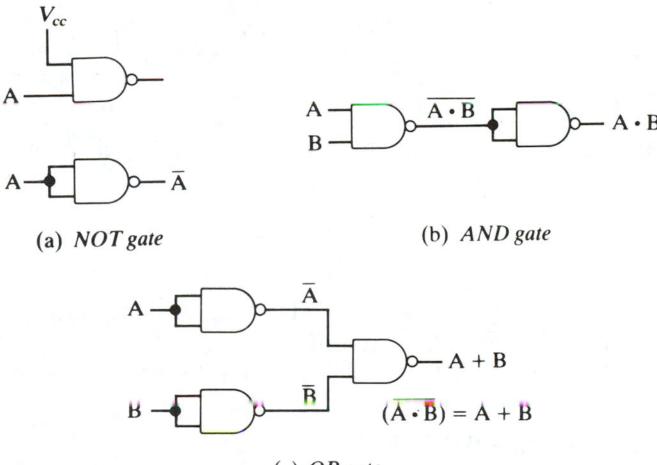

FIGURE 12.13 Realization of NOT, AND, and OR gates, using only NAND gates.

AND and OR gates because of the resulting lower power consumption, which can be of considerable importance in circuits containing hundreds of chips. The NAND and NOR gates are also faster than the AND and OR gates. The current requirements and "propagation delay" figures are given in Table 12.9 for the five simple gates we have just described.

TABLE 12.9. Power Consumption and Propagation Delay for Five Simple TTL Gates

	7400 NAND	*7402 NOR*	*7408 AND*	*7432 OR*	*7404 Inverter*
Current (mA)	12	12	16	19	12
Propagation delay[†] (ns)	10	10	15	12	10

[†]The propagation delay is the time delay between an input logic voltage change and the resultant output voltage change.

12.5.6 The Exclusive OR (XOR) Gate

The exclusive OR gate produces a 1 output if and only if *one* of the two inputs is a 1; otherwise the output is 0. The truth table and standard schematic symbol are shown in Fig. 12.14. A circle around the + OR

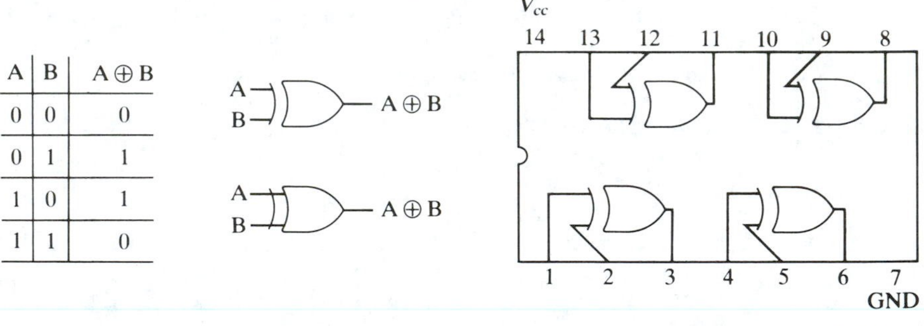

(a) *truth table* (b) *standard symbol* (c) *74LS86 or 7486 quad two-input XOR gate*

FIGURE 12.14 The XOR gate.

symbol indicates the XOR operation: $F = A \oplus B$. The output of an XOR gate is 1 only if the two inputs are different.

An XOR circuit can be made from only NAND gates or from only NOR gates, as shown in Fig. 12.15. From the truth table of the XOR gate, we can write the XOR function as

$$A \oplus B = A \cdot \bar{B} + \bar{A} \cdot B$$

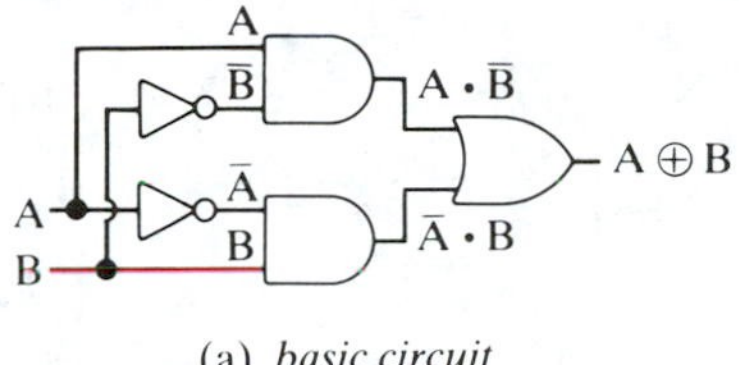

(a) *basic circuit*

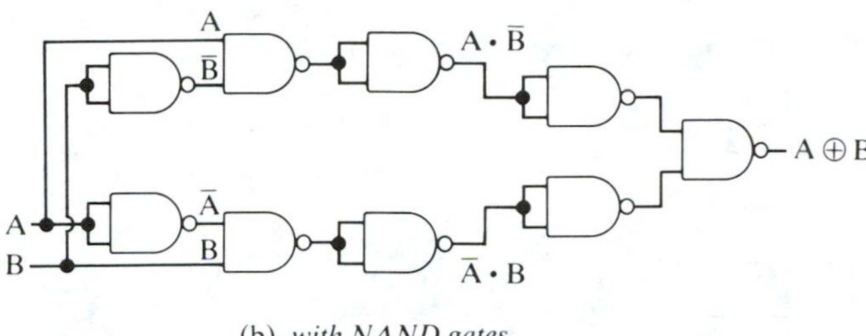

(b) *with NAND gates*

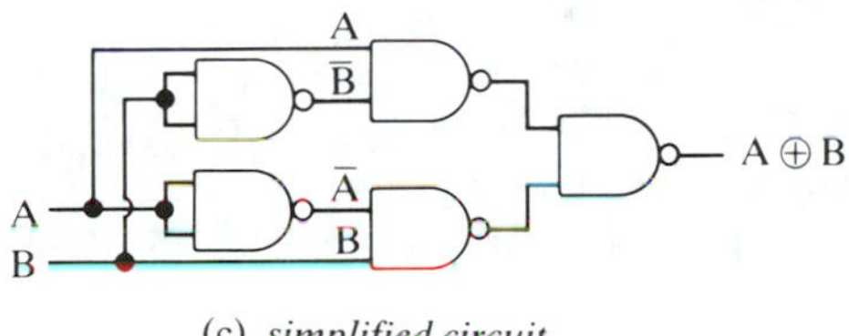

(c) *simplified circuit*

FIGURE 12.15 XOR gate using $A \oplus B = A \cdot \bar{B} + \bar{A} \cdot B$.

The NAND gate circuit of Fig. 12.15(a) is the realization of this equation. Notice how the inverters in series in Fig. 12.15(b) have been eliminated in 12.15(c). This follows from the fundamental theorem $\bar{\bar{A}} = A$.

From the truth table we can also write the XOR function as "A or B" and "NOT A AND B." Symbolically,

$$A \oplus B = (A + B) \cdot \overline{(A \cdot B)}$$

The circuit of Fig. 12.16(a) is the realization of this equation, and Fig. 12.16(b) shows the NAND gate version. Figures 12.15 and 12.16 illustrate how different connections of gates can be used to produce the same logical output function of the inputs.

The XOR gate can be used to invert a signal or to leave it unchanged, depending on the value of a "control" signal. From the XOR truth table we see that if B is the signal and A = 0, the XOR output equals B; but if A = 1, the output equals the complement or the inverse of B. The circuit is shown in Fig. 12.17.

Thus an XOR gate can be used to code ("scramble") and decode

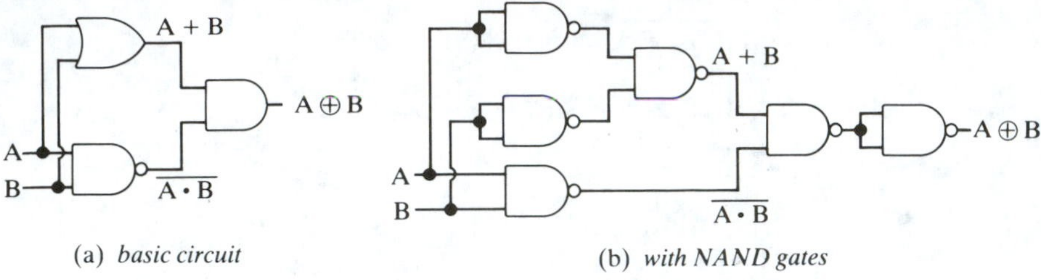

(a) *basic circuit* (b) *with NAND gates*

FIGURE 12.16 XOR gate using $A \oplus B = (A + B) \cdot \overline{(A \cdot B)}$.

A	B	A⊕B
0	0	0
0	1	1
1	0	1
1	1	0

A (control)
B
F = B if control = 0
$F = \overline{B}$ if control = 1

FIGURE 12.17 XOR inverter.

("unscramble") data. A known but apparently random series of 0s and 1s can be fed into the control input to produce coded or scrambled data D′ as shown in Fig. 12.18. At the receiver the coded data D′ and the *same* code

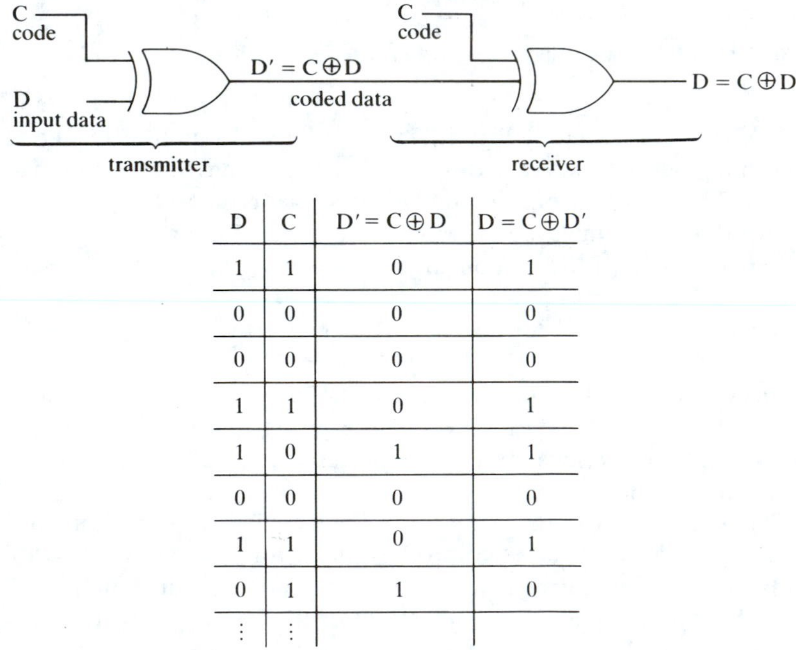

D	C	D′ = C⊕D	D = C⊕D′
1	1	0	1
0	0	0	0
0	0	0	0
1	1	0	1
1	0	1	1
0	0	0	0
1	1	0	1
0	1	1	0
⋮	⋮		

FIGURE 12.18 XOR gate coding/decoding.

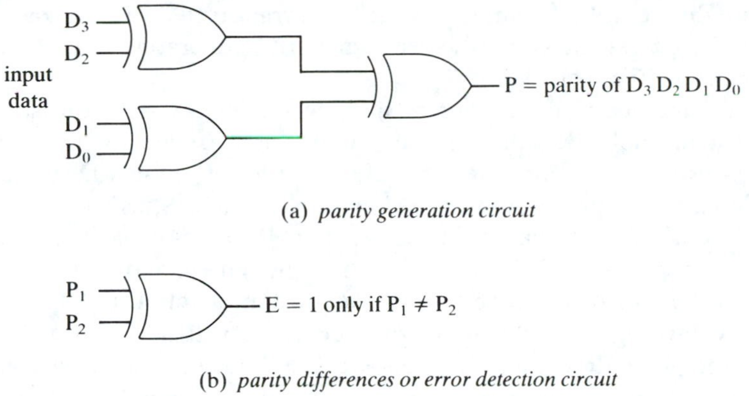

(a) *parity generation circuit*

(b) *parity differences or error detection circuit*

FIGURE 12.19 XOR parity circuits.

C are fed into another XOR gate. The output of the receiver XOR gate is then D, the correct data. Notice that the receiver must have a copy of the same code C used by the transmitter.

An XOR gate can also be used to generate and check parity bits, as shown in Fig. 12.19. In Fig. 12.19(a), the output P is the parity of the input data $D_3D_2D_1D_0$. In Fig. 12.19(b) output E is 1 only if the two parity bits P_1

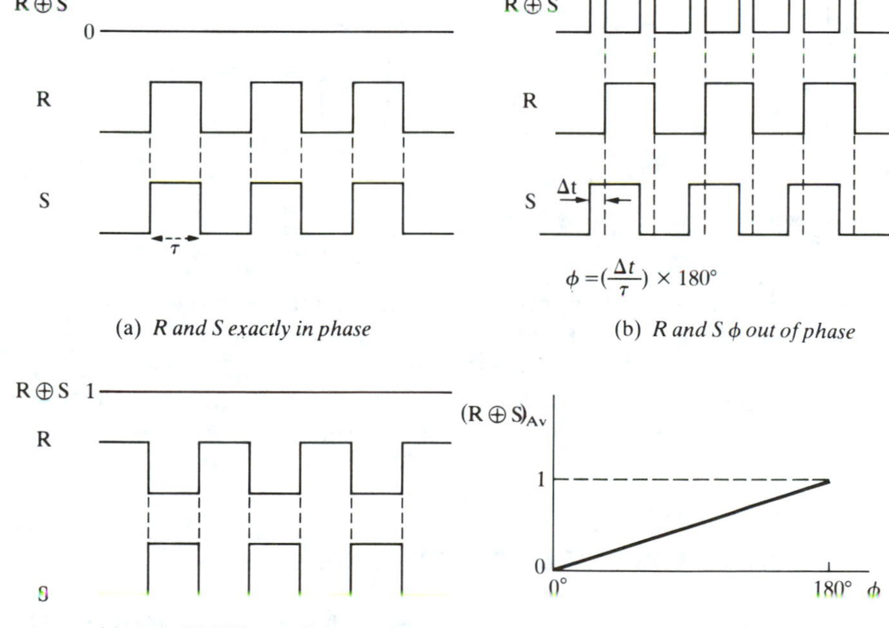

(a) *R and S exactly in phase*

(b) *R and S φ out of phase*

(c) *R and S 180° out of phase*

(d) *R⊕S average value vs. φ*

FIGURE 12.20 XOR phase detector.

and P_2 are different. P_1 would be the parity bit transmitted with the data and P_2 the parity bit calculated by the receiver parity-generating circuit. E = 1 would then signify a parity error.

If the reference R and the input signal S of a lock-in detector (see 8.8) are each unit amplitude square waves of the same frequency and are fed into an XOR gate, the output $R \oplus S$ will be a series of pulses whose width is proportional to the phase difference ϕ between R and S, as shown in Fig. 12.20. If R and S are exactly in phase, then $R \oplus S = 0$; if R and S are slightly out of phase, $R \oplus S$ is a series of narrow pulses; and if R and S are exactly 180° out of phase, then $R \oplus S$ is a constant level of unit amplitude. Thus, if we average $R \oplus S$, the average dc value will be linearly proportional to the phase difference between R and S. This can easily be done by feeding the $R \oplus S$ pulse train into a low-pass filter with the filter time constant $RC \gg \tau$ where τ is half the period of either R or S.

12.5.7 The Exclusive NOR (XNOR) Gate

The XOR gate followed by an inverter is called an *exclusive NOR* (XNOR) gate. Its truth table and diagram are shown in Fig. 12.21. Notice that the

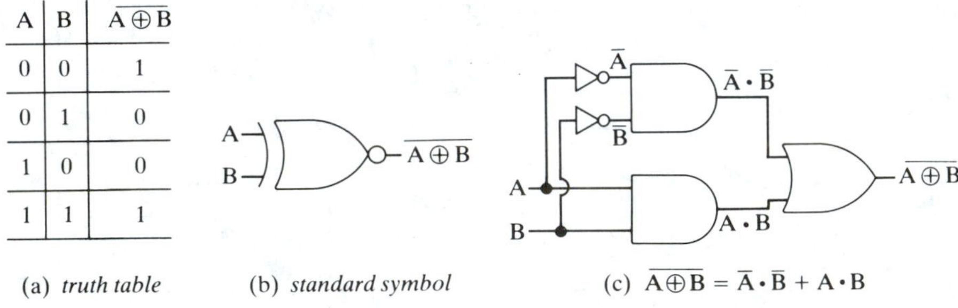

A	B	$\overline{A \oplus B}$
0	0	1
0	1	0
1	0	0
1	1	1

(a) *truth table* (b) *standard symbol* (c) $\overline{A \oplus B} = \bar{A} \cdot \bar{B} + A \cdot B$

FIGURE 12.21 The XNOR gate.

output is high only if both inputs are the same, either both 0 or both 1. The XNOR gate can be made from basic gates as shown in Fig. 12.21(c) by using the relationship $\overline{A \oplus B} = \bar{A} \cdot \bar{B} + A \cdot B$. The XNOR gate can also be used as a parity-error check; if the two inputs A and B are respectively the transmitted parity and the parity generated by the receiver, then an output of 0 indicates a parity error.

12.6 POSITIVE LOGIC AND NEGATIVE LOGIC

There are two stable electrical states for every logic output, whether TTL, CMOS, or any other family of hardware. These two states have distinctly different voltages (logical disaster occurs when the voltage differences become muddled!), and we have called them *high* or 1 and *low* or 0 up to now. A typical TTL high is +3.5 V, and a TTL low is 0.2 V. A typical CMOS high is 5 V, and a low is 0 V, if $V_{DD} = 5$ V. (CMOS is a low power family of gates.) We can assign the logic symbols 1 and 0 to either of the two voltage levels by simply choosing to do so.

If we choose to call the higher voltage level 1 and the lower voltage level 0, this system is called *positive logic* or *positive true* logic. If we choose to call the lower voltage level 1 and the higher voltage level 0, this system is called *negative logic* or *negative true* logic. Notice that negative logic does not necessarily mean negative voltages are involved.

We must always consistently use one system of logic or the other in analyzing any circuit. But it is useful to realize that for a given fixed circuit, it is sometimes useful to switch from positive logic to negative logic in our thinking processes.

The function of a particular gate circuit depends on whether we use positive logic or negative logic. Consider, for example, a TTL AND gate. Its "electrical" truth table is shown in Fig. 12.22(a). If we let 0 represent a

A (V)	B (V)	gate output F (V)
0.2	0.2	0.2
0.2	3.5	0.2
3.5	0.2	0.2
3.5	3.5	3.5

(a) *electrical truth table (TTL)*

A	B	F = A • B
0	0	0
0	1	0
1	0	0
1	1	1

(b) *positive logic truth table*

A	B	F = A + B
1	1	1
1	0	1
0	1	1
0	0	0

(c) *negative logic truth table*

FIGURE 12.22 AND gate truth tables.

voltage level of 0.2 V and 1 represent 3.5 V, then we get the positive-logic truth table of Fig. 12.22(b). In other words, the gate is an AND gate. If, however, we let 0 represent 3.5 V and 1 represent 0.2 V, we get the truth table of Fig. 12.22(c), which is for an OR gate. Thus the same electrical gate circuit is an AND gate with positive logic and an OR gate with negative logic. This equivalence is often useful in simplifying digital circuitry. Figure 12.23 shows the truth tables for a TTL NOR gate. Notice that the gate is a NOR gate with positive logic and a NAND gate with negative logic.

By convention the name of a gate is always for positive logic. That is, the gate whose electrical truth table is shown in Fig. 12.22(a) would be called an AND gate on the specification sheet. From now on we will use positive logic unless we explicitly specify negative logic.

Other positive-logic gates and their negative-logic gates are shown in

A (V)	B (V)	gate output F (V)
0.2	0.2	3.5
0.2	3.5	0.2
3.5	0.2	0.2
3.5	0.2	0.2

(a) *electrical truth table (TTL)*

A	B	$F = \overline{A + B}$
0	0	1
0	1	0
1	0	0
1	1	0

(b) *positive logic truth table*

A	B	$F = \overline{A \cdot B}$
1	1	0
1	0	1
0	1	1
0	0	1

(c) *negative logic truth table*

FIGURE 12.23 NOR gate truth tables.

Table 12.10. Notice that the schematic circuit symbols are different for the positive-logic and negative-logic gates, although the actual chip will be the same.

The bubble notation is equivalent to the superscript bar notation used in older literature to indicate whether a high voltage level or a low voltage level will perform some circuit function. For example, if a counter will reset to zero when its reset input terminal is at a low voltage (0.2 V for TTL or near 0 V for CMOS), the diagram of Fig. 12.24(a) would be used; the reset

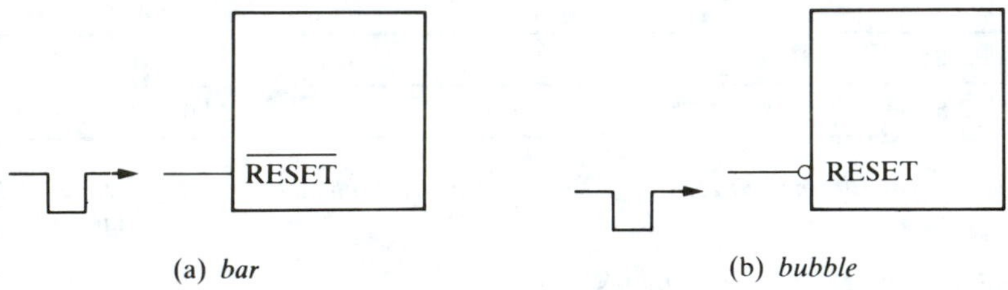

(a) *bar* (b) *bubble*

FIGURE 12.24 Bar and bubble notation for "low to RESET."

terminal is labeled $\overline{\text{Reset}}$. Exactly the same information is conveyed by the bubble notation of Fig. 12.24(b). With either the bubble or the superscript bar notation, the meaning is the same—a low voltage will "reset" the counter. The same notation is used for "Enable," "Set," "Reset," "Start," "Stop," and other terminals.

As an example, consider a specific counter that will be reset (R) by a low-voltage input at its reset terminal. Suppose we wish to reset the counter only when signal A is high and signal B is low. Then we wish R = 0 only when A = 1 and B = 0; the truth table is shown in Fig. 12.25(a). The straightforward way to realize this truth table with gates is shown in Fig. 12.25(b), because $R = \overline{(A \cdot \bar{B})}$ from inspection of the truth table. But the inverter at the output of the AND gate can be written as a bubble at the output of the AND gate, which means the gate is a NAND gate, as shown

TABLE 12.10. Positive and Negative Logic Truth Tables

POSITIVE LOGIC						NEGATIVE LOGIC					
Symbol	Gate Name	Equation	Truth Table A	B	F	Symbol	Gate Name	Equation	Truth Table A	B	F′
A, B — F	AND	$F = A \cdot B$	0 0 1 1	0 1 0 1	0 0 0 1	A, B — F′	OR	$F' = A + B$	1 1 0 0	1 0 1 0	1 1 1 0
A, B — F	NAND	$F = \overline{A \cdot B}$	0 0 1 1	0 1 0 1	1 1 1 0	A, B — F′	NOR	$F' = \overline{A + B}$	1 1 0 0	1 0 1 0	0 0 0 1
A, B — F	OR	$F = A + B$	0 0 1 1	0 1 0 1	0 1 1 1	A, B — F′	AND	$F' = A \cdot B$	1 1 0 0	1 0 1 0	1 0 0 0
A, B — F	NOR	$F = \overline{A + B}$	0 0 1 1	0 1 0 1	1 0 0 0	A, B — F′	NAND	$F' = \overline{A \cdot B}$	1 1 0 0	1 0 1 0	0 1 1 1
B — F	NOT	$F = \overline{A}$		0 1	1 0	B — F′	NOT	$F' = \overline{A}$		1 0	0 1
B — F	Buffer	$F = A$		0 1	0 1	B — F′	Buffer	$F' = A$		1 0	1 0

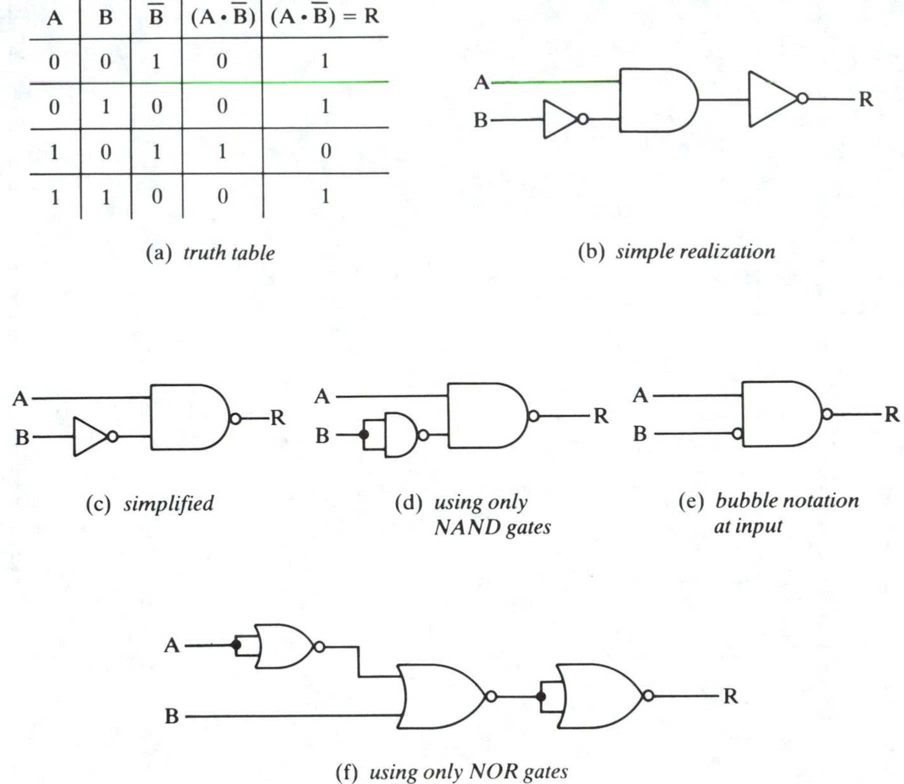

A	B	$\overline{B}$	$(A \cdot \overline{B})$	$\overline{(A \cdot \overline{B})} = R$
0	0	1	0	1
0	1	0	0	1
1	0	1	1	0
1	1	0	0	1

(a) *truth table*

(b) *simple realization*

(c) *simplified*

(d) *using only NAND gates*

(e) *bubble notation at input*

(f) *using only NOR gates*

FIGURE 12.25 Counter reset example.

in Fig. 12.25(c). And the inverter can be made from a NAND gate by connecting the two inputs, so the final circuit made entirely from NAND gates is shown in Fig. 12.25(d). The bubble notation could be used to represent the inverter for B as shown in Fig. 12.25(e). The truth table could also be realized with NOR gates, as shown in Fig. 12.25(f).

12.7 LOGIC FAMILIES

The two most important types or families of logic gates are the TTL family (transistor-transistor logic) and the CMOS family (complementary metal-oxide semiconductor). As you might guess, TTL gates are made mainly from bipolar transistors and CMOS gates mainly from MOSFETs. The TTL family includes three useful subfamilies: the regular TTL or 7400 series, the Schottky or 74S00 series, and the low-power Schottky or 74LS00 series. Most new TTL systems are being designed with the 74LS00 subfamily

TABLE 12.11. Logic Family Properties

	TTL (*7400*)	*LSTTL* (*74LS00*)	*Schottky TTL* (*74S00*)	*Fast TTL* (*74F00*)	*CMOS* *74C00*	*High-Speed CMOS* *74HC00*	*ECL*	*100 k ECL*
Propagation delay (ns)	10	9.5	3	2.7	50 @ 5 V 25 @ 10 V	8–10	2	0.75
Power per gate (mW)	10	2	19	4	10 nW	10 nW	25	40

because it has the same speed as the standard 7400 subfamily but uses only one-fifth the power. There are other, essentially obsolete, subfamilies of TTL: the 74L00, which is low power but very slow, and the 74H00, which is fast but uses much more power. A new "fast" subfamily, the 74F00 series, combines high speed with low power, but there is not yet available a wide variety of gates in the subfamily.

The 5400 TTL chips are a premium (military) version of the standard 7400 chips. 5400 chips will operate from −55° to +125°C, compared to 0° to +70°C for the 7400 chips. Most gates are available in all three TTL subfamilies; for example, the 7402 and the 74LS02 are each quad-two-input NOR chips in the TTL and low-power Schottky series, respectively, and the 74C02 is a CMOS quad-two-input NOR chip.

The CMOS family uses the least power but at the expense of significantly lower speed than TTL. CMOS is also more susceptible to permanent destruction from static electricity than TTL is. Most digital systems are made from TTL rather than CMOS unless very low power consumption is essential and the low speed of CMOS is acceptable. The 74HC00 family is a new high-speed CMOS family; it combines the speed of TTL with the low power consumption of CMOS.

Other logic families are the ECL (emitter coupled logic), for extremely high speeds, and the NMOS and PMOS families, which are mainly used in very large-scale integration (VLSI) systems where an extremely large number of gates per unit chip area is required. Table 12.11 summarizes the speed and power properties of the various families.

We will now discuss some of the more common logic gates in TTL and CMOS families.

12.7.1 The 7400 TTL NAND Gate

The 7400 TTL NAND gate is one of the most common and useful gates available in the TTL family. It is a quad-two-input NAND gate, which means there are four NAND gates (each with two inputs) in one package. Its schematic diagram is shown in Fig. 12.26 for one two-input NAND gate. The two inputs go to the two emitters of the npn transistor Q_1 (see Fig. 9.1), and thus a positive input will tend to turn off Q_1 and a negative input will tend to turn on Q_1. In the following arguments we will assume an "on" transistor has $V_{BE} = 0.7$ V and $V_{CE} = 0.2$ V.

Consider the state shown in Fig. 12.26(a) in which both inputs are positive enough ("high") to turn *off* both emitter-base junctions of Q_1. The base-collector junction of Q_1 is thus forward biased by the positive 5-V power supply connected to the base of Q_1 through the 4-kΩ resistance. Thus, the (positive) "collector" current I_1 of Q_1 flows into the base of Q_2 and turns on Q_2 strongly. The emitter current of Q_2 flowing through the 1-kΩ resistance thus forward biases the base-emitter junction of Q_3, and Q_3 conducts strongly also. Thus, the output voltage at the collector of Q_3 is the

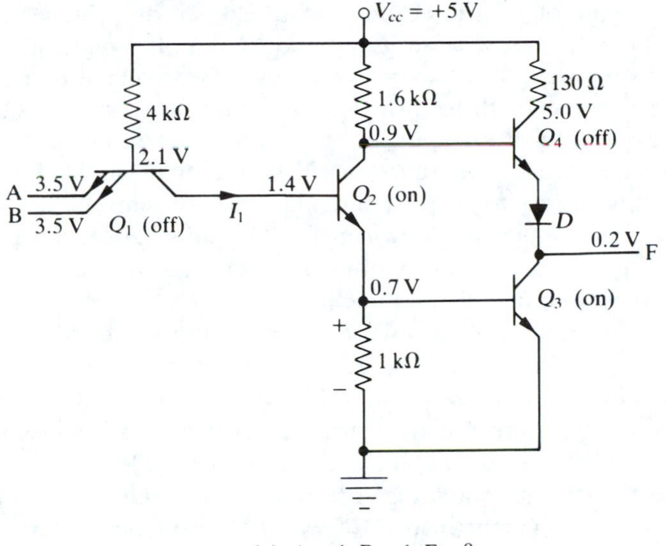

(a) A = 1, B = 1; F = 0

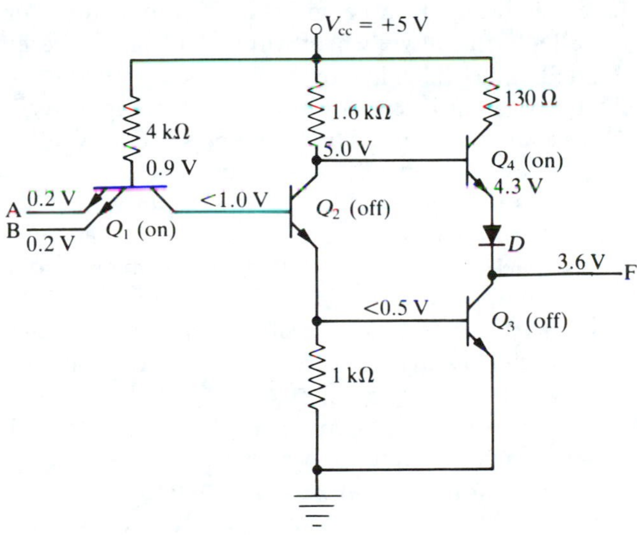

(b) A = 0, B = 0; F = 1

FIGURE 12.26 Two-input 7400 TTL NAND gate.

collector voltage of Q_3, which will be about 0.2 V if Q_3 is saturated. If Q_2 and Q_3 have base emitter voltages of 0.7 V each when they are strongly conducting or saturated, then the emitter of Q_2 will be 0.7 V (V_{BE} for Q_3), and the collector of Q_1 will be two base emitter drops above ground, or V_{C1} = 1.4 V. The base and collector of Q_1 act like a forward biased diode, so the base voltage of Q_1 will be one forward diode drop higher than the collector: $V_{B1} = 2.1$ V. Because Q_2 is saturated, its collector is 0.2 V higher than its emitter; thus $V_{C2} = 0.9$ V. The base voltage of Q_4 is thus 0.9 V, which is not enough to turn on Q_4 because the base of Q_4 would have to be *two* diode drops more positive than the output (0.2 V) to be on—one from the base-emitter junction of Q_4 and one from the diode. Thus the diode D is necessary to keep Q_4 off. The collector current of Q_4 is thus zero, and the voltages are as shown in Fig. 12.26(a). Also, if an input is *open*, the emitter current of Q_1 is obviously zero, so we see that *an open input is equivalent to a high input*.

Finally, let us consider how low the input must go to turn on Q_1 if the circuit is initially in the state of Fig. 12.26(a). Because the base of Q_1 is at 2.1 V, if *either input goes below about 1.4 V*, then the base-emitter junction of Q_1 will be forward biased and Q_1 will be turned on. We will now show that turning on Q_1 forces the output to go high (+3.6 V).

Consider the state shown in Fig. 12.26(b) in which either or both of the two inputs are low enough (below 1.4 V) to turn on one or both of the base-emitter junctions of Q_1. Q_1 can be considered as a common base amplifier with no phase inversion between the input at the emitter and the output at the collector. Thus, a negative-going input to the emitter of Q_1 will drive the collector negative, which will turn off Q_2. In terms of current, turning on Q_1 will cause the collector current of Q_1 to flow *out* of the base of Q_2, which "robs" Q_2 of its base current, thus turning off Q_2. The emitter current of Q_2 thus drops to zero, which removes the forward bias (the drop across the 1-kΩ resistor) from Q_3, thus turning off Q_3. The voltage at the collector of Q_2 rises to almost $V_{cc} = 5$ V, which turns on Q_4. The base of Q_4 is connected to the collector of Q_2, so $V_{B4} \cong 5$ V. The output voltage is two diode drops (one from V_{BE} of Q_4 and one from the diode) below the base of Q_4, so the output voltage will be 5 V − 1.4 V = 3.6 V.

To sum up, there are only two possible stable output states: (1) If both inputs are high so that Q_1 is off, then Q_3 will be on and Q_4 off, and the output voltage will be low, approximately $V_{CE3\,sat} \cong 0.2$ V. (2) If either input (or both) is low so that Q_1 is on, then Q_3 will be off and Q_4 on, and the output voltage will be high, approximately $V_{cc} - 2V_D \cong 3.6$. This is precisely the situation for a logical NAND relationship between the input and the output if we call a 3.6-V level a logic 1 and a 0.2-V level a logic 0. This is positive logic—the more positive voltage level being 1, and the more negative voltage level being 0.

Notice that Q_2 acts as a "phase splitter" in driving Q_3 and Q_4 because the emitter and collector voltages of Q_2 are always 180° out of phase. If Q_2 is on, it forces Q_3 on and Q_4 off, and if Q_2 is off, it forces Q_3 off and Q_4 on. Q_3 and Q_4 are always in *opposite* conditions in the two stable states: One of them is strongly on (saturated), and the other is off. This is usually called a *totem pole* output because one transistor (Q_4) "sits" on top of the other (Q_3) like the figures of an Indian totem pole.

We recall from Chapter 8 that when a transistor is saturated, it is difficult to turn it off quickly because there is essentially no electric field in the base region. Thus the charge carriers in the base (the electrons in the p-type base of the npn transistors of our NAND gate) have to *diffuse* out of the base, which is a slow process, much slower than their being swept out by an electric field. Thus, the saturated "on" transistor Q_3 is always turned off slower than the "off" transistor Q_4 is turned on. This means that for a brief time interval (approximately several ns) *both* output transistors Q_3 and Q_4 are on. Thus, a large spike of current briefly flows through Q_3 and Q_4. The 130-Ω resistance in the collector of Q_4 limits the amplitude of this current spike. To prevent the V_{cc} supply voltage from dropping abruptly due to the large current spike, we connect a bypass capacitor (typically 0.01 μF or 0.1 μF) from V_{cc} to ground for every two or three chips, or for every chip if the chip draws enough current. These bypass capacitors must have a very low reactance at high frequencies and are usually tantalum or ceramic. They are often called *despiking* capacitors. If the circuit changes state rapidly (e.g., 10^5 or 10^6 times per second), then the average current drawn from the power supply will be significantly increased over the steady-state current (12 mA for a 7400 chip or 1.5 mA for a 74LS00 chip).

Q_4 is essentially an active load for Q_3, thus producing a higher gain, enabling the output voltage to change quickly. There is always some capacitance C_L between the output terminal and ground, and this capacitance must be charged up for the output to go high and discharged for the output to go low. When Q_3 is turned off, Q_4 is turned on, and the load capacitance charges up from V_{cc} through the active transistor Q_4; this is much quicker than charging up C_L through a resistance. To force the output low, Q_3 is turned on and C_L discharges through Q_3 to ground, which again is much quicker than discharging through a resistance. To sum up, the TTL gates are fast because *the output load capacitance is charged and discharged through active transistors* rather than resistors.

A final note about the inputs: When the inputs are both high, Q_1 is off, and the input supplies no current because the emitter-base junction of Q_1 is reverse biased. But when either input is low, the emitter-base junction of Q_1 is forward biased and emitter current flows. In this case Q_1 is on, and Q_2 is off. Both the base-emitter and the base-collector junctions of Q_2 are reverse biased, so no current flows in the base of Q_2 (i.e., in the collector of Q_1), since $I_{B2} = I_{C1}$. Therefore, by the KCL with Q_1 on, $I_{B1} = I_{E1}$, and I_{B1} can be calculated from Ohm's law if we assume the low input is the standard TTL value of 0.2 V. $V_{B1} = 0.9$ V because the base-emitter junction of Q_1 is forward biased, so $I_{B1} = (5\text{ V} - 0.9\text{ V})/4\text{ k}\Omega = 1.025$ mA. Thus, $I_{E1} = 1.025$ mA; in other words, the low input must "sink" 1.025 mA of current when it has a voltage of 0.2 V. Thus, the low input must have a resistance of *no more than* 0.2 V/1.025 mA = 195 Ω (otherwise the input voltage would rise above 0.2 V). This restriction holds for *any* 7400 series TTL input circuit. (For the better 74LS00 TTL gate, a low input must sink only 0.4 mA, so the low input must have a resistance of no more than 0.2 V/0.4 mA = 500 Ω. For any CMOS gate a low input need sink essentially zero current, so a CMOS low input can have a high resistance.)

Any high output will be approximately 3.6 V and can drive a large number of TTL inputs high, because a high TTL input draws essentially no current (perhaps 50 μA). But the high output cannot supply much current to an external load. For example, suppose an LED is connected from the output to ground. The maximum current will flow through the LED when Q_4 is saturated, so $V_{CE4} = 0.2$ V. Thus, if there is 1.7 V across the LED,

$$V_{E4} = 1.7\text{ V} + V_D \cong 2.4\text{ V} \qquad V_{C4} = 2.6\text{ V}$$

and $I_{C4} = (5\text{ V} - 2.6\text{ V})/130\ \Omega = 18$ mA is the maximum output current. This is enough current to light one or two LEDs but no more.

12.7.2 The 7402 TTL NOR Gate

The 7402 TTL NOR gate is one of the most common and useful gates in the TTL family. It is a quad-two-input NOR gate, which means there are four NOR gates (each with two inputs) in one package. Its schematic diagram is shown in Fig. 12.27 for one two-input NOR gate. The two inputs are each connected to the emitter of a transistor in the common base configuration. Thus, there is a symmetry between the two inputs.

If either input goes high, as in Fig. 12.27(a), Q_3 and Q_4 will both be turned on, thus producing a low output. (Q_3 can be thought of as one npn transistor with two bases; if either base goes positive, Q_3 will turn on.) For example, if input A goes high, as in Fig. 12.27(a), Q_1 is turned off and its base-collector junction acts like a forward biased diode, which turns on (the left half of) Q_3. The emitter current of Q_3 flowing through the 1-kΩ resistance then turns on Q_4 strongly by making its base more positive, and thus the output voltage (the collector voltage of Q_4) drops to approximately $V_{CE4\,sat} = 0.2$ V. Q_3 is turned on strongly (saturated), so its collector emitter voltage is only 0.2 V; hence, the collector of Q_3 (the base of Q_5) is at approximately 0.9 V, which is not enough to turn on Q_5. Thus, Q_5 is off, and we have the familiar totem pole situation: Q_5 off, Q_4 on, and $V_{out} = 0.2$ V.

If, however, both inputs A and B are low, as in Fig. 12.27(b), then both Q_1 and Q_2 are conducting and rob (both sides of) Q_3 of base current; in other words, both sides of Q_3 are off. Thus, there is no Q_3 emitter current flowing through the 1-kΩ resistance to turn on Q_4, and thus Q_4 is off. But the collector of Q_3 (the base of Q_5) is now at approximately the positive supply voltage, so Q_5 is turned on strongly and the output is high. The output is approximately two diode drops lower than the base of Q_5 (one for the base-emitter junction of Q_5 and one for the diode); thus the output is at

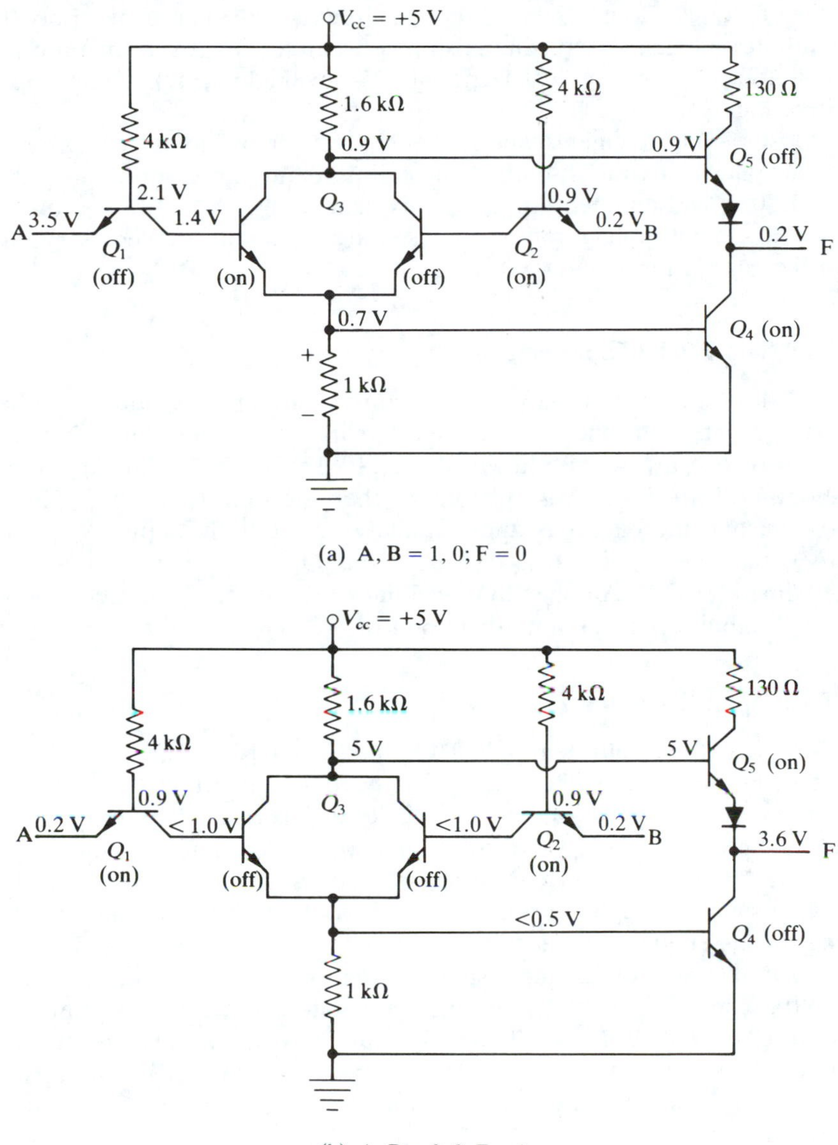

(a) A, B = 1, 0; F = 0

(b) A, B = 0, 0; F = 1

FIGURE 12.27 Two-input 7402 TTL NOR gate.

approximately 5 V − 1.4 V = 3.6 V. Again we have the familiar totem pole situation: Q_5 on, Q_4 off, and V_{out} = 3.6 V.

To sum up, there are only two stable states for the output: (1) If either input is high [Fig. 12.27(a)] so that either Q_1 or Q_2 is off (or both off), then Q_3 is on. Q_4 is then on and Q_5 is off, and the output will be low. (2) If both inputs are low [Fig. 12.27(b)], then both Q_1 and Q_2 are on and both sides of

Q_3 are off. Q_4 is off and Q_5 on, and the output is high. This is precisely the situation for a logical NOR relationship between the input and the output if we call a 3.6-V level a logical 1 and a 0.2-V level a logical 0. Again, this is positive logic.

Most of the comments about the NAND circuit apply to the NOR circuit. Specifically, the low input must sink about 1 mA current (0.4 mA for 74LS02), and an open input is equivalent to a high input. The output has the totem pole configuration, so despiking capacitors are necessary, just as in the case of the NAND gate.

12.7.3 The 7404 TTL Inverter

The 7404 TTL inverter is another common and useful TTL gate. It is a hex inverter, which means there are six inverters in one package. The schematic diagram of one inverter is shown in Fig. 12.28. It has the familiar totem pole output. If the input A is low (0.2 V), then Q_1 is on, Q_2 is off, Q_3 is off, Q_4 is on, and the output is approximately 3.6 V. If the input A is high (3.6 V), then Q_1 is off, Q_2 is on, Q_3 is on, Q_4 is off, and the output is approximately 0.2 V. An open input will mean Q_1 is off, which is equivalent to a high input. Thus the output will be low.

12.7.4 Open Collector TTL

The outputs of several totem pole TTL gates cannot be connected directly together because one of the gates might have a high output while another has a low output, as shown in Fig. 12.29(a). Then the high-output chip would supply a very large current I that would flow into the low-output chip; the two chips would be "fighting" one another, with the high chip trying to supply enough current to keep the output high, and the low chip trying to sink enough current to keep the output low. One solution is to use an *open collector* output gate, such as the NAND gate shown in Fig. 12.29(b). The output of any open collector gate can act *only* as a current sink (through Q_3). When Q_3 is on, the output is sinking current and the output voltage is low; when Q_3 is off, the output is high (+5 V). Notice that the external voltage supply to which R_L is connected can be different from V_{cc} for the gate. R_L is often called a *pullup* resistor.

To illustrate the use of open collector gates, let's construct a circuit to realize the function

$$Y = (\overline{A \cdot B}) \cdot (\overline{C \cdot D})$$

Our first naive solution would be to use two NAND gates and to AND their outputs as shown in Fig. 12.30(a). This would work fine, but a simpler solution (with fewer gates) is possible with open collector NAND gates, as shown in Fig. 12.30(b). Any low output from either the AB or the CD

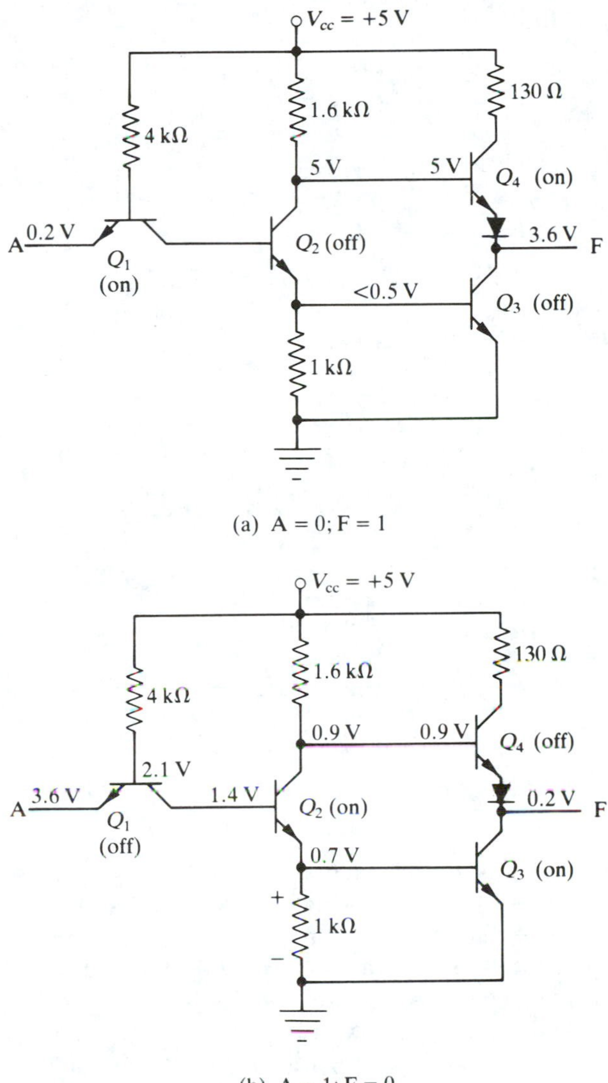

FIGURE 12.28 7404 TTL inverter.

NAND gate will produce a low output by sinking approximately 5 mA. A high output from one NAND gate and a low output from the other NAND gate will also produce a low output, because the high output from an open collector gate is merely an off transistor (Q_3) to ground. Thus a high open collector output does not fight a low open collector output. In other words, if the open collector device is in a high output state, Q_3 is off and acts as a high impedance, which means any other gate tied to the same output can drive the output low. This configuration is sometimes called the *wired OR*

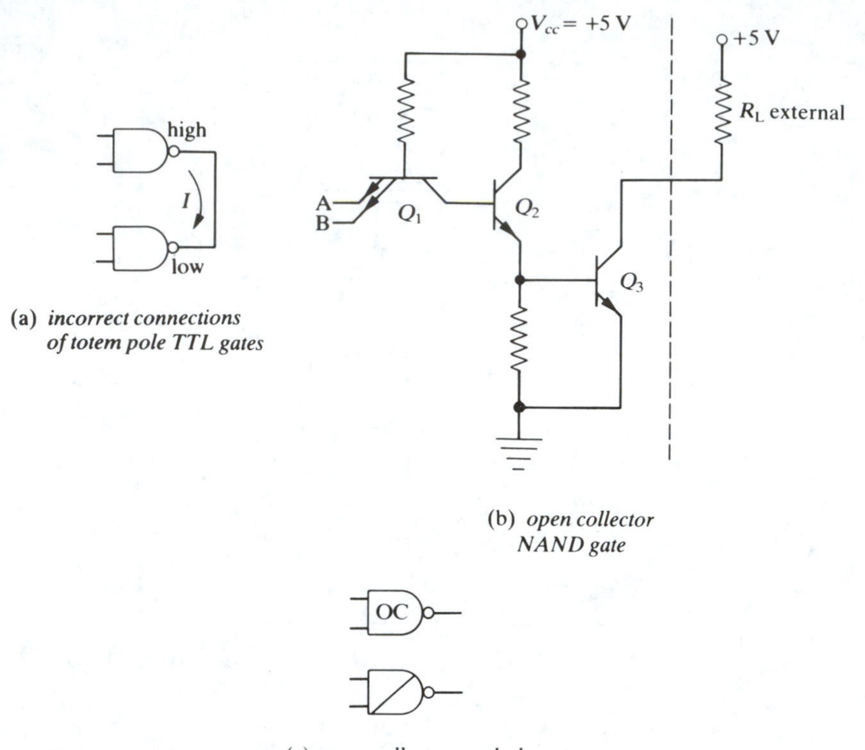

(a) *incorrect connections of totem pole TTL gates*

(b) *open collector NAND gate*

(c) *open collector symbols*

FIGURE 12.29 Open collector TTL NAND gate.

configuration, but this term is somewhat misleading because the overall logic function is AND with positive logic; that is, the output on the common line connecting all the open collector gates is high only when *all* the open collector outputs are high; *wired AND* would be a better name. Two ordinary totem pole NAND gates would *not* work in the circuit of Fig. 12.30(b) because the high and low outputs would fight each other.

By a similar argument, two open collector AND gates (but not totem pole gates) with outputs connected together would implement the function F = A · B · C · D. Notice also that the open collector configuration can be used to produce an output voltage greater or less than the 5-V supply to the open collector gates simply by changing the external voltage connected to the pullup resistor.

12.7.5 Three-State TTL Gates

The open collector gates just described have one disadvantage: They are much slower for a low-to-high output transition than the totem pole TTL gates. Consider what happens when the output goes from low to high. Q_3 is turned off, and the load capacitance C_L must charge to the high output

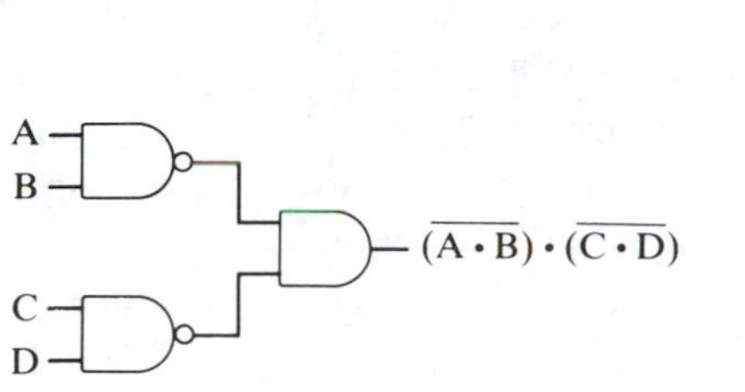

(a) *using conventional TTL gates*

(b) *using open collector NAND gates*

A	B	C	D	$\overline{A \cdot B}$	$\overline{C \cdot D}$	$\overline{(A \cdot B)} \cdot \overline{(C \cdot D)}$
0	0	0	0	1	1	1
0	0	0	1	1	1	1
0	0	1	0	1	1	1
0	0	1	1	1	0	0
0	1	0	0	1	1	1
0	1	0	1	1	1	1
0	1	1	0	1	1	1
0	1	1	1	1	0	0
1	0	0	0	1	1	1
1	0	0	1	1	1	1
1	0	1	0	1	1	1
1	0	1	1	1	0	0
1	1	0	0	0	1	0
1	1	0	1	0	1	0
1	1	1	0	0	1	0
1	1	1	1	0	0	0

(c) *truth table*

FIGURE 12.30 Open collector gate circuits.

voltage. With totem pole TTL the charging path is from V_{cc} through an active transistor Q_4, which has a very low impedance, typically less than 100 Ω, as shown in Fig. 12.31(a). But with an open collector gate, the charging path is from V_{cc} through the external pullup resistance, which is usually 1 kΩ or more, as shown in Fig. 12.31(b). Thus, the charging time constant is much larger for the open collector gate. The external pullup resistance cannot be made much smaller because it would then draw too much current from the power supply when the output is low. Totem pole and open collector gates have the same speed for a high-to-low output transition because in either case a transistor (Q_3) is being turned on and C_L discharges through Q_3 to ground. The three-state gate or tristate® (trademark of National Semiconductor) solves this problem by combining the speed of the totem pole TTL with the advantage of the open collector gate—allowing all the outputs to be connected together if desired. The solution, shown in Fig. 12.31(c), is called a three-state gate. There are three possible output states: (1) output high with Q_3 off and Q_4 on, (2) output low with Q_3 on and Q_4 off, and (3) the disabled state with *both* Q_3 and Q_4 off and the output presenting a high impedance. (This is equivalent to the open collector gate with a high output.) In the circuit of Fig. 12.31(c) the output is made a high impedance by a positive input C ("control" or "enable"), which makes the output $\bar{C}$ of the inverter C low (almost ground), which turns on both diodes D_2 and D_4. Turning on D_2 turns off Q_2 because it

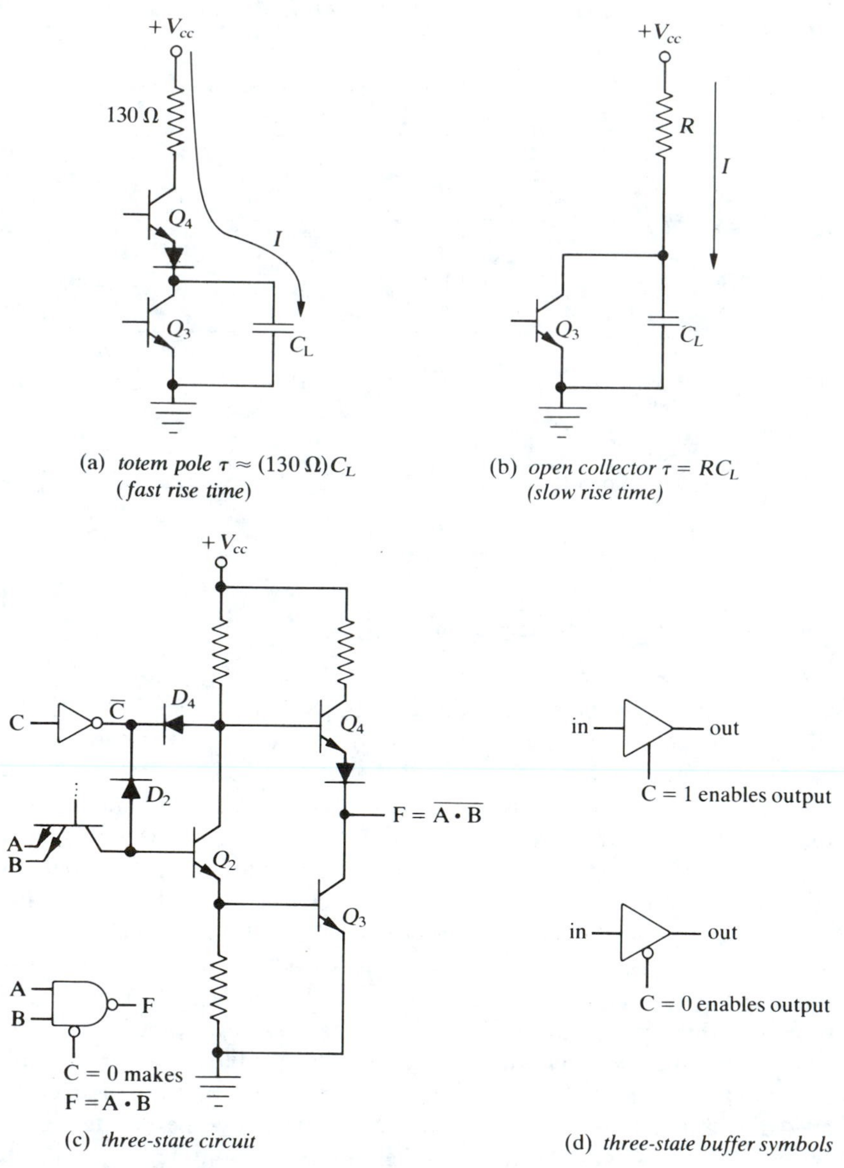

(a) *totem pole* $\tau \approx (130\ \Omega)C_L$ (*fast rise time*)

(b) *open collector* $\tau = RC_L$ (*slow rise time*)

(c) *three-state circuit*

(d) *three-state buffer symbols*

FIGURE 12.31 Three-state gate.

limits the base voltage of Q_2 to a maximum of about 0.7 V, which is not enough to turn on both Q_2 and Q_3. Turning on D_4 turns off Q_4 for the same reason. But when C is low, $\bar{C}$ is high, both diodes are off and the circuit works in the usual way. To sum up, when the control input is low, the gate operates as a normal totem pole gate with its high speed; but when the control input is high, both totem pole output transistors are turned off and the output of the gate is a high impedance to ground.

Three-state gates can be connected like open collector gates, with all their outputs connected together to make a wired AND circuit.

A buffer amplifier is also available in a tristate form; it is called a tristate® or three-state buffer, and its schematic symbol is shown in Fig. 12.31(d). It is basically a solid-state switch for low currents. When the enable or control input is high, the input and output are connected together; but when the enable or control input is low, there is a high impedance between the input and the output.

Open collector gates and three-state gates are often used to drive "buses" in logic systems. A *bus* here means a number of wires used to carry information (in the form of bits) between various parts of the system: for example, from data input lines to a storage memory, from the memory to a computation unit, or from a computation unit to an output device, such as a printer or a cathode tube display. Many such paths exist in digital systems, especially in computers. Thus, to minimize the wiring complexities, we want to use the same bus to carry different sets of information. The use of three-state or open collector gates allows us to do this by connecting three-state outputs from input devices, memory, computation units, and output devices all to the same bus.

Suppose we wish to transfer digital data to a computer memory and have the bus usable for other purposes when the data are not being actually transferred to the memory. One solution to this problem is shown in Fig. 12.32 with open collector gates. With the open collector NAND gates, the output of each gate is connected to one line of the bus, which goes to the computer memory. We assume here for simplicity that the data are in the form of four-bit words or nibbles, $D_3D_2D_1D_0$, and that we therefore have four lines, one for each bit. Each bus line is tied to a positive supply voltage through its own external pullup resistor. Each data bit comes in to one input of a NAND gate, and the other inputs are all tied together to a READ DATA line. If the READ DATA line is low, all the open collector gate outputs are high regardless of the data inputs; that is, each gate presents a relatively high impedance to the bus line, as shown in Fig. 12.32(b). Thus, the bus line can be used for other data transmission—any bus line can be driven low by the output of another open collector TTL gate connected to the same bus line. However, if the READ DATA line is high, then the gate outputs are the complement of the data input, as shown in Fig. 12.32(c). The point is simply that when the READ DATA line is low, all the open collector gate outputs are high (high impedance to ground) and the bus lines are unaffected by the data input to the gates. The bus can then be used

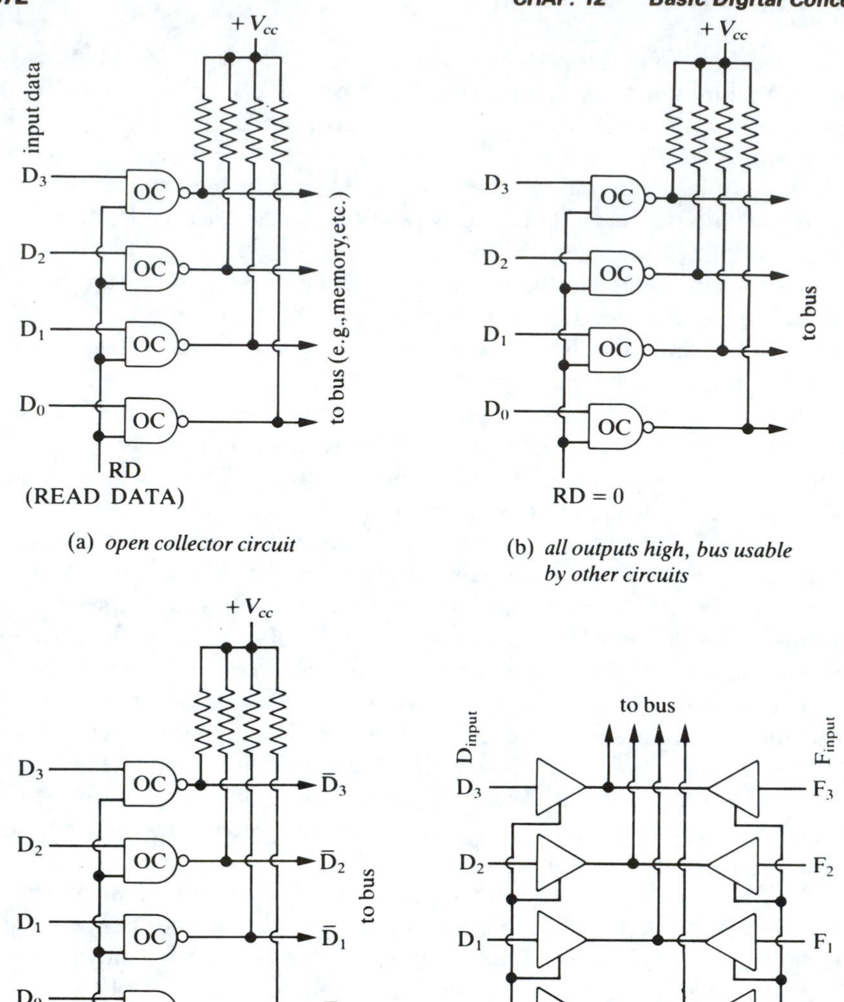

FIGURE 12.32 Open collector and three-state applications.

for transmission of other information. But when the READ DATA line is low, then the complement of the input data is fed onto the bus lines.

The data can also be fed onto the bus through three-state buffers, as shown in Fig. 12.32(d). If the ENABLE DATA (ED) line is low, the buffers act as open switches with a high output impedance and no data gets through to the bus. But if the ED line is high, the data is fed onto the bus. For example, if ED is high and EF is low, the D output will be fed onto the bus; if ED is low and EF is high, the F input will be fed onto the bus.

Three-state outputs are often built right into the output of some gates and more complicated devices, such as flip-flops and A/D converters. For example, microprocessor-compatible devices have three-state outputs that can usually be directly connected to a microcomputer bus.

If the data bus in a computer is bidirectional, then the gates and other circuitry connected to the bus *must* be open collector or three state. For example, connections to the older PDP-8/E and PDP-11 output lines (the *omnibus* and the *unibus*) must be open collector or three state. Most modern computers use three-state outputs.

12.7.6 General Comments on TTL

Inputs: A logic 0 or low input is ideally 0 V, but a typical TTL value is 0.2 V ($V_{CE\,sat}$ of the lower totem-pole transistor). A low input should be less than 0.8 V for reliable TTL operation. Each low input to TTL must sink approximately 1.6 mA (0.4 mA for LS TTL) of current because a low input turns on the input transistor, and the input is connected to the emitter. This 1.6-mA current is the emitter current of the input transistor of the TTL chip. Thus, the maximum impedance for the source of the low input is 0.8 V/1.6 mA = 500 Ω for TTL and 0.8 V/0.4 mA = 2 kΩ for LS TTL.

A logic 1 or high input is ideally the supply voltage (V_{cc} = 5 V for TTL), but a typical value is 3.6 V. A high input should be greater than 2.0 V for reliable TTL operation. The source of the high input need supply very little current, because a high input turns off the input transistor of the TTL chip. A typical current for a high input is 50 μA or less. A floating or unconnected TTL input will be equivalent to a high input because the input transistor is not conducting. Unused inputs should be tied either low (to ground) or high (to V_{cc} through a 1-kΩ resistance for reliable operation). A floating TTL input is not a reliable high because the open input voltage is usually around 1.4 V. Thus a small negative noise spike of 0.6 V or more would produce a low input.

Outputs: A logic 0 or low output is ideally 0 V, but a typical TTL value is 0.2 V, the collector emitter voltage across the saturated output transistor Q_3 between the output and ground. The resistance of this transistor is usually approximately 25 Ω. A low output can sink up to 16 mA in ordinary 7400 TTL (8 mA for LS TTL) and thus can drive ten TTL inputs (20 LS TTL inputs also) to a logical 0 or low level without the output rising above 0.4 V. The term *fanout* refers to how many similar gates can be driven by the output; we say the fanout of standard TTL is 10 (20 for LS TTL). The output of a standard TTL gate can sink up to 16 mA without its output voltage rising above 0.4 V. A TTL output can be shorted to ground but *not* to V_{cc}, because that would burn out the bottom transistor Q_3 in the totem pole output.

A logic 1 or high output is ideally V_{cc} = 5 V for TTL, but a typical value is 3.6 V, or approximately two diode drops below the supply voltage. The positive supply voltage is connected to the high output through a

130-Ω resistance, a saturated output transistor, and a conducting diode. Because a high input to a TTL gate requires almost zero driving current, a high TTL output can drive a large number of TTL gates high. The fanout limitation occurs when the output is low. Special TTL gates called *buffers* are available with extremely high fanouts to drive a large number of gates. The 7437 NAND gate, for example, has a fanout of 30. Fanout capability usually is not a problem in small systems, but it can occur if one chip is used to reset a large number of chips by driving their inputs low.

The *noise margin* is a measure of how much the signal voltage must change in order to transform a logical 0 to a logical 1 as interpreted by the TTL gates—the larger the noise margin, the better. TTL manufacturers guarantee that any input about +2.0 V will be interpreted by the gate as a good high input, and any input below 0.8 V will be interpreted as a good low input. They also guarantee that the minimum high TTL output from the gate will be +2.4 V, and that the maximum low TTL output will be +0.4 V. The noise margin for these worst cases is thus 0.4 V.

A typical high TTL output is +3.6 V, so a −2.8-V noise spike would be required to change it to a "sure" low input of 0.8 V for another TTL gate. Similarly, a typical low TTL output is +0.2 V, so a +1.8-V noise spike would be required to change it to a "sure" high input of 2.0 V for another TTL gate.

Notice that a high input held at +5 V has a very large noise margin; a −4.2-V noise spike would be necessary to change it to a "sure" low input of 0.8 V. On the other hand, a low input held at ground would need only a +2.0-V noise spike to be interpreted as a "sure" high input of 2.0 V.

This is one reason TTL circuits are usually designed with active *low* inputs (ENABLE, RESET, etc.). With an active *low* input the normal (unused) input condition is thus high—5 V, obtained by connecting the input pin to +5 V through a 1-kΩ resistor. Then a huge (and improbable) −4.2-V noise spike on the input would be required to enable the input accidentally, whereas if the input were active high, it normally would be near ground, obtained by connecting a 1-kΩ resistor to ground (you can't hard-wire it directly to ground because it would then be low *permanently*), and only a +2.0-V noise spike could accidentally enable it.

Active low inputs are also used because they save power. A pullup resistor to +5 V on a TTL input draws very low current (50 μA or less), but a pulldown resistor to ground would draw 1.4 mA for TTL or 0.4 mA for LS TTL.

If a low input drifts more positive, it would probably be interpreted as a high when it reaches approximately 1.4 V. But voltage levels between 0.4 V and 2.4 V should be avoided like the plague!

Consider a TTL gate with a high output; the bottom transistor Q_3 in the totem pole output is off. If the ground voltage goes negative from a noise spike, Q_3 may be turned on, thus producing a false low gate output. A negative spike of only −0.5 V on the ground lead can produce a false low output, so a reliable noise-free ground is absolutely essential.

A final comment concerns the V_{cc} power supply: All TTL gates are designed to run off $V_{cc} = 5\text{ V} \pm 0.25\text{ V}$. If V_{cc} rises to 7 or 8 V (for example, by short-circuit failure of the voltage regulator chip), all the TTL gates may be permanently destroyed in milliseconds. The 1-kΩ resistor usually used between a TTL input kept high and $+V_{cc}$ protects the chip in case V_{cc} increases.

12.7.7 Schottky and Low-Power Schottky TTL

The principal speed limitation in TTL logic is the delay caused in turning off the saturated "on" transistors. A saturated "on" transistor cannot be turned off until all the minority charges in the base region (electrons in the p-type base for the usual npn transistors) have *diffused* out of the base region, because in a saturated transistor there is essentially zero electric field in the base. In the Schottky and the low-power Schottky families all the transistors are prevented from ever saturating by connecting a Schottky diode between the base and the collector, as shown in Fig. 12.33(a). We recall that a Schottky diode turns on at approximately 0.3 V, and the transistor collector base voltage must drop to approximately 0.2 V for the transistor to saturate. Thus, the Schottky diode turns on before the transistor can saturate. In other words, the excess base current necessary to saturate the transistor is diverted around the transistor through the diode when the diode turns on, thereby preventing transistor saturation.

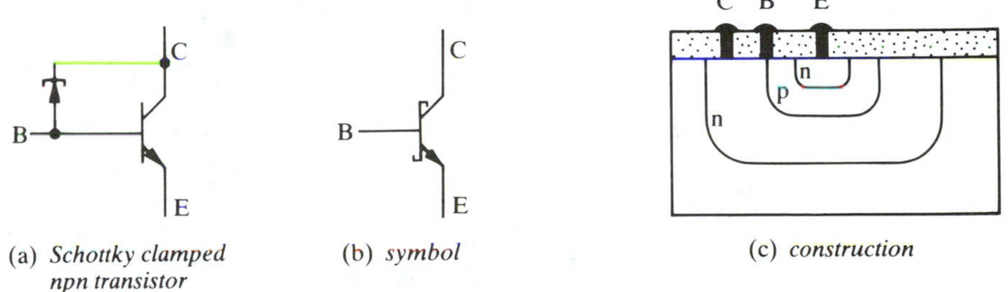

(a) *Schottky clamped npn transistor* (b) *symbol* (c) *construction*

FIGURE 12.33 Schottky transistors.

It is very easy to add a Schottky diode to an npn transistor, as shown in Fig. 12.33(c). The aluminum contact to the base is simply extended to cover part of the n-type collector. The aluminum-base junction acts like a Schottky diode.

The low-power Schottky TTL series gates draw approximately one fifth of the power of standard TTL gates and have essentially the same speed and price. Thus the LS series TTL gates have replaced the standard TTL gates in all new TTL designs.

12.7.8 CMOS

CMOS stands for complementary metal-oxide semiconductor. CMOS gates use much less power than TTL gates but are much slower. CMOS circuits use both n-channel and p-channel enhancement MOSFETs in "complementary" pairs. Extremely low power consumption is achieved because power is consumed only when the circuit is switching states—not while the circuit is in the steady state. Other advantages include larger noise margin, relative immunity to temperature fluctuations, operation from +3 to 18 V voltage supply, and a large fanout capability.

We recall that junction FETs or JFETs *always* must have the gate-channel junction *reverse* biased; if it is forward biased, the gate input will draw current and will be a low impedance. But with MOSFETs the gate is separated from the channel by an insulating layer of silicon dioxide, so the gate-channel voltage can be of *either* polarity and the gate input *always* has an extremely high input impedance. Plots of the drain current versus gate–source voltage for all four devices are shown in Fig. 12.34. Note that

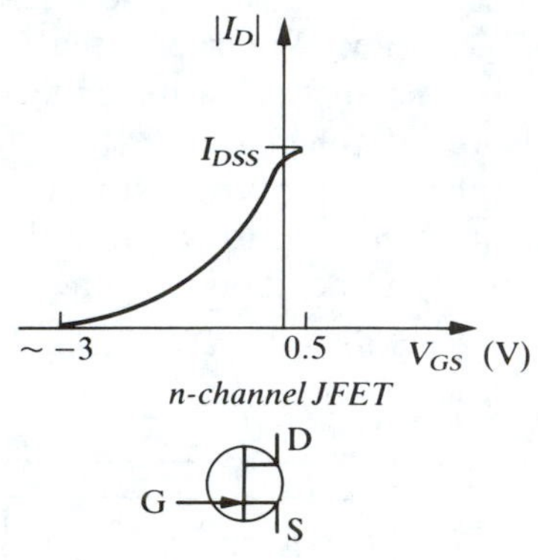

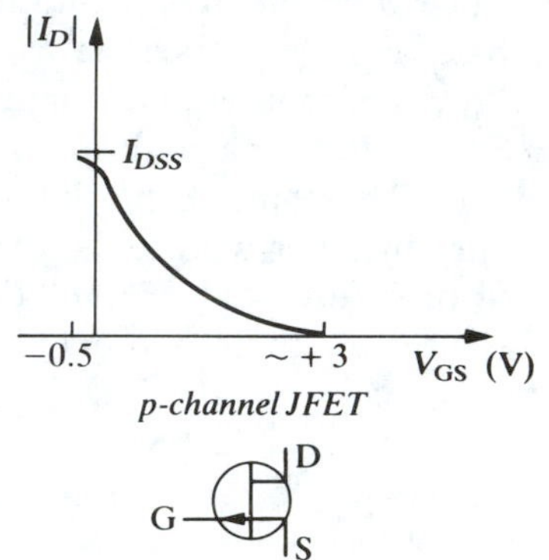

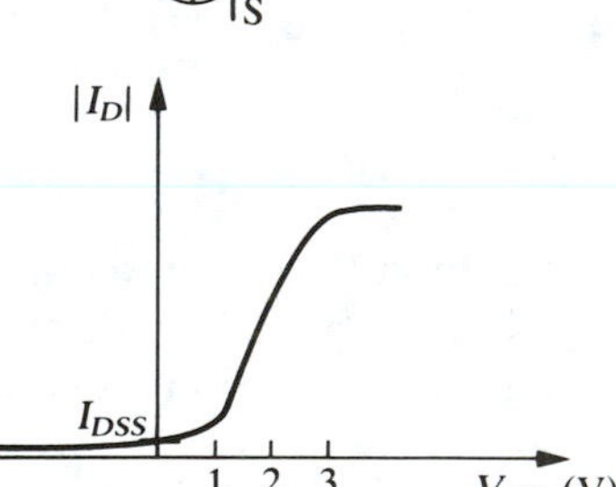

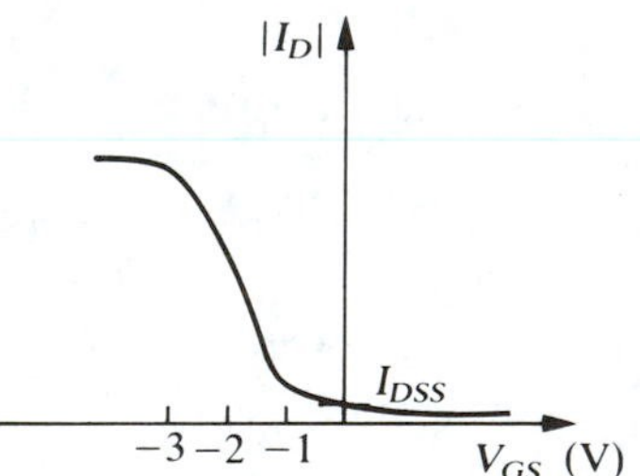

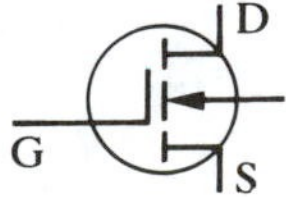

FIGURE 12.34 JFET and MOSFET curves.

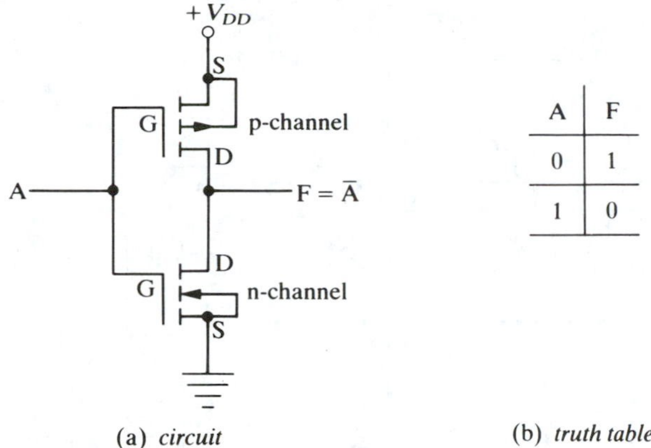

A	F
0	1
1	0

(a) *circuit* (b) *truth table*

FIGURE 12.35 CMOS inverter.

both types of enhancement MOSFETs need a nonzero gate–source voltage in order to conduct.

The basic operation of such complementary symmetry MOSFETs can be seen by considering the inverter circuit of Fig. 12.35. Notice that the output is taken from the two drains, which are connected together, and that the input is applied to the two gates, which are also connected together. Because the two enhancement MOSFETs are different types, one n-channel and one p-channel, it is impossible to forward bias *both* simultaneously so as to draw current from the power supply. Thus, no steady-state power is consumed (actually approximately 10 nW from leakage current) regardless of the logic state of the input. If the input is low (near 0 V), then the lower n-channel MOSFET is not conducting, but the upper p-channel MOSFET is conducting because its gate is negative with respect to its source. Thus, the output terminal is connected to the positive supply voltage through the upper (on, low-resistance) MOSFET, and the output is high, approximately the positive supply voltage. But, on the other hand, if the input is high (near the positive supply voltage), then the lower n-channel MOSFET is conducting because its gate is positive with respect to its source. The upper p-channel MOSFET is not conducting because its gate and source are at the same voltage. Thus, the output terminal is connected to ground through the lower (on, low-resistance) MOSFET, and the output is low, approximately 0 V.

A CMOS NAND gate and its truth table are shown in Fig. 12.36(a). Both Q_3 and Q_4 must be on to drive the output low; this output will be low only when inputs A and B are each high (near V_{cc}), which also turns off both Q_1 and Q_2. If both inputs are low (near ground), then Q_3 and Q_4 are both off and Q_1 and Q_2 are both on; thus the output is high. If A is high and B is low, then Q_1 is off and Q_3 is on from A, Q_2 is on, and Q_4 is off from B. The output is high. If A is low and B is high, then Q_1 is on and Q_3 is off from A. Thus Q_2 is off, and Q_4 is on from B, which makes the output high. A CMOS NOR gate is shown in Fig. 12.36(b).

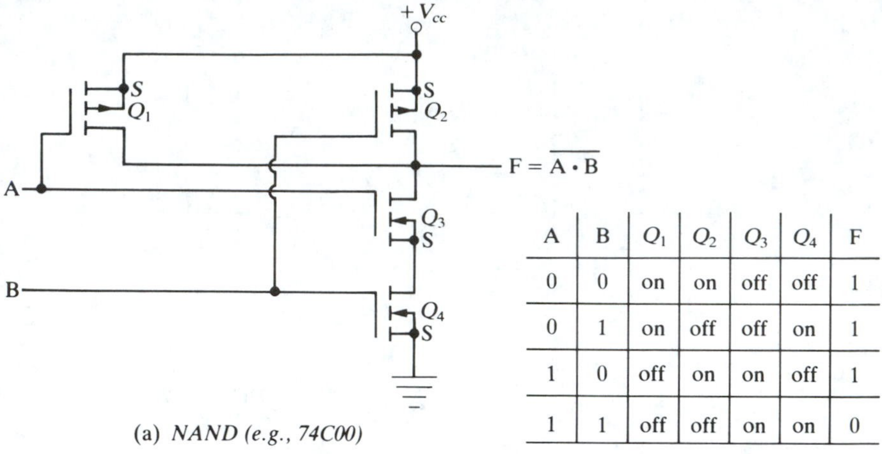

A	B	Q_1	Q_2	Q_3	Q_4	F
0	0	on	on	off	off	1
0	1	on	off	off	on	1
1	0	off	on	on	off	1
1	1	off	off	on	on	0

(a) *NAND (e.g., 74C00)*

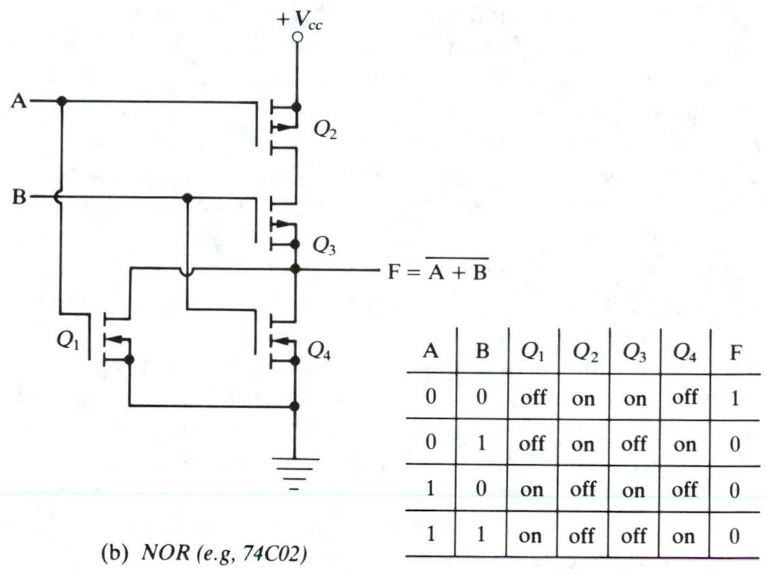

A	B	Q_1	Q_2	Q_3	Q_4	F
0	0	off	on	on	off	1
0	1	off	on	off	on	0
1	0	on	off	on	off	0
1	1	on	off	off	on	0

(b) *NOR (e.g, 74C02)*

FIGURE 12.36 CMOS NAND and NOR gates.

The following comments apply to all CMOS gates. A high state means a voltage level near the positive supply voltage V_{cc}, and a low state means a voltage near ground. The boundary between high and low is approximately *one-half* the supply voltage. The supply voltage can be from 3 to 18 V, with the speed increasing with increasing supply voltage. The input impedance is typically $10^{12}\,\Omega$ in parallel with 5 pF, and the input current drawn is essentially zero (10^{-11} A typically) because all inputs go to the gates of MOSFETs, which are insulated from the rest of the circuit by the silicon dioxide layer between the gate and the channel of the MOSFET. The high output state means the output is connected to the positive supply voltage through an "on" MOSFET, and a low output state means the output is connected to ground through two "on" MOSFETs in series. The minimum high input should be approximately $0.7V_{DD}$; the maximum low, $0.3V_{DD}$ for reliable operation, although $0.5V_{DD}$ marks the difference between high and low.

Unused inputs *must* be connected either to ground or to V_{cc}; they should not be left open. Because of their high impedance, the voltage of an open input may drift up and down, producing random 0 and 1 inputs. If an entire gate in a package is not used, *all* of its terminals should be grounded.

Because the input terminals have such a high impedance (high resistance and low capacitance) to ground, they are susceptible to damage from the buildup of static charge. The breakdown voltage of the SiO_2 CMOS gate is approximately 70 V, and once broken down the CMOS device is usually permanently destroyed. Most CMOS gates have several input protection diodes built into their inputs, as shown in Fig. 12.37(a). The

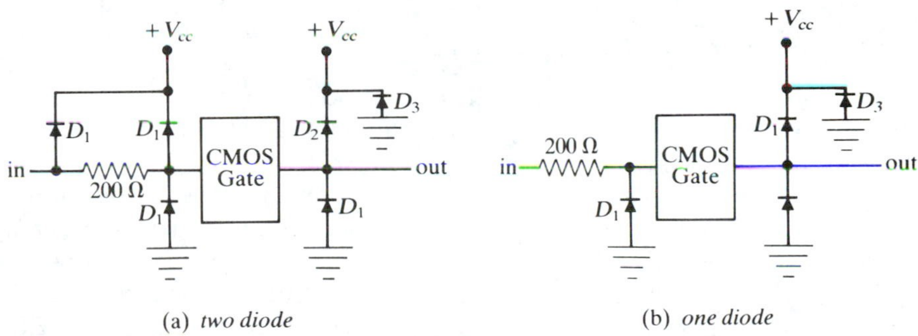

FIGURE 12.37 CMOS protection circuits.

breakdown voltage is usually 25 V for the D_1 diodes, 60 V for D_2, and 100 V for D_3. These diodes will conduct whenever the input voltage is more than one diode drop above V_{cc} or more than one diode drop below ground. It should be pointed out that some CMOS gates such as the 4049 inverter and the 4050 buffer contain only one input protection diode, as shown in Fig. 12.37(b).

Finally, although the steady-state power consumption is essentially zero because there is always a nonconducting enhancement MOSFET between V_{cc} and ground, a surge or spike of current will be drawn from the supply when the circuit is switching from one state to another, because for a brief time interval both of the MOSFETs are on: the longer the input rise time, the longer the current spike lasts and the greater the power dissipated. For moderate switching frequencies this represents a very low power consumption on the order of microwatts, but as the switching frequency goes up the average power consumption also goes up. Each time the output changes from 0 to 1, the output voltage changes from 0 to V_{cc}, and the output capacitance to ground must charge up to V_{cc}, requiring energy $CV_{cc}^2/2$. Thus the "switching" or "ac" power consumption is proportional to fCV_{cc}^2, where f is the frequency.

The fanout of CMOS gates is large because of the small input currents; one CMOS gate can typically drive 50 other CMOS gates.

The standard CMOS chips are numbered in the 4000 series; for example, 4011 is a quad-two-input NAND gate, 4001 is a quad-two-input NOR gate, and 4049 is a hex inverter. The 54C or 74C series is also CMOS and is typically 50% faster than the 4000 series; for example, a 74C00 is a quad-two-input NAND gate, and so on. Three-state CMOS gates are available, as in TTL. There are no open collector CMOS gates as there are in TTL, but the CMOS family does contain "analog switches" (which can conduct current in either direction), which are unavailable in TTL. One such analog switch is the 4051, which can be turned on or off by a voltage input that need supply essentially no current. Also, the switch input and output terminals can be interchanged just as for a mechanical switch.

12.7.9 Emitter-Coupled Logic (ECL)

The ECL family is the fastest, with propagation delays from 0.75 to 4 ns and operating frequencies up to several hundred megahertz. It is based on transistors in the common base configuration with the outputs taken off the emitters, hence the name. The transistors are prevented from ever saturating. The ECL family has a high fanout and operates from a negative supply voltage, $V_{ee} = -5.2$ V or -4.5 V. Its principal disadvantage is that the difference between the two logic levels is very small: A logic low is -1.7 V, and a logic high is -0.9 V. Many different gates are available; the ECL family chips are numbered in the 10000 and the 100000 series.

12.7.10 High-Speed (HC) CMOS Family

The newest version of CMOS is the high-speed CMOS family or *HC* series. A high-speed CMOS quad NAND gate would be denoted 74HC00 and is logically equivalent to a 74LS00 or 7400 chip. The 74HCXX series combines the speed of the 74LSXX low-power Schottky TTL family with the low power consumption of the 74CXX CMOS or 4000 CMOS families. At this writing, as many chip types are available in the HC family as in either the LS TTL or TTL families. Thus most new designs use the HC family.

The high speed of the HC family is made possible by the smaller size of the active area of the chip. Supply voltages can range from 2 to 6 V; 5 V is recommended. Both 54HCXX (military, −55° to 125°C) and 74HCXX (commercial, −40° to 85°C) versions are available. The 74HCXX noise margin is not quite as good as for standard CMOS: minimum 3.15 V for a high input, and maximum 0.90 V for a low input.

The minimum high output is (V_{cc} − 0.1 V), and the maximum low output is 0.10 V. HC D flip-flops can be clocked up to 30 MHz with a 15-pF load and 20 MHz with a 50-pF load, and the clock-to-Q output time is only approximately 25 ns.

As with any chip, the junction temperature T_J inside the chip active material is given by

$$T_J = T_A + P_D\theta$$

where T_A = the ambient temperature, P_D = the power dissipation in watts, and θ = the thermal resistance in degrees Celsius per watt of the chip. $\theta \cong 130°C/W$ for most HC chips in a plastic DIP, and $\theta \leqq 100°C/W$ for a ceramic DIP.

A few precautions are in order for (HC) CMOS.

1. The input signal should have a transition time of less than 500 ns (from $0.1V_{cc}$ to $0.9V_{cc}$).
2. Unused inputs should be connected to V_{cc} or to ground.
3. The V_{cc} power supply leads should be well bypassed.
4. Check the flip-flop "hold time" requirement—the data must remain stable for a certain number of ns even after the active edge.
5. The minimum high input required is approximately 3.1 to 3.5 V (compared to 2.0 V for TTL). This may be a problem when driving 74HCXX with 74LSXX if extra output current is drawn from the 74LSXX output. (The CMOS input requires only nA or less.) A pullup resistor may be necessary on the 74LSXX output or a driver/buffer between the 74LSXX output and the 74HCXX input.
6. As we can see from Fig. 12.37, the output voltage should never be more positive than V_{cc} plus one diode drop, or else the output diode will conduct. Similarly the output voltage should never be more negative than one diode drop below ground.

To avoid burning out either regular 74CXX CMOS or 74HCXX high-speed CMOS chips, observe the following rules:

1. Store in conducting (black) foam, never white foam.
2. Bare chips should be laid leads down on a *conducting* surface (metal).
3. Ground soldering iron tips.
4. Turn off the power *before* either inserting or removing chips.
5. Be sure unused inputs are either grounded or wired to V_{cc}. (Be careful of the high impedance "off" output of three-state outputs.)
6. In an extremely "noisy" electrical environment a 10-kΩ to 100-kΩ resistance in series with the input provides considerable protection at the expense of slightly lower speed.

12.7.11 Power Dissipation in CMOS Chips

The power dissipation is due to the steady dc leakage current (typically 1 nA or less) drawn from the V_{cc} supply plus the transient dissipation due to charging and discharging the internal and external capacitances as the circuit changes output states. The transient power dissipation depends linearly on the switching or signal frequencies, the rise and fall times of the input, V_{cc}^n ($n \cong 2$), and the internal and external capacitances.

It can easily be shown that charging up a capacitance C to a voltage V_{cc} dissipates $W_R = CV_{cc}^2/2$ energy in the charging resistance. The energy stored in C is also $CV_{cc}^2/2$ of course. Thus in charging C through R_L to a voltage V_{cc}, half the energy supplied by the input is stored in the electric field of the capacitance and half is converted into heat in R_L.

In discharging C through R_L the $CV_{cc}^2/2$ energy stored in the capacitance is all dissipated as heat in R_L. Thus, for each low-high-low transition, $2 \times CV_{cc}^2/2$ energy is dissipated as heat in R. Thus for a frequency f the power dissipated as heat in charging and discharging C is

$$P = f \times CV_{cc}^2$$

Table 12.12 contains a summary of the basic specifications of the various logic families.

12.8 INTERFACING

Connecting digital gates of different families to one another and to the "outside world," such as switches or lights, presents some special problems. These problems arise because of the different current and voltage requirements for the high and low logic states for the different families and the special requirements of the outside-world devices. For example, a low LS TTL input must sink approximately 0.4 mA of current, and a high TTL input need supply only $\approx$50 μA of current. Either a high or a low input to a CMOS gate need supply negligible current, but the CMOS high-voltage level can be substantially higher than for TTL. The LS TTL and CMOS input and output characteristics are given in Table 12.12.

12.8.1 CMOS to TTL

If a CMOS gate output is to drive a TTL input, the main difficulty is that the low CMOS output cannot sink enough current for the TTL input. This occurs because the low CMOS output is connected to ground through an "on" MOSFET that may have a resistance of up to 500 Ω. Thus, if the 1.6-mA sinking current for a good TTL low input flows into the CMOS output, the CMOS output voltage will rise to approximately (1.6 mA) × (500 Ω) = 0.8 V, which is the *maximum* voltage for a good low TTL input.

TABLE 12.12. Logic Gate Family Specifications

Family	*Propagation Delay (ns)*	*Max. Flip-Flop Frequency (MHz)*	*Power/Gate (mW)*	V_{cc} (V)	*Input Logic High* (V)	*Input Logic Low* (V)	*Output Logic High* (V)	*Output Logic Low* (V)	*Input* I_{sink} *for Low (mA)*	*Input* I_{source} *for High* (μA)	I_{out} *Source for High (mA)*	I_{out} *Sink for Low (mA)*	*Fanout*
Standard TTL (7400)	10	35	15	5	>2.0	<0.8	3.5 typ	0.2 typ	1.6	≈10	~2	16	10
Low-power Schottky (74LS00)	9.0	33 (74LS74)	2	5	>2.0	>0.8	>2.7 3.5 typ	<0.5 0.2 typ	0.4	20	0.4	8.0	20
Fast TTL (74F00)	3.5	125 (74F74)	5.5	5	>2.0	<0.8	>2.7	<0.5	0.6	20	1.0	20	33
CMOS (74C00)	25 @ 10 V 50 @ 5 V	10 @ 10 V 3.5 @ 5 V (74C74)	0.01	5 to 15	>3.5!!		V_{cc}	0	~0.01	~10	1.75	0.5	50
High-speed CMOS (74HC00)	8.0	40 (74HC74)	0.01	2 to 6	>3.5!!	<1.0	(V_{cc} − 0.1 V)	0.1	0.001	1	4	4	50 or 10 LS TTL loads
ECL	2	250	25	−5.2	>1.105	<1.475	~−0.885	~−1.750			~20		25
100 k ECL	0.75	500	40	−4.5	>−1.165	<1.475	~−0.955	~−1.705			~50		50-Ω line

In other words, an ordinary CMOS low output cannot reliably drive a standard TTL input. Because a 74LS00 series TTL input must sink only about 0.4 mA, an ordinary CMOS gate can drive several low-power Schottky TTL gates. The high CMOS output of 5 V is fine for supplying a high TTL input.

The solution is to use a CMOS buffer (with a 5-V supply) between the ordinary CMOS gates and the TTL gates, to be driven as shown in Fig. 12.38(a). The 4049 CMOS hex inverter, for example, will drive two 7400

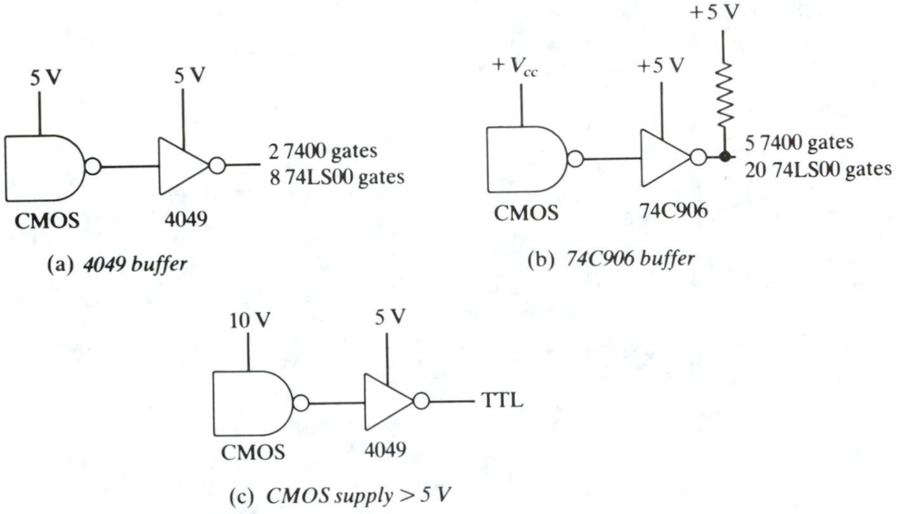

5 V
5 V
2 7400 gates
8 74LS00 gates
CMOS
4049

(a) *4049 buffer*

+5 V
$+V_{cc}$
+5 V
5 7400 gates
20 74LS00 gates
CMOS
74C906

(b) *74C906 buffer*

10 V
5 V
TTL
CMOS
4049

(c) *CMOS supply > 5 V*

FIGURE 12.38 CMOS to TTL interfacing.

series TTL gates or eight 74LS00 series gates. If a larger fanout is required, the 74C906/907 CMOS open drain buffer can be used. This is analogous to the open collector TTL gate and requires an external pullup resistor, as shown in Fig. 12.38(b). The 74C906 can sink 8 mA when operated on $V_{cc} = 5$ V.

If the CMOS power supply voltage is greater than 5 V, the high CMOS output voltage level must be reduced to avoid destroying the TTL input. The solution is to use a 4049 or a 4050 CMOS buffer running off a 5-V supply, as shown in Fig. 12.38(c). Unlike TTL gates, the CMOS input of the buffer is not destroyed if the input voltage is higher than the buffer supply voltage. The CMOS buffer fanout for a 4049 or a 4050 is again two 7400 series gates or eight 74LS00 series gates.

12.8.2 TTL to CMOS

If a TTL gate is used to drive a CMOS gate, the main difficulty is that the 3.5-V TTL output high is barely high enough for a good high CMOS input, which really should be 4 V or more if the CMOS supply is 5 V. (The

74HC00 high-speed CMOS series typically uses a 5-V supply.) If the CMOS supply is 10 V, then a good high CMOS input is about 8 V, and the 3.5-V TTL high is far too low.

For 5-V CMOS the solution is to use an external pullup resistor to 5 V on the output of the TTL, as shown in Fig. 12.39(a). This makes the high

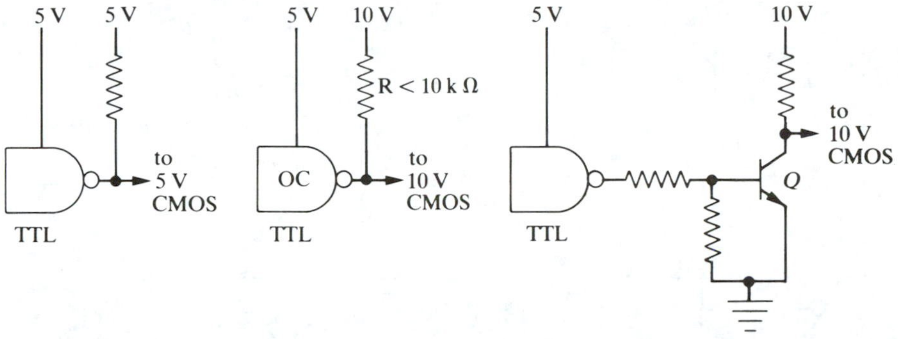

(a) *pullup resistor* (b) *open collector TTL* (c) *npn transistor driver*

FIGURE 12.39 TTL to CMOS interfacing.

TTL output 5 V. Either standard TTL or open collector TTL gates can be used in this way. For 10-V CMOS an open collector TTL gate can be used with the external pullup resistor going to a +10-V supply, as shown in Fig. 12.39(b). This makes the high TTL output 10 V. To get a good high output voltage equal to the external supply voltage, the pullup resistance should be small compared to the "off" resistance of the lower transistor in the output totem pole of the gate; this requirement is easy—any value under 10 kΩ is fine. To get a good low output voltage, the pullup resistance should be larger than the "on" resistance of the lower transistor, which is several hundred ohms at most. The larger R is, the larger the rise time will be, so values from 1 to 3 kΩ are used. Or an npn transistor can be used as shown in Fig. 12.39(c). If the TTL output is low, then Q is off and the input to the CMOS input is high (about 10 V). If the TTL output is high, then Q is on (saturated) and the CMOS input is low (about 0.2 V). Notice that the transistor inverts the output of the TTL gate; in other words, the transistor acts as an inverter.

12.8.3 TTL to Outside World

Because a TTL output low can sink 16 mA (7400 series) or 4 mA (74LS00 series) and a TTL output high can only supply 1 or 2 mA, it is generally best to turn on an outside-world device such as a lamp or a relay with a *low* TTL output. A TTL gate can turn on an LED, for example, with the circuit of Fig. 12.40(a). When the LED is on, the voltage drop across its terminals

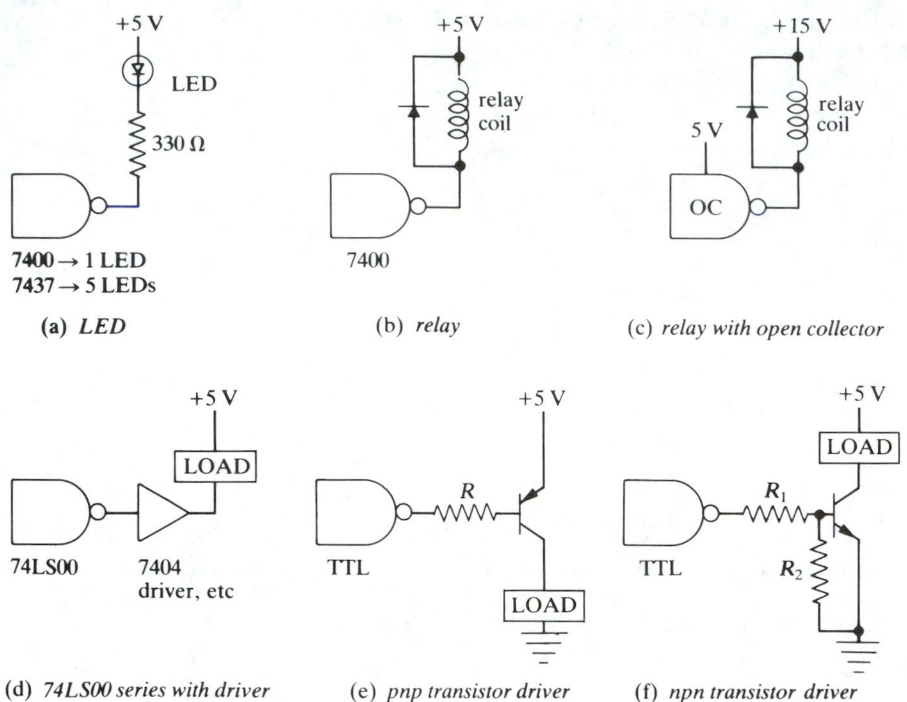

FIGURE 12.40 TTL to outside world devices.

is about 1.7 V, and its brightness depends on its current: a 10-mA current produces acceptable brightness. We shall consider a 10-mA current sufficient for the LED to be on, although 1 or 2 mA will produce a dimly lit LED that can be seen indoors but not in bright sunlight. The resistance in series with the LED can be calculated from Ohm's law: Assume the output low is the typical 0.2 V; then $R = (5\text{ V} - 1.7\text{ V} - 0.2\text{ V})/10\text{ mA} = 310\ \Omega$. If R is lower, the current is higher and the LED is slightly brighter; if R is larger, the current is lower and the LED is dimmer. Notice that because the TTL low output can only sink 16 mA, only *one* LED can be brightly lit. And a low-power Schottky TTL gate cannot light even one LED brightly because it can sink only 4 mA. If more sinking current capability is required (to light more than one LED, for example), then use a different gate such as the 7437 NAND buffer/clock driver, which can sink up to 48 mA, or else use an open collector gate such as the 7406 hex driver, which can sink up to 30 mA.

If a relay is to be driven by a TTL gate, it can be connected as shown in Fig. 12.40(b) provided the gate can sink the current required to turn on the relay. Small relays are available in a DIP package that operate on 5 V and draw only about 10 mA (the relay coil resistance is about 500 Ω). A standard TTL gate can operate one such relay, but a diode must be connected across the relay coil to prevent excessive ringing, which otherwise would occur when the current changed suddenly through the inductance of the relay coil. Without the diode the ringing can produce voltages larger than V_{cc}, which could destroy the gate. When the output of the TTL gate goes low, the relay draws current and the relay switch flips. Notice that a large current can either be turned on or off this way, depending on how the relay switch is connected. If the relay operates off a voltage higher than the 5-V TTL supply voltage, then an open collector TTL gate can be used as shown in Fig. 12.40(c), but the gate output must be able to withstand the higher supply voltage. Again, the diode is necessary to prevent ringing.

The low-power Schottky series of TTL gates can sink only up to 4 mA, so a higher current driver of some kind must be used between the LS gate and the load, as shown in Fig. 12.40(d). The 7404, 7406, or 7437 can be used.

To drive higher current loads (e.g., a large lamp) from TTL gates, we can use an external transistor, as shown in Fig. 12.40(e). The low TTL output will turn on the pnp transistor whose collector current is the load current. The resistance between the gate output and the transistor base is necessary because when the transistor is on, its base voltage will be about 4.4 V, which is considerably higher than the low gate output of only 0.2 V. Thus, the base current flowing through R must produce a drop of 4.2 V. If the load current with the transistor on is 100 mA, for example, and the transistor $h_{FE} = 50$, then the base current will be $I_B = I_C/h_{FE} = 100\text{ mA}/50 = 2\text{ mA}$. Thus, from Ohm's law $R = 4.2\text{ V}/2\text{ mA} = 2.1\text{ k}\Omega$. If R is larger than this value, the base current will be less and the load current will be too small. If R is smaller, the base and load currents will be larger, but too large a base current might exceed the current sinking capability of the low TTL output.

An npn transistor can also be used, as shown in Fig. 12.40(f). A high gate output will turn the transistor on, and current will flow through the load. The two resistances can be calculated when the load current is known. Notice, however, that the transistor base current can never be greater than the maximum current the high TTL output can supply, usually about 2 mA. Thus, for a transistor dc current gain of $h_{FE} = 50$, the maximum load current will be 100 mA. For such a case $R_1 = (3.5\text{ V} - 0.6\text{ V})/2\text{ mA} = 1.45\text{ k}\Omega$. The resistance R_2 is necessary for noise immunity. The low TTL output might rise up to 0.8 V, which would turn on the transistor if R_2 were much larger than R_1. Usually, $R_2 \cong R_1$, which robs the transistor of some base current drive but provides better noise immunity. A larger current load could be driven by using a Darlington configuration or a super beta transistor with exceptionally high dc current gain.

Large ac current loads can also be driven with a solid-state relay turned on by a TTL output. The solid-state relay is basically an optically coupled triac that can switch up to 40 A at 115 V ac. Whenever large currents are switched on and off suddenly, large voltage transients and ringing oscillations are produced that can often create havoc in adjacent circuitry. One of the best ways to prevent this is to use a *zero-crossing* relay to switch the currents. Such relays essentially wait until the 115-V ac current reaches zero before they open or close, thereby minimizing the transient effects.

12.8.4 CMOS to Outside World

A CMOS high output can supply or "source" only about 1 mA of current because the positive supply voltage is connected to the high output through an "on" MOSFET whose resistance is about 500 Ω. Thus, if 2 mA is drawn from a high CMOS output, the voltage will drop from 5 to 4 V. This severely limits what outside-world loads can be driven. A low CMOS output can sink only about 1 mA of current because the low output is connected to ground through an "on" MOSFET. Thus CMOS outputs can operate outside-world loads only if they draw or sink less than approximately 1 mA; this means, for example, that a CMOS output cannot drive even a single LED. It can turn on another MOSFET or a small transistor but not a large power transistor. The usual solution is to use a CMOS buffer such as the 4049 or 4050, which can source and sink more current, as shown in Fig. 12.41(a). The current that can be sourced or

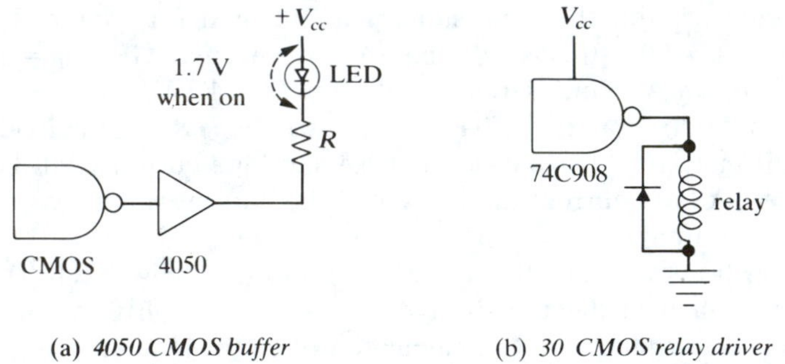

(a) *4050 CMOS buffer* (b) *30 CMOS relay driver*

FIGURE 12.41 CMOS to outside world devices.

sinked depends strongly on the supply voltage, the currents increasing with increasing supply voltage. For example, the 4050 hex noninverting buffer can sink 5 mA at 5 V, 12 mA at 10 V, and 40 mA at 15 V.

The resistance R is calculated from Ohm's law in the usual way. For example, for driving an LED, for a 10-mA current, $R = (V_{cc} - 1.7\text{ V})/10\text{ mA}$. If $V_{cc} = 10$ V, then $R = 830\ \Omega$.

In all CMOS interfaces we should remember that the output voltages from the 4049 or 4050 buffers are V_{cc} and 0 V for the logic high and low levels, respectively.

Special CMOS chips are available to drive larger loads. For example, the 74C908/918 is a dual CMOS 30-V relay driver and is shown in Fig. 12.41(b). It can source 250 mA at a 3.5-V output with $V_{cc} = 5$ V and can withstand up to 30 V across its output. It is useful in driving relays, lamps, and other devices.

12.8.5 LED and LCD Displays

It is often necessary to display numerical and verbal information so that it can be read by a human being. The general method for displaying a binary word (straight binary, or BCD, or ASCII, etc.) is shown in Fig. 12.42(a). A

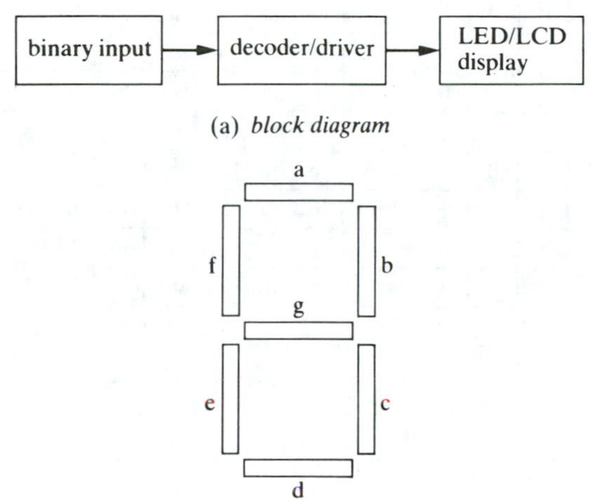

FIGURE 12.42 General display system.

decoder/driver circuit converts the binary input into the proper electrical signals required by the display. Both LED and liquid crystal displays (LCD) are popular. LEDs are best in dim ambient light, while LCDs are best in bright ambient light. LEDs are usually red but are available in green, blue, and yellow. LCD displays use considerably less power than LED displays and are ideally suited to CMOS circuitry in portable applications. To save power, it is common to *multiplex* the LED display, which means to energize one digit at a time in rapid sequence. The slow response of the human eye (dc to ~25 Hz) makes all the digits appear to be continuously lit.

A typical seven-segment LED is shown in Fig. 12.42(b). The seven segments are denoted by lower case letters, a . . . g. A popular TTL BCD to seven-segment LED decoder/driver chip is the 7447 which is shown in Fig. 12.43. When a particular 7447 output pin goes low (e.g., pin a), then the corresponding LED element lights up. For example, to display the number seven, the LED elements a, b, and c are lit up, and the other elements d, e,

Decimal	Inputs D	C	B	A	Outputs a	b	c	d	e	f	g	LED/Display
0	0	0	0	0	lo	lo	lo	lo	lo	lo	hi	0
1	0	0	0	1	hi	lo	lo	hi	hi	hi	hi	1
2	0	0	1	0	lo	lo	hi	lo	lo	hi	lo	2
3	0	0	1	1	lo	lo	lo	lo	hi	hi	lo	3
4	0	1	0	0	hi	lo	lo	hi	hi	lo	lo	4
5	0	1	0	1	lo	hi	lo	lo	hi	lo	lo	5
6	0	1	1	0	lo	hi	lo	lo	lo	lo	lo	6
7	0	1	1	1	lo	lo	lo	hi	hi	hi	hi	7
8	1	0	0	0	lo	lo	lo	lo	lo	lo	lo	8
9	1	0	0	1	lo	lo	lo	lo	hi	lo	lo	9
10	1	0	1	0	hi	hi	hi	lo	lo	hi	lo	c
11	1	0	1	1	hi	hi	lo	lo	hi	hi	lo	ɔ
12	1	1	0	0	hi	lo	hi	hi	hi	lo	lo	u
13	1	1	0	1	lo	hi	hi	lo	hi	lo	lo	⊆
14	1	1	1	0	lo	hi	hi	lo	lo	lo	hi	[
15	1	1	1	1	hi	hi	hi	hi	hi	hi	hi	blank

(a) *truth table*

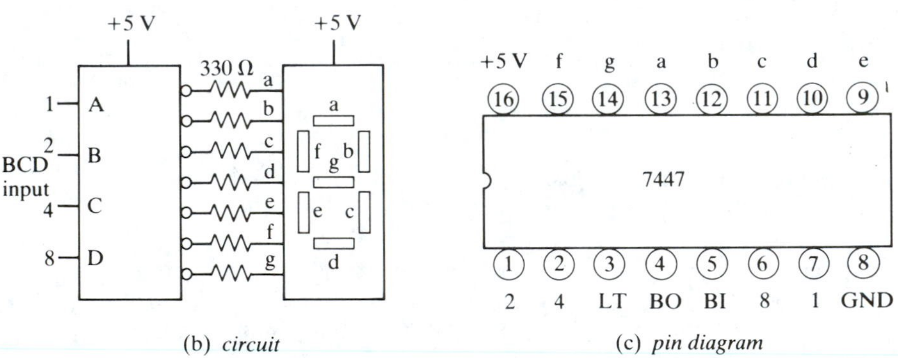

(b) *circuit*

(c) *pin diagram*

FIGURE 12.43 7447 truth table and pin diagram.

f, and g remain dark. Thus, for a BCD seven input, the 7447 outputs a, b, and c are low and the other outputs are high.

The LCD is ideally suited to CMOS circuitry because of its low power requirements. Although the LCD requires little power, it uses more complicated driving circuitry because it requires an ac voltage to depolarize the liquid crystal. Older CMOS driver chips such as the 4511 do not contain the oscillator, but newer chips are available that do contain the oscillator. One such example is the 74C945/947 which is a four-digit counter that can directly drive an LCD.

Other displays are also available such as the 5 × 7 dot LED display which can display many more characters than the seven-segment LED display. Many modern display units contain both the LED display and the decoder/driver circuitry. One such display is the HPDL-2416 which is an intelligent four-character alphanumeric (both letters and numbers) display in a DIP, including on-board CMOS memory, ASCII decoder, multiplexing, and driver circuitry.

12.8.6 Switch and Comparator Input to TTL and CMOS

We usually feed information into a digital circuit (or "input" the circuit) by typing on a keyboard or by having the analog information from a transducer (a thermistor or photocell, etc.) converted into digital form. The specifics of A/D conversion will be covered in Chapter 15.

High or low inputs can be supplied to a TTL gate or a CMOS gate with the pullup circuit of Fig. 12.44(a). When the switch is open, the input is

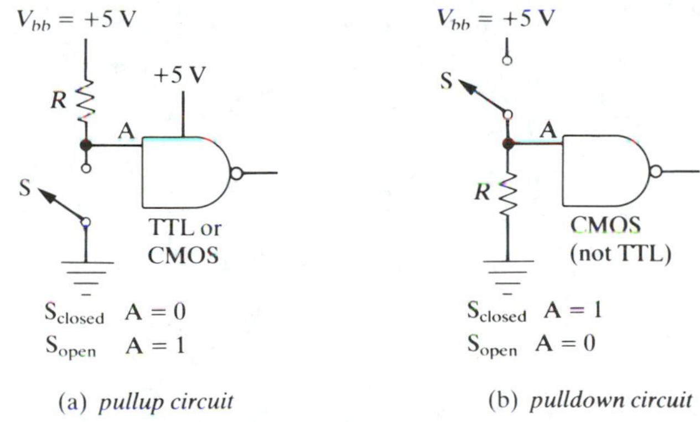

FIGURE 12.44 Inputs to TTL and CMOS gates.

5 V; when the switch is closed, the input is ground. The closed switch can obviously easily sink enough current for a good low TTL input. The value of the resistance R is not critical—several thousand ohms for TTL and perhaps 10 kΩ for CMOS. The V_{bb} voltage can be higher than 5 V if the CMOS gate runs off a voltage greater than 5 V. It is theoretically possible to use a pulldown resistor, as shown in Fig. 12.44(b), but there are three disadvantages. First, if the supply voltage ever exceeds 5 V for TTL, the TTL gates may be destroyed when the switch is closed. Second, the pulldown resistance must be fairly low to obtain a good low TTL input. For example, to keep the low input below 0.4 V, the resistance must be less than 250 Ω because the sinking current for a TTL low is 1.6 mA or less than 1 kΩ for LS TTL. Third, when the switch is closed to produce a high input,

a large current is drawn from the supply: $I = 5\ \text{V}/250\ \Omega = 20\ \text{mA}$. This pulldown circuit should be avoided for TTL inputs, but it can be used for CMOS. R can be much larger than 250 Ω in CMOS circuits.

Finally, mechanical switches are almost never used to supply high and low logic inputs to digital gates because they "bounce" open and shut (for approximately 10 ms) when they are closed. The remedy is the "bounceless" switch, described in Section 12.8.7.

A comparator can drive a TTL circuit, as shown in Fig. 12.45(a). The

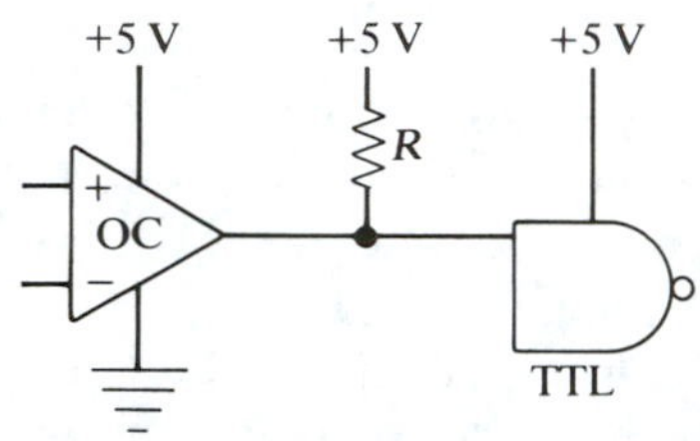

(a) *open collector comparator to TTL*

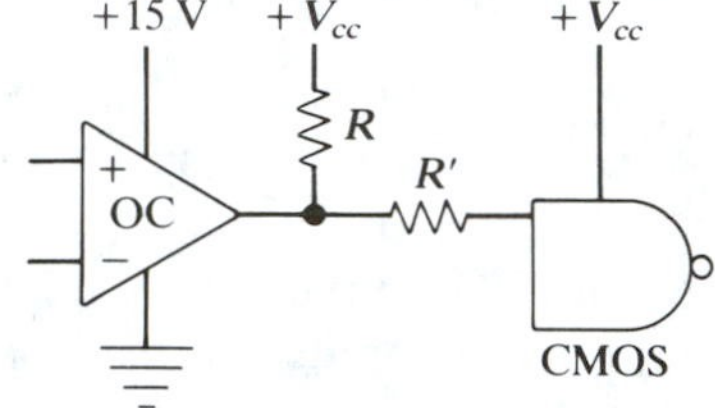

(b) *open collector comparator to CMOS*

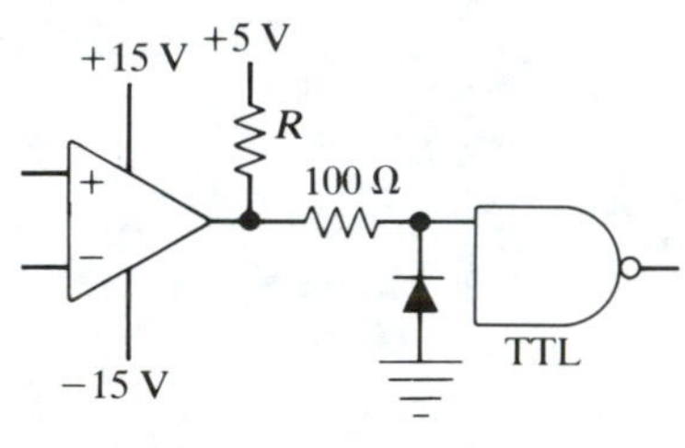

(c) *bipolar output from comparator*

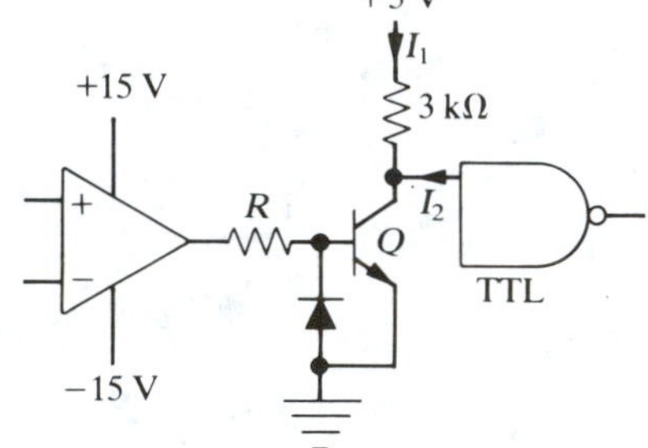

(d) *transistor drive*

FIGURE 12.45 Comparator/TTL–CMOS interfacing.

output of many comparators is of the open collector type, so a pullup resistor is used to +5 V for driving TTL and to V_{cc} for CMOS. In Fig. 12.45(b) the resistance R' is necessary to limit the current through the CMOS protection diodes. The circuits of Fig. 12.45(a) and (b) are suitable for a comparator that runs between a positive supply voltage and ground, because then the high comparator output provides a suitable TTL or CMOS high, and a low comparator output is essentially at ground, which provides a good low for both TTL and CMOS. But if a bipolar or CMOS op amp is used as a comparator, its output can swing from +13 V to +15 V high to −13 V to −15 V low, and the low output can destroy the TTL gate input circuitry. A protective diode with a 100-Ω current-limiting resistor [see Fig. 12.45(c)] prevents the low from going below approximately

−0.6 V. At this writing most experimentalists use comparators that swing from a positive voltage to near ground, so this is not a real problem. Another way of driving TTL is to use the comparator output to turn on an npn transistor, as shown in Fig. 12.45(d). A positive comparator output turns on Q and thus provides a low input of 0.2 V to the TTL gate. A negative comparator output turns Q off and provides a high TTL input of 5 V. Thus the transistor acts as an inverter. The protective diode is necessary to prevent the negative comparator output from exceeding the maximum base emitter breakdown voltage (typically 7 V) of Q. The resistance R limits the base drive to Q.

For example, for a good low, Q should be saturated and its collector voltage will be 0.2 V. Thus, $I_1 = (5\text{ V} - 0.2\text{ V})/3\text{ k}\Omega = 1.6\text{ mA}$, and $I_2 = 1.6\text{ mA}$ for the low sinking current for a good TTL low. Thus, $I_C = 3.2\text{ mA}$, and the base current is $I_B = I_C/h_{FE} = 3.2\text{ mA}/100 = 32\ \mu\text{A}$. When Q is saturated, its base voltage will be approximately 0.7 V and thus $R = (13\text{ V} - 0.7\text{ V})/0.032\text{ mA} = 384\text{ k}\Omega$. This is a maximum value for R. To ensure saturation of Q, we would use a smaller value for R.

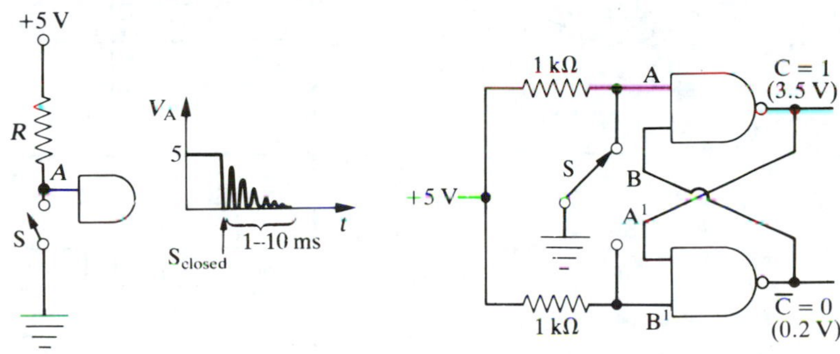

(a) *bouncing mechanical switch* (b) *NAND debounced switch*

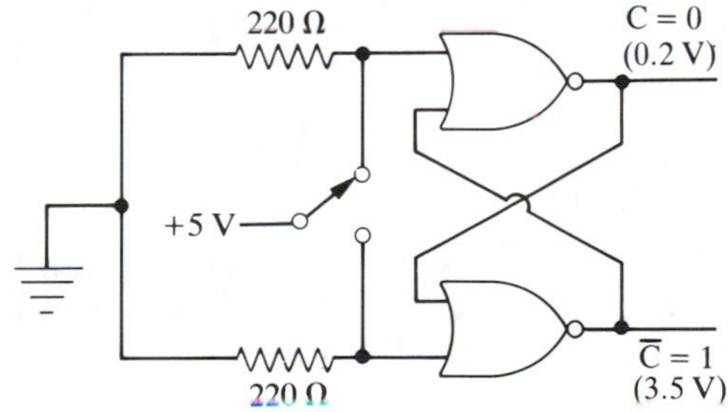

(c) *NOR debounced switch*

FIGURE 12.46 Bounceless switch.

12.8.7 The Bounceless Switch

A switch can be used to supply high or low logic inputs to digital gates. But most mechanical switches such as ordinary toggle switches and microswitches "bounce" when they are closed because of the mechanical bouncing of the switch contact inside the switch. The result is an erratic series of pulses approximately 1–10 ms long, as shown in Fig. 12.46(a). The low propagation time of TTL gates (typically 10 ns) can be used to make a "bounceless" switch, as shown in Fig. 12.46(b). The basis of this debouncing circuit is that the propagation time of the gate is much much less than the bounce period of the mechanical switch. The output of each NAND gate is fed back to the input of the other NAND gate. If the switch is flipped to the up position at $t = 0$, then A is set low at $t = 0$; and 10 ns later

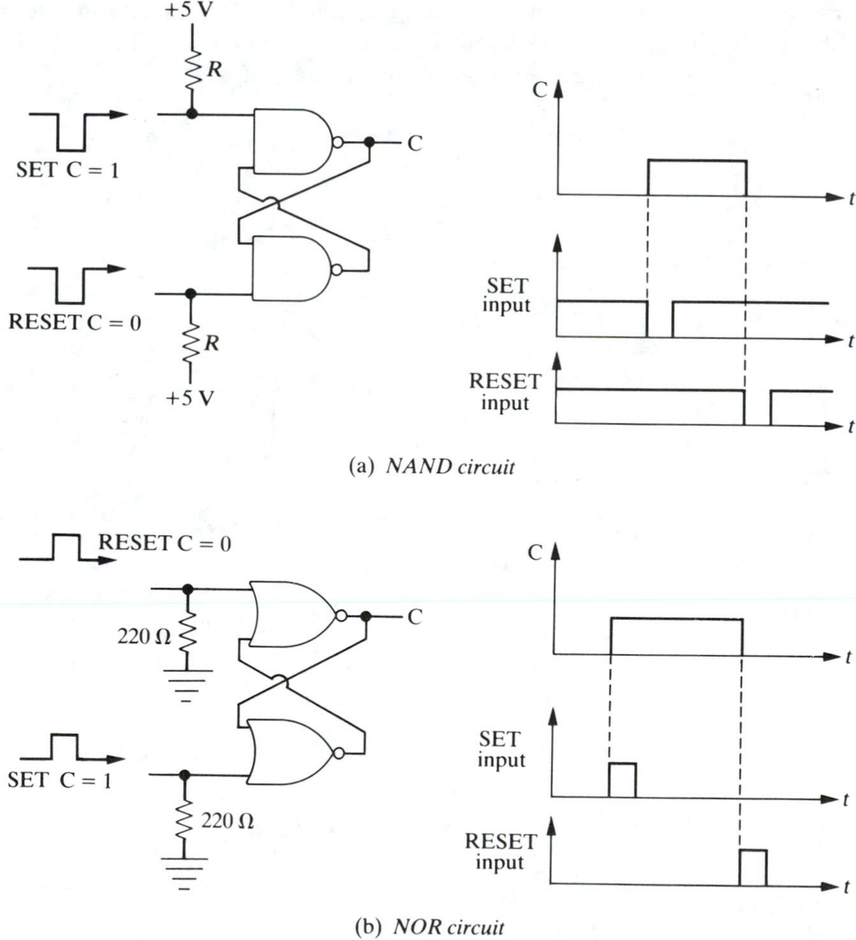

FIGURE 12.47 SET and RESET circuits.

the output C of the top NAND gates goes high, and the A′ input of the lower NAND gate also goes high at $t = 10$ ns. But the two inputs of the bottom NAND gate are both high now, so after 10 ns more have elapsed, the output $\bar{C}$ of the lower NAND gate goes low from the NAND truth table, which means the other (B) input to the top NAND gate goes low at $t = 20$ ns. If the switch bounces open (thus changing A from low to high), after $t = 20$ ns the output C of the top NAND gate will *remain* high because its B input is held low. Thus, if the switch mechanical *bounce period* is *greater than the two propagation times* (20 ns), the C output will go high and remain high even if the switch bounces open and shut. Basically all such switches in logical systems are debounced. NOR gates can be used, as shown in Fig. 12.46(c).

In digital systems the mechanical switch is often replaced by the output of a gate, as shown in Fig. 12.47(a) and (b). This technique is often used to set or reset a counter or other device with a low power pulse longer than 20 ns.

PROBLEMS

1. Count from 30_{10} to 50_{10} in (a) binary, (b) octal, and (c) hexadecimal.

2. $(10110)_2 = (\ \ )_{10}$; $(306)_8 = (\ \ )_{10}$; $(3F2)_{16} = (\ \ )_{10}$.

3. $(49)_{10} = (\ \ )_2$; $(49)_{10} = (\ \ )_8$; $(49)_{10} = (\ \ )_{16}$.

4. $(101011)_2 = (\ \ )_8$; $(37)_8 = (\ \ )_2$.

5. $(11010011)_2 = (\ \ )_{16}$; $(101011)_2 = (\ \ )_8$.

6. $(63)_{10} = (\ \ )_{BCD}$; $(10110111)_{BCD} = (\ \ )_{10}$.

7. The parity of 10110010 is __________. The parity of 10111010 is __________.

8. How many binary functions do two binary variables have? Three binary variables? Four binary variables?

9. Write the truth tables for the following functions of two binary variables: AND, NAND, OR, NOR, XOR, XNOR. Sketch the standard gate symbol for each function.

10. Diagram how you would implement the following functions using (a) only NAND gates, (b) only NOR gates.

$$F = A \cdot \bar{B} + \bar{A} \cdot B \qquad F = A + B$$

$$F = A \cdot B \qquad F = \bar{A}$$

11.

input coded data — [gate] — decoded data output
C

input data 1 0 1 0 1 1 0 0 1 1 0 1 1 0 0 1

C 0 1 0 1 0 1 0 1 0 1 0 1 0 1 0 1

output data:

12. Write the truth table for P in terms of D_3, D_2, D_1, D_0. What is P called?

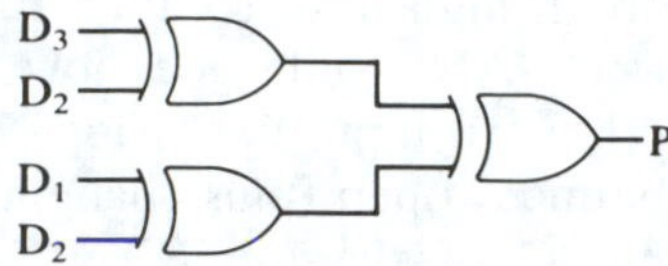

13. Write a circuit diagram to realize an XOR gate using only NAND gates. Use $A \oplus B = A \cdot \bar{B} + \bar{A} \cdot B$.
14. Repeat Problem 13, using only NOR gates.
15. Write a circuit diagram to realize an XOR gate using only NAND gates. Using $A \oplus B = (A + B) \cdot \overline{(A \cdot B)}$.
16. Repeat Problem 15, using only NOR gates.
17. Explain why a 7400 low input must sink current.
18. Explain why an open 7400 input acts like a high input.
19. Explain what is meant by a totem pole output.
20. Explain why a high output in a TTL totem pole circuit cannot supply more than about 10–20 mA of current to an LED.
21. Explain how much current a low output in a TTL totem pole circuit can sink.
22. Explain why a low TTL output is approximately 0.2 V and a high TTL output is approximately 3.5 V.
23. Explain what an open collector gate is.
24. Write the truth table and the circuit for the realization of the function $F = \overline{(A \cdot B)} \cdot \overline{(C \cdot D)}$, using only two open collector gates and a pullup resistor.
25. Repeat Problem 24 for the function $F = A \cdot B \cdot C \cdot D$.
26. Briefly summarize the high- and low-voltage requirements for TTL gates.
27. Discuss the problems and solutions for connecting a CMOS output to a TTL input.
28. Repeat Problem 24 for a TTL output to a CMOS input.
29. Discuss how TTL and CMOS outputs can drive high current "outside world" loads.

CHAPTER 13

Basic Digital Circuits

13.1 INTRODUCTION

In this chapter we consider some relatively simple digital circuits, many of which are available in a single chip. We consider flip-flops, which are the basic building blocks of all memory systems, counters, conversion from serial to parallel data, and vice versa; the transmission of digital data; arithmetic operations on binary numbers; and various other circuits. We assume positive logic: a high voltage level is a 1, and a low voltage level is a 0.

13.2 FLIP-FLOPS

A *flip-flop* is essentially any circuit with two output terminals and two stable voltage states (one higher than the other) for each terminal. It is thus ideal for storing binary digital information, either data or instructions. It usually has several input terminals, such as SET, RESET, clock, and DATA, as will be explained shortly. It can be switched very quickly from one stable state to the other with an external pulse.

13.2.1 The RS Flip-Flop

The basic idea of a flip-flop can be seen from the discrete component circuit of Fig. 13.1(a). There is exact symmetry between the two sides of the circuit, so one might conclude that each transistor will conduct equally (i.e., have the same collector current). We will now show that one transistor will always be on and the other always off because of the R_1 and R_2 resistor voltage dividers, which couple the collector of one transistor to the base of the other transistor. Suppose each transistor is conducting equally. There will always be a voltage fluctuation due to noise at the bases. Suppose the base of Q_2 goes slightly more positive. Then Q_2 will draw more collector current, and the voltage at the collector will decrease (i.e., become less positive). The fraction $R_1/(R_1 + R_2)$ of this *drop* in the collector voltage is applied to the base of Q_1, thus turning off Q_1 more. Q_1 draws less collector current, and its collector voltage rises (i.e., becomes more positive). A

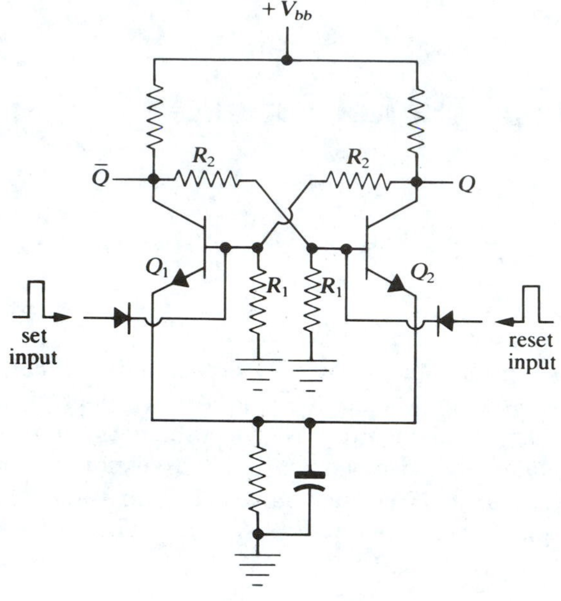

(a) *basic flip-flop with set and reset inputs*

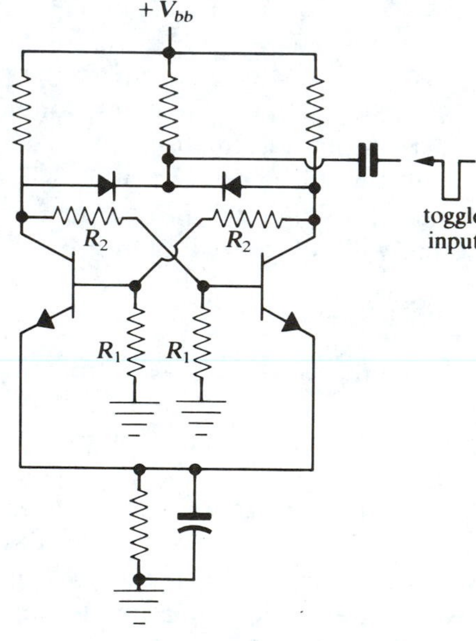

(b) *toggle flip-flop*

FIGURE 13.1 Discrete component flip-flop (obsolete).

fraction $R_1/(R_1 + R_2)$ of this positive-going voltage is applied to the base of Q_2, reinforcing the original positive-going fluctuation. The cycle starts over again, and the net result is to turn on Q_2 strongly and to cut off Q_1. The "on" and "off" transistors can be distinguished because the collector voltage of the "on" transistor will be lower than that of the "off" transistor. We usually denote the logic levels at the two collectors as Q and $\bar{Q}$, because one must be high and the other low. Thus, this circuit is sometimes called a *bistable multivibrator* or *flip-flop*. An external positive input fed into the base of the "off" (npn) transistor will quickly turn it on and turn off the other transistor. With two external inputs to the two bases, we thus can set or reset either transistor on or off. A positive set pulse input makes Q high; a positive reset input makes Q low. The flip-flop is usually called an *RS* (reset–set) flip-flop. Almost all flip-flops used today are on one chip as an integrated circuit, and a wide variety of flip-flops is available at low prices.

The circuit can be made to switch states or "toggle" for each single input pulse with the circuit of Fig. 13.1(b). The two "steering" diodes automatically steer the negative input toggle pulse to the base of the "on" transistor, thus turning it off and flipping the circuit to its other stable state. This is sometimes called a *toggle* circuit, and a little thought will show that if a series of pulses is used for the input, the output Q will be a series of pulses of exactly one-half the frequency of the input.

One of the simplest flip-flops is the RS flip-flop made from two cross-coupled NAND gates as shown in Fig. 13.2.

The two inputs are called $\bar{\text{S}}$ (SET) and $\bar{\text{R}}$ (RESET). The superscript bar means they are active low inputs: $\bar{\text{S}} = 0$ will set the output $\text{Q} = 1$, and $\bar{\text{R}} = 0$ will reset the output $\text{Q} = 0$. If both inputs are 1, then there are two possible stable states: $\text{Q} = 1$ and $\bar{\text{Q}} = 0$, or $\text{Q} = 0$ and $\bar{\text{Q}} = 1$. This is a "storage" mode, useful for storing data. This follows from the truth table for a NAND gate—any 0 in produces a 1 out. When the circuit is first turned on, the flip-flop may assume either state, depending on slight asymmetries in the circuit—the feedback from the output of one gate to the input of the other gate guarantees that one NAND gate output will be 1 and the other 0. To show this, just assume that the outputs of the two gates are equal at turn-on and that there is a small noise pulse at one output and use essentially the same argument that we used to describe the discrete component flip-flop.

Assume the flip-flop is in the state $\text{Q} = 1$, $\bar{\text{Q}} = 0$. If the $\bar{\text{R}}$ input now goes to 0, then the lower NAND gate has one 1 input (from Q) and one 0 input (from $\bar{\text{R}}$). Thus, approximately 10 ns after the $\bar{\text{R}}$ input goes to 0, the lower NAND gate output Q goes to 1, which means both inputs to the upper NAND gate are now 1, so the output of the upper NAND gate Q goes to 0 after another 10 ns. This $\text{Q} - 0$ output is fed to the top input of the lower NAND gate, which reinforces the high output $\bar{\text{Q}} = 1$ approximately 20 ns after the $\bar{\text{R}}$ input went to 0. This action holds $\bar{\text{Q}} = 1$ even if the $\bar{\text{R}}$ input returns to 1 at a later time. In other words, the $\bar{\text{R}}$ input need

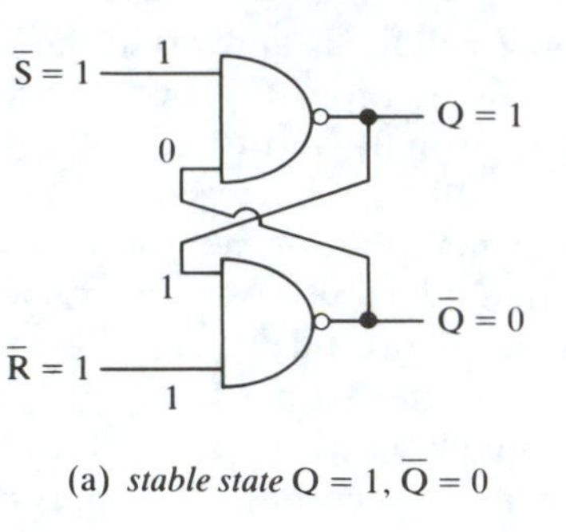

(a) *stable state* Q = 1, $\bar{Q} = 0$

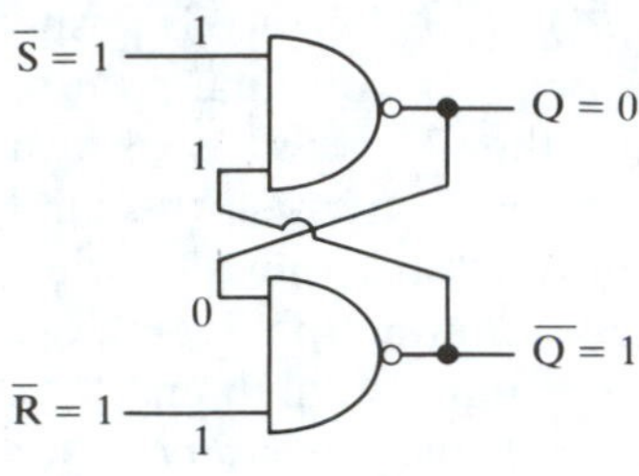

(b) *stable state* Q = 0, $\bar{Q} = 1$

$\bar{S}$	$\bar{R}$	Q	$\bar{Q}$
0	0	?	?
0	1	1	0
1	0	0	1
1	1	NC (data storage)	

NC = "no change"
? = indeterminate

(c) *truth table*

S Q
R $\bar{Q}$

(d) *schematic symbol*

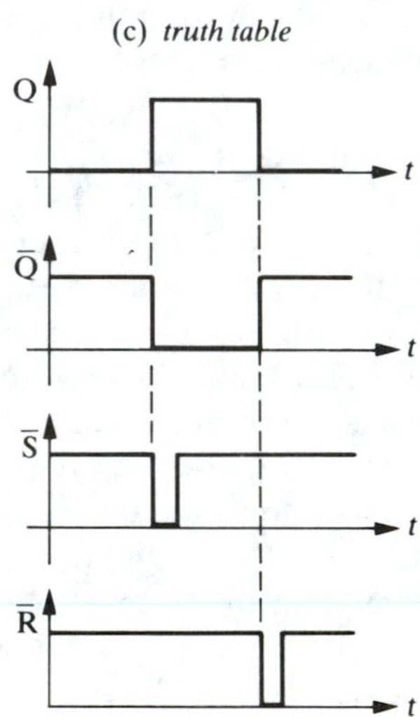

(e) *timing diagram*

FIGURE 13.2 The RS flip-flop with NAND gates.

remain 0 for only about 20 ns to cause the flip-flop to change states from Q = 1, $\bar{Q} = 0$, to Q = 0, $\bar{Q} = 1$. We can describe the $\bar{R}$ input as an *active low* reset, because a low $\bar{R}$ input resets the flip-flop output to Q = 0. The bar over $\bar{R}$ means a 0 input will reset the flip-flop.

If we assume the flip-flop is in the state Q = 0, $\bar{Q} = 1$, then a low $\bar{S}$

input will cause it to change or flip to the Q = 1, $\bar{Q}$ = 0 state. We can describe the $\bar{S}$ input as an *active low* set because a low $\bar{S}$ input sets the flip-flop output to Q = 1. The bar over $\bar{S}$ means a 0 input will set the flip-flop.

The important point is that a low $\bar{R}$ pulse (longer than 20 ns) will force the flip-flop outputs to the state Q = 0, $\bar{Q}$ = 1, and this output state will *persist* even after the $\bar{R}$ pulse returns to its normal high level. Similarly, a low $\bar{S}$ pulse will produce a Q = 1, $\bar{Q}$ = 0 output [see Fig. 13.2(e)]. We often say the output is *latched* because it persists in time even after the input $\bar{S}$ or $\bar{R}$ pulses return to their normal high states. Another name for the flip-flop is an *RS latch*.

The truth table for the RS flip-flop is shown in Fig. 13.2(c). We have just discussed the cases ($\bar{R}$, $\bar{S}$) = (1, 1) or (0, 1) or (1, 0). If both $\bar{R}$ and $\bar{S}$ are 1, the flip-flop output remains constant; that is, the flip-flop stores the state it had *before* the $\bar{R}$ and $\bar{S}$ inputs went high. This can be thought of as the *storage state*. But if both $\bar{R}$ and $\bar{S}$ go low, then *each* NAND gate has a low input, so each gate output goes high, and this condition persists as long as R and S are low. If $\bar{R}$ and $\bar{S}$ return to their normal high condition, the flip-flop output will be either Q = 1, $\bar{Q}$ = 0 or Q = 0, $\bar{Q}$ = 1, depending on the asymmetry of the two gates and on which input went high *first*. The input state $\bar{R}$ = $\bar{S}$ = 0 should be avoided like the plague for an RS flip-flop! The "?" in the truth table for $\bar{R}$ = $\bar{S}$ = 0 means the flip-flop state is indeterminate after $\bar{R}$ and $\bar{S}$ return to their normal high state.

The notation for the R and S inputs may be confusing at first glance. Why not, for example, call them R and S instead of $\bar{R}$ and $\bar{S}$? The word "reset" means to *reset to 0*; so when we reset a flip-flop, we would like its output Q to go to zero. By convention we consider the output of the flip-flop to be the condition of Q (not $\bar{Q}$). Thus, to "reset" means to make Q = 0. By analyzing the electrical behavior of the NAND gates, we see that the input labeled $\bar{R}$ must be low to make Q = 0. If $\bar{R}$ = 0, then obviously R = 1, so R stands for "reset," which is comforting. In other words, the symbol $\bar{R}$ means a *low* logic level will reset the flip-flop. This is sometimes indicated by drawing a small open circle or "bubble" at the reset input and labeling it "R," as shown in Fig. 13.2(d). Similarly, to "set" a flip-flop means to *set to 1*; so when $\bar{S}$ = 0 or S = 1, we make Q = 1. Either the overbar on the R and S or the bubble on the flip-flop symbol means a low logic level will reset or set; that is, these inputs are active low. An example of an $\bar{R}\bar{S}$ latch is the 74LS279, which contains four independent $\bar{R}\bar{S}$ latches on one 16-pin DIP.

An RS flip-flop can also be made from NOR gates, as shown in Fig. 13.3. An analysis of the NOR truth table shows that high inputs are necessary to reset and to set the flip-flop output; that is, these inputs are *active high*. The input state R = S = 1 should be avoided like the plague!

It is often desirable to set or to reset a flip-flop at a time indicated by another signal (the clock signal). The clocked RS flip-flop as shown in Fig.

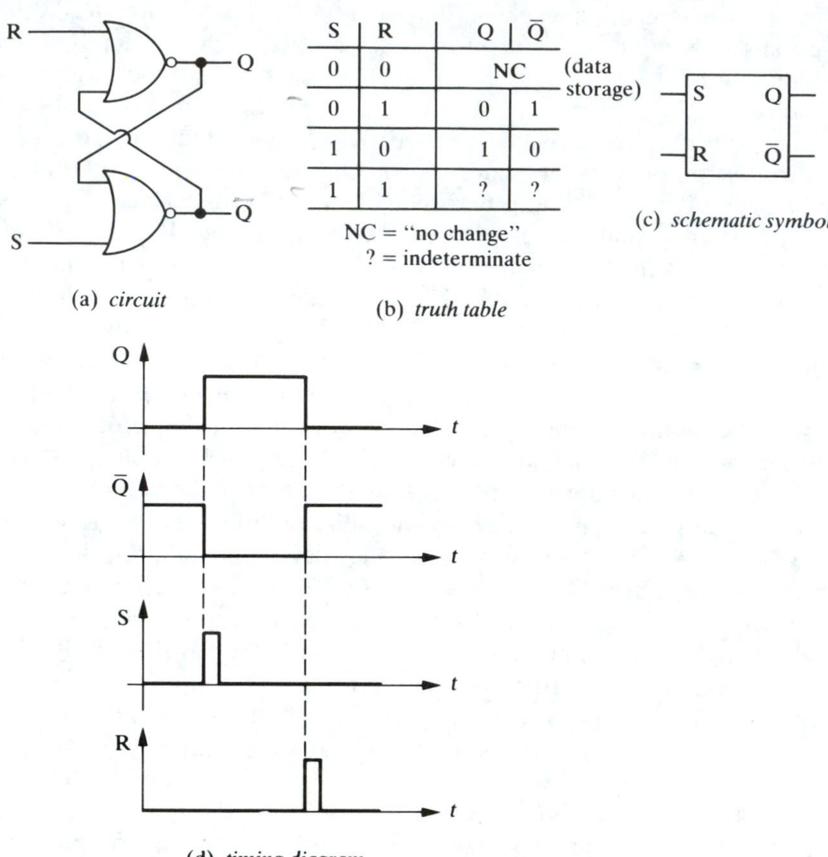

S	R	Q	$\bar{Q}$
0	0	NC	(data storage)
0	1	0	1
1	0	1	0
1	1	?	?

FIGURE 13.3 The RS flip-flop with NOR gates.

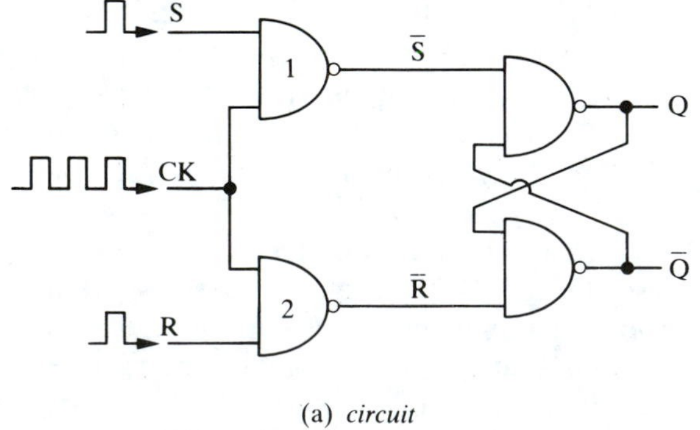

FIGURE 13.4 The clocked RS flip-flop.

CK	S	R	Q	$\bar{Q}$
0	0	0	NC	
0	0	1	NC	
0	1	0	NC	
0	1	1	NC	
1	0	0	NC	
1	0	1	0	1
1	1	0	1	0
1	1	1	?	?

(data storage)

(b) *truth table*

S Q
CK
R $\bar{Q}$

(c) *schematic symbol*

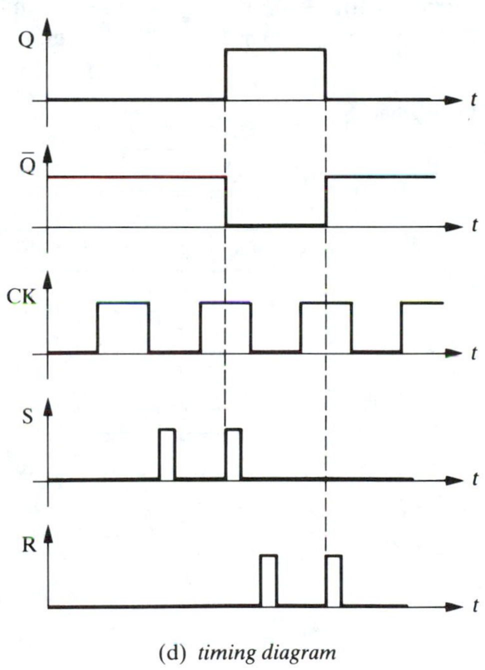

(d) *timing diagram*

FIGURE 13.4 Continued.

13.4 does just this. Its output assumes the state imposed by the set or reset input when the clock input is 1. If the clock input is 0, the flip-flop state is independent of the R and S inputs.

The explanation is simple. If the clock input is zero, then both NAND gates 1 and 2 have high outputs and $\bar{S} = \bar{R} = 1$, which means, from our previous discussion of the RS flip-flop, that the flip-flop outputs are latched or stable; the flip-flop can be thought of as storing information. But if the clock goes to 1, then S = 1, R = 0 implies $\bar{S} = 0$, $\bar{R} = 1$ and Q = 1, $\bar{Q} = 0$; and R = 1, S = 0 implies $\bar{R} = 0$, $\bar{S} = 1$ and Q = 0, $\bar{Q} = 1$, as shown in the truth table. Thus, we see that the SET and RESET inputs S and R are active *high*. We say that the clock *enables* the SET and RESET inputs, and the clock input is sometimes called an ENABLE input. Notice in the schematic symbol that there are no bubbles at the R and S inputs because they are active high. The SET and RESET inputs are called *synchronous* inputs because they affect the output only when the clock is high; that is, they affect the output at the same time (synchronously) as the clock.

Two other inputs that take effect regardless of the clock input are PRESET and CLEAR, which are *asynchronous*. In other words, PRESET and CLEAR override the clock. A clocked RS flip-flop with active low PRESET and CLEAR inputs as well as active high SET and RESET and clock inputs can be made with two 2-input and two 3-input NAND gates, as shown in Fig. 13.5. If the PRESET input $\overline{PR} = 0$, then Q = 1 immediately,

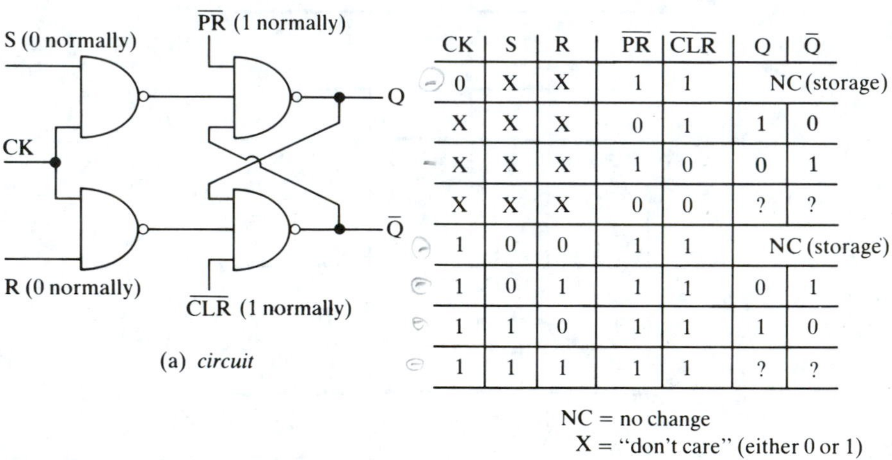

(a) *circuit*

CK	S	R	$\overline{PR}$	$\overline{CLR}$	Q	$\bar{Q}$
0	X	X	1	1	NC (storage)	
X	X	X	0	1	1	0
X	X	X	1	0	0	1
X	X	X	0	0	?	?
1	0	0	1	1	NC (storage)	
1	0	1	1	1	0	1
1	1	0	1	1	1	0
1	1	1	1	1	?	?

NC = no change
X = "don't care" (either 0 or 1)

(b) *truth table*

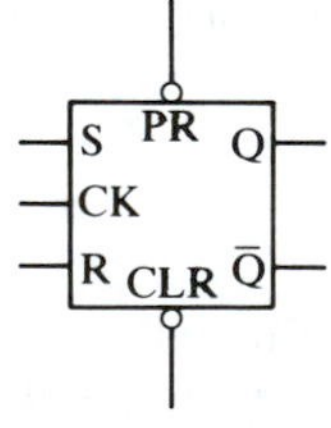

(c) *schematic symbol*

FIGURE 13.5 Clocked RS flip-flop with PRESET and CLEAR.

and if the clear input $\overline{\text{CLR}} = 0$, then Q = 0 immediately. Normally, both PRESET and CLEAR are held high, and both SET and RESET are held low. Thus we must avoid having R = S = 1 and $\overline{\text{PR}} = \overline{\text{CLR}} = 0$, because in either case $Q = \bar{Q} = 1$, and when the R,S or $\overline{\text{PR}}$,$\overline{\text{CLR}}$ inputs return to their normal states the outputs will be either $Q,\bar{Q} = 0,1$ or 1,0, depending on which input stays high (S,R) or low ($\overline{\text{PR}}$,$\overline{\text{CLR}}$) longer. In other words, the output will be either 0,1, or 1,0, depending on the *length* of the inputs.

13.2.2 The D Flip-Flop

The D flip-flop or the *data latch* has only one input for DATA and one for clock. It is shown in Fig. 13.6 with its truth table. The inverter forces R and

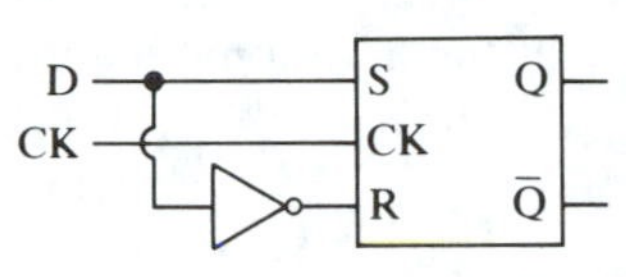

(a) *circuit using RS flip-flop*

CK	D	Q	$\bar{Q}$	
0	0	NC		data storage
0	1	NC		
1	0	0	1	
1	1	1	0	

NC = no change

(b) *truth table*

(c) *schematic symbol*

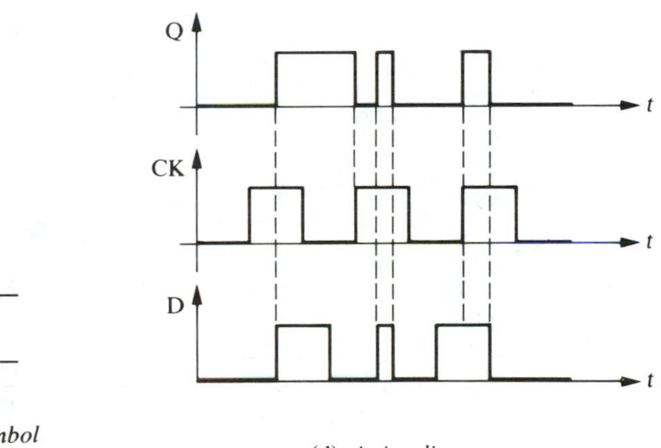

(d) *timing diagram*

FIGURE 13.6 The D flip-flop.

S to be different, thus avoiding the disastrous input condition R = S = 1, which could exist for the RS flip-flop. If the clock input is low, then the output is unaffected by the D input, and we think of the flip-flop as storing information or data, namely, the data that entered the flip-flop the last time the clock was high. If the clock input is high, then the flip-flop is enabled so that Q = S = D. The flip-flop output Q takes on the value of D as soon as the clock goes high. The flip-flop is said to be *transparent* because the data bit D is fed through the flip-flop to the output Q as long as the clock stays high. If D changes while the clock is high, Q will *follow* the changes in D, as shown in Fig. 13.6(d). This type of flip-flop is also called a *pulse-triggered* or *level-triggered* flip-flop because the output Q follows any input changes in D as long as the clock (voltage) level remains high. When the clock goes low, the Q output remains constant; that is, the previous D input is stored in the flip-flop.

D flip-flops are quite common in digital systems for data storage. One common type is the TTL 74LS75, which is a quad TTL D flip-flop or latch. Its pin diagram is shown in Fig. 13.7. Notice that flip-flops #1 and #2 are enabled by the same clock, and similarly for #3 and #4. The CMOS 4042 is another example of a quad D flip-flop or latch, with the additional feature

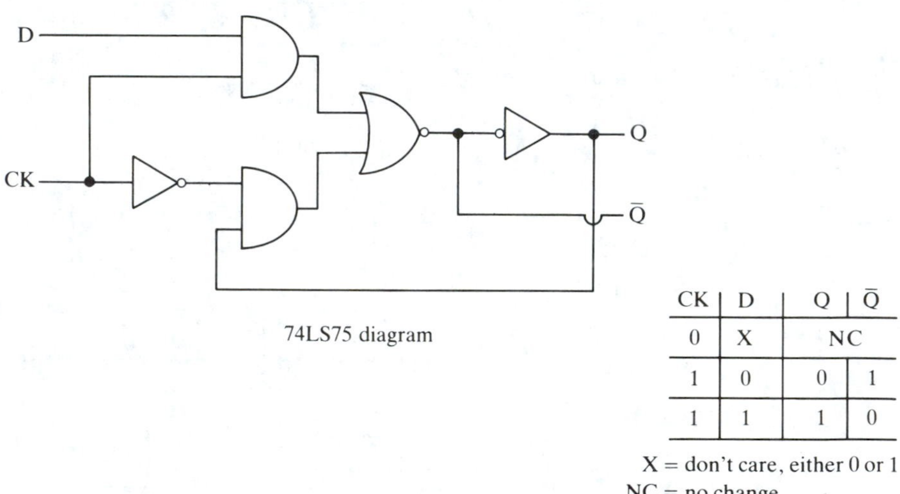

CK	D	Q	$\bar{Q}$
0	X	NC	
1	0	0	1
1	1	1	0

X = don't care, either 0 or 1
NC = no change

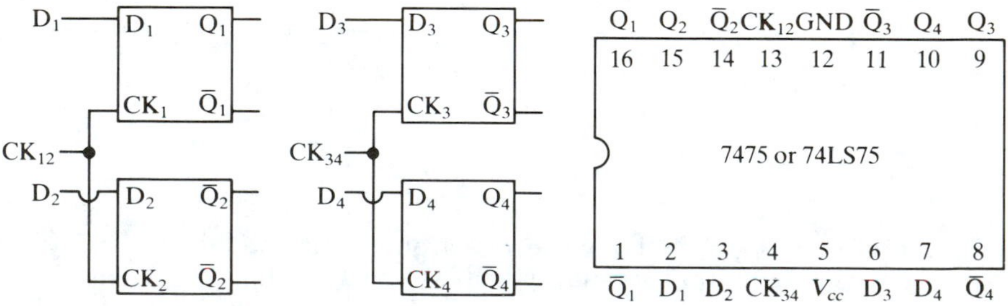

FIGURE 13.7 7475 quad D flip-flop (pulse triggered).

that the clock or ENABLE input can be either active high or active low, depending on the polarity of the "polarity" input: a low polarity produces an active low clock, and vice versa. One clock input enables all four flip-flops in the 4042.

13.2.3 Edge-Triggered Flip-Flops

In a pulse- or level-triggered flip-flop the flip-flop is transparent as long as the clock stays high; that is, changes in the data input will change the output state while the clock is high. But in an *edge-triggered* flip-flop, only the input present during the very short time (typically less than 10 ns) the clock is *changing* (during the "edge" of the clock pulse) will affect the output. Edge-triggered flip-flops can be made to trigger on either the rising edge or the falling edge of the clock pulse. A small triangle at the clock input indicates an edge-triggered flip-flop, or sometimes an arrow is drawn on the clock pulse, as shown in Fig. 13.8(b).

The circuit diagram for a D-type edge-triggered flip-flop is shown in Fig. 13.8(a). The logic levels are shown for a low clock input, a high DATA input, and the PRESET and CLEAR inputs (normally high). Either $Q = 1$, $\bar{Q} = 0$ or $Q = 0$, $\bar{Q} = 1$ is possible.

The basic point is that if the clock is low, the flip-flop is disabled; that is, the values of Q and $\bar{Q}$ cannot change in response to changes in D. This occurs because the clock input goes to gates #2 and #3, so a low clock forces both Q_2 and Q_3 high, which latches or preserves the values of Q and $\bar{Q}$, considering gates #5 and #6 as a simple RS latch or flip-flop. (Remember, $\overline{PR}$ and $\overline{CLR}$ each are normally high.)

For any clock level, high or low, a low $\overline{PR}$ will force $Q = 1$ and $\bar{Q} = 0$, and a low $\overline{CLR}$ will force $Q = 0$ and $\bar{Q} = 1$. These SET and RESET inputs are asynchronous; they override the clock because they are fed directly into the final two gates (#5 and #6). The truth table is shown in Fig. 13.8(c).

A timing diagram for the D-type positive edge-triggered flip-flop is shown in Fig. 13.8(d). Notice that the output changes all occur at the same time as the positive-going edge of the clock pulses, that is, at the leading edge of the clock pulses. Notice also what Q would be for a pulse-triggered flip-flop in response to the same clock and data input, as shown in Fig. 13.8(d).

One of the most popular flip-flops presently available is the 7474 or 74LS74, a dual positive edge-triggered flip-flop with active low asynchronous PRESET and CLEAR. Its pin diagram and some specifications are given in Fig. 13.9.

Edge-triggered flip-flops are available that trigger on the positive edge or on the negative edge of the clock pulse. Figure 13.10 shows the standard schematic symbols for the four types of D flip-flops. Edge-triggered flip-flops are generally preferred over pulse- or level-triggered flip-flops because they are more immune to noise.

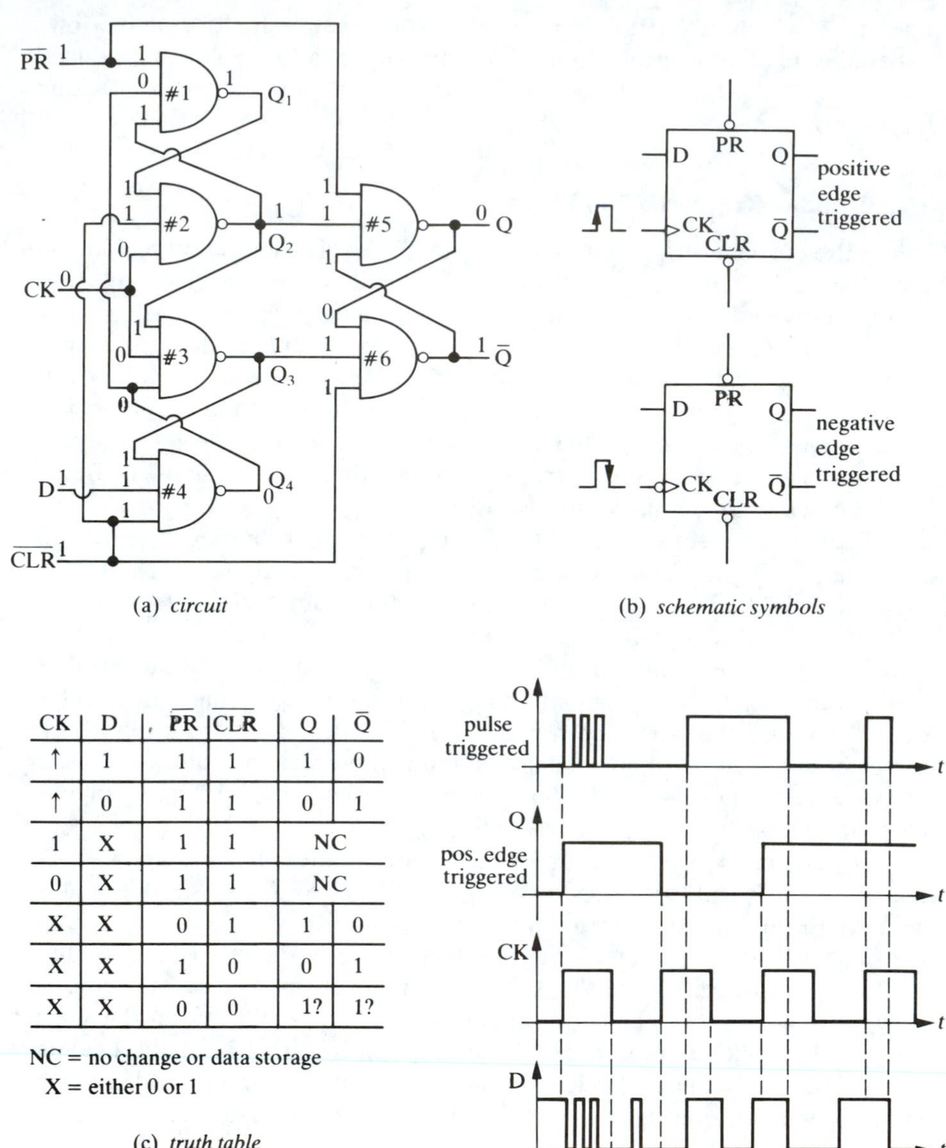

(a) *circuit*

(b) *schematic symbols*

CK	D	$\overline{PR}$	$\overline{CLR}$	Q	$\bar{Q}$
↑	1	1	1	1	0
↑	0	1	1	0	1
1	X	1	1	NC	
0	X	1	1	NC	
X	X	0	1	1	0
X	X	1	0	0	1
X	X	0	0	1?	1?

NC = no change or data storage
X = either 0 or 1

(c) *truth table*

(d) *timing diagram*

FIGURE 13.8 D edge-triggered flip-flop.

A toggle or divide-by-two circuit can easily be made from an edge-triggered D flip-flop (but not from a level- or a pulse-triggered flip-flop) by connecting the $\bar{Q}$ output to the D input, as shown in Fig. 13.11. Because of the external connection between the D input and the $\bar{Q}$ output, $D = \bar{Q}$ always, and when the clock goes high, Q takes on the value of $D = \bar{Q}$ at the

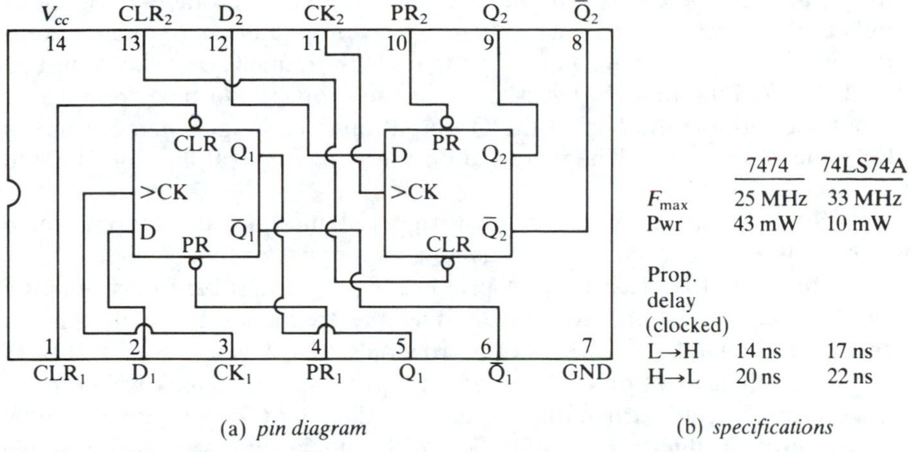

(a) *pin diagram*

	7474	74LS74A
F_{max}	25 MHz	33 MHz
Pwr	43 mW	10 mW
Prop. delay (clocked)		
L→H	14 ns	17 ns
H→L	20 ns	22 ns

(b) *specifications*

FIGURE 13.9 7474 dual positive edge-triggered flip-flop.

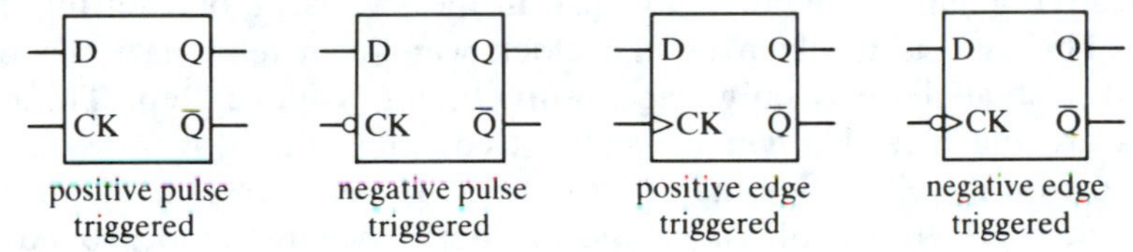

FIGURE 13.10 D flip-flop schematic symbols.

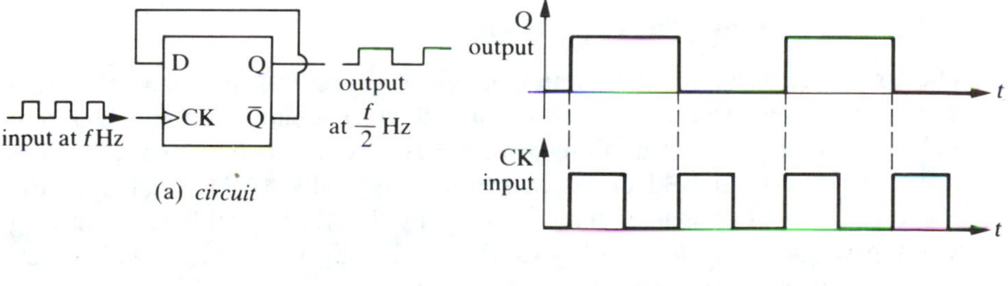

(a) *circuit*

(b) *square-wave input*

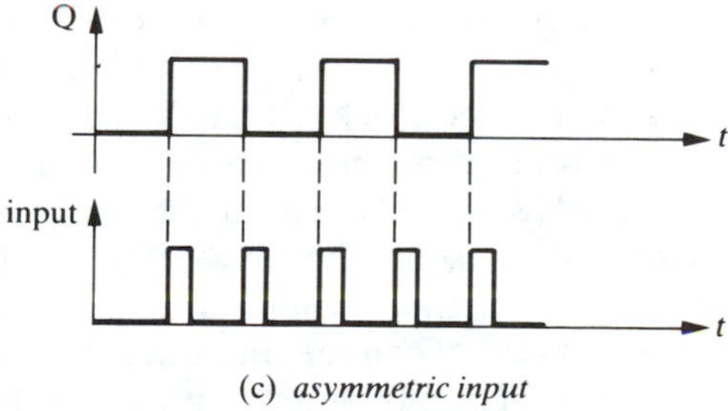

(c) *asymmetric input*

FIGURE 13.11 Toggle flip-flop (divide-by-two).

input; in other words, Q will *change*. Thus, for every positive-going clock pulse edge, the output Q will change; the output pulse frequency must therefore be exactly one half of the input clock frequency, as shown in Fig. 13.11(b). A little thought will show that the propagation time from the D input through the flip-flop to the Q output must be longer than the rise or fall time of the clock. The propagation time for a typical flip-flop is about 40 to 50 ns.

With a positive pulse- or level-triggered flip-flop, the output would rapidly oscillate between 1 and 0 as long as the clock remains high.

This circuit is often used to produce a perfectly square or symmetric clock waveform (at the cost of dividing the frequency by two) from an asymmetric input [e.g., a series of narrow pulses as shown in Fig. 13.11(c)].

A final comment on edge-triggered flip-flops: If the clock waveform is not "clean," erratic triggering or data reading may result. For example, with a positive edge-triggered flip-flop, if the clock waveform rises suddenly from 0.2 V and oscillates for a short while until settling down to a steady-state value of 3.6 V, then the flip-flop may trigger or read data on *several* of the positive-going oscillations, not just on the first large one starting from 0.2 V. The cure is to clean up the clock waveform (eliminate the oscillations) so that there is only *one* positive-going voltage step. The oscillations are much faster than the clock frequency, so a low-pass filter of some sort on the clock line may do the trick. In some cases a simple small capacitance (10 to 100 pF) from the flip-flop clock input to ground may suffice.

13.2.4 The Master/Slave Flip-Flop

The principal disadvantage of the pulse-triggered D flip-flop is that it is transparent; that is, the output Q will follow all changes in the D input, including any noise present along with the good data, *as long as* the clock is high. The Q output held or latched is the *last* value of D just before the clock goes low. The master/slave flip-flop avoids these problems by having two flip-flops—a *master* flip-flop and a *slave* flip-flop—as shown in Fig. 13.12(a). Because of inverter #2, $CK' = \overline{CK}$, and thus only one of these two flip-flops is enabled at any one time. The flip-flop is "opaque" in the sense that input data is never directly transferred to the output. The data is entered in the master flip-flop when the clock goes high; that is, when CK = 1, A will follow D, as long as the clock is high. When the clock goes low, the data bit A is then transferred to the slave flip-flop and it appears at the output. Specifically, when CK = 1, the master flip-flop is enabled and A = D. Inverter #2 makes CK′ = 0, so the slave flip-flop is "off" or disabled or latched. When CK = 0, the master flip-flop is disabled and CK′ = 1; the slave flip-flop is enabled. Then the output of the slave flip-flop Q = A when the clock goes low. The net result is that the data input is entered in the master flip-flop when the clock is high and is transferred to

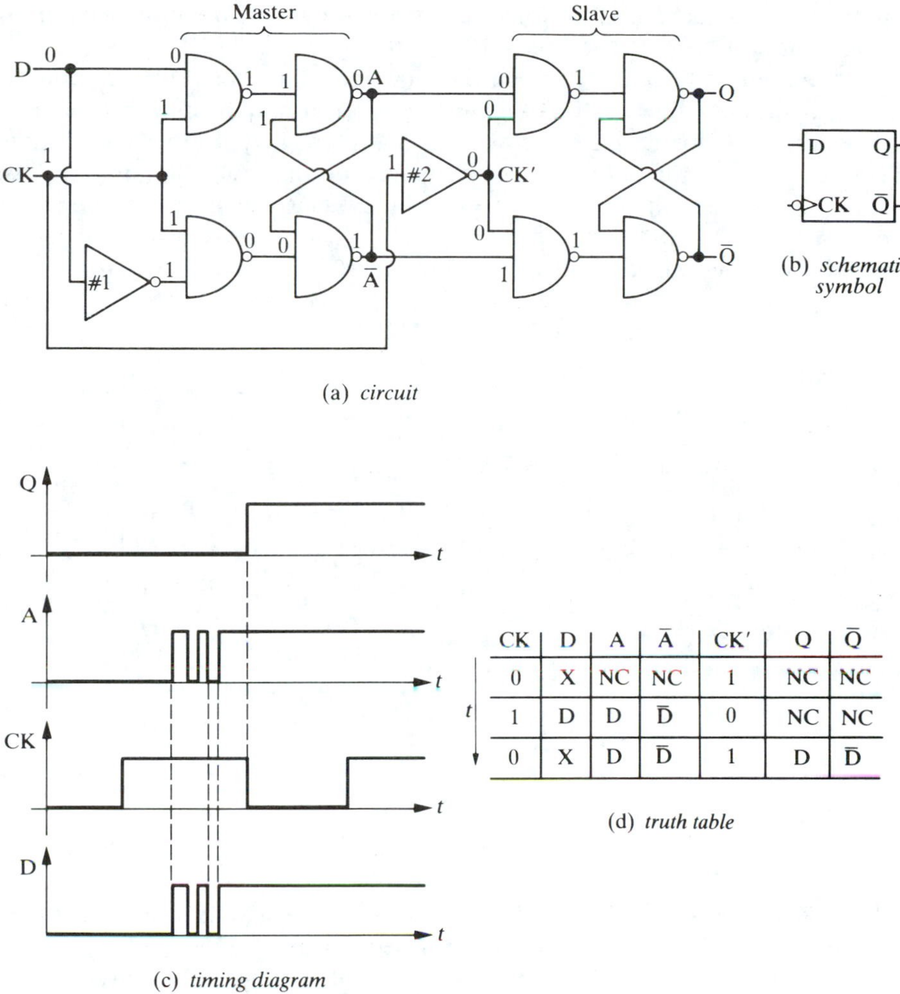

(a) *circuit*

(b) *schematic symbol*

(c) *timing diagram*

CK	D	A	$\bar{A}$	CK′	Q	$\bar{Q}$
0	X	NC	NC	1	NC	NC
1	D	D	$\bar{D}$	0	NC	NC
0	X	D	$\bar{D}$	1	D	$\bar{D}$

(d) *truth table*

FIGURE 13.12 D type master/slave flip-flop.

the slave flip-flop one-half clock period later when the clock goes low. The data input therefore appears at the output when the clock goes low. At no time are both master and slave flip-flops enabled together; in other words, the master flip-flop is never transparent from input to output. The circuit is approximately equivalent to a negative edge-triggered flip-flop.

The fact that the output Q = D when the clock goes *low* is symbolized by the small circle or bubble on the schematic diagram, as shown in Fig. 13.12(b). The timing diagram of Fig. 13.12(c) illustrates this; notice that Q does not change to equal D until the clock pulse goes low again. The truth table is shown in Fig. 13.12(d). X indicates *either* a 0 or a 1, and NC indicates "no change." The logic levels are shown on the circuit diagram [Fig. 13.12(a)] for the row CK = 1, D = 0.

The output Q equals A when the clock goes low, and Q stays equal to A until the next time the clock goes low when it may or may not change, depending on whether A has changed. The point is that Q will not follow *changes* in D as for a simple pulse-triggered flip-flop; Q will equal the *last* value of A = D when the clock goes low. This is similar to a negative edge-triggered flip-flop. An RS master/slave flip-flop can be made by simply omitting the first inverter of Fig. 13.12(a), as shown in Fig. 13.13.

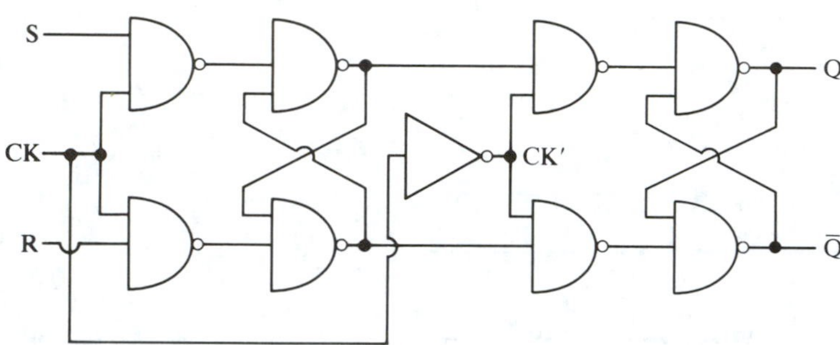

FIGURE 13.13 RS master/slave flip-flop.

In general, time edge-triggered flip-flops are preferred over master/slave flip-flops.

13.2.5 The JK Flip-Flop

The JK flip-flop is similar to the RS flip-flop except that all four possible input combinations of J and K are permitted, whereas in the RS flip-flop the input combination R = S = 1 is not permitted because it will produce an indeterminate output depending on the slight difference between the length of the R and S inputs. The truth table for the JK flip-flop is shown in Fig. 13.14(a), and the standard schematic symbol in 13.14(b). A toggle flip-flop can be made from a JK flip-flop by simply connecting both J and K inputs to a logic high level (through a 1-kΩ resistor to +5 V), as shown in Fig. 13.14(c). Then for each two complete clock pulses in, there will be exactly one output pulse.

A D-type flip-flop can be made from a JK flip-flop, as shown in Fig. 13.14(d), by simply adding an inverter so that $K = \bar{J}$.

A master/slave pulse-triggered JK flip-flop circuit, its truth table, and schematic symbol are shown in Fig. 13.15.

We recall that for either a two-input or a three-input NAND gate, any 0 input produces a 1 output; the output is 0 if and only if all the inputs are 1. Just as in the case of the D master/slave flip-flop, when the JK flip-flop

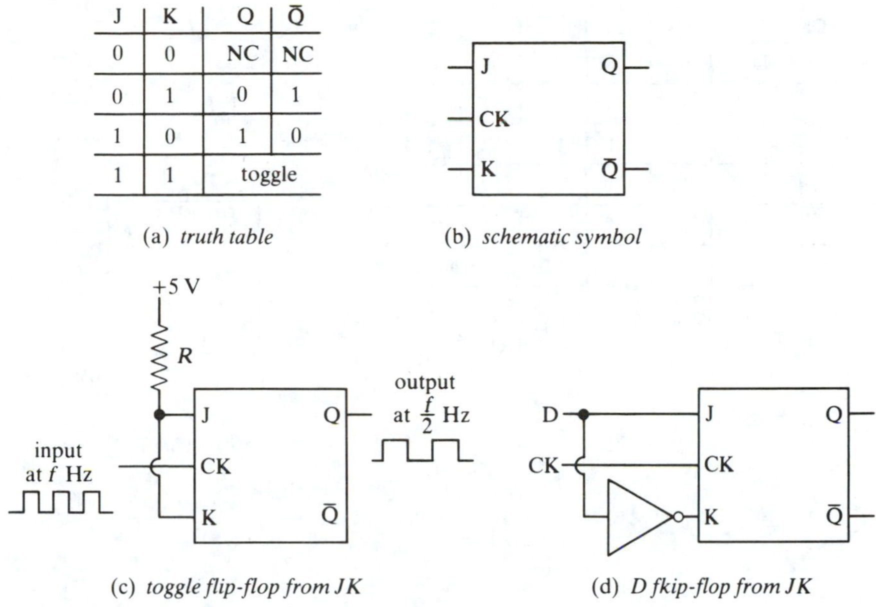

J	K	Q	$\bar{Q}$
0	0	NC	NC
0	1	0	1
1	0	1	0
1	1	toggle	

(a) *truth table*

(b) *schematic symbol*

(c) *toggle flip-flop from JK*

(d) *D fkip-flop from JK*

FIGURE 13.14 The JK flip-flop.

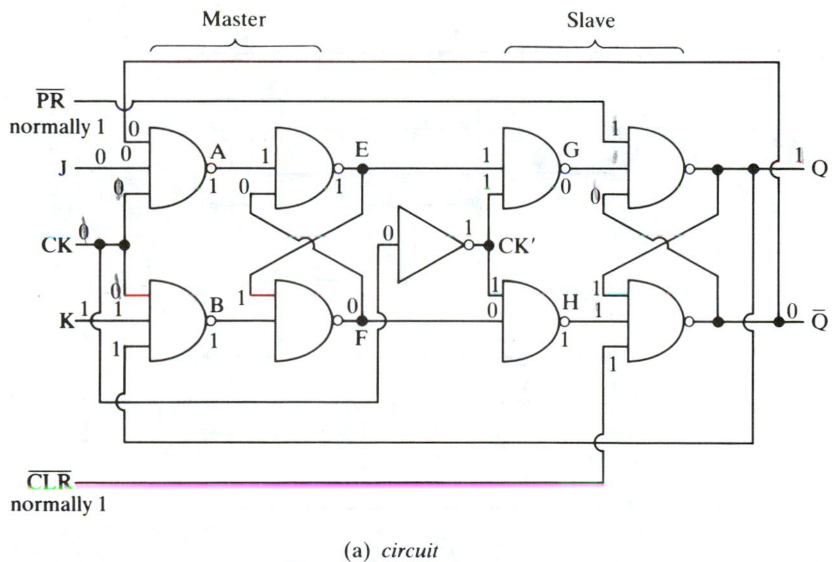

(a) *circuit*

FIGURE 13.15 JK master/slave flip-flop.

CK	J	K	$\overline{PR}$	$\overline{CLR}$	Q	$\overline{Q}$
X	X	X	1	0	0	1
X	X	X	0	1	1	0
⎍↓	0	0	1	1	NC	NC
⎍↓	0	1	1	1	0	1
⎍↓	1	0	1	1	1	0
⎍↓	1	1	1	1	toggle	

JK inputs transferred to Q $\overline{Q}$
outputs on falling edge of clock

(b) *truth table*

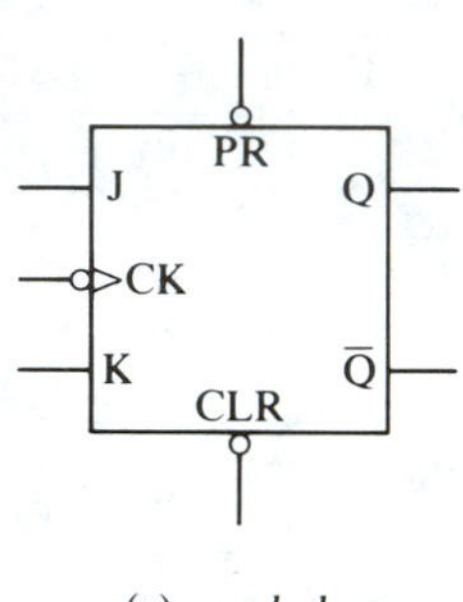

(c) *symbol*

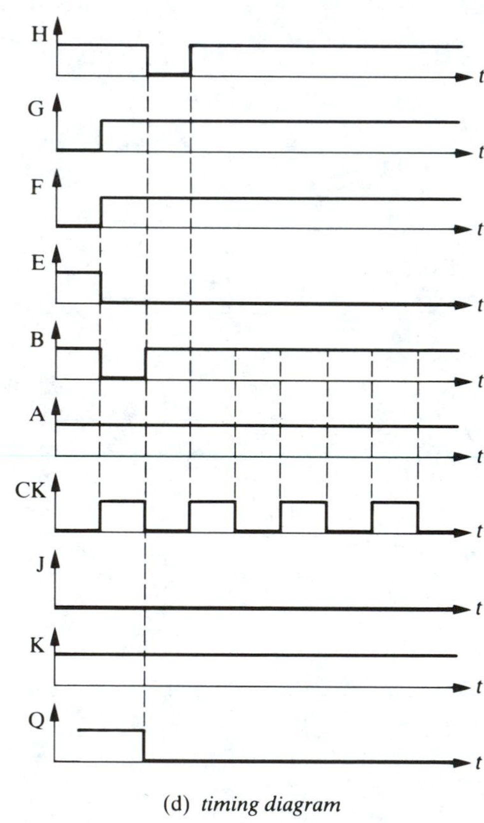

(d) *timing diagram*

FIGURE 13.15 Continued.

clock is high only the master flip-flop is enabled, and the J = 1 input sets E = 1. When the clock is low, CK′ = 1 and only the slave flip-flop is enabled; and the output is set to Q = E when the clock pulse goes low. The J and K inputs are *synchronous*, because they set or reset the output Q precisely at the trailing edge of the clock pulse. There often are two asynchronous inputs, called PRESET and CLEAR (sometimes SET and RESET); these inputs go directly to the last two NAND gates and override the clock; that is, they immediately change the output Q regardless of the clock level. The schematic symbol is shown in Fig. 13.15(c) and the timing diagram in Fig. 13.15(d).

13.2.6 Flip-Flop Setup and Hold Times

For any clocked device such as a flip-flop, the input should be stable for at least a *setup* time t_s before the clock pulse enables the device. This setup time varies from device to device and also depends slightly on whether the transition is $0 \rightarrow 1$ or $1 \rightarrow 0$. The setup time may be zero or up to 20 ns as for a 7474. The input should be stable for at least a ***hold*** time t_h after the clock pulse enables the device. For most devices the hold time is zero, but for the 7474 the hold time is 5 ns. Setup and hold times may differ for different family devices of the same type; for example, the 74LS74 has the same 20-ns setup time but a zero hold time compared to that of the 7474.

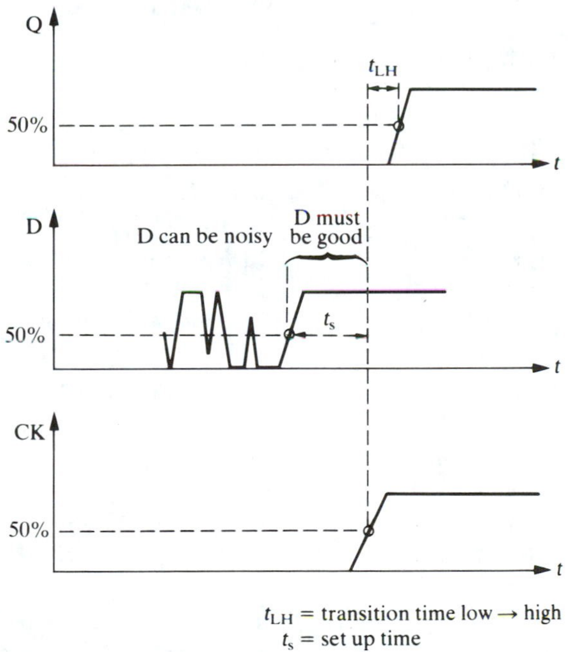

FIGURE 13.16 Setup and hold times.

For most pulse-triggered flip-flops enabled by a voltage *level*, the setup and hold times are zero; this is the case for the 7473 and the 7476 JK flip-flops. Figure 13.16 shows the setup time for a positive edge-triggered flip-flop.

There is also a *transition time* for flip-flops. This is the time for the flip-flop output to change from 0 to 1 or 1 to 0. Generally, the $1 \rightarrow 0$ transition is slightly faster than the $0 \rightarrow 1$ transition because it is faster to turn on a transistor than to turn one off.

13.2.7 Flip-Flop Summary

Three basic types of flip-flops are useful: (1) the simple RS flip-flop or latch, such as the 74LS279, which is a quad RS latch; (2) the D edge-triggered flip-flop, such as the 74LS74A, which is a dual D-type positive edge-triggered flip-flop with separate asynchronous PRESET and CLEAR inputs; (3) the JK flip-flop such as the 74LS107A dual JK master/slave negative edge-triggered flip-flop with separate CLEAR inputs (but no PRESET). Another example is the 74LS109, which is a dual positive edge-triggered JK flip-flop with separate PRESET and CLEAR inputs. Some useful flip-flops are given in Table 13.1.

TABLE 13.1. Useful Flip-Flops

Flip-Flop	*Type*	*No. Pins*	f_{max} *(MHz)*	t_s *(ns)*	t_h *(ns)*	*Power/FF (mW)*
7473	Dual JK pulse-triggered	14	20	0↑	0↓	50
74LS73A	Dual JK neg. edge-triggered	14	45	20↓	0↓	10
74LS279	Quad RS latch (RS FF)	16	(10-ns delay)			10
74LS75	Quad clocked latch (D FF)	16	(15-ns delay	20	0	160
7474	Dual D pos. edge-triggered	14	25	20↑	5↑	43
74LS74A	Dual D pos. edge-triggered	14	33	20↑	0↑	10
7476	Dual JK pulse-triggered	16	20	0↑	0↓	50
74LS76A	Dual JK neg. edge-triggered	16	45	20↓	0↓	10
74LS109A	Dual JK pos. edge-triggered	16	33	25↑	0↓	10
74LS107A	Dual JK master/slave	14	45	20↓	0↓	10

13.2.8 Simple Flip-Flop Applications

One of the most common uses of flip-flops is in data storage. All registers in microprocessors are basically flip-flops (e.g., eight flip-flops to store one byte). The circuit of Fig. 13.17 shows how incoming parallel four-bit data

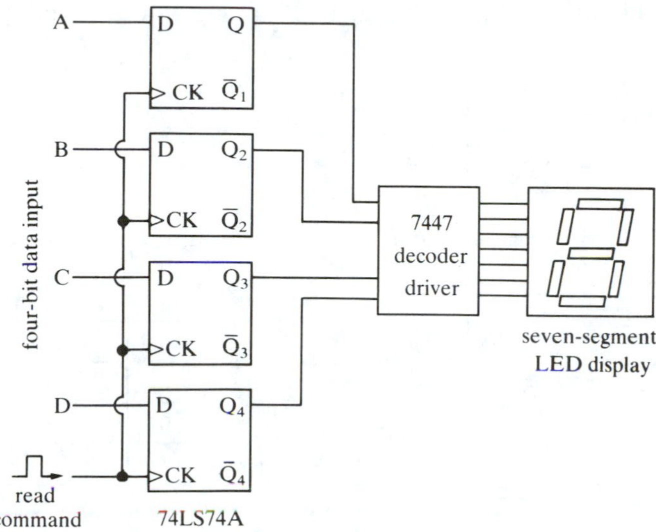

FIGURE 13.17 Simple flip-flop data latch circuit.

on four data lines A, B, C, and D can read out on seven-segment LEDs at a READ command pulse fed to all the flip-flop clocks. If the data bits are already present at the ABCD inputs, then, when the READ command pulse goes high, all four flip-flops are enabled simultaneously and each flip-flop output Q takes on the value of its data input. Thus, $Q_1 = A$, $Q_2 = B$, $Q_3 = C$, and $Q_4 = D$ when the READ pulse goes high. The decoder/driver then lights up the LEDs to display the data. When the READ pulse goes low, the flip-flop outputs are latched or frozen (Q does not change if D changes), so the LED display does not change until a new READ command comes along that will display the new ABCD data.

D flip-flops are also commonly used to store data in computers: Data is typically stored in one part of the computer and then at a later specific time transferred to another part of the computer over a *bus*, which is a group of wires, one for each bit. To minimize the number of buses, we usually use the same bus for data transmission in both directions, and for data transmission from several different flip-flop sources. In other words, for eight-bit words many different sets of eight D flip-flops will be connected to the same eight data lines or bus. The only way this can be accomplished without having different data on the same wire fighting each other is to use open collector outputs on the flip-flops, which was the case in the older PDP-8 and PDP-11 computers, or three-state outputs on the flip-flops. Most modern microprocessors use this type of data storage. Some flip-flop chips include the three-state output. For example, the 74LS374 is an octal D-type positive edge-triggered flip-flop with three-state output on a 20-pin chip. "Octal" means there are eight D flip-flops on the chip. A diagram of one of the eight D flip-flops is shown in Fig. 13.18, along with the truth table.

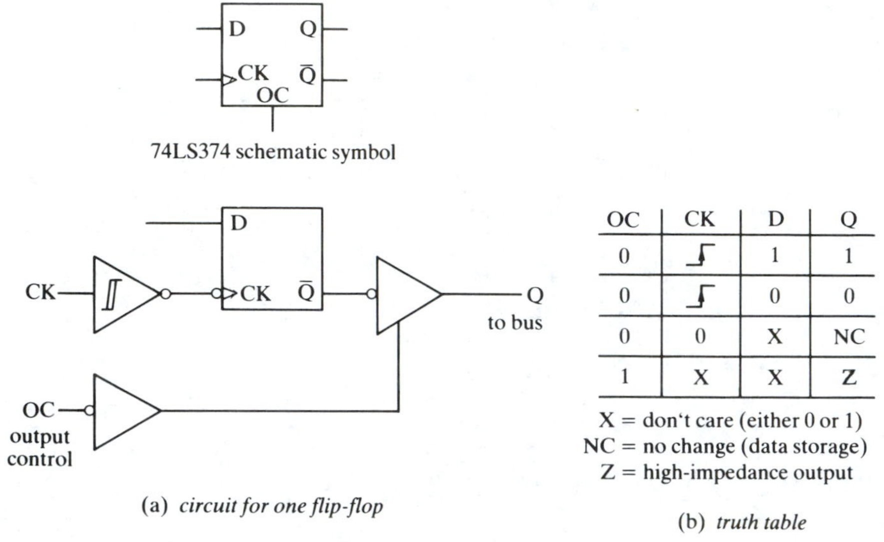

OC	CK	D	Q
0	↑	1	1
0	↑	0	0
0	0	X	NC
1	X	X	Z

X = don't care (either 0 or 1)
NC = no change (data storage)
Z = high-impedance output

(b) *truth table*

FIGURE 13.18 74LS374 octal three-state positive edge-triggered flip-flop.

On the positive edge of the clock, the output Q will be set equal to the data input D for a low "output control." When the clock is low, the outputs Q are latched or stored for a low output control. When the output control is high, the output Q is "off"; that is, a very high impedance is presented to the bus, and other flip-flops can feed data onto the same bus or read data off the bus. During the time the output control is high, new data can be fed into the flip-flops to set $\bar{Q}$, or else the flip-flops can store data as $\bar{Q}$. Notice that there is a Schmitt trigger inverter at the clock input for better noise rejection on the clock signal.

Edge-triggered flip-flops can be either positive edge triggered or negative edge triggered. Suppose we have a positive edge-triggered flip-flop and we wish to trigger it at the trailing edge of the clock pulse (i.e., when the clock goes from high to low). The solution is simply to invert the clock and use the inverted clock to trigger the flip-flop. If we wish to trigger a negative edge-triggered flip-flop on the leading edge of the clock, the same trick works—just invert the clock.

13.3 RIPPLE OR ASYNCHRONOUS COUNTERS

Counters of many different types can be made from flip-flops. The simplest one is a divide-by-two counter which can be made from one JK flip-flop or one edge-triggered flip-flop, as shown in Fig. 13.19(a).

A divide-by-2^N counter can be made by simply connecting N toggle

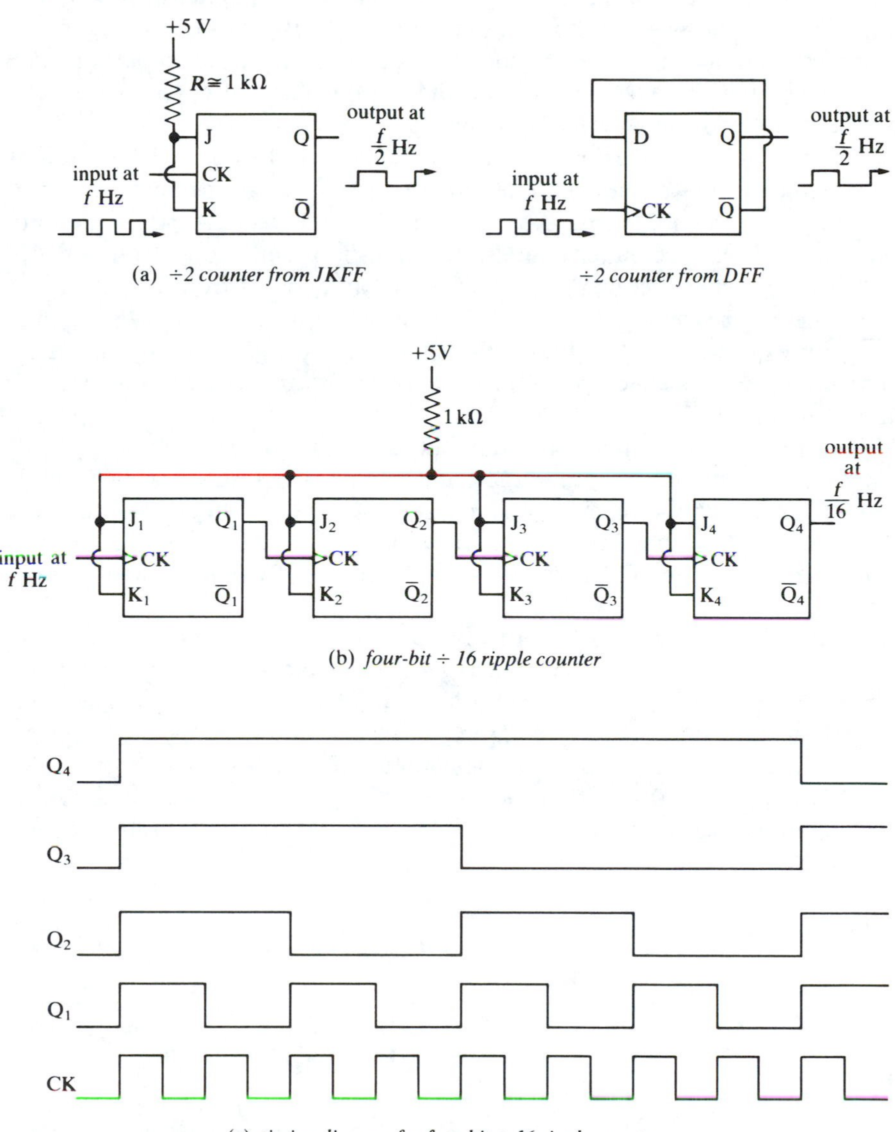

FIGURE 13.19 Asynchronous or ripple counters.

flip-flops in a chain, where the Q output of one flip-flop is the toggle input of the next flip-flop, as shown in Fig. 13.19(b) for $N = 4$. The timing diagram is shown in Fig. 13.19(c). Notice that the binary number $\bar{Q}_4\bar{Q}_3\bar{Q}_2\bar{Q}_1$ counts up from 0 to 15, and the binary number $Q_4Q_3Q_2Q_1$ counts down from 15 to 0, where Q_4 = MSB and Q_1 = LSB. For example, when 11_{10} input pulses have entered the counter, $\bar{Q}_1 = 1$, $\bar{Q}_2 = 1$, $\bar{Q}_3 = 0$, $\bar{Q}_4 = 1$.

Notice that all the flip-flop outputs do not change at exactly the same time because of the propagation delay (about 50 ns) in each flip-flop. The MSB is set N propagation times later than the LSB. We say that the count "ripples" through the counter, and such counters are called *ripple counters* or *asynchronous counters* because all the flip-flops do not change at exactly the same time as the clock input changes.

N flip-flops in a ripple counter can count from 0 to $2^N - 1$ before they repeat their count sequence. The name of the counter specifies the maximum count or number of flip-flops according to the name *N-bit counter*. For example, a four-bit counter contains four flip-flops and counts from 0 to 15; an eight-bit counter contains eight flip-flops and counts from 0 to 255.

The maximum toggle frequency is usually specified for a given flip-flop; for example, a 74LS74A dual positive edge-triggered D flip-flop has a maximum clock frequency of 33 MHz typically and is guaranteed to be more than 25 MHz.

If the complementary flip-flop output $\bar{Q}$ is connected to the toggle input of the next flip-flop, then the binary number $Q_4Q_3Q_2Q_1$ will count up from 0 to 15, and the binary number $\bar{Q}_4\bar{Q}_3\bar{Q}_2\bar{Q}_1$ will count down from 15 to 0. For example, a four-bit counter, cleared to $Q_4Q_3Q_2Q_1 = 0000$, after one input count will read 0001, after two input counts it will read 0010; and so on.

A major disadvantage of a ripple counter is due to the asynchronous nature of the counting. False or incorrect counts appear momentarily while the count is propagating or rippling down the chain of flip-flops. The worst case is when many bits are changing from one count to the next correct count. For example, when a four-bit counter changes count from $7_{10} = 0111$ to $8_{10} = 1000$, all four bits must change. If we assume a typical propagation time of 50 ns for a single flip-flop, then the count sequence will be as follows:

	Q_4	Q_3	Q_2	Q_1	
Correct 7_{10}	0	1	1	1	$t \leqq 0$ (initial condition)
Correct 7_{10}	0	1	1	1	$0 < t < 50$ ns
False 6_{10}	0	1	1	0	$50 < t < 100$ ns
False 4_{10}	0	1	0	0	$100 < t < 150$ ns
False 0_{10}	0	0	0	0	$150 < t < 200$ ns
Correct 8_{10}	1	0	0	0	200 ns $< t$

Thus, in the 200 ns it takes to change from 0111 to 1000, the counter briefly indicates a false 6_{10}, a false 4_{10}, and a false 0_{10}.

A ripple counter can be made to divide by any nonzero number with appropriate feedback to reset or clear the counter to zero at the final desired count. For example, a divide-by-five counter is shown in Fig. 13.20(a), and its count sequence in 13.20(b). The NAND gate clears the counter to 000 when the count of 101 is reached, because at the count of 101 the NAND gate output goes low, which clears all the flip-flops. (Remember the bubble at the flip-flop CLEAR terminal means a low CLEAR input will clear the flip-flop.)

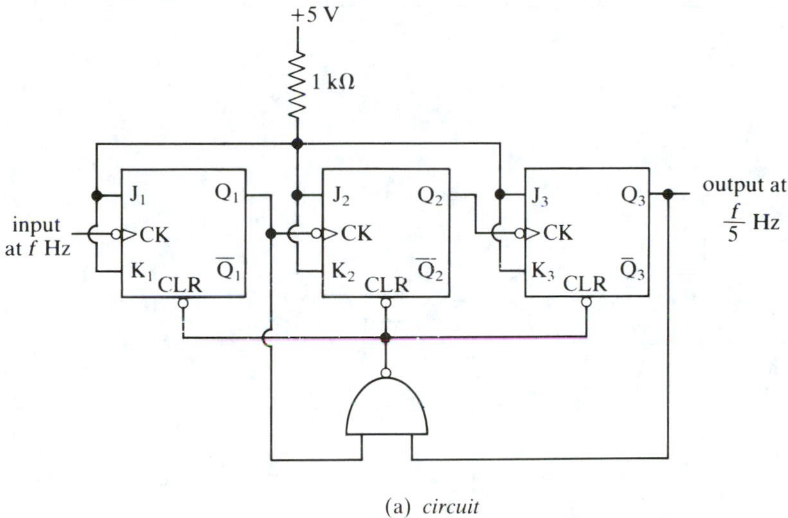

(a) *circuit*

Count	Q_3	Q_2	Q_1
0	0	0	0
1	0	0	1
2	0	1	0
3	0	1	1
4	1	0	0
5	1	0	1
6	0	0	0
7	0	0	1

transient count ≈ 50 ns long (count 5)

(b) *count sequence*

FIGURE 13.20 Divide-by-five ripple counter.

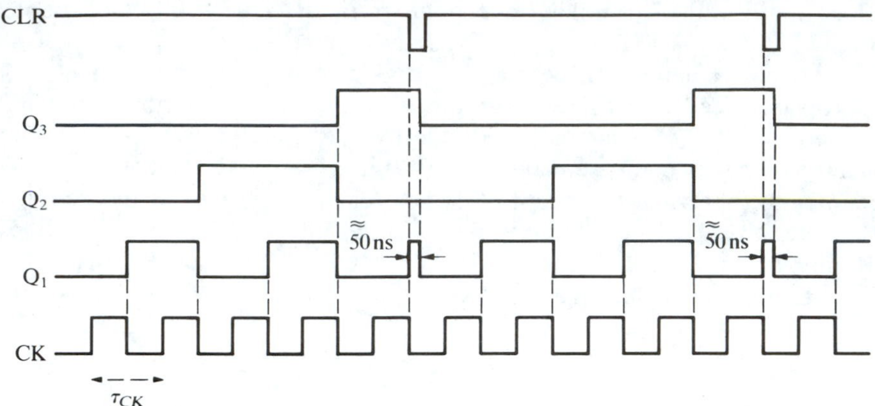

(c) *timing diagram*

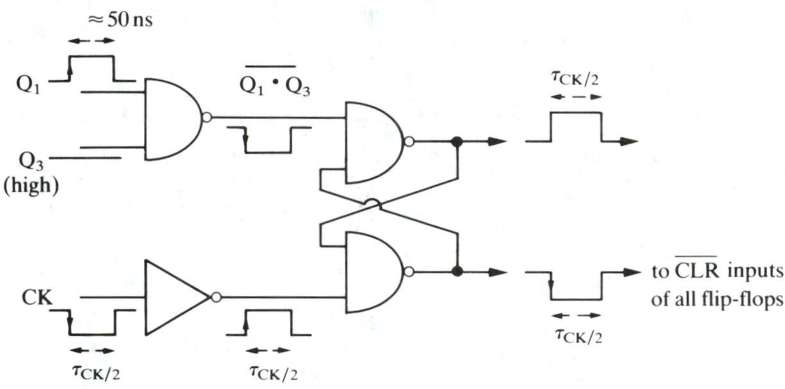

(d) *circuit to lengthen clear pulse*

FIGURE 13.20 Continued.

The condition 101, which is a false count of 5_{10}, exists for the very short time required to clear all the flip-flops, so the count sequence will be 000, 001, 010, 011, 100, 101 (briefly to clear), 000, 001, 010, 011, ... The timing diagram is shown in Fig. 13.20(c). If the brief 101 state is too short to provide a reliable clear for all three flip-flops, the circuit of Fig. 13.20(d) can be added to supply a CLEAR pulse one-half clock period long.

An asychronous or ripple up/down counter can be made that will count either up or down at an external command signal, as shown in Fig. 13.21. A = 1 will make Y = Q_1, and A′ = 0 will make Y′ = 0 (regardless of $\bar{Q}_1$),

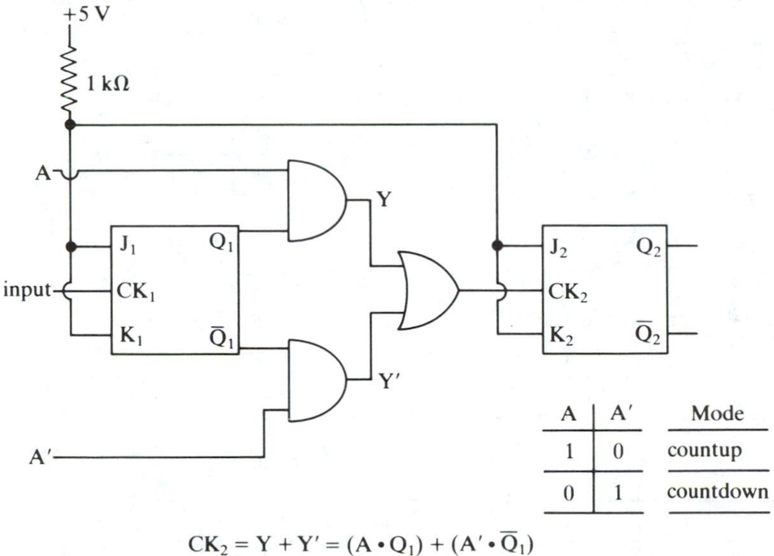

A	A′	Mode
1	0	countup
0	1	countdown

$$CK_2 = Y + Y' = (A \cdot Q_1) + (A' \cdot \overline{Q}_1)$$

FIGURE 13.21 Up/down counter.

which will form an up counter because $CK_2 = Q_1$. Similarly, a low A and a high A′ will make a down counter because $CK_2 = \overline{Q}_1$.

13.4 SYNCHRONOUS COUNTERS

A *synchronous counter* is one in which all the flip-flops change state at exactly the *same* time; that is, the output changes are synchronous with the clock. Thus, the same clock signal must be connected to all the flip-flop ENABLE or clock terminals, as shown in Fig. 13.22(a). Synchronous counters are usually made with JK flip-flops, and, for convenience, the JK flip-flop truth table is repeated in Fig. 13.22(b). If we wire $J = K = 1$ for the first flip-flop, it will divide by two in the usual way, but a little thought shows that we can't do this for all the flip-flops because each flip-flop would then toggle for each input pulse or count, and the output of *each* flip-flop would simply be half the input frequency. What we must do is to gate the JK input of each flip-flop so that it will change state only at the proper count. For example, to go from count 1_{10} to 2_{10} (001_2 to 010_2), Q_1 (the LSB) must change from 1 to 0, Q_2 must change from 0 to 1, but Q_3 must remain 0.

In Table 13.2 we list the JK inputs that must exist for the four possible flip-flop output transitions. Table 13.2 is merely another way of writing the JK truth table.

Let us now design a divide-by-five synchronous counter. Looking at

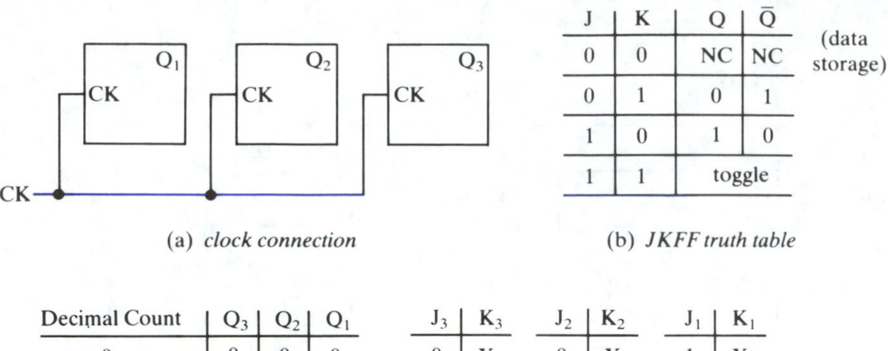

(a) *clock connection*

J	K	Q	$\bar{Q}$	
0	0	NC	NC	(data storage)
0	1	0	1	
1	0	1	0	
1	1	toggle		

(b) *JKFF truth table*

Decimal Count	Q_3	Q_2	Q_1
0	0	0	0
1	0	0	1
2	0	1	0
3	0	1	1
4	1	0	0
5	0	0	0
6	0	0	1
⋮	⋮	⋮	⋮

(c) *÷5 count sequence*

J_3	K_3	J_2	K_2	J_1	K_1
0	X	0	X	1	X
0	X	1	X	X	1
0	X	X	0	1	X
1	X	X	1	X	1
X	1	0	X	0	X
0	X	0	X	1	X
0	X	1	X	X	1
⋮	⋮	⋮	⋮	⋮	⋮

(d) *JK values for next count*

FIGURE 13.22 Synchronous counter.

the divide-by-five count sequence of Fig. 13.22(c) and Table 13.2, we can determine the required JK inputs for the three flip-flops. These JK inputs are shown in Fig. 13.22(d). For the first flip-flop we immediately see that $J_1 = \bar{Q}_3$ and $K_1 = 1$ will work for all the counts. For the second flip-flop $J_2 = K_2 = Q_1$ will work. For the third flip-flop $J_3 = Q_1 \cdot Q_2$ and $K_3 = 1$ or $K_3 = Q_3$. Thus, one AND gate and some wiring will produce the desired divide-by-five count sequence, as shown in Fig. 13.23(a). The timing diagram is shown in Fig. 13.23(b).

The RESET is not shown but should be included in an actual counter,

TABLE 13.2. Required JK Inputs for the Four Possible Flip-Flop Output Transitions

Q Transition	*J*	*K*
0 to 1	1	X
1 to 0	X	1
0 to 0	0	X
1 to 1	X	0

X = don't care (either 0 or 1).

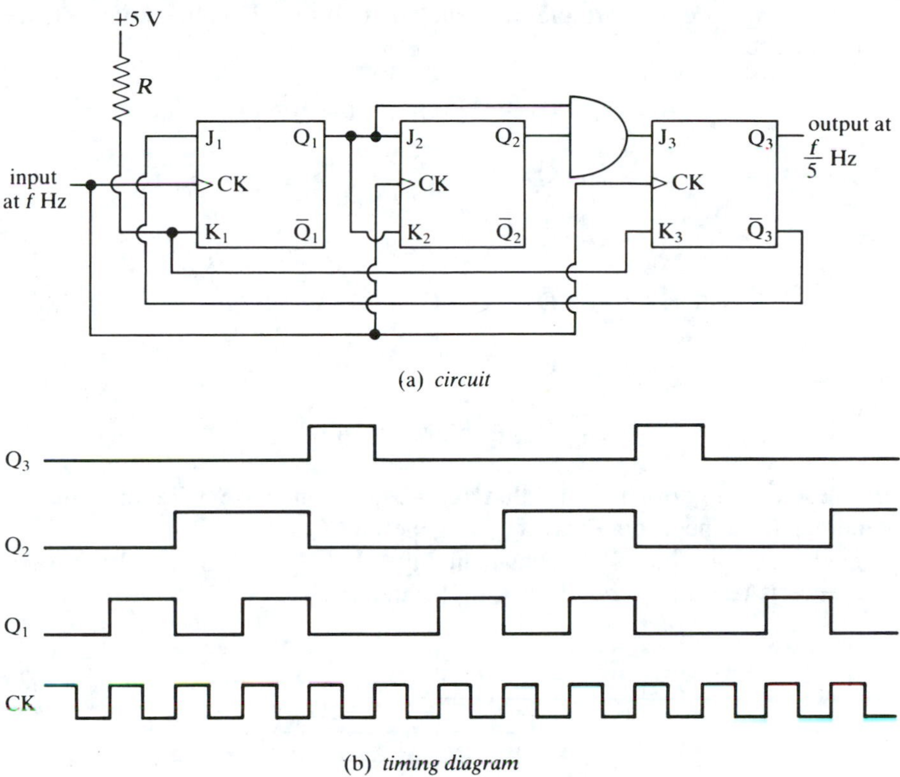

(a) *circuit*

(b) *timing diagram*

FIGURE 13.23 Divide-by-five synchronous counter.

because when the power is turned on, the flip-flops will assume random states, which might not even be in the desired count sequence.

We can make a synchronous counter that will count through an *arbitrary* sequence of counts in a fixed order. For example, for the Gray count sequence (in decimal notation) 0, 1, 3, 2, 6, 7, 5, 4, 0, . . . , the required JK outputs for the three flip-flops are as follows:

Decimal Count	Q_3	Q_2	Q_1	J_3K_3	J_2K_2	J_1K_1	Q	J	K
0	0	0	0	0X	0X	1X	0→0	0	X
1	0	0	1	0X	1X	X0	0→1	1	X
3	0	1	1	0X	X0	X1	1→0	X	1
2	0	1	0	1X	X0	0X	1→1	X	0
6	1	1	0	X0	X0	1X			
7	1	1	1	X0	X1	X0			
5	1	0	1	X0	0X	X1			
4	1	0	0	X1	0X	0X			
0	0	0	0	0X	0X	1X			
1	0	0	1	0X	1X	X0			

Using the preceding table, we conclude that the JK values for the three flip-flops are

$$J_1 = Q_2 \cdot Q_3 + \bar{Q}_2 \cdot \bar{Q}_3 \qquad (=\overline{Q_2 \oplus Q_3})$$

$$J_2 = Q_1 \cdot \bar{Q}_3$$

$$J_3 = \bar{Q}_1 \cdot Q_2$$

$$K_1 = Q_2 \cdot \bar{Q}_3 + \bar{Q}_2 \cdot Q_3 \qquad (=Q_2 \oplus Q_3)$$

$$K_2 = Q_1 \cdot Q_3$$

$$K_3 = \overline{Q_1 + Q_2} \quad \text{or} \quad K_3 = \bar{Q}_1 \cdot \bar{Q}_2$$

But Boolean algebra tells us that $K_1 = \bar{J}_1$, so once we wire up gates to generate J_1 we need only invert J_1 to generate K_1.

The counter circuit is shown in Fig. 13.24. The binary Gray count sequence is taken in parallel off Q_1, Q_2, and Q_3.

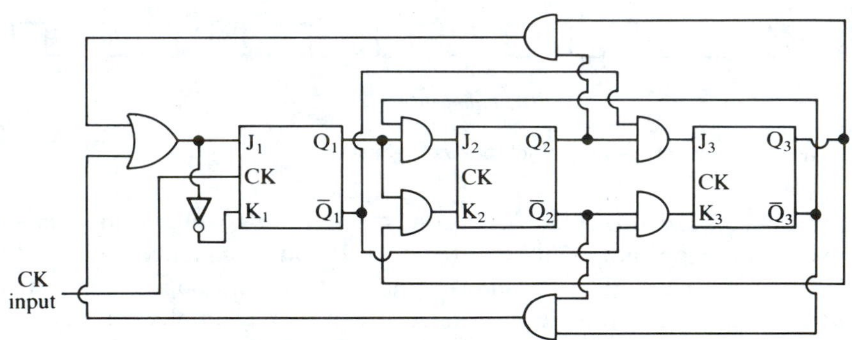

FIGURE 13.24 Gray synchronous counter; output = $Q_3Q_2Q_1$ (parallel).

There are many counter chips available that minimize counter design work. For example, the 74LS196 is a four-bit asynchronous ripple counter that contains a divide-by-two counter and a divide-by-five counter; connecting the output of the divide-by-two counter to the input of the divide-by-five counter gives a divide-by-ten or a decade counter. It will count up to 100 MHz. The 74LS197 contains a divide-by-two counter and a divide-by-eight counter. These two counter chips consist of four dc coupled master/slave flip-flops. They also have fully programmable inputs, which means the four flip-flop outputs can be preset to any desirable four values (the four data inputs) at the command of a count/low signal. They also have an asynchronous common CLEAR line that will reset all the flip-flops to zero regardless of the clock. Information is transferred from the input to the output of the flip-flops on the negative edge of the clock. The 74LS196 can also be used as a four-bit latch, so it is a versatile chip.

The 74LS163A is a four-bit fully programmable counter with a synchronous CLEAR. It consists of four dc coupled JK flip-flops and will count up to 25 MHz.

The 74LS160 is a fully programmable decade counter with an asynchronous or direct CLEAR and will count up to 25 MHz.

Modulo-N counters are also available; they will divide by N, where N is specified by a four-bit input; N can range from 2 to 15. The 8520 is one example. It can count up to 20 MHz and can be cascaded, so division by large numbers is possible; for example, two divide-by-15 counters in series will divide by $15^2 = 225$. It can also be used as a PISO (parallel in series out) shift register. (*PISO* is defined in Section 13.5.1.)

Many counters can be preset to a specific count (i.e., they have data inputs). There are two types of preset or loading: *synchronous* loading and *jam* or *asynchronous* loading. In synchronous loading the input data appear on the counter outputs only after a clock pulse. In other words, the data loaded in appear at the leading edge of the next clock pulse. In jam loading the input data appear on the counter output as soon as the LOAD command is activated, independent of the clock. The input data are loaded in a parallel fashion.

13.5 SERIAL/PARALLEL DATA CONVERSION

Binary data can be in either serial or parallel form. In serial form the bits come along on one wire at different times. One common example of this is ASCII code binary words from a keyboard. In parallel form the bits are simultaneously present on different wires; an eight-bit word would need eight separate wires, and so on. In general, serial data transmission is cheap but slow, and parallel data transmission is fast but expensive. Data words are invariably transmitted and processed within computers in parallel fashion but are often transmitted long distances in serial fashion (e.g., over phone lines). This section will briefly treat the conversion from serial form to parallel form, and vice versa.

13.5.1 The Shift Register

Serial to parallel and parallel to serial data conversions can be accomplished with a *shift register*, which is basically a series of flip-flops connected so that the output of one is the input of the next. For each clock pulse the data in one flip-flop is shifted over to the next flip-flop. A series of edge-triggered D flip-flops connected as shown in Fig. 13.25 is the basic shift register.

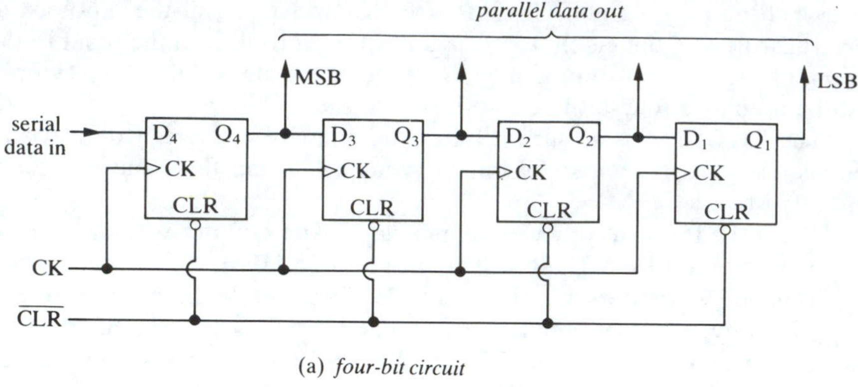

(a) *four-bit circuit*

No. clock pulses in	Q_4	Q_3	Q_2	Q_1
0	0	0	0	0
1	1	0	0	0
2	1	1	0	0
3	1	1	1	0
4	1	1	1	1

(b) *count sequence for* D = 1 *constant input*

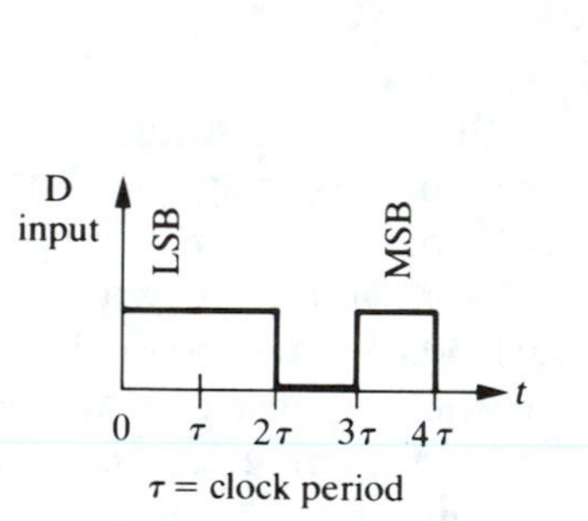

No. clock pulses in	MSB Q_4	Q_3	Q_2	LSB Q_1
0	0	0	0	0
1	1	0	0	0
2	1	1	0	0
3	0	1	1	0
4	1	0	1	1

(c) *count sequence for* D = 1011 *serial input*

FIGURE 13.25 The shift register.

Notice that the flip-flops must be *edge* triggered; if they were positive *pulse* triggered, the D input at the first flip-flop would "ripple" down the chain of flip-flops while the clock was high. The important requirement is that the rise time of the clock waveform be less than the propagation time for the signal through a flip-flop. If the flip-flops are initially cleared to zero and the D input is maintained at 1 for four clock pulses, then the sequence of flip-flop states would be as shown in Fig. 13.25(b).

We see that after each input clock pulse, the 1 data bit *moves one flip-flop down the chain* (i.e., the bit has been shifted). After four input clock pulses all four flip-flops have been set to 1. If we regard the flip-flop outputs Q_4, Q_3, Q_2, Q_1 as a four-bit binary word and the constant D = 1 input as a four-bit serial word 1111 and if we take the flip-flop outputs off the four separate wires, $Q_4Q_3Q_2Q_1$, we have converted the 1111 serial input to parallel form after four clock pulses. Similarly, if the D input is 1011, then the sequence of flip-flop outputs will be as shown in Fig. 13.25(c).

The input word 1011 is now available at the four flip-flop outputs in parallel form, with Q_4 being the MSB and Q_1 the LSB. This is an example of *serial in parallel out* (SIPO) data conversion. Notice that the parallel output data are available only *after* four clock pulses for a four-bit input serial word.

SIPO shift registers are available in chip form in many varieties. One is the 74LS164, which is an eight-bit SIPO shift register that can operate on a clock frequency of up to 25 MHz. It has an asynchronous CLEAR and shifts data on the positive edge of the clock. It also has two serial inputs; a high on either input enables the other input.

After more clock pulses the input word will appear in serial form at the output of the last flip-flop Q_1, being delayed by four clock pulses. This is called *serial in serial out* (SISO) data conversion.

To convert parallel data into serial form, we need only feed in the parallel data to the flip-flops through the PRESET or CLEAR inputs, as shown in Fig. 13.26. If the LOAD command is low (L = 0), then Y = Y′ = 1 and both the PRESET and CLEAR inputs are inactive; the flip-flop is unaffected. If L = 1, then $Y = \bar{D}$ and Y′ = D from the NAND gate truth table. Thus, if D = 0, Y = 1 and the PRESET is inactive; but Y′ = 0, which

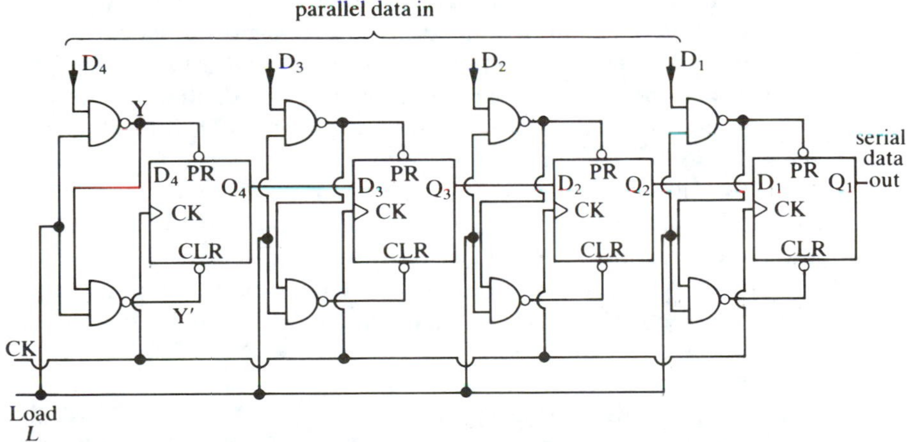

FIGURE 13.26 Shift register parallel load circuit.

clears the flip-flop to Q = 0. If D = 1, Y = 0, which presets the flip-flop to Q = 1, and Y′ = 1, which means the CLEAR is inactive. The net result is that if L = 1, then Q = D for all four flip-flops; if L = 0, Q is unaffected by D. Then each clock pulse will shift the data over one flip-flop; for four-bit data four clock pulses will cause the data to appear at Q_1 in serial form. This is called *parallel in serial out* (PISO) data conversion.

An example of an eight-bit PISO shift register is the 74LS166, which can operate on a clock frequency up to 25 MHz. It has an asynchronous CLEAR and a shift/load input S/L. When S/L = 1, serial data can be loaded and shifted serially with each clock pulse. When S/L = 0, parallel data can be loaded in synchronously (i.e., at the next positive-going clock pulse). There are two clock inputs; if either is low, the other clock is enabled; if either is high, the other clock is disabled. The CMOS 4021 is also an eight-bit PISO shift register.

There also exist *parallel in parallel out* (PIPO) shift registers, such as the 74LS195A and the CMOS 4035. Variable-length shift registers are also available, which have a variable number of bits from 1 to 64 controllable by a six-bit control input. An example is the CMOS 4557.

One of the more versatile shift registers is the *universal shift register*, which can shift the data either to the right or to the left and can be a SIPO, a SISO, a PISO, or a PIPO shift register. Examples are the 74194 and the CMOS 4034, which also has three-state outputs so that it can transfer parallel data between two buses in either direction.

13.5.2 Charge-Coupled-Device Shift Registers

A charge-coupled device (CCD) is a monolithic solid-state analog shift register consisting of a series of metallic "transfer" or "gate" electrodes deposited on a thin insulating layer of SiO_2 over an n-p silicon substrate, as shown in Fig. 13.27. If the CK is high (positive voltage) and CK′ is low, then a periodic potential well for electrons is produced, as shown in Fig. 13.27(a). These potential wells can be used to store or trap electrons. The shaded n-type regions produce a left-right asymmetry so that any trapped electrons will tend to move as far as possible to the right inside the well. When the next clock pulse comes along, making CK low and CK′ high, the potential wells will all be shifted to the right by a distance d, and any electrons trapped in a well will also be shifted along with the well a distance d. The new situation after one clock pulse is shown in Fig. 13.27(b).

Any number of electrons can be stored in each well, so the device is an *analog* shift register. Each pair of gate electrodes corresponds to one flip-flop in an ordinary shift register. A CCD shift register is approximately five times smaller than an MOS flip-flop shift register, but it does have the disadvantage that there is a small probability that a few electrons may be left behind in each shift from one well to the next well. This effect is cumulative—the more shifts, the worse the signal degradation.

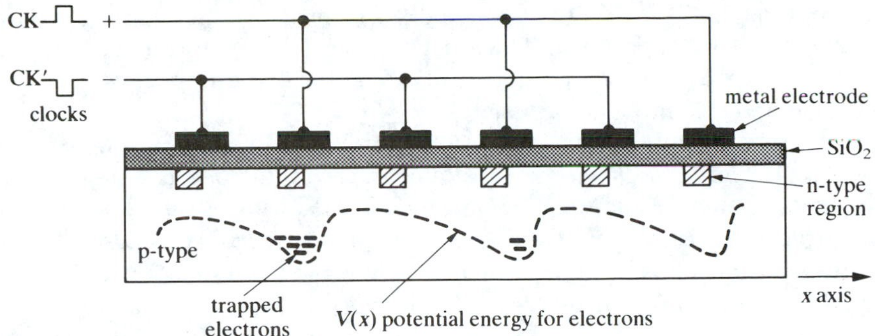

(a) *initial condition showing trapped electrons*

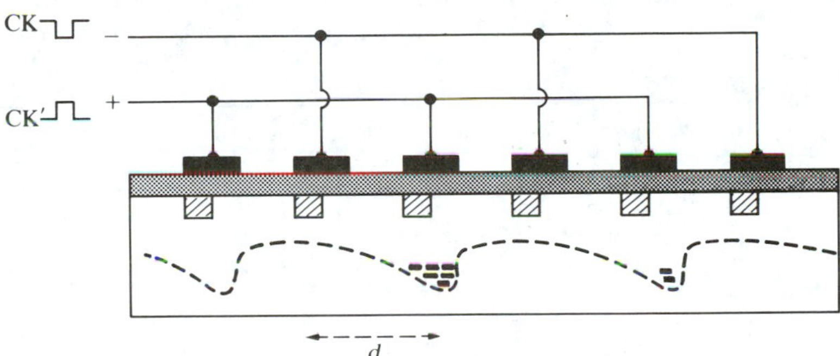

(b) *trapped electrons one clock pulse later*

FIGURE 13.27 CCD shift register.

The CCD can be loaded in parallel or series fashion. Typical shift rates are 5 to 10 MHz, and the output is taken from the last or end potential well. In other words, the output is serial. As a rule-of-thumb, each electron transferred produces approximately 1 or 2 μV at the output.

Applications of CCD devices include audio signal processing, memory, and optical scanning. An optical scanning device can be either one dimensional (linear imaging device or LID) or two dimensional (area imaging device or AID). An LID is a series of small photodiodes spaced approximately 10 μm apart in a monolithic array and parallel coupled to a CCD shift register. Each photodiode produces an electron charge linearly proportional to its incident light energy (400 to 1100 nm). When the LID is gated, each electron bunch is transferred to the potential well right next to the photodiode. The CCD is then clocked to feed out the charges serially from the last potential well. Thus, each charge pulse represents the total light energy that struck a particular photodiode. The output pulses can be digitized and fed into a computer.

An AID is simply a two-dimensional array of photodiodes coupled with CCD shift registers all on one substrate. Each photodiode is equivalent to one pixel (picture element). When an AID is used with an image intensifier, an extremely sensitive night vision device can be made. An AID is literally electronic film and is used in virtually every video camera.

AIDs are also used as film in reconnaissance satellite cameras. The light intensity from each pixel is digitized and radioed back to earth where the picture is reconstructed.

13.5.3 Shift Register Application: The Ring Counter

The *ring counter* is a shift register in which the serial output from the last flip-flop is fed back into the serial input of the first flip-flop, as shown in Fig. 13.28.

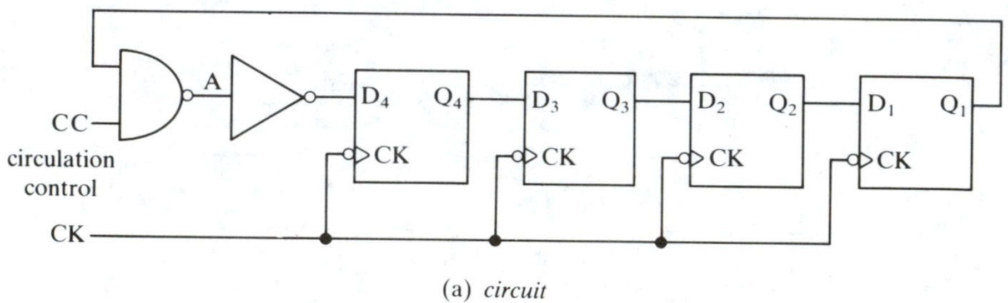

(a) *circuit*

No. clock pulses	Q_4	Q_3	Q_2	Q_1
0	1	0	0	0
1	0	1	0	0
2	0	0	1	0
3	0	0	0	1
4	1	0	0	0
5	0	1	0	0
6	0	0	1	0
7	0	0	0	1
⋮		⋮		

(b) *ring counter states if initial state* $Q_4\,Q_3\,Q_2\,Q_1 = 1000$

FIGURE 13.28 The ring counter.

The NAND gate and the inverter at the input act as a "circulation control." If the circulation control CC = 0, then A = 1 and $D_4 = 0$ regardless of Q_1. Thus, the flip-flops will be in the state 0000 after four or more clock pulses, because D_4 is held at 0. However, if the circulation control is 1, then $A = \bar{Q}_1$ and $D_4 = Q_1$; that is, the output Q_1 is fed back into the input of the first flip-flop. For example, if the word $Q_4Q_3Q_2Q_1 = 1000$ is loaded in through a separate parallel loading circuit (not shown in Fig. 13.28), when the circulation control goes to 1, the 1 will circulate indefinitely in the shift register, as indicated in Fig. 13.28(b).

13.5.4 Shift Register Application: The Johnson Counter

If the complement of the final flip-flop output is fed back to the input of the first flip-flop, then we have a *Johnson counter* or *twisted ring counter*, as shown in Fig. 13.29(a) for an eight-bit shift register.

If the flip-flops are all cleared to 00000000, then $\bar{Q}_1 = 1 = D_8$; and after the first clock pulse, Q_8 has been set to 1 and we have the state 10000000. Figure 13.29(b) shows the sequence of counter states.

If we add up all the 1's for $Q_8, Q_7, \ldots, Q_2$ in an analog summer, as shown in Fig. 13.29(c), we get an analog waveform that is a crude step approximation to a sine wave, as shown in Fig. 13.29(d). To get an odd number of steps we do not use Q_1. Obviously this waveform is only approximately sinusoidal in shape. The sine-wave frequency is one-sixteenth of the clock frequency for an eight-bit counter.

Let us now consider the problem of making a better approximation to a sine wave by varying the *amplitude* of the steps. Obviously, the more steps there are the smoother the waveform will be, but there will always be a discontinuity or abrupt step in the magnitude of the summer output for each clock pulse. These abrupt steps generate a large number of Fourier components. The waveform output from the summer can be made more nearly sinusoidal by making the steps near the bottom and top of the wave smaller than those near the center.

A Fourier analysis of the output from an eight-bit Johnson counter is

$$v(t) = \frac{a_0}{2} + \sum_{n=1}^{\infty} a_n \cos n\omega_0 t \qquad \omega_0 = \frac{2\pi}{16\tau} = \frac{\pi}{8\tau} \qquad \tau = \text{clock period}$$

We have omitted the sine terms because we have chosen $t = 0$ to make $v(t)$ even. The Fourier coefficients are

$$a_n = \frac{2}{16\tau}\int_{-8\tau}^{8\tau} v(t)\cos\frac{n\pi t}{8\tau}\,dt = \frac{1}{4\tau}\int_0^{8\tau} v(t)\cos\frac{n\pi t}{8\tau}\,dt$$

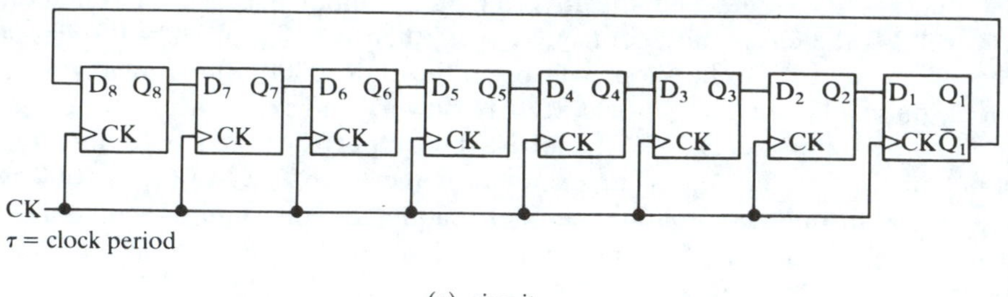

(a) *circuit*

after clock pulse	Q_8	Q_7	Q_6	Q_5	Q_4	Q_3	Q_2	Q_1
initial state	0	0	0	0	0	0	0	0
1	1	0	0	0	0	0	0	0
2	1	1	0	0	0	0	0	0
3	1	1	1	0	0	0	0	0
4	1	1	1	1	0	0	0	0
5	1	1	1	1	1	0	0	0
6	1	1	1	1	1	1	0	0
7	1	1	1	1	1	1	1	0
8	1	1	1	1	1	1	1	1
9	0	1	1	1	1	1	1	1
10	0	0	1	1	1	1	1	1
11	0	0	0	1	1	1	1	1
12	0	0	0	0	1	1	1	1
13	0	0	0	0	0	1	1	1
14	0	0	0	0	0	0	1	1
14	0	0	0	0	0	0	0	1
15	0	0	0	0	0	0	0	0
16	1	0	0	0	0	0	0	0
17	1	1	0	0	0	0	0	0
⋮								

$Q_8 = \overline{Q}_1$ of previous state

(b) *Johnson counter states*

FIGURE 13.29 The Johnson counter.

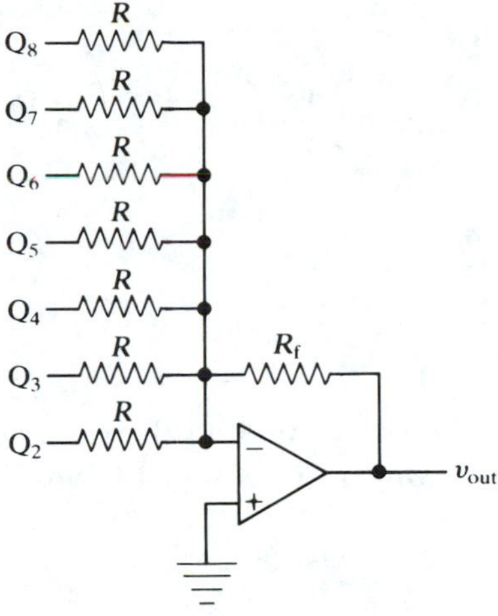

(c) *analog summer*

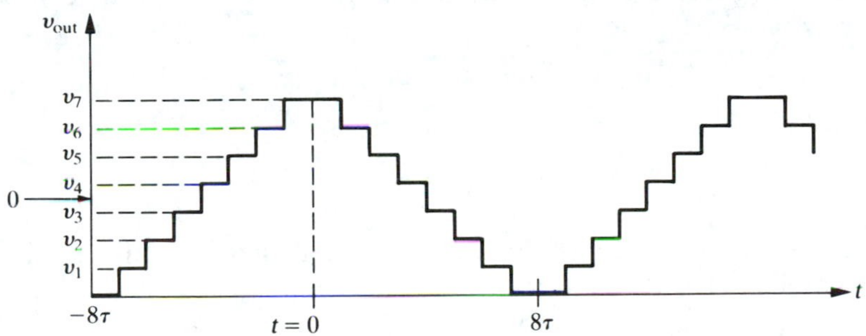

(d) v_{out} *waveform* (τ = *clock period*)

FIGURE 13.29 Continued.

Using

$$
v(t) = \begin{cases} v_7 & \text{for} \quad 0 < t \leqq \tau \\ v_6 & \quad \tau < t \leqq 2\tau \\ v_5 & \quad 2\tau < t \leqq 3\tau \\ v_4 & \quad 3\tau < t \leqq 4\tau \\ v_3 & \quad 4\tau < t \leqq 5\tau \\ v_2 & \quad 5\tau < t \leqq 6\tau \\ v_1 & \quad 6\tau < t \leqq 7\tau \\ 0 & \quad 7\tau < t \leqq 8\tau \end{cases}
$$

we obtain

$$a_n = \frac{2}{n\pi}\left[(v_7 - v_6)\sin\frac{n\pi}{8} + (v_6 - v_5)\sin\frac{2n\pi}{8} + (v_5 - v_4)\sin\frac{3n\pi}{8}\right.$$
$$+ (v_4 - v_3)\sin\frac{4n\pi}{8}$$
$$\left. + (v_3 - v_2)\sin\frac{5n\pi}{8} + (v_2 - v_1)\sin\frac{6n\pi}{8} + v_1\sin\frac{7n\pi}{8}\right] \tag{13.1}$$

With a little trigonometry it can be shown that all the even coefficients a_n are exactly zero. Thus the $v(t)$ step waveform of Fig. 13.29(d) is of the form

$$v(t) = a_1\frac{\cos \pi t}{8\tau} + a_3\frac{\cos 3\pi t}{8\tau} + a_5\frac{\cos 5\pi t}{8\tau} + \cdots$$

If we wish to make $v(t)$ close to a sine wave, we would like the amplitude a_1 to be nonzero (the fundamental) and all the other a_n's zero or as small as possible. We now will *vary* the step amplitudes $(v_7 - v_6)$, and so on, to reduce the higher-order harmonics $n > 1$. But first we note that from the symmetry of a sine wave

$$v_7 - v_6 = v_1$$
$$v_6 - v_5 = v_2 - v_1$$
$$v_5 - v_4 = v_3 - v_2$$

so instead of seven steps there are really only *four* steps. Let these four steps be denoted by A, B, C, D: Let

$$A = v_7 - v_6 = v_1$$
$$B = v_6 - v_5 = v_2 - v_1$$
$$C = v_5 - v_4 = v_3 - v_2$$
$$D = v_4 - v_3$$

Clearly D will be the largest step and A the smallest step to approximate a sine wave.

From (13.1) the first four Fourier components are

$$a_1 = \frac{2}{\pi}\left[2A\sin\frac{\pi}{8} + 2B\sin\frac{2\pi}{8} + 2C\sin\frac{3\pi}{8} + D\sin\frac{4\pi}{8}\right] \tag{13.2}$$

$$a_3 = \frac{2}{3\pi}\left[2A\sin\frac{3\pi}{8} + 2B\sin\frac{6\pi}{8} + 2C\sin\frac{9\pi}{8} + D\sin\frac{12\pi}{8}\right] \tag{13.3}$$

$$a_5 = \frac{2}{5\pi}\left[2A \sin\frac{5\pi}{8} + 2B \sin\frac{10\pi}{8} + 2C \sin\frac{15\pi}{8} + D \sin\frac{20\pi}{8}\right] \quad \textbf{(13.4)}$$

$$a_7 = \frac{2}{7\pi}\left[2A \sin\frac{7\pi}{8} + 2B \sin\frac{14\pi}{8} + 2C \sin\frac{21\pi}{8} + D \sin\frac{28\pi}{8}\right] \quad \textbf{(13.5)}$$

where we have used $\sin(n\pi/8) = \sin(7n\pi/8)$, and so on, for odd n.

If we now regard A, B, C, D as *variables*, we can *set* $a_3 = a_5 = a_7 = 0$ and solve (13.2)–(13.5) for A, B, C, D. For convenience we set $a_1 = 1$ because a_1 is the amplitude of the fundamental. In other words, the particular values for A, B, C, D obtained by solving (13.2)–(13.5) with $a_3 = a_5 = a_7 = 0$ will automatically produce a waveform with zero third, fifth, and seventh harmonics. The solution is tedious but straightforward with determinants. The results are

$$A = 0.150 \quad B = 0.278 \quad C = 0.363 \quad D = 0.392$$

and the resulting step waveform is shown in Fig. 13.30(a). These values for A, B, C, D also make $a_9 = a_{11} = a_{13} = 0$; that is, the ninth, eleventh, and thirteenth harmonics are also zero. Thus, after the fundamental the next nonzero harmonic is a_{15}!

We can now calculate the resistances in the summing circuit of Fig. 13.30(b) to produce the step amplitudes A, B, C, D that make a_3, a_5, a_7, a_9, a_{11}, a_{13} all equal 0. We choose $R_f = R_5 = 10\text{ k}\Omega$, which determines the amplitude of the largest (center) step. Then

$$\frac{A}{D} = \frac{0.150}{0.392} = 0.383 = \frac{\dfrac{R_f}{R_2}}{\dfrac{R_f}{R_5}} = \frac{R_5}{R_2} \qquad \therefore\ R_2 = \frac{R_5}{0.383} = 26.1\text{ k}\Omega$$

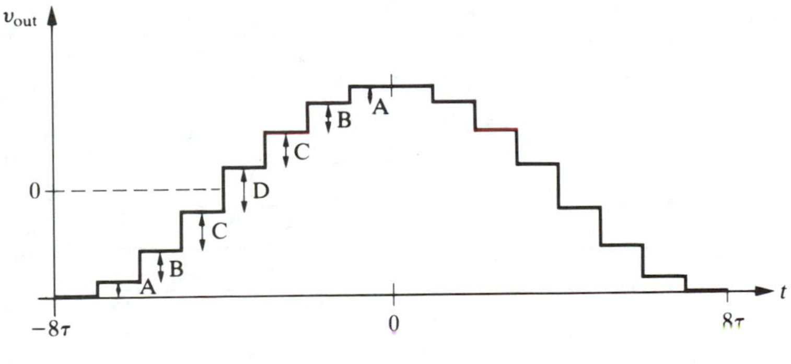

(a) *step waveform*

FIGURE 13.30 Sine-wave approximation with variable step amplitudes.

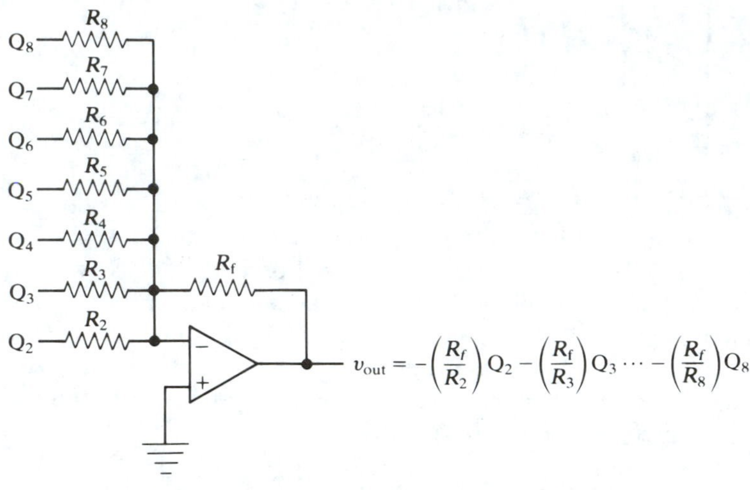

(b) *analog summer circuit*

Coefficient	Theoretical (dB)	experimental (dB)
a_1	0	0
a_2	$-\infty$	−60
a_3	$-\infty$	−70
a_4	$-\infty$	−70
a_5	$-\infty$	−63
a_6	$-\infty$	−70
a_7	$-\infty$	−65
a_8	$-\infty$	−70
a_9	$-\infty$	−68
a_{10}	$-\infty$	−70
a_{11}	$-\infty$	−68
a_{12}	$-\infty$	−70
a_{13}	$-\infty$	−70
a_{14}	$-\infty$	−60
a_{15}	−23.5	−21
a_{16}	−24.6	−22

(c) *Fourier coefficient of (a)*

FIGURE 13.30 Continued.

By symmetry $R_8 = R_2$, so $R_8 = 26.1\ \text{k}\Omega$.

$$\frac{B}{D} = \frac{0.278}{0.392} = 0.709 = \frac{\dfrac{R_f}{R_3}}{\dfrac{R_f}{R_5}} = \frac{R_5}{R_3} \qquad \therefore\ R_3 = \frac{R_5}{0.709} = 14.1\ \text{k}\Omega$$

By symmetry $R_7 = R_3$, so $R_7 = 14.1\ \text{k}\Omega$.

$$\frac{C}{D} = \frac{0.363}{0.392} = 0.926 = \frac{\dfrac{R_f}{R_4}}{\dfrac{R_f}{R_5}} = \frac{R_5}{R_4} \qquad \therefore\ R_4 = \frac{R_5}{0.926} = 10.8\ \text{k}\Omega$$

By symmetry $R_6 = R_4$, so $R_6 = 10.8\ \text{k}\Omega$.

These resistance values in the summer will produce an analog step approximation to a sine wave, with the $n = 15$ harmonic being the *first nonzero* harmonic above the fundamental! The theoretical and experimental values (from a student experiment) for some harmonics are shown in Fig. 13.30(c), and the agreement is excellent. Thus, a low-pass filter with a break frequency just above the fundamental can easily filter out the fifteenth and higher harmonics, and the filter output will be an analog sine wave (the $n = 1$ component) with extremely low harmonic distortion.

We will see in Section 15.3 that this shift register–summer circuit is an example of a *sampling* process—the 1-kHz sine wave is essentially sampled 16 times each period (i.e., at a sampling frequency of $f_s = 16$ kHz). The fundamental Nyquist sampling theorem states that in such a case the Fourier components of the sampled waveform (the 16-step approximation to the sine wave) contain the fundamental $f_0 = 1$ kHz, and the next highest nonzero harmonic is $f_s - f_0 = 16 - 1 = 15$ kHz.

13.5.5 Shift Register Application: The Pseudorandom Noise Generator

If the outputs of the last two flip-flops in a shift register are fed into the two inputs of an XOR gate, and the XOR gate output is fed back to the input of the first flip-flop, as shown in Fig. 13.31(a), then the shift register will step through different binary numbers in a specific sequence, where N is the number of flip-flops. In other words, whenever $Q_2 \neq Q_1$, then $D_4 = 1$, and a 1 is fed into the first flip-flop.

To be specific, suppose the word $Q_4Q_3Q_2Q_1 = 1000$ is loaded into the shift register. The sequence of states shown in Fig. 13.31(b) will occur. Notice that any initial four-bit word except 0000 could be used to start the sequence. Fifteen four-bit binary numbers are generated (not 0000) in a

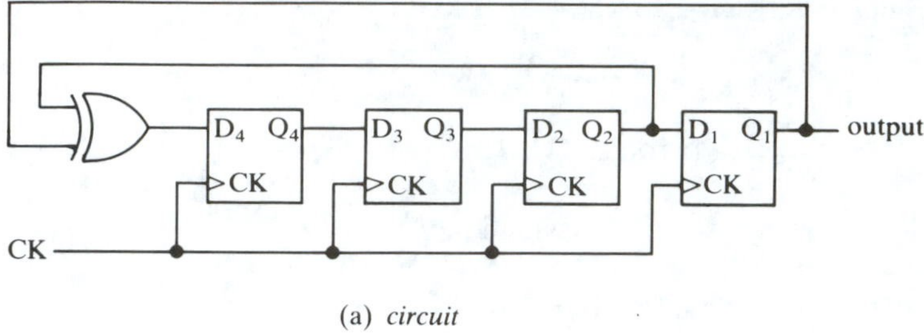

(a) *circuit*

Q_4	Q_3	Q_2	Q_1
1	0	0	0
0	1	0	0
0	0	1	0
1	0	0	1
1	1	0	0
0	1	1	0
1	0	1	1
0	1	0	1
1	0	1	0
1	1	0	1
1	1	1	0
1	1	1	1
0	1	1	1
0	0	1	1
0	0	0	1
1	0	0	0

(b) *state sequence*

FIGURE 13.31 Pseudorandom noise generator.

particular order. For N flip-flops $2^N - 1$ different N-bit binary numbers can be generated, one for each clock pulse, if the feedback is chosen properly. If we let τ be the clock period, then the $2^N - 1$ numbers will be generated in a time interval (the *cycle time*) of $(2^N - 1)\tau$; there will be no repetition of numbers in this time interval. After this time interval—that is, at the 2^N clock pulse—the sequence starts over again. Table 13.3 indicates how long this time interval is for various values of N for a 100-kHz clock, $\tau = 10\ \mu s$.

TABLE 13.3. Cycle Time for a 100-kHz Clock for a Pseudorandom Noise Generator

No. of flip-flops N	*No. of states $2^N - 1$*	*Cycle Time*
4	15	150 μs
8	255	2.55 ms
12	4,095	40.95 ms
16	65,535	655.35 ms
32	4.29×10^9	4.29×10^4 s $\simeq$ 11.9 hr
64	1.845×10^{19}	1.845×10^{14} s $\simeq 7 \times 10^6$ yr

$\tau \simeq 10\ \mu$s (100-kHz clock).
1 yr $\simeq 2.5 \times 10^7$ s.

For any shift register with a large number of bits, we obviously can't have a parallel output because it would take too many pins on the chip; so these noise generator chips are generally SISO shift registers. One of the "longest" shift registers is the 2533, which is a 1024-bit shift register made with large-scale integration techniques.

Because the output does not repeat itself in a time of $2^N - 1$ clock periods, we say the output is *pseudorandom* noise. For perfectly random noise the output would never repeat itself in any finite time interval, so the larger N is the more nearly random the output is. If we take the output off the last flip-flop Q_1, then the output consists of a series of 0s and 1s in a pseudo-random sense. A low-pass filter up to approximately $f_{CK}/4$ will produce a nice analog noise source.

The National MM5837 *digital noise source* is a 17-bit shift register on one eight-pin chip. It starts at a random state or count when power is applied and produces pseudorandom noise pulses sequentially on one pin with a flat power spectrum up to approximately 40 kHz (half-power point). It is useful as a broadband audio noise source. It requires two power supplies, typically −14 V and −27 V, and its output 0 and 1 levels are −13 V and −0.75 V, respectively.

A final comment on types of registers: The two general types are *static* and *dynamic*. A static register can store data indefinitely as long as the power is on, and all small-scale integration registers are static. However, dynamic registers must be clocked regularly to store data; most large MOS registers are dynamic.

13.5.6 The Modem (Modulator–Demodulator)

Digital data is often transmitted over existing telephone lines in serial form. The logic levels are converted to audio tones for transmission and are converted back into logic levels at the receiving end by a *modem*, which stands for "modulator-demodulator." The *acoustic* modem converts a logic 1 level into a short burst of 2225-Hz sine waves and a logic 0 level into a short burst of 2025-Hz sine waves for transmission. For reception 1270 Hz is a logic 1, and 1070 Hz is a logic 0. This technique is called *frequency shift keying* (FSK). Thus, the same phone line can be used simultaneously for transmission and reception (*full-duplex* operation) because the transmission and reception frequencies are so different and are all well within the usual telephone bandwidth of 300–3000 Hz. The modem at the other end of the line must be set to receive 2225 Hz as a 1 and to transmit 1270 Hz as a 1, and so on. A typical rate of data transmission for a modem on a phone line would be 300 bits/s, which is 300 *baud* for FSK. 1200 bits/s is the maximum for an ordinary ("voice" grade) phone line, but expensive data grade phone lines are available for 9600 bits/s. Physically, the modem is a cradle into which the telephone rests; the depression for the phone mouthpiece contains a small loudspeaker to transmit the audio tones, and the depression for the phone earpiece contains a small microphone to listen to the audio tones being transmitted over the phone line.

13.5.7 The Universal Asynchronous Receiver Transmitter (UART)

The UART was originally designed to transmit and receive serial data; the TR 1602 (on a 40-pin chip) was one of the first. Modern UARTs can accomplish both serial to parallel and parallel to serial data conversions.

Parallel data can also be converted to serial form with a PISO shift register, such as the 74165 for transmission along a single wire. At the receiving end of the wire a SIPO shift register, such as the 74164, can be used to convert the serial data back into parallel form. A typical source of serial data is the ASCII code from a keyboard.

The UART can be used for both ends of the data transmission (i.e., for transmission and reception). The UART contains two eight-bit shift registers, one for the receiver and one for the transmitter, each with a storage register for temporary data storage. The basic functions of the UART are indicated in Fig. 13.32.

The control logic basically determines how many bits the words contain, start and stop pulses, whether a parity bit is included, and whether the parity check is for even parity or for odd parity, and what type of data conversion is accomplished—for example, PIPO, SIPO, PISO, or SISO. No timing lines are required; the UART does it all. The "asynchronous" part of the name means that the data conversion is not linked to any clock.

The National MM5303 is a UART on a 40-pin chip. It is basically a programmable interface circuit between an asynchronous serial data channel and a parallel data channel. Parallel input data is converted into a serial form, including start and stop bits and a parity bit. And serial input data of the same format is converted into a parallel format. Both receiver and

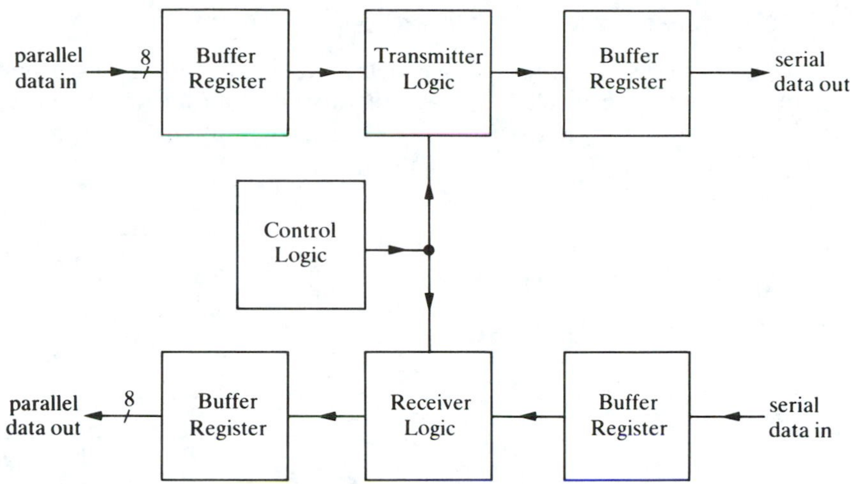

FIGURE 13.32 UART block diagram.

transmitter are buffered, and the outputs are three state. The voltage levels are TTL compatible. It requires +5 V and −12 V power supplies. The details of the UART are beyond the scope of this text, and the manufacturer's specification sheets should be consulted.

One common type of serial data transmission is specified by the IEEE RS-232C standard, which will be described in Chapter 16. Briefly, the serial output of the UART drives an external special driver chip (1488) for data transmission, and an external driver chip (1489) supplies the received data for the serial input of the UART.

Modern microprocessors generally use a *universal synchronous asynchronous receiver transmitter* (USART) to communicate with various peripheral drives such as printers via an RS-232C bus. The control is usually accomplished by the microprocessor. Each microprocessor CPU usually has its own particular USART. For example, the 8080/8085 CPU uses an 8251 USART. The Western Electric WD8250 is an asynchronous UART or *asynchronous communication element* (ACE) used in the IBM personal computer. It contains a three-state bidirectional eight-bit data bus. The 8530 is a dual USART for use with the 68000 microprocessor.

Data can be transmitted more rapidly using *phase shift keying* (PSK) instead of frequency shift keying. In PSK there are four possible types of phase changes (0°, 90°, 180°, 270°) that can occur, so one phase change can therefore encode *two* bits. The baud rate of a data line is the maximum possible number of phase changes or frequency changes. Thus, for FSK the bit rate equals the baud rate, but for PSK the bit rate equals two times the baud rate.

13.5.8 Multiplexers and Demultiplexers

A *multiplexer* (MUX) or *data selector* is the electrical analog of a rotary mechanical switch. It can select any one of several input lines and connect that input line to the single output terminal, as shown in Fig. 13.33(a) for a

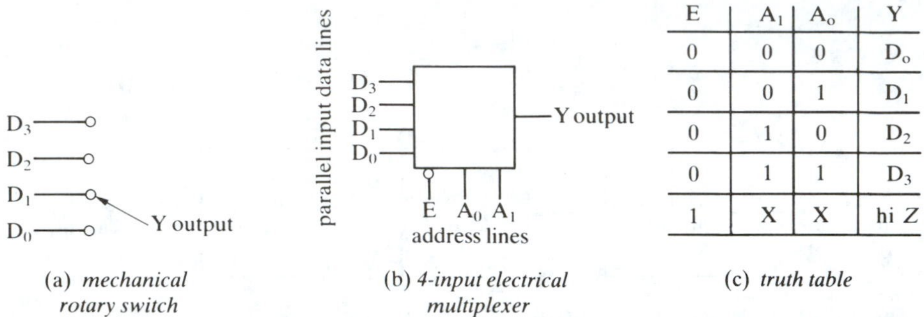

E	A_1	A_o	Y
0	0	0	D_o
0	0	1	D_1
0	1	0	D_2
0	1	1	D_3
1	X	X	hi Z

(a) *mechanical rotary switch* (b) *4-input electrical multiplexer* (c) *truth table*

FIGURE 13.33 Multiplexer or data selector (MUX).

four-input MUX. It is sometimes called an *N-input select gate*. The *digital* electrical multiplexer is a *unidirectional* device; logic levels at one of the inputs (usually called $D_0, D_1, \ldots$) appear at the output, but any electrical signal fed in at the output will not appear at the inputs. An *analog* multiplexer will conduct signals in either direction and will pass current in either direction. Which particular input appears at the output is selected by the rotary position of the mechanical switch or by the binary word "address" present at the address terminals. In a three-state multiplexer the output is connected to one of the input lines only when the STROBE or ENABLE input E is activated. Most multiplexers have active low STROBES, which means a low logic level at the STROBE will connect the output to the particular input selected by the address lines. When the STROBE is high, there is no connection between any of the input lines and the output; the output then looks like a (passive) high impedance to ground.

For a 16-input multiplexer or data selector, four address lines are necessary to select one of the 16 inputs. For an eight-input multiplexer only three address lines are needed. The 74LS151 is an example of a low-power Schottky TTL eight-input multiplexer with an 11.5-ns propagation delay from the input to the output and a 30-mW power consumption. It comes in a 16-pin chip—eight pins for the inputs, three for the address or data select lines, two for the output and its complement, one for V_{cc}, one for ground, and one for the STROBE or ENABLE line. The 74LS251 is an eight-input multiplexer with three-state outputs. The output presents a high impedance when the ENABLE or STROBE input is high.

One application of a MUX is to convert parallel input data into serial form. This is accomplished by connecting the parallel input data to the

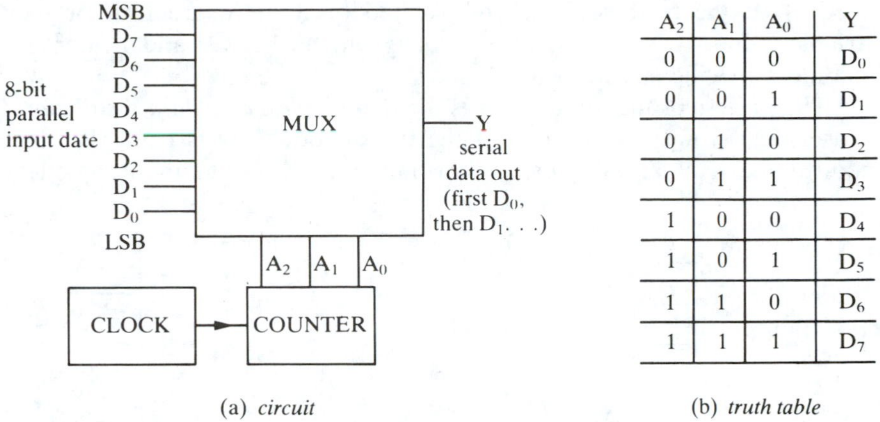

A_2	A_1	A_0	Y
0	0	0	D_0
0	0	1	D_1
0	1	0	D_2
0	1	1	D_3
1	0	0	D_4
1	0	1	D_5
1	1	0	D_6
1	1	1	D_7

(a) *circuit* (b) *truth table*

FIGURE 13.34 Parallel-to-serial data conversion with MUX.

input data lines and by connecting a counter to the address lines, as shown in Fig. 13.34. As the counter steps through its count sequence, the various data inputs will appear at the output in serial form.

Another application of a MUX is to realize or implement any arbitrary truth table—in other words, to generate any arbitrary Boolean function of N variables. We need one input for each possible output or row of the truth table. For example, for a function of three variables, A, B, and C, there are $2^3 = 8$ possible combinations of the three inputs. Thus, we would need an eight-input MUX, such as the 74151, with three address lines. The three address lines are the variables A, B, C, and the eight inputs are set or hard-wired to the values indicated by the truth table. For example, the truth table shown in Fig. 13.35(a) would be realized by the MUX circuit of Fig.

A	B	C	Y
0	0	0	0
0	0	1	0
0	1	0	1
0	1	1	0
1	0	0	1
1	0	1	0
1	1	0	1
1	1	1	1

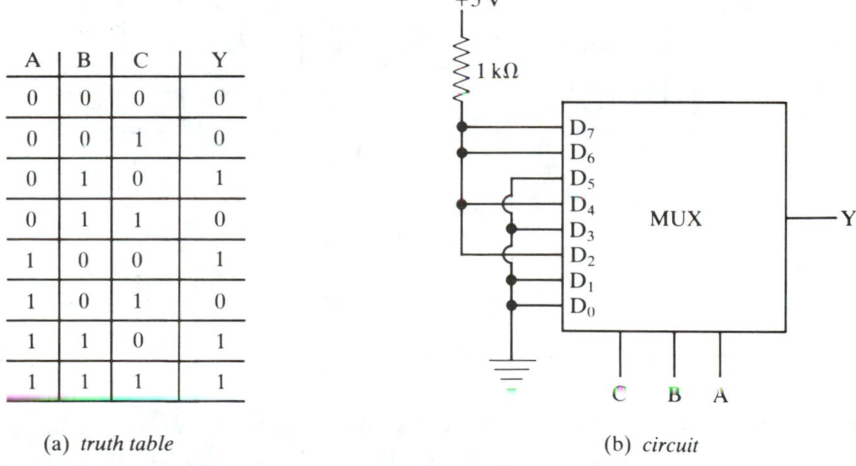

(a) *truth table* (b) *circuit*

FIGURE 13.35 MUX used to realize truth table.

13.35(b). If the truth table is complicated or if there are four or more variables, the MUX realization is usually much cheaper and neater than building a specific circuit of many gates.

The inverse of the multiplexer is the *demultiplexer* or *data distributor*. It is also like a rotary switch except the information flow is from the single input to one of N output lines, as shown in Fig. 13.36. The input logic level

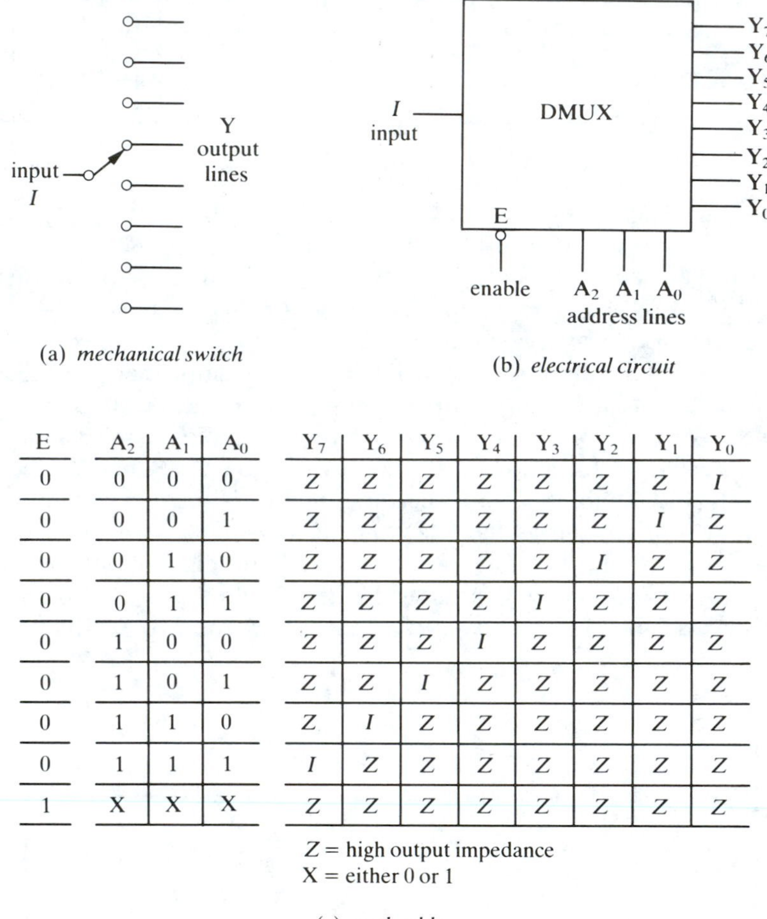

(a) *mechanical switch*

(b) *electrical circuit*

E	A_2	A_1	A_0	Y_7	Y_6	Y_5	Y_4	Y_3	Y_2	Y_1	Y_0
0	0	0	0	Z	Z	Z	Z	Z	Z	Z	I
0	0	0	1	Z	Z	Z	Z	Z	Z	I	Z
0	0	1	0	Z	Z	Z	Z	Z	I	Z	Z
0	0	1	1	Z	Z	Z	Z	I	Z	Z	Z
0	1	0	0	Z	Z	Z	I	Z	Z	Z	Z
0	1	0	1	Z	Z	I	Z	Z	Z	Z	Z
0	1	1	0	Z	I	Z	Z	Z	Z	Z	Z
0	1	1	1	I	Z	Z	Z	Z	Z	Z	Z
1	X	X	X	Z	Z	Z	Z	Z	Z	Z	Z

Z = high output impedance
X = either 0 or 1

(c) *truth table*

FIGURE 13.36 A demultiplexer.

appears at only one of the output lines, and the other unused output lines present a high impedance to ground. Which particular output line the input data appears on is determined by the address; thus for eight output lines we need three address lines, as shown in Fig. 13.36(b); for 16 output lines we need four address lines, and so on. (The word "distributor" is particularly appropriate because the distributor in a nondiesel automobile engine works the same way—the high-voltage spark input from the ignition coil is connected to the appropriate cylinder by the distributor.) There is also a STROBE or ENABLE input to activate the demultiplexer.

A *decoder* is a special case of a demultiplexer where there is no data input line; the address lines are the only "inputs." Thus, one of the *N* output lines goes high (or low, depending on the chip) for one particular input address: an input of 000 would make output line 0 go high; an input of 001 would make output line 1 go high; and so on. Thus, any demultiplexer can be used as a decoder by hard-wiring the input permanently high/low for a high/low decoded output.

The 7442 BCD-to-decimal decoder is one common example; only one selected output line is low, depending on the input address, and the other output lines are all high. The 7447 and the 7448 BCD-to-seven-segment decoder driver are other examples, although here several output lines are high for a particular input address.

Another type of multiplexer is the *analog* CMOS multiplexer, which is bidirectional; the signals (including currents) can pass in either direction. Thus, these devices are almost exactly equivalent to the mechanical rotary switches we discussed at the beginning of this section. They are digitally controlled analog switches. Thus, one circuit can act as either a multiplexer or a demultiplexer. The CMOS 4051 single eight-channel analog multiplexer/demultiplexer is one commonly used analog MUX on a 16-pin chip. It has a three-bit input address *ABC* to select one of the eight switch positions and an "inhibit" input INH. If INH = 1, the 4051 is off regardless of *ABC*. The truth table is shown in Fig. 13.37.

The "used" line or "on" line is usually driven low by a low input. A typical "on" resistance for the 4051 is 100–200 Ω (depending on the signal

(inhibit) INH	C	B	A	"On" Channel
0	0	0	0	0
0	0	0	1	1
0	0	1	0	2
0	0	1	1	3
0	1	0	0	4
0	1	0	1	5
0	1	1	0	6
0	1	1	1	7
1	X	X	X	none

FIGURE 13.37 4051 analog MUX/DMUX truth table.

voltage being switched), and the "off" resistance is extremely high—only a 10-pA leakage current is typical. A pullup resistor may be necessary on the output lines.

One application of an analog MUX is to assert digital control of the gain of an analog amplifier—that is, to build an automatic gain control circuit. We recall that the voltage gain of a standard inverting op amp circuit is $A_v = -R_2/R_1$, as shown in Fig. 13.38(a).

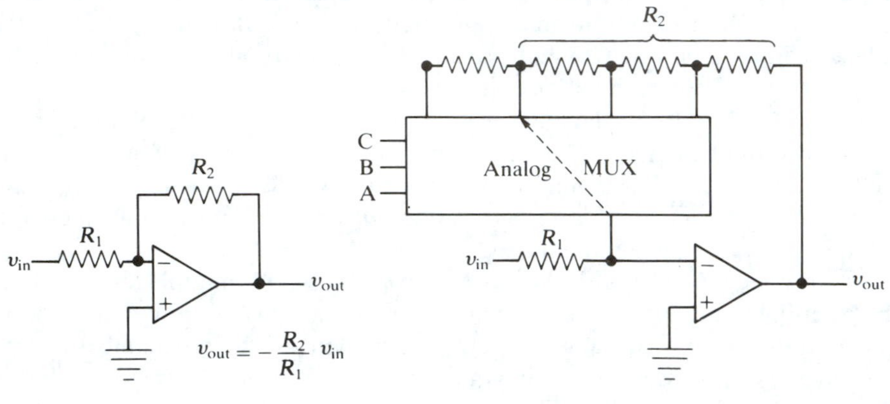

(a) *standard inverting amplifier*

(b) *with digital gain control (amplifier input impedance ≅ R_I for any gain)*

FIGURE 13.38 Digitally controlled amplifier gain.

The analog MUX is used to change the gain by changing R_2, as shown in Fig. 13.38(b). The digital word CBA on the address line determines the position of the switch and thus the R_2/R_1 ratio, which determines the gain. The input impedance remains R_1 regardless of the gain.

13.6 DIGITAL CLOCKS AND MONOSTABLE MULTIVIBRATORS (ONE-SHOTS)

In this section we will discuss clocks (square-wave oscillators or astable multivibrators) and monostable multivibrators or *one-shots* made either from basic digital gates or from specialized chips. Essentially all digital systems have one or more clocks, and monostable multivibrators are common either for producing single pulses or for introducing a time delay into a signal.

13.6.1 The 555 Timer

The 555 timer chip contains a combination of linear and digital circuits and is one of the most useful single chips available. It can function as an oscillator or as a monostable multivibrator ("timer"), depending on the configuration of a few simple external components. It can supply (either source or sink) up to 200 mA to an external load; it can generate single (one-shot) pulses from several μs to several hours in duration; and as a square-wave oscillator (up to several hundred kHz), it is stable in frequency against supply voltage variations to approximately 0.1%/V. It can operate over a supply voltage range from 5 to 16 V, so it is compatible with TTL or CMOS digital circuitry. In addition, it is inexpensive and available from many manufacturers.

The 555 circuit basically consists of two comparators, an RS flip-flop, and some linear driving circuitry. A simplified astable multivibrator circuit diagram is shown in Fig. 13.39 with the waveforms. The three 5-kΩ resistors form a voltage divider chain that keeps the negative input of comparator #2 at $2V_{cc}/3$ and the positive input of comparator #1 at $V_{cc}/3$. The discharge transistor Q_D is turned on and off by the flip-flop output.

The astable operation is as follows. Suppose the external capacitor C is initially uncharged. The output of comparator #2 is low (S = 0), and the output of comparator #1 is high (R = 1), which resets the flip-flop output Q to zero. The 555 output is thus high because of the inverter. The discharge transistor Q_D is held off by the low flip-flop output at its base. C will then charge toward V_{cc} through R_1 and R_2 with a time constant $(R_1 + R_2)C$.

When the capacitor voltage charges to $V_{cc}/3$, R goes low, but the flip-flop output remains low because S = R = 0. When the capacitor voltage charges to $2V_{cc}/3$, the output of comparator #2 goes high (S $\rightarrow$ 1) and the flip-flop output is set to Q = 1, which suddenly changes the output to a very low voltage. The discharge transistor is also turned on; this discharges the capacitor through R_1 and the transistor with a time constant of approximately R_1C. When the capacitor voltage starts to discharge, S goes low but the flip-flop output stays high because S = R = 0. When the capacitor voltage discharges to $V_{cc}/3$, R goes high and we have S = 0, R = 1, which resets the flip-flop output low, and the output goes high. Q_D is turned off, and C starts charging again through $R_1 + R_2$; the cycle then repeats itself, with the capacitor voltage charging and discharging between $V_{cc}/3$ and $2V_{cc}/3$, as shown in Fig. 13.39(b).

The output is thus a rectangular waveform with the output voltage high for a time of the order of $(R_1 + R_2)C$ (while C charges up), and the voltage is low for a time of the order of R_1C (while C discharges). If $R_1 \gg R_2$, the output waveform will be approximately symmetrical. The exact expression for the output frequency is $f = 1.44/(R_1 + R_2)C$. For an asymmetrical square wave the output is high for a time $T_1 = 0.693(R_1 + R_2)C$ and is low for a time $T_2 = 0.693(R_1C)$. Because the output is the complement of the flip-flop output, it will vary from approximately V_{cc} to ground. The maximum practical astable output frequency is approximately several hundred kHz, with the waveform rounding off as the frequency increases.

The 555 monostable multivibrator circuit is shown in Fig. 13.40 along

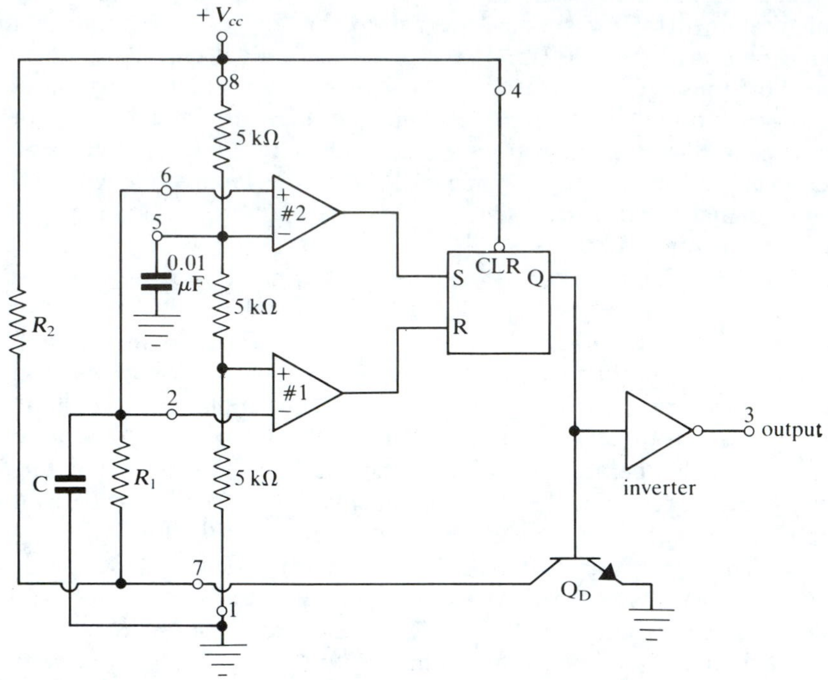

(a) *simplified circuit*

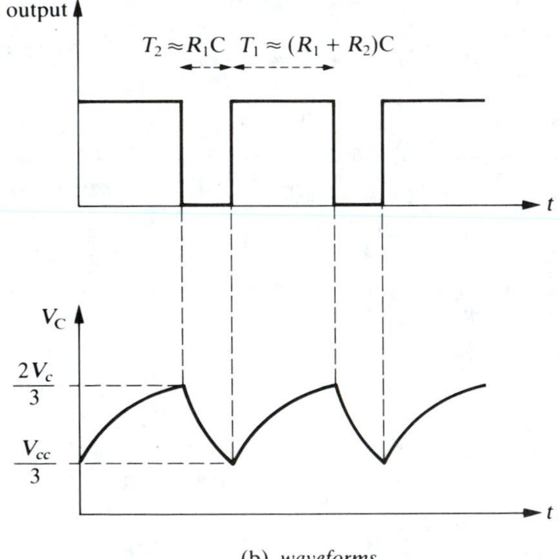

(b) *waveforms*

FIGURE 13.39 555 astable multivibrator.

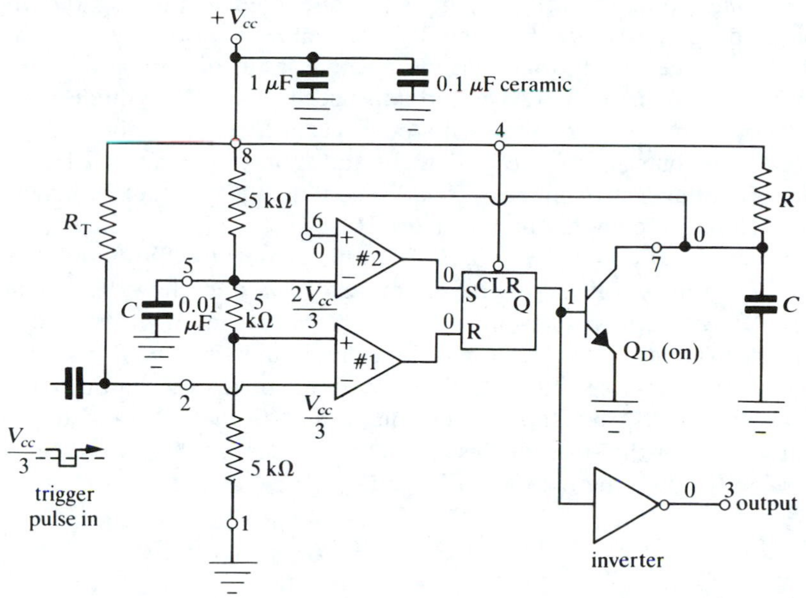

(a) *simplified circuit*

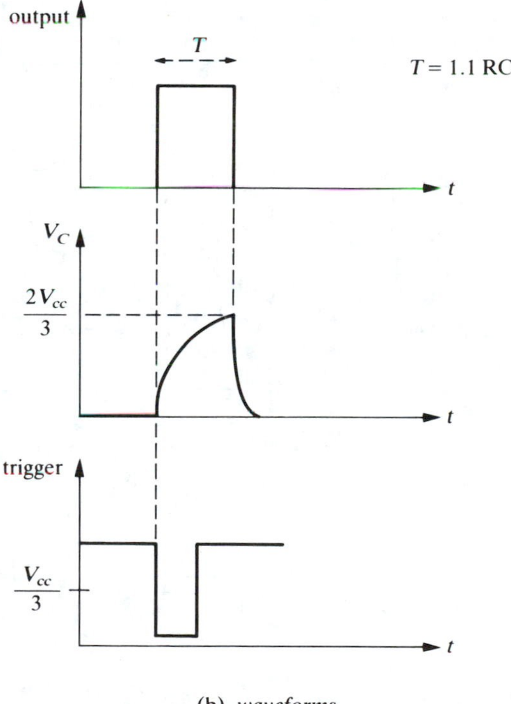

(b). *waveforms*

FIGURE 13.40 555 monostable multivibrator (one-shot).

with the waveforms. The trigger input is connected to the negative input terminal of comparator #1. The trigger input is normally held more positive than the $V_{cc}/3$ voltage at the positive input of comparator #1 by R_T; thus R is normally low. The flip-flop output is normally high (Q = 1), which holds the discharge transistor Q_D on, and the capacitor C is uncharged; the output voltage is low because of the inverter. The zero capacitor voltage also implies S = 0. When the trigger pulse drives the negative input of comparator #1 below $V_{cc}/3$, the output R of comparator #1 goes high and we have S = 0, R = 1, which resets the flip-flop output to Q = 0, which turns off the discharge transistor Q_D and allows C to charge through R with a time constant RC. When the capacitor voltage reaches $2V_{cc}/3$, the output of comparator #2 goes high and we have S = 1, R = 0, which sets the flip-flop output high again. This makes the output low and quickly discharges the capacitor through the discharge transistor. The output is high for the time it takes C to charge from 0 to $2V_{cc}/3$. The exact expression for the time the output is high is

$$T = 1.1RC$$

The stability of the pulse width of the monostable (and the period of

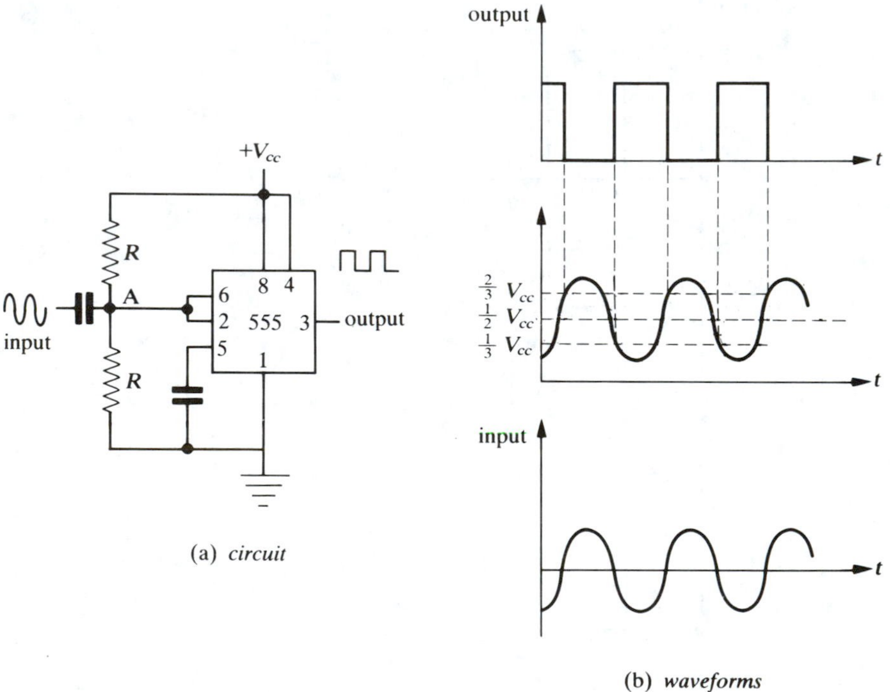

FIGURE 13.41 555 Schmitt trigger circuit.

the astable) multivibrator occurs because the *same* voltage V_{cc} supplies the capacitor charging current and also the reference voltages at the inputs to the two comparators. That is, if V_{cc} should increase for any reason (temperature change, line voltage fluctuation, etc.), the capacitor will charge up more quickly, but the $2V_{cc}/3$ threshold voltage at the input of comparator #2 will also increase, thus keeping the output pulse width constant.

Generally speaking, R should be less than 20 MΩ, and C should be a low-leakage tantalum electrolytic capacitor if a long pulse (large T) is desired. The 555 output will remain high as long as the trigger input is held low, so the trigger input must be shorter than the desired output pulse. Values of T as short as 10 μs and up to 10 min are easily obtainable. When the output is high, another trigger pulse will *not* affect the output. But a negative pulse at pin 4 (the flip-flop CLEAR) will drive the output low at any time in the cycle until a new low trigger pulse is applied to pin 2. The RESET (pin 4) should be held high (V_{cc}) if it is not used. The trigger pulse should be at least 1 μs long, in general, and the V_{cc} supply terminal should be bypassed to ground with a 1-μF electrolytic capacitor in parallel with a 0.1-μF ceramic capacitor.

There are many other applications of the 555: a frequency divider, pulse width modulator, pulse position modulator, linear ramp generator. A 555 Schmitt trigger is shown in Fig. 13.41.

13.6.2 The Inverter Clock

A simple reliable clock or square-wave oscillator can be made from two CMOS inverters, as shown in Fig. 13.42(a). The frequency or period is determined by the time constant RC. The period is approximately given by $T = 1.6RC$ but does depend upon the supply voltage, decreasing for increasing supply voltage by approximately 2%/V. This is not as stable as the 555, whose frequency or period changes by only about 0.1%/V, but the circuit is much simpler. The output amplitude is from 0 to very near V_{cc}, and the rise time is approximately 40 ns. The waveform is not very symmetric for frequencies above 100 kHz, with the output being high for a longer time than it is low: the lower the supply voltage V_{cc}, the more nearly symmetric the output. For $V_{cc} = 10$ V and $T = 5.7$ μs the output is low for approximately 26% of the period, and for $V_{cc} = 5$ V the output is low for 36%.

The oscillation is produced by the charging and discharging of C' through R. Assume that the voltages at points A, B, and C are all approximately $V_{cc}/2$ when the circuit is first turned on. Any slight noise voltage fluctuation at B will start the oscillation. Suppose a negative noise spike occurs at B. Then inverter #2 will cause the voltage at C to rise, which will be instantaneously coupled through C' to produce a rise at the input to inverter #1, which will produce a fall in voltage at B, which reinforces the original negative-going noise pulse at B.

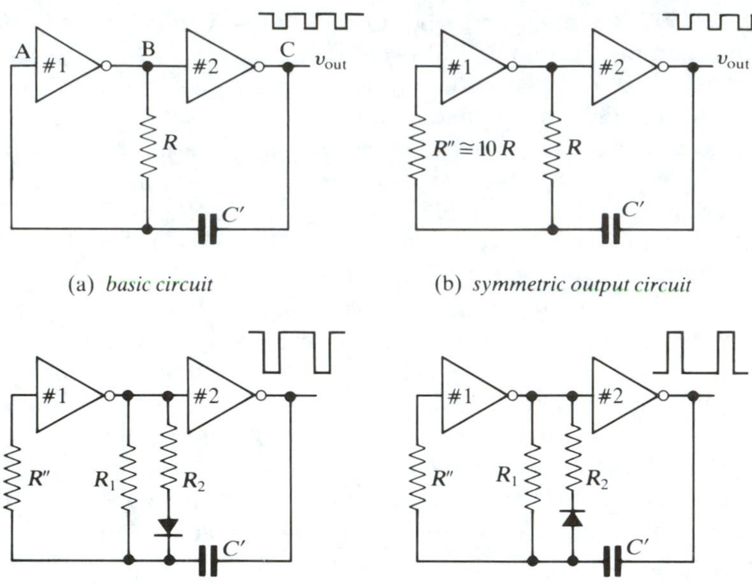

(a) *basic circuit* (b) *symmetric output circuit*

(c) *narrow-pulse output circuits*

FIGURE 13.42 Inverter clocks.

Thus the circuit will very quickly assume the state with $v_A = V_{cc}$, $v_B = 0$, and $v_C = V_{cc}$. Current will thus flow through R and charge up C' so that point A slowly falls in voltage. When v_A reaches $V_{cc}/2$, inverter #1 will flip its output from 0 to V_{cc}, and inverter #2 will thus immediately drive its output to 0. The output will stay at 0 until C' charges in the opposite sense; that is, v_A will rise. When v_A rises to $V_{cc}/2$, the circuit will flip back to $v_A = V_{cc}$, $v_B = 0$, $v_C = V_{cc}$. Thus the output will flip rapidly back and forth from V_{cc} to 0.

The asymmetry in the output waveform is caused by the asymmetric protection diodes usually built into the inverter inputs. If a more nearly symmetrical output is required, a resistance R'' (typically $R'' = 10R$) can be added in series with the input of inverter #1 at the expense of lowering the frequency and increasing the rise time. For $R''_2 = 68\ \text{k}\Omega$ and $R = 6.8\ \text{k}\Omega$ in the circuit of Fig. 13.42(b), the period is increased to 12.5 μs and the rise time is increased to approximately 100 ns, but the output waveform is nearly symmetric. The period is also less affected by changing supply voltage, approximately 0.4%/V.

An asymmetric output can be produced by using a diode to make the charging and discharging time constants different, as shown in Fig. 13.42(c). Either narrow positive or narrow negative pulses can be obtained by reversing the diode.

A perfectly symmetric square wave can be obtained by feeding the

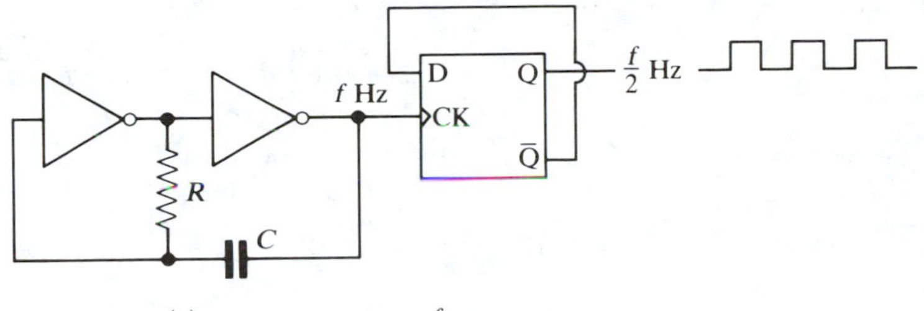

(a) *square output waveform*

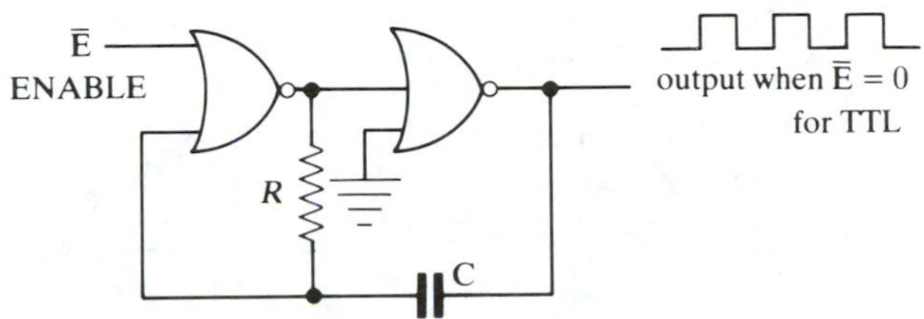

(b) *gated oscillator circuit (R< 220 Ω for TTL)*

FIGURE 13.43 Oscillator circuits.

output of the CMOS oscillator into a toggle flip-flop to divide the frequency by two, as shown in Fig. 13.43(a). The output rise time will be that of the flip-flop.

A gated oscillator can be constructed with two NOR gates, as shown in Fig. 13.43(b). When the gate or ENABLE input is low, each NOR gate acts as an inverter and the circuit oscillates. When the ENABLE input is high, the output is locked high and there is thus no oscillation. TTL inverters can be used if R is kept less than several hundred ohms, and the operating frequency will be higher for TTL than for CMOS gates, especially if LS TTL is used.

A simple CMOS square-wave oscillator using a Schmitt trigger is shown in Fig. 13.44(a). Another convenient CMOS square-wave oscillator uses a Schmitt trigger circuit as shown in Fig. 13.44(b). If the ENABLE input is high, the gate acts as an inverter and the circuit oscillates. If the ENABLE is low, the output is locked high. If v_2 and v_1 are the triggering levels for the Schmitt trigger, then with the output high at V_{cc}, C will charge up through R toward V_{cc}. When v_C reaches v_2, the output will flip low, and thus C will discharge through R toward ground. When v_C discharges to v_1, the output will flip high again and the cycle will repeat itself. For the 4093, $v_2 - v_1 = 1.5$ V for $V_{cc} = 5$ V, and 2.2 V for $V_{cc} =$ 10 V.

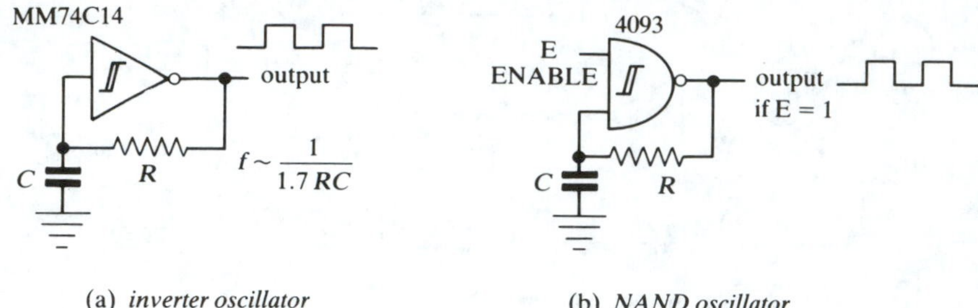

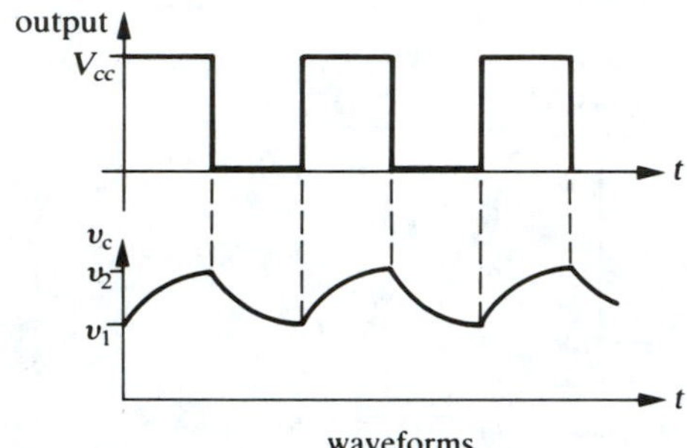

FIGURE 13.44 Schmitt trigger oscillators.

A simple Schmitt trigger can be made from two CMOS inverters, as shown in Fig. 13.45. A simple Ohm's law analysis shows that the input voltage is given by

$$v_{in} = \left(1 + \frac{R_1}{R_2}\right) v_A - \frac{R_1}{R_2} v_{out}$$

where v_A is the voltage at the input to the first inverter. For CMOS inverters the threshold input voltage is $V_{cc}/2$. So if we set $v_A = V_{cc}/2$ (the CMOS transition), we will obtain an expression for the threshold input voltage v'_{in} to the circuit:

$$v'_{in} = \left(1 + \frac{R_1}{R_2}\right) \frac{V_{cc}}{2} - \frac{R_1}{R_2} v_{out}$$

Thus we immediately see that the circuit input threshold depends on the state of the output. If the output $v_{out} = V_{cc}$, then v'_{in} is less than if $v_{out} = 0$. Specifically, for $v_{out} = V_{cc}$,

$$v'_{in} = \left(1 - \frac{R_1}{R_2}\right) \frac{V_{cc}}{2}$$

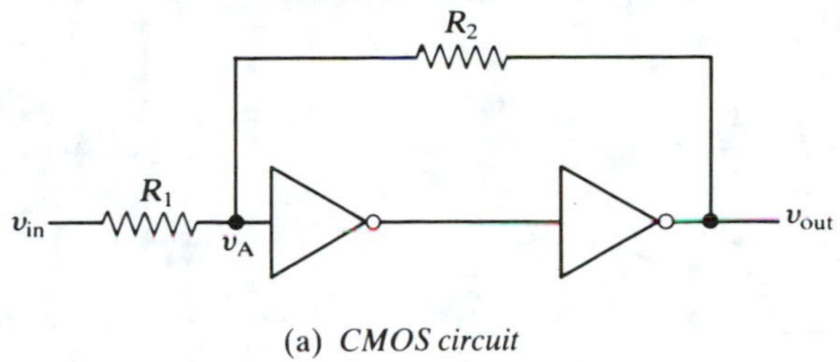

(a) *CMOS circuit*

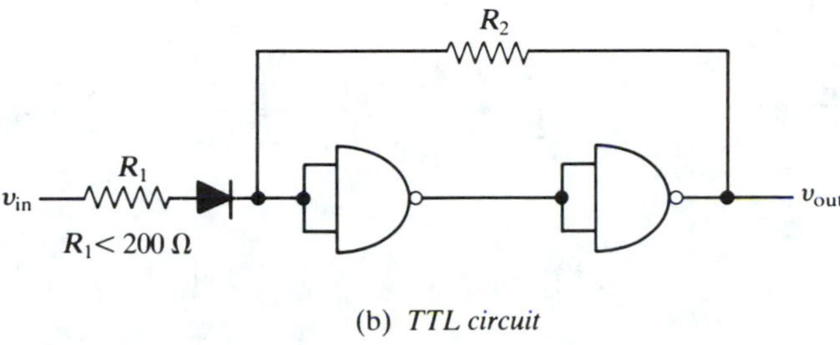

(b) *TTL circuit*

FIGURE 13.45 Simple Schmitt triggers.

and for $v_{out} = 0$,

$$v'_{in} = \left(1 + \frac{R_1}{R_2}\right)\frac{V_{cc}}{2}$$

Thus the difference Δ in the two input threshold voltages is

$$\Delta = \frac{R_1}{R_2} V_{cc}$$

For example, suppose we desire $\Delta = 0.5$ V. If $V_{cc} = 5$ V, then $R_1/R_2 = 1/10$. A TTL version of this is shown in Fig. 13.45(b), but the input signal source should have a low impedance to ground and R_1 should not exceed several hundred ohms.

13.6.3 Monostable Multivibrator from Digital Chips

A simple monstable multivibrator can be made from a single inverter or buffer or Schmitt NAND gate, as shown in Fig. 13.46. The *RC* differentiating circuit simply produces an input voltage that is high (or low) for a brief time interval of the order of *RC*. The output thus will have a pulse width of approximately *RC*. The circuits of Fig. 13.46(a) and (b) work best

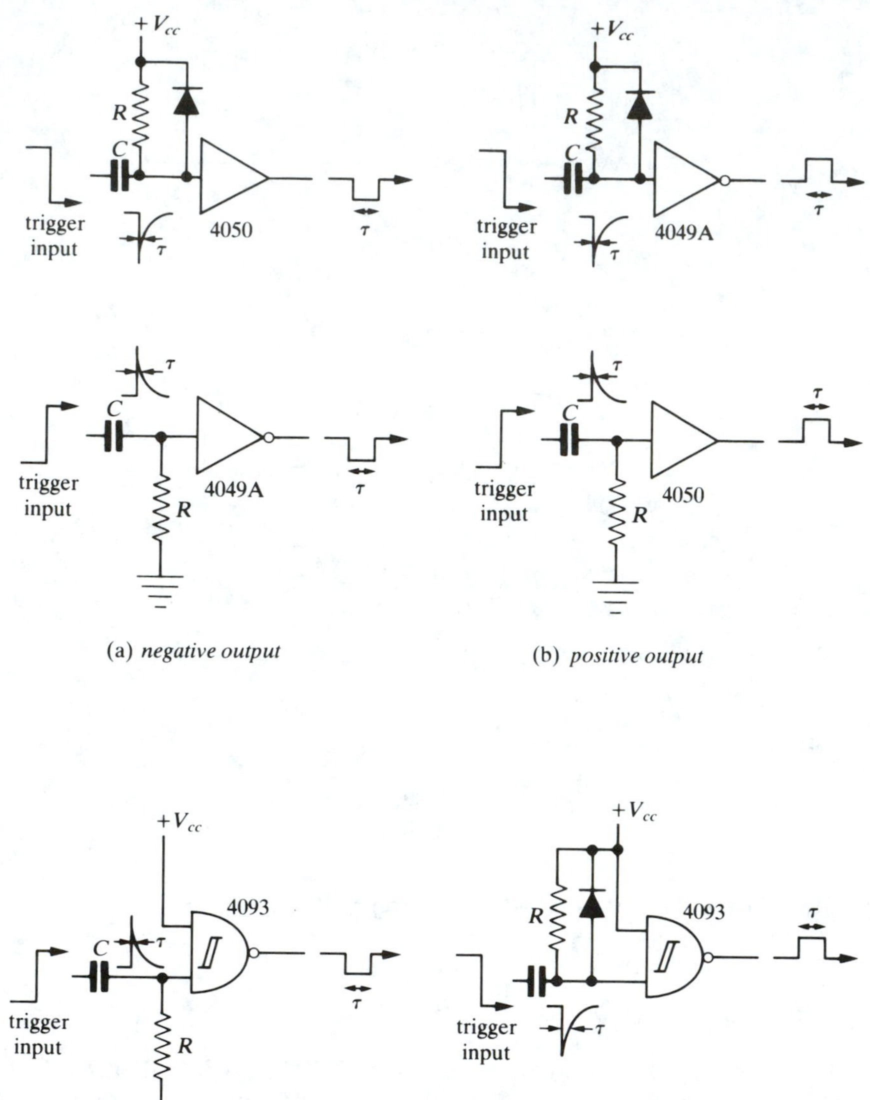

FIGURE 13.46 Simple one-shot circuits.

with CMOS chips because of their clean input threshold level (approximately $V_{cc}/2$) and their high input impedance. If TTL gates are used, R should be less than 200 Ω to get a good input low; the LS series is naturally better than the standard TTL series. If a Schmitt trigger is used as shown in Fig. 13.46(c), the output pulse will be slightly longer because of the hysteresis. In all these circuits the output pulse width will depend on the rise time and the amplitude of the input: the shorter the input rise time and the larger the input amplitude, the longer the output pulse. Thus the output pulse width cannot be synchronous with a system clock.

A monostable multivibrator can be made from a positive edge-triggered D flip-flop as shown in Fig. 13.47. The D input is wired to V_{cc}, and Q

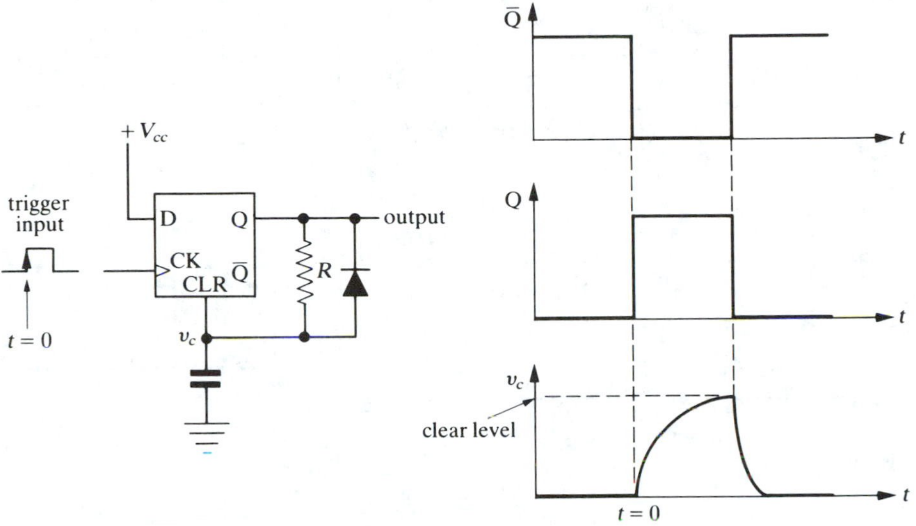

FIGURE 13.47 One-shot from D flip-flop.

is normally low because of R connecting Q to the CLEAR input. When a positive input pulse is applied to the clock input, Q goes high to V_{cc} and C starts to charge through R towards V_{cc} (with Q high, the diode is reversed biased). When v_C reaches the positive level necessary to clear the flip-flop, Q is driven low. Then C discharges quickly through the diode D. A negative output pulse is available at the $\bar{Q}$ output of the flip-flop.

Two other monostable multivibrator circuits are shown in Fig. 13.48. In the circuit of Fig. 13.48(a) the normal resting state is $v_{in} = 0$, $v_1 = 1$, $v_C = 1$, $v_{out} = 1$, and C charged. When an input trigger pulse goes high, the output v_{out} immediately goes low and v_1 goes low, which discharges C through R. But when v_{out} discharges to the low threshold for gate #2, the output v_{out} goes high. The output pulse is negative, and its width depends on RC. Generally, R must be less than several hundred ohms for TTL gates ($R < 800\ \Omega$ for LS TTL) to prevent the TTL 1.6-mA sinking current for a good TTL low input from raising the voltage v_1 too high. Notice that as soon as the input falls low, the output will go high; and the output pulse will go high when either v_C discharges or when the input returns low, whichever occurs first. Therefore, when $T_{in} > RC$, the output pulse width

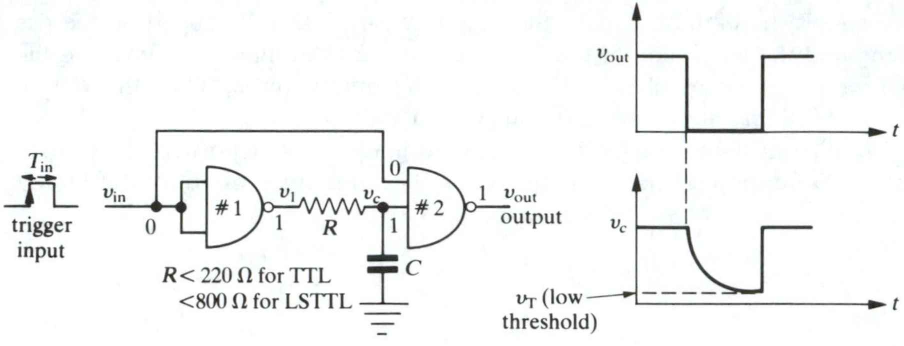

(a) *positive trigger (resting logic levels shown)*

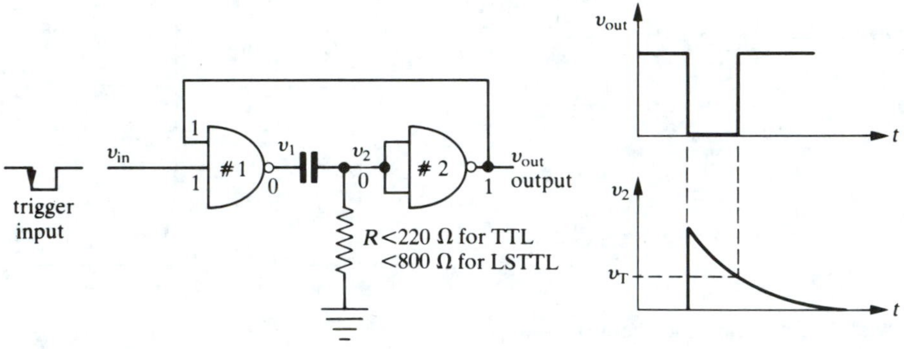

(b) *negative trigger (resting logic levels shown)*

FIGURE 13.48 One-shot circuits.

will depend on RC; if $T_{in} < RC$, the output pulse width will equal the input pulse width.

The circuit of Fig. 13.48(b) is probably more useful. The normal resting state is $v_{in} = 1$, $v_1 = v_2 = 0$, $v_{out} = 1$, and C uncharged. When an input trigger pulse goes low, v_1 goes high and the RC circuit acts like a differentiating circuit producing a positive spike v_2 at the input to gate #2, which exponentially decays toward ground; that is, the capacitor is charged by the high output of gate #1 with the charging current flowing through R. The positive spike v_2 drives the output low, and it remains low until v_2 falls to the threshold of gate #2, when the output then goes high again. The input trigger pulse can return high at *any* time without affecting the output because the low output of gate #2 fed back to the input of gate #1 keeps v_1 high. Thus the output pulse length is determined by the time constant RC, not by the length of the input trigger pulse.

Both of these circuits have the disadvantage that the pulse width will vary if the supply voltage varies (unlike the 555). Also, they are generally useful only for relatively short pulses.

If a precise, ultrareliable single pulse of variable width is needed, the best solution is to use a monostable multivibrator chip rather than make one from gates and other circuitry. Two such useful chips are the 74121 and 74123, shown in Fig. 13.49. The 74121 is *nonretriggerable*—that is, once it has been triggered, it will ignore any more trigger pulses during its output pulse. The 74123 is *retriggerable*; it will start a new timing cycle for each new input trigger pulse, even if it occurs during its output pulse. Output pulse widths from 50 ns to 10 ms can be obtained from the 74121; longer output pulses are probably best obtained from a CMOS (4528) monostable multivibrator or the 555. The 74121 contains a Schmitt trigger, which ensures reliable noise-free triggering on the *first* positive-going input trigger pulse.

Unlike the 74121, the 74123 contains a CLEAR input that overrides all other inputs and forces Q low. Such a CLEAR input is sometimes called a *jam CLEAR*.

A synchronous one-shot whose pulse width equals one clock period is shown in Fig. 13.50. It is synchronous because the rising edge of the output pulse occurs at the same time as the edge of the clock. Initially $Q_1 = Q_2 = 0$. If the positive trigger pulse overlaps the rise of the clock waveform, then, when the clock goes high, Q_1 goes high, but Q_2 does not go high because the propagation delay in flip-flop #1 is greater than the rise time of the clock pulse. Thus Q_2 goes high and $\bar{Q}_2$ goes low at the *next* rise of the clock τ s later. The output $Q = Q_1 \cdot \bar{Q}_2$ is thus a pulse exactly τ s long even if the trigger pulse is longer than one clock period! This output pulse will always have the same phase with respect to the clock, *even* if the clock frequency changes. With the usual one-shot the output pulse width depends on an *RC* time constant that is independent of the clock frequency. In such a case the circuit timing can get fouled up if the clock frequency changes significantly. Notice also that a delayed output pulse one clock period long can be obtained from a $\bar{Q}_1 \cdot Q_2$ output.

The CMOS 4060 chip contains an internal oscillator and a 14-stage ripple counter that can be used to generate *long* precise one-shot pulses, as shown in Fig. 13.51. The counter advances one count on the negative-going edge of each clock pulse and has an asynchronous RESET input that overrides the clock.

The 4013 flip-flop is normally in the state $Q = 0$, which means the 4060 counter reset is normally high, so the counter stages are all normally reset to 0. When a positive trigger pulse comes along, Q immediately goes high, which disables the 4060 counter RESET and allows the counter to count. After 2^{13} clock pulses into pin 9 of the 4060, Q_{14} goes high, which resets the 4013 flip-flop to $Q = 0$ and ends the output pulse. The 4060 reset is also forced high, which zeros all the 4060 chip counter flip-flops. The

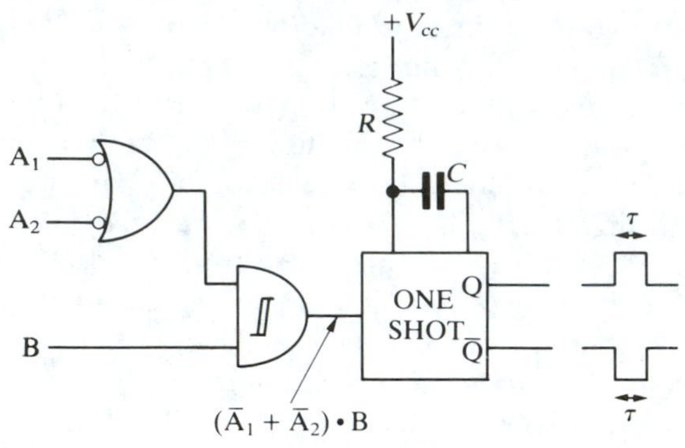

(a) *74121 nonretriggerable*

IC	Type	Retriggerable	Reset	
74121	TTL	No	No	Schmitt trigger input
74123	TTL	Yes	Yes	Schmitt trigger input
4528	CMOS	Yes	Yes	Triggerable on *either* a rising or a falling edge (A or B)

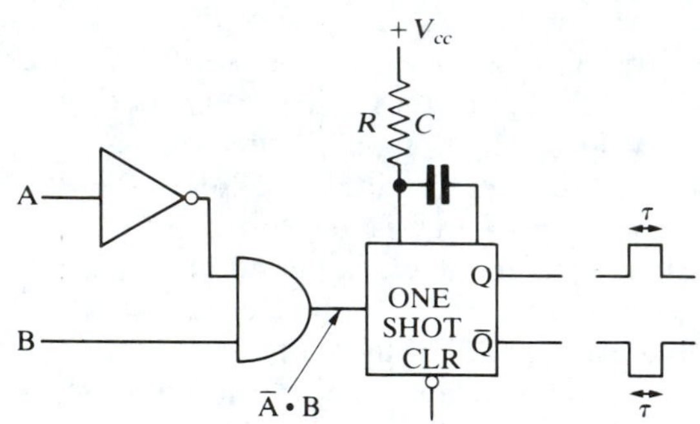

(b) *74123 retriggerable*

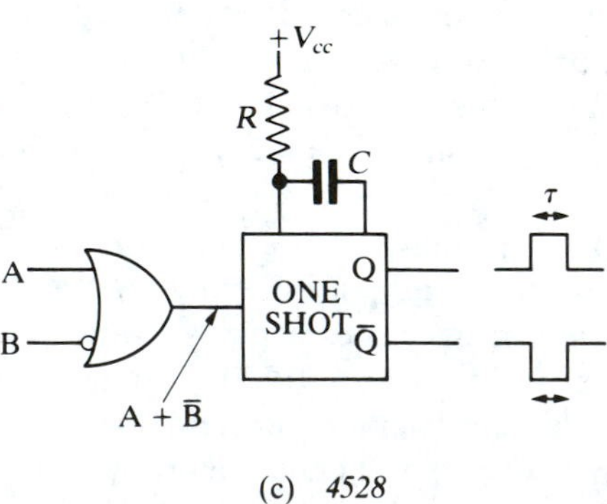

(c) *4528*

FIGURE 13.49 One-shot chips.

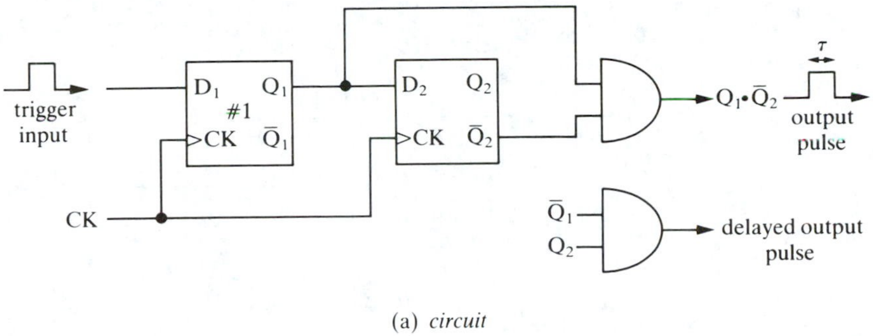

(a) *circuit*

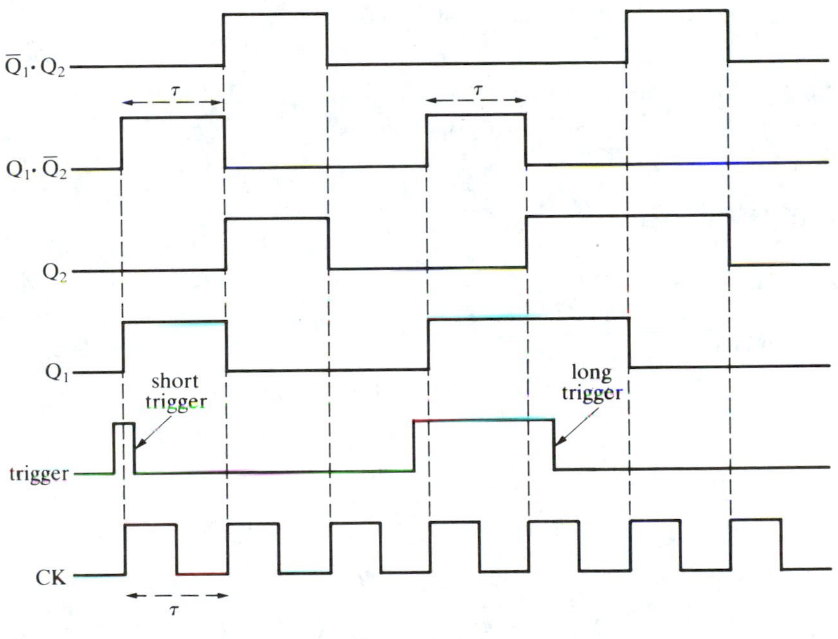

(b) *timing diagram*

FIGURE 13.50 Synchronous one-shot.

output pulse width thus equals 2^{13} t_c, where t_c is the period of the clock pulse. With the RESET input low, A is high and the NAND gate acts as an inverter; this circuit is similar to the oscillator circuit of Fig. 13.42, which we have already discussed. The oscillator period will be approximately $2R_1C$. Then if $R_1 = 1\ \text{k}\Omega$, $C = 0.01\ \mu\text{F}$, $t_c = 10\ \mu\text{s}$, the delay will be $T = 2^{13}t_c = 8192 \times 10\ \mu\text{s} = 81.92\ \text{ms}$. The time delay can be adjusted by varying R_1C or by taking the output off an earlier flip-flop in the counter chain. However, the internal clock period will depend slightly on the supply

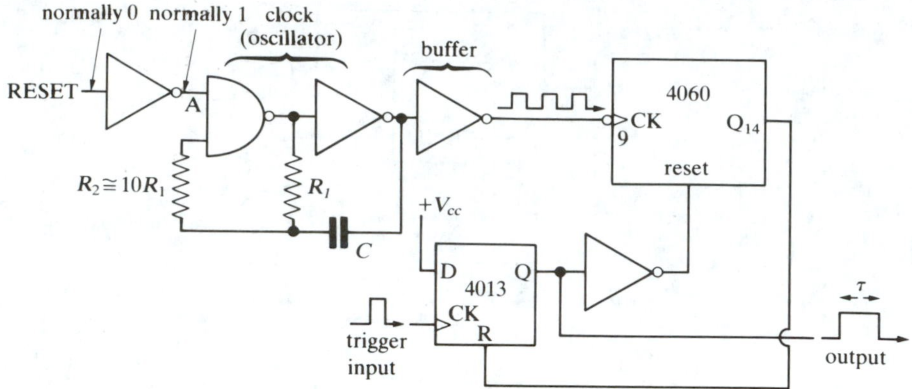

FIGURE 13.51 Counter monostable multivibrator.

voltage. The maximum clock frequency is only 3 MHz for $V_{DD} = 5$ V but increases to 10 MHz for $V_{DD} = 10$ V.

PROBLEMS

1. For the NOR gate RS flip-flop shown, write the truth table. (a) If Q = 1, $\bar{Q} = 0$, what can we infer about R and S? (b) If Q = 0, $\bar{Q} = 1$, what can we infer about R and S?

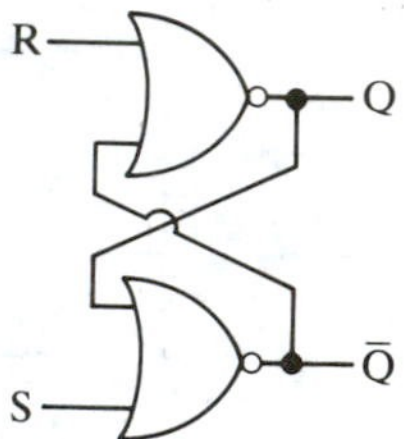

2. For the NAND gate RS flip-flop shown, write the truth table. (a) If Q = 1, $\bar{Q} = 0$, what can we infer about R and S? (b) If Q = 0, $\bar{Q} = 1$, what can we infer about R and S?

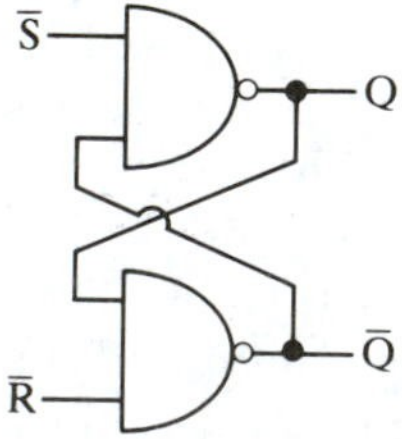

3. (a)If B = +5 V, what can we say about Y? (b) What happens to Y if B (normally +5 V) goes to 0 for 1 μs and then returns to +5 V? Assume Y is +5 V initially.

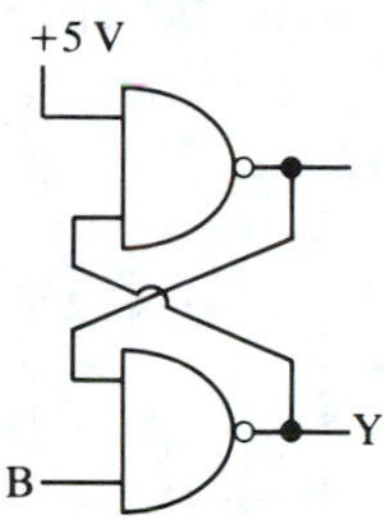

4. What is the relationship between B and Y?

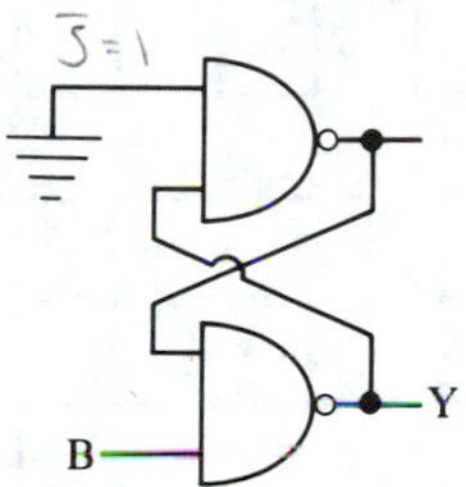

5. What is the relationship between B and Y?

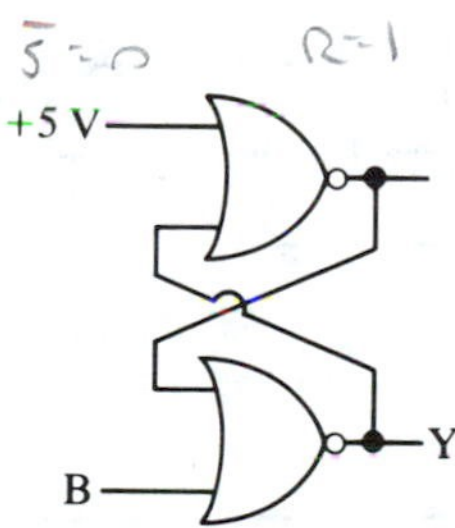

6. (a) If B = 0, what can we say about Y? (b) What happens to Y if B (normally 0 V) goes to +5 V for 1 μs and then returns to 0 V? Assume Y is initially +5 V.

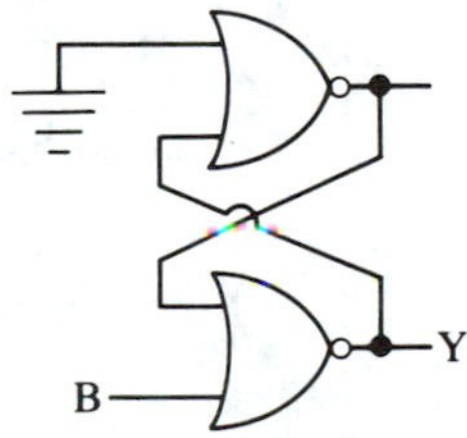

7. (a) Sketch a clocked RS flip-flop using NOR gates. (b) Add asynchronous PRESET and CLEAR inputs.

8. For the clocked RS flip-flop shown, with Q = 0, $\bar{Q}$ = 1 initially, sketch Q. If R is held at 0, sketch Q for the CK and S inputs shown.

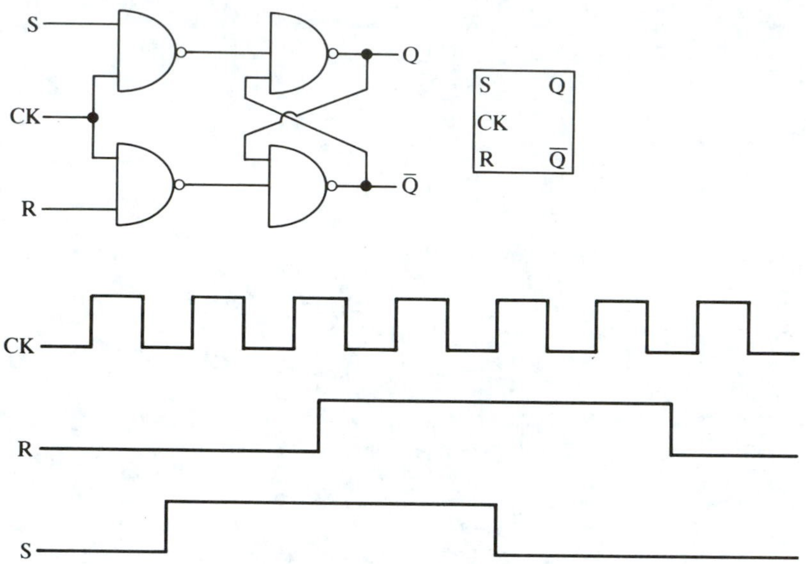

9. Sketch Q. The flip-flop PRESET and CLEAR are asynchronous.

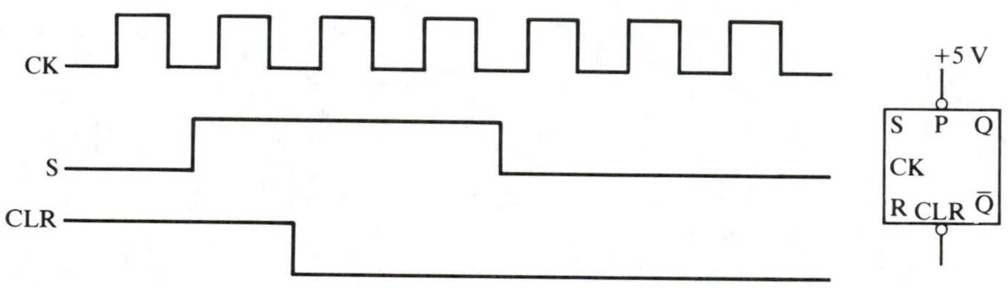

10. (a) Write the truth table. (b) Write the standard schematic diagram for this flip-flop. (c) This flip-flop is usually called a ______ flip-flop.

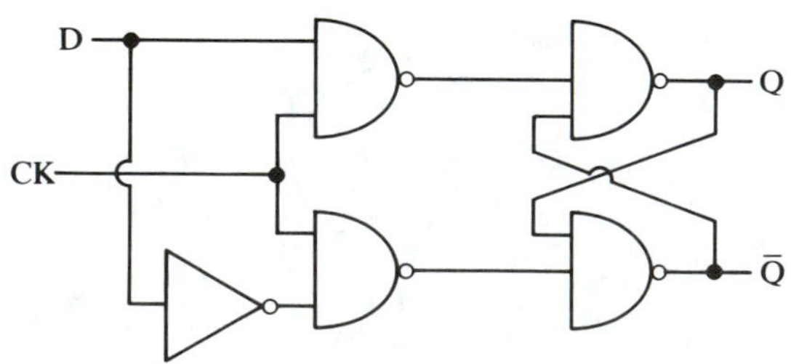

11. Complete the timing diagrams for the flip-flop of Problem 10. Assume Q = 0, $\bar{Q}$ = 1 initially.

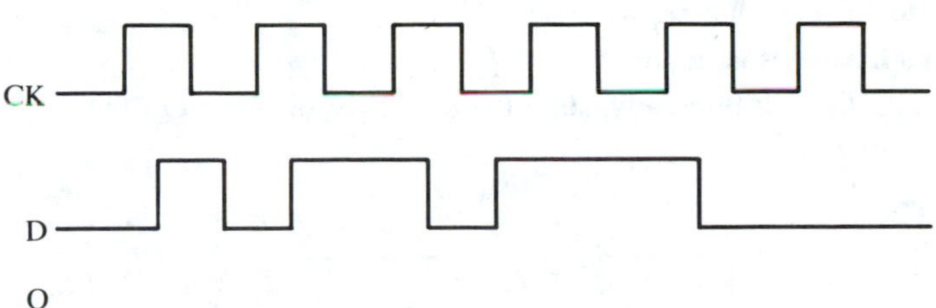

12. Sketch Q if the flip-flop is positive edge-triggered. D and CK are the same as in Problem 11.
13. Sketch Q if the flip-flop is a negative edge-triggered flip-flop. D and CK are the same as in Problem 11.
14. State the truth table for a JK flip-flop.
15. Sketch the block diagram for a master/slave RS flip-flop.
16. For the JK master/slave flip-flop of Fig. 13.15, if CK = 1, J = 1, K = 0, Q = 1, $\bar{Q}$ = 0, $\overline{PR}$ = 1, $\overline{CLR}$ = 1, find the logic levels at points A, B, E, F, G, H.
17. Repeat Problem 16 if CK = 1, J = 0, K = 1, Q = 0, $\bar{Q}$ = 1, $\overline{PR}$ = 1, $\overline{CLR}$ = 1.
18. Sketch the CK and Q waveforms for the two flip-flop circuits shown.

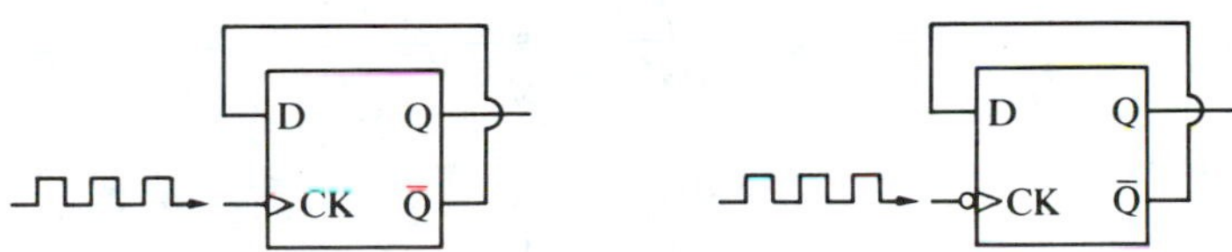

19. Sketch the CK and Q waveforms.

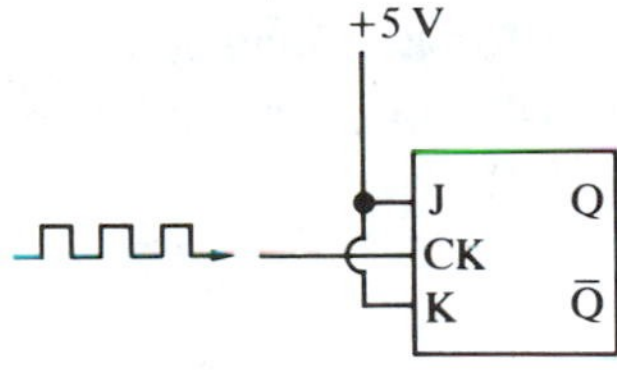

20. Design a divide-by-six ripple (asynchronous) counter and state its count sequence.
21. Design a synchronous divide-by-seven counter using JK flip-flops.
22. Design a synchronous counter that will count through the sequence 1, 3, 5, 7, 9, 1, 3, 5, 7, 9, . . . , using JK flip-flops.
23. For the four-bit shift register of Fig. 13.25, initially cleared to $Q_1 = Q_2 = Q_3 = Q_4 = 0$, state the count sequence for a serial D = 1 input for $0 < t < \tau$, D = 0 for $t \geqq \tau$, where τ = the clock period.

24. Repeat Problem 23 for D = 1 for $0 < t < \tau$, D = 0 for $\tau \leqq t \leqq 2\tau$, D = 1 for $2\tau < t < 3\tau$, D = 0 for $t \geqq 3\tau$.

25. Sketch a ring counter.

26. Sketch a Johnson counter.

27. If $Q_3Q_2Q_1 = 100$ initially, state the sequence of states $Q_3Q_2Q_1$.

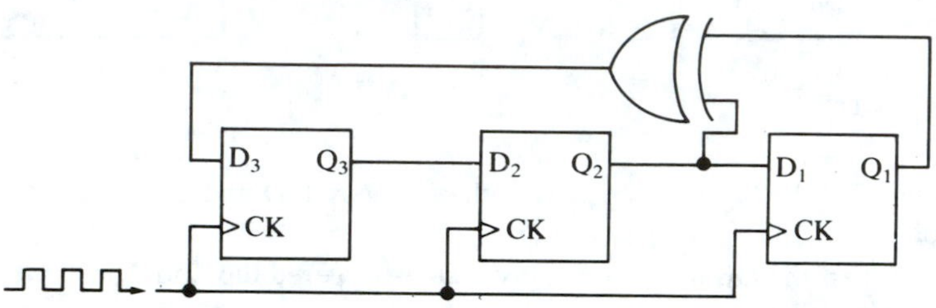

28. (a) Use an eight-input MUX to realize or implement the following truth table. (b) Use NAND gates to implement the truth table. Which is simpler?

ABC	Y
000	0
001	1
010	0
011	0
100	0
101	1
110	0
111	0

29. Sketch a block diagram of a 16-bit demultiplexer.

30. Q = S = R = 0 V initially. Sketch v_A versus t if S → 1 at $t = 0$.

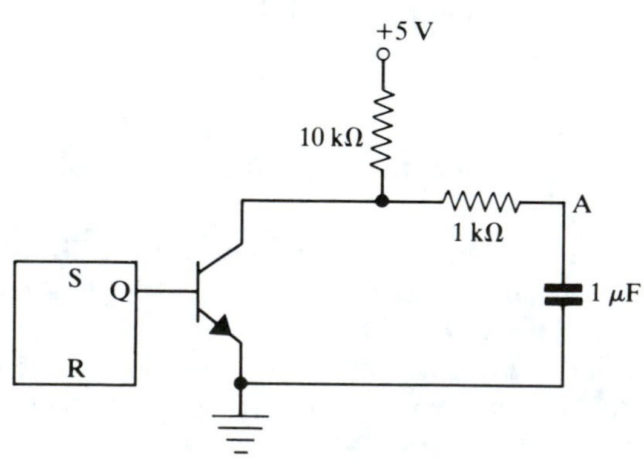

31. Sketch v_A and v_B versus t.

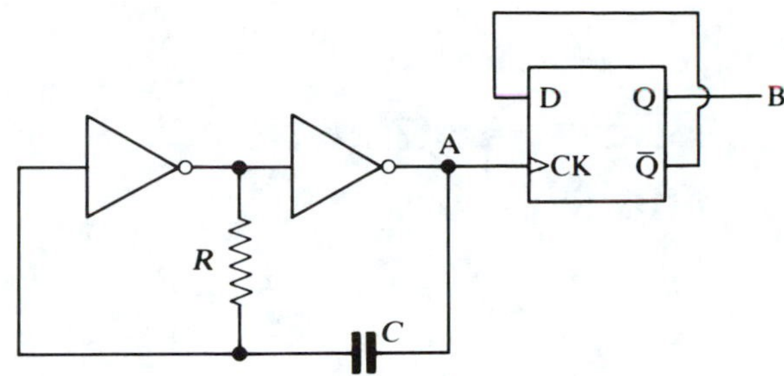

32. Sketch v_A and v_B versus t.

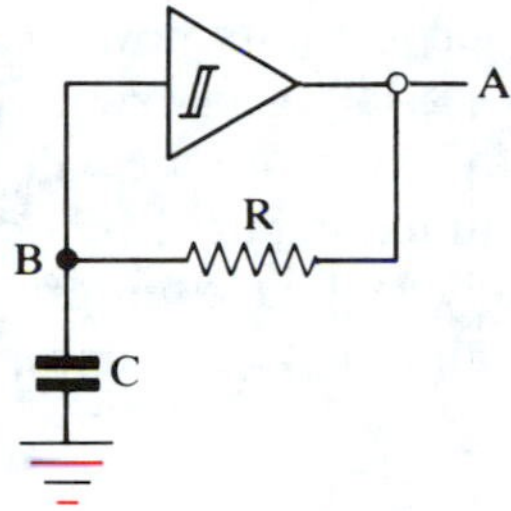

33. Sketch v_A and v_B versus t. How does the input rise time affect v_A?

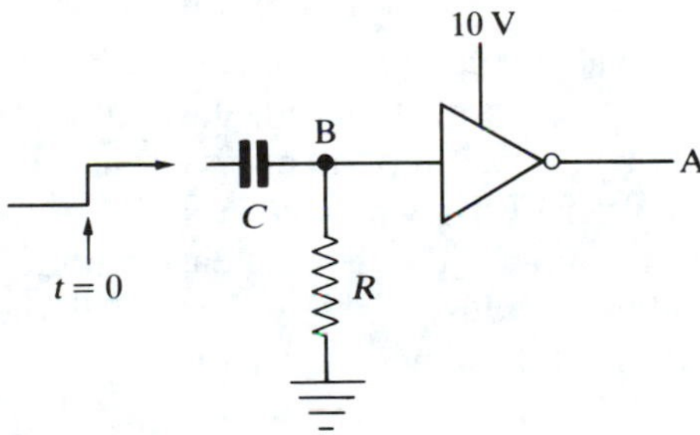

CHAPTER 14

Binary Arithmetic and Memory

14.1 INTRODUCTION

In this chapter we will briefly discuss how to add, subtract, multiply, and divide binary numbers. We also discuss some logical procedures to use in evaluating truth tables. In Chapter 16 we will see that binary arithmetic is useful not only in doing arithmetic but also in digital signal processing such as filtering. A few assembly language programs are given for arithmetic operations using the 8085 and the Z-80 microprocessors.

14.2 BINARY ADDITION

All modern digital computers are essentially very-high-speed adding machines and memory circuits designed to handle numbers in binary form. They also can perform subtraction, multiplication, and division, because these are basically special cases of addition. Two binary numbers can be added by a computer in two ways: in parallel or serially. In parallel addition the bits are added simultaneously, whereas in serial addition the bits are added one at a time, starting with the first or least significant bit.

Consider the addition A plus B = S. A is called the *addend*, B the *augend*, and S the *sum*. The fundamental arithmetic of binary addition is contained in four rules ("+" means "plus" not a logical OR):

1. $0 + 0 = 0$
2. $0 + 1 = 1$
3. $1 + 0 = 1$
4. $1 + 1 = 0$ but 1 must be carried over to the next higher (more significant) bit

For example, 00 + 01 = 01, according to rule 2 for the LSB and rule 1 for the MSB. (The LSB is at the right.) 01 + 10 = 11, according to rule 3 for the LSB and rule 2 for the MSB. 01 + 01 = 10, according to rule 4 for the LSB and rules 1 and 2 for the MSB. These four rules of binary addition can be expressed as the truth table of Fig. 14.1.

A	B	Sum Bit S	Carry Bit C
0	0	0	0
0	1	1	0
1	0	1	0
1	1	0	1

$S = A \oplus B$

$C = A \cdot B$

FIGURE 14.1 Binary addition truth table.

From inspection of the truth table, we see that the sum S is precisely A XOR B in each of the four possible cases, and that the carry C is A AND B. Thus, we see that to add two bits we need a circuit with two inputs (one for A and one for B) and two outputs (one for the sum bit S and one for the carry bit C). Such a circuit is shown in Fig. 14.2.

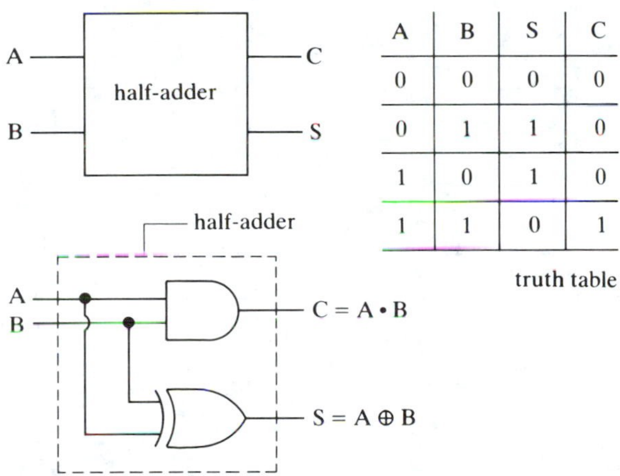

A	B	S	C
0	0	0	0
0	1	1	0
1	0	1	0
1	1	0	1

FIGURE 14.2 Half-adder circuit.

The adder circuit of Fig. 14.2 is often called a *half-adder* because two such circuits are necessary to complete an addition—one circuit to add the two bits, and another circuit to add in the carry bit from the next less significant bits. Thus, the half-adder circuit of Fig. 14.2 is suitable only for adding the least significant bits.

14.2.1 Parallel Addition

A *full-adder* can add two bits A_n and B_n and also add the carry bit C_{n-1} from the next less significant bit. A full-adder circuit is shown in Fig. 14.3.

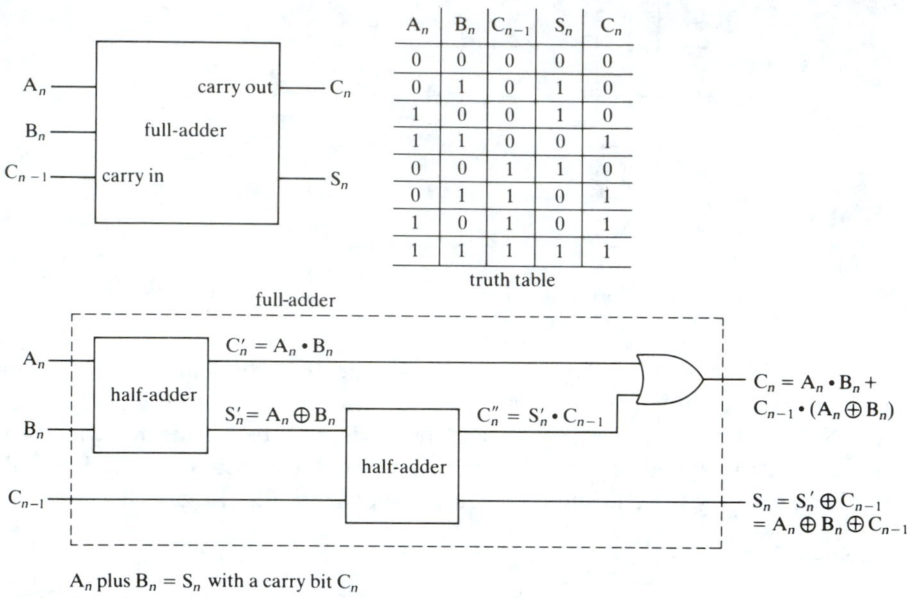

A_n	B_n	C_{n-1}	S_n	C_n
0	0	0	0	0
0	1	0	1	0
1	0	0	1	0
1	1	0	0	1
0	0	1	1	0
0	1	1	0	1
1	0	1	0	1
1	1	1	1	1

A_n plus $B_n = S_n$ with a carry bit C_n

C_{n-1} = carry bit from next less significant bit

FIGURE 14.3 Full-adder circuit.

C_{n-1}	A_n	B_n	S'_n	C'_n	C''_n	S_n	$C_n = C'_n + C''_n$
0	0	0	0	0	0	0	0
0	0	1	1	0	0	1	0
0	1	0	1	0	0	1	0
0	1	1	0	1	0	0	1
1	0	0	0	0	0	1	0
1	0	1	1	0	1	0	1
1	1	0	1	0	1	0	1
1	1	1	0	1	0	1	1

$C''_n = S'_n \cdot C_{n-1}$

$C'_n = A_n \cdot B_n$

$S'_n = A_n \oplus B_n$

$C_n = C'_n + (C_{n-1} \cdot S'_n)$

$S_n = S'_n \oplus C_{n-1}$

FIGURE 14.4 Complete full-adder truth table.

Inspection of the truth table for the full-adder in Fig. 14.4 shows that the sum bit S_n is

$$S_n = (A_n \oplus B_n) \oplus C_{n-1} = A_n \oplus B_n \oplus C_{n-1}$$

and the carry bit is

$$C_n = A_n \cdot B_n + C_{n-1} \cdot (A_n \oplus B_n)$$

One full-adder is required for each bit of a word to be added except for the least significant bit, which requires only a half-adder. For example, the parallel addition of two eight-bit words would require seven full-adders and one half-adder.

Figure 14.5 shows the circuit required for the parallel addition of two four-bit words.

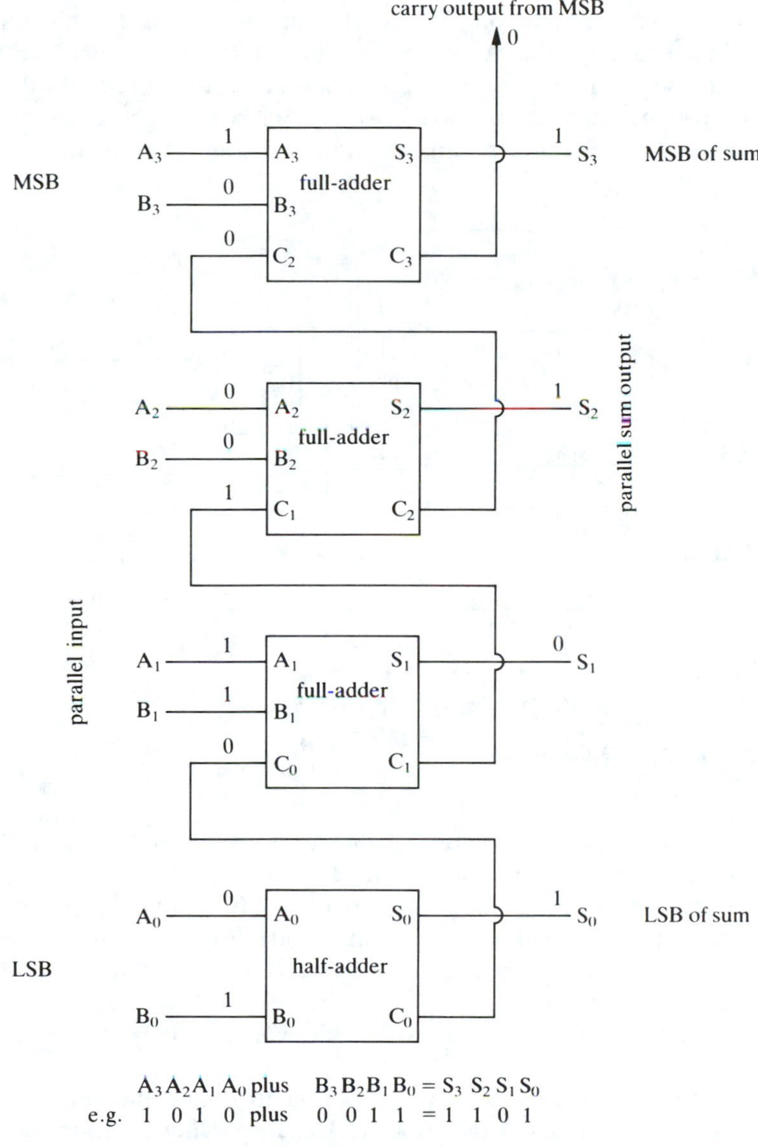

FIGURE 14.5 Parallel adder circuit for two four-bit words.

14.2.2 Serial Addition

The other method of addition of binary numbers, serial addition, first adds the least significant bits A_0 and B_0 and then, at a slightly later time, adds A_1 and B_1 plus any carry digit that might have resulted from the earlier addition of A_0 and B_0. Notice that the two binary numbers A and B must be synchronized in the form of a train of pulses such that the pulse for A_0 occurs at precisely the same time as the pulse for B_0, and so on.

The two numbers to be added are usually fed into separate shift registers that have a common clock so that the outputs of the shift registers are phased properly: the A_0 output occurs at the same time as the B_0 output, and so on. The carry bit from adding A_0 and B_0 is fed into the input of a D flip-flop, whose output (one clock pulse later) is the carry bit to be added to A_1 and B_1, and so on, as shown in Fig. 14.6. Notice that the purpose of the clock is to ensure that the A and B bits and the carry bit are fed into the full-adder at the proper times.

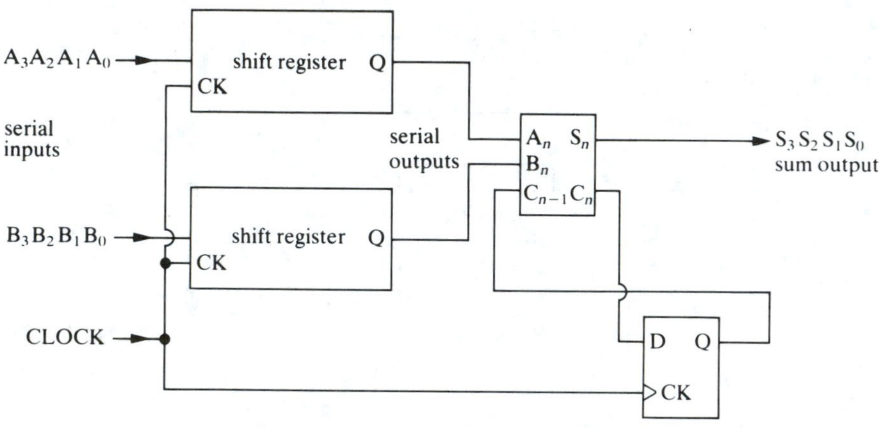

FIGURE 14.6 Serial adder circuit.

The train of pulses for the sum 1001 plus 0101 = 1110 is shown in Fig. 14.7. Notice that we read a binary number 001 from right to left, whereas the train of pulses representing 001 is read from left to right if time increases from left to right.

The time required to add two N-bit words serially is $N + 1$ clock periods, because only two bits A_n and B_n are added for each clock pulse, and we must wait a final extra clock period to obtain the carry bit from adding the most significant bits A_N and B_N. In parallel addition, however, the time required to add two N-bit words is normally the time delay in one

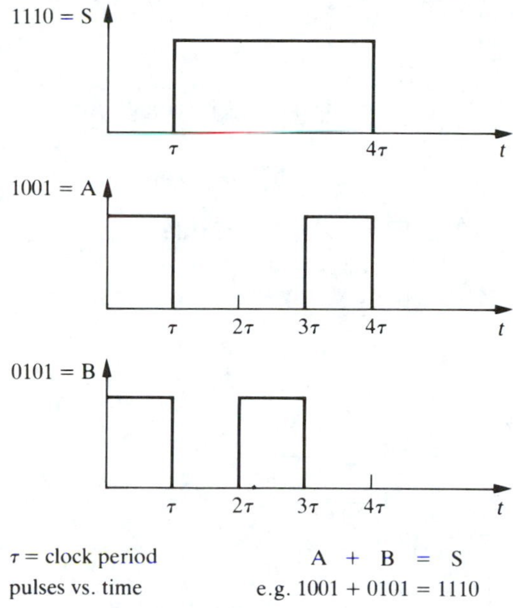

FIGURE 14.7 Pulses in serial addition of 1001 plus 0101 = 1110.

full-adder times the number of bits N. The reason is that the carry bit from adding two bits is generated after one full-adder delay and is then used in the addition of the next two more significant bits. For extremely fast parallel addition, a "look-ahead carry" technique is used, in which the carry bits are presented to all the full-adders simultaneously.

Adder circuits are available on a chip. The 74LS83A is a complete four-bit adder with look-ahead carry on one 16-pin chip. Two such chips can add two eight-bit words in approximately 25 ns with a power dissipation of 190 mW.

Addition may obviously be easily carried out with software on a computer using a high-level language. Addition may also be carried out using assembly language on a microprocessor such as the 8085 or the Z-80. For example, with the 8085 the instruction ADD B means the eight-bit number in register B will be added to the eight-bit number stored in register A (the accumulator) and the sum will be stored in the accumulator. The number in B is not affected, and the old number in A is "erased," i.e., it is replaced by the sum.

14.3 BINARY SUBTRACTION

Consider the subtraction B − A = D. B is called the *minuend*, A the *subtrahend*, and D the *difference*. The fundamental arithmetic of binary subtraction is contained in four rules:

1. $0 - 0 = 0$
2. $0 - 1 = 1$ and borrow 1 from the next more significant bit
3. $1 - 0 = 1$
4. $1 - 1 = 0$

For example, $11 - 01 = 10$, according to rule 4 for the LSB and rule 3 for the MSB. And $10 - 01 = 01$, according to rule 2 for the LSB and rule 1 for the MSB. These four rules are expressed in the truth table of Fig. 14.8,

B	A	D	B_r
0	0	0	0
0	1	1	1
1	0	1	0
1	1	0	0

$$D = B - A$$

$$D = A \oplus B$$

$$B_r = A \cdot \bar{B}$$

FIGURE 14.8 Binary subtraction truth table.

where B_r is the borrow bit. We immediately see from the truth table that the difference is the XOR $A \oplus B$ operation, and the borrow is $A \cdot \bar{B}$. Thus we see that to subtract two bits, we need a circuit with two inputs (B and A) and two outputs (D and B_n). Such a circuit is shown in Fig. 14.9 and is called a ***half-subtracter***. Notice that the half-subtracter circuit is very similar

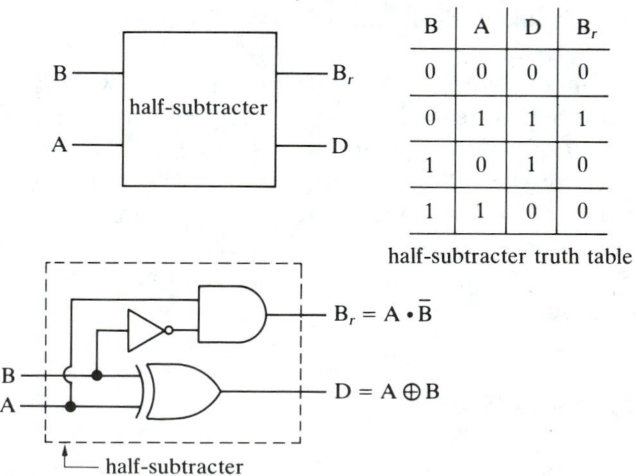

B	A	D	B_r
0	0	0	0
0	1	1	1
1	0	1	0
1	1	0	0

half-subtracter truth table

FIGURE 14.9 Half-subtracter circuit.

to to the half-adder circuit; the only difference is the presence of the inverter.

The half-subtracter circuit is suitable only for subtracting the LSB because, in general, two half-subtracter circuits are necessary to complete a subtraction—one circuit to subtract A_n from B_n, and one circuit to borrow from the next more significant bit if necessary. Two such half-subtracter circuits are called a *full-subtracter* and are shown in Fig. 14.10. The $B_{r_{n-1}}$

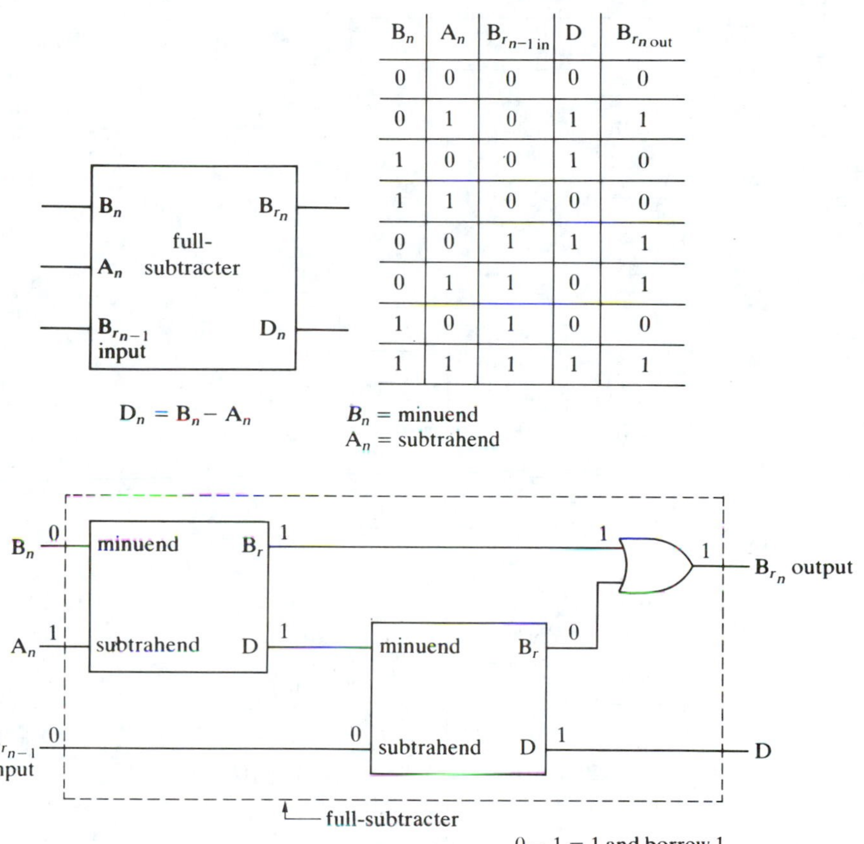

B_n	A_n	$B_{r_{n-1}\text{ in}}$	D	$B_{r_n\text{ out}}$
0	0	0	0	0
0	1	0	1	1
1	0	0	1	0
1	1	0	0	0
0	0	1	1	1
0	1	1	0	1
1	0	1	0	0
1	1	1	1	1

FIGURE 14.10 Full-subtracter circuit.

"input borrow" goes to the next *less* significant bit (the $n-1$ bit), and the B_{rn} "output borrow" goes to the *subtrahend* of the next *more* significant bit. This should become clear from inspection of Fig. 14.11, which shows the circuit required for the parallel subtraction of two four-bit words. The key to understanding the subtracter circuit is to realize that the output borrow bit must be the *subtrahend* for the next more significant bit; in other words, the output borrow must subtract 1 from the next more significant bit.

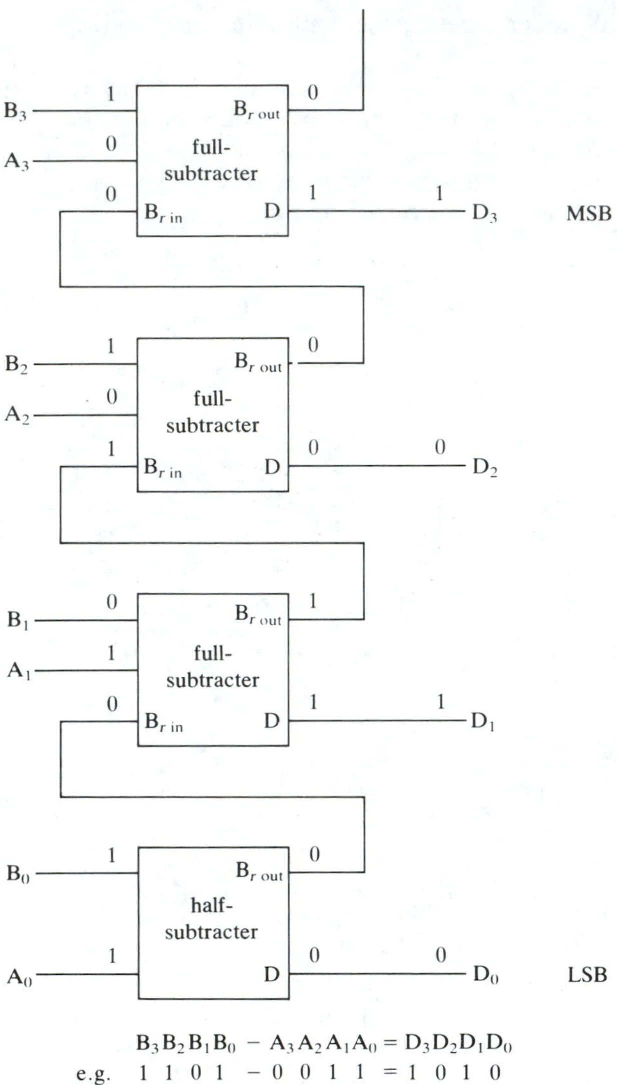

$$B_3B_2B_1B_0 - A_3A_2A_1A_0 = D_3D_2D_1D_0$$
e.g. 1 1 0 1 − 0 0 1 1 = 1 0 1 0

FIGURE 14.11 Parallel subtracter circuit for two four-bit words.

14.3.1 Twos Complement Subtraction

Subtraction can also be carried out by using the concept of the complement of a number. The *ones complement* of a binary number is defined as that number obtained by changing all 0s to 1s and all 1s to 0s. For example, the ones complement of 1011 is 0100. The *twos complement* of a binary number is defined as that number obtained by adding 1 to the ones complement. For example, the twos complement of 1011 is 0101. Another way to calculate the twos complement of a binary number is to leave (starting with the LSB) all the rightmost bits unchanged up to and including the first 1 and to complement the rest of the bits.

The twos complement is the "negative" of a binary number in the following sense. Let X be any binary number, and let X_1 be its ones complement and X_2 its twos complement. It is clear that $X + X_1 = 111 \cdots 111$ from the rules of binary addition. For example, if $X = 10011011$, $X_1 = 01100100$; and the sum is X plus $X_1 = 11111111$. Thus, it immediately follows that adding 1 (00000001) to $X + X_1$ will produce a sum of 00000000 with a carry bit of 1. For example, 11111111 plus 00000001 = (1)00000000. Thus, disregarding the (1) carry bit, we have shown that X plus X_2 equals zero with a carry bit of 1; this means that X_2 is the negative of X.

The basic idea of twos complement subtraction is that the binary subtraction B − A can be carried out by adding the minuend B to the *twos complement of the subtrahend* A, because adding the negative of A is the same as subtracting A. That is,

$$\text{B} \quad \text{minus} \quad \text{A} = \text{B} \quad \text{plus} \quad \text{A}_2$$

where A_2 is the twos complement of A.

Some examples of subtraction using the twos complement follow.

Case I: D = B − A with B > A

$$\begin{array}{r} 11_{10} \\ -10_{10} \\ \hline 01_{10} \end{array} \qquad \begin{array}{r} 1011 \\ -1010 \\ \hline \end{array} \xrightarrow{1c} \begin{array}{r} 1011 \\ +0101 \\ \hline \end{array} \xrightarrow{2c} \begin{array}{r} 1011 \\ +0110 \\ \hline (1)0001 \end{array} \quad \text{ans.}$$

$$\begin{array}{r} 147_{10} \\ -138_{10} \\ \hline 9_{10} \end{array} \qquad \begin{array}{r} 10010011 \\ -10001010 \\ \hline \end{array} \xrightarrow{1c} \begin{array}{r} 10010011 \\ +01110101 \\ \hline \end{array} \longrightarrow \begin{array}{r} 10010011 \\ +01110110 \\ \hline (1)00001001 \end{array} \quad \text{ans.}$$

Notice the carry bit is (1) if the difference is positive, as it is in the preceding examples.

Case II: D = B − A with B < A

$$\begin{array}{r} 3_{10} \\ -8_{10} \\ \hline -5_{10} \end{array} \qquad \begin{array}{r} 0011 \\ -1000 \\ \hline \end{array} \xrightarrow{1c} \begin{array}{r} 0011 \\ 0111 \\ \hline \end{array} \xrightarrow{2c} \begin{array}{r} 0011 \\ +1000 \\ \hline (0)1011 \end{array} \quad \text{ans. in twos complement}$$

$$\begin{array}{r} 87_{10} \\ -145_{10} \\ \hline -58_{10} \end{array} \qquad \begin{array}{r} 01010111 \\ -10010001 \\ \hline \end{array} \xrightarrow{1c} \begin{array}{r} 01010111 \\ 01101110 \\ \end{array} \xrightarrow{2c} \begin{array}{r} 01010111 \\ +01101111 \\ \hline (0)11000110 \end{array} \quad \begin{array}{l} \text{ans. in twos} \\ \text{complement} \end{array}$$

Notice the carry bit (0) is zero for B < A, and that the difference D in straight binary is obtained by taking the twos complement of $B + A_2$. If the difference is negative, we can also add the ones complement of the subtrahend to the minuend and take the ones complement of the sum. Notice that in both cases D is "automatically" expressed in twos complement notation.

All subtractions D = B − A can be performed using a twos complement notation for all numbers D, B, and A regardless of whether B is greater than or less than A. Both positive and negative numbers are represented in twos complement notation as shown in Fig. 14.12. Note that

Decimal	Straight Binary	Decimal	Twos complement
15	1111	7	0111
14	1110	6	0110
13	1101	5	0101
12	1100	4	0100
11	1011	3	0011
10	1010	2	0010
9	1001	1	0001
8	1000	0	0000
7	0111	−1	1111
6	0110	−2	1110
5	0101	−3	1101
4	0100	−4	1100
3	0011	−5	1011
2	0010	−6	1010
1	0001	−7	1001
0	0000	−8	1000

FIGURE 14.12 Twos complement number systems.

there is one more negative number than positive number in the twos complement system; also note that the MSB of any twos complement number is a sign bit: 0 for a nonnegative number, and 1 for a negative number.

With the addition of negative numbers addition is unchanged, and subtraction is merely the addition of the twos complement negative number. For example:

$$
\begin{aligned}
2_{10} + 3_{10} &= 5_{10} &&: 0010 + 0011 = 0101 \\
2_{10} - 3_{10} &= -1_{10} &&: 0010 + 1101 = 1111 \\
5_{10} - 1_{10} &= 4_{10} &&: 0101 + 1111 = 0100
\end{aligned}
$$

The hardware advantage of using the twos complement number system is that both addition and subtraction can be performed with adder circuits. No special subtracter circuits are required.

Subtraction may also be carried out using assembly language on a microprocessor, such as the 8085 or the Z-80. For example, with the 8085 microprocessor the instruction SUB B subtracts the eight-bit number in register B from the eight bit number in register A and places the difference in A. The number in B is not affected, and the old number in A is erased and replaced by the difference.

Subtraction may be easily carried out with software on a computer using a high-level language but at the expense of slower operation.

14.4 BINARY MULTIPLICATION

Binary multiplication is actually easier than addition or subtraction because there are no carry or borrow bits. The four rules for binary multiplication are given in Fig. 14.13. The truth table for A × B = P is shown in Fig.

1	$0 \times 0 = 0$
2	$0 \times 1 = 0$
3	$1 \times 0 = 0$
4	$1 \times 1 = 1$

(a) *rules*

A	× B	= P = A • B
0	0	0
0	1	0
1	0	0
1	1	1

(b) *truth table*

FIGURE 14.13 Binary multiplication.

14.13(b). A is called the *multiplicand*, B the *multiplier*, and P the *product*. To multiply two binary numbers, multiply the first bit (LSB) of the multiplier by each bit of the multiplicand, using the four rules of Fig. 14.13(a). This product is called the *first partial product*. The second bit of the multiplier is then multiplied by each bit of the multiplicand and the resulting product (the *second partial product*) is shifted over one bit (toward the MSB) and added to the first partial product. Each partial product is shifted over one bit toward the MSB before being added to the sum of the preceding partial products. This is the same procedure used in ordinary decimal multiplication, as shown in Fig. 14.14. Thus binary multiplication is simply a series of additions of the (shifted) partial products. The use of shift registers immediately comes to mind for binary multiplication because each partial product is shifted one bit to the left before it is added to the previous partial product. An additional register called the *accumulator* is necessary to temporarily store the partial products.

1100		12_{10}	
× 1101		$\times 13_{10}$	
1100	first partial product	36	first partial product
0000	second partial product	12	second partial product
1100	third partial product	156_{10}	product
1100	fourth partial product		
10011100	product		

(a) *in binary* (b) *in decimal*

FIGURE 14.14 Multiplication example.

Another way to multiply two numbers is to use repeated addition. If P = B × A, we simply add B to itself A times. For example, to calculate 4 × 2 we add 4 + 4; to calculate 4 × 3 we add 4 + 4 + 4. This technique is used with the popular 8085 and Z-80 microprocessors because there are no multiply instructions, but there are addition instructions. More advanced microprocessors, such as the 8086, contain multiply instructions.

In general, multiplication by addition of the shifted partial products is faster in real time than multiplication by repeated addition. The number of bits in the product equals the number of bits in the multiplicand plus the number of bits in the multiplier. For example, the product of an eight-bit number and an eight-bit number contains 16 bits. The maximum product of two eight-bit numbers is

$$(11111111)_2 \times (11111111)_2 = 11111110\ 00000001_2$$
$$FF_{16} \times FF_{16} = FE01_{16}$$

Multiplication is usually accomplished on a computer with software—by using the appropriate computer-language multiplication expression. For a high-level language the computer operator merely types in expressions like Y = A*B in BASIC. However, such high-level-language multiplication often takes too much time for certain applications, such as the *immediate* or *real-time* processing of data from laboratory instruments, and it also requires tying up the computer. For very rapid data processing from a specific instrument, it is sometimes better to use a specific large-scale-integration (LSI) multiplier chip. For example, a single LSI multiplier chip might take less than 1 μs to multiply two eight-bit numbers. For comparison, a software multiplication might take 100 μs, and a multiplication on a hand calculator might take several ms.

14.5 BINARY DIVISION

Division is basically repeated subtraction, just as multiplication is repeated addition. On a computer it is easily accomplished with a high-level languge instruction. If Q = B/A, B is called the *dividend*, A the *divisor*, and Q the *quotient*. If the division is not exact, there may be a *remainder*. For example, 7/2 = 3 with a remainder of 1: 7 is the divdend, 2 the divisor, and 3 the quotient.

Software division calculates the twos complement of the divisor A_2, then adds it to the dividend B, and examines the sum $B + A_2 = B - A$. If $B - A$ is greater than the divisor A, the process is repeated. If the sum is less than the divisor, the sum $B - A$ equals the remainder, and the division process is completed. The quotient is the number of times the divisor has been subtracted from the dividend.

14.6 MEMORY

Memory is needed not only to store data and instructions but also for temporary storage in number calculations—for example, to store the partial products in multiplication. Memory is an essential element of any computer or microprocessor system.

It is clear that an *N*-bit digital word or number can be stored in *N* flip-flops, and this is the preferred type of memory for most computers in the 1980s. Many of the flip-flops are formed on one chip by LSI techniques. In the early 1970s most memories were magnetic core, but these have largely been replaced by semiconductor, magnetic disk, and tape memories. Important memory characteristics are cost per bit stored, *access time* (the time required to extract the stored information), volatility (a *volatile* memory loses all its stored information when electrical power is removed, whereas a *nonvolatile* memory retains its information), and static or dynamic characteristics—a *dynamic* memory needs continual electrical "refreshing" to store data, but a *static* memory stores data as long as it has electrical power.

Several acronyms are used to describe different types of memory. A *random-access memory* (RAM) is a memory in which any particular word in the memory can be accessed (the word either written into or read from the memory) without looking at any other words in the memory. The location of the word memory is called the *address*. Reading a word in RAM does not erase the word; for example, a word in RAM can be read and transferred to another temporary storage memory called an accumulator, and the word will still be at its original address in the RAM. A dynamic random access memory is often called a DRAM.

A *read-only memory* (ROM) contains *permanently* stored information, usually put there in the manufacturing process. Such information is sometimes called *firmware*. Any word in a ROM can be read any number of times, but its contents cannot be changed. (Its contents are fixed at the time of manufacture.) A PROM or programmable ROM can be user programmed (electrically), and an EPROM can be completely erased (usually by UV light) and programmed (electrically) by the user. An EEPROM is an electrically erasable PROM whose contents can be selectively erased by the user with an electrical signal. A ROM or EPROM or EEPROM is random access like a RAM in the sense that any address can be selected to be read; you need not start at the first address, proceed to the second address, and so on.

A memory is organized into words that may be from one to eight bits long (or even longer), and each individual word can be read from the memory (RAM or ROM) or written into the memory (RAM only) one word at a time. For example, a 64-by-1-bit memory contains 64 one-bit words, and a 2048-by-eight-bit memory contains 2048 eight-bit words. The latter memory might be called a 2K memory, because one byte is an eight-bit word and 1K = 1024. Thus to select one particular word in the memory, we must have N address lines, where 2^N is the number of words in the memory, and M data lines, where M is the number of bits in each memory word. For example, the 64 × 1 memory would require six address lines ($2^6 = 64$) and one data line; the 2048 × 8 memory would require 11 address lines ($2^{11} = 2048$) and eight data lines.

A useful way to visualize a memory is shown in Fig. 14.15 in a memory

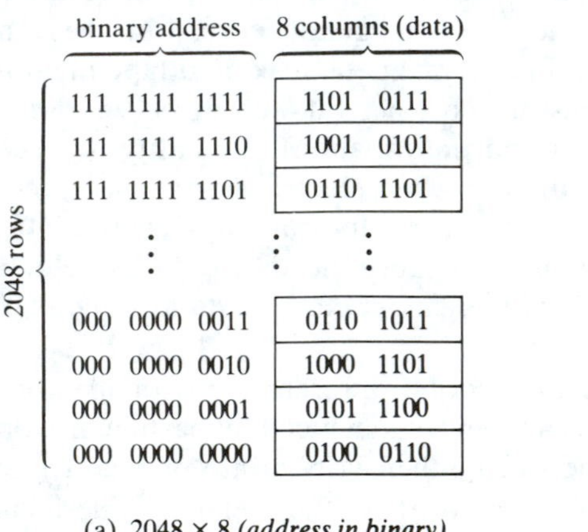

(a) 2048 × 8 *(address in binary)* (b) 256 × 8 *(address in hex)*

FIGURE 14.15 Memory map.

map with each row corresponding to a different address or location. The address refers to the location of the word, not to the word itself. For example, in Fig. 14.15(a), the data word stored at address 00000000010 is 10001101. The address may be also written in hex as in the 256 × 8 memory map in Fig. 14.15(b).

It is interesting to note that the definition of the information content of a message in information theory can be easily interpreted in terms of a memory address. The classical definition of information in a message is

$$I = \log_2\left(\frac{P_f}{P_i}\right) \qquad \textbf{(14.1)}$$

where I is the information (in bits) contained in the message expressed, P_i is the probability of finding an answer before receiving the message, and P_f is the probability of finding an answer after receiving the message. If the message gives us a hint as to where the answer is located, then $P_f > P_i$ and I is positive. If the message tells us *exactly* where to find the answer, $P_f = 1$. Suppose the answer sought is stored in one location of a memory containing 256 locations. If we know absolutely nothing about where the answer is located in the memory, the probability of finding it in one guess is $P_i = 1/256 = 2^{-8}$. If the message tells us exactly where to look, then $P_f = 1$; that is, we are certain to find the answer. The information of the message is thus

$$I = \log_2 \frac{1}{2^{-8}} = \log_2 2^8 = 8 \text{ bits} \qquad \textbf{(14.2)}$$

The information in the message is eight bits. In other words, the message contains or specifies the eight-bit address of the answer.

14.6.1 Magnetic Core Memories

Up until approximately 1974 most computer memories were made from tiny ferrite toroids or "doughnuts," several mm in diameter, that could be permanently magnetized into one of two different states (by passing a current through a wire inside the core), the states being either with the magnetic field (or flux) clockwise or counterclockwise, as shown in Fig. 14.16(a).

The B-H hysteresis curve for the ferrite material is shown in Fig. 14.16(b). Recalling that H is proportional to the magnetizing current I, we see that a current of magnitude $\pm I_1$ will cause a magnetic field of $\pm B_1$.

If the toroids are arranged in a two-dimensional array with a horizontal wire and a vertical wire going through each one, as shown in Fig. 14.16(c), then any one particular toroid will be magnetized only if *both* the horizontal wire and the vertical wire through it carry a current $I_1/2$. If only one wire carries a current $I_1/2$, only a very small B will be produced, as can be seen from the shape of the B-H curve in Fig. 14.16(b). If toroid #23 [in Fig. 14.16(c)] is to be magnetized counterclockwise as seen from the upper right, current should be sent left to right in wire x_2 and upward in wire y_3. Current in the opposite directions in wire x_2 and y_3 will produce a clockwise magnetization. Thus, we can selectively magnetize any one particular toroid in the array (i.e., "write" in the memory) by passing current through the two

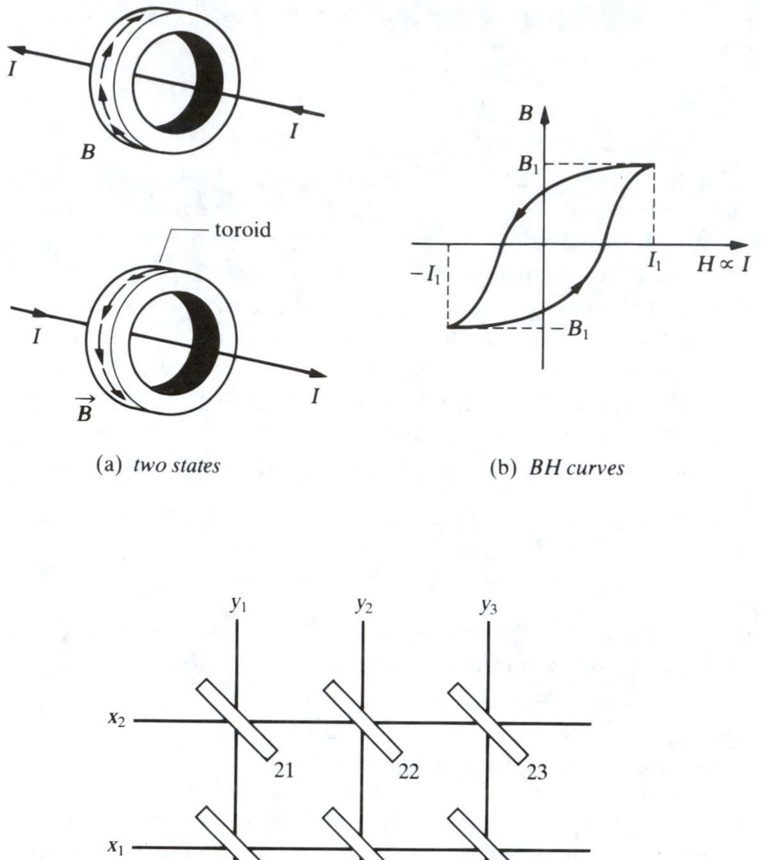

FIGURE 14.16 Magnetic memory core toroids.

wires passing through that particular toroid. To "read" the memory (i.e., to determine the state of magnetization of a particular toroid), we pass current $I_1/2$ through the two x and y wires that pass through the toroid to be read; if that toroid is already magnetized clockwise, nothing happens to it, but if it was magnetized counterclockwise, then it flips its magnetic state and a large change in the magnetic flux occurs, which induces a voltage in a "sense" wire (not shown). Thus the voltage pulse in the sense wire indicates the previous state of the toroid.

The early PDP-8 computers used magnetic core memories, and certain highly important military computers also use magnetic core memories because the information stored in them is preserved in the event of an electrical power failure. (The core memory is nonvolatile.) The construction of such cores is very tricky, requiring precise mechanical tolerances, and the resultant core is bulky and fragile, especially compared to modern solid-state monolithic memories. Also, a magnetic core memory uses a great deal of electrical power, because each time its state is changed from clockwise to counterclockwise magnetization, the B-H hysteresis curve is traversed. From electromagnetic theory we recall that the area inside the B-H curve represents the energy dissipated as heat in the magnetic material in the process of traversing the curve. Thus, every time the curve is traversed in the memory, an amount of energy equal to the area inside the B-H curve is dissipated. The net result is that magnetic core memories are essentially obsolete because of their size, mechanical fragility, expense, and power consumption, except for certain applications where their nonvolatile nature is essential.

14.6.2 Semiconductor Memories

There are three common types of solid-state or semiconductor memories: (1) bipolar, (2) MOS, and (3) CMOS. These types use, respectively, bipolar, metal-oxide-silicon, and complementary-metal-oxide-semiconductor technology. The principal tradeoff for semiconductor memories is between power and speed or access time as shown in Table 14.1; in other words, we can have short access times with high power or long access times with low power. At present, there are two fundamental limitations on memory density: the presence of extremely small amounts of radioactive alpha emitters in the chip material, and the bombardment of the chip by cosmic rays. In both cases the ionization in the chip from the radiation can change the bits stored.

TABLE 14.1. Semiconductor Memories

	Memory Type	*Approximate Access Time (ns)*	*Power Required*
Volatile	Bipolar (flip-flops)	10–50	High
Volatile	MOS (FET)	50–500	Medium
Volatile	CMOS	>500	Low

14.6.2.1 Bipolar Semiconductor Memories A bipolar semiconductor memory consists essentially of an array of bipolar transistor flip-flops with one flip-flop for each bit stored. If all the flip-flops are on one crystal, it is called a *monolithic* memory. Once the flip-flops are set corresponding to the information to be stored, they will retain the information for as long as electrical power is applied; in other words, they are *static* but volatile. At the present time bipolar memories are used only when their very high speed is required.

One example of a bipolar memory chip is the old-fashioned SN7489 64-bit read/write memory, shown in Fig. 14.17.

ME	WE	Operation	Outputs
0	0	Write	$\overline{\text{Data}}$ input
0	1	Read	Selected $\overline{\text{word}}$
1	0	Inhibit storage	$\overline{\text{Data}}$ input
1	1	Do nothing	High

33 ns access time
open collector outputs

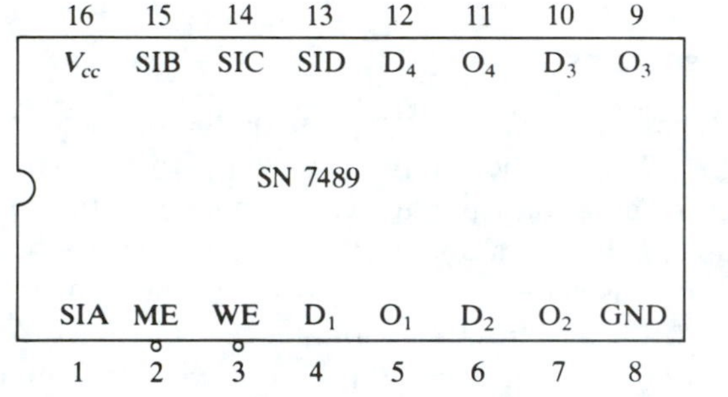

SI = sense input (address)
D = data input
O = sense output
WE = write ENABLE
ME = memory ENABLE

FIGURE 14.17 SN7489 bipolar memory (16 × 4).

14.6.2.2 MOS Semiconductor Memories The typical memory today is probably made from MOS devices. There are two general types: static and dynamic. A *static* memory stores information as long as it has power; a *dynamic* memory needs to have its information electrically "refreshed" or "topped off" every few milliseconds. A static MOS memory stores each bit in a flip-flop, as shown in Fig. 14.18(a), whereas a dynamic MOS memory stores each bit as charge (or voltage) on a capacitor, as shown in Fig. 14.18(b). The static MOS flip-flop is essentially a standard flip-flop constructed from integrated MOSFETs on a single silicon chip. Assume the row select line is high, which means the two MOSFET switches Q_1 and Q_2 are on. Then if the data in (write) line equals 1, $Q = 1$ and $\bar{Q} = 0$; whereas if the data in (write) line equals 0, $Q = 0$ and $\bar{Q} = 1$. The $Q\bar{Q}$ state of the flip-flop is read by measuring the Q and $\bar{Q}$ voltages. If the row select line is low, both Q_1 and Q_2 are off and the $Q\bar{Q}$ state is stable (i.e., data are stored). The two resistances R_1 and R_2 are made from MOSFETs with their gates connected to their sources.

The dynamic MOS storage element is a capacitance C_s charged up

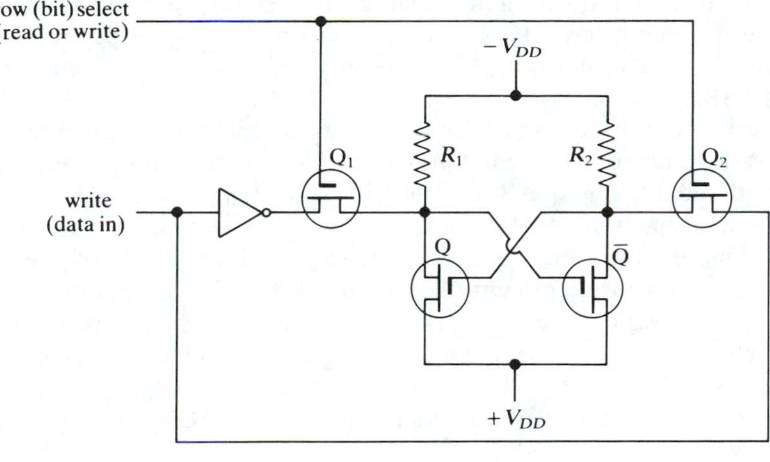

(a) *static (flip-flop)*

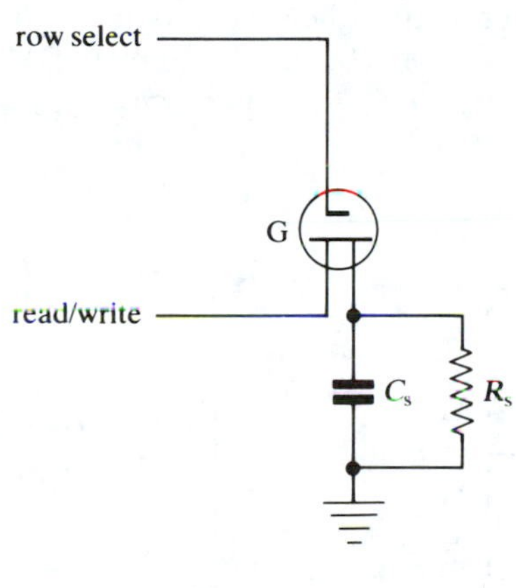

(b) *dynamic*

FIGURE 14.18 MOS memories.

through a MOSFET, as shown in Fig. 14.18(b). Because the charge tends to leak off with a time constant $R_s C_s$, where R_s is the leakage resistance (typically $C_s \approx 10\,\text{pF}$, $R_s \approx 10^9\,\Omega$), the information would be lost in approximately $R_s C_s \approx 10\,\text{ms}$. The remedy is to use special "refresh" circuitry [not shown in Fig. 14.18(b)] to recharge the capacitance approximately every ms. (Some dynamic memory chips have the refresh circuitry "on board" the chip.) Thus we immediately see that dynamic memory is inherently more complicated than static memory because of the refresh circuitry required. Also, the cost per bit of dynamic memory is larger than for static memories. But dynamic MOS memories typically have a density approximately four times higher than that of static MOS memories. For example, we might have a 1K × 8 static or a 4K × 8 dynamic memory on one chip. A typical dynamic MOS memory can have 3×10^4 bits on one 16-pin DIP.

We will not discuss the refresh circuitry in detail except to say that it involves a counter and a multiplexer to address the various rows to be refreshed and a timer to indicate when the refreshing is necessary.

To sum up, static MOS memories are easier and simpler to use, but they use more power and more chips are needed. Dynamic MOS memories require more elaborate circuitry because of their refresh requirements, but they require fewer chips and use less power. For both types of memories it is absolutely essential to use enough decoupling capacitors on the power supply leads. Type Z5U 0.1-μF ceramic capacitors are fine, one for each chip, and typically a 100-μF tantalum capacitor for the power supply input lead is needed.

14.6.2.3 CMOS Semiconductor Memories CMOS memories use the least power and are thus preferred for portable battery-powered systems. However, they are slower than either the bipolar or the MOS memories, with a typical access time of 700 ns. A simple CMOS 4096-bit static memory is shown in Fig. 14.19. A typical CMOS memory is the NMC6514,

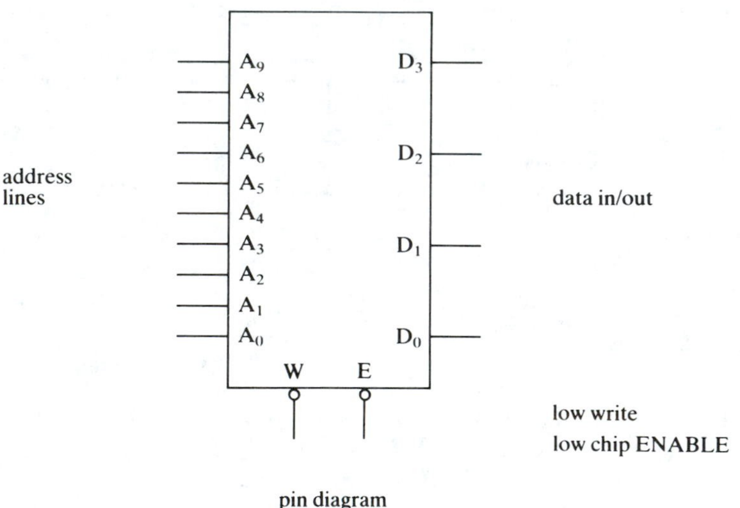

FIGURE 14.19 CMOS 4096-bit static memory (1024 × 4).

which is a 4096-bit (1024 × 4) static RAM in an 18-pin DIP. Its inputs and outputs are all TTL compatible, and it has three-state outputs for connection to a common bus. There are ten address lines ($2^{10} = 1024$), four data lines, a write line, a chip ENABLE line, and two pins for $V_{cc} = 5$ V and ground.

Some CMOS memories are available on cards with a built-in (on-board) lithium battery that makes them nonvolatile. The dual 06DULCMEM32 (@ \$660) is one such example; it contains 32K populated memory on a single S-100 board and has an access time of only 220 ns. It will store data for three to ten years with the regular power off. A large capacitor (1 F@6 V) can also be used as an emergency power source for a CMOS memory.

14.7 THE LOOKUP TABLE TECHNIQUE

The availability of inexpensive, fast RAMs has made practical a *lookup* table technique to map one function to another function. If an input function f (in digital form) is to be mapped into another desired function g, then the obvious solution or mapping technique is to multiply f by a conversion factor $K = g/f$ to obtain $g = Kf$. But this technique is impractical with analog electronics because of drift and stability problems and the difficulty of getting exactly the correct analog factor K. It is also often difficult to get one simple analytic expression for K, so even with a microcomputer the multiplication by K is difficult to program and slow to execute.

The lookup table technique is simply to store the desired values g at known locations or addresses in RAM and to *use the input f to address* the correct memory location where $g = Kf$ is stored. If $g_n = Kf_n$, then the number g_n is stored at the address f_n. Clearly there must be only one value of g corresponding to one value of f.

Consider a specific example. Suppose the input f is a nonlinear function of an independent experimental parameter x. For example, f could be the resistance $R(T)$ of a thermistor at a temperature $T = x$, as shown in Fig. 14.20. There is a one-to-one relationship between the resistance R and the temperature T, and we desire to convert this R value to the temperature C

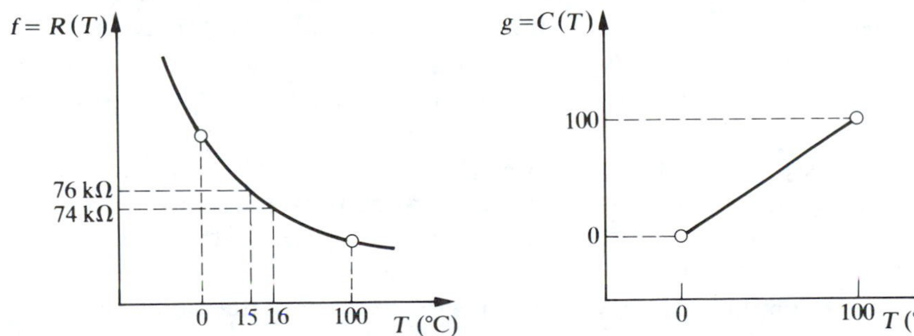

FIGURE 14.20 Lookup table example.

in degrees Celsius. Thus we must map the nonlinear function $R(T)$ into the linear function $C(T)$, where $C(T)$ is the (numerical) temperature in degrees Celsius.

First, we carefully measure the resistance at all the temperatures in the range of interest, using a standard accurately calibrated thermometer to measure T and an ohmmeter to measure R. How many measurements we take (e.g., every 1°C or every 0.1°C) depends on the desired precision and the amount of memory available. If a precision of ±1°C is sufficient, then for a temperature range from 0° to 100°C we would have a table of 101 values of T and the corresponding 101 values of R. The resistance value at any given temperature in this interval would determine the address of the corresponding temperature. For example, if $R = 76\ \text{k}\Omega$ at 15°C, then 15 would be stored in RAM at the address $76 = 4C_{16}$. If $R = 74\ \text{k}\Omega$ at 16°C, then 16 would be stored at the address $74 = 4A_{16}$.

The program required then would take the input digital word corresponding to $R(T)$, look up the memory contents at the address $R(T)$, and read out the memory contents to the output device, such as an LED readout. Then the program would again look at the $R(T)$ input, and so on. Such a program is very short and simple and can easily be executed by an inexpensive microprocessor; an expensive microcomputer is not required, although it would do, of course.

Actually, the analog resistance $R(T)$ would be digitized by an analog-to-digital converter (A/D), and the A/D output would be the address. For example, if an eight-bit A/D were used, there would be a maximum of 256 different digital outputs. Thus, if the correct temperature were loaded into each of the 256 locations of a 256 × 8 RAM, the A/D output would necessarily "point" at the correct temperature in RAM. This is a sort of "brute force" technique in that all 256 memory addresses are used. In the 8085 microprocessor the HL register pair can be used to point at a particular location in RAM; the contents of the HL pair is the address. Thus, an 8085 thermometer program would simply consist of (1) loading the A/D output into the HL register pair, (2) reading the correct temperature (using the instruction MOVA,M), which loads the memory contents at the address contained in the HL pair into the accumulator, and (3) reading the contents of the accumulator out on the LED display.

A second technique is slightly more sophisticated and uses less memory but more programming. The possible A/D outputs corresponding to unit temperature changes, not the temperatures, are loaded in the RAM. For example, for a 20° to 50°C range the A/D output for 20°C might be stored at address 2000, the A/D output for 21°C at address 2001, and so on. Then the program would take the A/D output and search the memory until it found (stored in memory) a value that just exceeded the A/D output. The *address* of this value would then be converted into the numerical temperature by some algorithm and displayed. In the previous example the temperature would be obtained by adding 20 to the two LSBs of the address. The address in this technique is a sort of index that specifies the temperature.

Such a mapping or transformation is now fairly common in linearizing the output of a transducer. This enables the transducer to produce a linear output as the variable x changes—the linear output being, of course, not the "raw" transducer output but the output from the contents of the lookup table in RAM. Without this linearization the transducer use might be restricted to a very narrow range of input values of x.

The second function need not be linear as in the preceding example; it could be any desired function.

14.8 READ-ONLY MEMORIES (ROMs)

A *read-only memory* contains permanently stored information. Like an encyclopedia in a library, it can be read but not written in. By definition a ROM is nonvolatile—that is, turning the power off does not erase the information in the ROM. ROMs are used in calculators to store various commonly used functions or algorithms for calculating such functions. A ROM could also be used to "translate" or "assemble" or "compile" a computer programming command in a high-level language, such as BASIC or FORTRAN, into machine language that can be understood by the computer. An ALU (arithmetic logic unit) chip is one example of a simple ROM. Another example of a ROM is the 7447 BCD-to-seven-segment LED decoder chip.

There are several types of ROMs: (1) the "plain" ROM, (2) the PROM and (3) the EPROM. In all three types any word stored can be selectively read (without erasing the word).

The ordinary ROM is usually programmed during the manufacture of the chip; such chips are invariably used only in high-volume applications such as calculators because it is so expensive to set up the manufacturing process.

The PROM is user programmable; that is, it comes completely blank. The user programs in binary data or instructions by applying large external voltages ($\approx$30 V) to the chip to vaporize some internal connections and thus create a specific information content in the PROM. Clearly, such a PROM can be programmed only once, but once programmed it is a permanent nonvolatile memory.

The EPROM is a PROM whose contents can be erased and reprogrammed. The programming is done electrically, as with a PROM, but the erasure is done by exposing the chip to ultraviolet light ($\lambda \cong 250$ nm) for from 10–30 min, depending on the strength of the UV lamp. The UV light penetrates a quartz window (which is part of the EPROM chip) and causes photoionization in the gate structures of the MOSFETs and thus discharges all the stored voltages, erasing the entire memory. When the EPROM is in use, all light must be kept from the window, because the small UV content of sunlight may cause partial erasure. Such chips are expensive, partly because of the expense of the quartz window. One such EPROM is the 27C16, which is a 16K (2048×8) 450-ns CMOS memory electrically programmable and erasable with UV light. It comes in a 24-pin DIP and costs approximately $10.

Just coming on the market are EPROMs that can be programmed and erased *electrically*. They are generally called *electrically erasable* PROMs or EEPROMs. Only one bit can be selectively erased with the electric erasure, which is an obvious advantage over UV erasure, which simultaneously erases the entire memory. The entire chip can be erased if desired.

14.9 MAGNETIC DISK MEMORY

Magnetic disk memory is the most popular technique for mass storage with microcomputers. The magnetic *floppy disk* is a flexible thin plastic disk with a magnetic coating approximately 8 inches in diameter that rotates at 360 rpm while inside its stationary paper jacket. The information is stored serially in a series of concentric magnetic tracks. The information is fed in and read out by read/write pickup heads or voice coils, which are moved radially to select the desired track. Because the disk is already spinning, only the radial movement is required to read any desired track. The maximum access time is the time for one complete revolution plus the time for any radial movement required. The radial movement is accomplished with stepping motors, and the numbers of steps are counted to indicate the desired track. An identification code is also permanently written on each track. Smaller, 5-in., mini floppy disks are used in many personal computers, such as the IBM and the Apple. Even smaller ($3\frac{1}{2}$-in.) micro floppy disks are available and are used in some personal computers.

There are two popular types of disk recording: single density and double density. In *single-density* recording there is a magnetic flux transition at the beginning of each clock period, and in the middle of the clock period a transition (either + or −) for a 1 bit and no transition for a 0 bit. The single-density clock transitions are 4 μs apart; each bit therefore occupies 4 μs.

In *double-density* recording, there is a transition (+ or −) for a 1 bit in the middle of the clock period and no transition for a 0 bit. But a clock pulse transition occurs only between two adjacent 0 bits. Each bit lasts only 2 μs; hence the name *double* density. The magnetic flux transitions are shown in Fig. 14.21.

Hard magnetic disks store many more data than do floppy disks. A hard disk consists of a thin magnetic coating on a rigid metal disk. The read/write heads are only approximately 10^{-5} cm above the magnetic coating and are essentially cushioned by a thin stream of air caused by the disk

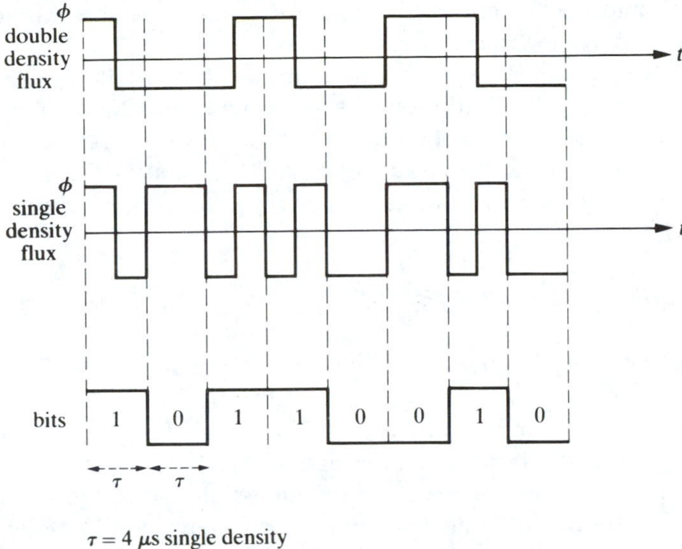

FIGURE 14.21 Single- and double-density magnetic disk recording.

rotation. Because of this close mechanical spacing, hard disks must be kept in a sealed dust-free environment. Also, such close mechanical tolerances mean that any repair must be done by a skilled service person from the manufacturer.

Hard disks are often called *Winchester* disks because one of the first such systems contained two platters or disks each of which stored 30MB (yes, 30 *mega*bytes!). This was naturally known as a 30-30 disk system,

TABLE 14.2. Magnetic Disk Properties

	Full-Size Floppy (8″ diam)	*Mini Floppy ($5\frac{1}{4}$″)*	*Micro Floppy ($3\frac{1}{3}$″)*	*($5\frac{1}{4}$″) Winchester*
Information storage capacity	246K	110KB (single-side density)	710KB	10MB
Data transfer rate	30KB/s	15KB/s	80B	5MB/s
Number of tracks	80	35		40 or 80
Rotational speed (rpm)	360	300	600	3000
Max access time (ms)	170	200	350	75
Approximate cost (1986) of disk drive		\$300	\$300	\$3000

B = byte.

which was informally named a Winchester disk after the famous 30-30 caliber Winchester lever-action rifle.

To appreciate how much information can be stored on a hard disk, assume one byte per letter of the English language and 2000 English letters per printed page for a typical book; a 20MB disk can store the equivalent of 10,000 pages of a book! In other words, a 20MB disk is the equivalent of forty 250-page books. Table 14.2 summarizes the properties of the magnetic disks.

14.10 MAGNETIC TAPE MEMORY

Magnetic tape used for digital data storage is similar to high-fidelity audio tape in that it comes in reels and consists of a thin layer of magnetic oxide on a thin plastic tape. But audio tape (in analog systems) is always run below magnetic saturation so that the signal amplitude is proportional to the degree of magnetization of the tape. Digital or computer tape is, however, deliberately saturated; the tape is either not magnetized at all (for a 0 bit) or fully saturated (for a 1 bit) or fully saturated in two different directions for the bits. The data are recorded on the tape by a magnetic recording head, which consists of a small electromagnet that magnetizes a tiny spot on the tape for a 1 bit. The data are read off the tape by a pickup head, which consists of a tiny pickup coil that produces a voltage pulse when a magnetized tape bit moves past it (à la Faraday's law).

14.11 MAGNETIC TAPE RECORDING TECHNIQUES

There are several ways to record data on magnetic tape; four of the most common are shown in Fig. 14.22. The tape magnetization ϕ is either positive, zero, or negative. In NRZ (non-return-to-zero) a positive ϕ means a 1 bit, and a negative ϕ a 0 bit. The main problem with NRZ is that if one change in ϕ is missed by the playback head, *all* the rest of the bits will be inverted. In NRZI (non-return-to-zero-inverted) a *transition* or *change* in ϕ in either direction (+ to − or − to +) means a 1 bit, and no transition or change means a 0 bit. In both NRZ and NRZI the transitions occur at the beginning of each clock pulse.

In Manchester or biphase M or single-density encoding, ϕ changes at *every* clock pulse and in the center of the clock pulse to indicate a 1 bit. The Manchester pulses can be twice as fast as for NRZ or NRZI; therefore twice the tape frequency response is necessary for Manchester. The Manchester encoding is seen to be a sort of frequency modulation with one cycle used per bit; a 1 bit is encoded as an up and down ϕ pulse, and a 0 bit as one clock period with ϕ either positive or negative.

In biphase L encoding there is no change in ϕ at the edge of the clock

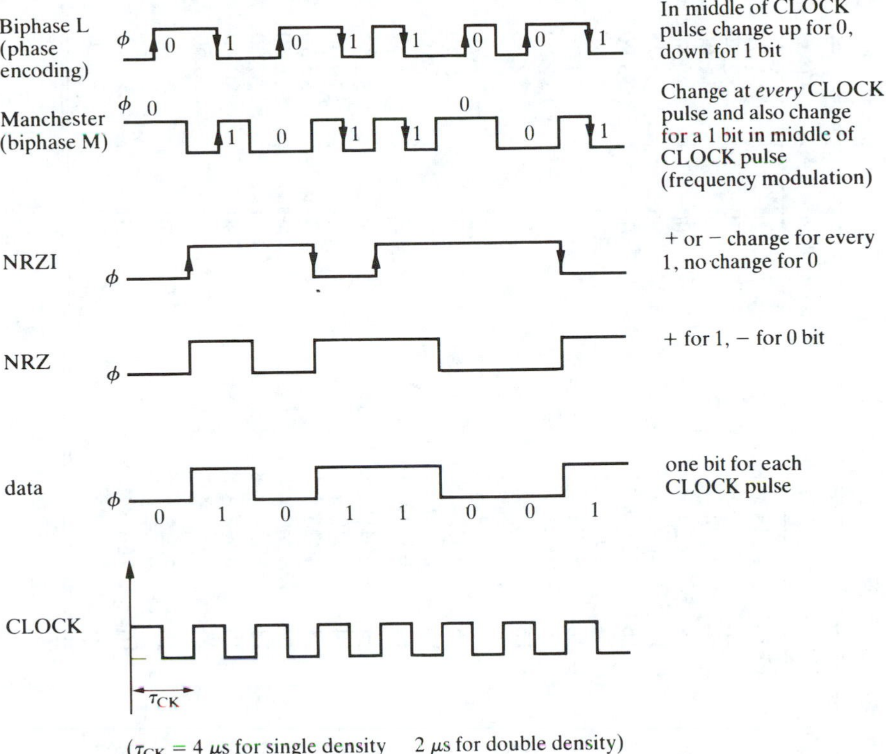

FIGURE 14.22 Magnetic tape data storage.

pulses, only in the middle of the clock pulses—down for a 1 bit, and up for a 0 bit.

The bits are thus recorded and read in *serial* fashion. Hence the access time may be very large large if you have to unwind most of the tape reel to access information. A typical bit density for modern tapes is 32 bits/mm (800 bits/in.), and typical data transfer rates may range from 30 to 240 bytes/s at $1\frac{7}{8}$ in./s or 3200 bytes/s at 20 in./s. At 800 bits/in. a 1000-ft tape can store 1.2×10^6 bytes, so magnetic tape is often the memory of choice for inexpensive long-term storage of large quantities of information if very rapid access is not important.

Cassette tape provides a low-cost, slow storage system, but professionals do not consider it really reliable. Up to 500KB can be stored on one C-60 cassette. If you must use cassette storage, one useful rule-of-thumb is to make *two* copies of all important files.

14.12 FUTURE STORAGE TECHNIQUES

Newer storage technologies not yet on the market include the magnetic bubble memory, which can store up to 1000MB on one chip, and the

TABLE 14.3. Typical Memory Chips

	Type	*Pins*	*Organization*	*Access time (ns)*	*Price*
Static RAM	TMM2016-12	24	2048 × 8	120	$1.70
	2111	18	256 × 4	450	$2.50
	2114N-2	18	1024 × 4	200	$1.00
	21C14 (CMOS)	18	1024 × 4	200	$0.50
	2147HN-3	18	4096 × 1	55	$4.25
	27LS00	16	256 × 1	80	$3.00
	6116P-4	24	2048 × 8	200	$1.90
	6264P-12 (CMOS)	28	8192 × 8	120	$4.75
	74S189	16	16 × 4	35	$2.00
	7489	16	16 × 4	50	$2.00
Dynamic RAM	1103 18	18	1024 × 1	300	$.75
	4116N-15	16	16384 × 1	150	$.50
	4164N-150	16	65536 × 1	150	$1.00
	50464-15	18	65536 × 4	150	$8.00
EPROM	2716	24	2048 × 8	450	$2.50
	27C16 (CMOS)	24	2048 × 8	450	$10.00
	2732	24	4096 × 8	450	$2.50
	27C32 (CMOS)	24	4096 × 8	450	$11.00
	2758	24	1024 × 8 (single 5 V)	450	$4.00
	2764-20	28	8192 × 8	200	$4.00
	27C64 (CMOS)	28	8192 × 8	450	$6.50
EEPROM	2816A	24	2048 × 8	350	$9.00
PROM	74S287 (3 state)	16	256 × 4	50	$1.75
	74S474 (3 state)	24	512 × 8	35	$3.50

optical laser disk, which can store up to 1000MB on one 12-in. disk. The popular 5-in. compact audio disc is a laser-read ROM containing approximately 400MB. At present, research is under way to enable the individual computer user to write on such a disc, i.e., to make it into a PROM.

PROBLEMS

1. Calculate the binary sum of 01100001 and 00010111.
2. Prove that (A + B) + C = A + (B + C). [*Hint*: Just write the truth table.]
3. Using half-adders and full-adders, sketch a block diagram of a parallel adder for adding two two-bit words.
4. Calculate 01100001 minus 00010111.
5. A half-adder can be used only to add the _______, whereas a half-subtracter can be used only to subtract the _______ (MSB or LSB).
6. Using half-subtracters and full-subtracters, sketch a block diagram of a parallel subtracter for subtracting two two-bit words.
7. Calculate the ones complement of 0110 and the twos complement of 0101.
8. Using twos complement arithmetic, calculate D = B − A, where B = 10010000 and A = 00101110.
9. Write the truth table for binary addition A + B = S.
10. Write the truth table for binary subtraction B − A = D.
11. Write the truth table for binary multiplication A × B = P.
12. Multiply 1001 × 0011.
13. How many address lines (*N*) and data lines (*M*) would the following memories require? (a) 1K × 4, (b) 4K × 8, (c) 1K × 1.
14. Distinguish between volatile and nonvolatile memories.
15. Distinguish between RAM and ROM. Which can the user write in? Read from?
16. What is a PROM? An EPROM? An EEPROM?
17. What type of memory has the fastest (shortest) access time?
18. Briefly explain the main advantages of magnetic disk memory.
19. Briefly explain the main advantages of magnetic tape memory.

CHAPTER 15

Analog/Digital Conversion

15.1 INTRODUCTION

Almost all of the physical variables we measure are intrinsically analog: voltage, current, force, position, speed, pressure, temperature, and many others. In order to use the powerful techniques of digital signal processing or computation, we must convert an analog input signal to digital form. The device that performs this conversion is called an *analog-to-digital* (*A/D*) *converter* or ADC. Analog data are often converted into digital form for long-distance transmission to avoid distortion and/or masking by noise or interference. If a laboratory experiment is to be controlled by a computer, the experimental parameters must be digitized so the computer can understand them. If the data are to be analyzed by a computer, the data must be converted into digital form.

Digital data must be converted into analog form to display it in a graphical form or to be easily comprehensible by human beings—few humans can interpret binary numbers quickly. The conversion device is called a *digital-to-analog* (*D/A*) *converter* or DAC. A D/A converter is also an essential component in several types of A/D converters.

Both A/D and D/A converters are now available for under $20 each on a single chip. A typical signal processing system is shown in Fig. 15.1.

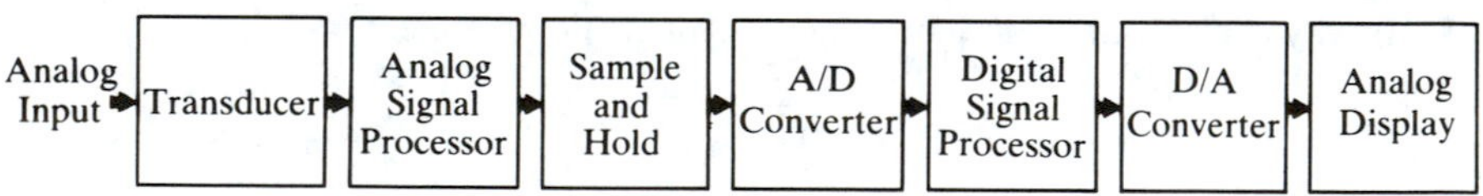

FIGURE 15.1 General signal processing system.

The transducer converts the analog input quantity (temperature, pressure, etc.) to be measured into a suitable voltage or current for the A/D converter. An analog signal processor (typically a filter) is sometimes necessary in front of the A/D converter to filter out noise or other undesirable components. The sample-and-hold circuit is the same as described in Chapter 10 and is necessary for fast inputs to "freeze" the analog input at a constant value while the A/D converter makes the conversion. Depending on the type of A/D converter used, the time required for a conversion may range from 100 ns to 100 ms.

Once the signal has been digitized, it can be manipulated by a microprocessor or computer in many different ways—the only limitation being the power of the computer. For example, the signal can be Fourier analyzed, various correlation functions can be calculated, the signal can be compared with other signals or voltage levels, and so on. One great advantage of this technique is that these calculations can be modified and/or corrected with changes in the software rather than with hardware changes, which are often more difficult, time consuming, and expensive.

The digital results of such calculations and operations then can be converted into analog form for controlling an experiment or for display in graphical form. The D/A converter changes the digital data into analog form. The display can be an X-Y recorder or plotter or a video terminal. The human eye and brain are extraordinarily good at rapidly interpreting large quantities of analog information in pictorial or graphical form. Examples are the operation of trained radar or sonar operators or physicians interpreting X-ray pictures or CAT scanner pictures. The human eye/brain combination is not nearly as good at interpreting binary numbers or large tables of decimal numbers.

15.2 THE DIGITAL-TO-ANALOG CONVERTER

15.2.1 Input Digital Codes and Analog Outputs

D/A converters can accept two different types of input digital numbers: unipolar and bipolar. For simplicity we will use four bits in all our examples. A *unipolar* input or unipolar code varies from 0000 to 111 for the analog output going from zero to maximum. A *bipolar* code means that the analog output runs from a negative value to a positive value.

The simplest unipolar code is to have 0000 input correspond to zero output and 1111 input correspond to the maximum, as shown in Table 15.1. Notice that for N bits there are 2^N different possible output levels and $2^N - 1$ output steps. For an eight-bit D/A converter $N = 8$, and there are $2^8 = 256$ different output levels and $2^8 - 1 = 255$ output steps.

There are three common ways to calibrate or set the magnitude of the analog output levels—in other words, to choose the output voltage corresponding to 1 LSB input. The first and most common calibration is to set the MSB of the digital input equal to a convenient voltage, usually 5.000 V. Thus a 1000 input will produce a 5.000-V output. Each LSB is then $5\text{ V}/2^{N-1} = 10\text{ V}/2^N$. For $N = 4$ each LSB $= 10\text{ V}/2^4 = 0.625$ V, and the maximum output is then $10\text{ V} - 1\text{ LSB} = 9.375$ V. For an eight-bit D/A

TABLE 15.1. Four-Bit Unipolar Codes

		Output Voltage		
Digital Input	*Output*	*MSB = 5.000* V	*LSB = 0.01* V	*MAX = 10.000* V
1111	15 LSB	9.375	0.150	10.000
1110	14 LSB	8.750	0.140	9.333
1101	13 LSB	8.175	0.130	8.667
1100	12 LSB	7.500	0.120	8.000
1011	11 LSB	6.875	0.110	7.333
1010	10 LSB	6.250	0.600	6.667
1001	9 LSB	5.625	0.090	6.000
1000	8 LSB	5.000	0.080	5.333
0111	7 LSB	4.375	0.070	4.667
0110	6 LSB	3.750	0.060	4.000
0101	5 LSB	3.125	0.050	3.333
0100	4 LSB	2.500	0.040	2.667
0011	3 LSB	1.875	0.030	2.000
0010	2 LSB	1.250	0.020	1.333
0001	1 LSB	0.625	0.010	0.667
0000	Zero	0.000	0.000	0.000

converter each LSB is $10\text{ V}/2^8 = 0.0391$ V, and full scale is 9.9609 V. The maximum output is always 1 LSB less than twice the midrange output (5.000 V).

The second common calibration is to choose the LSB to produce a convenient output level (e.g., 10 mV). Then full scale equals $(2^N - 1) \times 10\text{ mV} = 0.150$ V for $N = 4$ or 2.550 V for $N = 8$.

The third common calibration is to set the maximum output equal to a convenient value, usually 10.000 V. Then for $N = 4$ a 1111 input produces a 10.000-V output, and for $N = 8$, a 11111111 input produces a 10.000-V output. The size of the LSB is then $10\text{ V}/(2^N - 1) = 0.667$ V for $N = 4$, and 1 LSB = 0.0392 V for $N = 8$.

These three calibrations are shown in the three right-hand columns of Table 15.1.

The simplest bipolar code is the *offset binary* code in which the MSB produces a zero output; that is, for $N = 4$ a 1000 input produces a zero output, and for $N = 8$ a 10000000 produces a zero output. For $N = 4$ a 1001 input produces an output of +1 LSB, and a 0111 input a −1 LSB output. Notice that there is one more negative output level than positive output level; for $N = 4$ the output ranges from $-2^{N-1} \times \text{LSB} = -5.000$ V to $+(2^{N-1} - 1) \times \text{LSB} = +4.375$ V.

The twos complement code merely complements the MSB of the offset binary code. Both codes are shown in Table 15.2.

Table 15.3 shows the number of steps, step size, and resolution for various numbers of bits. The *resolution* is defined as the step size divided by

TABLE 15.2. Four-Bit Bipolar Codes

Offset Binary Code			*Twos Complement Code*		
	Output			*Output*	
Digital Input	*No. LSBs*	*Voltage* (V)	*Digital Input*	*No. LSBs*	*Voltage* (V)
1111	+7	+4.375	0111	+7	+4.375
1110	+6	+3.750	0110	+6	+3.750
1101	+5	+3.125	0101	+5	+3.125
1100	+4	+2.500	0100	+4	+2.500
1011	+3	+1.875	0011	+3	+1.875
1010	+2	+1.250	0010	+2	+1.250
1001	+1	+0.625	0001	+1	+0.625
1000	Zero	0.000	0000	Zero	0.000
0111	−1	−0.625	1111	−1	−0.625
0110	−2	−1.250	1110	−2	−1.250
0101	−3	−1.875	1101	−3	−1.875
0100	−4	−2.500	1100	−4	−2.500
0011	−5	−3.125	1011	−5	−3.125
0010	−6	−3.750	1010	−6	−3.750
0001	−7	−4.375	1001	−7	−4.375
0000	−8	−5.000	1000	−8	−5.000

TABLE 15.3. Step Size and Resolution (1 MSB = 5 V)

No. Input Bits	*Step Size* (*1 LSB*)	*No. Steps*	*Output Range* (V)	*Resolution* (%)
4	$5/2^3 = 0.625$	15	0–9.375	6.7
8	$5/2^7 = 0.0391$	225	0–9.9609	0.39
10	$5/2^9 = 0.00977$	1023	0–9.990	0.098
12	$5/2^{11} = 0.00244$	4095	0–9.99756	0.024
16	$5/2^{15} = 0.000153$	65535	0–9.999847	0.0015
N	$5\text{ V}/2^{N-1}$	$2^N - 1$	$0\text{–}\dfrac{(2^N - 1)5}{2^{N-1}}$	

the maximum output. A change of 1 LSB in the input produces a minimum percent output change equal to the resolution. For eight bits a 1-LSB change is a 0.39% change.

15.2.2 The Summing D/A Converter

The easiest type of D/A converter to understand is based on an op amp summing circuit as discussed in Chapter 10. The basic circuit is shown in Fig. 15.2(a) for a four-bit word input. The gain for the input v_0 is $-R_f/R_0$, and similarly for the other inputs. Thus, the gain for each of the inputs can

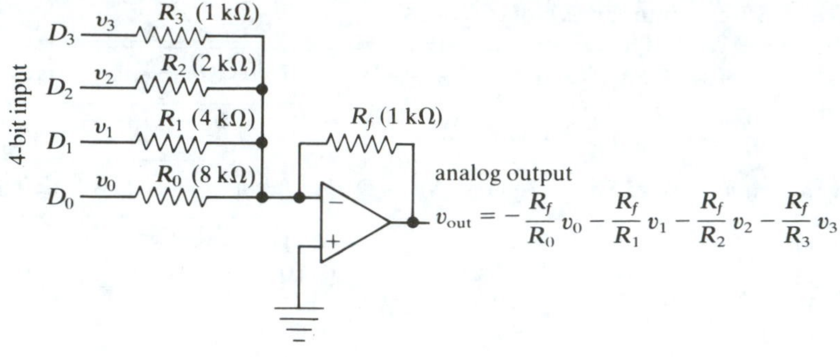

(a) *basic circuit*

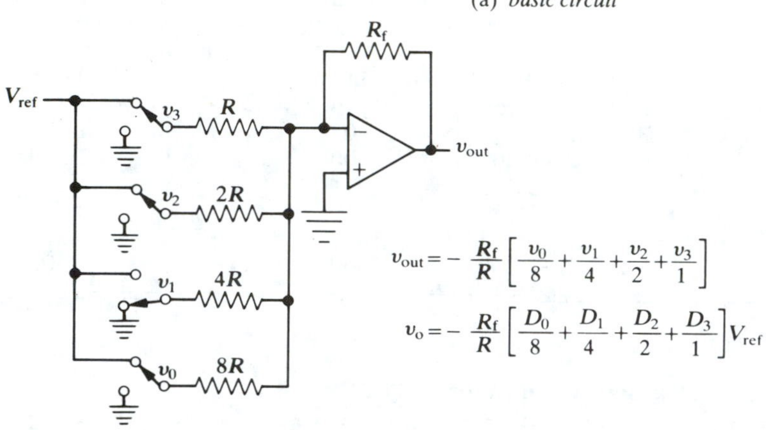

(b) *switches shown for input data* $D = D_3\,D_2\,D_1\,D_0 = 1101$

FIGURE 15.2 Summing D/A converter.

be adjusted by choosing the resistances R_0, R_1, R_2, R_3. If the inputs are the bits of a digital word, we can make the output equal to the analog equivalent of the digital input by making the gain for the MSB largest, the gain for the MSB one-half as large, and so on. For example, if v_0 is the LSB voltage and v_3 the MSB voltage, and if we arbitrarily choose $R_f = 1.0\,\text{k}\Omega$ and $R_3 = 1.0\,\text{k}\Omega$, we must have $R_2 = 2.0\,\text{k}\Omega$, $R_1 = 4.0\,\text{k}\Omega$, and $R_0 = 8.0\,\text{k}\Omega$. Then the gain for v_3 (the MSB) is 1, for v_2 is $\frac{1}{2}$, for v_1 is $\frac{1}{4}$, and for v_0 (the LSB) is $\frac{1}{8}$. Table 15.4 shows the output voltages for all the possible input four-bit words $D_3D_2D_1D_0$, assuming a 1 is 5 V and a 0 is 0 V. The analog dc output voltages are all negative and range from 0 to −9.375 V. The output can be made positive by adding an inverter to the output, so the final output would range from 0 to +9.375 V. The output voltage is "quantized" in steps of 1 LSB = 0.625 V = $\frac{5}{8}$ V, and there are 15 steps. The maximum output is one step less than twice 5 V, or 9.375 V.

TABLE 15.4. Inputs and Outputs for Four-Bit Summing D/A Converter of Fig. 15.2

	MSB D_3	D_2	D_1	LSB D_0	v_{out} (V)
	1	1	1	1	−75/8 = −9.375
	1	1	1	0	−70/8 = −8.750
	1	1	0	1	−65/8 = −8.125
	1	1	0	0	−60/8 = −7.500
	1	0	1	1	−55/8 = −6.375
	1	0	1	0	−50/8 = −6.250
	1	0	0	1	−45/8 = −5.625
0 = 0 V	1	0	0	0	−40/8 = −5.000
1 = 5 V	0	1	1	1	−35/8 = −4.375
	0	1	1	0	−30/8 = −3.750
	0	1	0	1	−25/8 = −3.125
	0	1	0	0	−20/8 = −2.500
	0	0	1	1	−15/8 = −1.875
	0	0	1	0	−10/8 = −1.250
	0	0	0	1	−5/8 = −0.625
	0	0	0	0	Zero

This summing D/A converter is conceptually simple and is easy to construct from a spare op amp if one is needed in a hurry. But it has three disadvantages that make it impractical for serious applications. First, the range of resistance values required is too large: from 100 kΩ down to 100 kΩ/2^{N-1} for N bits—for example, from 100 kΩ to 781.25 Ω for $N = 8$, and from 100 kΩ to 3.05176 Ω for $N = 16$. Each of these resistance values must be precisely chosen, which is very difficult to do (you can't just grab resistors from the bin). Also, for a large number of bits the low-value resistors will draw too much current from the source of the MSB. Also remember that large resistance values (above 1 MΩ) are noisy, expensive, and difficult to obtain in precise values. Second, the expected digital precision or resolution of ±1 bit means ±1 part out of $2^N - 1$, for example, ±0.39% for eight bits. To achieve this precision, we must have all the resistance values to ±0.39%/2 = ±0.2% (because the percent errors for R_f and R *add*) or better, and for more bits the requirements on the resistance values are even more stringent. Third, because the resistance in series with the LSB is the largest, the time constant associated with this resistance and any stray circuit capacitance can be large enough to slow down appreciably the conversion process. For, example, if $R_0 = 100$ kΩ for eight bits and if the stray capacitance is 5 pF, then the time constant $R_0C = 500$ ns, which is ten times longer than the time required for a typical flip-flop to change states and 50 times longer than the propagation time for a simple TTL gate. The op amp must also be very fast in terms of its bandwidth and slew rate.

Notice that the current through the feedback resistor R_f is proportional to the output voltage because the inverting input is a virtual ground. Thus, the output end of R_f acts like a current sink. Many commercial D/A converters output a current that is proportional to the digital input, although they do not use a summing op amp.

The input bits $D_0, \ldots, D_3$ are switched in with an electronic analog switch, as shown in Fig. 15.2(b). The switches are usually made from FETs rather than bipolar transistors to avoid the base-emitter offset voltage problems inherent in bipolar transistor switches.

15.2.3 The *R*/2*R* Ladder D/A Converter

The most common D/A converter technique uses an ingenious resistance "ladder" of values R and $2R$, as shown in Fig. 15.3 for a four-bit input. The

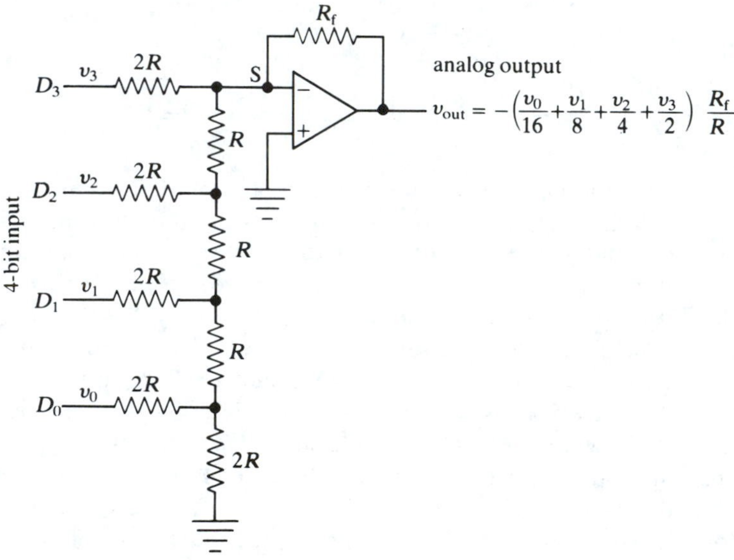

FIGURE 15.3 The *R*/2*R* ladder D/A converter.

digital input (either 5 V or 0 V) is connected by means of a high-speed analog switch to the $2R$ resistances.

The input v_3 has twice the gain of v_2, which has twice the gain of v_1, and so on. The argument is based upon a repeated application of Thevenin's theorem. To illustrate this, consider an input of 1 LSB: that is, $D_3D_2D_1D_0 = 0001$, $v_3 = v_2 = v_1 = 0$, and $v_0 = 5$ V. Neglecting the source resistance of the 5-V input, Thevenin's theorem applied to the 5-V source and the bottom two $2R$ resistances replaces them by a 2.5-V source in series with a single (Thevenin) resistance of R, as shown in Figs. 15.4(a) and (b).

Repeated application of Thevenin's theorem yields the circuit of Fig. 15.4(h)—a 0.3125-V source in series with a resistance R. Thus, the op amp input V_1 for the LSB input is $0.3125 \text{ V} = 5 \text{ V}/2^4$. A similar argument for an input $D_3D_2D_1D_0 = 0010$ yields $V_I = 0.625$ V. Thus, in general, the analog

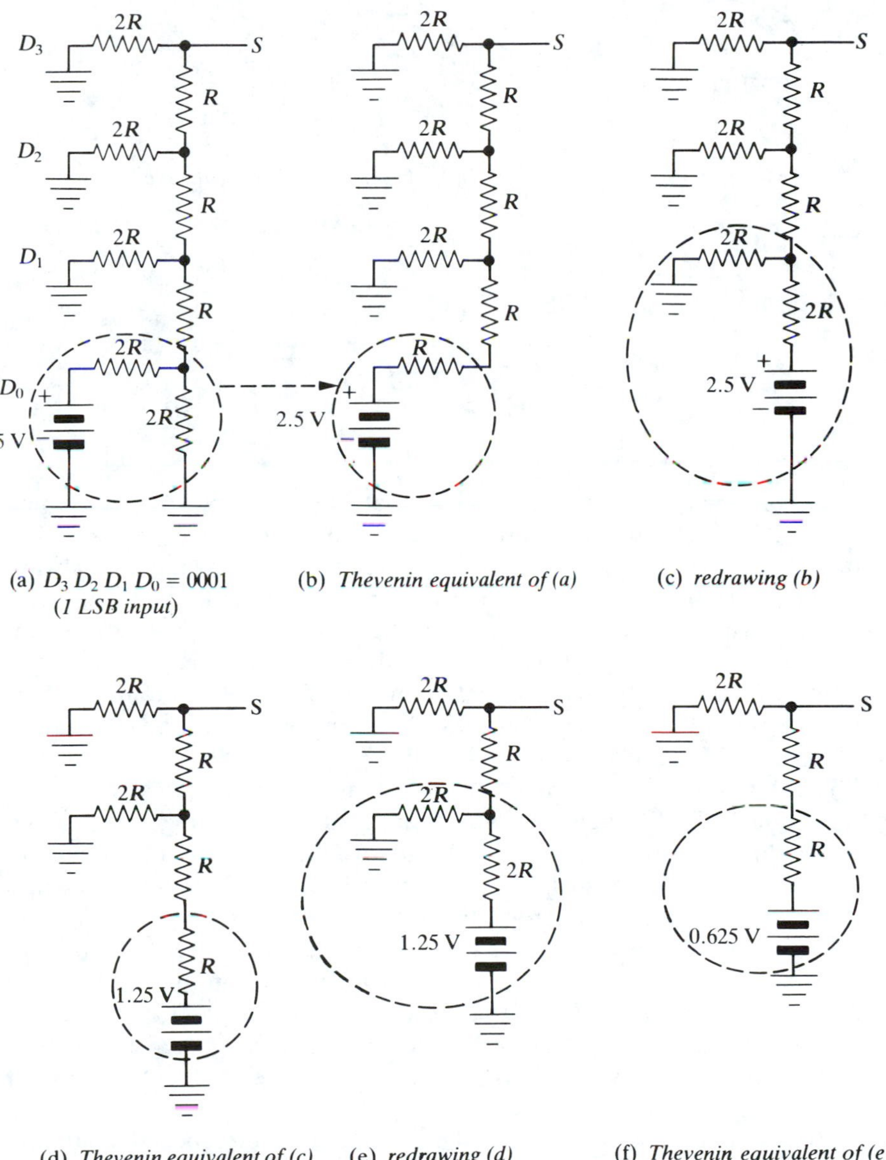

FIGURE 15.4 Thevenin analysis of $R/2R$ ladder.

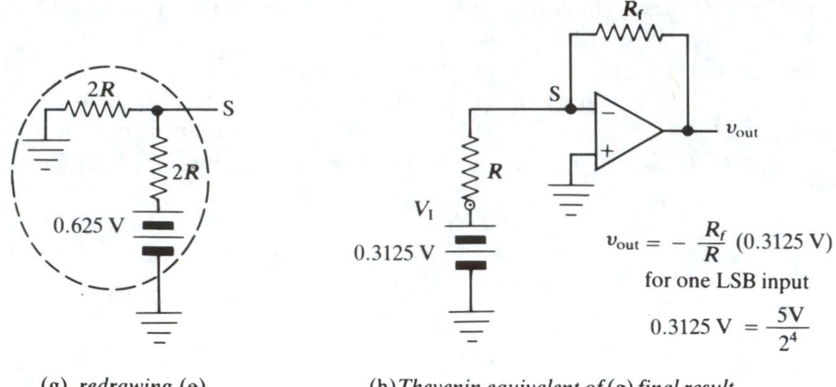

(g) *redrawing* (e)

(h) *Thevenin equivalent of* (g) *final result*

FIGURE 15.4 Continued.

output for a digital input $D_0D_1D_2D_3$ will be

$$v_{out} = -\frac{R_f}{R}\left(\frac{D_0}{16} + \frac{D_1}{8} + \frac{D_2}{4} + \frac{D_3}{2}\right)(5\text{ V})$$

where D_0, D_1, D_2, $D_3 = 0$ or 1 V.

The big advantage of this circuit is that only *two* values of resistance (R and $2R$) are required, even for many bits (e.g., $N = 16$). Another advantage is that the source impedance seen by the op amp is always constant ($=R$) regardless of the digital input, unlike the summing D/A converter circuit. R can be made low ($\sim 500\ \Omega$) to charge up the op amp input capacitance quickly to achieve a large bandwidth. These $R/2R$ ladders are now easily made directly on the chip with automatic laser trimming to produce the necessary resistance precision with a temperature coefficient of 1 ppm/°C or better.

The actual circuit for an $R/2R$ D/A converter is shown in Fig. 15.5, with the analog FET switches shown schematically. The voltage V_{ref} (5 V in the preceding explanation) can be either internal to the D/A chip or it can be supplied from an external source. It must be stable, of course, because the output analog is linearly proportional to V_{ref}.

The output of the D/A converter is seen to be linearly proportional to the digital input and also to the reference voltage V_{ref}. In other words, the analog output is equal to the *product* of the digital input and the reference voltage. Such D/A converters are known as *multiplying* converters. One advantage of a multiplying D/A converter is that the range of the output can be changed easily by merely changing the reference voltage. But in all cases the output stability is at best the same as the reference stability and usually worse because of instability in the op amp and resistor ladder.

Some commercial D/A converters have a current rather than a voltage

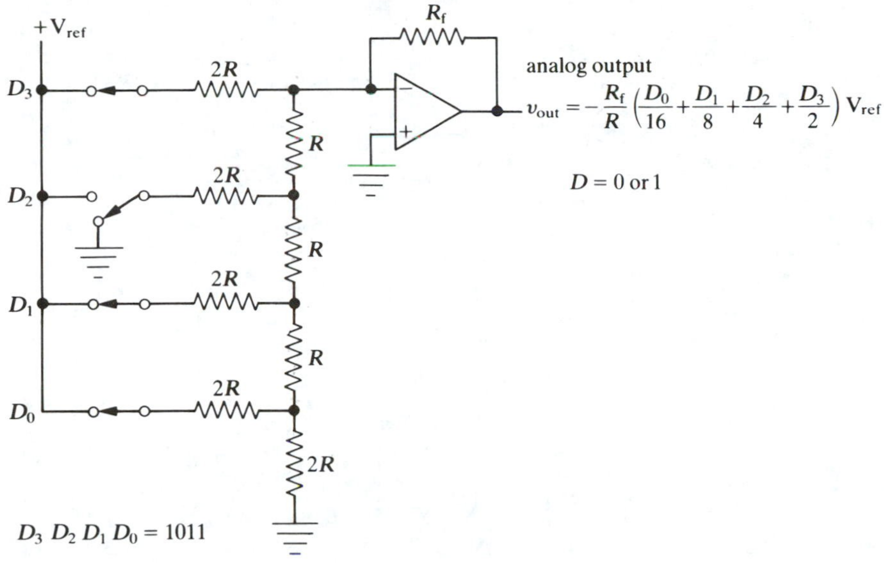

FIGURE 15.5 *R*/2*R* ladder four-bit D/A converter.

output; the output current is linearly proportional to the digital input. Such current D/A converters are generally faster than those with a voltage output. The current output can be converted into a voltage output by merely having it flow through a stable precise resistance, R_L, as shown in Fig. 15.6(a), or by feeding it into a current-to-voltage converter, as shown in Fig. 15.6(b). The maximum current output from the D/A converter is usually 2 mA; thus to obtain a 10-V maximum output, we must have $R_f = 5\ \text{k}\Omega$.

The capacitance C represents the output-to-ground capacitance of the D/A converter plus any stray wiring capacitance and is usually about 100 pF. Thus the response time for Fig. 15.6(a) will be $R_L C$. The op amp is often the slowest component in the circuit. The capacitance C' may be necessary to prevent oscillation, but it will limit the speed of the op amp converter—the op amp response time will be about $R_f C'$. In Fig. 15.6(b) the resistance R' may be necessary to reduce the output offset voltage. In some D/A converters, for the best accuracy and stability R_f is actually in the D/A chip.

The op amp offset voltage will be amplified by the gain factor $1 + R_f/R$ in the circuit of Fig. 15.6(a). Also, note that the op amp offset voltage typically varies with temperature by 10 μV/°C.

The reference voltage V_{ref} in the D/A converter can be positive or negative; in such cases the converter is called a *two-quadrant* converter. If the digital input can be positive or negative (with a sign bit), the converter is called a *four-quadrant* converter.

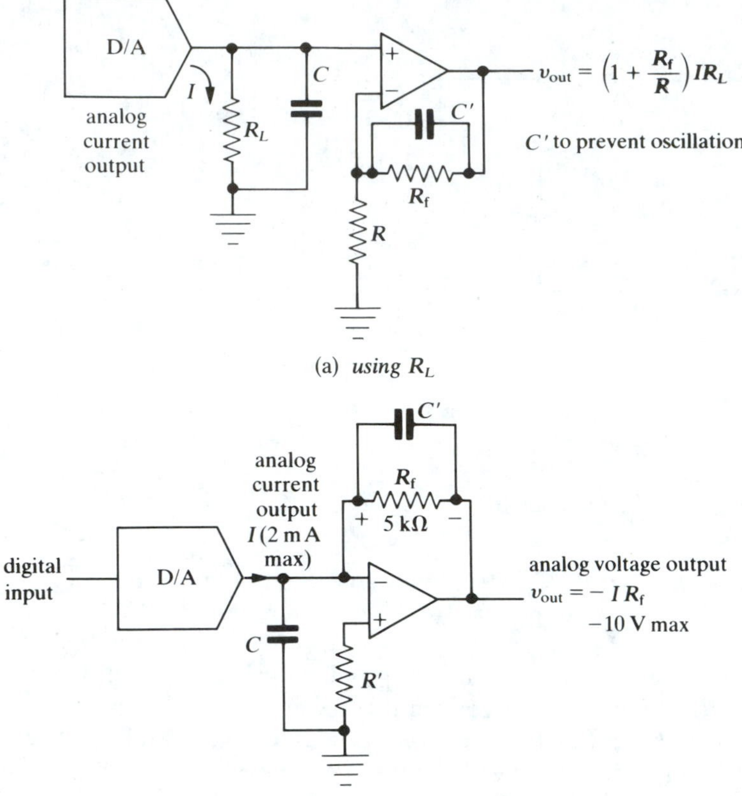

FIGURE 15.6 Conversion of D/A current output to voltage output.

As a general rule, the op amp in the output should be as stable as possible against temperature changes and changes in its supply voltage. The op amp bandwidth should also be as small as possible to pass the expected changes in the digital input. Any excess bandwidth merely adds noise to the output. Also, it may be necessary to add a stable resistance in series with the noninverting input to minimize the dc output offset voltage.

15.2.4 D/A Converter Specifications

All D/A converters have the following parameters:

1. *Resolution or precision*. The maximum output resolution is one bit in $(2^N - 1)$ —for example, 1 part in 255 (0.39%) for an eight-bit converter, 0.024% for 12 bits.

2. *Accuracy*. The percentage error in the voltage output—usually how much the full-scale output differs from a certain value (e.g., 10 V).
3. *Linearity*. The deviation of the stepwise output from the best straight linear line—usually, within $\pm\frac{1}{2}$ LSB of the best straight line. ±0.1% to ±0.5% is typical.
4. *Monotonicity*. The analog output always increases for each one-bit increase in the digital input. The *maximum* possible error is $+\frac{1}{2}$ LSB for a monotonic output. A D/A can be monotonic at 25°C but nonmonotonic (e.g., ±1 LSB) over a 0° to 50°C range.
5. *Settling time*. The minimum time from the instant that the correct digital input appears to the instant that the output settles down to within $\pm\frac{1}{2}$ LSB of the final equilibrium output value. From 10 μs to 100 ns is typical. The settling time can vary by a factor of 2 or more with changes in the load resistance for a current output D/A and with changes in the feedback (gain) resistor for a voltage output D/A. The settling time is also largely determined by the speed of the op amp (bandwidth and slew rate). An externally compensated op amp may be necessary for the fastest settling time. The fastest settling time is usually obtained with a current output D/A converter driving a simple film resistor (no op amp). The settling time is usually worst when the maximum number of input bits changes (e.g., from 0111 to 1000).
6. *Output range*. The range of output analog voltage (typically 0 to 10 V) for the maximum input digital swing (e.g., from 0000 to 1111 for a four-bit input).
7. *Input digital code*. The type of input digital code the D/A converter will operate on: for example, straight binary, BCD, twos complement, or offset binary.
8. *Offset*. The analog dc output (in mV or μV or fraction of LSB) with the digital input for zero applied.
9. *Feedthrough and quadrature error*. In multiplying D/A converters the changing digital input will slightly differentiate the reference voltage. Thus the output will contain a small component (typically mV) proportional to the derivative of the varying reference voltage. The amplitude increases linearly with the frequency of the reference.
10. *Reference*. An *internal* reference means the reference voltage circuitry is contained in the D/A converter; an *external* reference must be supplied by the user. In some units there is included an internal reference connected to a REF OUT pin. To use this internal reference, the REF OUT pin must be externally connected (by the user) to the REF IN pin. Or an external reference can be connected to the REF IN pin.

Some D/A converters produce "glitches" or false analog output spikes when large numbers of bits change. For example, in an eight-bit converter when the input changes from 01111111 to 10000000, all eight bits must change and thus all eight analog switches must "flip." The time for the switch to turn on is usually slightly different from the time to turn off, so for a brief interval all the switches may be on or all may be off. If they are all off for a few microseconds, the output briefly drops to near zero before increasing to a value 1 LSB larger. Some D/A converters contain deglitching circuits that are fast sample-and-hold circuits that hold the output constant until all the switches have flipped. Figure 15.7 illustrates some

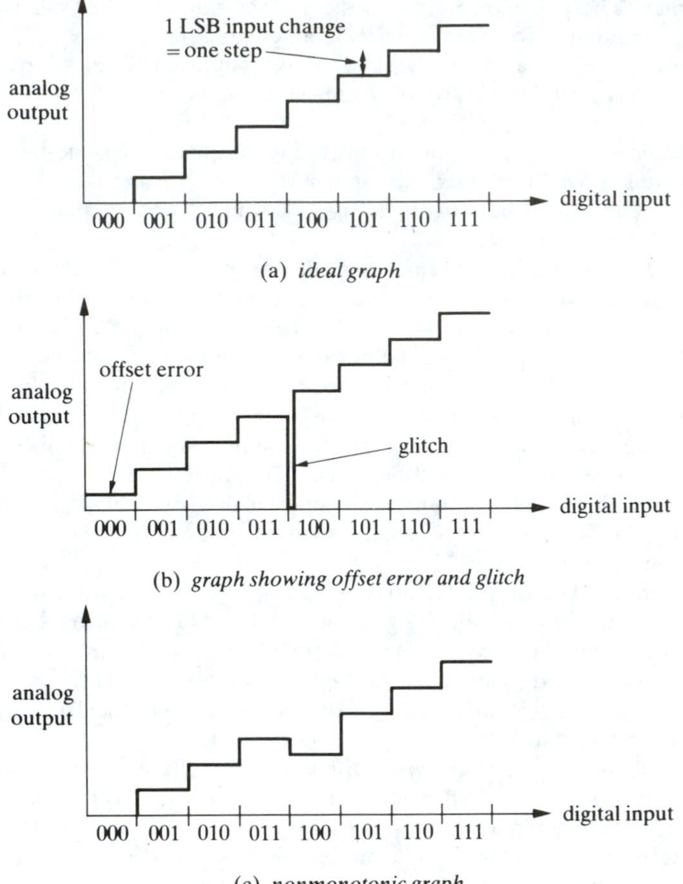

(a) *ideal graph*

(b) *graph showing offset error and glitch*

(c) *nonmonotonic graph*

FIGURE 15.7 D/A converter graphs.

D/A parameters and errors. Some typical D/A converters are shown in Table 15.5. The use of an external voltage reference and the conversion of the current output to a voltage output for a DAC 0808 is shown in Fig. 15.8. Notice that the op amp used (LF351) is one with a high slew rate (13 V/μs) and low input bias and offset currents. The high slew rate and large bandwidth are necessary to match the fast settling time (150 ns) of the 0808 D/A converter. It is also desirable to choose an op amp with minimum dc offset drifts.

The concepts of "left-justified" and "right-justified" data should also be mentioned. If a 12-bit D/A converter is used in connection with an eight-bit microprocessor, the D/A converter must be loaded in two bytes. The two types of loading are *right-justified* and *left-justified* and are shown below. In left-justified data one byte contains the eight *most* significant bits and the

TABLE 15.5. D/A Converter Specifications

Device	*No. bits*	*Nonlinearity*	*Settling Time*	*Reference*	*Output*	*Comments*
PMI						
DAC-10	10	±0.05%	85 ns	External	0–4 mA	
DAC-100	10	±0.05%	375 ns	Internal	0–2 mA	$20 ea., popular
DAC-80	12	±0.08%	300 ns (*I*) 3 μs (*V*)	Internal	±1 mA 0–2 mA ±5 V, ±10 V 0–+5 V	$16 ea.
National						
DAC 0800LCN	8	±0.195%	100 ns	External	*I*	2-quadrant, complementary output
DAC0808	8	±0.195%	150 ns	External	*I*	
DAC 1000	10	±0.05%	500 ns	External	*I*	microprocessor compatible, 4-quadrant, double buffered
Analog Devices						
AD7533LN	10	$\pm\frac{1}{2}$ LSB		External	*I*	$9 ea.
AD567JD	12	$\pm\frac{1}{2}$ LSB_{max}	500 ns	Internal	*I*	double buffered, $23 ea.
AD9768SD	8		5 ns	Internal	*I*, up to 20 mA	ultrafast, $35 ea., ECL input
DAC1146	18	±0.00076% full scale	6 μs	Internal 10 V	*I* or *V*	400 V/μs, $32
AD7543JN	12	±1 LSB	2 μs	External	$0 - V_{ref}$	(*serial* input) right justified data, $21 ea.
AD7226KN	8	±1 LSB	5 μs + 20 μs −	External	0–5 mA	quad $19/100s
AD558JN	8	$\pm\frac{1}{2}$LSB (±0.195%)	1 μs	Internal	0–5 mA or 0–10 V 2.56 V	$6/100s, microprocessor compatible
AD DAC-08AD	8	±0.1% ($\frac{1}{4}$ LSB)	85 ns	External	0–2 mA	popular, $10 ea.

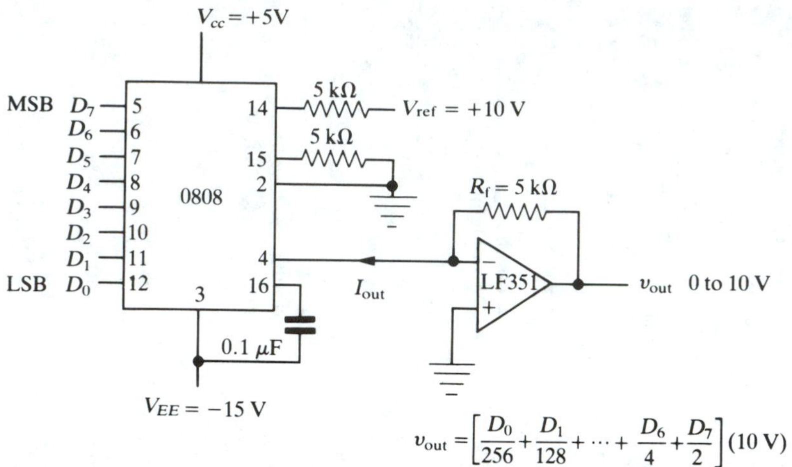

FIGURE 15.8 D/A converter circuit with voltage output.

other byte the four least significant bits. In right-justified data one byte contains the eight *least* significant bits and the other byte the four most significant bits.

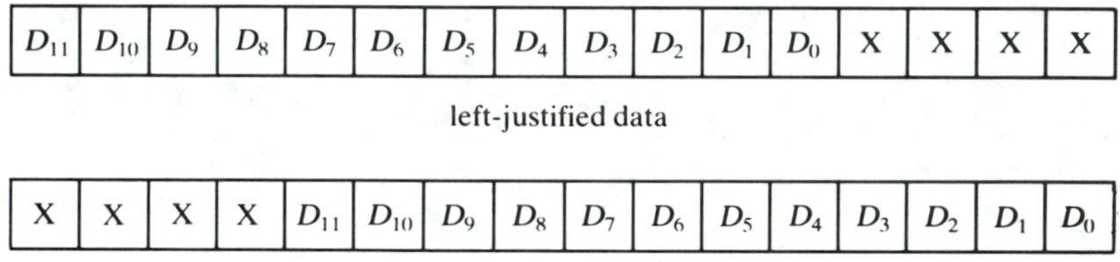

Some D/A converters have latched or three-state digital inputs, which means the D/A converter produces an analog output from the latched or stored digital input only when it is clocked or enabled. Thus, the digital input to the chip can be changing (when the chip is not clocked) without changing the analog output; hence the digital input terminals can be connected directly to the data lines from a microprocessor or computer. *Microprocessor compatible* applied to a D/A (or an A/D) converter simply means three-state input (or output) that can be directly connected to a microprocessor.

15.2.5 The Frequency-to-Voltage D/A Converter

Both the preceding D/A converters produce an analog output that is proportional to an essentially instantaneous value of the digital input. A circuit that converts an input train of standard-sized pulses into an analog output proportional to the frequency of the input pulses (or equivalently to the number of input pulses in a fixed time interval) can be considered to be a type of D/A converter. The simplest way to get an analog output is to feed the pulse train into a low-pass filter. The disadvantage of this technique is that to eliminate the ripple (at the pulse frequency) a long-time-constant filter must be used, which is slow to respond to changes in the input pulse train.

Specific frequency-to-voltage converter chips are available, such as the Analog Devices AD537JD, which accepts a variety of input waveforms (sine, square, etc.) up to 100 kHz and outputs currents up to 20 mA, which can drive a relay or meter.

15.3 THE SAMPLING THEOREM

Before we discuss specific A/D converters, let us consider some of the fundamental concepts and problems of A/D conversion.

The fundamental idea of converting an analog signal to digital form is to divide the continuous analog waveform into a set of discrete points and to represent each voltage point as a digital number. In other words, the analog signal must be *sampled* and each sampled value must be converted to digital form. Let t_s be the time between sampled points; the sampling frequency is thus $f_s = 1/t_s$. In Fig. 15.9(a) the input analog signal v_a and the sampled points v_s are shown. Obviously, the more often we sample the signal the better; otherwise we might miss some rapidly changing details of the signal.

Figure 15.9(b) shows how we would miss a detail of the signal if we sampled too slowly. The sharp spike simply doesn't show up in the sampled points. We can immediately conclude that we must sample fast enough (large f_s, small t_s) to get several sample points during the short interval of time when the spike occurs. But there is a limit, as we shall soon see; in other words, the faster we sample the better, but only up to a certain sampling frequency.

Consider a rectangular pulse of width T in the analog input, as shown in Fig. 15.10(a). If $t_s = T$, we might get one sample point in the pulse, but we might miss the pulse completely if the two sample points nearest the pulse lie at each edge of the pulse, as shown in Fig. 15.10(b). Thus we see that to guarantee *at least one* sample point inside the pulse, we must have $t_s < T$. And a little thought will show that if $t_s = T/2$, then we are bound to get *two* sample points inside the pulse, as shown in Fig. 15.10(c), and thus resolve the flat top of the pulse. (With $t_s = T/2$, there is only one way to get only one sample point inside the pulse as shown in Fig. 15.10(d), and the probability of this is negligible—the single sample point would have to lie *exactly* at the center of the pulse.) In conclusion, we see that to resolve or

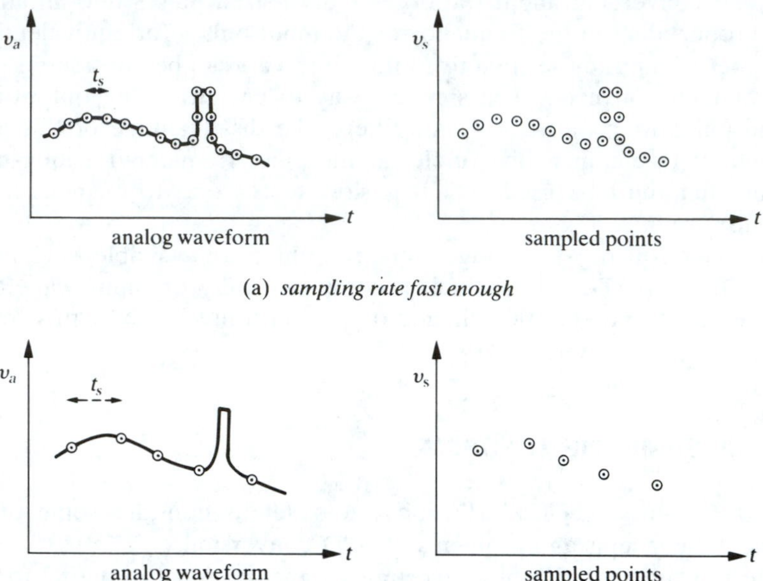

(a) *sampling rate fast enough*

(b) *sampling rate too slow; spike is not resolved*

FIGURE 15.9 Sampling an analog waveform.

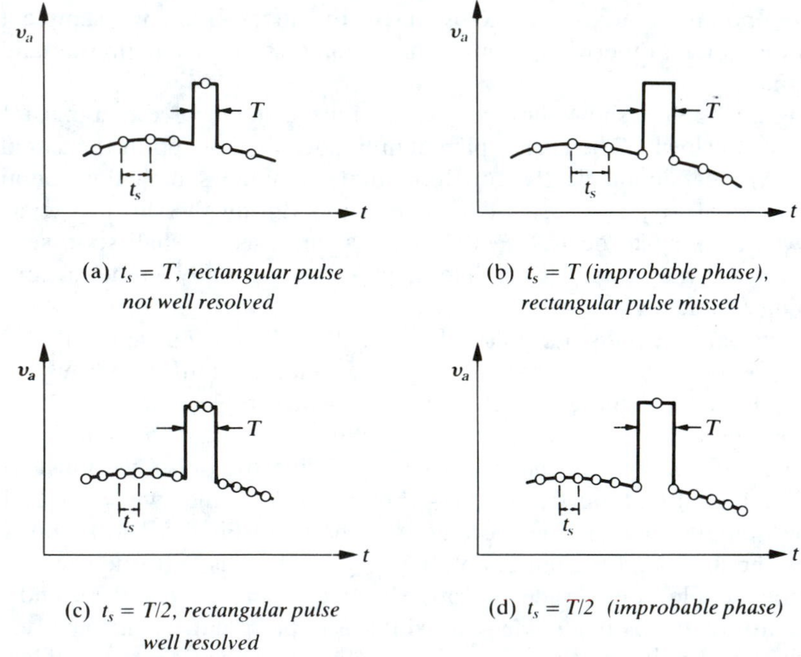

(a) $t_s = T$, *rectangular pulse not well resolved*

(b) $t_s = T$ *(improbable phase), rectangular pulse missed*

(c) $t_s = T/2$, *rectangular pulse well resolved*

(d) $t_s = T/2$ *(improbable phase)*

FIGURE 15.10 Sampling an analog waveform containing a rectangular pulse.

detect the rectangular pulse with two or more sample points inside the pulse we must have

$$t_s < \frac{T}{2} \quad \text{or} \quad f_s > \frac{2}{T} \tag{15.1}$$

From Chapter 3 we recall that for a rectangular pulse of width T, the major Fourier components range from zero frequency (dc) up to a maximum frequency of $f_{\max} = 1/T$. Thus, considering an arbitrary analog input to consist of a series of rectangular pulses, we should have the sampling frequency f_s greater than twice the maximum Fourier component frequency of the *shortest* analog input pulse, and (15.1) becomes

$$f_s > 2f_{\max} \tag{15.2}$$

It is important to note that $f_{\max}$ is the *highest* frequency component of the analog input signal. For example, in sampling an audio signal the highest Fourier component audible to the human ear is $f_{\max} = 20$ kHz. Thus, we must use a sampling frequency of at least 40 kHz. It is also important to note that we cannot have $f_s = 2f_{\max}$; we must have the *inequality*.

Equation (15.2) is called the *Nyquist sampling theorem*, and we will now derive it mathematically. In the following treatment the subscript a refers to the analog signal and the subscript s to the sampled waveform.

The sampled waveform can be thought of as a series of very narrow spikes; the analog signal $v_a(t)$ and the sampled signal $v_s(t)$ are shown in Figs. 15.11(a) and (b). Let us now resolve the analog signal input $v_a(t)$ into its Fourier components by using the technique of the Fourier transform as discussed in Chapter 3. Letting $g_a(\omega)$ be the Fourier transform of $v_a(t)$, we then have

$$v_a(t) = \frac{1}{\sqrt{2\pi}}\int_{-\infty}^{\infty} g_a(\omega)e^{j\omega t}\,d\omega \quad \text{and} \quad g_a(\omega) = \frac{1}{\sqrt{2\pi}}\int_{-\infty}^{\infty} v_a(t)e^{-j\omega t}\,dt \tag{15.3}$$

and for the sampled waveform we have

$$v_s(t) = \frac{1}{\sqrt{2\pi}}\int_{-\infty}^{\infty} g_s(\omega)e^{j\omega t}\,d\omega \quad \text{and} \quad g_s(\omega) = \frac{1}{\sqrt{2\pi}}\int_{-\infty}^{\infty} v_s(t)e^{-j\omega t}\,dt \tag{15.3'}$$

The Fourier components $g_a(\omega)$ depend on the shape of $v_a(t)$. For a particular analog signal $v_a(t)$ its Fourier transform $g_a(\omega)$ might look like Fig. 15.11(c), for example, with component frequencies from dc to a maximum $\omega_{\max}$.

Now let us Fourier analyze the sampled waveform $v_s(t)$—that is, calculate $g_s(\omega)$. Because $v_s(t)$ is a *periodic* waveform with period t_s, we can

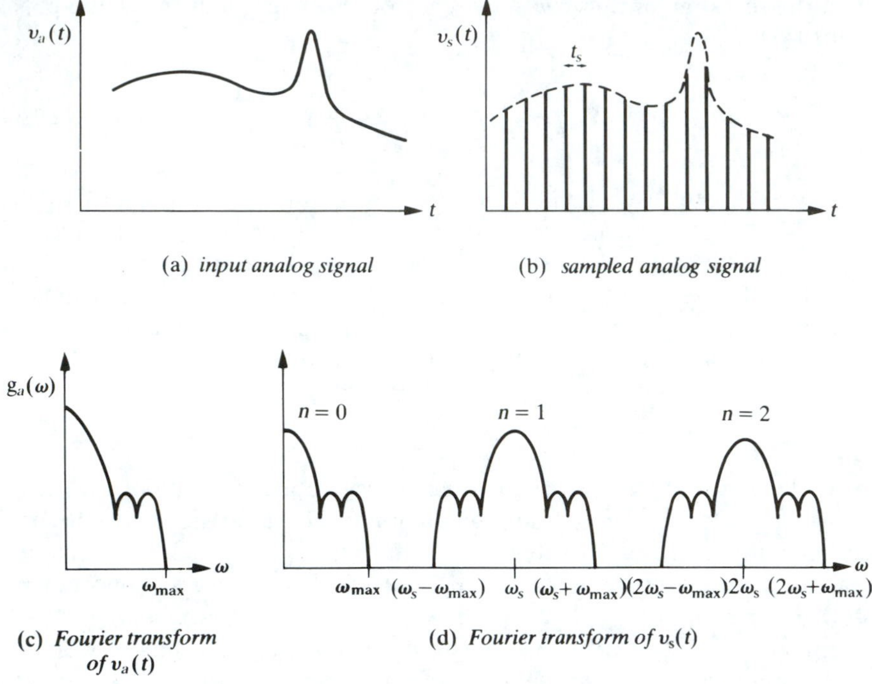

FIGURE 15.11 Analog input and sampled waveforms and Fourier components.

formally write it as a (sum of) delta functions with each delta function having the amplitude of the analog waveforms at the instants of time $t = Kt_s$ when the sampling occurs and zero for $t \neq Kt_s$.

$$v_s(t) = \sum_{K=-\infty}^{\infty} v_a(t)\beta\,\delta(t - Kt_s) \tag{15.4}$$

where β is a constant having the dimensions of time because $\delta(t - Kt_s)$ has the dimensions of inverse time.

We are ultimately interested in the Fourier components of the sampled waveform because we might reasonably expect the sampling process to generate many new, false Fourier components. In other words, we know that the input *analog* signal contains Fourier components from dc out to ω_{max}, but what are the Fourier components of the *sampled* waveform? That is, what is the form of $g_s(\omega)$?

From (15.3′)

$$g_s(\omega) = \frac{1}{\sqrt{2\pi}}\int_{-\infty}^{\infty} v_s(t)e^{-j\omega t}\,dt = \frac{1}{\sqrt{2\pi}}\int_{-\infty}^{\infty} v_a(t)\sum_K \beta\,\delta(t - Kt_s)e^{-j\omega t}\,dt \tag{15.5}$$

Because the delta function is *periodic* (with period t_s), it can be expanded in a *discrete* Fourier series

$$\sum_K \beta\,\delta(t - Kt_s) = \sum_{n=-\infty}^{\infty} a_n e^{jn\omega_s t} \quad \text{where } \omega_s = \frac{2\pi}{t_s} \tag{15.6}$$

Substituting (15.6) in (15.5) then yields

$$g_s(\omega) = \frac{1}{\sqrt{2\pi}} \int_{-\infty}^{\infty} v_a(t) \left[\sum_{n=-\infty}^{\infty} a_n e^{jn\omega_s t} \right] e^{-j\omega t}\, dt$$

or

$$g_s(\omega) = \sum_{n=-\infty}^{\infty} a_n \frac{1}{\sqrt{2\pi}} \int_{-\infty}^{\infty} v_a(t) e^{-j(\omega - n\omega_s)t}\, dt \tag{15.7}$$

The integral expression in (15.7) is very similar to the Fourier transform (15.3) of the *analog* input; comparing (15.7) and (15.3), we conclude that

$$g_s(\omega) = \sum_{n=-\infty}^{\infty} a_n g_a(\omega - n\omega_s) \tag{15.8}$$

It can be shown that $a_n = \beta / t_s$ for any value of n, so

$$g_s(\omega) = \beta / t_s \sum_{n=-\infty}^{\infty} g_a(\omega - n\omega_s) \tag{15.9}$$

Equation (15.9) says that the Fourier transform $g_s(\omega)$ of the *sampled* waveform is a multiple of the *sum* of the Fourier transforms of the original analog input *shifted in frequency by* $n\omega_s$ (i.e., shifted by an integral multiple n of the (angular) sampling frequency ω_s. That is,

$$g_s(\omega) = \cdots + g_a(\omega - \omega_s) + g_a(\omega) + g_a(\omega + \omega_s) + g_a(\omega + 2\omega_s) + \cdots \tag{15.10}$$

Equation (15.10) implies that the Fourier spectrum $g_s(\omega)$ of the sampled waveform is as shown in Fig. 15.11(d).

The important point is that the Fourier spectrum $g_s(\omega)$ of the sampled waveform $v_s(t)$ *repeats itself* every ω_s. The spectra for $n \neq 0$ are called *images*. To reconstruct the original analog waveform from the Fourier spectrum $g_s(\omega)$, *we want only the* $n = 0$ *image*. The $n = 0$ image can be obtained without any of the higher images by simply feeding the sampled waveform into a low-pass filter with a bandpass from 0 (dc) up to a cutoff frequency greater than $\omega_{\max}$ [but less than $(\omega_s - \omega_{\max})$], *provided* that the $n = 0$ and the $n = 1$ images *do not overlap*! The $n = 0$ and the $n = 1$ images will not overlap if

$$\omega_{\max} < (\omega_s - \omega_{\max}) \tag{15.11}$$

But (15.11) can be rewritten as

$$\omega_s > 2\omega_{\text{max}} \quad \text{or} \quad f_s > 2f_{\text{max}} \tag{15.12}$$

which we recognize (with a sigh of relief) as the Nyquist sampling theorem: the sampling frequency must be *at least* twice as high as the highest Fourier component frequency of the input analog signal. (The equality $f_s = 2f_{\text{max}}$ will not work!) In other words, if the Nyquist sampling theorem is satisfied, then the different images in Fig. 15.11(d) will *not overlap* in the frequency domain.

The practical implications of all this are as follows: (1) With a *fixed* sampling frequency f_s we must use a low-pass prefilter at the A/D converter input that passes input analog signal components from dc to less than $f_s/2$. (2) If we can vary the sampling frequency f_s, it should be chosen high enough so that f_s is greater than $2f_{\text{max}}$, twice the highest-frequency Fourier component of the signal. Thus, in digitizing audio signals, from $f = 20$ Hz to 20 kHz, we would choose a sampling frequency $f_s > 40$ kHz and put in a low-pass filter to pass only input signal components up to 20 kHz *before* the input signal is sampled. In making modern compact audio discs, the audio signals are usually sampled at approximately $f_s = 44$ kHz to ensure that the low-pass filter attenuation is good at 20 kHz. A larger f_s allows the use of a cheaper, simpler low-pass filter with a rolloff that is not as steep.

The Fourier spectrum of the sampled signal, $g_s(\omega)$ can also be described by using the language of amplitude modulation (AM). We recall from Chapter 3 that an AM sine wave has a Fourier spectrum consisting of the carrier frequency ω_c and sidebands above and below the carrier as shown in Fig. 15.12 for a pure sine-wave modulation of angular frequency ω_m. The sampled waveform $v_s(t)$ in Fig. 15.11(b) can be thought of as a carrier of frequency f_s whose amplitude is modulated by the input analog

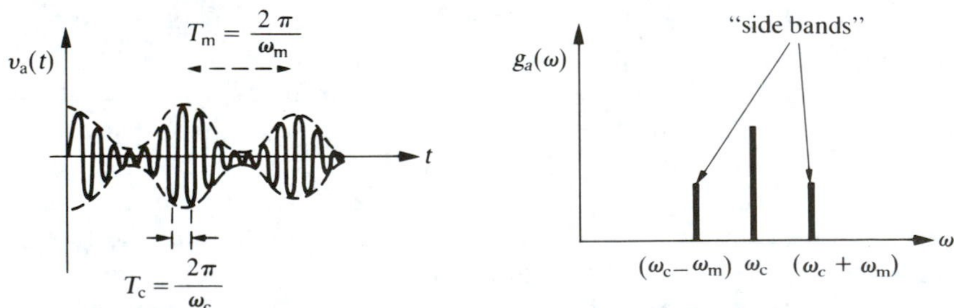

(a) *AM carrier with pure sinusoidal modulation at* $f_m = 1/T_m$

(b) *Fourier components of AM carrier in (a)*

FIGURE 15.12 Amplitude-modulated carrier and Fourier spectrum.

signal $v_a(t)$ except that $v_s(t)$ has an infinite number of Fourier components itself because it is a series of very *narrow* pulses, not a sine wave of one frequency. Thus, we expect the "carrier" frequency f_s to have AM sidebands above and below it in frequency from $(\omega_s - \omega_{max})$ to $(\omega_s + \omega_{max})$. This is what equation (15.9) says; the sidebands above and below ω_s are caused by the amplitude modulation of the analog signal $v_a(t)$. And each of the carrier frequencies, $\omega_s = 0, \omega_s, 2\omega_s, 3\omega_s, \ldots,$ has these sidebands.

A numerical example showing some overlapping of the $n = 0$ and the $n = 1$ sidebands due to too low a sampling frequency is the case of an analog signal with its fundamental at 4 kHz and a small second harmonic at $f_{max} = 8$ kHz, sampled at a rate of only 15 kHz: $f_{max} = 8$ kHz, $f_s = 15$ kHz. The input analog signal thus contains only two Fourier components, 4 kHz and 8 kHz, as shown in Fig. 15.13(a). The frequency spectrum of the sampled waveform is shown in Fig. 15.13(b). Notice that it contains the fundamental $f_0 = 4$ kHz, $f_{max} = 8$ kHz, $f_s - f_{max} = 7$ kHz, $f_s - f_0 = 11$ kHz, $f_s = 15$ kHz, $f_s + f_0 = 19$ kHz, and $f_s + f_{max} = 23$ kHz as well as the higher images or sidebands (not shown) for $n = 2, 3, \ldots$.

In this case the sampling frequency (15 kHz) is certainly more than twice the input analog fundamental (4 kHz) but not twice the second harmonic at $f_{max} = 8$ kHz. Thus, there is a 1-kHz overlap of the $n = 0$ and the $n = 1$ sideband structure—the $(f_s - f_{max}) = 7$-kHz $(n = 1)$ component is below the $f_{max} = 8$-kHz $(n = 0)$ component. Therefore, if we try to recover the original analog input (4 kHz and 8 kHz) by putting in a low-pass filter from dc to 9 kHz on the sampled output, we will get a *distorted* analog output because of the presence of the 7-kHz component.

If we had fed the analog input into a low-pass filter from dc to $f_s/2 = 15$ kHz$/2 = 7.5$ kHz, we would measure the 4-kHz component but totally miss the 8-kHz second harmonic of the input and retain the false 7-kHz $= (f_s - f_{max})$ component.

To avoid all these problems, we should use a sampling frequency $f_s = 18$ kHz which is greater than $2f_{max} = 16$ kHz and an input low-pass prefilter from dc to 9 kHz to make absolutely sure no input frequency component is greater than $f_s/2 = 9$ kHz. Then an output low-pass filter from dc to 9 kHz will pass the correct two components at 4 kHz and 8 kHz but no component from the $n = 1$ image at $f_s = 18$ kHz or $(f_s - f_{max}) = 18 - 8 = 10$ kHz or $(f_s + f_{max}) = 18 + 8 = 26$ kHz. The higher images for $n \geqq 2$ will also not be passed, of course. Figure 15.13(c) shows the spectrum.

We can now see why the shift register step approximation to a sine wave in 13.5.4 produces a 1-kHz fundamental and the first nonzero (distortion) harmonic at 15 kHz. The sampling frequency is $f_s = 16$ kHz because there were 16 steps or samples in each period of the 1-kHz wave. Thus the $n = 1$ image consists of $(f_s - f_0) = 16 - 1 = 15$ kHz, $f_s = 16$ kHz, and $(f_s + f_0) = 16 + 1 = 17$ kHz.

Thus, the general block diagram for an A/D converter circuit is that of

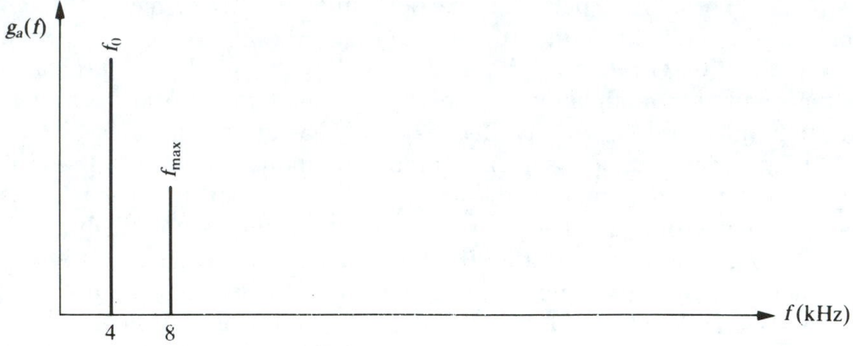

(a) *spectrum of analog input*

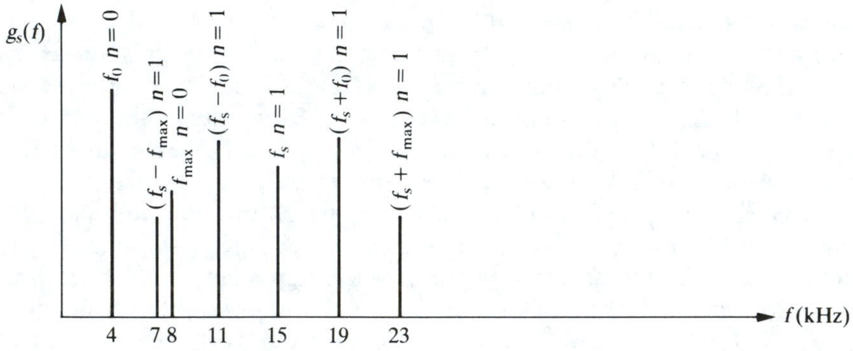

(b) *specrum of analog input sampled at* $f_s = 15$ kHz. $f_s < 2 f_{max}$ (bad!)
$n \geqq 2$ *images omitted*

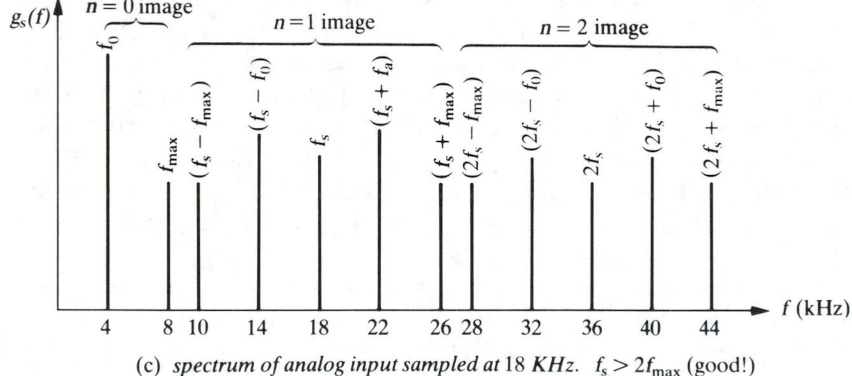

(c) *spectrum of analog input sampled at* 18 *KHz.* $f_s > 2f_{max}$ (good!)

FIGURE 15.13 Overlapping sidebands due to low sampling frequency.

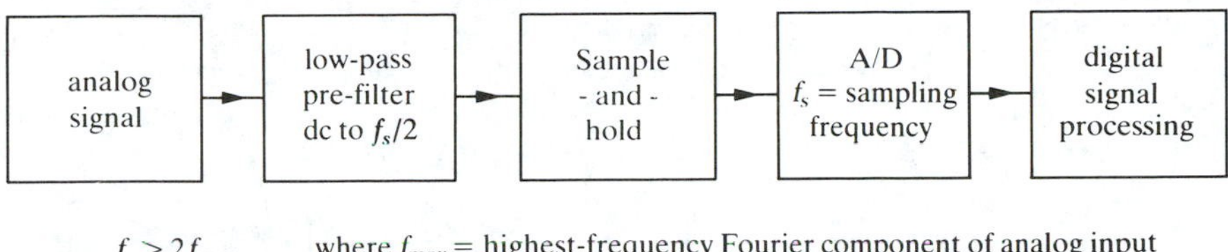

$f_s > 2f_{max}$ where f_{max} = highest-frequency Fourier component of analog input

FIGURE 15.14 General A/D converter.

Fig. 15.14, but remember this will not distort the analog input only if the maximum Fourier component frequency is less than one-half of the sampling frequency.

Another way of looking at the requirement that the sampling frequency f_s be more than twice the highest signal component frequency is to look at Fig. 15.15. Here, the signal frequency is a pure sine wave of frequency $f_0 = 30$ kHz, and the sampling frequency is only $f_s = 40$ kHz, which means samples are taken every $t_s = 1/f_s = 25\ \mu$s. The sampled points are shown as circles in Fig. 15.15(a). Notice that the sampling frequency is too low—it should be greater than $2f_0 = 60$ kHz. Notice also that *exactly the same* sampled points would be obtained from an input analog sine wave at 10 kHz. This 10-kHz signal is shown as a dashed line and is called an *image* or false frequency. A careful sketch will show that input analog sine waves of frequencies 10, 30, 50, 70, 90, 110, . . . kHz will each produce the same sampled points. Except for the 30-kHz input sine wave, the other frequencies (10, 50, 70, . . .) are all false image frequencies. The 50-kHz image is shown in Fig. 15.15(b). The image frequencies f_n are spaced $f_s/2$ apart and are given by $f_n = ((2n - 1)/4)f_s$ for $n = 1, 2, 3, \ldots$. In this case f_2 is the actual input.

These image frequencies are precisely what we would predict from Fig. 15.11 and the Nyquist analysis. In words, the Fourier spectrum of the sampled waveform repeats itself every f_s Hz in frequency. This is shown for $f_0 = 30$ kHz, $f_s = 40$ kHz in Fig. 15.15(c), because 10 kHz $= (f_s - f_0)$, 50 kHz $= 2f_s - f_0$, 70 kHz $= f_s + f_0$, 90 kHz $= 3f_s - f_0$, 110 kHz $= 23f_s + f_0$, and so forth. (There are no components at $f_s, 2f_s, \ldots$ because there is no dc component in the analog input.)

The usual way to recover the input analog waveform is to use a low-pass filter from 0 to $f_s/2$, in this case from 0 to 20 kHz. But this filter output would contain only the false image at 10 kHz, not the true input at 30 kHz.

The lowest image frequency $f_s - f_0$ is familiar to anyone who has seen a moving picture in which the rotating wheel of a vehicle appears to be rotating very slowly. In this case the sampling frequency is only slightly different from the wheel rotation frequency f_0, and what we see is a very low frequency $f_s - f_0$. For example, if $f_s = 24$ Hz (24 frames/s) and $f_0 =$

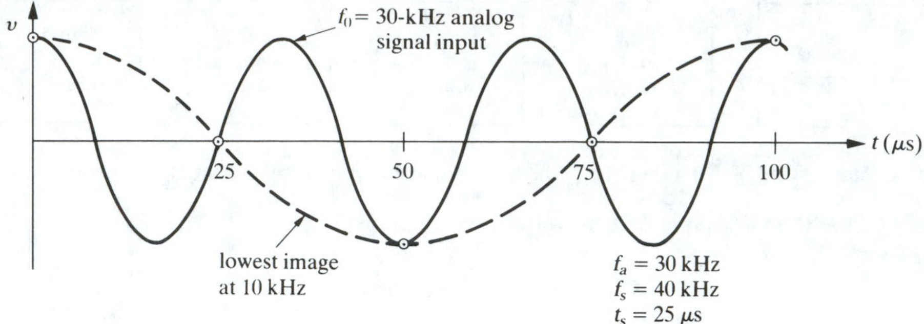

(a) 10-*kHz image* $(f_s - f_0) = 40\text{-}30 = 10$ *kHz*

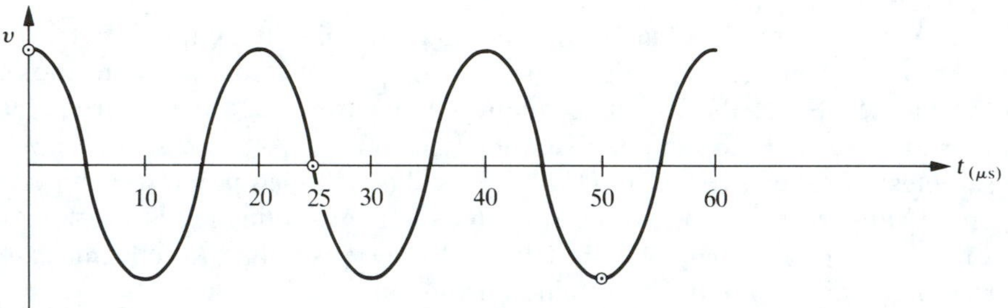

(b) 50-*kHz image*

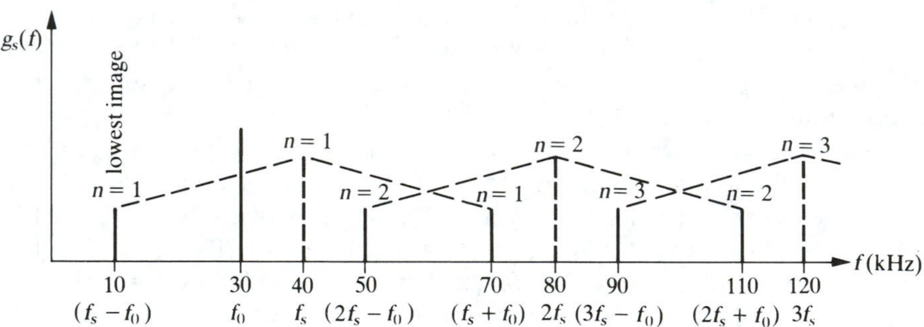

(c) *complete Fourier spectrum of sampled waveform*

FIGURE 15.15 Image frequencies (f_s too low).

22 Hz (22 revolutions/s), then we see the rotating wheel image of frequency $f_s - f_0 = 24 - 22 = 2$ Hz. The same effect also occurs in viewing a rotating wheel with a stroboscopic light. The reverse wheel rotation occurs when $f_s < f_0$.

These false low frequencies are called "image" frequencies, and the process by which they appear is often called "aliasing."

15.4 A/D CONVERTER CIRCUITS

We will consider five types of A/D converters: the staircase or counter type, the voltage-to-frequency A/D converter, the single- and dual-slope integrating types, the successive approximation type, and the flash or parallel type.

For any A/D converter, the *quantization error* is ±1 count; so there should clearly be many counts in the measurement to minimize the relative quantization error.

15.4.1 The Staircase or Counter A/D Converter

The basic circuit of the counter A/D converter is shown in Fig. 15.16. The counter is reset to zero to start the conversion, which sets $v_A = 0$, and a constant frequency clock is fed into a counter with parallel outputs. The

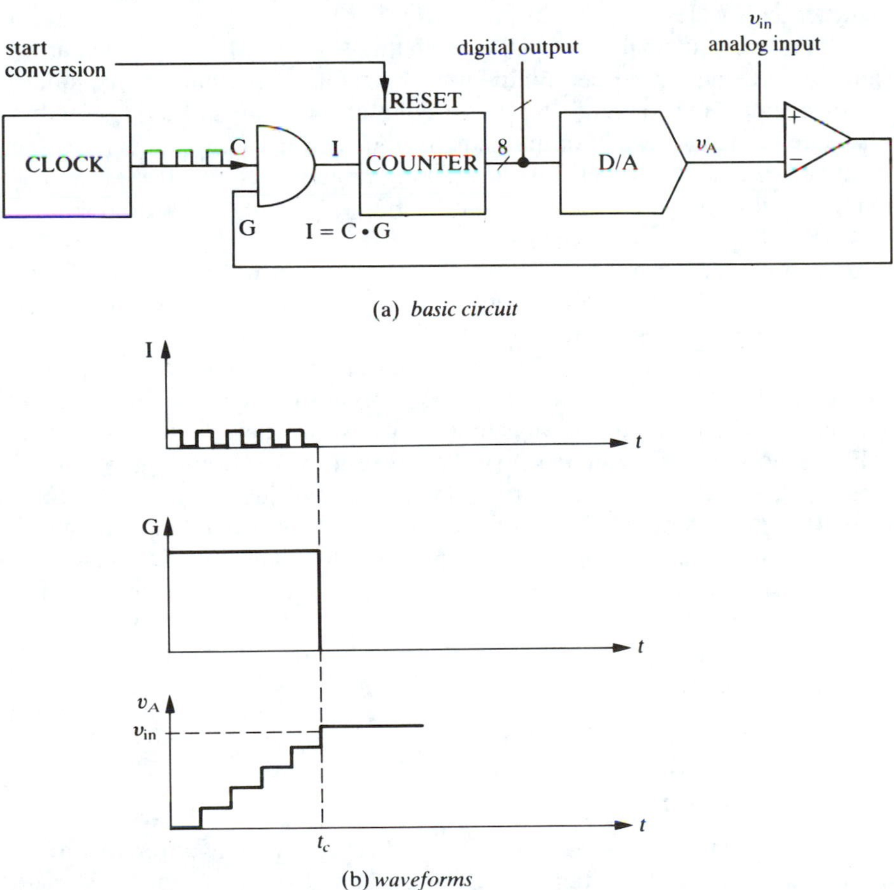

(a) *basic circuit*

(b) *waveforms*

FIGURE 15.16 The staircase or counter A/D converter.

digital counter parallel output is converted to an analog form where it is compared to the analog input in the comparator. The comparator output is initially high because the D/A converter output v_A is less than v_{in}; thus the gate is on because $G = 1$. When the analog output of the D/A converter slightly exceeds the analog input, the comparator output goes low ($G = 0$), which makes $I = 0$, and no more clock pulses are input to the counter, thus freezing or latching the counter at the digital value of the analog input. The final latched count is the digital value of the analog input. For example, if each count or step corresponds to 0.01 V at the output of the D/A converter and if the counter stops at 397 counts, we know that the analog input is between 3.97 V and 3.98 V. The resolution of the A/D converter is thus equal to the resolution of the D/A converter; for example, an eight-bit D/A converter would give a resolution of one part in 255 or 0.039%. Notice also that the settling time of the D/A converter must be less than the clock period because the D/A output has to settle down before a new count comes along from the next clock pulse. This is usually no problem, because even inexpensive D/A converters have settling times of several hundred nanoseconds or less.

The two principal disadvantages of this type of A/D converter are its slow speed and its susceptibility to noise spikes: the larger the analog voltage input, the longer it takes the counter to count up high enough so that the counter digital output equals the analog input. The maximum conversion time thus equals the maximum number of counts (for a full-scale input) times the clock period. For example, if the clock frequency is 100 kHz, the clock period $T = 10\ \mu s$, so for an eight-bit D/A converter, the maximum conversion time is $255T = (255)(10\ \mu s) = 2550\ \mu s = 2.55$ ms. For higher resolution, with a 12-bit counter, the conversion time would be $4095T = 40.85$ ms. A faster clock would help, of course.

The main disadvantage of this counter type of A/D converter is its susceptibility to noise. This type of converter measures the last instantaneous value of the analog input, just before the counter is stopped. Thus, a noise spike occurring just near the end of the counting period will greatly distort the count; a positive noise spike produces too large a count, and a negative noise spike produces too small a count by stopping the counter too soon. For both of these reasons this type of A/D converter is seldom used in practice, although it is perhaps the easiest one to understand conceptually.

The accuracy also depends on the stability of the comparator and on the stability of the D/A converter analog output.

15.4.2 The Voltage-to-Frequency A/D Converter

The (integrating) voltage-to-frequency A/D converter differs from the preceding staircase/counter converter by actually integrating the analog input voltage. This means that any short spikes in the input do not affect the output as much. You might think that the way to do this would be to compare the integrated analog input with a reference in a comparator and to use the comparator to gate a counter driven by a clock of fixed frequency, something like the staircase counter. But this won't work because the output of an integrator is given by

$$v_{\mathrm{I}} = -\frac{1}{RC}\int_0^{\Delta T} v_{\mathrm{in}}\, dt = -\overline{v_{\mathrm{in}}}\frac{\Delta T}{RC}$$

where $\overline{v_{\mathrm{in}}}$ is the average input and ΔT is the time of integration. Thus, the larger the input voltage, the shorter the time interval ΔT, and the counter reading would be inversely proportional to the magnitude of the input voltage. (You can't get around this problem even by using a "down" counter.)

The way around this difficulty is to make the time interval ΔT equal to the period of an output waveform: in other words, to make a voltage-to-frequency converter or a voltage controlled oscillator (VCO) and then to count the output frequency. The frequency can be counted with a computer using the appropriate software, and an inexpensive 555 chip can be used. Computer game "paddles" are often a variable resistance that controls the frequency of the VCO.

The general circuit and waveforms are shown in Fig. 15.17. The switch S is opened to start the conversion; the integrator output v_{I} then ramps down linearly until it reaches the negative reference at the other input of the comparator. When this happens, the comparator output goes high, which triggers the one-shot to output a standard pulse to the counter. The comparator output going high also briefly closes the switch S (which is an electronic switch), which then sets v_{I} to zero. This causes the comparator output to go low again, and the conversion process starts again. The waveforms are shown in Fig. 15.17(b), and we see that a series of standard pulses comes out of the one-shot every ΔT seconds; their frequency or repetition rate is $f = 1/\Delta T$. But ΔT is the time for v_{I} to reach $-V_{\mathrm{ref}}$. So

$$v_{\mathrm{I}} = \frac{1}{RC}\int_0^{\Delta T} v_{\mathrm{in}}\, dt = -\frac{1}{RC}\overline{v_{\mathrm{in}}}\,\Delta T = -V_{\mathrm{ref}}$$

Thus $$\Delta T = \frac{RCV_{\mathrm{ref}}}{\overline{v_{\mathrm{in}}}} \quad \text{or} \quad f = \frac{1}{\Delta T} = \frac{\overline{v_{\mathrm{in}}}}{RCV_{\mathrm{ref}}}$$

The output frequency is linearly proportional to the average input voltage. The number N of (digital) counts on the frequency counter is just fT_{G}, where T_{G} is the fixed time the counter is gated "on." Thus, the counter

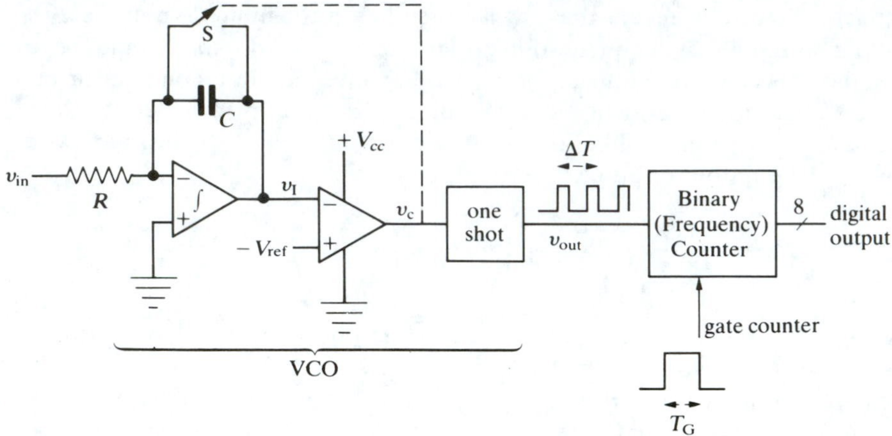

(a) *basic circuit*

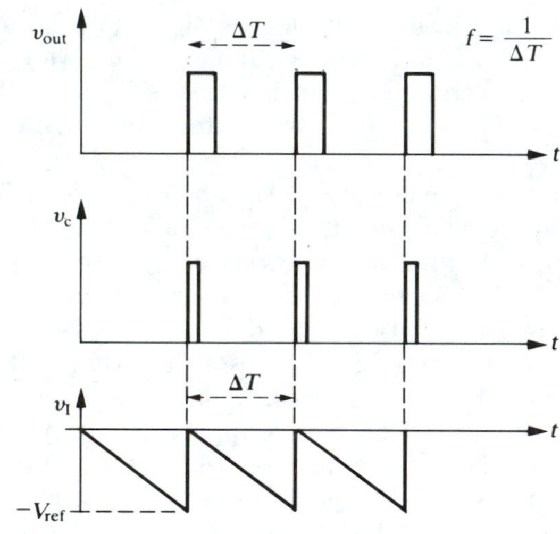

(b) *waveforms*

FIGURE 15.17 Voltage-to-frequency A/D converter.

reading is

$$N = \frac{\overline{v_{in}}}{RCV_{ref}} T_G$$

But the accuracy of this technique is limited by the stability and accuracy of the integrator, RC, and the reference voltage.

If v_{in} is proportional to V_{ref}—that is, if V_{ref} powers the input transducer—then drifts in V_{ref} will not affect N. Notice also that the reference voltage must be less than the maximum integrator output (so v_I can reach V_{ref}), and the output unit is proportional to the time average of the input, thus eliminating large errors due to short spikes in the input. If we wish a new reading every 100 ms, $T_G = 100\text{ ms} = 10^{-1}\text{ s}$. Then for an eight-bit counter, the maximum count will be given by

$$N_{max} = 11111111 = f_{max} T_G$$

or because $f_{max} = 1/RC$ (for $v_{in\,max} = V_{ref}$)

$$255 = \frac{T_G}{RC}$$

Thus $RC = T_G/255 = 10^{-1}\text{ s}/255 = 390\ \mu\text{s}$: for example, if $R = 39\text{ k}\Omega$, then $C = 0.01\ \mu\text{F}$. One advantage of this type of A/D converter is that the pulse train output from the one-shot can be easily transmitted over a simple coaxial conductor cable with good noise immunity. Thus the A/D converter can be placed at a "hostile" site (where the transducer produces v_{in}) and the pulse train output transmitted long distances to a "friendly" laboratory environment where the counter and computers are located.

Specific voltage-to-frequency converter chips are available, such as the Analog Devices AD537, which can accept ±30-V inputs and which outputs a full-scale frequency of up to 100 kHz, depending on an external RC network. Using the new 10-ns LT1016 high-speed comparator, a 1-Hz–10-MHz V/F converter can be made for a 0–12-V input.

15.4.3 The Single-Slope Integrating A/D Converter

The basic idea of the single-slope integrating A/D converter is to integrate a fixed (negative) reference voltage until it produces a value that equals the analog input voltage and to count a constant frequency clock waveform for this time of integration. The circuit and waveforms are shown in Fig. 15.18. The counter is reset to zero and the switch S is opened at $t = 0$. The integrator output v_I is

$$v_I = -\frac{1}{RC}\int_0^T (-V_{ref})\,dt = \frac{V_{ref}T}{RC}$$

Let T be the time when $v_I = v_{in}$; then $v_{in} = V_{ref}T/RC$. The comparator output goes low and then shuts off the counter input. The counter output is

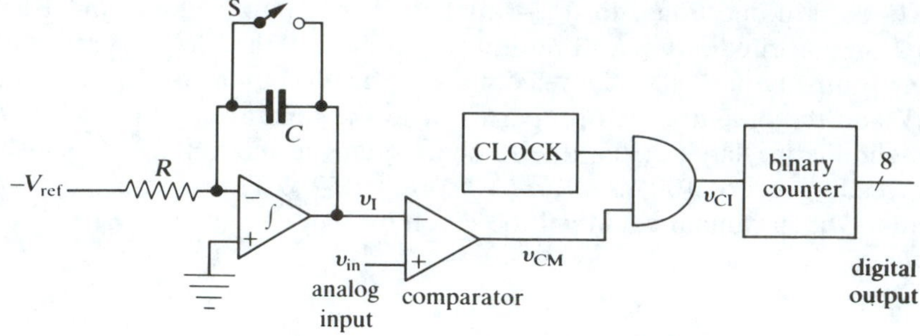

(a) *basic circuit*

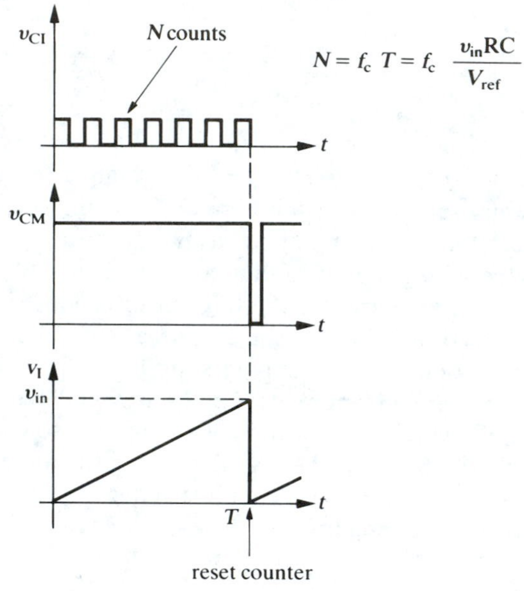

(b) *waveforms*

FIGURE 15.18 Single-slope integrating A/D converter.

$N = f_c T$, where f_c is the constant clock frequency. Thus,

$$N = f_c T = f_c \frac{v_{in} RC}{V_{ref}}$$

The counter count N is linearly proportional to the analog input voltage; that is, an A/D conversion has been achieved. If v_{in} is proportional to V_{ref}, then drifts in V_{ref} will not affect N.

This type of A/D converter is simple but slow (especially for large inputs), and both RC and the clock frequency tend to drift with temperature, which affects the calibration.

15.4.4 The Dual-Slope Integrating A/D Converter

The most popular slow integrating A/D conversion technique is the dual-slope technique, whose basic circuit is shown in Fig. 15.19. In this tech-

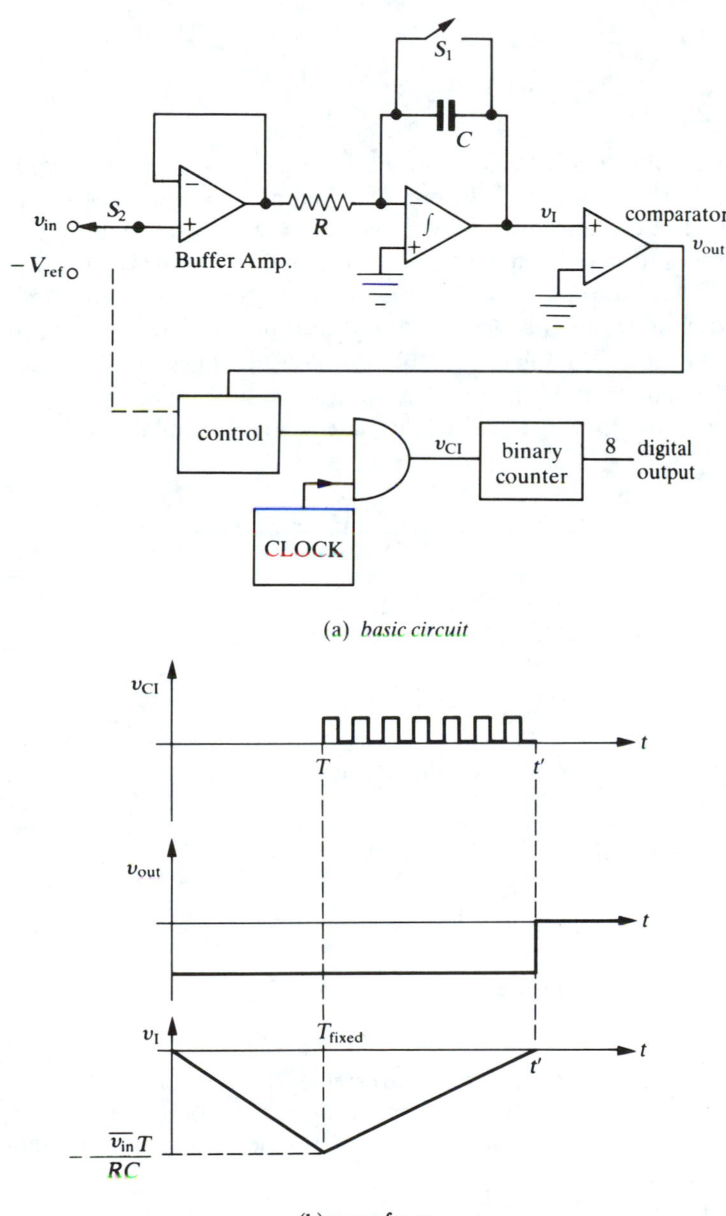

FIGURE 15.19 Dual-slope A/D converter.

nique the accuracy does not depend on either the integrating time constant RC or the clock frequency but only on the *ratio* of the analog input voltage to the internal reference voltage.

The switch S_1 is opened, and S_2 is connected to the analog input to start the conversion at $t = 0$, so v_I starts at 0 and ramps negative according to

$$v_I = -\frac{1}{RC}\int_0^T v_{in}\,dt$$

The control circuitry allows this integration to take place for a *fixed time* T; thus at $t = T$, $v_I = -\overline{v_{in}}T/RC$, where $\overline{v_{in}}$ is the average value of the input from 0 to T. The control circuitry then switches S_2, so the integrator input is a constant negative reference voltage $-V_{ref}$ and simultaneously gates the counter on. Thus, starting at $t = T$, the counter begins to count and the integrator output v_I begins to ramp upward towards 0, starting from $-\overline{v_{in}}T/RC$, as shown in Fig. 15.19(b). When v_I reaches 0 at $t = t'$, the comparator output goes high, which stops the counter. The counter then reads a count that is proportional to the time interval from $t = T$ to $t = t'$, i.e. $(t' - T)$.

$$v_I(t') - v_I(T) = -\frac{1}{RC}\int_T^{t'} (-V_{ref})\,dt$$

$$0 - \left(\frac{\overline{v_{in}}T}{RC}\right) = +\frac{1}{RC}V_{ref}(t' - T)$$

or

$$\overline{v_{in}}T = V_{ref}(t' - T)$$

Thus the time $(t' - T)$ over which the counter counts is

$$t' - T = \frac{\overline{v_{in}}}{V_{ref}}T \quad \text{and} \quad \overline{v_{in}} = \frac{V_{ref}(t' - T)}{T}$$

and the counter count N is

$$N = f_c(t' - T) = \frac{\overline{v_{in}}}{V_{ref}}Tf_c$$

which is *independent* of RC; thus an ultrastable RC is unnecessary.

The time interval $(t' - T) = Nt_c$, where N is the counter reading and t_c is the clock period. For example, if the clock frequency is 1.0 MHz, then N is the time interval $(t' - T)$ expressed in μs, and

$$\overline{v_{in}} = \frac{V_{ref}Nt_c}{T} = \frac{V_{ref}10^{-6}N}{T}$$

The quantization error of the counter is ±1 count, so for $\overline{v_{in}}$ to be measured to ±0.1%, ±1 count/$N = 10^{-3}$ or $N = 10^3$ counts. For example, if $\overline{v_{in}} = V_{ref}$, then $T = 1$ ms. The total time for the measurement will be from 0 to T (1 ms) plus from T to t' (1 ms), or a total of 2 ms.

Thus to get a high-precision measurement, we need a large count and hence a long measurement time. This process is somewhat slow, but it is independent of RC and is an integrating process, which means it is relatively immune to any short spikes in the input. It is widely used in laboratory instruments that are read by human eyeballs (e.g., digital voltmeters, etc.) because it is noise free and accurate, although slow. The human eye/brain cannot perceive a digital counter change if it changes more often than about every 50 ms, so the slow speed is no disadvantage if the instrument is to be read by a human rather than an electronic instrument.

If the time interval T is measured by taking a fixed number n of clock pulses, $T = nt_c$, then

$$\overline{v_{in}} = \frac{V_{ref}Nt_c}{nt_c} = V_{ref}\frac{N}{n}$$

which means the measurement does not depend upon the clock period! The measurement accuracy depends only on the stability and accuracy of the reference voltage.

If the analog input is derived from a transducer (such as a strain gauge) that must be supplied with a dc voltage, then if the A/D reference voltage *also supplies* the transducer, drifts in the reference voltage will not offset the accuracy, at least to first order. The reason is that if V_{ref} increases, so does v_{in}, and the output count being proportional to the ratio $\overline{v_{in}}/V_{ref}$ does not change. This is an example of a *ratiometric* method of measurement and is widely used to minimize long-term drift errors. It can be used whenever the output reading of an instrument (N in this case) is proportional to a ratio (v_{in}/V_{ref}), provided the v_{in} source is supplied by the same voltage reference.

15.4.5 The Successive Approximation A/D Converter

The successive approximation A/D converter is probably the most useful single type of converter because it combines high accuracy with reasonably high speed and moderate price. Conversion times from 1 to 100 μs are typical. The circuit is shown in Fig. 15.20, and the basic idea of operation is quite simple. A control register or successive approximation register (SAR) prepares a digital number (an approximation to the input) that is converted to an analog value by a fast D/A converter whose output is then compared to the analog input by a comparator. Depending on whether the digital

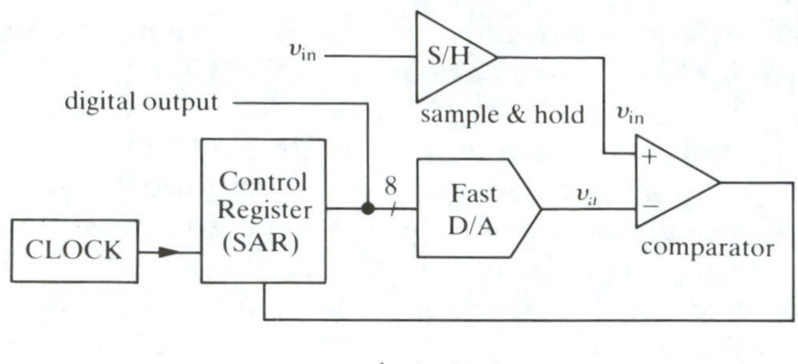

FIGURE 15.20 Successive approximation A/D converter.

number (converted to analog) is larger or smaller than the input, the control register then prepares a second digital number (approximation) and repeats the comparison process until the best possible digital approximation to the analog input is obtained. The final digital input to the D/A converter is the output of the A/D converter and is present in parallel form. A sample-and-hold amplifier at the input holds the comparator input constant while the control register operates.

To be specific, consider an eight-bit successive approximation A/D converter. With the analog input sampled and held and the eight bits of the SAR all cleared to 0, the first step is to set the eight-bit SAR output to one half of its maximum output, namely, 10000000. The D/A converter then converts this digital number to analog v_a form and compares it to the analog input. If $v_a < v_{in}$, the second step is for the signal output to be set to three fourths of its maximum output, namely, 11000000, and the D/A converter converts this number to analog form and again compares it to the input. If $v_a > v_{in}$, the second step is for the register output v_a to be set to one fourth of its maximum output, and so on. The "search" is a binary one; each approximation is in the center of the interval where the input is known to lie. When all eight bits are set to get the best eight-bit approximation to the input, the conversion is completed and the eight-bit input to the D/A converter is the output of the A/D converter. The time for the conversion is the time for the eight D/A conversions and comparisons; it is essentially the sum of all eight D/A converter settling times. This is the time for an A/D conversion *regardless* of the magnitude of the analog input—unlike the staircase or integrating A/D techniques where the larger the input, the longer the conversion time. For example, for an eight-bit A/D converter with a 1-μs settling time for the D/A converter, it would take approximately 8 μs to make an A/D conversion with the successive approximation technique but 255 μs for a maximum amplitude input to a staircase A/D converter with a 1-MHz clock.

The analog input cannot change appreciably during the conversion time, or else a significant error will result; this is why the sample-and-hold circuit is necessary. Also, any noise spike on the input will cause an incorrect bit to be set. Thus, for noisy inputs it may be necessary to add an analog low-pass filter to the input.

The worst errors will be produced by noise spikes in the input during the early part of the conversion time, when the more significant bits are being set. Thus, an input analog filter with a break frequency approximately equal to the reciprocal of the conversion time will prevent the input from changing appreciably during the conversion time.

The resolution is ±1 bit out of N bits or $\pm 1/2^N$: 0.39% for eight bits, ±0.1% for ten bits, etc.

Extremely inexpensive eight-bit successive approximation A/D converters are now available. One example is the ADC 0809 made by several manufacturers. It is an eight-bit successive approximation CMOS converter with 100-μs conversion time, including eight multiplexed input lines for about \$5. Any of eight analog inputs can be selected by a three-bit address

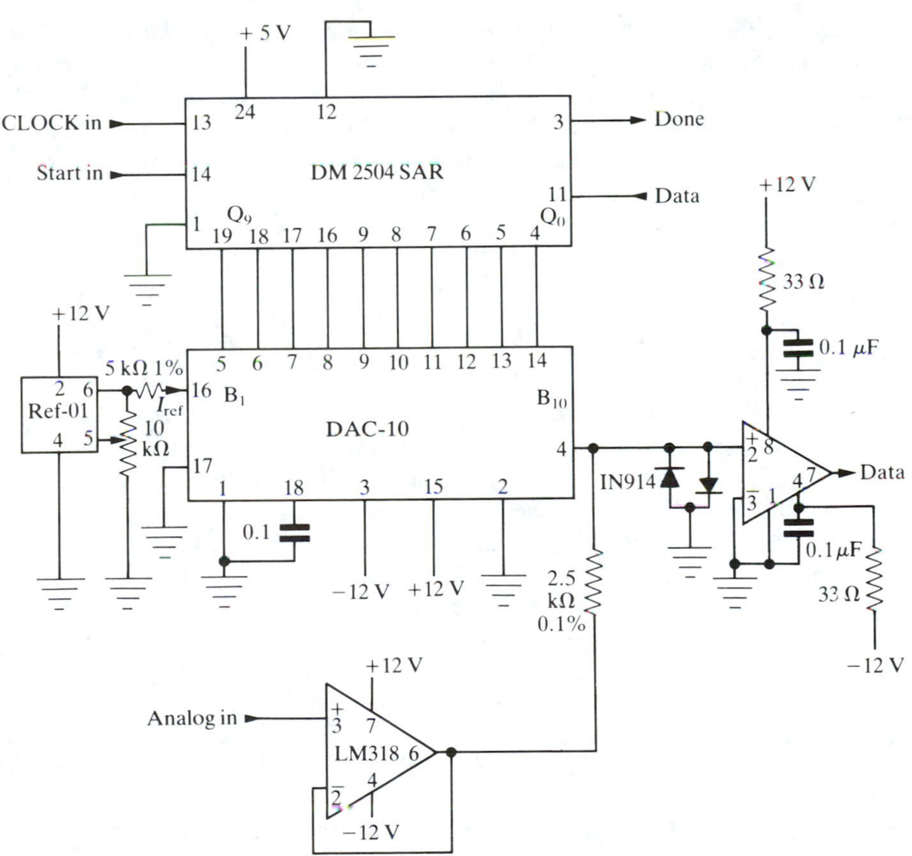

FIGURE 15.21 Ten-bit successive approximation A/D converter.

line. It needs only one 5-V supply and has latched three-state outputs for easy interfacing to microprocessors. The ADC 1210 is a 10-bit or 12-bit successive approximation converter with a 12-bit conversion time of 100 μs and a 10-bit conversion time of 30 μs.

In general, faster successive approximation A/D converters can be made by using a separate SAR chip along with a separate fast D/A converter, rather than a one-chip A/D converter. The SAR and the D/A converter plus a few simple components make up the A/D converter circuit. The DM2502 (8-bit) and the DM2504 (12-bit) are examples of separate SARs. Conversion times of less than 1 μs can be achieved with this technique.

A specific ten-bit A/D circuit is shown in Fig. 15.21. The DAC-10 D/A converter has an 84-ns settling time, so the conversion time of the circuit of Fig. 15.21 is about 1 μs.

15.4.6 The Flash or Parallel A/D Converter

The fastest technique for A/D conversion is the *flash* or *parallel* encoding technique in which a resistor ladder divides a stable analog precision reference voltage into many steps, and the input voltage is *simultaneously* compared with each of the voltage steps. A *priority encoder* then produces a digital output, depending on where the input voltage falls in relation to the steps. In other words, all the output bits are determined at essentially the same time. Conversion times of less than 100 ns are available. Often no sample-and-hold amplifier is required because of the high speed. The digital output usually cannot be interfaced directly to a microcomputer because the computer is too slow. Some sort of direct memory access (DMA) and a high-speed buffer latch are usually required for a computer interface. The basic circuit is shown in Fig. 15.22 for a three-bit converter.

The output of the comparators is sometimes called a *thermometer code*. The resistor ladder divides the reference voltage into eight ranges, and the digital output for each of the eight ranges is shown in Fig. 15.22(b). The priority encoder is simply an array of gates to produce the desired digital output, which can be either straight binary, as shown, or BCD.

For an N-bit output we need $2^N - 1$ comparators; for example, for $N = 8$ bits, we need 255 comparators, which requires a very large power supply as well as LSI to get so many comparators on a single chip. Also, because the input is directly connected to each of the $2^N - 1$ comparators, there is a large input-to-ground capacitance; for example, if each comparator has an input-to-ground capacitance of only 5 pF, an eight-bit flash comparator would have $255 \times 5\text{ pF} = 1275\text{ pF}$. An input buffer with a low output impedance to drive all the comparator inputs helps enormously, but the flash conversion technique is not very practical for high resolution (large N).

The 74LS148 is a one-chip priority encoder that encodes eight input

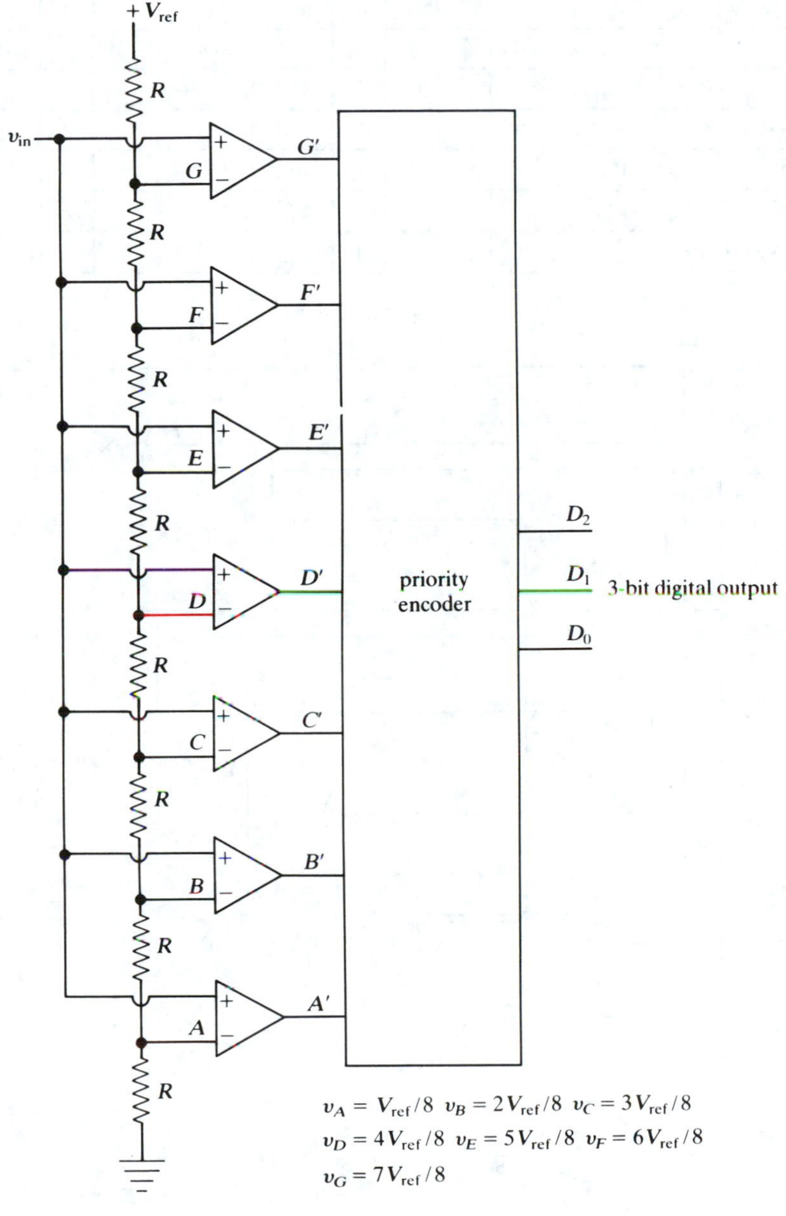

(a) *basic circuit*

FIGURE 15.22 Three-bit flash A/D converter.

Input	G'	F'	E'	D'	C'	B'	A'	D_2	D_1	D_0
$0 < v_{in} \leqq V_{ref}/8$	0	0	0	0	0	0	0	0	0	0
$V_{ref}/8 < v_{in} \leqq 2V_{ref}/8$	0	0	0	0	0	0	1	0	0	1
$2V_{ref}/8 < v_{in} \leqq 3V_{ref}/8$	0	0	0	0	0	1	1	0	1	0
$3V_{ref}/8 < v_{in} \leqq 4V_{ref}/8$	0	0	0	0	1	1	1	0	1	1
$4V_{ref}/8 < v_{in} \leqq 5V_{ref}/8$	0	0	0	1	1	1	1	1	0	0
$5V_{ref}/8 < v_{in} \leqq 6V_{ref}/8$	0	0	1	1	1	1	1	1	0	1
$6V_{ref}/8 < v_{in} \leqq 7V_{ref}/8$	0	1	1	1	1	1	1	1	1	0
$7V_{ref}/8 < v_{in}$	1	1	1	1	1	1	1	1	1	1

(b) *truth table*

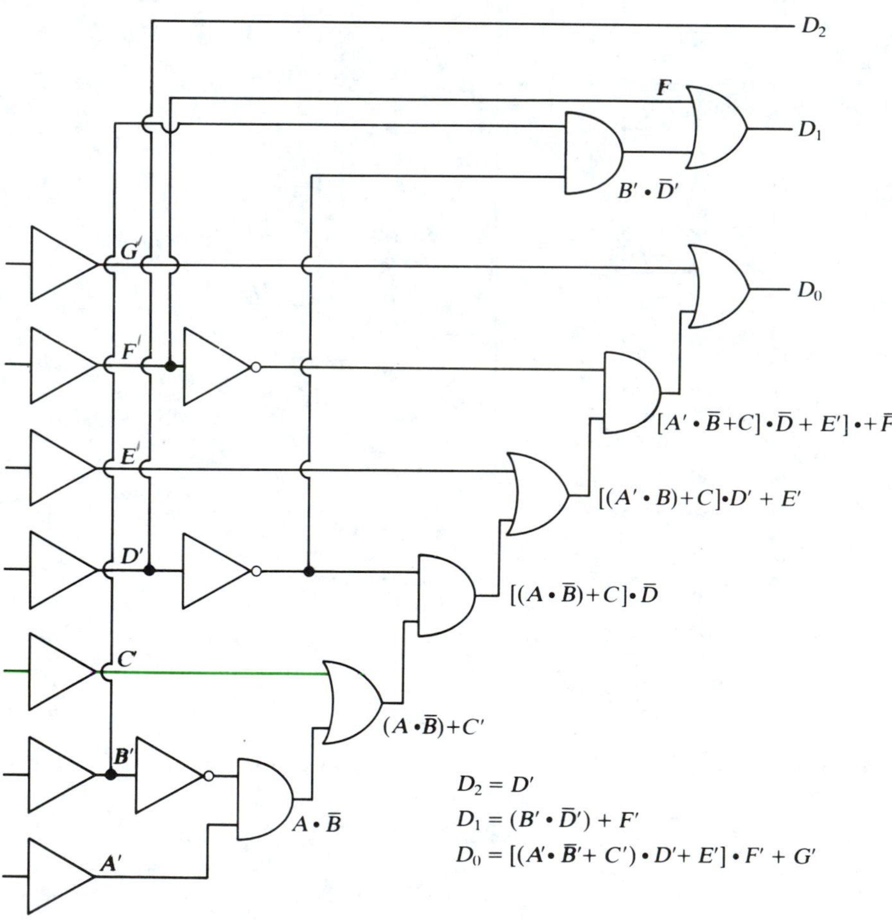

(c) *three-bit priority encoder* (74LS148)

FIGURE 15.22 Continued.

lines into three output bits in straight binary. The three-bit priority encoder is shown in Fig. 15.22(c).

15.4.7 The Parallel-Series A/D Converter

The one-pass flash converter discussed in 15.4.6 needs too many comparators to be practical for high resolution. The *parallel-series* or *subranging* A/D converter is a *two-pass* conversion technique that obtains high resolution and high speed, although not quite as high speed as that of the one-pass flash converter. The basic idea is to digitize the input analog signal in one pass with a "coarse" four-bit flash converter, convert the digital output to analog form with a D/A converter, subtract the D/A output from the analog input, and convert this difference or "residue" with a second "fine" flash four-bit converter. The digital output of the two four-bit flash A/D converters is then the eight-bit digital output. In other words, the four most significant bits are obtained from the first pass, the four least significant bits from the second pass. The basic circuit is shown in Fig. 15.23.

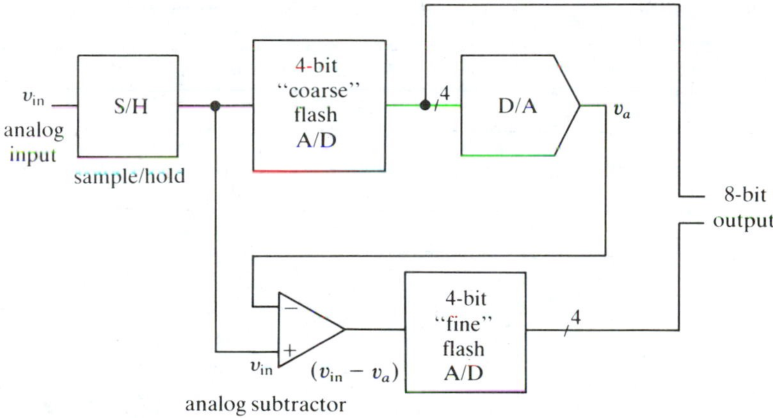

FIGURE 15.23 Eight-bit parallel-series A/D converter.

We immediately see that the number of comparators is much reduced compared to a flash converter. One eight-bit flash converter contains 255 comparators, whereas the two four-bit flash converters contain only 30 comparators. But the parallel-series converter does contain a high-speed D/A converter, a fast sample-and-hold input amplifier, which must hold the input constant during the time required for these two passes, and a fast analog subtracter, which must have an accuracy and stability corresponding to less than $\frac{1}{2}$ LSB.

Consider the specific case of making an eight-bit parallel-series A/D

Analog Input (mV)	8-Bit Output							
$2550 \leqq v_{in}$	1	1	1	1	1	1	1	1
$2540 \leqq v_{in} < 2550$	1	1	1	1	1	1	1	0
$2530 \leqq v_{in} < 2540$	1	1	1	1	1	1	0	1
⋮	⋮							
$1300 \leqq v_{in} < 1310$	1	0	0	0	0	0	1	0
$1290 \leqq v_{in} < 1300$	1	0	0	0	0	0	0	1
$1280 \leqq v_{in} < 1290$	1	0	0	0	0	0	0	0 midscale
$1270 \leqq v_{in} < 1280$	0	1	1	1	1	1	1	1
$1260 \leqq v_{in} < 1270$	0	1	1	1	1	1	1	0
⋮	⋮							
$170 \leqq v_{in} < 180$	0	0	0	1	0	0	0	1
$160 \leqq v_{in} < 170$	0	0	0	1	0	0	0	0
$150 \leqq v_{in} < 160$	0	0	0	0	1	1	1	1
⋮	⋮							
$90 \leqq v_{in} < 100$	0	0	0	0	1	0	0	1
$80 \leqq v_{in} < 90$	0	0	0	0	1	0	0	0
$70 \leqq v_{in} < 80$	0	0	0	0	0	1	1	1
⋮	⋮							
$20 \leqq v_{in} < 30$	0	0	0	0	0	0	1	0
$10 \leqq v_{in} < 20$	0	0	0	0	0	0	0	1
$0 \leqq v_{in} < 10$	0	0	0	0	0	0	0	0

(a) *input/output*

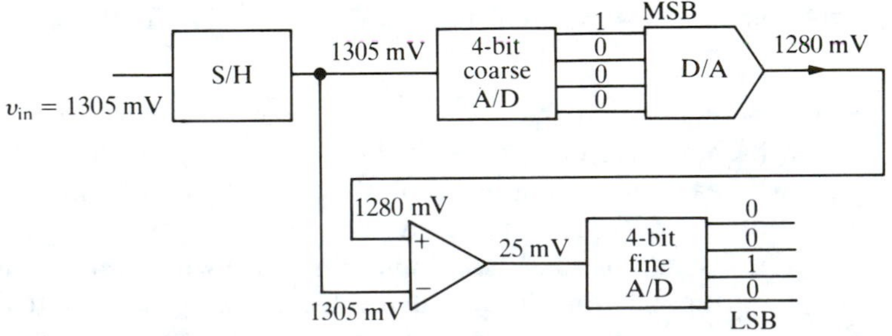

(b) v_{in} = 1305 *mV*, *digital output* =10000010

FIGURE 15.24 Eight-bit parallel-series A/D example.

converter. Let us assume the least significant of the eight output bits represents an analog input of 10 mV. Then, we desire an A/D relationship as shown in Fig. 15.24(a). The fine four-bit converter then should have its LSB equivalent to 10 mV and its MSB equivalent to 80 mV. The coarse four-bit converter should have its LSB equivalent to 160 mV and its MSB equivalent to 1280 mV. For a 1305-mV input, for example, the first coarse converter output should be 1000 (1280 mV), and the second fine converter output 0010 (20 mV). These two outputs taken as one eight-bit output then are 10000010, which is the correct output. The analog voltage levels and the bits are shown in Fig. 15.24(b) for a 1305-mV input.

The new Hewlett-Packard 5180A waveform recorder contains a 10-bit parallel-series A/D converter that can make 2×10^7 conversions per second (a sampling frequency of 20 MHz). Thus, its A/D conversion time is only 50 ns, and according to the sampling theorem it can therefore analyze an input signal with Fourier components up to 10 MHz. It contains a sample-and-hold input amplifier that settles in only 5 ns, a five-bit D/A converter with a settling time of only 15 ns, and several amplifiers with slew rates of 1000 V/μs.

15.4.8 Sample/Track Hold Amplifiers

The conversion time of an A/D converter is the time required to make the conversion of a (steady) analog input to a stable digital output. Clearly, the conversion time must be small compared to the time in which the input analog signal changes appreciably. If not, the digital output will be in error. For slow analog signals and short conversion times this is no problem. However, if the analog input changes appreciably during the conversion time, then the input must be sampled over a short time interval and then held constant while the A/D conversion is performed. The sample-and-hold amplier does just this.

The basic circuit of a sample-and-hold amplifier is shown in Fig. 15.25 along with a waveform diagram. When the "sample" command is given, the switch S is closed and the capacitor C is charged up to the input voltage with a time constant RC where R is the effective series resistance from the first op amp and the switch. The *acquisition* time is the time required for the capacitor to charge to within 0.1% of the input voltage. The sample command must clearly be longer than the acquisition time. A typical acquisition time for an inexpensive sample-and-hold amplifier, such as the AD583, is approximately 4 μs; it can be as low as several hundred ns for more expensive types. The input signal should not change appreciably during the acquisition time. When the hold command is given, the switch S is opened, and the capacitor stores the input voltage while the A/D converter performs the conversion. The time delay between the beginning of the hold command and the opening of the switch (during which time the

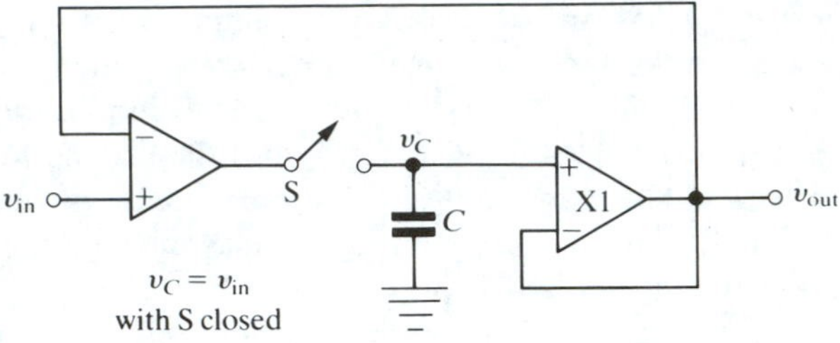

(a) *basic circuit (switch S is closed in "sample" mode, open in "hold" mode)*

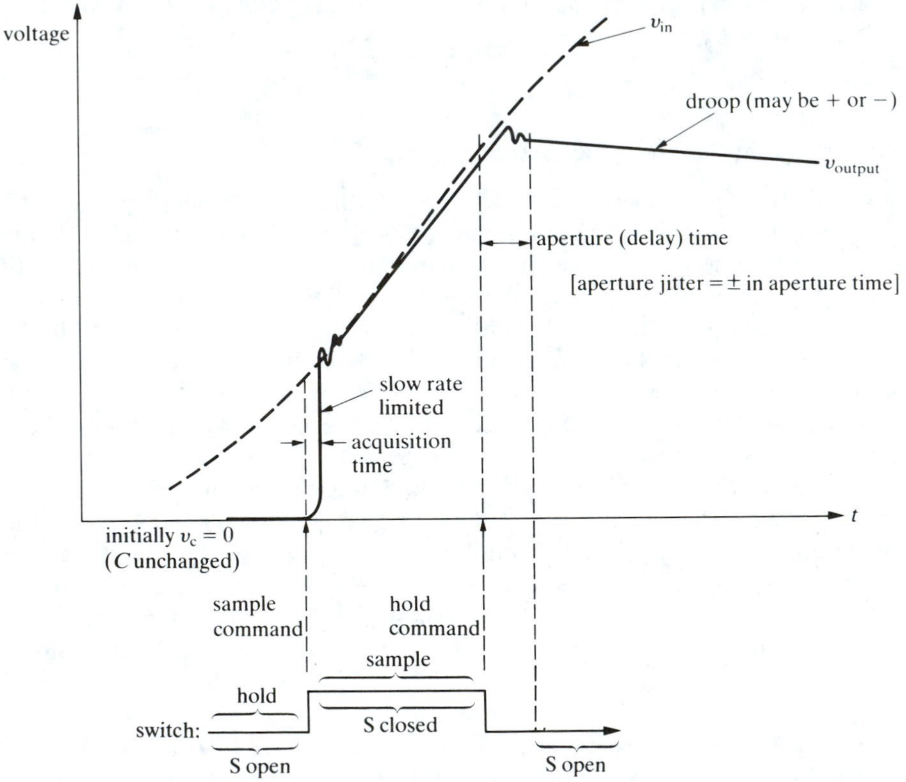

voltage

v_{in}

droop (may be + or −)

v_{output}

aperture (delay) time

[aperture jitter = ± in aperture time]

slow rate limited

acquisition time

t

initially $v_c = 0$ (C unchanged)

sample command

hold command

sample

hold

switch:

S closed

S open

S open

(b) *waveforms*

FIGURE 15.25 Sample/track hold amplifier.

input may continue to change) is called the *aperture time* of the sample-and-hold amplifier. A typical aperture time is about 50 ns. The *aperture jitter* is the ± uncertainty in the aperture time. The *droop* of the sample-and-hold amplifier refers to the slow decrease in the capacitor voltage (due to leakage, etc.) once the switch is open (i.e., during the hold command).

A change in the analog input that will result in an output bit error depends on the precision of the A/D conversion (i.e., on the number of bits of the A/D converter). For example, a change of 1 LSB in the output of an eight-bit A/D converter corresponds to a 0.39% change in the analog input. Thus, to avoid an error in the LSB of the A/D output, the input should not change by an analog amount equivalent to 1 LSB during the sample-and-hold acquisition time. For example, if for a sinusoidal input $V = V_0 \cos \omega t$ the A/D gain is set so that a $2V_0$ analog input corresponds to a maximum digital output, then for an eight-bit converter 1 LSB corresponds to an analog input voltage change of $2V_0/255 = 0.00784V_0$. The maximum voltage change per second for the input is $(dV/dt)_{\max} = \omega V_0$, so the requirement on the effective sample-and-hold acquisition time T_c is

$$(\omega V_0)T_c < 0.00784V_0$$

or

$$\omega T_c < 0.00784 \quad \text{and} \quad T_c < \frac{0.00125}{f}$$

The higher the input frequency, the shorter the acquisition time must be. If $f = 1$ kHz, the acquisition time must be less than 1.25 μs. The AD HAS-1201 is a 12-bit A/D converter with a 1-μs conversion time and an on-board track/hold amplifier.

15.4.9 A/D Specifications

All A/D converters have the following specifications:

1. *Accuracy*. The sum of all the errors due to temperature changes, and so on. There is always the possibility of a one-bit "quantization" error in *any* digitizing process: $\pm\frac{1}{2}$ LSB. The accuracy is often expressed as a percentage of full scale.
2. *Linearity*. The difference between the actual staircase output versus the best straight-line approximation; $\pm\frac{1}{2}$ LSB is the best possible linearity. *Monotonicity* means the digital output code steadily increases as the analog input increases. The output can be monotonic but quite nonlinear, but a nonmonotonic output cannot be linear. Missing output digital codes are a common cause of nonlinearity; for example, the output may skip from 0010 to 0100 as the input increases. Noise spikes can also cause nonlinearity or nonmonotonicity.
3. *Resolution*. In any N-bit A/D converter there are 2^N different output levels and $2^N - 1$ steps. The resolution is $1/(2^N - 1) \cong 1/2^N$—that is, the maximum fractional error in the output if a one-bit quantization error is assumed.

No. bits, N	*No. Steps* $= 2^N - 1$	*LSB/F.S.** (%)	*LSB* (*F.S.* = 10 V) (mV)	*D.R.*† (dB)
8	255	0.39	39.2	48
10	1023	0.098	9.78	60
12	4095	0.024	2.44	72
14	16383	0.0061	0.61	84
16	65535	0.0015	0.15	96

*F.S. = full scale.
†D.R. = dynamic range.

4. *Dynamic range*. The ratio of the largest to the smallest output, usually expressed in dB.

$$\text{D.R.} = 20 \log_{10} \frac{2^N - 1}{1} \cong 20 \log_{10} 2^N$$

Modern digital audio processing uses 16-bit resolution and thus has a 96-dB dynamic range, much larger than the typical 60 dB for conventional analog recording.
5. *Offset*. The output when there is zero input. Can be set equal to zero by user in most units but will drift with temperature.
6. *Conversion time*. The time required to convert a stable analog input to a stable digital output. The time between the start conversion (SC) and the end-of-conversion (EOC) pulse. Usually maximum for a full-scale output. Often assumes the input sample-and-hold amplifier has settled down and the A/D has been reset to zero. (The aperture time is similar but refers only to the time required to open the switch in the sample-and-hold circuit—the time to go from the hold to the sample condition. Aperture jitter refers to the time-to-time variation in the aperture time. Modern sample/hold amplifiers are available with a jitter as low as ±2 ps.)

   A quick numerical illustration might be useful. Consider an eight-bit A/D converter with a 10-μs conversion time (but with no sample/hold amplifier) sampling a sinusoidal 10-V peak-to-peak input. We clearly wish the input to change by less than 1 LSB in the output during the conversion time. This means that for a 10-V full-scale output from the A/D converter, we require

$$\frac{10\text{ V}}{2^N - 1} \geqq \left(\frac{dv_{\text{in}}}{dt}\right) T_c$$

where T_c is the conversion time. For a sine wave $V = V_0 \sin \omega t$, and $(dV/dt)_{\max} = \omega V_0$. Thus

$$\frac{10\text{ V}}{2^N - 1} \geqq \omega V_0 T_c$$

or
$$\omega \leqq \frac{10\text{ V}}{(2^N - 1)V_0 T_c}$$

Thus for our $N = 8$-bit A/D converter with a 10-μs conversion time, and with a $V_0 = 5$-V input,

$$\omega \leqq \frac{10\text{ V}}{(2^8 - 1)(5\text{ V})(10\ \mu\text{s})} = 784\text{ rad/s} = 2\pi f$$

or
$$f \leqq 1280\text{ Hz}$$

In other words, if the input sinusoidal signal frequency is greater than 1280 Hz, the digital output will be in error by more than 1 LSB! If a sample/hold (S/H) amplifier is used, the effective aperture time should be used for T_c.

7. *Throughput*. Refers to the maximum rate at which stable digital outputs can be produced. It depends on the time to reset the A/D plus the conversion time plus any other time delay in the unit.
8. *Buffer amplifier*. Some A/D converters may need a buffer amplifier at the input. The AD670 is an eight-bit 10-μs successive approximation A/D with a built-in instrumentation amplifier.
9. *S/H circuit*. Necessary for flash successive approximation A/D; not necessary for any integration A/D. Input should change by less than 1 LSB during the effective acquisition time.

A summary of the types of A/D converters is given in Table 15.6, and some specific A/D converters are given in Table 15.7.

TABLE 15.6. A/D Types

	Succ. Approx.	*Dual-Slope Int.*	*Flash*	*V → F*
Speed	med.	low	high	low
Accuracy (no. bits)	high-med.	high	low	med.
Cost	low	low	high	low

TABLE 15.7. Some A/D Converters

A/D type	*No. Bits*	*Accuracy*	*Conv. Time (μs)*	*Comments*
ADC0809	8	±1 LSB	100	CMOS, 8-input MUX, ~$5
ADC0801	8	$\pm\frac{1}{4}$ LSB	50	8080, 8085 compatible
AD670	8		10	On-chip inst. amp., ~$10
AD7575	8	±1 LSB	5	On-chip track/hold
AD573	10		30	On-chip reference, clock
AD579	10		1.8	On-chip reference, clock
AD574A	12	$\pm\frac{1}{2}$ LSB	35	3-state output, $35
AD578	12	$\pm\frac{1}{2}$ LSB	3	
AD7578	12	±1 LSB	100	3-state output
APC1140	16		35	

Source: *Data Acquisition Databook Update and Selection Guide*, 1986 Analog Devices.

15.5 SERIAL AND PARALLEL INTERFACING

Data or instructions can be transmitted either in parallel or in serial fashion. Parallel data transmission requires one wire for each bit [e.g., eight data wires and one ground wire for eight-bit data (a data *bus*)], and the bits appear at the receiving end of the bus essentially simultaneously. Serial data transmission, in contrast, requires only two wires, one for the data and one for ground, but the data arrive bit by bit at the receiving end. Parallel data transmission is rapid but expensive, whereas serial data transmission is slow but cheap. Generally speaking, parallel data transmission is limited to short distances (e.g., from a transducer to a microprocessor controller in the laboratory, a few feet), and serial data transmission is required for longer distances.

Parallel data transmission usually requires two data buses—one for the input and one for the output. For asynchronous transmission (no clock) there must also be *handshaking* information exchanges between the transmitter and the receiver to ensure the receiver is ready to listen before the transmitter sends the data. But, surprisingly, there is no one standardized format for parallel data transmission; each computer has its own special format. However, there is a standard serial data transmission, namely, RS-232.

15.5.1 The RS-232 Serial Interface

The Electronics Industries Association (EIA) RS-232C standard is the generally accepted standard for serial data transmission. Virtually all microcomputers contain an RS-232 serial interface port: the Zenith, the TRS-80, the Apple, the IBM PC, etc. Also, most peripheral devices (plotters, printers, or keyboard terminals) come with an RS-232 port. Teletype, tape, and disk peripheral devices require serial communication.

The complete RS-232 serial interface standard consists of 25 data lines or wires. The standard D25 connector, female at one end and male at the other end of the link, is used. Table 15.8 shows the pin designations for all 25 wires.

Fortunately, in most cases, only a few of the 25 wires are necessary to use. Figure 15.26 shows the D25 pin connections.

The main thing to remember about the RS-232C interface is that a logical 0 is represented by a positive voltage anywhere from +5 to +15 V, and a logical 1 is represented by a negative voltage anywhere from −5 to −15 V. (The older RS-232 interface used +5 to +25 V for a 0 and −5 to −25 V for a 1.) And it is comforting to realize that any two lines can be connected together without damage. RS-232 is the complement of the NRZ magnetic tape recording technique discussed in Chapter 14.

In a sense the RS-232C voltage levels are inverted from the usual TTL levels, where the higher voltage (3.5 V) is a 1 and the lower voltage (0.2 V)

TABLE 15.8. RS-232C Pin Designations for D25 Connector

Pin	Function
1	Protective ground
2	Transmit data (TXD)
3	Receive data (RXD)
4	Request to send (RTS)
5	Clear to send (CTS)
6	Data set ready (DSR)
7	Signal ground
8	(Carrier detect)
9	Reserved for data testing
10	Reserved for data testing
11	Uncommitted
12	Sec. received line signal detector
13	Sec. clear to send
14	Sec. transmittal data
15	Transmission signal timing (for ≧ 1200 baud)
16	Secondary received data
17	Receiver signal timing (for ≧ 1200 baud)
18	Uncommitted
19	Secondary request to send
20	Data terminal ready (DTR)
21	Signal quality detector
22	Ring indicator
23	Data signal rate selector
24	Transient signal timing (for ≧ 1200 baud)
25	Uncommitted

is a 0. The two RS-232C levels need not be symmetric about 0 V; for example, +12 V could be a 0 and −5 V could be a 1. And, ±3 V will work fine in most cases, but it is best to use ±5 V or larger swings. The large voltage swing between the 1 and the 0 means there is a much greater noise immunity than for the ordinary TTL logic levels; thus RS-232 signals can be sent over cables up to 100 ft long.

There are a few other electrical requirements for RS-232C data transmission. The maximum data transfer rate is 20,000 bits/s. The load impedance at the receiver should be between 3 and 7 kΩ, and the *maximum* slew rate for the data pulses should be 30 V/μs. Larger slew rates (e.g., rise and fall times that are too short) may produce too much crosstalk between wires; the short rise times produce more high-frequency Fourier components that radiate energy more effectively than lower-frequency components of pulses with longer rise times.

The rate at which data are transmitted is called the *baud* rate; 1 baud normally equals 1 bit/s. Typical baud rates are 110 (teletype), 300 (slow telephone line), 1200 (fast telephone line), 2400, 4800, 9600, and 19,200. Crosstalk is usually no problem below 9600 baud.

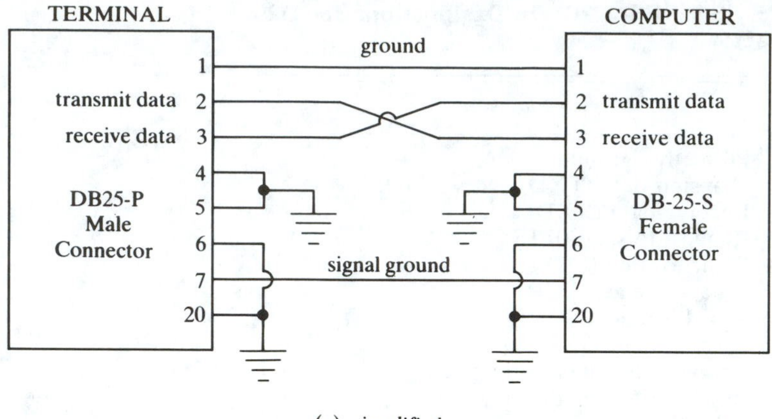

(a) *simplified*

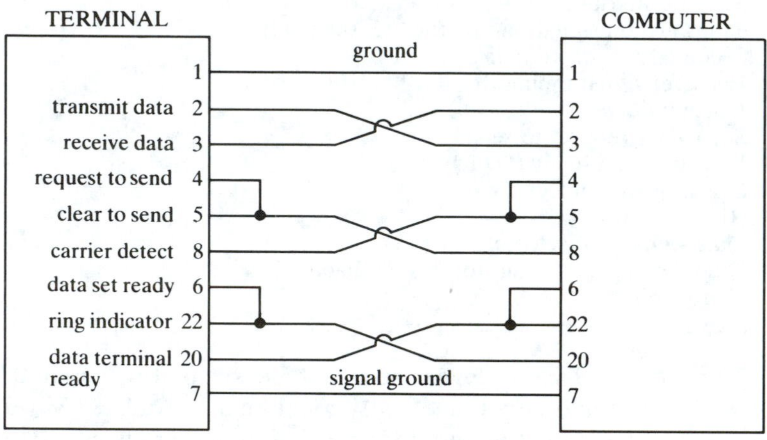

(b) *with handshaking protocol*

FIGURE 15.26 DB25 connectors for RS-232C serial data transmission.

The data are usually transmitted with the standard ASCII code plus one low (0) start bit and one high (1) stop bit, as shown in Fig. 15.27. For example, if the baud rate is 300 and each character takes eight bits (seven bit data plus one parity bit), then to transmit one character requires ten bits: one start bit, eight character bits, and one stop bit. The number of characters transmitted per second is

$$\frac{300 \text{ bits/s}}{10 \text{ bits/character}} = 30 \text{ character/s}$$

Notice that two of the ten bits (the start and stop bits) are essentially wasted

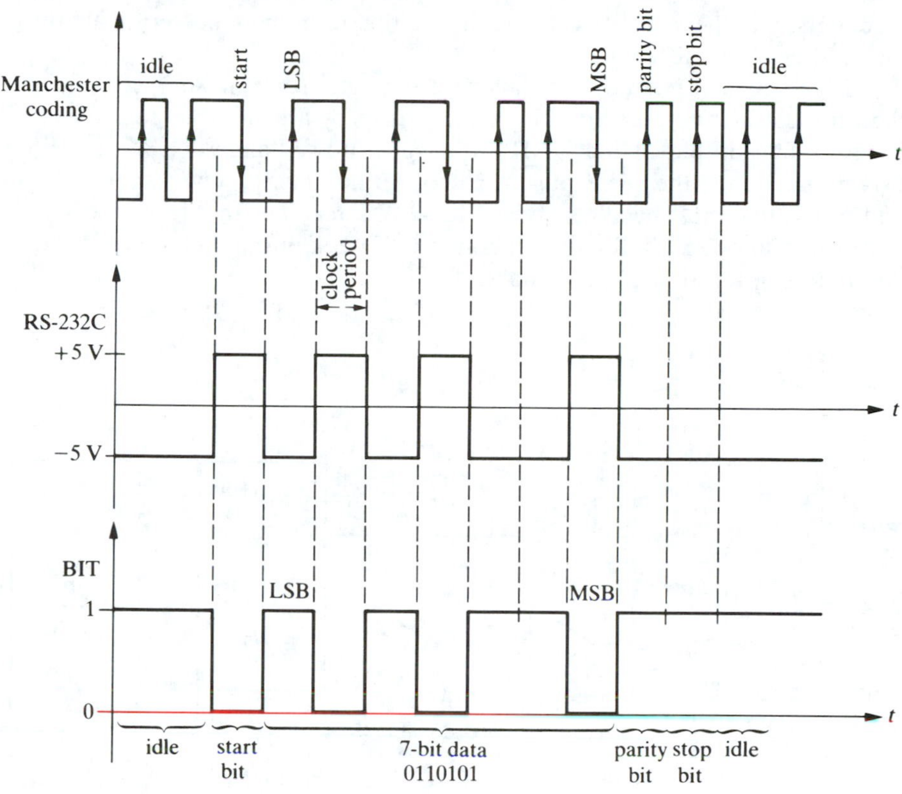

FIGURE 15.27 Serial data in RS-232C and Manchester format (5 V = 0, −5 V = 1).

in terms of information. For TTY (teletype) one start bit, eight character bits, and two stop bits are used, so each character uses 11 bits. Also notice that dc coupling is required because of the long down time (1) when no data are transmitted. This type of serial data coding is sometimes called *non-returnable zero* (NRZ).

Another type of serial coding is *Manchester* or *biphase M* coding, which is self-clocked—a transition occurs at the beginning of each bit period, and a positive *transition* at the center of a bit cell indicates a 1 and no *transition* indicates a 0. Notice that the line need be only ac coupled (not dc) because of the idle pulses. *Biphase L* encoding is also self-clocked—a transition occurs at the center of each bit period: an upward transition for a 0, downward for a 1.

There are two types of data transmission: *full-duplex* and *half-duplex*. In the full-duplex mode outgoing and incoming data are simultaneously present on the line; in the half-duplex mode the outgoing and incoming data are present at different times. An ordinary phone conversation is full duplex. In half-duplex a terminal can transmit data to a computer, but not vice versa; if the computer transmits to the terminal, the terminal must "keep quiet" and listen.

If the data originate in parallel form (e.g., from an A/D converter), then it must first be converted into serial form by a PISO shift register or a UART. The output of the UART then goes to a standard RS-232C line driver that sends the serial pulses over the line. The receiver at the other end of the line must have a standard RS-232C receiver that feeds the serial pulses into another UART, which converts it back into parallel form for the computer input. Figure 15.28 shows this.

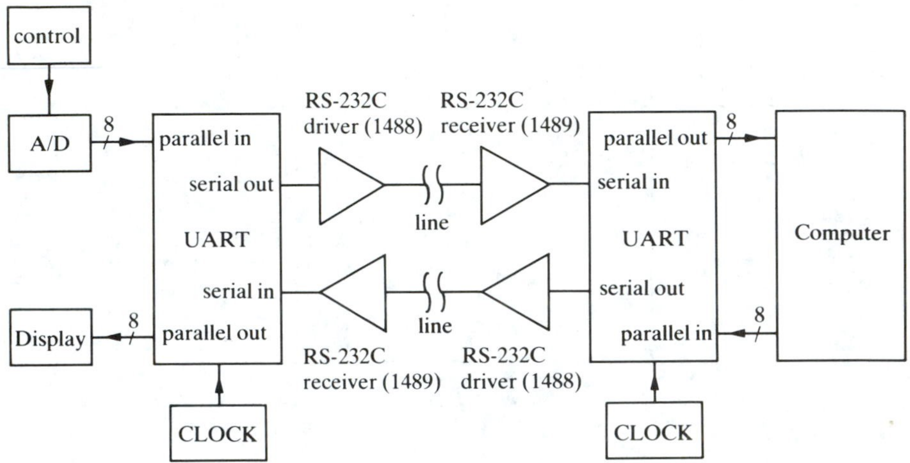

FIGURE 15.28 Parallel-to-serial-to-parallel data transmission.

The standard chips for RS-232C line drivers and receivers are the 1488 (driver) and the 1489 (receiver). Their pin diagrams and the 1489 transfer characteristic are shown in Fig. 15.29. The 1488 driver requires a ±9-V supply typically and yields ±7-V outputs: +7 V for a 0 input and −7 V for a 1 input. For the output slew rate to be below 30 V/μs, the total output capacitance seen by the 1488 must be approximately 330 pF or *larger*.

The 1489 receiver requires a +5-V supply, an external resistance R_T, and a threshold voltage V_{TH} to control the hysteresis thresholds. The typical ($R_T \to \infty$) turn-on voltage is 1.9 V, and the turn-off voltage is 0.8 V for a 1.1-V hysteresis. With $V_{TH} = -5$ V the turn-on–turn-off voltages are higher, typically +4 V + 2.9 V with $R_T = \infty$; and with V_{TH} positive the turn-on–turn-off voltages are lower, typically −1.5 V − 2.6 V. An open or grounded input will be interpreted as a negative voltage input (i.e., a 1 input).

An external capacitance C_T can be used to filter short high-frequency

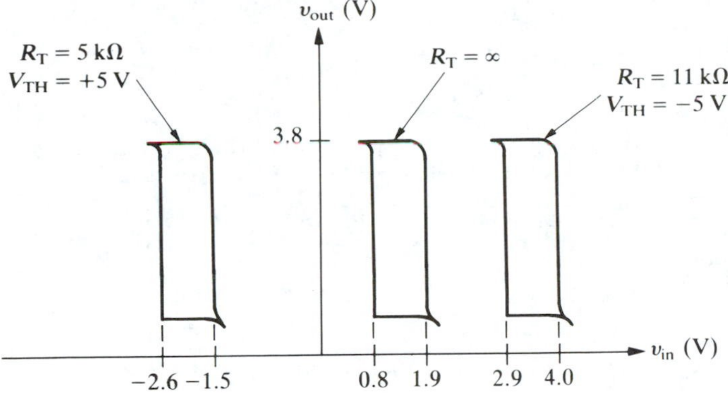

1489 transfer characteristics

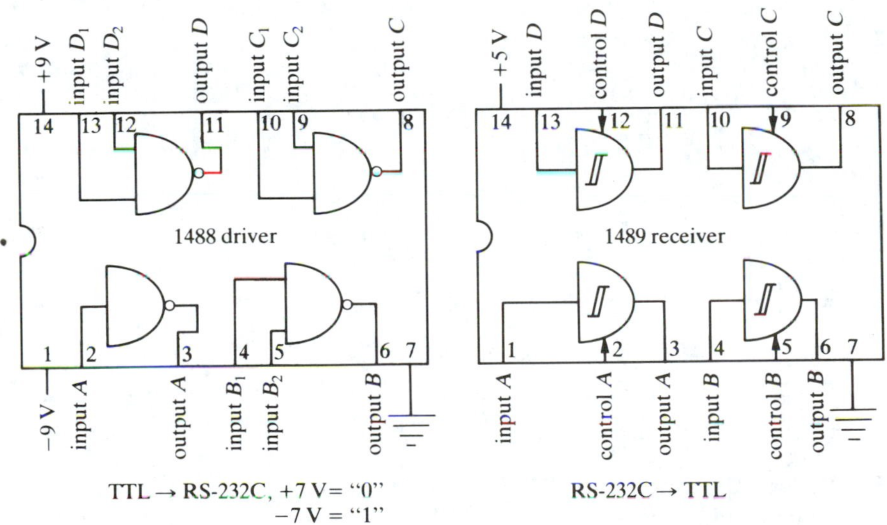

FIGURE 15.29 1488/1489 RS-232C line driver/receiver.

noise pulses; for $C_T = 500$ pF, 6-V noise pulses shorter than approximately 850 ns will be eliminated. The value of C_T must be chosen for the cable length—a large C_T for a short cable, and vice versa. This filtering plus the hysteresis largely eliminate many noise pulses and contribute to reliable operation for line lengths up to approximately 100 ft.

The 9636/9637 is another driver/receiver pair of chips for RS-232C signals; each is a *dual* driver/receiver and comes in an eight-pin DIP. The slew rate of the 9636 driver can be controlled with an external resistance.

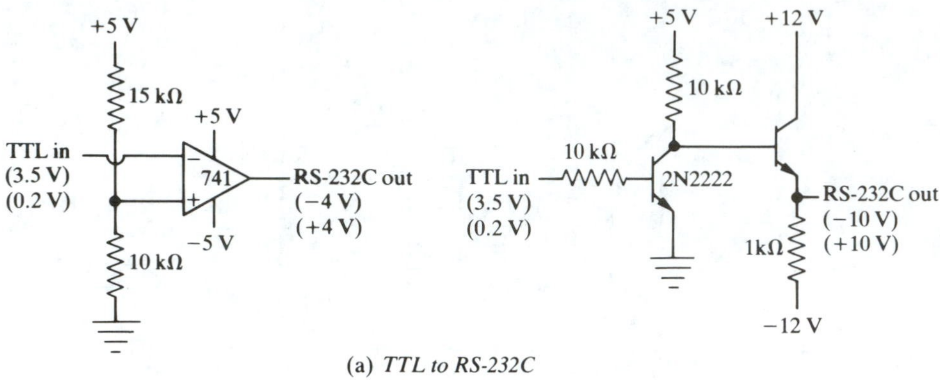

(a) *TTL to RS-232C*

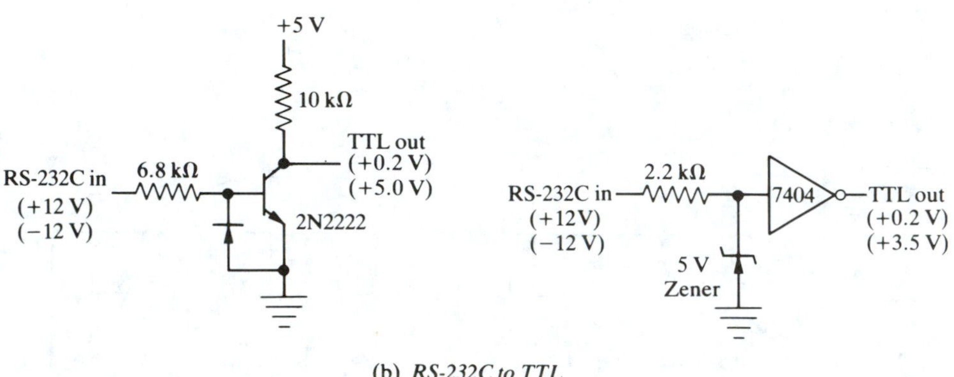

(b) *RS-232C to TTL*

FIGURE 15.30 TTL/RS-232C conversion circuits.

Conversion from TTL to RS-232 voltage levels and vice versa can also be made with homemade circuits, as shown in Fig. 15.30.

15.5.2 The Current Loop (TTY) Serial Interface

The older teletypes used a current loop (usually 20 mA) to transmit serial data. A 20-mA current flowing represented a 1, and no current represented a 0. This technique is useful over very long distances with a constant current source turned on and off by the logic signals. One convenient way to do this is with optical isolators at both ends of the line, as shown in Fig. 15.31. A logic 1 input from the 8251 USART is inverted and turns on the photodiode in the optical isolator. The photodiode light then turns on the phototransistor in the isolator, which causes a 20-mA current to flow in the line. This current causes the receiver photodiode to emit light, which turns on the receiver phototransistor. The phototransistor provides a logic 0 to the inverter, which then produces a 1 output. Notice that there is complete

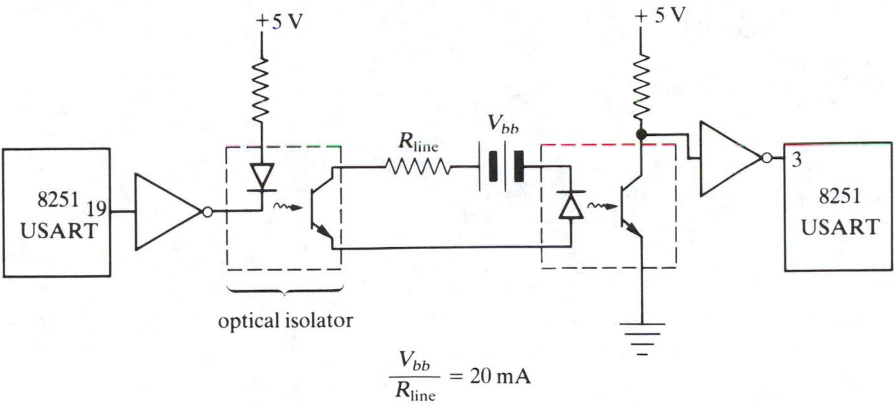

FIGURE 15.31 Current-loop data transmission.

electrical isolation between the transmitter and the receiver due to the optical isolators at both ends of the line. The 20-mA current was required to actuate a relay and print one character in the teletype.

15.5.3 Example

A specific example of the connection of a parallel data source to the serial 232 port of a computer using a UART is shown in Fig. 15.32, where an A/D converter provides the data input to the UART, which inputs the computer. The computer sends a serial RS-232 word to the UART via pin 2 of the computer DB25 female connector. The 1489 receiver converts this word into TTL voltage levels and feeds the input to pin 20, the serial input (SI) terminal of the UART. This word selects one of the eight analog inputs to the A/D converter by settling the CBA address inputs to the A/D converter. The UART DAV (DATA AVAILABLE) pin 19 goes high, which sets both the ALE (ADDRESS LATCH ENABLE) pin 22 and the SC pin 6 of the A/D converter high. The high ALE pin allows the A/D to read the CBA address and choose one of the A/D analog inputs. The high SC (START CONVERSION) pin resets the A/D output to zero and starts the A/D conversion process. The positive DAV pulse is also converted into a negative pulse with a slight delay caused by the *RC* integrating network, and this negative pulse is the RDAV (RESET DATA AVAILABLE) signal to the UART; that is, the UART DAV pin is forced low again. Approximately 100 clock periods later the conversion process is finished, and the A/D indicates this by setting the EOC pin high.

When the EOC goes high, the DS (DATA SEND) pin of the UART also goes high, which starts the parallel-to-serial conversion in the UART and sends the serial TTL result out via the SO (SERIAL OUTPUT) pin 25 of the UART. This TTL output is converted into an RS-232 format by the

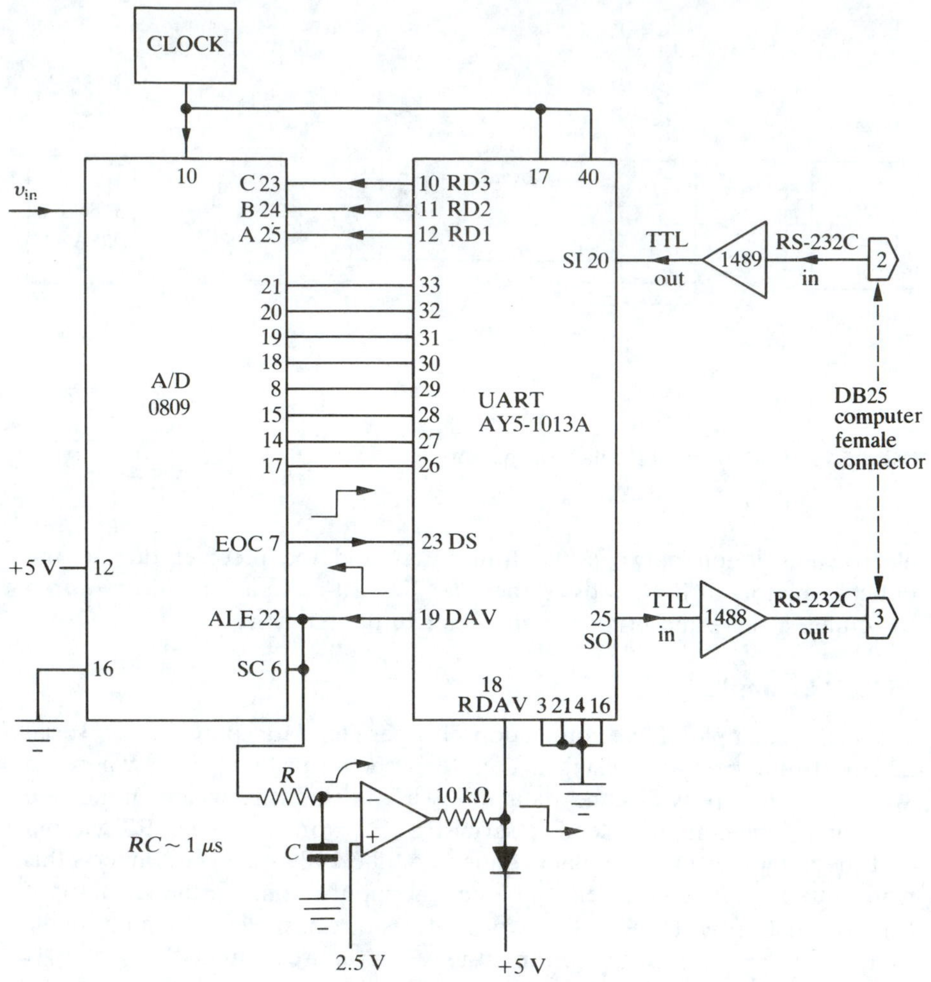

FIGURE 15.32 Parallel data to computer RS-232C. (Courtesy Robert Tinker, Technical Education Research Center, 8 Eliot Street, Cambridge, Massachusetts 02138).

1488 chip, whose output is fed into the serial input to the computer on pin 3 of the computer's DB25 connector.

For a nice article on how to interface the 0809 A/D converter to the VIC-20 microcomputer, see *Hands On*! Vol. 6, No. (2) (Summer 1983), 7 by Technical Education Research Centers Inc., 8 Eliot St., Cambridge, Mass. 02138.

15.5.4 Parallel Interfacing

As we have said, there is no one standard technique for parallel interfacing between computers, or from terminals to computers, and so on. But it is useful to know that there are available one-chip parallel interface chips for some of the more popular microprocessors. For example, two of the most common one-chip parallel I/O (input output) ports are the 6821 PIA (parallel interface adapter) for use with the 6800 microprocessor, and the 8255 PPI (programmable peripheral interface) for the 8080 family of microprocessors. Both data and control information enter the PIA (or PPI) through data bus buffers and are sent to data output or control registers under the control of the computer program. The registers are addressed by the computer as if they were memory locations in the computer.

A simplified block diagram of the 8255 is shown in Fig. 15.33.

chip select $\overline{CS}$	A_1	A_0	$\overline{RD}$	$\overline{WR}$	Operation
0	0	0	0	1	μP reads port A to data bus
0	0	1	0	1	μP reads port B to data bus
0	1	0	0	1	μP reads port C to data bus
0	1	1	0	1	illegal
0	0	0	1	0	μP writes data to port A
0	0	1	1	0	μP writes data to port B
0	1	0	1	0	μP writes data to port C
0	1	1	1	0	μP writes data to control
1					D bus floating (disabled)

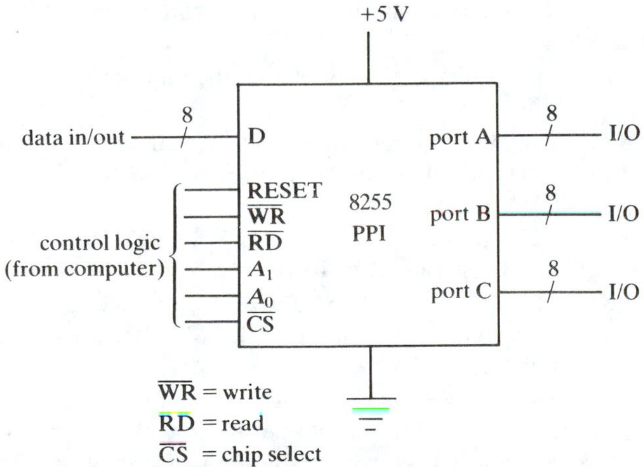

FIGURE 15.33 8255 Programmable peripheral interface.

PROBLEMS

1. If the MSB of an eight-bit A/D converter equals 5 V (simple unipolar code), what analog input will produce an output of 10100101? What will the digital output be for an 8.75-V input?
2. If the MSB of an eight-bit D/A converter equals 5 V (simple unipolar code), what is the analog output for a 01100101 input? For a 10010000 input?
3. In a D/A converter with an offset binary code, what would the output be for an input of 11111111? Of 00000001?
4. Repeat Problem 3 for a twos complement code.
5. Explain briefly the speed disadvantage of the summing D/A converter.
6. Calculate v_{out} for D_0D_1 = 00, 01, 10, 11, where 0 = 0 V and 1 = 3.5 V. What source impedance does the op amp see? What is this device called?

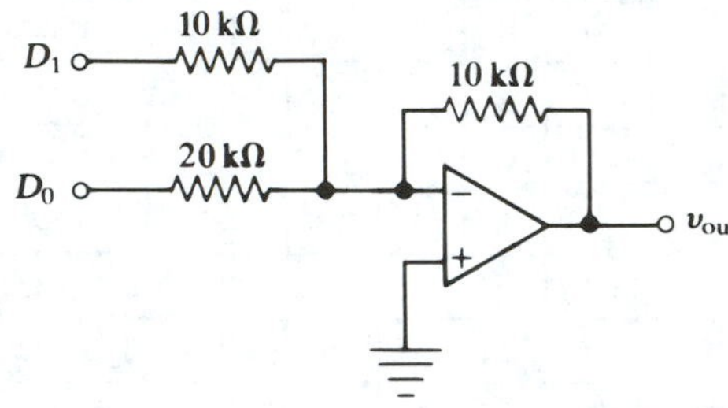

8. If the A/D sampling frequency is 1 MHz and the input analog signal has Fourier components from dc to 800 kHz, what sort of prefilter is necessary before the sampling takes place? Why?
9. Sketch the Fourier spectrum of the sampled waveform if the sampling frequency is 10 kHz and the input analog signal has Fourier components at 2 kHz, 4 kHz, and 6 kHz.
10. What are the two main disadvantages of the staircase or counter A/D converter?
11. What would be the time required for a ten-bit staircase A/D converter to convert a 7.5-V input signal to digital form if 1 MSB = 5 V and the clock frequency is 200 kHz?
12. In an eight-bit single-slope A/D converter with V_{ref} = 5 V, f_{ck} = 1 MHz, RC = 100 μs, what will the digital output be for v_{in} = 5 V? For RC = 1 ms?
13. In an eight-bit dual-slope A/D converter with V_{ref} = 5 V, f_{ck} =100 kHz, T = 1 ms, what will the digital output be for v_{in} = 2.5 V? How could you make the output = 0100000 for v_{in} = 2.5 V?
14. Briefly explain what determines the conversion time for a successive approximation A/D converter. What would be the conversion time for a ten-bit successive approximation A/D converter whose D/A settling time is 2 μs?
15. Briefly explain the advantages and disadvantages of the flash A/D converter.

What would the input capacitance be for a four-bit flash A/D converter if each comparator has a 3-pF capacitance to ground?

16. Sketch a six-bit series-parallel A/D converter block diagram using two three-bit flash A/D converters.

17. What are the RS-232C voltage levels for a 0 and a 1?

18. What are the advantages and disadvantages of serial RS-232C serial data transmission?

19. Sketch a simple op amp circuit to convert standard TTL logic levels to RS-232C levels, and vice versa.

CHAPTER 16

Microprocessors and Microcomputers

16.1 INTRODUCTION

In this chapter we will discuss the microprocessor and its applications and briefly compare it with the microcomputer. We will consider the architecture, programming, languages, speed of operation, and applications to various problems to illustrate the advantages and disadvantages of microprocessors. Finally, we will discuss briefly some of the more common interface buses, such as the S-100 bus and the GPIB (IEEE-488) or general-purpose interface bus.

16.2 THE MICROPROCESSOR VERSUS THE MICROCOMPUTER

Let us consider the general organization or *architecture* of a digital computer. First, there is an *input device* that feeds information into the computer; this input device could be a keyboard terminal producing an ASCII serial digital input or an A/D converter producing a parallel digital input.

Second, there is an *arithmetic logic unit* (ALU) that performs all the various mathematical and logical operations on the input information. The ALU consists of various gates, adders, subtracters, and other circuits.

Third, there is a *control unit* that sends the information to the proper gates. For example, in the case of multiplication the partial products must be sent to the proper terminals of the adder. The control unit also regulates the input and output flows of information.

Fourth, there is a *memory unit* to store both data and instructions. The read-only memory (ROM) stores the monitor program [including input/output (I/O) routines, arithmetic algorithms, etc.] and other instructions (subroutines) that are used often. The random-access memory (RAM) stores both data and instructions. Microprocessor-associated RAM is usually approximately 1 kbyte, whereas microcomputer RAM is usually 64 kbytes or larger. Larger disk memories can, of course, be used with microcomputers, and additional memory chips can be used with microprocessors.

Fifth, there is an *output device*, such as a video display, printer, or plotter to display or convert the output digital information to a form comprehensible to the human operator. In control applications where the computer is actually running apparatus or a machine, the output will be the appropriate analog output for the particular machine used, the output being generated by a D/A converter and appropriate amplification. Much apparatus is now available with digital control inputs, so the digital computer output can be used for control.

The general block diagram of such a digital computer is shown in Fig. 16.1. Generally speaking, the ALU and the control unit are referred to as a

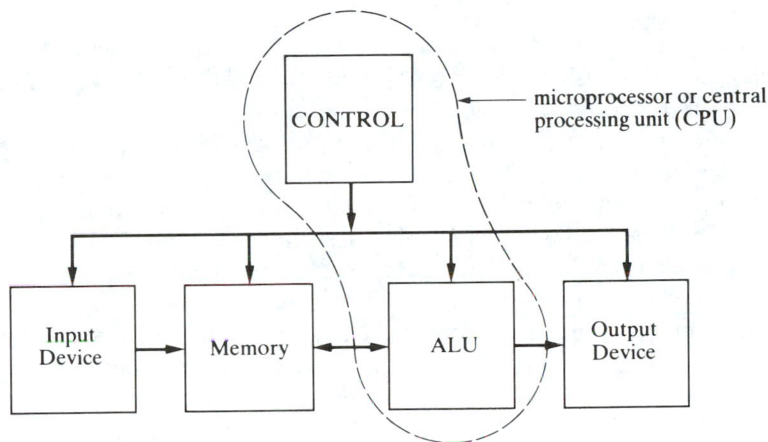

FIGURE 16.1 General digital computer architecture.

microprocessor if they are on the same chip. In any case, whether on one chip or several, they are called the *central processing unit* (CPU). The CPU is the intellectual "heart" or "brains" of the computer and controls all the operations. A computer can thus be thought of as an input device, a CPU, a memory, and an output device, as shown in Fig. 16.2. The CPU generally fetches or retrieves instructions from memory, decodes them, and executes them. A microprocessor is thus a *part* of a computer.

A *microcomputer* is a computer in which the CPU consists of only one (or a few) LSI chips. It typically contains a keyboard, several hundred kbytes of RAM, a video display, and the ability to access one or more small disk memories. Examples are the TRS-80, the Apple, and the IBM personal computer, which have as CPUs the Z-80, the 6502, and the 8086, respectively. They usually cost under $3000 each and can be programmed in a high-level language such as BASIC or FORTRAN.

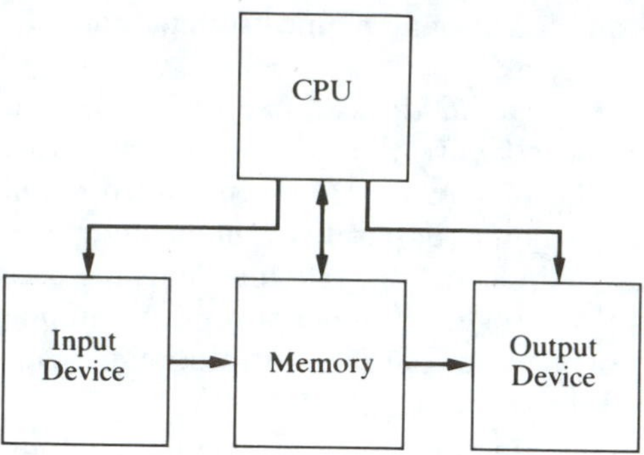

FIGURE 16.2 Computer architecture.

One-chip microcomputers have recently become available. One example is the MC68701, which contains 2 kbytes of EPROM, 124 bytes RAM, a UART, a 6801 CPU, a timer, and a baud-rate generator.

A microprocessor, on the other hand, does not contain the input or output devices or large RAM, and usually must be programmed in a more primitive (but faster in real-time application) lower-level language called *assembly* language, which will be discussed in 16.3. Some one-chip microprocessors, such as the Z-80, are available with BASIC in the chip ROM.

The CPU of a computer can, of course, understand only binary language, but it is clearly impractical in almost all cases to program in binary or machine language. Consider, for example, the memorization and debugging problems involved in writing binary instructions like 001110100101000000100000. These 24 bits mean load the contents of memory address 2050 into the accumulator register for the 8085 microprocessor.

In assembly language the same instruction would be written by the programmer as LDA 2050. The machine cannot understand LDA 2050 because it is not written in terms of 1s and 0s. But LDA 2050 is easy for the programmer to understand and remember—LDA stands for "load accumulator A," and 2050 is the memory address in hexadecimal. Terms such as LDA are often called *mnemonics* because they sound like the instruction expressed in words and thus are easier to remember.

An *assembler* is a fixed computer program on a disk or in ROM that translates the assembly language program, step by step, into binary machine language; for example, the assembler would translate LDA 2050 into 001110100101000000100000.

In some primitive microprocessor systems there is no assembler program in ROM, and the assembly language programs must be hand assembled into machine language, using hexadecimal rather than binary notation. For example, the programmer would look up LDA = $3A_{16}$. Thus the instruction originally written as LDA 2050 would be typed in hexadecimal as 3A, 50, 20 for the 8085. Notice that for the Intel 8085 microprocessor, the lower-order memory address byte (50) is written first. For other microprocessors, such as with the higher-order address byte first, the instruction LDA 2050 would be typed in as 3A, 20, 50. Notice also that two hexadecimal characters represent one byte: for example, $3A_{16} = 00111010$.

In a microcomputer, of course, the programming is usually done in a higher-level language such as FORTRAN or PASCAL. A *compiler* is a fixed computer program on a disk that translates the program instructions such as Z = X*Y into binary machine language: a FORTRAN compiler is used for FORTRAN, a PASCAL compiler for PASCAL.

With the addition of I/O devices and possibly additional memory, a microprocessor system can do many of the functions of a microcomputer at much less hardware cost. However, the price paid is that the programming must usually be done in assembly language, which is much more tedious than using a higher-level language. But programs written in assembly language almost always run much faster than those written in a higher-level language. It should also be noted that recently developed chips can execute FORTRAN and BASIC commands very efficiently, so microprocessor systems can be programmed in higher-level languages.

Typical microprocessor applications are control of instruments or apparatus in which the answer must be calculated immediately (in real time) to control an instrument. Examples are control of automobile ignition systems, washing machine cycles, digital TV tuners, cash registers, machine tools such as lathes or milling machines, analysis of radar or sonar signals, signals from spectrophotometers or NMR spectrometers. Also, microprocessors can be used in programmable calculators, TV games, scientific test equipment, telephone equipment, and aircraft instruments. In other words, whenever some fancy calculation must be made quickly over and over again, the microprocessor is appropriate. Generally speaking, whatever a microprocessor can do, a microcomputer can also do with greater programming versatility but with slower real-time speed and with greater expense. The microprocessor is faster and cheaper but more difficult to program in assembly language. Again, remember that some microprocessors have BASIC and can be programmed in BASIC at the price of slower speed.

A complete microprocessor system containing a microprocessor, several kilobytes of RAM, and buffers can be purchased on a standard bus card from several manufacturers. One such board is the STD Bus 8085A Processor Card 7805 made by Pro-Log. It contains an 8085A microprocessor, 2 kbytes of CMOS RAM, and sockets for another 2 kbytes of RAM and up to 8 kbytes of EPROM or ROM (2716). It comes on a standard bus 56-pin card that can be plugged into a standard female $\frac{1}{2}$-in. card slot. It runs on a single 5-V power supply and typically draws 460 mA (higher current with EPROM). Another, larger, system is made by Cubit and uses the 80186 with 64 kbytes of EPROM and 128 kbytes of RAM.

A collection of hard-wired gates can usually be made to perform like a microprocessor, but, of course, once the gates are hard-wired, their operation is essentially fixed and cannot be modified without rewiring. Thus the gates are immune to software tampering. Such a circuit is sometimes called a *sequencer* in control applications. It is often easier to teach someone to reprogram a microprocessor than to rewire a card without making bad solder joints or wiring mistakes. Thus, we see that the microprocessor system lies somewhere between the microcomputer and a hard-wired collection of gates. As a general rule-of-thumb, if more than two dozen gates are required to perform some operation or calculation, it is probably better to use a microprocessor system rather than a hard-wired circuit. On the other hand, if a great deal of modification involving extensive programming will probably be called for, a microcomputer might be best because of its easy-to-program higher-level languages, provided it is fast enough in real-time calculations. In some cases, when using a microcomputer, only one part of the program slows down the calculation, and to achieve higher speed this part can be written in assembly language, and the rest of the program can be written in a higher-level language.

Table 16.1 summarizes the essential features of hard-wired circuits, microprocessor systems, and microcomputer systems.

TABLE 16.1. Three General Types of Digital Systems

Hard-Wired System	*Microprocessor System*	*Microcomputer*
All hardware, no software changes possible (immune to software tampering)	Easy to change function by changing software in assembly language	Change software in high-level language
Dedicated to one application, usually control	More versatile (many applications), but usually dedicated to one application	General-purpose, very versatile, including data processing
Very fast ($\approx\mu$s) real-time control	Fast ($\approx$100 μs) real-time control	Slower (ms)
Little or no memory	Up to 1 kbyte RAM, specific I/O configuration	Up to 500 kbytes RAM general I/O: can operate printer plotter, video display, etc.

16.2.1 The Input Device

The basic function of the input device is to translate the input information, which usually comes in analog form, into some form of language the CPU can understand, usually binary language. One example is the standard

ASCII-coded keyboard, in which pressing a key generates a standard serial binary word output that can be fed into a serial RS-232 interface. Another example is an A/D converter that converts an analog input voltage (from a transducer, such as a strain gauge or thermometer or electrode) into a parallel binary output. In a microprocessor system the same "port" or set of wires can usually be programmed to be either an input or an output port.

16.2.2 The ALU

The ALU is part of the CPU and performs all the numerical calculations and logical operations. A *register* is a collection of D or RS master/slave flip-flops (eight for an eight-bit machine) in which binary numbers can be stored in parallel fashion, one bit in each flip-flop. The ALU also contains a number of registers for temporary storage. The combination of the adder, subtracter, registers, and other hardware and the proper programming instructions allows the microprocessor system to add, subtract, multiply, divide, shift bits, and perform Boolean logic operations. The ALU also generates and stores certain *flag* bits that specify a number of situations—for example, the presence of a carry bit from addition, a borrow bit from subtraction, the sign or the parity of a number, and whether a number is zero or nonzero. For example, if the result of an operation is not zero, then the zero flag is not set (i.e., it is 0), but if the result is zero, then the zero flag is set (i.e., it is 1). In programming, the presence of flags is essential for branching; for example, a program may jump to a certain subroutine if the zero flag is set. Most, but not all, operations set or alter all these flags.

In most microprocessor systems *all* ALU operations involve one register called register A or simply the *accumulator*. For an eight-bit microprocessor such as the 8085 or the Z-80, the accumulator stores eight bits (or one byte). Parentheses are used to denote the digital contents of a register; thus A means register A or the accumulator, but (A) means the contents of A.

A microprocessor typically contains six or more registers; they are used for temporary storage in most operations: for example, to store the partial products in a multiplication operation, or the minuend and subtrahend in a subtraction operation. Registers are also used as a *memory pointer* to specify a memory address in the RAM; usually, two 8-bit registers are used together to form a 16-bit memory address register. (A 16-bit register can address 64K of RAM because $2^{16} = 65{,}536$.) In the 8085 microprocessor, for example, the two eight-bit registers used for this purpose are called the H (high) and L (low) registers, representing the high and low orders of the memory address. Registers are also used as a *stack pointer* to designate the address of a location in the portion of the memory designated as the stack, where information is temporarily stored.

16.2.3 The Control Device

The control device uses the clock to ensure that the various operations occur in the proper sequence. For example, to calculate the difference (A) − (B), the minuend and subtrahend must first be loaded into registers A and B, respectively, and then the difference must be taken. When a register is loaded, the new contents are usually written directly "over" the old contents of the register, in other words, the old contents are erased. The control device also regulates the INTERRUPT and WAIT operations. The INTERRUPT occurs, for example, when the main program is halted and the JUMP operation switches the program to a subroutine that is executed before the return to the main program. The WAIT operation occurs when a slow I/O device asks the computer to wait until it has data ready to send or is ready to receive.

The control unit contains the *program counter*, which is merely a special register consisting of JK flip-flops and containing the memory location (address) of the next instruction in the program. When one step of the program is executed, the program counter is automatically incremented by 1 to indicate the next instruction of the program.

Branching instructions or jump instructions are often found in programs, and they affect the program counter. If instruction 4 is a JUMP ON ZERO, then if the result (in the accumulator) of instruction 3 is not zero, the program counter advances to instruction 5. If, however, the result of instruction 3 is zero, then the program counter is advanced to the instruction number contained in the JUMP instruction.

The control unit also includes the INTERRUPT routine. If an active logic level is placed on the INTERRUPT lead, the program counter is set to a new value and the program then proceeds along a new branch. The value of the program counter plus the contents of the registers are saved until the new INTERRUPT branch is executed, and the old program is resumed. The part of the memory used for this storage is called the *stack* and is specified by a stack pointer register. The INTERRUPT routine starts by "pushing" the contents of the accumulator and other registers onto the stack. When the INTERRUPT routine is finished, the last few INTERRUPT routine instructions "pop" the saved accumulator and register contents back from the stack into the registers. This process is called *popping* data back from the stack to the registers. The original contents of the program counter are also popped back from the stack into the program counter, and then the program proceeds from where it was just before the INTERRUPT.

There are usually two kinds of INTERRUPTS: *maskable* and *nonmaskable*. The nonmaskable INTERRUPT occurs whenever the INTERRUPT line is made active, and the program executes the INTERRUPT subroutine and then returns to the main program. But maskable INTERRUPTS can be controlled by the program—an instruction called the

INTERRUPT DISABLE causes the microprocessor to ignore any maskable INTERRUPT request, and the instruction INTERRUPT ENABLE causes the microprocessor to execute the INTERRUPT request. There are usually different external leads for the maskable and the nonmaskable INTERRUPTS.

16.2.4 The Memory Unit

A microprocessor usually contains ROM but little or no RAM; additional ROM and RAM can be added with additional chips. The typical general-purpose microprocessor contains anywhere from 6 to 32 registers, including at least one accumulator register through which *all* arithmetic operations have to flow. In addition, external memory can be addressed by a microprocessor. Eight bits can address $2^8 = 256$ memory locations, and the typical microprocessor has a 16-bit memory address, so $2^{16} \cong 64K$ memory locations can be addressed. The 8086 16-bit microprocessor can use a 20-bit memory address, so $2^{20} = 1.024MB$ locations can be addressed.

The CPU ROM contains the often-used instructions for the various arithmetic and logical operations and cannot, of course, be changed by the user. Typical chip ROM instructions display data on seven-segment LEDs,

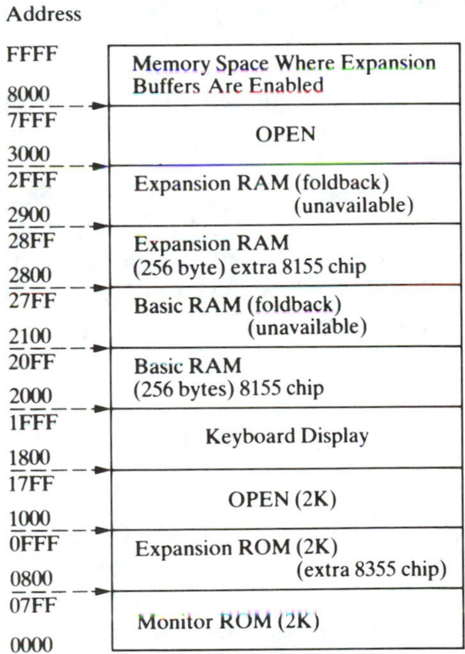

FIGURE 16.3 SDK-85 memory map (8085 microprocessor kit).

input and output data to and from the RAM, CPU registers, and other locations. Separate EPROM chips can usually be used if desired.

The RAM can contain either data or the user-written program. However, there must be a specific way for the microprocessor to distinguish between data and instructions in its memory. In the 8085, for example, pressing the GO key means the following address will contain instructions to be executed and not data.

Figure 16.3 shows the memory map of the 8085 microprocessor. Notice both ROM and RAM are shown in the map. The 256 bytes of RAM are located from addresses 2000 to 20FF, and data or instructions can be entered by the user in this section of memory. If the user attempts to enter data at a memory address in the ROM, the LED display will usually display an ERROR message.

16.2.5 Specific Microprocessors

Microprocessors basically come in two general types: control-oriented and data-processing-oriented microprocessors. *Control-oriented* microprocessors generally have fewer bits, less memory, and limited capability for arithmetic operations but are easily interfaced with a variety of I/O devices. *Data-processing-oriented* microprocessors, on the other hand, tend to have more bits and a much more versatile instruction set, including many arithmetic operations and twos complement arithmetic. Some of the 16-bit microprocessors have multiply and divide instructions, but few of the 8-bit microprocessors do. (The MC68701, MC6801, and 8051 have 8-bit × 8-bit multiply instructions.) The newer microprocessors usually require only one supply voltage (usually 5 V) and dissipate approximately 1 W. But the HD630X family series of CMOS microprocessors dissipate less than 20 mW.

Some of the more common microprocessors are listed in Table 16.2.

Figure 16.4(a) shows the architecture of the 8085 A microprocessor. A typical microprocessor is contained in a 40-pin dual-in-line package. The pin diagram for the Intel 8085 A is shown in Fig. 16.4(b). Notice that only one +5-V power supply is needed.

16.3 MICROPROCESSOR INSTRUCTIONS AND PROGRAMMING

Microprocessors are generally programmed in assembly language rather than a higher-level language such as BASIC or FORTRAN. The following treatment will use the Intel SDK-85 microprocessor kit which contains an 8085, support chips, a hex keypad, and seven-segment LED readout. In any language the programmer's introductions must eventually be translated into binary machine language before the computer can run the program. This

TABLE 16.2. Some Microprocessors

Microprocessor	*Word Size (bits)*	*Application*	*Usable RAM (bytes)*	*Comments*
TMS 1000	4	Control, calculations	124 nibbles	Inexpensive, no external ROM or RAM
Intel 8080	8	Control, some EDP*	64K	1973, many support chips, widely used GP†
Intel 8085	8	Control/EDP	64K	Improved 8080, built-in clock and controller, single 5-V supply, GP
Zilog Z-80	8	Control/EDP	64K	Improved 8080, more registers, single 5-V supply, used in TRS-80, GP
Intel 80C31	8	Control/EDP	128K	Low-power CMOS, single 5-V supply, fast 1-μs/cycle
Intel 8086	16	Control/EDP	1024K	Much-improved 8080, both 8-bit and 16-bit multiply, single 5-V supply, external clock, used in IBM PC
Zilog Z8000	16	Control/EDP	64K up to 48MB	Improved Z-80, multiply and divide, instructions *different* from those of Z-80
6800	8	Control/EDP	64K	Similar to PDP-11, GP
6809	8	Control/EDP	64K	Improved 6800, multiply and divide
6502	8	Control/EDP	64K	Popular GP, used in KIM, Apple
68000	16	Control/EDP	16.8MB	Used in Macintosh

* EDP = electronic data processing.

†GP = general purpose.

translation into machine language is usually done by the computer using either an assembler or a compiler. For example, if the program is written in BASIC, the computer must contain a BASIC assembler that translates the BASIC program into machine language in a line-by-line sense. In FORTRAN the FORTRAN compiler translates the FORTRAN program into machine language. The assembler or compiler is present in some form of ROM or disk. The price paid for the higher-level language is slower speed and the additional memory space required for the assembler or compiler.

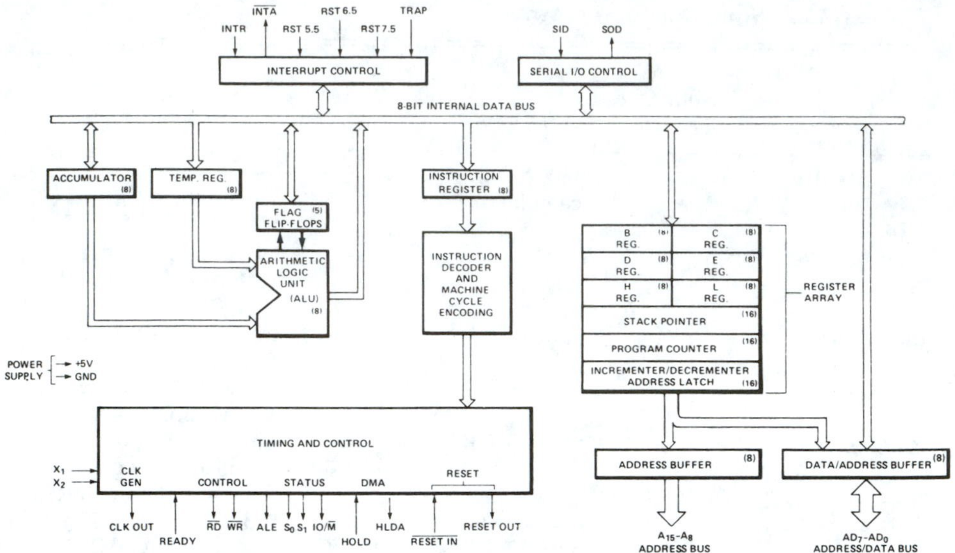

(a) *functional block diagram*

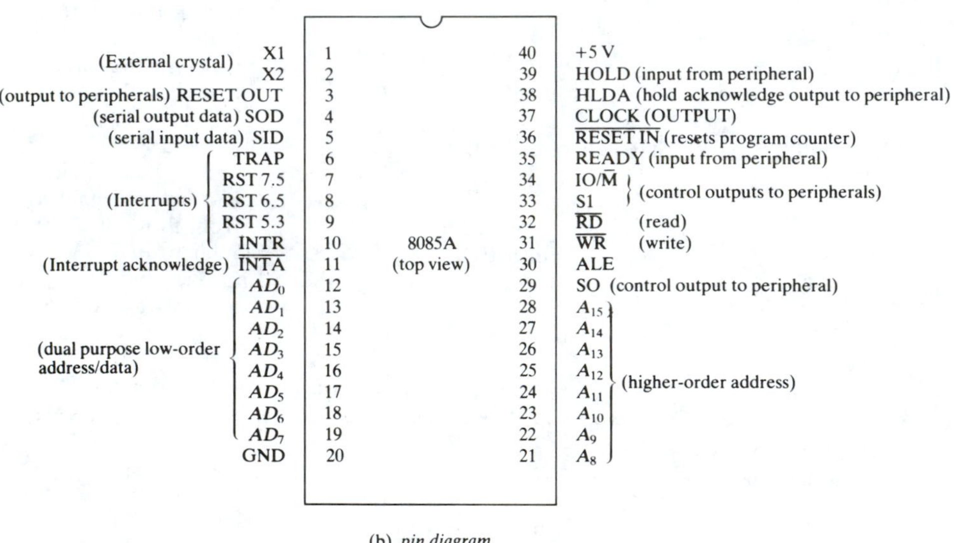

(b) *pin diagram*

FIGURE 16.4 The 8085A microprocessor. (Courtesy Intel Corporation)

Generally, microprocessor systems (as opposed to microcomputers) have limited memories, and thus higher-level languages are not used.

In assembly language the programmer writes short alphanumeric commands called mnemonics. One example would be LDAM, which stands for load accumulator A with the contents of memory M. Also, in assembly language, variables can be referred to by six alphanumeric characters instead of by memory location.

In a high-level language the programmer simply writes an alphanumeric expression or formula. In FORTRAN, for example, the expression $b = (x + y)z$ would be evaluated for $x = 5$, $y = 6$, $z = 9$ by the program

```
X = 5
Y = 6
Z = 9
B = (X + Y)Z
STOP
END
```

The contrast between assembly language and a higher-level language can be simply summarized by saying that assembly language is executed faster than a higher-level language but is harder to program.

When assembly language is compiled, it produces no error messages to help in debugging a program. It is also much more concrete and less abstract than a higher-level language in that its instructions actually specify the physical place (the particular register) where bytes are stored and the actual numerical addresses of data in RAM.

16.3.1 Sample Addition Program

To be specific, let us consider the simple program of adding two numbers X and Y in assembly language for the 8085 microprocessor. The available RAM in the 8085 is from addresses 2000 to 20FF, with all addresses in hexadecimal. The simplest way to perform this addition requires an 11-byte assembly language program. The two numbers to be added must first be entered into two memory locations in the available RAM. We will assume X has already been entered at 2050 and Y at 2051; hence, (2050) = X, (2051) = Y. [Remember that (M) means the *contents* of memory location M.] Let us place the sum at 2052. Notice that X + Y means X "plus" Y not X "OR" Y. The program is the following:

```
LDA 2050    ; load X = (2050) into A
LXIH 2051   ; load 2051 into the HL register pair; that is, (H) = 20 and
              (L) = 51. This "points" HL at Y. Note (2051) = Y.
ADD M       ; add the contents of A to Y, which is at the address stored
              in HL, and place the results in A. Thus (A) = X + Y
STA 2052    ; store contents of A at 2052. Thus (2052) = X + Y
RST         ; reset (return to monitor program)
```

The actual machine language program would be the hex numbers or op codes for the various steps like LDA 2050. The complete program is given in Program 16.1.

$address_{16}$	mnemonic	op $code_{16}$	documentation
2000	LDA,2050	3A	;load X = (2050) into A
2001		50	;
2002		20	;
2003	LXIH2051	21	;load 2051 into HL "point" HL at Y
2004		51	;Y = (2051)
2005		20	;
2006	ADD M	86	;add (A) to Y, which is at (HL), sum in A
2007	STA 2052	32	;store (A) at 2052
2008		52	;
2009		20	;
200A	RST	CF	;return to monitor program

PROGRAM 16.1 8085 addition program: (2050) plus (2051) = (2052).

An assembly language program generally has four parts or *fields*: (1) the label, (2) the mnemonic, (3) the operand, and (4) the documentation or comments. The *label* is simply a word chosen to be meaningful to the programmer; it is ignored by the microprocessor unless a JUMP instruction in the program jumps the program back to that point. In assembly language the actual address must be used, e.g., JUMP to 204A. Usually only important program steps are labeled. The *mnemonic* is a descriptive abbreviation of the program step; there is one menomic for each program step. Each microprocessor has its own set of copyrighted mnemonics, although there is considerable similarity among eight-bit microprocessors. For example, in Program 16.1 LDA is the mnemonic standing for "load accumulator A," and LXIH stands for "load the HL register pair" with the two bytes immediately following: 2051 in Program 16.1. The *operand* is the data or raw material necessary for the operation specified by the mnemonic. For example, in Program 16.1, 2050 is the operand for the LDA operation; 2050 is the 16-bit address (two bytes) of the number loaded into the accumulator. Similarly, 2051 is the operand for the LXIH operation; 20 is loaded into the H register, and 51 into the L register. The mnemonic LXIH is slightly misleading because it loads *both* H and L registers—it might better have been called LXIHL. The *documentation* is simply an explanation of the program step for the programmer in simple English. It is written to the right of a semicolon, and the microprocessor ignores anything after a semicolon. The documentation is absolutely essential to enable other human beings to understand the program, and even the original programmer may need documentation to understand the program several months after writing it. A corollary of Murphy's law is that no program ever has clear enough documentation.

The program written depends on the instructions for the particular microprocessor used. Within a family of microprocessors made by one company, there is often a common subset of instructions that will work on all microprocessors in that family. For example, all the Intel 8080 instructions will work on the Intel 8085, and all the 8085 instructions will work on the 8086. And, all the 8080 instructions will work on the Zilog Z-80, but the Z-80 and the Z8000 instructions are almost completely incompatible.

In the SDK-85 microcomputer, which is based on the 8085 microprocessor, the programmer first writes the program on paper, using appropriate mnemonics. But the microprocessor can understand only binary language. The programmer must look up the hexadecimal op code for each mnemonic and hand key the op code on the keypad of the SDK-85. The SDK-85 then internally converts the hex op code into binary language. For example, the programmer writes LDA as a program step, looks up the op code, and finds it to be 3A. The hex number 3A must then be keyed in by hand. This procedure, which must be repeated for each mnemonic of the program, is called *hand assembly* of the program. Hand assembly is obviously a slow and tedious process. It is possible to get a software package to translate directly from the mnemonic to the binary version—that is, to "assemble" the program. But such an assembler typically requires 10 kbytes of memory and is not normally a part of a small microprocessor system such as the SDK-85. An assembler is obviously part of larger systems, such as personal computers. And a simple BASIC assembler is even available in some of the newer microprocessors.

Let's return to Program 16.1. In the 8085, 3A is machine language (in hex) for LDA, 21 is LXIH, 86 is ADD M, 32 is STA, and CF is RST. The first step, LDA 2050, loads the contents of memory address or location 2050 into the accumulator A. It is a three-byte instruction; the first byte is the load command, the second byte is the low-order memory address 50_{16}, and the third byte is the high-order memory address 20_{16}. Notice that one byte is specified by two hexadecimal numbers, and that the lower-order address is specified before the higher-order address. In many microprocessors the high-order address byte comes first, and the low-order address byte comes second, so be careful. In the Intel 8085, LDA 2178 would load the contents of 2178 into the accumulator and would be entered

LDA
78
21

In a microprocessor with the higher-order memory address first (e.g., the

Motorola 6800 family), the instruction LDA 2178 would be entered

LDA
21
78

The second step in Program 16.1 loads the two-byte number (16 bits) 2051 into the HL register pair. Notice that only the high register symbol H appears in this mnemonic; the L is implied. After this step (HL) = 2051. This is also a three-byte instruction.

The third step adds the contents of the accumulator A to the contents of memory, whose address is the contents of the HL register pair. Since (2051) = Y, this instruction adds the contents of A to Y and places the sum in the accumulator. The previous number in the accumulator is "lost" or written over by the sum. This is a one-byte instruction.

The fourth step transfers or stores the contents of the accumulator into the memory at address 2052. This is a three-byte instruction.

The fifth and final step returns to the monitor so that a new program can be executed, or new data entered, and so on. This step essentially tells the computer that the program is over; if this were not done, the computer would try to execute the instruction contained at the next location in memory (i.e., at 200A), then the instruction at the next location (200B), and so on, and chaos would probably result.

Newer microprocessors operate "in parallel"—on several instructions simultaneously, thus increasing the speed.

A similar 8085 program to add (2050) to (2051) and store the results at 2052 is given in Program 16.2.

no. states	$address_{16}$	mnemonic	documentation
10	2000,2001,2002	LXIH 2050	;set (HL) = 2050
7	2003	MOV A,M	;load contents of memory at address = (HL) into A; i.e., load (2050) into A
6	2004	INX H	;increment (HL); i.e., set (HL) = 2051
7	2005	ADD M	;add contents of memory at address = (HL) to A & place sum in A, i.e., add (2051) to (A)
6	2006	INX H	;increment (HL); i.e., set (HL) = 2052
7	2007	MOV M,A	;move contents of A to memory location 2052
12	2008	RST	;reset

55 total × 322 ns = 17.710 ns = 17.71 μs

PROGRAM 16.2 8085 addition program: (2050) plus (2051) = (2052).

In this program the HL register is used to "point" at various RAM memory locations. The first step points HL at 2050; the MOV A,M step moves the contents of the memory at (HL) into the accumulator. Notice that the information flow is from *right to left* in the MOV A,M instruction, i.e., from the memory M to the accumulator A.

The hex "op code" for each mnemonic must be looked up (hand assembled) using the 8085 instruction set. The SDK-85 kit contains a chip to convert the hex op code into binary code which the 8085 can "understand."

This program occupies only nine bytes of memory: 2000 through 2008. The actual machine language program is

address_{16}	mnemonic	op code (hex)
2000	LXIH	21
2001		50
2002		20
2003	MOV A,M	7E
2004	INX H	23
2005	ADD M	86
2006	INX H	23
2007	MOV M,A	77
2008	RST	CF

Let us now briefly consider the speed of the preceding two programs. The number of states or clock cycles is listed in the 8085 user's manual for each instruction. For both Program 16.1 and Program 16.2 the total number of states is 55. With a 3.11-MHz clock frequency the clock period is 322 ns, and each program therefore takes 55×322 ns $= 17.71$ μs to run.

If the two numbers were added using FORTRAN, the program would be simply

$$B = X + Y$$

and would take hundreds of μs to run, depending on the data. Thus the assembly language approach, although tedious to write, is faster than the higher-level-language approach, and thus is preferred for fast real-time control or data processing.

16.3.2 Addressing Modes

There are five different methods of getting data into the accumulator for the 8085 (and for most eight-bit microprocessors): these methods are called the *addressing modes* of the microprocessor.

1. *Immediate* addressing means that the data or number is contained in the instruction itself (i.e., in the program). LXIH 2051 is an example: the

(hex) number 2051 is loaded into the HL register pair. In other words, the data immediately follow the instruction right in the program.

2. *Direct* addressing means that the addresses of the data needed are actually part of the instruction. For example, LDA 2050 loads the data at address 2050 into the accumulator. Notice that the correct data must already have been loaded into memory location 2050. Also, three bytes are required for each instruction—one for the mnemonic LDA and two more for the 16-bit address. Direct addressing tends to be somewhat slow.

3. *Register* addressing means that the data are already in one of the registers. For example, ADD C means to add the number already in register C to the number in register A and place the sum in A. The number in C is not affected. Register addressing is generally very rapid because no fetch from memory is involved.

4. *Register indirect* addressing means that the data are already in the memory, and a register pair (usually the HL pair) is used to point at the memory. In other words, the content of the HL pair specifies the address of the data. For example, the operation MOV A,M means to move the data from memory to the accumulator A with the address of the data in memory being the contents of the HL register pair. The H register contains the high-order byte and the L register the low-order byte of the memory address. The HL pair is used in a sense as a memory pointer. Notice no address is necessary in the instruction itself; the address is already in the HL pair.

5. *Implied* addressing means that the instruction itself implies the location of the data. For example, STC (SET CARRY FLAG) means only the carry flag is used.

Table 16.3 summarizes the data location for the various addressing modes.

TABLE 16.3. 8085 Addressing Modes

Addressing Mode	*Data Location*	*Comments*
Immediate	In instruction itself	Fast
Direct	In memory, address in instruction	Slow
Register	In register specified in instruction	Fast
Register indirect	In memory, address in HL pair	Slow
Implied	Implied by instruction	

16.4 THE 8085 MICROPROCESSOR INSTRUCTION SET

Programs 16.1 and 16.2 have illustrated a few of the 100 instructions for the 8085. All the instructions comprise the *instruction set*. The 8085 instruction set is given in Appendix A. The instructions fall into five broad areas: (1) data transfer, (2) arithmetic, (3) logic, (4) branching group, and (5) stack, input/output, and control.

1. *Data transfer* These instructions move data between registers or from memory to registers, and vice versa. One example is MOV A,M, which moves the contents of the memory location whose address is in the HL register pair to the accumulator. Another example is MVI A,3E, which moves the actual number 3E into the accumulator. The flags are not set or altered by these instructions.
2. *Arithmetic* These instructions perform arithmetic operations such as addition and subtraction on data in registers or memory. The Zero, Carry, Sign, and Parity flags are usually all set by these instructions. One example is ADD M, which adds the contents of the memory location whose address is in the HL register pair to the number already in the accumulator; the sum appears in the accumulator.
3. *Logic* These instructions perform Boolean logical operations on data in registers and memory and on flags. They usually set the flags. Examples are ANA M, which ANDs the contents of the accumulator A with the contents of the memory whose address is in the HL register pair, and RLC, which rotates the contents of the accumulator one bit to the left. The MSB becomes the new Carry flag bit, and the new LSB is the old MSB.
4. *Branching* (*Jumping*) These instructions allow the computer to make logical decisions; they do not set any flags. There are two kinds of branches; unconditional and conditional. An *unconditional* branch simply changes the program counter. For example, JMP 2064 jumps to address 2064 regardless of the flags. A *conditional* branch changes the program counter only if some condition is met (e.g., if the Zero flag is set). JZ jumps when the Zero flag is set. For example, the instruction JZ 2041 at addresses 2010, 2011, and 2012 jumps to address 2041 if the Zero flag is set; if the Zero flag is not set, the next instruction at 2013 is executed.
5. *Stack, Input/Output, and Control* These instructions perform I/O operations, push data from a register into the stack, pop data from the stack back to a register, and control the interrupts. This group includes the "no operation" (NOP) instruction (whose op code is 00_{16}); this is useful to "waste" time or to delay an operation. Each NOP instruction lasts four clock periods or 1.3 μs in the 8085 on the SDK-85.

This is merely a brief overview of 8085 programming; the 8085 User's Manual or Programming Manual should be consulted for the complete instruction set.

Useful references for assembly languages are given at the end of the book; the book by Leventhal and Walsh is particularly recommended for the SDK-85 kit and the 8085 microprocessor.

16.5 WAVEFORM GENERATION

The microprocessor-based computer can be used to generate almost any type of analog voltage waveform, and the waveform can be changed by changing the programming or software without making hardware changes. This feature is, of course, a considerable advantage in many applications.

We will now consider a number of specific waveforms to illustrate the general techniques used.

In most microprocessor systems one hardware port (e.g., eight terminals for an eight-bit parallel port) can be programmed to be either input or output. Such ports are referred to as parallel I/O ports.

Thus, before we even choose the waveform to be generated, we must "initialize" the I/O ports. That is, a particular I/O port must be programmed to be either an input or an output port. With the SDK-85 microcomputer, for example, the two I/O ports are in the 8355 support chip and are labeled port 0 and port 1. Whether a port is an input or an output depends on the value (either 0 or 1) of that port's data direction register (DDR) according to the following table:

Contents of DDR	*Type of Port*
0	input
1	output

Port 2 is the DDR for port 0, and port 3 is the DDR for port 1. The command OUT 02 transfers the contents of the accumulator to port 2, and OUT 03 transfers the contents of the accumulator to port 3. Table 16.4 shows assembly language programs for the four possible combinations.

TABLE 16.4. Initialization of I/O Ports

Port 0 Output	*Port 0 Input*	*Port 1 Output*	*Port 1 Input*
MVI A,FF OUT 02	MVI A,00 OUT 02	MVI A,FF OUT 03	MVI A,00 OUT 03

MVI A,FF sets all eight bits of the accumulator equal to 1, and OUT 02 transfers these 1 bits to port 02, thus making all eight bits of port 0 an output. (Ports 0 and 1 are automatically set to be input ports in the SDK-85 unless otherwise specified.)

Individual bits of the two I/O ports can also be set to be either input or output. MVI A,03 (03_{16} = 00000011) and OUT 02 would make pins 0 and 1 outputs and pins 2 through 7 inputs for port 0, and MVI A,07 (07_{16} = 00000111) and OUT 02 would make pins 0, 1, and 2 outputs, and pins 3 through 7 inputs.

Program 16.3 outputs the eight-bit number n to port 0.

$address_{16}$	mnemonic	op $code_{16}$	documentation
2000	MVIA,FF	3E	;set (A) = 11111111
2001		FF	;
2002	OUT,02	D3	;make all 8 bits of port 0 outputs
2003		02	;
2004	MVI,n	3E	;set (A) = n
2005		n	;n = 1 byte
2006	OUT,00	D3	;output (A) = n to port 0
2007		00	;
2008	Reset	CF	;return to monitor

PROGRAM 16.3 Output the number n to port 0.

For example, if n = 80_{16} = 10000000, pins 0 through 6 of port 0 will be low and pin 7 will be high. The voltage levels on port 0 can be checked with a voltmeter, or LEDs and resistors can be connected to the pins of port 0.

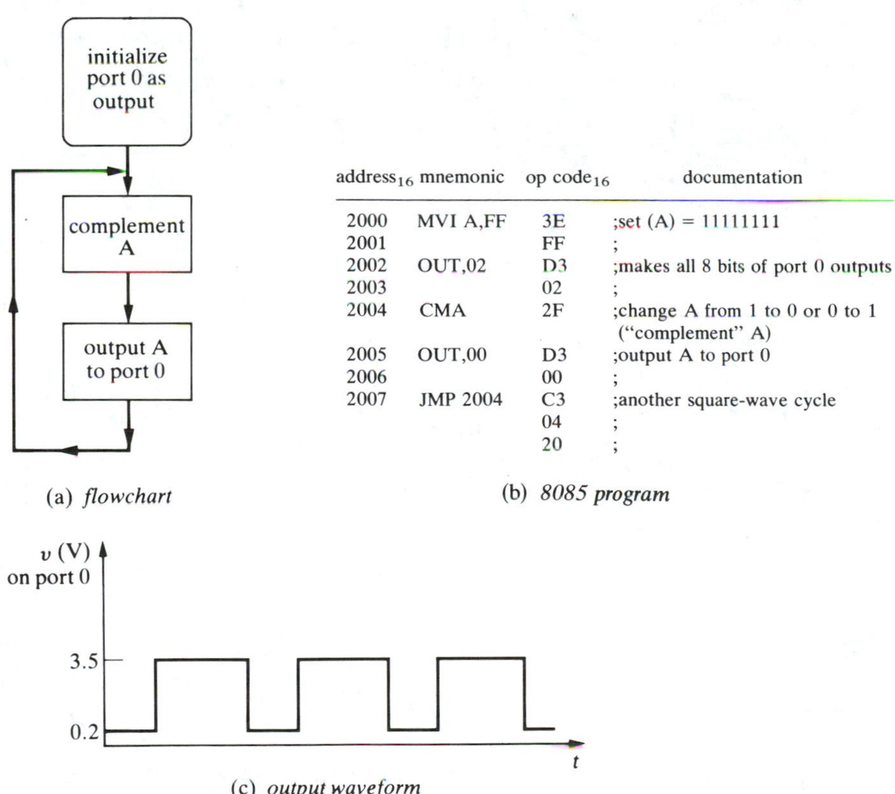

$address_{16}$	mnemonic	op $code_{16}$	documentation
2000	MVI A,FF	3E	;set (A) = 11111111
2001		FF	;
2002	OUT,02	D3	;makes all 8 bits of port 0 outputs
2003		02	;
2004	CMA	2F	;change A from 1 to 0 or 0 to 1 ("complement" A)
2005	OUT,00	D3	;output A to port 0
2006		00	;
2007	JMP 2004	C3	;another square-wave cycle
		04	;
		20	;

(a) *flowchart* (b) *8085 program*

(c) *output waveform*

FIGURE 16.5 Fast 8085 square-wave program.

Consider the problem of generating a repeating square wave—in other words, set an output port alternately high (3.5 V) and low (0.2 V). A simple 8085 program to do this on all eight pins of port 0 is given in Fig. 16.5(b).

When the program is running, an "E" is shown on the SDK-85 LED display. This is the fastest square wave that can be generated by such a program. For the 8085 the period of the fastest square wave is approximately 24 μs high and 14 μs low.

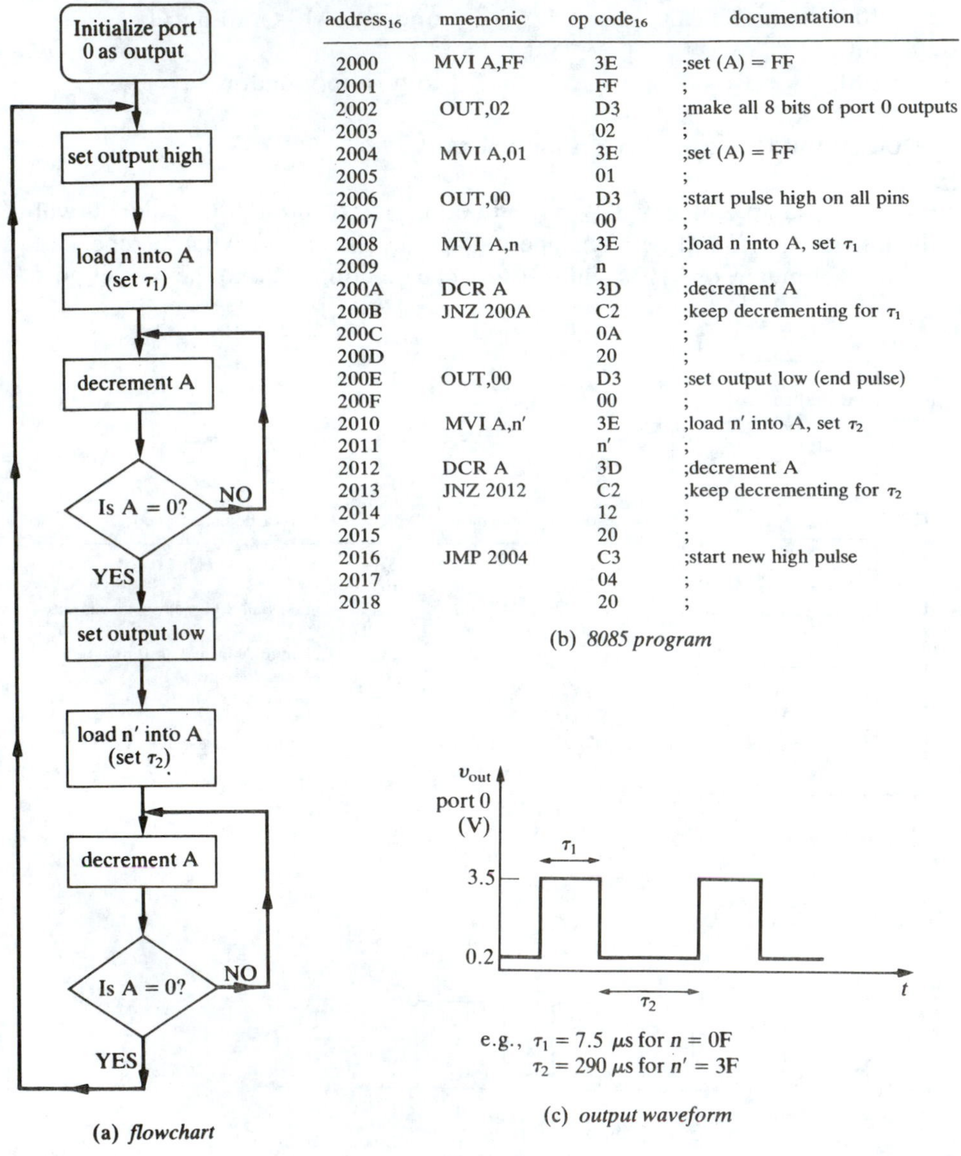

address$_{16}$	mnemonic	op code$_{16}$	documentation
2000	MVI A,FF	3E	;set (A) = FF
2001		FF	;
2002	OUT,02	D3	;make all 8 bits of port 0 outputs
2003		02	;
2004	MVI A,01	3E	;set (A) = FF
2005		01	;
2006	OUT,00	D3	;start pulse high on all pins
2007		00	;
2008	MVI A,n	3E	;load n into A, set τ_1
2009		n	;
200A	DCR A	3D	;decrement A
200B	JNZ 200A	C2	;keep decrementing for τ_1
200C		0A	;
200D		20	;
200E	OUT,00	D3	;set output low (end pulse)
200F		00	;
2010	MVI A,n′	3E	;load n′ into A, set τ_2
2011		n′	;
2012	DCR A	3D	;decrement A
2013	JNZ 2012	C2	;keep decrementing for τ_2
2014		12	;
2015		20	;
2016	JMP 2004	C3	;start new high pulse
2017		04	;
2018		20	;

(b) *8085 program*

FIGURE 16.6 8085 square-wave program (with variable high and low pulse length).

A square-wave output with variable duty cycle can be produced by waiting a time τ_1 with a high output and a time τ_2 with a low output and then beginning the program again. A simple square-wave program for the 8085 program is given in Fig. 16.6(b). The basic idea is first to set the output high. Then a number n is loaded into A and decremented until A = 0. (To decrement A means to subtract 1 from A.) Then the output is set low. The time the output is high (τ_1) thus depends on n. A similar technique determines τ_2.

The length of the output pulse can be also increased by using a 16-bit register instead of the single 8-bit register A. In such case a 16-bit number *N* would be loaded into a register pair consisting of two 8-bit registers and then decremented as a pair.

An 8085 one-shot program is given in Fig. 16.7(b). This outputs a

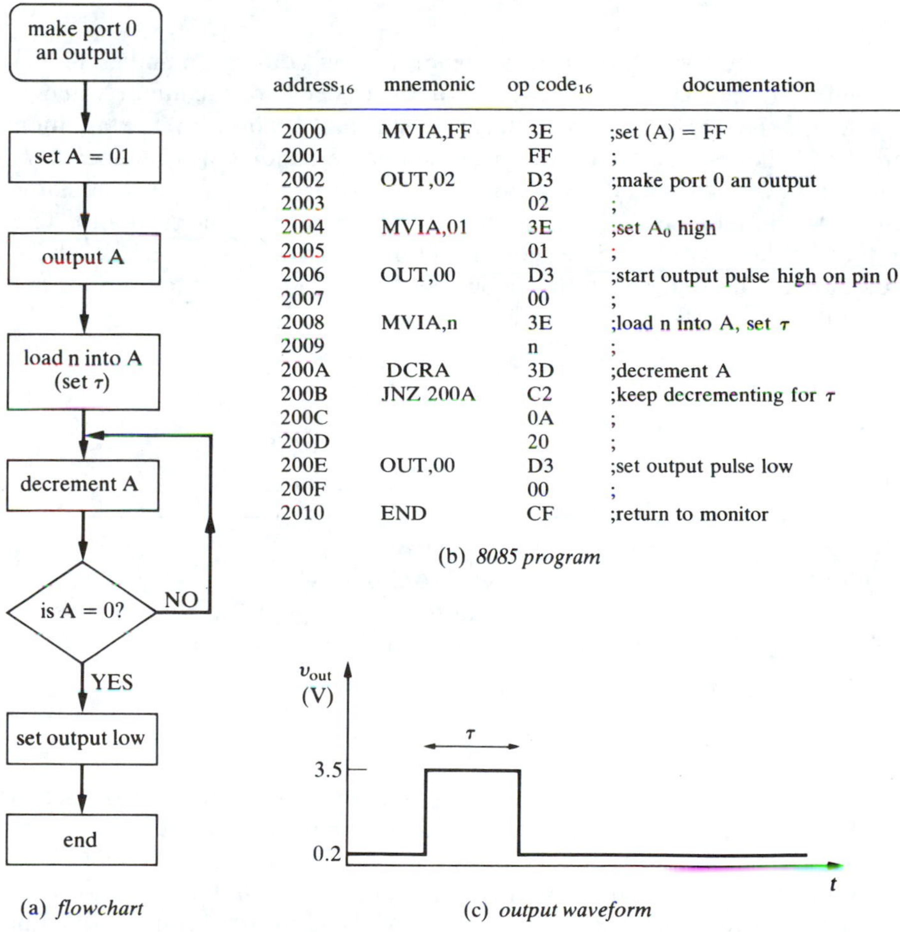

address$_{16}$	mnemonic	op code$_{16}$	documentation
2000	MVIA,FF	3E	;set (A) = FF
2001		FF	;
2002	OUT,02	D3	;make port 0 an output
2003		02	;
2004	MVIA,01	3E	;set A_0 high
2005		01	;
2006	OUT,00	D3	;start output pulse high on pin 0
2007		00	;
2008	MVIA,n	3E	;load n into A, set τ
2009		n	;
200A	DCRA	3D	;decrement A
200B	JNZ 200A	C2	;keep decrementing for τ
200C		0A	;
200D		20	;
200E	OUT,00	D3	;set output pulse low
200F		00	;
2010	END	CF	;return to monitor

(a) *flowchart* (b) *8085 program* (c) *output waveform*

FIGURE 16.7 8085 one-shot program.

single pulse of duration τ on pin 0 of port 0. τ is determined by the number n in A which is decremented.

A one-shot program for the Z-80 is Program 16.4.

```
        LDB,n           ;load integer n into register B
        LDA,1           ;set accumulator A = 1
        OUT (PORT),A    ;set port = A = 1, out pulse starts
HERE:   DJNZ HERE       ;decrement B and jump to HERE if
                         B ≠ 0, next step when B = 0
        XORA            ;set accumulator A = 0
        OUT (PORT),A    ;set port = A = 0 (end of pulse)
```

PROGRAM 16.4 Z-80 one-shot program.

A sawtooth waveform can be generated by adding a number to the contents of the accumulator register until a specified maximum value is reached, then zeroing the register and starting again with the addition process. When A reaches FF (its maximum value), adding any number to A will set A = 0; that is, the sawtooth will return to zero. The accumulator contents will then be a binary stepwise approximation to the sawtooth. The output of the register can be fed into the input of a D/A converter whose output is the desired sawtooth analog waveform. The 8085 program for the sawtooth waveform is given in Fig. 16.8(b).

The Z-80 program for the sawtooth waveform is given in Program 16.5.

label	operation	documentation
	XOR A	;set A = 0
LOOP:	OUT (DAC),A	;set DAC port = A
	ADD A,n	;add n to A (A = 0 if overflow occurs)
	JP LOOP	;jump to LOOP

PROGRAM 16.5 Z-80 sawtooth program.

A triangular waveform can be generated by incrementing the contents of the accumulator register until a specified maximum value is reached, decrementing the register until zero is reached, and then starting over. The output of the register can be fed into a D/A converter and filtered to obtain a smooth sawtooth if desired. The triangular waveform program is given in Fig. 16.9(b).

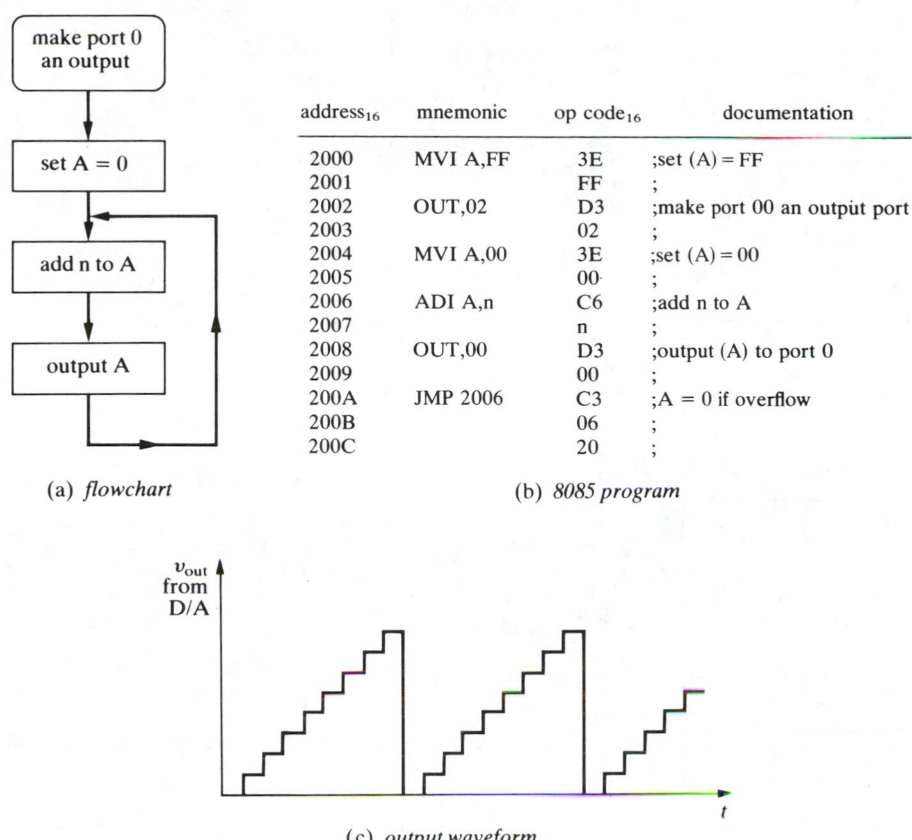

(a) *flowchart*

address$_{16}$	mnemonic	op code$_{16}$	documentation
2000	MVI A,FF	3E	;set (A) = FF
2001		FF	;
2002	OUT,02	D3	;make port 00 an output port
2003		02	;
2004	MVI A,00	3E	;set (A) = 00
2005		00	;
2006	ADI A,n	C6	;add n to A
2007		n	;
2008	OUT,00	D3	;output (A) to port 0
2009		00	;
200A	JMP 2006	C3	;A = 0 if overflow
200B		06	;
200C		20	;

(b) *8085 program*

(c) *output waveform*

FIGURE 16.8 8085 sawtooth program.

A Z-80 program is given in Program 16.6.

label	mnemonic	documentation
	XOR A	;set A = 0
L1:	OUT(DAC), A	;set DAC port = A
	LDB,A	;set B = A = 0
	INC A	;increment A (until A = 255)
	DJNZ L1	;decrement B and jump to L1 if B $\neq$ 0, next step when B = 0
	LDA, FF$_{16}$	;load A to 255$_{10}$
L2:	OUT(DAC), A	;set DAC port = A
	DEC A	;decrement A
	DJNZ L2	;decrement B

PROGRAM 16.6 Z-80 triangular waveform program.

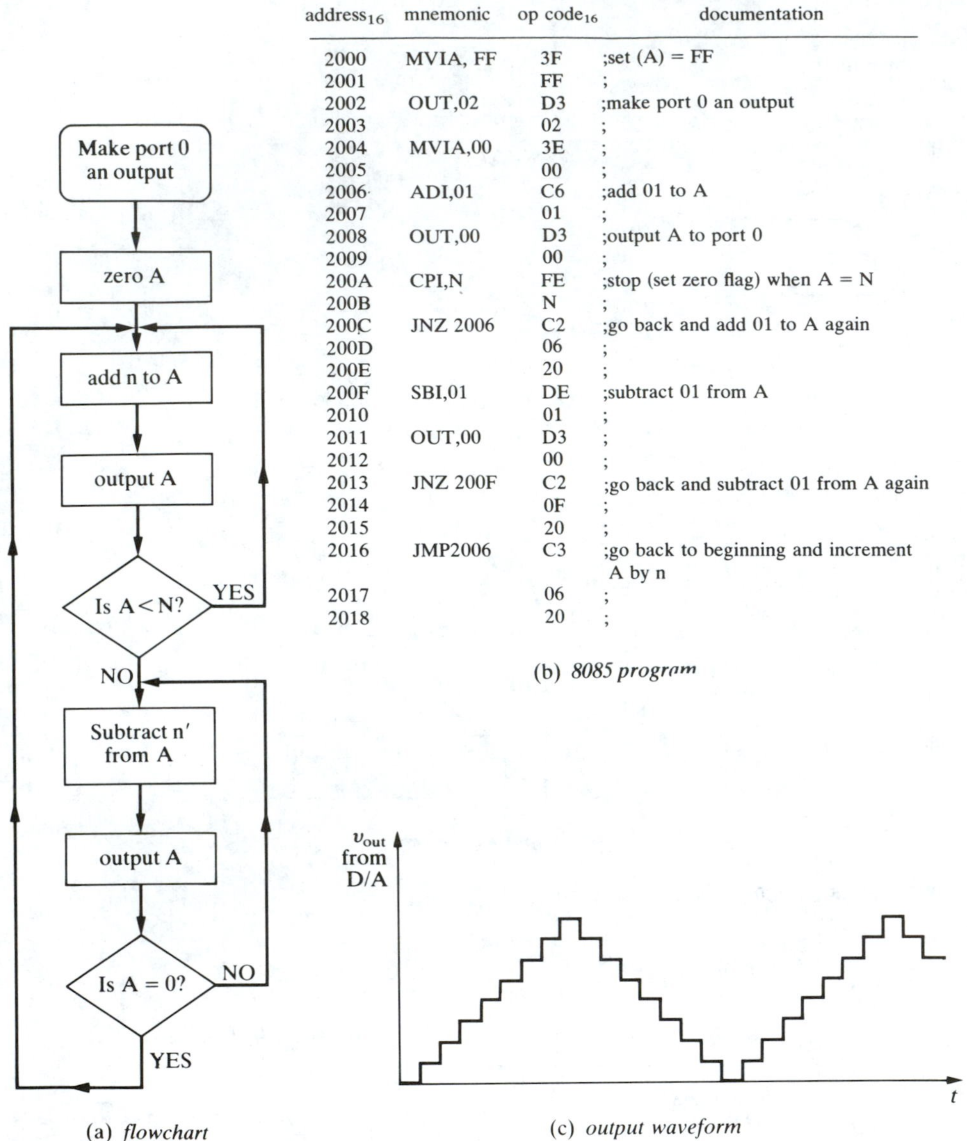

address$_{16}$	mnemonic	op code$_{16}$	documentation
2000	MVIA, FF	3F	;set (A) = FF
2001		FF	;
2002	OUT,02	D3	;make port 0 an output
2003		02	;
2004	MVIA,00	3E	;
2005		00	;
2006	ADI,01	C6	;add 01 to A
2007		01	;
2008	OUT,00	D3	;output A to port 0
2009		00	;
200A	CPI,N	FE	;stop (set zero flag) when A = N
200B		N	;
200C	JNZ 2006	C2	;go back and add 01 to A again
200D		06	;
200E		20	;
200F	SBI,01	DE	;subtract 01 from A
2010		01	;
2011	OUT,00	D3	;
2012		00	;
2013	JNZ 200F	C2	;go back and subtract 01 from A again
2014		0F	;
2015		20	;
2016	JMP2006	C3	;go back to beginning and increment A by n
2017		06	;
2018		20	;

(a) *flowchart* (b) *8085 program* (c) *output waveform*

FIGURE 16.9 8085 triangular waveform program.

An arbitrary waveform can be generated by storing the waveform amplitude values (as binary numbers) in a read/write memory (RAM), and then reading the memory contents out at regular intervals to the D/A converter whose output will be the stepwise approximation to the desired wave. For an 8-bit microcomputer 256 memory locations can be specified or addressed by one byte, so a 256-byte memory is convenient.

An 8085 program is given in Fig. 16.10(b).

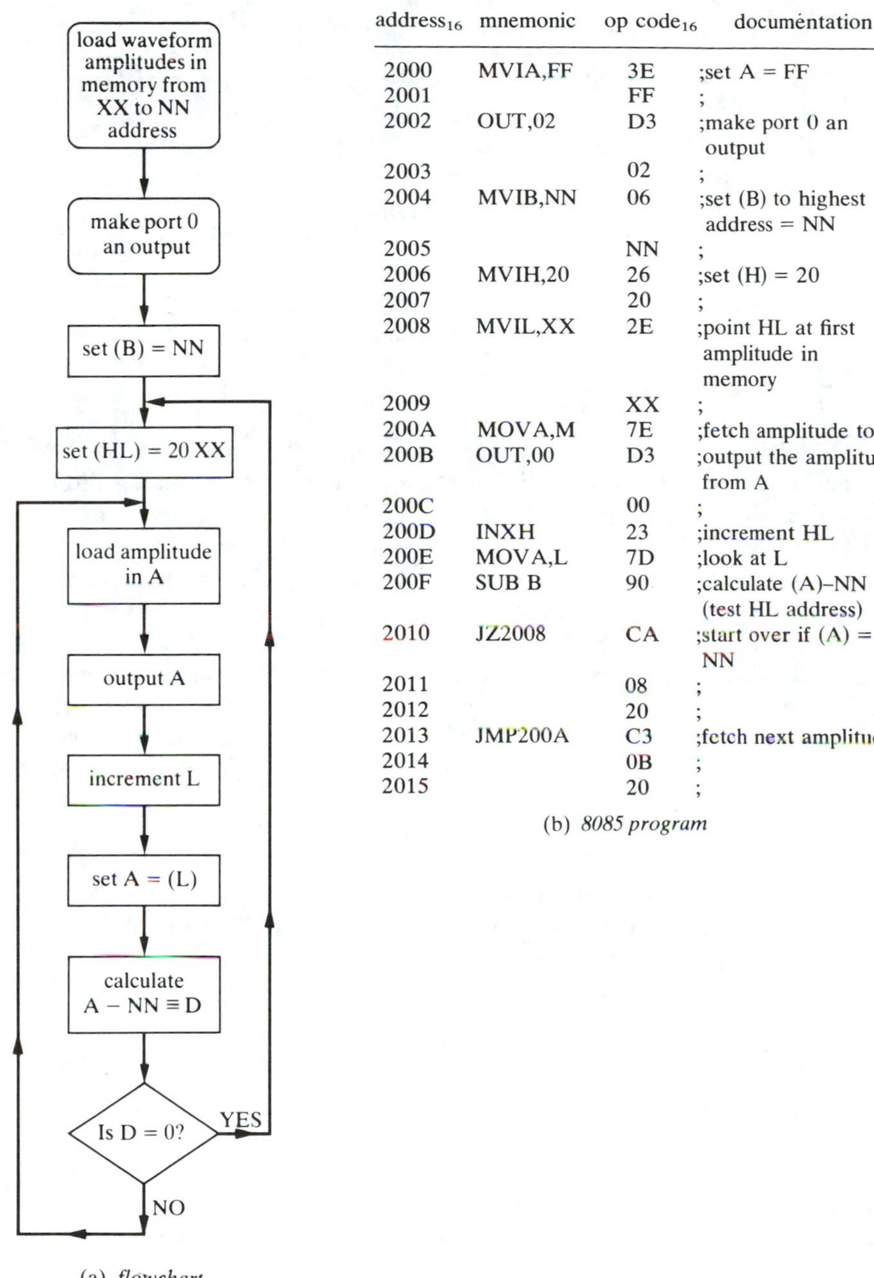

(a) *flowchart*

address$_{16}$	mnemonic	op code$_{16}$	documentation
2000	MVIA,FF	3E	;set A = FF
2001		FF	;
2002	OUT,02	D3	;make port 0 an output
2003		02	;
2004	MVIB,NN	06	;set (B) to highest address = NN
2005		NN	;
2006	MVIH,20	26	;set (H) = 20
2007		20	;
2008	MVIL,XX	2E	;point HL at first amplitude in memory
2009		XX	;
200A	MOVA,M	7E	;fetch amplitude to A
200B	OUT,00	D3	;output the amplitude from A
200C		00	;
200D	INXH	23	;increment HL
200E	MOVA,L	7D	;look at L
200F	SUB B	90	;calculate (A)–NN (test HL address)
2010	JZ2008	CA	;start over if (A) = NN
2011		08	;
2012		20	;
2013	JMP200A	C3	;fetch next amplitude
2014		0B	;
2015		20	;

(b) *8085 program*

FIGURE 16.10 8085 arbitrary waveform program.

A Z-80 program is given in Program 16.7.

label	operation	documentation
	LD HL, TABLE	;load first address in HL register pair
LOOP:	LDA,(HL)	;load (HL) into A
	OUT (DAC),A	;set output port = (A)
	INC L	;increment L by one—go to next memory address
	JP LOOP	;

PROGRAM 16.7 Z-80 arbitrary waveform program.

16.6 DIGITAL FILTERING

Digital filtering is one of the most important types of digital signal processing. The main advantage of a digital filter compared to an analog filter is that the filter characteristics (bandpass, center frequency, etc.) can be changed merely by changing the software—that is, by changing the computer program. The disadvantages are that the digital filter characteristics change with changes in the clock frequency, and the mathematics is rather formidable in some cases.

In all digital filters the input analog waveform is sampled at regular times by an A/D converter to produce a series of amplitude values that are converted to binary form and stored in memory. If the time between successive samples is t_s, then the sampling frequency is $1/t_s$. For example, if the input is sampled every 5 μs, $t_s = 5\ \mu$s, and the sampling frequency $f_s = 200$ kHz.

For the most general digital filter the output of the filter is an amplitude value, y_n, in binary form, and it may depend on many input values x_n and possibly also on some of the *output* values. That is, in the general case the digital filter output is

$$y_n = \sum_{k=-N}^{N} a_k x_{n-k} + \sum_{k=-M}^{M} b_k y_{n-k} \qquad \textbf{(16.1)}$$

where a_k and b_k are constants. We will shortly see that the choices of a_k and b_k determine the type of filter (e.g., low pass, high pass, etc). The $a_k x_{n-k}$ terms represent constant multiples of the inputs x_{n-k}, and the $b_k y_{n-k}$ terms represent constant multiples of the outputs.

We must now distinguish between the "past" and the "future." Remembering that the input is sampled every t_s, we let each integral value of the index n represent one additional time interval t_s. That is, if the input

x is sinusoidal with amplitude A and angular frequency ω, then

$$x = Ae^{j\omega t} \tag{16.2}$$

and the sampled input will be

$$x_n = Ae^{j\omega n t_s} \tag{16.3}$$

Thus if x_n is the *present* input, then x_{n-1} is the input that occurred t_s seconds *earlier*, and x_{n+1} is the input that will occur t_s seconds in the *future*.

If we are dealing with a real-time or "causal" filter, the present output y_n can depend only on the present input x_n and *earlier* inputs x_{n-k} $(k > 0)$ and on *earlier* outputs y_{n-k} $(k > 0)$; it cannot depend on inputs and outputs in the *future*. The general *causal* digital filter output is then

$$y_n = \sum_{k=0}^{N} a_k x_{n-k} + \sum_{k=1}^{M} b_k y_{n-k} \tag{16.4}$$

If, however, all the sampled inputs x_n have been stored in some memory, then we can, in a sense, look into the future in calculating the filter output, and (16.1) is the general filter output expression.

There are two general types of digital filters: recursive and non-recursive. The *nonrecursive* filter is the simpler type and is one in which the output y_n depends only on the *inputs* x_n, not on any of the outputs. In other words, for a nonrecursive causal filter, all the b_k coefficients are zero, and the output is related to the inputs x_n by

$$y_n = \sum_{k=0}^{N} a_k x_{n-k} \tag{16.5}$$

Another way of looking at a nonrecursive filter is that there is no "feedback" from past outputs to the present output y_n. Other names for a nonrecursive filter are *finite impulse response* (FIR) filter, *transversal* filter, *tapped delay line* filter, and a *moving average* filter.

The name "moving average" filter is particularly meaningful. Suppose $N = 2$ in (16.5), which means the output y_n equals a linear combination of the present input x_n, the input x_{n-1} taken t_s earlier, and the input x_{n-2} taken $2t_s$ earlier:

$$y_n = a_0 x_n + a_1 x_{n-1} + a_2 x_{n-2} \tag{16.6}$$

For example, if $a_0 = a_1 = a_2 = \frac{1}{3}$, then the output is simply the arithmetic average of the present input and the previous two inputs:

$$y_n = \frac{x_n + x_{n-1} + x_{n-2}}{3} \tag{16.7}$$

This might be termed a "smoothing by threes" average. It should be clear that such an averaging process tends to smooth out any rapid fluctuations in the input x and thus would act something like a *low*-pass filter.

A simple analog way to realize a nonrecursive causal filter is with a discrete delay line, in which each section of the delay line delays the incoming signal by t_s s. A summing circuit then sums together all the signals at the junctions between the delay elements, as shown in Fig. 16.11.

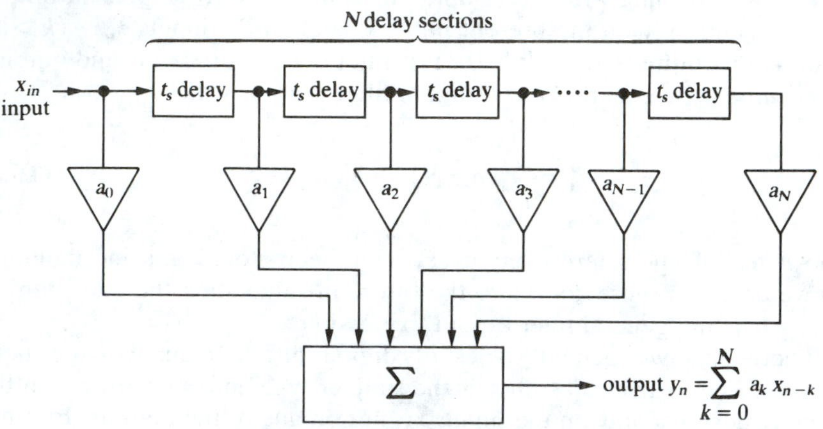

FIGURE 16.11 Nonrecursive causal digital filter realization.

In a *recursive* filter the output depends on both the inputs x_n and the outputs y_n, and the general relationship (16.1) applies. For a *causal* recursive filter

$$y_n = \sum_{k=0}^{N} a_k x_{n-k} + \sum_{k=1}^{N} b_k y_{n-k} \tag{16.8}$$

Other names for a recursive filter are ***infinite impulse response*** (IIR) filter, ***ladder*** filter, ***lattice*** filter, ***wave digital*** filter, ***autoregressive moving average*** (ARMA) filter, and ***autoregressive integrated moving*** filter. A recursive digital filter is shown in Fig. 16.12. Recursive filters can produce steeper filter skirts than nonrecursive filters, but they can oscillate. Nonrecursive filters cannot oscillate.

16.6.1 The Flanger or Comb Filter

We will now consider three simple nonrecursive digital filters. Our first example is the simplest possible nonrecursive filter; the output y_n contains only two terms in the $a_k x_{n-k}$ sum:

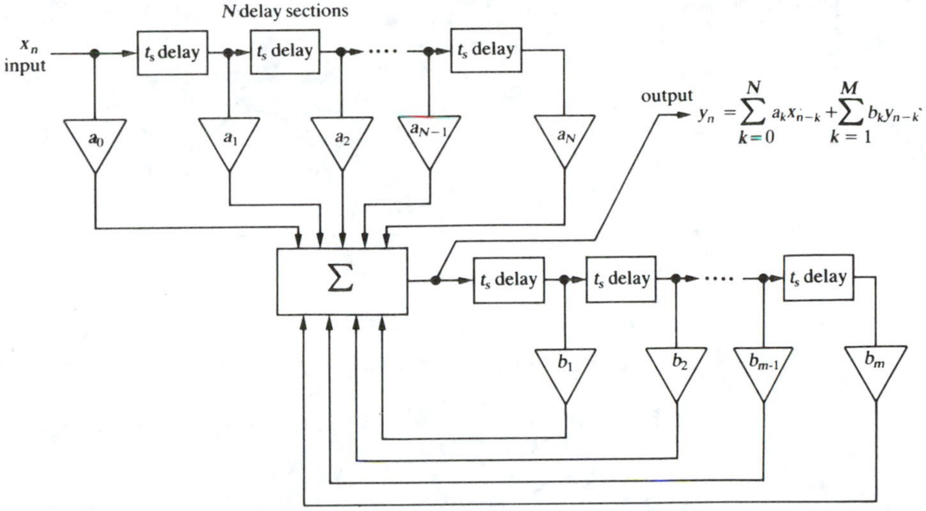

FIGURE 16.12 Recursive digital filter realization.

$$y_n = a_0 x_n + a_J x_{n-J} \tag{16.9}$$

This filter is sometimes called a *flanger filter* or a *comb filter*. In words, the output equals a constant, a_0, times the most recent input x_n plus another constant, a_J, times the input that occurred Jt_s earlier. J is usually greater than 1 in most applications. One way of thinking of this filter is that it calculates a running average of the input by adding the present input with an earlier input. Another way is to think of the filter as adding an input waveform to an earlier version (Jt_s seconds earlier) of the input.

If an analog sine wave of frequency f and period T is fed into such a filter, an A/D converter must first digitize the input to form the x_n values every t_s seconds, where t_s is the time required for one A/D conversion. t_s is determined by the clock that drives the A/D converter. If the delay Jt_s equals one period of the input, then the output will be a larger amplitude sine wave of the same frequency. The output amplitude will be twice the input amplitude, and the filter gain will be 6 dB. This will be true if Jt_s equals any integral multiple m of the input period:

$$Jt_s = mT \quad \text{or} \quad f = \frac{mf_s}{J} \qquad m = 1, 2, 3, \ldots \tag{16.10}$$

Thus the filter gain will be maximum at the frequencies $f = mf_s/J = f_s/J$, $2f_s/J, 3f_s/J, \ldots$, as shown in Fig. 16.13. However, if $Jt_s = T/2$ or $f = \frac{1}{2}(f_s/J)$, then the output will be exactly zero, because x_n and x_{n-1} will be equal in magnitude and opposite in sign; that is, they will be two sine waves 180° out

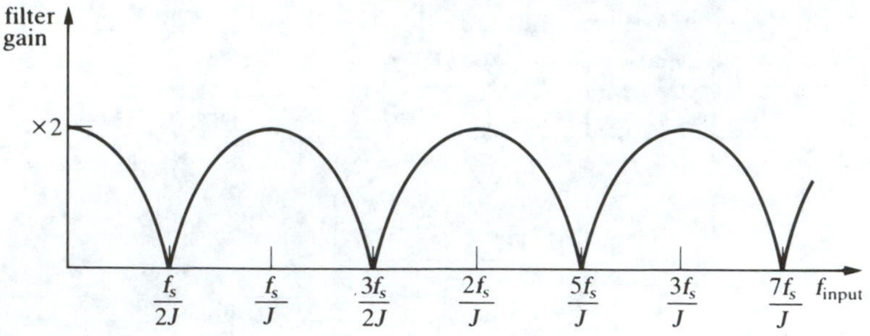

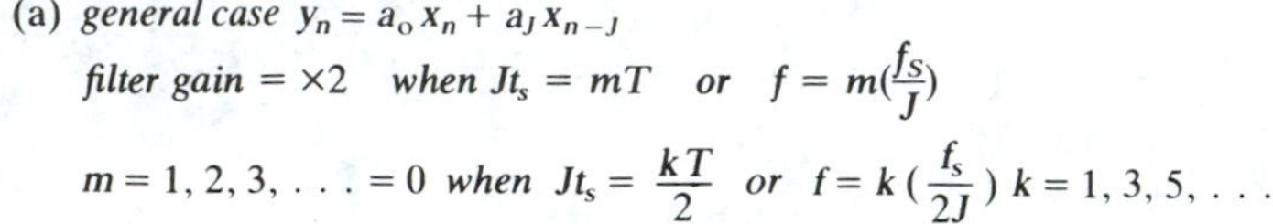
(a) *general case* $y_n = a_o x_n + a_J x_{n-J}$

filter gain = ×2 *when* $Jt_s = mT$ *or* $f = m(\frac{f_S}{J})$

$m = 1, 2, 3, \ldots = 0$ *when* $Jt_s = \frac{kT}{2}$ *or* $f = k(\frac{f_s}{2J})$ $k = 1, 3, 5, \ldots$

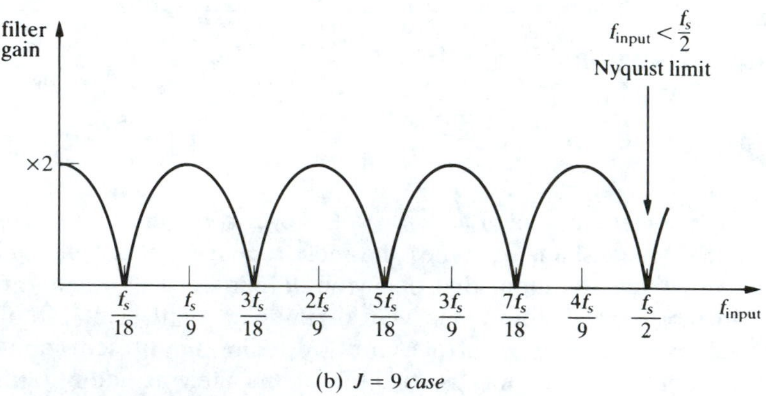

(b) $J = 9$ *case*

FIGURE 16.13 Flanger or comb filter gain.

of phase. The filter gain will be zero for any input frequency for which x_n and x_{n-J} are exactly 180° out of phase. Thus, if x_{n-1} is delayed $T/2$ or $3T/2$ or $5T/2$—that is, if $Jt_s = kT/2$, where k is an odd number—the filter output will be zero. In terms of input frequency, the filter output will be zero if

$$f = k\left(\frac{f_s}{2J}\right) \qquad k = 1, 3, 5, \ldots \tag{16.11}$$

A similar analysis for intermediate values of delay shows that the gain is between 0 and 1, as shown in Fig. 16.13(a). Notice that for $m = 0$ (which means practically that $t_s \ll T$—that is, the sampling frequency is much larger than the input frequency) the filter gain is 6 dB. Such a filter is sometimes called a comb filter because its gain versus frequency graph looks like the teeth of a comb. Such a comb filter is useful in detecting a repetitive waveform whose fundamental frequency is f_s/J, because it has peak gain at $m(f_s/J)$. For example, a sonar pulse whose basic frequency is $f_s/J = 2$ kHz will produce an echo with Fourier components at $m(f_s/J)$—that is, 2, 4, 6, 8, . . . kHz along with noise at other frequencies. The comb filter will clearly tend to pass all the Fourier components of the echo and to attenuate the noise at other frequencies. If the same clock frequency is used to generate the transmitted pulse frequency and also to determine the sampling frequency, then the filter comb will "track" any frequency changes in the transmitted pulse regardless of the cause of the frequency drift.

The Nyquist sampling theorem states that to recover an accurate replica of the input analog signal, the sampling frequency f_s must be more than twice as high as the highest Fourier component of the input. Thus, in the comb filter response curve the input frequency should always be less than $f_s/2$. The specific case $J = 9$ is shown in Fig. 16.13(b), and we see that we should use a prefilter to limit the input frequencies to less than $f_s/2$. For the 2-kHz sonar pulse mentioned previously, we would set the peak of the first tooth at 2 kHz—that is, $f_s/9 = 2$ kHz, or $f_s = 18$ kHz—and use a low-pass prefilter with a bandwidth out to $f_s/2 = 9$ kHz. Actually, because of the presence of $1/f$ noise, we would probably use a bandpass filter from approximately 1 kHz to 7 or 8 kHz.

The hardware realization of such a filter is shown in Fig. 16.14. The 0809 A/D output is the binary number x_n, which occurs every t_s seconds. This number is fed into the parallel (input) port 00 of the 8085 microprocessor. The 3-MHz clock of the 8085 is too fast for the 0809 A/D converter, so it must be divided down before it is fed into the 0809. The 8085 output on port 2B is used to start the A/D conversion.

The microprocessor is programmed to store an x_n input amplitude in memory at some address and then to store the next input amplitude value x_{n+1} t_s later at the next memory address. Then it adds these two values and outputs the sum to the D/A converter from its output port 01. The 0806 D/A current output is then converted to a voltage and displayed on an oscilloscope. The 8085 program is given in Program 16.8.

The first eight steps of the program are initialization and are executed only once at the beginning of the program. The next two steps are the main part of the program and continue to recycle as long as the program is running (i.e., as long as there is an analog input to the filter). The value x_n is read in from the A/D converter and stored at the address equal to the contents of the HL register pair. Registers L and E are incremented and ANDed with the hex number 1F. This limits to 32 the number of memory locations used. The datum x_n at the memory location specified by the HL register pair is then loaded into the accumulator A; DE and HL are exchanged, and the x_n content of A is added to the contents at the location specified by the HL pair (formerly the DE pair). The sum is output to the output port. A positive output pulse at output port 2B is sent to the A/D

	address$_{16}$	mnemonic	op code$_{16}$	documentation
Initialization	2800	LXI H,L 200A	21	;load HL with 200A
	2801		0A	
	2802		20	
	2803	LXI D,E 2000	11	;load DE with 2000
	2804		00	
	2805		20	
	2806	MVI A,00	3E	;set A = 00
	2807		00	
	2808	OUT,02	D3	;make port 00 an input
	2809		02	
	280A	MVI, A,FF	3E	;set (A) = FF
	280B		FF	
	280C	OUT,03	D3	;make port 01 an output
	280D		03	
	280E	MVI A,0F	3E	;
	280F		0F	
	2810	OUT,20	D3	;make ports A and B outputs
	2811		20	
Calculation	2812	IN 00	DB	;load x_n data into A from port 00 (from A/D)
	2813		00	
	2814	MOV M,A	77	;store x_n data at (HL) 200A, etc.
	2815	XCHG	EB	;HL $\rightleftarrows$ DE
	2816	ADD M	86	;calculate $x_n + x_{n-k}$
	2817	XCHG	EB	;HL $\rightleftarrows$ DE
	2818	OUT,01	D3	;sum output to D/A on port 01
	2819		01	

PROGRAM 16.8 Flanger filter program.

	address$_{16}$	mnemonic	op code$_{16}$	documentation
test C	281A	INR L	2C	; increment L
	281B	MOV A,L	7D	;
	281C	ANI 1F	E6	; test L to see if it's = 1F = 00011111
	281D		1F	
	281E	MOV L,A	6F	;
test E	281F	INR E	1C	; increment E.
	2820	MOV A,E	7B	; test E to see if it's = 1F = 00011111
	2821	ANI 1F	E6	;
	2822		1F	
	2823	MOV E,A	5F	;
Start next A/D conversion	2824	MVI A,FF	3E	;set (A) = FF
	2825		FF	
	2826	OUT,20	D3	;sends a 1 out on *all* of the 2B port wires to start next ADC (to SC)
	2827		20	
	2828	MVI A,00	3E	;set (A) = 00
	2829		00	
	282A	OUT,2B	D3	;sends a 0 out on 2B
	282B		2B	
Delay	282C	MVI A,0F	3E	;
	282D		0F	
	282E	DCR A	3D	; delay to allow A/D to settle
	282F	(NOP)^{10}	00	;
	2839	JNZ 28 2E	C2	
	283A		2E	
	283B		28	
	283C	JMP 2812	C3	;load next data
	283D		12	
	283E		28	

PROGRAM 16.8 Continued.

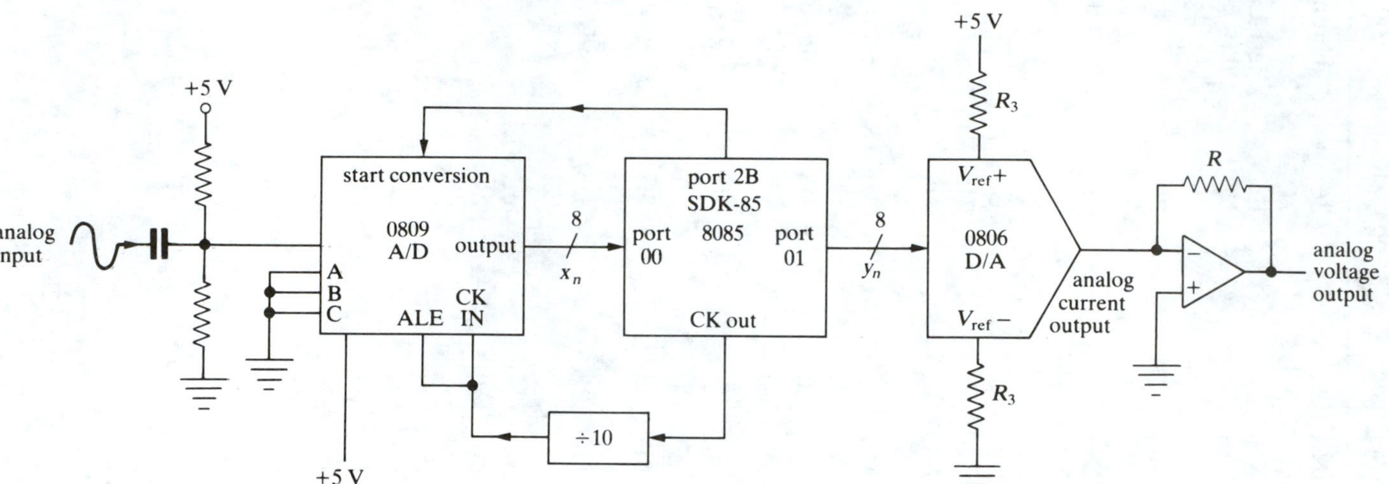

FIGURE 16.14 Nonrecursive flanger or comb filter circuit.

converter to start another conversion. The NOP and DCR steps form a delay in the program to ensure that the A/D output settles down before it is read into the accumulator.

16.6.2 The Look-up Table Thermometer

Figure 16.15 shows a simple look-up table thermometer with an 0809 A/D converter, an 8085 microprocessor SDK-85 kit, and a thermister temperature sensor. The thermistor resistance $R(T)$ decreases with increasing temperature, so the analog voltage input v to the 0809 increases as the temperature rises. The 8085 starts the A/D conversion by setting the 0809 START pin high and then low. (When the 0809 START pin goes high, its SAR is reset, and the conversion is started when the START pin goes low.) The 0809 EOC pin goes low about 8 clock periods after the conversion starts, so the program first checks EOC to make sure it is low, and then checks when EOC is high to determine when the conversion is completed. The 8-bit 0809 digital output is then enabled and fed into the SDK-85 on port 1 and loaded into the L register. The HL register pair then is used to point at the memory address containing the correct temperature. The correct temperature numbers (as determined by an independent thermometer) must, of course, be loaded into the proper memory locations 2800 . . . before the program is run. The instruction MOV A,M transfers the temperature number from the memory location whose address = (HL) to the accumulator A. The contents of A are then displayed on the LED readout, and the temperature is read again.

16.6.3 More Digital Filter Theory

Many different types of filters can be made by including more than two terms in the $a_k x_{n-k}$ sum and by varying the coefficients a_k. For example, consider a "smoothing by fives" nonrecursive filter:

$$y_n = \sum_{k=0}^{4} a_k x_{n-1} = a_0 x_n + a_1 x_{n-1} + a_2 x_{n-2} + a_3 x_{n-3} + a_4 x_{n-4} \quad \textbf{(16.12)}$$

The output y_n is a linear combination of the most recent input x_n and the four previous inputs, x_{n-1} at t_s earlier, and so on. We will now show that by varying the coefficients a_k, we can obtain a variety of different filters.

Suppose the input is sinusoidal of amplitude A and angular frequency ω.

$$x = Ae^{j\omega t} \quad \textbf{(16.13)}$$

With a sampling frequency of $f_s = 1/t_s$ the sampled input x_n will then be

$$x_n = Ae^{j\omega n t_s} \quad n = 1, 2, 3, \ldots \quad \textbf{(16.14)}$$

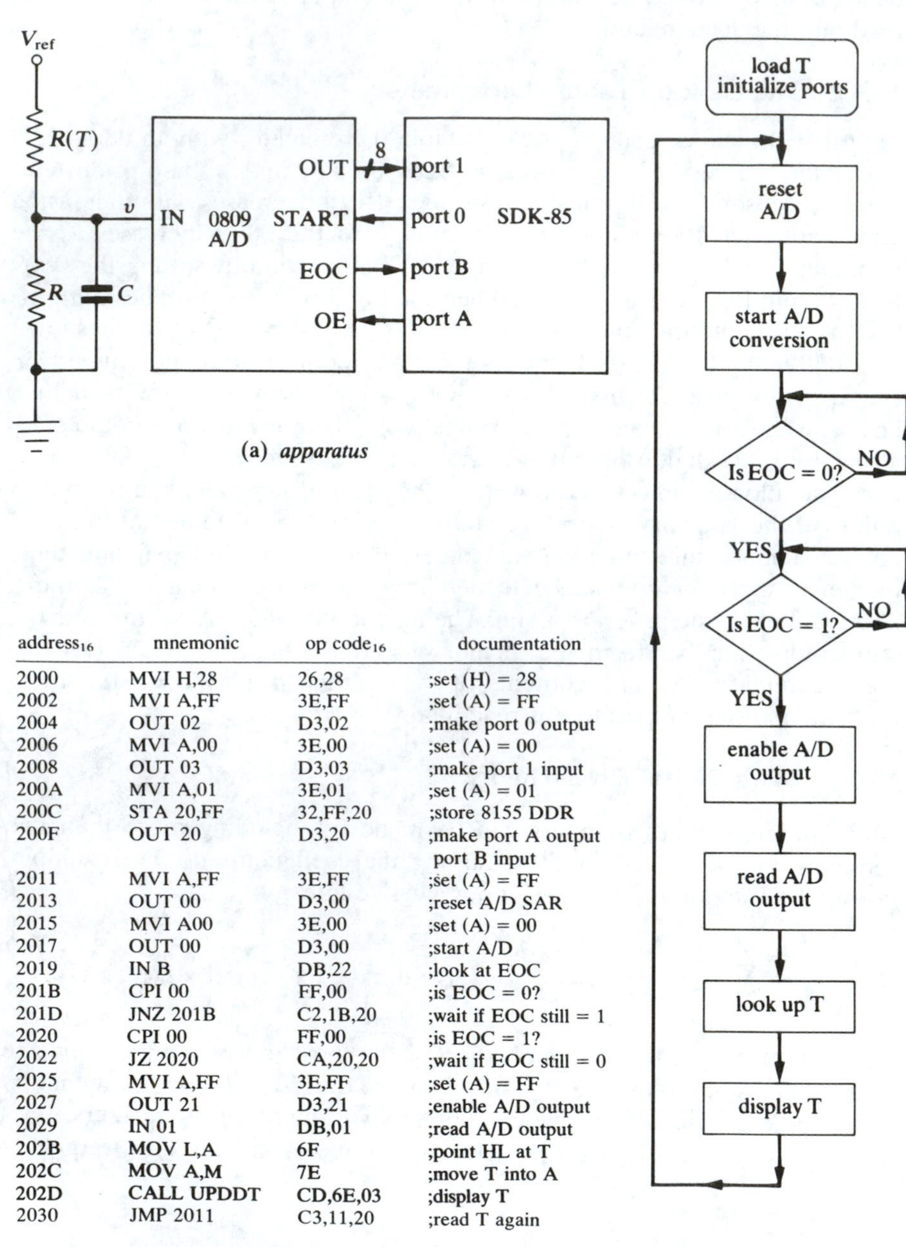

address$_{16}$	mnemonic	op code$_{16}$	documentation
2000	MVI H,28	26,28	;set (H) = 28
2002	MVI A,FF	3E,FF	;set (A) = FF
2004	OUT 02	D3,02	;make port 0 output
2006	MVI A,00	3E,00	;set (A) = 00
2008	OUT 03	D3,03	;make port 1 input
200A	MVI A,01	3E,01	;set (A) = 01
200C	STA 20,FF	32,FF,20	;store 8155 DDR
200F	OUT 20	D3,20	;make port A output port B input
2011	MVI A,FF	3E,FF	;set (A) = FF
2013	OUT 00	D3,00	;reset A/D SAR
2015	MVI A00	3E,00	;set (A) = 00
2017	OUT 00	D3,00	;start A/D
2019	IN B	DB,22	;look at EOC
201B	CPI 00	FF,00	;is EOC = 0?
201D	JNZ 201B	C2,1B,20	;wait if EOC still = 1
2020	CPI 00	FF,00	;is EOC = 1?
2022	JZ 2020	CA,20,20	;wait if EOC still = 0
2025	MVI A,FF	3E,FF	;set (A) = FF
2027	OUT 21	D3,21	;enable A/D output
2029	IN 01	DB,01	;read A/D output
202B	MOV L,A	6F	;point HL at T
202C	MOV A,M	7E	;move T into A
202D	CALL UPDDT	CD,6E,03	;display T
2030	JMP 2011	C3,11,20	;read T again

(c) *8085 program*

(b) *flow chart*

FIGURE 16.15 Look-up table thermometer.

The filter output y_n will be

$$y_n = a_0 A e^{j\omega n t_s} + a_1 A e^{j\omega(n-1)t_s} + a_2 A e^{j\omega(n-2)t_s} + a_3 A e^{j\omega(n-3)t_s} + a_4 A e^{j\omega(n-4)t_s} \tag{16.15}$$

Factoring out the common $Ae^{j\omega n t_s}$ term, we have

$$y_n = Ae^{j\omega n t_s}[a_0 + a_1 e^{-j\omega t_s} + a_2 e^{-2j\omega t_s} + a_3 e^{-3j\omega t_s} + a_4 e^{-4j\omega t_s}]$$

Recalling the Euler identity $2\cos\theta = e^{j\theta} + e^{-j\theta}$, we can create terms like $e^{j\theta}$ and $e^{-j\theta}$ by factoring out $e^{-2\omega t_s}$:

$$y_n = Ae^{j\omega(n-2)t_s}[a_0 e^{2j\omega t_s} + a_1 e^{j\omega t_s} + a_2 + a_3 e^{-j\omega t_s} + a_4 e^{-2j\omega t_s}]$$

To group the $e^{j\theta}$ and $e^{-j\theta}$ terms together, we let $a_0 = a_4 \equiv a$, $a_1 = a_3 \equiv b$, and $a_2 \equiv c$. The filter output then becomes

$$y_n = Ae^{j\omega(n-2)t_s}[a(e^{2j\omega t_s} + e^{-2j\omega t_s}) + b(e^{j\omega t_s} + e^{-j\omega t_s}) + c]$$

or

$$y_n = Ae^{j\omega(n-2)t_s}[2a\cos 2\omega t_s + 2b\cos\omega t_s + c] \tag{16.16}$$

which can be rewritten as

$$y_n = H(\omega)x_{n-2} \tag{16.17}$$

where

$$H(\omega) \equiv 2a\cos 2\omega t_s + 2b\cos\omega t_s + c \tag{16.18}$$

and

$$x_{n-2} = Ae^{j\omega(n-2)t_s} \tag{16.19}$$

Because t_s is constant, $H(\omega)$ depends only on ω, the frequency of the input, and the constants a, b, and c, which we are free to specify. *$H(\omega)$ is the filter gain as a function of the input frequency.*

Equation (16.17) says that the filter output at any time equals the filter gain $H(\omega)$ times the input two sampling periods ($2t_s$) earlier at $t - 2t_s$.

A little thought will show that for every two more $a_k x_{n-k}$ terms included in (16.12), another cosine term will appear in $H(\omega)$. For example, if $y_n = a_0 x_n + \cdots + a_6 x_{n-6}$, then $H(\omega) = 2a_0\cos 3\omega t_s + 2a_1\cos 2\omega t_s + 2a_2\cos\omega t_s + a_3$. In general, $H(\omega)$ will always be a linear combination of $\cos n\omega t_s$ terms:

$$H(\omega) = \sum_{n=0}^{N} c_n \cos n\omega t_s \tag{16.20}$$

From Fourier analysis we recall that *any* even function $H(\omega)$ can be analyzed into an *infinite* sum ($N \to \infty$) of terms of the form $c_n \cos n\omega t_s$ if the fundamental period of $H(\omega)$ is t_s. In other words, the infinite set of terms $c_n \cos n\omega t_s$ forms a *complete* set for the expansion of any even function $H(\omega)$. For a *finite* number N of $c_n \cos n\omega t_s$ terms, the sum (16.20) can *approximate* any desired $H(\omega)$ for a low-pass or high-pass filter by proper choice of the constants c_n.

Suppose we wish to make a low-pass filter, as shown in Fig. 16.16. We

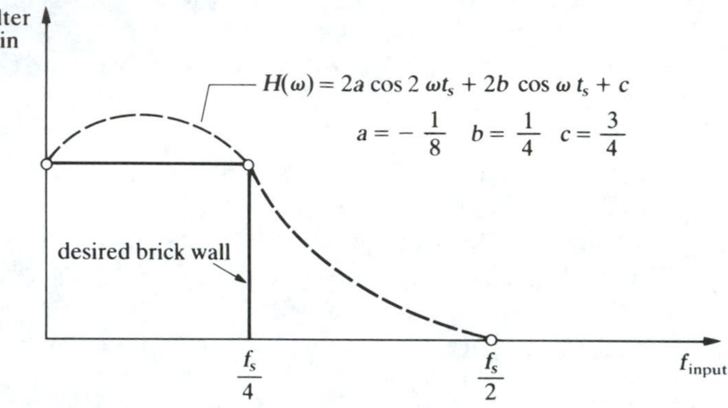

FIGURE 16.16 Low-pass filter approximation.

wish the gain $H(\omega)$ to be 1.0 from dc out to $f_s/4$ and 0 from $f_s/4$ out to $f_s/2$. (Remember there is a prefilter that limits the input frequencies to the range 0 to $f_s/2$ to satisfy the Nyquist sampling theorem.) It would clearly take an infinite number of terms like $\cos n\omega t_s$ to produce this "brick-wall" filter, but we can approximate the desired gain by the finite sum of

$$H(\omega) = 2a \cos 2\omega t_s + 2b \cos \omega t_s + c \tag{16.21}$$

So we impose the following three conditions on a, b, and c:

$$\text{when } f = 0 \ (\omega = 0) \qquad H(0) = 1 \quad \text{or} \quad 1 = 2a + 2b + c \tag{16.22}$$

$$\text{when } f = \frac{f_s}{4} \ \left(\omega = \frac{\pi}{2t_s}\right) \qquad H\left(\frac{\pi}{2t_s}\right) = 1 \quad \text{or} \quad 1 = -2a + 0 + c \tag{16.23}$$

$$\text{when } f = \frac{f_s}{2} \ \left(\omega = \frac{\pi}{t_s}\right) \qquad H\left(\frac{\pi}{t_s}\right) = 0 \quad \text{or} \quad 0 = 2a - 2b + c \tag{16.24}$$

We now simply solve (16.22)–(16.24) for a, b, and c. The results are

$$a = -\frac{1}{8} \qquad b = \frac{1}{4} \qquad c = \frac{3}{4} \tag{16.25}$$

$$H(\omega) = -\frac{1}{4}\cos 2\omega t_s + \frac{1}{2}\cos \omega t_s + \frac{3}{4}$$

$H(\omega)$ for these values of a, b, and c is shown in Fig. 16.16 and we can see that it is indeed an approximation to the desired low-pass filter.

Clearly, the more $\cos n\omega t_s$ terms we include in $H(\omega)$, the more accurately we can approximate any desired filter shape. The constants $2a$, $2b$, c $(a_0, a_1, a_2, a_3, a_4)$ are the Fourier coefficients of the desired filter gain $H(\omega)$.

If we had desired a high-pass filter, we would have set $H(0) = 0$, $H(\pi/2t_s) = 1$, and $H(\pi/t_s) = 1$. The constants work out to be $a = -\frac{1}{4}$, $b = -\frac{1}{8}$, and $c = \frac{3}{4}$.

Different conditions can be imposed on $H(\omega)$. For example, to obtain a flatter frequency response near dc for a low-pass filter, we might impose the three conditions

$$H(0) = 1 \qquad \left.\frac{dH}{d\omega}\right|_{\omega=0} = 0 \quad \text{and} \quad H\left(\frac{\pi}{t_s}\right) = 0$$

We must always remember that the number of imposed conditions cannot exceed the number of constants in the $H(\omega)$ expression—otherwise we could not solve for the constants.

The set of constants a_k is usually called the *window* of the filter, because the inputs x_n are multiplied by a_k to produce the output y_n; in other words, the inputs x_n are "looked at" through the a_k constants.

A final comment about the shape of the desired filter gain $H(\omega)$ approximated by the finite sum (16.21): The finite sum always will have "wiggles" or oscillations, as shown in Fig. 16.17, especially near discontinuities, such as a discontinuous drop in $H(\omega)$ for a brick-wall filter. The period of such oscillations is always the period of the highest term $\cos N\omega t_s$. Thus the oscillations in $H(\omega)$ can be greatly reduced by averaging

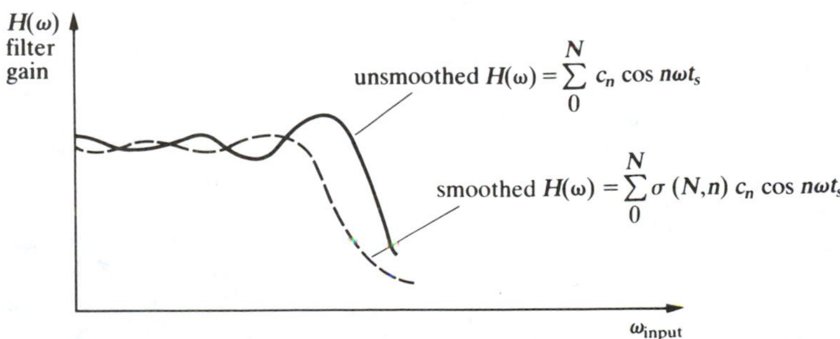

FIGURE 16.17 Effect of smoothing.

over this period. The net effect is to multiply each term $\cos n\omega t_s$ by a "smoothing" filter $\sigma(N, n)$, where

$$\sigma(N, n) = \frac{\sin \dfrac{\pi n}{N}}{\dfrac{\pi n}{N}} \tag{16.26}$$

Figure 16.17 shows the smoothing effect. The price paid for the smoothing is that the filter shift is less steep. The oscillations are usually called the *Gibbs phenomenon*, and the above smoothing factor is called the *Lanczos smoothing factor*. For further details consult the excellent monograph on digital filters by Hamming (second edition).

16.6.4 The Switched Capacitor Filter

A switched arrangement of two capacitors can yield a hardware realization of simple *recursive* digital filter in which the output depends on both the present input and the previous output. Consider two capacitors and two switches, as shown in Fig. 16.18. Let the two switches be driven out of phase at some clock frequency; that is, S_1 is on and S_2 is off for half of the clock period; and S_1 is off and S_2 is on for the other half of the clock period. When S_1 is closed (and S_2 open), as in Fig. 16.18(a), C_1 will charge up to the input voltage v_{in}. When S_1 is open and S_2 is closed, as in Fig. 16.18(b), the two capacitors will be in parallel and will thus have the same voltage. Therefore

$$v_{out} = \frac{Q_{total}}{C_1 + C_2}$$

where Q_{total} is the total charge on both C_1 and C_2. But the charge contributed by C_1 is just $Q_1 = C_1 v_{in}$, and the charge contributed by C_2 is $Q_2 = C_2 v'_{out}$, where v'_{out} is the *previous* voltage at the output. Thus,

$$v_{out} = \frac{Q_1 + Q_2}{C_1 + C_2} = \frac{C_1 v_{in} + C_2 v'_{out}}{C_1 + C_2}$$

or

$$v_{out} = \frac{C_1}{(C_1 + C_2)} v_{in} + \frac{C_2}{(C_1 + C_2) v'_{out}}$$

which is of the form

$$y_n = Ax_n + (1 - A) y_{n-1}$$

where

$$A = \frac{C_1}{C_1 + C_2} \quad \text{and} \quad 1 - A = \frac{C_2}{C_1 + C_2}$$

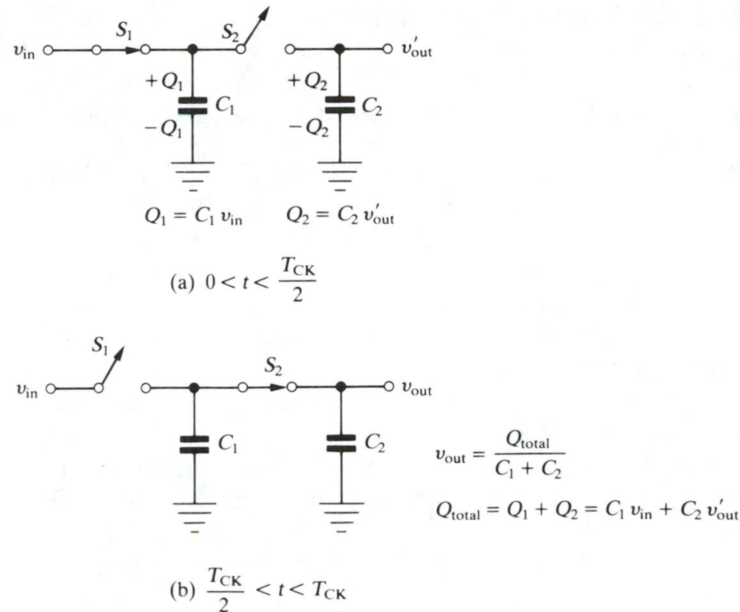

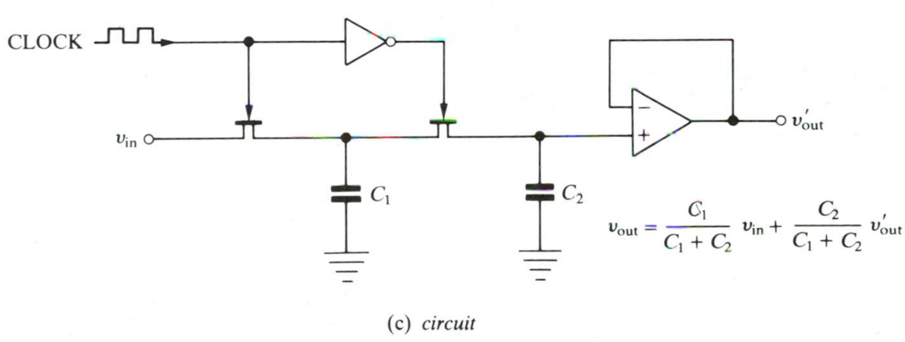

(c) *circuit*

FIGURE 16.18 Switched capacitor recursive low-pass filter.

This is a simple example of a recursive filter. Remember that x_n is the present input for the present output y_n, and y_{n-1} is the immediately previous output (one clock period earlier). The switches are usually FETs, and the output is buffered as shown in Fig. 16.18(c).

It can be shown that this filter is a low-pass filter with a breakpoint of

$$\omega_B = f_{ck} \ln\left(1 + \frac{C_2}{C_1}\right)$$

Thus by changing the clock frequency, we can change the break frequency; the filter is electrically tunable.

16.7 MICROPROCESSOR-CONTROLLED MEASUREMENT

Microprocessor-controlled measurement of physical quantities is becoming more and more common. The basic idea is to use dedicated hardware (transducers) to measure real-time physical phenomena and to convert the analog data to digital code and manipulate it with software for everything else. For fast data processing, assembly language is usually used, and for more leisurely processing, a more convenient easier-to-program high-level language is used. In this section we will describe one simple example in which the angular position of a light source is read out on the 8085 SDK microcomputer LED readout.

A visible light source is allowed to move anywhere on a circle in a horizontal plane, and the object is to read the angular position of the light source on the LED display of the 8085 SDK microcomputer. The general apparatus is shown in Fig. 16.19. Three photoresistors (CdS cells) are

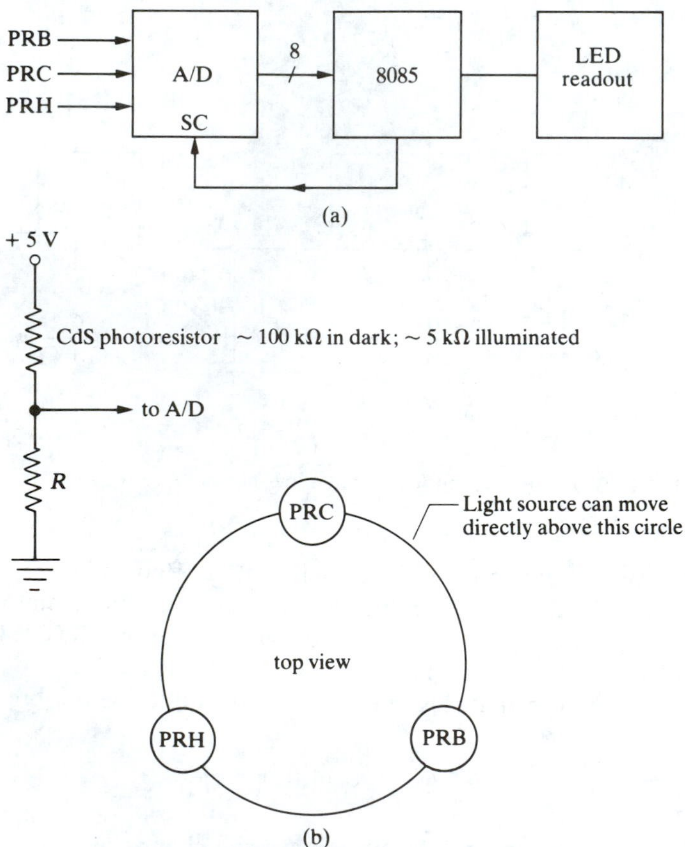

FIGURE 16.19 Light-position measuring apparatus.

located 120° apart on a horizontal circle and are illuminated by the light source, which moves in a circle just above the photoresistors. The photoresistor closest to the light source will obviously be more strongly illuminated and will have the lowest resistance. Each photoresistor is biased or powered with 5 V dc, as shown in Fig. 16.19(b), and thus puts out a dc analog voltage approximately proportional to the illumination. The analog output is converted into digital form by the A/D converter and fed into the 8085 microcomputer. Only one A/D is used; it is multiplexed from one photoresistor to the next by the microprocessor. The program or software then analyzes the digital light inputs and converts the inputs into one hex number, which is a measure of the angular position of the light source.

The photoresistor angular positions are shown in Fig. 16.20. The three

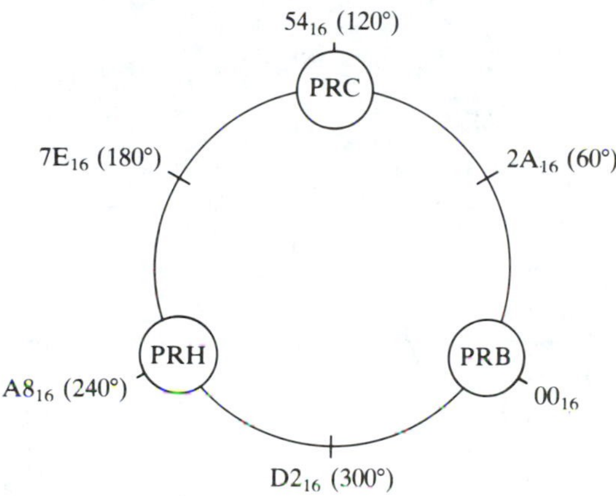

FIGURE 16.20 Photoresistor angular position.

photoresistors are labeled PRB, PRC, and PRH because their outputs are stored, respectively, in registers B, C, and H in the 8085 microprocessor. The angular reference position on the circle starts at 00_{16} at photoresistor B and increases linearly counterclockwise around the circle to 54_{16} at photoresistor C, to $A8_{16}$ at photoresistor H, and to $FD_{16} = 253_{10}$ just before photoresistor B. Thus there are 254_{10} angular markings on the circle, and each hex digit corresponds to 1.42°.

The basic idea of the program is simple. The digitized output signals from the three photoresistors are compared with each other by a series of "compare" program statements to determine in which 60° segment the light source lies. Figure 16.21 illustrates this procedure. For example, if the signal from photoresistor B is less than that from photoresistor C, the light source must be somewhere between 2A and A8 on the circle, and so on. If the signal from photoresistor B is also greater than the signal from photoresistor H, the light source must be somewhere in the 60° arc from 2A to

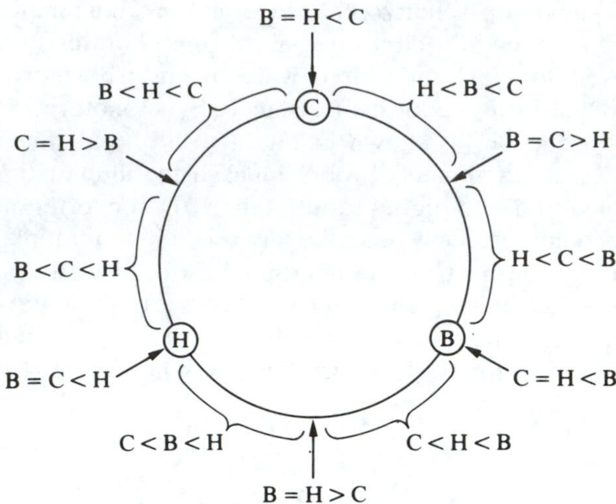

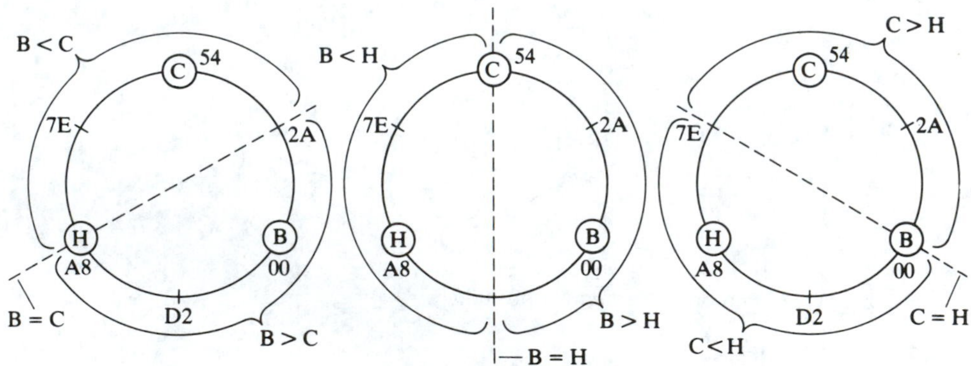

FIGURE 16.21 Angular light positions.

54—between C and B but closer to C. Assuming that the light signals are approximately linearly proportional to the angular position of the light source, we can calculate approximately the numerical angular position of the light source (in the sector between C and B) from the formula 2A + (C − B). Some analog signal processing would make this relationship more accurate. Or a look-up table technique could be used to convert to degrees. Similar arguments are used for the other comparisons. The flowchart of Fig. 16.22 shows the program logic, and the program itself is given in Program 16.9.

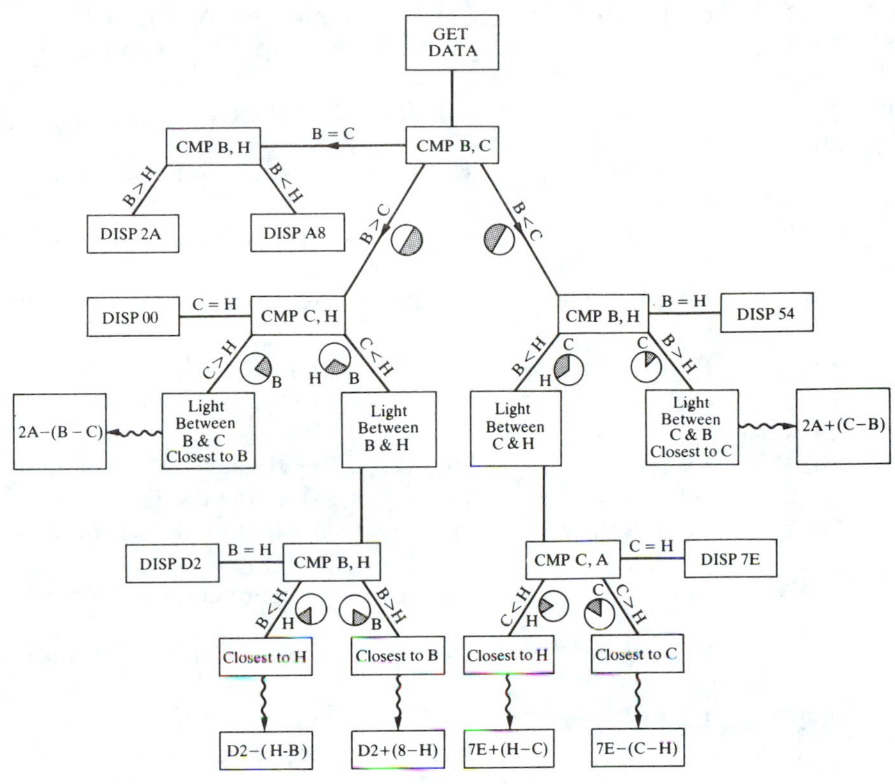

FIGURE 16.22 Flowchart for angular light-position measurement.

$address_{16}$	mnemonic	op $code_{16}$	documentation
2000	MVI A,00	3E,00	;make port 0 input
2002	OUT,02	D3,02	
2004	MVI A,03	3E,03	;make port 1 output (pins 0 and 1)
2006	OUT,03	D3,03	
2008	MVI A,00	3E,00	;look at photoresistor 00
200A	OUT,01	D3,01	
200C	LXI SP, 20D0	31,D0,20	;define stack pointer
200D	RNC	D0	;definition
200E	RIM	20	;call instruction
200F	MVI D,1F	16,1F	;prepare countdown register D
2011	MVI E,00	1E,00	;prepare countdown register E

PROGRAM 16.9 WHERE IS LIGHT?

$address_{16}$	mnemonic	op $code_{16}$	documentation
2013	CALL DELAY	CD	;time for A/D to settle
2014		F1	;ROM subroutine DE → 0
2015		05	
2016	IN 00	DB,00	;read data
2018	MOV B,A	47	;store data in B
2019	MVI A,01	3E,01	;look at photoresistor 01
201B	OUT,01	D3,01	
201D	MVI D,1F	16,1F	;prepare countdown register D
201F	MVI E,00	1E,00	;prepare countdown register E
2021	CALL DELAY	CD	;time for A/D to settle
2022		F1	;ROM subroutine DE → 0
2023		05	
2024	IN 00	DB,00	;read data
2026	MOV C,A	4F	;store data in C
2027	MVI A,02	3E,02	;look at photoresistor 02
2029	OUT,01	D3,01	
202B	MVI D,1F	16,1F	;prepare countdown register D
202D	MVI E,00	1E,00	;prepare countdown register E
202F	CALL DELAY	CD	;time for A/D to settle
2030		F1	;ROM subroutine DE → 0
2031		05	
2032	IN 00	DB,00	;read data
2034	MOV A,B	78	
2035	CMPC	B9	
2036	JZ 2050	CA,50,20	;B = C
2039	JC 2050	DA,50,20	;B < C
203C	MOV A,C	79	;B > C
203D	CMP H	BC	
203E	JZ 2069	CA,69,20	;C = H
2041	JC 2071	DA,71,20	;C < H
2044	MOV A,B	78	;H < C
2045	SUB C	91	;(B − C)
2046	MOV L,A	6F	
2047	MVI A,2A	3E,2A	
2049	SUB L	95	;2A − (B − C)
204A	CALL UPDDT	CD,6E,03	;display light position
204D	JMP 2008	C3,08,20	;look at photoresistors again
2050	MVI A,2A	3E,2A	;B = C from 2036
2052	CALL UPDDT	CD,6E,03	;display light position

PROGRAM 16.9 Continued.

address$_{16}$	mnemonic	op code$_{16}$	documentation
2055	JMP 2008	C3,08,20	;look at photoresistors again
2058	CMP H	BC	;B < C from 2039
2059	JZ 207F	CA,7F,20	;B = H
205C	JC 2087	DA,87,20	;B < H
205F	MOV A,C	79	;H < B
2060	SUB B	90	;C − B
2061	ADI A,2A	C6,2A	
2063	CALL UPDDT	CD,6E,03	;display light position
2066	JMP 2008	C3,0820	;look at photoresistors again
2069	MVI A,00	3E,00	;C = H from 203E
206B	CALL UPDDT	CD,6E,03	;display light position
206E	JMP 2008	C3,08,20	;look at photoresistors again
2071	MOV A,B	78	;C < H from 2041
2072	CMP H	BC	
2073	JC 2095	DA,95,20	;B < H
2076	SUB H	94	;(B − H) B > H
2077	ADI A,D2	C6,D2	;D2 + (B − H)
2079	CALL UPDDT	CD,6E,03	;display light position
207C	JMP 2008	C3,08,20	;look at photoresistors again
207F	MVI A,54	3E,54	;B = H from 2059
2081	CALL UPDDT	CD,6E,03	;display light position
2084	JMP 2008	C3,08,20	;look at photoresistors again
2087	MOV A,H	7C	;B < H from 205C
2088	CMP C	B9	
2089	JC 20A1	DA,A1,20	;H < C
208C	SUB C	91	;(H − C) C < H
208D	ADI A,7E	C6,7E	;FE + (H − C)
208F	CALL UPDDT	CD,6E,03	;display light position
2092	JMP 2008	C3,08,20	;look at photoresistors again
2095	MOV A,H	7C	;B < C from 2073
2096	SUB B	90	;H − B
2097	MOV L,A	6F	
2098	MVI A,D2	3E,D2	
209A	SUB L	95	;D2-(H − B)
209B	CALL UPDDT	CD,6E,03	;display light position
209E	JMP 2008	C3,08,20	;look at photoresistors again
20A1	MOV A,C	79	;H < C from 2089
20A2	SUB H	94	;C − H
20A3	MOV L,A	6F	
20A4	MVI A,7E	3E,7E	
20A6	SUB L	95	;7E − (C − H)
20A7	CALL UPDDT	CD,6E,03	;display light location
20AA	JMP 2008	C3,08,20	;look at photoresistors again

PROGRAM 16.9 Continued.

For a more elegant program, wherever CALL UPDDT, JMP 2008 appears, we might write CALL SUBROUTINE, where SUBROUTINE = CALL UPDDT, JMP 2008.

16.8 MICROPROCESSOR CONTROL OF EXPERIMENTAL VARIABLES

In this section we shall briefly describe how temeprature can be controlled; in other words, we shall construct a "thermostat" with the microprocessor. We assume that we have a heater that can be turned on or off to control the temperature and some sort of temperature sensor (e.g., a thermometer or thermistor) that senses the temperature. The temperature sensor is connected to an input port of the microprocessor, and the acquired temperature value is then logically compared with a fixed temperature value

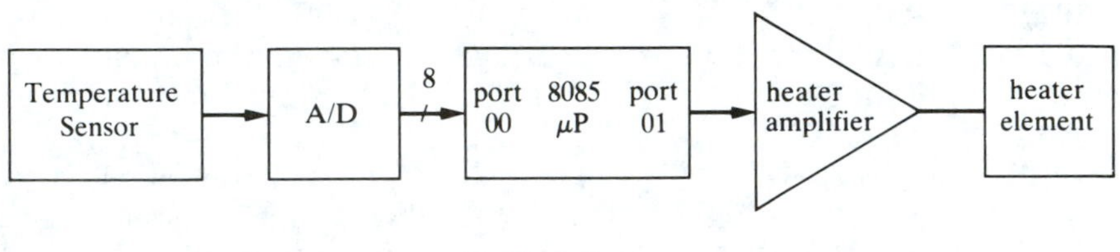

(a) *block diagram*

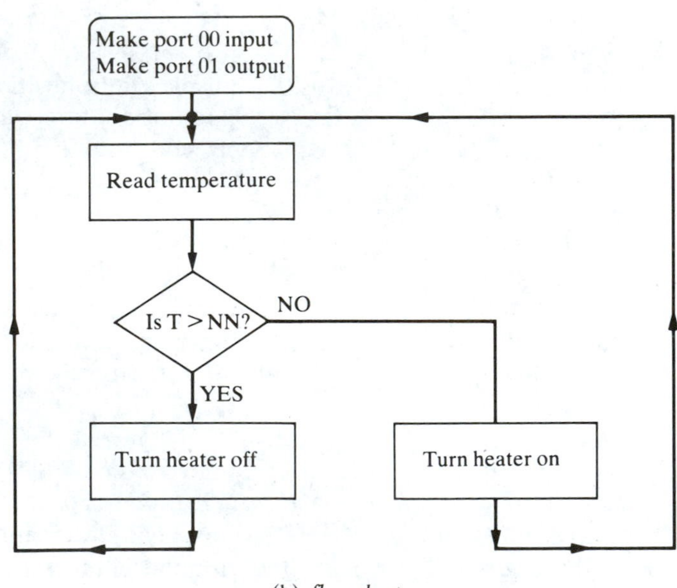

(b) *flow chart*

FIGURE 16.23 Temperature regulator block diagram.

specified in the program. If the experimental temperature is greater than the specified temperature, the heater is turned off; if it is less, the heater is turned on. The microprocessor then reads the temperature over and over again, turning the heater on or off as desired.

The basic apparatus and flow chart are shown in Fig. 16.23. The analog output of the temperature transducer must be set to cover a range appropriate for the A/D input (e.g., from 0 to 10 V). The A/D output is connected to an input port of the microprocessor, and the output port of the microprocessor is connected to the heater control. The low-power microprocessor output must obviously be amplified by using one of the interface techniques of Chapter 13 before it can control a high-wattage heater.

The program for the 8085 is given in Program 16.10. We assume that the temperature reading (the output of the A/D converter) is connected to port 00 of the microprocessor and that port 01 is connected to the heater. We also assume that a 00 output will turn off the heater and that an FF output will turn on the heater.

$address_{16}$	mnemonic	op $code_{16}$	documentation
2000	MVI A,00	3E,00	;set (A) = 00
2002	OUT,02	D3,02	;make port 00 an input
2004	MVI A,FF	3E,FF	;set (A) = FF
2006	OUT,03	D3,03	;make port 01 an output
2008	IN,00	DB,00	;read temperature T from port 00
200A	CPI NN	FE,NN	;calculate T − NN, NN = desired temperature
200C	JC 2016	DA,19,20	;jump to 2016 if T < NN CY = 1 if T < NN CY = 0 if T ≥ NN
200F	MVI A,00	3E,00	; } turn off heater
2011	OUT,01	D3,01	; }
2013	JMP 2008	C3,08,20	;read temperature again
2016	MVI A,FF	3E,FF	; } turn on heater
2018	OUT,01	D2,01	; }
201A	JMP 2008	C3,08,20	;read temperature again

PROGRAM 16.10 8085 temperature control program.

Steps 2000 through 2006 are initialization; they merely define the input and output ports. Then the temperature input from the A/D converter output is read into the accumulator, and $T - NN$ is calculated, where T is the temperature input and NN is a one-byte number that is the desired temperature chosen by the programmer. The subtraction is done with the compare immediate (CPI) instruction, which subtracts the number NN from the contents of the accumulator. [The difference (A) − *NN* is not accessible; it is stored in an internal register. (A) is unaffected.] If the difference is zero or positive, the Carry flag is not set (CY = 0), but if the difference is negative—that is, if $T < NN$—then the Carry flag is set (CY = 1). Thus the Carry flag tells us whether the temperature T is greater than, equal to, or less than *NN*, which in turn tells us whether to turn the heater off or on. If the Carry flag is set to 1, the temperature is too low and the JUMP ON CARRY instruction (JC) jumps to step 2016, which turns on the heater. If the Carry flag is not set, the temperature is either just right (equal to *NN*) or too high. In this case steps 200F through 2014 are executed, which turn off the heater. Then the temperature is read again, and the program repeats itself.

The preceding program will read the temperature and turn the heater off or on, as required, in approximately several hundred microseconds. This is much too fast for any macroscopic heater, because a large heater is inherently a very slow device with a thermal time constant of the order of seconds or even minutes. Thus, it is not necessary to keep on reading the temperature and controlling the heater power. In other words, there is no need to tie up the CPU continuously when controlling such a slow device. This comment applies to *any slow* external device controlled by a microprocessor.

Another disadvantage of the preceding program is that if the temperature sensor is very sensitive and has a fast response (short time constant), the microprocessor may turn the heater on and off very rapidly. Such a rapidly changing current in the heater element will produce broadband electromagnetic interference and is obviously undesirable. One way to prevent this rapid on and off heater operation is to program in hysteresis. If the temperature sensor reads a temperature above T_2, the heater will be turned off, but only when the temperature sensor reads a temperature below T_1 ($T_1 < T_2$; e.g., $T_2 = 37°\text{C}$, $T_1 = 36°\text{C}$) will the heater be turned on.

Another solution is to use the interrupt capability of the microprocessor to check the temperature every second or every minute. An INTERRUPT is a direct input to the CPU that tells it of the occurrence of some external event (e.g., the temperature being too low), or the INTERRUPT could be activated every second or every minute by a real-time clock.

When the INTERRUPT line goes low, the current value of the program counter is pushed onto the stack automatically, the contents of the registers are (usually) pushed onto the stack and stored (in RAM), and the program counter is set to point to some location where the INTERRUPT routine or "service routine" is stored. (The stack pointer must, of course, first be initialized.) After the INTERRUPT routine or program is executed, the program counter and the register contents are popped back, and the regular operation of the microprocessor continues as it did before the INTERRUPT.

There are two basic types of INTERRUPTS: maskable and nonmaskable. A *maskable* INTERRUPT can be changed or overridden by software, but a *nonmaskable* INTERRUPT cannot. Once a nonmaskable INTERRUPT is activated, then the CPU always executes the INTERRUPT process or operation. A nonmaskable INTERRUPT would be used only for an extremely important function, such as a power failure when there is a need to switch over to an auxiliary power source or to store valuable data before power is fully lost.

In the case of our heater control, the INTERRUPT could be activated every second by an external real-time clock signal. A real-time clock circuit is shown in Fig. 16.24. The advantage of this INTERRUPT scheme is that

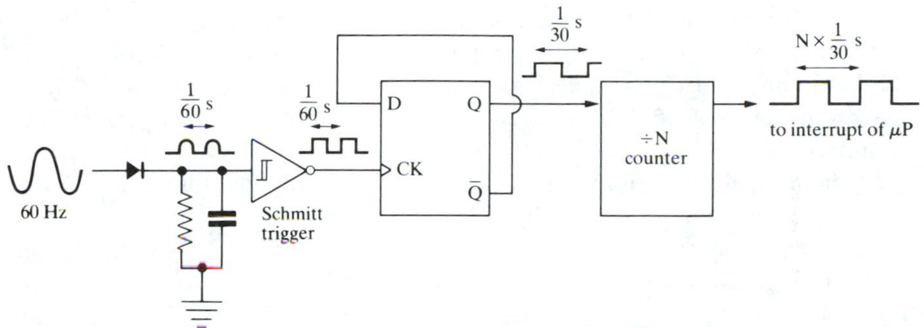

FIGURE 16.24 Real-time clock.

the temperature control routine to turn the heater on or off takes only approximately 200 μs, and if the INTERRUPT is activated every second, only this 200 μs of each second is taken up with the temperature control problem task. The rest of the second (999,800 μs) can be utilized for various calculations. In other words, the microprocessor is not completely tied up with the task of temperature control; it can be "simultaneously" used for other control or calculations. This is an example of *time* multiplexing. The data in the various registers used during the 999,800 μs must obviously be kept separate from the temperature program. The disadvantage is that the external hardware of the clock and counter is required.

Another example of microprocessor control is the electronic control of the internal combustion engine. A microprocessor (usually called an "electronic brain" by the mechanics) measures several input variables and controls several outputs as well as providing on-board diagnostic information for various malfunctions. The five inputs usually measured are (1) the engine coolant temperature, (2) the engine rpm, (3) the intake manifold pressure, (4) the throttle position, and (5) the oxygen concentration in the exhaust gas. The only unusual transducer or sensor is that used to measure the oxygen concentration in the exhaust gas. It usually consists of a ceramic

ZrO_2 or TiO_2 sensor whose electrical resistance is a strong function of oxygen concentration. The purpose of this measurement is to control the air/fuel ratio, because this ratio determines in large part the fuel economy, drivability, and amount and type of exhaust pollutants. The oxygen concentration is an extremely sensitive function of the air/fuel ratio, being very low for a "rich" mixture (excess fuel) and orders of magnitude higher for a "lean" mixture (excess air). The desired stoichiometric air/fuel ratio is approximately 14.6/1 for most gasoline; this is the ratio for which there is just enough oxygen in the air to completely oxidize all the gasoline hydrocarbons with no oxygen left over. The oxygen concentration will increase greatly if the air/fuel ratio is raised from 14.0/1 to 15.0/1. The sensor output may change from 1.0 V to 50 mV as the mixture changes from rich to lean. This dc signal can be compared to a reference signal and used to control the amount of fuel injected into the cylinders. The efficiency of the catalytic converter (in the exhaust line) is high for eliminating both NO_x and unburned CH and CO *only in a very narrow range* around 14.6/1, so a *precise* control of the air/fuel ratio enables the engine to operate with minimum pollution.

Note that if the engine rpm is taken from the speedometer, then a broken speedometer cable will disable the engine!

The eight outputs usually controlled are (1) the air/fuel ratio, (2) the spark timing, (3) the idle speed, (4) the transmission torque converter clutch, (5) the exhaust gas recirculation, (6) the fuel preheating, (7) the purging of the adsorbed fuel vapor from a charcoal cannister into the intake manifold, and (8) the secondary airflow to the catalytic converter.

For each start the microprocessor assumes the previous values for the operating conditions. Another interesting feature is the capability for on-board diagnosis of malfunctions. If a system malfunctions, an error LED can be lit on the dash and an error message *stored* in the RAM. Even if the malfunction is transient, the error message will remain stored and can be read by a mechanic at a later time in the garage. The same basic unit can be used for a line of cars, with different plug-in ROMs for the different models. The electronics are usually located under the dash to avoid the hostile environment (varying temperatures, dust, water, etc). of the engine compartment, although this means more cabling and connectors are required than if the unit were located right next to the engine. There is redundant hardware in the crucial spark and air/fuel mixture control circuits so that a malfunction will not make the vehicle undrivable; if a malfunction occurs, the backup hardware is automatically switched on. There is also redundant software. For example, if the engine coolant temperature sensor fails, the software will automatically assume a hot engine temperature, thus ensuring enough circulation of cooling fluid to avoid engine damage. All of these features tend to be found on larger cars because the smaller models often can meet legislated economy and pollution limits without such sophistication. For this reason few Japanese cars have such systems.

Future extensions of microprocessor control could include antiskid braking control, collision avoidance in connection with an on-board short-range radar transmitter/receiver, cruise control, and a tire-pressure monitor.

16.9 MICROPROCESSOR BUSES

In this section we will discuss some of the more popular buses used to transfer information between a microprocessor or microcomputer and other devices, such as A/D converters, printers, and others. We will discuss only parallel buses. The most common serial technique is the RS-232 standard, which we have already covered in Chapter 15. The advantage of parallel communication over serial communication is, of course, speed.

The typical bus contains eight data lines (which are usually bidirectional—they can transmit information in either direction), 16 address lines which can address up to 64K locations in memory, power supply and ground lines, and a number of control lines.

There are basically two general approaches to the organization of bus lines: by logical function and by hardware criteria. Using logical function would mean placing all 16 address lines in sequence on pins 0 to 15, all eight data lines in sequence, and so on. Using hardware criteria would mean placing the lines in the best locations to minimize crosstalk from one line to another (for example, from a clock line to a DATA line) or to minimize the damage to the circuit if the card were inserted incorrectly (for example, if a 12-V pin accidentally touched the 8-V unregulated input to the 5-V regulator chips, all the 5-V regulator chips and possibly many of the chips would be destroyed. This is a real danger in the S-100 bus if it is put in "backwards.")

16.9.1 The S-100 Bus (IEEE-696)

The most popular bus is probably the S-100 bus, which originated in 1976 in the Altair microcomputer, which was built around the popular 8080 microprocessor. It contains 100 pins, as the name implies, and is not organized according to either logical function or hardware criteria! It more or less evolved from the original design in 1976, with few changes, but it is now standardized. It can be used in connection with other microprocessors—the Z-80, 6800, 6502, among others.

It contains eight DATA IN lines, eight DATA OUT lines, which are often hard-wired together on the peripheral boards to make an on-board bidirectional eight-data-line bus. However, these 16 data lines can obviously be used for the newer 16-bit microprocessors, so this is an advantage. It contains 16 address lines plus 7 more "extended address" lines, making a possible total of 23 address lines that can address 16.8MB of memory ($2^{23} = 16.8 \times 10^6$). It contains two unregulated 8-V lines, which must be converted to a regulated 5 V by an on-board regulator chip on each card, and unregulated +16-V and −16-V lines for ±12-V regulators. The control lines include eight INTERRUPT lines and 39 other types of control lines. Finally, there are two unspecified lines reserved for future use.

The reason there are so many lines is that the S-100 bus was developed before the development of a system controller for the 8080 microprocessor, so separate lines had to be allocated for each control function.

Literally hundreds of different devices (A/D converters, memory, etc.) are available on S-100 cards, and the IEEE has standardized the S-100 bus, so it should be around for a long time.

16.9.2 The STD Bus

The STD bus is very popular and uses relatively small ($4\frac{1}{2}'' \times 6\frac{1}{2}''$) edge-connector cards that plug into a 56-pin socket. The socket contains 28 actual pins, with the two sides of each pin electrically separate. The STD bus is generally used for control rather than for data processing and is usually restricted to eight-bit systems.

There are four sections in the STD bus. (1) The power distribution section includes two pins for +5-V inputs, two pins for ground, two for −5-V inputs, two for ±12-V inputs, and two for an auxiliary ground. Thus, both digital (5 V) and analog (12 V) systems can be accommodated. (2) The data section provides eight data pins for bidirectional data flow, both in and out. All eight data lines are three-state buffered. (3) The address section includes 16 address pins with the low-order byte (A_0 to A_7) on one side of the card and the high-order byte (A_8 to A_{15}) on the other side. All 16 address lines are three-state buffered. (4) The control bus section provides the usual read/write comments, address commands, bus acknowledge, interrupt, and so on.

Each card plugs into a standard bused *motherboard* or *cardrack* or *backplane* that supplies the ±5 V, ±12 V, and ground. Any card can be plugged into any slot in the motherboard.

Many different types of cards are available: A/D and D/A converters; CPU or microprocessor cards for the 8085, Z-80, 6502, and 6800; RAM, PROM, and EPROM memory cards; and output cards capable of driving motors or relays. A complete microcomputer system can be assembled with STD bus cards. One of the more recent STD bus cards by Cubit contains an 80186 16-bit CPU, 64K of EPROM, 128K of RAM, and serial and parallel I/O all on one STD bus card. Two of the main suppliers of STD bus cards are Pro-Log Corp. and Mostek Corp.

16.9.3 The SS50 Bus

The SS50 bus was developed for 6800 microprocessor systems by Southwest Technical Products. It contains 50 lines for memory and the processor and a 30-line subset for peripherals. There is no standard numbering system for the lines; the names of the lines are printed directly on the motherboard. It contains the usual 8 data lines, 16 address lines, power supply and ground lines, and control lines. Its 8-V and ±12-V lines are unregulated.

16.9.4 The LSI-11 Bus (the Q-bus)

The LSI-11 bus is based on the Digital Equipment PDP-11 computer and is an asynchronous 16-bit bus. Because it is asynchronous, data transfer is based on handshaking protocol rather than controlled by a master clock. The power supply pins supply regulated +5 V and ±12 V.

16.9.5 TRS-80 Bus

The TRS-80 bus is made for the popular TRS-80 microcomputers by Radio Shack. It is designed for a Z-80 microprocessor system and contains 40 pins, but the TRS-80 III bus has been expanded to 50 pins. It contains the usual 8 data lines, 16 address lines, power supply, and control lines.

16.9.6 Apple II Bus

The Apple II bus is designed for the popular Apple II microcomputer, which uses the 6502 microprocessor. It contains 50 pins for external peripheral devices. It contains the usual 8 data lines, 16 address lines, power supply and control lines. It supplies ±5 V and ±12 V.

16.9.7 The GPIB or HPIB (IEEE 488)

The GPIB (general-purpose interface bus) or HPIB was developed by Hewlett-Packard and is sometimes known as the Hewlett-Packard interface bus. It is an asynchronous or handshaking bus intended mainly for interfacing instruments to one another and to a controller or computer. It is available on the Commodore Pet microcomputer. It connects three types of instruments: (1) a "talker" instrument, such as an A/D converter or counter that inputs data, (2) a "listener" instrument, such as a pulse generator which is controlled (sent data) by the controller, and (3) a talk/listen instrument that can either talk or listen, such as a digital multimeter. These three types of instruments are connected to a controller (usually a microprocessor or microcomputer) with 16 lines.

The lines are eight bidirectional data lines, three byte transfer control lines for the handshaking, and five general control lines. Notice that there are no separate address lines; the data lines must be used.

The three transfer lines are DAV (data available), NDAC (not data accepted), and NRFD (not ready for data). When the data are valid, DAV is high. When NDAC is low, the data have been accepted. When NRFD = 0, the listener is ready for data. The basic idea of data transfer is done with a handshaking protocol. The unit having the data (the talker) asks the proposed listener, "Are you ready to receive data"; the listener replies, "Yes, I'm ready"; the talker then sends the data, and the listener replies "OK, data received."

The five control lines are ATN (attention), IFC (interface clear), REN (remote enable), SRQ (service request), and EOI (end or identify).

Two chips made to implement the GPIB operation are the 8291 talker/listener and the 8292 controller.

16.9.8 The CAMAC System (IEEE 582)

The CAMAC (computer-automated measurement and control) system is more than just a bus; it includes standards for power supplies and other hardware as well as for the bus. It was developed by the European Standard of Nuclear Electronics and is intended to be a standardized general system usable with any type of computer. It specifies the mechanical dimensions of all the system modules. A standard CAMAC "crate" fits in a 19-in. rack and accepts up to 25 modules, each 17.2 mm × 200 mm × 306 mm. Many different modules are available, each with 86 pins that fit in the 86-pin sockets in the crate.

The CAMAC system is used primarily for nuclear physics experimentation in which the data come in very rapidly; it can handle up to 24 Mbits/s. The standard CAMAC power supply includes ±6-V, ±12-V, and ±24-V power supplies, 3 control lines, 33 command lines, 5 address lines (one to specify the plug-in module, the other four to specify the location within the module), 24 READ DATA lines (which can handle three bytes in parallel), 24 WRITE DATA lines (also three bytes in parallel), 2 timing lines, and 27 status lines.

PROBLEMS

1. Briefly describe the function of the microprocessor.
2. How many memory locations can be addressed with an 8-bit address? With a 16-bit address?
3. What are the usual four parts or "fields" of an assembly language program?
4. Write a brief 8085 assembly language program to add the contents of memory location 2010 to the contents of 2011 and place the sum in 2012.
5. Sketch a block diagram for a microprocessor system (including RAM) to generate an arbitrarily shaped repetitive waveform.
6. Write a brief 8085 assembly language program to make: (a) port 0 an input, (b) port 0 an output.
7. Write a brief 8085 assembly language program to generate a repetitive series of pulses with an output on port 1 high for τ and low for 5τ. How can τ be varied?
8. Describe the difference between a recursive and a nonrecursive digital filter.

9. Sketch a graph of the digital filter output versus input frequency for $x_n = a_0 x_n + a_1 x_{n-10}$.
10. Repeat Problem 9 for $x_n = a_0 x_n + a_1 x_{n-3}$.
11. For a simple nonrecursive filter $y_n = \sum_{k=0}^{6} a_k x_{n-k}$, show that the filter gain

$$H(\omega) = 2a_0 \cos 3\omega t_s + 2a_1 \cos 2\omega t_s + 2a_2 \cos \omega t_s + a_3$$

12. Prove (16.25).
13. Prove $a = -\frac{1}{8}$, $b = -\frac{1}{4}$, $c = \frac{3}{4}$ for a high-pass nonrecursive filter $y_n = \sum_{k=0}^{4} a_k y_{n-k}$ with $H(0) = 0$, $H(f = f_s/4) = 1$, $H(f = f_s/2) = 1$.
14. Calculate $H(\omega)$ for $y_n = \sum_{k-0}^{8} a_k x_{n-k}$.
15. Write a temperature control program with hysteresis.

APPENDIX A

Components: Resistors, Capacitors, Inductors, and Transformers

A.1 RESISTORS

The principal types of resistors along with their characteristics and applications are given below.

TYPE	*CHARACTERISTICS*
Carbon	Inexpensive, common. Useful up to hundreds of MHz. Wide range of resistance values. 1 Ω–22 MΩ, $\frac{1}{8}$ W–2 W.
Metal Film	Stable, high voltage rating, available up to 10 kV. Low noise, low temperature coefficient: 25–100 ppm/°C. Very small inductance due to helix shape of film. Low distributed capacitance causes drop off of impedance at higher frequencies; for example, 10% drop at 300 kHz for 10 MΩ, and 10% drop at 3 MHz for 1 MΩ at 30 MHz for 100 kΩ at 100 MHz for 10 kΩ at 300 MHz for 1 kΩ
Wirewound	High power capabilities available—up to 1000 W. Low temperature coefficient: 20–100 ppm/°C. Maximum voltage 500–1000 V. Relatively high inductance; thus useful only at frequencies below several hundred Hz.
Potentiometers (variable resistors)	*Wirewound*—For low frequency; ac impedance tends to rise well above dc resistance at tens of MHz (high precision). *Nonwirewound*—Better high-frequency performance (less inductance), lower precision. *Metal Film*—Useful up to 50 MHz.

Dual-in-line resistor packages are available for printed circuit boards; they are especially useful for pullup resistors for open collector TTL chips. Square resistor packages are also available for printed circuit boards.

Chip or surface mount resistors and capacitors are also available with no leads for high-density circuits.

A.2 CAPACITORS

The principal types of capacitors and their characteristics are given below.

TYPE	*CHARACTERISTICS*
Paper dielectric	Primarily for frequencies less than 10 MHz. Inexpensive. 0.001–10 μF, 100–2000 V max. WVdc (working voltage dc).
Polyester dielectric	0.001 –0.5 μF. Compact, inexpensive. Resistance greater than 10^{10} Ω. Less than 1% power factor at 1 kHz. Temperature coefficient: 150 ± 50 ppm/°C, −40° to +75°C, 500 WVdc. Constant capacitance up to 20 kHz (1% low at ~20 kHz).
Polystyrene film dielectric	Resistance greater than 10^{14} Ω, 0.001–0.5 μF. −40° to +85°C. Less than 0.005% power factor. Temperature coefficient: 150 ± 50 ppm/°C. 500 WVdc. Long-term stability ±0.2%.
Ceramic	1 pF–1 μF. High capacitance per unit volume. Many different temperature coefficients available.

Temperature-Compensated Type	*Temperature Coefficient (ppm/°C)*
P100	+100
N750	−750
NPO	0
N030	−30

TYPE	*CHARACTERISTICS*
	General-purpose "high K." Inexpensive, wide capacitance tolerance, large temperature coefficient.
	Special types available to carry up to 10 A rf current at frequencies up to several hundred MHz.
	Disc ceramic bypass capacitors form a series resonant

circuit due to lead inductance. The following table gives the self-resonant frequency for various values assuming $\frac{1}{2}$-in. lead length.

Capacitance Value (μF)	*Self-Resonant Frequency* (MHz)*
0.01	165
0.001	55
0.0001	165

* From *Radio Amateur's Handbook*, courtesy of American Radio Relay League.

Such bypass capacitors act properly only at frequencies approximately equal to or less than their self-resonant frequency.

Electrolytic (Tantalum) — High capacitance per unit volume. Polarized. 0.01–5000 μF. Capacitance falls off and power loss increases appreciably at 10 kHz and higher. Special types available for rf bypass applications up to 10 MHz. Approximately 300 WVdc max.

Electrolytic (Aluminum) — 1–10,000 μF. Extremely high capacitance values per unit volume. Polarized. Generally good only at frequencies less than 50 kHz. Commonly used as filters in power supplies. 450 WVdc max.

Mica — 5 pF–0.005 μF. Stable, high Q, close capacitance tolerance. Good for tuned circuits. 500 WVdc max.

A new development is the *supercap* double-layer capacitor, which has from 10 to 25 times the capacitance per unit volume of an ordinary aluminum electrolytic capacitor, combined with an extraordinarily low leakage current. Capacitance values up to 1 F (1,000,000 μF) at 5 V and 10 V are available at prices under $10 each. These capacitors are +80% to −20% in capacitance value like most electrolytics, but, unlike electrolytics, the supercaps are nonpolar. They can be used as standby power sources for CMOS memories and microcomputers for up to several weeks during a power outage. An NEC FZ OH105Z has 1.0-F capacitance in a cylindrical case 1.1 in. in diameter and 0.98 in. high for $9.65 each in small quantities.

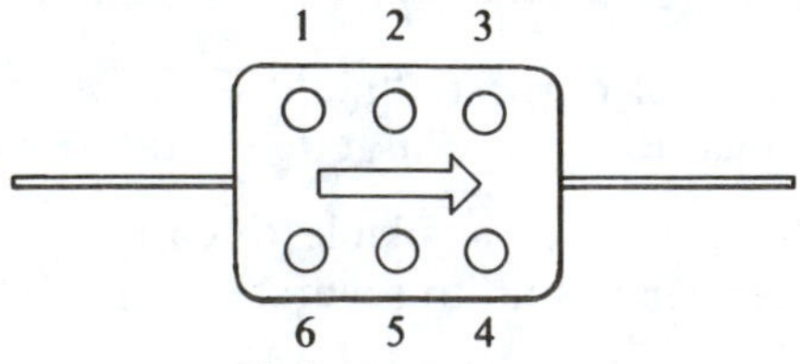

FIGURE A1.1 Molded mica capacitor color code.

C in pF from colors in (2), (3), and (4) positions per the resistor color code; for example, (2) = yellow, (3) = violet, and (4) = brown means C = 470 pF. (1) = white for commercial—black for mil. spec. (5) = tolerance; black ±20%, silver ±10%, gold ±5%, and brown ±1%. (6) = brown ±500 ppm/°C and ±3% drift; yellow −20 to +100 ppm/°C and ±0.1% drift; and green 0 to 70 ppm/°C and ±0.05% drift.

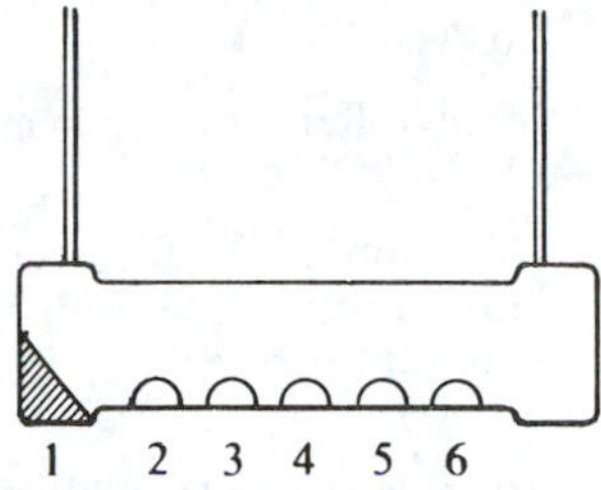

FIGURE A1.2 Ceramic tubular capacitor color code.

(1) and (2) determine temperature coefficient.

(1) Black "NPO" zero	(1) Silver "Y5D" ±3.3%
(2) missing temp. coeff.	(2) Brown −30° to +85°C

(1) Brown "Z5U" +22%–56% capacitance variation +10°C to +85°C
(2) Gray

C in pF from colors in (3), (4), and (5) positions by the resistor color code; for example, (3) = orange, (4) = orange, and (5) = red means C = 3300 pF. (6) = black ± 20%, white ±10%, and green ±5%.

A.3 INDUCTORS (RF CHOKES)

Type	*Self-Resonant Freq. (MHz)*	*Max. dc Current (A)*	*Q*	*dc Resistance* (Ω)
25-mH ferrite core 3-section (Miller 6308)	0.47	0.065	102	82
2.5-mH iron core 3-section (Miller 4666)	1.7	0.160	80	15
1-mH phenolic core 3-section (Miller 4652)	3.7	0.160	59	19
100-μH iron core single layer (Miller 4632)	12	0.400	107	3
10-μH iron core single layer (Miller 4622)	40	1.5	69	0.11
1-μH phenolic core single layer (Miller 4602)	190	2.0	60	0.05
0.1-μH phenolic core single layer (Miller 4580)	500	3.0	68	0.017

rf chokes should be used at frequencies less than their self-resonant frequency.

(Courtesy of J. W. Miller Inc.)

A.4 TRANSFORMERS

Type	*Characteristics*
Laminated iron core	Used up to several hundred hertz. Higher frequencies require smaller core. Special audio transformers available for 20 Hz–20 kHz.
Powdered iron core	Many different types available for use from 60 Hz–250 MHz.
Ferrite iron core (ceramic ferromagnetic)	High core resistivity, stable, low eddy current loss. Used from 1 kHz–1 GHz. Constant permeability available up to 30 MHz.
Air core	10 MHz–450 MHz typical. At 450 MHz only approximately one turn needed per winding. For higher frequencies microwave cavity and waveguide techniques are used.

APPENDIX B

Batteries

Any two different metals (called *electrodes*) placed in a solution containing ions (called the *electrolyte*) will generate a voltage difference between the two metals. For example, carbon and zinc electrodes in a solution of ammonium chloride will generate a voltage difference of 1.5 V. This is the familiar flashlight battery. Such an arrangement of two electrodes in an electrolyte is called a *cell*. Several cells mounted in one package are called a *battery*; for example, four 1.5-V carbon–zinc cells connected in series form a 6-V battery. Strictly speaking, a 1.5-V carbon–zinc flashlight battery is really a cell. In everyday usage, however, the word "battery" refers to either a cell or a battery. There are six basic types of batteries whose main characteristics and applications are listed below.

Battery Type	*Volts per Cell*	*Characteristics*	*Application*
Carbon–zinc (flashlight battery)	1.5	Inexpensive, widely available in many sizes. Voltage falls off gradually with use. Poor shelf life, especially at temperatures above 90°F. Poor performance below 32°F.	Inexpensive equipment, toys.
Alkaline–manganese	1.5	Good for relatively high current applications, but no better than carbon–zinc for low current applications. Good low temperature performance. Up to twice as much energy as carbon–zinc. Voltage falls off gradually with use. Good shelf life.	Radios, toys, movie cameras, electronic flash.

Battery Type	*Volts per Cell*	*Characteristics*	*Application*
Mercury	1.35	Good at high temperatures (up to 130°F). Poor below 40°F. Voltage remains constant with use until a sudden fall off when "used up." Excellent shelf life. Constant ampere-hour capacity for different load currents.	Portable scientific equipment, TVs, radios, not exposed to low temperatures.
Silver oxide	1.5	Good at low temperatures. Voltage remains constant with use. Mainly for low current applications. Excellent shelf life. Good shelf life at 10°F or cooler.	Portable instruments, hearing aids, electric watches.
Nickel–cadmium	1.25	Good at both high and low temperatures, −4°F to +113°F. Voltage remains constant with use—typically from 1.25–1.1 V per cell. May be recharged separately or in series. For parallel recharging, each cell or battery must have its own current-limiting resistor. Charging current should equal ampere-hour capacity ÷ 10.	Critical portable equipment—radiation detectors, radios, satellites, alarms.
Lead–acid (automobile battery)	2.6	Capable of supplying high currents (up to 100 A). Very heavy. Rechargeable. Exudes flammable hydrogen gas when recharged. Corrosive sulfuric acid electrolyte.	Automobiles, portable high-power equipment.

(Courtesy of Union Carbide Corporation.)

A no-load or open-circuit measurement of the battery terminal voltage may or may not identify a bad battery. If the battery terminal voltage is low, then the battery is certainly bad; if high then the battery may be good or bad. As a general rule, batteries should be tested by measuring the terminal voltage with a load that draws approximately one half the recommended current; for example, if the recommended current is 100 mA or less, the battery terminal voltage should be measured with a load drawing 50 mA from the battery. A low terminal voltage indicates a bad battery.

A final comment: Always use a voltmeter *not an ammeter* to test a battery. Connecting an ammeter across a battery will usually destroy the ammeter and damage the battery. For more details see various battery application data manuals from the battery manufacturers.

APPENDIX C

Measuring Instruments

C.1 METERS

Digital multimeters (DMM) are gradually replacing analog meters. They cost approximately the same as the analog meters with comparable scales but are available in more precise (and more expensive) versions. They are mechanically sturdier because they do not contain the delicate analog meter movement. They are usually specified as "$3\frac{1}{2}$"- or "$4\frac{1}{2}$"-digit meters, which requires an explanation. The "$\frac{1}{2}$" means the most significant digit can be only a 0 or a 1. The other digits can be anything from 0 to 9. Thus, a $3\frac{1}{2}$-digit meter can read a maximum of 1999, a $4\frac{1}{2}$-digit meter a maximum of

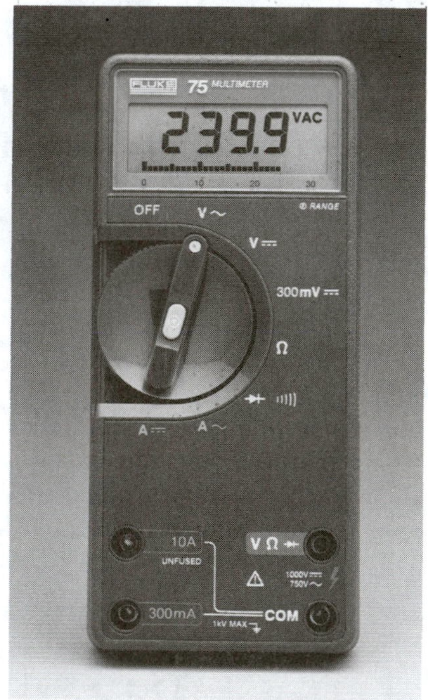

FIGURE C1.1 Digital multimeter. (Reproduced by permission of John Fluke Mfg. Co., Inc.)

19,999. If the most significant digit can be either 0 or 1 or 2 (plus three other ordinary digits), it is called a "$3\frac{3}{4}$"-digit meter. Thus, the maximum reading for a $5\frac{3}{4}$-digit meter is 29,999. The decimal point can usually be anywhere.

A small hand-held $3\frac{1}{2}$-digit DMM typically costs \$100 to \$200 and measures 0.2 to 1000 V (ac and dc), 2 mA to 2 A, and 200 Ω to 20 MΩ. The input impedance is usually 10 MΩ for all the voltage ranges. One such meter is the Fluke 75, shown in Fig. C1.1.

A large benchtop DMM typically costs \$500 for $4\frac{1}{2}$ or $5\frac{1}{2}$ digits and measures 0.02 to 1000 V, 20 Ω to 20 MΩ, and 0.2 mA to 2 A with a 10-MΩ input impedance and a 0.03% accuracy.

Digital capacitance meters that measure from 199.9 pF to 1999 μF full-scale with a 0.1% accuracy for a typical $3\frac{1}{2}$-digit readout are also available.

Digital thermometers that measure from −65° to +150°C with 0.1°C resolution with an 8-s response time are available.

C.2 OSCILLOSCOPES

The cathode ray oscilloscope or scope is still probably the single most useful electrical instrument. It can measure dc or ac voltages directly, currents indirectly, and most commonly is used to display the actual waveform of a voltage—that is, to plot a graph of voltage versus time on the face of the cathode ray tube.

A simplified block diagram of the oscilloscope is given in Fig. C1.2. The basis of the oscilloscope is the cathode ray tube, which is essentially a TV picture tube. The interior is highly evacuated, and the inside of the broad face is coated with a phosphor that emits light when struck by electrons. A hot filament emits electrons, and an accelerating voltage and series of focusing electrodes produce a thin beam of electrons aimed from the neck of the tube toward the broad face. Thus, a small light spot would be seen in the center of the screen if nothing deflects the electron beam. The signal voltage to be observed is fed into the vertical input where it is amplified and applied to the vertical deflection plates. If the signal voltage is positive, the upper deflection plate is made positive with respect to the lower plate, and the electron beam is deflected upward. On the face of the cathode ray tube the bright spot moves upward. If the signal voltage is negative, then the lower deflection plate is made positive with respect to the upper plate, and the spot moves downward.

At the same time the electron beam is moving up and down in response to the signal voltage, it is also being swept horizontally by a sawtooth voltage generated within the oscilloscope and applied to the horizontal deflection plates. As the sawtooth increases linearly, the electron beam is swept horizontally across the face of the cathode ray tube from left to right

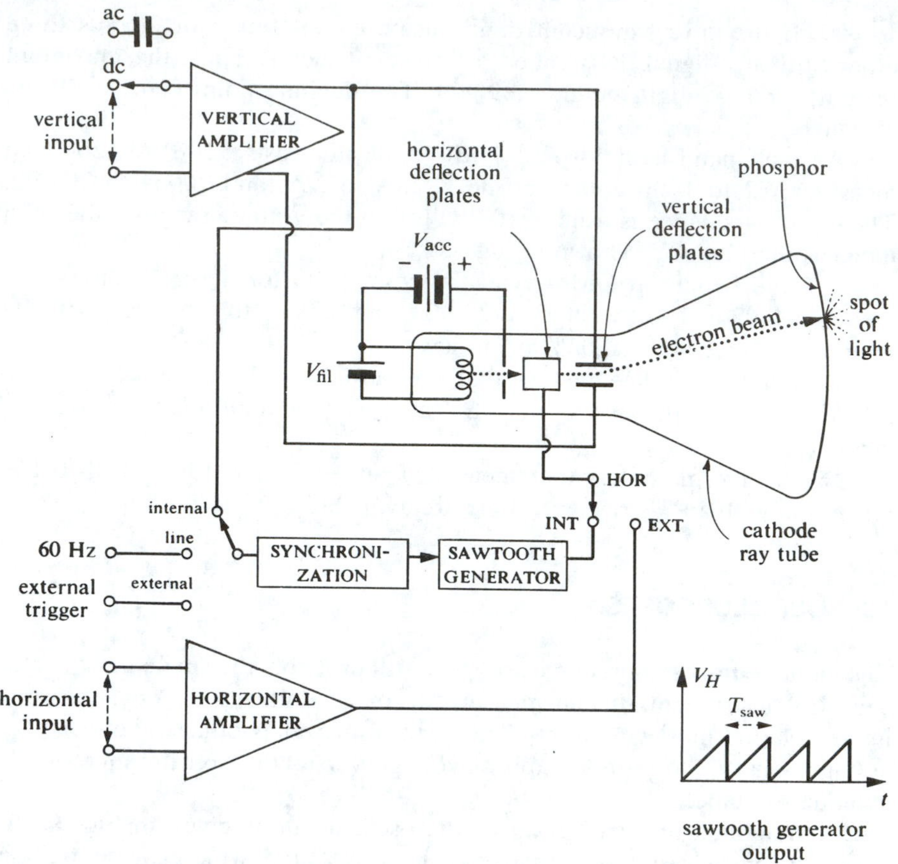

FIGURE C1.2 Cathode ray oscilloscope block diagram.

(as seen by the observer) at a constant speed. This speed is read off the knob labeled "sweep speed" or "time/div.," etc., depending on the particular scope used. The speed is usually expressed in time for the beam to move one horizontal division on the CRT—for example, 50 μs/div rather than in cm/s: the steeper the sawtooth voltage (the shorter the period T_{saw}) the faster the sweep speed and the smaller the time/div. When the sawtooth horizontal sweep drops rapidly to zero again, the electron beam is quickly deflected back to the left. During this short time, the tube is usually blanked out; that is, the electron beam is actually stopped during this time, so the observer cannot see the rapid right-to-left traverse of the beam. It is instructive for the beginner to set the horizontal sweep speed to a very low value, say, 0.5 s/div with no vertical signal input, and watch the beam steadily move across the face of the cathode ray tube at a constant speed of 2 div/s from left to right. Be sure to set the intensity or brightness control to a low enough value so that the phosphor is not burned by the beam spot. For higher sweep speeds, say, 10 ms/div or 50 μs/div, the beam spot moves so rapidly that the observer sees a smooth horizontal line because the phosphor glows for approximately 0.1–0.5 s after the electron beam strikes it. Thus a new sweep comes along to excite the phosphor before the light from the preceding sweep has completely disappeared.

With a sinusoidal input of frequency f and period $T = 1/f$, if the sawtooth period T_{saw} equals T, then one sine wave will be displayed; if T_{saw} equals $2T$ (a slower sweep), then two sine waves will be displayed, and so on. The purpose of the synchronization or sync or trigger controls is to ensure that the horizontal sweep starts (is triggered) at the *same point* on the input signal voltage for each sweep. Thus, successive sweeps will lie on top of one another, yielding one stationary display on the face of the cathode ray tube. If the horizontal sweep starts at different points on the input signal voltage for different sweeps, then the waveforms displayed will not lie on top of one another and a blurred picture will be produced. The scope is then said to be *out of sync* or not triggering properly.

The simplest type of synchronization or triggering is the *internal* or *automatic* sweep mode in which the horizontal sawtooth is automatically started when the vertical signal input reaches a certain low-voltage value. No external adjustments are required, and this usually results in a stable display unless the signal is extremely low in amplitude and/or high in frequency. Usually there are two internal triggering modes, + and −, meaning that the sweep starts when the vertical signal input goes positive or negative, respectively. The *external* triggering mode means that the horizontal sweep sawtooth does not start unless a signal voltage from outside the scope is fed into the "ext. sync." jack. On many scopes there is a "line" triggering mode, which means that the horizontal sweep sawtooth is started in phase with the line voltage (usually 110 V, 60 Hz). This triggering is useful in displaying waveforms that are phase coherent with the 110-V line voltage. For example, if a high-frequency signal appears blurred when the conventional internal triggering is used due to the presence of a 60-Hz amplitude modulation, then switching to line triggering will lock in the 60-Hz envelope of the signal or, in fact, any envelope that is phase coherent with the line (e.g., 120 Hz), and so forth. Often this is the case when the power supply voltage is insufficiently filtered, and a strong 60-Hz or 120-Hz ripple voltage appears superimposed on the dc supply voltage.

C.3 BRIDGES

One general class of measuring instruments that does not require a calibrated meter is the bridge. Bridges can be used to measure resistance, capacitance, and inductance as well as other parameters. In a bridge the meter is used as a null indicator; that is, when the bridge is properly adjusted, the meter reads zero and the value of the parameter being measured (e.g., resistance) is determined by the values of the bridge components.

The simplest bridge is perhaps the Wheatstone bridge, which we discussed in Chapter 1. If an ac signal source is used instead of a battery, then we have an ac bridge that can be used to measure L or C. An ac capacitance bridge is shown in Fig. C1.3 for relatively high-Q capacitors

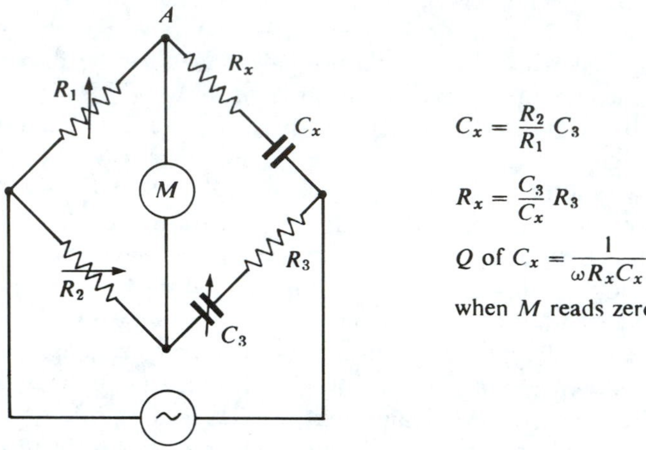

FIGURE C1.3 An ac capacitance bridge.

C_x, $Q \geq 1$. Value R_x is the series resistance of the capacitor C_x (R_x is zero for an ideal capacitor). The null indicator may be an ac meter or a sensitive oscilloscope. The ac source can be at any frequency; hence the variation of capacitance with frequency can be studied. When the bridge is balanced so that the meter or scope reads zero, then $V_A = V_B$ and the unknown capacitance is given by $C_x = (R_2/R_1)C_3$.

An ac inductance bridge is shown in Fig. C1.4 for high-Q inductors, $Q > 1$. Notice that R_x is the effective *parallel* resistance of L_x. Thus the higher R_x is, the higher Q is: $Q = R_x/\omega L_x$. In making capacitance and

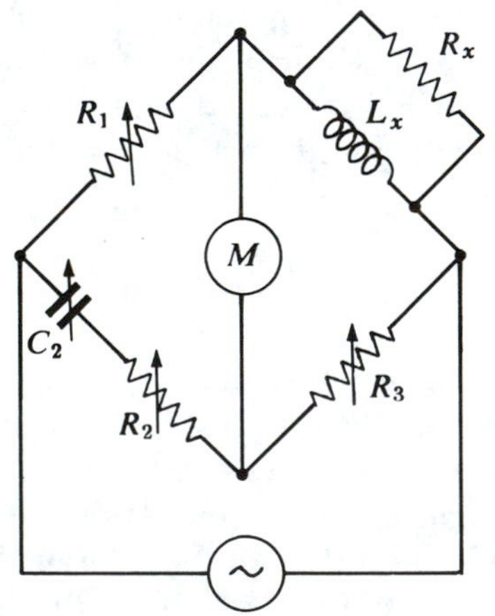

$$L_x = R_1 R_3 C_2$$

$$Q = \frac{R_x}{\omega L_x} \omega R_2 C_2$$

when M reads zero

FIGURE C1.4 An ac inductance bridge.

inductance measurements, remember that often the impedance of the component being measured is not purely capacitive or purely inductive. However, the impedance of *any* real (lossy) component can always be written as

$$Z = a + jb$$

where a and b are real numbers. And the equivalent circuit for any real (lossy) component can be written as *either* a series or a parallel combination of a resistance and a reactance.

TABLE C1.1. Component Impedances

Component	a	b	Z
Pure resistance	Positive	Zero	R
Pure capacitance	Zero	Negative	$j\left(-\frac{1}{\omega C}\right)$
Pure inductance	Zero	Positive	$j(\omega L)$
Lossy capacitance	Positive	Negative	$R_{ser} + j\left(-\frac{1}{\omega C_{ser}}\right)$ or $\frac{R_{par}}{1 + j\omega R_{par} C_{par}}$
Lossy inductance	Positive	Positive	$R_{ser} + j(\omega L_{ser})$ or $\frac{R_{par}(j\omega L_{par})}{R_{par} + j\omega L_{par}} = \frac{R_{par}\omega^2 L_{par}^2 + j\omega L R_{par}^2}{R_{par}^2 + \omega^2 L_{par}^2}$

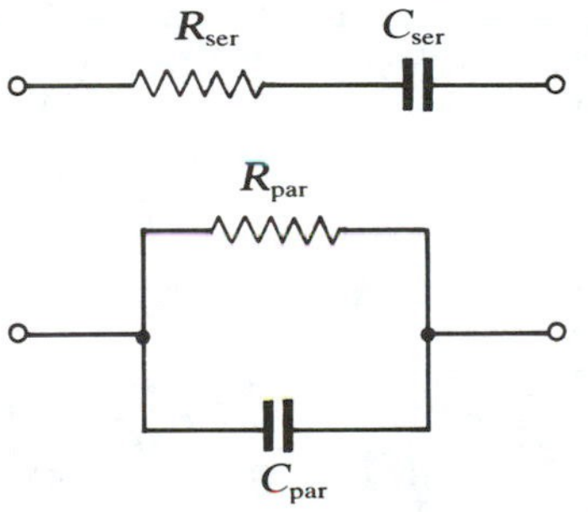

R_{ser} C_{ser}

R_{par} C_{par}

$$\frac{R_{par}\left(\frac{1}{j\omega C}\right)}{R_{par} + \frac{1}{j\omega C}} = \frac{R_{par} + j(-\omega R_{par}^2 C)}{1 + \omega^2 R_{par}^2 C^2}$$

(a) *lossy capacitance*

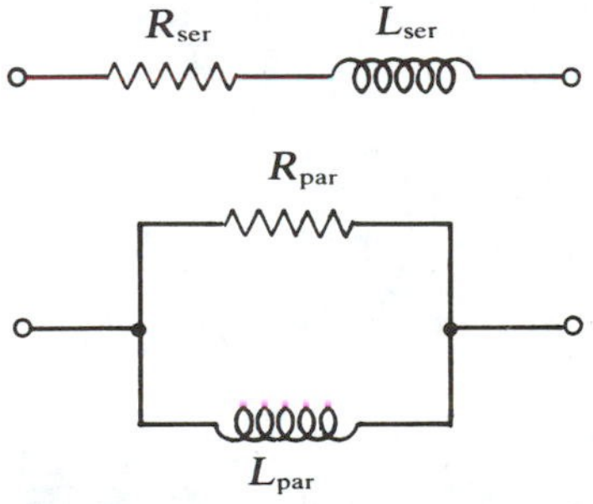

R_{ser} L_{ser}

R_{par} L_{par}

(b) *lossy inductance*

Commercial impedance bridges are available that incorporate dc and ac resistance bridges and ac inductance capacitance bridges in one instrument. The bridge circuits in this section were taken from the General Radio Type 1650 impedance bridge and were used by the kind permission of the General Radio Company.

Modern autoranging digital impedance meters are available which can measure L, C, R, and G (conductance) and dissipation factor D at a 1-kHz test frequency. One such meter is the Model 253 impedance meter from Electro Scientific Industries, Inc. Although a bit expensive, it is worth its weight in gold in the laboratory.

APPENDIX D

Cables and Connectors

Many cables used in scientific apparatus and quality electrical equipment are of the *coaxial* type; an outer grounded conducting sheath completely surrounds one or more inner conductors that carry the desired electrical signals. The principal advantages of such a configuration are shielding and safety. The outer shielding is usually a grounded single sheath of braided copper but may consist of two sheaths for extra shielding for use in noisy environments or where little extraneous pickup can be tolerated. Different types of insulating jackets are available, the most common being black polyethylene (type 111a), which gives good long-term stability and moisture and abrasion resistance from −55°C to +85°C. Polyvinylchloride is, in general, inferior. Teflon jackets are available where high temperatures and/or corrosive chemicals may be encountered. Table D1.1 lists nine of the more commonly used single-center-conductor, coaxial cables along with their mechanical and electrical characteristics.

A variety of coaxial connectors is available to allow one to connect cables to and disconnect cables from other cables and equipment. Connectors come in various sizes to go with the different diameter cables and also with various characteristic impedances to match the cable impedance. Connectors can also be classified according to the type of coupling, whether threaded or bayonet (push-and-twist), and the maximum voltage rating. Threaded connectors are much more resistant to mechanical vibration than are the bayonet type, whereas the bayonet type is most convenient for laboratory use. Usually the connector is chosen to go with the cable. Most of the power loss usually occurs in the cable rather than in the connector, but a bad cable-connector connection can result in large reflections and signal loss.

Many connectors, including BNC, are available in several types with drastically different electrical and mechanical properties. For example, the regular BNC connector line is useful only up to 150 MHz, whereas the improved BNC is useful up to 10 GHz. The standard solder connection between the connector and the cable braid will withstand approximately a 30-lb pull, whereas the BNC crimp connection will withstand a 90-lb pull. Some of the most useful connectors are described in Table D1.2 and are illustrated in Fig. D1.1.

For frequencies below 150 MHz, BNC connectors and RG58/U cable

TABLE D1.1. Cables

Cable Type	*Inner Conductor*	*Outer Diameter*	*Dielectric*	*Characteristic Impedance* (Ω)	*Maximum Voltage* (*V*)	*Attenuation* (*dB*/100 *ft*)		*Capacitance* (*pF*/*ft*)	*Use*
						100 MHz	1000 MHz		
RG58/U	1 copper	0.200″	PE†	53.5	1900	4.1	14	28.5	small g. p.*
RG55B/U	1 silver-plated copper	0.206″	PE	53.5	1900	4.8	17	28.5	double-braided shield, g.p.
RG59/U	1 copper weld	0.240″	PE	73	2300	3.4		21	small g. p.
RG8/U	7 copper	0.405″	PE	50	5000	1.8		26	med. diam., g. p.
RG9/U	7 silver-plated copper	0.420″	PE	51	5000	1.9	6.5	30.0	microwave double-braided shield
RG174/U	7 copper weld	0.100″	PE	50		8.8 (at 400 MHz)		30.0	very small diam.
RG188/AU	7 copper weld	0.105″	Teflon	50				27.5	high temp.
RG62/U	1 copper weld	0.242″	Air	93	750	3.1	1.1	13.5	small diam. low cap.
RG63B/U	1 copper weld	0.415″	Air	125	1000	2.0	6.5	10.0	med. diam. low cap.

* g. p. = general purpose.
† PE = polyethylene.

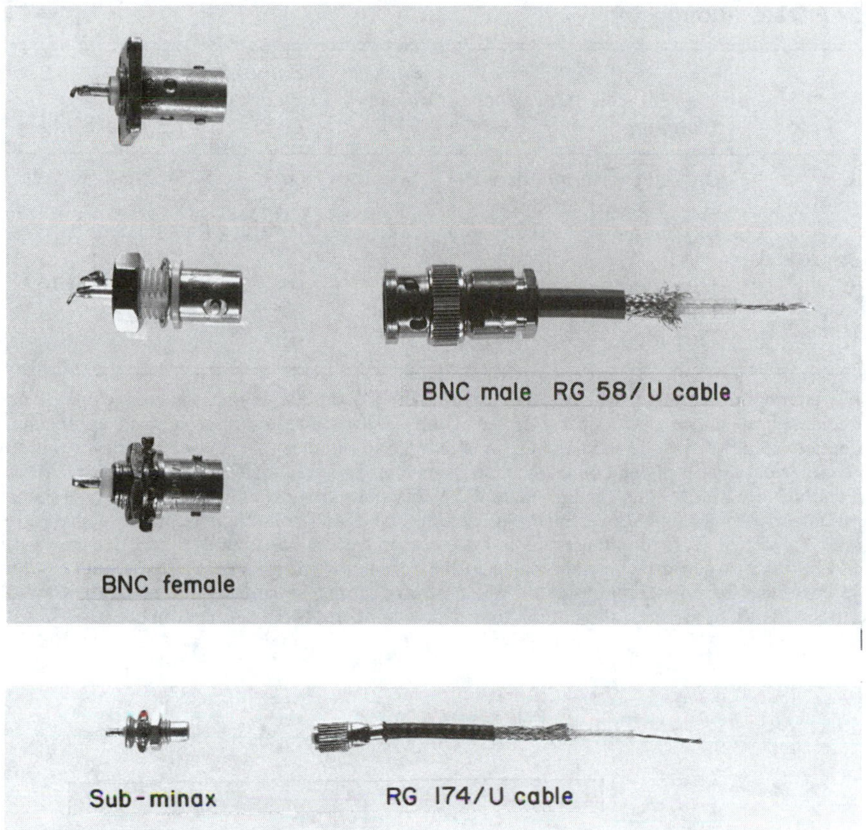

FIGURE D1.1 Coaxial connection and cable.

are good general-purpose choices. For higher frequencies great care should be given to the cable–connector interface, and improved BNC connectors should be used. Type UHF connectors should not be used much above 300 MHz. There are many other types of connectors for high-voltage applications, extreme mechanical strength, and so on. For example, the type APC connector is a precision-threaded 7-mm 50-Ω type for use up to 18 GHz with rigid air dielectric lines. The type SMA connector is a subminiature threaded 3-mm 50-Ω type for use up to 18 GHz.

Mention should also be made of *ribbon* connecters and cables. Ribbon cable can be obtained with 10, 16, 25, 34, 44, or 56 wires (all No. 28/AWG wire) with each wire a different color. It is especially useful for parallel bus lines in digital systems. Crimp ribbon connectors are used for rapid connection.

TABLE D1.2. Connectors

Type	*Coupling*	*Impedance* (Ω)	*Peak Voltage* (*V*)	*Maximum Frequency* (*GHz*)	*Typical Cable*
BNC (regular)	Bayonet	nonconstant	500	0.3	RG–55,58,59,162
BNC (improved)	Bayonet	50	500	10*	RG–55,58,59,162
TNC	Threaded	50	500	10*	RG–55,58,59,162
N	Threaded	50,70	500	10*	RG–8,9
UHF	Threaded	—	500	0.4	RG–8,9

* The maximum frequency of operation really depends on the connector *and* cable. If an appreciable impedance discontinuity exists at the cable–connector interface, an appreciable percentage of power will be reflected and a high VSWR will exist. As a general rule-of-thumb, if the electrical length of the connector is $\lambda/50$ or less, an impedance mismatch is not serious. For a BNC connector pair (male and female) this corresponds to frequencies of less than approximately 150 MHz. Above this frequency the cable and connector impedance should be carefully matched (e.g., don't use a 50-Ω connector with a 75-Ω cable), and extreme care should be taken when attaching the cable to the connector: no gap should exist between the cable core and the connector insulator, all shoulders should be square, and so forth. Careful attention should be given to the cutting and soldering of the cable to the connector.

APPENDIX E

Complex Numbers

E.1 DEFINITION

Any complex number z means simply $z = a + jb$, where $j^2 = -1$ ($j = \sqrt{-1}$), and a and b are real numbers (positive or negative). a is called the *real* part of z, and b is called the *imaginary* part of z. Both a and b must have the same dimensions, e.g., both ohms or both volts and so forth. The $a + jb$ form for a complex number is often called the *Cartesian* form. Examples: $z = 2 + 3j$, $z = -4 + 2j$, $z = R + j\omega L$, $z = R - j/\omega C$.

E.2 COMPLEX CONJUGATE

The *complex conjugate* of z, written as z^*, means $z^* = a - jb$. To obtain the complex conjugate of any complex number, no matter how complicated, merely replace j by $-j$ and $-j$ by j. Examples: If $z = 2 + 3j$, $z^* = 2 - 3j$. If $z = re^{j\theta}$, $z^* = re^{-j\theta}$.

E.3 GEOMETRIC REPRESENTATION

We can represent any complex number by a single point in two-dimensional space by plotting the imaginary part along the vertical axis and the real part along the horizontal axis (Fig. E1.1). The *absolute value* of z or the

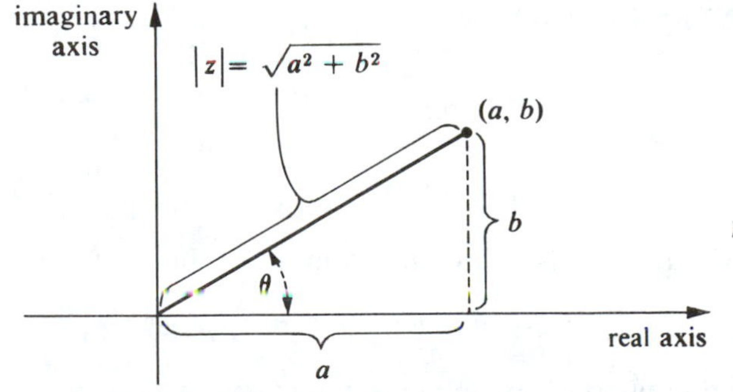

FIGURE E1.1 Geometric representation of a complex number.

magnitude of z, written $|z|$, means the distance from the origin to the point (a, b): $|z| = +\sqrt{a^2 + b^2}$. Notice $|z|$ is always real and positive; and z can represent an impedance, a voltage, or a current.

E.4 POLAR FORM

Any complex number z can also be written in polar form as $z = |z|e^{j\theta}$. Using the trigonometric identity $e^{j\theta} = \cos\theta + j\sin\theta$, we see that $z = |z|\cos\theta + j|z|\sin\theta$. Thus, the real part of z is given by $a = |z|\cos\theta$, and the imaginary part by $b = |z|\sin\theta$. Also $|z| = +\sqrt{a^2 + b^2}$ and $\theta = \tan^{-1}(b/a)$. In polar form the complex conjugate of z is written $z^* = |z|e^{-j\theta}$.

E.5 EQUALITY

Two complex numbers $z_1 = a_1 + jb_1$ and $z_2 = a_2 + jb_2$ are equal if and only if $a_1 = a_2$ and $b_1 = b_2$. In words, their real parts must be equal, *and* their imaginary parts must be equal. If z_1 and z_2 are written in polar form, then $z_1 = z_2$ if and only if $|z_1|e^{j\theta_1} = |z_2|e^{j\theta_2}$: $|z_1|\cos\theta_1 = |z_2|\cos\theta_2$ *and* also $|z_1|\sin\theta_1 = |z_2|\sin\theta_2$.

E.6 ADDITION

Two complex numbers $z_1 = a_1 + jb_1$ and $z_2 = a_2 + jb_2$ can be added or subtracted by the following rule:

$$(z_1 \pm z_2) = (a_1 \pm a_2) + j(b_1 \pm b_2)$$

It is very inconvenient to add two complex numbers in polar form but very easy in the Cartesian form $a + jb$. Addition commutes: $z_1 + z_2 = z_2 + z_1$. Example: $(2 + 3j) + (4 + 5j) = 6 + 8j$.

E.7 MULTIPLICATION

Two complex numbers $z_1 = a_1 + jb_1$ and $z_2 = a_2 + jb_2$ can be multiplied by the following rule:

$$z_1 z_2 = (a_1 + jb_1)(a_2 + jb_2) = (a_1 a_2 - b_1 b_2) + j(a_1 b_2 + a_2 b_1)$$

If z_1 and z_2 are expressed in polar form, multiplication is very easy:

$$z_1 z_2 = |z_1|e^{j\theta_1}|z_2|e^{j\theta_2} = |z_1||z_2|e^{j(\theta_1+\theta_2)}$$

Multiplication of complex numbers commutes: $z_1 z_2 = z_2 z_1$.

E.8 DIVISION

Two complex numbers can be divided by the following rule:

$$\frac{z_1}{z_2} = \frac{a_1 + jb_1}{a_2 + jb_2} = \frac{a_1 + jb_1}{a_2 + jb_2} \times \frac{a_2 - jb_2}{a_2 - jb_2} = \frac{a_1 a_2 + b_1 b_2}{a_2^2 + b_2^2} + j\frac{a_2 b_1 - a_1 b_2}{a_2^2 + b_2^2}$$

where we have multiplied and divided z_1/z_2 by $(a_2 - jb_2)$ to get z_1/z_2 in Cartesian form. Division is much easier if z_1 and z_2 are in polar form:

$$\frac{z_1}{z_2} = \frac{|z_1|e^{j\theta_1}}{|z_2|e^{j\theta_2}} = \frac{|z_1|}{|z_2|}\, e^{j(\theta_1 - \theta_2)}$$

E.9 MISCELLANEOUS

It is often desired to calculate quickly the magnitude of a complex number. Perhaps the most useful technique is to multiply the number by its complex conjugate and to take the square root of the product:

$$zz^* = (a + jb)(a - jb) = a^2 + b^2 = |z|^2$$

Therefore $|z| = \sqrt{zz^*}$.

Some common ac RLC circuits and their complex impedances are given below.

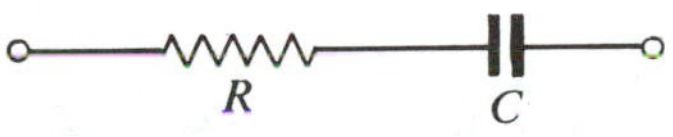

R C (series)

$$z = R + \frac{1}{j\omega C} = R - \frac{j}{\omega C}$$

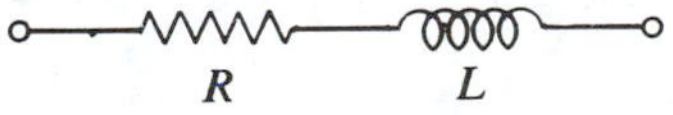

R L (series)

$$z = R + j\omega L$$

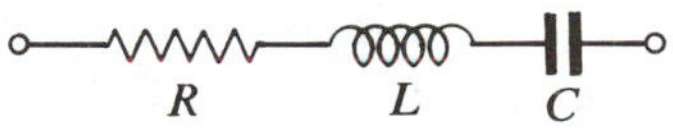

R L C (series)

$$z = R + j\omega L + \frac{1}{j\omega C} = R + j\left(\omega L - \frac{1}{\omega C}\right)$$

C, L (parallel)

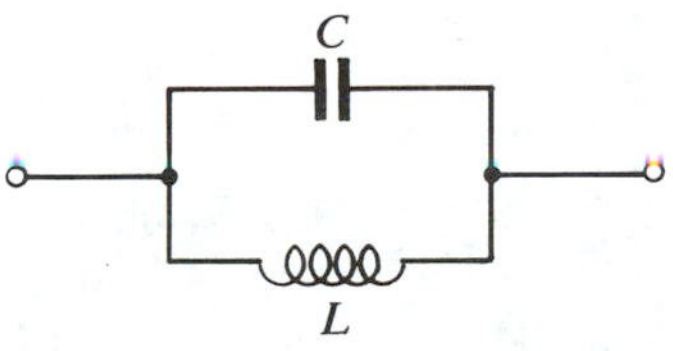

$$z = \frac{(j\omega L)(1/j\omega C)}{j\omega L + 1/j\omega C} = \frac{L/C}{j(\omega L - 1/\omega C)}$$

$$= \frac{-j/\omega C}{1 - 1/\omega^2 LC}$$

APPENDIX F

Best-Loved Transistors

Type	*Case*	BV_{CEO} (*V*)	$I_{C\,max}$ (*mA*)	h_{FE}	f_T (*MHz*)	*Application*
2N3904 npn	P*	40	200	100	300	GP
2N3906 pnp	P	40	200	100	300	GP
2N2222A npn	M†	40	800	100–300	300	GP
2N2907A pnp	M	60	600	100–300	200	GP
2N2723 npn	M	60	40	2,000	100	Small-signal Darlington
2N6427 npn	P	40	500	10,000		Small-signal Darlington
TIP101 npn	P		8A	1,000–20,000		Power Darlington
TIP106 pnp	P		8A	1,000–20,000		Power Darlington
2N3055 npn	M	60	15A	20–70		GP power
2N5879 pnp	M	60	15A	20–100		GP power
2N5088 npn	P	30	300	>100	50	Small signal, low noise
2N5086 pnp	P	50	150	>100	40	Small signal, low noise
LI4G3		45	1.2 @ 1 mW/cm^2, 100 μA dark, 5 μs			Phototransistor
2N5780 npn		40	8 @ 1 mW/cm^2, 100 μA dark, 75 μs			Photodarlington
2N2369A npn	M	15	200	40	12 ns t_{on}	Fast switching
2N4208 pnp	M	12	200	30	15 ns t_{on}	Fast switching
2N918 npn	M	13	50	20	600	Up to 600 MHz
2N4260 pnp	M	15		30		Up to 2 GHz

2N5459 n-channel	JFET	25 V_{GSS}	$I_{DSS} > 4$ mA	2–6 mS		GP
2N5462 p-channel	JFET	40 V_{GSS}	$I_{DSS} > 4$ mA	2–6 mS		GP
40673 n-channel	MOSFET	20 V		12 mS		Dual gate, 400 MHz
VN10KE n-channel	MOSFET	60 V_{DSS}	200	$g_{fs} = 100$ mS††	5 Ω $R_{DS(on)}$	GP enhancement
IRF132 n-channel	MOSFET	100 V_{DSS}	12 A	$g_{fs} = 4$ S††	0.25 Ω $R_{DS(on)}$	Power

* P = plastic.
† M = metal.
††1 mS = 1000 μmhos.

Best-Loved Op Amps ("Typical" Parameters)

Type	V_{io} (*mV*)	I_{io} (*nA*)	I_B (*nA*)	*S.R.* (*V/μs*)	f_T (*MHz*)	$I_{out\,max}$ (*mA*)	V_{cc} (*V*)	*Application*
741	1	20	80	0.5	1.2	20	±22	GP
741C	2	20	80	0.5	1.2	20	±18	
1741S	1	30	200	10			±22	D/A converter (high slew rate)
714E	0.03	0.5	1.2	0.17	0.6		±18	Ultra low offset
4136C	0.5	5	140	1.7	3	20		GP medium speed
324	2	5	45	0.5	1.0	20	32	Single supply
LM11	0.1	0.5	25	0.3	0.5	2	±20	Precision, slow
318	4	30	150	70	15	10	±20	High speed, high slew rate
218	2	6	120	70	15	10	±20	−25° to +85°C, lower V_{io}
TL071C	3	0.005	0.03	13	3	10	±18	BIFET low noise
MC34022A	0.3	0.015	0.04	13	4		±18	Precision BIFET
0P-01	0.3	0.5	18	12	2.5			GP
0P-27	10 μV	7	10	2.8	8	±17		Low noise GP
CA3140A	2	0.0005	0.01	9	4.5	±10 −1	±22	MOSFET input

APPENDIX G

The 8085 Instruction Set

DATA TRANSFER GROUP									ARITHMETIC AND LOGICAL GROUP								
	Move			Move (cont)			Move Immediate			Add*			Increment**			Logical	
MOV	A,A	7F	MOV	E,A	5F	MVI	A, byte	3E	ADD	A	87	INR	A	3C	ANA	A	A7
	A,B	78		E,B	58		B, byte	06		B	80		B	04		B	A0
	A,C	79		E,C	59		C, byte	0E		C	81		C	0C		C	A1
	A,D	7A		E,D	5A		D, byte	16		D	82		D	14		D	A2
	A,E	7B		E,E	5B		E, byte	1E		E	83		E	1C		E	A3
	A,H	7C		E,H	5C		H, byte	26		H	84		H	24		H	A4
	A,L	7D		E,L	5D		L, byte	2E		L	85		L	2C		L	A5
	A,M	7E		E,M	5E		M,byte	36		M	86		M	34		M	A6
MOV	B,A	47	MOV	H,A	67				ADC	A	8F	INX	B	03	XRA	A	AF
	B,B	40		H,B	60		Load Immediate			B	88		D	13		B	A8
	B,C	41		H,C	61					C	89		H	23		C	A9
	B,D	42		H,D	62					D	8A		SP	33		D	AA
	B,E	43		H,E	63	LXI	B, dble	01		E	8B					E	AB
	B,H	44		H,H	64		D, dble	11		H	8C					H	AC
	B,L	45		H,L	65		H, dble	21		L	8D					L	AD
	B,M	46		H,M	66		SP, dble	31		M	8E					M	AE

Instruction	Operand	Code
MOV	C,A	4F
MOV	C,B	48
MOV	C,C	49
MOV	C,D	4A
MOV	C,E	4B
MOV	C,H	4C
MOV	C,L	4D
MOV	C,M	4E
MOV	D,A	57
MOV	D,B	50
MOV	D,C	51
MOV	D,D	52
MOV	D,E	53
MOV	D,H	54
MOV	D,L	55
MOV	D,M	56
MOV	L,A	6F
MOV	L,B	68
MOV	L,C	69
MOV	L,D	6A
MOV	L,E	6B
MOV	L,H	6C
MOV	L,L	6D
MOV	L,M	6E
MOV	M,A	77
MOV	M,B	70
MOV	M,C	71
MOV	M,D	72
MOV	M,E	73
MOV	M,H	74
MOV	M,L	75
XCHG		EB

Load/Store

Instruction	Code
LDAX B	0A
LDAX D	1A
LHLD adr	2A
LDA adr	3A
STAX B	02
STAX D	12
SHLD adr	22
STA adr	32

Subtract*

Instruction	Operand	Code
SUB	A	97
SUB	B	90
SUB	C	91
SUB	D	92
SUB	E	93
SUB	H	94
SUB	L	95
SUB	M	96
SBB	A	9F
SBB	B	98
SBB	C	99
SBB	D	9A
SBB	E	9B
SBB	H	9C
SBB	L	9D
SBB	M	9E

Double Add†

Instruction	Operand	Code
DAD	B	09
DAD	D	19
DAD	H	29
DAD	SP	39

Decrement**

Instruction	Operand	Code
DCR	A	3D
DCR	B	05
DCR	C	0D
DCR	D	15
DCR	E	1D
DCR	H	25
DCR	L	2D
DCR	M	35
DCX	B	0B
DCX	D	1B
DCX	H	2B
DCX	SP	3B

Specials

Instruction	Code
DAA*	27
CMA	2F
STC†	37
CMC†	3F

Rotate†

Instruction	Code
RLC	07
RRC	0F
RAL	17
RAR	1F

Instruction	Operand	Code
ORA	A	B7
ORA	B	B0
ORA	C	B1
ORA	D	B2
ORA	E	B3
ORA	H	B4
ORA	L	B5
ORA	M	B6
CMP	A	BF
CMP	B	B8
CMP	C	B9
CMP	D	BA
CMP	E	BB
CMP	H	BC
CMP	L	BD
CMP	M	BE

Arith & Logical Immediate

Instruction	Code
ADI byte	C6
ACI byte	CE
SUI byte	D6
SBI byte	DE
ANI byte	E6
XRI byte	EE
ORI byte	F6
CPI byte	FE

byte = constant or logical/arithmetic expression that evaluates to an 8-bit data quantity (Second byte of 2-byte instructions)

dble = constant, or logical/arithmetic expression that evaluates to a 16-bit data quantity (Second and Third bytes of 3-byte instructions)

adr = 16-bit address (Second and Third bytes of 3-byte instructions)

* = all flags (C, Z, S, P, AC) affected

** = all flags except CARRY affected (exception INX and DCX affect no flags)

† = only CARRY affected

8085 Instruction Set (continued)

BRANCH CONTROL GROUP

Jump

Instruction	Code
JMP adr	C3
JNZ adr	C2
JZ adr	CA
JNC adr	D2
JC adr	DA
JPO adr	E2
JPE adr	EA
JP adr	F2
JM adr	FA
PCHL	E9

Call

Instruction	Code
CALL adr	CD
CNZ adr	C4
CZ adr	CC
CNC adr	D4
CC adr	DC
CPO adr	E4
CPE adr	EC
CP adr	F4
CM adr	FC

Return

Instruction	Code
RET	C9
RNZ	C0
RZ	C8
RNC	D0
RC	D8
RPO	E0
RPE	E8
RP	F0
RM	F8

Restart

Instruction	Operand	Code
RST	0	C7
	1	CF
	2	D7
	3	DF
	4	E7
	5	EF
	6	F7
	7	FF

I/O AND MACHINE CONTROL

Stack Ops

Instruction	Operand	Code
PUSH	B	C5
	D	D5
	H	E5
	PSW	F5
POP	B	C1
	D	D1
	H	E1
	PSW*	F1

Instruction	Code
XTHL	E3
SPHL	F9

Input/Output

Instruction	Code
OUT byte	D3
IN byte	DB

Control

Instruction	Code
DI	F3
EI	FB
NOP	00
HLT	76

New Instructions (8085 Only)

Instruction	Code
RIM	20
SIM	30

ASSEMBLER REFERENCE

Operators

NUL
LOW, HIGH
W, MOD, SHL, SHR
?, –
NOT
AND
OR, XOR

ASSEMBLER REFERENCE (Cont.)

Pseudo Instruction

General:
ORG
END
EQU
SET
DS
DB
DW

Macros:
MACRO
ENDM
LOCAL
REPT
IRP
IRPC
EXITM

Relocation:

ASEG	NAME
DSEG	STKLN
CSEG	STACK
PUBLIC	MEMORY
EXTRN	

Conditional Assembly:
IF
ELSE
ENDIF

Constant Definition

Constants	Type
0BDH, 1AH	Hex
105D, 105	Decimal
72O, 72Q	Octal
11011B, 00110B	Binary
'TEST', 'A' 'B'	ASCII

APPENDIX H

Suggested Laboratory Experiments

The following experiments take approximately one laboratory period each unless otherwise noted. The experiments gradually become more sophisticated as the semester progresses, and the material covered in the lectures becomes more advanced. For every experiment, however, the student is deliberately given the minimum possible instruction in writing. I feel it is essential to require the student to calculate the values of the various resistors and capacitors in the circuits. Otherwise, the experiments quickly degenerate into "cookbook" exercises that require little or no original thinking. In this way the student also quickly gains a feeling for the approximate magnitudes of various components and voltages.

It is essential that the instructor be available for most of the laboratory period to answer questions. It is also essential that the instructor already have done the experiment rather than just have looked at the apparatus and instruction sheet. Numerous small practical questions arise in students' minds, particularly in the first few weeks, which can be answered quickly only by someone who has actually performed the experiment himself. Therefore, if a student is used to supervise part of the laboratory period, he/she should have a detailed practical background in the subject matter. Indeed, he/she should have taken the course if at all possible. A good senior physics major, for example, could probably perform well as a laboratory instructor if he/she did well in the course the previous year. However, a faculty member still should give the lectures and be available for the first hour of each laboratory period.

Students typically work in groups of two at a laboratory bench that contains one commercial oscilloscope with probe (a 1× probe is sufficient); one homemade, variable-voltage, short-circuit-protected, dc power supply; one power supply with fixed 1-A short-circuit-protected outputs of 5 V, ±15 V; a homemade "digital" board containing a homemade variable-frequency square-wave oscillator (clock), debounced high (+5 V) and low (0 V) terminals, eight buffered LEDs, and a $2\frac{1}{4}'' \times 6\frac{1}{2}''$ plug-in socket for DIP chips and No. 22 solid wire; one commercial broadband sine/square-wave generator; and a homemade circuit breadboard vertically mounted on a table-top-size relay rack. A photograph of a bench setup with two bread-

FIGURE H1.1 Laboratory bench for two students.

boards is shown in Fig. H1.1. There is a common supply of resistors, transistors, capacitors, chips, and other small parts centrally located in the laboratory. Each bench also has its own set of two BNC–BNC cables, banana plugs, clip leads, and a dozen or so spring-clip solderless connectors that fit snugly into the holes in the breadboard.

The breadboard consists of a 4.80″ × 16.99″ × 0.06″ piece of phenolic punched terminal board (Vectorboard No. 64A18 @ $1.50) mounted in the vertical plane on the front of the table model relay rack. This board contains 1152 0.091-in.-diameter holes on 0.265-in. centers in a rectangular grid. The spring-clip push-in terminals (Vector No. T30N–2 @ $7.00/100) fit snugly in the holes and provide solderless tie points for circuit wiring. The phenolic board is securely screwed to aluminum U-channels on the top and bottom, which are in turn screwed to the relay rack. A vertical $\frac{1}{8}''$ thick aluminum angle is at each end of the phenolic board on which are mounted female banana plugs for power supply leads and several female BNC chassis connectors for feeding signals in and out of the breadboard. A horizontal $\frac{1}{16}''$ thick aluminum plate 2″ wide is mounted just above the breadboard resting on the tops of the $\frac{1}{8}''$ angle. Several $\frac{3}{8}''$ diameter holes in this plate provide convenient mounting for potentiometers. Hundreds of different electronic experiments can be easily performed at this bench.

For integrated circuit experiments a small ($2\frac{1}{4}'' \times 6\frac{1}{2}''$) plug-in terminal board (SK–10 "Universal Component Socket") is used, @ $18. Dual-in-line integrated circuits plug directly into the SK–10 socket along with various resistors and capacitors. Number 20 or 22 hook-up wire also fits into the socket.

The short-circuit-protected, variable-voltage dc power supply shown in Fig. H1.2 has been used for most of the analog experiments. Normally D_5 and D_6 do not conduct unless a large enough current is drawn through transistor T_1 so that $V_{EB} + I_E R_2$ exceeds the turn-on voltages of D_5 and D_6. When D_5 and D_6 conduct, the output current from the supply is limited. With $R_2 = 1\ \Omega$ the output current will be limited to a maximum of about 600 mA. A smaller value of R_2 would result in a higher limiting value of the output current; for example, if $R_2 = 0.5\ \Omega$, then the maximum output current would be about 1.2 A.

The +5-V, ±15-V power supply is shown in Fig. H1.3. The 7800 series regulator chips are internally protected against short circuits, and the supplies have performed reliably for seven years at UNH in the student labs.

To the Student

In all of your lab write-ups include:

1. date of experiment
2. complete schematic diagrams
3. careful sketches of all waveforms
4. data in tables
5. sample calculations

EXPERIMENT 1—KIRCHHOFF'S LAWS

I. Design and construct two circuits using the dc power supply and three or more resistors (1 to 10 kΩ) to demonstrate the Kirchhoff voltage law (KVL) and the Kirchhoff current law (KCL). Calculate the dc voltages at all points in the circuit

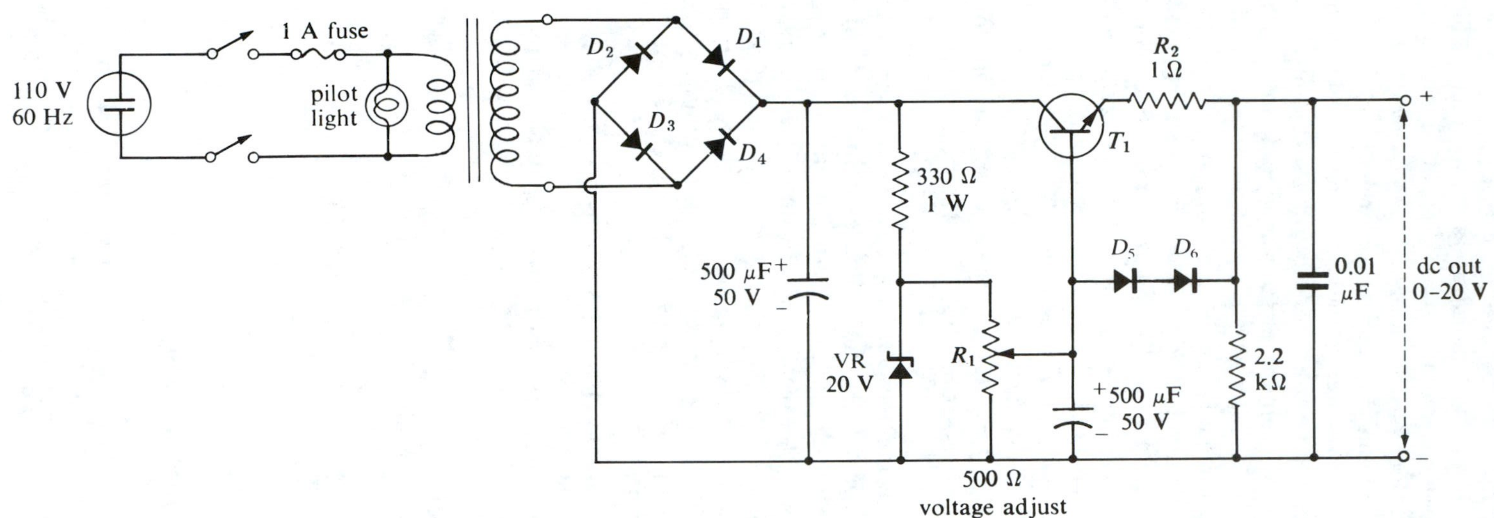

$D_1 - D_6$ = 1N2484 or 1N4004 silicon rectifier diode
VR = 20 V zener diode 1 W
T_1 = 2N 3055 or equivalent silicon power transistor

FIGURE H1.2 Variable-voltage dc short-circuit-protected power supply.

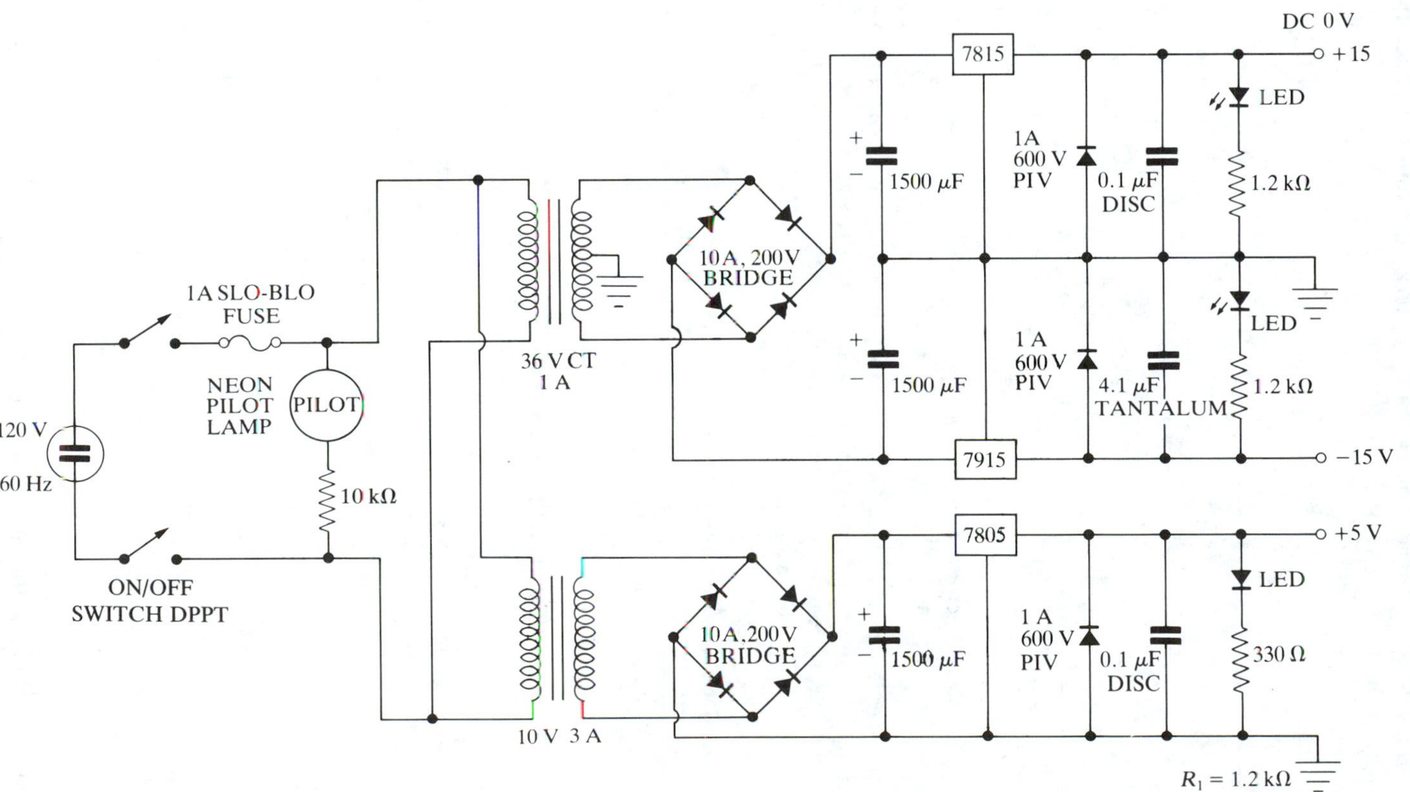

FIGURE H1.3 ±15 V, +5 V regulated power supplies.

using the KVL and KCL and compare with your measured values. Measure the dc voltages with the oscilloscope (on dc coupling).

One such circuit might be:

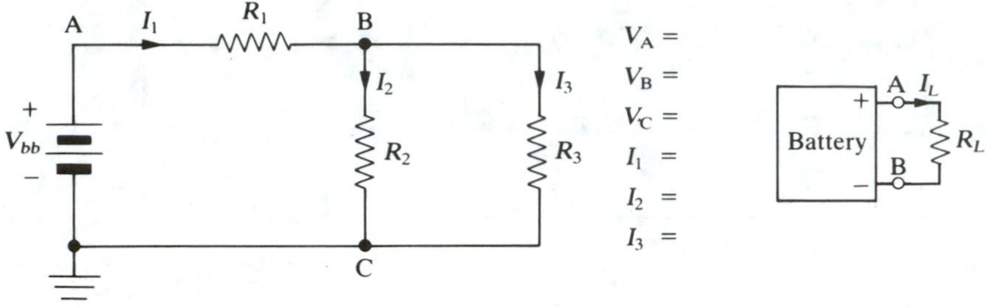

Check with the instructor the circuits you propose to use *before* you construct them. Use volts, mA, and kΩ.

II. Measure the terminal voltage V_{AB} and the current I_L for a battery or a low-current dc power supply. Graph on linear–linear paper. Start with $R_L = 1$ kΩ and decrease R_L. Measure V_{AB} with a digital voltmeter.

QUESTIONS

1. Express the KCL in words. What conservation law is involved?

2. Express the KVL in words. What conservation law is involved?

3. What is the slope of your graph in Part II?

EXPERIMENT 2—THEVENIN'S THEOREM

I. Design and construct a circuit using the fixed 15-V power supply and several resistors to demonstrate Thevenin's theorem.

One such circuit might be:

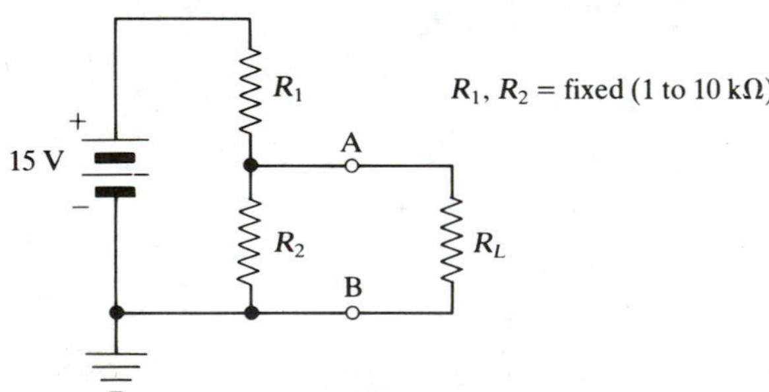

II. Calculate the Thevenin equivalent (*e* and *r*) of your circuit in Part I and construct it with the variable-voltage power supply and one resistor. Measure its terminal voltage V_{AB} for the same values of R_L you used in Part I and compare with Part I.

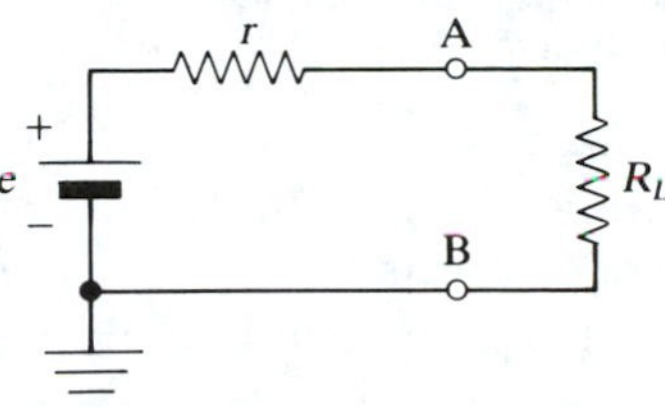

III. Construct the circuit below with R at least $1\ \text{M}\Omega$. Measure V_{AB} and I_L for several values of R_L: $R_L = 1\ \text{k}\Omega$, $10\ \text{k}\Omega$, $100\ \text{k}\Omega$. What can we say about I_L as long as $R_L \ll R$? In other words, the 5-V battery and R act like a constant ________ source.

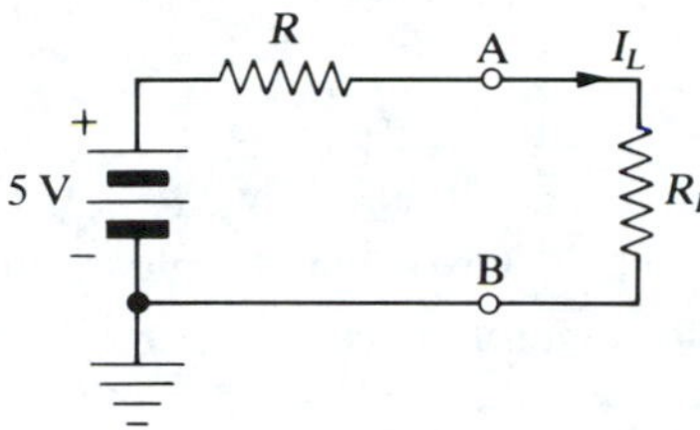

QUESTIONS

1. As far as R_L is concerned, is there any difference between your original circuit in Part I and your Thevenin equivalent in Part II?
2. In terms of e and r in II, what is the open-circuit voltage (V_{oc}) ($R_L \to \infty$) between A and B?
3. In terms of e and r, what is the short-circuit current (I_{sc}) ($R_L \to 0$)?
4. How is r related to V_{oc} and I_{sc}?

EXPERIMENT 3—*RC* LOW-PASS FILTER

I. Design and construct a low-pass RC filter with a breakpoint (frequency) between 1 kHz and 50 kHz. Measure the voltage gain $|A_v|$ as a function of frequency for a sinusoidal input from the variable-frequency oscillator. C must be nonpolar. Measure the input and output voltages with the oscilloscope for about eight different frequencies.

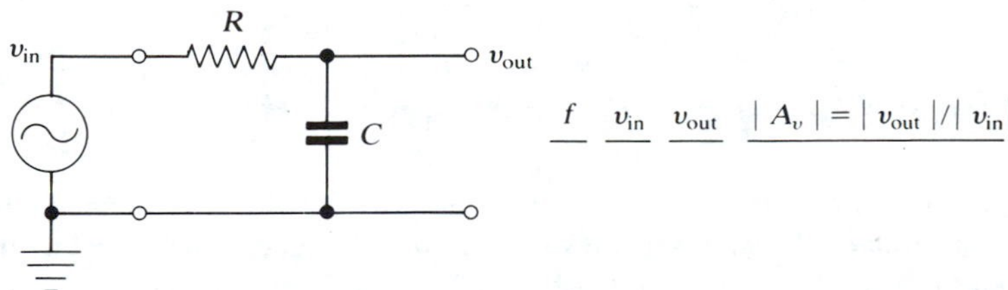

Use a range of frequency from well below the break frequency to well above the break frequency. Negligible current should be drawn from the output terminal (i.e., the filter should not be "loaded").

Graph log $|A_v|$ versus log f.

QUESTIONS

1. Are v_{in} and v_{out} in phase? Explain.
2. Does $|A_v|$ depend upon the phases of v_{in} and v_{out}?
3. Compare the theoretical and experimentally measured break frequencies.
4. Calculate expressions for A_v and $|A_v|$ as functions of f, R, and C.

II. Measure the phase shift between v_{in} and v_{out} by using the trigger input of the oscilloscope. (Connect the input v_{in} to the external trigger input.) Use an input frequency near the break frequency. You can use a dual-trace scope, if available, to measure the phase shift, or you can use Lissajous figures. Carefully sketch v_{in} and v_{out}.

QUESTIONS

1. Compare the theoretical and experimental phase shift θ.
2. Sketch the phasor diagram for v_{in}, v_{out}, v_R, and v_C. Place the current i along the real voltage axis.
3. What is θ (a) for a dc input, (b) for a very-high-frequency input?
4. Calculate an expression for θ as a function of f, R, and C.
5. Sketch a graph of θ versus f.

EXPERIMENT 4—*RC* HIGH-PASS FILTER

I. Repeat Experiment 3 using a *high*-pass filter (C must be nonpolar).

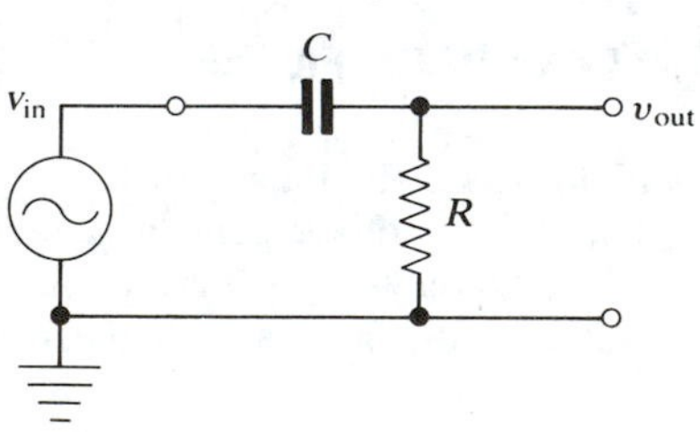

EXPERIMENT 5—*LC* RESONANT CIRCUITS

I. Design and construct a parallel LC resonant circuit with a resonant frequency between 50 and 200 kHz and measure the voltage gain $|A_v|$ as a function of frequency.

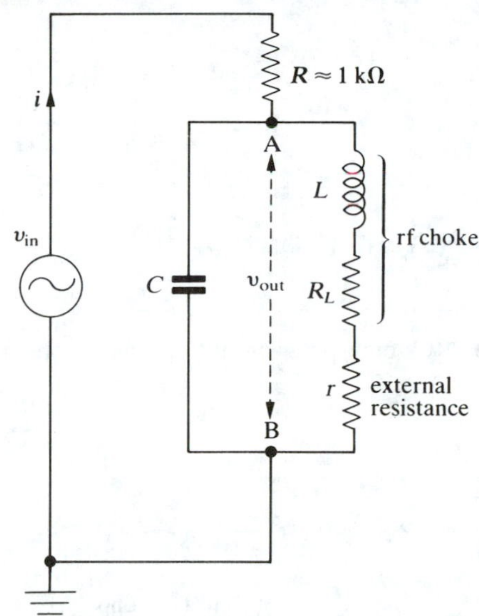

$$|A_v| = \frac{|v_{out}|}{|v_{in}|}$$

$L = 10\ \mu\text{H}$ or $100\ \mu\text{H}$ rf choke

R_L = inherent resistance of rf choke

(C must be nonpolar)

Graph $|A_v| = |v_{out}|/|v_{in}|$ versus frequency for $r = 0$.
Repeat with $r \cong R_L$.
Compare the theoretical and experimental values of the resonant frequency.
Calculate the Q for each case and compare with the theoretical Q.

QUESTIONS

1. Show that ($r = 0$)

$$Q = \frac{1}{R_L}\sqrt{\frac{L}{C}} = \frac{1}{\omega_0 R_L C}$$

2. Show that $Z_{AB} \cong Q\omega_0 L$ at resonance.

3. If we want $v_{out} \cong v_{in}$ at resonance, what can we say about Z_{AB} and R?

II. Design and construct a series LRC resonant circuit. Graph $|A_v| = |v_{out}|/|v_{in}|$ versus frequency.

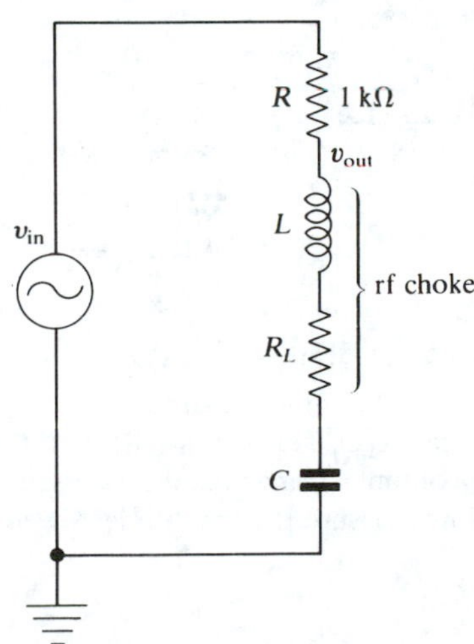

QUESTIONS

1. What is the current i at resonance?
2. What is the current at dc?
3. What is the current as $f \to \infty$?

EXPERIMENT 6—PULSES AND *RC* FILTERS (2 lab periods)

I. Hook up a pulse generator to an *RC* high-pass circuit as shown below (differentiating circuit).

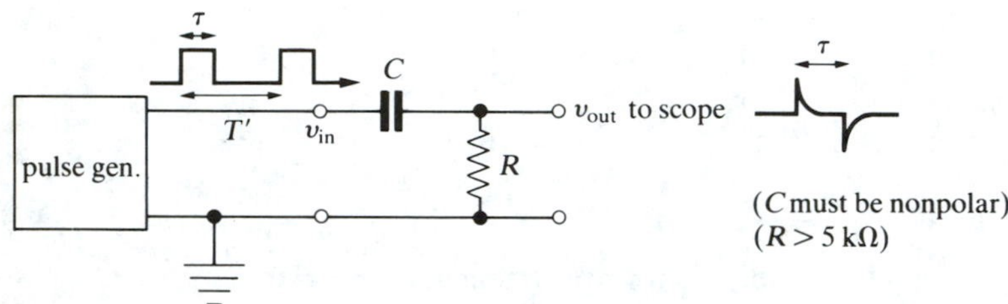

Measure v_{in} and v_{out} for (a) $RC \cong \tau/10$, (b) $RC = \tau$, (c) $RC \cong 10\tau$. [Adjust the pulse width τ on the pulse width generator to achieve (a), (b), and (c).] Sketch v_{out} and v_{in} for all three cases. (Label axes quantitatively.)

What happens if v_{in} is made twice as large?

What happens if you double R and make C half as large?

QUESTIONS

1. Calculate the v_{out} waveform expected theoretically for $RC = \tau$ and $\tau/10$ and compare with the experimentally observed v_{out}. (You may assume $\tau \ll T'$; i.e., each pulse of the input can be treated as an individual pulse.)

II. Repeat I using an *RC* low-pass filter (integrating circuit).

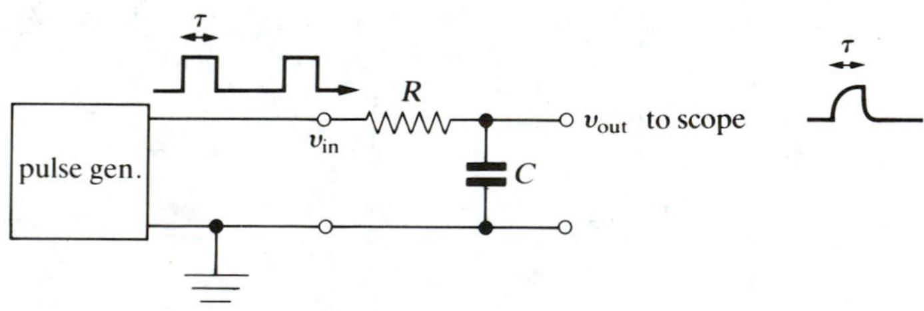

III. Measure the amplitude V_{01} and rise time τ of the pulse generator output pulse v_{in} with the scope (no R_1C_1 network at all). Set $R_1C_1 = \tau$ = time constant for the pulse generator output. (It may be necessary to put $C \approx 0.01\ \mu$F in to make τ long enough.) Now measure the amplitude V_{02} and rise time τ' of v_{out}.

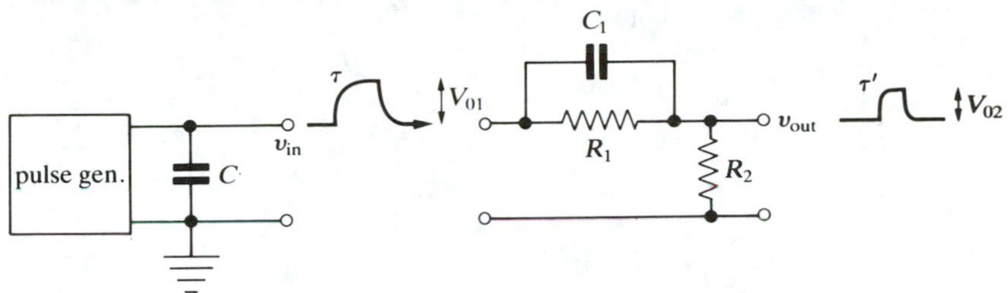

The rise of the pulse generator output v_{in} should be

$$v_{in}(t) = V_{01}(1 - e^{-t/\tau})$$

where τ = the "characteristic" time (rise time = 2.2τ) $\cong R_{out}C$ where R_{out} = output resistance of the pulse generator, and C = the total capacitance to ground.

The output of the $R_1C_1R_2$ divider circuit should be (across R_2)

$$v_{out}(t) = V_{02}(1 - e^{-t/\tau'})$$

where τ' is the characteristic time and $\tau' < \tau$. It can be shown that $\tau' = R_{12}C_1 < \tau = R_1C_1$.

Compare your measured value of τ' to $R_{12}C_1$.

QUESTIONS

1. Qualitatively, why is τ' less than τ; that is, why does the output v_{out} rise more quickly than the input?
2. What "price" is paid for decreasing the characteristic time; that is, in what respect is the output v_{out} worse than the input?

EXPERIMENT 7—THE *Q* (QUALITY FACTOR) OF A PARALLEL *LC* RESONANT CIRCUIT (2 lab periods)

Construct a parallel LC circuit with a resonant frequency of approximately 100 kHz. Measure the Q by three methods:

I. With the variable-frequency sine-wave generator circuit, plot $|v_{out}/v_{in}|$ versus frequency and use

$$Q = \frac{f_0}{\Delta f} \tag{1}$$

where f_0 = the resonant frequency and Δf = the full frequency width at half the maximum output.

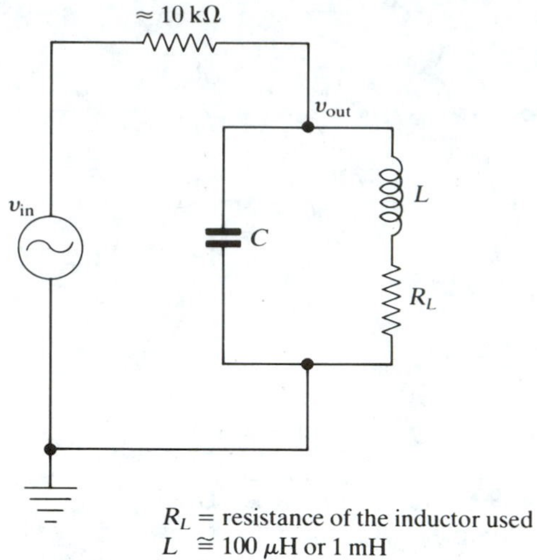

II. Measure R_L, the dc resistance of the inductor used, and use

$$Q_0 = \frac{\omega_0 L}{R_L} \qquad \text{where } \omega_0 = 2\pi f_0 \tag{2}$$

III. With the pulse generator circuit above measure the "ringing" or damped sinusoidal oscillations in v_{out}. The damped oscillations should be at f_0, and their envelope should decay in time according to

$$e^{-R_l t/2L} = e^{-\omega_0 t/2Q_0} \tag{3}$$

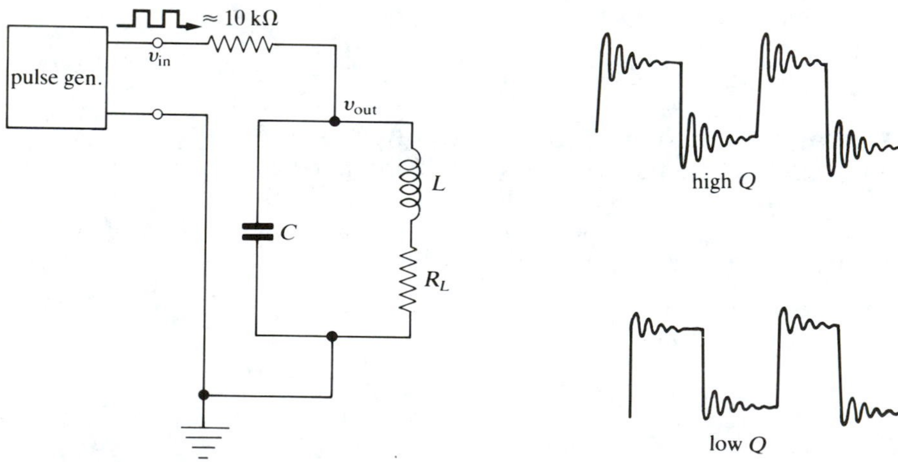

Measure the envelope and thereby determine Q_0.

QUESTIONS

1. What happens to Q if R_L is increased?
2. In terms of energy, what role do R_L, L, and C play?
3. Derive (3) by solving the differential equation for the damped harmonic oscillator.

EXPERIMENT 8—THE DIODE

The object is to measure the current–voltage curve of a silicon diode [e.g., 1N4004 (rectifier diode), 1N4148, or 1N914 (signal diode)].

I. Reverse current direction. Construct the following circuit and measure the diode reverse current I_r and the voltage across the diode for several supply voltages.

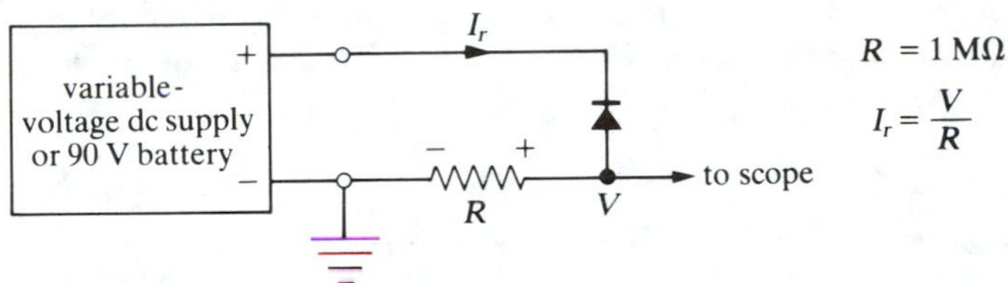

The reverse current I_r will be very small (less than 1 μA), so the voltage drop V across R will be small. For maximum sensitivity R can be the dc input resistance (1 MΩ for most scopes) of the scope itself with the scope on dc coupling. If you connect a scope probe in parallel with R, remember that $R_{\text{total}} = R \| R_{\text{scope}}$. A digital voltmeter can also be used to measure V.

QUESTIONS

1. Why can't you use an old-fashioned analog VOM (50 μA/V) to measure V?
2. The reverse biased diode acts like a constant ________ source.

II. Forward current direction. Construct the following circuit and measure the diode forward current I_f and the voltage V_D across the diode. (Measure V_D for $I_f \cong 0.1, 1, 2, 4, 10$ mA.) With the ground between the diode and R, the scope lead can be put on A to measure V_D and on B to measure I_f with the scope on dc coupling.

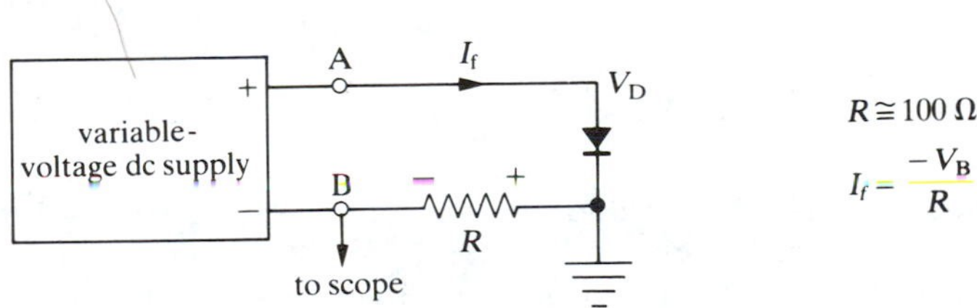

III. Plot the diode current vertically versus the diode voltage horizontally. Use an expanded scale for the reverse current.

IV. Repeat for a germanium diode, if available (1N34). You should measure a smaller turn-on voltage and a larger reverse current.

QUESTIONS

1. A forward biased diode has a very ________ resistance. A reverse biased diode has a very ________ resistance.
2. For your diode the approximate dc forward resistance is ________ for a forward current of ________. The approximate dc reverse resistance is ________ for a reverse voltage of ________.
3. For your silicon diode the approximate turn-on voltage is ________.

EXPERIMENT 9—DIODE CIRCUITS

I. Construct the following circuit, using a signal silicon diode (e.g., 1N4148, 1N914, etc.). Carefully sketch $v_A = V_0 \cos 2\pi f t$ and v_B for $V_0 = 0.1$ V, 1.0 V, 2.0 V. Use $f \cong 1$ kHz.

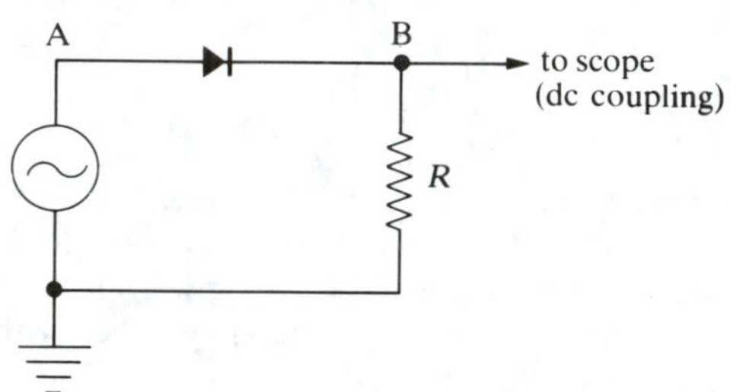

II. Construct the following circuit.

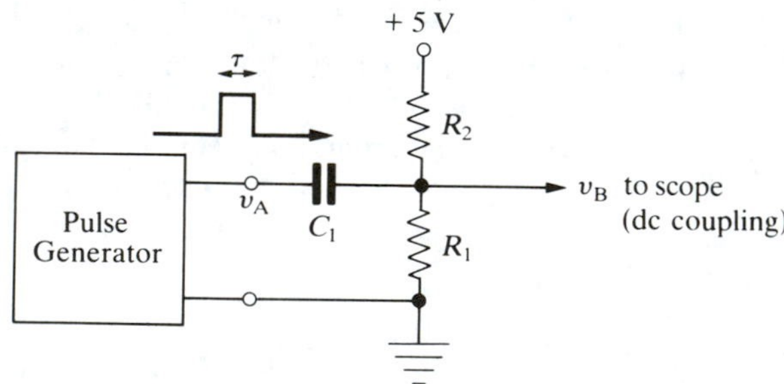

Choose R_1 and R_2 so that (a) the current drain from the +5-V dc supply is less than 1.0 mA, and (b) the dc voltage $v_B = 3$ V (with no pulse input).

Choose C_1 and τ so that $R_{12}C_1 \ll \tau$—that is, so that $R_{12}C_1$ forms a differentiating circuit. Measure v_A and v_B on the scope. Sketch v_B and v_A. Remember $v_{out} \cong RC(dv_{in}/dt)$ for a differentiating circuit.

Now add a diode to your circuit to prevent v_B from rising much above +5 V; in other words, the diode should "clip" off the positive spikes. Sketch your diode circuit. Measure v_A and v_B on the scope and sketch.

QUESTION

1. Explain your results in Parts I and II in terms of the diode turn-on voltage.

EXPERIMENT 10—SIMPLE UNREGULATED POWER SUPPLIES

Construct the full-wave circuit in Part III and one or more of the other power supply circuits. Use a 120-V primary, 24-V secondary transformer and silicon rectifier diodes (1N4004). ***Rectify*** means to change from ac to dc.

CAUTION: 120-V LINES ARE VERY DANGEROUS. TAPE ALL 120-V CONNECTIONS BEFORE PLUGGING IN THE TRANSFORMER. HAVE YOUR INSTRUCTOR CAREFULLY CHECK YOUR CIRCUIT BEFORE YOU PLUG IN THE TRANSFORMER OR TURN IT ON. WHEN YOUR CIRCUIT IS TURNED ON, USE ONLY ONE HAND TO ATTACH THE SCOPE PROBE TO MAKE VOLTAGE MEASUREMENTS. KEEP THE OTHER HAND IN YOUR POCKET!

I. Half-Wave Rectifier Circuit:

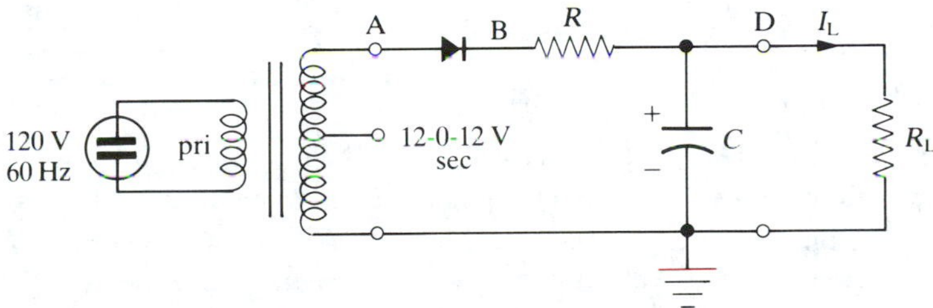

II. Full-Wave Rectifier Circuit:

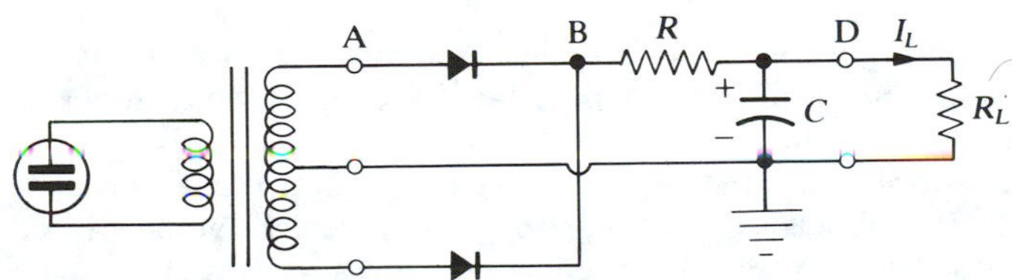

III. Full-Wave Rectifier Bridge Circuit:

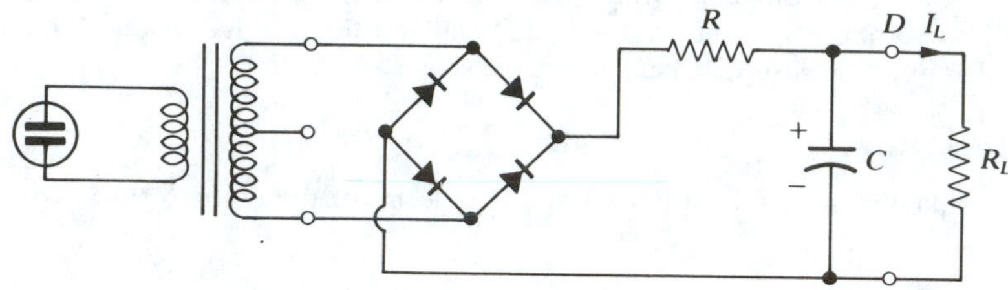

For all these circuits C will be electrolytic and $C > 100\ \mu\text{F}$.

R is chosen to make the breakpoint of the RC low-pass filter less than 60 Hz. R should be less than 100 Ω (the larger C and the smaller R, the better).

Measure V_A, V_B, and V_D on the scope for R_L from approximately 50 to 1000 Ω. Calculate the dc I^2R_L power dissipated for each value of R_L. If any resistor is too hot to touch (gently) with a finger, its power rating is too small.

Tabulate the dc output voltage and the ripple amplitude and frequency for several load currents:

Circuit	$V_{\text{out dc}} = V_D$	I_L	V_{ripple}	*Ripple Frequency*

QUESTIONS

1. Why not use a large R and a small C to obtain a small ripple?
2. Explain why the ripple amplitude increases as I_L increases.
3. By making a rough graph of $V_{\text{out dc}}$ versus I_L, estimate the effective output resistance of the entire power supply. *Hint*: Calculate the Thevenin equivalent circuit of your power supply by drawing an approximate straight line through your data points. Remember that the voltage intercept at $I_L = 0$ is the Thevenin voltage and the magnitude of the slope is the Thevenin (output) resistance of the power supply.

EXPERIMENT 11—THE ZENER DIODE

The zener or regulator diode is designed to be run "backwards" (with reverse polarity) and will regulate its terminal voltage against variations in either the unregulated input voltage or the load current. Its current–voltage curve is shown below. For forward bias (where it is never run) it acts like an ordinary diode. For reverse bias it conducts very little current until the reverse voltage reaches a specific value V_Z (the "zener voltage"), whereupon it draws current in the reverse direction to keep the voltage across the diode at almost exactly V_Z. In other words, the normal range of operation of the zener or regulator diode is from A to B on the curve below.

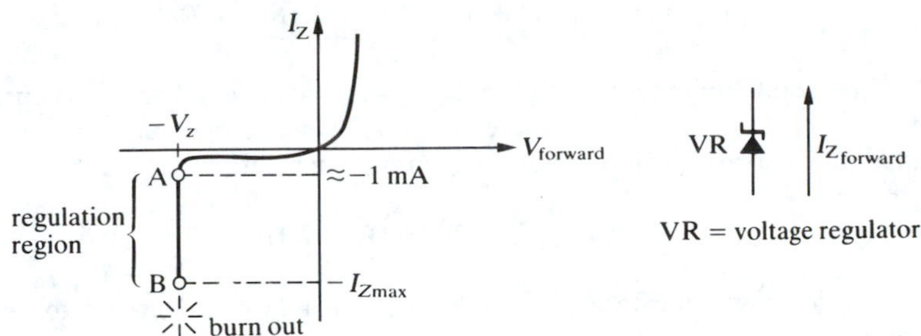

Each zener diode is rated for a maximum dc power dissipation $P_{max} = I_{Z\,max} V_Z$ corresponding to just below B on the curve. Thus for any particular diode in regulation between A and B

$$\sim 1 \text{ mA} < |I_Z| < I_{Z\,max} \tag{1}$$

For $P_{max} = 1$ W, $V_Z = 12$ V, $I_{Z\,max} = P_{max}/V_Z = 1\text{ W}/12\text{ V} = 83$ mA.

I. Calculate $I_{Z\,max}$ for your zener.

Connect your zener diode as shown below so that it is conducting in the reverse direction. Adjust V_1 and R so that (a) $I_Z \cong 1$ mA (use $R = 1$ kΩ), (b) $I_Z \cong I_{Z\,max}/2$ (use smaller R).

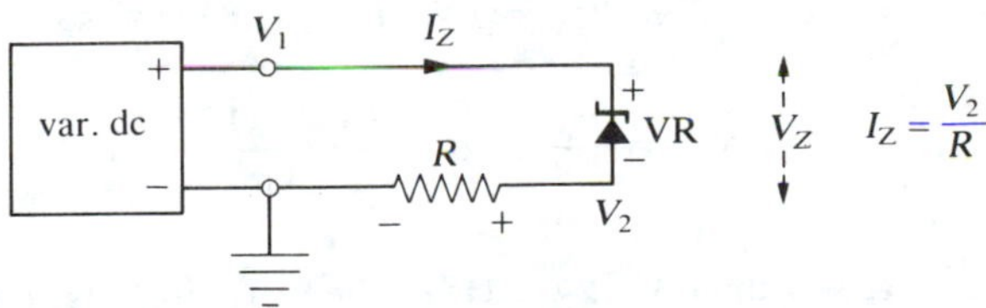

Carefully measure V_Z for each current I_Z (to high precision with a DMM).

Calculate an approximate ac or dynamic resistance for your zener between A and B: $R_{ac} \cong \Delta V_Z/\Delta I_Z$.

II. Connect your zener as a voltage regulator as shown below.

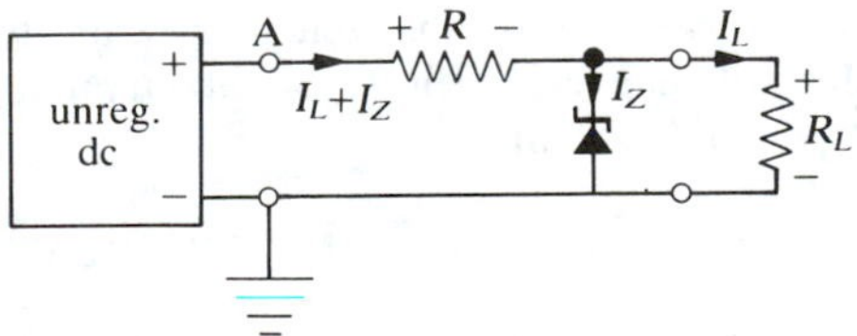

The unregulated dc voltage supply in practical circuits is usually the filtered output of a full-wave bridge rectifier as in circuit III of Experiment 10, but it can be any voltage supply at least a few volts more than V_Z for this experiment (e.g., $V_A = 8$ V if $V_Z = 6$ V).

From the KVL

$$V_A = (I_Z + I_L)R + V_Z = (I_Z + I_L)R + I_L R_L \quad \textbf{(2)}$$

The regulation action of the zener occurs because it changes its current I_Z to keep V_Z constant. If V_A increases, for example, due to a positive-going ripple pulse, then I_Z will increase to keep V_Z constant:

$$V_Z = V_A - (I_Z + I_L)R \quad \textbf{(3)}$$

Or if I_L increases due to a decrease in R_L, then I_Z will decrease to keep V_Z constant.

But there are limits to how much I_Z can change; it must always be more than ~1 mA and less than the maximum burnout current (83 mA for a 12-V, 1-W zener); that is, the zener must always be between A and B on the current–voltage curve. If I_Z drops much below 1 mA, the zener stops regulating and V_Z decreases substantially. If I_Z exceeds $I_{Z\,max}$, the zener burns out.

Consider the possibility of burning out the diode. This will happen when I_L is *minimum* ($I_L \to 0$); that is, when R_L is completely removed ($R_L \to \infty$). Then

$$V_A = I_Z R + V_Z$$

and

$$R = \frac{V_A - V_Z}{I_Z}$$

For $V_A = 15$ V, $V_Z = 12$ V, $P_{max} = 1$ W, $I_{Z\,max} = 83$ mA, so

$$R_{min} = \frac{V_A - V_Z}{I_{Z\,max}} = \frac{15\text{ V} - 12\text{ V}}{83\text{ mA}} = 36\ \Omega$$

Any value of R *less* than 36 Ω will cause the zener to draw *more* than 83 mA and burn out (when $R_L \to \infty$) For a conservative design we might choose R such that $I_{Z\max}$ is only 40 mA when $R_l \to \infty$. In such a case

$$R = \frac{V_A - V_Z}{I_{Z\,max}} = \frac{15\text{ V} - 12\text{ V}}{40\text{ mA}} = 75\ \Omega$$

Consider the possibility of the zener drawing too little current (i.e., $I_Z < 1$ mA). This will happen when I_L is *maximum* (i.e., when R_L gets too small). From (1) when $I_Z = 1$ mA

$$15\text{ V} = (1\text{ mA} + I_{L\,max})R + 12\text{ V}$$

with $R = 75\ \Omega$

$$I_{L\,max} = 39\text{ mA}$$

which corresponds to

$$R_L = \frac{V_Z}{I_L} = \frac{12\text{ V}}{39\text{ mA}} = 308\ \Omega$$

Any R_L less than 308 Ω will cause I_Z to drop below 1 mA.

To sum up, if we have a 12-V, 1-W zener, $V_A = 15$ V, $R = 75\ \Omega$, then the output voltage V_Z across R_L will remain pegged at 12 V for any R_L between ∞ and 308 Ω. If R_L drops below 308 Ω, the zener will lose regulation and the voltage across R_L will fall below 12 V.

III. Calculate a practical value for R for your zener diode for $I_L = 20$ mA or some other value given by your instructor. Measure V_Z for $I_L = 1$, 10, 20 mA. Tabulate your data, including the ripple.

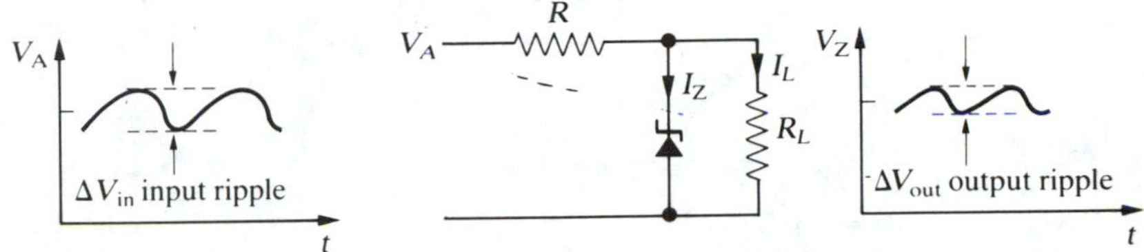

With an unregulated input voltage V_A containing a ripple ΔV_{in} superimposed on a dc component, the zener ac or dynamic resistance can be measured by

$$R_{ac} \cong \frac{\Delta V_{out}}{\Delta I_Z} = \frac{\Delta V_{out}}{\dfrac{\Delta V_{in} - \Delta V_{out}}{R}} \tag{4}$$

where ΔV_{out} equals the output ripple amplitude across the zener diode.

A capacitor is often wired in parallel with the zener to reduce the high-frequency noise.

QUESTIONS

1. Derive

$$I_L = \frac{V_A - V_Z}{R} - I_Z$$

2. What is the advantage of using a smaller value for R [but not less than $(V_A - V_Z)/I_{Z\,max}$]?
3. Derive (4) or explain it carefully in words. Is

$$R_{ac} \cong \frac{\Delta V_{out}}{\Delta V_{in}} R$$

a good approximation?
4. What happens if $R_L \to 0$, that is, the output is "shorted"? Is the zener diode damaged?

EXPERIMENT 12—THE ONE-CHIP REGULATOR

All practical power supplies are now made with *one-chip* regulators. The regulator chip is usually a three-terminal chip that takes an unregulated dc input voltage containing ripple and produces a highly regulated dc output voltage with very low ripple. The chips usually have built-in short-circuit protection and are available in a wide range of voltage outputs.

Construct the following one-chip regulator power supply circuit.

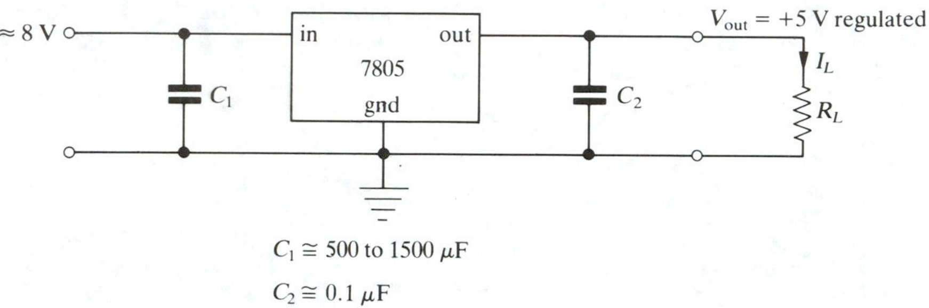

$C_1 \cong 500$ to $1500\ \mu F$

$C_2 \cong 0.1\ \mu F$

The chip must be bolted to a good heat sink to safely dissipate its maximum rated power. The unregulated input voltage must be at least 2 or 3 V more than the regulated output voltage for virtually any regulator chip. The unregulated dc input voltage usually comes from the output of a full-wave bridge rectifier circuit as in Experiment 10.

Measure the input and output dc and ripple voltages for various values of I_L. Determine the maximum useful range of operation.

EXPERIMENT 13—DC TRANSISTOR CURVES

Choose an inexpensive low-power silicon transistor (e.g., 2N2222, 2N3904, 2N3906, etc.) and look up its maximum voltage, current, and power ratings.

The most dangerous thing for transistors is too large a voltage—more than approximately 30 V applied between the collector and emitter or more than only 5 V between the base and emitter will usually quickly (in milliseconds) and permanently destroy the transistor.

The second most dangerous thing for transistors is heat. Dissipating too much power in a transistor will destroy it from excessive heat—the higher the power, the quicker the destruction. For example, if the maximum power dissipation for your transistor is 200 mW = 0.2 W and the collector emitter voltage is 10 V, then the *maximum* collector current is 20 mA; a smaller current would be desirable to allow a safety margin.

With these thoughts in mind, hook up one of the following circuits with your transistor. Use a variable dc supply for V_{cc}, and $R_E \cong 100\ \Omega$. V_{bb} can be a small battery or dc supply.

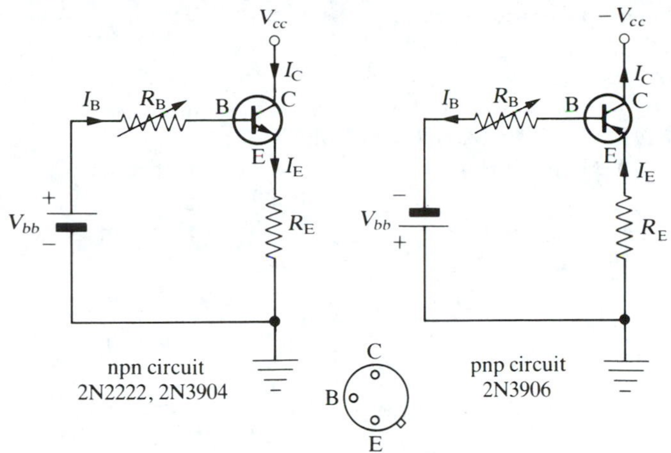

bottom view 2N2222, 2N3904, 2N3906

Measure I_E for various values of V_{CE} by varying V_{cc} for a fixed value of I_B.

$$I_B \cong \frac{V_{bb} - V_{BE}}{R_B}$$

Repeat for $I_B \cong 2\ \mu A$, $5\ \mu A$, $10\ \mu A$. For example, if $V_{bb} = 1.5$ V and you desire $I_B = 5\ \mu A$, then $R_B \cong 0.9\ V/5\ \mu A = 180\ k\Omega$. Remember to keep $I_C V_{CE} < P_{max}$ for your transistor.

Measure V_B and V_E for each value of I_B and V_{cc}.

Graph I_C versus V_{CE} for the three values of I_B. Remember $I_E = I_B + I_C \cong I_C$.

$$h_{FE} I_B = I_C \qquad (1 + h_{FE}) I_B = I_E \qquad I_C = \alpha I_E \qquad h_{FE} = \frac{\alpha}{1-\alpha}$$

The I_C-versus-V_{CE} curves can be displayed on a scope by using a ~10-V sawtooth waveform for V_{CE} and the scope horizontal input. The scope vertical input should be taken across R_E.

Draw the $P_{max} = I_C V_{CE}$ curve on your graph and lightly shade in the forbidden region corresponding to $P > P_{max}$.

QUESTIONS

1. What is h_{FE} or β for your transistor?
2. What is α for your transistor?
3. The base must always be more _______ than the emitter for an npn transistor for the transistor to conduct I_C. The base must always be more _______ than the emitter for a pnp transistor for the transistor to conduct I_C.
4. For an npn or a pnp transistor to conduct I_C, the base-emitter junction must always be _______ biased.
5. For an npn or a pnp transistor the collector-base junction must be _______ biased.

6. For a silicon npn transistor $V_{BE} \cong$ _______ for $I_C > 0.1$ mA. For a silicon pnp transistor $V_{BE} \cong$ _______ for $I_C > 0.1$ mA.

EXPERIMENT 14—THE COMMON EMITTER AMPLIFIER (2 lab periods)

I. Construct a common emitter amplifier circuit as shown below. Use an inexpensive 2N2222, 2N3904, or 2N3906.

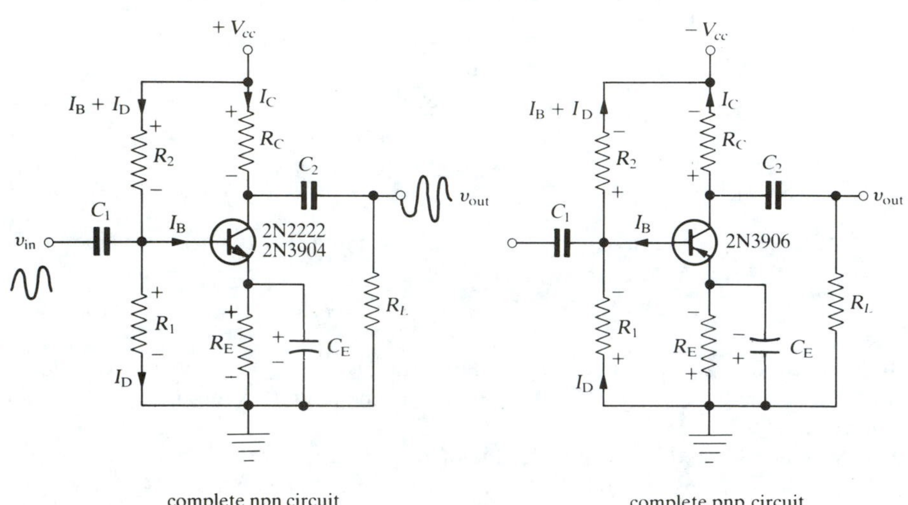

complete npn circuit   complete pnp circuit

Most transistors amplify well only when $I_C > 0.5$ mA, although there are special transistors (e.g., 2N5086) designed to amplify well with $I_C \cong 100\ \mu A$ or less.

Our job is now to calculate all the R and C values. The constraints for the dc design are the following:

1. $I_C > 0.5$ mA dc with no input signal.
2. $V_{CE} \cong V_{cc}/2$ (V_{CE} always must be > 2 V for good gain).
3. $V_{cc} < V_{CE\,max}$ for your particular transistor.
4. $V_{BE} \cong 0.6$ V forward bias for any silicon transistor.
5. $I_D < 1$ mA to avoid drawing too much current from the V_{cc} power supply. $I_D \cong 10I_B$ is usually a good choice.

Notice that we can ignore C_1, C_2, and R_L for the *dc* design.

1. Draw the I_C-versus-V_{CE} curves in your lab book from Experiment 13 or from the transistor specifications. Note the value of h_{FE} or β for your transistor.
2. Draw the load line desired on the I_C-versus-V_{CE} curves. A flat load line will produce a high voltage gain; a steep load line, a high current gain.

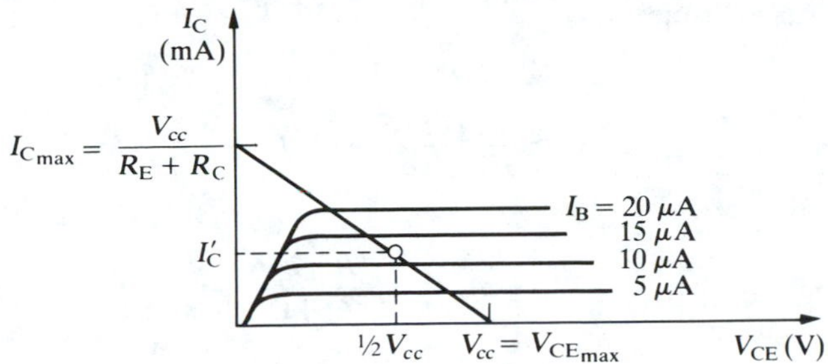

The two endpoints of the load line fix V_{cc} and $R_C + R_E$ because the KVL implies

$$V_{cc} = I_C R_C + V_{CE} + I_E R_E$$

Using $I_E \cong I_C$ gives

$$I_C = \frac{V_{cc} - V_{CE}}{R_C + R_E} \tag{1}$$

(1) is called the load-line equation.

Thus $\quad I_{C\max} = \dfrac{V_{cc}}{R_C + R_E} \quad$ when $V_{CE} = 0$

that is, when the transistor is conducting so strongly it is essentially a short circuit (zero voltage drop across it).

And $\quad V_{CE\max} = V_{cc} \quad$ when $I_C = 0$

that is, when the transistor is completely nonconducting or cut off.

3. Now choose a dc operating point I'_C (the ***quiescent*** point). If the amplifier must amplify both positive and negative input voltages (e.g., a sinusoidal input), then a good general-purpose choice is $V_{CE} = V_{cc}/2$—that is, at the center of the load line. I'_C is now determined as well as the quiescent base current $I'_B = I'_C/\beta$. If the amplifier must amplify only positive or only negative input pulses, then the operating point should be near one end of the load line; for example, $V_{CE} \cong V_{cc}$ for an npn transistor to amplify positive input pulses, and V_{CE} near 0 for negative pulses.
4. Using the simplified h parameter ac equivalent circuit shown below ($h_{re} = 0$, $h_{oe} \to \infty$), we can easily calculate the voltage gain:

$$A_v = \frac{v_{out}}{v_{in}} \cong \frac{-\beta i_b R_{CL}}{i_b h_{ie}} = -\beta \frac{R_{CL}}{h_{ie}} \tag{2}$$

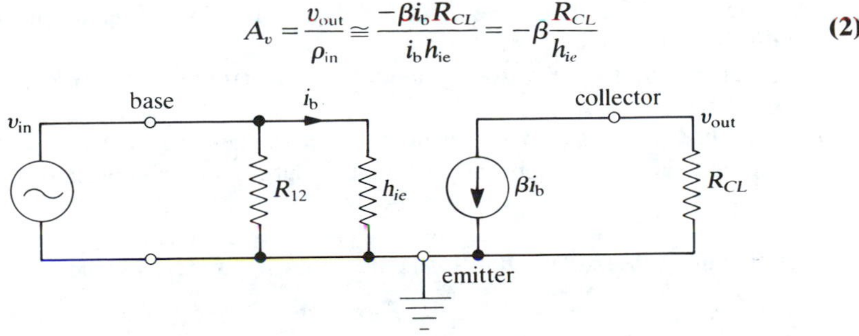

And from Chapter 5

$$h_{ie} \cong \frac{2.6\ \text{k}\Omega}{I_C} \qquad \text{with } I_C \text{ in mA} \tag{3}$$

Thus,

$$A_v \cong -\beta \frac{R_{CL} I_C}{2.6\ \text{k}\Omega} \tag{4}$$

Notice that the voltage gain is approximately equal to the β of the transistor, or at most two or three times larger, depending on R_{CL} and I_C.

Knowing I'_C from the chosen operating point, β from the transistor type, and $R_C + R_E$ from the load line, we can get any reasonable required gain from (4) by choosing R_C. The higher R_C the higher the gain, but for stability reasons R_E shouldn't be too small. Try $R_E = 220\ \Omega$ or so. Then R_C is determined from (4) if we see A_v equal to the desired gain. Calculate R_C and A_v. Use an oscilloscope for R_L so that $R_{CL} \cong R_C$.

5. Now we must calculate R_1 and R_2 to set the dc operating point.
   Calculate V_E from $V_E = I_E R_E \cong I_C R_E$.
   Calculate V_B from $V_B = V_E + 0.6$ V (npn).

$$I_B = \frac{I'_C}{\beta} \qquad I_D \cong 10 I_B$$

$$V_B \cong I_D R_1 \tag{5}$$

$$I_D = \frac{V_{cc}}{R_1 + R_2} \tag{6}$$

   Calculate R_1 and R_2 from (5) and (6).
   Check to make sure $(R_1 \| R_2)/R_E \sim 20$ for thermal stability (not too large). If $(R_1 \| R_2)/R_E$ is too large, simply lower R_1 and R_2.
6. Calculate C_E such that $1/\omega C_E \ll R_E$ at the *lowest* expected signal frequency.
7. Calculate C_1 such that the $R_{12}C_1$ low-pass filter has a breakpoint below the lowest expected signal frequency.
8. Calculate C_2 such that the $R_L C_2$ high-pass filter has a breakpoint below the lowest expected signal frequency.
9. Measure all the dc voltage values in your circuit and compare with the predicted values.
10. Measure A_v versus f over a wide range of frequencies. Be sure to use a small enough sinusoidal input v_1 so that the output v_2 remains undistorted (not clipped). Compare A_v with the $\beta R_{CL}/h_{ie}$ predicted value.
11. Measure R_{in} and R_{out} at an intermediate frequency (usually ~ 1 kHz) where the gain A_v versus f is flat. Compare with the predicted values.

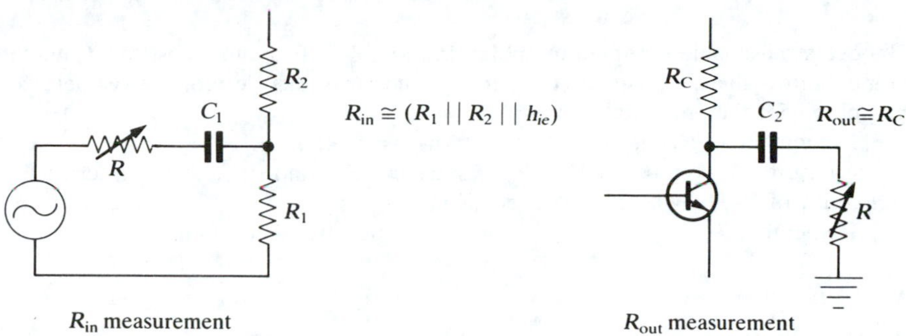

R_{in} measurement

R_{out} measurement

12. Remove C_E and remeasure A_v versus f, and R_{in}.

QUESTIONS

1. Why (qualitatively) does a flat load line produce a high voltage gain and a steep load line a high current gain?
2. What would be a good operating point for an npn common emitter amplifier used to amplify negative pulses? Positive pulses?
3. Repeat Question 2 for a pnp common emitter amplifier.
4. Suppose the transistor is burned out and has infinite resistance between collector and emitter. What will V_C be? V_E? I_C?
5. Suppose the transistor is burned out and is a short—that is, it has zero resistance between collector and emitter. What will V_C be? V_E? I_C?
6. What happens to R_{in} if I_D is increased from $10I_B$ to $50I_B$?
7. Sketch the ac equivalent circuit of the common emitter amplifier ($X_C \sim 0$ for all C).
8. Why does removing C_E reduce the ac voltage gain (qualitatively)?

EXPERIMENT 15—THE EMITTER FOLLOWER OR COMMON COLLECTOR AMPLIFIER

Construct either emitter follower shown below.

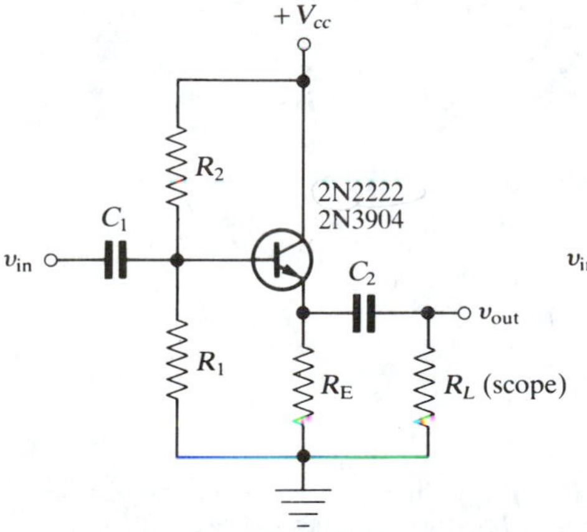

$+V_{cc}$ | R_2 | C_1 | v_{in} | 2N2222 2N3904 | C_2 | v_{out} | R_1 | R_E | R_L (scope)

npn circuit

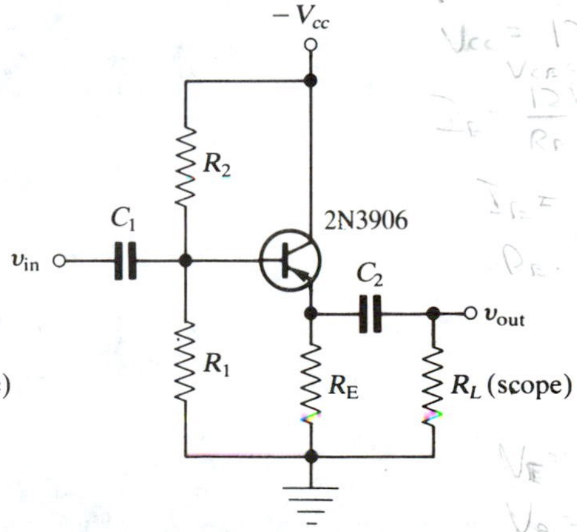

$-V_{cc}$ | R_2 | C_1 | v_{in} | 2N3906 | C_2 | v_{out} | R_1 | R_E | R_L (scope)

pnp circuit

Choose a reasonable dc operating point and load line for your transistor. Calculate component values, construct the circuit, measure the dc voltage values, and compare with the predicted voltages.

Remember $C_1 R_{12}$ and $C_2 R_L$ form high-pass filters.

Measure A_v versus f, R_{in}, R_{out}. Measure R_{in} and R_{out} at an intermediate frequency of 1 kHz, where A_v versus f is flat.

Remember $I_C > 1$ mA, $V_{CE} > 2$ V to get good transistor gain.

QUESTIONS

1. Why isn't there a capacitor in parallel with R_E?
2. Explain qualitatively why the ac voltage gain is approximately equal to 1.
3. Sketch an approximate graph of I_E versus V_{BE} and explain why a larger I_E will produce a slightly larger ac voltage gain with a maximum gain of 1.
4. Is there any phase change between input and output?
5. Sketch the output voltage waveform for a 0.1-V positive input pulse.

EXPERIMENT 16—MISCELLANEOUS TRANSISTOR CIRCUITS

Construct several of the following transistor circuits. Remember, in all silicon transistor circuits if the transistor is conducting, $V_{BE} = 0.6$ V forward bias, and V_{CE} must be at least several volts to obtain a decent gain. In all circuits it may be necessary to bypass the power supply lead to ground with a 0.1-μF capacitor (ceramic is OK).

I. Darlington Emitter Follower

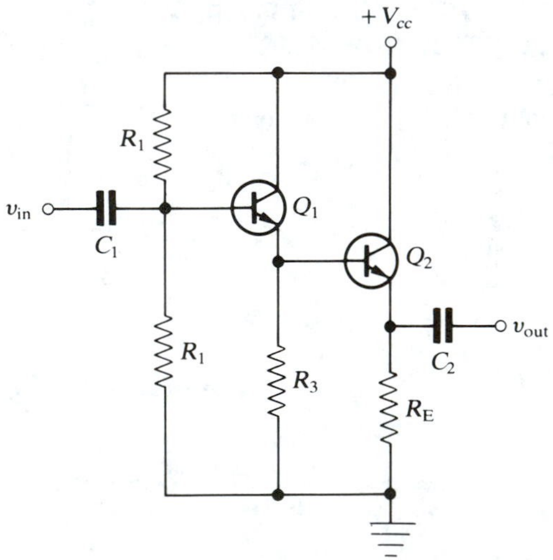

Measure A_v versus f and R_{out}. $\beta_{eff} \cong \beta_1\beta_2$. Q_2 should generally be a higher current transistor because $I_{E1} \cong I_{B2}$: for example, Q_1 = 2N5088, Q_2 =

2N2222 (npn) or Q_1 = 2N5086, Q_2 = 2N2907 (pnp). Try $R_3 \cong 47$ kΩ. Try a TIP 101 high-power npn Darlington.

II. Voltage Regulator

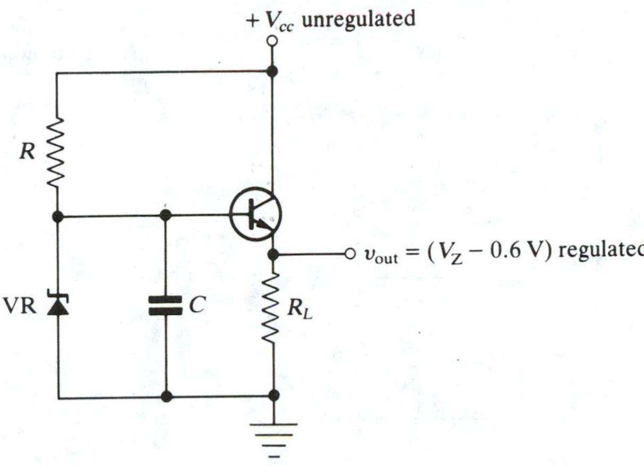

Measure the input and output voltages and ripple and (by varying R_L) the range of I_L over which the circuit regulates.

III. Current Regulator

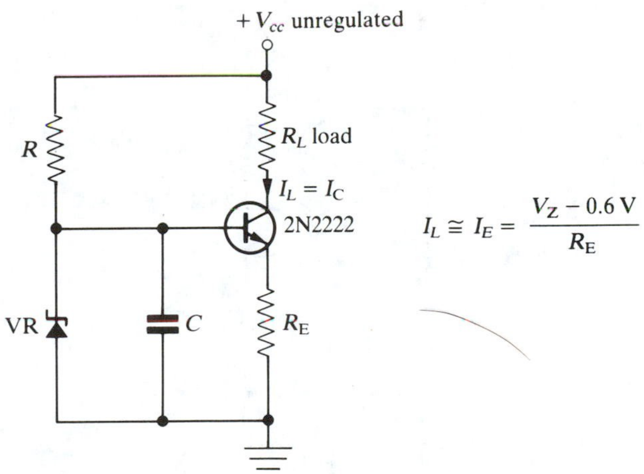

Measure I_L (dc and ripple) for various values of R_L.
Measure the ripple in V_{cc}.

IV. Short-Circuit-Protected Supply

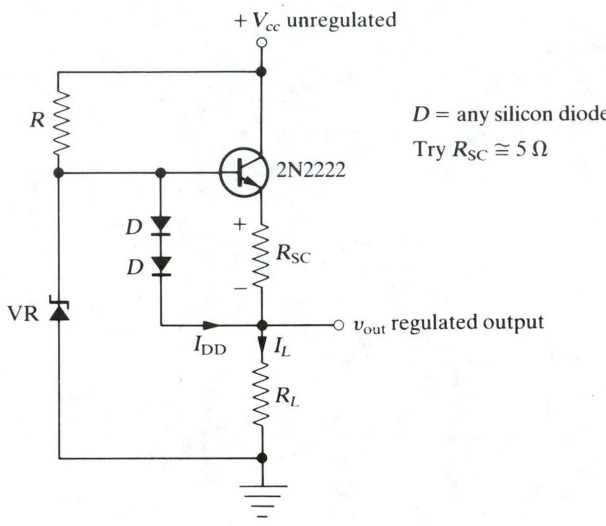

$I_{DD} = 0$ normally. Why?

The maximum output current is $I_{L\,max} \cong 0.6\ \text{V}/R_{sc}$. Why?

V. Temperature-Regulated Current supply

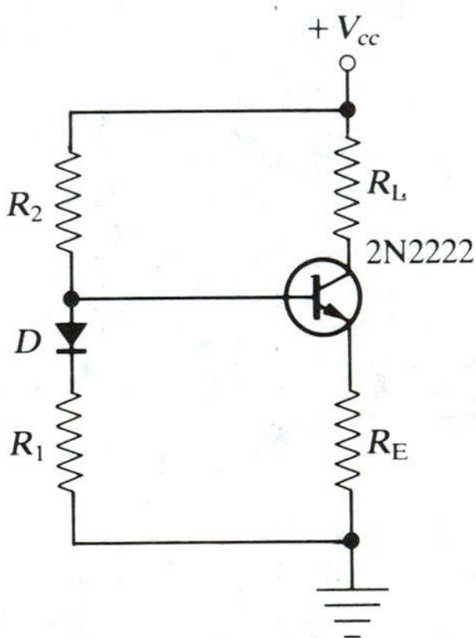

As the transistor temperature increases (try a hot-air gun or a match cautiously applied), the diode turn-on voltage decreases (at ~2.5 mV/°C), which tends to turn off the transistor.

VI. Push-Pull Amplifier

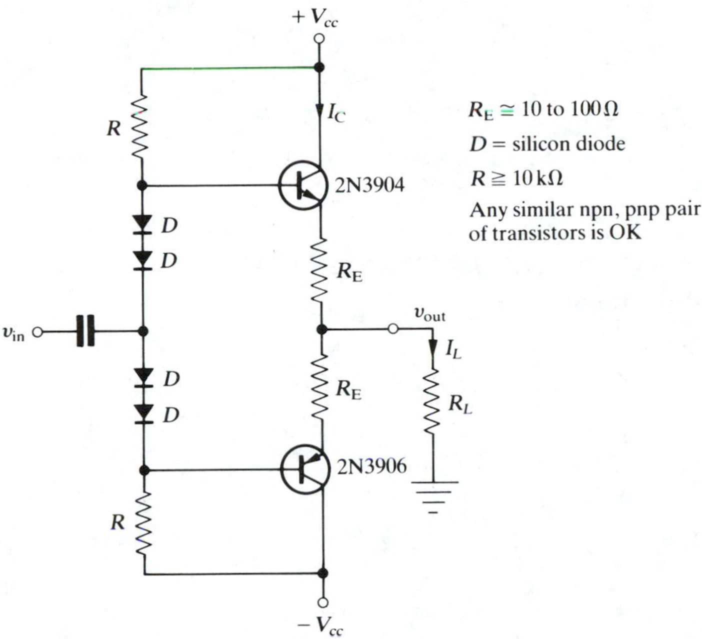

The larger the collector current with no input (I_{CO}), the less output crossover distortion, but the more dc power dissipated in Q_1 and Q_2. $I_{CO} \cong 0.6 \text{ V}/R_E$. Why? Adjust R, R_E, so $I_{CO} \cong 0$.

How much current I_L can be supplied to a low-resistance load such as an 8-Ω loudspeaker?

VII. Negative-Feedback Amplifier

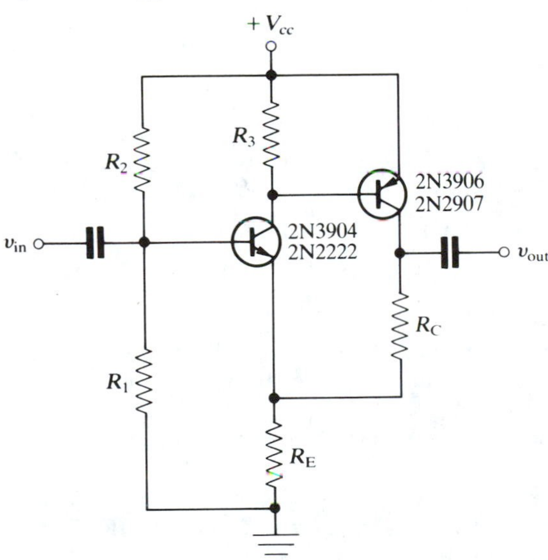

The feedback ratio $B = R_E/(R_E + R_C)$, so the expected gain is

$$A_v = \frac{1}{B} = \frac{R_E + R_C}{R_E} + 1 + \frac{R_C}{R_E}$$

Try $R_E \approx 200\ \Omega$.

Measure the gain A_v versus f, R_{in}, and R_{out}.

EXPERIMENT 17—THE JUNCTION FIELD-EFFECT TRANSISTOR (JFET)

I.

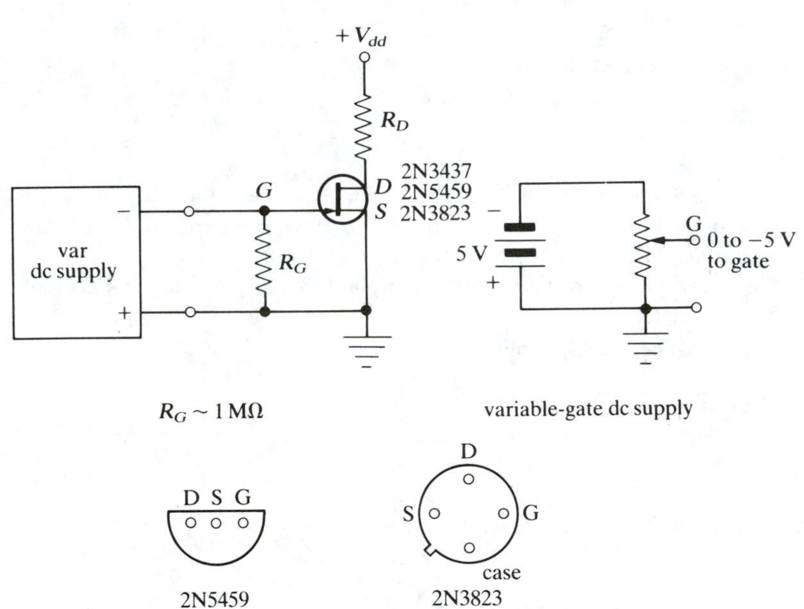

Look up the y parameters of your JFET. $I_{DSS} = I_{D\,\text{max}}$ when $V_{GS} = 0$.

For a fixed V_{GS}, plot $I_D = I_S$ versus V_{DS} by varying V_{dd}. Repeat for several values of V_{GS}. Try $|V_{GS}| \cong 0$ V, 0.5 V, 1.0 V, 2.0 V.

Remember that the gate must always be negative with respect to the source for an n-channel JFET.

Plot I_D versus V_{GS}.

Calculate $y_{fs} = y_{21} = g_m = \Delta I_D/\Delta V_{GS}$ for your JFET.

Does y_{fs} depend on I_D?

EXPERIMENT 18—THE JFET COMMON SOURCE AMPLIFIER

Construct the following amplifier circuit.

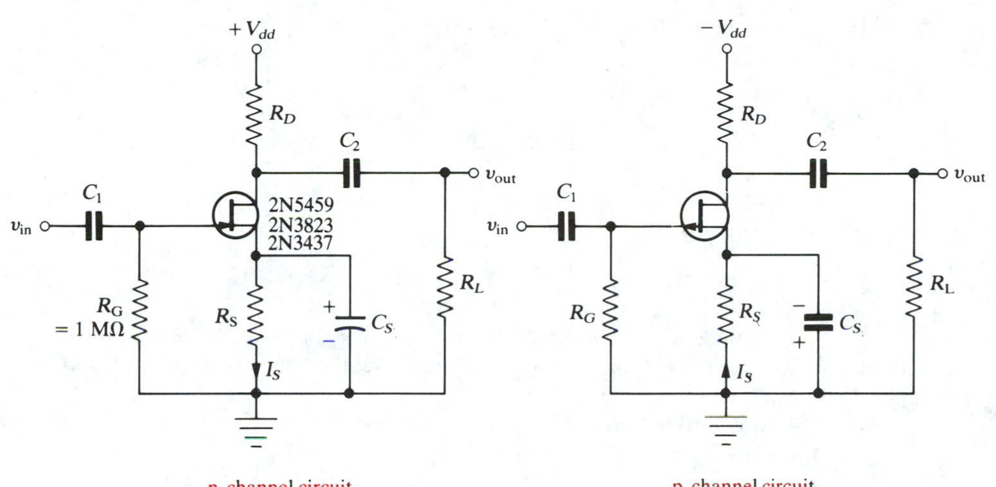

n channel circuit p channel circuit

Use the I_D-versus-V_{DS} curve from Experiment 17 or from the JFET specifications to set the dc operating point.

Note: $I_S R_S = V_{GS}$ because $I_G = 0$ and there is no voltage drop across R_G.

Note: C_1 and R_G form an input high-pass filter, and C_2 and R_2 form an output high-pass filter.

Choose C_S such that $1/\omega C_S \ll R_S$ at the lowest operating frequency.

Calculate the expected voltage gain from the approximate y parameter equivalent circuit. (Use $y_{12} = 0$, $y_{11} \sim 100\ \text{M}\Omega$, $y_{21} = y_{fs} = g_m$, $y_{22} = 0$.)

Measure A_v versus f, and R_{in} and R_{out} at 1 kHz. Compare the measured A_v with the predicted A_v.

Increase R_G by a factor of 10. Does anything change?

QUESTIONS

1. What would happen if C_S were too small?
2. Why can we use a very small value of C_1 and still get a good low-frequency response with this circuit?
3. How does this circuit compare with the common emitter amplifier?

EXPERIMENT 19—THE JFET SOURCE FOLLOWER

Construct the source follower shown below.

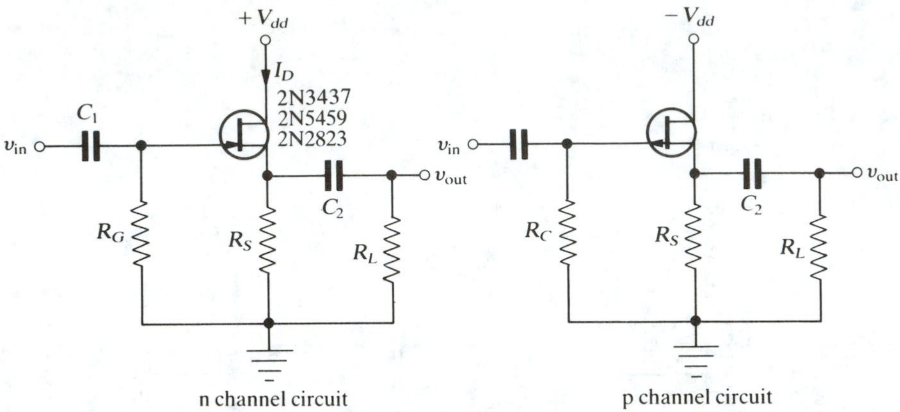

Choose a reasonable dc operating point and load line for your JFET. Calculate component values, construct the circuit, and measure the dc voltage values and compare with the predicted voltages.

Measure A_v versus f and R_{in} and R_{out} at 1 kHz.

Repeat for a larger value of I_D (change R_S).

QUESTION

1. How does this circuit compare with the emitter follower circuit?

EXPERIMENT 20—THE EFFECTS OF TEMPERATURE ON DIODES

Construct the following circuit. R_1, R_2, R_E, and Q form a constant current generator that drives a current I_C through diode D.

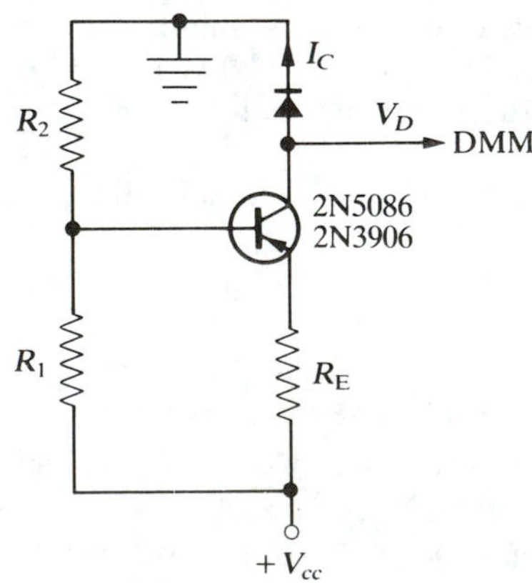

$$I_E = \frac{\frac{V_{cc}R_1}{R_1 + R_2} - 0.6\text{ V}}{R_E} \cong I_C \tag{1}$$

Choose R_1, R_2, and R_E so that $I_C \cong 10$ to $100\ \mu$A so that the self-heating of the diode is negligible. The power dissipated in the diode is

$$P_D = I_C V_D \qquad V_D \cong 0.6\text{ V} \quad \text{for any silicon diode} \tag{2}$$

Measure the voltage V_D across the diode with a digital multimeter (DMM) for various temperatures. You should find

$$\frac{\Delta V_D}{\Delta T} \cong -2.5\text{ mV/°C} \tag{3}$$

Repeat with a different diode of the same type.

EXPERIMENT 21—SATURATED AND UNSATURATED TRANSISTOR SWITCHES

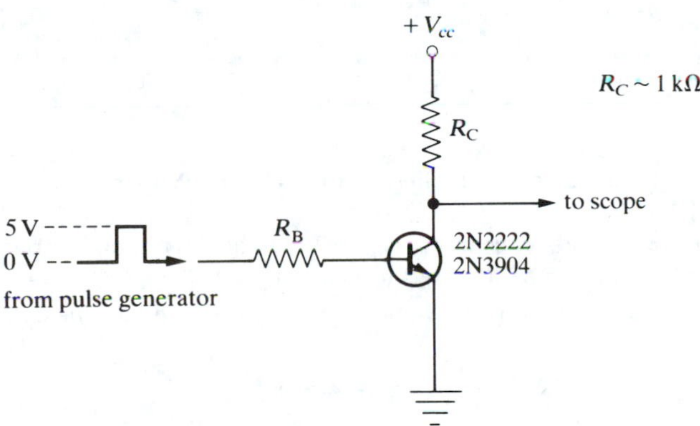

Look up $V_{CE\,sat}$ and β for your transistor. Measure the rise and fall times for your pulse generator. Carefully measure the collector voltage waveform (including rise and fall times) for positive input pulses that go from 0 to +5 V as a function of R_B. $1\text{ k}\Omega < R_B < 50\text{ k}\Omega$. Be sure to do two cases:

Case I: Unsaturated, $I_B = I_C/\beta$ at all times (large R_B).

Case II: Saturated, $I_B > I_C/\beta$ (smaller R_B).

QUESTION

1. Which case gives the shorter rise and fall times for the collector voltage? Why?

EXPERIMENT 22—THE DIFFERENTIAL AMPLIFIER

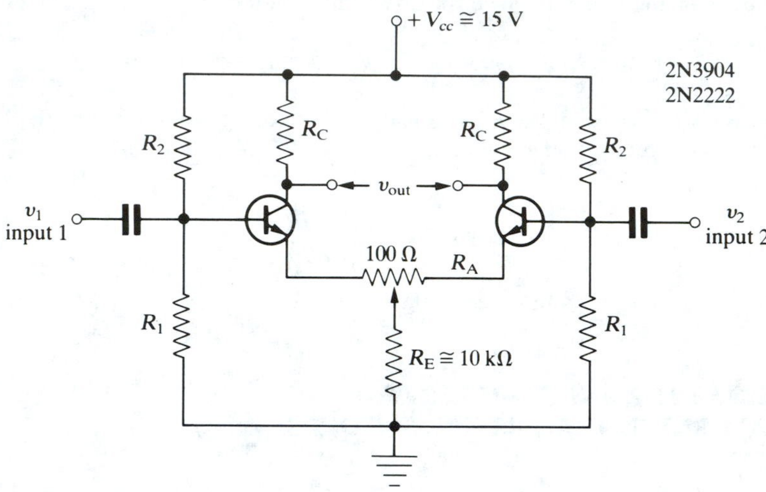

Construct the differential amplifier shown above. With no input, make $I_{C1} = I_{C2} \cong 0.1$ to 0.2 mA.

Try to make R_1 and R_2 as large as possible to achieve a high input impedance. With $v_1 = v_2$, R_A can be adjusted to make $v_{out} = 0$.

Measure the ac voltage gain A and show that it is truly a ***differential*** amplifier:

$$v_{out} = A(v_2 - v_1) \tag{1}$$

Measure the common mode gain when $v_1 = v_2$. (This should be very small.)

$$A_{cm} \equiv \frac{v_{out}}{v_1} = \frac{v_{out}}{v_2} \tag{2}$$

Calculate the common mode rejection ratio in dB.

$$\text{CMRR} \equiv 20 \log_{10}\left(\frac{A}{A_{cm}}\right) \tag{3}$$

EXPERIMENT 23—INTRODUCTION TO OP AMPS: INVERTING AMPLIFIER (2 lab periods)

Construct the inverting amplifier shown below. It is good practice to wire in a 0.1-μF ceramic bypass capacitor from each op amp power supply lead to ground.

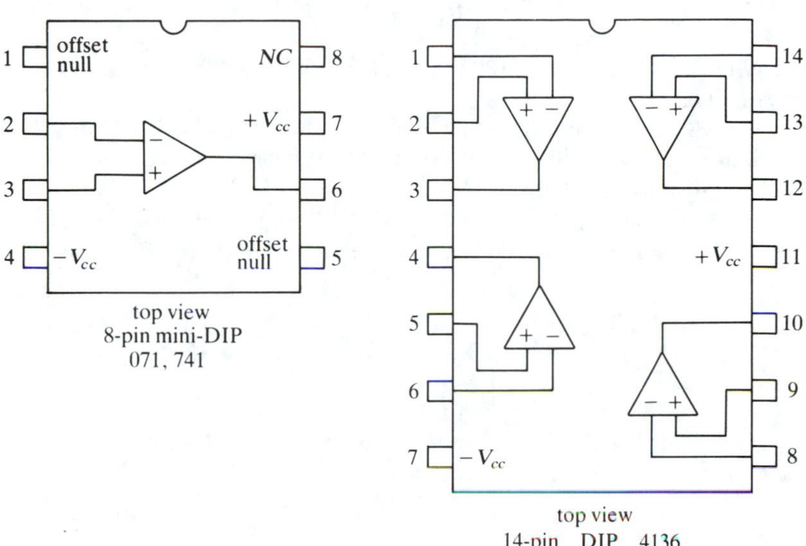

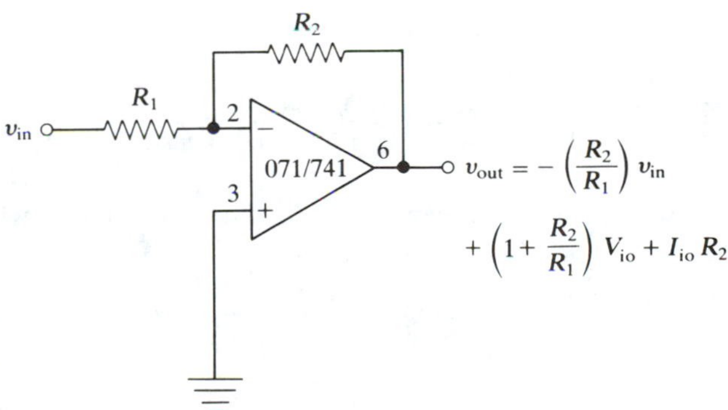

I. Using $R_1 = 1\ \text{k}\Omega$, $R_2 = 10\ \text{k}\Omega$, measure the voltage gain A_v versus f from 100 Hz to 2 MHz.
Repeat for $R_1 = 1\ \text{k}\Omega$, $R_2 = 100\ \text{k}\Omega$.
Repeat for $R_1 = 1\ \text{k}\Omega$, $R_2 = 1\ \text{M}\Omega$.
Graph A_v in dB versus log f for all three cases.

II. Measure R_{in} for the ×10 and ×100 amplifier at ~100 Hz and ~10 kHz.

III. Measure R_{out} for the $\times 10$ and $\times 100$ amplifier at ~ 100 Hz and ~ 10 kHz, being careful to use a very small input amplitude so that the output v_{out} remains undistorted.

IV. The output is

$$v_{out} = -\frac{R_2}{R_1} V_1 + \left(1 + \frac{R_2}{R_1}\right) V_{io} + I_B R_2$$

With $R_1 = 1\text{ k}\Omega$, $R_2 = 100\text{ k}\Omega$, and $v_{in} = 0$ (ground the input), measure $v_{out}(\text{dc})$.

Measure R_1 and R_2 with a precision ohmmeter to at least three significant figures. Use a sensitive digital voltmeter to measure v_{out}.

Measure $v_{out}(\text{dc})$ with $R_1 = 10\text{ k}\Omega$, $R_2 = 1\text{ M}\Omega$, and $v_{in} = 0$.

Calculate V_{io} and I_B. *Hint*: Subtract the two values of v_{out} to get I_B.

V. With $R_1 = 1\text{ k}\Omega$, $R_2 = 100\text{ k}\Omega$, add a resistance $R_3 = R_1 \| R_2$, as shown below.

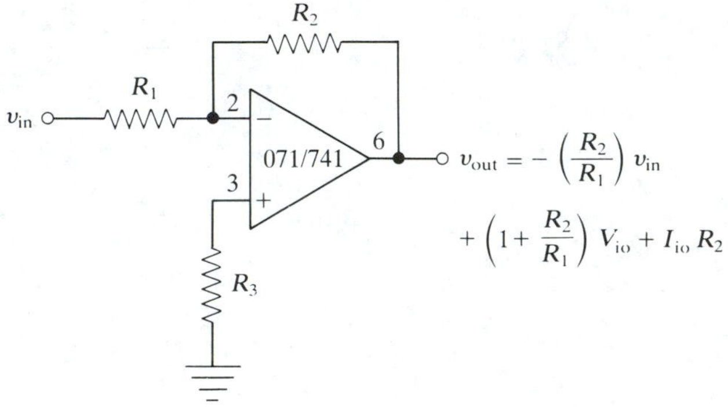

Measure the output $v_{out}(\text{dc})$ with $v_{in} = 0$.

Change R_1 to $10\text{ k}\Omega$ and R_2 to $1\text{ M}\Omega$ and measure v_{out} again. Calculate I_{io}.

VI. Add the offset null circuit recommended by the op amp manufacturer.

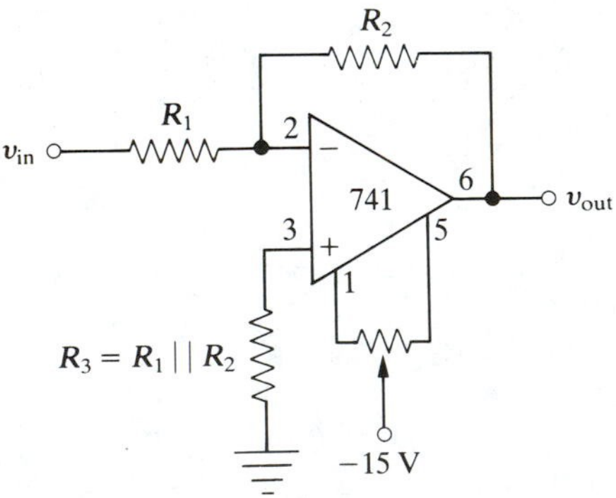

Adjust the pot to minimize v_{out} with $v_{in} = 0$. Estimate the output noise.

VII. Add a capacitor C_2 in parallel with R_2 and measure A_v versus f. Choose C_2 such that $1/\omega C_2 \sim R_2$ at 10 kHz.

Add a capacitor C_1 in series with R_1 and measure A_v versus f. Choose C_1 such that $1/\omega C_1 \sim R_1$ at 1 kHz.

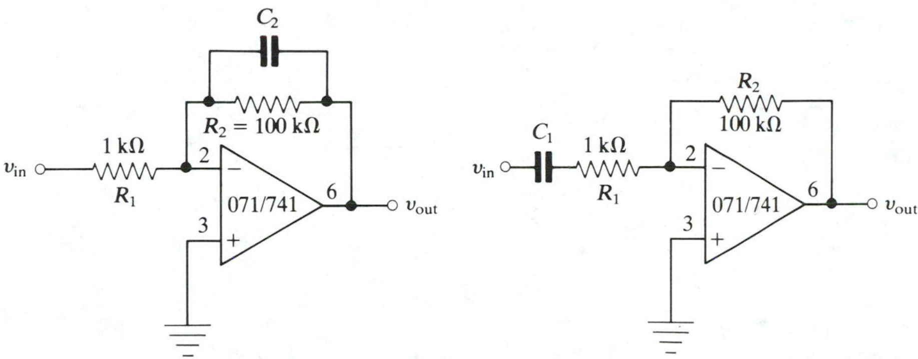

VIII. Observe the slew rate limitation by feeding in a large enough amplitude sine wave v_{in} so that the output is triangular in shape. To measure the slew rate, feed in a rectangular pulse with a short rise time and observe the output.

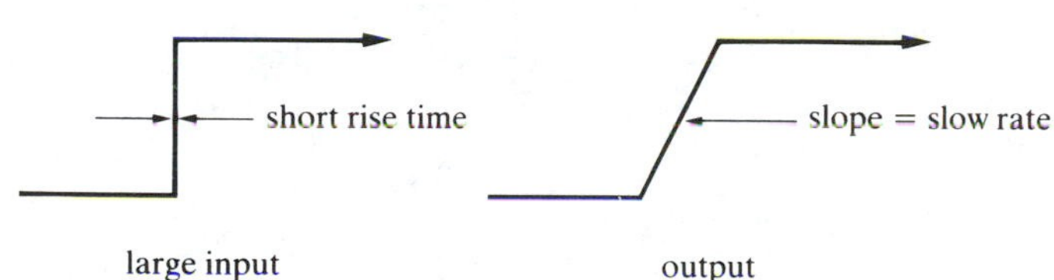

IX. Measure the power supply rejection ratio (PSRR) by changing V_{cc} (with $v_{in} = 0$) and by measuring the change in v_{out}.

X. Measure the output noise rms voltage approximately with $v_{in} = 0$.

QUESTIONS

1. From your graphs in Part I calculate the gain bandwidth product for your op amp.
2. What happens to R_{in} and R_{out} as the frequency increases?
3. Explain why the addition of R_3 helps to minimize the dc output offset voltage.
4. Compare I_B and I_{io}.
5. What is the dc output voltage v_{out} in Parts V and VI due to?

EXPERIMENT 24—INTRODUCTION TO OP AMPS: NONINVERTING AMPLIFIER

Construct the noninverting amplifier shown below.

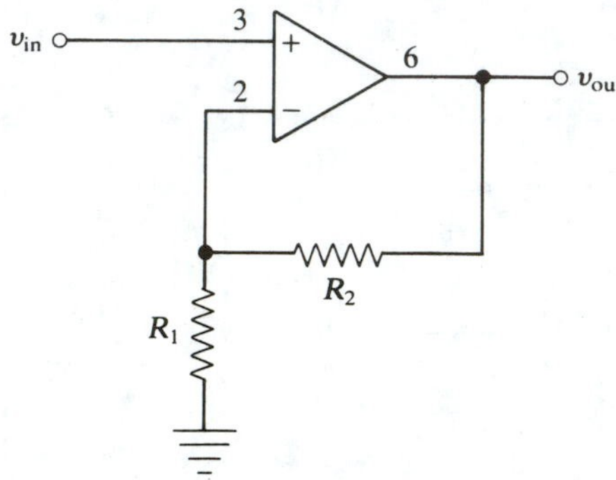

Repeat the procedures of Experiment 23 except that resistor $R_3 = R_1 \| R_2$ in Part V goes in series with the noninverting input:

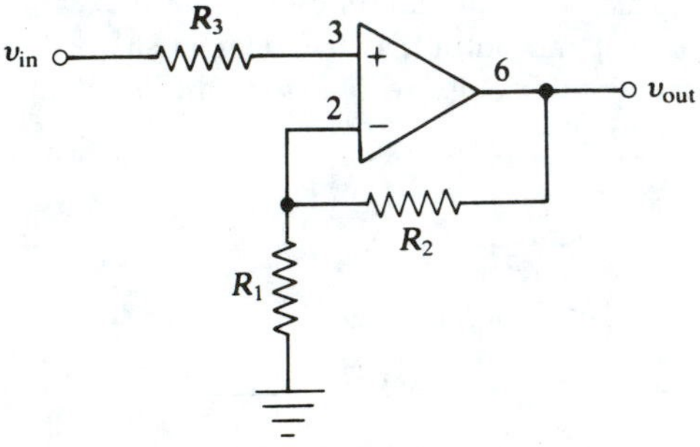

In Part VII use

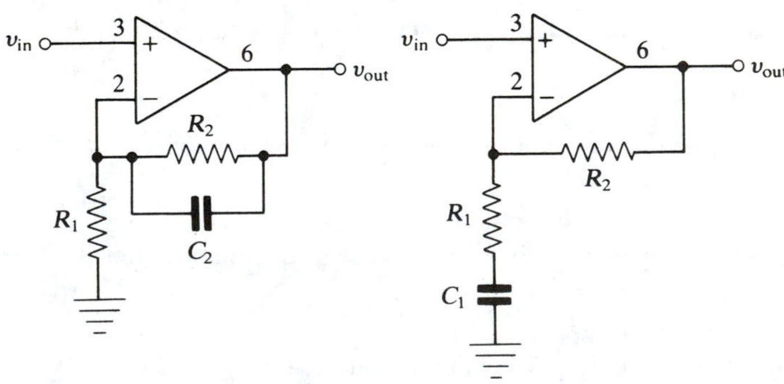

EXPERIMENT 25—OP AMP APPLICATIONS

Construct several of the following op amp circuits and determine the range of operation with respect to input frequency, voltage, and output load (or current) as appropriate. In all cases it is probably a good idea to connect 0.1-μF bypass capacitors between the $\pm V_{cc}$ op amp terminals and ground directly at the op amp chip.

I. Voltage Follower. Why is this circuit often called a *buffer* amplifier? What is R_{in}?

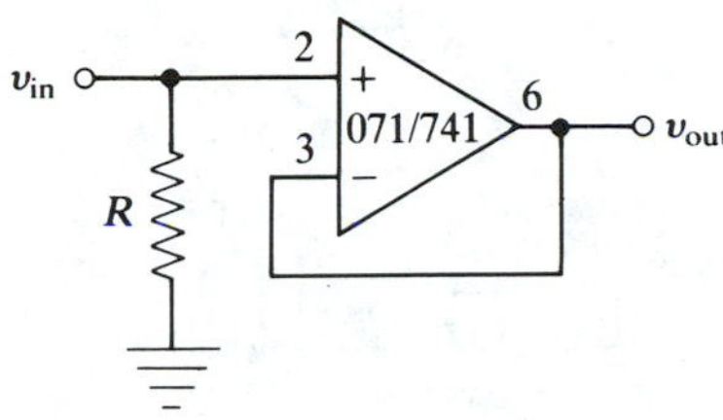

II. Power Booster or Amplifier

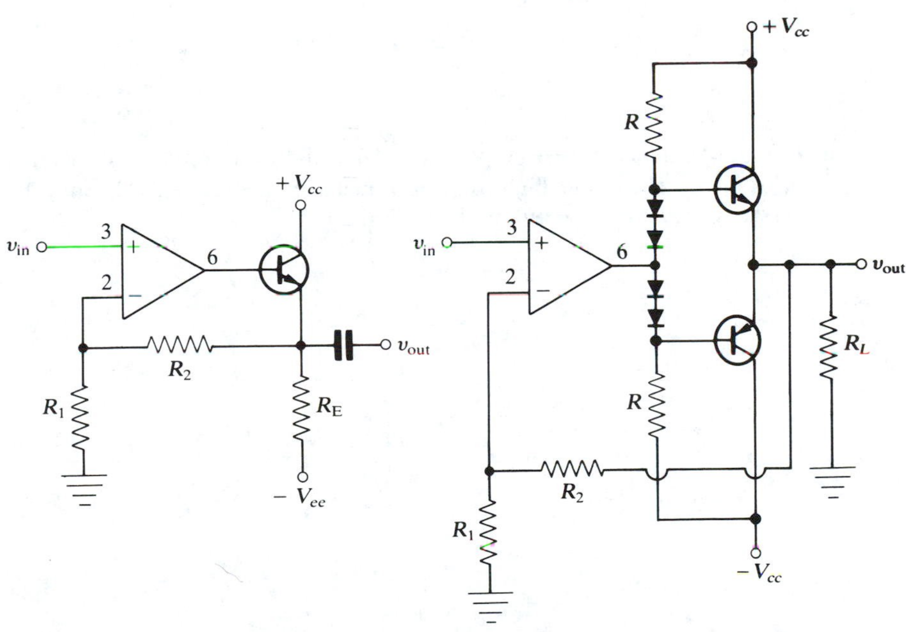

III. Difference Amplifier. $v_{out} = R_2/R_1(v_B - v_A)$. How could you make the input resistances seen by v_A and v_B equal?

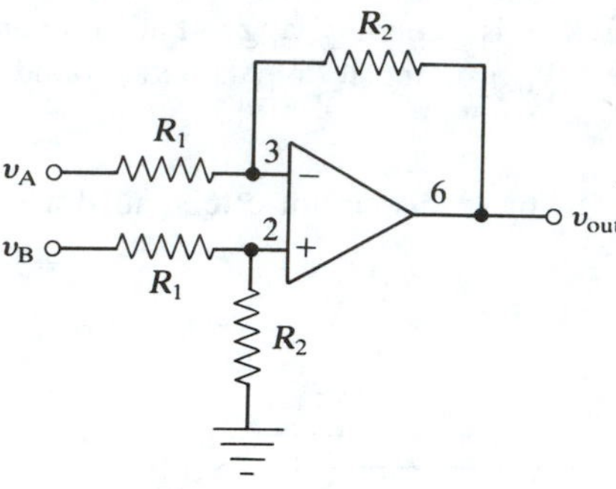

IV. Summing Amplifier

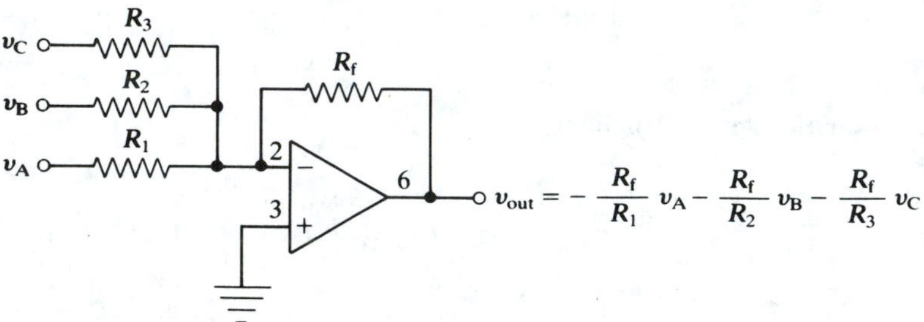

V. Current-to-Voltage Converter. You may have to put a capacitor in parallel with R to decrease the high-frequency noise in the output. R_B may be necessary to minimize the output dc offset.

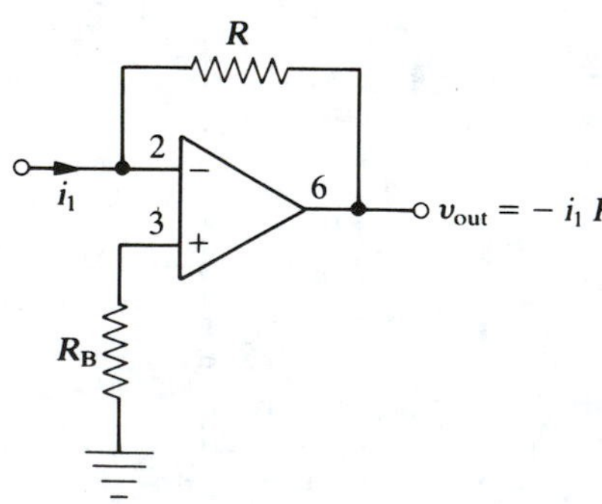

VI. Voltage-to-Current Converter.

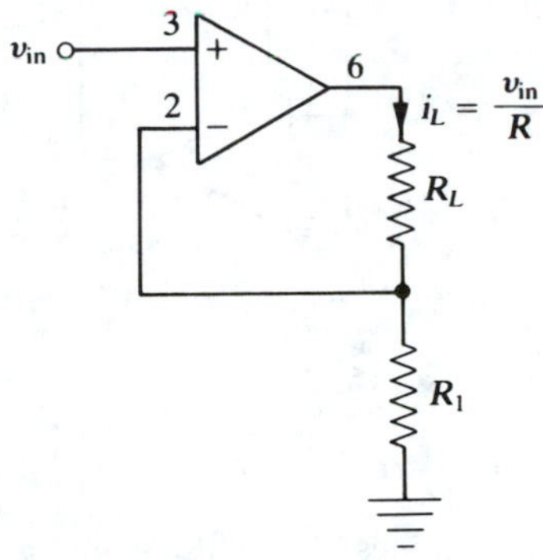

$i_L = v_{in}/R_1$ independent of R_L. $i_{L\,max}$ ~20 mA for most op amps.

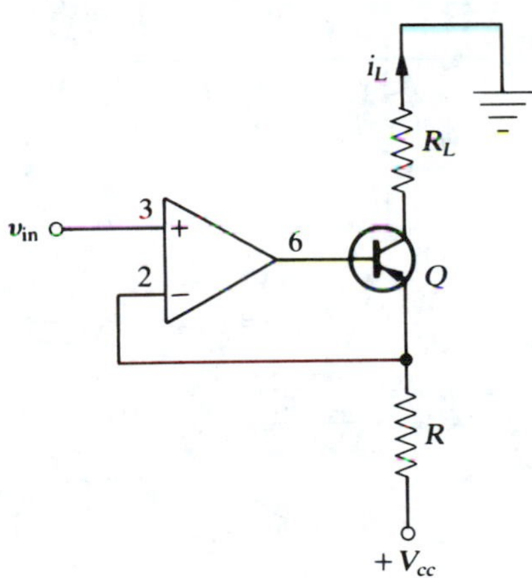

$i_L = V_{cc} - v_{in}/R$ independent of R_L. $i_{L\,max}$ is limited by Q (~50 to 100 mA for a small transistor). R is fixed. The voltage v_{in} controls the current through R_L regardless of changes in R_L.

VII. Logarithmic Amplifier. $v_{out} \propto \ln v_{in}$. A low bias-current op amp is best (e.g., 071, 3140). $v_{out\,max} \cong 0.6$ V, $R \cong 100$ kΩ. $C \simeq 0.1\ \mu$F may be necessary to prevent oscillation. R_B may be necessary to decrease the dc output drift.

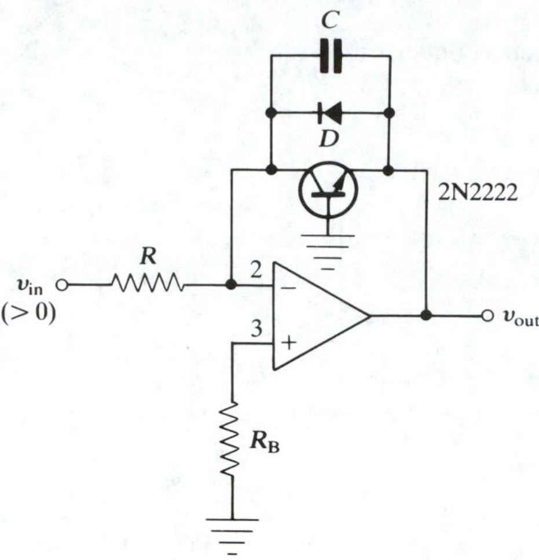

VIII. Phase Shifter. RC determines the phase shift ϕ. v_{out} lags v_{in} by ϕ. Use $v_{in} \sim 1$ kHz.

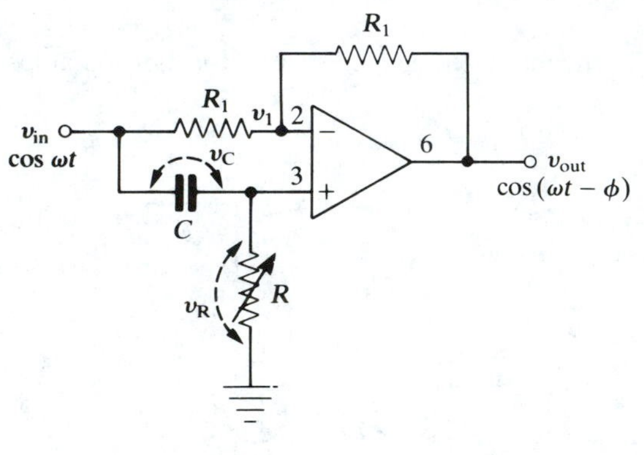

$$v_{in} = v_R + v_C$$

$$\text{OACR} \rightarrow \frac{v_{in} - v_1}{R_1} = \frac{v_1 - v_{out}}{R_1}$$

$$\text{OAVR} \rightarrow v_1 = v_R$$

$$v_{out} = v_R - v_C$$

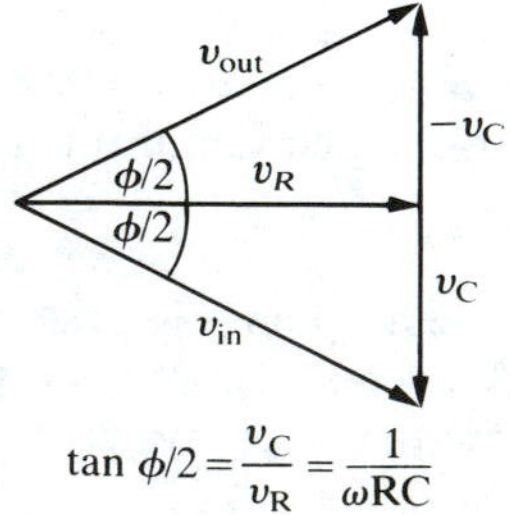

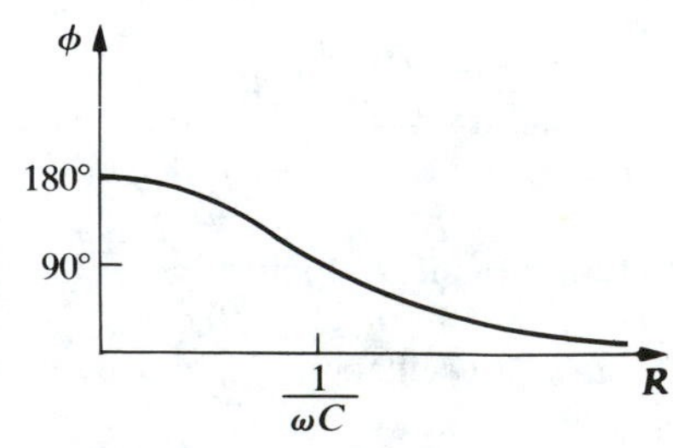

Note that v_{out} is constant in magnitude for $0 \leqq \phi \leqq 180°$ and that $\phi = 90°$ when $\omega = 1/RC$.

EXPERIMENT 26—OP AMP DIODE CIRCUITS

I. Ideal Diode Half-Wave Rectifier

(a) Slow, simple circuit. D = signal diode. Try f = 100 Hz, 1 kHz, 10 kHz, 100 kHz. Sketch the output carefully and consider the op amp slew rate. Use a 5-V pk-pk input.

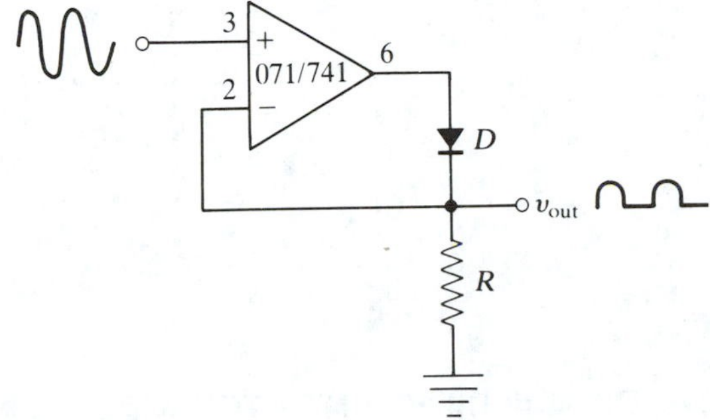

(b) Faster circuit. $R \sim 1$ to 10 kΩ. Try 100 Hz, 1 kHz, 10 kHz, 100 kHz and sketch the output carefully. Why is this circuit faster? *Hint*: What is the op amp dc output voltage when the input swings negative as compared with (a)?

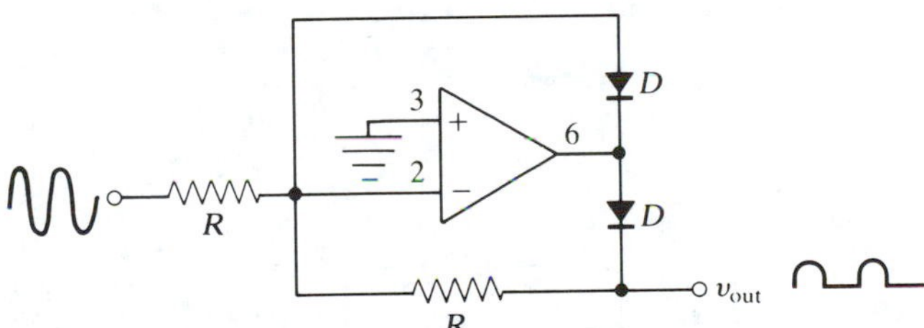

II. Ideal Diode Full-Wave Rectifier. Use a good quad op amp (4136). D = signal diode.

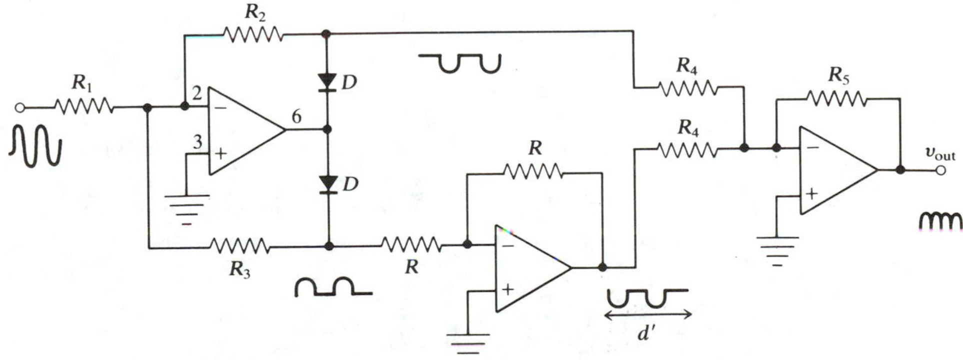

III. Diode Clamp. D = signal diode. $R \ll R_L$. $v_{out} = V_{ref}$ if $v_{in} < V_{ref}$. $v_{out} = v_{in}$ if $v_{in} > V_{ref}$. When is the diode on? When is it off?

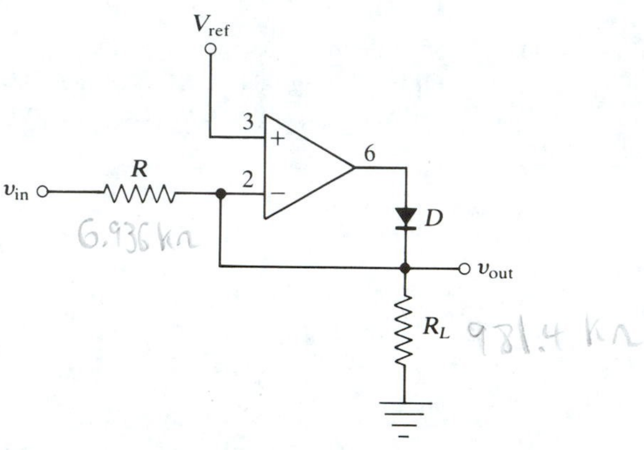

EXPERIMENT 27—OP AMP DIFFERENTIATOR AND INTEGRATOR

I. Differentiator. $R \sim 10\ \text{k}\Omega$, $C \sim 0.1\ \mu\text{F}$. Try a sinusoidal input at various frequencies, triangular, and a square-wave input.

$$v_{out} = -RC\frac{dv_{in}}{dt}$$

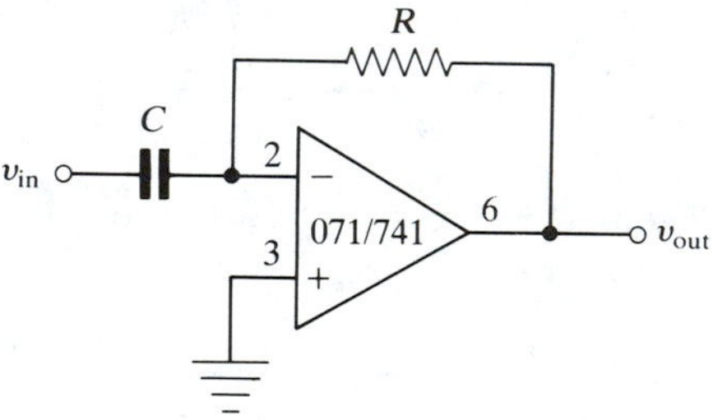

(A small capacitor, C_2, in parallel with R may be necessary to prevent oscillation.)

II. Integrator. $R \sim 100\ \text{k}\Omega$. $C \sim 1\ \mu\text{F}$, low leakage (e.g., polystyrene or mylar). R_2 and R_3 may be necessary (>1 MΩ) to eliminate output dc drift with a 741.

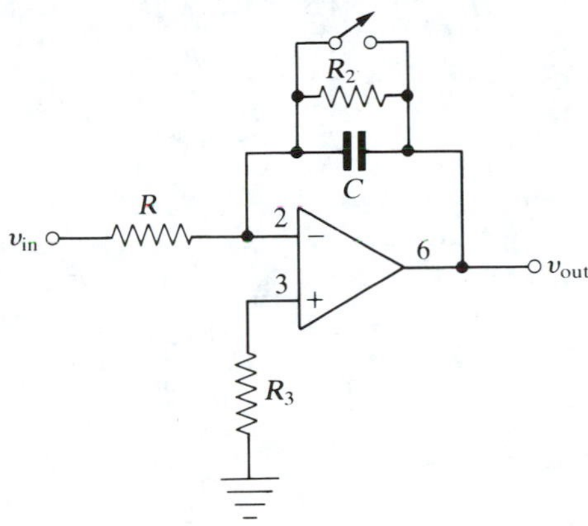

R_2 and R_3 may not be necessary with a lower bias-current op amp such as an 071 or a 3140.

$$v_{out} = -\frac{1}{RC}\int v_{in}\, dt$$

III. Charge-Sensitive Amplifier. $R_2C < \tau$. (R_2 discharges C.) Try a sinusoidal input for various frequencies. Try a square-wave input.

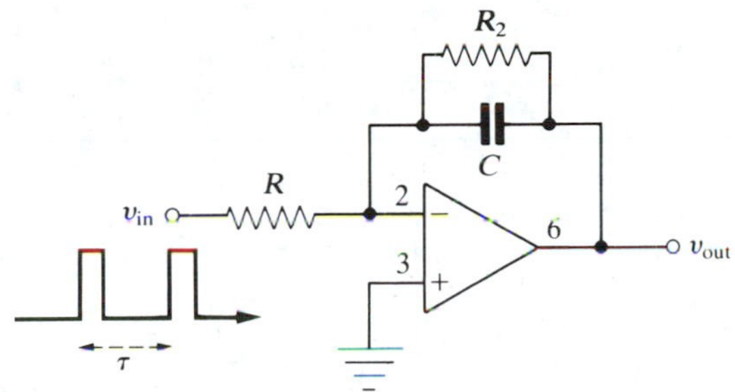

EXPERIMENT 28—SAMPLING CIRCUITS

I. Peak Detector. D = signal diode. C = low leakage (polystyrene or mica). $v_{out} \rightarrow$ peak of v_{in} and holds there.

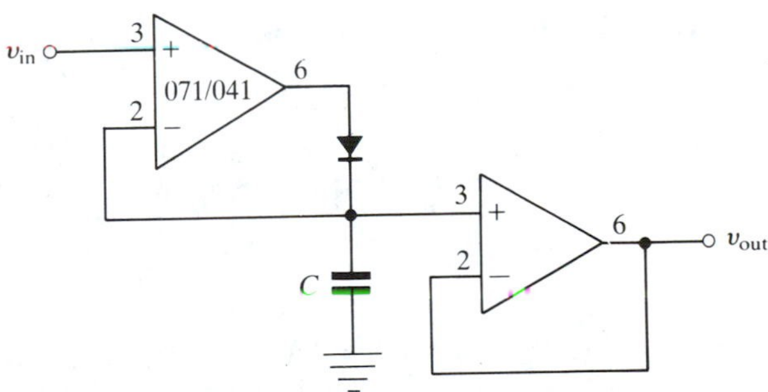

II. Sample and Hold

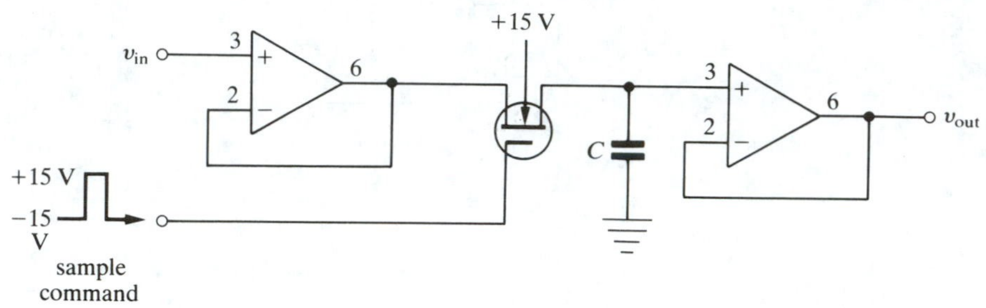

EXPERIMENT 29—OP AMP COMPARATOR

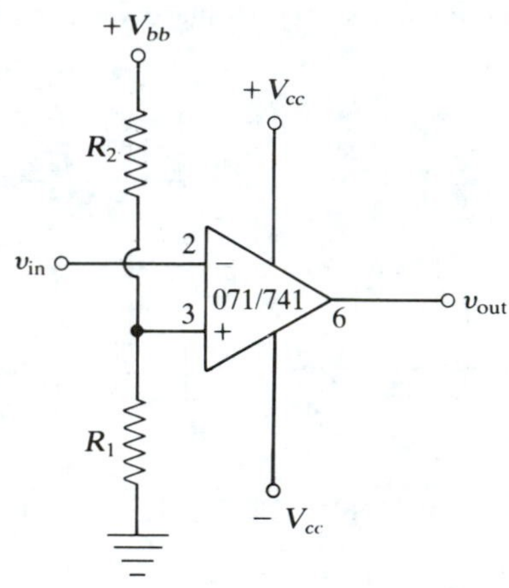

$$v_{out} \cong +V_{cc} \qquad \text{if } v_{in} < \frac{R_1}{R_1 + R_2} V_{bb}$$

$$v_{out} \cong -V_{cc} \qquad \text{if } v_{in} > \frac{R_1}{R_1 + R_2} V_{bb}$$

V_{bb} can be either + or −.

Measure the "threshold" or "trip" voltages by varying v_{in} slowly. Try a sinusoidal input of varying amplitude. Sketch the input and output waveforms.

EXPERIMENT 30—THE SCHMITT TRIGGER

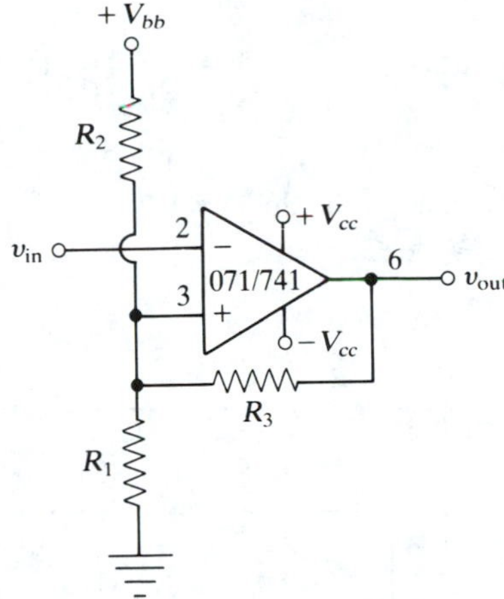

Derive the threshold or trip levels:

$$v_{out} \cong +V_{cc} \qquad \text{if } v_{in} < \_\_\_\_\_\_.$$

$$v_{out} \cong -V_{cc} \qquad \text{if } v_{in} > \_\_\_\_\_\_.$$

Vary v_{in} slowly and measure v_{out}. Sketch the hysteresis loop of v_{out} versus v_{in}. Try a sinusoidal low-frequency input. Sketch the input and output waveforms. Try a noisy input signal.

EXPERIMENT 31—OP AMP OSCILLATORS

Carefully sketch the output waveforms.

I. Square Wave Oscillator. A low bias current op amp is needed here.

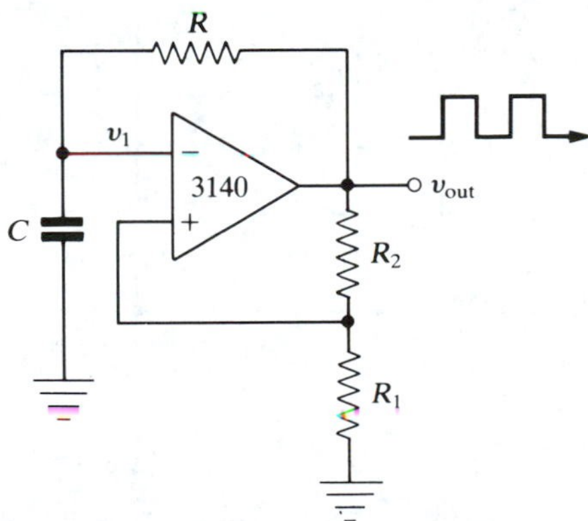

$$v_{out} \cong +V_{cc} \qquad \text{if } 0 < v_1 < \frac{R_1}{R_1 + R_2} V_{cc}$$

$$v_{out} \cong -V_{cc} \qquad \text{if } -\frac{R_1}{R_1 + R_2} V_{cc} < v_1 < 0$$

Try $R_1 \cong R_2$.

How could you change the duty cycle?

II. Ramp or Triangle-Wave Generator

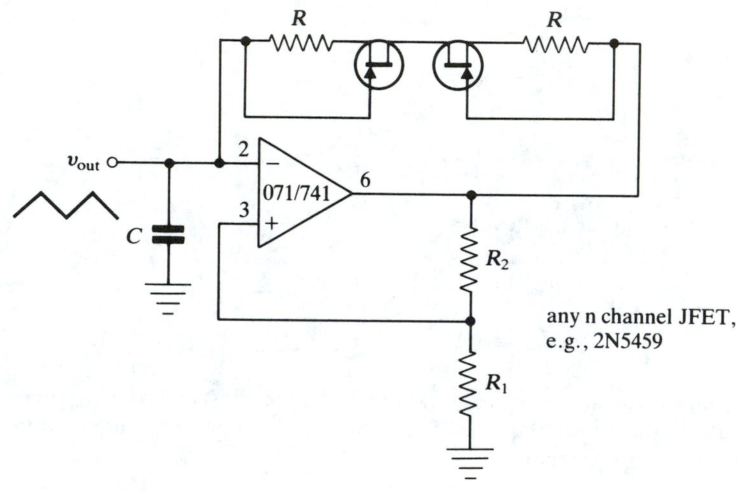

III. Monostable Multivibrator

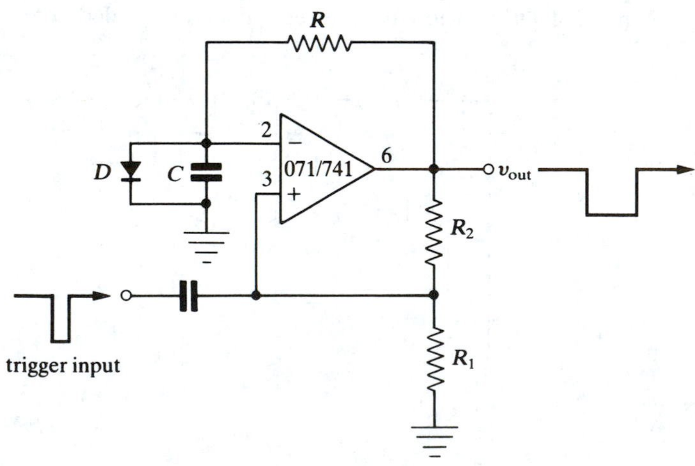

IV. Colpitts Oscillator (Sinusoidal). The C_2/C_1 ratio determines the amount of feedback. Try $L = 100\ \mu\text{H}$ and $C_1 \cong C_2 = 1000\ \text{pF}$.

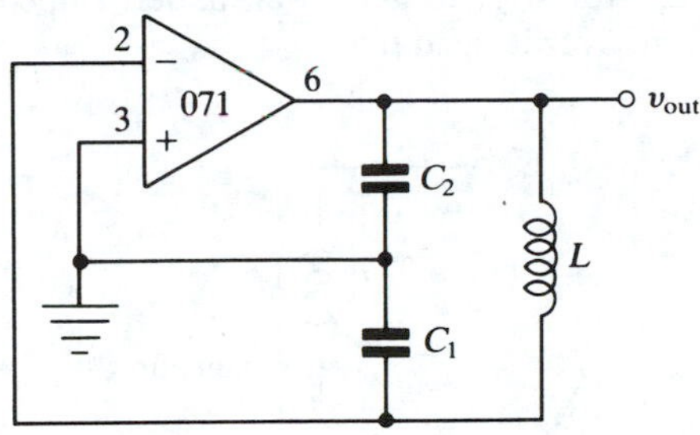

V. Hartley Oscillator (Sinusoidal). The position of the tap on L determines the amount of feedback (L_2/L_1).

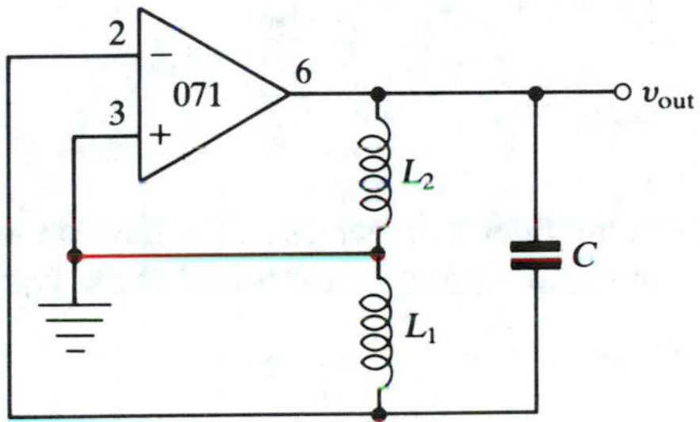

VI. Crystal-Controlled Oscillator (Sinusoidal). Crystal frequency $\cong$ 100 kHz. You may have to vary R_2 to obtain oscillation.

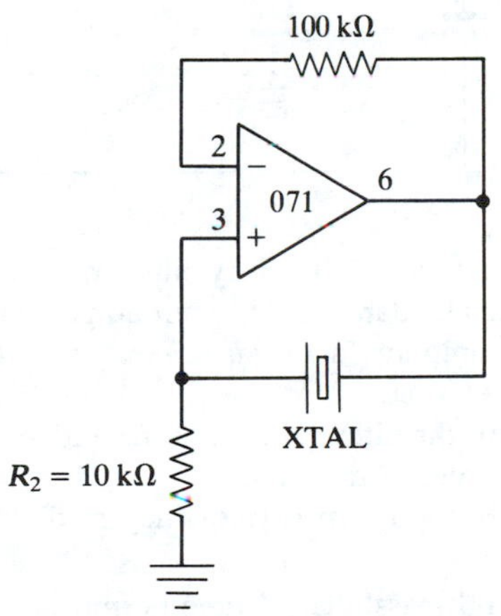

EXPERIMENT 32—ACTIVE LOW-PASS FILTERS

I. Construct a passive low-pass *RLC* filter with a break frequency of approximately 1 kHz. Adjust R to get (a) an underdamped filter, (b) a critically damped filter, (c) an overdamped filter.

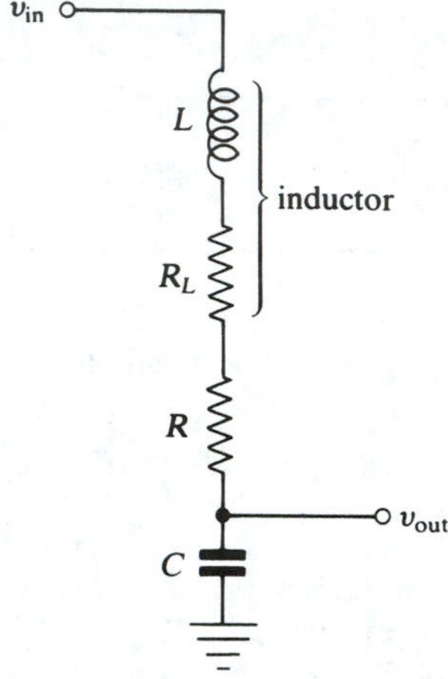

Measure and graph the filter gain versus frequency (sinusoidal input) and sketch the transient response (to a pulse input) for all three types of damping.

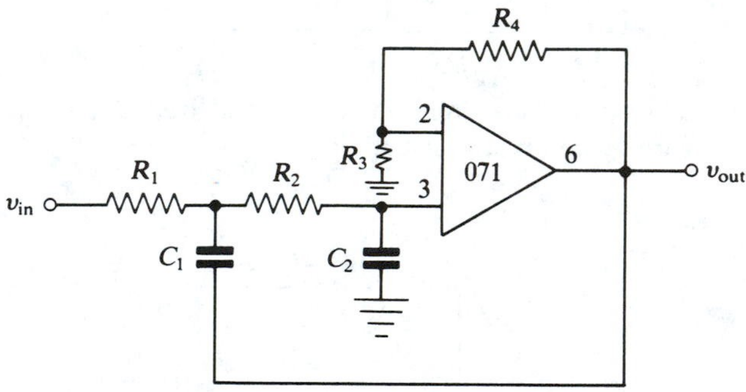

II. Construct a 1-kHz low-pass Sallen Key filter with variable gain for the (a) underdamped, (b) critically damped, (c) overdamped case. Use a pot for R_3 and R_4. Remember the damping factor $d_0 = 3 - A = 3 - (1 + R_4/R_3) = 2 - R_4/R_3$. Use $R_1 = R_2$ and $C_1 = C_2$. $\omega_0 = 1/R_C$.

Measure and graph the filter gain versus frequency and sketch the transient response for all three types of damping.

Measure the gain rolloff versus frequency in dB/octave or dB/decade for $f > 1$ kHz.

III. Repeat Part II for a high-pass filter, if time permits.

EXPERIMENT 33—ACTIVE BANDPASS FILTER

I. Construct the following bandpass filter. Design for $f_0 \cong 10$ to 100 kHz.

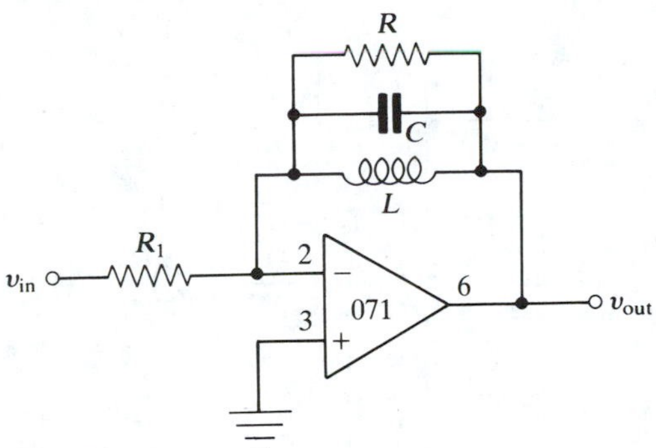

Measure and graph the gain versus frequency. Measure the Q. How does R affect the Q? What value of R gives the highest Q?

II. Construct the following twin T bandstop filter. Design for $f_0 \cong 10$ to 100 kHz. Be sure to carefully match the component values (R, C, $R/2$, $2C$) in the twin T section.

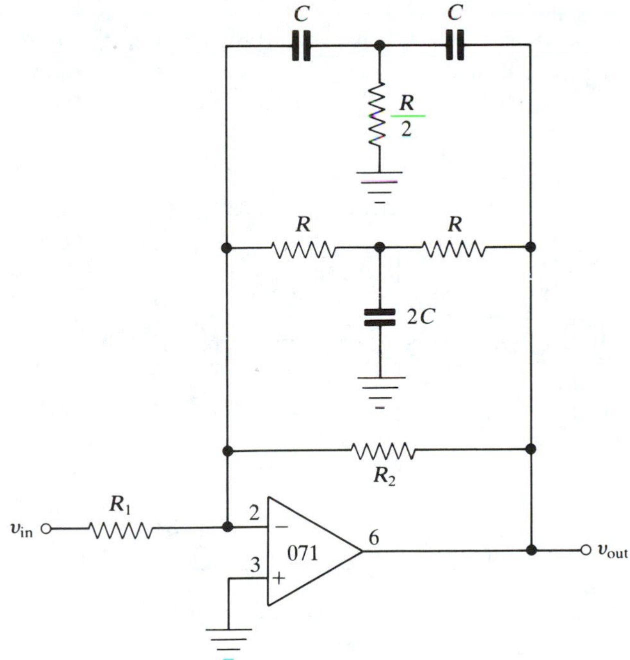

Measure and graph the gain versus frequency. Measure the Q. Compare the theoretical and experimental values for f_0.

EXPERIMENT 34—OP AMP POWER SUPPLIES

I. Simple Circuit (no short-circuit protection). $v_{out} = (1 + R_2/R_1)V_{ref}$ (regulated). VR = zener diode.

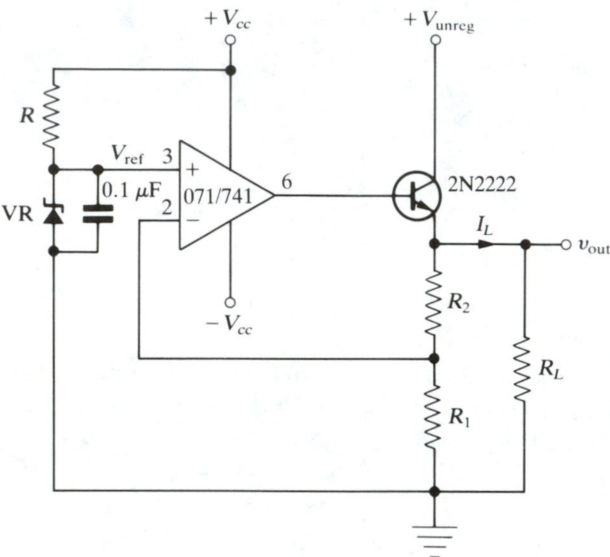

Measure the unregulated and output voltages (dc and ripple) for various I_L.

II. Short-Circuit-Protected Circuit. D = silicon diode. $I_{E\,max} \cong 0.6\text{ V}/R_{sc}$. $v_{out} = (1 + R_2/R_1)V_{ref}$ (regulated).

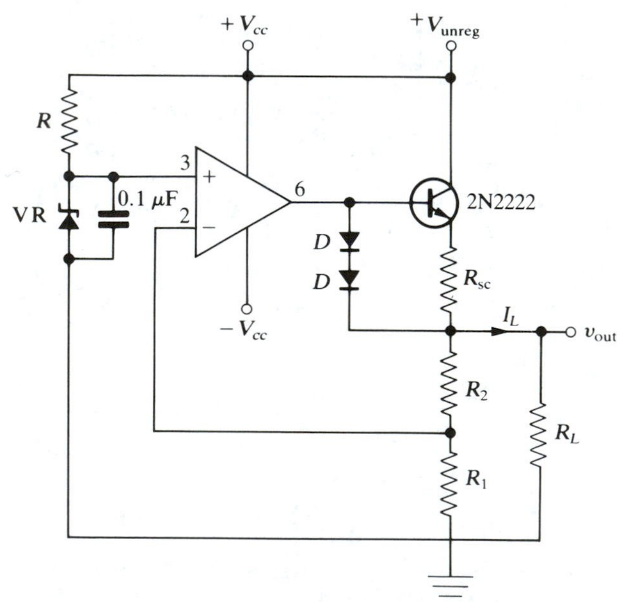

Measure the unregulated and output voltages (dc and ripple) for various I_L. Calculate R_{sc} so $I_{E\,max}$ doesn't blow out the transistor. Short the output to ground and measure the maximum I_E.

In both circuits V_{CE} should never fall below 2 or 3 V to ensure good output regulation. R_1 and R_2 should be chosen so that v_{out} is at most 2 V less than V_{cc}. Why? R is chosen in the usual way to allow the zener diode to regulate.

III. Constant Current Source ($I_{L\,max}$ = max op amp output current). Show $I_L = V_{ref}/R_3$ regardless of R_L.

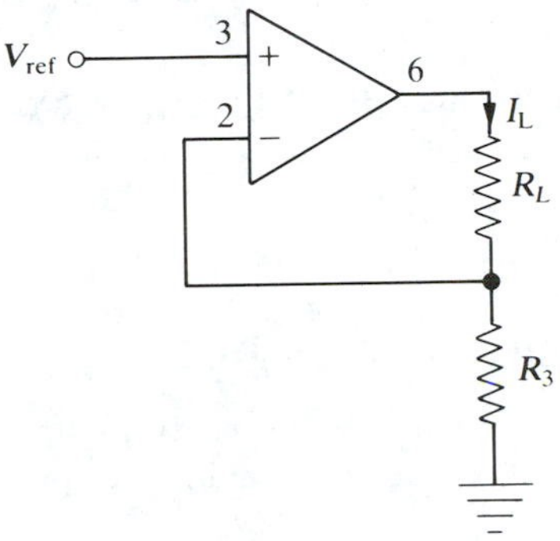

IV. Constant Current Source with Current Booster. Show $I_L = V_{ref}/R_3$ regardless of R_L.

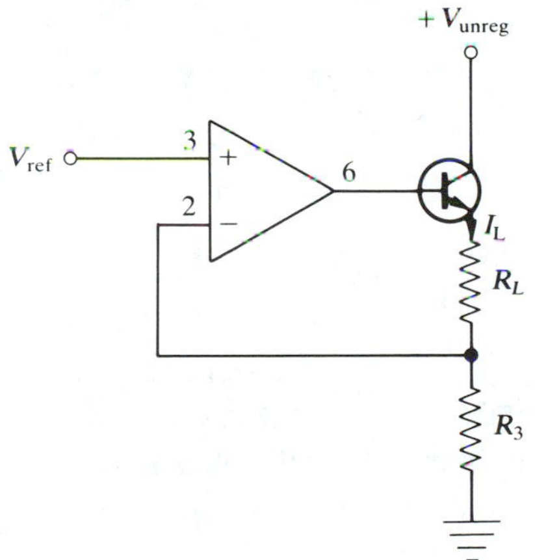

EXPERIMENT 35—THE "QUICK AND DIRTY" PRACTICAL POWER SUPPLY

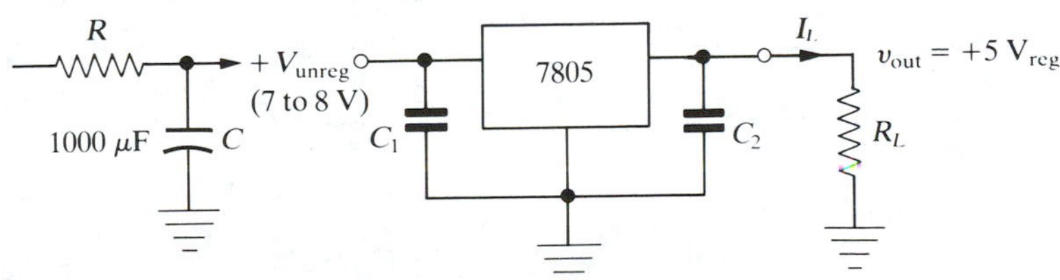

$+V_{unreg}$ usually comes from a four-diode full-wave bridge rectifier with a simple low-pass RC filter with $C \sim 1000\ \mu F$. $C_1 \cong 0.2\ \mu F$, $C_2 \cong 1\ \mu F$ (tantalum if possible).

Measure the output voltage v_{out} (dc and ripple) for various values of I_L. Calculate the dc output resistance of the supply. Omit C_1 and C_2 and observe v_{out}.

The 7800 series regulator chips are available in many voltages (e.g., 7815 for +15 V, etc.). The 7900 series chips produce negative voltages (e.g., 7915 output = −15 V). The chip must be bolted to a good heat sink (metal chassis, metal fins, etc.) if I_L is large. The power dissipated in the regulator chip is $P = (V_{unreg} - V_{reg})I_L$.

EXPERIMENT 36—BASIC NAND AND NOR GATE LOGIC

Use a 7400 (TTL), 74LS00 (TTL), 74C00 (CMOS), 4011 (CMOS), or 74HC00 (CMOS) two-input NAND gate for all parts of this experiment. Use V_{cc} = +5 V for TTL, or V_{cc} = +5 V or +10 V for CMOS gates.

I. Verify the truth table for your gate by measuring the actual electrical voltages at A, B, and F for the four possible combinations of S_1 and S_2.

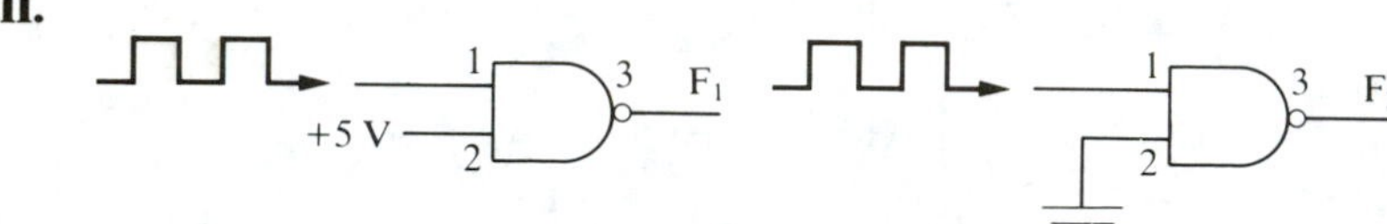

Write an "electrical" truth table, a positive-logic truth table, and a negative-logic truth table for your gate.

II.

Experimentally determine F_1 and F_2 waveforms.

III. Repeat Part I for a NOR gate: 7402, 74LS02, 4001, 74C02, or 74HC02.

IV. Construct an inverter from your NOR gate and from your NAND gate.

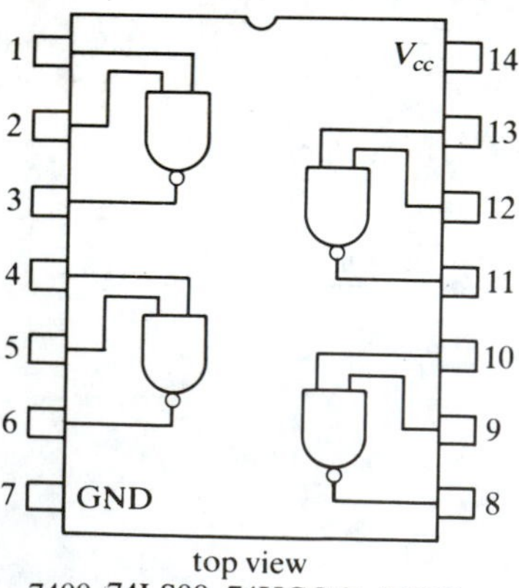

top view
7400, 74LS00, 74HCOO, 74C00

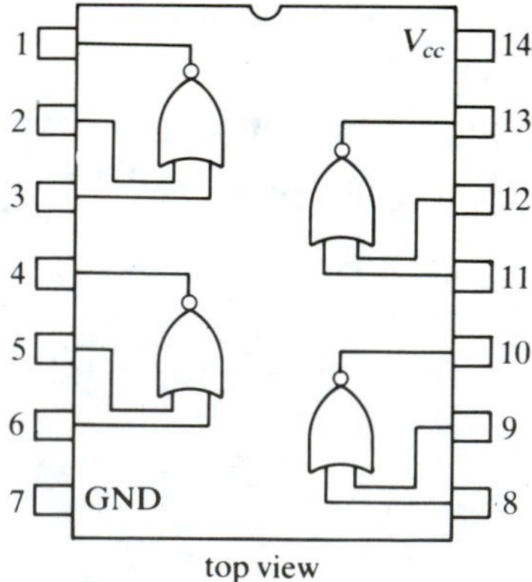

top view
7402, 74LS02, 74C02

EXPERIMENT 37—GOOD TTL LOGIC HIGHS AND LOWS

Use a 7400 or 74LS00 TTL gate. A good TTL high is >2.0 V, and a good TTL low is <0.8 V.

I. Verify that F = $\bar{A}$ if the two NAND inputs are tied together, by slowly varying V_A from 0 to V_{cc} and measuring V_F with an oscilloscope or high-impedance voltmeter. Graph V_F versus V_A.

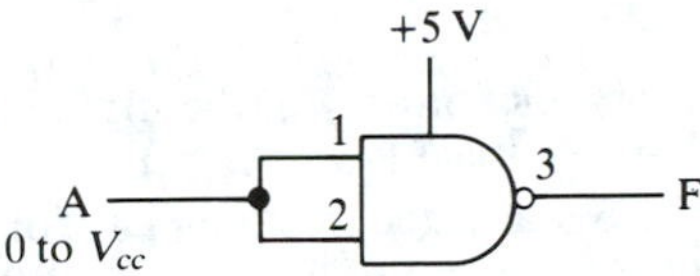

V_A can be the output of a variable-voltage power supply, but V_A should not exceed $V_{cc} = 5$ V. Graph V_F versus V_A. What is the maximum V_A for which $V_F > 2.4$ V? What is the minimum V_A for which $V_F < 0.4$ V?

II. High Output Source Current. Measure V_{out} as a function of I_{out} by varying R. Plot V_{out} versus I_{out}. (Try $R = 1$ MΩ, 10 kΩ, 1 kΩ, 470 Ω, 220 Ω, 100 Ω.)

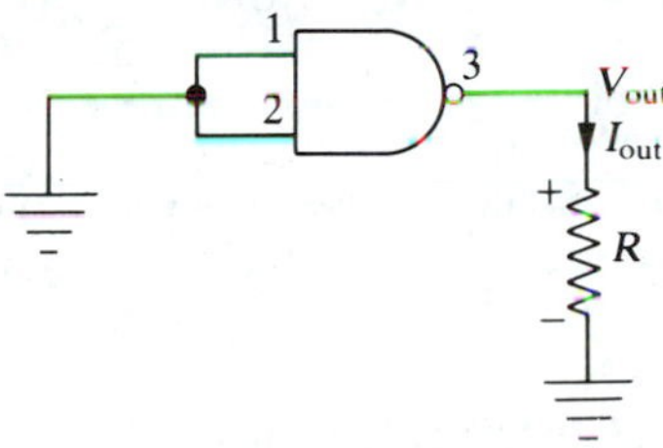

What is the maximum I_{out} ("source" current) for which $V_{out} > 2.0$ V (a good TTL logic high)?

III. Low Output Sink Current. Measure V_{out} as a function of I_s (the sink current) by varying R. Plot V_{out} versus I_s. (Try $R = 1$ MΩ, 10 kΩ, 1 kΩ, 100 Ω.) What is the maximum current I_s the output can sink and keep $V_{out} < 0.4$ V (a good TTL logic low)?

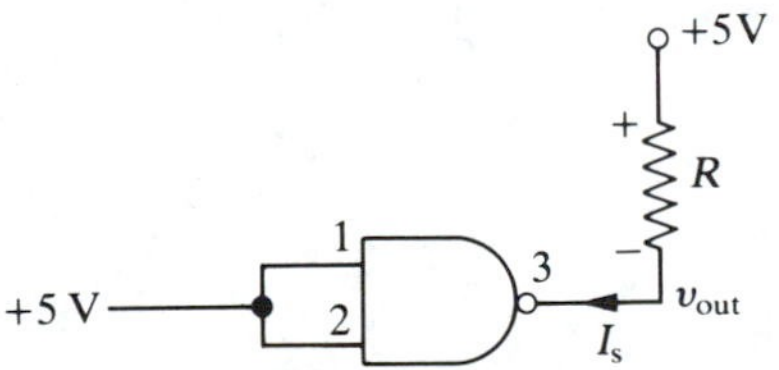

IV. B = 1. A = ? Note: If A and B are both good logic highs, then F is low.

Measure V_A and V_{out} as functions of I_{in} by varying R. [The lower R, the larger I_{in} and the better the logic high at A.]

Plot V_{out} versus I_{in}. What is the minimum I_{in} for A to be a good logic high (i.e., to keep V_{out} less than 0.4 V)? Try R = 1 MΩ, 100 kΩ, 10 kΩ, 1 kΩ.

V. B = 1. A = ? Note: If A and B are both good logic lows, then F is high.

Measure V_A and V_{out} as functions of I_{si} (the input sink current) by varying R. (The smaller R, the better a logic low is at A.)

Plot V_{out} versus I_{si}. What is the maximum I_{si} for A to be a good logic low (i.e., to keep V_{out} more than 2.0 V)? Try R = 100 kΩ, 10 kΩ, 1 kΩ, 470 Ω, 100 Ω.

VI. Floating (Unconnected) Inputs. Measure V_{out} and infer what A must be: 0 or 1? A floating TTL input acts like a _______.

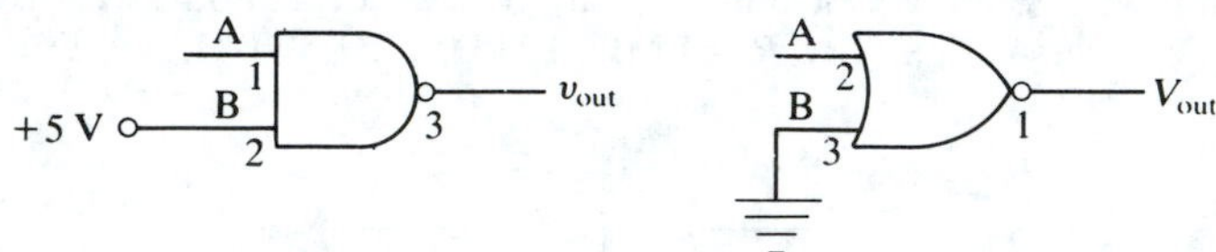

Beware: In actual circuits, *never* leave an unused input floating. Either ground it or connect it to $+V_{cc}$ through a 1-kΩ resistor.

EXPERIMENT 38—LEDS AND DECODER

I. The LED.

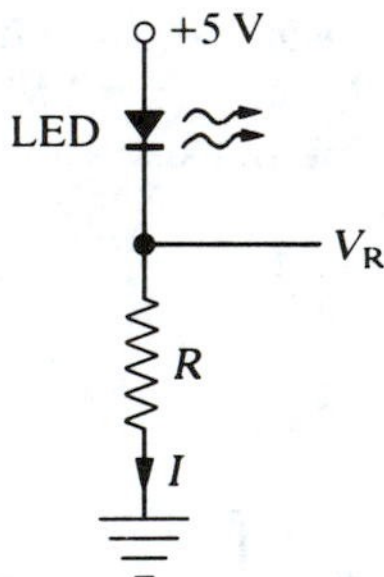

Observe the LED light output as a function of current by varying R and measuring V_R. Approximately how much current I is necessary to make the LED appear bright in a lighted room? Approximately what is the voltage drop across the LED when it is lit? Is this voltage essentially constant as long as the LED is lit? Try R = 330 Ω.

II.

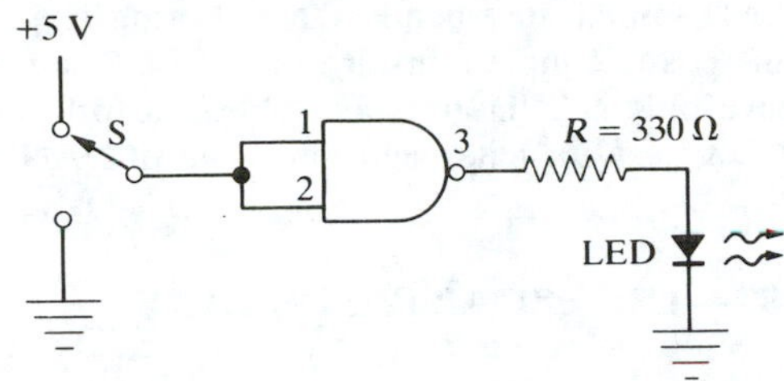

For what input is the LED on?
Can a single 7400 (74LS00) gate light one LED? Two LEDs?

III.

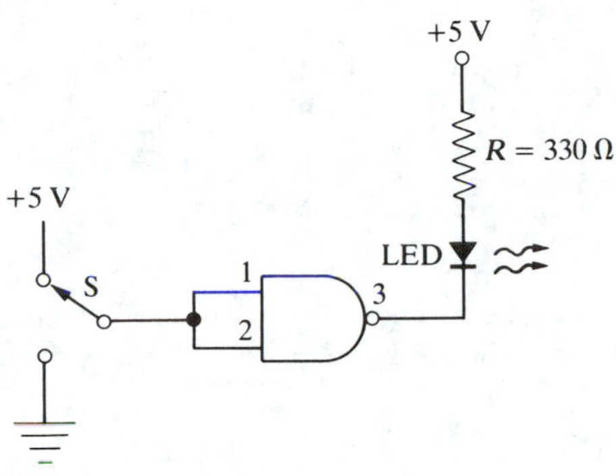

For what input is the LED on?
Can a single 7400 (74LS00) gate light one LED? Two LEDs?

IV.

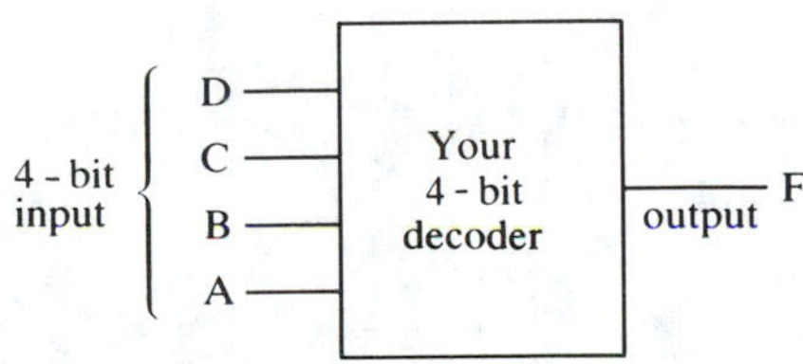

Your instructor will give you a four-bit input (e.g., ABCD = 1110). Design and construct a combination of gates (your four-bit decoder) that will make F = 1 only for ABCD = 1110. For example, the decoder for ABCD = 1110 would be

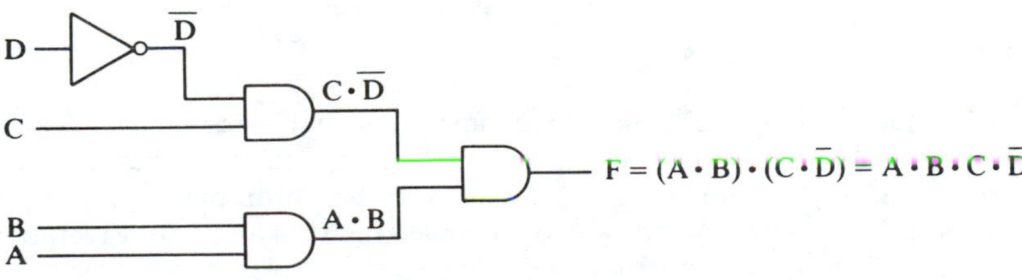

Use an LED to indicate when F = 1.

How many possible input four-bit words ABCD are there?

If you have only NAND gates available, you will probably have to make inverters, AND gates, OR gates, and so on, out of NAND gates.

EXPERIMENT 39—THE EXCLUSIVE OR GATE

I. Verify the truth table for an XOR gate (7486 or 74LS86), using switches for the inputs and an LED for the output. Write an electrical truth table, a positive-logic truth table, and a negative-logic truth table.

II. Construct an XOR gate, using only NAND (or NOR) gates.

III.

A 1 2 7486 3 F = ?

A 1 +5 V 2 3 F = ?

How could you make a "controllable inverter" from an XOR gate? Sketch the input, output, and control waveforms or "timing diagram." Assume a square-wave input.

IV. Phase Detector

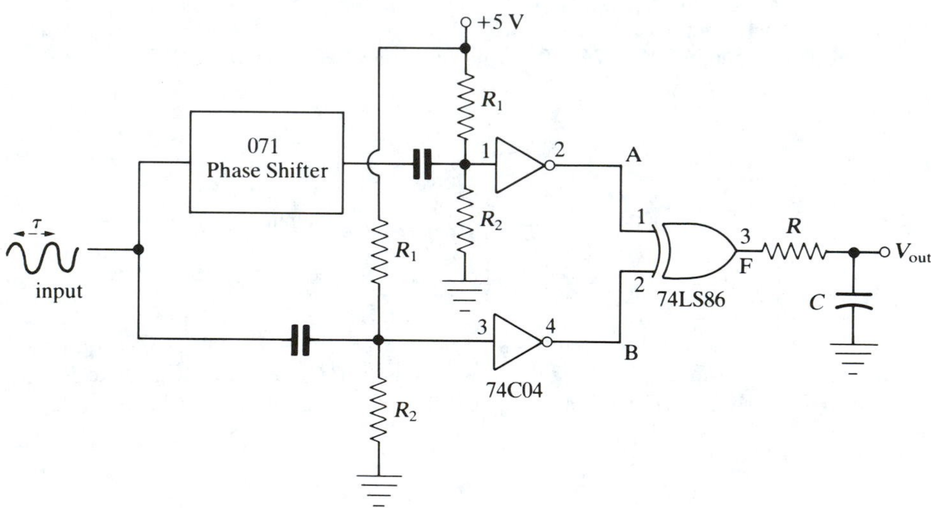

Construct a phase shifter (see Experiment 25) and measure the phase shift ϕ between A and B on a dual-beam scope. Pick R_1 and R_2 so that the waveforms at A and B are rectangular. Also measure the dc output V_{out} and plot V_{out} versus ϕ. Pick $RC \gg \tau$ for a good low-pass filter. Sketch the waveforms at A, B, and F.

V_{out} should be proportional to the phase difference between the signals at ________ and ________.

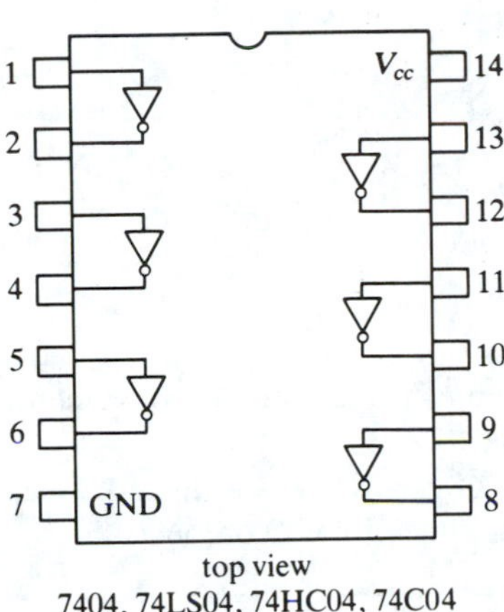

top view
7404, 74LS04, 74HC04, 74C04

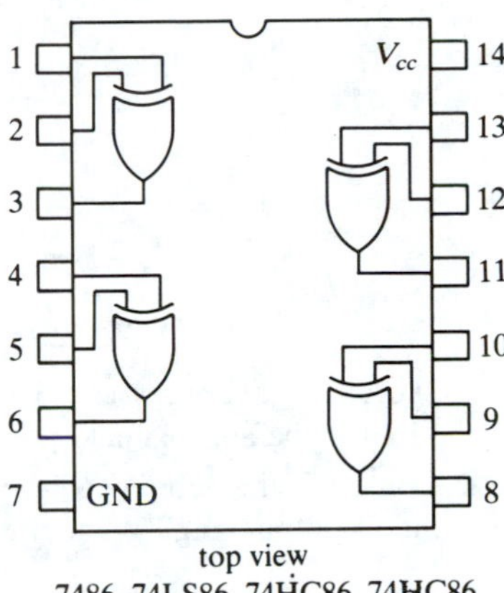

top view
7486, 74LS86, 74HC86, 74HC86

EXPERIMENT 40—THE RS FLIP-FLOP

I. Construct the two flip-flops shown and determine their truth tables.

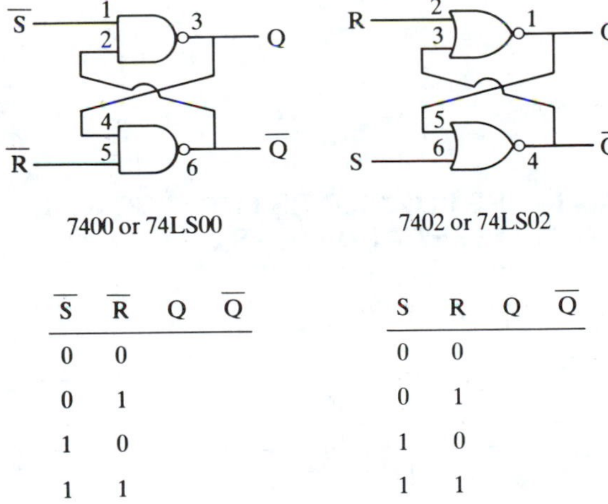

7400 or 74LS00

7402 or 74LS02

$\overline{S}$	$\overline{R}$	Q	$\overline{Q}$
0	0		
0	1		
1	0		
1	1		

S	R	Q	$\overline{Q}$
0	0		
0	1		
1	0		
1	1		

$\bar{S} = \bar{R} = 0$ is undesirable. Why? $S = R = 1$ is also undesirable. Why?

II. Construct the bouncing mechanical switch circuit shown.

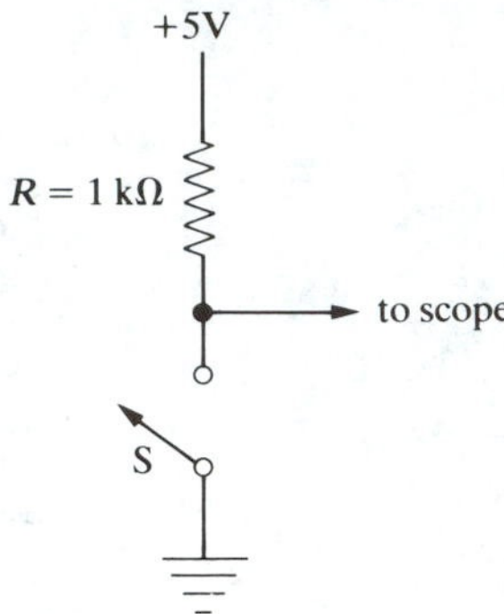

Sketch V_{out} versus t as S is closed. A memory or storage scope is convenient. What is the approximate period of the bouncing?

III. Construct the debounced switch. Sketch Q and $\bar{Q}$ versus t as S is flipped. Does the Q output ring? What is the approximate rise time of the Q output?

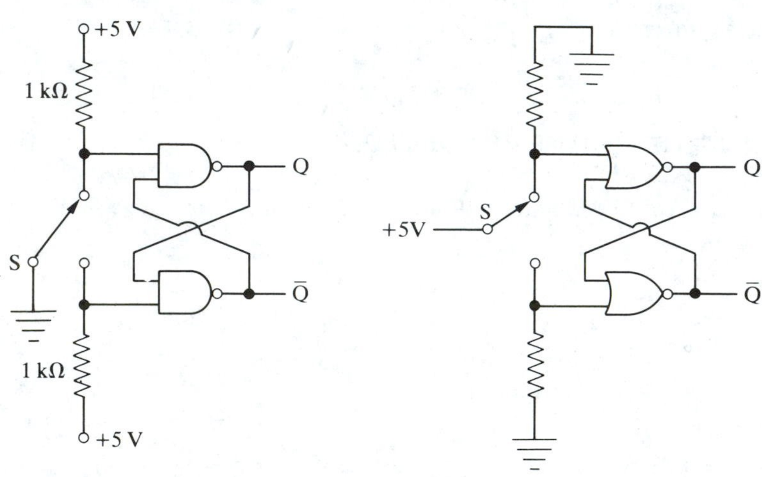

EXPERIMENT 41—THE CLOCKED RS FLIP-FLOP WITH ASYNCHRONOUS PRESET AND CLEAR

I. Construct the clocked RS flip-flop shown.

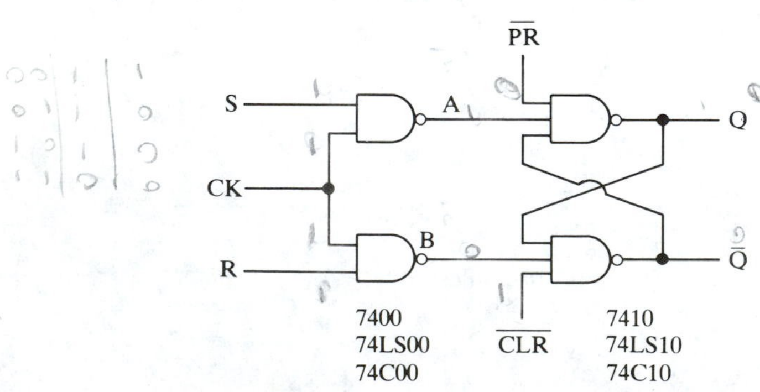

a
b
c
F

a b c	$F = \overline{abc}$
0 0 0	1
0 0 1	1
0 1 0	1
0 1 1	1
1 0 0	1
1 0 1	1
1 1 0	1
1 1 1	0

The R, S, CK, $\overline{PR}$, and $\overline{CLR}$ inputs can all be switches. Determine the truth table. X = "don't care" (either 0 or 1).

$\overline{PR}$	$\overline{CLR}$	CK	S	R	A	B	Q	$\bar{Q}$
1	1	0	0	0				
1	1	0	0	1				
1	1	0	1	0				
1	1	0	1	1				
1	1	1	0	0				
1	1	1	0	1				
1	1	1	1	0				
1	1	1	1	1				
0	1	X	X	X				
1	0	X	X	X				
0	0	X	X	X	X	X		

Notice that the $\overline{\text{PRESET}}$ and $\overline{\text{CLEAR}}$ inputs are active low inputs because they are inputs to a NAND gate. Thus, they are normally held ________ when they are not used.

Why are the $\overline{\text{PRESET}}$ and $\overline{\text{CLEAR}}$ inputs called asynchonous? *Hint*: "Synchonous" means controlled or activated by the clock, and the prefix "a," grammatically speaking, means "not."

What happens if both $\overline{PR}$ and $\overline{CLR}$ = 0? Is this a good situation?

What should R and S be to store information (as Q and $\bar{Q}$)?

II. Design (do not build) a similar clocked RS flip-flop using only NOR gates, and state the truth table. The PRESET and CLEAR inputs are active ________ and thus are normally held ________ when they are not used.

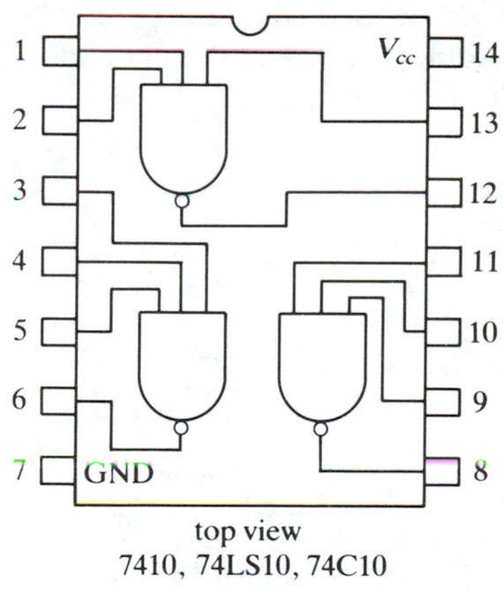

top view
7410, 74LS10, 74C10

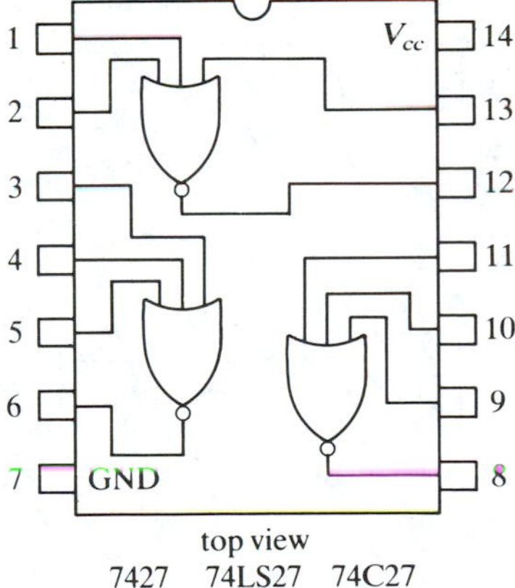

top view
7427 74LS27 74C27

EXPERIMENT 42—TIMER CIRCUIT

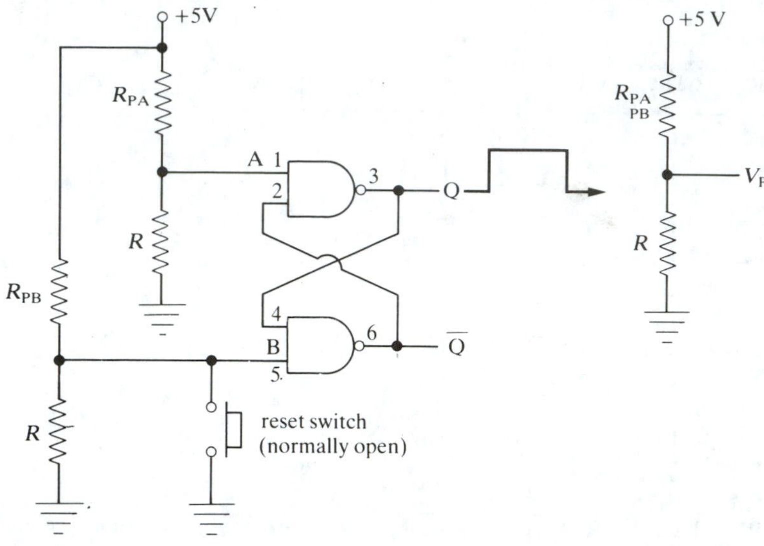

Wire two photocells R_P for two NAND gates as above. The photocell resistance is high ($\sim 100\ k\Omega$) when it's in the dark and low ($\sim 10\ k\Omega$) when it's illuminated. Choose R such that the voltage V_p is a good logic low (~ 0.5 V or less) when the photocell is dark, and V_p is a good logic high (~ 2.5 V or more) when the photocell is illuminated.

With the two photocells illuminated, A = B = 1. Then the switch will reset Q = 0 (and $\bar{Q}$ = 1) when it is momentarily closed. Now when photocell PA is momentarily darkened (something blocks its light), Q should go high about 10 ns after the light to photocell A is decreased.

When the other photocell (PB) is momentarily darkened, $\bar{Q}$ should go high and Q should go low. Thus, the *length* of time Q is high is a measure of the time interval between the two events that darken the two photocells. Thus the Q output level can be used to measure the time intervals between the events darkening the photocells. For example, if the clock is 1 kHz, the counter reading will be the time interval in milliseconds.

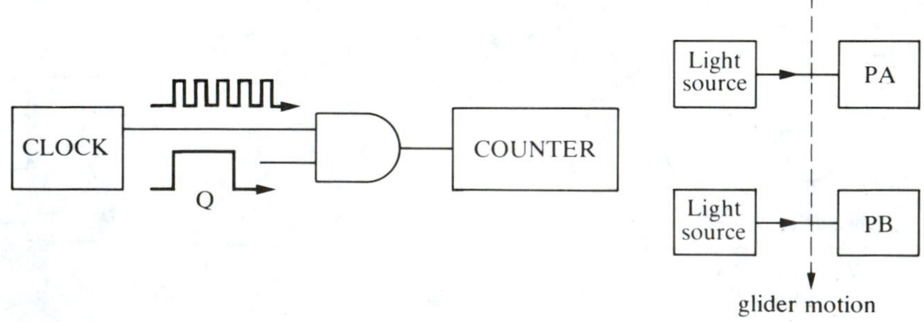

Use the circuit to measure the speed of a glider on an airtrack, if available, or the speed of one's hand moving past the two photocells.

The speed of the circuit can be improved by using a phototransistor in place of the photocells.

The use of a 74LS00 instead of a 7400 will improve things because the LS TTL gate must sink less current for a good logic low input. Why does this help? A CMOS NAND gate (74C00) would be even better because its input currents are so small.

EXPERIMENT 43—THE D FLIP-FLOP

I. Add an inverter to your clocked RS flip-flop to make a D (data) flip-flop.

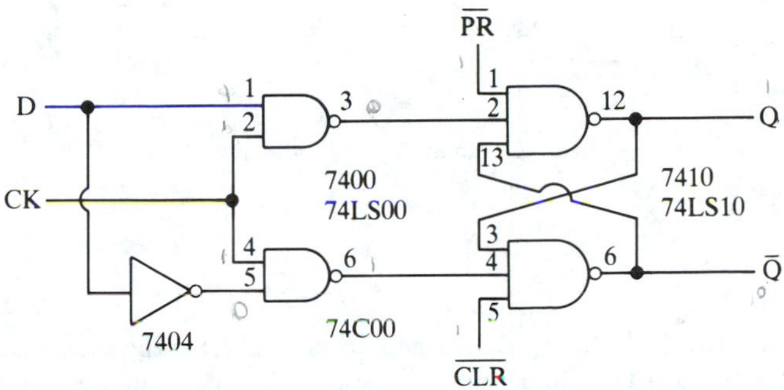

Determine the truth table. All the inputs can be switches. Notice that $S = \bar{R}$ always because of the inverter. Thus, we never can have RS = 00 or 11.

$\overline{PR}$	$\overline{CLR}$	CK	D	Q	$\bar{Q}$
1	1	0	0		
1	1	0	1		
1	1	1	0		
1	1	1	1		
0	1	X	X		
1	0	X	X		
0	0	X	X		

Notice that as long as CK = 1, Q = D; that is, if D changes while the CK is high, Q will follow the changes in D. This is an example of a pulse-triggered or level-triggered D flip-flop. An example is the 7475 or 74LS75, a quad clocked latch—that is, four level-triggered D flip-flops on one chip.

Notice also that the PRESET ($\overline{PR}$) and the CLEAR ($\overline{CLR}$) inputs are *asynchronous*—that is, they "override" the clock input. Question: What are $\overline{PR}$ and $\overline{CLR}$ normally?

II. As explained in Chapter 13, there are two types of flip-flop triggering: *level* or *pulse* triggering, and *edge* triggering. In an edge-triggered D flip-flop the D input present during the short time the clock is *changing* (typically 20 ns) will affect the Q output. In some cases the D input must be stable for a setup time (typically 20 ns) before the clock edge, and also for a hold time (typically 0 ns but 5 ns for some flip-flops) after the clock edge. The clock edge can be either the rising (positive edge triggering) or the falling (negative edge triggering) edge of the clock pulse. A 7474 is a very popular positive edge-triggered flip-flop.

Construct a divide-by-two counter with a 7474 or 74LS74A flip-flop. Carefully sketch the CK input and the Q output waveforms.

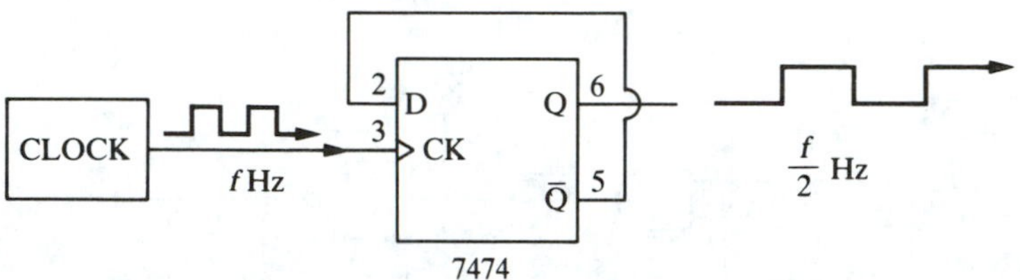

How fast will the flip-flop toggle? *Toggle* means to *change* the output state once for each complete input clock pulse. Measure the rise and fall times of the flip-flop output.

Repeat with a CMOS D edge-triggered flip-flop such as the 74C74 or 4013.

Do for V_{cc} = 5 V and 10 V. The flip-flop should be faster with V_{cc} = 10 V.

III. Make a real-time clock with a square waveform at 30 Hz.

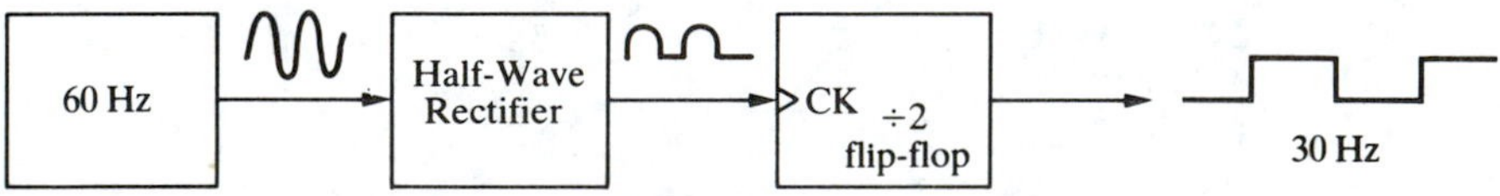

Use a low-voltage transformer secondary winding (6 V, etc.) for the 60-Hz sinusoidal signal. The input to the DFF divide-by-2 circuit should have a *maximum* peak amplitude of +5 V. The frequency stability of this clock depends on the power company keeping the ac line voltage at precisely 60 Hz (which it does quite well). This circuit also illustrates that the triggering waveform input to the CK terminal of the DFF need not be square.

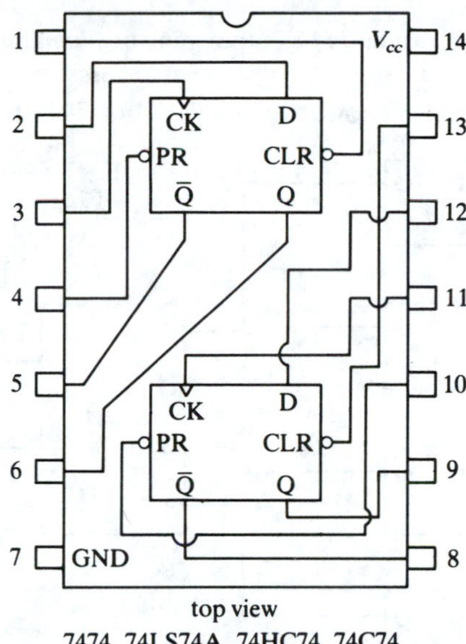

top view
7474, 74LS74A, 74HC74, 74C74

EXPERIMENT 44—THE 7447 SEVEN-SEGMENT LED DRIVER AND A DATA LATCH

I.

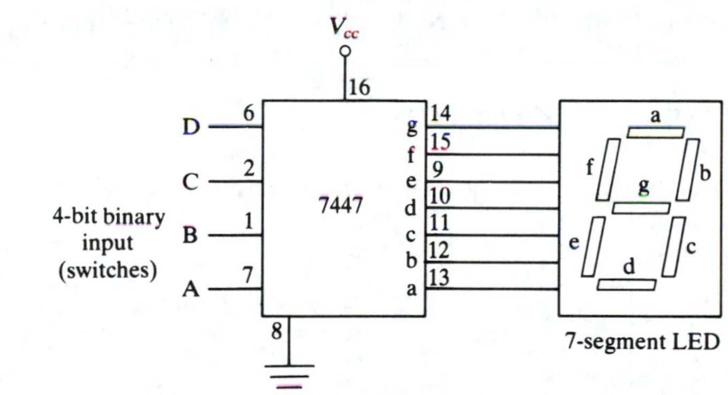

The 7447 is a widely used decoder that converts an ordinary four-bit binary input into an output to light up the appropriate segments in a seven-segment LED. For example:

Input	LED Display
DCBA = 0101 (D = MSB A = LSB)	5 (five)

Verify that the appropriate 7447 output pins are set low for the various binary DCBA inputs.

II. Construct a data latch as shown with the 7474 or 7477 four-bit bistable latch.

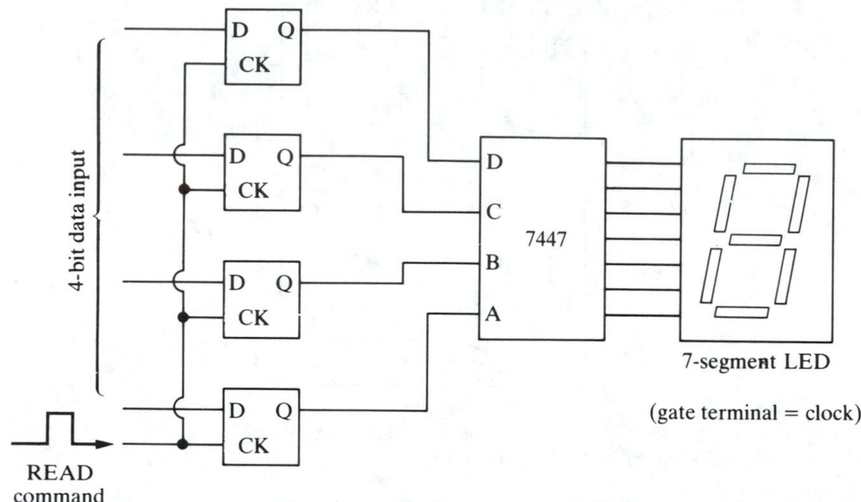

Set various four-bit inputs with the switches and observe that the decimal equivalent appears on the LED display only after the READ command of the gate terminal enables the flip-flops. Notice also that the display terminal remains constant after the READ pulse is over.

EXPERIMENT 45—THE JK FLIP-FLOP

I. Wire up the JK flip-flop shown and determine the truth table.

74LS76A:

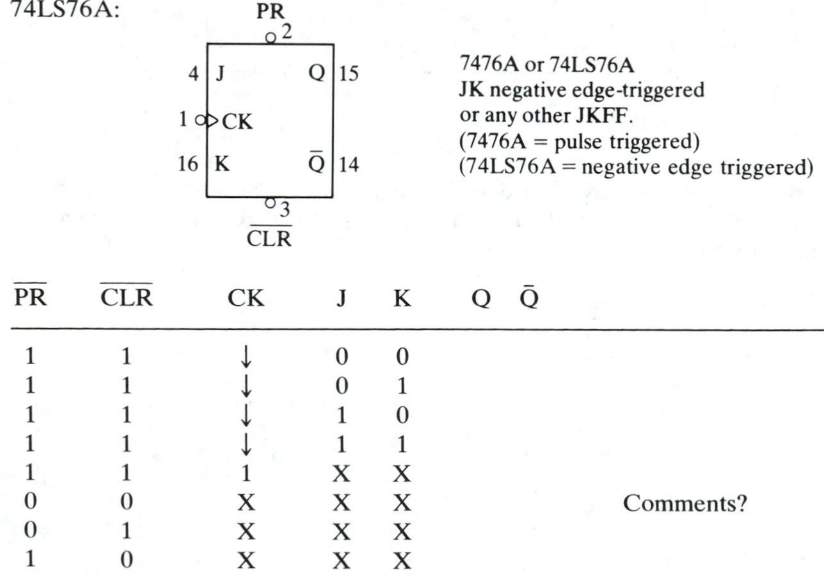

$\overline{PR}$	$\overline{CLR}$	CK	J	K	Q	$\bar{Q}$
1	1	↓	0	0		
1	1	↓	0	1		
1	1	↓	1	0		
1	1	↓	1	1		
1	1	1	X	X		
0	0	X	X	X		
0	1	X	X	X		
1	0	X	X	X		

Comments?

The ↓ in the truth table means negative edge triggering.

7476A:

PR	CLR	CK	J	K	Q	$\bar{Q}$	
1	1	⎍	0	0			
1	1	⎍	0	1			
1	1	⎍	1	0			
1	1	⎍	1	1			
0	0	X	X	X			Comments?
0	1	X	X	X			
1	0	X	X	X			

The ⎍ in the truth table means positive pulse (level) triggering.

II. JK Toggle Flip-Flop. Construct a toggle flip-flop from your JKFF and verify that it divides by two by observing (and sketching) the clock and the output (Q) waveforms.

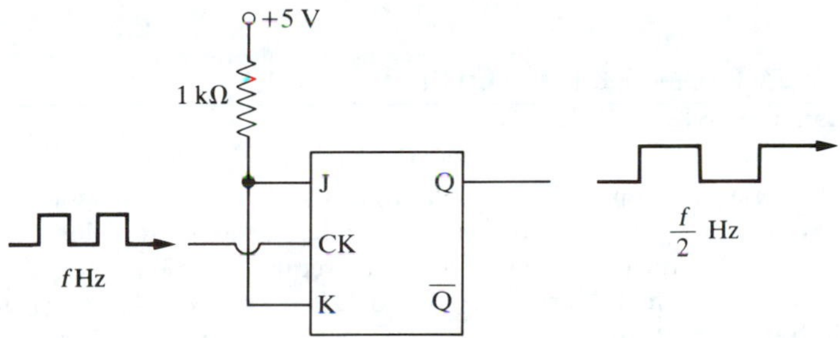

The 1-kΩ resistor is optional. R = 0 Ω will work, but the 1 kΩ will protect the FF against voltage spikes, and so on, on the V_{cc} line.

III. Sketch how you would make a DFF from a JKFF.

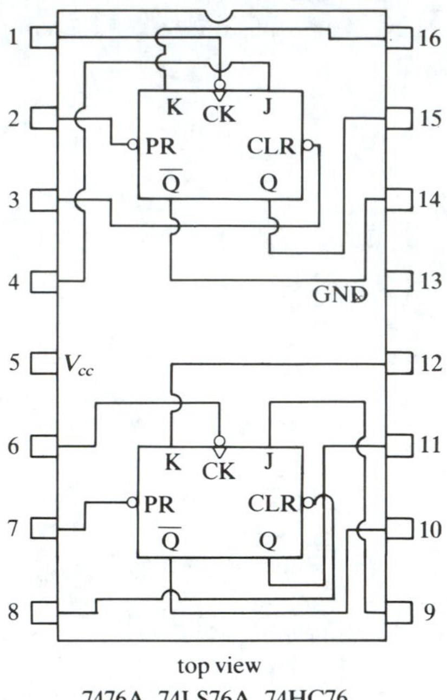

top view
7476A, 74LS76A, 74HC76

EXPERIMENT 46—THE RIPPLE OR ASYNCHRONOUS COUNTER

I. If N toggle flip-flops are connected in "series" (see illustration on page 911)—that is, with the Q output of one providing the toggle (clock) input of the next—then the final flip-flop Q output frequency will be the input clock frequency (to the first flip-flop) divided by 2^N. Using four flip-flops, construct a divide-by-16 ripple counter. You can use one 7475 or 74LS75, which is a quad DFF with each DFF connected as a toggle FF (see Experiment 43). Or you can use two 7476A or 74LS76A, each of which is a dual JKFF, with each JKFF connected as a toggle FF (see Experiment 45). Many other flip-flops could be used—you just need four toggle-connected flip-flops. Wire the unused active low PR and CLR inputs to +5 V.

What is the propagation delay for each flip-flop?

Do all the flip-flop outputs change state at precisely the same time?

What is the approximate total delay between the input clock change and the change in Q_4 for the divide-by-16 counter?

Sketch the timing diagram.

Write the count sequence.

Hook up a buffered LED to each Q and use a low-frequency (~1-Hz) input. Observe the LEDs blinking. Notice that all the Q outputs are the bits of the count in binary form, with Q_1 = LSB and Q_4 = MSB.

II. Construct a divide-by-3 or a divide-by-6 or a divide-by-7 or a divide-by-9 or a divide-by-11 ripple counter, but clear all the flip-flops at the final desired count.

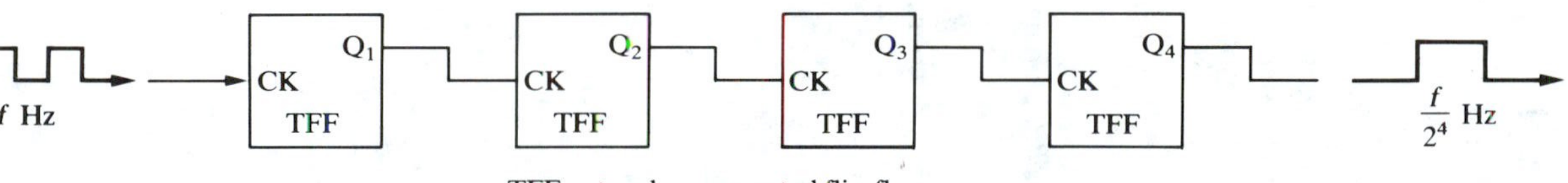

Illustration for Exp. 46.I.

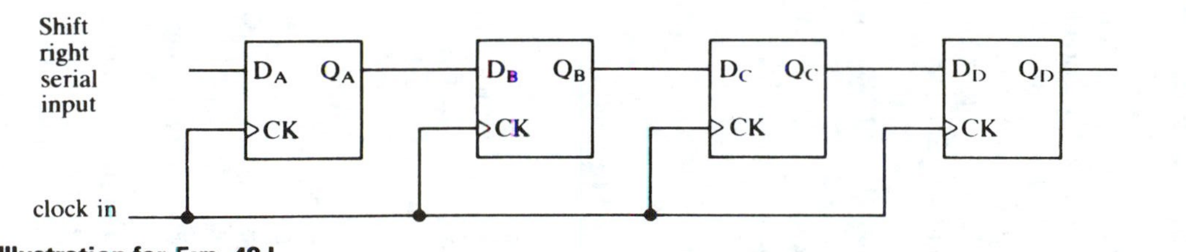

Illustration for Exp. 48.I.

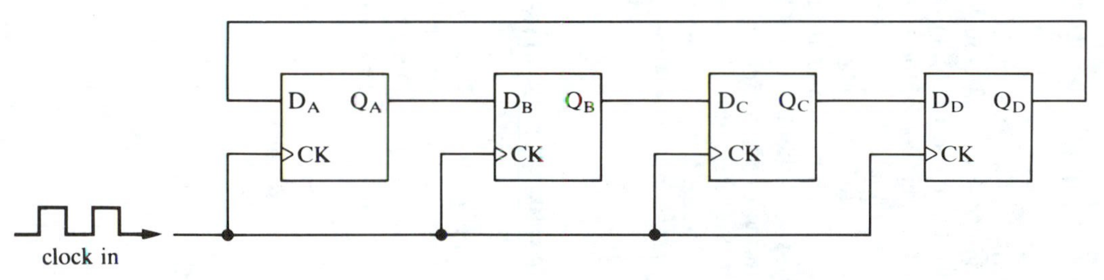

Illustration for Exp. 48.II.

See Section 13.3 for a divide-by-5 ripple counter. Sketch the circuit.
Sketch the timing diagram.
Write the count sequence.
Observe the false transient count(s) on the scope and sketch.

EXPERIMENT 47—THE SYNCHRONOUS COUNTER

Using JK flip-flops (e.g., 74LS76A, 74LS109A), design and construct a synchronous divide-by-7 or divide-by 9 counter. See 13.4 for a divide-by-6 synchronous counter. Wire the unused $\overline{PR}$ and $\overline{CLR}$ inputs to +5 V. Connect a seven-segment display to the four flip-flop outputs. Write the count sequence.

Calculate the logic expressions for the various JK inputs (e.g., $J_1 = Q_2 \cdot Q_3 + \bar{Q}_2 \cdot Q_3$, etc., for the divide-by-6 counter in Section 13.4).

Sketch the timing diagram.

Do all the flip-flop outputs change state at essentially the same time? Compare with the ripple counter.

Are there any false transient counts as in the ripple counter?

Notice that all the flip-flop Q outputs contain the binary numbers in the count sequence in *parallel* form.

EXPERIMENT 48—THE SHIFT REGISTER

I. The 74194 or 74LS194A is a four-bit, bidirectional or *universal* shift register with an asynchronous CLEAR input. (See illustration on page 911.) A series of edge-triggered D flip-flops can be used instead of the 74194.

With the $\overline{\text{CLEAR}}$ input at 1, when the clock goes high:

Control inputs	S_1	S_0	Function
	0	0	Inhibit or no change, $Q_AQ_BQ_CQ_D$ remain unchanged
	0	1	Shift Right $Q_A \rightarrow$ input, $Q_B \rightarrow Q_A$, $Q_C \rightarrow Q_B$, $Q_D \rightarrow Q_C$
	1	0	Shift Left $Q_A \rightarrow Q_B$, $Q_B \rightarrow Q_C$, $Q_C \rightarrow Q_D$, $Q_D \rightarrow$ input
	1	1	Parallel Load $Q_A \rightarrow$ a, $Q_B \rightarrow$ b, $Q_C \rightarrow$ c, $Q_D \rightarrow$ d abcd = bits preset at ABCD inputs.

A. Shift Right Operation:

Set ABCD = 1000 and load these data in. Then with a slow square-wave clock input (~1 Hz), note the sequence of $Q_AQ_BQ_CQ_D$.

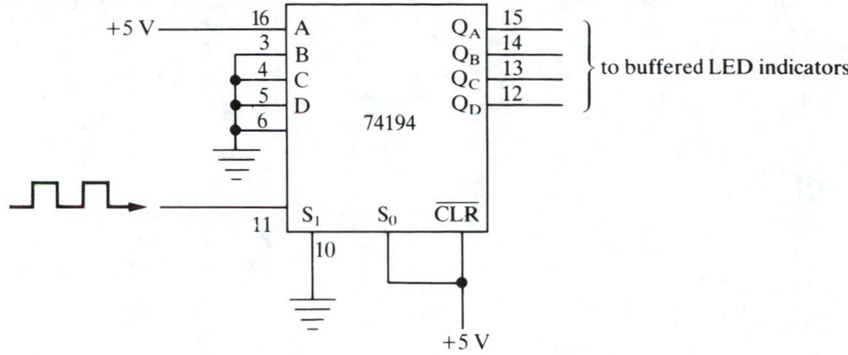

B. Repeat for the shift left mode of operation, loading in ABCD = 0001. Remember $S_1 = 1$ and $S_0 = 0$ for shift left.

The two preceding modes are a parallel to serial data conversion or parallel in serial out (PISO). Where is the serial output?

If the data are fed in serially and then taken out from the $Q_AQ_B \ldots$ flip-flop outputs, we have a serial in parallel out (SIPO) shift register. The 74164 or 74LS164A is an eight-bit SIPO shift register.

II. Ring Counter. (See illustration on page 911.)

The output of the last flip-flop is connected to the input of the first flip-flop:

Load in $Q_AQ_BQ_CQ_D$ = 1000 and then observe the Q outputs (on the LED indicators) as the slow clock input comes in. Write the $Q_AQ_BQ_CQ_D$ sequence.

Load in $Q_AQ_BQ_CQ_D$ = 1100 and repeat.

EXPERIMENT 49—THE JOHNSON COUNTER AND SINE-WAVE SYNTHESIS

I. Using an eight-bit SIPO shift register (74164, 74LS164A), construct a Johnson counter by connecting the $\bar{Q}_H$ output to the A input (see illustration on page 914). Load in $Q_AQ_BQ_CQ_DQ_EQ_FQ_GQ_H = 00000000$ by momentarily grounding the CLEAR input. Notice, there is no parallel load input on the 74164, unlike the four-bit 74194.

Observe the $Q_A \ldots Q_H$ states on the LED indicators as the slow clock input comes in.

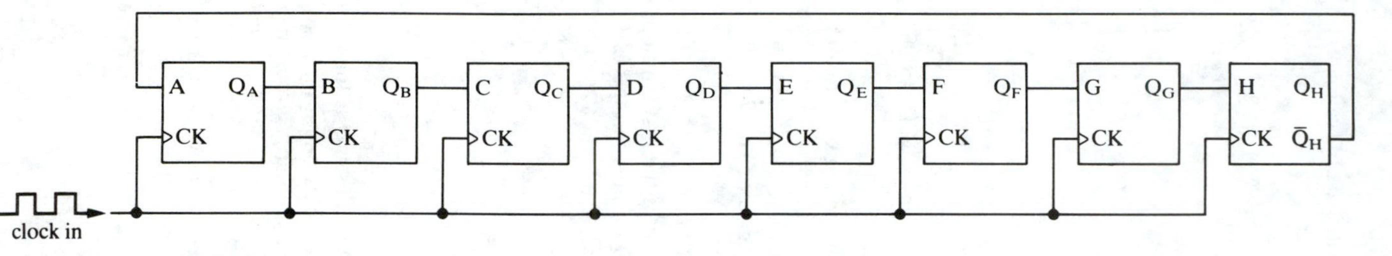

Illustration for Exp. 49.I.

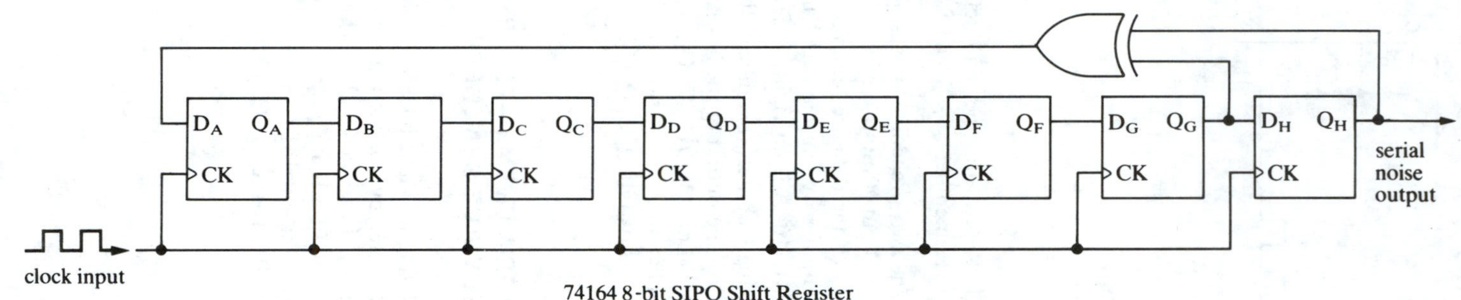

Illustration for Exp. 50.

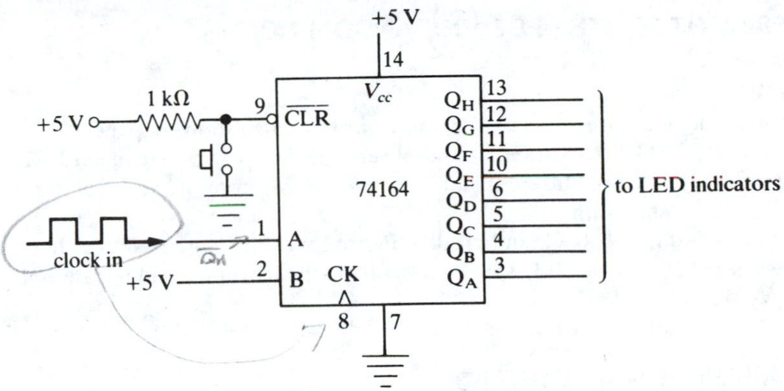

Write the $Q_A \ldots Q_H$ count sequence.

II. Connect the $Q_B \ldots Q_H$ outputs to a summing amplifier and display the analog waveform V_{out} on a scope. This should be a crude stepwise approximation to a sine wave. What is the relationship between the clock period and the period of this waveform?

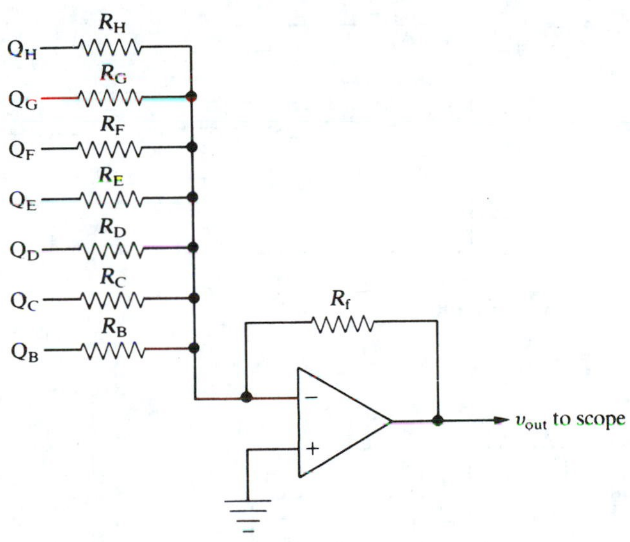

As explained in Section 13.5.3, calculate values for R_B, R_C, R_D, R_E, R_F, R_G, R_H, and R_f to make the 2nd through the 14th harmonic zero amplitude. Use these values in your circuit and observe v_{out}. Carefully sketch it.

Measure the harmonic distortion of v_{out} on a harmonic analyzer such as the Tektronix 5L4N, if available. The $n = 1$ fundamental should be large, the $n = 2$ through 14 should be very small, and the $n = 15$ harmonic should be the first significant one after the fundamental.

EXPERIMENT 50—THE PSEUDORANDOM NOISE GENERATOR

Construct a digital noise generator as shown in the illustration on page 914. Load in $Q_A \ldots Q_H = 10000000$ and predict and observe the $Q_A \ldots Q_H$ states on LEDs with a slow clock. How many different states are there? Observe the serial output at Q_H; what is its noise spectrum?

Try a different feedback and predict and observe the states. Notice that you do not always get $2^N - 1$ possible states. What happens if you load in a state not in the sequence?

EXPERIMENT 51—THE ANALOG MULTIPLEXER/DEMULTIPLEXER

The 4051 is an analog (signals can flow in both directions) eight-channel multiplexer/demultiplexer. It is essentially an eight-position switch with three digital inputs ABC that determine which position the switch takes. Note: INH must equal 0.

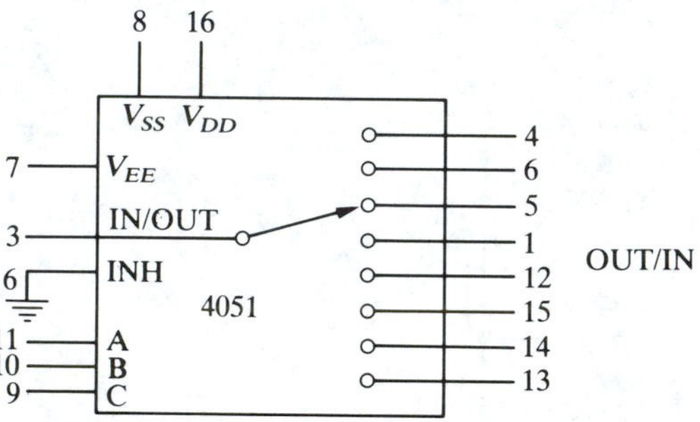

I. A simple divide-by-8 counter:

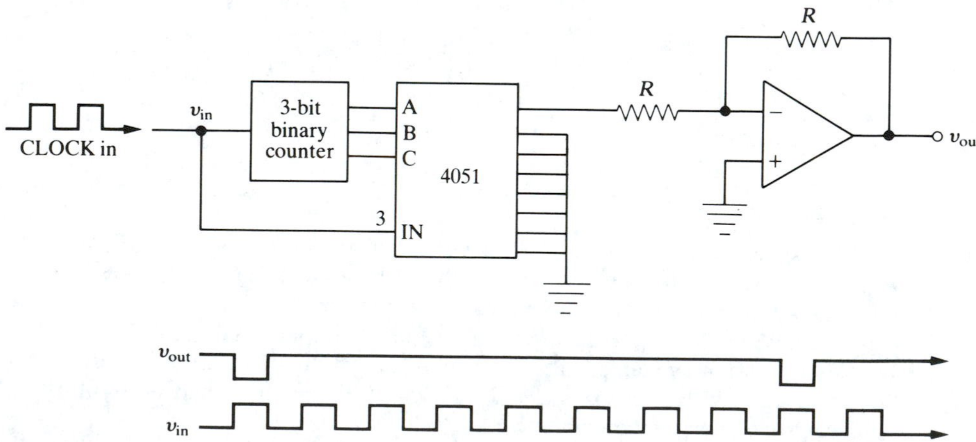

By rewiring the 4051, we can divide by any number from 2 through 8. Or we can produce output waveforms like

Choose a frequency division N and construct the circuit.

II. A digital signal can also control the gain of an analog amplifier, as shown below.

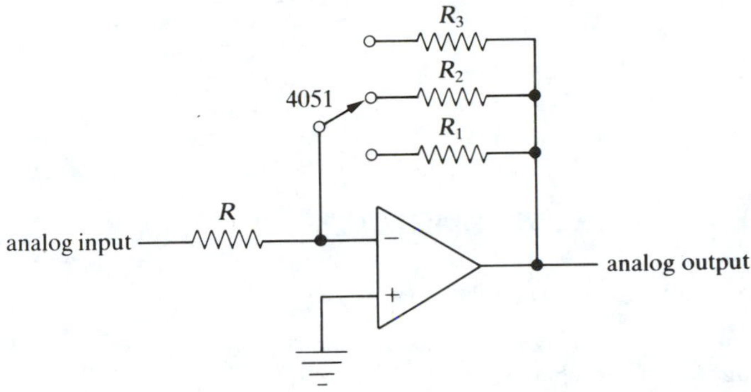

III. A sinusoidal input can be rectified with a 4051 used as a "chopper." Its "on" resistance is typically 80 Ω.

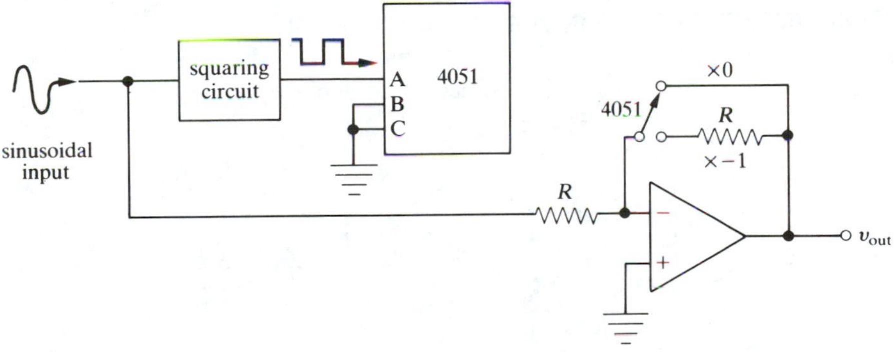

The output will be an inverted half-wave rectified version of the input, provided the input and the A square-wave are in phase.

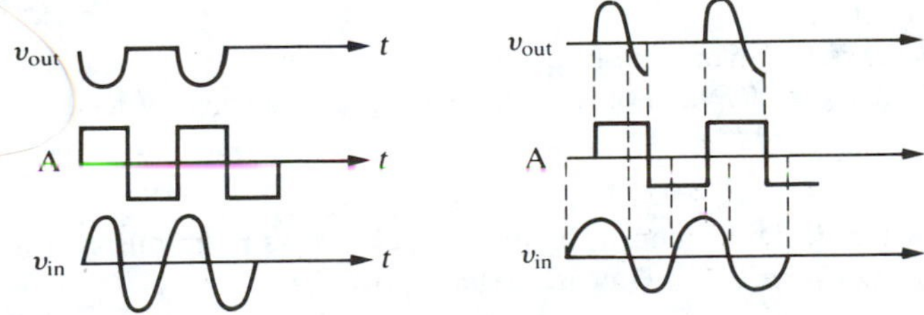

If there is a phase difference between V_{in} and A, then the dc average of the output v_{out} will be proportional to the phase shift ϕ. See 8.8. If $\phi = 90°$, then the dc average of V_{out} will be zero.

A simple analog bilateral switch such as the 4066 will do the same job.

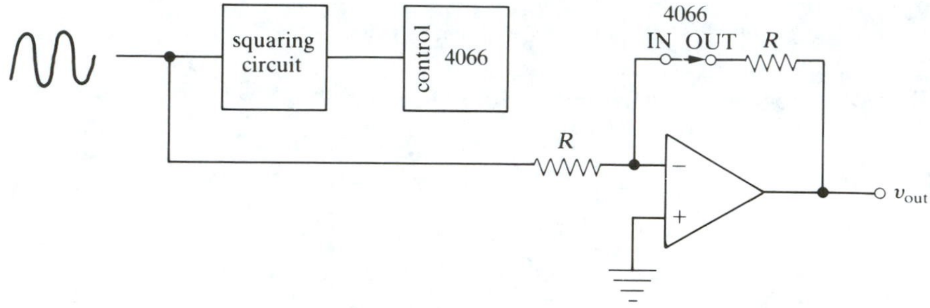

EXPERIMENT 52—THE 555 TIMER CHIP

Using a 555 chip, construct the following:

I. Square-wave oscillator. Measure the output frequency change with respect to power supply voltage changes.

II. Monostable multivibrator or one-shot. What is the shortest output pulse you can get? Measure the rise and fall times.

EXPERIMENT 53—THE INVERTER OSCILLATOR

I. Using two CMOS inverters, construct the oscillator shown below.

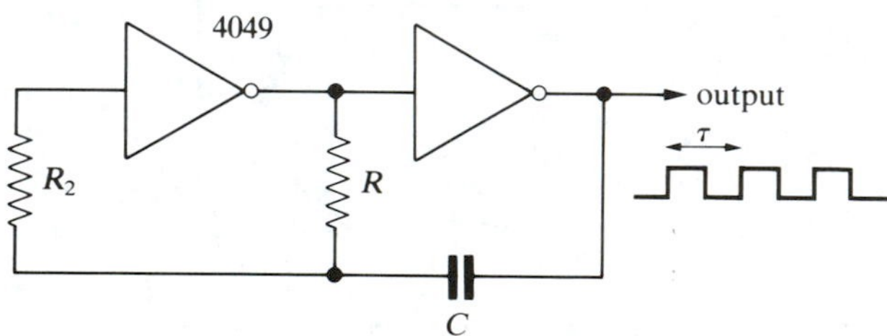

With $R_2 = 0$, measure the output period τ and its relation to R and C. How can you adjust τ?

What is the maximum frequency of oscillation?

Add $R_2 \cong 10R$ and observe the effect on the waveform.

II. Modify the circuit to produce an asymmetric output waveform (see Section 13.6.2).

III. Add a toggle D flip-flop to produce a perfectly square output wave. Carefully sketch the waveforms. How is τ related to RC?

IV. Construct a gated oscillator circuit, using two input NOR gates.

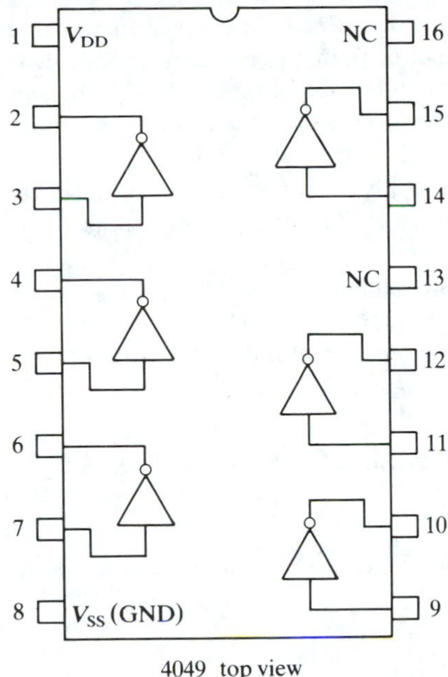

4049 top view

EXPERIMENT 54—THE MONOSTABLE MULTIVIBRATOR OR ONE-SHOT

Construct several one-shots as shown in Section 13.6.3. Construct at least one whose output depends *not* on an *RC* time constant but on a clock period (i.e., a synchronous one-shot). Measure the rise and fall times of the output pulse.

EXPERIMENT 55—A 64-BIT RANDOM-ACCESS MEMORY

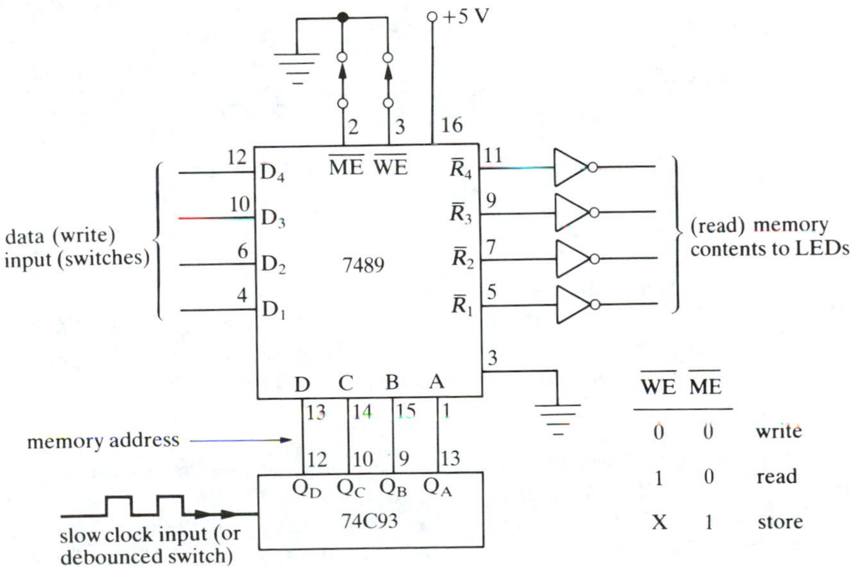

I. Hook up a 7489 RAM to a 7493 four-bit counter as shown. Load in 16 four-bit numbers by grounding both the memory enable ($\overline{ME}$) and the write enable ($\overline{WE}$) (pins 2 and 3) and selecting the address by the 7493 counter outputs. With a slow clock input to the counter, verify that the loading has been successful by stepping through all 16 memory locations.

II. Turn off the 5-V supply and then turn it on again. Now step through all 16 memory locations. Are the contents the same? Is this memory volatile or nonvolatile?

III. How could you automatically "branch" from one memory location (e.g., 0111) back to 0000? [*Hint*: RESET the 7493 counter to 0000 when the 7489 output is 0111. Use a 7442 decoder with the 7489 output as the 7442 input. Thus only locations 0000 through 0111 will be stepped through.]

EXPERIMENT 56—THE DIGITAL-TO-ANALOG (D/A) CONVERTER (DAC)

Hook up the DAC0800 eight-bit D/A converter after studying its spec sheet.

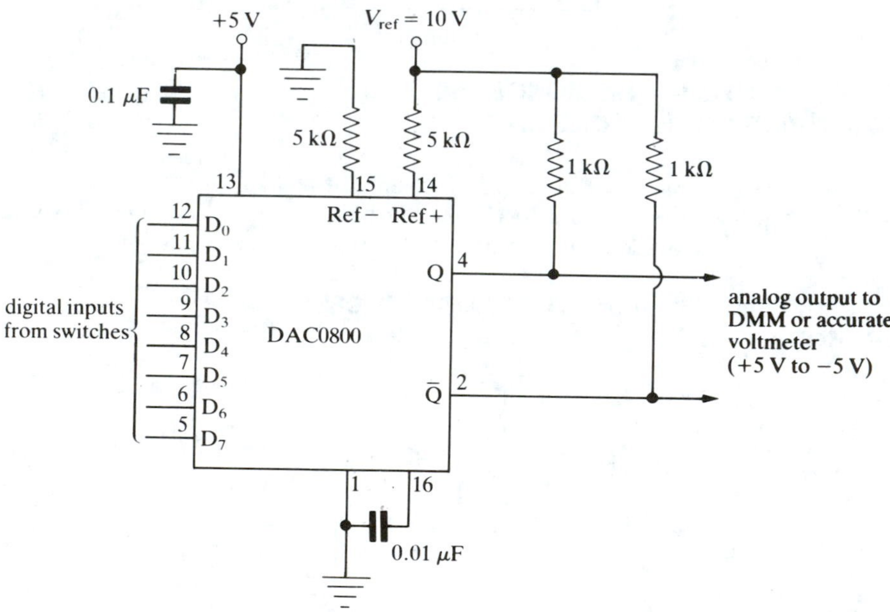

I. Vary the digital input by using the switches and measure the analog output. Wire the MSB and the next MSB inputs to ground so that there are only $2^6 = 64$ possible inputs. Plot V_{out} vertically and the digital input horizontally.

Does the output increase monotonically with an input increase?

What is the output for a 000001 input (one LSB)?

What is the accuracy?

Remove the MSB and the next MSB grounds and measure the output for 11111110 and 11111111 inputs. Is the difference one LSB?

Check the linearity by putting in the digital numbers 00000001, 00000010, 00000100, 00001000, 00010000, 00100000, 01000000, and 10000000 and measuring the outputs.

II. Hook up a 7493 four-bit counter and observe the output on a scope.

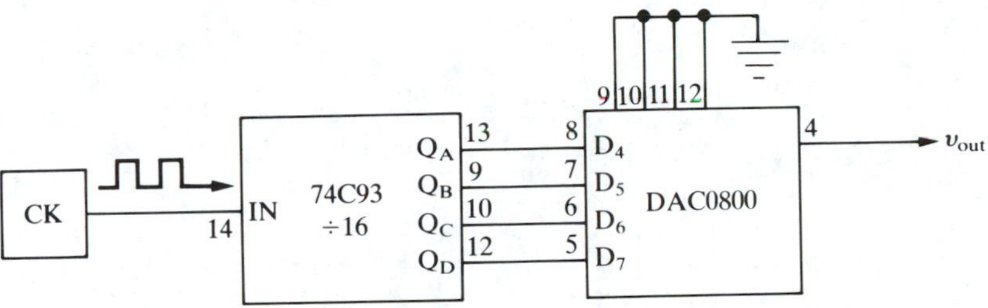

The output should be a "staircase"—a digital approximation to a triangular wave. Increase the input clock frequency and observe the output. At high-frequency inputs the settling time of the op amp will distort the analog output.

III. You can make an arbitrary function generator by storing the function values in a memory (e.g., 7489) and stepping through the memory locations at a constant rate. The analog function output is obtained by connecting the 7489 memory output to the input of the D/A converter. (See the illustration on page 922.)

IV. Digital Control of an Analog Signal. Connect an analog sine wave to the $+V_{ref}$ terminal. The output will then be the *product* of the eight-bit digital input and the analog sine wave. Be sure the amplitude of the V_{ref} sine wave is not too large.

EXPERIMENT 57—THE ANALOG-TO-DIGITAL (A/D) CONVERTER (ADC)

Hook up the 0809 successive approximation A/D converter after carefully studying its spec sheet.

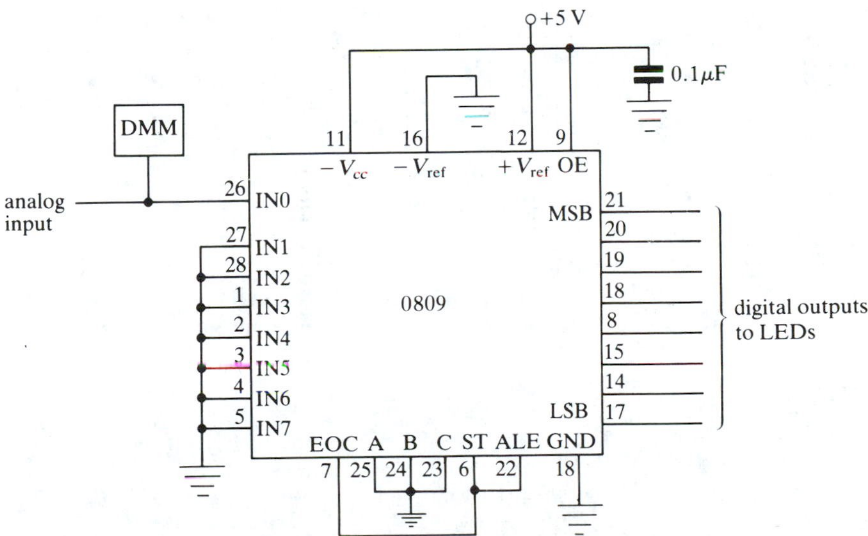

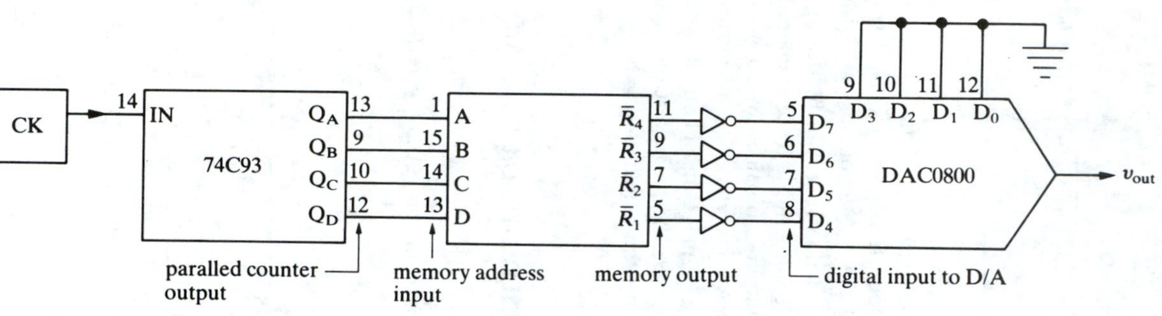

Illustration for Exp. 56.III.

I. Vary the analog input voltage and note the digital outputs. What analog input corresponds to 1 LSB?

Check the linearity by putting in 1, 2, 3, 4, 5 V and noting the digital output.

II. Hook up a positive square-wave input (0 to approximately 3 V amplitude) and short rise and fall times, and note the MSB output on a scope. Now increase the square-wave frequency. When does the MSB output stop faithfully following the input? Briefly explain, using the 0809 spec sheet. [*Hint*: The conversion time is 100 μs.]

For the next four experiments, familiarize yourself with the 8085 microprocessor kit by reading:

1. *Microcomputer Experimentation with the Intel SDK-85* by Lowenthal and Walsh, Prentice-Hall, © 1980. Working through this book is the best way to become familiar with using the SDK-85 microprocessor kit.
2. *The MCS-80185 Family User's Manual* © Intel, especially Chapter 5 on the instruction set.
3. *The 8080/8085 Assembly Language Programmer's Manual* © Intel, especially Chapter 3 on the instruction set.
4. *SDK-85 System Design Kit User's Manual* © Intel. This comes with the SDK-85 kit and tells you how to assemble the kit, and gives brief descriptions of all the chips, operating instructions, and some short sample programs.

EXPERIMENT 58—SIMPLE ARITHMETIC WITH THE 8085 MICROPROCESSOR

Learn to add and subtract numbers (in hex) by using the 8085 SDK-85. Your lab instructor will give you two numbers and ask you to write a program to add and subtract them. Check your results by performing the additions and subtractions by hand. Be sure to review your hexadecimal arithmetic first! For example:

$$06_{16} + 0C_{16} = 12_{16}$$

$$1E_{16} + A3_{16} = C1_{16}$$

EXPERIMENT 59—DIGITAL WAVEFORM GENERATION

I. Write a simple program (in assembly language) to generate a series of rectangular pulses at an output port.

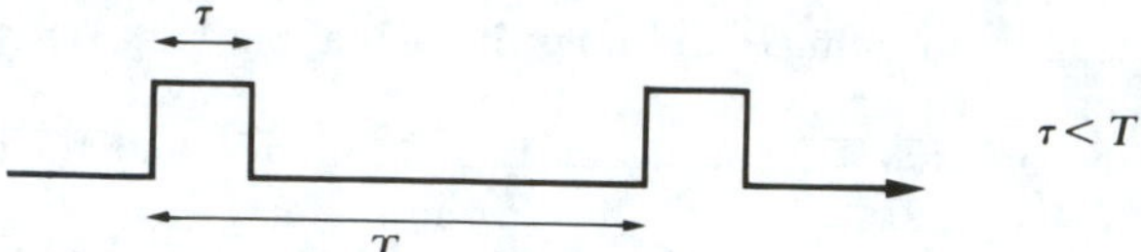

The amplitude can simply be the analog voltage level for a logic 1, but the pulse width τ and the period T should be determined by the program. Notice that by changing the program you can change either τ or T, but τ must always be less than T.

Remember to initialize the output port as an *output*.

What are the limitations on τ and T? How are these limitations related to the clock frequency?

II. Generate a sawtooth waveform.

EXPERIMENT 60—THE 8085 MICROPROCESSOR AS A LOOKUP TABLE

Use the 8085 microprocessor and memory as a lookup table to measure temperature with a thermistor and have the readout in degrees Celsius.

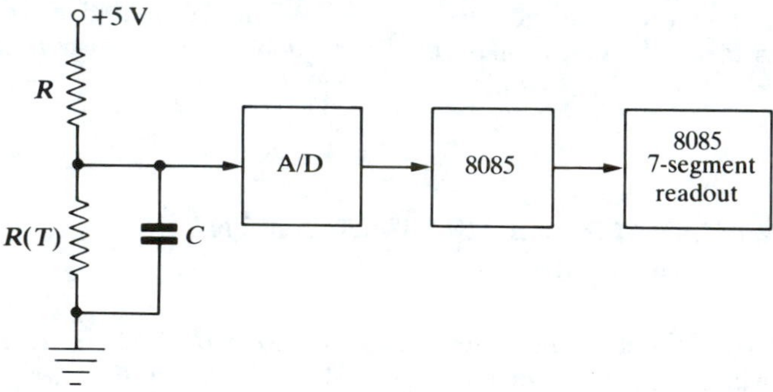

You will have to calibrate the thermistor $R(T)$ first by putting it in a water bath of known temperature, measured with an ordinary thermometer. The A/D output for 20°C will then be the address where 20 is stored in memory, and so on.

The capacitor C may be necessary to reduce noise at the input.

Be sure the thermistor leads are not shorted by the water bath.

Any other transducer can be used in place of the thermistor.

EXPERIMENT 61—DIGITAL FILTERS (at least three or four lab periods)

Program your SDK-85 to act as a simple nonrecursive digital filter. You may choose either a low-pass or a high-pass filter. To keep things simple, use only three terms in your filter.

$$y_n = C_1 x_n + C_2 x_{n-1} + C_3 x_{n-2}$$

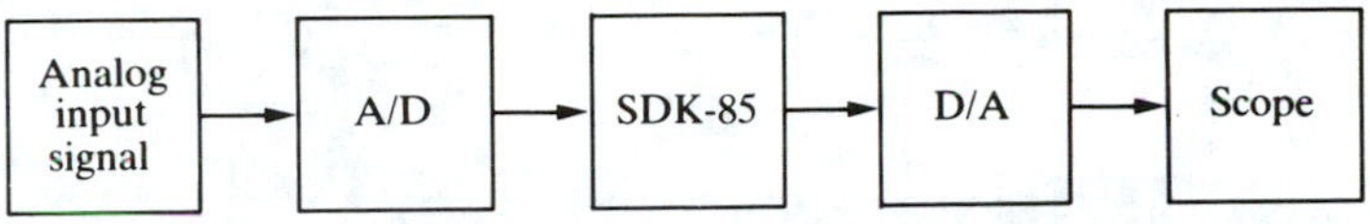

Analog input signal → A/D → SDK-85 → D/A → Scope

Measure the gain versus frequency for your filter and compare with your theory.

A swept-frequency analog input is convenient but not necessary.

Remember the Nyquist sampling theorem.

APPENDIX I

References

ELEMENTARY REFERENCES

Basic Electronics, Revised ed. Prepared by the U.S. Navy Bureau of Naval Personnel, Dover Publications, Inc., New York, 1973 (republication of *Basic Electronics Volume I*, NAVPERS 10087-C).

Second Level Basic Electronics. 22841-X, Dover Publications, Inc., New York (reproduction of *Basic Electronics Volume II*, NAVPERS 10087-C).

The ABCs of DMMs, 1984. Hohn Fluke Mfg. Co., Inc. Short booklet.

COLLEGE TEXTS

Conductors and Semiconductors, A. Holden. Bell Telephone Labs, 1964. Elementary level.

Solar Cells, M. A. Green. Prentice-Hall, 1982.

The Physics of Semiconductor Devices, 3rd ed., D. A. Fraser. Clarendon Press, 1983. Intermediate Level.

Basic Electronics for Scientists, 4th ed., J. J. Brophy. McGraw-Hill, 1983. Basic text.

Principles of Electronic Instrumentation, 2nd ed. J. A. Diefendorfer. W. B. Saunders, 1979. Basic text.

Electronics with Digital and Analog Integrated Circuits, R. J. Higgins. Prentice-Hall, 1983. Covers digital first, then analog.

Electronics and Instrumentation for Scientists, H. V. Malmstadt, C. G. Enke, and S. R. Crouch. Benjamin/Cummings, 1981.

Microelectronics, J. Millman. McGraw-Hill, 1979. Oriented toward electrical engineers; lots of circuit analysis of discrete component circuits.

Digital Circuits with Microprocessor Applications, P. M. Chirlian. Matrix Publishers, Inc., 1982. Oriented toward electrical engineers.

The Art of Electronics, P. Horowitz and W. Hill. Cambridge University Press, 1980. Advanced treatment, little on circuit analysis.

Basic Electric Circuit Analysis, D. E. Johnson, J. L. Hilburn, and J. R. Johnson. Prentice-Hall, 1984.

INTERMEDIATE AND ADVANCED REFERENCES

Data Conversion Handbook, D. B. Bruck. Hybrid Systems Corp., 1974. Good general reference on A/D, D/A conversion.

Handbook of Microcircuit Design and Application, D. F. Stout and M. Kaufman. McGraw-Hill, 1980. Lots of circuits.

CMOS Cookbook, D. Lancaster. Howard W. Sams & Co. Inc., 1977.

TTL Cookbook, D. Lancaster. Howard W. Sams & Co. Inc., 1974.

Microcomputer Interfacing, B. A. Artwick. Prentice-Hall, 1980. Good practical book from industrial viewpoint.

Analog-Digital Conversion Notes, ed., D. H. Sheingold. Analog Devices Inc., 1980.

IC Op Amp Cookbook, 2nd ed. W. Jung. Howard W. Sams & Co. Inc., 1980.

Noise in Electronic Devices & Systems, M. J. Buckingham. Halsted Press, John Wiley, 1983.

Measurements Detection of Radiation, N. Tsoulfonidis. McGraw-Hill, 1983. Covers all types of nuclear solid-state detectors.

Digital Filters, 2nd ed. R. W. Hamming. Prentice-Hall, 1983. Excellent treatment.

Design of Phase Locked Loop Circuits with Experiments, H. M. Berlin, Howard W. Sams & Co. Inc., 1978.

Active Filter Cookbook, D. Lancaster. Howard W. Sams & Co. Inc., 1975.

Designing Microprocessor-Based Instrumentation, J. J. Carr. Reston/Prentice-Hall, 1982. Covers Z-80 and 6502 plus general material on I/O ports, transducers, A/D/A conversion, and so on.

Semiconductor and Integrated Circuit Fabrication Techniques, P. E. Gise and R. Blanchard. Reston/Prentice-Hall, 1979. How the chips are actually made—oxidation, photomasking, epitaxial deposition, and so on.

Digital Filters and their Applications, V. Cappellini, A. G. Constantinides, and P. Emiliani. Academic Press, 1978.

"Interfacing Solid-State Sensors with Digital Systems," J. E. Brignall, in *J. Physics E. Sci. Instrum.*, Vol. 18 (1985), p. 559.

"Sensors in Microprocessor Based Applications," J. E. Brignall and A. P. Dorey, in *J. Physics E. Instrum.*, Vol. 16 (1983), p. 952.

MANUFACTURERS' DATA BOOKS (Partial Listing)

National Semiconductor
2900 Semiconductor Dr.
Santa Clara, Calif. 95051

High-Speed CMOS Family Data Booklet
Voltage Regulator Handbook
Interface Data Book
Transistor Data Book
Data Acquisition Handbook
Linear Data Book
Logic Data Book
CMOS Data Book
Special Functions Data Book

Motorola Box 20912 Phoenix, Arizona 85036	*Linear Integrated Circuits* *Interface Circuits* *CMOS Data*
Texas Instruments P.O. Box 5012, MS308 Dallas, Texas 75222	*The TTL Data Book* *The Linear Circuits Data Book* *The Optoelectronics Data Book* *High-Speed CMOS Logic Data Book*
Siliconix P.O. Box 4777 Santa Clara, Calif. 95054	*MOSPOWER Design Catalog* *Small-Signal FET Design Catalog* *Analog Switch and IC Product Databook*
PMI (Precision Monolithics, Inc.) 1500 Space Park Dr. Santa Clara, Calif. 95050	*Linear Integrated Circuits* *Linear and Conversion Products*
Fairchild MOS Memory/Logic Division 101 Bernal Rd. Santa Clara, Calif. 95050	*High-Speed CMOS Data Book*
Analog Devices Two Technology Way P.O. Box 280 Norwood, Mass. 02062	*1984 Data Book, Vol. 1, Integrated Circuits* *1984 Data Book, Vol. 2, Modules Subsystems*

ELECTRONIC SUPPLIERS

Newark Electronics, 500 N. Pulaski Rd., Chicago, Ill. 60624, huge catalog, branches everywhere

Jameco Electronics, 1355 Shoreway Rd., Belmont, Calif. 94002, short catalog

Digi-Key Corporation, Highway 32 South, P.O. Box 677, Thief River Falls, Minn. 56701, short catalog

Mouser Electronics, 11433 Woodside Ave., Santee, Calif. 92071, short catalog

Gerber Electronics, 128 Carnegie Row, Norwood, Mass. 02062, large catalog

COMPREHENSIVE DATA BOOKS

IC Master (two volumes), Hearst Business Communications, Inc., 645 Stewart Ave., Garden City, N.Y. 11530, $95.

eem Master Catalog, Hearst Business Communications, Inc., 645 Stewart Ave., Garden City, N.Y. 11530. Components, free to qualified persons.

D.A.T.A. Book (seven volumes), 45 U.S. Highway No. 46, Pine Brook, N.J. 07058. Specifications on *all* digital ICs, linear JCs, transistors, diodes, and so on. Very useful.

MAGAZINES (Partial Listing)

Electronics. Broad general-interest coverage including computers.

Electronic Design News. General coverage plus specific circuits.

Spectrum IEEE. Useful review articles, high level.

Byte. Computer oriented, state-of-the-art ads.

Index

A

B

D

E

F

G

M

N

Q

R

U

V

W

X

Y

Z